FOR THE
IB DIPLO
PROGRAM

THIRD EDITION

Chemistry

Christopher Talbot
Chris Davison
David Fairley

Dedication

To my late Father, David Talbot, who encouraged my love of science (Chris Talbot)

Author acknowledgements

We would like to thank and recognise Dr David Fairley for his invaluable knowledge of chemistry employed in his meticulous reviews and for his contributions in writing a number of the linking questions and toolkit boxes. Thank you also to David Allen for his help with writing toolkit boxes, the online Tools and Inquiry reference guide and Internal Assessment chapter.

We thank Professor Norman Billingham, University of Sussex, for his detailed advice and reviews of the physical chemistry chapters, Dr Jurica Bauer, Maastricht University, for advice on kinetics, and Professor Mike Williamson, University of Sheffield, for advice on spectroscopy.

IB advisor: The Publishers would like to thank Roula Barghout and John Green for their advice and support in the development of this project.

Acknowledgements

Every effort has been made to trace all copyright holders, but if any have been inadvertently overlooked, the Publishers will be pleased to make the necessary arrangements at the first opportunity.

Orders: please contact Hachette UK Distribution, Hely Hutchinson Centre, Milton Road, Didcot, Oxfordshire, OX11 7HH. Telephone: +44 (0)1235 827827. Email education@hachette.co.uk Lines are open from 9 a.m. to 5 p.m., Monday to Friday. You can also order through our website: www.hoddereducation.com.

ISBN: 978 1 3983 6990 0

First edition published in 2012
Second edition published in 2014
This edition published in 2023 by
Hodder Education,
An Hachette UK Company
Carmelite House
50 Victoria Embankment
London EC4Y 0DZ

www.hoddereducation.com

Impression number 10 9 8 7 6 5 4 3 2

Year 2027 2026 2025 2024 2023

Cover photo © berkay08/stock.adobe.com

Illustrations by Aptara, Inc.

Typeset in Times New Roman 10/14pt by DC Graphic Design Limited, Hextable, Kent.

Produced by DZS Grafik, Printed in Slovenia

A catalogue record for this title is available from the British Library.

Contents

Free online content

Go to our website www.hoddereducation.com/ib-extras for free access to the following:

- Practice exam-style questions for each chapter
- Glossary
- Answers to self-assessment questions and practice exam-style questions
- Answers to linking questions
- Tools and Inquiries reference guide
- Internal Assessment – the scientific investigation

Introduction

Welcome to *Chemistry for the IB Diploma Third Edition*, updated and designed to meet the criteria of the new International Baccalaureate (IB) Diploma Programme Chemistry Guide. This coursebook provides complete coverage of the new IB Chemistry Diploma syllabus, with first teaching from 2023.

The aim of this syllabus is to integrate concepts, topic content and the nature of science through inquiry. Differentiated content for SL and HL students is clearly identified throughout with a blue banner.

Key concepts

The six themes that underpin the IB Chemistry Diploma course are integrated into the conceptual understanding of all units, to ensure that a conceptual thread is woven throughout the course. Conceptual understanding enhances your overall understanding of the course, making the subject more meaningful. This helps you develop clear evidence of synthesis and evaluation in your responses to assessment questions. Concepts are explored in context and can be found interspersed in the chapter.

About the authors

Christopher Talbot has taught IB Chemistry, TOK (Theory of Knowledge), and IB Biology and a range of science courses, including IGCSE and MYP, in a number of local and international schools in Singapore since 1995. Chris studied at the University of Sussex, UK, to gain a BSc (Honours) degree in biochemistry. He also holds master's degrees in life sciences (chemistry) and science education from NIE in Singapore.

Christopher Davison studied at the University of Nottingham, UK, and gained an MSci degree in chemistry and a PhD in organic chemistry. He moved into teaching in 2014, and he works at Wellington College, UK, teaching IB Chemistry.

IB advisors

David Fairley studied at the University of Canterbury, New Zealand, and gained a BSc degree (Honours) and a PhD in chemistry. He taught IB Chemistry, MYP Science, IGCSE Co-ordinated Sciences and TOK at Overseas Family School, Singapore.

John Sprague has been teaching TOK for 20 years, in the UK, Switzerland and Singapore. Previously Director of IB at Sevenoaks School in the UK, he now teaches philosophy and TOK at Tanglin Trust School, Singapore.

The 'In cooperation with IB' logo signifies that this coursebook has been rigorously reviewed by the IB to ensure it fully aligns with the current IB curriculum and offers high-quality guidance and support for IB teaching and learning.

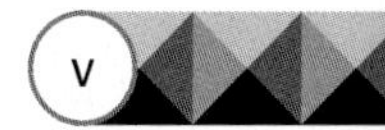

How to use this book

The following features will help you consolidate and develop your understanding of chemistry, through concept-based learning.

Guiding questions

- There are guiding questions at the start of every chapter, as signposts for inquiry.
- These questions will help you to view the content of the syllabus through the conceptual lenses of the themes.

SYLLABUS CONTENT

- This coursebook follows the order of the contents of the IB Chemistry Diploma syllabus.
- Syllabus understandings are introduced naturally throughout each topic.

Key terms

◆ Definitions appear throughout the margins to provide context and help you understand the language of chemistry. There is also a glossary of all key terms online at www.hoddereducation.com/ib-extras.

Tools

The Tools features explore the skills and techniques that you require and are integrated into the chemistry content to be practised in context. These skills can be assessed through internal and external assessment.

Inquiry process

The application and development of the Inquiry process is supported in close association with the Tools.

Common mistake

These detail some common misunderstandings and typical errors made by students, so that you can avoid making the same mistakes yourself.

Nature of science

Nature of science (NOS) explores conceptual understandings related to the purpose, features and impact of scientific knowledge. It can be examined in chemistry papers. NOS explores the scientific process itself, and how science is represented and understood by the general public. NOS covers 11 aspects: Observations, Patterns and trends, Hypotheses, Experiments, Measurements, Models, Evidence, Theories, Falsification, Science as a shared endeavour, and The global impact of science. It also examines the way in which science is the basis for technological developments and how these modern technologies, in turn, drive developments in science.

ATL ACTIVITY

Approaches to learning (ATL) activities, including learning through inquiry, are integral to IB pedagogy. These activities are designed to get you to think about real-world applications of chemistry.

Top tip!

This feature includes advice relating to the content being discussed and tips to help you retain the knowledge you need.

WORKED EXAMPLE

These provide a step-by-step guide showing you how to answer the kind of quantitative questions that you might encounter in your studies and in the assessment.

Going further

Written for students interested in further study, this optional feature contains material that goes beyond the IB Diploma Chemistry Guide.

Links

Due to the conceptual nature of chemistry, many topics are connected. The Links feature states where relevant material is covered elsewhere in the coursebook. They may also help you to start creating your own linking questions.

Self-assessment questions (SAQs) appear throughout the chapters, phrased to assist comprehension and recall and also to help familiarize you with the assessment implications of the command terms. These command terms are defined in the online glossary. Practice exam-style questions for each chapter allow you to check your understanding and prepare for the assessments. The questions are in the style of those in the examination so that you get practise seeing the command terms and the weight of the answers with the mark scheme. Practice exam-style questions and their answers, together with self-assessment answers, are on the accompanying website, IB Extras: **www.hoddereducation.com/ib-extras**

Links to Theory of Knowledge (TOK) allow you to develop critical thinking skills and deepen scientific understanding by bringing discussions about the subject beyond the scope of the content of the curriculum.

LINKING QUESTIONS

These questions are introduced throughout each topic and are for all students to attempt (apart from those labelled as HL only or in HL-only sections of the book). They are designed to strengthen your understanding by making connections across the themes. The linking questions encourage you to apply broad, integrating and discipline-specific concepts from one topic to another, ideally networking your knowledge. Practise answering the linking questions first, on your own or in groups. Sample answers and structures are provided online at **www.hoddereducation.com/ib-extras**. The list in this coursebook is not exhaustive; you may encounter other connections between concepts, leading you to create your own linking questions.

International mindedness is indicated with this icon. It explores how the exchange of information and ideas across national boundaries has been essential to the progress of science and illustrates the international aspects of chemistry.

The **IB learner profile** icon indicates material that is particularly useful to help you towards developing the following attributes: to be inquirers, knowledgeable, thinkers, communicators, principled, open-minded, caring, risk-takers, balanced and reflective. When you see the icon, think about what learner profile attribute you might be demonstrating – it could be more than one.

Tools and Inquiry

Skills in the study of chemistry

The skills and techniques you must experience through this chemistry course are encompassed within the tools. These support the application and development of the inquiry process in the delivery of the course.

Tools

- **Tool 1:** Experimental techniques
- **Tool 2:** Technology
- **Tool 3:** Mathematics

Inquiry process

- **Inquiry 1:** Exploring and designing
- **Inquiry 2:** Collecting and processing data
- **Inquiry 3:** Concluding and evaluating

Throughout the programme, you will be given opportunities to encounter and practise the skills; they will be integrated into the teaching of the syllabus when they are relevant to the syllabus topics being covered.

You can see what the Tools and Inquiry boxes look like in the *How to use this book* section on page vi.

The skills in the study of chemistry can be assessed through internal and external assessment. The approaches to learning provide the framework for the development of these skills.

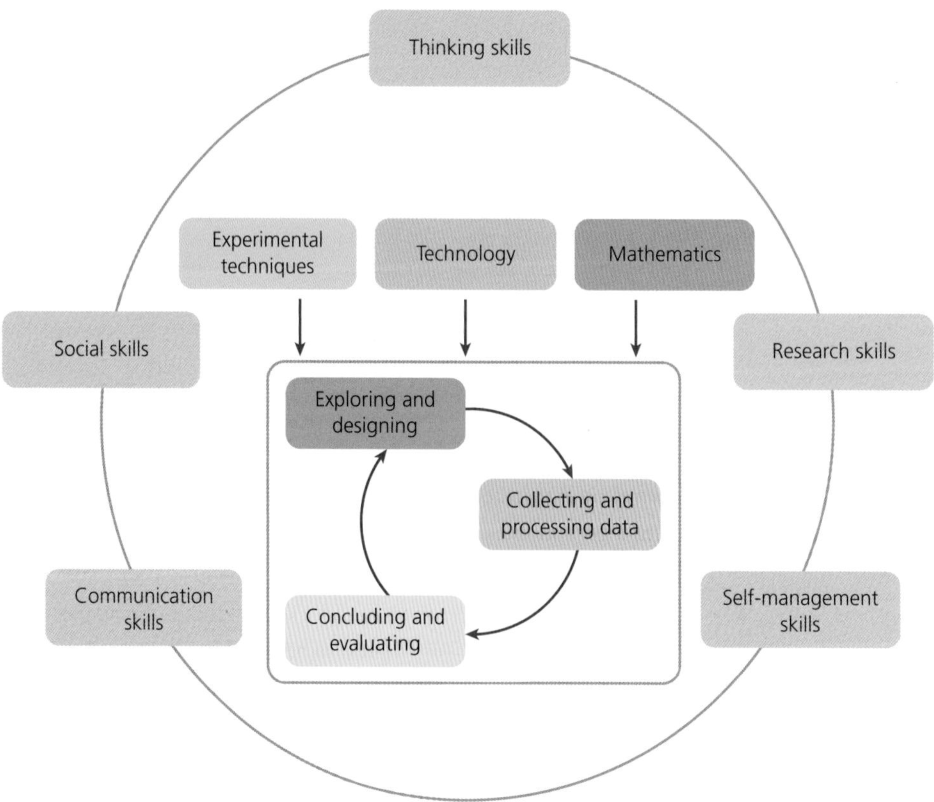

■ Skills for chemistry

IB Diploma Programme Chemistry Guide page 28

Visit the link in the QR code or this website to view the online Tools and Inquiries reference guide: **www.hoddereducation.com/ib-extras**

Tool 1: Experimental techniques

Skill	Description
Addressing safety of self, others and the environment	Recognize and address relevant safety, ethical or environmental issues in an investigation.
Measuring variables	Understand how to accurately measure the following to an appropriate level of precision. • Mass • Volume • Time • Temperature • Length • pH of a solution • Electric current • Electric potential difference
Applying techniques	Show awareness of the purpose and practice of: • preparing a standard solution • carrying out dilutions • drying to constant mass • distillation and reflux • paper or thin layer chromatography • separation of mixtures • calorimetry • acid–base and redox titration • electrochemical cells • colorimetry or spectrophotometry • physical and digital molecular modelling • recrystallization • melting point determination.

Tool 2: Technology

Skill	Description
Applying technology to collect data	• Use sensors • Identify and extract data from databases • Generate data from models and simulations
Applying technology to process data	• Use spreadsheets to manipulate data • Represent data in a graphical form • Use computer modelling

Tool 3: Mathematics

Skill	Description
Applying general mathematics	• Use basic arithmetic and algebraic calculations to solve problems. • Carry out calculations involving decimals, fractions, percentages, ratios, reciprocals and exponents. • Carry out calculations involving logarithmic functions. • Carry out calculations involving exponential functions (additional higher level). • Determine rates of change from tabulated data. • Calculate mean and range. • Use and interpret scientific notation (e.g. 3.5×10^6). • Use approximation and estimation. • Appreciate when some effects can be ignored and why this is useful. • Compare and quote values to the nearest order of magnitude. • Understand direct and inverse proportionality, as well as positive and negative correlations between variables. • Calculate and interpret percentage change and percentage difference. • Calculate and interpret percentage error and percentage uncertainty. • Distinguish between continuous and discrete variables.
Using units, symbols and numerical values	• Apply and use International System of Units (SI) prefixes and units. • Identify and use symbols stated in the guide and the data booklet. • Express quantities and uncertainties to an appropriate number of significant figures or decimal places.
Processing uncertainties	• Understand the significance of uncertainties in raw and processed data. • Record uncertainties in measurements as a range (±) to an appropriate level of precision. • Propagate uncertainties in processed data, in calculations involving addition, subtraction, multiplication, division and (HL only) exponents. • Express measurement and processed uncertainties—absolute, fractional (relative), percentage—to an appropriate number of significant figures or level of precision. • Apply the coefficient of determination (R^2) to evaluate the fit of a trend line or curve.
Graphing	• Sketch graphs, with labelled but unscaled axes, to qualitatively describe trends. • Construct and interpret tables, charts and graphs for raw and processed data including bar charts, histograms, scatter graphs and line and curve graphs. • Plot linear and non-linear graphs showing the relationship between two variables with appropriate scales and axes. • Draw lines or curves of best fit. • Interpret features of graphs including gradient, changes in gradient, intercepts, maxima and minima, and areas. • Draw and interpret uncertainty bars. • Extrapolate and interpolate graphs.

Inquiry process

Inquiry 1: Exploring and designing

Skill	Description
Exploring	• Demonstrate independent thinking, initiative, and insight. • Consult a variety of sources. • Select sufficient and relevant sources of information. • Formulate research questions and hypotheses. • State and explain predictions using scientific understanding.
Designing	• Demonstrate creativity in the designing, implementation and presentation of the investigation. • Develop investigations that involve hands-on laboratory experiments, databases, simulations, modelling. • Identify and justify the choice of dependent, independent and control variables. • Justify the range and quantity of measurements. • Design and explain a valid methodology. • Pilot methodologies.
Controlling variables	Appreciate when and how to: • calibrate measuring apparatus • maintain constant environmental conditions of systems • insulate against heat loss or gain.

Inquiry 2: Collecting and processing data

Skill	Description
Collecting data	• Identify and record relevant qualitative observations. • Collect and record sufficient relevant quantitative data. • Identify and address issues that arise during data collection.
Processing data	• Carry out relevant and accurate data processing.
Interpreting results	• Interpret qualitative and quantitative data. • Interpret diagrams, graphs and charts. • Identify, describe and explain patterns, trends and relationships. • Identify and justify the removal or inclusion of outliers in data (no mathematical processing is required). • Assess accuracy, precision, reliability and validity.

Inquiry 3: Concluding and evaluating

Skill	Description
Concluding	• Interpret processed data and analysis to draw and justify conclusions. • Compare the outcomes of an investigation to the accepted scientific context. • Relate the outcomes of an investigation to the stated research question or hypothesis. • Discuss the impact of uncertainties on the conclusions.
Evaluating	• Evaluate hypotheses. • Identify and discuss sources and impacts of random and systematic errors. • Evaluate the implications of methodological weaknesses, limitations and assumptions on conclusions. • Explain realistic and relevant improvements to an investigation.

Source: Tables from IB Diploma Programme Chemistry Guide, pages 28–32.

S1.1 Introduction to the particulate nature of matter

Guiding question

- How can we model the particulate nature of matter?

SYLLABUS CONTENT

By the end of this chapter, you should understand that:

- elements are the primary constituents of matter, which cannot be chemically broken down into simpler substances
- compounds consist of atoms of different elements chemically bonded together in a fixed ratio
- mixtures contain more than one element or compound in no fixed ratio, which are not chemically bonded and so can be separated by physical methods
- the kinetic molecular theory is a model to explain physical properties of matter (solids, liquids and gases) and changes of state
- temperature (K) is a measure of average kinetic energy (E_k) of particles.

By the end of this chapter you should know how to:

- distinguish between the properties of elements, compounds and mixtures
- distinguish the different states of matter
- use state symbols (s, l, g and aq) in chemical equations
- interpret observable changes in physical properties and temperature during changes of state
- convert between values in the Celsius and Kelvin scales.

Note: There is no higher-level only content in S1.1.

◆ **Matter**: Includes atoms, ions and molecules which have mass and volume.

◆ **Energy**: The ability to move matter and do work.

◆ **Mixture**: A mixture has the properties of its components, and can be separated by physical processes.

◆ **Macroscopic level**: Direct observation and measurement of physical properties.

Introduction

Chemistry is the study of the composition and reactions of substances (**matter**). Matter has mass and occupies space (has volume). Pure substances can exist as a solid, liquid or gas (states of matter) and are made up of particles (atoms, ions or molecules).

Matter is associated with **energy** which exists in various forms, such as heat, light, electrical, nuclear and chemical energy (within bonds and intermolecular forces).

Matter can be classified into two main groups: pure substances (elements and compounds) and impure substances (**mixtures**). Chemists need to separate mixtures of substances in order to identify pure substances and find their structures.

Chemistry can be understood at three levels. At one level, changes can be observed: for example, when magnesium burns in oxygen and releases energy as light and heat energy (Figure S1.1) to form magnesium oxide.

■ **Figure S1.1** A chemical reaction: the burning of magnesium in oxygen to form magnesium oxide

This level is the **macroscopic level**, because it deals with the bulk properties of substances, such as density and colour. This interpretation is derived from observations (especially visual) at the macroscopic level, and associated quantitative measurements involving weighing, that establish oxygen and magnesium as the reactants and magnesium oxide (not magnesium peroxide) as the product.

◆ **Particle level**: Interpretation in terms of atoms, molecules, ions and subatomic particles, e.g. electrons.

◆ **Symbolic level**: Representations of chemical phenomena using algebra, symbols, shapes and diagrams.

◆ **State symbol**: Used in chemical equations to describe the physical state of a reactant or product.

At the **particle level**, chemistry interprets these phenomena as changes involving the bonding between atoms, ions and molecules. Magnesium atoms combine with oxygen molecules to form magnesium oxide.

The third level is the **symbolic level**, the expression of chemical phenomena in terms of chemical equations using symbols and equations such as $2Mg(s) + O_2(g) \rightarrow 2MgO(s)$. The **state symbols** (s), (l), (g) and (aq) represent pure solid, pure liquid, pure gas and aqueous solution, respectively.

A chemist thinks at the particle level, carries out experiments at the macroscopic level (chemicals), and represents both symbolically. These three aspects of chemistry can be mapped as a triangle (Figure S1.2).

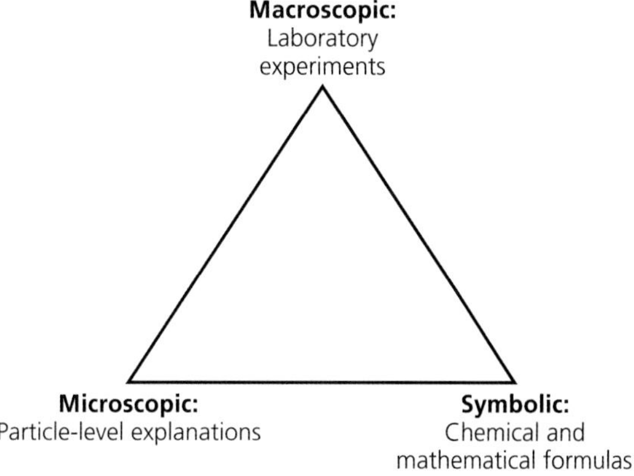
Macroscopic:
Laboratory experiments
Microscopic:
Particle-level explanations
Symbolic:
Chemical and mathematical formulas

■ **Figure S1.2** This triangle illustrates the three modes of scientific inquiry used in chemistry: macroscopic, microscopic and symbolic

At the macroscopic level, matter can be classified into mixtures or pure substances. These can be further subdivided as shown in Figure S1.3.

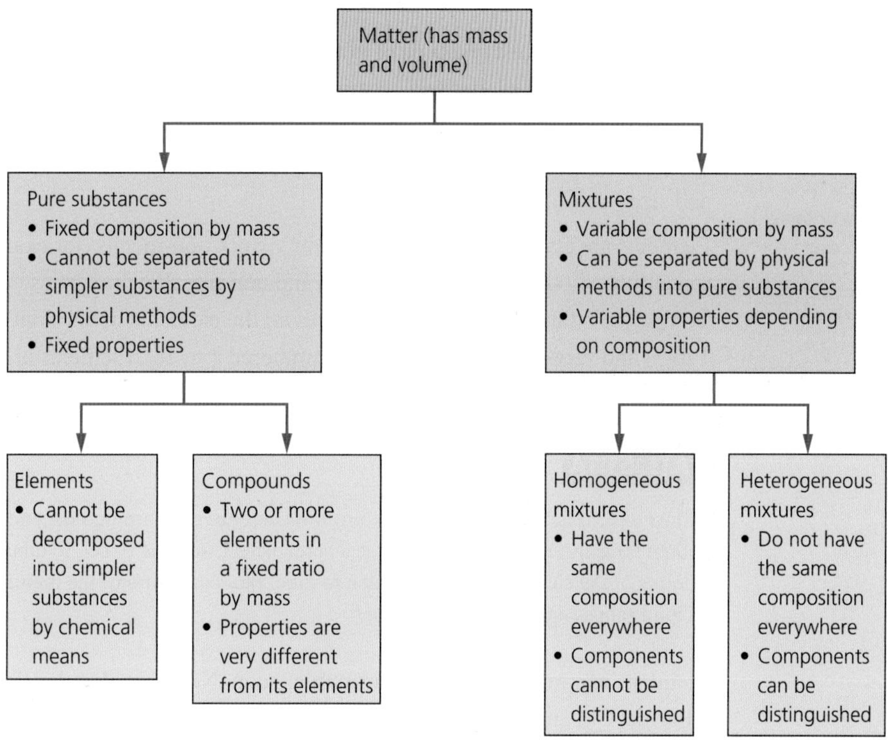
Matter (has mass and volume)
Pure substances
• Fixed composition by mass
• Cannot be separated into simpler substances by physical methods
• Fixed properties
Mixtures
• Variable composition by mass
• Can be separated by physical methods into pure substances
• Variable properties depending on composition
Elements
• Cannot be decomposed into simpler substances by chemical means
Compounds
• Two or more elements in a fixed ratio by mass
• Properties are very different from its elements
Homogeneous mixtures
• Have the same composition everywhere
• Components cannot be distinguished
Heterogeneous mixtures
• Do not have the same composition everywhere
• Components can be distinguished

■ **Figure S1.3** Classification of matter

TOK

What is the role of inductive and deductive reasoning in scientific inquiry, prediction and explanation?

The natural sciences, including chemistry, aim to understand the natural world through observation, via sense perception, and reasoning (logic). Science begins with observations; therefore, much of science is purely descriptive. However, science moves beyond pure observation through the use of both deductive and inductive reasoning.

Deductive reasoning uses general principles to make specific predictions and experiments, then produce observations to confirm whether the general principles are reliable enough to make predictions. A general example might be: 'If it is true that X causes Y, then when I design an experiment to investigate X, I should also then observe Y.'

Inductive reasoning works in the opposite direction: it uses a number of specific observations to develop general conclusions. A general example here would be: 'In all experiments when I see X, I also see Y; therefore, I can conclude that X and Y have a common cause.'

Scientists use both these forms of reasoning in the scientific method to gain understanding of the natural world. One form follows the sequence: observation, formulation of a hypothesis (scientific explanation for an observation), prediction, experimentation and then conclusion.

Elements and compounds

◆ **Element**: A substance (made of one type of atom) that cannot be decomposed into simpler, stable substances by a chemical change.

◆ **Atom**: The smallest particle of an element that can exist.

◆ **Molecule**: Two or more different atoms bonded (covalently).

◆ **Diatomic**: Made of two atoms bonded together (covalently).

Elements are the simplest substances and cannot be broken down or decomposed into simpler, more stable substances by a chemical reaction. They can be regarded as the simplest chemical 'building blocks'.

Elements contain just one type of **atom** (which all have the same number of protons). Elements are represented by a letter (e.g. uranium, U) or two letters (e.g. cobalt, Co). Although the atoms of a particular element may differ slightly in mass (isotopes), they all have essentially identical chemical reactions. The atoms of each element differ in mass (measured by relative atomic mass) and size (measured by atomic radius).

Elements are classified as metals (such as silver), non-metals (for example, nitrogen) or metalloids (including silicon) which have metallic and non-metallic properties or intermediate properties. The periodic table (Figure S1.4) arranges all the elements in horizontal rows (periods, numbered 1 to 7) and vertical columns (groups, numbered 1 to 18). Elements are ordered by atomic (proton) number and elements in the same group have similar chemical properties.

ATL S1.1A

Most of the elements are solids but bromine and mercury are liquids, and eleven elements are gases at room temperature and pressure. Use a spreadsheet, such as Excel, to tabulate the melting and boiling points of the elements from hydrogen to zinc. Plot bar charts of the melting points, boiling points and liquid ranges against atomic number.

Elements can exist as atoms (for example, neon, Ne), or **molecules** (for example, oxygen, O_2, chlorine, Cl_2, and sulfur, S_8). Molecules of elements are groups of two or more atoms that are chemically bonded together by covalent bonds. **Diatomic** molecules are formed by the covalent bonding of two atoms, for example, hydrogen (H_2).

Key: non-metals; metals; metalloids

Atomic number **Element** Relative atomic mass

	1	2	3	4	5	6	7	8	9	10	11	12	13	14	15	16	17	18
1	1 **H** 1.01																	2 **He** 4.00
2	3 **Li** 6.94	4 **Be** 9.01											5 **B** 10.81	6 **C** 12.01	7 **N** 14.01	8 **O** 16.00	9 **F** 19.00	10 **Ne** 20.18
3	11 **Na** 22.99	12 **Mg** 24.31											13 **Al** 26.98	14 **Si** 28.09	15 **P** 30.97	16 **S** 32.07	17 **Cl** 35.45	18 **Ar** 39.95
4	19 **K** 39.10	20 **Ca** 40.08	21 **Sc** 44.96	22 **Ti** 47.87	23 **V** 50.94	24 **Cr** 52.00	25 **Mn** 54.94	26 **Fe** 55.85	27 **Co** 58.93	28 **Ni** 58.69	29 **Cu** 63.55	30 **Zn** 65.38	31 **Ga** 69.72	32 **Ge** 72.63	33 **As** 74.92	34 **Se** 78.96	35 **Br** 79.90	36 **Kr** 83.80
5	37 **Rb** 85.47	38 **Sr** 87.62	39 **Y** 88.91	40 **Zr** 91.22	41 **Nb** 92.91	42 **Mo** 95.96	43 **Tc** (98)	44 **Ru** 101.07	45 **Rh** 102.91	46 **Pd** 106.42	47 **Ag** 107.87	48 **Cd** 112.41	49 **In** 114.82	50 **Sn** 118.71	51 **Sb** 121.76	52 **Te** 127.60	53 **I** 126.90	54 **Xe** 131.29
6	55 **Cs** 132.91	56 **Ba** 137.33	57 **La †** 138.91	72 **Hf** 178.49	73 **Ta** 180.95	74 **W** 183.84	75 **Re** 186.21	76 **Os** 190.23	77 **Ir** 192.22	78 **Pt** 195.08	79 **Au** 196.97	80 **Hg** 200.59	81 **Tl** 204.38	82 **Pb** 207.20	83 **Bi** 208.98	84 **Po** (209)	85 **At** (210)	86 **Rn** (222)
7	87 **Fr** (223)	88 **Ra** (226)	89 **Ac ‡** (227)	104 **Rf** (267)	105 **Db** (268)	106 **Sg** (269)	107 **Bh** (270)	108 **Hs** (269)	109 **Mt** (278)	110 **Ds** (281)	111 **Rg** (281)	112 **Cn** (285)	113 **Nh** (286)	114 **Fl** (289)	115 **Mc** (288)	116 **Lv** (293)	117 **Ts** (294)	118 **Og** (294)

†	58 **Ce** 140.12	59 **Pr** 140.91	60 **Nd** 144.24	61 **Pm** (145)	62 **Sm** 150.36	63 **Eu** 151.96	64 **Gd** 157.25	65 **Tb** 158.93	66 **Dy** 162.50	67 **Ho** 164.93	68 **Er** 167.26	69 **Tm** 168.93	70 **Yb** 173.05	71 **Lu** 174.97
‡	90 **Th** 232.04	91 **Pa** 231.04	92 **U** 238.03	93 **Np** (237)	94 **Pu** (244)	95 **Am** (243)	96 **Cm** (247)	97 **Bk** (247)	98 **Cf** (251)	99 **Es** (252)	100 **Fm** (257)	101 **Md** (258)	102 **No** (259)	103 **Lr** (262)

■ **Figure S1.4** Periodic table of elements (from the IB *Chemistry data booklet* section 7)

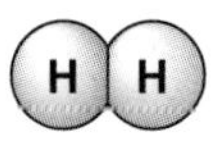

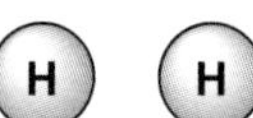

■ **Figure S1.5** A molecule of hydrogen, H_2, and two hydrogen atoms, 2H

H_2 represents a molecule of hydrogen; the subscript $_2$ represents two individual atoms of hydrogen, H (Figure S1.5). Writing the 2 as a suffix implies there are two atoms in the formula (chemically bonded). Hence, H_2, and not 2H, is the molecular formula of hydrogen (Figure S1.5).

Each element is represented by a chemical symbol. The symbol consists of either one or two letters. The first letter is always a capital (upper-case) letter and the second letter is always small (lower-case). The symbol H is derived from the first letter of the English name, hydrogen.

Top tip!

When writing symbols, you must be very careful to ensure your capital letters cannot be mistaken for small letters and vice versa.

ATL S1.1B

A number of elements have symbols that are not related to their name in English. Choose two of the elements and symbols listed below and present a short history of the element's discovery, its current and historical uses, and the linguistic origin of the name from which the symbol is derived. Sodium, Na; potassium, K; iron, Fe; copper, Cu; silver, Ag; tin, Sn; tungsten, W; mercury, Hg; lead, Pb.

O=O

oxygen molecule O_2

N≡N

nitrogen molecule N_2

H—H

hydrogen molecule H_2

Cl—Cl

chlorine molecule Cl_2

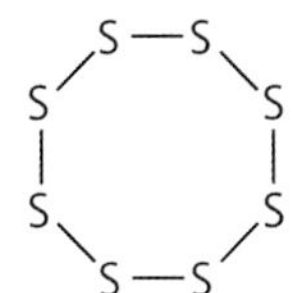

sulfur molecule S_8

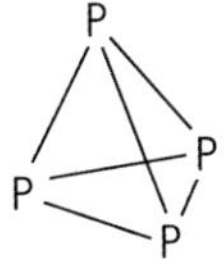

phosphorus molecule P_4

■ **Figure S1.6** Diagram of oxygen, nitrogen, hydrogen, chlorine, sulfur and phosphorus molecules

◆ **Compound:** A substance containing two or more different elements chemically bonded in a fixed ratio.

Top tip!

Water can be decomposed to its elements by passing an electric current through it. This is a chemical reaction known as electrolysis (see R3.2 Electron transfer reactions). This reaction can be described by the following word and balanced symbol equations with the correct formulas: water → oxygen + hydrogen, and $2H_2O \rightarrow O_2 + 2H_2$

Examples of elements that exist as molecules include the diatomic molecules oxygen, O_2, chlorine, Cl_2, and nitrogen, N_2, and the polyatomic molecules (more than two atoms) phosphorus, P_4, ozone, O_3, and sulfur, S_8 (Figure S1.6).

ATL S1.1C

A number of elements, including oxygen, carbon and phosphorus, exist as allotropes. Find out about the concept of allotropy and the two different forms of the element oxygen.

Prepare notes on the chemical and physical properties of dioxygen (oxygen) and trioxygen (ozone) to illustrate the concept of allotropes.

TOK

How is it that scientific knowledge is often shared by large, geographically spread and culturally diverse groups?

People from all cultures have contributed to science, as science is part of the social and cultural traditions of many human societies. The development of scientific knowledge relies on observations, experimental evidence and logical arguments. However, scientific ideas are affected by the social and historical setting.

Modern chemistry has its origins in alchemy (200BCE–AD1500), which was a series of theories and experiments designed to 'transmute' (change) base metals into gold. The word chemistry is derived from the Egyptian word *chēmeia*, which referred to *qemi*, the Black Land—a reference to the dark fertile soil of the Nile River. Hence, chemistry originally meant the Art of the Black Land (Black Arts).

The development and rise of alchemy were influenced by the Greek philosophers: for example, the theory of the four elements, which was the cornerstone of alchemy, was originally described by Empedocles and developed by Aristotle (around 350BCE). These four elements are fire, earth, water and air, which alchemists believed could be transformed into one another. Alchemists were concerned with the transmutation (conversion) of base (cheap) metals into gold and silver.

Between the eleventh and thirteenth centuries, much of this Arabic knowledge, including earlier Greek works of science, medicine and philosophy, was translated into Latin and transmitted to European centres of learning.

The Muslim chemist Al-Jabir Ibn Hayyan (died AD815) discovered 19 elements and perfected chemical separation techniques such as distillation, crystallization and sublimation. He also suggested that all matter can be traced to a simple, basic particle composed of charge and fire.

Compounds

Compounds are pure substances made up from the atoms of two or more elements, bonded together chemically. The ratio of elements in a particular compound is fixed and is given by its chemical formula (Figure S1.7). The physical and chemical properties of a compound are always different from those of the elements that make it up.

■ **Figure S1.7** A model showing the structure of the compound calcium carbonate, $CaCO_3$ (black spheres represent carbon; red, oxygen; and white, calcium)

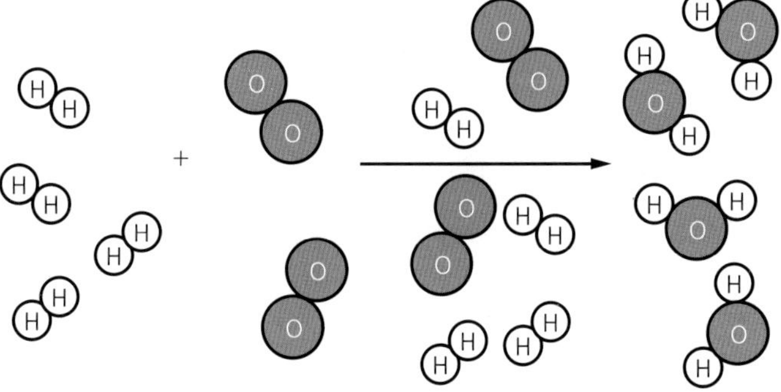
Hydrogen: a pure element
Oxygen: a pure element
Hydrogen and oxygen mixed together
Water: a pure compound formed from hydrogen burning in oxygen
H
O
+

Water (H_2O) is a compound. It is made up of two atoms of the element hydrogen and one atom of the element oxygen. Only two hydrogen atoms and one oxygen atom chemically bonded together can form a water molecule.

The ratio of hydrogen atoms to oxygen atoms in water is always 2 : 1 (Figure S1.8). A different ratio of these two elements will give a different compound, which will have different chemical properties. For example, adding one more oxygen atom gives a ratio of 2 : 2 and the compound hydrogen peroxide (H_2O_2) which has properties very different from those of water.

■ **Figure S1.8** The ratio of hydrogen to oxygen atoms in water molecules is always 2 : 1

Nature of science: Models

Scientific models

A scientific model represents or imitates many of the features of entities (for example, atoms), phenomena (for example, changes of state) and chemical reactions. Models can be communicated by mathematical formulas, graphs, equations and drawings: for example, atoms can be modelled as hard indivisible spheres that can connect (bond) to other atoms. All representations of a model have limitations: for example, atoms are not hard spheres and have their own internal structure. Scientists change and refine their models as new experimental data becomes available.

TOK

How do models limit scientific knowledge?

Models are not reality – so the decision to model our understanding of the world in one way as opposed to another will have an impact on our knowledge about the world. As with maps of the world, different depictions emphasize different aspects, leading our understanding in certain directions, downplaying or simply not conveying other facts.

◆ **Ion**: An electrically charged atom or group of atoms containing different numbers of protons and neutrons; formed by the loss or gain of electrons from an atom or group of bonded atoms.

A compound may be made up of atoms, molecules or **ions**. Ions are electrically charged particles formed from atoms. An ion has either a positive or a negative charge. Water is made of molecules but sodium chloride is made of ions (Figure S1.9).

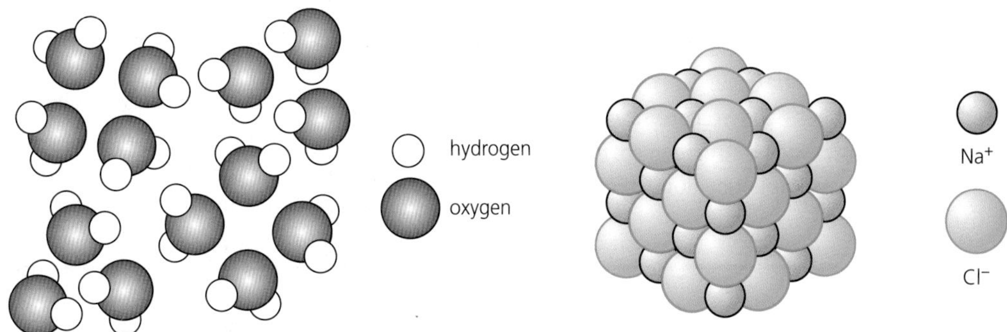
hydrogen
oxygen
Na⁺
Cl⁻
Water is a compound made up of water molecules
Sodium chloride is a compound made up of positive sodium ions and negative chloride ions (in a 1 : 1 ratio)

■ **Figure S1.9** The particles present in water and sodium chloride

◆ **Chemical change**: A process in which one or more substances are converted into new substances.

The formation of a compound is a **chemical change** (or chemical reaction) and the chemical and physical properties of the compound are different from those of the elements from which it is formed.

For example, iron sulfide (FeS) can be formed by heating iron with sulfur (Figure S1.10). This is a chemical change and can be described by the chemical equation with state symbols:

$Fe(s) + S(l) \rightarrow FeS(s)$

■ **Figure S1.10** Elemental iron and elemental sulfur; the compound iron(II) sulfide and a mixture of elemental iron and elemental sulfur (from left to right)

Chemical reactions always involve the formation of new chemical substances. Sodium ions and chloride ions are formed from sodium atoms and chlorine molecules, and heat and light energy are released during this chemical change.

◆ **Dissolving**: The interaction between a solute and water molecules to form a solution.

◆ **Physical change**: A reversible change in the physical properties of a substance.

If sodium chloride is mixed with water, an aqueous solution is formed containing sodium and chloride ions surrounded by water molecules. The formation of the solution involves a process known as **dissolving**. The dissolving of sodium chloride (and many other substances, for example, sucrose) is a physical process (**physical change**) and no new substances are formed. Changes of state such as boiling and freezing are also physical changes. Physical changes are reversible and heat energy can be released or absorbed in a physical change.

Going further

Energy

Physical changes and chemical reactions usually involve energy changes. Energy appears in different but related forms.

Consider a ball rolling across a table and colliding with a second, stationary, ball. The rolling ball has kinetic energy, due to its motion. When it collides with the other ball it does work, resulting in the transfer of kinetic energy from one ball to the other, which then begins to roll across the table. A similar process happens at the molecular level in a sample of gaseous molecules, which exchange kinetic energy when they collide.

If a ball is raised above the surface of the table, it gains gravitational potential energy. If you compress a spring or stretch a rubber band you are storing elastic potential energy. Compressed or stretched springs can be made to do work (and work must be done to compress, stretch or lift them).

The energy contained in a beaker of hot water is **internal energy**, which includes kinetic energy (present in the constant random motion of the water molecules) as well as the energy stored in the bonds and that of all the atoms' electrons and the protons and neutrons in their nuclei. Heat is the transfer of thermal energy (molecular kinetic energy) from places or objects at a high temperature to those at a lower temperature. Heat can be made to do work through the action of machines such as steam engines, and work can be converted to heat through friction.

Chemical energy is a form of potential energy associated with chemical bonds, including ionic bonds, and (weaker) intermolecular forces. During a chemical reaction, the chemical bonds and/or intermolecular forces change; the chemical energy may be converted into thermal energy (or vice versa). Chemical energy can be made to do work when, for example, a gaseous mixture of petrol and oxygen explodes and its expansion moves the piston down in an engine.

Another form of energy is electrical energy, associated with the energy of moving electrons. An electrical cell (battery) does work and transfers energy since the potential difference (voltage) between its terminals (which results from a chemical reaction) can 'push' electrons around a circuit. If that circuit is a simple resistance (for example, a filament in a light bulb) then the chemical energy change in the battery is converted into thermal energy (heat). An electric motor can convert at least some of the chemical energy in the battery into work.

Another important form of energy is electromagnetic radiation. When you lie in the sun on a warm day, electromagnetic radiation from the sun is converted into thermal energy in your skin. Photosynthesis in plants is an example of the conversion of electromagnetic energy into chemical energy, and a photocell converts electromagnetic energy into electrical energy.

The law of conservation of energy states that energy cannot be created or destroyed, but can only be transferred from one form to another and/or transferred from one object or particle to another. This law tracks energy changes, implying that if energy appears to be lost somewhere, it must have appeared somewhere else in another form.

◆ **Internal energy**: The total energy a system has; it is the sum of the total kinetic and potential energy of all the molecules in the system.

◆ **Solvation**: The interactions between solute particles and solvent molecules in a solution.

◆ **Solute**: The substance that dissolves in a liquid solvent.

◆ **Solvent**: The liquid in which a solute dissolves.

◆ **Solution**: A solute dissolved in a solvent.

◆ **Soluble**: A substance that dissolves in a solvent.

Solvation

When a compound dissolves, the individual ions or molecules interact with the solvent molecules. This is a physical process called **solvation** and the ions or molecules are said to be solvated. Hydration is a specific type of solvation with water as the solvent.

For example, sodium chloride (**solute**) dissolves in water (**solvent**) to form a **solution** of hydrated aqueous ions. Sodium chloride is said to be **soluble** in water. When no more sodium chloride will dissolve in the water, the solution is said to be saturated.

The interaction between positive and negative ions and water molecules is due to ion–dipole forces. They involve electrostatic attraction between ions and the charged ends of the water molecules (Figure S1.11). This means that ionic substances are more soluble in water than other solvents.

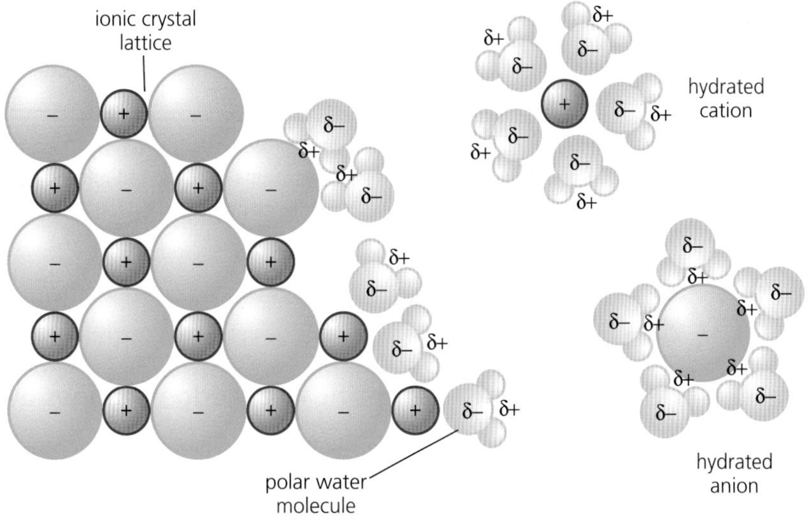

■ **Figure S1.11** An aqueous solution of sodium chloride showing hydrated ions

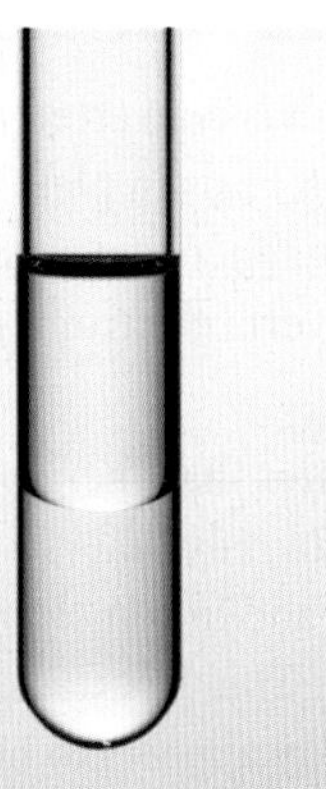

■ **Figure S1.13** Cyclohexane (lower density and chemically similar to oil) floating on top of water (higher density)

Going further

Solvation

The solvent molecules around an ion are known as the solvation shell; if water is the solvent, they are known as the hydration sphere. For ions, the greater the charge density, the larger the hydration sphere. Small, highly charged ions like Fe^{3+} or Al^{3+} typically have more water molecules associated with them than large, low charged ions such as K^{+}. Small and highly charged ions are described as having a high charge density: the large charge is located over a small volume.

Typically, polar molecules interact strongly with polar solvents through dipole–dipole interactions and/or hydrogen bonds. This means that polar molecules tend to dissolve in polar solvents. Methanol, ethanol (Figure S1.12) and water all contain –O–H groups that can form hydrogen bonds with ionic substances and polar molecules.

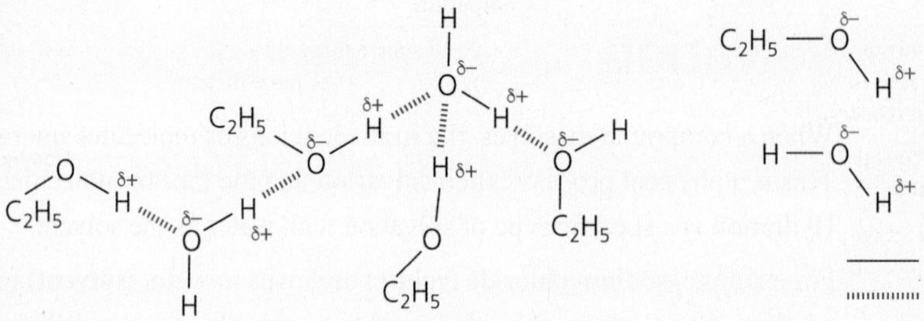

C_2H_5—O ethanol molecule
H—O water molecule
—— covalent bond
|||||||| hydrogen bond

■ **Figure S1.12** Interactions in a solution of ethanol in water

◆ **Insoluble**: A substance that does not dissolve in a solvent.

◆ **Miscible**: Liquids that mix together.

◆ **Immiscible**: Liquids that do not mix but separate into layers.

If two liquids are **miscible** (such as water and ethanol) they mix to form a homogeneous mixture: a single liquid of uniform composition. If two liquids do not mix (for example, water and oil) they are described as **immiscible** and two layers form with the less dense liquid floating on top (Figure S1.13). The resulting mixture is described as heterogeneous. Some pairs of liquids exhibit partial miscibility.

Top tip!

The term **insoluble** does not mean that none of the substance will dissolve; it only means that the solubility of that compound is so low that it can be ignored for most practical purposes. Dissolving and crystallization are reversible reactions and, at equilibrium, the rate of dissolving equals the rate of crystallization leading to no obvious macroscopic change.

ATL S1.1D

Go to https://lab.concord.org/embeddable.html#interactives/interactions/dissolving-experimental.json

Run the simulation with different combinations of polar and non-polar molecules to explore how the polarity of molecules affects how they mix or do not mix. Does this model illustrate the 'like dissolves like' principle?

TOK

How can it be that scientific knowledge changes over time?

Chemistry is not a body of unchanging facts, but a process of generating new chemical knowledge, theories and laws, using the scientific method. You might therefore consider 'chemistry' as what people who are interested in this type of knowledge *do*. Being a chemist, in other words, is being able to do the sorts of things that people in this community of knowers do. Robert Boyle (1627–1691) was perhaps one of the most important early chemists and regarded as the father of modern chemistry. He believed in the importance of elements of what we call the scientific method: experimental science, claiming that theory should conform to observation, and the publication of experimental results. Through experiments involving pure chemicals, he falsified the ancient Greek theory of four elements and proposed a concept of elements close to the one we have today. He believed that all matter is composed of tiny particles and the universe works like a complex machine.

LINKING QUESTION

How do intermolecular forces influence the type of mixture that forms between two substances?

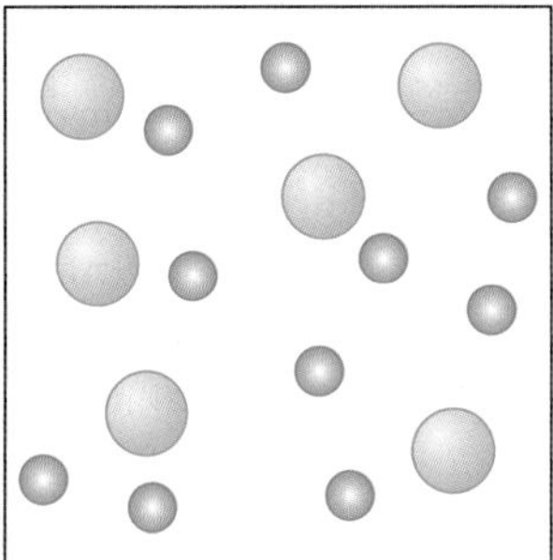

a mixture of two elements existing as atoms

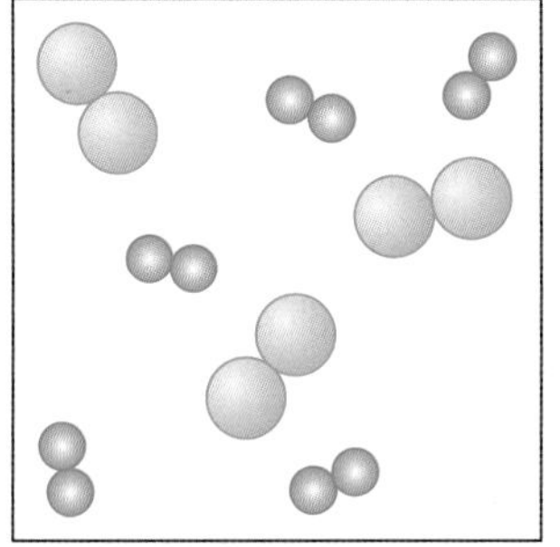

b mixture of two elements existing as molecules

■ **Figure S1.14** Particle representation of a mixture of atoms and a mixture of molecules

Mixtures

Mixtures (Figure S1.14) consist of more than one compound or element, mixed but not chemically combined by chemical bonds. The components can be mixed in any proportion and the properties of a mixture are often the sum of, or the average of, the properties of the individual components.

The major differences between elements, compounds and mixtures are summarized in Table S1.1.

■ **Table S1.1** The major differences between elements, compounds and mixtures

	Element	**Compound**	**Mixture**
Definition	A pure substance that cannot be decomposed into simpler substances by a chemical reaction.	A pure substance formed by the combination of two or more different elements in a chemical reaction.	A combination of two or more substances that are mixed but not chemically bonded.
Separation	Cannot be separated (from itself) by chemical or physical processes.	Cannot be separated into its component elements by physical means.	Can be separated by a physical method.
Composition	Made up of only one type of atom, which may be present as molecules.	Made up of different types of atoms, chemically combined and present as molecules or ions. Proportion of elements (by mass and number) is fixed.	Proportion of constituent substances (components) may vary. The components may be elements and/or compounds.
Chemical properties	The properties are determined by its location in the periodic table, especially the group (column).	The chemical properties of the compound are typically very different from those of the elements from which it is made.	Mixture shows all the chemical properties of its constituent substances.
Melting and boiling points	Fixed and sharp.	Fixed and sharp.	Usually melts and boils over a range of temperatures.

◆ **Mixture:** A mixture has the properties of its components, and can be separated by physical processes.

1 State whether each diagram shows an element, a compound or a mixture.

a 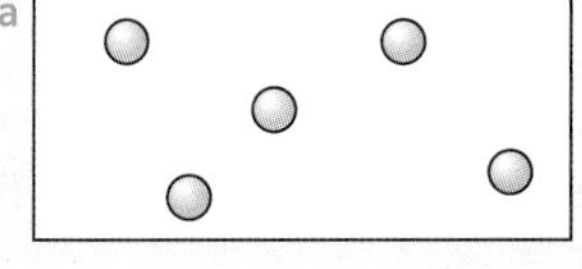

b 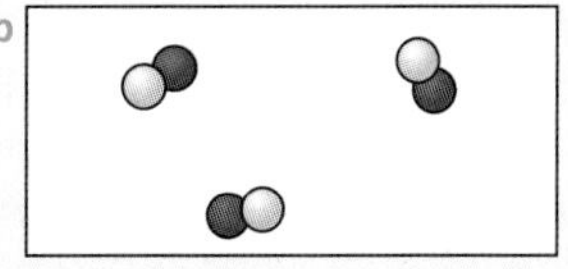

c 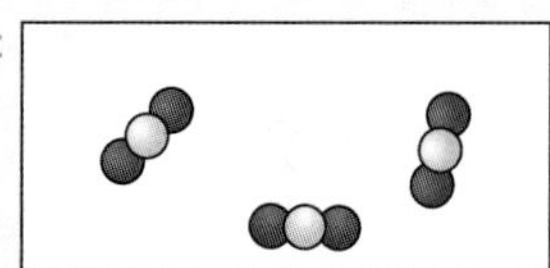

d 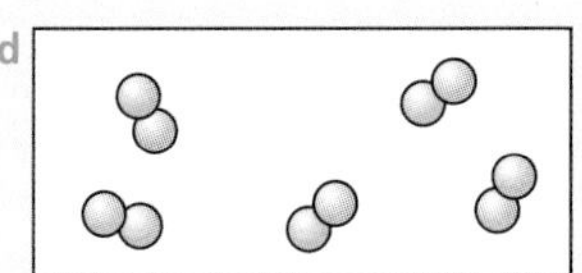

e 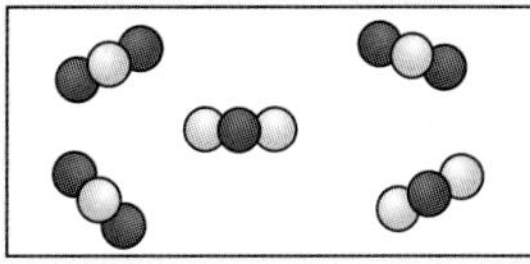

f 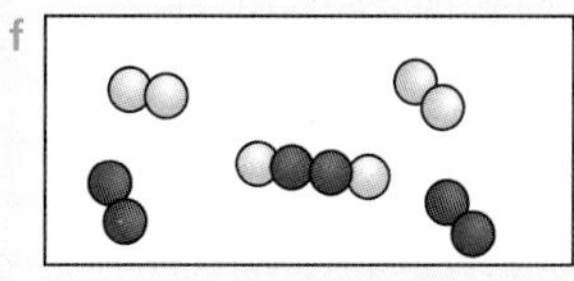

ATL S1.1E

Watch the Royal Society of Chemistry video at www.youtube.com/watch?v=Y8WWWvtakgI that shows the formation of iron(II) sulfide from its elements (starting at 07:20). This type of reaction is known as a synthesis reaction and is also an example of a redox reaction.

- Record your observations in the data table.
- Explain how you know that heating the iron and sulfur together in the test tube results in a new compound composed of chemically combined iron and sulfur.
- Interpret your results in terms of particle theory assuming iron and sulfur exist as atoms (from the decomposition of S_8 molecules) and iron sulfide has a 1 : 1 ratio of ions.

Physical properties	**sulfur**	**iron**	**iron–sulfur before heating**	**iron–sulfur after heating**
Colour				
Effect of magnet				

◆ **Heterogeneous**: A mixture that does not have a uniform composition and properties.

◆ **Homogeneous**: A mixture with a uniform composition and properties.

◆ **Solubility**: The amount of solute needed to form a saturated solution.

◆ **Separating funnel**: A glass vessel with a tap at the bottom, used to separate immiscible liquids.

Types of mixtures

Mixtures can be further classified into **heterogeneous** and **homogeneous** mixtures. Homogeneous mixtures, such as air and solutions, have a uniform distribution of the different particles. Solids can also form homogeneous mixtures such as copper–nickel alloys.

A heterogeneous mixture is a mixture in which the composition is not uniform throughout the mixture.

- A suspension is a heterogeneous mixture of a solid and a liquid: for example, fine mud in water.
- An emulsion (a type of colloid) is a heterogeneous mixture of different liquids which cannot be mixed homogeneously: for example, milk, which is a stable mixture of small drops of oil in water (stabilized by protein).
- Aerosols are heterogeneous mixtures of solids or liquids in gases: for example, smoke or fog.
- Sols are heterogeneous mixtures of two or more solids or liquids. A sulfur sol is formed when sodium thiosulfate reacts with dilute hydrochloric acid. It consists of tiny particles of sulfur molecules that scatter light.

LINKING QUESTION

Why are alloys generally considered to be mixtures, even though they often contain metallic bonding?

ATL S1.1F

Work collaboratively with a partner on the following activity.

Classify the mixtures in the list below as homogeneous or heterogeneous and identify the major components present in sea water, blood, air and solder. Use the internet to identify the appearance of any substances you are not familiar with, such as magnesium, iodine and solder.

- vinegar (ethanoic acid in water)
- a mixture of helium and hydrogen gases
- sea water
- electrical solder (a low-melting-point alloy)
- cooking oil and water (left to reach equilibrium)
- magnesium powder and iodine crystals
- blood
- air

Present your findings to the class and show how the concept of phase (research this on the internet) is important in homogeneous or heterogeneous mixtures.

Top tip!

The production of metallic alloys is possible because of the non-directional nature of metallic bonding (all cations are attracted to the 'sea' of delocalized electrons) and because the lattice of the alloy can accommodate cations of different sizes. This means alloys can have variable proportions of metals and retain the chemical properties of the components.

Separation techniques

Chemists have developed many different methods of separation (Table S1.2), especially for separating components from complex mixtures.

■ **Table S1.2** Common separation techniques

Type of mixture	Name of separation technique	Physical difference	Examples of mixtures separated
insoluble solid and liquid	filtration	**solubility**	sand and water; calcium carbonate (chalk) and water
two miscible liquids	distillation (simple and fractional)	boiling point	vinegar (ethanoic acid and water) and crude oil (a mixture of hydrocarbon liquids)
soluble solid and liquid (solution)	crystallization or evaporation	volatility	sodium chloride (salt) and water
soluble solids	paper chromatography of solution and recrystallization	solubility (partitioning)	food colourings; plant pigments
immiscible liquids	**separating funnel**	insolubility (polarity)	water and petrol or water and oil

2 State the technique(s) that could be used to separate the following mixtures:

a ethanol from water

b insoluble magnesium carbonate from a suspension in water

c cyclohexane (an insoluble hydrocarbon) from water

d food colourings in a sweet (to determine if the colouring is the result of a single dye or a mixture of dyes)

e water from potassium chloride solution

f copper(II) sulfate crystals from its aqueous solution.

Tool 1: Experimental techniques

What factors are considered in choosing a method to separate the components of a mixture?

The choice of separation technique (Figure S1.15) depends on what is in the mixture and the physical and chemical properties of the substances present. It also depends on whether the substances to be separated are solids, liquids or gases.

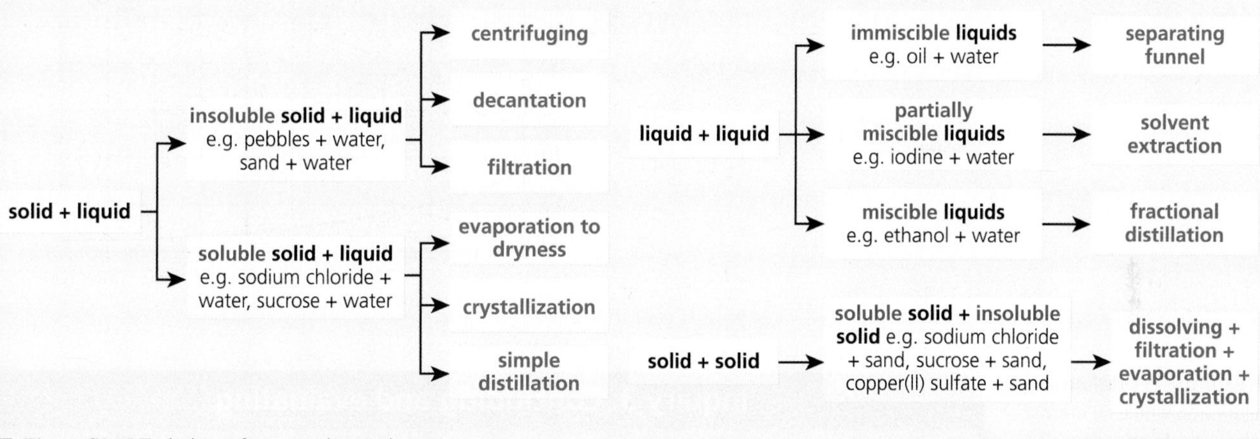

■ **Figure S1.15** Techniques for separating a mixture

WORKED EXAMPLE S1.1

Some skin ointments contain a white solid, camphor (a hydrocarbon). Before incorporation into the ointment, a natural sample of camphor was found to be contaminated with table salt (sodium chloride) and rust (hydrated iron(III) oxide). The effect of three liquids on these substances is shown in the table below.

	Liquid		
	water	***ethanol***	***dilute hydrochloric acid***
table salt	dissolves to produce a colourless solution	no effect	dissolves to produce a colourless solution
camphor	no effect	dissolves to produce a colourless solution	no effect
rust	no effect	no effect	reacts to give a brown solution

Describe how you would obtain pure and dry camphor from the impure sample of camphor (without using sublimation).

Answer

Add ethanol, stir to dissolve. Filter; the **filtrate** will be camphor solution. Heat gently using a water bath to evaporate off all ethanol and obtain dry camphor. Alternatively, add dilute hydrochloric acid to dissolve both salt and rust (via a chemical reaction), then filter to obtain solid camphor as a **residue**. Dry the camphor in the air or in a desiccator or oven.

◆ **Filtrate**: The solution that passes through the filter paper during filtration.

◆ **Residue**: The solid left on the filter paper during filtration.

ATL S1.1G

The solubility of a solute is often expressed as the mass of solute that will dissolve in 100 g (or 100 cm³) of water at a particular temperature; for example, 5 g in 100 cm³ of water. The solubility over a range of temperatures can be shown as a solubility curve like the one here.

- Deduce which compound is most soluble at 20 °C, which is least soluble at 40 °C and which is least soluble at 10 °C.
- A mass of 80 g of KNO_3 is dissolved in 100 g of water at 50 °C. The solution is heated to 70 °C. Determine how many more grams of potassium nitrate must be added to make the solution **saturated**.
- Find out how the solubility of gases that do not react with water varies with temperature. Outline your findings.

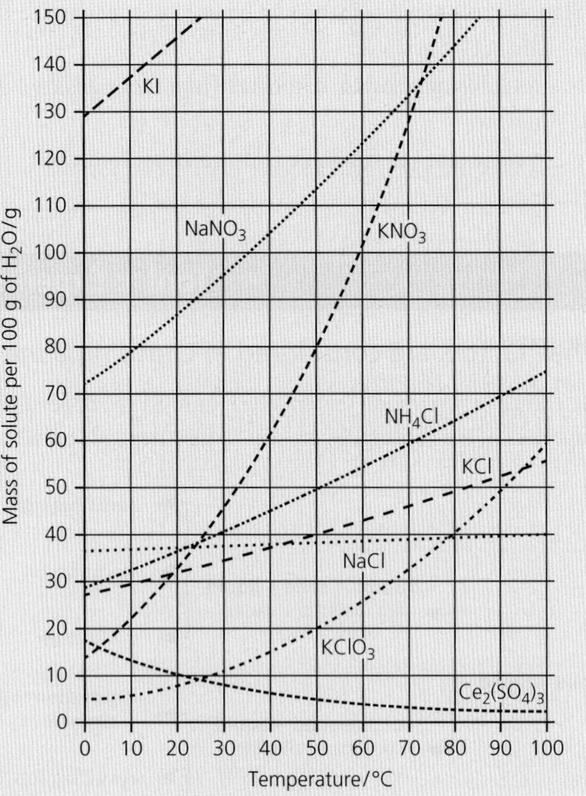

◆ **Saturated (solution)**: A solution in which no more solute can dissolve at a particular temperature.

Inquiry 3: Concluding and evaluating

Compare the outcomes of your own investigation to the accepted scientific context

The results of your investigation (practical and/or simulation) should be explained using accurate and relevant chemical concepts, principles and facts.

You should compare the results of your investigation with what would be expected with reference to published data or chemical concepts. Compare the conclusions with published research or with the general scientific consensus among chemists about the research question. Do your conclusions conform to the consensus or are they unexpected?

It is not necessary to find an identical investigation with the exact same results: it is possible to compare findings with another investigation that is different but with results that either support or falsify those of your investigation.

Tool 1: Experimental techniques

How can the products of a reaction be purified?

Many organic reactions are reversible (the backward and forward reactions occur together) and do not go to completion, so the reaction mixture will contain the reactants, the products and the solvent. A good synthesis reaction will give a yield of at least 70–80% product but small amounts of other organic materials may be formed as by-products.

- If the desired product is a solid present in large amounts, it may be crystallized by slow cooling and then filtering from the solution and further purified by recrystallization (see page XXX).
- If the desired product is a liquid, it may be removed by distillation. However, some organic liquids of high molar mass will decompose before reaching the boiling point. Vacuum distillation involves carrying out the distillation at reduced pressure; this lowers the boiling points of liquids making the process faster and reducing the risk of decomposition.
- If the product is a soluble solid, or present in a small amount in a mixture, it may be possible to remove and purify it by column chromatography (Figure S1.16).
- If the products are immiscible liquids, they can be separated using a separating funnel, or preferentially extracted into a specific liquid.

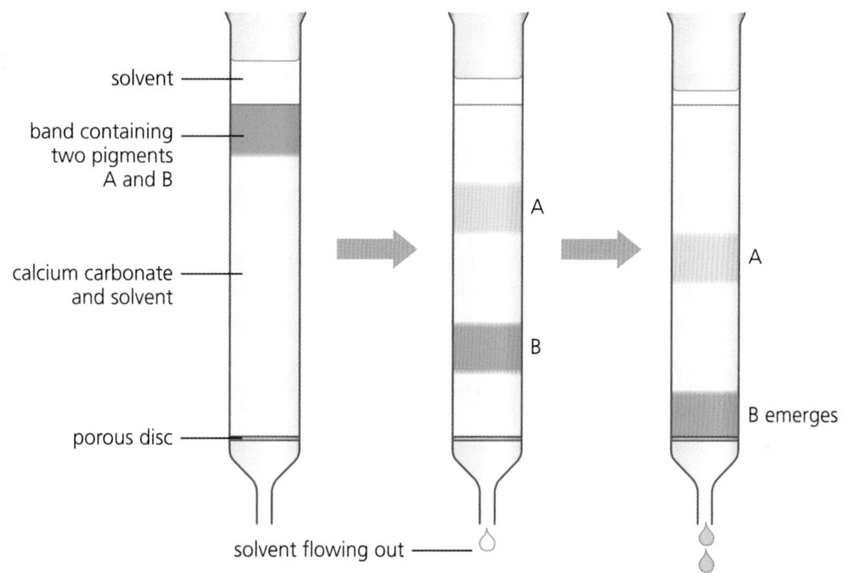

■ **Figure S1.16** Separation of the components in plant pigments during column chromatography

Going further

Solvent extraction

If a solute is added to two immiscible liquids and the solute is soluble in both, then some of the solute will dissolve in both liquids. It will distribute itself in a definite ratio in a process known as partitioning.

For example, when a small quantity of iodine (non-polar) is shaken with a mixture of water (polar) and hexane (non-polar), some iodine will dissolve in both liquids, but very little in the water. The mixture is allowed to separate into two layers and left until equilibrium is established (Figure S1.17). At equilibrium, the rate at which iodine dissolves in the water is balanced by the rate at which iodine dissolves in the hexane. $I_2(aq) \leftrightharpoons I_2(hexane)$. The partition coefficient is the ratio of concentrations of iodine in the two phases.

Solvent extraction is a practical example of using partitioning and is the process used to purify a compound by using its different solubility between two solvents.

Solvents are used to selectively separate out the components of a mixture. The two components of the mixture must have different values of solubility in the solvents A and B. The two solvents will form two separate layers. The lower layer is run off and separated from the top layer and the extraction process is repeated. When all the top layers, for example, are combined and the solvent evaporated, nearly pure solute is obtained. The quantity of solute obtained depends upon the position of equilibrium and how many times the process is repeated. A larger number of smaller extractions will yield more solute than one large extraction.

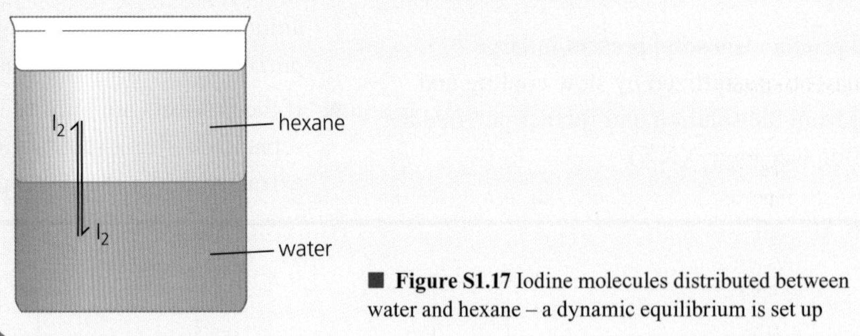

■ **Figure S1.17** Iodine molecules distributed between water and hexane – a dynamic equilibrium is set up

Kinetic molecular theory

States of matter

◆ **States of matter**: Solids, liquids and gases are the three states of matter, depending on the conditions of temperature and pressure.

The **states of matter**, solid, liquid and gas, can be described in terms of their physical properties. Each state of matter has its own properties and these are shown in Table S1.3.

■ **Table S1.3** The properties of the three states of matter

Property of matter	Solid	Liquid	Gas
shape	fixed	not fixed (takes the shape of the container it occupies)	not fixed (takes the shape of the container it occupies)
volume	fixed	fixed	not fixed (occupies the volume of its container)
ability to flow or spread	does not flow	flows easily	diffuses easily (from high concentration to low concentration)
compressibility	limited compressibility	limited compressibility	highly compressible
density	much higher than gases and usually slightly higher than the liquids	usually slightly greater than solids, but much less than liquids	low

Top tip!

A fluid can flow easily and takes the shape of its container. Gases and liquids are fluids.

Chemistry in zero gravity

The International Space Station (ISS) is the largest space project so far undertaken and is the largest structure ever to orbit the Earth. It is a research facility 400 km above the Earth and has been in operation since 2000.

It has permanent accommodation for six people and short-term accommodation for up to 15 people when a space vehicle visits. On the ISS, components of which were built in 15 countries, scientists conduct research and experiments that are impossible to conduct on Earth—for example, precipitation reactions involving metal silicates—in a zero-gravity environment.

ATL S1.1H

Find a video on the internet that shows the behaviour of water on a space station where there is 'zero gravity' (for example, www.youtube.com/watch?v=5GqsE09uUuw).

- Compare and contrast the properties of the water with those it exhibits on Earth in the presence of a gravitational field. Research this phenomenon that occurs at the surface of a liquid.
- Oil and water mix in zero gravity. Find out how they can be separated in these conditions.

The differences between the physical properties of the three states of matter can be explained by kinetic molecular theory as shown in Table S1.4. This model describes all substances as being made up of moving particles and explains how the arrangement of particles (atoms, ions or molecules) relates to the physical properties of solids, liquids and gases.

Table S1.4 Kinetic molecular theory

State of matter	Solid	Liquid	Gas
Distances between particles (interparticle distances)	The particles are often arranged as a crystal with a regular repeating arrangement (lattice). Solids cannot be compressed very much because the particles are closely packed and are touching. They will repel if brought closer.	The particles are not arranged in a lattice and are slightly further apart than in solids. Liquids cannot be easily compressed as the particles are close together and there is little space between them.	The particles are far apart. A gas can be easily compressed because there are large distances between particles which are present at low concentration.
Forces acting between particles	There are balanced forces (of attraction and repulsion) between the particles which hold them in fixed positions. The particles can only vibrate about their fixed positions within the lattice. This explains why a solid has a fixed shape and a fixed volume. The strong attractive forces prevent the particles from leaving their positions. The repulsive forces act when the particles are brought closer than their equilibrium positions and this limits compressibility. When a solid is heated, the particles gain kinetic energy and vibrate. The separation between particles increases slightly and the solid expands.	The particles can vibrate, rotate and move freely within the body of the liquid. There are attractive forces operating between the particles but they are not held in a fixed position and move past each other throughout the liquid. The attractive forces operating between the particles make it difficult for particles to leave (except at the surface by evaporation), hence liquids have a definite volume. When a liquid is heated the particles vibrate more and move more quickly. The liquid expands slightly and the amount in the gas above it increases.	The particles move randomly at high speed, colliding with other gas particles and with the walls of the container. The intermolecular forces act only at moments of collision. Between collisions the particles are sufficiently far apart that the intermolecular forces are negligible. Therefore, a gas is free to fill its container completely.

Tool 2: Technology

Use computer modelling

There is a wide variety of software designed to simulate chemical processes and illustrate key chemical concepts. Some of these programs are interactive: it is possible for the user to change the value of variables and observe the effects on the simulated system.

Computer models can be deterministic or stochastic. Deterministic models are calculated with fixed probabilities—they always change in the same way from chosen values of variables. Stochastic models use a random number generator to create a model with variable outcomes—for diffusion, for example. Every time you run the model you are likely to get different results, even with the same values of variables.

ATL S1.1I

Visit the simulation on gas laws at https://phet.colorado.edu/sims/html/states-of-matter-basics/latest/states-of-matter-basics_en.html Go to **States of Matter**: **Basics** > **States**.

- Observe water as a solid (ice), liquid and gas (water vapour), then complete the following table.

State	Temperature in °C	Particle (molecule) speed (slow, medium, fast)	How the particles (molecules) move
Solid			
Liquid			
Gas			

- State and explain how particle (molecule) motion changes when you add heat.
- State and explain how particle (molecule) motion changes when you remove heat.
- Outline one limitation of this model.

Now choose **Phase Changes** > **Oxygen**. Set the temperature to above 90 K (the boiling point of oxygen). Use your finger to press down on the chamber slowly. State what happens to the:

- pressure
- temperature
- average velocity of the gas particles.

Reset the simulation. Add more gas particles using the bike pump.

- State what happens to the temperature when you add more gas particles.

◆ **Changes of state**: The interconversion (via physical changes) of a substance between the solid, liquid and gaseous states.

◆ **Melting**: The change of a solid into a liquid at constant temperature.

◆ **Boiling**: The change of a liquid into a gas at constant temperature (at a pressure equal to the surrounding pressure).

◆ **Condensing**: The change of a gas into a liquid (at constant temperature).

◆ **Freezing**: The change of a liquid into a solid (at constant temperature).

◆ **Sublimation**: A change of state from solid to gas (at constant temperature) without melting occurring.

◆ **Deposition**: The formation of a solid on a surface from a gas (at constant temperature).

Figure S1.18 shows the relationship between the states of matter and the arrangement (simplified and in two dimensions only) of their particles (ions, atoms or molecules). The arrows represent physical changes known as **changes of state**.

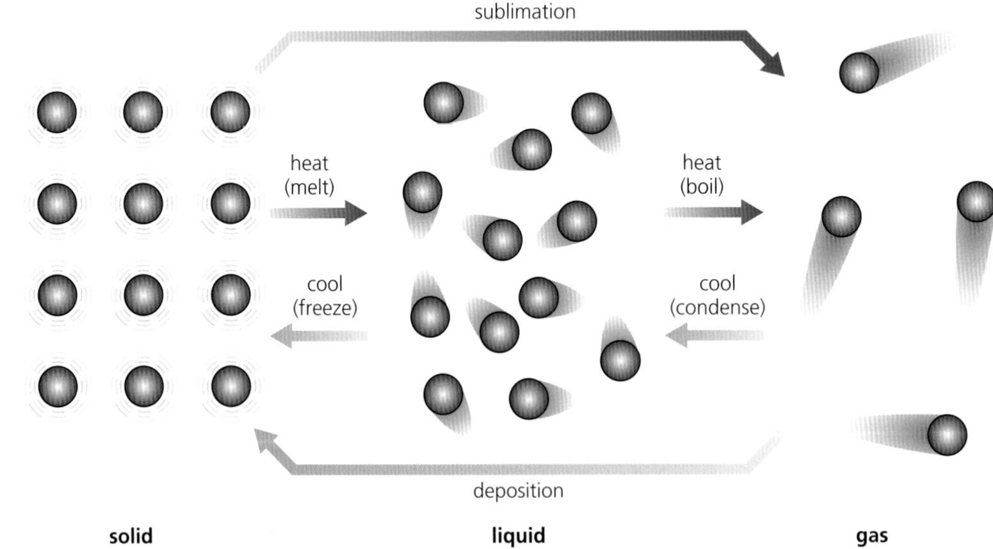

■ **Figure S1.18** The three states of matter and their interconversion

Top tip!

Nearly all solids are crystalline and have a regular pattern of particles known as a lattice.

Changes of state

During **melting** and **boiling**, heat is absorbed from the surroundings. During **condensing** and **freezing**, heat is released to the surroundings. The heat supplied during melting and boiling is used to 'break' the attractive forces between particles by increasing their potential energy. The heat released during condensing and freezing is from the decrease in the potential energy of the particles.

When heated, some pure solids, such as $CO_2(s)$ and $I_2(s)$, can undergo **sublimation** and change directly to a gas without passing through the liquid state. The molecules leave the solid's surface with enough kinetic energy to exist as gas particles. If the temperature is lowered, the gas particles slow down and re-form the solid without passing through the liquid state in a process known as **deposition**.

◆ **Evaporation**: When the particles from the surface of a liquid form a gas below its boiling point.

Some particles at the surface of a liquid have enough kinetic energy to overcome the forces of attraction between themselves and they escape to form a gas. This process is **evaporation** and it takes place at all temperatures below the boiling point of the liquid. If the temperature is lowered, the reverse process, known as condensation, occurs. The gas particles move more slowly and enter the surface of the liquid.

At the boiling point, the particles are trying to escape from the liquid so quickly that bubbles of gas form inside the bulk of the liquid and the pressure of the gas created above the liquid equals that of the air (atmospheric pressure). Liquids with higher boiling points have stronger bonds (intermolecular forces) between their particles.

Top tip!

The heat released or absorbed during a change in state is known as the latent heat.

When a gas is cooled, the average kinetic energy (and hence speed) of the particles decreases, the particles move closer and their average separation decreases. The forces of attraction become significant and, if the temperature is lowered to the condensation point, the gas will condense to form a liquid. When a liquid is cooled to its freezing point (equal in value to the melting point), it changes to a solid. During condensing and freezing, heat energy is released.

Tool 3: Mathematics

Interpret features of graphs including gradient, changes in gradient, intercepts, maxima and minima, and areas

Many investigations in science are concerned with finding relationships between continuous variables. After collecting a set of data, the data points for two variables can be plotted on a graph, and then a line drawn that best expresses the apparent relationship suggested by the data. This is called a line or curve of best fit.

For a linear relationship, the gradient at any point along the line is the same, but on a curve, the gradient varies at different points along the curve. On a graph that shows a change in a variable over time, the steepness of the line represents how fast the change is happening. In other words, the gradient of the line represents a rate of change. Since this is the rate of change at a particular instant in time, it is called an instantaneous rate of change.

The x-intercept is the point where a line or curve crosses the x-axis, and the y-intercept is the point where a line or curve crosses the y-axis. It may have some chemical significance depending on the graph. An intercept may also be determined by extrapolation.

Some lines or curves may appear to increase indefinitely. Graphs such as those showing the concentration or amount of a product in a reaction over time may level out as they reach a particular value whereas others, including those showing how reactant concentrations vary, may tend to a steady minimum value. Other curves, such as those showing how chemical potential energy between particles changes with separation, may have a turning point (maximum or minimum) that can be interpreted in a way that aids understanding of a process or interaction.

In some cases the area under the graph may have some chemical significance. For example, the area under the Maxwell–Boltzmann distribution is related to the total number of molecules in the gas being described.

ATL S1.1J

When a soluble solid, such as sodium chloride or citric acid, is added to water, its freezing point is lowered below 0 °C.

Plan an investigation on the effect of a range of salts (for example, NaCl, KCl and $CaCl_2$) of different concentrations on the depression of distilled water's freezing point.

Carry out a risk assessment and perform the investigation following approval from your teacher.

Present your findings alongside research into the concept of colligative properties.

Top tip!

A substance is a solid at temperatures below its melting point, a liquid at temperatures between its melting and boiling points and a gas at temperatures above its boiling point.

Common mistake

Misconception: *Since changes of state, such as freezing and sublimation, occur at constant temperature, they do not involve energy changes.*

Scientific concept: Remember that physical processes and chemical changes both involve energy changes and these can involve both kinetic and potential energy.

Top tip!

Changes of state are physical changes: no new chemical substances are formed. Ice, water and steam all contain molecules with the formula H_2O. Whenever melting, freezing, boiling or condensing occurs, the temperature remains constant during the change in state.

ATL S1.1K

A heating curve shows the changes of state occurring when the temperature of a solid is gradually increased until it melts and boils. Similarly, a cooling curve shows the changes of state when a gas or liquid is cooled gradually until it forms a solid.

Heating

Time / s ± 1 s	Temperature / °C ± 0.1 °C
0	30.0
30	33.2
60	35.5
90	37.5
120	39.0
150	41.0
180	42.0
210	42.6
240	43.0
270	43.4
300	43.6
330	43.7
360	43.8
390	44.0
420	44.0
440	44.1
470	44.2
500	44.5
530	45.2
560	46.0
590	47.5
620	49.0
650	51.4

Cooling

Time / s ± 1 s	Temperature / °C ± 0.1 °C
0	55.0
30	52.6
60	49.5
90	45.5
120	44.3
150	44.1
180	44.0
210	44.0
240	44.0
270	44.0
300	44.0
330	44.0
360	44.0
390	44.0
420	44.0
440	43.7
470	43.5
500	43.3
530	43.0
560	42.6
590	42.3
620	41.9
650	41.5

- Use a spreadsheet to plot the data for the heating and cooling curves on a single graph of temperature versus time.
- Use your graph to determine and compare the melting and freezing points of the substance (dodecanoic acid, $C_{12}H_{24}O_2$).
- Use the kinetic molecular theory to explain what is happening at all diagonal and flat parts of the two curves.

LINKING QUESTION

Why are some changes of state endothermic and some exothermic?

LINKING QUESTION

Why are some substances solid while others are fluid under standard conditions?

◆ **Temperature**: Measure of the average kinetic energy of particles in a substance.

◆ **Absolute zero**: A temperature (−273.15 °C) which corresponds to the complete absence of heat, and hence no particle vibrations.

◆ **Absolute temperature**: A temperature expressed on the thermodynamic temperature scale (with units of kelvin).

◆ **Celsius scale**: Scale of temperature based on a one-hundred degree range between the melting point of pure ice and the boiling point of water (under standard conditions).

Top tip!

The distribution of particle kinetic energies in a substance is given by the Maxwell–Boltzmann distribution. This is used in the collision theory of reaction rates to explain the effects of temperature changes on rates of reaction.

Top tip!

A change in temperature has the same numerical value on both scales.

LINKING QUESTION

What is the graphical distribution of kinetic energy values of particles in a sample at a fixed temperature?

LINKING QUESTION

What must happen to particles for a chemical reaction to occur?

Temperature

Temperature is measured by a thermometer and is directly related to the random motion of particles.

At temperatures above **absolute zero** (0 K or −273.15 °C), all particles in matter have kinetic energy. Absolute zero is the lowest possible temperature and is the absence of particle vibration (internal energy). Whenever a substance's temperature increases, the average kinetic energy of its particles has increased.

The kelvin (K) is the unit of **absolute** or thermodynamic **temperature** (Figure S1.19). The kelvin is the SI unit of temperature and should always be used in calculations involving temperature, such as those involving gas laws or thermodynamics.

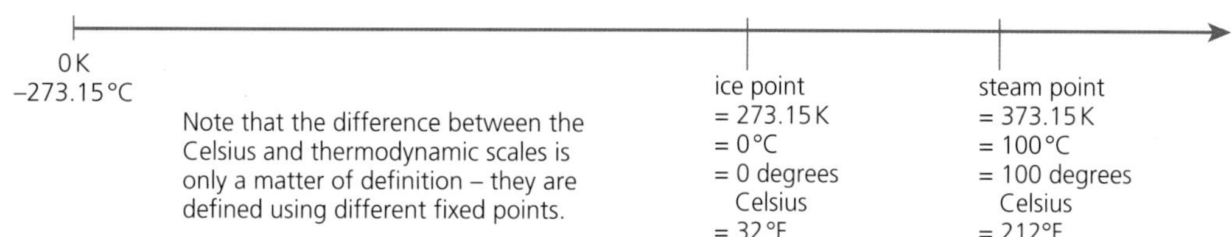

■ **Figure S1.19** The defining temperatures on the absolute and Celsius scales

However, you will usually measure temperature using a thermometer that measures in degrees Celsius with a scale based on two fixed points: the freezing and boiling points of water.

You can convert temperatures from the **Celsius scale** to the absolute scale by adding 273.15. Hence, the boiling point of water is approximately 100 °C + 273.15 = 373.15 K. Note that the kelvin unit does not have a degree symbol, °. Negative temperatures are possible on the Celsius scale but not on the Kelvin scale.

Figure S1.20 shows the difference between heat and temperature. The large and small beakers contain water at the same temperature. However, the masses of the water are different and when the same amount of heat is transferred to the water in the small beaker and in the large beaker, the temperature rises are different. The smaller mass of water has a larger increase in temperature.

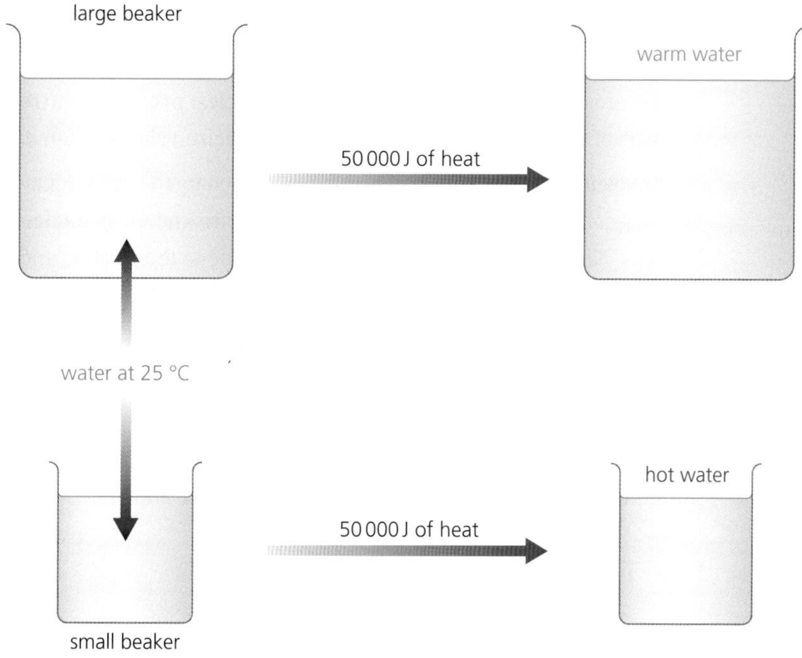

■ **Figure S1.20** The difference between heat and temperature

S1.2 The nuclear atom

Guiding question

- How do the nuclei of atoms differ?

SYLLABUS CONTENT

By the end of this chapter, you should understand that:

- atoms contain a positively charged, dense nucleus composed of protons and neutrons (nucleons)
- negatively charged electrons occupy the space outside the nucleus
- subatomic particles have different masses and charges
- isotopes are atoms of the same element with different numbers of neutrons
- mass spectra are used to determine the relative atomic masses of elements from their isotopic composition (HL only).

By the end of this chapter you should know how to:

- use the nuclear symbol $^{A}_{Z}X$ to deduce the number of protons, neutrons and electrons in atoms and ions
- perform calculations involving non-integer relative atomic masses and abundances of isotopes from given data
- interpret mass spectra in terms of identity and relative abundance of isotopes (HL only).

Atomic structure

◆ **Subatomic particles**: The particles which atoms are composed of.

◆ **Proton**: Positively charged particle found inside the nucleus of an atom.

◆ **Neutron**: Neutral particle found inside the nucleus of an atom.

◆ **Electron**: Negatively charged particle found in energy levels (shells) outside the nucleus inside atoms.

◆ **Nucleus**: Small, dense region at the centre of atoms, containing protons and neutrons (nucleons).

◆ **Nucleons**: Particles in a nucleus, either protons or neutrons.

◆ **Energy level (shell)**: The region an electron occupies outside the nucleus inside an atom.

Atoms are the smallest particles that can take part in a chemical reaction and are fundamental units of matter. Dalton's atomic theory viewed the atom as a solid ball but the discovery of the three fundamental **subatomic particles** completely changed the scientific view of atoms. New theories and models had to be developed to account for the fact that the atom is composed of smaller particles.

Atoms are composed of three subatomic particles: **protons**, **neutrons** and **electrons**. Each atom consists of two regions: the nucleus and the electron shells (main energy levels).

The **nucleus** is a very small, dense, positively charged region located at the centre of the atom. The nucleus contains protons and neutrons. These are known as **nucleons** since they are found in the nucleus. Nearly all of the mass of an atom is due to the protons and neutrons.

The electrons occupy the empty space around the nucleus and are arranged in **shells** (main **energy levels**). Each shell can hold a specific maximum number of electrons. Electrons in different shells have different amounts of energy. The first shell can hold a maximum of two electrons and the second shell can hold up to eight electrons.

This simple nuclear model of the atom is illustrated in Figure S1.21 for hydrogen, helium and lithium atoms with atomic (proton) numbers of 1, 2 and 3.

Top tip!

Atoms have to be electrically neutral and must contain equal numbers of protons and electrons.

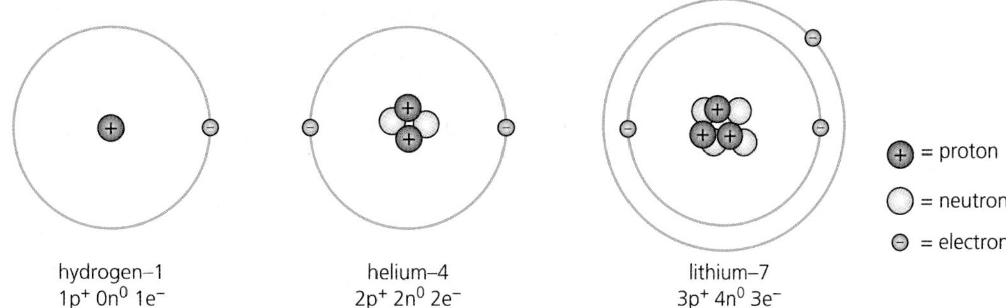

Figure S1.21 A simple model of hydrogen, helium and lithium atoms

Subatomic particles

◆ **Neutral**: A particle with zero charge or no overall charge.

◆ **Attraction**: A force pulling two (particles with) opposite charges together.

Protons have a positive charge; neutrons have no electrical charge and are electrically **neutral**. These particles have almost the same mass.

Electrons have a negative charge and a very small mass – almost negligible compared with the mass of protons or neutrons. Atoms are electrically neutral because they contain equal numbers of protons and electrons. The opposite charges of the proton and electron hold the atom together through electrostatic forces of **attraction**.

A summary of the characteristics of the subatomic particles is given in Table S1.5. The charge is measured relative to that of a proton; the mass is measured relative to that of the proton or neutron (as they have nearly the same mass).

Table S1.5 Characteristics of protons, neutrons and electrons

Subatomic particle	Symbol	Nuclide notation	Approximate relative mass	Mass/kg	Exact relative charge	Charge/C
proton	p	$^{1}_{1}p$	1	$1.672\,622 \times 10^{-27}$	+1	$1.602\,189 \times 10^{-19}$
neutron	n	$^{1}_{0}n$	1	$1.674\,927 \times 10^{-27}$	0	0
electron	e	$^{0}_{-1}e$	5×10^{-4} (negligible)	$9.109\,383 \times 10^{-31}$	−1	$-1.602\,189 \times 10^{-19}$

Tool 3: Mathematics

Use and interpret scientific notation

Scientific notation is a way of writing numbers that are too large or too small to be conveniently written in decimal form. In scientific notation every number is expressed in the form $a \times 10^b$, where a is a decimal number larger than 1 and less than 10, and b is an integer called the exponent.

Scientific notation is useful for making the number of significant figures clear. It is also used for entering and displaying large and small numbers on calculators.

$\times 10^x$ or the letter E is often used on calculators to represent 'times ten to the power of'. For example, 4.62E3 represents 4.62×10^3, or 4620.

Table S1.6 Some numbers in decimal and scientific notation

Decimal notation	Scientific notation
3	3×10^0
400	4×10^2
−53 000	-5.3×10^4
6 730 000 000	6.73×10^9
0.000 000 007 52	7.52×10^{-9}

Top tip!

Remember that: **p**rotons are **p**ositive, **n**eutrons are **n**eutral leaving electro**n**s as **n**egative.

Common mistake

Misconception: *Electrons have no mass but have a charge.*

Electrons have a small mass and a charge that is equal in size but opposite in sign to the much more massive proton.

3 Refer to the table of physical constants in section 2 of the IB *Chemistry data booklet.* Calculate the charge (in coulombs) of one mole of electrons.

4 An atom has a radius of 0.1 nm and the nucleus has a radius of 10^{-15} m. The formula for the volume of a sphere is given by the expression $\frac{4}{3}\pi r^3$, where r is the radius.

a Calculate the volume in m^3 of:

i the atom

ii the nucleus.

b Deduce the percentage of the atom that is occupied by the nucleus.

5 Use the data in the IB *Chemistry data booklet* to calculate how many times heavier a hydrogen-1 atom (one proton and one electron) is than an electron.

6 Use the data in the IB *Chemistry data booklet* to calculate the mass of a carbon-12 atom. Give your answer in kg to 2 s.f.

7 To 3 s.f., the mass of a proton is 1.67×10^{-27} kg and the radius of a proton is 8.41×10^{-16} m. Calculate the density of the nucleus of a hydrogen atom giving the value to 3 s.f. in $kg\,m^{-3}$.

Common mistake

Misconception: *An atom is neutral because it contains equal numbers of protons and neutrons.*

An atom is neutral because it contains equal numbers of protons and electrons.

◆ **Coulomb**: The SI unit of electrical charge.

◆ **Elementary charge**: The size of the charge carried by a proton or electron.

◆ **Electric field**: A region of space where a force acts on electric charges.

Charge is measured in **coulombs**, symbol C. An electron has a charge of $-1.602\,189 \times 10^{-19}$ C and a proton has a charge of $+1.602\,189 \times 10^{-19}$ C. This charge is known as the **elementary charge**. The size of the charges is the same but they have opposite signs.

Electric charges are surrounded by an **electric field**. Like charges repel, so two electrons repel and two protons also repel. Opposite charges attract, so a proton and an electron attract and experience electrostatic attraction for each other. The force of attraction between a proton and an electron decreases as the distance between the two charged particles increases.

ATL S1.2A

All types of chemical bonds and intermolecular forces can be explained in terms of electrostatic attraction (and repulsion).

Go to https://reader.activelylearn.com/authoring/preview/2841734/notes

Choose the **Macro Scale** screen to visualize the electrostatic force that two charges exert on each other. Adjust the charge magnitude and separation to see how these factors affect the force.

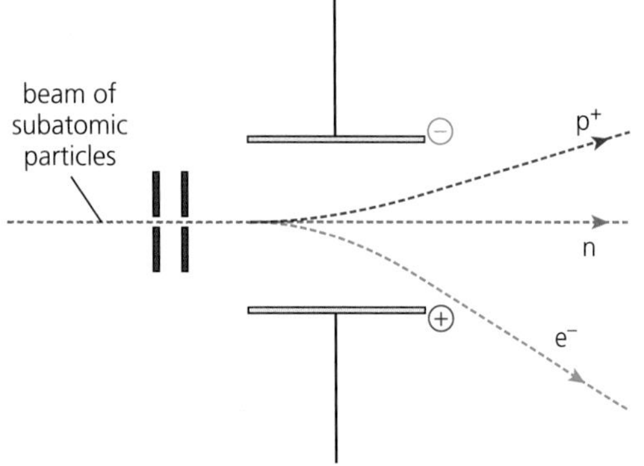

■ **Figure S1.22** The behaviour of protons, neutrons and electrons in an electric field

Moving charged particles such as protons, electrons and ions are deflected by electric (Figure S1.22) and magnetic fields in an evacuated vacuum tube. The deflection of positive ions in a magnetic field is a key feature of the mass spectrometer used to study gaseous atoms and molecules.

Ions that enter a magnetic field are deflected from a straight line to follow a circular path, the radius of which depends on their mass-to-charge ratio, m/z.

Why are many of the laws in the natural sciences stated using the language of mathematics?

Chemists use mathematics to describe and explain the physical world. They believe that mathematical relationships reflect real aspects of the physical world. Science relies on the assumption that we live in an ordered universe that is subject to precise mathematical laws. The laws of physics, the most fundamental of the sciences, are all expressed as mathematical equations, such as Coulomb's law, which describes the attraction between charged particles as a function of charge and distance.

Mathematical concepts, such as differential equations and logarithms, which were developed for purely abstract reasons, turn out to explain real chemical phenomena. Their usefulness, as physicist Eugene Wigner once wrote, 'is a wonderful gift which we neither understand nor deserve.' Mathematics is a powerful tool for describing physical laws and especially making predictions. For example, James Clerk Maxwell's equations of electromagnetism predicted the existence of radio waves two decades before the German physicist Hertz detected them. Radio waves are now widely used in MRI scanners in hospitals and in mobile phones.

What is the difference between a scientific theory and a law?

Scientists will often use language in specific ways. The term 'law' (as in Coulomb's law) is one such example. A law in science is a generalized description of what is happening. The mathematical formulation of Coulomb's law shows that, no matter what fundamental entities are being considered, the general properties of the forces between them can always be described in this manner.

In science, a law simply *describes* a phenomenon whereas a 'theory' is a series of beliefs and ideas which *explain* why some things happen in the way they do.

Confusion may arise when terms such as law or theory are used in the context of a different area of knowledge, or outside a community that uses them in some particular way. When a non-scientist uses the word theory, it might be a way of talking about a sort of educated guess or highlighting that the real facts have not yet been established. However, when a chemist uses atomic theory to explain chemical reactions, there is no suggestion that their explanation is merely an educated guess!

8 a Complete the table below, showing the relative masses and charges of subatomic particles.

Particle	Mass relative to mass of proton	Charge relative to charge of proton
proton		
neutron		
electron		

b State where these particles are found in an atom.

9 State the type of electrostatic force (if any) between the following pairs of particles:

a two protons
b two neutrons
c an electron and a proton.

10 The diagram shows a carbon atom. Deduce the numbers of protons, neutrons, electrons and nucleons.

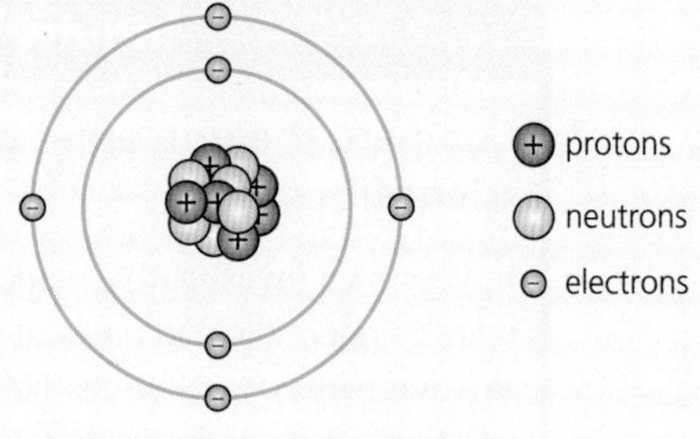

Nuclear forces

◆ **Strong nuclear force**: A force that acts between nucleons in a nucleus to keep it stable.

◆ **Repulsion**: A force pushing apart two like charges (or two particles with the same charge).

Inside the nucleus, the **strong nuclear force** binds the protons and neutrons together. It is an attractive force between protons and neutrons (nucleons). At very small distances, the attractive strong nuclear force prevents the electrical **repulsion** between protons from breaking the nucleus apart. However, as the nucleus becomes larger, and as the average distance between protons increases, the magnitude of the force of repulsion between protons becomes larger than that of the strong nuclear force (force of attraction between nucleons). This can make a nucleus unstable and, therefore, the nucleus of a large atom, such as uranium, may undergo radioactive decay to make itself stable.

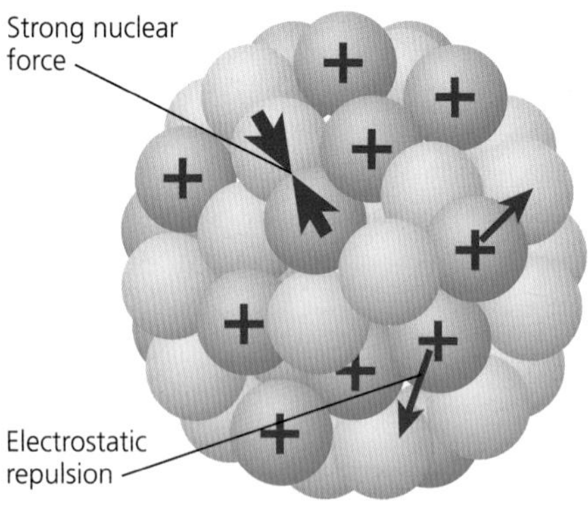

■ **Figure S1.23** The forces acting in a nucleus

Although electrons are the smallest of the three subatomic particles, they control the chemical properties of the chemical elements. Different elements have different chemical properties because they have different numbers of electrons and hence different arrangements in their electron shells.

Common mistake

Misconception: *The nucleus contains an equal number of protons and neutrons because the neutrons neutralize the charge on the protons.*

There is no relationship between the numbers of protons and neutrons. The neutrons bind the protons together but the charges of the protons are neutralized by the electrons. In light atoms the numbers of protons and neutrons are often equal, but as the atomic number increases, the number of neutrons tends to increase more than the number of protons.

Inquiry 3: Concluding and evaluating

Compare the outcomes of your own investigation to the accepted scientific context

In your own investigative work, especially your internal assessment and extended essay, you must compare the outcomes of your own investigation to the accepted scientific context (the existing paradigm). Use the published studies, such as review papers, you have referenced in your background introduction to establish whether your own results agree with previous work in the field.

Nature of science: Observations

The discovery of the nucleus

In 1909, Geiger and Marsden found a method of probing the inside of atoms using alpha particles emitted from radioactive radon as 'bullets' (Figure S1.24). Alpha particles are helium nuclei and consist of two protons bound to two neutrons. When alpha particles were fired at very thin sheets of gold foil, most of them passed straight through it. But some of the alpha particles were deflected by the gold foil and a few of them even bounced back from it.

In order to explain these observations, Rutherford proposed that each atom had a small dense positive nucleus, orbited by tiny negative electrons. During Geiger and Marsden's experiments, nearly all of the positive alpha particles passed straight through the large empty spaces. However, a few alpha particles passed close to a positive nucleus and were deflected (Figure S1.25). A small proportion of the alpha particles approached a nucleus head-on. When this happened, the positive alpha particle was repelled by the positive nucleus and bounced back.

Five years before his discovery of the nucleus, Rutherford had observed that alpha particles beamed through a hole onto a photographic plate would make a sharp-edged picture, while alpha particles beamed through a thin sheet of mica (a silicate mineral) only 20 micrometres (0.002 cm) thick would make an impression with blurry edges. Remembering those results allowed Rutherford and his students to refine the gold-foil experiment and use very thin gold foils only 0.00004 cm thick.

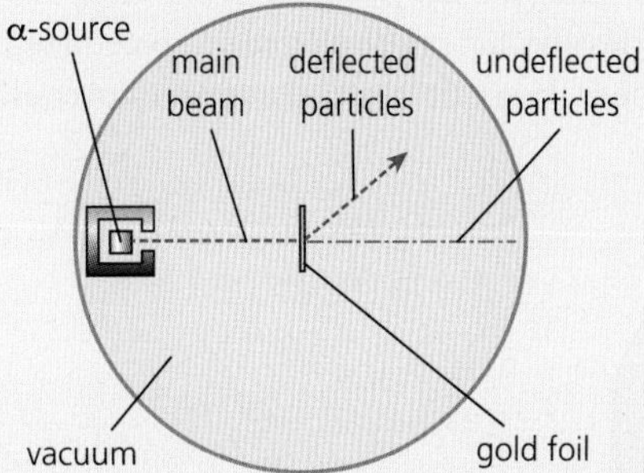

■ **Figure S1.24** Alpha particle scattering experiment

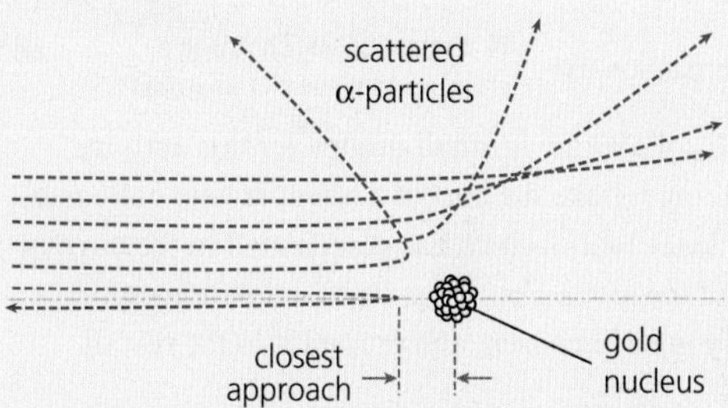

■ **Figure S1.25** Alpha particles approaching a gold nucleus in Rutherford's gold-foil experiment

Inquiry 1: Exploring and designing

Consult a variety of sources

You should consult a wide variety of sources when selecting topics for a chemistry investigation. The sources that might be consulted include:

- chemistry text books and practical books
- journals and periodicals such as:
  - *Chemistry Review*, published by Hodder Education
  - *The Australian Journal of Education in Chemistry*
  - *Journal of Chemical Education* (JCE), published by the American Chemical Society (ACS)
  - *School Science Review* (SSR), published by the Association for Science Education (ASE) (print and online)
- websites and other online resources.

ATL S1.2B

The development of ideas about atomic structure illustrates how scientific models and theories develop over time. When new discoveries are made, existing models (such as that shown in Figure S1.26) and theories may have to be altered or, sometimes, completely replaced if they do not fit in with the new discoveries.

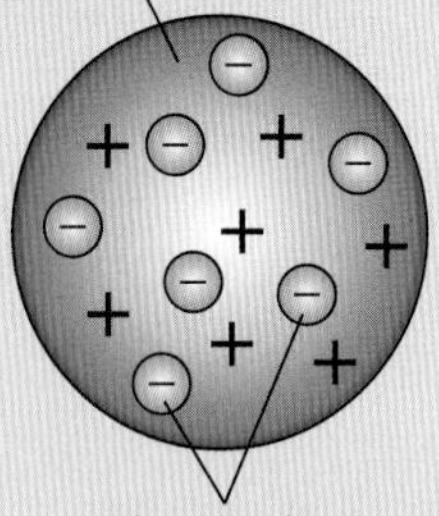

■ **Figure S1.26** Thomson's plum-pudding model of atoms

You will need to collaborate in small groups for this activity. Each group is going to take the part of a scientist who collected some of the evidence (data) which has contributed to the current or past models of the atom. You are going to prepare a report on the scientist you have chosen, and present it to the rest of the class.

Your presentation should be 5 minutes long, illustrated with slides (produced using, for example, PowerPoint) and covering the following points:

- Who you are and what country you are from.
- When you did the research you are about to describe (the year or years in brackets).
- What you already knew about the atom.
- What conclusions you drew from the results of your experimentation.
- An outline of your model of the atom.
- What applications arose from your experimentation, or what theories or models were developed.
- What aspects of this case suggest the knowledge created is more or less reliable.
- A clear bibliography.

Use suitable textbooks, magazine articles or the internet to help you find the information you need.

The scientists are:

Pierre Gassendi (1660)	Henry Moseley (1913)
John Dalton (1803)	Niels Bohr (1913)
Joseph J. Thomson (1897–1899)	Erwin Schrödinger (1926)
Ernest Rutherford (1909)	James Chadwick (1932)

Electron arrangement

◆ **Shells**: The main energy levels of an atom where the electrons are located.

◆ **Atomic number**: The number of protons in the nucleus of an atom or ion.

◆ **Electron arrangement**: Describes the arrangement of electrons in the shells (main energy levels and sub-levels) of atoms and ions.

The electrons in atoms are arranged in **shells**. The hydrogen atom has an **atomic number** of 1 and therefore its nucleus contains one proton. The atomic number is the number of protons in the nucleus and determines the number of electrons. There is one electron for electrical neutrality in a hydrogen atom. This electron enters the first shell nearest the nucleus.

The first shell ($n = 1$) can hold a maximum of two electrons so, in the lithium atom (atomic number 3), the third electron has to enter the second shell (second main energy level). The second shell ($n = 2$) can hold a maximum of eight electrons.

The maximum number of electrons each shell can hold is given by the expression $2n^2$, where n is the shell number. The first, second, third and fourth shells can hold up to $2 \times 1^2 = 2$, $2 \times 2^2 = 8$, $2 \times 3^2 = 18$ and $2 \times 4^2 = 32$ electrons, respectively. The second and subsequent shells are divided into a number of sublevels.

Chemists often use a shorthand notation to describe the arrangement of electrons in shells. This is known as the **electron arrangement**. Hydrogen has an electron arrangement of 1; lithium, which has three electrons, has an electron arrangement of 2.1; and sodium, which has 11 electrons, has an electron arrangement of 2.8.1. Figure S1.27 shows the electron shells (shell structures) for atoms of these and selected other elements. Table S1.7 lists electron arrangements for atoms of the first 20 elements.

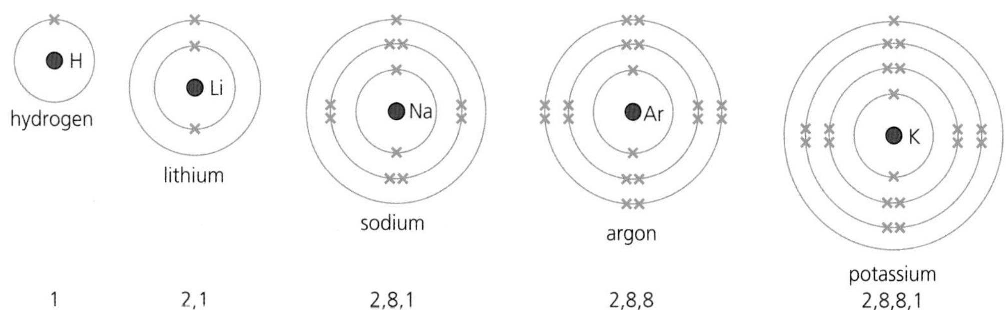

■ **Figure S1.27** Electron arrangements of hydrogen, lithium, sodium, argon and potassium atoms shown as shell structures

■ **Table S1.7** Electron arrangements for the first 20 chemical elements

		Shell (main energy level)			
Name	**Atomic number**	***1st***	***2nd***	***3rd***	***4th***
hydrogen	1	1			
helium	2	2			
lithium	3	2	1		
beryllium	4	2	2		
boron	5	2	3		
carbon	6	2	4		
nitrogen	7	2	5		
oxygen	8	2	6		
fluorine	9	2	7		
neon	10	2	8		

		Shell (main energy level)			
Name	**Atomic number**	***1st***	***2nd***	***3rd***	***4th***
sodium	11	2	8	1	
magnesium	12	2	8	2	
aluminium	13	2	8	3	
silicon	14	2	8	4	
phosphorus	15	2	8	5	
sulfur	16	2	8	6	
chlorine	17	2	8	7	
argon	18	2	8	8	
potassium	19	2	8	8	1
calcium	20	2	8	8	2

◆ **Valence electron**: An electron in the outer shell (main energy level) of an atom that takes part in bond formation.

Top tip!

The third shell can hold a maximum of 18 electrons. However, when there are eight electrons in the third shell, there is a degree of stability and the next two electrons enter the fourth shell. For the transition metals beyond calcium, the additional electrons enter the third shell until it contains a maximum of 18 electrons.

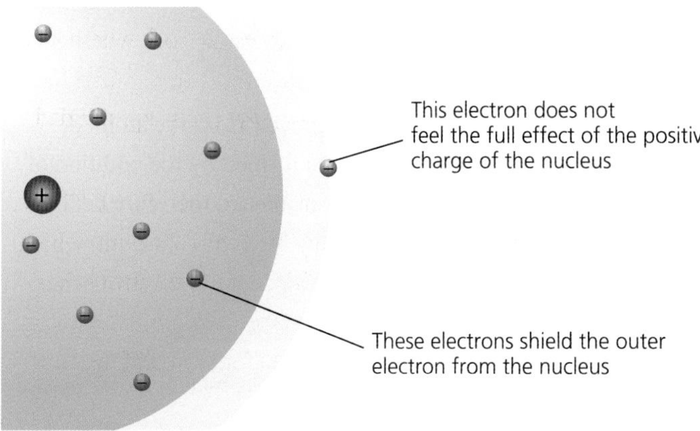

■ **Figure S1.28** Electron shielding

The **valence electrons**, found in the shell furthest from the nucleus, are the most important since it is these electrons which are involved in the formation of chemical bonds. For example, a potassium atom (2.8.8.**1**) has a single valence electron in the fourth shell.

Another important concept is electron shielding (Figure S1.28). The electrons in the different shells experience different attractive forces from the nucleus due to the repulsive forces from other electrons. The outer electrons experience the most shielding since electrons in inner shells approach the nucleus more closely and so partially neutralize and reduce its charge.

TOK

Why is scientific knowledge falsifiable?

Theories in chemistry can never be verified or proved true; chemical knowledge is always provisional. Theories can only be shown to be false (refuted) by experimental results. So, chemists only accept those laws, models and theories which have been extensively tested and, *so far*, have not been falsified. According to Karl Popper, the difference between science and non-science (or pseudoscience) is that scientific statements are open to being falsified by experimental data. Often facts are offered which look like scientific facts in the sense that they rely only on confirmation. For example: 'Holding this crystal will make you feel better.' If you happen to feel better while holding the crystal, you might think that the claim has been tested and confirmed. But there may be any number of other reasons for feeling better. Thus, for Popper, this means nothing: it is pseudoscience that only looks like science.

Nuclide notation

The atomic number or proton number (symbol Z) is the same for every atom of a particular element and no two different elements have the same atomic number. In the periodic table, the elements are arranged in increasing order of atomic (proton) number.

◆ **Mass number**: The number of protons and neutrons in an atom or ion.

◆ **Nuclide**: One particular type of atom (or ion), as defined by the number of protons and neutrons in its nucleus.

◆ **Nuclear symbol**: The general notation for a nuclide, $^{A}_{Z}X$, where X is the symbol of the atom or element, Z is the atomic or proton number and A is the nucleon or mass number.

The total number of protons and neutrons is called the **mass number** (symbol A).

Nuclides of a chemical element are described by their atomic and mass numbers using **nuclear symbols** $^{A}_{Z}X$ where X represents the symbol of the chemical element, Z represents the atomic number (number of protons) and A represents the mass number or nucleon number (number of protons and neutrons). Here is how this notation works for aluminium-27:

mass number, $A \rightarrow$ 27
atomic number, $Z \rightarrow$ 13 Al

$$^{27}_{13}Al$$

The number of neutrons in an atom can be found from the following relationship:

number of neutrons = mass number (A) − atomic number (Z)

So an atom of aluminium-27 contains 13 protons, 13 electrons and (27 − 13) = 14 neutrons.

Common mistake

Misconception: *No two nuclides may have the same mass or nucleon number.*

Different nuclides may have the same nucleon number, for example: $^{40}_{18}Ar$, $^{40}_{19}K$ and $^{40}_{20}Ca$.

Top tip!

Not all of the atoms in naturally occurring samples of some chemical elements are identical. Atoms of the same element that have different mass numbers are called isotopes.

An element forms an ion when one or more electrons are added or removed from the atom. A positive ion is formed by the removal of electrons and a negative ion is formed by the addition of electrons. The number of protons, neutrons and electrons in an atom or ion can therefore be calculated from the mass number, atomic number and charge.

Common mistake

Misconception: *The elements in the periodic table are arranged in order of increasing relative atomic mass.*

The elements in the periodic table are arranged in order of increasing atomic (proton) number which usually correlates with increasing relative atomic mass.

Top tip!

Positive ions have more protons than electrons. Negative ions have more electrons than protons. This is because electrons are negatively charged.

LINKING QUESTION

What determines the different chemical properties of atoms?

LINKING QUESTION

How does the atomic number relate to the position of an element in the periodic table?

11 Deduce and fill in the missing data in the table below.

Element and nuclide symbol	Atomic number	Mass number	Number of protons	Number of neutrons	Number of electrons
	8			8	
		56		28	
				21	20
	11	23			
Fe		57			

12 Deduce the number of electrons, protons and neutrons in a $^{31}_{15}P^{3-}$ and b $^{24}_{12}Mg^{2+}$. Explain your answers.

Common mistake

Misconception: *Ionic bonding occurs between metals and non-metals, while covalent bonds form between non-metals.*

The true nature of a chemical bond is determined using the differences in electronegativities. Bonds have 'ionic character' or 'covalent character' along a bonding continuum.

Isotopes

◆ **Isotopes**: Atoms with the same number of protons but a different number of neutrons.

◆ **Radioisotopes**: Atoms of an unstable element that release radiation from the nucleus.

Many elements exist as a mixture of **isotopes** (this was established by the invention of the mass spectrometer). Figure S1.29 shows the stable isotopes of carbon, chlorine and hydrogen as nuclide symbols and Figure S1.30 shows the subatomic particles of the three isotopes of hydrogen.

Carbon:	**Chlorine:**	**Hydrogen:**
Carbon-12 ($^{12}_{6}C$)	Chlorine-35 ($^{35}_{17}Cl$)	Hydrogen-1 ($^{1}_{1}H$)
Carbon-13 ($^{13}_{6}C$)	Chlorine-37 ($^{37}_{17}Cl$)	Hydrogen-2 ($^{2}_{1}H$)
		Hydrogen-3 ($^{3}_{1}H$)

■ **Figure S1.29** Stable isotopes of carbon, chlorine and hydrogen

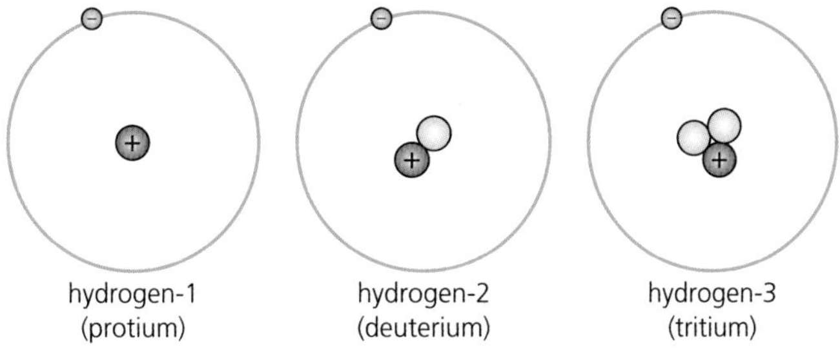

■ **Figure S1.30** The three isotopes of hydrogen: protium (^{1}H), deuterium (^{2}H) and tritium (^{3}H)

Common mistake

Misconception: *For all nuclides, the proton number is always less than the mass or nucleon number.*

Protium (hydrogen-1) has a proton number equal to the mass or nucleon number.

Isotopes of the same chemical element have identical chemical properties but slightly different physical properties due to the different masses. For example, samples of $^{20}_{10}Ne$ and $^{22}_{10}Ne$ have different densities, different melting points and different boiling points. The lighter isotope, $^{20}_{10}Ne$, will diffuse more rapidly than the heavier isotope, $^{22}_{10}Ne$, at the same temperature. The lighter isotope, neon-20, has a slightly lower melting point, boiling point and density than the heavier isotope, neon-22.

Some isotopes of some elements are unstable and are known as **radioisotopes**. Their nuclei break down spontaneously, in a process known as radioactive decay; these isotopes are described as radioactive. They emit rays and particles known as nuclear radiation. For some radioisotopes, nuclear decay occurs quickly; for others the decay is very slow.

ATL S1.2C

Investigate how the balance of neutrons (N) and protons (Z) in the nucleus influences the stability of an atom by using the simulation at https://phet.colorado.edu/sims/html/build-an-atom/latest/build-an-atom_en.html to build the atoms shown in the table. Explore which atoms are stable and which are unstable and release radiation.

Isotope	N	Z	N/Z ratio	Stable?
hydrogen-1				
hydrogen-2				
hydrogen-3				
hydrogen-4				
helium-2				
helium-3				
helium-4				
helium-5				
lithium-5				
lithium-6				
lithium-7				
lithium-8				
beryllium-7				
beryllium-8				
beryllium-9				
beryllium-10				
boron-9				
boron-10				
boron-11				
boron-12				

Isotope	N	Z	N/Z ratio	Stable?
carbon-11				
carbon-12				
carbon-13				
carbon-14				
nitrogen-13				
nitrogen-14				
nitrogen-15				
nitrogen-16				
oxygen-15				
oxygen-16				
oxygen-17				
oxygen-18				
fluorine-17				
fluorine-18				
fluorine-19				
fluorine-20				
neon-18				
neon-19				
neon-20				
neon-21				

- Generate a graph of atomic number (Z) on the x-axis and number of neutrons (N) on the y-axis. Plot the unstable isotopes using a different colour.

On a graph like this, stable isotopes are found in an area known as the band of stability.

- Deduce and explain the observed relationship between N and Z for atoms that lie in this region.
- Explain how you could use this plot to predict if an isotope is stable or unstable.

◆ **Isotopic abundance:** The proportions, often expressed as percentages, of the different isotopes in a naturally occurring sample of an element.

A few elements, such as fluorine and aluminium, have been found to consist of only one isotope. However, most elements exist as a mixture of two or more isotopes. The **isotopic abundance** (relative abundance) of an isotope in a sample of the element can be expressed as a fraction or percentage (Table S1.8).

■ **Table S1.8** Isotopes of chromium

Isotope	Relative abundance / %	Fractional abundance
chromium-50	4	0.04
chromium-52	84	0.84
chromium-53	10	0.10
chromium-54	2	0.02

13 Deduce the number of protons, neutrons and electrons in the following hydrogen species:

a $^{3}_{1}H^{-}$ b $^{3}_{1}H^{+}$ c $^{2}_{1}H^{-}$ d $^{2}_{1}H^{+}$

Going further

Nuclear power

Nuclear power plants depend on uranium-235 nuclei undergoing fission (splitting) when bombarded with slow moving neutrons. During fission, several neutrons are emitted from the nucleus and a large amount of energy is released. These neutrons bombard more uranium-235 nuclei and the process continues as a chain reaction with the release of more energy. However, natural uranium only has a very small percentage of U-235 with over 99% of the uranium atoms being U-238 which do not undergo fission.

To increase the percentage of uranium-235 in a sample of uranium, a process called enrichment is carried out. The uranium is reacted with fluorine gas, converting it to uranium hexafluoride, UF_6, which is volatile (easily vaporized). The UF_6 is allowed to diffuse through a series of membranes with small holes.

Because the lighter 235-UF_6 molecules travel about 0.4% faster than the heavier 238-UF_6 molecules, the gas that first passes through the membrane has a higher concentration of uranium-235. When this vapour passes through another membrane, the uranium-235 vapour becomes further concentrated. It takes many of these effusion stages (Figure S1.31) to produce enriched uranium suitable for use in nuclear power stations.

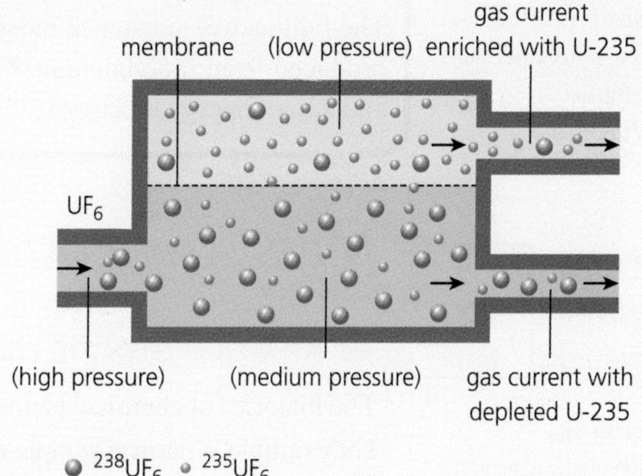

■ **Figure S1.31** Separation of uranium isotopes by effusion

TOK: Scope

How might developments in scientific knowledge trigger political controversies or controversies in other areas of knowledge?

The relationship between science and politics can be described as the coupling of two interdependent developments: the 'scientification' of politics and the politicization of science.

A number of political problems were first recognized and described by scientists: for example, environmental pollution only entered the political agenda after scientists discovered the insecticide DDT in the food chain. The greater the extent to which science becomes part of the political process, the greater its role becomes in defining the problems that it is then asked to solve.

The coupling of scientific knowledge with politics drives the politicization of science. As it enters the public arena, scientific knowledge is assessed and evaluated by society. The expert's position in a controversy is therefore viewed as determined as much by politics as by scientific knowledge.

ATL S1.2D

Nuclear power stations produce large amounts of electrical energy with small amounts of fuel and very low greenhouse gas emissions. This can enable industrialized countries to become less dependent on imported fossil fuels. Opponents will highlight the accident risks of reactors, proliferation concerns (nuclear proliferation refers to the spread of nuclear weapons or fissile material to countries or terrorist groups that do not have them) and storage of nuclear waste. Present and defend an ethical argument (using ethical theories) involving the concepts of sustainability and intergenerational justice for or against the use of nuclear power.

Going further

Nuclear medicine

Nuclear medicine involves the use of radioactive substances to detect (or treat) abnormalities, especially cancer, in the function of particular organs in the body. Substances introduced into the body for this purpose are called tracers. This is because they emit radiation which can be detected and allow the substance to be traced. They may be injected or ingested (eaten or drunk).

The radioactive substance most commonly used is technetium-99m. This is an excited atom produced from molybdenum-99 by beta decay (which involves a proton forming a neutron and a high energy electron).

ATL S1.2E

Research the properties of technetium-99m and explain why it is often chosen as a source of radiation for use in medical diagnosis.

HL ONLY

Mass spectrometry

Relative atomic mass

Atomic number	10
Element	**Ne**
Relative atomic mass	20.18

■ **Figure S1.32** The representation of elements in the periodic table of the IB *Chemistry data booklet*. 20.18 is the relative atomic mass of neon. The nuclide notation for neon-20 is ${}^{20}_{10}\text{Ne}$.

The majority of chemical elements in nature exist as a mixture of isotopes in fixed proportions. For example, a natural sample of chlorine atoms consists of 75% chlorine-35 and 25% chlorine-37. This is why the atomic mass may not be a whole number: instead of showing the mass of a single nuclide it shows the relative atomic mass (Figure S1.32).

The **relative atomic mass,** A_r, of an element is the weighted average mass of the atoms of its isotopes compared to one-twelfth of the mass of one atom of carbon-12:

$$\text{relative atomic mass} = \frac{\text{weighted average mass of the isotopes of the chemical element}}{\frac{1}{12}\text{ the mass of one atom of carbon-12}}$$

However, since one-twelfth the mass of one atom of carbon-12 is 1, then the relative atomic mass of a chemical element is effectively the weighted average isotopic mass divided by 1.

Relative atomic masses can be calculated from a mass spectrum of a chemical element by multiplying the relative isotopic mass of each isotope by its fractional abundance and adding all the values together.

Conversely, the use of simple algebra allows us to calculate the percentage abundance of one isotope given the relative atomic mass of the chemical element and the mass numbers of both isotopes.

◆ **Relative atomic mass:** The ratio of the average mass per atom of the naturally occurring form of an element to one-twelfth of the mass of a carbon-12 atom.

Link

Using mass spectrometry as a technique to determine an accurate value for the molar mass of an organic compound (HL only) is covered in Chapter S3.2, page 290.

Top tip!

Relative atomic mass is not the same as mass number. Mass numbers are integers because they are the number of protons plus the number of neutrons. Relative atomic mass is an average mass that takes into account all the isotopes of an element and so is not usually an integer.

When you calculate relative atomic mass, the answer should have a value somewhere between the masses of the lightest isotope and the heaviest isotope.

14 Rubidium exists as a mixture of two isotopes, ${}^{85}\text{Rb}$ and ${}^{87}\text{Rb}$. The percentage abundances are 72.1% and 27.9%, respectively. Calculate the relative atomic mass of rubidium.

15 The relative atomic mass of gallium is 69.7. Gallium is composed of two isotopes: gallium-69 and gallium-71. Let %Ga-69, the percentage abundance of gallium-69, equal x. Calculate x.

ATL S1.2F

Find out how mass spectrometry was used in the discovery of carbon-60, carbon-70 and other fullerenes. Outline how the clusters of carbon were generated and analysed in a mass spectrum and how their structures gained greater support from 'shrink wrapping' experiments.

LINKING QUESTION

How can isotope tracers provide evidence for a reaction mechanism?

Mass spectrometers

A **mass spectrometer** (Figure S1.33) allows chemists to determine accurately the relative atomic masses of atoms. The instrument produces gaseous positive ions from vaporized atoms, separates them and detects and measures their abundances. **Mass spectrometry** can also be used to determine the relative molecular masses of molecular compounds and establish their structure.

◆ **Mass spectrometer**: An analytical instrument that separates gaseous ions according to their mass to charge ratio.

◆ **Mass spectrometry**: Study of substances from gas-phase ions using a mass spectrometer.

◆ **Mass spectrum**: A bar graph of ion abundance against the mass/charge ratio.

◆ **Radical cation**: A cation containing an unpaired electron (in its ground state).

◆ **Molecular ion**: An ion formed by the removal of an electron from a molecule.

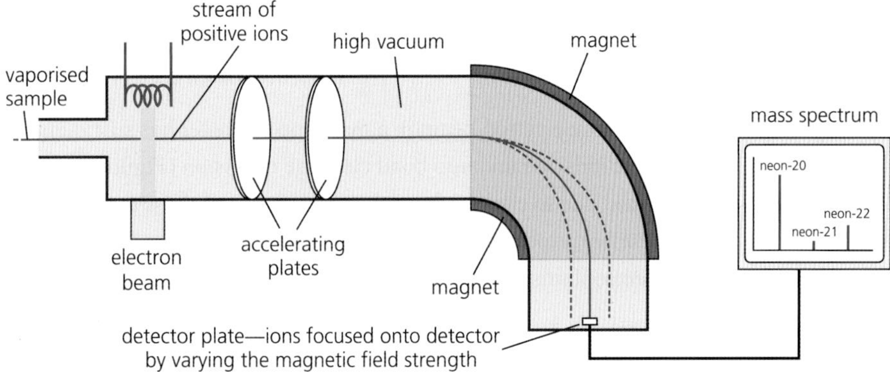

■ **Figure S1.33** A mass spectrometer in cross section

A **mass spectrum** for chlorine atoms is shown in Figure S1.34. The two peaks indicate the detection of $^{35}Cl^+$ and $^{37}Cl^+$ ions. The mass spectrum shows that chlorine is composed of two isotopes, chlorine-35 and chlorine-37, in a 3 : 1 or 75% : 25% ratio by abundance.

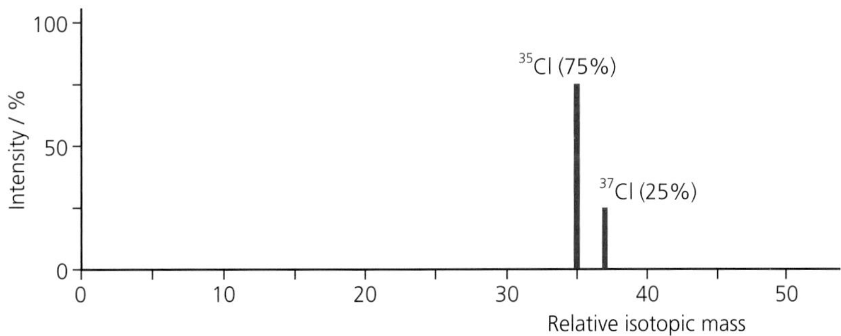

■ **Figure S1.34** Mass spectrum of a sample of naturally occurring chlorine atoms

LINKING QUESTION

How does the fragmentation pattern of a compound in the mass spectrometer help in the determination of its structure?

Top tip!

The horizontal axis for a mass spectrum is actually mass-to-charge ratio (*m* / *z*). However, since all the ions are unipositive, the scale is equivalent to mass since the charge on all the ions is +1.

The mass spectrometer (Figure S1.35) uses a beam of electrons with a high kinetic energy to ionize molecules. An outer electron is removed from the molecule, leaving an unpaired electron and forming a **radical cation**: a positive ion with an unpaired electron. Figure S1.36 shows the formation of a **molecular ion** (radical cation) from propanone.

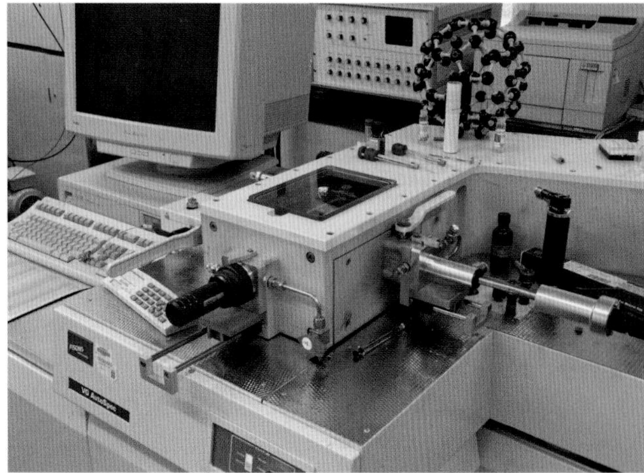

■ **Figure S1.35** A mass spectrometer

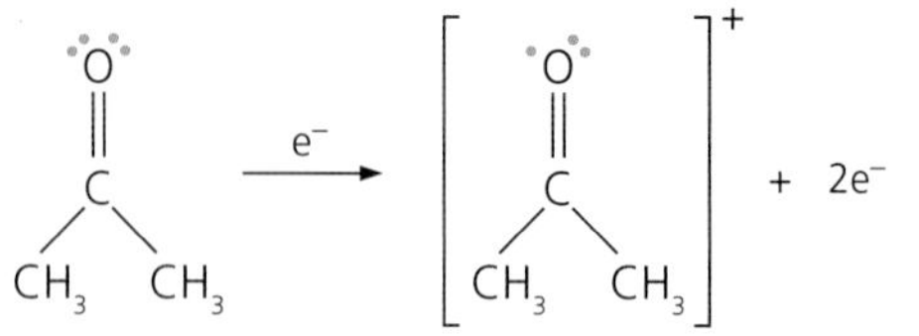

$$\mathrm{CH_3COCH_3} \xrightarrow{e^-} [\mathrm{CH_3COCH_3}]^{+} + 2e^-$$

■ **Figure S1.36** Ionization of a propanone molecule (with lone pairs of electrons shown as green dots) to form a unipositive radical cation

If the ionizing electron beam in a mass spectrometer has sufficient kinetic energy, the propanone molecular ions can undergo bond cleavage or fission (Figure S1.37) and molecular fragments are formed due to electron bombardment. Some of these fragments will carry a positive charge and therefore appear as further peaks in the mass spectrum. The mass spectrum of propanone therefore contains peaks at $m/z = 15$ and 43 (Figure S1.38), as well as the molecular ion peak at 58.

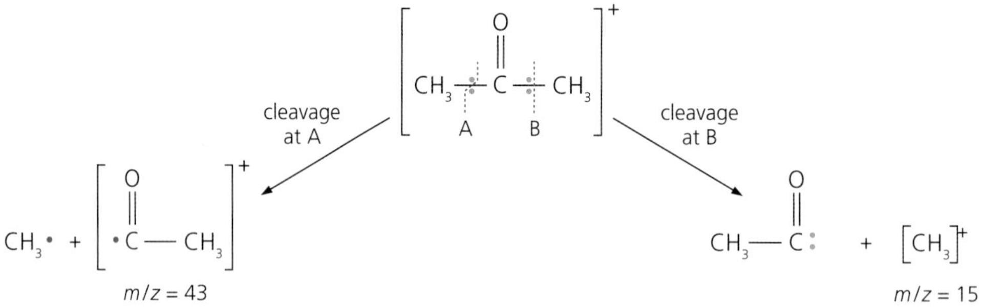

$[\mathrm{CH_3}\text{–}\mathrm{CO}\text{–}\mathrm{CH_3}]^+$ (bonds marked A and B)

cleavage at A: $\mathrm{CH_3}\bullet + [\bullet\mathrm{C(O)}\text{—}\mathrm{CH_3}]^+$ ($m/z = 43$)

cleavage at B: $\mathrm{CH_3}\text{—}\mathrm{C(O)}\colon + [\mathrm{CH_3}]^+$ ($m/z = 15$)

■ **Figure S1.37** Fragment ions produced from the propanone molecule

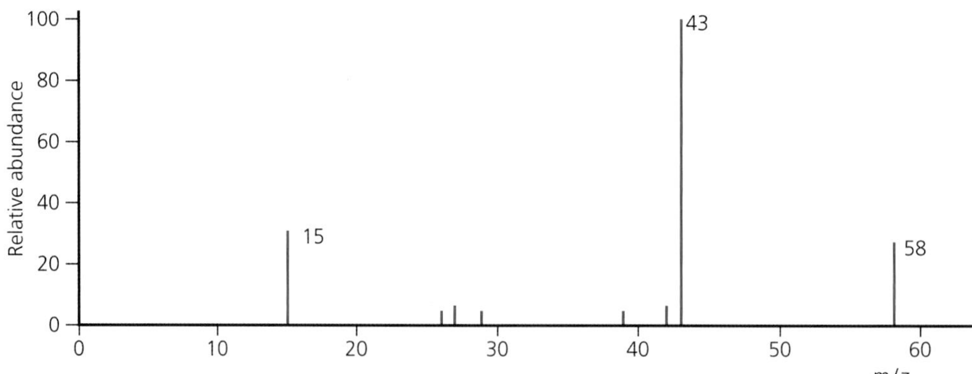

■ **Figure S1.38** Mass spectrum of propanone

$\mathrm{R}\text{—}\overset{+}{\mathrm{C}}\text{=}\mathrm{O} \longleftrightarrow \mathrm{R}\text{—}\mathrm{C}\equiv\overset{+}{\mathrm{O}}$ the acylium ion

$\mathrm{R_3C^+}$ a tertiary carbocation

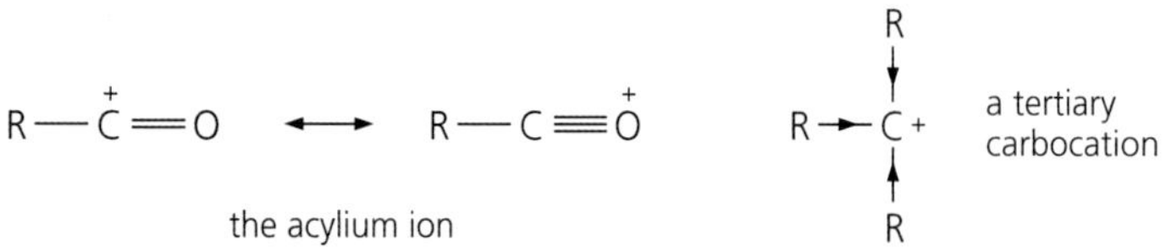

■ **Figure S1.39** The acylium ion (resonance stabilized) and tertiary carbocation; R is an alkyl group such as $-CH_3$ or $-CH_2CH_3$

Top tip!

The more stable an ion, the more likely it is to form and the greater its peak height (relative abundance). Small, highly charged ions, such as H^+, are unstable. Fragments with ions or tertiary carbocations (Figure S1.39) are made more stable by the inductive effect, which pushes electron density toward the positively charged carbon.

■ **Table S1.9** Common peaks in mass spectra

Mass lost	Fragment lost
15	CH_3
17	OH
18	H_2O
28	$CH_2{=}CH_2$, C=O
29	CH_3CH_2, CHO
31	CH_3O
45	COOH

Table S1.9 lists peaks that are often seen in the fragmentation patterns of mass spectra. They provide useful clues for determining the structure of an organic molecule.

WORKED EXAMPLE S1.2A

The diagram shows the mass spectra of two compounds with the molecular formula $C_2H_4O_2$. One compound is methyl methanoate and the other is ethanoic acid. Decide from the major fragments which mass spectrum corresponds to which substance.

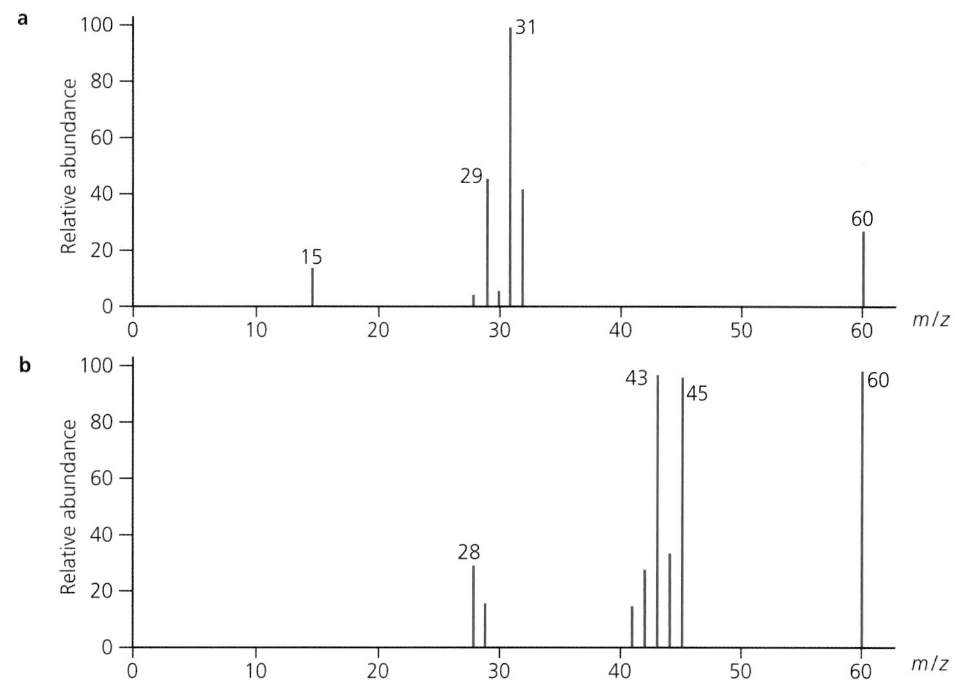

Answer

The major peaks in mass spectrum **a** are due to: CH_3^+ ($m/z = 15$); CHO^+ ($m/z = 29$) and CH_3O^+ ($m/z = 31$). These fragments can be readily generated from methyl methanoate.

$$\left[H-\overset{\overset{\displaystyle O}{\|}}{C} \vdots O \vdots CH_3 \right]^+$$

A B

cleavage at A: $H-\overset{\overset{\displaystyle O}{\|}}{C}^+$ ($m/z = 29$) or $^+O-CH_3$ ($m/z = 31$)

cleavage at B: $H-\overset{\overset{\displaystyle O}{\|}}{C}-O$ ($m/z = 45$) (absent, because it does not form an ion) + CH_3^+ ($m/z = 15$)

The peak at $m/z = 31$ can only be generated from methyl methanoate, and not from ethanoic acid. The major peaks in mass spectrum **b** are due to: CO^+ ($m/z = 28$), CH_3CO^+ ($m/z = 43$) and $COOH^+$ ($m/z = 45$). These fragments can be readily generated from ethanoic acid.

$[CH_3 - C(=O) - O - H]^+$ (A: bond between CH_3 and C; B: bond between C and O)

cleavage at A → CH_3 ($m/z = 15$, absent) + ^+COOH ($m/z = 45$)

cleavage at A and B → $^+C{=}O$ ($m/z = 28$)

cleavage at B → $CH_3{-}^+C{=}O$ ($m/z = 43$) + OH ($m/z = 17$, absent)

Isotope patterns for organic compounds containing one chlorine atom or one bromine atom can be easily recognized.

Chlorine has two isotopes, ^{35}Cl and ^{37}Cl, and these are present in the ratio $^{35}Cl : ^{37}Cl = 3:1$. Therefore, any molecule or fragment that contains one chlorine atom will give rise to two peaks, separated by two mass units and with a 3 : 1 height ratio.

Figure S1.40 shows the mass spectrum of chloroethane, C_2H_5Cl. There are two molecular ion peaks, at masses of 64 and 66, with heights in the ratio 3 : 1. These are due to the formation of the ions $CH_3CH_2{}^{35}Cl^+$ and $CH_3CH_2{}^{37}Cl^+$, respectively. There are also peaks at masses of 49 and 51. These also have heights in the ratio 3 : 1. These peaks represent the loss of 15 units of mass, or a CH_3 group, from the molecule leaving $CH_2{}^{35}Cl^+$ and $CH_2{}^{37}Cl^+$ ions. The peak at mass 29 corresponds to loss of chlorine and formation of $[CH_3CH_2]^+$.

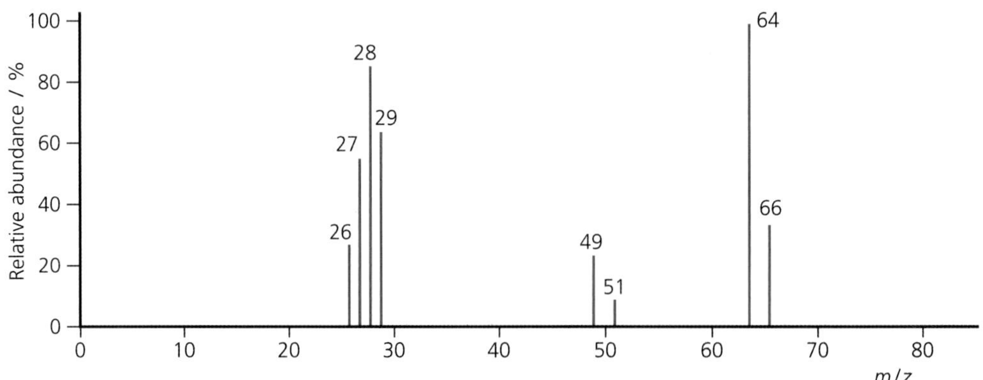

■ **Figure S1.40** The mass spectrum of chloroethane

Top tip!

A molecule containing one bromine atom will show a molecular ion with two peaks separated by two mass units in an approximate 1 : 1 ratio. This is because the ratio of the two isotopes ^{79}Br and ^{81}Br is approximately 1 : 1.

S1.3 Electron configurations

Guiding question

- How can we model the energy states of electrons in atoms?

SYLLABUS CONTENT

By the end of this chapter, you should understand that:

- emission spectra are produced by atoms emitting photons when electrons in excited states return to lower energy levels
- the line emission spectrum of hydrogen provides evidence for the existence of electrons in discrete energy levels, which converge at higher energies
- the main energy level is given an integer number, n, and can hold a maximum of $2n^2$ electrons
- a more detailed model of the atom describes the division of the main energy levels into s, p, d and f sublevels of successively higher energy
- each orbital has a defined energy state for a given electron configuration and chemical environment, and can hold two electrons of opposite spin
- sublevels contain a fixed number of orbitals, regions of space where there is a high probability of finding electrons
- in an emission spectrum, the limit of convergence at higher frequency corresponds to ionization (HL only)
- successive ionization (IE) data for an element give information about its electron configuration (HL only).

By the end of this chapter you should know how to:

- qualitatively describe the relationship between colour, wavelength, frequency and energy across the electromagnetic spectrum
- distinguish between a continuous spectrum and a line spectrum
- describe the emission spectrum of a hydrogen atom, including the relationship between the lines and energy transitions to the first, second and third energy levels
- deduce the maximum number of electrons that can occupy each energy level
- recognize the shape and orientation of an s atomic orbital and the three p atomic orbitals
- apply the Aufbau principle, Hund's rule and the Pauli exclusion principle to deduce electron configurations for atoms and ions up to $Z = 36$
- explain the trends and discontinuities in first ionization energy (IE) across a period and down a group (HL only)
- calculate the value of the first IE from spectral data that gives the wavelength or frequency of the convergence limit (HL only)
- deduce the group of an element from its successive ionization data (HL only).

◆ **Bohr model**: A model of the hydrogen atom (and other one-electron systems) with a proton in the nucleus and an electron moving in circular orbits, with each orbit corresponding to a specific fixed energy. A similar model with electrons in main energy levels can also be used to describe other atoms (in a more approximate manner) with more than one electron.

Introduction

In Chapter S1.2, electrons were described as particles in electron shells (main energy levels) around the nucleus. Experimental evidence has shown that electrons can behave as particles and waves and are found in regions within atoms known as atomic orbitals. The electrons in atoms can, under certain conditions, undergo transitions between orbitals and release or absorb light.

Figure S1.41 shows the change from a simple **Bohr model** of a sodium atom with electrons orbiting in circular shells (main energy levels) of different energies (2,8,1), to a quantum-mechanical model of a sodium atom based on atomic orbitals of different energies, shapes and orientations forming sublevels within the shells ($1s^2$ $2s^2$ $2p^6$ $3s^1$).

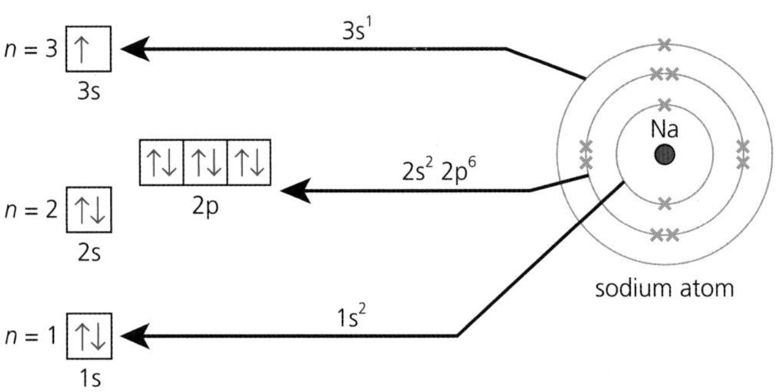

■ **Figure S1.41** Quantum-mechanical and Bohr models of a sodium atom

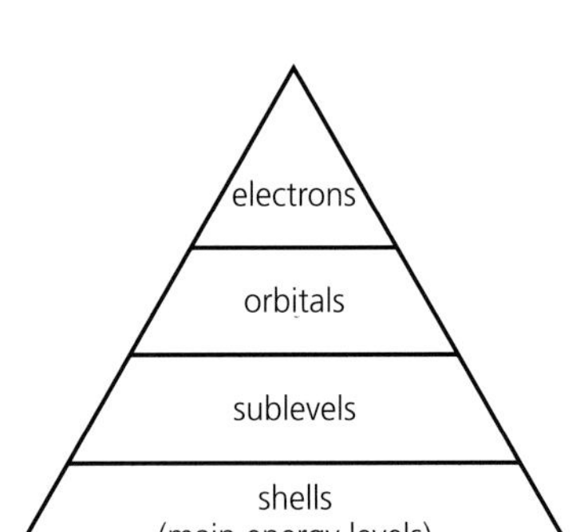

■ **Figure S1.42** Arrangement of electrons in atoms (and ions)

Each orbital has a label, for example, 1s. The number indicates the electron shell in which the orbital is located. The 1s orbital is in the first shell, closest to the nucleus. The letters *s* and *p* indicate the shape of the orbitals: spherical and dumbbell.

Orbitals with the same shape and in the same shell have the same energy and are grouped together to form sublevels (Figure S1.42). Sublevels of similar energy, for example, 2s and 2p, are grouped together into main energy levels (electron shells).

Light

◆ **Wave**: Oscillations that carry energy.

◆ **Photon**: A particle consisting of a quantum of electromagnetic radiation.

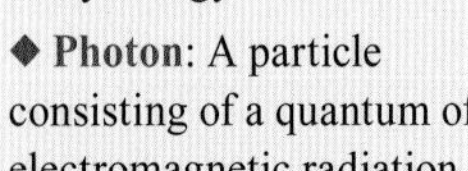

◆ **Quantum**: The minimum amount of energy by which the energy of a system can change.

◆ **Planck's equation**: $E = hf$; the energy of a photon is proportional to its frequency.

◆ **Frequency**: Number of waves that pass a point in one second.

◆ **Hertz**: SI unit of frequency equal to one cycle per second.

◆ **Planck's constant**: A fundamental constant equal to the ratio of the energy of a quantum of energy to its frequency, $h = 6.63 \times 10^{-34}$ Js.

◆ **Wavelength**: Distance between two neighbouring crests or troughs of a wave.

◆ **Wave equation**: The velocity of a wave is given by the product of the wavelength, λ, and the frequency (v), $c = \lambda \times f$.

Light can be described as an electromagnetic **wave** or by a particle model which views light as a stream of **photons** or tiny 'packets' or **quanta** (singular: **quantum**) of light energy (Figure S1.43).

- Light waves and other electromagnetic waves are described by **wavelength** (Figure S1.44) and **frequency**.

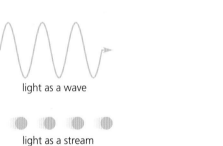

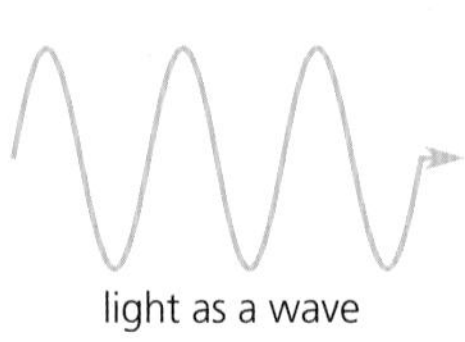

light as a stream of photons (quanta: packets of energy)

■ **Figure S1.43** Wave and particle nature of light

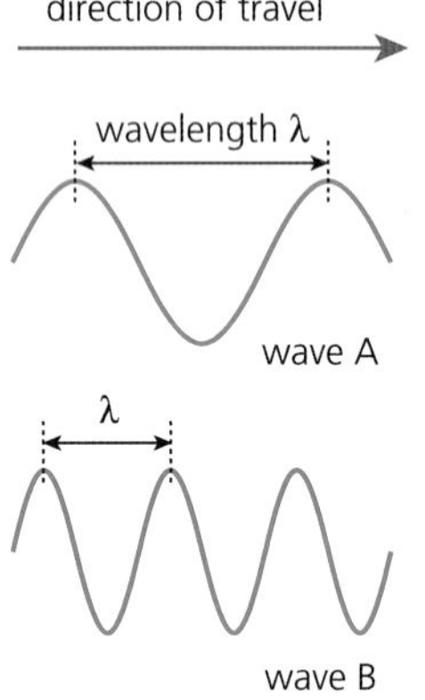

■ **Figure S1.44** Wave A has twice the wavelength of wave B

Top tip!

Each quantum has a very small but fixed amount of energy that varies with the colour of the light.

- The two models of light are linked by **Planck's equation**: $E = h \times f$, where E is the energy of a photon (in joules), f is the frequency of the light (in **hertz**, Hz, or s^{-1}) and h represents **Planck's constant** (6.63×10^{-34} J s). The frequency, f, and wavelength, λ, are related to the speed by the **wave equation**: $c = f \times \lambda$, or $f = c / \lambda$ where c is the velocity of light.

HL ONLY

◆ **Wavenumber:** The reciprocal of the wavelength (in cm); the number of waves per cm.

HL ONLY

Another unit of frequency often used by spectroscopists is **wavenumber** (reciprocal of wavelength).

Light and other electromagnetic waves travel in a vacuum at a speed of $3.00 \times 10^8\,m\,s^{-1}$.

16 The yellow light emitted by a sodium streetlight has a wavelength of 589 nm. Calculate:

a the frequency of the light (in Hz or s^{-1})

b the energy (in J) carried by one of these photons

c the energy (in $kJ\,mol^{-1}$) carried by one mole of these photons ($N_A = 6.02 \times 10^{23}\,mol^{-1}$).

Tool 3: Mathematics

Apply and use SI prefixes and units

Most quantities in science consist of a number and a unit. A measured quantity without units is usually meaningless. However, there are some derived quantities in chemistry that do not have units (for example, pH and absorbance) because they are based on ratios or logarithmic functions.

The Système Internationale (SI) is a system of units used throughout the scientific world. All SI units are derived from the units of seven base quantities (Table S1.10 shows five of these base quantities).

■ **Table S1.10** Selected SI base units

Property measured	Name of unit	Abbreviated unit
mass	kilogram	kg
length	metre	m
temperature	kelvin	K
electric current	ampere	A
amount of substance	mole	mol

Volumes in chemistry are usually measured in cubic centimetres (cm^3) or cubic decimetres (dm^3) and masses are usually measured in grams (g). SI units should be used for other quantities where possible and appropriate.

Express compound units using negative powers rather than a solidus (/): for example, write the speed of light as $3 \times 10^8\,m\,s^{-1}$ rather than $3 \times 10^8\,m/s$.

Use prefixes to indicate multiples of 10^3, so that numbers are kept between 0.1 and 1000.

Treat a combination of a prefix and a symbol as a single symbol.

■ **Table S1.11** Standard SI prefixes

Prefix (word)	Prefix (symbol)	Power
giga	G	10^9
mega	M	10^6
kilo	k	10^3
deci	d	10^{-1}
centi	c	10^{-2}
milli	m	10^{-3}
micro	μ	10^{-6}
nano	n	10^{-9}

Express very small or large numbers in scientific notation: for example, $6.02 \times 10^{23}\,mol^{-1}$.

◆ **Spectrum:** The range of electromagnetic energies arrayed in order of increasing or decreasing wavelength or frequency.

Different colours in visible light correspond to electromagnetic waves of different wavelengths, frequencies and energies, with red light having the longest wavelength and hence the lowest energy and frequency. These different colours are separated out in a **spectrum** (Figure S1.45).

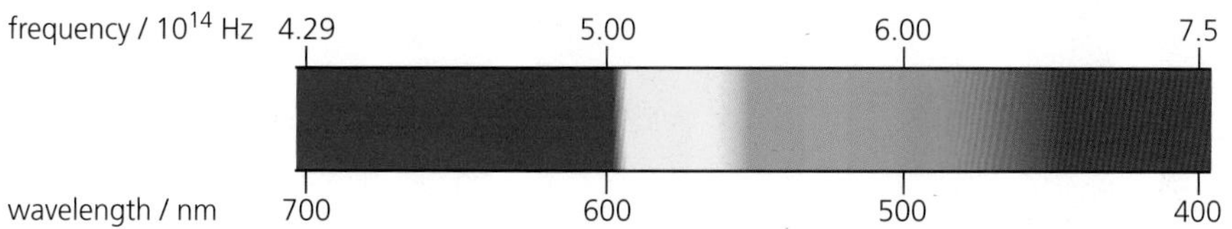

■ **Figure S1.45** Continuous emission spectrum of white light separated into all its merging colours

◆ **Electromagnetic spectrum:** The range of wavelengths over which electromagnetic radiation extends.

◆ **Continuous spectrum:** A spectrum containing all frequencies over a wide range.

◆ **Electron excitation:** A process by which an electron gains energy that moves it to a higher energy level.

Figure S1.46 shows the **electromagnetic spectrum**. It is an arrangement of all the types of electromagnetic radiation in increasing order of wavelength or decreasing order of frequency.

Figure S1.46 also shows the visible region of light as a **continuous spectrum** ('rainbow') ranging from red to violet with an infinite number of colours.

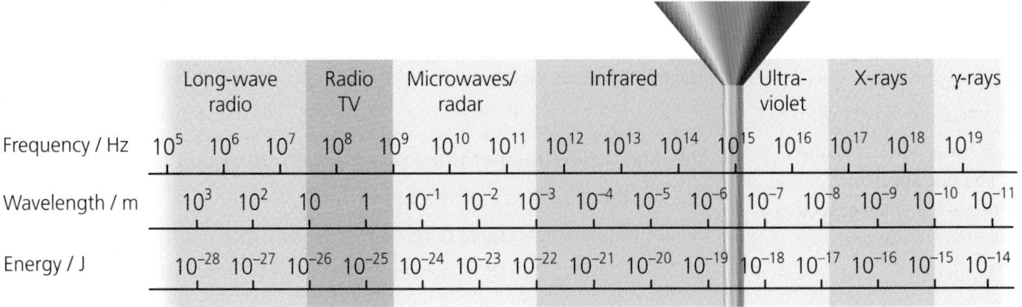

■ **Figure S1.46** The electromagnetic spectrum

17 Identify the region of the electromagnetic spectrum where the following will be found:

a an absorption with a wavelength of 5.00×10^4 m

b an absorption with a wavelength of 5.00×10^{-7} m

c an absorption with a frequency of 5×10^{15} s^{-1}

18 Compare the relative frequencies, wavelengths, energies and wavenumbers of ultraviolet and infrared radiations.

WORKED EXAMPLE S1.3A (HL ONLY)

Astronomical observations, signals observed from the distant stars, are generally weak. If a photon detector receives a total of 3.15×10^{-18} J from radiation of 600 nm, calculate the number of photons (to the nearest integer) received by the detector.

Answer

Energy of one photon, $E = hf = \frac{hc}{\lambda}$

$$= \frac{6.63 \times 10^{-34}\,\text{J s} \times 3.00 \times 10^{8}\,\text{m s}^{-1}}{600 \times 10^{-9}\,\text{m}}$$

$$= 3.32 \times 10^{-19}\,\text{J}$$

Therefore number of photons received, $n = \frac{3.15 \times 10^{-18}}{3.32 \times 10^{-19}} = 10$

Electron excitation

If gaseous atoms are excited, they emit light of certain frequencies. The process of **electron excitation** may be thermal or electrical. Thermal excitation occurs when a substance is vaporized and a flame is formed (Figure S1.47).

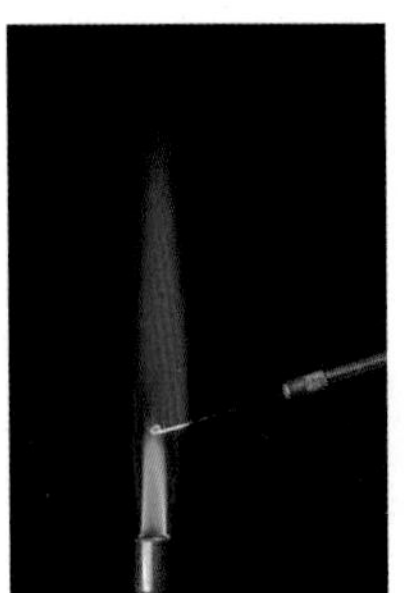 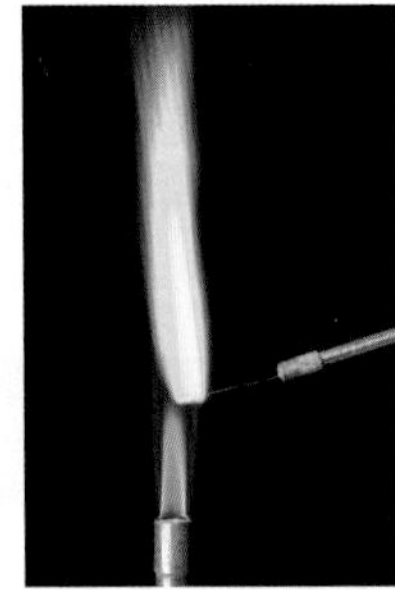 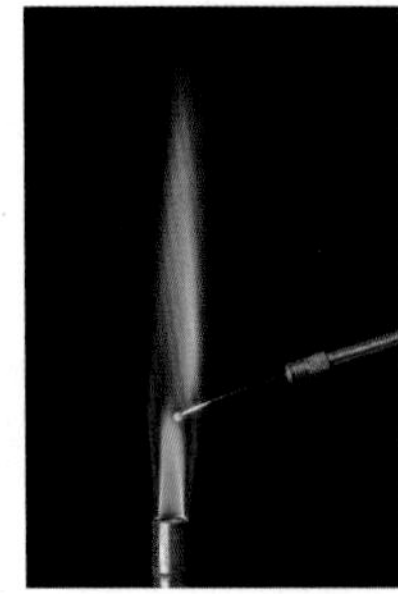 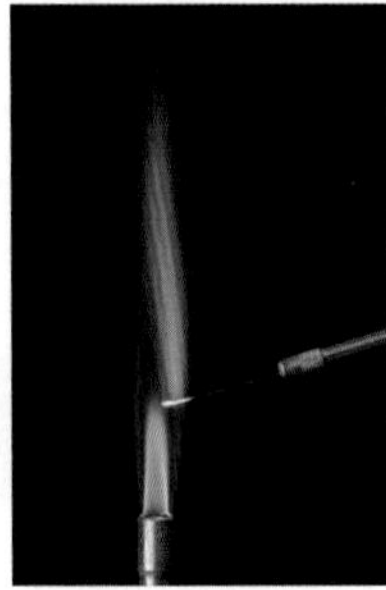 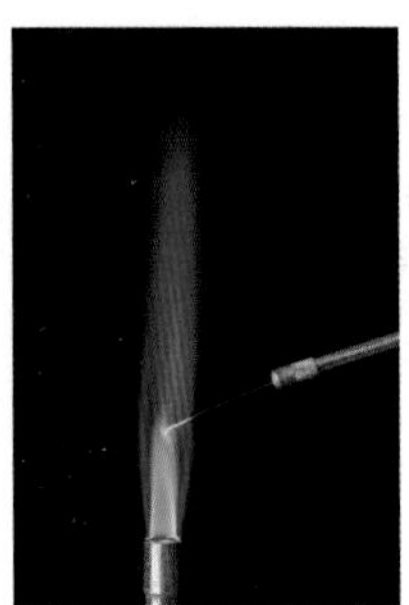 

■ **Figure S1.47** The flames generated by lithium, sodium, potassium, calcium, strontium and barium

◆ **Discharge tube**: A glass tube, containing gas of an element (or compound) at low pressure, with a cathode and an anode; when a high voltage is applied to the electrodes, the gas glows and emits an emission spectrum consisting of narrow lines.

◆ **Emission spectrum**: The line spectrum produced when gaseous atoms are excited.

◆ **Spectral series**: A group of related lines in an emission (or absorption) spectrum produced by transitions between one specific energy level and a set of different energy levels.

◆ **Orbit**: The path of an electron as it travels around the nucleus of an atom.

◆ **Energy level**: A fixed energy that an electron, an atom or a molecule can have.

Top tip!

Balmer and later Rydberg found simple mathematical relationships involving spectral lines in the emission spectrum of hydrogen.

Top tip!

'Discrete energy levels' means an electron can exist at a distinct number of levels. It cannot be between these levels. A ladder has discrete potential energy levels when it is being climbed.

Electrical excitation also occurs when a high voltage is passed across a tube containing a gaseous sample of the element at low pressure. Molecules will be dissociated by the high voltage and excited atoms formed. Advertising signs using noble gases in high-voltage **discharge tubes** and flame tests are both examples of electron excitation.

If the light from gaseous hydrogen atoms with excited electrons is passed through a prism (or diffraction grating), the visible region of the **emission spectrum** can be viewed. This spectrum consists of four narrow coloured lines (red, aqua, blue and violet) on a black background. The lines converge (get progressively closer) as the frequency or energy of the emission lines increases.

Figure S1.48 shows the complete emission spectrum of hydrogen atoms including the similar sets of lines (**spectral series**) that are observed in the ultraviolet and infrared regions of the electromagnetic spectrum.

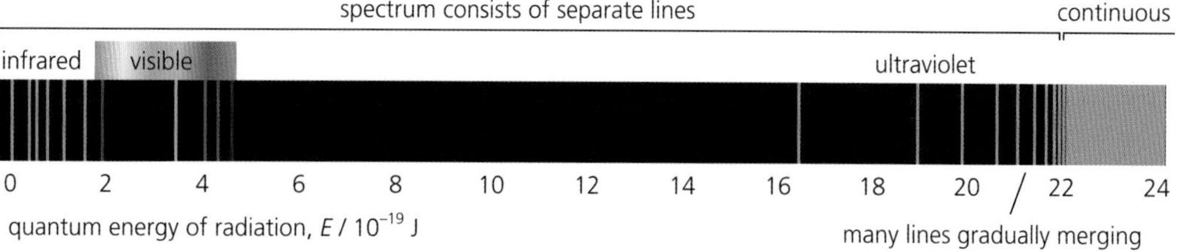

■ **Figure S1.48** The complete emission spectrum of atomic hydrogen

ATL S1.3A

Go to **https://teachchemistry.org/classroom-resources/exciting-electrons-simulation**

Use this simulation to explore what happens when electrons in an atom are excited from their ground state. You will see that when an electron moves from an excited state to its ground state, energy is released in the form of electromagnetic radiation (in the form of a wave).

Now go to **https://interactives.ck12.org/simulations/chemistry/bohr-model-of-electron/app/**

Use this second simulation to explore how electrons create coloured light in a hydrogen gas discharge tube. Relate what you see in the simulation to Bohr's model, spectral lines of hydrogen, and atomic emission spectra.

Energy levels and spectra

The spectrum of the simplest atom, hydrogen, shows several series of lines at different energies (frequencies and wavelengths). This suggests that the hydrogen atom can lose different amounts of energy, which in turn suggests that its electrons can exist with a range of energies.

The Danish physicist Niels Bohr explained the many lines in the emission spectrum of hydrogen and other atoms by suggesting that electron **orbits** were related to discrete **energy levels**—each orbit corresponded to a fixed amount of energy. The further the orbit was away from the nucleus, the greater the amount of energy the electron had.

According to Bohr's theory, an electron moving in an orbit does not emit energy. To move to an orbit further away from the nucleus, the electron must absorb energy (electrical or thermal energy) to do work against the attraction of the positively charged nucleus. The atom or electron would then be in an unstable **excited state**.

An emission (line) spectrum is formed when excited electrons move from orbits of high energy to orbits of lower energy. During this transition, the electrons emit light of a particular frequency (Figure S1.49). The energy of the emitted photon of light is equal to the difference between the two energy levels (ΔE) and the frequency of the light is related to the energy difference ($\Delta E = hf$).

19 Calculate the energy difference (in joules) corresponding to an electron transition that emits a photon of light with a wavelength of 600 nm.

◆ **Excited state:** An atom (ion or molecule) with an electron in an energy level higher than its ground state.

◆ **Lyman series:** A series of lines in the emission spectrum of hydrogen caused by the transition of electrons from the second energy level ($n = 2$) or higher to the ground state ($n = 1$), emitting ultraviolet radiation.

◆ **Ground state:** The lowest stable energy state of an atom (or ion or molecule).

◆ **Balmer series:** A series of lines in the emission spectrum of visible light emitted by excited hydrogen atoms which correspond to the electrons undergoing a transition from the third ($n = 3$) or higher energy level to the second-lowest energy level ($n = 2$), emitting visible light.

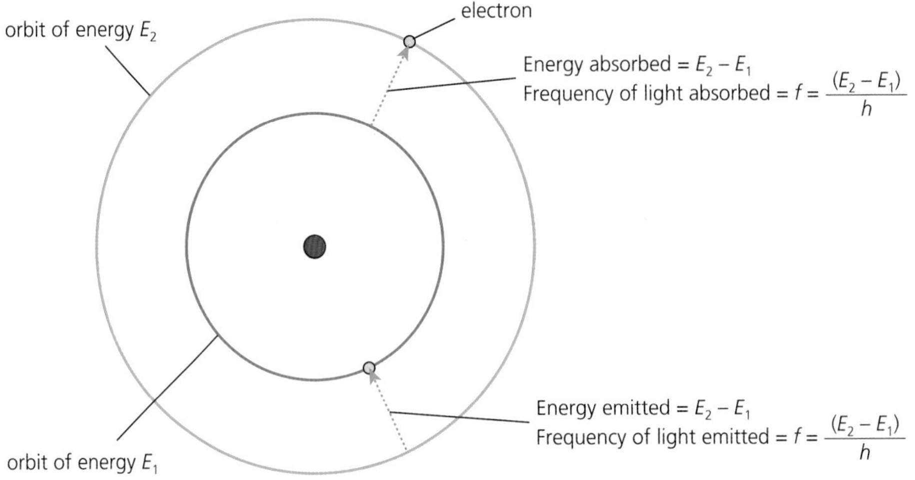

■ **Figure S1.49** The origin of spectral lines in the emission spectrum of gaseous atoms

Figure S1.50 shows how the Bohr model can be used to explain the origin of the **Lyman series** (in the ultraviolet) in the spectrum of hydrogen atoms. The circles represent the electron energy levels. The distances between them represent the energy differences between the energy levels.

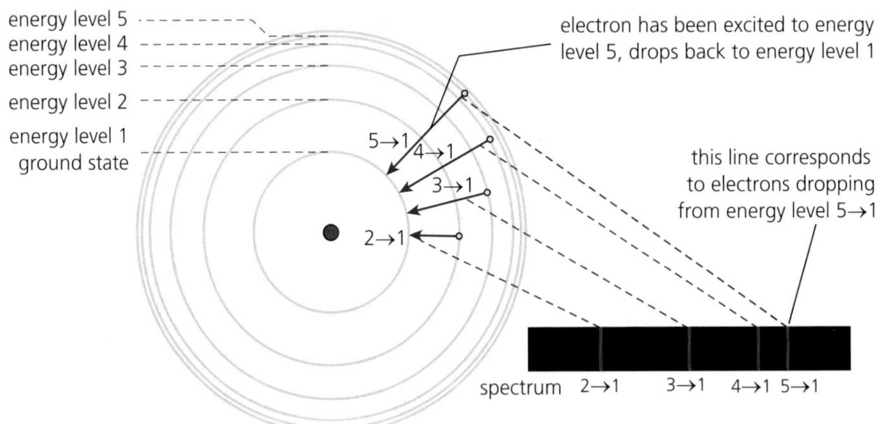

■ **Figure S1.50** How the discrete energy levels in the hydrogen atom give rise to the emission lines of the Lyman series in the ultraviolet region of the electromagnetic spectrum

The energy levels or orbits are labelled as $n = 1$, $n = 2$, $n = 3$ starting from the nucleus. An electron in the lowest energy level (nearest the nucleus) is in its **ground state** ($n = 1$), the most stable state for a hydrogen atom. The energy levels or orbits correspond to electron shells (main energy levels).

The **Balmer series** of lines, in the visible region, is formed when excited electrons transition from higher energy levels to the second energy level ($n = 2$).

Top tip!

All atoms have the converging arrangement of energy levels seen in the hydrogen atom but the spacing, and hence energy differences, are unique to each atom.

◆ **Convergence limit**: The frequency (or wavelength) at which spectral lines converge.

◆ **Ionization energy**: The minimum energy required to remove a mole of electrons from a mole of gaseous atoms or ions to form a mole of unipositive gaseous ions.

Figure S1.50 shows that the energy levels become more closely spaced until they converge at high potential energy. This is known as the **convergence limit** and the difference in energy between the convergence limit and the ground state is the **ionization energy**, the minimum energy required to remove a mole of electrons from a mole of gaseous atoms to form a mole of unipositive gaseous ions. The converging energy levels are directly responsible for the convergence of emission lines in the Lyman and other spectral series.

Figure S1.51 is similar to Figure S1.50 except the energy levels have been drawn as straight lines, rather than as circles. In addition, the electron transitions that give rise to the other spectral series have been added.

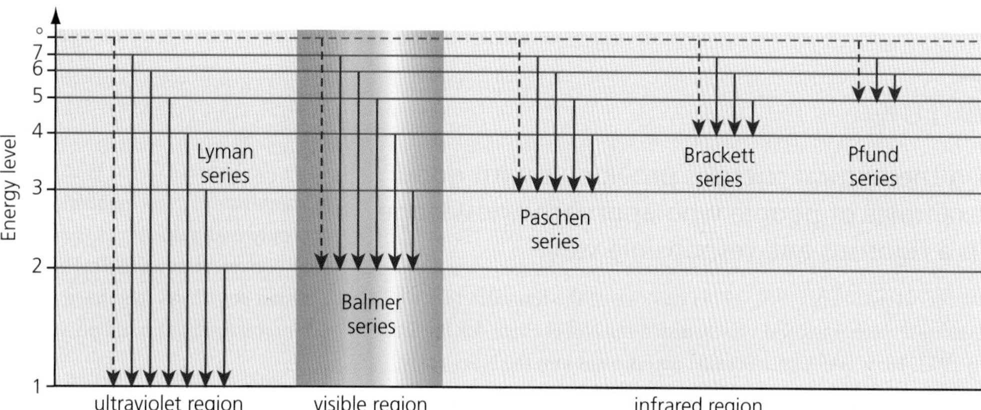

■ **Figure S1.51** The origin of all the major spectral series in the atomic hydrogen emission spectrum

Top tip!

You do not need to remember the names of the spectral series, but you need to know that all electron transitions to $n = 2$ from a higher energy level will lead to the release of light in the visible region; all electron transitions to $n = 1$ from a higher energy level will lead to the release of light in the ultraviolet region; and all electron transitions to $n = 3$ from a higher energy level will lead to the release of light in the infrared region.

Figure S1.51 above is a simplified diagram showing the origin of the visible emission spectrum of atomic hydrogen (gaseous hydrogen atoms at low pressure).

20 Identify three physical quantities about light that could be represented by values on the horizontal axis.

21 State whether these quantities are increasing or decreasing from left to right.

22 Explain why the emission spectrum of hydrogen consists of a series of spectral lines that converge.

23 State which energy level the electron transitions corresponding to the visible lines in the emission spectrum of hydrogen relate to.

24 Find out why the emission spectrum of an element is sometimes compared to the fingerprint of a criminal.

If white light is passed through a sample of gaseous atoms, the transmitted light will be missing certain wavelengths of light due to absorption. The absorbed energy has caused electron excitation; electrons have undergone a transition to a higher energy level. These absorptions are observed as black lines against the coloured background of the visible spectrum.

◆ **Absorption spectrum**: A spectrum of electromagnetic radiation transmitted through a substance, showing dark lines due to absorptions at specific frequencies.

Figure S1.52 shows the relationship between the **absorption spectrum** and the highest energy series of the emission spectrum of an excited atom of the same element.

■ **Figure S1.52** Relationship between an absorption spectrum and an emission spectrum of atoms of the same element

The study of absorption spectra of the Sun and other stars reveals the presence of hydrogen, helium and other elements depending on the type of star, its age and its temperature.

TOK

Do the natural sciences provide us with good examples of people who approach knowledge in a rigorous and responsible way?

Werner Heisenberg (1901–1976) was one of the founders of quantum mechanics and was awarded the Nobel Prize for Physics in 1932. He was not anti-Semitic and did not join the Nazi party, but he was a dedicated German nationalist.

Interestingly, during the Second World War (in September 1941 when German victory in Europe seemed likely), Heisenberg visited Niels Bohr (1885–1962) in Denmark. There has been much speculation about this meeting over the years.

Did Heisenberg try to find out more about the feasibility of nuclear fission from Bohr in order to help develop nuclear weapons for Nazi Germany? Or was he discussing the ethical objections to developing nuclear weapons?

Heisenberg was involved in Nazi Germany's secret war-time project to develop atomic weapons. Some historians believe he participated in the project to deliberately slow progress and prevent Hitler from having nuclear weapons; other historians believe he tried hard to make an atomic bomb but failed because he did not understand the physics properly.

After the Second World War, Bohr and Heisenberg managed to re-establish their personal and professional relationship, but it was only maintained by their tacit agreement to disagree about their conversation in 1941. Heisenberg always maintained that the German physicists 'had been spared the difficult moral decision of whether we should build an atomic bomb.'

The issue at stake is whether scientists should develop knowledge responsibly—not in terms of following the scientific method, but in considering whether the knowledge gained will be used as a force for good or evil. There is little doubt that scientific knowledge about how the world works is beneficial: without it we could not harness energy and there would be no medicine. However, neither would there be nuclear and chemical weapons or dying environmental systems. Should scientists seek knowledge (about the energy stored in atoms and how to release it, for example) knowing that the primary application is as a weapon? Is that knowledge valuable in and of itself, or valuable enough to risk that it will not be used as a force for good?

For a full discussion of what might have occurred during the conversations between Bohr and Heisenberg, see **www.nww2m.com/2011/09/the-mysterious-meeting-between-niels-bohr-and-werner-heisenberg/**

LINKING QUESTION

What qualitative and quantitative data can be collected from instruments such as gas discharge tubes and prisms in the study of emission spectra from gaseous elements and from light?

HL ONLY

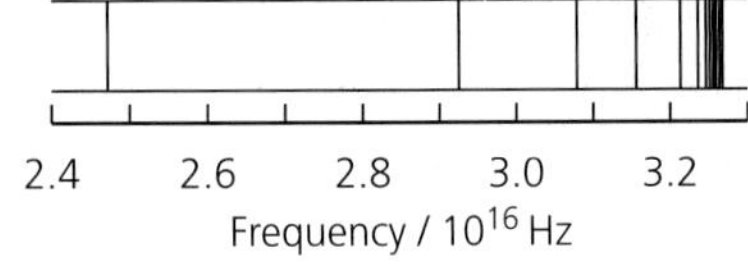

■ **Figure S1.53** The Lyman series in the emission spectrum of atomic hydrogen

Determination of ionization energy from an atomic emission spectrum

The ionization energy of hydrogen can be calculated from the point in the atomic emission spectrum where the lines of the Lyman series (Figure S1.53) converge. This is the minimum amount of energy required to remove one mole of electrons from one mole of gaseous hydrogen atoms and can be represented by the equation $H(g) \rightarrow H^{+}(g) + e^{-}$.

Top tip!

If an electron receives enough energy to remove it completely from the attraction of the nucleus, the atom is ionized. The energy required to ionize a mole of gaseous atoms is known as the ionization energy. It is equivalent to the electron transition $n = 1$ to $n = \infty$ (infinity).

WORKED EXAMPLE S1.3B

For the atomic hydrogen emission spectrum, the convergence limit occurs at a frequency of 3.27×10^{15} Hz. Calculate the ionization energy for a single hydrogen atom and for a mole of hydrogen atoms.

Answer

Planck's relationship, $E = hf$, is used to convert the frequency to energy:

$$E = 3.27 \times 10^{15}\,\text{s}^{-1} \times 6.63 \times 10^{-34}\,\text{J s} = 2.367 \times 10^{-18}\,\text{J}$$

To convert to ionization energy in kilojoules per mole, multiply by the Avogadro constant:

$$2.367 \times 10^{-18}\,\text{J} \times 6.02 \times 10^{23}\,\text{mol}^{-1} = 1\,424\,879.82\,\text{J mol}^{-1} = 1420\,\text{kJ mol}^{-1} \text{ (to 3 s.f.)}$$

25 For the helium atom emission spectrum, the convergence limit occurs at a wavelength of 5.04×10^{-7} m. Calculate the ionization energy for a single helium atom and a mole of helium atoms.

26 Explain why the convergence limit in the visible spectrum is not used to calculate the ionization energy.

Going further

Balmer and Rydberg equations

The Swiss mathematician Johann Balmer (1825–1898) examined the five visible lines in the spectrum of the hydrogen atom (Figure S1.54); their wavelengths are 397 nm, 410 nm, 434 nm, 486 nm and 656 nm.

In 1885 he derived an empirical (experimentally determined) relationship for the wavelengths observed in the visible region of the hydrogen atom emission spectrum:

$$\lambda = h\frac{m^2}{m^2 - n^2}$$

where λ is the wavelength of a specific emission line, $n = 2$ and $m = 3, 4, 5, 6$ etc., and $h = 3.6456 \times 10^{-7}$ m (the Balmer constant). Here, $n = 2$ is the lower level reached by the electron and $m = 3, 4, 5\ldots$ represents the higher energy level from which the electron has undergone a transition.

Balmer's formula was later found to be a special case of the Rydberg formula derived by Swedish physicist Johannes Rydberg in 1888. Rydberg studied the wavelengths of alkali metals and found he could simplify his calculations using the wavenumber.

■ **Figure S1.54** Balmer series of hydrogen

$$\frac{1}{\lambda} = R_H\left(\frac{1}{n_1^2} - \frac{1}{n_2^2}\right)$$

where R_H is the Rydberg constant for hydrogen ($\approx 1.096\,78 \times 10^7\ \text{m}^{-1}$), $n_1 = 2$ for Balmer's formula and $n_2 > n_1$. A full explanation of why the Balmer and Rydberg formulas work was not forthcoming until Neils Bohr (1885–1962) developed his atomic model in 1913.

A value for the Rydberg constant can be determined by plotting a graph of reciprocal wavelength (in m^{-1}) for the lines from a chosen spectral series (Lyman, Balmer and so on) against $\frac{1}{n_2^2}$. The Rydberg constant is given by the gradient of the plot (Figure S1.55).

n2	lambda	(1/n2)^2	1/lambda
3	6.56E-07	0.1111111	1.52E+06
4	4.86E-07	0.0625	2.06E+06
5	4.34E-07	0.04	2.30E+06
6	4.10E-07	0.0277778	2.44E+06
7	3.97E-07	0.0204082	2.52E+06

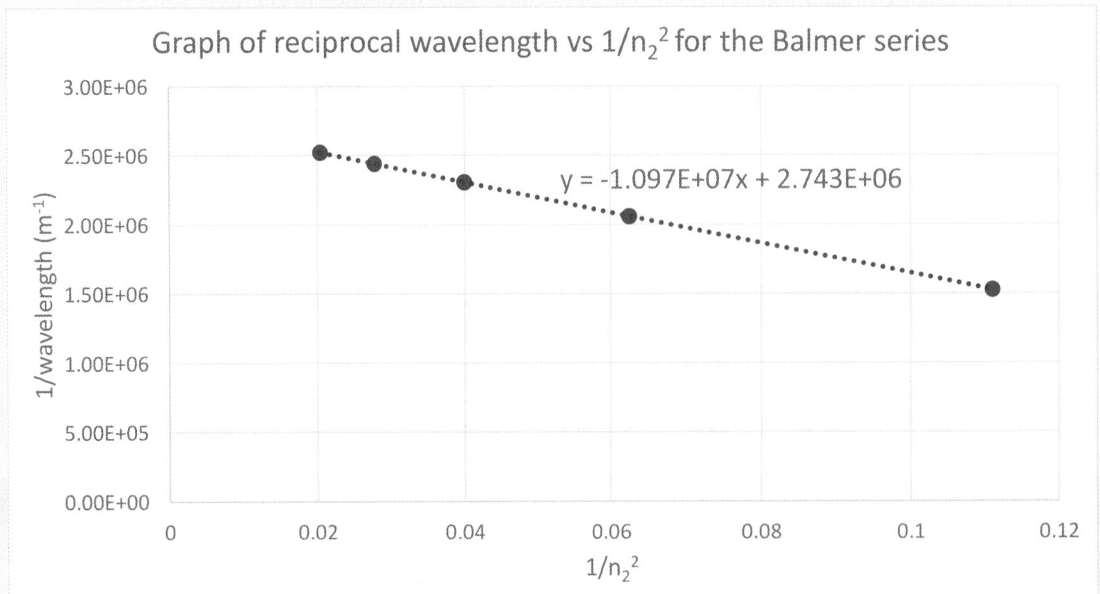

■ **Figure S1.55** A graph of the reciprocal wavelength (in m^{-1}) for lines from the Balmer series against $\frac{1}{n_2^2}$

Tool 2: Technology

Use spreadsheets to manipulate data

Calculators and spreadsheets, such as Excel, are widely used for processing and displaying information. Spreadsheets provide a very effective way of processing raw data, particularly when there is a lot of it. They can be used to sort data, carry out calculations, generate graphs, perform statistics and even create simple simulations using formulas in cells.

- (HL only) Construct a spreadsheet with formulas to calculate theoretical wavelengths using the Balmer formula.

LINKING QUESTION

How do emission spectra provide evidence for the existence of different elements?

What knowledge, if any, is likely to always remain beyond the capabilities of science to investigate or verify?

It is perhaps unwise to explicitly suggest that there will be some physical phenomenon that cannot be investigated by the scientific method. The history of science has many examples of intuition or a fresh perspective leading to new and unexpected scientific knowledge.

August Comte (1798–1857) was a French scientist and philosopher who formulated the doctrine of positivism which, at a simple level, can be described as generating scientific knowledge using empirical data from experimentation.

'We shall never be able to study, by any method, the chemical composition of stars,' Comte asserted. He was unaware of the faint radiation from stars that can now be studied spectroscopically to detect the elements present in their atmospheres (Figure S1.56).

However, what are the known unknowns, those things which we believe will remain forever unknown? Are there any? They may include the following:

- Did time have a beginning and does it have an end?
- We have evidence for the Big Bang but can we ever know what happened before?
- Will we ever understand consciousness?

Some of the answers to these questions might be unknowable because (like Comte) we simply cannot imagine that we will ever have the technology or understanding of the universe to provide an answer. Maybe there are answers, but we are not smart enough or skilled enough to find them.

Other questions might be of the sort that simply cannot be answered with the methods we use. For example, our scientific method begins with observations of the world. Without observations there can be no experiments, no scientific hypotheses, no falsifiable predictions. However, this means that phenomena which are not observable pose questions that science cannot answer.

- What happens after death?—a state which, by definition, we cannot observe—is not a scientific question.
- What happened before the beginning of time and space? also refers to something which, by definition, is not observable.
- Why are the scientific laws the way they are and not some other way? is a question which the scientific method is not equipped to answer.

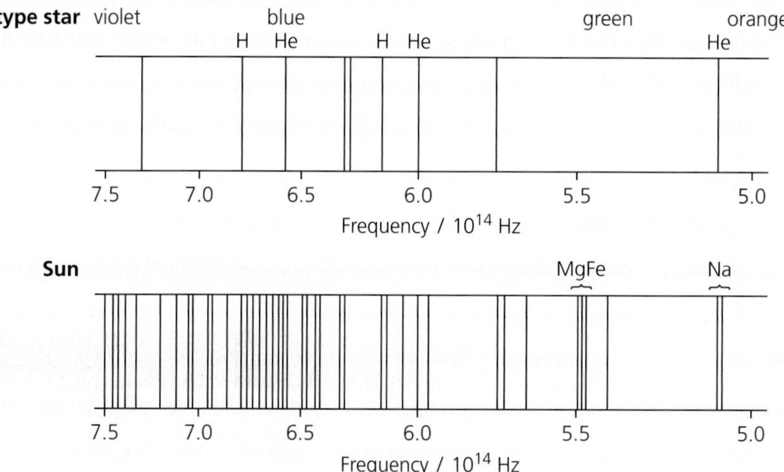

■ **Figure S1.56** The absorption spectra of a B-type (very luminous and blue) star and the Sun (luminous and yellow)

◆ **Principal quantum number** (*n*): Indicates the main energy levels and has values of 1, 2, 3 and so on; the higher the number, the further the electron is from the nucleus.

◆ **Atomic orbital**: A region in space where there is a high probability that an electron of a specific energy may be found. It represents the three-dimensional motion of an electron described by a mathematical wave function.

Orbitals and energy levels

The maximum number of electrons that a main energy level can have is $2n^2$ where n is the **principal quantum number** (shell number or main energy level).

Electrons can behave as waves and if their behaviour in atoms is modelled mathematically, the result is a series of three-dimensional shapes or **atomic orbitals**. An orbital is the *volume* of space around the nucleus of an atom in which there is a very high probability of locating an electron.

■ **Table S1.12** Structure of the first four main energy levels (shells)

Main energy level (shell) / principal quantum number, *n*	**Total number of electrons ($2n^2$)**	**Type and number of orbitals**	**Total number of orbitals**
1	2	s (one)	1
2	8	s (one) p (three)	4
3	18	s (one) p (three) d (five)	9
4	32	s (one) p (three) d (five) f (seven)	16

Spin

Nature of science: Experiments

The Stern-Gerlach experiment

Experiments by Otto Stern and Walther Gerlach, in Germany in the 1920s, showed that an electron has a magnetic dipole moment. In the Stern–Gerlach experiment, a beam of hot atoms was passed through a non-uniform magnetic field. The electrons were deflected either up or down by a constant amount, in roughly equal numbers.

Electromagnetic theory predicts that a spinning electrically charged sphere produces a magnetic dipole: it acts like a tiny bar magnet, with a north and a south pole. The results of the Stern–Gerlach experiment can be explained if the electron is spinning on its axis and the direction of spin is quantized—able to take one of two distinct values: clockwise or anticlockwise. These two directions of spin produce magnetic moments in opposite directions, often described as 'up' (given the symbol ↑) and 'down' (given the symbol ↓).

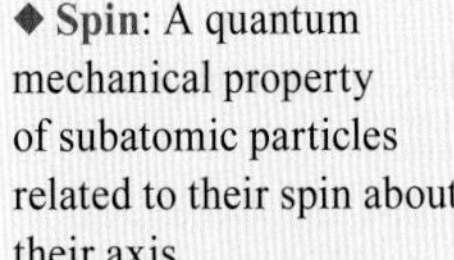

◆ **Spin**: A quantum mechanical property of subatomic particles related to their spin about their axis.

◆ **Spin pair**: Two electrons with opposite spins, occupying the same orbital.

◆ **Sublevel**: A subdivision (s, p, d or f) of a main energy level (electron shell).

Electrons (considered as particles) **spin** and generate a magnetic field. They have two spin states: 'clockwise' and 'anticlockwise' (Figure S1.57), represented by arrows pointing up and down.

In any orbital, the electron spins must be opposite: a single atomic orbital can hold a maximum number of two electrons as a **spin pair**.

Consider the helium atom. The two electrons in the 1s orbital must have opposite spins: one clockwise and one anticlockwise (Figure S1.58).

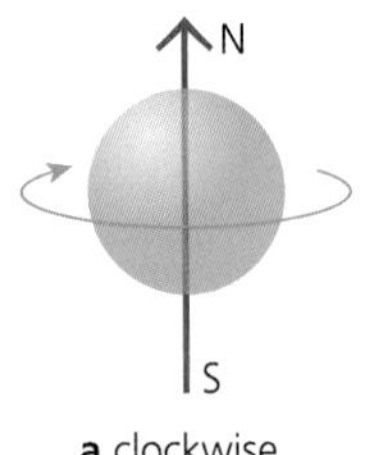
N
S
a clockwise

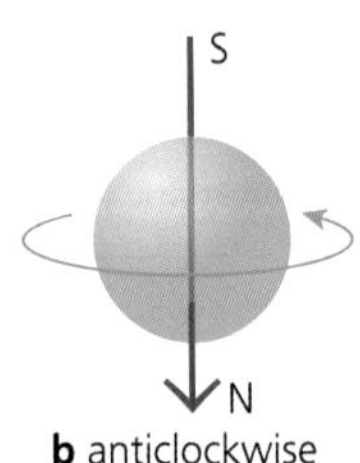
S
N
b anticlockwise

■ **Figure S1.57** Electron spin

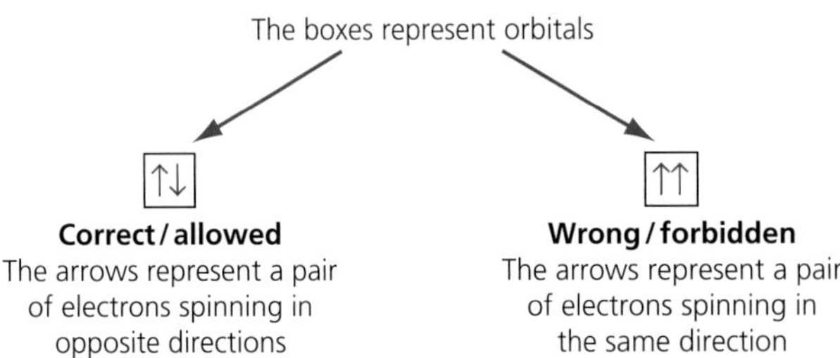
The boxes represent orbitals
↑↓
Correct / allowed
The arrows represent a pair of electrons spinning in opposite directions
↑↑
Wrong / forbidden
The arrows represent a pair of electrons spinning in the same direction

■ **Figure S1.58** Permitted and forbidden electron configurations for a helium atom

This type of 'electrons in boxes' diagram is often used to show the arrangement of electrons.

Types of orbital

There are four types of orbitals (in atoms in the ground state of the first seven periods) – s, p, d and f – which have different shapes and energies, and form **sublevels** within each shell. They can be visualized as groups of boxes, as in Figure S1.59, and each box can hold up to two electrons. An s sublevel holds two electrons, a p sublevel holds six electrons, a d sublevel holds 10 electrons and an f sublevel holds 14 electrons.

1s 2p 3d 4f

■ **Figure S1.59** 1s, 2p, 3d and 4f sublevels, each containing an unpaired electron, shown as boxes

The s orbitals are spherical and exist individually. Figure S1.60 shows the shapes of the 1s, 2s and 3s orbitals. The 3s orbital is larger and of higher energy than the 2s orbital which, in turn, is larger and of higher energy than the 1s orbital. Each s orbital can hold up to two electrons.

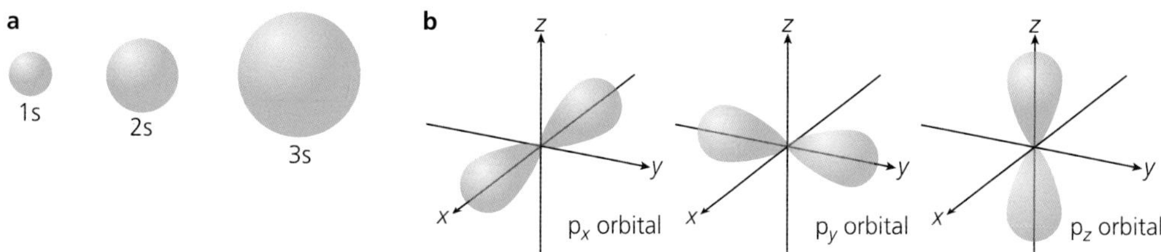

◆ **Lobe**: A region of electron density of an atomic orbital pointing away from the nucleus.

◆ **Degenerate**: Quantum states with the same energy.

■ **Figure S1.60 a** Shapes of s orbitals and **b** Shapes of three p orbitals from the same sublevel

Top tip!

For clarity, the three different p orbitals have been drawn separately. In reality, it is assumed that they are all superimposed on top of each other.

p orbitals have a dumbbell shape (with two **lobes**) and exist in groups of three arranged at right angles to each other. They are labelled according to the axis along which they lie. Figure S1.60 shows the shapes and arrangements of three p orbitals from the same sublevel.

Each single p orbital can hold up to two electrons. The three p orbitals all have the same energy—the orbitals are said to be **degenerate**. The 3p orbitals have the same shape as the 2p orbitals but are larger with a lower electron density.

Atoms may also have electrons in d orbitals (from $n = 3$ onwards) and in f orbitals (from $n = 4$ onwards). d orbitals exist in groups of five and f orbitals exist in groups of seven.

27 Deduce the total number of electrons in the fifth main energy level ($n = 5$) and the number of orbitals in the 5s, 5p and 5d sublevels.

The five 3d orbitals have five different shapes. Four of these orbitals ($3d_{xy}$, $3d_{xz}$, $3d_{yz}$ and $3d_{x^2-y^2}$) consist of four lobes in the same plane as one another and pointing mutually at right angles. The fifth ($3d_{z^2}$) is best represented as a two-lobed orbital surrounded by a 'doughnut' of electron density around the nodal plane (Figure S1.61).

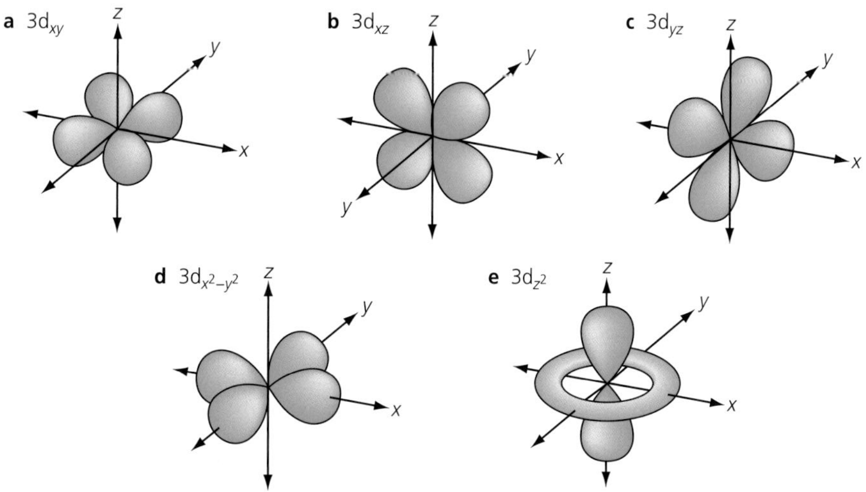

■ **Figure S1.61** The shapes of the 3d orbitals

Top tip!

The shapes of the d orbitals do not need to be known but d orbitals are important in the bonding of transition metal complexes. Their shapes and the orientations of their lobes are used to explain the splitting patterns of the d orbitals in complexes.

The most important orbitals are those in the outer or valence shells, which are involved in the formation of chemical bonds.

Covalent bonds are formed when orbitals overlap and merge to form molecular orbitals. The electron pair in a single bond occupies a molecular orbital formed by the overlap of an atomic orbital on each atom (Figure S1.62). The atomic orbitals can be s orbitals, p orbitals, or a combination of s and p orbitals.

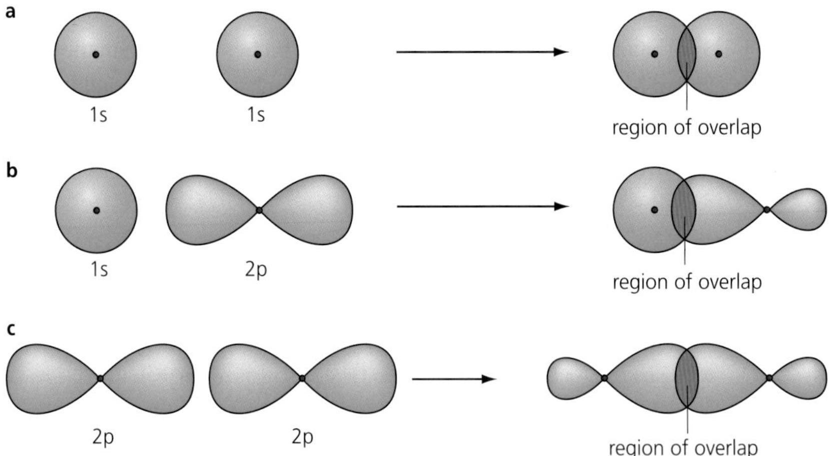

■ **Figure S1.62** Molecular orbitals formed by the overlap of s and p atomic orbitals to form sigma bonds where the electron density (shown by the overlap) will be located between the two nuclei

Going further

Penetration and shielding

The electron density of the 1s orbital is shown in Figure S1.63. Notice how the electron density is highest at the nucleus and falls off exponentially with increasing distance from the nucleus (in all directions).

Like the 1s orbital, the 2s orbital is a sphere but its electron density is not a simple exponential function. The 2s orbital has a smaller amount of electron density close to the nucleus. Most of the electron density is farther away, beyond a region of zero electron density called a node (Figure S1.63). Because most of the 2s electron density is farther from the nucleus than the 1s, the 2s orbital is higher in energy. This means that the 2s orbital is less penetrating and has a smaller shielding effect than the 1s orbital. Shielding describes the amount of screening from nuclear charge that an electron can do compared to its neighbouring electrons. Electrons that are more penetrating can get closer to the nucleus and effectively block out the charge from electrons that are further away.

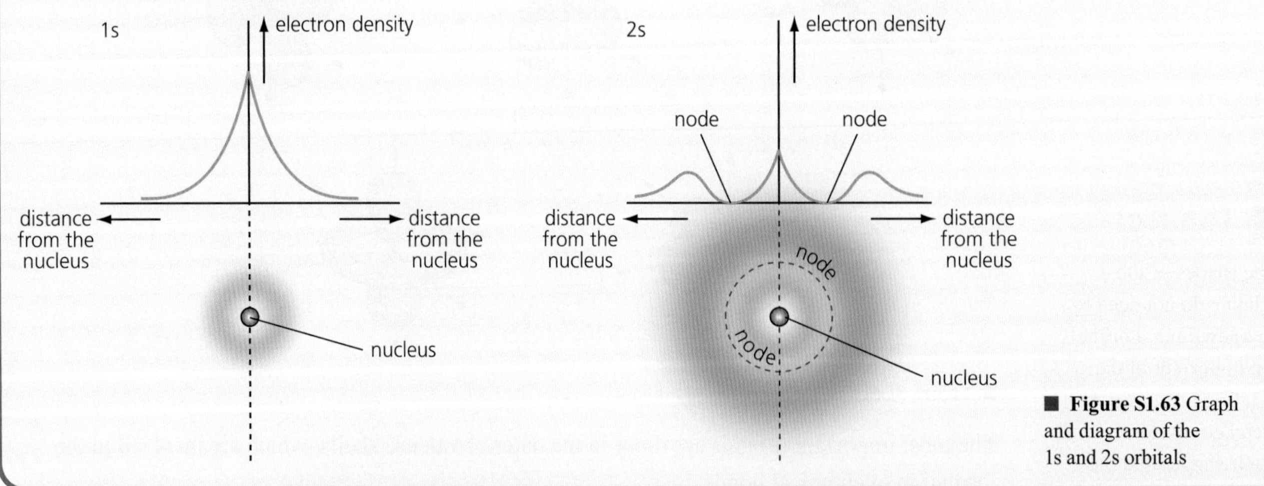

■ **Figure S1.63** Graph and diagram of the 1s and 2s orbitals

LINKING QUESTION

What is the relationship between energy sublevels and the block nature of the periodic table?

◆ **Aufbau principle**: A principle that gives the order in which atoms' orbitals are filled from lowest to highest energy in successive elements in the periodic table.

◆ **Electron configuration (full)**: The arrangement of all the electrons within the atomic orbitals of an atom.

◆ **Electron configuration (condensed)**: Includes the symbol for the nearest noble gas in square brackets and then the symbols for the occupied orbitals that contain valence electrons.

Top tip!

Exceptions to the Aufbau principle may occur when orbitals have similar energies and electron repulsion favour the presence of a single electron in each orbital.

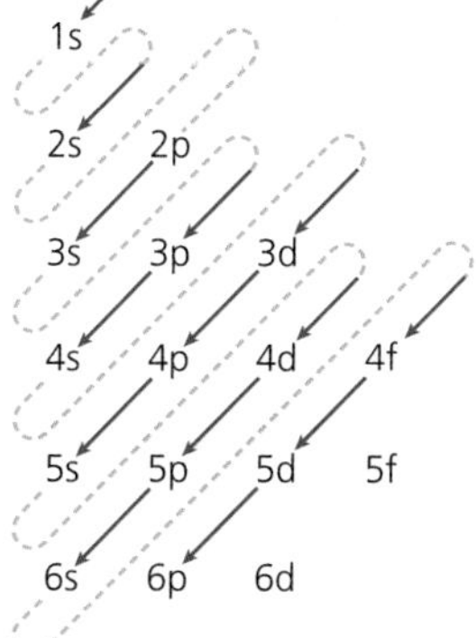

■ **Figure S1.65** Electrons in energy levels or orbitals to show the application of the Aufbau principle

Filling atomic orbitals

Figure S1.64 shows the arrangement of s, p, d and f atomic orbitals as boxes, where *n* represents the main energy level. The higher up the diagram, the greater the energy the electrons have and the further they are from the nucleus (on average).

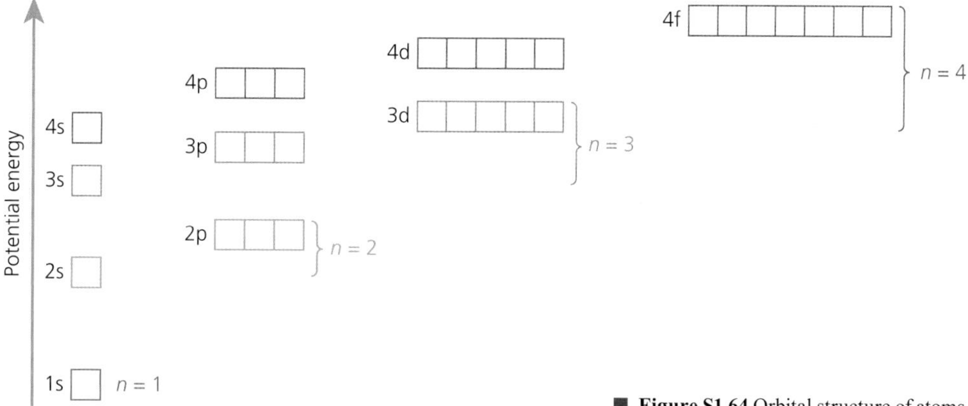

■ **Figure S1.64** Orbital structure of atoms

Figure S1.65 shows the order in which atomic orbitals are filled with electrons. Electrons fill the lowest-energy orbitals until these are full and then fill the next lowest levels. This rule is known as the **Aufbau principle** (*Aufbau* is German for 'building up').

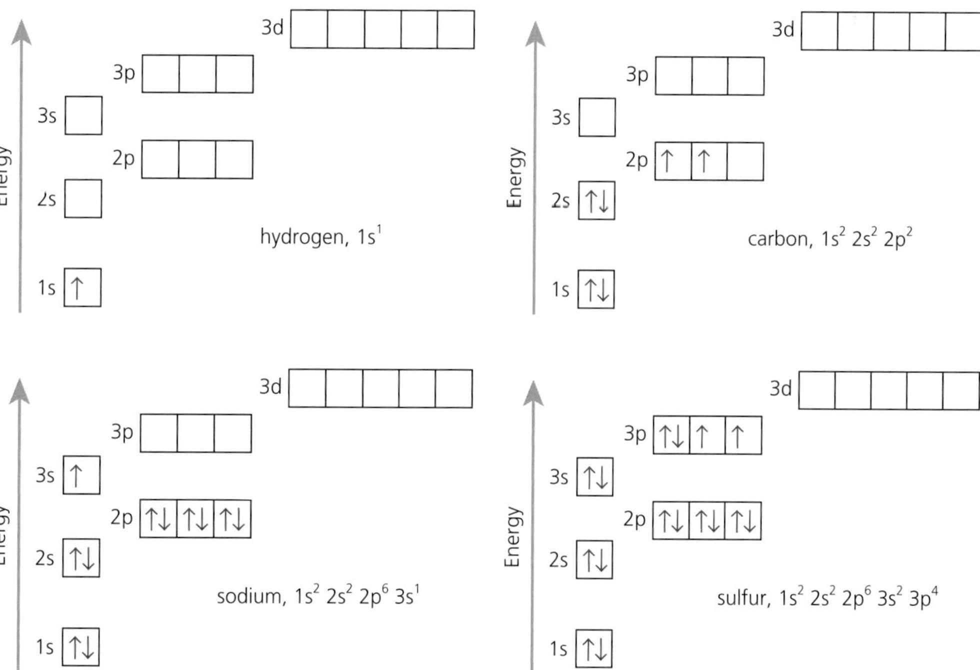

■ **Figure S1.66** The Aufbau principle for filling atomic orbitals with electrons

Figure S1.66 shows the Aufbau principle applied to four atoms: hydrogen, sodium, carbon and sulfur. Electrons are added to their atomic orbitals to produce the most stable **electron configuration** (this is the ground state). The diagram also shows how the detailed electron configuration of an element can be written on a single line.

Condensed electron configurations can be written by replacing the inner electron configuration with a noble gas symbol. For example, the full configuration of the lithium atom is $1s^2\ 2s^1$ and the condensed configuration is [He] $2s^1$.

28 Draw electrons in atomic orbitals for a silicon atom with the orbitals arranged in a linear sequence from lowest to highest energy. Label the sublevels with s, p, d and f, as appropriate.

As the principal quantum number (n) increases, the energy levels of the sublevels come close to each other. As a result, overlapping of sublevels occurs. The first overlap occurs between the 3d and 4s sublevels (Figure S1.67). The 3d orbitals are at a slightly higher energy level than the 4s orbitals and so the 4s orbital fills first. The 3d orbitals are filled in the first row of the transition metals: scandium to zinc.

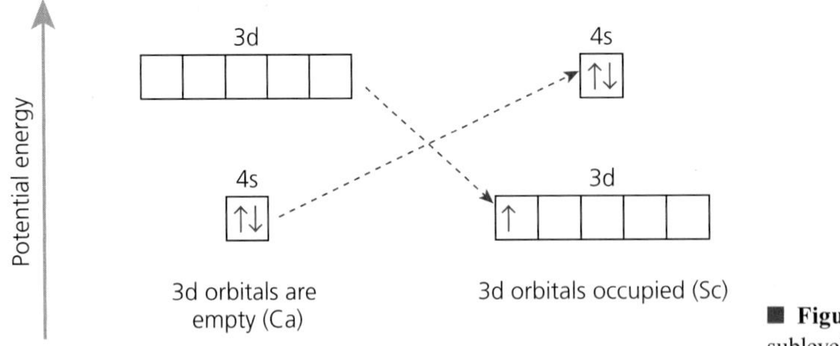

■ **Figure S1.67** The 3d–4s sublevel overlap

◆ **Pauli exclusion principle**: A maximum of two electrons (with opposite spins) can occupy an atomic orbital.

◆ **Hund's rule**: Every orbital in a sublevel is singly occupied with one electron before any one orbital is doubly occupied, and all electrons in singly occupied orbitals have the same spin.

◆ **Quantum mechanics**: A branch of physics that describes the behaviour of light, molecules, atoms and subatomic particles.

Rules for filling atomic orbitals

1 Aufbau principle: electrons enter the lowest available energy level according to the energy levels.

2 **Pauli exclusion principle**: two electrons of opposite spin enter each orbital.

3 **Hund's rule**: the orbitals of a sublevel (p, d or f) must be occupied singly by electrons (with parallel spins) before they can be occupied in spin pairs.

Consider two electrons in the orbitals of a 2p sublevel (Figure S1.68). The permitted configuration has two electrons with the same spin.

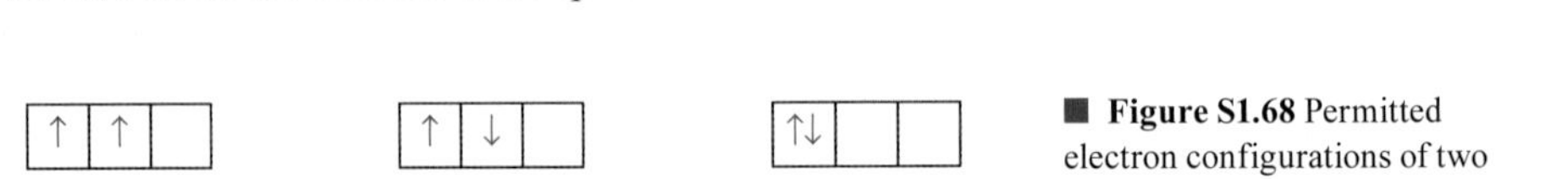

■ **Figure S1.68** Permitted electron configurations of two electrons in a 2p sublevel

Going further

Pauli exclusion principle

The Pauli exclusion principle is one of the fundamental principles of **quantum mechanics**. It can be tested by a simple experiment with liquid helium.

If the two electrons in the 1s orbital of a helium atom (in the ground state) had the same (parallel) spins (↑↑ or ↓↓), their overall magnetic fields would reinforce each other (Figure S1.69). Such an arrangement would make helium gas paramagnetic. Paramagnetic substances are those that contain unpaired electrons and are weakly attracted by a strong magnetic field. However, if the electron spins are paired and antiparallel to each other (↓↑ or ↑↓), the magnetic effects cancel out and the gas is diamagnetic.

Diamagnetic substances do not contain any unpaired electrons and are slightly repelled by a strong magnetic field. Experiments have shown that helium is diamagnetic and therefore the 1s orbital has a spin pair with paired or antiparallel spins.

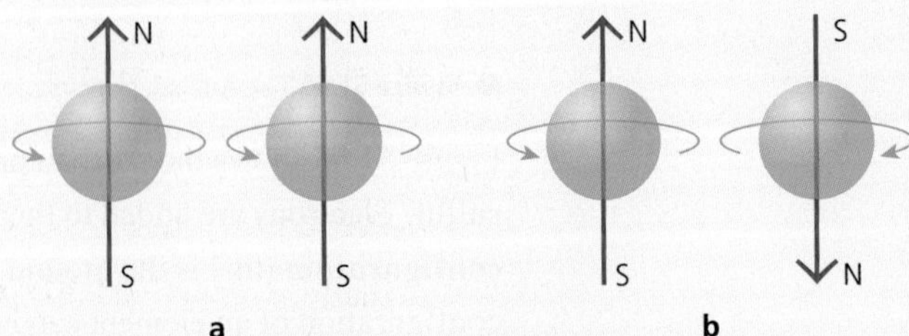

■ **Figure S1.69** The **a** parallel and **b** antiparallel spins of two electrons. In **a** the two magnetic fields reinforce each other and in **b** the two magnetic fields cancel each other.

When an electron is added to an atom, a negative ion is formed. The electron is added to the empty orbital of the lowest energy, for example: F $1s^2\ 2s^2\ \mathbf{2p^5}$; F^- $1s^2\ 2s^2\ \mathbf{2p^6}$. When an electron is removed from a metal atom, a positive ion is formed. The electron is removed from the occupied orbital with the highest energy, for example: Mg $1s^2\ 2s^2\ 2p^6\ \mathbf{3s^2}$; Mg^{2+} $1s^2\ 2s^2\ 2p^6$.

When one or more electrons absorb thermal or electrical energy, they are promoted into higher energy orbitals. An excited sodium atom is shown in Figure S1.70. The return of the excited electron to the ground state will give rise to emission of electromagnetic radiation corresponding to a specific line in the emission spectrum of gaseous sodium atoms.

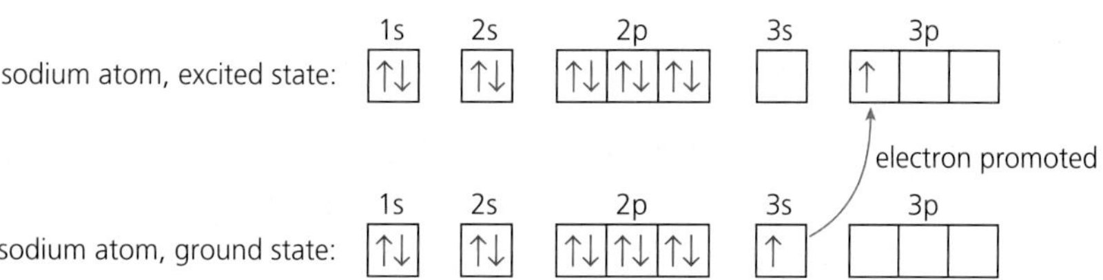

■ **Figure S1.70** Orbital notation for sodium atoms in ground and excited states

Transition elements

The elements from scandium to zinc form the first row of the d-block (Table S1.13). This block contains ten metallic elements, because the 3d sublevel contains five 3d orbitals, each able to hold two electrons (a spin pair).

■ **Table S1.13** Outer electron configurations of the first-row d-block metals ([Ar] represents the configuration of the noble gas argon). Electron configurations are condensed configurations and orbital diagrams have arrows representing electrons

		Electron configurations								
Element	**Atomic number**				***3d***					***4s***
Sc	21	[Ar]	$3d^1\ 4s^2$	[Ar]	↑					↑↓
Ti	22	[Ar]	$3d^2\ 4s^2$	[Ar]	↑	↑				↑↓
V	23	[Ar]	$3d^3\ 4s^2$	[Ar]	↑	↑	↑			↑↓
Cr	24	[Ar]	$3d^5\ 4s^1$	[Ar]	↑	↑	↑	↑	↑	↑
Mn	25	[Ar]	$3d^5\ 4s^2$	[Ar]	↑	↑	↑	↑	↑	↑↓
Fe	26	[Ar]	$3d^6\ 4s^2$	[Ar]	↑↓	↑	↑	↑	↑	↑↓
Co	27	[Ar]	$3d^7\ 4s^2$	[Ar]	↑↓	↑↓	↑	↑	↑	↑↓
Ni	28	[Ar]	$3d^8\ 4s^2$	[Ar]	↑↓	↑↓	↑↓	↑	↑	↑↓
Cu	29	[Ar]	$3d^{10}\ 4s^1$	[Ar]	↑↓	↑↓	↑↓	↑↓	↑↓	↑
Zn	30	[Ar]	$3d^{10}\ 4s^2$	[Ar]	↑↓	↑↓	↑↓	↑↓	↑↓	↑↓

Top tip!

A simplified explanation for these observations is that a half-filled or completely filled 3d sublevel is a particularly stable electron configuration. These configurations can be more fully explained using the concept of exchange energy (a quantum-mechanical effect).

Note that chromium and copper have one 4s electron, not two. This is to allow either a half-filled or a filled d sublevel to be formed: the chromium atom has a $3d^5\ 4s^1$ configuration and a copper atom is $3d^{10}\ 4s^1$.

WORKED EXAMPLE S1.3C

Deduce the total number of electrons in d orbitals in a tin atom.

Answer

A tin atom has the following electron configuration: $1s^2\ 2s^2\ 2p^6\ 3s^2\ 3p^6\ 3d^{10}\ 4s^2\ 4p^6\ 4d^{10}\ 5s^2\ 5p^2$, so it has 20 d electrons.

The d-block elements form positive ions as they are metals (most are transition metals). This means electrons are removed from the atom and a cation is formed. The 4s electrons are removed before the 3d electrons—this can be seen in, for example, the iron(II) ion, [Ar] $3d^6$ and the iron(III) ion, [Ar] $3d^5$; and the copper(I) ion, [Ar] $3d^{10}$ and the copper(II) ion, [Ar] $3d^9$.

29 Write full and condensed electron configurations for the following:

Ti^{3+} Cr^{2+} Cu P^{3-} Cu Ga Mg As Sr K^+

Assume all atoms and ions are gaseous and in the ground state.

Quantum numbers specify the properties of the atomic orbitals and the electrons in those orbitals. An electron in an atom has four quantum numbers to describe its state, including its energy and spin. Research and make a presentation to your class on the four quantum numbers and apply them to the atoms of the first ten elements.

Does the precision of the language used in the natural sciences successfully eliminate all ambiguity?

A traditional view of language in science is that it plays a passive role: it is simply a 'vehicle' conveying meaning and information from one person to another. Attempting to express a new scientific idea becomes simply 'trying to find the right words'. Such an attitude is an extension of the assumption that the essential role of language is to transport a cargo which has meaning or content.

However, many scientists and philosophers, including Niels Bohr, consider that language plays a more active role than is implied by this model. Bohr made the point that scientific investigations do not just employ mathematical analysis; they also include discussions centred around concepts that are described by words. Quoting Bohr:

> *Quantum mechanics provides us with a striking illustration of the fact that though we can fully understand a connection we can only speak of it in images and parables. We must be clear that when it comes to atoms, language can only be used as in poetry. The poet, too, is not so nearly concerned with describing facts as with creating images and establishing mental connections.*

During a paradigm shift—a change in scientific perspective, such as the move from classical mechanics to quantum mechanics—the meaning of words often changes. Like a map, the availability and choice of certain concepts and terms may constrain the knowledge being sought. Bohr argued that some words (such as *position, momentum, space* and *time*) refer to classical concepts which are irrelevant to quantum theory. He also believed that it was not possible to easily modify or change the meanings of words used to describe classical concepts. This is the challenge faced by emerging paradigms of how to describe the natural world: until the new language, or the new understanding of old concepts, is developed and accepted, it is difficult to shift our descriptions of the world.

One example is the term 'spin'. In the classical paradigm, spin is defined as the number of rotations a particle must make to regain its original state and 'look' exactly 'the same'. For example, an asymmetrical object like a book requires a complete rotation of 360° in order to appear as it did before it was turned round. It has a spin of 1: it is back in the same state after a full rotation.

However, the spin of a particle is a purely quantum-mechanical quantity. Electrons are assigned a spin number of $\pm\frac{1}{2}$. A spin $-\frac{1}{2}$ particle needs two full rotations ($2 \times 360° = 720°$) in order to return to the same state. There is nothing in our classical, macroscopic world which has a symmetry like that.

Ionization energies

Ionization energy (sometimes abbreviated to IE) is a measure of the amount of energy needed to remove one mole of electrons from gaseous atoms. As electrons are negatively charged and protons in the nucleus are positively charged, there is electrostatic attraction between them.

The greater the electrostatic 'pull' of the nucleus, the harder it is to remove an electron from an atom. The first ionization energy is the minimum energy required to convert one mole of gaseous atoms into gaseous ions with a single positive charge.

Ionization energies are always endothermic: energy has to be absorbed and work done so that the negatively charged electron can be removed from the influence of the positively charged nucleus.

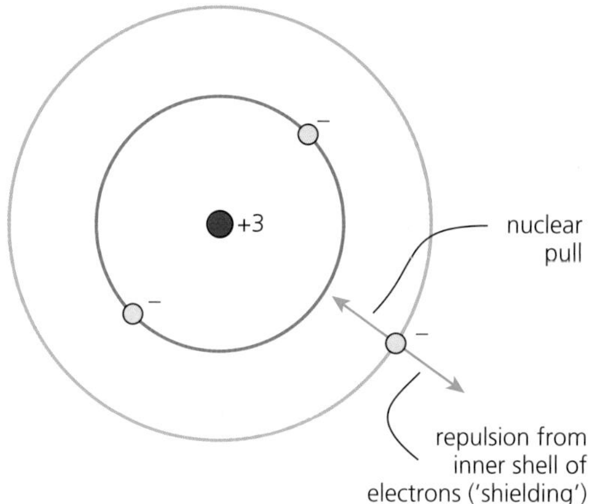
+3
nuclear pull
repulsion from inner shell of electrons ('shielding')

■ **Figure S1.71** Electrostatic forces operating on the outer or valence electron in a lithium atom

◆ **Shielding**: For atoms or ions with more than one electron, the effect that electrons have on each other in terms of shielding and hence reducing the nuclear charge experienced by the valence (outer) electrons.

For example, the first ionization energy of chlorine is the energy required to bring about the process $Cl(g) \rightarrow Cl^+(g) + e^-$. The electron is removed from the outer 3p sublevel of the chlorine atom. Figure S1.71 shows the removal of the outer 2s electron of a lithium atom (Li) and the formation of its cation (Li^+).

Top tip!

Ionization energies are always endothermic, even for reactive group 1 metals. Ionization energies are measured for isolated gas-phase atoms. Ions are more stable, and hence lower in energy, when in a lattice or hydrated in aqueous solution.

Common mistake

Misconception: *A low value of ionization energy means it is harder to remove an electron from an atom.*

The lower the value of the ionization energy, the *easier* it is to remove an electron from an atom since less energy is required.

Factors that affect ionization energy

Atomic radius

As the distance of the outer electrons from the nucleus increases, the attraction of the positive nucleus for the negatively charged electrons falls. This causes the ionization energy to decrease.

Nuclear charge

When the nuclear charge becomes more positive (due to the presence of additional protons), its attraction on all the electrons increases. This causes the ionization energy to increase.

Shielding effect

The outer or valence electrons are repelled by all the other electrons in the atom in addition to being attracted by the positively charged nucleus. The outer valence electrons are **shielded** from the attraction of the nucleus (Figure S1.72) due to the presence of the inner electrons (core electrons).

In general, the shielding effect is greatest when the electrons are close to the nucleus. Consequently, electrons in the first shell, where there is high electron density, have a stronger shielding effect than electrons in the second shell, which have a stronger shielding effect than electrons in the third shell. Electrons added to the same shell have a relatively small shielding effect. The simplest model of electron shielding is that each inner electron shields one unit of nuclear charge irrespective of the energy level it is in.

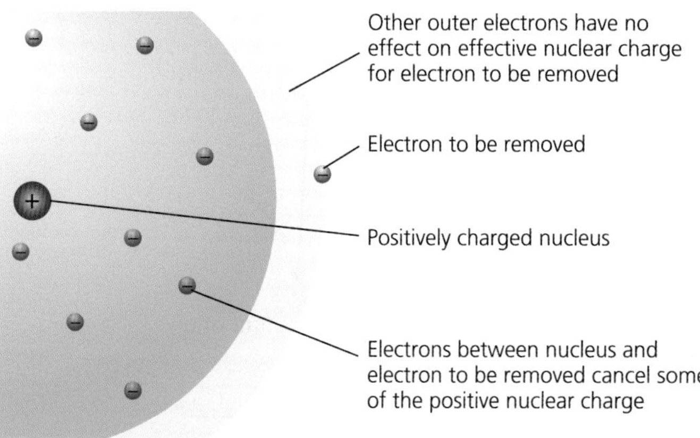
Other outer electrons have no effect on effective nuclear charge for electron to be removed
Electron to be removed
Positively charged nucleus
Electrons between nucleus and electron to be removed cancel some of the positive nuclear charge

■ **Figure S1.72** The shielding effect

Common mistake

Misconception: *The electrons in an inner shell of an atom set up a barrier to the attractive force between the nucleus and electrons in an outer shell.*

The inner (core) electrons of an atom approach more closely to the nucleus and so partially neutralize its charge.

Going further

Effective nuclear charges

The attractive force experienced by valence electrons is known as the effective nuclear charge (ENC). It can be approximated by the expression:

ENC = nuclear charge (proton number) − number of shielding (inner) electrons

The greater the ENC, the greater the amount of energy needed to remove the electron (Table S1.14). The simple ENC calculations suggest sodium will have the lowest IE and chlorine will have the highest IE. ENC increases across a period of elements and decreases down a group of elements.

■ **Table S1.14** Calculation of ENC for selected atoms

Element and electron arrangement	Nuclear charge	Number of core electrons	ENC
Na (**2,8**,1)	+11	10	+1
Mg (**2,8**,2)	+12	10	+2
P (**2,8**,5)	+15	10	+5
Cl (**2,8**,7)	+17	10	+7

Ionization energies down a group and across a period

Figure S1.73 shows a plot of the first ionization energies for atoms of the first forty elements. Some clear trends can be observed for the elements before the transition metals (d-block).

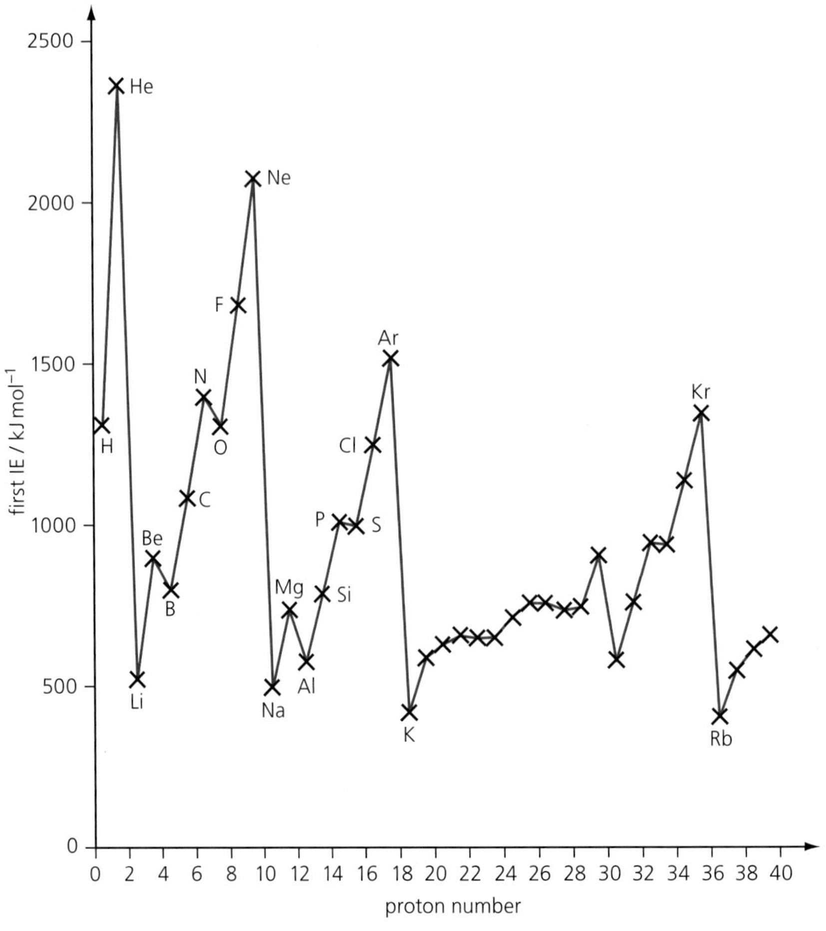

■ **Figure S1.73** First ionization energies for atoms of elements 1 to 40

Changes down a group

Figure S1.73 shows that down a particular group, for example from helium to neon to argon, or from lithium to sodium to potassium, ionization energies decrease. The larger the atom, the less energy is needed to remove an electron from it and the lower the value of the first ionization energy.

The larger the radius of the atom, the larger the distance between the outer electron and the nucleus, so the smaller the electrostatic attraction between them. The increase in nuclear charge (due to an increase in proton number) down a group is cancelled by electron shielding, with more core electrons located between the nucleus and the valence electrons, reducing the charge on the nucleus.

30 Explain why the hydrogen atom, H(g), has a smaller value of first ionization energy than the helium ion, $He^{+}(g)$.

Changes across a period

Ionization energies tend to increase across a period: atoms of the group 1 elements (the alkali metals) have the lowest ionization energy within each period, and atoms of the noble gases in group 18 have the highest values. However, the increase is not uniform. For example, in periods 2 and 3 there are dips between groups 2 and 13 and between groups 15 and 16.

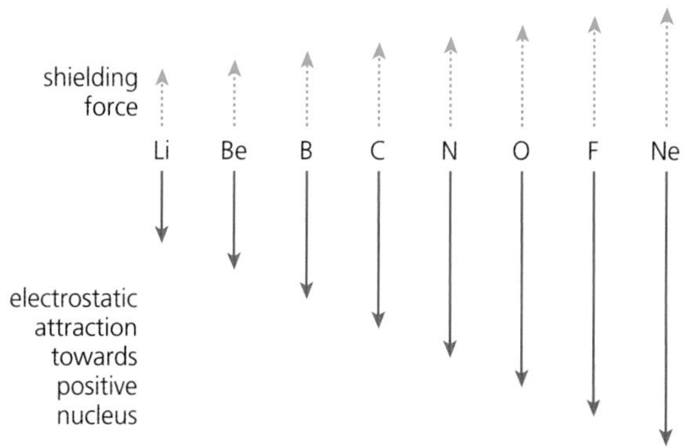

■ **Figure S1.74** A diagram illustrating how the balance between electron shielding and nuclear charge changes across period 2

Going across a period, the atom of each element has an extra proton in the nucleus and an extra electron in the outermost (valence) shell, increasing the attractive force between electrons and the nucleus. The additional valence electron enters the same shell (main energy level). Because electrons in the same shell are at (approximately) the same distance from the nucleus, the increase in shielding is very small (Figure S1.74). Thus, the effective nuclear charge increases, the atomic radius is reduced, and the increased electrostatic attraction results in a greater ionization energy across the period.

The first decrease in each period is due to a change in the sublevel from which the electron is lost. The change in electron shielding that results from this has a greater effect than the increase in nuclear charge and decrease in atomic radius.

In period 2, the first decrease occurs between beryllium ($1s^2\,2s^2$) and boron ($1s^2\,2s^2\,2p^1$). The outermost (valence) electron in a boron atom is in a 2p orbital (Figure S1.75). The average distance of a 2p orbital from the nucleus is slightly greater than the average distance of a 2s orbital from the nucleus and electrons in the 2p orbital are partially shielded by those in the 2s orbital (Figure S1.76). The outer electron in the boron atom therefore experiences a smaller electrostatic force of attraction than the outer 2s electron in the beryllium atom and so is easier to remove. This means the ionization energy of boron is less than that of beryllium.

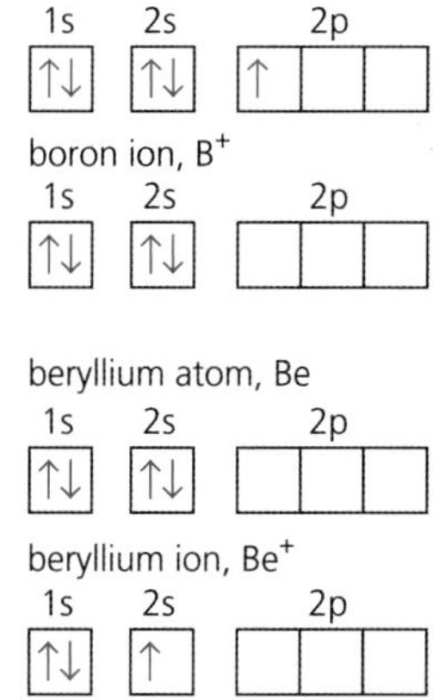

■ **Figure S1.75** Orbital notations for boron and beryllium atoms and their unipositive ions

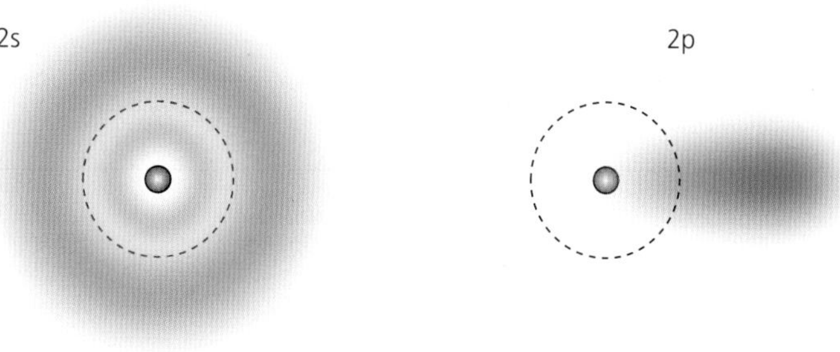

■ **Figure S1.76** Electron density clouds of the 2s and 2p orbitals (only one lobe shown). The dotted line shows the extent of the 1s orbital; the 2s electrons can partially penetrate the 1s orbital, increasing its stability

The decrease in first ionization energy from magnesium ($1s^2\,2s^2\,2p^6\,3s^2$) to aluminium ($1s^2\,2s^2\,2p^6\,3s^2\,3p^1$) can be explained in a similar way: the electrons in the filled 3s orbital are more effective at shielding the electron in the 3p orbital than they are at shielding each other.

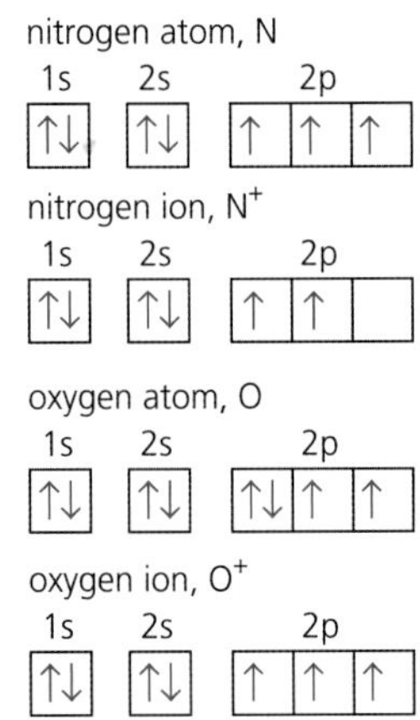

■ **Figure S1.77** Orbital notation for nitrogen and oxygen atoms and their unipositive ions

In period 2, the second dip in first ionization energy occurs between nitrogen ($1s^2 2s^2 2p^3$) and oxygen ($1s^2 2s^2 2p^4$). Due to Hund's rule, the nitrogen atom has the maximum number of unpaired electrons in the 2p sublevel: the full electronic configuration is $1s^2 2s^2 2p_x^1 2p_y^1 2p_z^1$. This is the lowest energy, most stable state with the least amount of electrostatic repulsion.

The additional electron in the oxygen atom has to enter an orbital that already contains an electron and, by the Pauli principle, form a spin pair (Figure S1.77). The two electrons sharing the same orbital repel each other relatively strongly. This electron–electron repulsion is larger than the extra attraction the new electron experiences from the additional proton in the nucleus. Thus, the energy needed to remove a 2p electron from the oxygen atom is slightly less than that needed to remove a 2p electron from a nitrogen atom.

In period 3, the first ionization energy of sulfur ($1s^2 2s^2 2p^6 3s^2 3p_x^2 3p_y^1 3p_z^1$) is also less than that of phosphorus ($1s^2 2s^2 2p^6 3s^2 3p_x^1 3p_y^1 3p_z^1$) because less energy is required to remove an electron from the 3p orbitals of sulfur than from the half-filled 3p orbitals of phosphorus.

Similar inter-electron repulsions are experienced by the additional electrons in the fluorine and neon atoms, so the ionization energy of each of these elements is, like oxygen's, about $430\,kJ\,mol^{-1}$ less than expected (Figure S1.78).

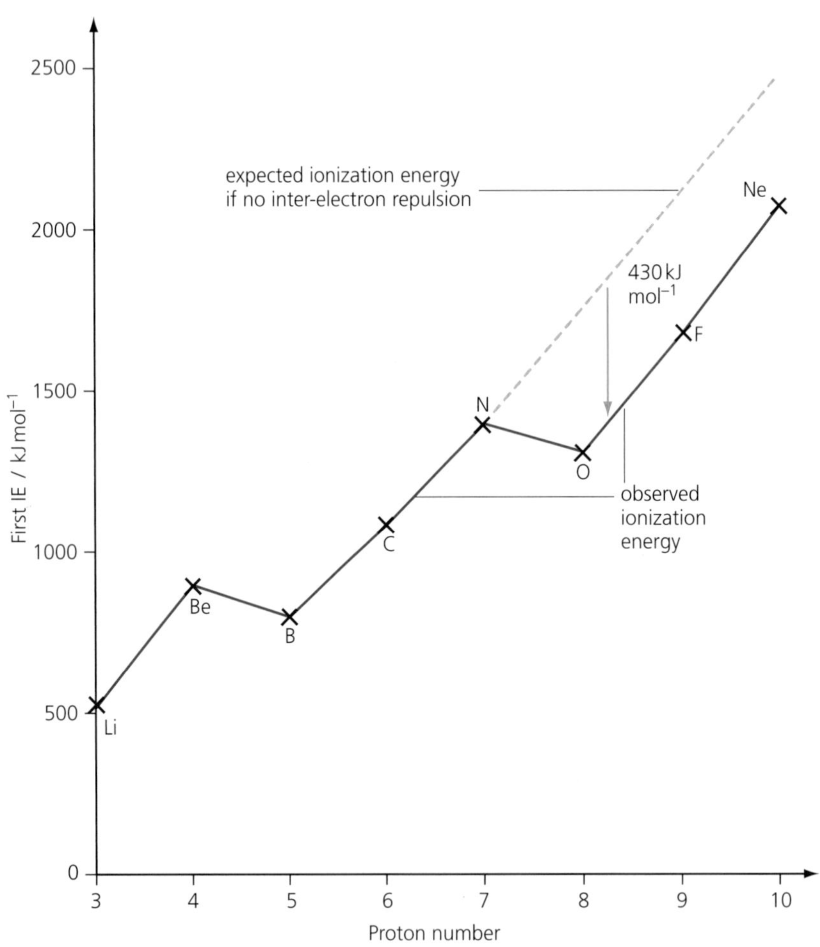

■ **Figure S1.78** A graph of first ionization energy against proton number for period 2

The 2-3-3 pattern of first ionization energy across periods 2 and 3 (Figure S1.79) is identical except the first ionization energies for period 2 are higher. This is because the electrons being removed are in a second shell closer to the nucleus, rather than in a third shell for the period 3 elements.

The outer electrons in period 2 atoms experience a higher effective nuclear charge (due to lower shielding) than those in period 3.

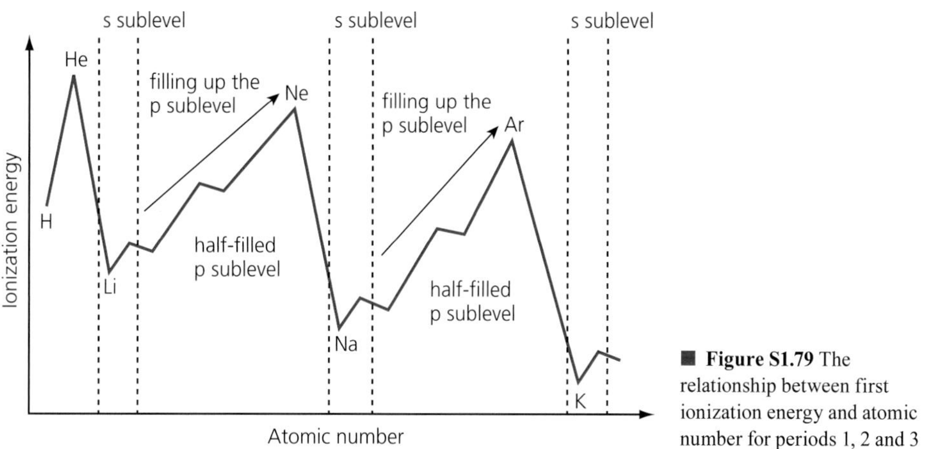

■ **Figure S1.79** The relationship between first ionization energy and atomic number for periods 1, 2 and 3

Table S1.15 summarizes trends in first ionization energy across a period and down a group of the periodic table.

■ **Table S1.15** Summary of trends in first ionization energy

Across a period: ionization energy increases	Down a group: ionization energy decreases
Increase in nuclear charge (number of protons).	Increase in nuclear charge (number of protons).
The number of main energy levels (shells) is the same; the average distance of the valence electrons from the nucleus decreases.	Increase in the number of main energy levels (shells); the average distance of the valence electrons from the nucleus increases.
Shielding increases slightly.	Shielding increases significantly.
Decrease in atomic / ionic radius.	Increase in atomic / ionic radius.
All the electrons including the valence electrons are held closer to the nucleus with stronger forces of electrostatic attraction.	All the electrons including the valence electrons are held further away from the nucleus with weaker forces of electrostatic attraction.

Ionization energies and metallic character

The key physical and chemical properties of metals and non-metals are summarized in Table S1.16. Most elements are metals; some are non-metals and others are classified as metalloids because they have some properties in common with both metals and non-metals.

■ **Table S1.16** Selected properties of metals and non-metals

Property	Metals	Non-metals
Thermal and electrical conductivity	Excellent conductors of heat and electricity.	Poor conductors of heat and electricity.
Ions formed	Usually form cations (positive ions).	Usually form anions (negative ions).
Bonding in compounds	Usually ionic.	Ionic and covalent.
Nature of oxides	Most metals form basic oxides that are ionic (but transition metals can form acidic oxides in their higher oxidation states).	Most non-metals form acidic oxides that are covalent.

The most important chemical property of metals is their electropositivity, their ability to lose valence electrons and form positively charged ions.

In general, metals (especially the alkali metals in group 1) have low ionization energies, but non-metals (especially the halogens in group 17) have high ionization energies. This means that group 1 metals are chemically reactive, since their atoms readily lose electrons to form cations, but halogen atoms have little tendency to lose electrons and form cations.

HL ONLY

Due to their lower ionization energies, metals tend to lose electrons and act as reducing agents, whereas non-metals, with higher ionization energies, tend to gain electrons, filling their outer shell and acting as oxidizing agents.

The extent to which an element demonstrates metallic behaviour can be compared with changes in first ionization energies as we move around the periodic table:

- The first ionization energy of an element generally increases across a period and decreases down a group. This results in less tendency for cation formation, and so less metallic behaviour, across the period and up a group.
- Atomic radius increases down a group with an increasing number of electron shells (main energy levels). The valence electrons are less firmly held by the nucleus and increasingly shielded from it by the inner electron shells. The ionization energy decreases (Table S1.17), so the tendency for cation formation, and so metallic behaviour, increases.

■ **Table S1.17** First ionization energies of group 1 elements (alkali metals)

Group 1 metal	First ionization energy / $kJ\,mol^{-1}$
lithium, Li	520
sodium, Na	496
potassium, K	419
rubidium, Rb	403
caesium, Cs	376

LINKING QUESTION

How does the trend in ionization energy values across a period and down a group explain the trends in properties of metals and non-metals?

Successive ionization energies

Table S1.18 shows the equations describing the first, second and third ionization energies of a potassium atom (using data from WebElements).

■ **Table S1.18** Successive ionization energies for potassium atoms (in a mass spectrometer from electron bombardment)

Ionization energy	Ionization equation	Ionization energy/ $kJ\,mol^{-1}$
first	$K(g) \rightarrow K^+(g) + e^-$	+419
second	$K^+(g) \rightarrow K^{2+}(g) + e^-$	+3051
third	$K^{2+}(g) \rightarrow K^{3+}(g) + e^-$	+4419

◆ **Successive ionization energies:** Energies involved in the successive removal of molar quantities of electrons from gaseous atoms.

The second ionization energy of any atom is always larger than the first ionization energy because more energy is required to remove an electron from a unipositive ion than a neutral atom. Further **successive ionization energies** increase because the electrons are being removed from increasingly positive ions and so the electrostatic forces on the remaining electrons are greater. There are also fewer remaining electrons to provide shielding.

Figure S1.80 provides strong experimental evidence that the electron configuration of a potassium atom is 2,8,8,1 as the greatest increases in ionization energy occur after 1, 9 and 17 electrons are removed (1 + 8 = 9 and 1 + 8 + 8 = 17). These large increases in ionization energy correspond to removal of core electrons from inner shells whose electrons are nearer to the nucleus. A logarithmic scale (to the base 10) is used because the values of successive ionization energies have such a large range.

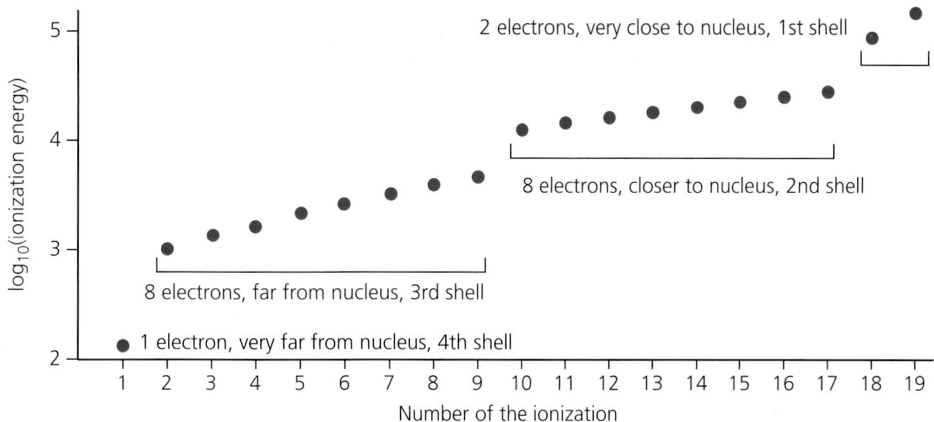

■ **Figure S1.80** Successive ionization energies for potassium

Identifying an element from successive ionization energies

Considering successive ionization energies for individual elements allows us to identify an element's group in the periodic table.

Consider the successive ionization energies of aluminium ($1s^2\ 2s^2\ 2p^6\ 3s^2\ 3p^1$):

- The first ionization energy is 578 kJ mol^{-1}. This is relatively low because the 3p electron is shielded from the nucleus by all the other electrons.
- The second and third ionization energies (1817 kJ mol^{-1} and 2745 kJ mol^{-1}) are significantly higher than the first because the 3s electrons are being removed from increasingly positive ions and are not as shielded as the 3p electron.
- There is a large increase to the fourth ionization energy, 11 578 kJ mol^{-1}, since the 2p electron being removed is in a shell (main energy level) closer to the nucleus and the shielding has decreased.
- The next two ionization energies are 14 831 kJ mol^{-1} and 18 378 kJ mol^{-1}, so we again have relatively small increases.

Since three valence electrons are removed before a large jump in the ionization energy, this suggests the element is in group 13.

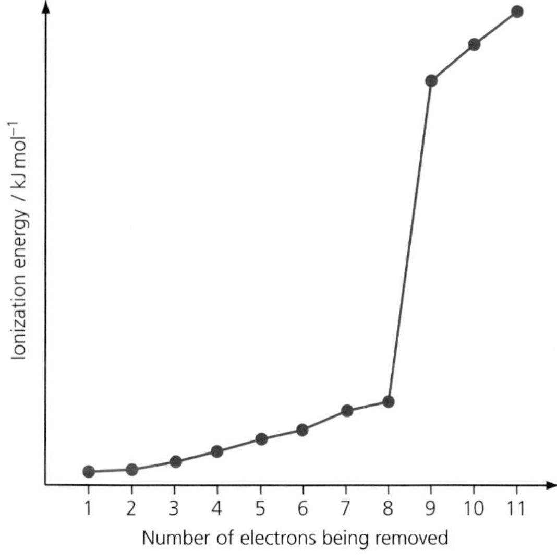

■ **Figure S1.81** Eleven successive ionization energies of an argon atom

Evidence for electron pairing from successive ionization energies

Successive ionization energy data can also be used to provide evidence for sublevel electron pairing in atoms.

Consider the successive ionization energies of the argon atom (Figure S1.81):

- The relatively large increase in ionization energy between electrons 8 and 9 occurs because up to electron number 8, the electrons were removed from the third shell but electron 9 is removed from the second shell.
- The slight increase in ionization energy between electrons 6 and 7 is due to a change in sublevel from 3p to 3s.
- The slight increase in ionization energy for electrons 4, 5 and 6 relative to electrons 1, 2 and 3 can be accounted for by enhanced electron–electron repulsion.

HL ONLY

Tool 3: Mathematics

Logarithms

Logarithms, or 'logs', express one number in terms of a base number that is raised to a power. For example, $100 = 10^2$, where 10 is the base and 2 is the power or index. This expression can be written in terms of logs as $\log_{10}100 = 2$. This is read as 'log to base 10 of 100 is 2'.

Logarithms are a powerful way to reduce a set of numbers that range over many orders of magnitude to a smaller more manageable scale.

Presentation of data on a logarithmic scale can be helpful when drawing graphs.

Logarithmic scales are used when the data covers a large range of values, such as values of pH and successive ionization energies.

Logarithmic scales are also used when the data contains exponential or power laws since they will then be displayed as linear relationships.

Rate graphs and Arrhenius equation graphs are important logarithmic graphs in chemical kinetics.

LINKING QUESTION

Why are log scales useful when discussing $[H^+]$ and ionization energies?

LINKING QUESTION

How do patterns of successive IEs of transition elements help to explain the variable oxidation states of these elements?

LINKING QUESTION

Why are data on successive ionization energies of a transition element often not useful in assigning the group number of the element?

Going further

Transition elements (except scandium) cannot form ions with a noble gas configuration. For example, consider a cobalt atom ($1s^2\ 2s^2\ 2p^6\ 3s^2\ 3p^6\ 3d^7\ 4s^2$ or 2,8,15,2). It is not energetically favourable for cobalt to lose nine electrons and form Co^{9+} ($1s^2\ 2s^2\ 2p^6\ 3s^2\ 3p^6$ or 2,8,8) with the noble gas configuration of argon.

However, a number of transition metals (Table S1.19) form cations with an 18-electron configuration, sometimes referred to as a pseudo noble gas configuration. This corresponds to the removal of all the 4s or 5s and the 4p and 5p valence electrons to leave the inner 'core' electrons. Tin and lead in group 14 form +4 ions with a pseudo noble gas configuration.

■ **Table S1.19** Ions with a pseudo noble gas configuration

Transition metal ion	Electron configuration	Electron arrangement
Cu^+	$1s^2\ 2s^2\ 2p^6\ 3s^2\ 3p^6\ 3d^{10}$	2,8,18
Zn^{2+}	$1s^2\ 2s^2\ 2p^6\ 3s^2\ 3p^6\ 3d^{10}$	2,8,18
Ag^+	$1s^2\ 2s^2\ 2p^6\ 3s^2\ 3p^6\ 4s^2\ 3d^{10}\ 4p^6\ 4d^{10}$	2,8,18,18

S1.4 Counting particles by mass: the mole

Guiding question

- How do we quantify matter on the atomic scale?

SYLLABUS CONTENT

By the end of this chapter, you should understand that:

- the mole (mol) is the SI unit of amount of substance
- one mole contains exactly the number of elementary entities given by the Avogadro constant
- masses of atoms are compared on a scale relative to ^{12}C and are expressed as relative atomic mass (A_r) and relative formula mass (M_r)
- the empirical formula of a compound gives the simplest ratio of atoms of each element present in that compound
- the molecular formula gives the actual number of atoms of each element present in a molecule
- the molar concentration is determined by the amount of solute and the volume of solution
- Avogadro's law states that equal volumes of all gases measured under the same conditions of temperature and pressure contain equal numbers of molecules
- Avogadro's law applies to ideal gases.

By the end of this chapter you should know how to:

- convert the amount of substance, n, to the number of specified elementary entities
- calculate the molar masses of atoms, ions, molecules and formula units
- solve problems involving the relationships between the number of particles, the amount of substance in moles and the mass in grams
- use molar masses with chemical equations to determine the masses of the products of a reaction
- interconvert the percentage composition by mass and the empirical formula
- determine the molecular formula of a compound from its empirical formula and molar mass
- use experimental data on mass changes in combustion reactions to derive empirical formulas
- solve problems involving the molar concentration, amount of solute and volume of solution
- interconvert molar concentration and concentration in $g\,dm^{-3}$
- solve problems involving the mole ratio of reactants and/or products and the volume of gases.

Note: There is no higher-level only content in S1.4

The mole

When chemists measure how much of a particular chemical reacts, they measure the mass in grams or, if the reactant is a gas, they may measure its volume. However, chemists prefer to use an SI unit called the **mole** which can be found from the mass of a pure substance or volume of a pure gas and is directly related to the number of particles.

◆ **Mole**: One mole of a substance contains 6.02×10^{23} elementary entities of the substance.

◆ **Amount**: A measure of the number of specified elementary entities (units of mol).

Particles react in simple numbers but different particles have different masses. Since we normally measure reacting masses, we need a way of converting mass to number of particles.

The mole, symbol mol, is the SI unit of **amount** of substance and the kilogram (kg) is the SI unit of mass. The mole is the chemist's counting unit and is used to measure the amount of a substance. A pair (2), a dozen (12) and a ream of writing paper (500 sheets) are all familiar examples of counting units (Figure S1.82).

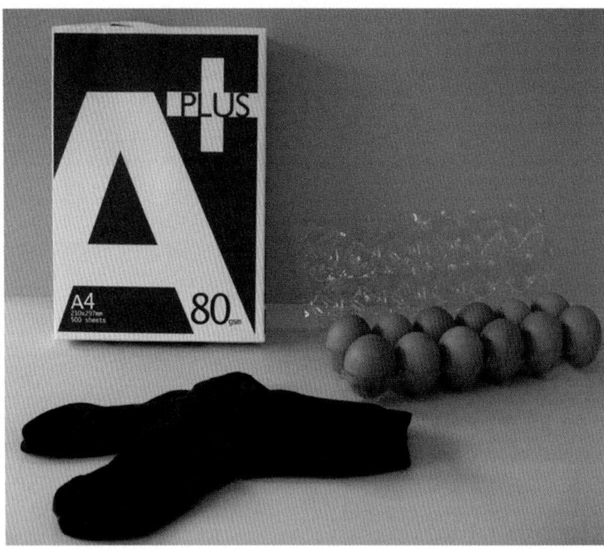

■ **Figure S1.82** Counting units: from left to right, a pair of socks, a ream of paper and a dozen eggs

◆ **Elementary entity:** An atom, molecule, ion, electron, other particle or group of particles.

◆ **Formula unit:** The empirical formula of any ionic or giant molecular compound used as an elementary entity.

◆ **Avogadro constant:** The number of atoms in exactly 12 grams of carbon-12 atoms. It has the value $6.02 \times 10^{23}\,mol^{-1}$.

The mole is a quantity that chemists use to count **elementary entities**. These may be atoms (for example, Ar); molecules (for example, S_8); ions (for example, Na^+); **formula units** for ionic substances (for example, NaCl) or giant molecular compounds (for example, SiO_2); or electrons. The elementary entity being considered should be specified, for example: 1 mole of chlorine atoms (Cl), 1 mole of chloride ions (Cl^-) or 1 mole of chlorine molecules (Cl_2).

6.02×10^{23} atoms	6.02×10^{23} molecules	6.02×10^{23} ions	6.02×10^{23} formula units	6.02×10^{23} electrons

one mole of particles

■ **Figure S1.83** The mole concept applied to different elementary entities

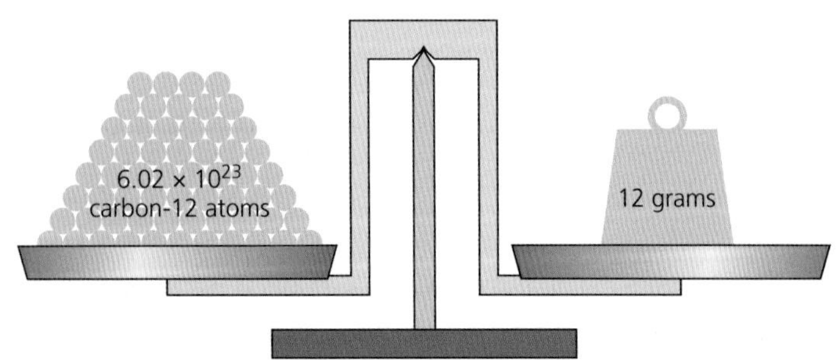

■ **Figure S1.84** An illustration of the Avogadro constant

The mole is defined as the amount of substance that contains 6.02×10^{23} elementary entities of a specific substance. Exactly 12 g (0.012 kg) of carbon-12 contains 1 mole or 6.02×10^{23} carbon-12 atoms (Figure S1.84). The number of particles per mole is the **Avogadro constant** (symbol N_A) which has the value of 6.02×10^{23} with units of per mol.

The equation below describes the relationship between the amount of a substance and the number of elementary entities:

$$\text{amount of substance (mol)} = \frac{\text{number of particles}}{6.02 \times 10^{23}\,\text{mol}^{-1}}$$

The formula may be rearranged to make the number of particles the subject.

Top tip!

Avogadro's constant provides an important connection between the macroscopic view of matter (viewed on a mole-by-mole or gram-by-gram basis) and the particulate view of matter (viewed on an atom-by-atom or molecule-by-molecule basis).

ATL S1.4A

The Avogadro constant is an experimentally determined value that has been measured using a number of methods, including electrolysis and X-ray crystallography. Use the internet to find a video describing a simple experiment involving the spreading out of an oil to form a monolayer on the surface of water, whose results can be used to estimate the Avogadro constant.

Perform a risk assessment for a similar activity. When your teacher has approved your method, carry out the experiment and evaluate your results.

WORKED EXAMPLE S1.4A

Calculate the amount (mol) of propan-1-ol, C_3H_7OH, that contains 1.20×10^{24} atoms of carbon.

Answer

$C_3H_7OH \rightarrow 3C + 8H + O$

Each molecule of propanol has three carbon atoms so $\frac{1.20 \times 10^{24}}{3}$ converts the number of atoms of carbon into the number of molecules.

Division by the Avogadro constant converts the number of molecules into an amount (in mol):

$$\left(\frac{1.20 \times 10^{24}}{3}\right) \div 6.02 \times 10^{23} = 0.66445 \text{ mol} \approx 0.664 \text{ mol}$$

Top tip!

Read questions carefully to identify the correct elementary entity, for example: a specific atom, or all atoms; a molecule; a specific ion or all ions. It is often helpful to write equations dissociating molecules into atoms: for example, $HNO_3 \rightarrow H + N + 3O$.

31 Calculate the number of molecules of water in 0.0100 mol of water, H_2O.

32 a Calculate the amount (in mol) of nitric(V) acid, HNO_3, that contains 9.0×10^{23} molecules.

b Calculate the number of oxygen atoms present in 9.0×10^{23} molecules of nitric(V) acid, HNO_3.

Tool 1: Experimental techniques

Measuring mass

The mass of a substance can be measured in the laboratory with an electronic balance. The SI unit for mass is the kilogram (kg) and the smaller masses used in chemistry are measured in grams (g), and milligrams (mg) (1000 mg = 1 g). In the chemical industry, large quantities of substances are used. Large masses can be measured in tonnes (t): 1 tonne = 1000 kg.

For many experimental methods, masses—such as the mass of chemical added—need to be measured accurately. If a specific mass is required for an experiment:

- Put a container onto a top-pan balance and zero (tare) the balance. This sets the mass reading to zero.
- Add the chemical using a spatula (if solid) until the mass required is shown.
- Transfer the chemical to the experiment.
- Return the container to the top-pan balance to check that it still reads zero and no chemical has remained in it.

If the mass to be used is not specific but still needs to be measured accurately, the method used involves measuring by difference.

- Weigh the mass of the chemical and its container.
- Add the chemical to the experiment.
- Weigh the mass of the (now empty) container.

The difference between the two masses involves the mass of chemical added to the experiment. This accounts for any chemical remaining in the container.

ATL S1.4B

Student A claims that 3 moles of carbon dioxide, CO_2, contain 3.01×10^{24} atoms of carbon dioxide. Student B claims that there is one error in their statement. Do you agree with either student? Justify your answer, presenting your argument and calculation to a classmate.

ATL S1.4C

How many atoms in my signature?

The focused research question for this practical investigation is: How many carbon atoms are there in your pencil signature on a piece of paper? Plan, carry out and evaluate a methodology based on mass measurements.

Guidance

- You will need access to an analytical electronic balance with a high sensitivity.
- Find out how pencils are classified into different grades using the letters H and B and choose a pencil that has the highest percentage of carbon in the form of graphite.
- It will take a large number of signatures on a single sheet of paper to get a measurable change in mass.
- Use the concept of ratios to scale down 12.01 g (mass of one mole of carbon atoms and equivalent to the relative atomic mass of 12.01) and $6.02 \times 10^{23}\,\text{g mol}^{-1}$ to the mass of a single signature.

Tool 3: Mathematics

Calculations involving decimals

Adding and subtracting decimals

When adding and subtracting decimals, the final answer should have the same number of decimal places as the minimum number of decimal places used to get the final answer.

For example: 23.82 g (2 decimal places) of magnesium was mixed with 18.7 g (1 decimal place) of hydrated copper sulfate. The final mass is 23.82 g + 18.7 g = 42.52 g = 42.5 g to 1 decimal place.

Multiplying and dividing decimals

When multiplying and dividing decimals, the number of decimal places is irrelevant. The final answer should be given to the smallest number of significant figures used in the calculation.

For example, 50.00 g of water was raised in temperature by 7.1 °C. The energy absorbed by the water is given by $Q = mc\Delta T$ where c, the specific heat capacity of water, is $4.18\,\text{J g}^{-1}\,\text{K}^{-1}$.

$Q = 50.00 \times 4.18 \times 7.1 = 1483.9\,\text{J}$

The smallest number of significant figures is 2 and comes from the temperature change. The final answer is therefore written as 1500 J to 2 s.f.

Hint: Always state the number of decimal places or significant figures you have used to obtain your answer.

ATL S1.4D

Find out about the research of the German chemist Johann Josef Loschmidt and the concept later known as the Loschmidt number. State the definition of the Loschmidt number, outline how it was developed from theory and experiment, and explain how it is related to the Avogadro constant.

Nature of science: Measurement

The kilogram

Since 1889, the SI unit of mass, the kilogram, has been defined as being equal to the mass of the international prototype kilogram (IPK). The IPK is a cylinder of platinum–iridium alloy located in the International Bureau of Weights and Measures (BIPM), near Paris.

■ **Figure S1.85** The international prototype kilogram at the BIPM

When the IPK was created in the 1880s, so were other identical prototype cylinders, which were given to various countries. However, over the years, the IPK has lost mass when compared with these other prototypes and this created small systematic errors in weighing.

The uncertainty that this mass change has created also impacted the mole which, in 1971, was defined in relation to the kilogram as the amount of substance of a system which contains as many elementary entities as there are atoms in 0.012 kg of carbon-12.

However, since May 2019, the kilogram has been defined with respect to Planck's constant (a fundamental physical constant in quantum mechanics) and the mole has been defined as being an amount of elementary entities equal to Avogadro's constant.

What do you think are the benefits of changing the definition of a kilogram from one based on the platinum–iridium alloy to one based on Planck's constant? What problems are avoided by introducing this new definition? What problems arise about how knowledge is established and defined in the natural sciences?

The importance of the mole

Imagine baking a cake. The recipe will tell you the mass or number of each ingredient that is needed to bake the cake. For example: 100 g of flour, 2 eggs, 20 g of sugar and 3 g of baking powder.

◆ **Coefficients**: The numbers in a balanced equation that represent the moles of reactants and products.

In the same way, in chemistry, the 'ingredients' (reactants) need to be mixed and reacted in the correct quantities (proportions). Chemists calculate those quantities from the **coefficients** in a balanced equation: for example, $2Na(s) + Cl_2(g) \rightarrow 2NaCl(s)$.

This equation indicates that 2 moles ($2 \times 6.02 \times 10^{23}$) of sodium atoms react with 1 mole (6.02×10^{23}) of chlorine molecules to form 2 moles ($2 \times 6.02 \times 10^{23}$) of sodium chloride formula units.

To a chemist, this means that if you have 45.98 g (2 mol) of sodium metal and you react it with 70.90 g (1 mol) of chlorine gas, you will form 116.88 g (2 mol) of sodium chloride. The total mass of the reactants will be equal to the total mass of the products. Amounts (in moles) can be converted to masses (in grams) using molar masses.

Top tip!

If a chemical reaction produces a gas (for example, baking a cake or decomposition of copper(II) carbonate: $CuCO_3(s) \rightarrow CuO(s) + CO_2(g)$) then the mass of the solid product will be less than the mass of the reactant. However, if the gas is collected and its mass weighed, then it can be seen that no change in mass has occurred.

Tool 3: Mathematics

Understand the significance of uncertainties in raw and processed data

No recorded measurement, or value calculated from such measurements, can be exact, that is, known with infinite precision. Random errors (uncertainties) due to the imprecision of measuring devices result in readings being above or below the true value, causing one measurement to differ slightly from the next. In raw data, uncertainties are expressed as a range within which the true value lies.

Knowing uncertainties allows you to evaluate the reliability of experimental data. Random errors in raw data propagate through a calculation to give a value for the error in the final result. The greater the size of this final error, the less reliable the conclusions are in answering the research question.

◆ **Stoichiometry**: The ratio in which elements chemically combine in a compound, or the ratio in which substances react in a chemical reaction.

◆ **Stoichiometric**: When amounts of reactants are reacted together so that they are all consumed at the same time and no reactant is left over.

A balanced chemical equation shows the number of particles of a substance which react with another substance. The coefficients in the balanced equation show the ratios (in moles) in which these substances react. This is known as the **stoichiometry** of the reaction.

For example, in $N_2(g) + 3H_2(g) \rightarrow 2NH_3(g)$, the coefficients are the reacting mole ratios: 1 mole of N_2 reacts with 3 moles of H_2 to produce 2 moles of NH_3. So 0.05 mol of N_2 would react with 0.15 mol of H_2 using the $1N_2:3H_2$ ratio. Similarly, if 4 moles of NH_3 were to be produced, we would need 6 moles of H_2 using the $2NH_3:3H_2$ ratio. All these reactions are known as **stoichiometric** reactions.

Common mistake

Misconception: *Mass is not conserved during a chemical reaction. The products of chemical reactions need not have the same mass as the reactants.*

Mass is always conserved during a chemical reaction. The products of chemical reactions must have the same mass as the reactants since atoms cannot be created or destroyed during a chemical reaction.

Chemical equations

Balanced symbol equations (Figure S1.86) are more useful than word equations because they indicate the reacting proportions of the reactants.

Word equation:	nitrogen	+	hydrogen	→	ammonia
Symbol equation:	$N_2(g)$	+	$3H_2(g)$	→	$2NH_3(g)$
Particulate equation:					

N_2	+	$3H_2$	→	$2NH_3$
1 molecule of N_2		3 molecules of H_2		2 molecules of NH_3
2 N atoms		6 H atoms		2 N atoms and 6 H atoms

■ **Figure S1.86** Balanced particulate equation for the synthesis of ammonia: two nitrogen atoms and six hydrogen atoms on each side of the equation.

Balancing equations

For an equation to be balanced, there must be an equal number of each type of atom on both sides of the reaction. The formulas of the reactants and products cannot be changed: the only numbers that can be changed are the coefficients in front of the formulas. The coefficients are usually whole numbers.

Top tip!

It is allowable, and sometimes preferable, to use fractional coefficients in the balancing process: for example, $C_2H_6(g) + \frac{7}{2}O_2(g) \rightarrow 2CO_2(g) + 3H_2O(l)$. Generally, the fractional coefficient is not retained in the final answer. In this example, multiplying the coefficients through by two removes the fraction: $2C_2H_6(g) + 7O_2(g) \rightarrow 4CO_2(g) + 6H_2O(l)$. However, if an equation represents the standard molar enthalpy of combustion, then fractional coefficients have to be used. For example, $C_2H_6(g) + \frac{7}{2}O_2(g) \rightarrow 2CO_2(g) + 3H_2O(l)$ represents the standard molar enthalpy of combustion of ethane.

WORKED EXAMPLE S1.4B

Balance the following hydrolysis reaction between phosphorus(V) chloride (a solid) and water to form phosphoric(V) acid and hydrochloric acid. Include state symbols.

Answer

Word equation: phosphorus(V) chloride + water → phosphoric(V) acid + hydrochloric acid

Unbalanced symbol equation: $PCl_5 + H_2O \rightarrow H_3PO_4 + HCl$

Balance chlorine: $PCl_5 + H_2O \rightarrow H_3PO_4 + 5HCl$

Balance hydrogen and oxygen: $PCl_5 + 4H_2O \rightarrow H_3PO_4 + 5HCl$

Add state symbols: $PCl_5(s) + 4H_2O(l) \rightarrow H_3PO_4(aq) + 5HCl(aq)$

	reactants	**products**
phosphorus atoms	1 × 1 = 1	1 × 1 = 1
hydrogen atoms	4 × 2 = 8	(3 × 1) + (5 × 1) = 8
oxygen atoms	4 × 1 = 4	4 × 1 = 4
chlorine atoms	5 × 1 = 5	5 × 1 = 5

In reactions dealing only with ions, leave the polyatomic ions (such as sulfate, SO_4^{2-}, and carbonate, CO_3^{2-}) as groups. They are usually collected in brackets to make balancing easier. 'Atom accounting' makes this easier by using a table, as shown in Worked example S1.4B. Start with all coefficients of one and total the number of each type of atom or species. The more atoms in a given molecule, the larger the effect it has on balancing, so begin with these.

End with molecules or atoms that consist of only one type, since the number can be changed independently of the other atom types. If a coefficient comes out to a fraction, multiply all coefficients by the fraction denominator to result in all whole-number coefficients.

WORKED EXAMPLE S1.4C

Balance the equation for the precipitation reaction between barium chloride solution and aluminium sulfate solution to form solid barium sulfate and aluminium chloride solution.

Answer

Word equation: barium chloride solution + aluminium sulfate solution → barium sulfate solid + aluminium chloride solution

Unbalanced symbol equation: $BaCl_2(aq) + Al_2(SO_4)_3(aq) \rightarrow BaSO_4(s) + AlCl_3(aq)$

Balance sulfate ions: $BaCl_2(aq) + Al_2(SO_4)_3(aq) \rightarrow 3BaSO_4(s) + AlCl_3(aq)$

Balance barium: $3BaCl_2(aq) + Al_2(SO_4)_3(aq) \rightarrow 3BaSO_4(s) + AlCl_3(aq)$

Balance aluminium: $3BaCl_2(aq) + Al_2(SO_4)_3(aq) \rightarrow 3BaSO_4(s) + 2AlCl_3(aq)$

	reactants	**products**
barium	3 × 1 = 3	3 × 1 = 3
chlorine	3 × 2 = 6	2 × 3 = 6
aluminium	2 × 1 = 2	2 × 1 = 2
sulfur	3 × 1 = 3	3 × 1 = 3
oxygen	3 × 4 = 12	3 × 4 = 12

33 Balance the following equations by putting the appropriate coefficients in front of the formulas.

a $H_2SO_4 + NaOH \rightarrow Na_2SO_4 + H_2O$
b $Fe + Cl_2 \rightarrow FeCl_3$
c $CuSO_4 + KOH \rightarrow Cu(OH)_2 + K_2SO_4$
d $CuO + HNO_3 \rightarrow Cu(NO_3)_2 + H_2O$
e $CuCO_3 + HCl \rightarrow CuCl_2 + H_2O + CO_2$
f $Pb(NO_3)_2 + KI \rightarrow PbI_2 + KNO_3$
g $Al_2O_3 \rightarrow Al + O_2$
h $C_3H_8 + O_2 \rightarrow CO_2 + H_2O$
i $Al(NO_3)_3 + NaOH \rightarrow Al(OH)_3 + NaNO_3$
j $CaCl_2 + Na_3PO_4 \rightarrow Ca_3(PO_4)_2 + NaCl$
k $Fe + O_2 \rightarrow Fe_2O_3$
l $Ca(OH)_2 + HCl \rightarrow CaCl_2 + H_2O$
m $HCl + MnO_2 \rightarrow MnCl_2 + Cl_2 + H_2O$

Measuring moles

When you take a large bag of the same coins to the bank, the cashier will not waste their time by counting every coin in the bag. Instead they will weigh the bag of coins to determine the number of coins inside. This is possible because the cashier knows the unique mass of each coin. For example, a coin might weigh 1 g so a bag of coins that weighs 100 g has 100 coins inside.

In a similar way, you can weigh out one mole of an element if you know the relative atomic mass of that element from the periodic table. The relative atomic mass is a number with no units but when units of g are added it is the mass of one mole of atoms of that element. For example, carbon has a relative atomic mass of 12.01. Therefore one mole of carbon atoms has a mass of 12.01 g.

Relative atomic mass

The key to working with amounts of atoms measured in moles is to know the relative masses of the different atoms. This helps in the 'counting' of atoms when a large sample is weighed.

Hydrogen-1 atoms have the smallest relative atomic mass of 1. A carbon-12 atom has a relative atomic mass of 12, which means a carbon-12 atom is twelve times heavier than a hydrogen-1 atom. These masses are known as **relative isotopic masses**.

◆ **Relative isotopic mass:** The mass of a particular isotope compared to the mass of $\frac{1}{12}$ of a carbon-12 atom.

Tool 3: Mathematics

Fractions

The division sign ÷ is rarely used in scientific reports: division is normally expressed in terms of a fraction.

A fraction consists of three parts: the numerator is the number on top; the denominator is the number on the bottom; and the third component—the line (vinculum)—represents division.

$$\text{fraction} = \frac{\text{numerator}}{\text{denominator}}$$

So we can represent dividing 3 by 4 as $\frac{3}{4}$ rather than writing 3 ÷ 4.

If we multiply the numerator and the denominator by the same quantity, the fraction remains unchanged. For example: $\frac{1}{3} = \frac{1 \times 5}{3 \times 5} = \frac{5}{15}$ so $\frac{1}{3}$ and $\frac{5}{15}$ are equivalent fractions.

You can only add and subtract fractions if they have the same denominator so these operations rely on this concept of equivalent fractions. For example: $\frac{1}{2} + \frac{1}{4} = \frac{2}{4} + \frac{1}{4} = \frac{3}{4}$.

The reciprocal of a fraction is simply the fraction inverted. A reciprocal is usually denoted by the superscript index −1 so the reciprocal of $\frac{1}{3} = \left(\frac{1}{3}\right)^{-1} = \frac{3}{1} = 3$.

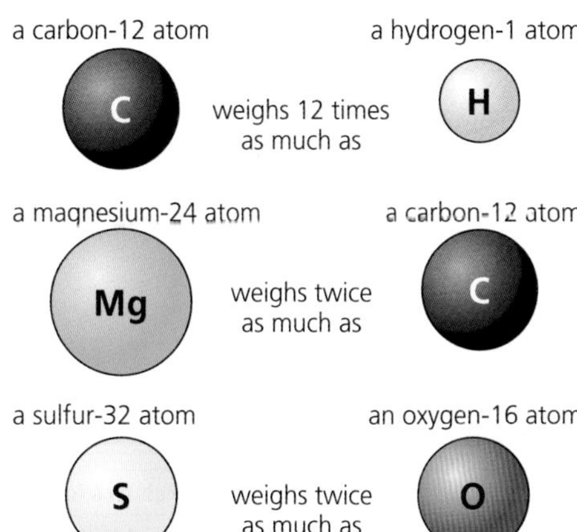

a carbon-12 atom
C
weighs 12 times as much as
a hydrogen-1 atom
H
a magnesium-24 atom
Mg
weighs twice as much as
a carbon-12 atom
C
a sulfur-32 atom
S
weighs twice as much as
an oxygen-16 atom
O

■ **Figure S1.87** Comparing the relative isotopic masses of abundant specific nuclides where the numbers represent the nucleon or mass numbers (the sum of the protons and neutrons in the nucleus)

However, most naturally occurring elements exist as a mixture of stable isotopes and so the mass of a sample of a pure element needs to reflect the percentages of the different isotopes. The relative masses of elements are known as relative atomic masses.

For example, the relative isotopic mass of a carbon-12 atom is exactly 12, but the relative atomic mass (A_r) of carbon is 12.01. This is because in a sample of carbon there are carbon-12 atoms (98.89%) and also small amounts of carbon-13 atoms (1.11%).

$$\text{relative atomic mass, } A_r\text{, of carbon} = \left(\frac{98.89}{100} \times 12\right) + \left(\frac{1.11}{100} \times 13\right) = 12.01$$

The relative atomic mass is the ratio of the weighted average mass of one atom of an element to one twelfth of the mass of an atom of carbon-12. So, for example, a magnesium-24 atom has the same relative mass as two carbon-12 atoms; and one carbon-12 atom has the same relative mass as three helium-4 atoms (Figure S1.88). In effect, this scale expresses the mass as a multiple of that of the lightest atom: hydrogen-1.

Top tip!

The accurate method for determining relative atomic masses involves the use of a mass spectrometer.

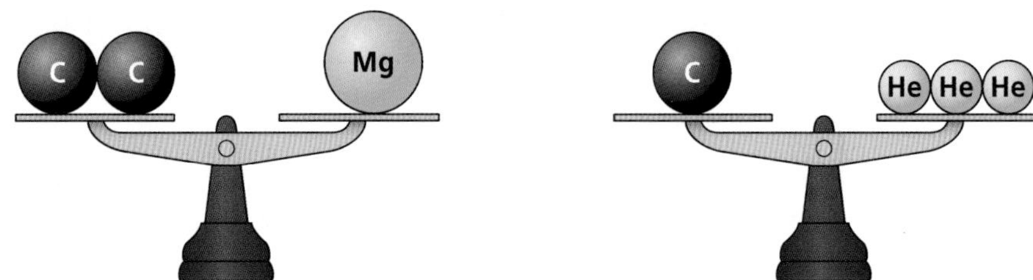

■ **Figure S1.88** The concept of relative atomic mass applied to carbon-12, magnesium-24 and helium-4 atoms

ATL S1.4E

Use the isotope simulation at **https://phet.colorado.edu/en/simulations/isotopes-and-atomic-mass/about** to build nuclides of different isotopes of the first twenty elements.

Tabulate the atomic mass, mass (nucleon) number, relative abundance and stability of the nuclides you build.

ATL S1.4F

Relative atomic masses can, in principle, use any nuclide as a reference mass. Outline the problems that arose between chemists using an element (oxygen) as a reference mass and physicists using a common isotope (oxygen-16) as the reference mass.

Relative formula mass

◆ **Relative formula mass**: The sum of the relative atomic masses of all the atoms shown in the formula unit or molecular formula.

◆ **Molar mass**: The mass in grams of one mole of molecules or the formula unit of an ionic compound.

The **relative formula mass** of a compound is the sum of the atomic masses of all the atoms in a molecular formula or the formula unit of an ionic compound. It is, therefore, the weighted average mass of one molecule or formula unit compared to $\frac{1}{12}$th the mass of an atom of carbon-12. Since it is a relative mass, it has no units.

The relative formula mass of a carbon dioxide molecule (44.01) is approximately the same as the combined relative atomic mass of 44 hydrogen atoms, each with a relative atomic mass of 1.01.

■ **Table S1.20** Examples of relative formula mass calculations

Compound	Chemical formula	Relative formula mass
water	H_2O	$(2 \times 1.01) + 16.00 = 18.02$
ethanol	C_2H_6O	$(2 \times 12.01) + (1.01 \times 6) + 16.00 = 46.08$
sodium oxide	Na_2O [$2Na^+$ O^{2-}]	$(2 \times 22.99) + 16.00 = 61.98$
ammonium chloride	NH_4Cl [NH_4^+ Cl^-]	$14.01 + (4 \times 1.01) + 35.45 = 53.50$

Molar mass

Top tip!

Molar mass must be distinguished from relative atomic and relative formula masses, which are ratios and hence have no units, although both have the same numerical value.

The **molar mass** (symbol M) is the mass of one mole of any substance (atoms, molecules, ions or formula units) where the carbon-12 isotope is assigned a value of *exactly* 12 $g\,mol^{-1}$. It is a particularly useful concept since it can be applied to any elementary entity (Table S1.21). It has units of grams per mol ($g\,mol^{-1}$).

■ **Table S1.21** Some molar masses

Formula	Molar mass / $g\,mol^{-1}$	Number of particles	Type of particles
C	12.01	6.02×10^{23}	atoms
H_2O	18.02	6.02×10^{23}	molecules
$CaCO_3$ [Ca^{2+} CO_3^{2-}]	100.09	$(2 \times 6.02 \times 10^{23})$	ions

Nature of science: Measurement

Relative atomic mass scale

Hydrogen was initially chosen as the standard for relative masses because chemists knew that the element had the lightest atoms, which could therefore be assigned a mass of one. Later, when more accurate values for atomic masses were obtained, chemists knew that an element could contain atoms of different masses, known as isotopes. It then became necessary to choose a single isotope as the international reference standard for relative atomic masses. In 1961, carbon-12 was chosen as the new standard.

LINKING QUESTION

Atoms increase in mass as groups are descended in the periodic table. What properties might be related to this trend?

ATL S1.4G

Molar volume is the atomic mass of an element divided by the density of its solid form:

$$\text{molar volume } (\text{cm}^3\,\text{mol}^{-1}) = \frac{\text{molar mass } (\text{g}\,\text{mol}^{-1})}{\rho\ (\text{g}\,\text{cm}^{-3})}$$

For elements which are liquids or gases at room temperature, the density of the liquid element at its boiling point is used.

Julius Lothar Meyer (1830–95) plotted molar volume against relative atomic mass. He noted there were periodic maxima and minima and that similar elements occupied similar positions on the graph. In doing this, he also drew attention to the idea of arranging the elements in order of increasing relative atomic mass.

The table below shows the atomic volumes (in $\text{cm}^3\,\text{mol}^{-1}$) of the first 18 elements (to 2 s.f.).

H	14		O	14		Al	10
He	27		N	17		Si	12
Li	13		F	13		P	17
Be	4.9		Ne	17		S	16
B	4.7		Na	24		Cl	23
C	5.3		Mg	14		Ar	29

- Use a spreadsheet to plot a graph of atomic volume against atomic number for these elements.
- Describe and account for the appearance of the graph.
- State some of the reasons why it is difficult to compare the elements using this approach.

■ **Figure S1.89** 0.1 mol of water, H_2O (M = 18.02 g mol^{-1}); potassium dichromate(VI), $K_2Cr_2O_7$ (M = 294.2 g mol^{-1}); and copper(II) sulfate-5-water, $CuSO_4.5H_2O$ (M = 249.72 g mol^{-1}) weighed to one decimal place

Calculating quantities

The equation relating the amount of substance in moles (n), mass (m) and molar mass (M) can be used to calculate any quantity given the values of the other two.

$$\text{amount of substance (mol)} = \frac{\text{mass (g)}}{\text{molar mass } (\text{g}\,\text{mol}^{-1})}$$

or, in symbols, $n = \dfrac{m}{M}$.

Figure S1.90 summarizes the interconversions between amount (mol), mass (g) and number of particles via the molar mass and Avogadro's constant.

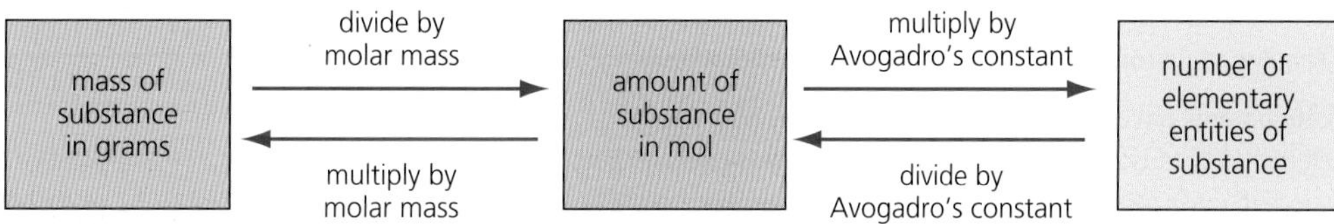

■ **Figure S1.90** Interconversion between mass, amount and number of elementary entities

WORKED EXAMPLE S1.4D

Calculate the number of iron atoms (to 2 s.f.) present in an iron rod of length 50.0 cm and diameter 2.0 cm.

Density of iron = $7.84\,g\,cm^{-3}$

The volume of a cylinder is given by the formula $\pi r^2 l$, where r is the radius and l is the length.

Answer

$$\text{volume of iron rod} = \pi \times \left(\frac{2.0\,\text{cm}}{2}\right)^2 \times 50.0\,\text{cm} = 157\,\text{cm}^3$$

$$\text{mass of iron rod} = 7.84\,\text{g cm}^{-3} \times 157\,\text{cm}^3 = 1231\,\text{g}$$

$$\text{amount of iron atoms} = \frac{1231\,\text{g}}{55.85\,\text{g mol}^{-1}} = 22.1\,\text{mol}$$

$$\text{number of iron atoms} = 22.1 \times 6.02 \times 10^{23} = 1.3 \times 10^{25}\,\text{atoms}$$

Determining the amount from mass and molar mass

34 Calculate the amount of water present in 54.00 g of water, H_2O.

35 Calculate the amount of calcium present in 0.500 kg of calcium.

36 Calculate the amount of water present in a drop with a mass of 180 mg.

Determining the mass from amount and molar mass

37 Calculate the mass of 0.400 mol of calcium carbonate, $CaCO_3$.

Determining the molar mass from mass and amount

38 0.002 00 mol of a substance has a mass of 1.00 g. Calculate the molar mass of the substance.

Calculating the numbers of atoms of an element in a given mass of a compound

39 Calculate the number of carbon and hydrogen atoms and the total number of atoms in 22.055 g of propane, C_3H_8 (molar mass = $44.11\,g\,mol^{-1}$).

Top tip!

Convert the mass into an amount (mol) of molecules and then a number (using the Avogadro constant). The formula shows the relationship between the number of atoms of each element in one molecule.

Nature of science: Science as a shared endeavour

Law of definite and multiple proportions

The law of definite proportions applies to two or more samples of the *same compound* and states that the ratio of one element to the other is always the same.

For example, the decomposition of 18.02 g of water results in 16.00 g of oxygen and 2.02 g of hydrogen, or an oxygen-to-hydrogen mass ratio close to 8 : 1. This ratio holds for any sample of pure water, regardless of its origin (synthesis or purification). The law of definite proportions applies to all compounds.

Compounds have definite proportions of their constituent elements because the atoms that compose them, each with its own specific mass, occur in a definite ratio. Because the ratio of atoms is the same for all samples of a particular compound, the ratio of masses is also the same.

The law of multiple proportions applies to two *different compounds* containing the same two elements (A and B) and states that the masses of B that combine with 1 g of A are always related to each other as a small whole-number ratio (Table S1.22). An understanding of these two laws provided the basis for Dalton's atomic theory.

■ **Table S1.22** Law of multiple proportions applied to oxides of carbon

Compound		Mass of oxygen that combines with 1.00 g of carbon / g
carbon dioxide	O=C=O	2.67
carbon monoxide	C≡O	1.33

The ratio of these two masses is a whole number:

$$\frac{\text{mass of oxygen to 1 g of carbon in carbon dioxide}}{\text{mass of oxygen to 1 g of carbon in carbon monoxide}} = \frac{2.67}{1.33} = 2$$

Using the molecular models, we can see why the ratio is 2 : 1—carbon dioxide contains two oxygen atoms to every carbon atom, while carbon monoxide contains only one.

Do human rights exist in the same way that the laws of gravity exist?

It can be argued that human rights exist in the same way that laws exist in a country. Laws are a social construct that have been formulated by politicians and enacted in legal documents. Declarations of human rights are living documents in that they are open to, and subject to, change as societies change. Human rights do not exist in a physical sense but, like other ideas and concepts, exist in a more abstract manner.

The laws of physics are also abstract objects, generalizations from empirical data (observations and measurements) which do exist. The laws of physics have no matter or energy and they do not occupy regions of space. However, they exist as mathematical concepts due to the behaviour of objects—which have matter, contain energy and occupy regions of space.

Determining the masses of the products of a reaction

Chemical equations can be interpreted in terms of numbers of molecules, amounts in moles and masses in grams.

For example, in the equation in Table S1.23, one mole of nitrogen (N_2) molecules reacts with three moles of hydrogen (H_2) molecules to form two moles of ammonia (NH_3) molecules.

Amounts in moles can be converted to masses via the molar mass.

■ **Table S1.23** Balanced equations give ratios in which substances react by number, amount and mass

N_2	+	$3H_2$	→	$2NH_3$
1 molecule of N_2	reacts with	3 molecules of H_2	to make	2 molecules of NH_3
100 molecules of N_2	react with	300 molecules of H_2	to make	200 molecules of NH_3
602 molecules of N_2	react with	1806 molecules of H_2	to make	1204 molecules of NH_3
6.02×10^{23} molecules of N_2	react with	18.06×10^{23} molecules of H_2	to make	12.04×10^{23} molecules of NH_3
1 mole of molecules of N_2	reacts with	3 moles of molecules of H_2	to make	2 moles of molecules of NH_3
28.02 g of N_2	reacts with	6.06 g of H_2	to make	34.08 g of NH_3

LINKING QUESTION

How can molar masses be used with chemical equations to determine the masses of the products of a reaction?

Steps for solving stoichiometry problems

1. Write a balanced equation for the reaction.
2. Convert the given amount of reactants or products (in grams or other units) to amounts (in mol) using the molar mass.
3. Use the mole ratio from the balanced equation to calculate the amount (in mol) of the required products or reactants.
4. Convert the amount of products or reactants to grams or other units.

WORKED EXAMPLE S1.4E

Hydrogen reacts with oxygen to form water. Calculate the mass of water formed if 8.08 g of hydrogen reacts with excess oxygen to form water.

Answer

1 $2H_2(g) + O_2(g) \rightarrow 2H_2O(g)$

2 $\text{amount of } H_2 = \dfrac{\text{mass (g)}}{\text{molar mass (g mol}^{-1})} = \dfrac{8.08\,\text{g}}{2.02\,\text{g mol}^{-1}} = 4.00\,\text{mol}$

3 2 mol of H_2 forms 2 mol of H_2O; the reactant and product are in a 1 : 1 ratio by moles hence amount of water formed = 4.00 mol

4 mass of water = amount (mol) × molar mass (g mol^{-1})

$= 4.00\,\text{mol} \times 18.02\,(\text{g mol}^{-1}) = 72.08\,\text{g}$

Common mistake

Misconception: *The coefficients in the balanced equation represent the numbers of atoms of the elements in the substance.*

The coefficients in the balanced equation represent the *ratios* of the numbers of atoms, ions, molecules, formula units or electrons that are involved in the chemical reaction.

40 The pollutant sulfur dioxide, SO_2, can be removed from the air by the following chemical reaction:

$$2CaCO_3(s) + 2SO_2(g) + O_2(g) \rightarrow 2CaSO_4(s) + 2CO_2(g)$$

Determine the mass of calcium carbonate needed to remove 10.00 g of sulfur dioxide.

Molecular formula and empirical formula

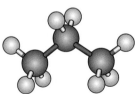

The **molecular formula** shows the number of atoms in one molecule of the element or molecular compound. The **empirical formula** shows the simplest whole-number ratio of the atoms of different elements present.

◆ **Molecular formula:** A formula that shows the actual number of atoms of each element in a molecule.

◆ **Empirical formula:** A formula showing the simplest whole-number ratio of atoms or ions in a molecule or formula unit.

Top tip!

Empirical means from experimental data; the molecular formula can only be determined if the molar mass is known (or found via another experiment). The empirical formula may or may not be the same as the molecular formula (Figure S1.91).

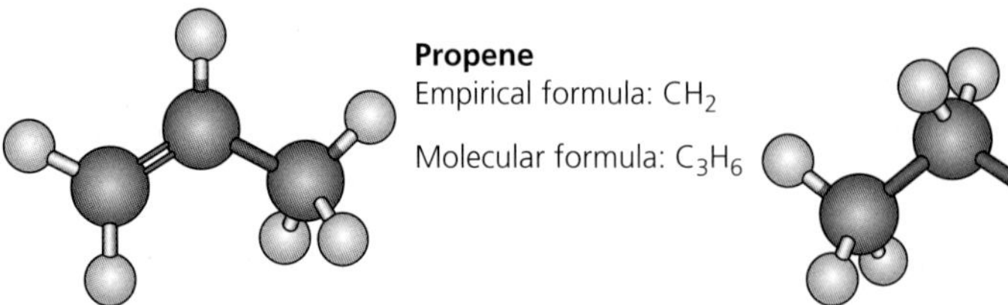

■ **Figure S1.91** Molecular and empirical formulas and three-dimensional structures of propene and propane

Top tip!

For simple molecular compounds, there are many possible multiples of the empirical formula. For example, CH_2 is the empirical formula of C_2H_4, C_3H_6, C_4H_{10} and many other compounds. However, for ionic compounds, such as NaCl or CaO, or giant molecular compounds, such as SiO_2, the empirical formula is the chemical formula of the compound.

Percentage composition by mass

The mole ratio of elements in a compound is the ratio of the amounts of these elements in the compound. For example, one mole of nitrogen(III) oxide, N_2O_3, contains two moles of nitrogen atoms and three moles of oxygen atoms. Hence, the mole ratio of nitrogen atoms to oxygen atoms in nitrogen(III) oxide is 2 : 3. The mole ratio can be used to calculate the **percentage composition by mass** of a compound from its formula.

◆ **Percentage composition by mass:** The percentage by mass of each element in a compound.

Tool 3: Mathematics

Calculations involving percentages

Percentages are often used in chemistry. Consider, for example, atom economy, % yield, % composition by mass, % ionization and % isotopic abundance. These are all concepts covered elsewhere in this book.

In experimental work, we also can calculate % error. This will help us to decide whether systematic errors or random errors are more relevant to the collection of data.

The % error compares your calculated value with the literature value and is calculated from:

$$\%\ \text{error} = 100 \times \frac{\text{theoretical value} - \text{experimental value}}{\text{theoretical value}}$$

A student calculated the enthalpy of combustion of hexane to be $-399\,\text{kJ}\,\text{mol}^{-1}$. The IB *Chemistry data booklet* (section 14) quotes this as $-4163\,\text{kJ}\,\text{mol}^{-1}$.

The % error is: $100 \times \dfrac{-4163-(-399)}{-4163} = -90.0\%$

WORKED EXAMPLE S1.4F

Calculate the percentage composition by mass of nitrogen and oxygen in nitrogen(III) oxide, N_2O_3.

Answer

Let $n(N_2O_3) = 1\,\text{mol}$, then: $m(\text{N}) = 2\,\text{mol} \times 14.01\,\text{g}\,\text{mol}^{-1} = 28.02\,\text{g}$

$m(\text{O}) = 3\,\text{mol} \times 16.00\,\text{g}\,\text{mol}^{-1} = 48.00\,\text{g}$

$m(N_2O_3) = 1\,\text{mol} \times 76.02\,\text{g}\,\text{mol}^{-1} = 76.02\,\text{g}$

$$\%\ \text{by mass of nitrogen} = \frac{28.02\,\text{g}}{76.02\,\text{g}} \times 100 = 36.86\%$$

$$\%\ \text{by mass of oxygen} = \frac{48.00\,\text{g}}{76.02\,\text{g}} \times 100 = 63.14\%$$

Determination of empirical formula

To determine the empirical formula of a compound, first calculate the amount of atoms of each substance present in a sample (it is usually helpful to assume a 100 g sample) then calculate the simplest whole-number ratio of the amounts. The data will be presented as mass of elements or % by mass of elements.

Table S1.24 shows how to determine an empirical formula from masses or percentage by mass of constituents:

■ **Table S1.24** Determining an empirical formula

Step 1	Write down the mass of each element in grams or the mass of each element in 100 g of compound if you are given % by mass.
Step 2	Write down the molar mass of each element (refer to section 7 in the IB *Chemistry data booklet* to find the relative atomic mass if needed).
Step 3	Find the amount of each element by dividing the mass of each element by the respective molar mass.
Step 4	Find the relative amount of each element by dividing the amount of each element by the smallest amount.
Step 5	Express the relative amount as a whole number.
Step 6	Write the formula.

Top tip!

If the ratio has .5, multiply by 2; if it has .33 or .66 multiply by 3; if it has .25, multiply by 4.

WORKED EXAMPLE S1.4G

Determine the empirical formula for the compound with the following composition by mass: lead 8.32 g, sulfur 1.28 g, oxygen 2.56 g.

Answer

1 mass of each element: Pb = 8.32 g S = 1.28 g O = 2.56 g

2 molar mass of each element: Pb = 207.20 g mol⁻¹ S = 32.07 g mol⁻¹ O = 16.00 g mol⁻¹

3 divide by the molar mass:

$$\text{Pb: } \frac{8.32\,\text{g}}{207.20\,\text{g mol}^{-1}} = 0.0402\,\text{mol}$$

$$\text{S: } \frac{1.28\,\text{g}}{32.07\,\text{g mol}^{-1}} = 0.0399\,\text{mol}$$

$$\text{O: } \frac{2.56\,\text{g}}{16.00\,\text{g mol}^{-1}} = 0.160\,\text{mol}$$

4 divide by the smallest amount:

$$\text{Pb: } \frac{0.0402\,\text{mol}}{0.0399\,\text{mol}} = 1.008$$

$$\text{S: } \frac{0.0399\,\text{mol}}{0.0399\,\text{mol}} = 1$$

$$\text{O: } \frac{0.160\,\text{mol}}{0.0399\,\text{mol}} = 4.010$$

5 whole number ratio: 1 : 1 : 4

6 empirical formula: $PbSO_4$

ATL S1.4H

The online program at https://chemcollective.org/chem/curriculum/stoich/empform.php gives the elemental analysis of a random mineral. Reloading the page gives a new set of data. Use this activity to gain confidence in calculating empirical formulas. Step-by-step support and feedback are provided if you need additional help.

Inquiry 3: Concluding and evaluating

Identify and discuss sources and impacts of random and systematic errors

Systematic error arises from a flaw in the methodology, apparatus or instrumentation and can be detected and corrected. This type of error leads to inaccurate measurements of the true value. The best way to check for systematic error is to use different methods to perform the same measurement.

Random uncertainty is always present and cannot be corrected; it is connected with the precision of repeated measurements.

- Read the article about the direct synthesis of tin(IV) iodide at www.scienceinschool.org/article/2019/classic-chemistry-finding-empirical-formula/
- Outline the sources of errors in this procedure, their effects, and whether they are systematic or random.

ATL S1.4I

Significant figures (s.f.) are all the digits used in data to carry meaning, whether they are before or after a decimal point (this includes zeros).

Imagine you have to teach this topic to your peers.

Using software such as PowerPoint, develop a presentation that defines the concept of s.f. and outlines the rules for determining the number of s.f. in a measurement and the number of s.f. that should be given in answers to calculations. Include your own examples of rounding numbers to a given number of s.f. You can use an s.f. calculator (for example, www.omnicalculator.com/math/sig-fig) to check your answers.

LINKING QUESTION

What is the importance of approximation in the determination of an empirical formula?

WORKED EXAMPLE S1.4H

A metal (M) forms an oxide with the formula MO. The oxide contains 39.70% O by mass. Determine the identity of M. Assume 100 g of the compound is present.

Answer

mass of M = 100 − 39.70 = 60.30 g

$$\text{amount of oxygen atoms} = \frac{39.70\,\text{g}}{16.00\,\text{g mol}^{-1}} = 2.48\,\text{mol}$$

The M : O molar ratio is 1 : 1, so amount of M atoms = 2.48 mol.

$$\text{molar mass} = \frac{60.30\,\text{g}}{2.48\,\text{mol}} = 24.3145\,\text{g mol}^{-1} = 24.3\,\text{g mol}^{-1}$$

M is magnesium

WORKED EXAMPLE S1.4I

An iron-containing protein has a molar mass of 136 000 g mol^{-1} and 0.33% by mass is iron. Calculate the number of iron atoms present in one molecule of the protein.

Answer

molar mass of iron is 55.85 g mol^{-1}

mass of iron in 1 mole of protein = 0.0033 × 136 000 = 448.8 g mol^{-1}

$$\text{number of moles of iron} = \frac{448.8\,\text{g mol}^{-1}}{55.85\,\text{g mol}^{-1}} = 8 \text{ (to 1 s.f.)}$$

There are eight iron atoms in each molecule of the protein.

Determination of molecular formula

The molecular formula is a simple multiple (n) of the empirical formula where:

$$n = \frac{\text{molecular formula mass}}{\text{empirical formula mass}}$$

We can use this to find a molecular formula from an empirical formula as follows:

1 Write down the empirical formula of the compound.
2 Calculate the molar mass of the empirical formula.
3 Find the multiple n.
4 Multiply subscripts in the empirical formula by n to obtain the molecular formula.

WORKED EXAMPLE S1.4J

0.035 g of nitrogen forms 0.115 g of an oxide of nitrogen. Calculate **(i)** the empirical formula of the compound formed and **(ii)** the molecular formula given that the molar mass of the compound is $92\,\text{g mol}^{-1}$.

Answer

elements:	nitrogen	oxygen
masses:	0.035 g	0.115 − 0.035 g
moles:	$\frac{0.035\,\text{g}}{14.01\,\text{g mol}^{-1}} = 0.0025$	$\frac{0.080\,\text{g}}{16.00\,\text{g mol}^{-1}} = 0.0050$
ratio:	1	2

empirical formula: NO_2

empirical formula mass = 46

since molecular formula mass = 92, $n = \frac{92}{46} = 2$

so the molecular formula is $2 \times (NO_2) = N_2O_4$

LINKING QUESTION

How can experimental data on mass changes in combustion reactions be used to derive empirical formulas?

◆ **Hydrated salt**: An ionic salt that contains water of crystallization.

◆ **Water of crystallization**: Water molecules chemically bonded (in a fixed ratio) into the lattice of an ionic salt following crystallization.

◆ **Anhydrous salt**: An ionic salt that does not have water of crystallization.

A similar approach can be used to determine the empirical formula of a **hydrated salt** whose **water of crystallization** can be removed, by heating, without the **anhydrous salt** undergoing decomposition. In the calculation, the water and anhydrous salt are treated as formula units and divided by their molar masses. The mass of water of crystallization is the difference between the mass of the hydrated salt and the anhydrous salt.

■ **Figure S1.92** Blue hydrated copper(II) sulfate crystals and white anhydrous copper(II) sulfate crystals

Tool 1: Experimental techniques

Drying to constant mass

Drying to constant mass involves periodic monitoring of the mass of the residue of a filtration or a solution being crystallized. When the mass does not change between two readings, all the remaining liquid has been driven off. The substance may be dried by:

- heating in a crucible
- heating in an oven
- placing in a desiccator which contains a drying agent (Figure S1.93).

The appropriate method depends on the volume of liquid to be removed and whether the substance is likely to decompose if heated strongly. In practice, one or more of the methods may be used in the sequence shown. The residue can be transferred to a watch glass and heated in an oven to dry the solid.

■ **Figure S1.93** A desiccator used to store bottles in a dry atmosphere created by the desiccant silica gel

WORKED EXAMPLE S1.4K

Use the data below to determine the mass of water lost and the value of *x* in the formula for hydrated cobalt(II) chloride, $CoCl_2.xH_2O(s)$.

- Mass of clean, dry empty crucible = 10.45 g
- Mass of crucible and hydrated pink cobalt(II) chloride crystals = 12.83 g
- Mass of crucible and anhydrous blue cobalt(II) chloride crystals = 11.75 g

Answer

mass of hydrated salt = 2.38 g

mass of anhydrous salt = 1.30 g

mass of water = 2.38 g − 1.30 g = 1.08 g

$$\text{amount of water} = \frac{1.08\,\text{g}}{18.02\,\text{g mol}^{-1}} = 0.0599\,\text{mol}$$

$$\text{amount of anhydrous salt} = \frac{1.30\,\text{g}}{129.83\,\text{g mol}^{-1}} = 0.0100\,\text{mol}$$

Hence $x = \frac{0.0599\,\text{mol}}{0.0100\,\text{mol}} = 6$ (to 1 s.f.) and the formula is $CoCl_2.6H_2O(s)$.

41 12.3 g of hydrated magnesium sulfate, $MgSO_4.xH_2O$, gives 6.0 g of anhydrous magnesium sulfate, $MgSO_4$, on heating to constant mass. Deduce the value of *x* and explain how, experimentally, it may be possible to obtain values of *x* significantly less or more than the value you have calculated.

Determining the percentage by mass of an element in a compound of known formula

The experimentally determined percentage composition by mass of a compound is used to calculate the empirical formula of a compound. The reverse process can also be applied and the percentage by mass of a specific element in a compound of known formula can be calculated.

The method may be divided into three steps:

1 Determine the molar mass of the compound from its formula.
2 Write down the fraction by molar mass of each element (or water of crystallization) and convert this to a percentage.
3 Check to ensure that the percentages sum to 100.

WORKED EXAMPLE S1.4L

Calculate the percentage composition by mass of hydrated sodium sulfate, $Na_2SO_4.10H_2O$ ($M = 322\ g\ mol^{-1}$).

Answer

$$\text{percentage by mass of sodium} = \frac{2 \times 22.99}{322} \times 100 = 14.28\%$$

$$\text{percentage by mass of sulfur} = \frac{32.07}{322} \times 100 = 9.96\%$$

$$\text{percentage by mass of oxygen} = \frac{4 \times 16.00}{322} \times 100 = 19.88\%$$

Note that this does not include the oxygen in the water.

$$\text{percentage by mass of water} = \frac{10 \times ((2 \times 1.01) + 16.00)}{322} \times 100 = 55.96\%$$

$$\text{sum of percentages by mass} = 14.28 + 9.96 + 19.88 + 55.96 = 100.08 \approx 100\%$$

The minor discrepancy is due to rounding during the calculations.

Deducing the coefficients in an equation from reacting masses

The balancing numbers in a chemical equation can be calculated by calculating the moles of the substances in the reaction. In order to do this:

1. Calculate the amount of each substance using $n = \frac{m}{M}$.
2. Find the simplest whole-number ratio of these values of amounts by dividing all the values of amounts by the smallest value.
3. If this does not give a whole-number ratio, multiply up. Use a factor of 2 where there is a value ending in approximately 0.5; a factor of 3 where there is a value ending in approximately 0.33 or 0.67; or a factor of 4 where there is a value ending in approximately 0.25 or 0.75.

◆ **Gravimetric analysis:** A technique for determining the composition and formulas based on accurate measurement of the masses of reactants and products.

WORKED EXAMPLE S1.4M

2.431 g of magnesium reacts with 1.600 g of oxygen to form 4.031 g of magnesium oxide (Figure S1.94). Use this information to deduce the equation for the synthesis reaction.

Answer

Substance	***magnesium, Mg***	***oxygen, O_2***	***magnesium oxide, MgO***
Amount of each substance / mol	$\frac{2.431\ g}{24.31\ g\ mol^{-1}}$ = 0.1000 mol	$\frac{1.600\ g}{32.00\ g\ mol^{-1}}$ = 0.05000 mol	$\frac{4.031\ g}{40.31\ g\ mol^{-1}}$ = 0.1000 mol
Whole-number ratio	$\frac{0.1000}{0.05000} = 2$	$\frac{0.05000}{0.05000} = 1$	$\frac{0.1000}{0.05000} = 2$

Therefore the reacting ratio is 2 : 1 : 2 and so the balanced equation is:

$2Mg(s) + O_2(g) \rightarrow 2MgO(s)$

This is an example of a process called **gravimetric analysis**.

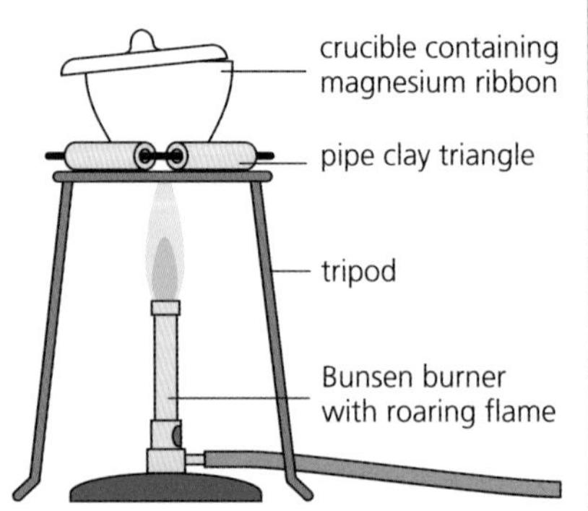

■ **Figure S1.94** Apparatus for determining the empirical formula of magnesium oxide by gravimetric analysis

Molar and mass concentration

◆ **Molar concentration:** The amount (mol) of a solute in 1 dm^3 of a solution.

◆ **Mass concentration:** The mass (g) of a solute in 1 dm^3 of a solution.

The **molar concentration** states the amount (mol) of a solute dissolved in solvent to form a solution. A 1.00 $mol\,dm^{-3}$ solution contains 1.00 mole of solute per cubic decimetre of solution. For example, a 1.00 $mol\,dm^{-3}$ solution of sulfuric acid contains 98.09 g of sulfuric acid in 1 dm^3 or 1000 cm^3 of solution.

You often need to calculate the amount of solute dissolved in a certain volume of solution.

$$\text{amount (mol)} = \text{volume (dm}^3) \times \text{molar concentration (mol dm}^{-3})$$

Top tip!

Molar concentrations are often represented using square brackets: for example, $[H_2SO_4(aq)] = 1.00\,mol\,dm^{-3}$. The symbol M is also used to represent molar concentration (the context, coming after a number, will stop you confusing it with M being used to represent molar mass).

WORKED EXAMPLE S1.4N

Calculate the mass (g) of potassium hydroxide in 25.00 cm^3 of a 0.125 $mol\,dm^{-3}$ solution.

Answer

$$\text{amount of KOH} = \text{volume (dm}^3) \times \text{molar concentration (mol dm}^{-3})$$

$$\text{amount of KOH} = \frac{25.00}{1000}\,dm^3 \times 0.125\,mol\,dm^{-3} = 0.003\,13\,mol$$

$$\text{mass of KOH} = 0.003\,13\,mol \times 56.11\,g\,mol^{-1} = 0.176\,g$$

When soluble ionic solids dissolve in water, they dissociate into separate ions. This can lead to the concentration of ions differing from the concentration of the solute.

If 9.521 g (0.1 mol) of magnesium chloride ($MgCl_2$) is dissolved to form 1 dm^3 of solution, then the concentration of magnesium chloride solution, $MgCl_2(aq)$, is 0.1 $mol\,dm^{-3}$. However, the 0.1 mol magnesium chloride would be dissociated to form 0.1 mol of magnesium ions and 0.2 mol of chloride ions. The concentration of magnesium ions is therefore 0.1 $mol\,dm^{-3}$ and the concentration of chloride ions is 0.2 $mol\,dm^{-3}$.

$$MgCl_2(s) + H_2O(l) \rightarrow Mg^{2+}(aq) + 2Cl^-(aq) + H_2O(l)$$

0.1 mol　　　　0.1 mol　0.2 mol

The concentration of a solution can also be measured in terms of mass of solute per volume of solution (Figure S1.95). A molar concentration can be converted to a **mass concentration** (grams of solute per cubic decimetre) by using the following expression:

$$\text{mass concentration (g dm}^{-3}) = \text{molar mass (g mol}^{-1}) \times \text{molar concentration (mol dm}^{-3})$$

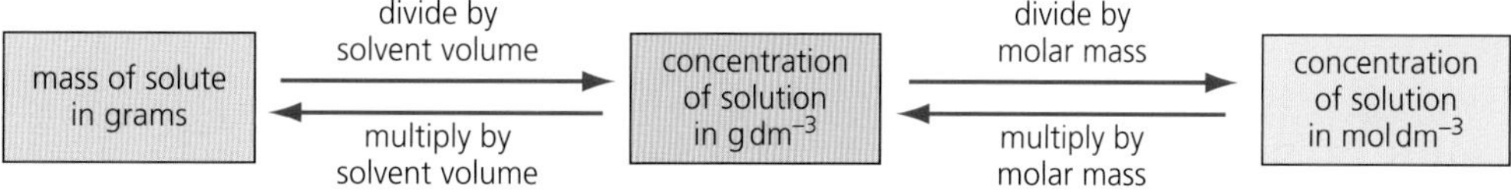

■ **Figure S1.95** Summary of the interconversion between concentration in moles per cubic decimetre and concentration in grams per cubic decimetre of solution

LINKING QUESTION

How can a calibration curve be used to determine the concentration of a solution?

LINKING QUESTION

What are the considerations in the choice of glassware used in preparing a standard solution and a serial dilution?

◆ **Titration**: A technique in which a standard solution is used to analyse another and determine its concentration.

◆ **Standard solution**: A solution of accurately known concentration for use in titrations.

◆ **Primary standard**: A reagent of high purity and stability used to prepare a standard solution.

Titrations

The equation of some reactions can be found using an acid–base **titration** to find the reacting volumes of **standard solutions** in the presence of an indicator. Standard solutions are solutions of known concentration that have been made from a **primary standard**.

Top tip!

If the solution to be titrated is concentrated then it may need to be diluted before titration. A dilution factor then needs to be applied in the calculation.

Tool 1: Experimental techniques

Preparing a standard solution

A volumetric flask is calibrated to contain a precise volume of solution at a particular temperature (usually 25 °C).

To prepare a standard solution, an accurately known mass of a solid must be dissolved in distilled (deionized) water and then made up to an accurate volume using a volumetric flask.

The primary standard should be stable (low reactivity with water and oxygen) and have a relatively high molar mass so that random errors in mass measurement are minimized.

Some common substances are not stable in air: for example, sodium hydroxide absorbs both water and carbon dioxide from air and is therefore unsuitable as a primary standard.

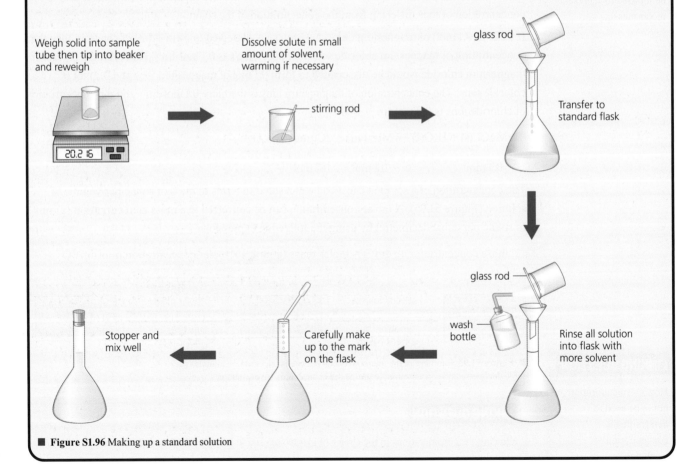

■ **Figure S1.96** Making up a standard solution

Tool 1: Experimental techniques

Acid–base titration

The volume of one solution is accurately measured using a burette as it is added to another solution, the volume of which has been accurately measured by pipette into a conical flask (on a white tile).

◆ **End-point**: The point in a titration when the reaction is complete as shown by the change in the colour of the indicator.

There is normally a visual method using an indicator (acid–base or redox indicator) of knowing when the reaction is complete (the **end-point**). In some titrations, the pH or temperature may be measured instead.

In titrations, the concentrations are usually between about 0.05 and 0.5 mol dm^{-3}, but they need to be adjusted by dilution to give a suitable titre value. A titre value of around 25 cm^3 is desirable as it has a relatively low uncertainty. As the titre value falls below 25 cm^3 the uncertainty increases accordingly.

The initial and final burette readings should both be recorded. The difference between the two measurements can then be used to calculate the titre delivered in cm^3. The measurement is repeated until at least two concordant titres (usually ±0.1 cm^3) are obtained. Any non-concordant results are reported in the raw data table but marked as anomalous and excluded from the calculation of the mean titre. Another approach is to average all values, apart from obvious outliers.

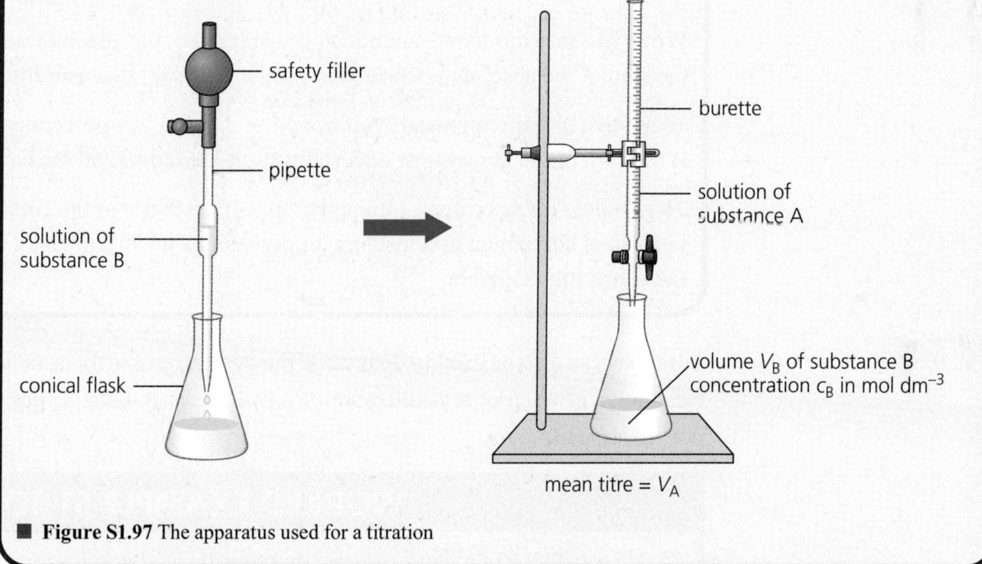

■ **Figure S1.97** The apparatus used for a titration

42 17.50 cm^3 of 0.150 mol dm^{-3} potassium hydroxide solution react with 20.00 cm^3 of phosphoric(V) acid, H_3PO_4, of concentration 0.0656 mol dm^{-3}. Deduce the equation for the reaction.

Tool 3: Mathematics

Express, record and propagate uncertainties

Each type of apparatus has an absolute uncertainty that can be expressed as a range. For example, an electronic balance has an uncertainty of ±0.001 g (if using a 4 d.p. balance) and a burette has an uncertainty of $\pm 0.05\,cm^3$ for single readings. The random error (absolute uncertainty) for the balance takes into account the random error (absolute uncertainty) in taring the balance to zero (±0.0005 g) and the random error (absolute uncertainty) in the reading when the mass is added (±0.0005 g).

The fractional uncertainty is given by the expression:

$$\frac{\text{absolute uncertainty}}{\text{measurement}}$$

The percentage uncertainty is given by the expression:

$$\frac{\text{absolute uncertainty}}{\text{measurement}} \times 100$$

In general, if the absolute uncertainty is not indicated on apparatus, the following assumptions are made:

- For an analogue scale, the uncertainty of a reading (one reading recorded) is ±0.5 of the smallest scale reading.
- If the apparatus has a digital display, the random uncertainty is ± the smallest measurement.

When two measurements are added or subtracted, the absolute uncertainty in the result is the sum of the absolute uncertainties of the individual measurements.

When two measurements are multiplied or divided, the percentage uncertainty in the result is the sum of the percentage uncertainties in the individual measurements.

To calculate the maximum total apparatus uncertainty in the final result, add all the individual equipment and instrument percentage uncertainties of readings that are used in the calculation together.

Titrations can also be used to determine the percentage purity or molar mass of an acid or base, or the amount of water of crystallization in a hydrated salt. Other types of titration include redox and precipitation titrations.

WORKED EXAMPLE S1.40

(2.355 ± 0.002) g of a weak organic monoprotic acid was dissolved in $(250.00 \pm 0.30)\,cm^3$ distilled water using a volumetric flask. $(20.00 \pm 0.06)\,cm^3$ of the acid solution was then titrated with $(0.100 \pm 0.005)\,mol\,dm^{-3}$ sodium hydroxide standard solution and the results are given in the following table. Determine the molar mass of the organic monoprotic acid and show the error propagation steps.

Titration number	1	2	3
Initial volume of aqueous NaOH / cm^3 (± 0.05)	0.00	0.00	0.00
Final volume of aqueous NaOH / cm^3 (± 0.05)	25.60	25.40	25.45
Volume of aqueous NaOH required / cm^3 (± 0.10)	25.60	25.40	25.45
Titration used	trial	yes	yes

Answer

Note that reliable readings (consistent readings) are used to calculate the mean instead of using all readings. Values are kept to 3, 4 or even 5 s.f. during the calculation and rounded at the end.

Absolute value calculation	Percentage uncertainty determination
mass of weak organic acid = (2.355 ± 0.002) g	% uncertainty in mass of weak organic acid = $\frac{0.002}{2.355} \times 100\% = 0.08493\%$
concentration of NaOH = (0.100 ± 0.005) mol dm^{-3}	% uncertainty in concentration of NaOH = $\frac{0.005}{0.100} \times 100\% = 5.00\%$
average volume of NaOH used = $\frac{25.40 + 25.45}{2} = (25.43 \pm 0.10)\,\text{cm}^3$	% uncertainty in volume of aqueous NaOH = $\frac{0.10}{25.43} \times 100\% = 0.3932\%$
volume of organic acid used (via pipette) = (20.00 ± 0.06) cm^3	% uncertainty in volume of organic monoprotic acid in pipette = $\frac{0.06}{20.00} \times 100\% = 0.3000\%$
concentration of acid solution = $\frac{25.43}{20.00} \times 0.100 = 0.127\,\text{mol dm}^{-3}$	% uncertainty in concentration of acid = % uncertainty in concentration of NaOH + % uncertainty in volume of NaOH + % uncertainty in volume of acid = 5.00% + 0.3932% + 0.3000% = 5.6932% absolute uncertainty in concentration of acid = 0.127 × 5.6932% = 0.007 mol dm^{-3}
volume of acid solution prepared using 250 cm^3 volumetric flask = (250.00 ± 0.30) cm^3	% uncertainty in volume of acid solution = $\frac{0.30}{250.00} \times 100\% = 0.12000\%$
amount of acid used in preparing 250 cm^3 of stock solution = $0.127 \times \frac{250.00}{1000.00} = 0.0318\,\text{mol}$	% uncertainty in amount of acid in 250 cm^3 of stock solution = 0.12000% + 5.6932% = 5.8132% absolute uncertainty in number of moles of acid in 250 cm^3 stock solution = 0.0318 × 5.8132% = 0.0019 mol
mass of 0.0318 mol of acid = 2.355 g molar mass of acid = $\frac{2.355}{0.0318} = 74.1\,\text{g mol}^{-1}$	% uncertainty in molar mass = 0.08493% + 5.8132% ≈ 6% (rounded up) absolute uncertainty of molar mass = 74.1 × 6% = 4.4 g mol^{-1}

So, molar mass of acid = (74.1 ± 4.5) g mol^{-1}.

43 A 50.00 cm^3 (0.05 dm^3) sample of concentrated sulfuric acid was diluted to 1.00 dm^3 (a dilution factor of 20). A sample of the diluted sulfuric acid was analysed by titrating with aqueous sodium hydroxide. In the titration, 25.00 cm^3 of 1.00 mol dm^{-3} aqueous sodium hydroxide required 20.00 cm^3 of the diluted sulfuric acid for neutralization.

Determine the concentration of the original concentrated sulfuric acid solution using the steps below.

a Construct the equation for the complete neutralization of sulfuric acid by sodium hydroxide.

b Calculate the amount of sodium hydroxide that was used in the titration.

c Calculate the concentration of the diluted sulfuric acid.

d Calculate the concentration of the original concentrated sulfuric acid solution.

Tool 3: Mathematics

Use approximation and estimation

Do you trust your calculator? Your calculator can only calculate the numbers it is given, but what happens if you make a mistake? For example, multiplying instead of dividing (the answer to 3.4×2 is very different to $3.4/2$) or mistyping a number (1.89 is very different to 8.19). A decimal point in the wrong place can also have a big effect.

Approximation and estimation are useful skills when it comes to checking if a calculation answer seems correct. They should let you know if a mistake has been made. For example, if you have $50.3/11.6$ you should be able to approximate the answer to $48/12 = 4$. If you have an answer of 6.7, it is clearly incorrect. Delve a little deeper with this calculation and you should also be able to see that your number will be slightly more than 4, so an answer of 3.9 which, on the surface, may seem correct would be wrong.

Titrations can take a long time and calculations often involve many steps. This makes being able to estimate the volume of solution you might need to deliver from the burette and the result you expect particularly useful. For example, consider the reaction between potassium hydroxide and sulfuric acid. Two moles of the base are needed to react with 1 mole of the acid. If the solutions are equimolar, the volume of base will be double that of the sulfuric acid. This is a quick way to estimate whether the volume of the solution of unknown concentration is more or less than that of the substance it is titrated with.

◆ **Back titration:** Where an excess amount of reagent is added to the solution to be analysed. The unreacted amount of the added reagent is then determined by titration, allowing the amount of substance in the original solution to be calculated.

Back titration

In a **back titration**, a known excess of one reagent A is allowed to react with an unknown amount of a reagent B. At the end of the reaction, the amount of A that remains unreacted is found by titration. A simple calculation gives the amount of A that has reacted with B and also the amount of B that has reacted.

In a typical acid–base back titration, a quantity of a base is added to an excess of an acid (or vice versa). All the base and some of the acid react. The acid remaining is then titrated with a standard alkali and its amount determined. From the results, the amount of acid which has reacted with the base can be found and the amount of base can then be calculated. The principle of this type of titration is shown in Figure S1.98.

Amount of standard acid (**calculated** from its volume and concentration)	
Amount of acid reacting with sample (**unknown**)	Amount of acid reacting with the standard solution of alkali used in the titration (**calculated** from its volume and concentration)

■ **Figure S1.98** The principle of an acid–base back titration

Top tip!

Back titrations are usually used when the determination of the amount of a substance poses some difficulty in the direct titration method: for example, for insoluble solid substances where the end-point is difficult to detect and for volatile substances where inaccuracy arises due to loss of substance during titration.

WORKED EXAMPLE S1.4P

An antacid tablet with the active ingredient sodium hydrogencarbonate, $NaHCO_3$, with a mass of 0.300 g was added to 25.00 cm^3 of 0.125 mol dm^{-3} hydrochloric acid.

After the reaction was completed, the excess hydrochloric acid required 3.55 cm^3 of 0.200 mol dm^{-3} KOH to reach the equivalence point in a titration.

Calculate the percentage of sodium hydrogencarbonate, $NaHCO_3$, in the tablet.

Answer

$HCl(aq) + NaHCO_3(s) \rightarrow NaCl(aq) + H_2O(l) + CO_2(g)$, then

$KOH(aq) + HCl(aq) \rightarrow KCl(aq) + H_2O(l)$

amount of KOH $= \frac{3.55}{1000} dm^3 \times 0.200\,mol\,dm^{-3} = 0.000\,710\,mol$

amount of excess HCl = 0.000 710 mol

amount of HCl before reaction with $NaHCO_3$ =
$\frac{25.00}{1000} dm^3 \times 0.125\,mol\,dm^{-3} = 0.003\,13\,mol$

amount of HCl that reacted with $NaHCO_3$ =
0.003 125 mol − 0.000 71 mol = 0.002 415 mol

amount of $NaHCO_3$ that reacted with HCl = 0.002 415 mol

mass of $NaHCO_3$ = 0.002 415 mol × 84.01 g mol^{-1} = 0.203 g

percentage purity of $NaHCO_3 = \frac{0.0203\,g}{0.300\,g} \times 100 = 67.7\%$

Inquiry 2: Collecting and processing data

Carry out relevant and accurate data processing

Data processing involves combining and manipulating raw numerical data to produce meaningful information. Then you can draw conclusions based on it.

Processing data could comprise using simple operations, for example, addition or multiplication, or might be more complicated such as calculating weighted averages, absolute or percentage changes, rates and standard deviation. You need to choose and carry out suitable calculations to simplify your data or to make comparisons between data sets, for example, finding the mean of a series of results, or the percentage gain or loss.

Remember to include metric or SI units for final processed quantities, which should be expressed to the correct and same number of significant figures.

You should calculate a percentage error if there is an accepted value with which you may compare your results. If it is appropriate, display the averaged data (after omitting outliers) in the form of a graph or chart.

44 Magnesium oxide is not very soluble in water, so it is difficult to titrate directly. Its purity can be determined by use of a back titration method. 4.08 g of impure magnesium oxide was completely dissolved in 100.0 cm^3 of 2.00 mol dm^{-3} aqueous hydrochloric acid. The excess acid required 19.70 cm^3 of 0.200 mol dm^{-3} aqueous sodium hydroxide for neutralization.

Determine the percentage purity of the impure magnesium oxide.

Guidance:

- Construct equations for the two neutralization reactions.
- Calculate the amount of hydrochloric acid added to the magnesium oxide.
- Calculate the amount of excess hydrochloric acid titrated.
- Calculate the amount of hydrochloric acid reacting with the magnesium oxide.
- Calculate the mass of magnesium oxide that reacted with the initial hydrochloric acid, and hence determine the percentage purity of the magnesium oxide.

Going further

Parts by mass and percent by mass

It is often convenient to quote a concentration as a ratio of masses. A parts by mass concentration is the ratio of the mass of the solute to the mass of the solution, all multiplied by a multiplication factor:

$$\frac{\text{mass of solute}}{\text{mass of solution}} \times \text{multiplication factor}$$

The particular parts by mass unit we use, which determines the size of the multiplication factor, depends on the concentration of the solution. For example, for percent by mass the multiplication factor is 100:

$$\text{percentage by mass} = \frac{\text{mass of solute}}{\text{mass of solution}} \times 100$$

Percent means 'per hundred'; a solution with a concentration of 5% by mass contains 5 g of solute per 100 g of solution. Vinegar is approximately 5% by mass of ethanoic acid and its concentration can be determined by titration with freshly prepared NaOH(aq).

For more dilute solutions, we can use parts per million (ppm), which has a multiplication factor of 10^6:

$$\text{ppm} = \frac{\text{mass of solute}}{\text{mass of solution}} \times 10^6$$

A solution with a concentration of 25 ppm by mass, for example, contains 25 g of solute per 10^6 g of solution. The units of ppm are used in determination of BOD (biological oxygen demand) of polluted water.

For dilute aqueous solutions near room temperature, the units of ppm are equivalent to milligrams of solute per cubic decimetre of solution. This is because the density of a dilute aqueous solution near room temperature is $1.0\,\text{g}\,\text{cm}^3$, so that $1\,\text{dm}^3$ has a mass of almost 1000 g.

Top tip!

A good way to remember the relative scale of a concentration in ppm is to remember that 1 ppm is equivalent to $1\,\text{mg}\,\text{dm}^{-3}$.

WORKED EXAMPLE S1.4Q

$150\,\text{cm}^3$ of an aqueous sodium chloride solution contains 0.0045 g NaCl. Calculate the concentration of NaCl in parts per million (ppm).

Answer

$$\text{ppm} = \frac{\text{mass of solute (mg)}}{\text{volume of solution (dm}^3\text{)}}$$

$$\text{mass of NaCl} = 0.0045\,\text{g} = (0.0045 \times 1000)\,\text{mg} = 4.5\,\text{mg}$$

$$\text{volume of solution} = 150\,\text{cm}^3 = \frac{150}{1000}\,\text{dm}^3 = 0.150\,\text{dm}^3$$

$$\text{concentration of NaCl} = \frac{4.5\,\text{mg}}{0.150\,\text{dm}^3} = 30\,\text{mg}\,\text{dm}^{-3} = 30\,\text{ppm}$$

Clean water

According to the United Nations (UN), 3.6 billion people today live in areas where water can be scarce at least one month per year and this number could increase to 5 billion by 2050. Water has a major role in maintaining peace and political stability across borders. Transboundary cooperation between countries on shared water resources is vital to securing lasting peace and sustainable development. Oxidizing agents such as chlorine and ozone are often used to purify drinking water by killing bacteria and viruses. The presence of toxic nitrates from the use of artificial fertilizers in freshwater is another common problem.

◆ **Avogadro's law**: Equal volumes of all gases (when behaving ideally) contain equal numbers of molecules at the same temperature and pressure.

◆ **Molar gas volume**: The volume occupied by one mole of an ideal gas, usually under standard conditions.

◆ **STP**: Standard temperature and pressure: conditions used for comparing gases (273.15 K and 100 kPa).

Avogadro's law

If the pressure and temperature of a sample of gas are fixed, then the volume of gas depends on the amount of gas, which is a measure of the number of molecules. This is **Avogadro's law** and can be easily demonstrated using balloons (Figure S1.99).

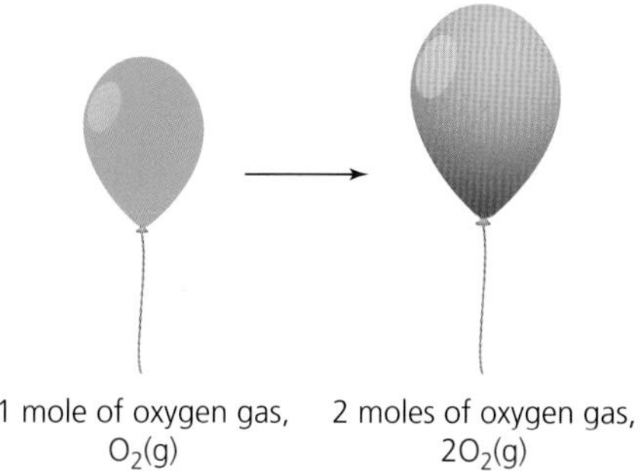

■ **Figure S1.99** The volume of the balloon is increased by a factor of 8 when the amount of the gas is doubled (at constant pressure and temperature)

Top tip!

Avogadro's law is a consequence of large intermolecular distances (a gas is mainly empty space) and so is only exactly obeyed by an ideal gas. An ideal gas can be visualized as having particles that do not attract or repel each other and whose total volume is insignificant compared to the volume of the container the gas is in. Many gases approach ideal behaviour when the pressure is low and the temperature is high.

A more formal statement of Avogadro's law is that equal volumes of gases at the same temperature and pressure contain equal numbers of molecules.

Since one mole of any (ideal) gas (measured at the same temperature and pressure) occupies the same volume (the **molar gas volume**, 22.7 dm^3 at standard room temperature and pressure, **STP**), equal volumes of different gases will contain the same amount of gas particles (in moles).

Mathematically, Avogadro's law can be expressed as: $V \alpha n$, where V represents the volume of gas and n represents the amount of gas (in moles). For example, if the number of molecules of a mass of gas is doubled, then the volume (at constant temperature and pressure) of the gas is doubled (Figure S1.100).

Common mistake

Misconception: *One mole of all substances (pure solids, liquids and gases) has a volume of 22.7 cubic decimetres.*

One mole of an ideal gas at STP has a volume of 22.7 cubic decimetres (dm^3) at 273.15 K and 100 kPa.

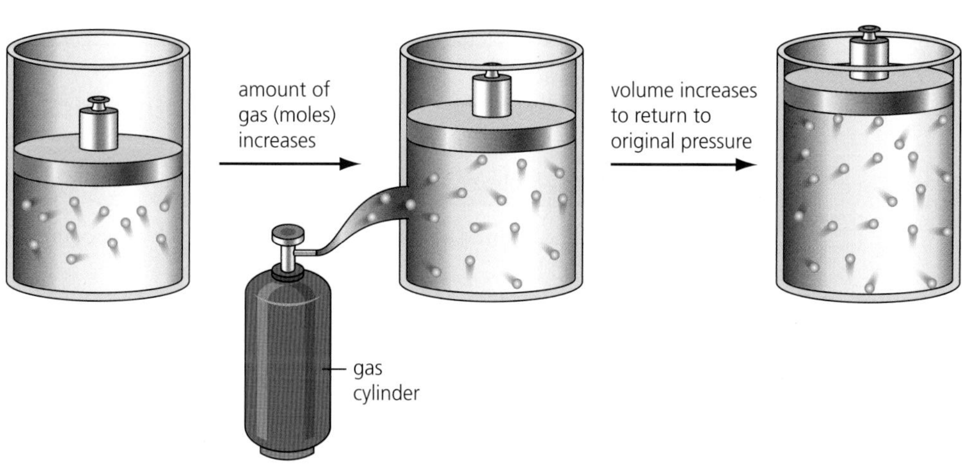

■ **Figure S1.100** The effect of increasing the amount of gas particles (at constant temperature and pressure)

◆ **Gay-Lussac's law:** The ratio of the volumes of gases consumed or produced in a chemical reaction is equal to the ratio of simple integers.

A consequence of Avogadro's law is that the ratio of the volumes of gases consumed or produced in a chemical reaction is equal to a ratio of simple whole numbers (**Gay-Lussac's law**).

Using Avogadro's law, measurements on the formation of hydrogen chloride by direct synthesis (Figure S1.101) can be interpreted as shown below:

1 volume of hydrogen	+	1 volume of chlorine	→	2 volumes of hydrogen chloride
1 mol of hydrogen molecules	+	1 mol of chlorine molecules	→	2 mol of hydrogen chloride molecules
1 chlorine molecule	+	1 hydrogen molecule	→	2 hydrogen chloride molecules

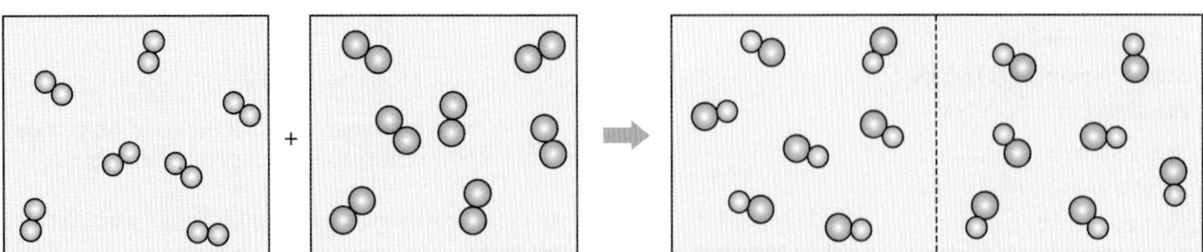

■ **Figure S1.101** Avogadro's law applied to the formation of hydrogen chloride from hydrogen and chlorine

Calculating the volumes of gaseous reactants used

WORKED EXAMPLE S1.4R

Calculate the volume of oxygen needed to burn 200 cm^3 of propane and, therefore, the volume of air needed to burn 200 cm^3 of propane. (Air is 20% oxygen by volume.)

Answer

$$C_3H_8(g) + 5O_2(g) \rightarrow 3CO_2(g) + 4H_2O(l)$$

This equation indicates that one mole of propane reacts with five moles of oxygen to produce three moles of carbon dioxide and four moles of water. Applying Avogadro's law means that one volume of propane gas reacts with five volumes of oxygen gas to produce three volumes of carbon dioxide gas and four volumes of water.

The volumes of propane and oxygen must be in a 1 : 5 ratio, hence the volume of oxygen needed is $5 \times 200\,cm^3 = 1000\,cm^3$.

The volume of oxygen required is 1000 cm^3. However, air is only 20% oxygen and so you need five times more air than oxygen. Hence, the volume of air required is $5 \times 1000\,cm^3 = 5000\,cm^3$.

Calculating the volumes of gaseous products

WORKED EXAMPLE S1.4S

Calculate the volume of carbon dioxide produced by the complete combustion of 0.500 dm^3 of butane, C_4H_{10}.

Answer

$$2C_4H_{10}(g) + 13O_2(g) \rightarrow 8CO_2(g) + 10H_2O(l)$$

This equation indicates that 2 volumes of butane react with 13 volumes of oxygen to form 8 volumes of carbon dioxide.

The amounts of butane and carbon dioxide are in a 2 : 8 ratio, hence the volume of carbon dioxide formed is $0.500\,dm^3 \times 4 = 2.00\,dm^3$.

Deducing the molecular formula

WORKED EXAMPLE S1.4T

When 20 cm^3 of a gaseous hydrocarbon is reacted with excess oxygen, the gaseous products consist of 80 cm^3 of carbon dioxide, CO_2, and 80 cm^3 of steam, H_2O, measured under the same conditions of pressure and temperature (above 100 °C).

Deduce the molecular formula of the hydrocarbon.

Answer

20 cm^3 hydrocarbon + excess oxygen → 80 cm^3 CO_2 + 80 cm^3 H_2O

1 molecule of hydrocarbon → 4 molecules of CO_2 + 4 molecules of H_2O

Each molecule of the hydrocarbon must contain four carbon atoms and eight hydrogen atoms.

The molecular formula is therefore C_4H_8.

◆ **Eudiometry**: Measurement and analysis of gases with a eudiometer.

◆ **Eudiometer**: Apparatus for measuring changes in volumes of gases during chemical reactions.

Going further

Eudiometry

The molecular formula of a hydrocarbon can be determined by **eudiometry**. The hydrocarbon is completely burnt in oxygen in a **eudiometer** (a glass tube with a volume scale, closed at one end) and the volumes of gaseous mixtures are measured at various stages. Carbon dioxide can be removed by reaction with a strong base, and any excess oxygen can be removed by reaction with pyrogallol. The measurements are done with cooled gases so the water is a liquid.

1 Write the balanced generalized equation for the combustion of one mole of hydrocarbon.

$$C_xH_y(g) + (x + \frac{y}{4})O_2(g) \rightarrow xCO_2(g) + \frac{y}{2}H_2O(l)$$

2 Using the volumes of the hydrocarbon and carbon dioxide from the experimental data, determine the value of x using the volume and mole ratio:

$$\frac{\text{volume of carbon dioxide}}{\text{volume of hydrocarbon}} = \frac{x}{1}$$

This follows from Avogadro's law: equal volumes of gases at the same temperature and pressure contain the same number of particles.

3 Determine the volume of O_2 used and the value of y.

volume of O_2 used = initial volume of O_2 − volume of unused O_2

$$\frac{\text{volume of oxygen used}}{\text{volume of hydrocarbon}} = \frac{x + \frac{y}{4}}{1}$$

WORKED EXAMPLE S1.4U

Determine the molecular formula of an unknown hydrocarbon. 15 cm^3 of a gaseous hydrocarbon was completely combusted with 95 cm^3 (an excess) of oxygen. After cooling to room temperature, the residual gaseous products (carbon dioxide and excess oxygen) occupied a volume of 80 cm^3. On adding aqueous sodium hydroxide to remove CO_2, the volume decreased to 50 cm^3. Determine the molecular formula of the hydrocarbon.

Step 1 Write the balanced generalized equation for the combustion of one mole of hydrocarbon.

$$C_xH_y(g) + (x + \frac{y}{4})O_2(g) \rightarrow xCO_2(g) + \frac{y}{2}H_2O(l)$$

Step 2 Using the volumes of the hydrocarbon and carbon dioxide from the experimental data, determine the value of x using the volume and mole ratio:

$$\frac{\text{volume of carbon dioxide}}{\text{volume of hydrocarbon}} = \frac{x}{1}$$

The volume of (acidic) CO_2 produced is equal to the volume of gas that is removed upon addition of the aqueous sodium hydroxide (a base) according to the equations below:

$$2NaOH(aq) + CO_2(g) \rightarrow Na_2CO_3(aq) + H_2O(l)$$

$$2NaOH(aq) + 2CO_2(g) \rightarrow 2NaHCO_3(aq)$$

Step 3 Determine the volume of O_2 used and the value of y. Volume of O_2 used = initial volume of O_2 – volume of unused O_2

$$\frac{\text{volume of oxygen used}}{\text{volume of hydrocarbon}} = \frac{x + \frac{y}{4}}{1}$$

	$C_xH_y(g)$	$(x + \frac{y}{4})O_2(g)$	$xCO_2(g)$	$\frac{y}{2}H_2O(l)$
Initial gas volume/cm^3	15	95	0	0
Final gas volume/cm^3	0	50	30	-
Change in gas volume/cm^3	–15	$-(95 - 50) = -45$	$80 - 50 = +30$	-

For $CO_2(g)$ and $C_xH_y(g)$: $\frac{1}{x} = \frac{15}{30}$; $x = 2$

For $O_2(g)$ and $C_xH_y(g)$: $\frac{1}{x + \frac{y}{4}} = \frac{15}{45}$; $y = 4$

Hence the molecular formula of the hydrocarbon is C_2H_4.

■ **Figure S1.102** Amedeo Avogadro

Nature of science: Theories

Developing atomic theory

Amedeo Avogadro (1776–1856) trained and practised as a lawyer, but later became Professor of Physics at Turin University. His research was based on the careful experimental work by Gay-Lussac and John Dalton's atomic theory. As a tribute to him, the number of particles in one mole of a substance is known as the Avogadro constant (formerly the Avogadro number).

In 1808, Dalton published his atomic theory proposing that all matter is made of atoms. He further stated that all atoms of an element are identical and the atoms of different elements have different masses. However, he was incorrect about the way elements reacted to form compounds. For example, he thought water was made of one hydrogen atom and one oxygen atom and wrote it with the formula HO.

In 1809, Joseph Gay-Lussac published his law of combining gas volumes. He had noticed that when two volumes of hydrogen gas reacted with one volume of oxygen gas, they formed two volumes of water vapour. All gases that he reacted seemed to react in simple volume ratios. Avogadro reconciled Dalton's theory and Gay-Lussac's observations by proposing that elements could exist as molecules rather than as individual atoms.

LINKING QUESTION

Avogadro's law applies to ideal gases. Under what conditions might the behaviour of a real gas deviate most from an ideal gas?

ATL S1.4J

Extend this beginning of a concept map on a large piece of paper to illustrate all aspects of the mole concept introduced in this chapter.

- Include number of particles, mass, gas volumes, the molar gas volume and concentrations (by mass and molarity).
- Use arrows to indicate names and values of the unit conversions.
- Include worked examples for all the conversions shown on your concept map.

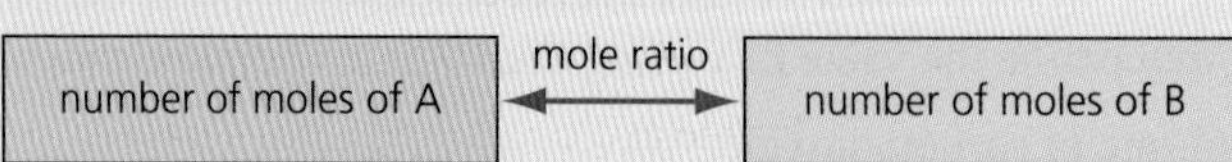

■ **Figure S1.103** Start of a concept map for the mole

Ideal gases

Guiding question

- How does the model of ideal gas behaviour help us to predict the behaviour of real gases?

SYLLABUS CONTENT

By the end of this chapter, you should understand that:

- an ideal gas consists of moving particles with negligible volume and no intermolecular forces; all collisions between particles are considered elastic
- real gases deviate from the ideal gas model, particularly at low temperature and high pressure
- the molar volume of an ideal gas is constant at a specific temperature and pressure
- the relationship between the pressure, volume, temperature and amount of an ideal gas is shown in the ideal gas equation $PV = nRT$ and the combined gas law.

By the end of this chapter you should know how to:

- recognize the key assumptions in the ideal gas model
- explain the limitations in the ideal gas model
- investigate the relationship between temperature, pressure and volume for a fixed mass of an ideal gas and analyse graphs relating these variables
- solve problems relating to the ideal gas equation
- use the ideal gas equation to calculate the molar mass of a gas from experimental data.

Note: There is no higher-level only content in S1.5

◆ **Gas**: A state of matter which occupies the full volume of a container regardless of its quantity.

◆ **Pascal**: SI unit of pressure equal to one newton per square metre; $1\,Pa = 1\,N\,m^{-2}$.

◆ **Elastic collision**: A collision in which the total kinetic energy of the colliding bodies after collision is equal to their total kinetic energy before collision.

Common mistake

Misconception: *Gas pressure is caused by gas molecules colliding with each other inside the container.*

Gas pressure is caused by gas molecules colliding with the walls of the container.

Top tip!

Gas expanding to fill a container by diffusion is a spontaneous process due to the increase in entropy.

Introduction

Gases under conditions of low temperature and pressure are simpler to understand than liquids and solids. The overall molecular motion in gases is totally random and the forces of attraction between gas molecules are so small that each molecule moves freely and essentially independently of other molecules.

The physical behaviour of gases can be predicted. The laws that control this behaviour have played an important role in the development of the atomic theory of matter and the kinetic molecular theory of gases.

Gas pressure can be measured with a calibrated pressure gauge or digital gas-pressure sensor. $\text{Pressure} = \frac{\text{force}}{\text{area}}$ and so the SI unit of pressure is $N\,m^{-2}$ which is also known as the **Pascal** (Pa).

Ideal gases

The behaviour of gases is explained by the molecular kinetic theory of gases, which makes the following assumptions about the properties of particles in a gas:

- The total volume occupied by all of the molecules is negligible compared to the volume of the container (Figure S1.104).
- The intermolecular interactions between the molecules, and between the molecules and walls of the container, are negligible and can be ignored. The collisions of the molecules or atoms with each other and the walls of the container are perfectly elastic and give rise to gas pressure (Figure S1.105). During these **elastic collisions**, gas molecules do not lose kinetic energy.

- The mean kinetic energy of the molecules or atoms is directly proportional to the absolute temperature of the gas. This means gas particles of lower molar mass move more quickly (on average) than particles of higher molar mass at the same temperature.

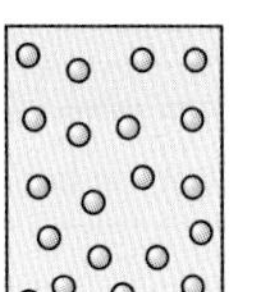

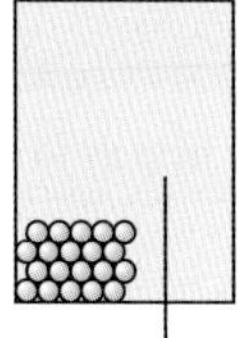

■ **Figure S1.104** Concept of an ideal gas

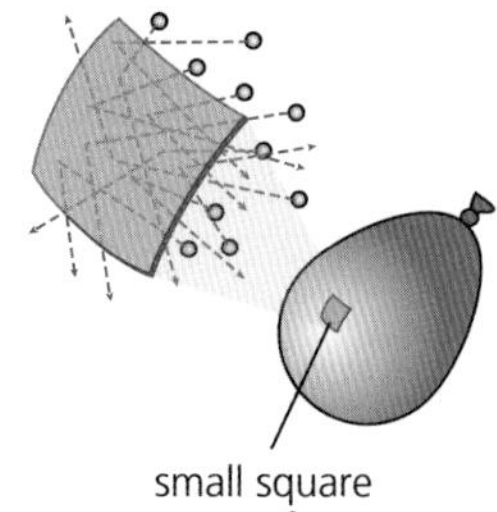

■ **Figure S1.105** The generation of gas pressure on the inner surface of a balloon

Top tip!

The average kinetic energy of gas molecules is constant as long as the temperature is constant.

Top tip!

Absolute zero (0 K) (see Chapter S1.1, page 21) is the temperature at which atoms and molecules in the solid state stop vibrating – there is no internal energy. It is the lowest temperature possible on the absolute temperature scale and, due to the laws of thermodynamics, cannot be reached but scientists have cooled objects to temperatures very close to absolute zero (−273.15 °C).

◆ **Ideal gas**: A hypothetical gas that obeys the ideal gas equation and the gas laws exactly and would have molecules that occupy negligible space and have negligible forces between them.

◆ **Diffusion**: The overall movement of particles from a region of high concentration to a region of low concentration.

This is the **ideal gas** model and it is a good mathematical and chemical model of the behaviour of most gases, as long as the temperature is high enough and/or the pressure low enough for the assumptions to hold. However, an ideal gas is a *hypothetical* state since the first two assumptions cannot be precisely true. No real gas behaves exactly as an ideal gas, but real gases can behave almost ideally under certain conditions.

TOK

What is the role of imagination and intuition in the creation of hypotheses in the natural sciences?

'What are substances made from?' is a question that has been asked for thousands of years—since the time of the Ancient Greek philosophers. Gradually, scientists began to believe that all substances are made up of particles (later demonstrated to be atoms, molecules and ions). They are too small to be seen, but many simple experiments and observations suggest that matter must be made of very small particles.

Observations (empirical data) using sense perception involving the **diffusion** of coloured gases, such as bromine (Figure S1.106), can only be explained if the substances are assumed to be composed of large numbers of moving particles (Figure S1.107).

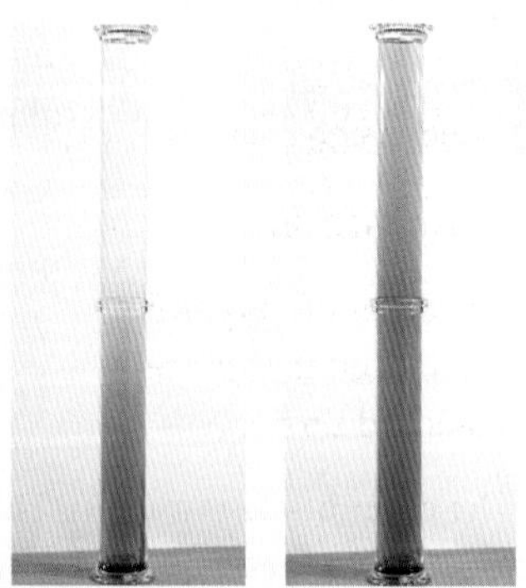

■ **Figure S1.106** After 24 hours the orange bromine fumes have diffused throughout both gas jars

The rate of diffusion of bromine molecules is slow, even though the speed of the gas molecules is very high (typically hundreds of metres per second), because there are billions of collisions between bromine molecules and oxygen and nitrogen molecules in the air.

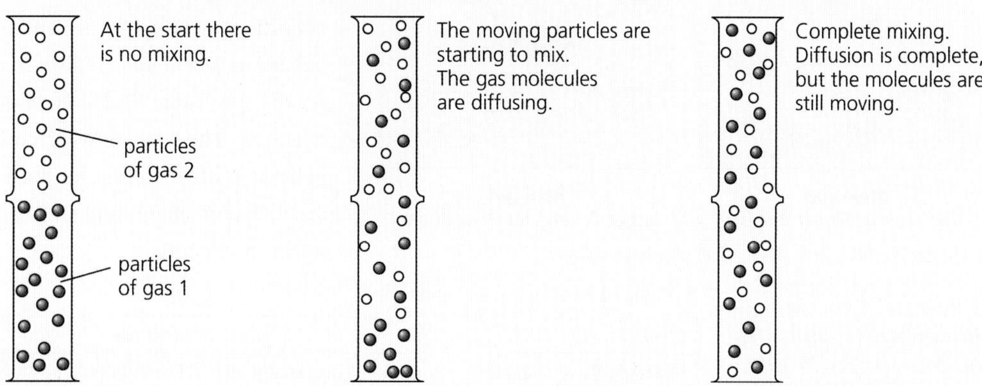
At the start there is no mixing.
particles of gas 2
particles of gas 1
The moving particles are starting to mix.
The gas molecules are diffusing.
Complete mixing.
Diffusion is complete, but the molecules are still moving.

■ **Figure S1.107** A particle description of diffusion in two gases that diffuse into each other

The early development of the kinetic molecular theory was a direct consequence of the intuition, imagination and reasoning (logic) of various scientists involved in the development of the kinetic theory. The later scientists included Einstein, who quantitatively explained gas motion in terms of a random (or drunken) walk model (Figure S1.108), with particles travelling in straight lines between collisions. The distance travelled by a bromine molecule between collisions (the so-called mean free path) is very short—in the order of tens of nanometres. Each molecule travels only a few centimetres per second in the direction of diffusion.

Imagination and intuition play an important part in the construction of scientific knowledge, but the results of the intuitions (hypotheses or other possible explanations) mean nothing to a scientist unless they can be confirmed or falsified through scientific experiments.

When discussing the role of imagination or intuition (or any other of the sources of knowledge), many students simply point out that it happens and perhaps illustrate it through examples. The truly interesting point, however, will be to explore the impact of these sources on the knowledge that is constructed. Showing what role imagination and intuition play in the scientific method is more important than merely stating that they matter: doing so provides you with an opportunity to further develop and explore the scientific method itself.

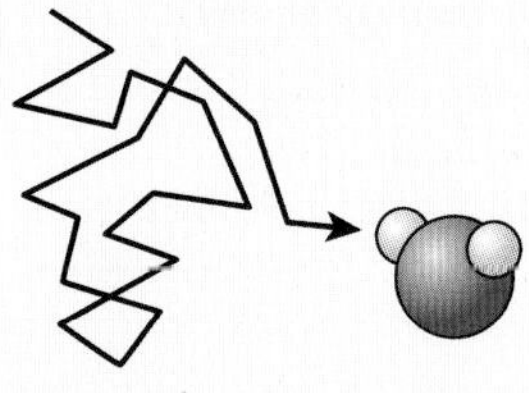

■ **Figure S1.108** Drunken walk trajectory of a single gas molecule

Tool 3: Mathematics

Compare and quote values to the nearest order of magnitude

The order of magnitude of a number is the smallest power of ten used to represent the number in scientific notation (standard form).

The order of magnitude of 10 is 1. The order of magnitude of 10 000 is 4 and the order of magnitude of 0.3 (3×10^{-1}) is −1.

Two numbers can be said to have the same order of magnitude if the large one divided by the small one is less than 10. This means that 57 and 19 have the same order of magnitude, but 570 and 19 do not.

We can compare orders of magnitude easily using standard form. The diameter of a hydrogen atom is 2.55×10^{-11} m. The diameter of the Earth is approximately 1.2×10^{7} m.

We can compare these two diameters by dividing the larger power of ten by the smaller one: $10^{7} \div 10^{-11} = 10^{18}$. The diameter of the Earth is therefore approximately 1 000 000 000 000 000 000 times larger than that of a hydrogen atom.

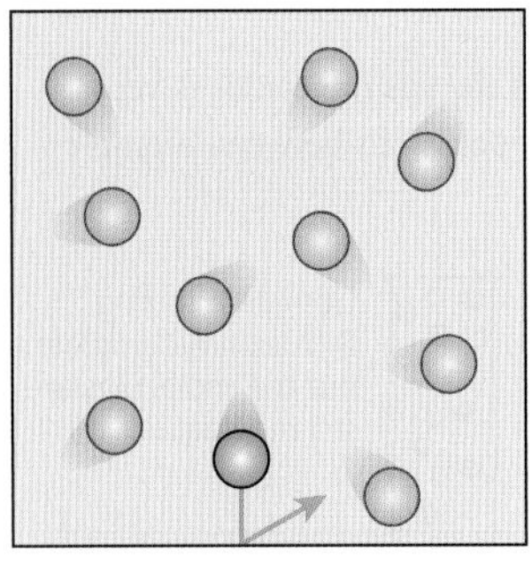

Ideal gas
(no intermolecular forces)

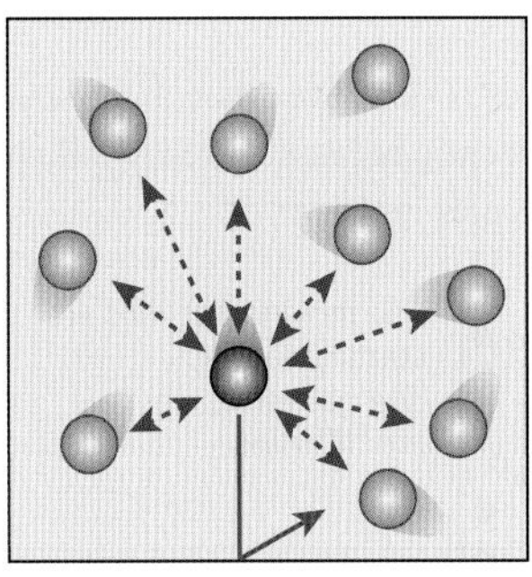

Real gas
(attractive and repulsive forces)

■ **Figure S1.109** Comparison of an ideal and real gas

Real gases

You should not think that a gas always behaves as an ideal gas. Every gas will behave almost ideally under certain conditions of temperature and pressure. The same gas will behave as a **real gas** (Figure S1.109) under other sets of conditions (Table S1.25) and not exactly obey the ideal gas equation. The ways in which real gases deviate from ideal gas behaviour are different for each gas because their molecules are different shapes and sizes and their intermolecular forces differ in strength.

◆ **Real gas:** A gas that does not behave ideally and does not obey the ideal gas equation.

Common mistake

Misconception: *Gases are divided into real and ideal gases.*

All gases are real gases but can approach ideal behaviour under certain conditions.

■ **Table S1.25** Ideal and real gases

Ideal gas	Real gas
Molecules have negligible total volume compared to the volume of the container they are in.	The total volume of the molecules is a significant fraction of that of the container at high pressure.
Molecules act in an independent manner. They show no attraction (or repulsion) for each other. Intermolecular forces are assumed to be absent.	Molecules are beginning to associate together, as they start to form a liquid. Intermolecular attractions are significant when molecules are close or slow moving.
Collisions between molecules are perfectly elastic (no loss of total kinetic energy).	Collisions may be inelastic due to transfer of energy to internal vibration and rotation of molecules.

Effect of intermolecular forces

The extent to which a gas deviates from ideal behaviour depends on the polarity and size of the molecule since these factors determine the strength of the intermolecular forces. Deviation is greatest for polar molecules and large molecules, which experience strong intermolecular forces of attraction; it is smallest for small non-polar molecules or noble gas atoms.

For example, in decreasing order of deviation: $Cl_2 > NH_3 > HCl > O_2 > N_2 > H_2 > He$. Although the chlorine molecule is non-polar, its deviation from ideal behaviour is greater than that of ammonia molecules because the London (dispersion) forces are collectively stronger than the hydrogen bonding between ammonia molecules.

London (dispersion) forces, which operate between molecules, especially at high pressure and low temperatures, increase with the number of electrons and, hence, the molar mass. The shape of the molecule also plays a role in determining the strength of these forces.

The strongest intermolecular forces are in small polar molecules with hydrogen bonding, such as NH_3, H_2O and HF.

The effect of intermolecular forces and size of particles on deviation from ideal behaviour over a range of temperatures can be assessed via plots of $\frac{PV}{RT}$ against P (or from the values of a and b in the van der Waals' equation).

WORKED EXAMPLE S1.5A

List the gases He, Ne, O_2, H_2O, NH_3 and C_2H_6 in order of increasing deviation from ideal gas behaviour. Explain the order.

Answer

In order of increasing deviation: He, Ne, O_2, C_2H_6, NH_3 and H_2O.

H_2O deviates more from ideal behaviour than NH_3. NH_3 forms fewer hydrogen bonds per molecule than H_2O and these bonds are weaker. H_2O and NH_3 deviate the most from ideal behaviour due to the presence of hydrogen bonding, the strongest of the intermolecular forces.

Deviation increases in the order He < Ne < O_2 < C_2H_6. The strength of the London (dispersion) forces depends on the number of electrons present in the molecule and its shape. The greater the number of electrons a molecule or atom has, the more polarizable the electron cloud is, and hence the stronger the intermolecular or interatomic interactions.

Top tip!

Polarizability is a measure of how easily an electron cloud is distorted by an electric field.

LINKING QUESTION

Under comparable conditions, why do some gases deviate more from ideal behaviour than others?

Molar volume of a gas

Link

How to calculate moles and Avogadro's law are also covered in Chapter S1.4, from page 65.

It is not easy to measure the mass of a sample of gas. Gases have low densities and they rapidly diffuse. A large container is needed to hold a measurable mass of gas, and buoyancy effects in air become important. However, it is easy to measure the volume of a pure gas using a sealed gas syringe, then convert this to an amount (mol).

It follows from Avogadro's law that the volume occupied by one mole of molecules must be the same for all gases, provided they approach ideal behaviour. This volume is known as the molar gas volume. At 273.15 K and 100 kPa—conditions known as standard temperature and pressure (STP)—the molar volume has an approximate value of 22.7 dm^3, which is roughly the same as the volume of a cube of side 28.3 cm.

Amounts and volumes of gas at STP can be interconverted via the molar gas volume (Figure S1.110).

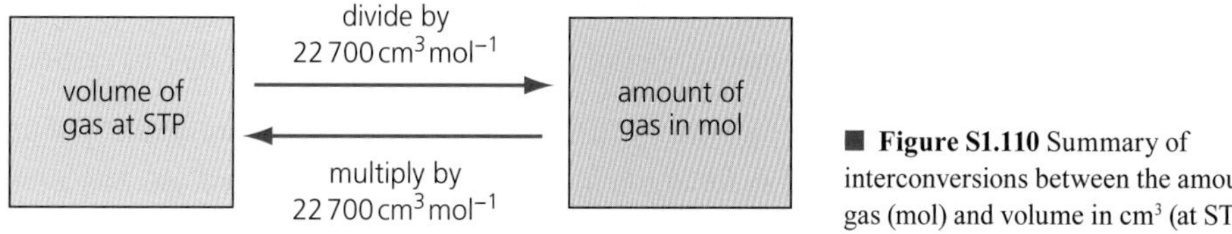

■ **Figure S1.110** Summary of interconversions between the amount of gas (mol) and volume in cm^3 (at STP)

The molar gas volume at STP together with Avogadro's constant ($6.02 \times 10^{23} mol^{-1}$) can be used to solve a range of stoichiometry problems (Figure S1.111).

Top tip!

The molar gas volume varies with temperature and pressure (see **https://planetcalc.com/7918/**). At 100 kPa and 25 °C the molar gas volume is 24.87 $dm^3 mol^{-1}$.

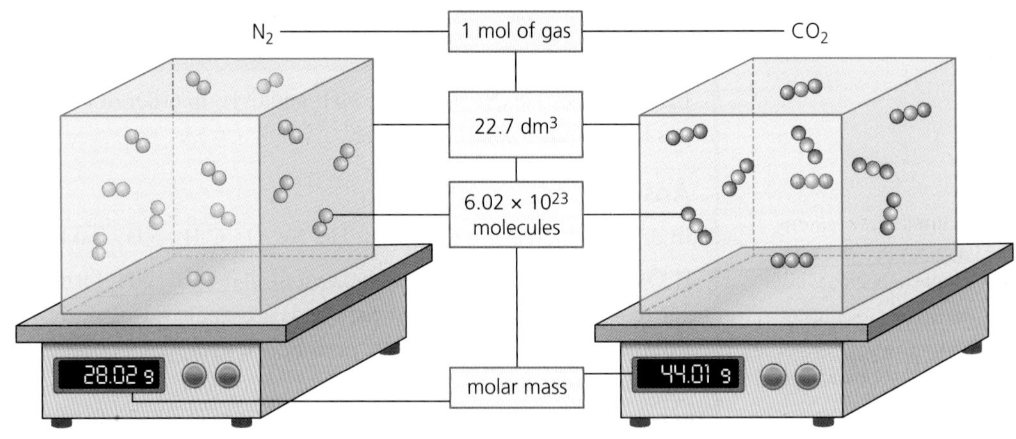

■ **Figure S1.111** Both samples of gas contain the same number of particles (atoms or molecules) if the gas approaches ideal behaviour

Inquiry 3: Concluding and evaluating

Evaluating a hypothesis

A hypothesis is a scientific explanation, often based on a model, of the event or process that caused the physical phenomenon that was observed. The hypothesis will identify which variables are involved in the cause (independent) and effect (dependent), and which variables are not involved (the controlled variables).

For example, Avogadro's law proposes that equal volumes of gases at the same temperature and pressure contain equal numbers of particles. His law was based upon a clear distinction between atoms and molecules. Avogadro's law predicts that if the number of particles of a gas is doubled (at constant temperature and pressure), then the volume is also doubled. His law assumes that the number of atoms or molecules in a specific volume of gas is independent of their size or molar mass.

A hypothesis allows a prediction to be made if the independent variable is changed. If the results of the investigation do not support the hypothesis, a new hypothesis must be developed that takes into account the data after evaluating the methodology.

Top tip!

Gas volumes are usually compared at STP.

LINKING QUESTION

How can the ideal gas law be used to calculate the molar mass of a gas from experimental data?

45 Calculate the volume of oxygen in dm^3 at STP that contains 1.35 mol of molecules.

46 Calculate the amount in moles of hydrogen molecules in 175 cm^3 of hydrogen gas at STP.

47 Calculate the number of molecules present in 2.85 dm^3 of carbon dioxide at STP.

48 Calculate the density (in grams per cubic decimetre, $g\,dm^{-3}$) of argon gas at STP. The relative atomic mass for argon is 39.95 (so 22.7 dm^3 of argon gas has a mass of 39.95 g).

49 20.8 g of a gas occupies 7.44 dm^3 at STP. Determine the molar mass of the gas.

50 When 3.06 g of potassium chlorate(V), $KClO_3$, is heated, it produces 840 cm^3 of oxygen at STP and leaves a residue of potassium chloride, KCl. Deduce the balanced equation. (The molar mass of potassium chlorate(V) is 122.5 $g\,mol^{-1}$.)

The gas laws

A sample of gas can be described in terms of four variables: pressure, volume, temperature and amount (number of moles). The four variables are interdependent and there are various laws that describe the relationship between them.

◆ **Gas laws:** Laws relating the temperature, pressure and volume of an ideal gas.

◆ **Atmospheric pressure:** The pressure due to the Earth's atmosphere at a given altitude on a given day.

There are three **gas laws** (with temperature, pressure and volume), each of which states the effect of one variable on another when the third is kept constant. They assume ideal behaviour and can be explained by the kinetic molecular theory.

Top tip!

The temperature scale used in the gas laws is the absolute or thermodynamic (Kelvin) scale. Conversion between the Celsius and thermodynamic scales is via the equation $T(°C) = T(K) - 273.15$, where $T(°C)$ is the temperature in Celsius and $T(K)$ is the absolute temperature.

■ **Table S1.26** The four gas laws

Gas law	Formula	Description	Explanation
Boyle's law	$P_1V_1 = P_2V_2$	For a fixed amount of gas at constant temperature, as volume decreases, pressure increases by the same proportion.	If a gas is compressed without changing its temperature, the average kinetic energy (speed) of the gas particles is constant. The particles travel from one end of the container to the other in a shorter period of time. This means that they hit the walls more often. Any increase in the frequency of collisions with the walls leads to an increase in the pressure of the gas.
Charles's law	$\frac{V_1}{T_1} = \frac{V_2}{T_2}$	For a fixed amount of gas at constant pressure, as temperature increases, volume increases by the same proportion.	Because the mass of individual gas particles is constant, the particles must move more quickly as the gas becomes warmer. This means the particles exert a greater force on the container each time they hit the walls, which leads to an increase in the pressure of the gas. Since the atmospheric pressure is constant, the increased pressure causes the volume of the gas to expand until its pressure becomes equal to the **atmospheric pressure**.
Gay-Lussac's law (pressure law)	$\frac{P_1}{T_1} = \frac{P_2}{T_2}$	For a fixed amount of gas at constant volume, as temperature increases, pressure increases by the same proportion.	Because the mass of these particles is constant, their kinetic energy can only increase if the average velocity of the particles increases. The faster these particles are moving when they hit the wall, the greater the force they exert on the wall. Since the force per collision becomes larger as the temperature increases, the pressure of the gas must increase as well.
Avogadro's law	$\frac{V_1}{n_1} = \frac{V_2}{n_2}$	For a given mass of gas, the volume and amount (mol) of the gas are directly proportional if the temperature and pressure are constant.	As the number of gas particles increases, the frequency of collisions with the walls of the container increases and so the volume increases to maintain a constant pressure.

Tool 3: Mathematics

Identify and use symbols stated in the guide and the data booklet

Although the IB *Chemistry data booklet* lists a number of key equations, you need to remember what the symbols mean – and the appropriate units for the quantities the symbols represent – in order to use them correctly. For example, in the equation:

$$n = CV$$

- n represents the amount of dissolved solute in moles (mol)
- C represents the concentration in $mol\,dm^{-3}$
- V represents the volume of solution in dm^3.

Physical constants are usually represented by a consistent symbol: for example, c is used across physics and chemistry to represent the speed of light in a vacuum, $3.0 \times 10^8\,m\,s^{-1}$.

Be aware of the differences between STP (standard temperature and pressure), SATP (standard ambient temperature and pressure), standard conditions (⊖) and standard states:

- STP is 273.15 K and 100 kPa
- SATP is 298.15 K and 100 kPa.
- The plimsoll symbol (⊖) indicates standard thermodynamic conditions of 298 K and 100 kPa for gases and 1 mol dm^{-3} for aqueous solutions. These are an essential part of the definitions of thermochemical equations describing standard enthalpy changes and electrode potentials.
- The standard states of elements are the forms that they exist in under standard conditions.

Top tip!

The names of specific gas laws will not be assessed but the four relationships need to be known.

WORKED EXAMPLE S1.5B

A gas tank contains 50 dm^3 of helium at 597 kPa. Calculate how many party balloons can be filled from this tank if each balloon has a volume of 2500 cm^3 at a pressure of 101 kPa.

Answer

$P_1V_1 = P_2V_2$

$597\,\text{kPa} \times 50\,\text{dm}^3 = 101\,\text{kPa} \times V_2$

$V_2 = 296\,\text{dm}^3$

$$\text{Number of balloons} = \frac{296\,000\,\text{cm}^3}{2500\,\text{cm}^3} = 118$$

Common mistake

Misconception: *Different units for pressure and volume can be used in a Boyle's law calculation.*

Although it does not matter what units are used for pressure and volume in a Boyle's law calculation, they must be the same on both sides of the equation.

ATL S1.5A

Visit the gas laws simulation at https://phet.colorado.edu/sims/html/gas-properties/latest/gas-properties_en.html. Choose the **Explore** option.

- Adjust the width of the container to 5.0 nm and add 100 heavy particles. Note the temperature (in K) and pressure (in kPa). These are the control conditions.
- Compare the conditions in the table below with the control conditions to see how the simulation illustrates the gas laws.

Real-life scenario	Changes from control conditions	Question
Scuba diving	Double the volume, then heat or cool so temperature returns to original value.	Does the new pressure suggest V and P are inversely proportional?
Hot air balloons	Double the volume then heat or cool so pressure returns to original value.	Does the new temperature suggest V and T are directly proportional?
Air-tight oven	Heat or cool so temperature is double original value.	Does the new pressure suggest P and T are directly proportional?
Balloon	Double the number of particles then adjust the volume until the pressure is at the original value (adjust the temperature too, if it strays from the original value).	Does the new volume suggest n and V are directly proportional?

- Discover whether the gas laws depend on whether you use heavy gas particles or light gas particles. (Hint: Turn on the collision counter, measure the control setup with heavy particles and then repeat using light particles.)
- Do the relationships between P and V and P and T remain unchanged if you double the temperature in Celsius?
- Starting with the control setup, start the timer and quickly open the top of the chamber using the handle.
- How long does it take for half of the heavy particles to leave the chamber (100 → 50)?
- How long does it take for half of the light particles to leave the chamber (100 → 50)?

Does scientific language have a primarily descriptive, explanatory or interpretative function?

The aim of the natural sciences is to go beyond merely describing the physical world to explaining and interpreting phenomena. To paraphrase Immanuel Kant's famous saying, explanations without facts (descriptions) are empty, and facts without explanations are blind.

According to this distinction, explanation is different from description. Description tells us what is there—it identifies the phenomenon—whereas explanations explore why it is there or why it is the way it is. The natural sciences explain phenomena by theories which are derived from empirical data (obtained via sense perception and instrumentation). Description is what happens and an explanation is why it happens.

However, an opposing view, advocated by radical empiricists, is that explanation is the same as description: explanations are just elaborate descriptions. The idea here is that describing and identifying what is happening is, in some important respect, the same process as explaining why it is the case. One example would be when describing the nature of gas: describing the behaviour of gas (thereby developing laws) is also to explain why the gas behaves the way that it does (thereby developing theories).

Interpretation refers to the idea that scientific language 'interprets' or 'gives an interpretation to' the observations that scientists make, or the perceptual experiences they have when they perform experiments. That might be a reference to a perspective known as 'the theory-ladenness of observation', popular since the work of Thomas Kuhn and Paul Feyerabend in the 1960s.

Theory-ladenness of observation holds that everything scientists observe is interpreted through a prior understanding of other theories and concepts. Whenever a scientist describes observations, they are constantly using terms and measurements that the scientific community has adopted. Therefore, it would be impossible for someone else to understand these observations if they are unfamiliar with, or disagree with, the theories that these terms come from.

This is why scientific paradigm shifts are so important: they represent a moment where the new phenomena being observed simply cannot be made sense of using the terms and concepts (or theories) that have been used in the past. A new theory—and all the accompanying concepts and principles—needs to be developed to account for the new observations.

Common mistake

Misconception: *Calculations involving gas law problems can be solved using temperatures in Celsius.*

All gas law problems that involve temperature in calculations must use the absolute scale (with temperatures in kelvin).

51 A 4.50 dm^3 sample of gas is warmed at constant pressure from 300 K to 350 K. Calculate its final volume.

52 A sample of gas collected in a 350 cm^3 container exerts a pressure of 103 kPa. Calculate the volume of this gas at a pressure of 150 kPa. (Assume that the temperature remains constant.)

53 10 dm^3 of a gas is found to have a pressure of 97 000 Pa at 25.0 °C. Determine the temperature (in degrees Celsius) that would be required to change the pressure to 101 325 Pa.

54 The temperature of a monatomic gas sample is changed from 27 °C to 2727 °C at constant volume. Determine the ratio of the final pressure to the initial pressure.

ATL S1.5B

A sealed plastic syringe was connected to a pressure sensor and data logger. The following results were obtained at constant temperature.

- Plot pressure (y-axis) against volume (x-axis).
- Describe the relationship between the two variables and explain it using the kinetic molecular theory.
- Explain why the temperature needs to be kept constant.

Volume of air in gas syringe / cm^3	Pressure / kPa
20	101.7
18	115.0
16	128.0
14	147.0
12	168.0
10	196.0
8	246.0

55 At 60.0 °C and 1.05×10^5 Pa the volume of a sample of gas collected is $60.0\,cm^3$. Determine the volume of the gas at STP.

◆ **Combined gas law**: States that the ratio of the product of pressure and volume to the absolute temperature of a fixed amount of gas is constant under different conditions.

◆ **Gas constant**: (Also known as the universal gas constant.) the constant of proportionality that appears in the ideal gas equation. It has a value of $8.31\,J\,K^{-1}\,mol^{-1}$.

◆ **Ideal gas equation**: An equation relating the absolute temperature (*T*), pressure (*P*), volume (*V*) and amount (*n*) of an ideal gas; $PV = nRT$.

Combined gas law

The **combined gas law** is formed by combining $P \times V$ = constant and $\frac{V}{T}$ = constant to give $\frac{PV}{T}$ = constant, for a fixed amount of gas.

(Note that the constants in the three expressions will all be different.)

This relationship is often written as $\frac{P_1V_1}{T_1} = \frac{P_2V_2}{T_2}$

This relationship can be used to calculate the new pressure, volume or temperature (P_2, V_2 or T_2) of a fixed mass of gas if two of the variables change, as long as all the initial conditions (P_1, V_1 and T_1) are known.

ATL S1.5C

Watch the video at **https://wonders.physics.wisc.edu/collapsing-can/** which demonstrates the collapsing of a metal can. Use your knowledge of molecular kinetic theory to explain how air pressure caused this to happen.

The ideal gas equation

Since one mole of any gas at STP occupies the molar volume, V_m, the quantity $\frac{PV_m}{T}$ is the same for all gases. It is called the **gas constant** and given the symbol *R*. The equation defining it can be more easily written as $PV_m = RT$.

A more general form of this relationship is $PV = nRT$, where *n* is the amount of gas in moles. This equation, which describes how a given amount of gas behaves as its pressure, volume and temperature change, is known as the **ideal gas equation**.

Each of the individual gas laws (Table S1.26) can be derived from this. For example, when *n* and *T* are held constant, the product *nRT* contains three constants and so must also be a constant:

$$PV = nRT = \text{constant, so } PV = \text{constant}$$

Top tip!

It is vital that volumes expressed in dm^3 and cm^3 are converted to cubic metres, and that pressure is expressed in pascals, if the value of the gas constant, *R*, given above is used.

$1\,dm^3 = 0.001\,m^3 = 10^{-3}\,m^3$ and $1\,cm^3 = 0.000\,001\,m^3 = 10^{-6}\,m^3$.

56 At 273 K and 101 325 Pa, 12.64 g of a gas occupy $4.00\,dm^3$. Calculate the relative formula mass of the gas.

Top tip!

Density of a gas is the mass per unit volume.

Density (ρ) = $\frac{\text{mass } (m)}{\text{volume } (V)}$.

The ideal gas equation can be used to determine the relative formula mass of gases or the vapour of volatile liquids using the apparatus shown in Figure S1.112. A known mass of a volatile liquid is injected into the gas syringe. It is vaporized and the syringe is pushed out.

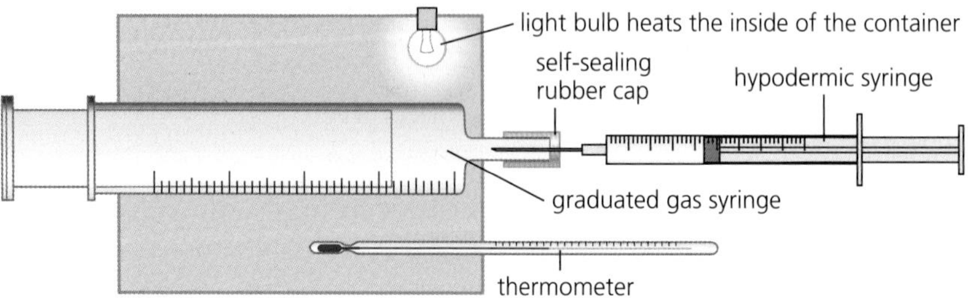

■ **Figure S1.112** Apparatus used to determine the relative formula mass of a volatile liquid

The ideal gas equation can also be modified and used to determine the molar mass ($g\,mol^{-1}$), density ($g\,m^{-3}$) and concentration ($mol\,dm^{-3}$) of an ideal gas (Table S1.27).

57 Calculate the relative formula mass of a gas which has a density of $2.615\,g\,dm^{-3}$ at 298 K and 101 325 Pa.

■ **Table S1.27** Using the ideal gas equation to derive other quantities

To determine	combine $PV = nRT$ with	to get
molar mass, M	$n = \frac{m}{M}$	$M = \frac{mRT}{PV}$
density, ρ	$\rho = \frac{m}{V}$	$\rho = \frac{PM}{RT}$
concentration, c	$c = \frac{n}{V}$	$c = \frac{P}{RT}$

Nature of science: Theories

Gas laws

The various gas laws were developed at the end of the 18th century, when scientists began to realize that relationships between the pressure, volume and temperature of a sample of gas could be obtained which would describe the behaviour of all gases. (Scientific laws are simply descriptions of what seems to happen in all cases.) Gases and mixtures of gases behave in a similar way over a wide variety of physical conditions because (to a good approximation) they all consist of molecules or atoms which are widely spaced: a gas is mainly empty space. The ideal gas equation can be derived from kinetic molecular theory. (Theories are explanations of *why* the laws are the way they are.) The gas laws are derivable by simple manipulation of the ideal gas equation, with one or more of the variables (pressure, volume and absolute temperature) held constant.

Inquiry 1: Exploring and designing

Creativity in designing, implementation and presentation of investigations

Chemistry is about being creative – be that in the method used to collect data, the interpretation of it or the way that it is presented. An open mind in all three of these respects can reveal things that others may miss and may help others understand things they might not have been able to.

For example, if you are looking at electroplating, why not take the idea of a calibration curve and apply this to your method? Electroplate for known amounts of time and measure the mass of the metal deposited. Can you produce a calibration curve to show the amount of metal deposited? Could it be used to predict how long you need to electrolyse for to collect a fixed amount of metal?

Investigations do not need to be overly complex or use complicated or expensive equipment. Keep them simple but elegant. You will find out about optical isomers in this course and how they rotate plane-polarized light – but by how much? A simple, yet elegant investigation could be used to investigate this. You may need to speak to your physics department to borrow a polarimeter, but the investigation itself would be simple enough to carry out.

When it comes to displaying your results, do you need to stick to 2D graphs? Spreadsheets provide some powerful graphing solutions. For example, if you are investigating how both temperature and pressure affect the volume of a gas, this can be displayed in 3D in one graph, rather than providing two graphs (of volume versus pressure and volume versus temperature).

A real-life example of creativity is the Mpemba effect, named after Tanzanian schoolboy Erasto Bartholomeo Mpemba who, in 1963 at the age of 14, tested the hypothesis that hot water freezes faster than cool water. To date, we still do not have an explanation for this phenomenon.

ATL S1.5D

You are asked to investigate a set of gas lighters. The manufacturer states that the gases in their lighters should belong to the alkanes homologous series of gases. Molecules of the alkanes all contain only carbon and hydrogen atoms and the general formula for alkanes is C_nH_{2n+2}, starting with methane, CH_4.

You find that the combined mass of the gas lighter and the gas is 11.62 g and that when some of the gas was released into an upside down measuring cylinder full of water, it occupied a volume of 4.20 dm^3 at a temperature of 16.0 °C. When the lighter was re-weighed on an electronic balance it had a mass of 8.78 g.

Present a report outlining and explaining your calculation, with any assumptions. Evaluate your method and comment on the calculated molar mass of the gas inside the gas lighter.

Determining the gas constant, *R*

The gas constant, *R*, can be experimentally determined using simple laboratory equipment. Magnesium and hydrochloric acid can be used to generate hydrogen gas. The gas can be collected in a eudiometer (Figure S1.113), a glass tube with a volume scale, closed at one end. Insoluble hydrogen will be collected in the closed end of the tube by water displacement.

Once collected, the volume of the gas, its temperature, the amount of gas and its pressure can be directly measured or indirectly determined. By substituting these values into the ideal gas law and solving for *R*, an experimental value for the gas constant can be determined.

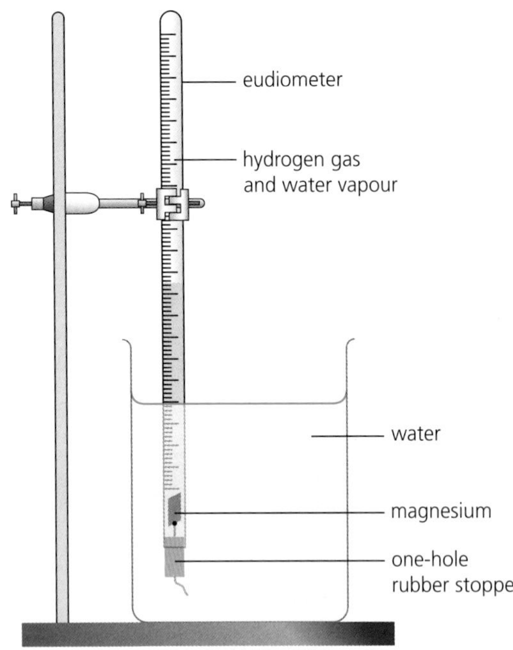

Figure S1.113 Apparatus for collecting and measuring hydrogen gas in a eudiometer

58 0.0147 g of pure magnesium was reacted with excess hydrochloric acid in the apparatus shown in Figure S1.113. 13.06 cm^3 of hydrogen gas was collected and the temperature of the room was 24.3 °C. The pressure of the hydrogen gas was corrected for water vapour and calculated to be 101 305 Pa. Calculate the gas constant, *R*, and compare with the literature value of 8.31 J K^{-1} mol^{-1}.

Common mistake

Misconception: *The ideal gas equation can be applied to describe pure liquids.*

The ideal gas equation can only be applied to describe gases (or mixtures of gases) that behave ideally.

Solving problems involving two connected vessels

Start by using the ideal gas law for the individual vessels to determine the amount of gas in each. The sum of the amounts of gas in each vessel is the same before and after they are connected. When the vessels are connected, some of the gas will move from one to the other to equalize the pressure. The linked vessels can then be considered as a single system containing gas which will obey the ideal gas law.

WORKED EXAMPLE S1.5C

The two containers shown in Figure S1.114 are connected by a valve. Calculate the total pressure after the valve is opened and the two gas samples are allowed to mix at constant temperature.

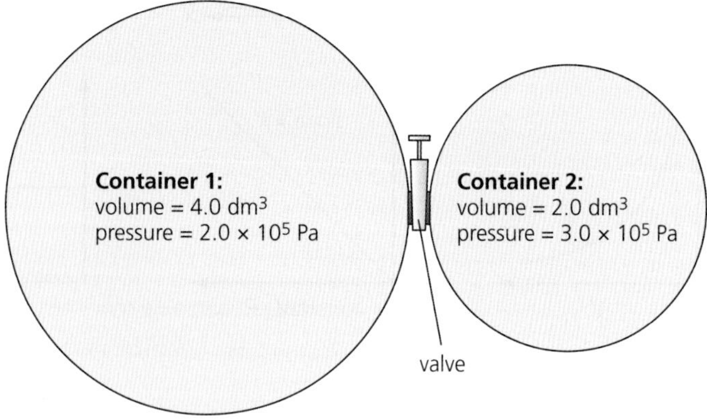

■ **Figure S1.114** Two vessels containing ideal gases with different volumes and pressures connected by a valve

Answer

container 1: $n_1 = \frac{P_1V_1}{RT} = \frac{8}{RT}$

container 2: $n_2 = \frac{P_2V_2}{RT} = \frac{6}{RT}$

When the gas valve is opened: total amount, $n_F = \frac{14}{RT}$ and total volume, $V_F = 6$

Applying the ideal gas law: $P_F = \frac{n_FRT}{V_F} = \frac{14}{RT} \times \frac{RT}{6} = \frac{14}{6} = 2.3$

Reinstating the exponent gives the final pressure $P_F = 2.3 \times 10^5$ Pa

(The question is asking for pressure to be determined so the units for volume, provided they are consistent, are not relevant.)

Graphical plots of ideal gas behaviour

Tool 3: Mathematics

Sketch graphs with labelled but unscaled axes to qualitatively describe trends

When drawing graphs from data, the y-axis is usually used to show values of a dependent variable and the x-axis shows the values of the independent variable. The dependent variable is the variable that is measured after the independent variable is changed.

Sketched graphs have labelled but unscaled axes and are used to show qualitative trends, such as variables that are proportional or inversely proportional. A sketch does not generally need to show units, only the variables.

The ideal gas equation can be used to sketch graphs that show the behaviour of a fixed amount of ideal gas. These graphs will have one of the forms shown in Figure S1.115, depending on the variables that are plotted.

For example, all six graphs in Figure S1.116 show the relationship between pressure and volume of a fixed amount of gas at constant temperature (Boyle's law: PV = constant).

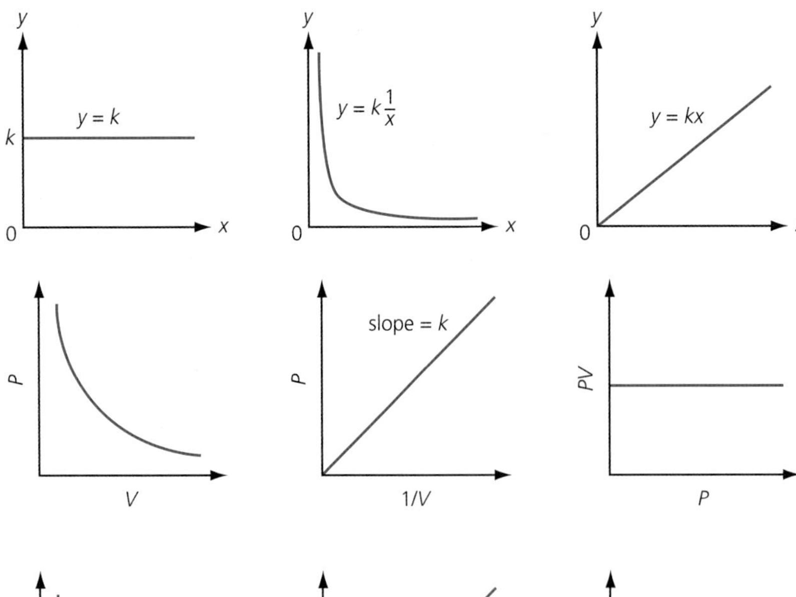

■ **Figure S1.115** Three main types of graphs related to the ideal gas equation

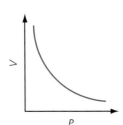

■ **Figure S1.116** Relationships between pressure and volume for an ideal gas (at constant temperature)

Top tip!

$y = mx + c$ is the equation of a straight line where m is the gradient and c is the intercept on the y-axis.

WORKED EXAMPLE S1.5D

Sketch a graph to show how **a** the volume and **b** the density of a fixed mass of ideal gas vary with temperature at constant pressure.

Answer

a Rearranging $PV = nRT$ gives $V = \frac{nRT}{P}$

Since n, R and P are constants, $V \propto T$.

The plot of V against T is a straight line passing through the origin.

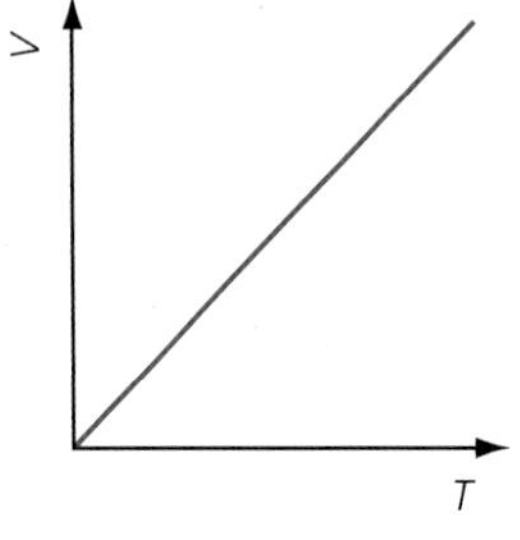

b Modifying the ideal gas equation to include a density term (ρ) by substituting $\rho = \frac{M}{V}$ where M is the mass of gas, gives $P = \rho \times \frac{RT}{M}$ and hence $\rho = \frac{PM}{RT}$.

Since P, M and R are constants, $\rho \propto \frac{1}{T}$.

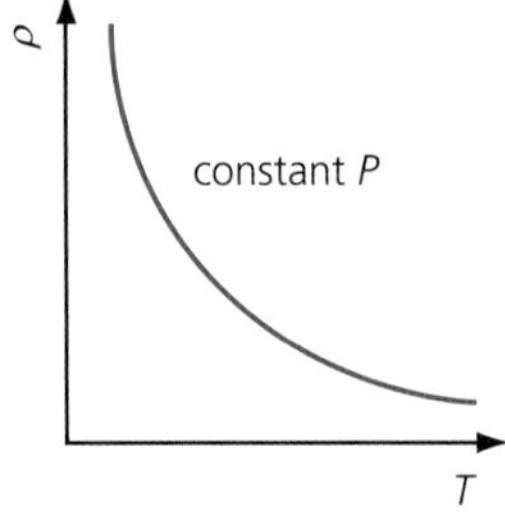

59 For a fixed amount of ideal gas, sketch the following graphs to show its ideal behaviour. Justify your answers with appropriate algebra.

a pV against V

b c (molar concentration) against $\frac{1}{T}$

c ρ (density) against $\frac{1}{T}$

d c (molar concentration) against P

Top tip!

If n, R and T are constants, when P increases, V must decrease to maintain the equality in the ideal gas equation.

Deviation from ideal behaviour

Effect of temperature and pressure

The deviation from ideal gas behaviour can be shown by plotting $\frac{PV}{RT}$ against pressure, P. For a gas behaving ideally, these plots would give straight lines (Figure S1.117).

The greatest deviation from ideal behaviour occurs when a gas is at a low temperature and high pressure (Figure S1.118).

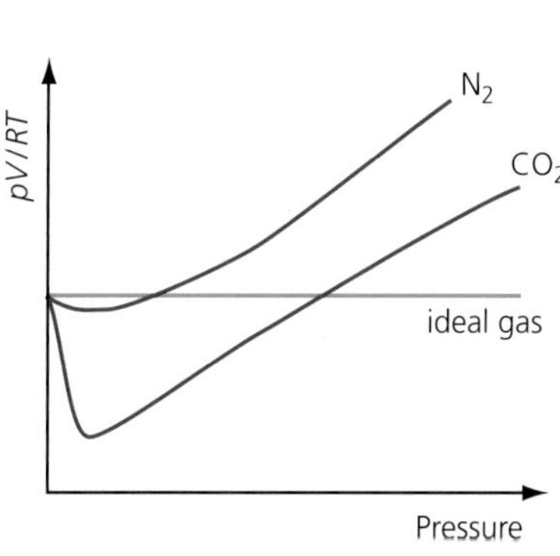

■ **Figure S1.117** Deviation from ideal behaviour at high and low pressures by two real gases (nitrogen and carbon dioxide)

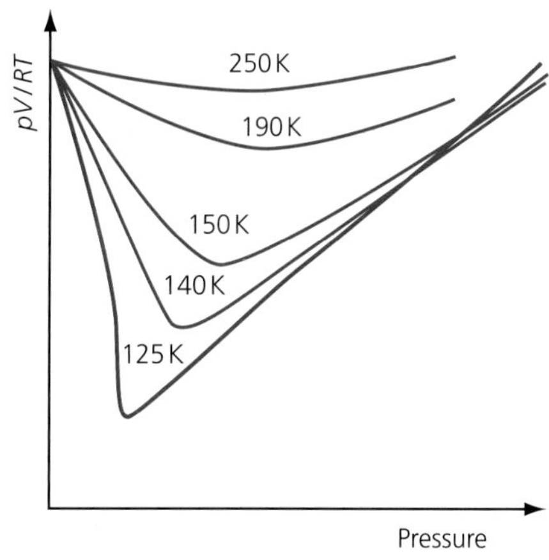

■ **Figure S1.118** Deviation from ideal behaviour at low temperature

Top tip!

An ideal gas is a hypothetical state; gases do not behave ideally and all will condense to form a liquid or freeze to form a solid before absolute zero.

A real gas deviates from ideal behaviour considerably at high pressures. This is because when gases are compressed, the molecules or atoms come sufficiently close together for attractive intermolecular forces to operate. The ideal gas model also assumes that the volume of molecules is negligible compared with the volume occupied by the gas. This is no longer true in a highly compressed gas where the actual volume of the gas molecules becomes significant.

At low temperatures, deviation from ideal behaviour occurs because the molecules are moving slowly, which significantly strengthens the intermolecular forces operating between neighbouring molecules or atoms.

LINKING QUESTION

Graphs can be presented as sketches or as accurately plotted data points. What are the advantages and limitations of each representation?

Going further

Gas mixtures

Dalton's law (Figure S1.119) states that the total pressure of a mixture of ideal gases that do not react is the sum of the partial pressures of the constituent gases in the mixture. The partial pressures, P_1, P_2 and so on, are the pressures each constituent gas would exert if it alone were present in the container. P_{Total} is the sum of the partial pressures. Dalton's law assumes ideal behaviour: the nature of the gas molecules does not affect the properties of the gas.

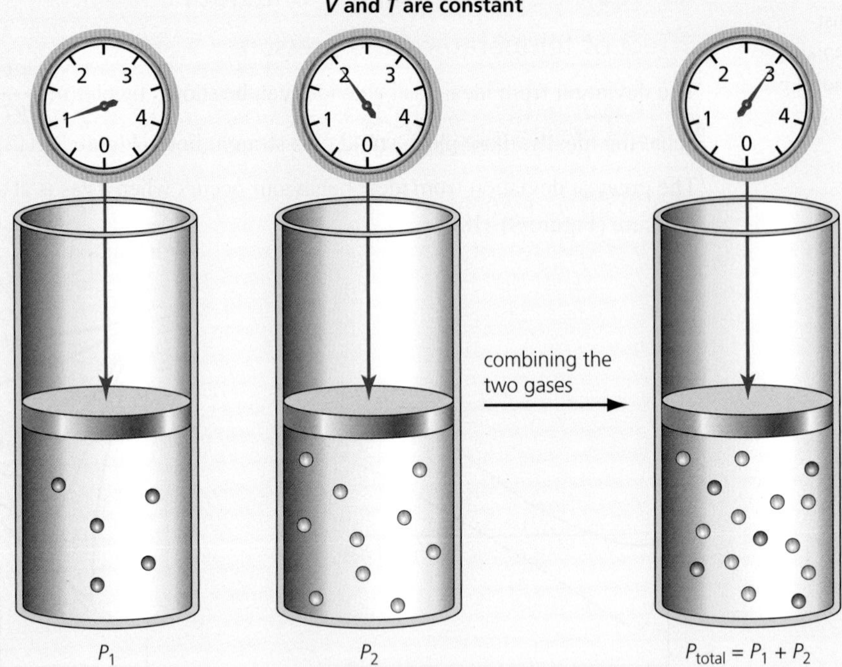

■ **Figure S1.119** Dalton's law of partial pressures

Dalton's law of partial pressures can be written as follows: $P_{total} = P_1 + P_2 + P_3 \ldots$

The partial pressure of an individual gas is equal to the total pressure multiplied by the mole fraction of that gas. The mole fraction is the number of moles of one particular gas divided by the total number of moles of gas in that mixture.

WORKED EXAMPLE S1.5E

4.0 dm^3 of nitrogen at a pressure of 400 kPa and 1.0 dm^3 of argon at a pressure of 200 kPa are introduced into a container of volume 2.0 dm^3. Calculate the partial pressures of the two gases and the total pressure in the container.

Answer

From Boyle's law, the partial pressure of nitrogen, $P_{N2} = 400 \times \frac{4.0}{2.0} = 800\,\text{kPa}$.

Similarly, the partial pressure of argon, $P_{Ar} = 200 \times \frac{1.0}{2.0} = 100\,\text{kPa}$.

Total pressure $= P_{N2} + P_{Ar} = 800\,\text{kPa} + 100\,\text{kPa} = 900\,\text{kPa}$.

Going further

Collecting gases over water

When the product of a chemical reaction is an insoluble and unreactive gas, it is often collected by the displacement of water (preferably in a eudiometer). However, the collected gas will not be pure as it contains water vapour. This is because some water molecules evaporate and mix by diffusion with the gas molecules.

If a gas is collected by displacement of water, then the gas will be saturated with water vapour. A gas is normally collected at atmospheric pressure and Dalton's law of partial pressures can be used to calculate the actual pressure of the gas: $P_{\text{atmosphere}} = P_{\text{water vapour}} + P_{\text{gas}}$.

Going further

Solubility of gases

Gases dissolve reversibly in liquids in direct proportion to their partial pressure above the liquid. This is called Henry's law and assumes ideal behaviour and no chemical reaction between the gas and the solvent.

Carbon dioxide is dissolved in carbonated (fizzy) drinks under high pressure during manufacture. If there is a head space above the drink, the gas and liquid in the bottle can be described as an equilibrium system: $CO_2(g) \rightleftharpoons CO_2(aq)$. When the bottle or can is opened, the pressure is reduced and the carbon dioxide comes out of solution. This releases bubbles so the drink fizzes. This is a simple description of the reaction between carbon dioxide and water and ignores the reversible chemical reaction with water to form carbonic acid.

Is there a single 'scientific method'?

A belief commonly held in the non-scientific community is that science operates via a single set of processes called the scientific method.

The idea is perpetuated by science textbooks which suggest an ordered procedure: observe; formulate a hypothesis and state its predictions; collect raw data and process it to test the hypothesis; draw a conclusion; if the hypothesis is not supported, modify the hypothesis.

This formulation, suggesting science proceeds by the generation of hypotheses from which predictions that can be tested by experiment are deduced, is known as hypothetico-deductivism.

While the steps listed above are used to communicate research findings, they do not necessarily describe the research process itself—the actual methods used in science. Research processes are highly varied and there is no universal methodology that all scientists use.

S2.1 The ionic model

Guiding question

- What determines the ionic nature and properties of a compound?

SYLLABUS CONTENT

By the end of this chapter, you should understand that:

- when metal atoms lose electrons, they form positive ions called cations
- when non-metal atoms gain electrons, they form negative ions called anions
- the ionic bond is formed by electrostatic attractions between oppositely charged ions
- binary ionic compounds are named with the cation first followed by the anion. The anion adopts the suffix *'ide'*
- ionic compounds exist as three-dimensional lattice structures, represented by empirical formulas
- lattice enthalpy is a measure of the strength of the ionic bond in different compounds, influenced by ion radius and charge.

By the end of this chapter you should know how to:

- predict the charge of an ion from the electron configuration of the atom
- deduce the formula and name of an ionic compound from its component ions, including polyatomic ions
- interconvert names and formulas of binary ionic compounds
- explain the physical properties of ionic compounds to include volatility, electrical conductivity and solubility.

Note: There is no higher-level only content in S2.1

◆ **Ionic bonding**: Electrostatic attraction between oppositely charged ions.

◆ **Cation**: A positively charged ion.

◆ **Anion**: A negatively charged ion.

◆ **Non-directional bonding (ionic)**: Attractive electrostatic forces operate in all directions on all ions of opposite charge.

Top tip!

Ionic bonding is described as **non-directional** since each ion is attracted to every other ion of opposite charge, though the attraction decreases with distance. Specifically, the attracting force between ions at distance d decays as $\frac{1}{d^2}$ (Coulomb's law). In contrast, covalent bonding involves the sharing of electron pairs between atoms and is directional.

Introduction to ionic bonding

Ionic bonding occurs when one or more electrons are transferred from the valence or outer shell (main energy level) of one atom to the valence or outer shell (main energy level) of another atom.

The atom losing an electron or electrons forms a positively charged ion (**cation**) and the atom gaining an electron or electrons forms a negatively charged ion (**anion**). The loss of electrons by atoms to form cations is oxidation; the gain of electrons by atoms to form anions is reduction.

Ionic bonding is the electrostatic attraction between oppositely charged ions (Figure S2.1): cations and anions.

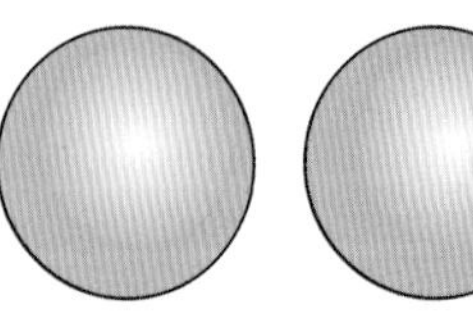

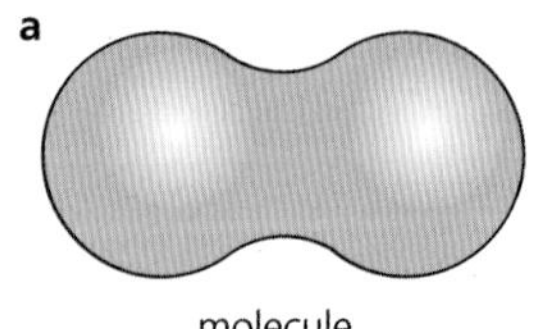

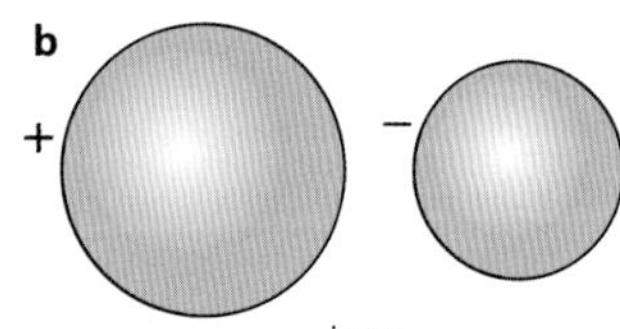

■ **Figure S2.1** Changes during the formation of **a** covalent bonds and **b** ionic bonds

Common mistake

Misconception: *Ionic bonding involves a sharing of electrons.*

Ionic bonding involves transfer of valence electrons.

Formation of ions and lattice in sodium chloride

Top tip!

The chemically active electrons are the valence electrons in the outer shell (in Na, for example, 2,8,**1**) and the core electrons are those strongly bound near the nucleus and not chemically active (**2**,**8**,1 in Na).

The formation of an ionic compound typically involves the reaction between a metal and a non-metal, for example, the reaction between sodium and chlorine to form sodium chloride.

The sodium atom has the configuration $1s^2\ 2s^2\ 2p^6\ 3s^1$ and the chlorine atom is $1s^2\ 2s^2\ 2p^6\ 3s^2\ 3p^5$.

In the formation of sodium chloride, the electron in the 3s orbital of a sodium atom is transferred to a 3p orbital of a chlorine atom.

The sodium cation has the configuration $1s^2\ 2s^2\ 2p^6$ and the chloride anion has the configuration $1s^2\ 2s^2\ 2p^6\ 3s^2\ 3p^6$.

Lewis formulas showing the valence electrons (bonding and non-bonding pairs) can be used to represent the transfer of electrons that occurs during the formation of ionic bonds. For example, the reaction between sodium and chlorine atoms is described using Lewis formulas in Figure S2.2.

Common mistake

Misconception: *The atoms of sodium and chlorine attract each other and form sodium chloride, NaCl.*

Ionic bonding involves cations and anions.

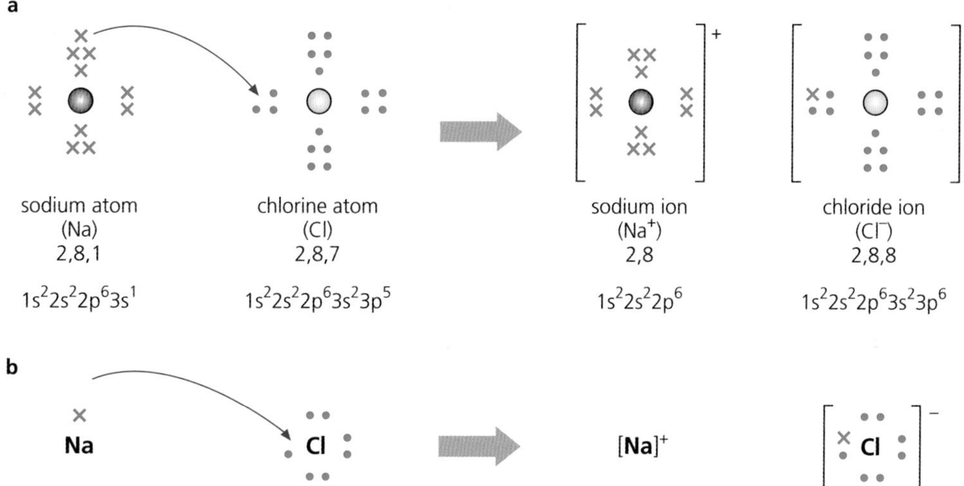

■ **Figure S2.2** Ionic bonding in sodium chloride, NaCl, showing **a** all electrons and **b** only the valence electrons. The arrow indicates the transfer of a single electron from the sodium atom to the chlorine atom.

1 Describe the formation of the following in terms of electron configurations of the atoms involved and the ions formed after electron transfer:
   a calcium fluoride
   b sodium oxide
   c aluminium oxide

LINKING QUESTION

Why is the formation of an ionic compound from its elements a redox reaction?

◆ **Giant structure:** A lattice with a very large number of particles that extends in all directions.

◆ **Lattice:** A regular, repeating three-dimensional arrangement of atoms, ions or molecules in a crystal.

The sodium cations and chloride anions will be arranged into a regular arrangement, a **giant structure** known as a **lattice** (Figure S2.3). Oppositely charged ions attract and ions of the same charge repel each other. However, there is a net (overall) attractive force because the oppositely charged ions are the nearest neighbours while other ions with the same charge are the next-nearest neighbours. As a result, the repulsive forces are weaker than the attractive forces.

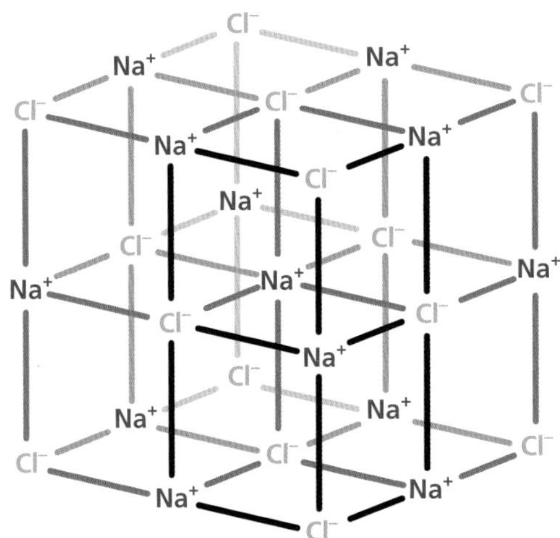

■ **Figure S2.3** Ionic lattice for sodium chloride (simple cubic lattice)

Top tip!

Ionic compounds form because the energy released when attractive electrostatic forces bring ions close together in an ionic crystal more than outweighs the energy required to form the ions.

The strength of an ionic lattice is measured by its lattice enthalpy, $\Delta H^{\ominus}_{lattice}$, which is the energy required to break up one mole of an ionic lattice into gaseous ions under standard conditions.

For sodium chloride this change is represented by the equation:

$NaCl(s) \rightarrow Na^+(g) + Cl^-(g) \quad \Delta H^{\ominus}_{lattice} = 790\,kJ\,mol^{-1}$

This shows that one mole of sodium chloride is converted into gaseous ions an infinite distance apart when 790 kJ of energy is absorbed.

ATL S2.1A

Use the simulation at https://contrib.pbslearningmedia.org/WGBH/arct15/SimBucket/Simulations/chemthink-ionicbonding/content/index.html to investigate how the transfer of electrons between atoms forms cations and anions and how the electrostatic forces of attraction lead to ionic bonding. You can also explore trends in the periodic table and how the structure of an ionic compound is related to its formula.

Naming ionic compounds

TOK

What is the role of language in a community of knowers?

Whether or not one is part of a knowledge community depends on whether or not one can 'speak the language' of that community. People outside the community of chemists might not have any idea what chemists are talking about when they use language and names like those in this book. Part of joining any community of knowers, then, is learning the language of the community and the conventions it uses to create new names. Problems often arise when people outside a particular community use specialist language improperly.

◆ **Polyatomic ion**: An ion that contains two or more covalently bonded atoms.

Link

Resonance structures (HL only) are described in more detail in Chapter S2.2, page 174.

Polyatomic ions

Many ions contain more than one atom. These types of ion are known as **polyatomic ions**. Many polyatomic ions—such as sulfate, carbonate and phosphate—may be described by two or more structures (resonance structures).

Table S2.1 summarizes the names, formulas and structures of commonly encountered polyatomic ions. A number of these ions are stabilized by resonance (delocalization of pi electrons).

■ **Table S2.1** List of common polyatomic ions

Name of ion	Formula	Structure of polyatomic ion	Example of compound
ammonium	NH_4^+	[H — N(→H above) — H, with H below]$^+$	ammonium chloride, NH_4Cl
hydronium	H_3O^+	[H — O(→H above) — H]$^+$	hydrochloric acid, H_3O^+, Cl^-
sulfate(VI)	SO_4^{2-}	$^-$:O: — S(=O above, =O below) — :O:$^-$	magnesium sulfate, $MgSO_4$
hydrogencarbonate	HCO_3^-	$^-$:O: — C(=O) — :O: — H	potassium hydrogen-carbonate, $KHCO_3$
nitrate(V)	NO_3^-	:O: = N^+ (— :O:$^-$, — :O:$^-$)	silver nitrate(V), $AgNO_3$
phosphate(V)	PO_4^{3-}	$^-$:O: — P(=O above, — :O:$^-$ below) — :O:$^-$	potassium phosphate(V), K_3PO_4
hydroxide	OH^-	$^-$:O: — H	sodium hydroxide, NaOH
carbonate	CO_3^{2-}	:O: = C (— :O:$^-$, — :O:$^-$)	sodium carbonate, Na_2CO_3

There are two common polyatomic cations: the ammonium ion, NH_4^+, and the hydronium ion, H_3O^+.

LINKING QUESTION (HL ONLY)

Polyatomic anions are conjugate bases of common acids. What is the relationship between their stability (with regard to the formation of the conjugate acid) and the conjugate acid's dissociation constant, K_a?

Naming the cation

The name of the cation is usually placed before that of the negative ion. If the cation is a metal, it takes the name of the element that is ionized. If the element forms more than one positive ion then a number showing the oxidation state is also required. Ammonium cations, NH_4^+, and hydroxonium cations, H_3O^+, are cations that are not formed from metal atoms.

The alkali metals (group 1 metals) and the alkaline earth metals (group 2 metals) form only one cation. For example, sodium always forms a unipositive ion, Na^+.

However, transition metals and p-block metals often form a range of stable cations. For example, copper forms two positive ions: copper(I), Cu^+, and copper(II), Cu^{2+}. Hence 'copper' is ambiguous when used in the name of a compound and these names should start with either copper(I) or copper(II): including the Roman numeral shows the oxidation state (which is related to the charge) of copper in the compound.

Naming the anion

If the anion is monatomic (has one atom), it is named by changing the ending of the element name to *—ide*. This leads to, for example, chloride, Cl^-, oxide, O^{2-}, and nitride, N^{3-}.

There are a few negative ions that consist of more than one atom whose names also end in *—ide,* including, for example, cyanide, CN^-, and hydroxide, OH^-.

Examples of compounds containing these ions are potassium oxide, K_2O [$2K^+$ O^{2-}]; potassium nitride, K_3N [$3K^+$ N^{3-}]; potassium cyanide, KCN [K^+ CN^-]; and potassium hydroxide, KOH [K^+ OH^-]. Ammonium chloride has the formula NH_4Cl [NH_4^+ Cl^-].

WORKED EXAMPLE S2.1A

What are the names of the compounds with the formulas $CoCl_2$ and $BaCl_2$?

Answer

Cobalt is a transition metal and likely to form two or more stable positive ions. Hence the name is cobalt(II) chloride.

Since barium is in group 2, the charge of the cation in its compounds is always 2+, so there is no need to specify the oxidation state. The compound is simply barium chloride.

Oxyanions

Oxyanions are polyatomic ions consisting of an atom of an element plus a number of oxygen atoms covalently bonded to it. The name of the oxyanion is given by the name of the element with the ending —*ate*. If the element has a number of oxidation states, it is necessary to add a Roman number to show which is present. Examples of this nomenclature include sodium nitrate(V), $NaNO_3$ [Na^+ NO_3^-] and sodium nitrate(III), $NaNO_2$ [Na^+ NO_2^-] (formerly known as sodium nitrite); and potassium sulfate(IV), K_2SO_3 [$2K^+$ SO_3^{2-}] (formerly known as potassium sulfite) and potassium sulfate(VI), K_2SO_4 [$2K^+$ SO_4^{2-}].

The charge on the nitrate (NO_3^-) ion is −1 not −3. The charge on the ammonium (NH_4^+) ion is +1 not +4.

Many other oxyanions are commonly found in compounds. These include chromate(VI), CrO_4^{2-}, dichromate(VI), $Cr_2O_7^{2-}$, ethanoate, CH_3COO^- and peroxide, O_2^{2-} ($^-O–O^-$).

Table S2.2 gives examples of ionic compounds including those with polyatomic ions and with elements showing variable oxidation states. These examples will allow you to deduce formulas and names from the ions present.

Table S2.2 Examples of ionic compounds including those with polyatomic ions

Name	Formula	Ions	Ions present
sodium oxide	Na_2O	[$2Na^+$ O^{2-}]	sodium and oxide
cobalt(III) fluoride	CoF_3	[Co^{3+} $3F^-$]	cobalt(III) and fluoride
magnesium nitride	Mg_3N_2	[$3Mg^{2+}$ $2N^{3-}$]	magnesium and nitride
potassium ethanoate	CH_3COOK	[CH_3COO^- K^+]	ethanoate and potassium
potassium manganate(VII)	$KMnO_4$	[K^+ MnO_4^-]	potassium and manganate(VII)
potassium dichromate(VI)	$K_2Cr_2O_7$	[$2K^+$ $Cr_2O_7^{2-}$]	potassium and dichromate(VI)
calcium phosphate(V)	$Ca_3(PO_4)_2$	[$3Ca^{2+}$ $2PO_4^{3-}$]	calcium and phosphate(V)
calcium hydrogencarbonate	$Ca(HCO_3)_2$	[Ca^{2+} $2HCO_3^-$]	calcium and hydrogencarbonate
potassium chlorate(VII)	$KClO_4$	[K^+ ClO_4^-]	potassium and chlorate(VII)
sodium chlorate(I)	NaClO	[Na^+ ClO^-]	sodium and chlorate(I)

Top tip!

All of the ionic formulas shown in Table S2.2 are effectively empirical formulas since they show the simplest integer ratio of ions in an ionic compound. For example, the formula of sodium chlorate(I), NaClO, indicates that the compound is composed of an equal number of sodium ions, Na^+, and chlorate(I) ions, ClO^-.

Ionic formulas

The formulas of many ionic compounds can be deduced using the list of ions shown in Table S2.3.

■ **Table S2.3** List of common ions

Positive ions	
Simple ions	***Formula***
sodium	Na^+
potassium	K^+
hydrogen	H^+
copper(II)	Cu^{2+}
iron(II)	Fe^{2+}
magnesium	Mg^{2+}
calcium	Ca^{2+}
iron(III)	Fe^{3+}
aluminium	Al^{3+}
Polyatomic ions	***Formula***
ammonium	NH_4^+

Negative ions	
Simple ions	***Formula***
chloride	Cl^-
bromide	Br^-
iodide	I^-
oxide	O^{2-}
sulfide	S^{2-}
nitride	N^{3-}
Polyatomic ions	***Formula***
hydrogencarbonate	HCO_3^-
hydroxide	OH^-
nitrate(V) or nitrate	NO_3^-
nitrate(III)	NO_2^-
sulfate(VI) or sulfate	SO_4^{2-}
sulfate(IV)	SO_3^{2-}
carbonate	CO_3^{2-}
phosphate(V)	PO_4^{3-}

Common mistake

Misconception: *The reason an ionic bond is formed between chloride and sodium ions is because an electron has been transferred between them.*

The reason an ionic bond is formed between chloride ions and sodium ions is because they have opposite electrostatic charges.

In forming compounds, the number of ions used is such that the number of positive charges is equal to the number of negative charges. Ionic compounds are electrically neutral.

Top tip!

The subscript number after a bracket, as in $(OH)_2$ in the formula for magnesium hydroxide, $Mg(OH)_2$, multiplies the entire polyatomic ion inside the bracket.

WORKED EXAMPLE S2.1B

What is the formula of lead(II) hydroxide?

Answer

Lead(II) hydroxide is composed of lead(II) ions, Pb^{2+}, and hydroxide ions, OH^-.

Twice as many hydroxide ions as lead(II) ions are necessary in order to achieve electrical neutrality. Hence, the formula of lead(II) hydroxide is $Pb(OH)_2$ [Pb^{2+} $2OH^-$].

2 Deduce the formulas of the following:

a iron(II) phosphate
b ammonium iodide
c aluminium nitrate
d calcium bromide
e iron(III) oxide
f potassium chloride
g sodium carbonate
h iron(III) hydroxide
i potassium sulfate(IV)
j magnesium sulfide

Ion charges and electron configurations

◆ **Octet rule**: Atoms with an atomic number greater than five fill their valence shell with eight electrons (an octet) when they form compounds.

Atoms (except for those of noble gases and some non-metals, for example boron and carbon) form simple ions by the gain or loss of electrons. Many simple ions follow the **octet rule** and have four pairs of valence electrons. Atoms of groups 1, 2, 13, 15, 16 and 17 will usually lose or gain valence electrons to attain the electron configuration of a noble gas (ns^2 np^6).

Consider the sodium atom with the full electron configuration of $1s^2\ 2s^2\ 2p^6\ \mathbf{3s^1}$, where the valence electron is highlighted. The sodium ion, Na^+, will have the electron configuration $1s^2\ 2s^2\ 2p^6$, the same as neon. All group 1 cations (+1) have the electron configuration of the previous noble gas.

Consider the phosphorus atom with the full electron configuration of $1s^2\ 2s^2\ 2p^6\ \mathbf{3s^2\ 3p^3}$, where the valence electrons are highlighted. The phosphide ion, P^{3-}, will have the electron configuration $1s^2\ 2s^2\ 2p^6\ 3s^2\ 3p^6$. All group 15 anions (−3) have the electron configuration of the next noble gas.

Top tip!

The first two elements in group 14, carbon and silicon, have outer shells (main energy levels) that contain 4 electrons. These two elements generally do not form simple ions but instead form covalent bonds. However, carbon reacts with metals to form a number of metal carbides such as CaC_2 $[Ca^{2+}\ C_2^{2-}]$. Magnesium silicide has the formula Mg_2Si and there is evidence that it contains Mg^{2+} and Si^{4-}.

Table S2.4 shows the electron arrangements of the atoms and simple ions of the elements in period 3 of the periodic table.

■ **Table S2.4** Electron arrangements of the atoms and simple ions of the elements in period 3

Group	1	2	13	14	15	16	17	18
Element	sodium	magnesium	aluminium	silicon	phosphorus	sulfur	chlorine	argon
Electron arrangement and configuration of atom	2,8,1 $1s^2\ 2s^2\ 2p^6\ 3s^1$	2,8,2 $1s^2\ 2s^2\ 2p^6\ 3s^2$	2,8,3 $1s^2\ 2s^2\ 2p^6\ 3s^2\ 3p^1$	2,8,4 $1s^2\ 2s^2\ 2p^6\ 3s^2\ 3p^2$	2,8,5 $1s^2\ 2s^2\ 2p^6\ 3s^2\ 3p^3$	2,8,6 $1s^2\ 2s^2\ 2p^6\ 3s^2\ 3p^4$	2,8,7 $1s^2\ 2s^2\ 2p^6\ 3s^2\ 3p^5$	2,8,8 $1s^2\ 2s^2\ 2p^6\ 3s^2\ 3p^6$
Number of electrons in outer (valence) shell	1	2	3	4	5	6	7	8
Common simple ion	Na^+ cation	Mg^{2+} cation	Al^{3+} cation	—	P^{3-} (phosphide) anion	S^{2-} (sulfide) anion	Cl^- (chloride) anion	—
Electron arrangement and configuration of ion	2,8 $1s^2\ 2s^2\ 2p^6$	2,8 $1s^2\ 2s^2\ 2p^6$	2,8 $1s^2\ 2s^2\ 2p^6$	—	2,8,8 $1s^2\ 2s^2\ 2p^6\ 3s^2\ 3p^6$	2,8,8 $1s^2\ 2s^2\ 2p^6\ 3s^2\ 3p^6$	2,8,8 $1s^2\ 2s^2\ 2p^6\ 3s^2\ 3p^6$	—

3 Prepare a similar table for the atoms of elements in period 2.

LINKING QUESTION

How does the position of an element in the periodic table relate to the charge of its ion(s)?

Ions of the transition elements

Most of the transition elements form more than one stable simple cation (Table S2.5). For example: copper forms copper(I), Cu^+, and copper(II), Cu^{2+}; and iron forms iron(II), Fe^{2+}, and iron(III), Fe^{3+}. The Roman number indicates the oxidation state of the transition metal. Charges of 2+ and 3+ (or oxidation states of +2 and +3) are the most common charges on a simple transition metal ion.

For elements in groups up to 8, the highest charge (oxidation state) is equal to the group number. For example, manganese is in group 7 of the periodic table and forms the strongly oxidizing ion, MnO_4^- $[Mn^{7+}\ 4O^{2-}]$, where the oxidation state of manganese is +7.

■ **Table S2.5** Charges of selected transition metal ions

Transition metal	**Simple positive ions**
silver	Ag^+
iron	Fe^{2+}, Fe^{3+}
copper	Cu^+, Cu^{2+}
manganese	Mn^{2+}, Mn^{3+}, Mn^{4+}
chromium	Cr^{3+}, Cr^{2+} (not stable in air)

Common mistake

Misconception: *An ionic bond is when one atom donates an electron to another atom so that they both have full outer shells.*

The ionic bond is the attractive electrostatic force operating between ions of opposite charge.

Iron, which forms either the Fe^{2+} or Fe^{3+} ion, loses electrons.

$$\underset{[Ar]\,3d^6\,4s^2}{Fe} \rightarrow \underset{[Ar]\,3d^6}{Fe^{2+}} + 2e^-$$

$$\underset{[Ar]\,3d^6\,4s^2}{Fe} \rightarrow \underset{[Ar]\,3d^5}{Fe^{3+}} + 3e^-$$

Iron can also form the polyatomic ferrate(VI) ion, FeO_4^{2-}.

According to the Aufbau principle, electrons fill the 4s sublevel before beginning to enter the 3d sublevel. These outermost 4s electrons are always the first to be removed in the formation of simple cations. A common cation of most transition elements is, therefore, +2.

A half-filled d sublevel (d^5) is particularly stable (due to exchange energy) and this is formed as the result of an iron atom losing a third electron.

Several other simple ions do not follow the octet rule (Table S2.6).

■ **Table S2.6** Selected simple ions that do not follow the octet rule

Name of ion	Formula of ion	Electron configuration
proton / hydrogen ion	H^+	nil
hydride ion	H^-	$1s^2$
tin(II) ion	Sn^{2+}	$[Kr]\,5s^2\,4d^{10}$
lead(II)	Pb^{2+}	$[Xe]\,4f^{14}\,5d^{10}\,6s^2$
gallium(III) ion	Ga^{3+}	$[Ar]\,3d^{10}$
bismuth(III) ion	Bi^{3+}	$[Xe]\,4f^{14}\,5d^{10}\,6s^2$
thallium(I) ion	Tl^+	$[Xe]\,4f^{14}\,5d^{10}\,6s^2$

WORKED EXAMPLE S2.1C

Ytterbium (Yb) is an f-block metal. The formula for ytterbium phosphate is $Yb_3(PO_4)_2$. Deduce the formula of ytterbium nitride.

Answer

$Yb_3(PO_4)_2$ is composed of the following ions: $[3Yb^{2+}\ 2PO_4^{3-}]$. It therefore contains ytterbium(II) cations.

Hence, ytterbium nitride is composed of the following ions: $[3Yb^{2+}\ 2N^{3-}]$. It has the formula Yb_3N_2.

TOK

How can it be that scientific knowledge changes over time?

The English chemist and physicist Michael Faraday (1791–1867) first demonstrated that solutions of certain compounds in water could conduct electricity. He proposed that the electricity caused these compounds to break up into charged particles (ions) which were responsible for the electrical conductivity. In his early experiments, he appears to have considered that the ions might be present in the compound itself. It was Faraday who introduced the term ion and, recognizing that the ions had electrical charges, he also introduced the terms cation and anion. A hypothesis, such as the postulated existence of ions in certain classes of compounds, serves two related functions: it classifies and explains chemical observations and it leads to predictions which will suggest new experiments to find data supporting (or falsifying) the idea of ions as charged particles. In this way, the amount of scientific knowledge—in the form of new or improved models, theories and hypotheses—increases.

Link

Electronegativity is discussed in more detail in Chapter S3.1, page 234.

Ionic bonding and electronegativity

Ionic bonding is favoured if the metal and non-metal elements are reactive.

The electronegativity of an atom is the ability of the atom to attract shared pairs of electrons to itself. The greater the electronegativity of an atom, the greater its ability to attract shared pairs of electrons. Figure S2.4 shows that the most electronegative elements are highly reactive non-metals and the least electronegative elements are the reactive metals. (This information can also be found in section 9 in the IB *Chemistry data booklet*.) Ionic bonding is, therefore, most likely when there is a large difference in electronegativity values between the two elements.

					H 2.2												He
Li 1.0	Be 1.6											B 2.0	C 2.6	N 3.0	O 3.4	F 4.0	Ne
Na 0.9	Mg 1.3											Al 1.6	Si 1.9	P 2.2	S 2.6	Cl 3.2	Ar
K 0.8	Ca 1.0	Sc 1.4	Ti 1.5	V 1.6	Cr 1.7	Min 1.6	Fe 1.8	Co 1.9	Ni 1.9	Cu 1.9	Zn 1.6	Ga 1.8	Ge 2.0	As 2.2	Se 2.6	Br 3.0	Kr
Rb 0.8	Sr 1.0	Y 1.2	Zr 1.3	Nb 1.6	Mo 2.2	Tc 2.1	Ru 2.2	Rh 2.3	Pd 2.2	Ag 1.9	Cd 1.7	In 1.8	Sn 2.0	Sb 2.0	Te 2.1	I 2.7	Xe
Cs 0.8	Ba 0.9	La 1.1	Hf 1.3	Ta 1.5	W 1.7	Re 1.9	Os 2.2	Ir 2.2	Pt 2.2	Au 2.4	Hg 1.9	Tl 1.8	Pb 1.8	Bi 1.9	Po 2.0	At 2.2	Rn
Fr 0.7	Ra 0.9	Ac 1.1															

■ **Figure S2.4** Electronegativity values (Pauling scale)

There are some simple rules for predicting the type of chemical bond based upon the electronegativity differences:

- If the difference in electronegativity values is greater than 1.7, then the bond is likely to be ionic. For example, francium fluoride (FrF) would be highly ionic.
- If the difference in electronegativity values is less than 0.5, then the bond is essentially non-polar covalent (for example, the C–H bond).
- If the difference in electronegativity values is greater than 0.5 but less than 1.7, then the bond is likely to be polar covalent.

Ionic and covalent bonding are extremes; polar bonds are an intermediate form. They are covalent bonds with ionic character (partial electron transfer). The larger the difference in electronegativity between the atoms, the greater the polarity of the bond and the greater the ionic character (Figure S2.5).

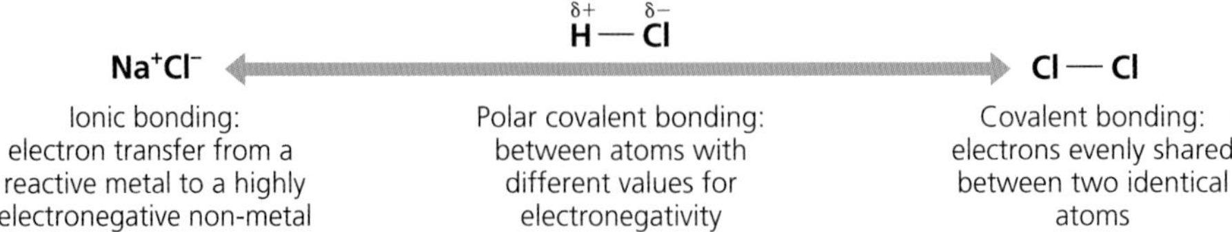
δ+ δ−
H — Cl
Na⁺Cl⁻
Cl — Cl
Ionic bonding: electron transfer from a reactive metal to a highly electronegative non-metal
Polar covalent bonding: between atoms with different values for electronegativity
Covalent bonding: electrons evenly shared between two identical atoms

■ **Figure S2.5** The spectrum of bonding from ionic to covalent via polar covalent. The delta symbols shown for polar covalent bonding represent fractional charges on the two atoms

Charge-shift bonds

Israeli chemist Sason Shaik and French chemist Philippe Hiberty collaborated to study molecules using computer simulations based on Linus Pauling's valence bond theory. One of the molecules they studied was the fluorine molecule, F_2, which they found was best described by including an ionic form, $F^+ F^-$, with the familiar covalent form, F–F. The calculated bond energy was then in closer agreement with the experimental bond enthalpy. They found other cases similar to fluorine and named such bonds charge-shift bonds.

WORKED EXAMPLE S2.1D

Using electronegativity values from Figure S2.4, predict the type of bonding in chlorine molecules (Cl_2), hydrogen iodide (HI) and lithium fluoride (LiF).

Answer

Using the values given:

Cl_2 difference in electronegativity = (3.2 − 3.2) = 0
non-polar covalent bond, Cl–Cl

HI difference in electronegativity = (2.7 − 2.2) = 0.5
polar covalent bond, $^{\delta+}H–I^{\delta-}$

LiF difference in electronegativity = (4.0 − 1.0) = 3.0
ionic bond, Li^+F^-

Top tip!

In compounds such as potassium sulfate(VI), K_2SO_4 or $[2K^+ SO_4^{2-}]$, the bonding within the polyatomic sulfate(VI) ions is (polar) covalent, but the bonding between the sulfate(VI) and potassium ions is ionic.

LINKING QUESTION (HL ONLY)

How is formal charge used to predict the preferred structure of sulfate?

4 Deduce the difference in electronegativity in the following bonds and predict the type of bonding:
 a chlorine–chlorine
 b phosphorus–hydrogen
 c carbon–chlorine
 d beryllium–fluorine
 e carbon–hydrogen

TOK

Is prediction the primary purpose of scientific knowledge?

The primary goal of the natural sciences is to generate new scientific knowledge via the scientific method. It could be argued that prediction is the primary and ultimate purpose of the natural sciences. Chemists aim to improve theories that explain and predict chemical behaviour to develop better analytical methods and more effective control over chemical reactions. To some degree, when chemists understand a chemical reaction or physical process at some level, especially at the molecular level, they can predict the occurrence of that event. Prediction may also permit a substantial degree of control. However, prediction should not be fully equated with understanding.

Properties of ionic compounds

Ionic compounds are hard crystalline solids at room temperature and pressure. This is due to the presence of very strong attractive forces between oppositely charged ions in a lattice.

Consequently, large amounts of thermal energy are required to overcome the strong electrostatic forces of attraction. The strong ionic bonding in the solid and liquid states means that ionic substances have low volatility and hence have high melting and boiling points. They undergo negligible evaporation at temperatures below their boiling point.

The brittle nature of an ionic crystal, and its ability to be cleaved along planes, results from the mutual repulsion between layers when ions are displaced (moved) as seen in Figure S2.6. This occurs because ions are forced into new positions, in which ions of the same charge experience repulsive forces that weaken the crystal along the plane of shearing.

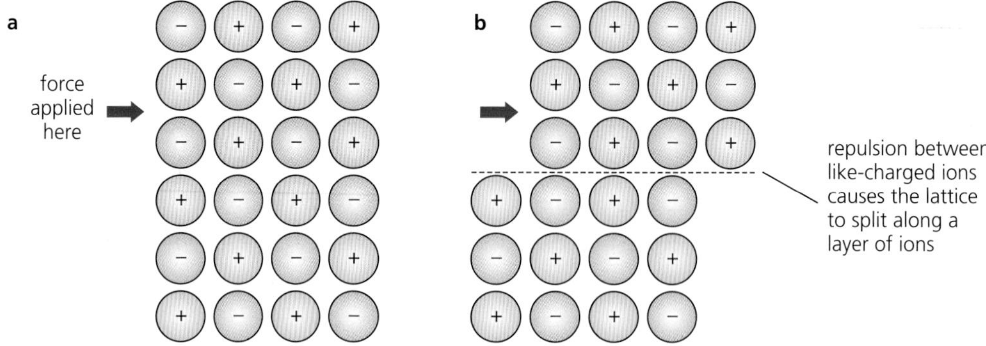

■ **Figure S2.6** Cleavage in ionic solids: a layer of ions **a** before and **b** after cleavage

Common mistake

Misconception: *Sodium chloride conducts electricity in the solid and molten states and in aqueous solution due to the presence of mobile electrons that act as charge carriers.*

Sodium chloride conducts when melted or dissolved in water to form a solution due to the presence of mobile ions that act as charge carriers; these ions cannot move in the crystalline solid state.

Since the ions are held rigidly in the lattice by electrostatic forces of attraction, ionic solids cannot conduct electricity when a voltage is applied. However, when molten or dissolved in water to form an aqueous solution (if the ionic compound is soluble), the lattice is broken up and the hydrated ions are free to move and act as charge carriers.

Tool 1: Experimental techniques

Electric current

To measure the size of an electric current, an ammeter is used. The ammeter must be connected in the circuit in series. If the circuit consists of one loop, the same current flows through the circuit and it does not matter where the ammeter is placed. However, since an ammeter is a polarized device, the negative terminal of the power supply should be connected to the negative terminal of the ammeter (directly or indirectly) and, likewise, positive should be connected to positive.

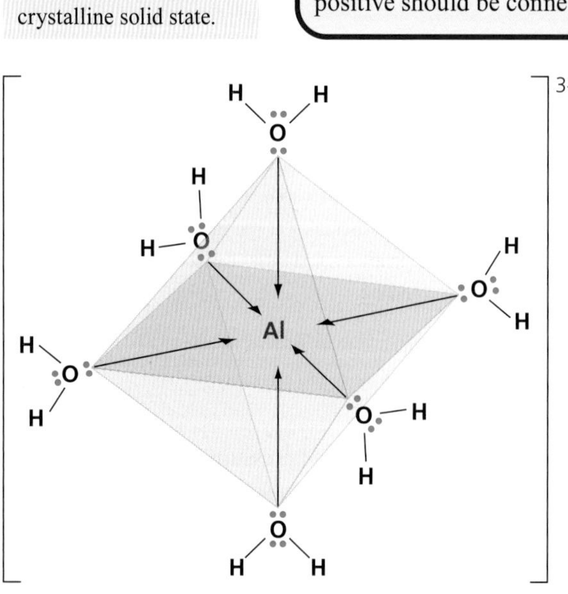

■ **Figure S2.7** Structure of the hexaaquaaluminium ion, $[Al(H_2O)_6]^{3+}$

When an ionic solid dissolves in water, the ions become hydrated by the polar water molecules. The solubility of an ionic solid will decrease as the polarity of the solvent molecules decreases. This is partly due to the decrease in strength of the interaction between the ions and the solvent molecules.

The electrostatic forces of attraction between an ion and the oppositely charged region of a water molecule are ion–dipole forces.

If a metal ion has empty low-energy 3d and 4s orbitals, it can form a coordination bond with the lone pair of electrons on the oxygen atom in a water molecule, creating a complex ion. Transition metals and metals from period 3 onwards (for example, lead(II) ions, aluminium ions and tin(II) ions) can all form complex ions with water molecules (Figure S2.7).

Top tip!

Many ionic substances, but not all, dissolve in water.

Tool 2: Technology

Using sensors

Data-logging is an electronic method of recording physical measurements: sensors produce electrical signals which are calibrated and recorded by a computer system.

■ **Table S2.7** Possible uses of some common sensors

Sensor	Possible investigation
conductivity sensor	ionic precipitations, hydrolysis reactions
temperature sensor	enthalpy changes in endothermic and exothermic reactions such as neutralization
balance	loss in mass due to gas production
light sensor	reactions that result in a colour change, chemiluminescence
pH	titrations and other reactions involving a change in pH
pressure sensor	reactions that produce a gas, gas laws
dissolved gas sensor	solubility of gases

The main advantage of data-logging is that it facilitates the processes of analysing and interpreting the raw data from practical work.

ATL S2.1B

Ionic liquids are salts that are liquids at room temperature. Find out about the history, structure and properties of some well-studied ionic liquids. Outline their potential uses and commercial applications. Summarize your findings in a presentation using, for example, PowerPoint.

Going further

Lattices

The exact arrangement of the ions in an ionic lattice varies according to the ions. The cations and anions may have similar sizes or be different in size. The geometry of the crystalline structure adopted by the compound partly depends on the relative sizes of the ions (the ionic radii ratio).

In sodium chloride, each Na^+ ion is surrounded by six Cl^- ions and vice versa. This is known as 6:6 coordination (Figure S2.8). The sodium chloride lattice structure has a simple cubic arrangement and this structure is usually adopted by ionic compounds in which the anion is larger than the cation (radii ratios 0.6 to 0.4). The ionic radius of Na^+ is 95 pm and that of Cl^- is 181 pm so the radii ratio is 0.525.

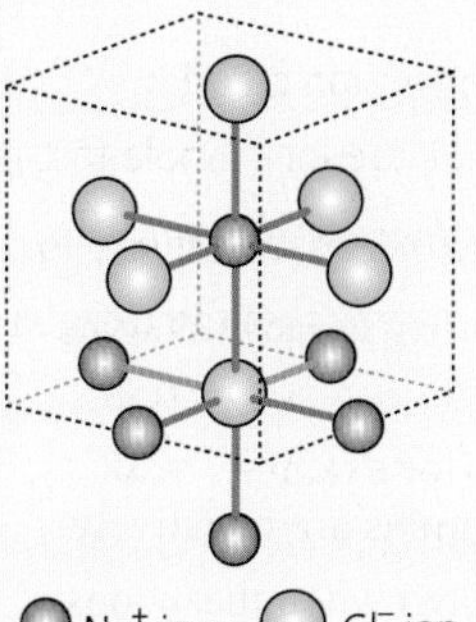

■ **Figure S2.8** The sodium chloride lattice showing 6:6 coordination

Many ionic compounds adopt the sodium chloride structure because the radius ratio commonly falls within the range 0.4–0.6 for which this is the most stable structure when the anion is larger than the cation. The optimum arrangement is a compromise between maximum attraction of oppositely charged ions and minimum repulsion of ions with the same charge.

When the positive and negative ions have similar sizes, the crystalline structure adopted is more likely to be that of caesium chloride (Figure S2.9), which has 8:8 coordination. The ionic radius for Cs^+ is 174 pm, which is fairly similar to that of Cl^- (181 pm).

■ **Figure S2.9** A model of the caesium chloride lattice: red balls represent caesium ions; green balls represent chloride ions

LINKING QUESTION

What experimental data demonstrate the physical properties of ionic compounds?

Common mistake

Misconception: *Each molecule of sodium chloride, NaCl, contains one sodium ion and one chloride ion.*

There are no molecules in sodium chloride, just a giant structure of ions in a 1 : 1 ratio.

Common mistake

Misconception: *A sodium atom can only form one ionic bond because it only has one electron in its outer shell to donate.*

A sodium ion can strongly bond to as many chloride ions as can effectively pack around it in the regular crystal lattice. In sodium chloride, NaCl, there are six chloride anions strongly ionically bonded to each sodium cation in a simple cubic lattice.

ATL S2.1C

Use the internet to find the structures of the following ionic substances:

- NaCl (rock salt structure)
- CsCl and ZnS (zinc blende and wurtzite structures)
- CaF_2 and $CaTiO_3$ (perovskite structure).

Use commercial modelling kits (for example, Molymod) or other materials (such as polystyrene balls, wooden sticks and glue) to construct the unit cells (smallest repeating unit) of each structure.

Find out about the technique of X-ray crystallography which is used to establish the structures of ionic lattices.

Lattice enthalpy

The formation of an ionic compound from elements is usually a very exothermic (favourable) reaction, with heat being released to the surroundings.

For example, when a mole of potassium chloride forms from the elements potassium and chlorine (under standard conditions), 436.7 kJ of heat is released in the following exothermic reaction:

$$K(s) + \frac{1}{2}Cl_2(g) \rightarrow KCl(s) \quad \Delta H^{\ominus} = -436.7\,kJ\,mol^{-1}$$

This equation describes the enthalpy of formation of potassium chloride. This is the enthalpy (energy) change when one mole of a compound is formed from its elements.

This energy is not just from the tendency of metals to lose electrons and non-metals to gain electrons. In fact, the transfer of an electron from a potassium atom to a chlorine atom to form ions actually *absorbs* energy (it is an energetically unfavourable endothermic process). The first ionization energy of potassium is $+419\,kJ\,mol^{-1}$ and the first electron affinity of chlorine is only $-349\,kJ\,mol^{-1}$. Based only on these energies, the reaction for the formation of potassium chloride should be endothermic (unfavourable) by $+70\,kJ\,mol^{-1}$.

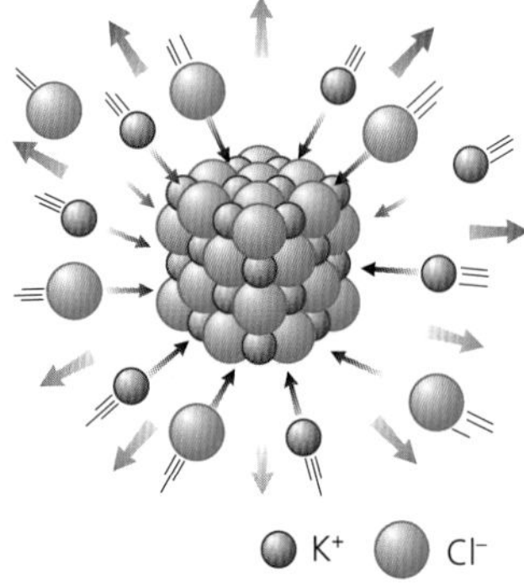

■ Figure S2.10 The reverse of lattice enthalpy is the energy which would be released to the surroundings (short red arrows) if one mole of an ionic compound could form directly from infinitely free gaseous ions rushing together (black arrows) and forming a lattice

Common mistake

Misconception: *A potassium cation is only bonded to the chloride anion it donated its electron to.*

Each potassium cation is strongly ionically bonded to each of the neighbouring negative chloride ions. It is irrelevant how the ions came to be charged.

Since the potassium cations are positively charged and the chloride anions are negatively charged, the potential energy decreases (as described by Coulomb's law) when these ions pack together to form a lattice. That energy is emitted as heat when the lattice forms, as shown in Figure S2.10.

◆ **Lattice enthalpy**: The energy needed to convert one mole of ionic solid into gaseous ions infinitely far apart under standard conditions.

Ionic compounds form because of the reverse of **lattice enthalpy**: the energy associated with the formation of a crystalline lattice of cations and anions from the gaseous ions.

The lattice enthalpy of an ionic crystal is the heat energy absorbed (at constant pressure) when one mole of solid ionic compound is broken up to form gaseous ions separated to an infinite distance from each other (under standard thermodynamic conditions).

For example, the lattice enthalpy of potassium chloride is the enthalpy change for the reaction:

$$KCl(s) \rightarrow K^+(g) + Cl^-(g) \quad \Delta H^{\ominus}_{lattice} = +720\,kJ\,mol^{-1}$$

This reaction is illustrated in Figure S2.11.

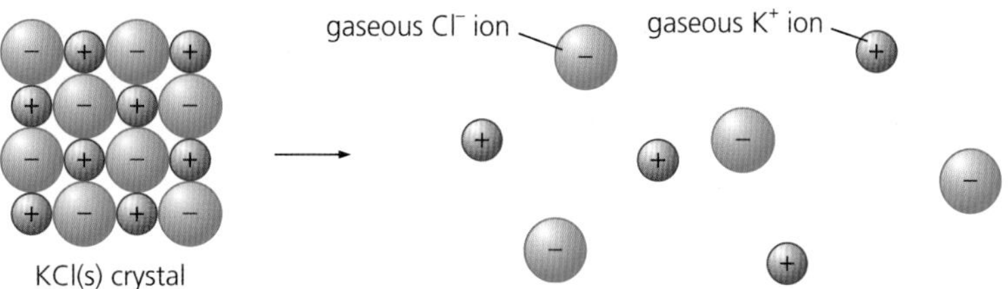

■ **Figure S2.11** Lattice enthalpy for potassium chloride

Lattice energies are a measure of the stability of an ionic crystal. The greater the value of the lattice enthalpy, the more energetically stable the lattice. This results in higher melting and boiling points. The size of the lattice enthalpy also has an effect on the solubility of an ionic salt. The size of the lattice enthalpy is controlled by the charges on the ions, their ionic radii and the packing arrangement of the ions (type of lattice).

The exact value of the lattice enthalpy, however, is not simple to determine because it involves a large number of interactions among many charged ions. The easiest way to calculate it is with the Born–Haber cycle based on other experimental data.

In an ionic compound, the lattice enthalpy depends on the attractive electrostatic forces operating between oppositely charged ions in the crystal. The larger the electrostatic forces of attraction, the larger the size of the lattice enthalpy.

According to Coulomb's law, the attractive force F operating between two adjacent oppositely charged ions can be expressed as follows:

$$F \propto \frac{\text{charge on positive ion} \times \text{charge on negative ion}}{d^2}$$

where d represents the distance between the nuclei of the two ions (Figure S2.12).

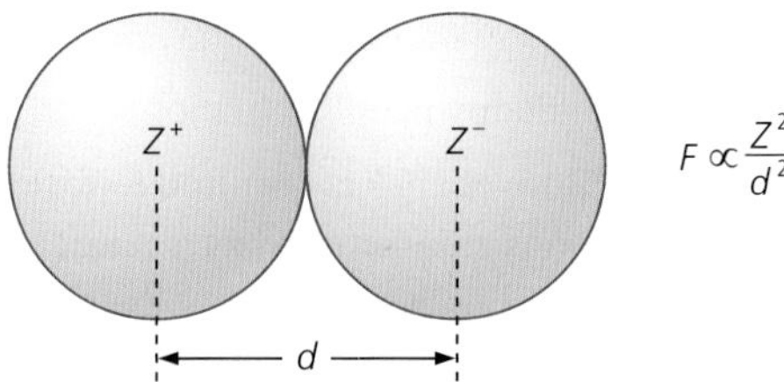

■ **Figure S2.12** Ions with a charge of Z^+ and Z^- separated by a distance d

Ionic charge, therefore, has a large effect on the size of the lattice enthalpy. For example, the lattice enthalpy of magnesium oxide, MgO, is about four times greater than the lattice enthalpy of sodium fluoride, NaF.

$$\Delta H^{\ominus}_{\text{lattice}}[\text{MgO(s)}] = +3889\,\text{kJ}\,\text{mol}^{-1}$$

$$\Delta H^{\ominus}_{\text{lattice}}[\text{NaF(s)}] = +902\,\text{kJ}\,\text{mol}^{-1}$$

This is largely due to the doubling of the charges of the ions. The sum of ionic radii and the lattice structures of magnesium oxide and sodium fluoride are similar.

The effect of ionic radius can be clearly seen in the sodium halides (Table S2.8). The smaller the distance between the centres of ions (sum of ionic radii), the larger the value of lattice enthalpy. Smaller ions can approach each other more closely leading to strong electrostatic forces of attraction.

■ **Table S2.8** The effect of ionic radius on lattice enthalpy

Sodium halide	Radius of halide ion / nm	Distance between centres of ions / nm	Lattice enthalpy / kJ mol^{-1}
Na^+ F^-	0.133	0.235	+930
Na^+ Cl^-	0.181	0.283	+790
Na^+ Br^-	0.196	0.298	+754
Na^+ I^-	0.220	0.322	+705

Going further

Lattice enthalpy and formulas

Because lattice enthalpy is a measure of the energetic stability of ionic compounds relative to their gaseous ions, its value can help explain the formulas of ionic compounds.

Consider magnesium chloride, $MgCl_2$, as an example. The ionization energy of an element increases rapidly as successive electrons are removed from its atom. The first ionization energy of magnesium is 738 kJ mol^{-1}. The second ionization energy is 1450 kJ mol^{-1}, which is almost twice the first. All these values are endothermic, meaning energy needs to be absorbed and work done to remove the electrons.

So, from the viewpoint of energetics, why does magnesium not form the unipositive ion, Mg^+, in its compounds leading to magnesium chloride with the formula, MgCl? No energy would be needed for the second ionization. Calculations based on the ions as charged spheres suggest that MgCl would have a lattice enthalpy near 753 kJ mol^{-1}.

More energy is required to form Mg^{2+}, but the resulting ion is smaller and has a greater charge than Mg^+. The resulting higher charge density leads to a higher value of lattice enthalpy (2540 kJ mol^{-1}) which is more than enough to compensate for the energy needed to remove both 3s valence electrons from a magnesium atom.

Nature of science: Observations

The existence of ions

Scientists obtain a great deal of the evidence they use by collecting and producing empirical results from observations and measurements. Perhaps the most convincing empirical evidence for the existence of ions is the movement of coloured ions in electrolysis (Figure S2.13). For example, electrolysis of green aqueous copper(II) chromate(VI) gives rise to a yellow colour (chromate(VI) ions) at the anode and a blue colour (copper(II) ions) at the cathode. The Royal Society of Chemistry has a video demonstration of this at **https://edu.rsc.org/exhibition-chemistry/the-movement-of-ions-bringing-electrolysis-to-life/4014567.article**.

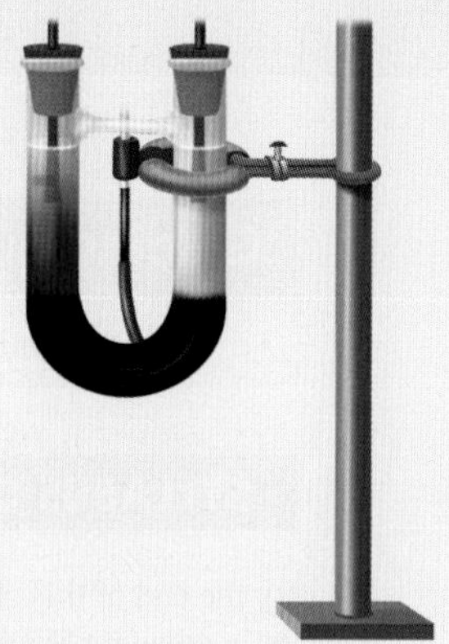

■ **Figure S2.13** Electrolysis of aqueous copper(II) chromate

LINKING QUESTION

How can lattice enthalpies and the bonding continuum explain the trend in melting points of metal chlorides across period 3?

S2.2 The covalent model

Guiding question

- What determines the covalent nature and properties of a substance?

SYLLABUS CONTENT

By the end of this chapter, you should understand that:

- a covalent bond is formed by the electrostatic attraction between a shared pair of electrons and the positively charged nuclei
- the octet rule refers to the tendency of atoms to gain a valence shell with a total of eight electrons
- single, double and triple bonds involve one, two or three shared pairs of electrons respectively
- a coordination bond is a covalent bond in which both electrons of the shared pair originate from the same atom
- the valence shell electron-pair repulsion (VSEPR) model enables the shapes of molecules to be predicted from the repulsion of electron domains around a central atom
- bond polarity results from the difference in electronegativities of the bonded atoms
- molecular polarity depends on both bond polarity and molecular geometry
- carbon and silicon form covalent network structures
- the nature of the force that exists between molecules is determined by the size and polarity of the molecules; intermolecular forces include London (dispersion), dipole–induced dipole, dipole–dipole and hydrogen bonding
- given comparable molar mass, the relative strengths of intermolecular forces are generally: London (dispersion) forces < dipole–dipole forces < hydrogen bonding
- chromatography is a technique used to separate the components of a mixture based on their relative attractions involving intermolecular forces to mobile and stationary phases
- resonance structures occur when there is more than one possible position for a double bond in a molecule (HL only)
- benzene, C_6H_6, is an important example of a molecule that has resonance (HL only)
- some atoms can form molecules in which they have an expanded octet of electrons (HL only)
- formal charge values can be calculated for each atom in a species and used to determine which of several possible Lewis formulas is preferred (HL only)
- sigma bonds (σ) form by the head-on combination of atomic orbitals where the electron density is concentrated along the bond axis (HL only)
- pi bonds (π) form by the lateral combination of p orbitals where the electron density is concentrated on opposite sides of the bond axis (HL only)
- hybridization is the concept of mixing atomic orbitals to form new hybrid orbitals for bonding (HL only).

By the end of this chapter you should know how to:

- deduce the Lewis formula of molecules and ions
- explain the relationship between the number of bonds, bond length and bond strength
- identify coordination bonds in compounds
- predict the electron domain geometry and the molecular geometry for species with up to four electron domains
- deduce the polar nature of a covalent bond from electronegativity values
- deduce the net dipole of a molecule or ion by considering bond polarity and geometry
- describe the structures and explain the properties of silicon, silicon dioxide and carbon's allotropes: diamond, graphite, fullerenes and graphene
- deduce the types of intermolecular force present from the structural features of covalent molecules
- explain the physical properties of covalent substances to include volatility, electrical conductivity and solubility in terms of their structure
- explain, calculate and interpret retardation factor, R_F, values
- deduce resonance structures of molecules and ions (HL only)

- discuss the structure of benzene from physical and chemical evidence (HL only)
- represent Lewis formulas for species with five and six electron domains around the central atom (HL only)
- deduce the electron domain geometry and the molecular geometry for these species using the VSEPR model (HL only)
- apply formal charge to determine a preferred Lewis formula from different Lewis formulas for a species (HL only)
- deduce the presence of sigma bonds and pi bonds in molecules and ions (HL only)
- analyse the hybridization and bond formation in molecules and ions (HL only)
- identify the relationships between Lewis formulas, electron domains, molecular geometry and type of hybridization (HL only)
- predict the geometry around an atom from its hybridization, and vice versa (HL only).

Covalent bond formation and notation

When two non-metal atoms react to form molecules, they share one or more pairs of electrons. A shared pair of electrons is a covalent bond.

The two hydrogen atoms in a molecule are bonded because their nuclei (protons) are both attracted to the shared electron pair. The atoms are identical so the electrons are shared equally. A single non-polar covalent bond is formed (Figure S2.14).

◆ **Polar covalent bond**: A covalent bond with partial charges due to a difference in electronegativity.

However, often the two atoms bonded will have different values of electronegativity. In iodine monochloride, ICl, the more electronegative chlorine atom will attract the shared pair(s) of electrons more strongly and a single **polar covalent bond** is formed (Figure S2.15).

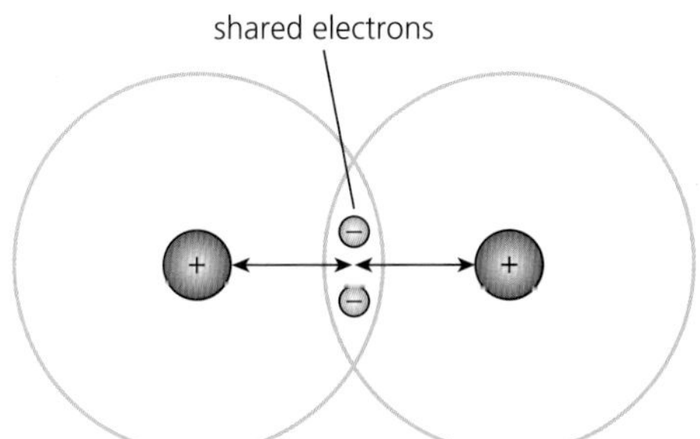
shared electrons
Both nuclei are attracted to the same pair of shared electrons. This holds the nuclei together.

■ **Figure S2.14** A simple electrostatic model of the covalent bond in the hydrogen molecule, H_2

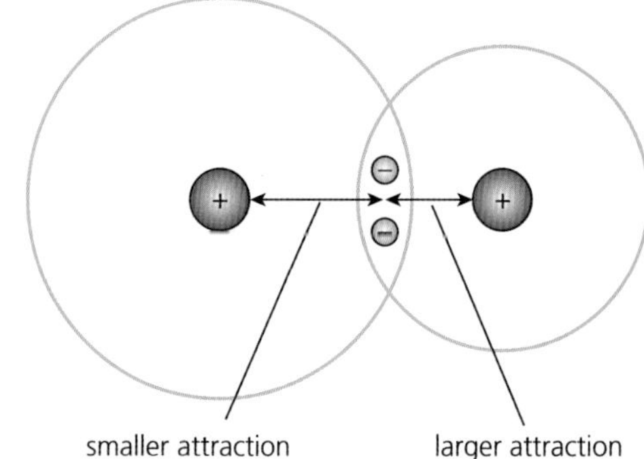
smaller attraction
larger attraction

■ **Figure S2.15** The unequal sharing of the electron pair in a polar covalent bond such as that in I–Cl

In some cases, a stable electronic arrangement may only be achieved if more than one pair of electrons is shared between the two nuclei. The oxygen molecule, O_2, contains a double bond made of two shared pairs. The nitrogen molecule, N_2, contains a triple bond consisting of three shared pairs. Figure S2.16 shows simple electrostatic models of the oxygen and nitrogen molecules.

LINKING QUESTION

Why do ionic bonds only form between different elements while covalent bonds can form between atoms of the same element?

a

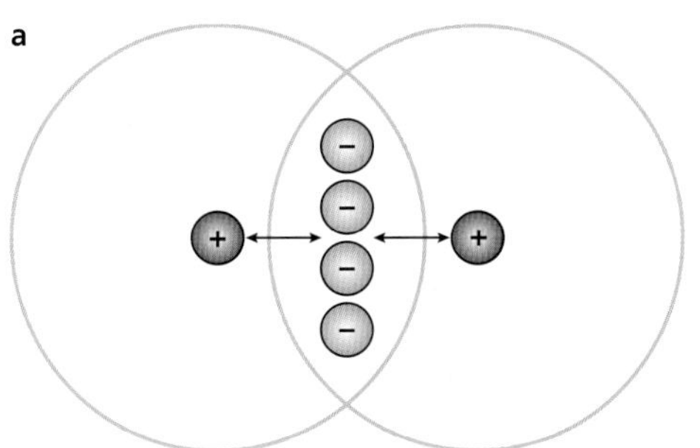

b

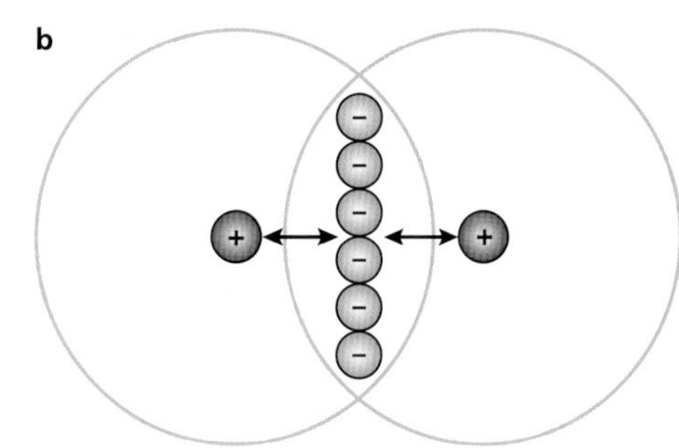

■ **Figure S2.16** A simple electrostatic model of the covalent bonds in **a** the oxygen molecule, O=O, and **b** the nitrogen molecule, N≡N.

Octet rule

Noble gases in group 18 (with the exception of helium) have four pairs of electrons—a stable octet—in the valence shells of their atoms. When an atom forms a chemical bond by gaining, losing or sharing valence electrons, its electronic configuration often becomes the same as that of a noble gas.

The formation of chemical bonds to achieve a stable noble gas configuration is known as the octet rule (Figure S2.17). It applies to ionic compounds and to many molecules, but there are exceptions: for example, hydrogen follows a duplet rule, since the first main energy level holds a maximum of two electrons.

However, note that some molecules (for example, BCl_3) have fewer than eight valence electrons, others (for example, PCl_5) have more than eight valence electrons, and some species (for example, a chlorine atom) have unpaired electrons (they are radicals).

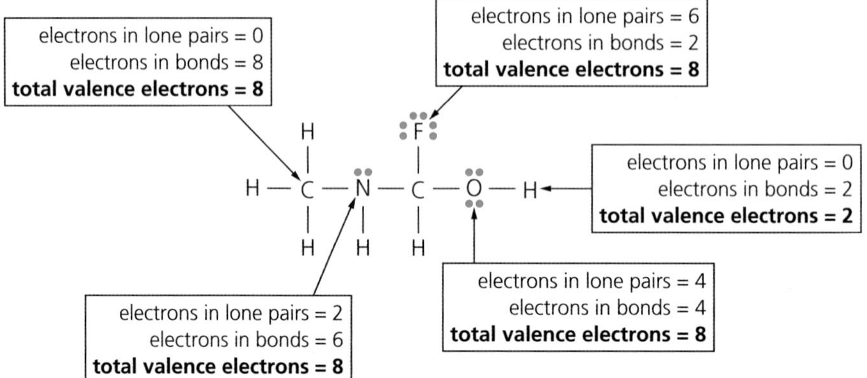

■ **Figure S2.17** A molecule showing atoms obeying the octet rule and hydrogen obeying the duplet rule

LINKING QUESTION

What are some of the limitations of the octet rule?

ATL S2.2A

Some of the heavier noble gases, especially xenon, form a range of molecular compounds.

Research how Neil Bartlett (1932–2008) synthesized the first solid noble gas compound, $XePtF_6$, in 1962. What was his theoretical justification for planning and carrying out the experiment? Outline your findings to the class, presenting the syntheses, structures (Lewis formulas) and shapes of the fluorides of xenon.

LINKING QUESTION

Why do noble gases form covalent bonds less readily than other elements?

Going further

Covalency

The number of covalent bonds formed by an atom depends on the number of electrons available for bonding and the number of valence shell orbitals available. For atoms of many elements, the number of bonds formed is a fixed quantity related to the group number and known as the covalency of the element.

Atoms of elements in groups 15 to 17 in period 3 (the third row of the periodic table) and beyond can use a variable number of electrons in bonding and have more than one covalency. Table S2.9 shows the most usual covalencies of some common elements. The covalency is often equal to the number of electrons that need to be gained, or lost, to achieve an octet (duplet for hydrogen) structure.

■ **Table S2.9** Covalencies of some common elements

Element	Symbol	Group	Covalency
hydrogen	H	1	1
beryllium	Be	2	2
boron	B	13	3
carbon	C	14	4
nitrogen	N	15	3
oxygen	O	16	2
fluorine	F	17	1
aluminium	Al	13	3
silicon	Si	14	4
phosphorus	P	15	3 or 5
sulfur	S	16	4 or 6
chlorine	Cl	17	1 or 3 or 5
bromine	Br	17	1 or 3 or 5
iodine	I	17	1 or 3 or 5 or 7

Figure S2.18 shows how to work out the formula of some binary molecular compounds from covalencies using the cross-over method.

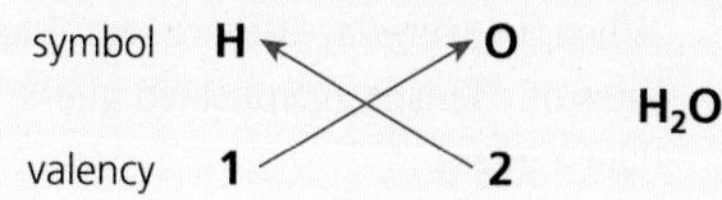

symbol C H
valency 4 1
CH_4

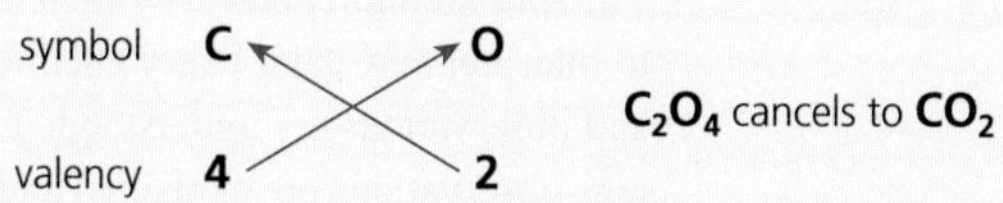

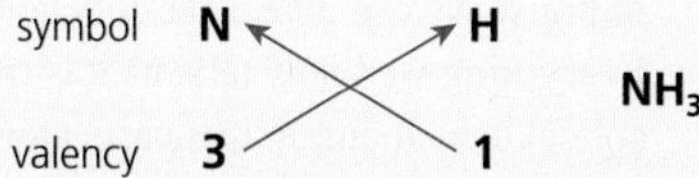

■ **Figure S2.18** Cross-over method for deducing the formulas of water, methane, carbon dioxide and ammonia molecules

Top tip!

The higher valencies of heavier atoms such as sulfur and phosphorus are due to there being more space around larger atoms, their large polarizability and, most importantly, the availability of empty low-energy 3d and 4s orbitals.

The importance of theories in the natural sciences

One of the most important terms used in science is 'theory'. Understanding just what this term means in a scientific community is crucial to understanding the scope of science because the term is used quite differently outside that community.

When, in everyday life, we say, 'I have a theory' or 'That's just a theory', we mean something along the lines of, 'This is an educated guess; an idea that makes sense of the data, but which is really no more than a guess.'

However, scientists use the term very differently. A theory in science is a set of ideas and explanations that have developed over a long period of time to make sense of the available data. A theory will have been tested and challenged but will have withstood the test of time. A theory will interpret facts and data by putting those facts and data in a wider framework in order to explain them.

Atomic theory, for example, is not just a speculative guess about the nature and structure of atoms; it is a comprehensive framework of ideas explaining why atoms are the way they are, given the data available. Nearly everything in this book is about the atomic theory.

Theories are more, however, than the summation of all the available data; theories also tell scientists how to interpret new data. Every chemical experiment uses atomic theory as a framework in which the data and observations are understood. The theory is both the result of observations and the lens through which new observations are explained and understood.

Were someone to point to atomic theory (more correctly described as an atomic paradigm) and say, 'Sure, but it's just a theory!' the chemist would reply: 'Yes, but a "theory" in our community is the highest form of praise we can give a set of beliefs! It means that there is virtually nothing that speaks against its truth. Everything we know tells us it is true and accepting its truth makes sense of everything we see when we collect data about, in this case, atoms.'

Lewis formulas

◆ **Lewis formula:** Diagram of a molecule, atom or ion showing the valence electrons (bonding and non-bonding pairs).

Lewis formulas show the valence electrons in an atom, a simple ion or each atom of a molecule or polyatomic ion. They highlight the importance of electron pairs in molecules and how these pairs lead to the constituent atoms having a stable noble gas arrangement of electrons.

■ **Table S2.10** The Lewis formulas of atoms of elements (when not bonded to other atoms)

Group	Electronic configuration	Lewis formula	Electronic configuration	Lewis formula
1	Li $1s^2\ 2s^1$	Li•	Na [Ne] $3s^1$	Na•
2	Be $1s^2\ 2s^2$	•Be•	Mg [Ne] $3s^2$	•Mg•
13	B $1s^2\ 2s^2\ 2p^1$	•Ḃ•	Al [Ne] $3s^2 3p^1$	•Ȧl•
14	C $1s^2\ 2s^2\ 2p^2$	•Ċ• (one dot above, one below)	Si [Ne] $3s^2 3p^2$	•Ṡi• (one dot above, one below)
15	N $1s^2\ 2s^2\ 2p^3$	•N̈•	P [Ne] $3s^2 3p^3$	•P̈•
16	O $1s^2\ 2s^2\ 2p^4$	•Ö• (pair below)	S [Ne] $3s^2 3p^4$	•S̈• (pair below)
17	F $1s^2\ 2s^2\ 2p^5$	:F̈• (pair below)	Cl [Ne] $3s^2 3p^5$	:C̈l• (pair below)

◆ **Lone pair**: A pair of electrons in the atomic orbital of a valence shell that are not used to form covalent bonds within a molecule.

◆ **Coordination bond**: A covalent bond where one atom supplies the bonding pair. (Also known as a coordinate bond.)

A pair of electrons can be represented by dots, crosses or dashes. Figure S2.19 shows different versions of the Lewis formula for the chlorine molecule, Cl_2.

■ **Figure S2.19** Lewis formulas for the chlorine molecule, Cl_2

Top tip!

The use of a combination of dots and crosses makes it clear which atom contributed the shared or bonding electrons. Dashes can be used to represent bonding pairs, as well as **lone pairs** (pairs not involved in bonding), which must be represented in a Lewis formula.

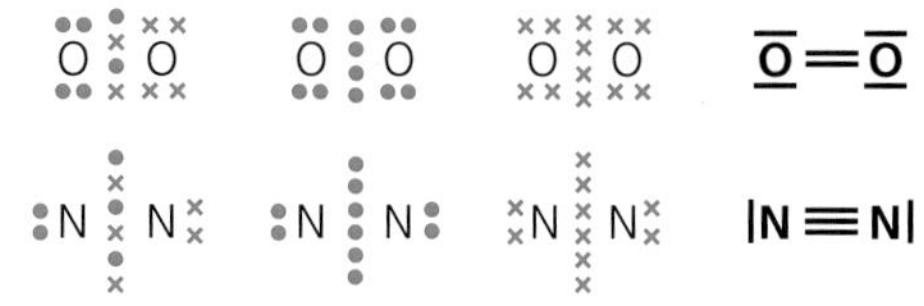

■ **Figure S2.20** Lewis formulas for the oxygen and nitrogen molecules

Oxygen atoms in a diatomic oxygen molecule, O_2, share two pairs of electrons to achieve an octet. The nitrogen molecule, N_2, has a triple bond formed by sharing three electron pairs (Figure S2.20).

The diagrams in Figure S2.21 show the Lewis formulas for several molecules in which the bonded atoms have achieved the electron arrangement of a noble gas.

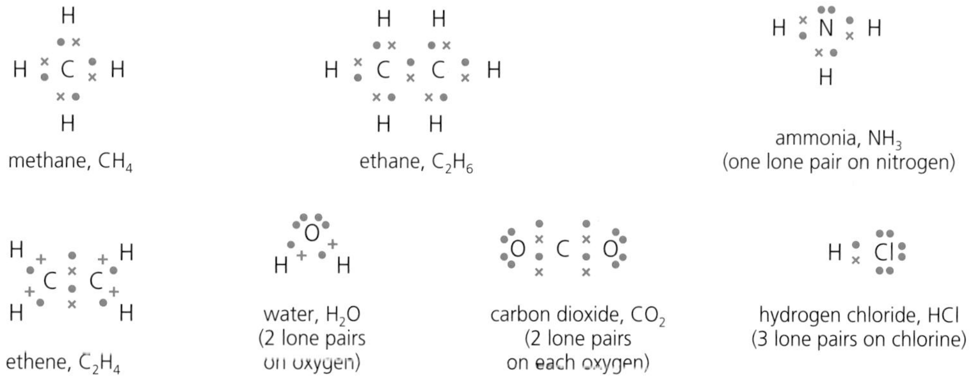

methane, CH_4

ethane, C_2H_6

ammonia, NH_3 (one lone pair on nitrogen)

ethene, C_2H_4

water, H_2O (2 lone pairs on oxygen)

carbon dioxide, CO_2 (2 lone pairs on each oxygen)

hydrogen chloride, HCl (3 lone pairs on chlorine)

■ **Figure S2.21** Lewis formulas for a selection of molecules of simple covalent compounds

Top tip!

Lewis formulas may also be drawn with the dots and crosses as part of a Venn diagram (Figure S2.22). This makes it easy to check exactly which bonds the electrons are in.

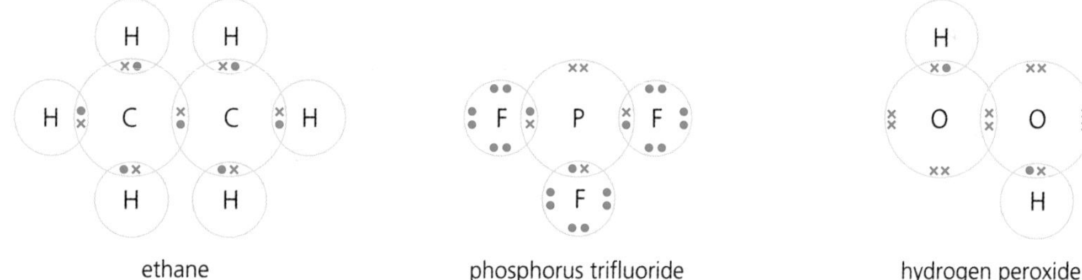

■ **Figure S2.22** Lewis formulas including Venn diagrams for the ethane, phosphorus trifluoride and hydrogen peroxide molecules

Coordination bonds

In some molecules and polyatomic ions, both electrons to be shared come from the same atom. The **coordination bond** formed (also known as a coordinate bond) is often shown by an arrow pointing from the atom which donates the lone pair to the atom which receives it. One atom needs a lone pair of electrons and a second atom needs an empty orbital to accept the lone pair.

For example, the carbon monoxide molecule, CO (Figure S2.23), contains one coordination bond. The coordination bond is indistinguishable from the other two single covalent bonds.

◆ **Adduct**: Compound formed by coordination between a Lewis acid electron (acceptor) and a Lewis base electron (donor).

Coordination bonding may also be found in molecular addition compounds (or **adducts**), such as boron trifluoride ammonia, $BF_3.NH_3$ (Figure S2.24). The boron atom in boron trifluoride has only six electrons in its outer shell and so can accept a lone pair to fill the shell, obeying the octet rule. The nitrogen atom on the ammonia molecule donates its lone pair of electrons to form the coordination bond between the nitrogen and boron atoms.

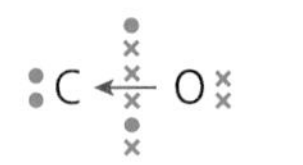

■ **Figure S2.23** Lewis formula for the carbon monoxide molecule, CO

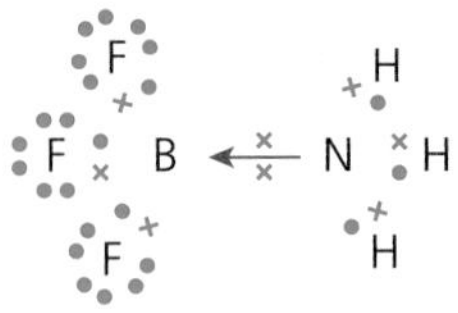

■ **Figure S2.24** Lewis formula for the boron trifluoride ammonia molecule, $BF_3.NH_3$

The formation of coordination bonds between a pair of reacting chemical species is the basis of a theory of acidity known as Lewis theory. Coordination bonds are also involved in the formation of transition metal complex ions and are often part of organic reaction mechanisms. Aqueous solutions of acids contain the oxonium cation, H_3O^+, which has a coordination bond.

Coordination bonding is also present in some common polyatomic ions. Three examples are shown in Figure S2.25.

- The gases ammonia and hydrogen chloride react together rapidly to form ammonium chloride, a white ionic solid, NH_4Cl [NH_4^+ Cl^-]. The ammonium ion is formed when an ammonia molecule accepts a proton.
- When a fluoride ion shares a lone pair with the boron atom in boron trifluoride, a tetrafluoroborate ion, BF_4^-, is formed.
- In the nitrate(V) ion, NO_3^-, the nitrogen atom achieves an octet by forming a coordination bond with one of the oxygen atoms.

(HL only – this ion exists as a hybrid of three equivalent resonance structures.)

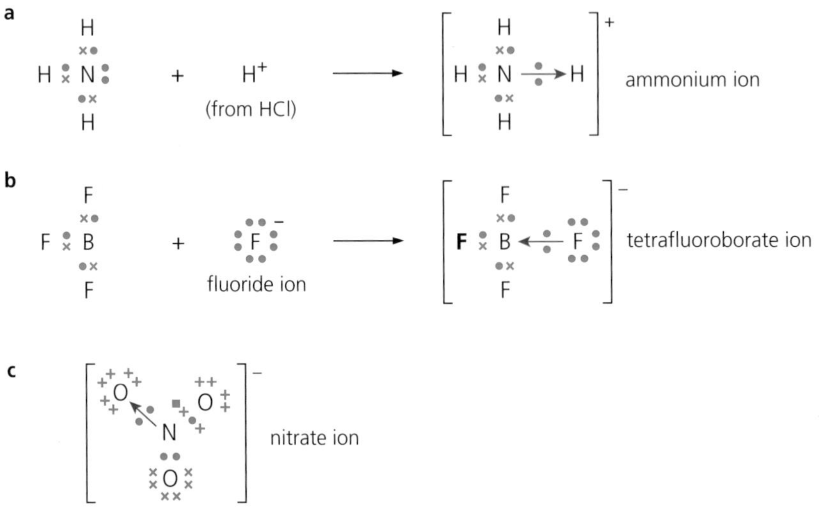

■ **Figure S2.25** Formation of **a** ammonium and **b** tetrafluoroborate ions; **c** structure of the nitrate(V) ion

Going further

Polymeric covalent substances and back bonding

Beryllium chloride is simple molecular in the gas phase but when cooled and condensed it forms a polymeric solid, $(BeCl_2)_n$. It forms long-chain molecules (linear polymers) with each beryllium atom accepting two lone pairs of electrons from two chlorine atoms (Figure S2.26). The arrangement of bonds around each beryllium is tetrahedral.

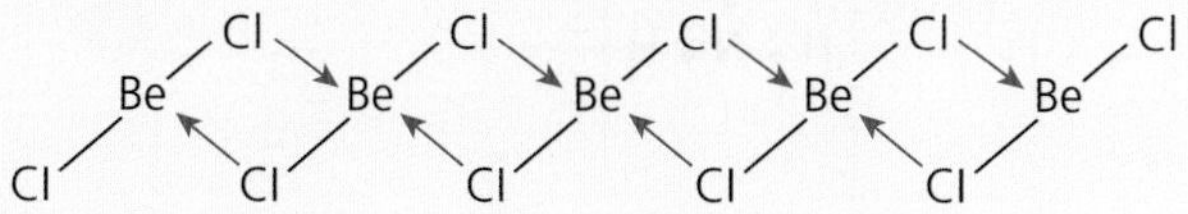

■ **Figure S2.26** Structure of beryllium chloride (in the solid state) with the arrows representing coordination bonds

The boron trifluoride molecule, BF_3, is also **electron deficient** but exhibits a form of coordination bond known as back bonding. Boron has an empty p orbital, while fluorine has a lone pair of electrons in its p orbital.

Fluorine atoms donate a lone pair of electrons to boron atoms (Figure S2.27). Boron is an electron-pair acceptor (Lewis acid), while fluorine is an electron-pair donor (Lewis base).

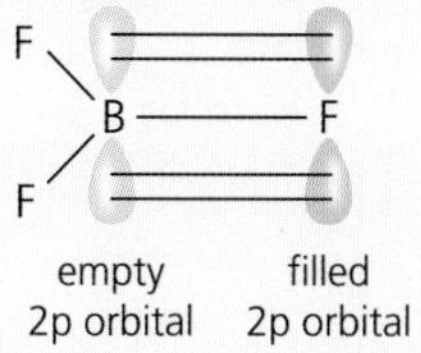

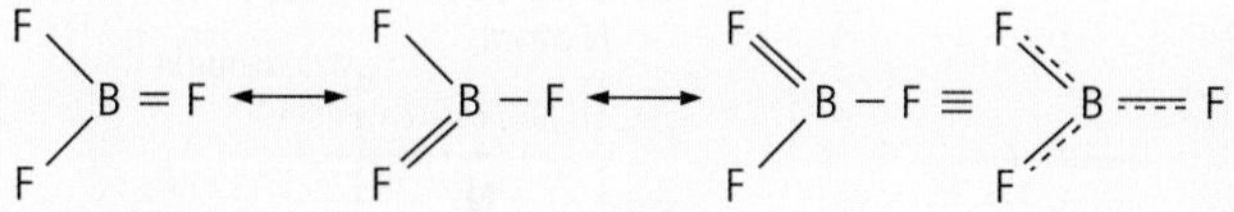

■ **Figure S2.27** Back bonding in boron trifluoride

◆ **Electron deficient**: A molecule in which there is an insufficient number of electrons to complete the octet of the central atom.

LINKING QUESTION (HL ONLY)

Why do Lewis acid–base reactions lead to the formation of coordination bonds?

Drawing Lewis formulas for molecules and ions

The general steps for drawing Lewis formulas for molecules and polyatomic ions are as follows:

1. Identify the central atom based on the number of bonds it can form. For example, hydrogen can only form one bond and cannot be the central atom. The central atom is usually the atom with the lowest subscript in the molecular formula and the atom that can form the most bonds. If all of the atoms usually form the same number of bonds, the least electronegative atom is usually the central atom. Arrange the other atoms (the terminal atoms) around the central atom.
2. For polyatomic ions, draw a square bracket around the Lewis formula and write the overall charge outside the bracket at the top right-hand corner.
3. If the ion is negatively charged, add the corresponding number of extra electrons to the more electronegative atom. The extra charge should be spread out: if there is, for example, a 2^- charge, distribute the two extra electrons to two different atoms.
4. If the ion is positively charged, remove the corresponding number of electrons from the less electronegative atom.
5. Work from the terminal atoms, ensuring that the terminal atoms achieve a noble gas configuration by forming the necessary number of bonds with the central atom.
6. Add the remaining electrons (total number of valence electrons − number of electrons involved in bonding with the terminal atoms) as lone pairs to the central atom.

7 Check the number of electrons around the central atom, remembering that an octet is not always necessary.

- If the central atom is from period 2, it cannot have more than eight electrons around it. If it does, consider changing a double bond to a coordination bond.
- If the central atom is from period 3 or beyond, it can have more than eight electrons around it.

Figure S2.28 follows the steps laid out above to determine the Lewis diagram of NH_2^-.

1. N is the central atom
2. Brackets and charge added

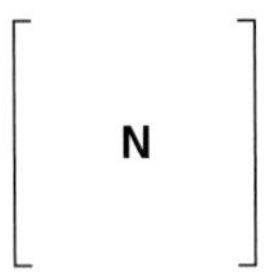

3. Extra electron is added to more electronegative N atom

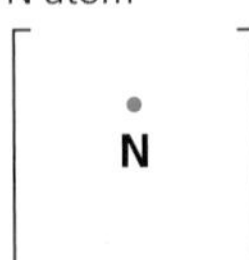

4. No positive charge on this species
5. Add in the terminal H atoms

H N H

6. Add in the three remaining valence electrons of N

H N H

■ **Figure S2.28** The construction of the Lewis formula for NH_2^-

ATL S2.2B

The widget at **www.stolaf.edu/depts/chemistry/courses/toolkits/121/js/lewis/** (hosted by St Olaf College) allows you to construct Lewis formulas by clicking on the bond or atom you want to modify. It recommends a procedure to follow and will check your structure. Work through the atoms and ions available, using the widget to deduce and draw Lewis formulas for all the ions and molecules included on the website.

5 Draw Lewis formulas showing the formation of the phosphonium cation, PH_4^+, from the phosphine molecule, PH_3, and a proton, H^+.

6 At high temperatures aluminium chloride exists as molecules (monomers) with the formula $AlCl_3$. This molecule is electron deficient; it still needs two electrons to complete the outer shell of the aluminium atom.

At lower temperatures two molecules of $AlCl_3$ combine to form a molecule with the formula Al_2Cl_6 (dimer). The $AlCl_3$ molecules are able to combine because lone pairs of electrons on two of the chlorine atoms form coordination bonds with the aluminium atoms.

Draw Lewis formulas to show the dimerization of two $AlCl_3$ molecules to form a molecule of Al_2Cl_6.

7 Draw a Lewis formula for the dinitrogen monoxide, N_2O (NNO) molecule.

Bond lengths and enthalpies

◆ **Bond enthalpy**: The minimum amount of energy required to break a mole of gaseous covalent bonds and form gaseous atoms under standard conditions.

For the same pair of atoms, double bonds are stronger than single bonds (because there are more pairs of shared electrons between the two nuclei) and triple bonds are stronger than double bonds. This observation can be accounted for using a simple electrostatic model of covalent bonding: the attraction by the nuclei increases with each shared pair of electrons.

The **bond enthalpy** is the energy needed to break a mole of gaseous covalent bonds to form gaseous atoms. Bond strengths, as measured by bond enthalpies, generally *increase* and bond lengths *decrease* from single to double to triple bonds between the same pair of atoms. This is illustrated for carbon in Table S2.11.

■ **Table S2.11** Bond enthalpies and lengths of carbon–carbon bonds

Bond type	Bond enthalpy / $kJ\,mol^{-1}$	Bond length / nm
single (C–C)	348	0.154
double (C=C)	612	0.134
triple (C≡C)	837	0.120

Ethanoic acid, CH_3COOH, contains two carbon–oxygen bonds, one single and one double. The double bond, C=O (0.122 nm) is significantly shorter than the single bond, C–O (0.143 nm).

Bond lengths in resonance structures

Molecules with resonance structures have bond lengths and bond strengths with values intermediate between those of the related single, double and triple bonds. For example, the carbon–carbon bonds in benzene have a bond enthalpy of $507\,kJ\,mol^{-1}$ and a bond length of 140 pm. The carbon–carbon bonds in benzene are 'half way' between single and double bonds and said to have a bond order of 1.5.

Top tip!

Bond enthalpies need to be used with some caution as they are often average values for many different types of molecules having the same bond. When there are differences in oxidation states, the bond enthalpies can be significantly different: for example, the P–Cl bond enthalpy is $319\,kJ\,mol^{-1}$ in PCl_3 and $264\,kJ\,mol^{-1}$ in PCl_5.

HL ONLY

WORKED EXAMPLE S2.2A

Deduce which compound has the shortest carbon to oxygen bond: C_2H_5CHO; C_2H_5COOH; CO_2, or CO.

Answer

The shortest carbon to oxygen bond is formed when there is the highest bond multiplicity. CO has a triple bond; CO_2 has two double bonds; C_2H_5CHO has a double bond; and C_2H_5COOH has one single bond and one double bond. Hence, CO has the shortest (and strongest) bond.

8 The table below lists bond lengths and bond enthalpies of some hydrogen halides.

Hydrogen halide	Bond length / 10^{-12} m	Bond enthalpy / $kJ\,mol^{-1}$
H–Cl	128	431
H–Br	141	366
H–I	160	298

a State the relationship between the bond length and the bond enthalpy for these hydrogen halides.

b Suggest why the bond enthalpy values decrease in the order HCl > HBr > HI.

c Suggest a value for the bond enthalpy in hydrogen fluoride, HF.

LINKING QUESTION

How does the presence of double and triple bonds in molecules influence their reactivity?

Going further

Reactivity of fluorine

The F–F bond length in F_2 is 142 pm, which is 22 pm longer than twice the covalent radius of the fluorine atom (60 pm). The F–F bond is also relatively weak (bond dissociation enthalpy = 159 kJ mol^{-1}) relative to the Cl–Cl bond (242 kJ mol^{-1}).

The reason for this is that the F–F bond is 'stretched' by repulsion of the lone pairs on the two fluorine atoms. This crowding is caused by the fact that the [He] 1s^2 core orbital and the valence orbitals of the fluorine atoms are contracted by the high nuclear charge.

A similar lone pair repulsion effect explains the anomalously long and weak N–N and O–O single bonds in hydrazine, $H_2N–NH_2$, and hydrogen peroxide, HO–OH, (Figure S2.29) which are both highly reactive molecules.

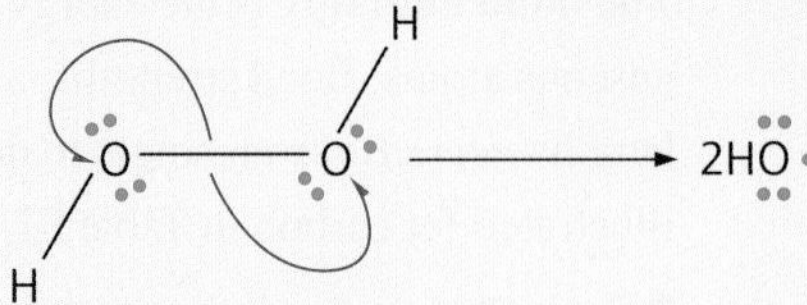

■ **Figure S2.29** Lone pair repulsion in H_2O_2 and formation of hydroxyl radicals where single headed arrows show movement of a single electron from head to tip

TOK

What are the uses and limitations of models?

It would be impossible to develop useful knowledge about the world without the use of models. Models are general representations of events and objects in the world which are meant to ignore the differences and only focus on what makes them similar. Without models, our knowledge of the world would be limited to knowledge of specific events and objects in the world. You might develop knowledge about some of the very large numbers of cells in your body, but without models we would only know about those particular cells. This textbook is full of examples of models—every diagram representing an atom in this book is a model.

However, there might be a danger here. Were you to look at the representation of the model in Figure S2.29 and form beliefs about features of the atoms that are not intended, you would be in trouble. No chemical reaction, obviously, has little arrows and numbers. This point is so obvious as to be comical. So, part of the responsibility of the knower is to understand which elements of a model are intended or unintended.

The flip side of this is that models are only useful when the information they provide answers the question being asked. As you use each model in this textbook, it is worth considering what the model is being used for, what features of the model are therefore relevant, and which elements of the model are not meant to convey knowledge.

The word 'model' can also be used as a 'style' of explaining or understanding an observation. The discussion of VSEPR theory uses both the term 'theory' (an established explanatory framework) and the term 'model' to refer to a way of making sense of the bonding of atoms.

HL ONLY

Complex ions

◆ **Ligand**: A molecule or negative ion that donates a pair (or pairs) of electrons to a central metal ion to form a coordinate covalent bond.

Link

The formation of complex ions is covered in Chapter S3.1, page 254 and Chapter R3.4, page 650.

A complex ion has an ion of a transition metal or a metal from groups 13 or 14 (for example, aluminium, tin or lead) at its centre, with a number of other molecules or anions bonded around it. These molecules and/or anions are known as **ligands**. The coordination number is the number of positions on the metal ion where ligands are bonded.

Ligands have one or more available lone pairs of electrons in the outer energy level and these are used to form coordination bonds with the metal ion. All ligands are electron-pair donors and hence function as Lewis bases. The central metal ion functions as a Lewis acid (electron-pair acceptor).

The electronic configuration of copper(II) ions is [1s^2 2s^2 2p^6 3s^2 3p^6] **3d^9**. When bright blue copper(II) sulfate solution reacts with aqueous ammonia solution, the ammonia first acts as a base to produce a pale blue precipitate of copper hydroxide. When it is added in excess, a ligand exchange reaction occurs and the dark blue tetraamminediaquacopper(II) ion, $[Cu(NH_3)_4(H_2O)_2]^{2+}$, is formed (Figure S2.30).

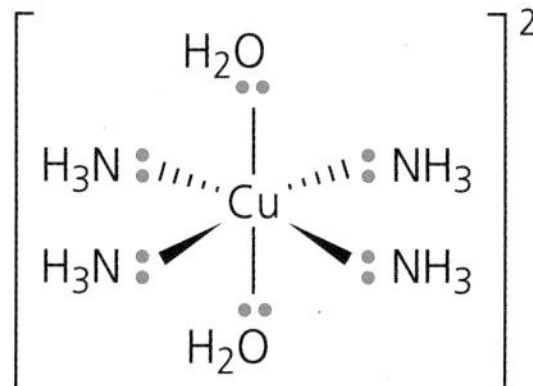

■ **Figure S2.30** Structure and shape of tetraamminediaquacopper(II) ion

Top tip!

Many transition cations exist as complex ions, which are often coloured in aqueous solution.

HL ONLY

Two water molecules (the 'aqua' in the name) and four ammonia molecules ('ammine') act as ligands. The lone pairs of electrons on the oxygen and nitrogen atoms of the water and ammonia molecules, respectively, are accommodated in the empty orbitals of the copper(II) ion forming coordination bonds.

The central copper has the electronic configuration $[1s^2 2s^2 2p^6 3s^2 3p^6]$ $\mathbf{4s^2 3d^{10} 4p^6 4d^3}$. It has not achieved the electronic configuration of a noble gas but this can be explained using more complex molecular orbital theory.

Top tip!

A double salt such as iron(II) ammonium sulfate, $Fe(NH_4)_2(SO_4)_2.6H_2O(s)$, will release iron(II) ions, $Fe^{2+}(aq)$, ammonium ions, $NH_4^+(aq)$, and sulfate ions, $SO_4^{2-}(aq)$, in water. These are all simple ions. A complex ion, such as $[Fe(CN)_6]^{4-}(aq)$, does not dissociate to release iron(II) ions, $Fe^{2+}(aq)$, and cyanide ions, $CN^-(aq)$.

Figure S2.31 below summarizes the structure of a complex, its complex ion and its counter ion.

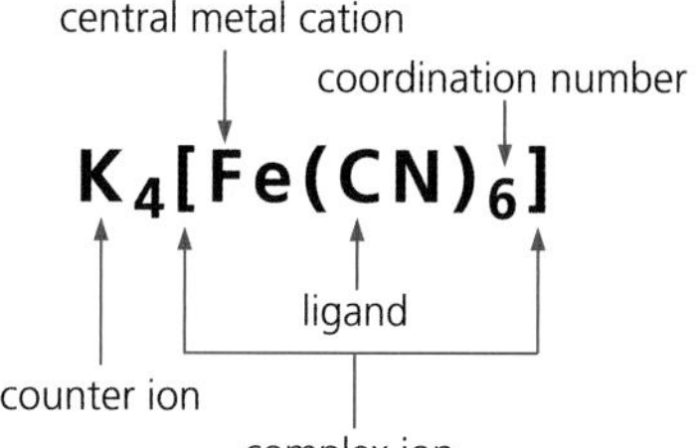

■ **Figure S2.31** Structure of $[4K^+ [Fe(CN)_6]^{4-}]$

The shapes of molecules

Nature of science: Models

Developing models of molecules

One of the major advances in chemistry occurred in 1874, when the Dutch chemist Jacobus van't Hoff (1852–1911) suggested that molecules possessed a definite, unique, three-dimensional shape defined by fixed bond angles. He suggested that the four single bonds around a carbon atom were equivalent and arranged tetrahedrally. This was a paradigm shift in chemical thinking and was used to explain the existence of enantiomers (pairs of molecules which are mirror images of each other and affect plane polarized light in opposite ways). He was responsible for developing a new area of chemistry known as stereochemistry, which deals with the effects produced by the organization of atoms and functional groups in space.

TOK

What is a scientific paradigm?

A paradigm is a web of interconnected ideas and explanations that are developed from the data and used to make sense of new data. Paradigm shifts are crucial in the understanding of how scientific knowledge has developed over time.

In his 1962 book, *Structure of Scientific Revolutions*, the philosopher Thomas Kuhn suggested that rather than developing bit by bit, with new knowledge gradually replacing older knowledge, science sometimes undergoes a huge shift in understanding. When old ways of thinking about the world and explaining it cannot fully explain enough newer observations, a fundamentally new way of understanding the data is introduced. If the new paradigm is more efficient, comprehensive and informative in how it explains the old and new data, it might become the new norm.

VSEPR theory

◆ **Electron domain:** A lone pair or bond location around a particular atom in a molecule.

◆ **VSEPR theory:** A model which states that the shape of bonds around the central atom of a molecule is determined by minimizing the repulsion between the electron domains.

Because of their identical (negative) charge, electron pairs in molecules repel each other and will experience the least electrostatic repulsion when they are as far apart from one another as possible. This applies to bonded pairs and to non-bonded (lone) pairs. Bonding and lone pairs are termed **electron domains**.

The angles between the pairs depend on the number of electron domains around the central atom. This is the **valence shell electron-pair repulsion theory** (or **VSEPR** for short).

To deduce the shape of a simple molecule or polyatomic ion with one central atom:

1 Draw the Lewis formula.

2 Count the number of electron pairs around each atom.

3 Deduce the arrangement that positions these pairs as far apart from one another as possible.

Table S2.12 shows the arrangement of the electron pairs (electron domains) that results in minimum repulsion and the basic shapes of the molecules.

■ **Table S2.12** Basic molecular shapes for species with two, three and four electron domains

Molecule shape	Number of electron domains	Description
	2	linear
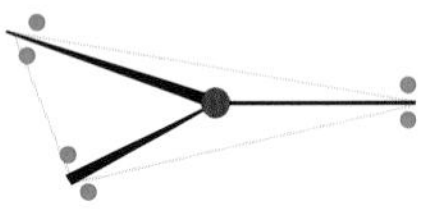	3	trigonal planar
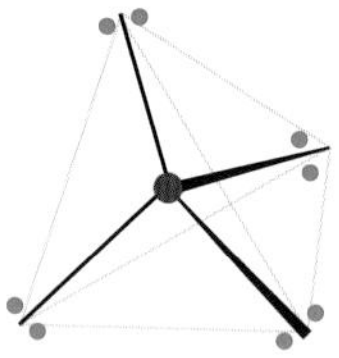	4	tetrahedral

Impact of lone pairs

An orbital containing a lone pair is larger than an orbital containing a bonded pair and occupies more space around the central atom. It therefore repels the other electron pairs that surround the central atom more strongly.

This causes the angle between two lone pairs to be slightly larger than the angle between a lone pair and a bonded pair, which in turn is larger than the angle between two bonded pairs:

lone pair–lone pair repulsion > lone pair–bond pair repulsion > bond pair–bond pair repulsion

strongest **weakest**

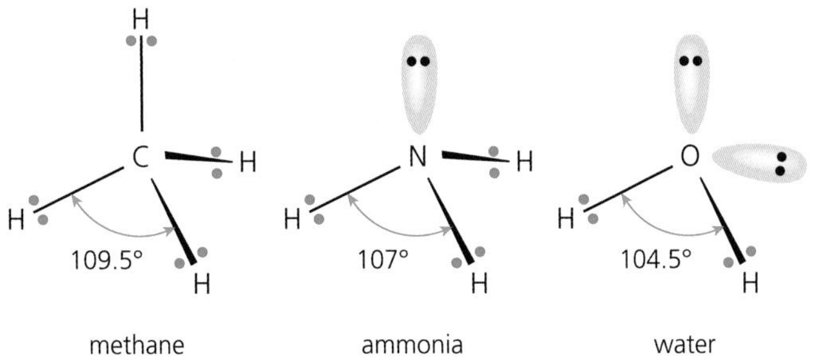

■ **Figure S2.32** The shapes of the methane, ammonia and water molecules. Lone pairs repel other electron pairs more strongly than bonding pairs do, so the bond angles in the ammonia and water molecules are slightly smaller than the tetrahedral angle of 109.5°.

The lone pairs play a critical role in determining the electron domain geometry, but they are not included in describing the shape of a molecule or ion.

For example, the molecules CH_4, NH_3 and H_2O all have four pairs of electrons (four electron domains) in the valence shell of the central atom (Figure S2.32). These four pairs arrange themselves in a tetrahedron, or a close approximation to one. But only the methane molecule is described as having a tetrahedral shape. The ammonia molecule, which has only three bonding pairs, is trigonal pyramidal and water, with two bonding pairs, is described as bent, V-shaped or non-linear.

Cl — Be — Cl

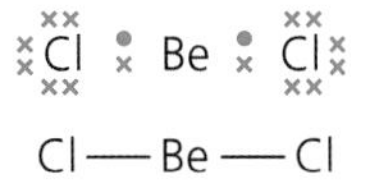

beryllium chloride

■ **Figure S2.33** Lewis formula and molecular shape of beryllium chloride showing the bond angle of 180°

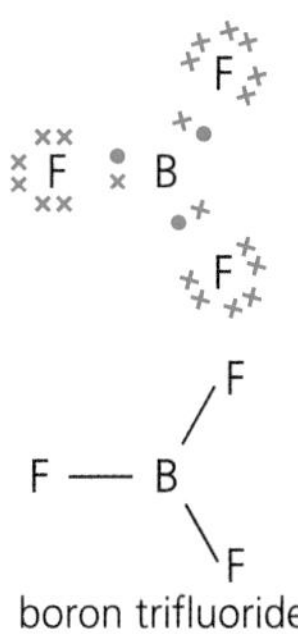

boron trifluoride

■ **Figure S2.34** Lewis formula and molecular shape of boron trifluoride showing three bond angles of 120°

WORKED EXAMPLE S2.2B

Deduce the shapes of the beryllium chloride molecule, $BeCl_2(g)$, and the boron trifluoride molecule, BF_3.

Answer

The beryllium chloride molecule, $BeCl_2(g)$, is linear with a bond angle of 180°, because it has two electron domains. The electron pairs are farthest apart when they are on opposite sides of the central beryllium atom (Figure S2.33).

The Lewis formula of the boron trifluoride molecule, BF_3 (Figure S2.34), shows there are three electron domains in the valence shell of the boron atom. These three bonding pairs repel each other equally so the boron trifluoride molecule is a trigonal planar (flat) molecule. The three boron–fluorine bonds point toward the three corners of an equilateral triangle and the bond angles are all 120°.

9 Arrange the following in decreasing order of bond angle (largest one first) and explain your reasoning: PH_2^-, PH_3, PH_4^+.

Shapes of molecules with multiple bonds

A double or triple bond has the same repulsive effect as a single bond. It is counted as one electron domain because all the bonding pairs of electrons are located between the two atoms.

The carbon dioxide molecule has a linear structure (Figure S2.35) and the ethene molecule is trigonal planar around each of the two carbon atoms (Figure S2.36).

O = C = O

carbon dioxide

■ **Figure S2.35** Lewis formula and molecular shape of the carbon dioxide molecule

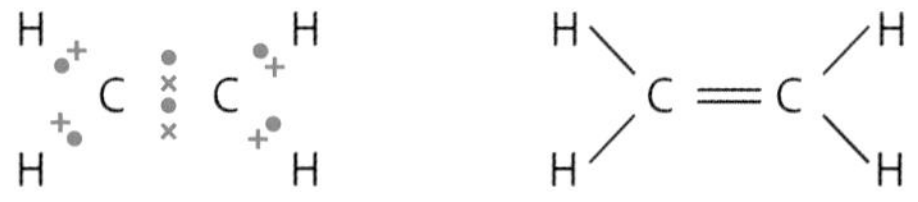

ethene

■ **Figure S2.36** Lewis formula and molecular shape of the ethene molecule

Isoelectronic means having the same number of electrons and can be applied to species which have the same total number of electrons or have the same number of electrons in their valence shells.

Because the arrangement of electrons in the valence shell of the central atoms determines the shape of a molecule, species that are isoelectronic have the same basic shape and often have identical bond angles too.

For example: the oxonium ion, H_3O^+, and the ammonia molecule, NH_3, are both trigonal pyramidal with bond angles close to 107°.

ATL S2.2C

- State and explain how adding an atom or lone pair affects the position of existing atoms or lone pairs.
- State and explain what happens to the bond angle when you add or remove an electron domain.
- Use the simulation at https://phet.colorado.edu/sims/html/molecule-shapes/latest/molecule-shapes_en.html to construct the molecules mentioned in this chapter (CH_4, NH_3, H_2O, BF_3, $BeCl_2$, CO_2, SO_2, SO_3 and so on). Draw them in three dimensions, showing the bond angles, and state their electron domain geometry and molecular shape.

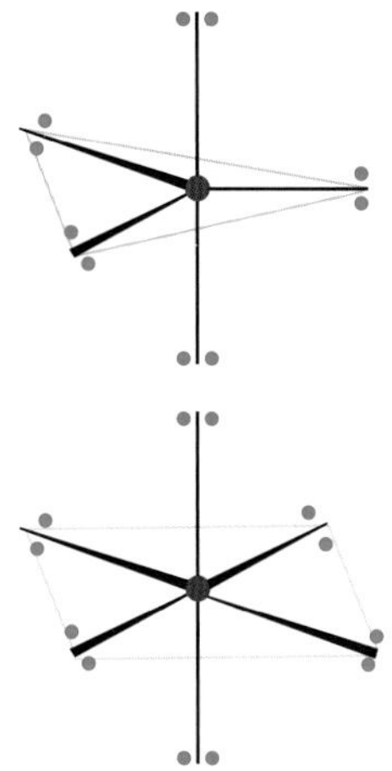

■ **Figure S2.37** The basic shapes for molecules with five and six electron pairs (electron domains)

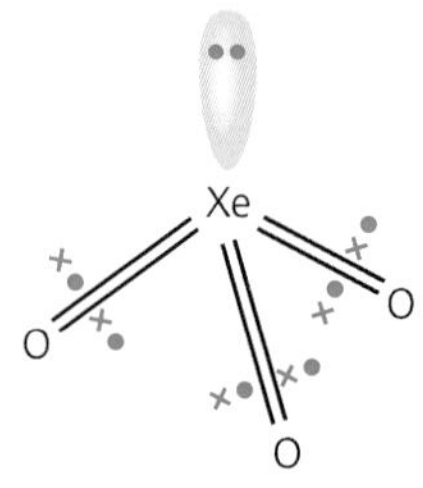

■ **Figure S2.38** Structure and shape of the XeO_3 molecule

Shapes of molecules with expanded octets

VSEPR theory can also be applied to molecules and ions that have five electron domains (PCl_5, for example) or six electron domains (such as SF_6). The basic shapes adopted by these molecules to minimize the repulsion between electron pairs in the valence shell are trigonal bipyramidal and octahedral, respectively (Figure S2.37). The trigonal bipyramid has three equatorial bonds and two axial bonds; the octahedron has four equatorial bonds and two axial bonds.

A multiple bond is still treated as if it is a single electron pair. For example, the xenon trioxide molecule has a pyramidal shape (Figure S2.38). The valence shell of the xenon atom in xenon trioxide contains 14 electrons: eight from the xenon and two each from the three oxygen atoms.

Top tip!

For species with five electron domains, there are alternative positions for any lone pairs of electrons. The favoured positions will be those where the lone pairs are located furthest apart, thus minimizing the repulsive forces in the molecule. Consequently, lone pairs usually occupy equatorial positions.

Tool 1: Experimental techniques

Physical modelling

Physical models can be more helpful than two-dimensional representations when trying to visualize three-dimensional structures.

Balloons blown to the same size can be used to represent an electron domain. The balloons can be taped or tied together at the tied-off ends as if surrounding a central atom. When connected, the balloons naturally adopt the lowest energy arrangement predicted by the VSEPR model.

You could use two different colours to distinguish between a bond and a non-bonding lone pair.

■ **Table S2.13** Molecules that can be modelled using balloons

Molecule	One balloon colour (bonding pairs)	Another balloon colour (lone pairs)
sulfur hexafluoride	6	0
boron trichloride	3	0
carbon dioxide	2	0
arsenic trihydride	3	1
hydrogen sulfide	2	2
tin(II) bromide	2	1
chlorine tribromide	3	2
carbon tetrachloride	4	0
phosphorus pentabromide	5	0
selenium tetrabromide	4	1
iodine tetrafluoride ion	4	2
bromine pentachloride	5	1

LINKING QUESTION

How does the ability of some atoms to expand their octet relate to their position in the periodic table?

WORKED EXAMPLE S2.2C

Deduce the shapes of:

a PF_5 b SF_6 c $XeCl_4$.

Answer

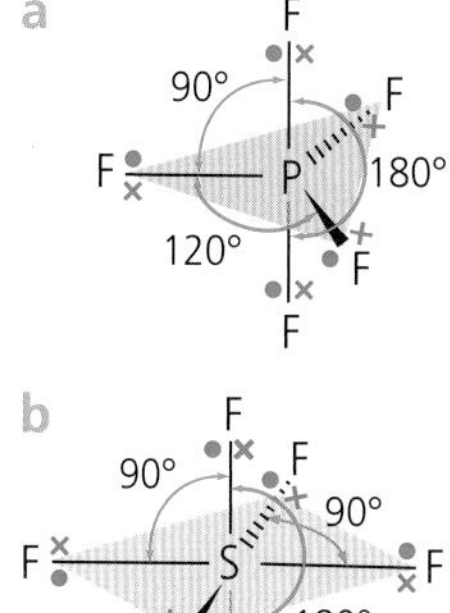

a The valence shell of the phosphorus atom in phosphorus(V) fluoride contains ten electrons: five from the phosphorus and one each from the five fluorine atoms. The shape will be a trigonal bipyramid with bond angles of 120°, 180° and 90°.

b The valence shell of the sulfur atom in the sulfur(VI) fluoride molecule contains 12 electrons: six from the sulfur and one each from the six fluorine atoms. The shape will be an octahedron with bond angles of 90° and 180°.

c The valence shell of the xenon atom contains 12 electrons: eight from the xenon and one each from the four chlorine atoms. There are four bonding pairs and two lone pairs. The basic shape adopted by the molecule is octahedral. However, there are two possible arrangements for the lone pairs. The first structure, square planar, minimizes the repulsion (the lone pairs are at 180° to each other) and is hence adopted as the molecular shape. As a general rule, for a molecule where the electron domains adopt an octahedral structure, any lone pairs will occupy axial positions (opposite each other).

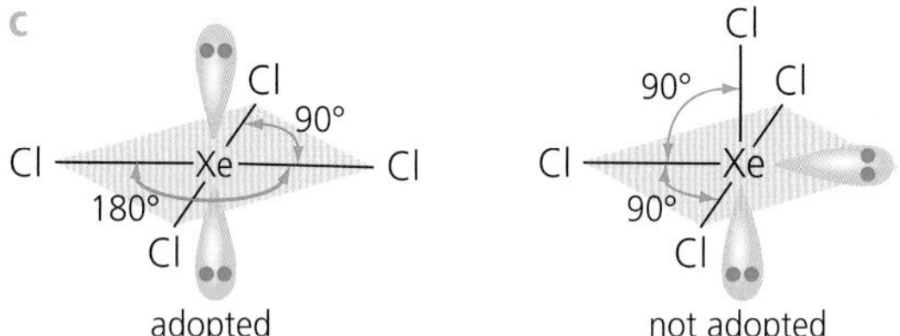

10 Deduce and draw the shapes of:

a SF_4
b ClF_3
c ICl_2^-
d I_3^-.

Table S2.14 summarizes how the numbers of bonded and non-bonded electron pairs determine the shapes of molecules.

■ **Table S2.14** The shapes of molecules as determined by the numbers of bonding and non-bonding electron pairs

Total number of electron domains	Number of bonding pairs	Number of lone pairs	Shape	Example
2	2	0	linear	BeF_2 and I_3^-
3	3	0	trigonal planar	BF_3
3	2	1	bent	GeF_2
4	1	3	linear	HF
4	2	2	bent	H_2O
4	3	1	trigonal pyramidal	NH_3
4	4	0	tetrahedral	CH_4
5	2	3	linear	XeF_2
5	3	2	T-shaped	ClF_3
5	4	1	'see-saw' / 'saw horse'	SF_4
5	5	0	trigonal bipyramidal	PF_5
6	4	2	square planar	XeF_4
6	5	1	square-based pyramidal	ClF_5
6	6	0	octahedral	SF_6

LINKING QUESTION

How useful is the VSEPR model at predicting molecular geometry?

Molecular polarization

Bond polarity

When the electronegativity values of the two atoms forming a covalent bond are the same, the bonding pair of electrons is equally shared and a non-polar covalent bond is formed. Examples include hydrogen, H_2, nitrogen, N_2, and hydrogen astatide, H–At$^-$ (when considered to 1 d.p., hydrogen and astatine both have an electronegativity value of 2.2). Phosphorus and hydrogen also have the same values of electronegativity (to 1 d.p.), so the P–H bond is non-polar.

When two atoms with different electronegativity values form a covalent bond, the shared pair(s) of bonding electrons are attracted more strongly by the more electronegative element. This results in an asymmetrical distribution of the bonding electrons.

For example, in the hydrogen chloride molecule, the more electronegative chlorine atom attracts the bonding pair more strongly than the hydrogen atom. Consequently, the chlorine atom has a partial negative charge and the hydrogen atom has a partial positive charge. The hydrogen chloride molecule is polar (Figure S2.39) and the bond is polar covalent.

δ+ δ–
H — Cl

H — Cl

hydrogen chloride

■ **Figure S2.39** Bond polarity in the hydrogen chloride molecule

The arrow always points from the positive charge to the negative charge and shows the **dipole** (pair of equal and opposite, separated charges: δ^+ and δ^-). The arrow shows the shift in electron density toward the more electronegative chlorine atom. The crossed end of the arrow represents a plus sign that designates the positive end (the less electronegative atom).

The strength of a dipole in a covalent bond is measured by its **dipole moment**, which is equal to the product of charge and the distance between the charges. Dipole moments can be measured experimentally for molecules with a single bond and molecules with two or more bonds and they can also be calculated knowing the shapes of the molecule. Table S2.15 lists the dipole moments of some common molecules with dipole moments expressed in units of Debye, D (1 D = 3.336×10^{-30} coulomb metres).

■ **Table S2.15** The dipole moments of some common molecules

Molecule	Dipole moment / D
HF	1.91
HCl	1.05
HBr	0.80
HI	0.42
H_2O	1.84
NH_3	1.48

◆ **Dipole**: A pair of separated equal and opposite electrical charges located on a pair of atoms within a molecule.

◆ **Dipole moment**: The product of the charge on a dipole and the distance between the ends; a measure of the polarity of a bond.

◆ **Non-polar molecule**: A molecule that has a symmetric distribution of charge and whose individual bond dipoles sum to zero or cancel.

◆ **Polar molecule**: A molecule that has an asymmetric distribution of charge: the individual bond dipoles do not sum to zero or cancel.

Molecular polarity

In molecules containing more than two atoms, the bond polarity and the arrangement of bonds (molecular geometry) must both be considered. Depending on the angles between the bonds, the individual bond dipoles can either reinforce or cancel each other. Molecules with large dipole moments can be formed when the dipoles reinforce each other.

If cancellation is complete, as in, for example, carbon dioxide (Figure S2.40), the resulting molecule will have no molecular dipole moment and so will be **non-polar**.

Fluoromethane, CH_3F, is a **polar molecule**. The combined effect of the dipole moments of the three C–H bonds is not cancelled out by the polarity of the single highly polar C–F bond because the C–H bond is virtually non-polar. The molecule therefore has a net dipole moment with the negative end centred toward the fluorine atoms (Figure S2.41).

The tetrafluoromethane molecule, CF_4, has four polar C–F bonds pointing toward the four corners of a regular tetrahedron (Figure S2.42). The dipole moments in each bond cancel, the overall dipole moment is zero and the molecule is non-polar.

δ– δ+ δ+ δ–
O=C=O

■ **Figure S2.40** The molecular geometry of the carbon dioxide molecule showing how the bond dipoles cancel

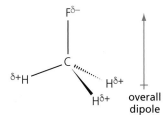

Fδ–
C
δ+H
Hδ+
Hδ+
overall dipole

■ **Figure S2.41** The molecular geometry of the CH_3F molecule

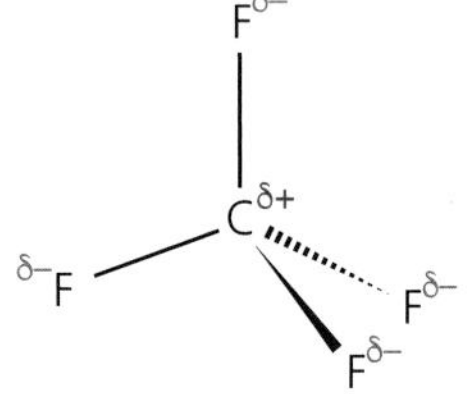

■ **Figure S2.42** The molecular geometry of the CF_4 molecule

11 Deduce the shapes of the following molecules and state which of them are polar.

a C_2H_6
b C_6H_6 (benzene)
c CH_3OH
d SF_4
e SCl_6
f PCl_5
g BCl_3

WORKED EXAMPLE S2.2D

Deduce which of the difluorobenzene molecules shown is non-polar (has zero dipole moment).

A B C

Answer

Compound C, 1,4-difluorobenzene.

In compound C, the C–F dipoles are on opposite sides of the benzene ring so their two equal dipoles oppose each other and cancel.

In molecules A and B, the C–F dipoles are, more or less, on the same side of the benzene ring and do not cancel.

Going further

Testing for polarity (in bulk)

When polar molecules are placed in an electric field, the electrostatic forces cause the molecules to align with the electric field (Figure S2.43).

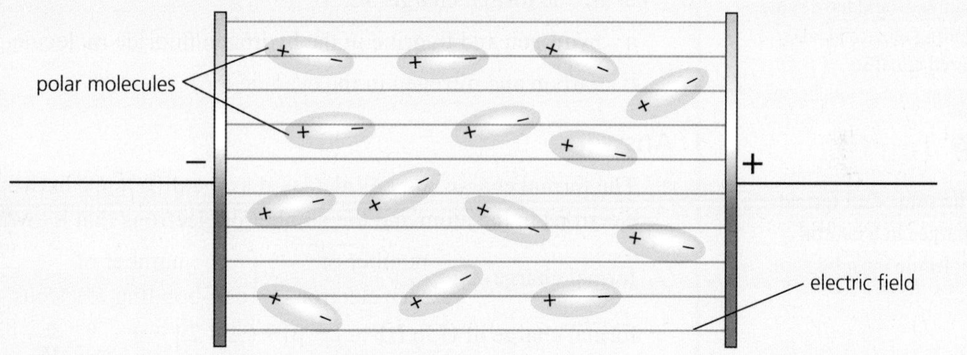

■ **Figure S2.43** Polar molecules in an electric field

When a charged plastic rod is brought close to the stream of a liquid running from the jet of a burette, a polar liquid will be deflected from its vertical path toward the charged rod but a non-polar liquid will not be affected. The greater the polarity of the liquid, the greater the deflection (under the same experimental conditions). The deflection is observed with both a positively and a negatively charged rod.

LINKING QUESTION

What properties of ionic compounds might be expected in compounds with polar covalent bonding?

ATL S2.2D

Run a dry plastic comb through dry hair or rub it with a dry cloth. Electrons are transferred from nylon to the cloth. Slowly bring the comb near a thin stream of water from a tap.

- Record and explain your observations using annotated diagrams.
- Predict the results if the experiment were repeated with ethanol (less polar than water) and cyclohexane (non-polar).
- Explain why poor results are obtained in humid conditions.

TOK

What kinds of explanations do natural scientists offer?

Physicists and chemists are generally interested in the level of analysis emphasizing atomic particles and subatomic particles, especially electrons in chemistry. Atoms chemically combine via formation of covalent bonds to form molecules. The questions usually asked by chemists, therefore, deal with small molecules or polymers as the unit of analysis. Chemical theories, such as kinetic molecular theory or transition state theory, are often described as being reductionist because they aim to explain complex behaviours in terms of relatively simple structures and functions.

HL ONLY

LINKING QUESTION

What features of a molecule make it 'infrared active'? (Infrared active refers to certain covalent bonds that absorb infrared radiation and cause a change in the molecule's dipole moment.)

◆ **Formal charge**: The charge an atom in a Lewis formula would have if the bonding electrons were shared equally.

Top tip!

The sum of all formal charges in a neutral molecule must be zero.

Top tip!

In an ion, the sum of all formal charges must equal the charge of the ion.

Formal charge

In polar covalent bonds, the partial charges on the bonded atoms are real. **Formal charges** provide a method for keeping track of electrons, but they may or may not correspond to real charges.

A formal charge is an imaginary charge placed on each atom in a Lewis formula that distinguishes the different Lewis formulas that can be drawn for a molecule or polyatomic ion. The concept of formal charge helps us determine which atoms bear most of the charge in a charged molecule. It also helps us to identify charged atoms in molecules that are neutral overall.

In most cases, if the Lewis formula shows that an atom has a formal charge, it actually has at least part of that charge.

WORKED EXAMPLE S2.2E

What are the formal charges of:

a hydrogen and fluorine in the hydrogen fluoride molecule

b carbon and nitrogen in the cyanide ion?

Answer

a The formal charge can be calculated as the difference between the number of valence electrons in the atom and the number of electrons that it 'owns' in a Lewis formula.

$$\text{formal charge} = \text{number of valence electrons} - \left(\text{number of non-bonding electrons} + \tfrac{1}{2}\,\text{number of bonding electrons}\right)$$

formal charge of H in HF $= 1 - [0 + (\frac{1}{2} \times 2)] = 0$

formal charge of F in HF $= 7 - [6 + (\frac{1}{2} \times 2)] = 0$

$\overset{0}{\text{H}}\,:\overset{0}{\underset{\cdot\cdot}{\overset{\cdot\cdot}{\text{F}}}}:$

b The cyanide ion has the Lewis structure:

$[:\text{C}\equiv\text{N}:]^-$

The carbon atom has two non-bonding electrons (one lone pair) and three electrons from the six in the triple bond, giving a total of five. The number of valence electrons on a neutral carbon atom is 4, so $4 - 5 = -1$.

Nitrogen has two non-bonding electrons and three electrons from the triple bond. Because the number of valence electrons on a neutral nitrogen atom is 5, its formal charge is $5 - 5 = 0$. Hence, the formal charges on the atoms in the Lewis structure of CN^- are:

$[:\overset{-1}{\text{C}}\equiv\overset{0}{\text{N}}:]^-$

HL ONLY

The concept of formal charge can be used to identify the correct Lewis formula for a molecule or the most important resonance structure for a resonance stabilized molecule or ion.

Determine the preferred Lewis formula using the following rules:

- Small (or zero) formal charges on individual atoms are better than large ones.
- Formal charges of same sign on adjacent atoms make an unfavourable structure.
- For better stability, more electronegative atoms should bear negative formal charge.

LINKING QUESTION

What are the different assumptions made in the calculation of formal charge and of oxidation states for atoms in a species?

12 The following are three possible Lewis formulas for the thiocyanate ion, NCS^-:

$[:\ddot{N}-C\equiv S:]^-$ $[\ddot{N}=C=\ddot{S}]^-$ $[:N\equiv C-\ddot{S}:]^-$

Determine the formal charges of the atoms in each Lewis formula.

Neutral nitrogen, carbon and sulfur atoms have five, four and six valence electrons, respectively.

Covalent network structures

◆ **Covalent network structure**: A regular arrangement, usually three-dimensional, of covalently bonded atoms that extends throughout the substance.

◆ **Allotrope**: One of the different structural forms of an element.

Covalent network structures usually consist of a three-dimensional lattice of covalently bonded atoms. These atoms can be all of the same type, as in silicon and carbon (diamond and graphite), or of two or more different elements, as in silicon dioxide (silicon(IV) oxide). Covalent network structures may be two-dimensional, as in graphite and graphene, or three-dimensional, as in, for example, diamond and silicon dioxide.

Carbon

Allotropes are crystalline forms of the same element in which the atoms (or molecules) are bonded differently. Pure carbon exists in several allotropic forms: diamond, graphite, graphene and a family of related molecules known as the fullerenes.

Graphite is the most thermodynamically stable allotrope of carbon under standard conditions.

The differences in physical properties of diamond, graphite, graphene and the fullerene C_{60}, summarized in Table S2.16, are due to the large differences in the bonding between the carbon atoms in the allotropes.

■ **Table S2.16** The physical properties of diamond, graphite, graphene and carbon-60

Allotrope	*diamond*	*graphite*	*graphene*	C_{60}
Colour	colourless and transparent	black and opaque	nearly transparent	black (in large quantities)
Hardness	very hard	very soft and slippery	stronger and stiffer than diamond	soft
Electrical conductivity	very poor—a good insulator	good along the plane of the layers	higher than graphite	very poor—a good insulator
Density	3.51 g cm^{-3}	2.23 g cm^{-3}	2.27 g cm^{-3}	1.72 g cm^{-3}
Melting point / K	3823	sublimation point 3925–3970	sublimation point 5100–5300	sublimation point 800
Boiling point / K	5100			

Diamond

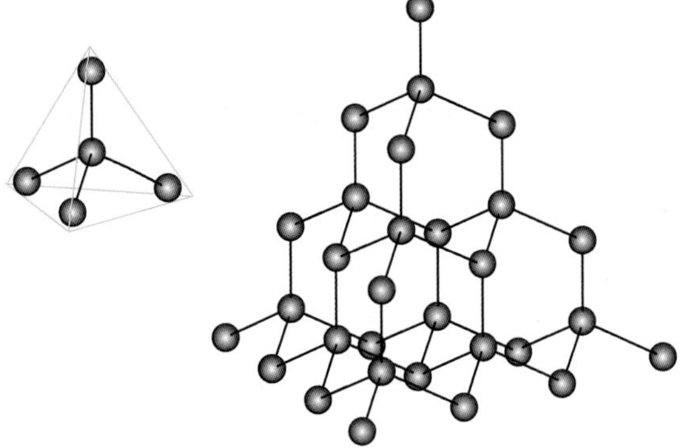

■ **Figure S2.44** Structure of diamond: tetrahedral unit and lattice

In diamond, each carbon atom is tetrahedrally bonded to four other carbon atoms by localized single covalent bonds (Figure S2.44). A very rigid three-dimensional covalent network structure is formed. The bond angles are 109.5°. Each carbon atom is sp^3 hybridized and has a coordination number of four because there are four neighbouring carbon atoms near to it.

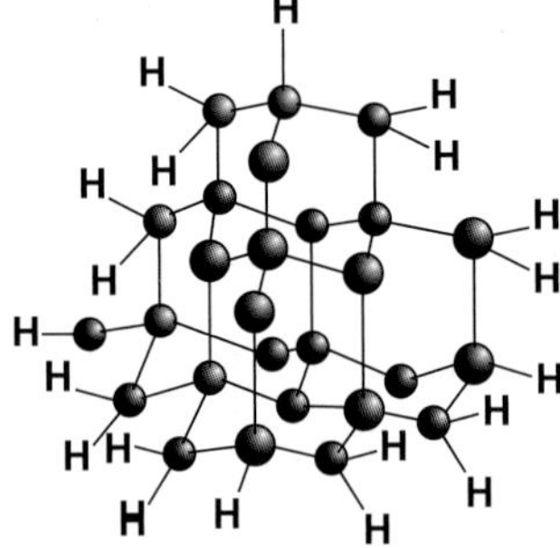

■ **Figure S2.45** The carbon atoms on the edges of a diamond crystal have hydrogen atoms bonded to them

Going further

Although the carbon atoms inside a diamond crystal are all bonded to four other carbon atoms, those on the flat surface of the side of a crystal have only three carbon atoms bonded to them.

When very clean diamond surfaces were studied with an electron microscope, it was discovered that the surface of a diamond crystal is covered with hydrogen atoms, each hydrogen atom singly bonded to a carbon atom (Figure S2.45). The carbon atoms on corners have two hydrogen atoms.

Graphite

In graphite, the carbon atoms are sp^2 hybridized and arranged in planar (flat) layers which are held together by weak London (dispersion) forces. Within the layers, the carbon atoms are arranged in regular hexagons. Each carbon atom is bonded to three other carbon atoms by strong covalent bonds. The fourth electron of each carbon atom occupies a p orbital. These p orbitals on every carbon atom in each planar layer overlap above and below the plane. Each layer has an extended delocalized pi (π) molecular orbital containing mobile electrons (Figure S2.46).

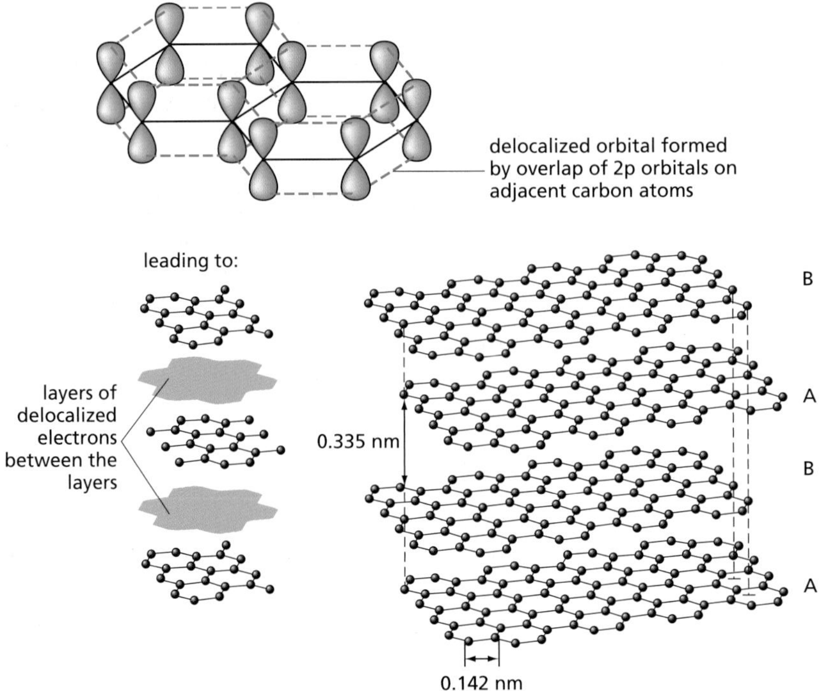

■ **Figure S2.46** The layered structure of graphite

The physical properties of graphite are directly related to its structure.

- High melting and sublimation points: there is strong covalent bonding throughout the layers of carbon atoms. A large amount of thermal energy is needed to overcome these strong bonds. The carbon–carbon bond length is in between that of a single and a double carbon–carbon bond, suggesting that there is a partial double bond character between carbon atoms within the layers.
- Softness: graphite is easily scratched because the London (dispersion) forces between the layers of carbon atoms (known as graphene) are weak. The layers of graphite can slide over each other and flake off when a force is applied. This is why graphite is used in pencils and as a lubricant.
- It is a good conductor of electricity along the planes: when a voltage is applied, the delocalized electrons (mobile electrons) can move along the layers.

Fullerenes

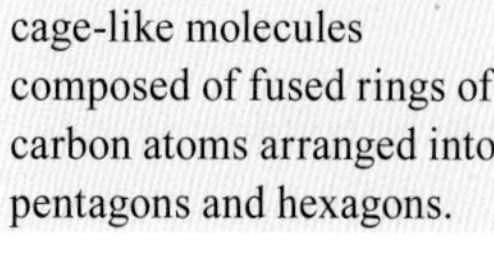
◆ **Fullerenes**: Closed cage-like molecules composed of fused rings of carbon atoms arranged into pentagons and hexagons.

C_{60} (buckminsterfullerene) was the first of a family of simple molecular forms of carbon known as **fullerenes** to be isolated. It was first prepared by very rapidly condensing vapour produced from graphite using a high-power laser in an inert atmosphere of helium maintained at low pressure. Mass spectrometry (see page 34) confirmed the presence of C_{60}, C_{70} and other fullerenes in the vapour. Fullerenes with 72, 76, 84 and up to 100 carbon atoms are commonly obtained. The smallest possible fullerene is C_{20}.

The atoms in a molecule of C_{60} are arranged into the shape of a truncated icosahedron: a football or soccer ball (Figure S2.47). This symmetrical structure has 60 vertices (corners) and 32 faces (12 pentagons and 20 hexagons). All the larger fullerenes have 12 five-membered rings which are 'isolated'—no pentagons are adjacent—and a variable number of six-membered rings. Fullerenes with fewer than 60 carbon atoms do not obey the isolated pentagon rule.

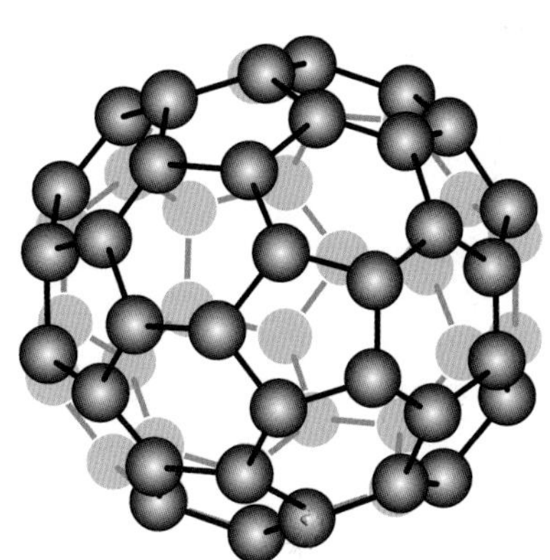
■ **Figure S2.47** Shape of the C_{60} molecule

The bonding in C_{60} is a series of alternating carbon–carbon double and single bonds (Figure S2.48). This arrangement of bonds is known as a conjugated system and would be expected to give C_{60} similar chemical properties to benzene.

However, the p orbital overlap inside and outside the curved surface is poor (Figure S2.49). Inside, the orbital lobes are too close and repulsion occurs; outside, the orbital lobes are too far away from each other for effective overlap. The molecule's carbon–carbon double bonds therefore behave like those of an alkene and it undergoes a variety of addition reactions. The π electrons cannot be transferred from one molecule to another when a potential difference is applied to a sample of C_{60}, and hence C_{60} is non-conducting. C_{60} also has a relatively low boiling point as there are only London (dispersion) forces to overcome.

C_{70} is another well-studied fullerene. It takes the shape of a rugby ball (Figure S2.50).

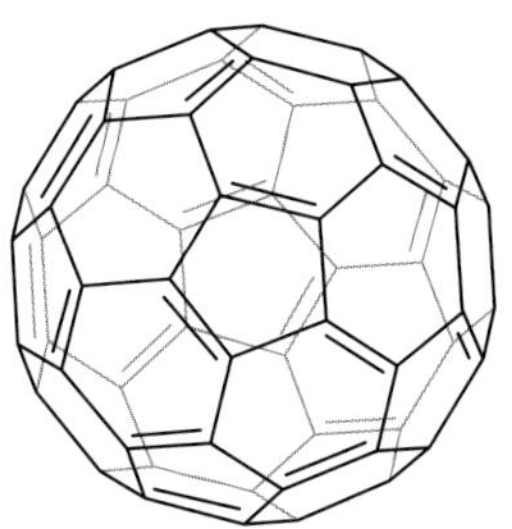
■ **Figure S2.48** The alternating single and double carbon–carbon bonds in C_{60}

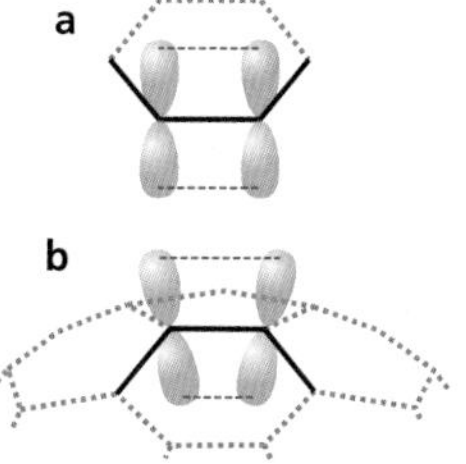

■ **Figure S2.49** p orbital overlap in **a** benzene (planar) and **b** carbon-60 (non-planar)

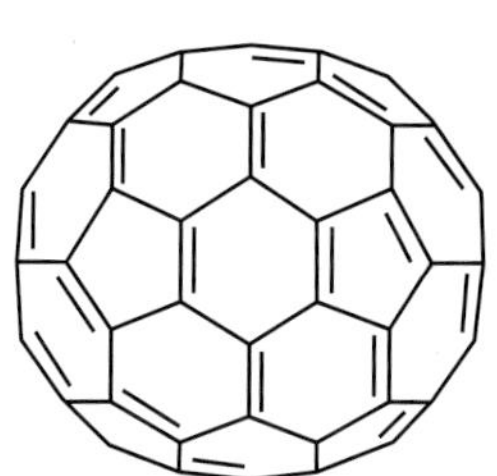
■ **Figure S2.50** Structure of C_{70}

Does competition between scientists help or hinder the production of knowledge?

The initial synthesis and detection of C_{60} by British chemist Sir Harry Kroto (1939–2016) and American physicist Richard Smalley (1943–2005) involved intense and very productive collaboration between two scientists with different but complementary scientific skills and areas of expertise. Their work, with that of Robert Curl, was recognized by the Nobel Prize in Chemistry in 1996.

However, a bitter personal dispute arose between Smalley and Kroto over the discovery of C_{60} and the events leading to the proposal of a structure for it. They stopped collaborating and instead worked independently with their own research groups.

Kroto likened the situation to the Japanese story *Rashomon*, in which witnesses to a crime all have different recollections of the events. Shared absolute truth for the group becomes the subjective truth of the individuals.

Graphene

Graphene is a form of carbon that consists of a monolayer of graphite and can be considered to be the structural element of fullerenes and nanotubes. It can also be viewed as a very large aromatic molecule formed from many fused benzene molecules (Figure S2.51).

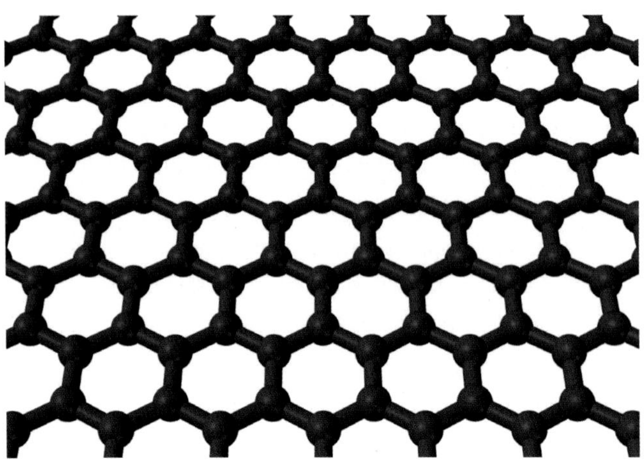

■ **Figure S2.51** Graphene, showing the atomic-scale honeycomb lattice

Graphene shares some of the properties of graphite but these properties are often enhanced due to quantum effects. For example:

- Graphene is the most chemically reactive form of carbon.
- Single sheets of graphene burn at very low temperatures and are much more reactive than graphite due to the dangling bonds of carbon atoms on the edges.
- Graphene is extremely strong for its mass.
- For a given amount of material, graphene conducts electricity and heat much better than graphite.
- Graphene is transparent, unlike graphite.

Potential applications of graphene include miniaturized electrical circuits and components (including transistors), touchscreens, solar cells and energy storage devices.

Common mistake

Misconception: *All covalent substances are electrical insulators because the valence electrons are either shared or present as lone pairs.*

Graphite and graphene are exceptions. They conduct along the layers because of delocalized π electrons.

13 a State the number of bonding pairs around each carbon atom in diamond.
   b State the coordination number of each carbon atom.
   c Draw a sketch showing how each carbon atom in a diamond structure is related to its immediate neighbours. Include bond angles.

14 a Sketch part of a single layer of the graphite structure, in plan view.
   b State the coordination number of each atom.
   c Compare the interatomic distance within a layer with the interatomic distance between layers, and comment on the difference.

15 The C–C bond length in a graphite layer is 0.142 nm, while in diamond it is 0.154 nm and in a typical double-bond compound such as ethene, C_2H_4, it is 0.134 nm. Comment on this data.

16 State the type of bonding and describe the lattice structure of solid carbon-60.

Going further

Nanotubes

Following the discovery of C_{60}, a family of structurally related carbon nanotubes was discovered. These resemble a rolled-up sheet of graphene, with the carbon atoms arranged in repeating hexagons (Figure S2.52). If pentagons are present in the structure, they may be closed at either end. Nanotubes can be single-walled or multi-walled and of different lengths and radii.

Bundles of carbon nanotubes have tensile strengths up to 100 times greater than iron because of the strong covalent bonding within the walls of the nanotube. Some nanotubes are conductors and some are semiconductors because the behaviour of electrons is very sensitive to the dimensions of the tube.

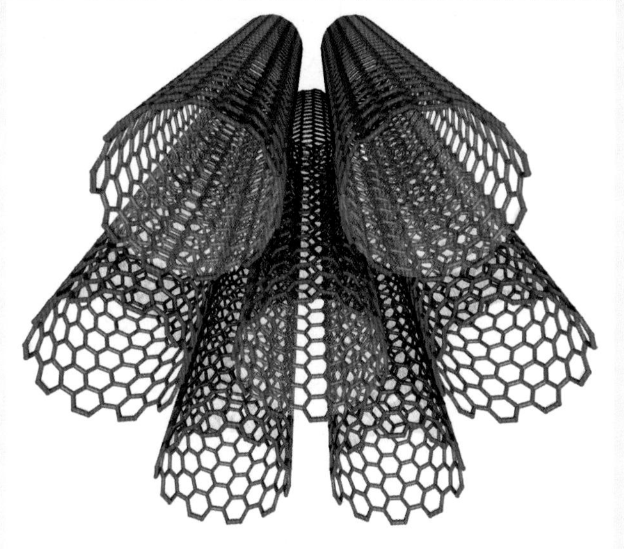

■ **Figure S2.52** Carbon nanotubes

Top tip!

Silicon has a structure similar to diamond based on interconnected tetrahedra.

Silicon and its compounds

Silicon

Silicon is a metalloid element that exists as a giant three-dimensional covalent network structure. In bulk, it is a shiny silver solid that resembles a metal. It is hard but brittle and does not feel cold to the touch, as a metal does, since it is a poor thermal conductor. It has low density, which is a typical property of non-metals. The electrical conductivity of silicon is poor under standard conditions but can be greatly improved by adding small amounts of gallium or arsenic.

ATL S2.2E

Silicon carbide (SiC) has a covalent network structure based on diamond.

- Use the internet to find its structure and draw an annotated diagram.
- Predict the properties of silicon carbide and check against published data.
- Find out about the uses of silicon carbide and relate them to its structure and bonding.

Silicon dioxide

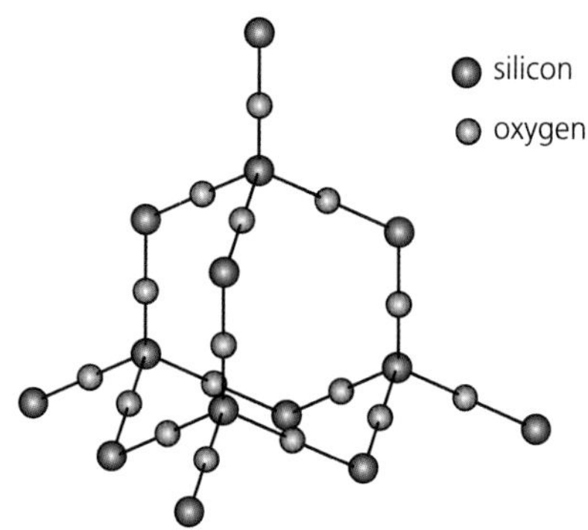

■ **Figure S2.53** Structure of quartz (silicon dioxide)

The most common form of silicon dioxide (silica) is quartz, which has a structure similar to diamond in which tetrahedral SiO_4 groups are bonded together by Si–O–Si bonds (Figure S2.53). Silicon dioxide has physical properties that are very similar to diamond. It is hard, transparent and has high melting and boiling points. It is the basis of a wide range of ceramics. A common impure form of silicon dioxide is sand, which is coloured yellow by the presence of iron(III) oxide.

17 Describe the structure and bonding in silicon dioxide and explain why it is suitable for making bricks for the inside of a furnace.

Going further

Silicates

A silicon atom can form four single covalent bonds (σ bonds) to form the silicate unit, SiO_4^{4-}, which has a tetrahedral distribution of bonds (Figure S2.54).

Each of the four oxygen atoms in the SiO_4^{4-} ion has an unshared electron which it can use to bond with other SiO_4^{4-} units. Anions such as $Si_2O_7^{2-}$, $Si_3O_9^{6-}$ (Figure S2.55) and $Si_6O_{18}^{12-}$ occur in many silicate-based minerals.

Silicates may have long chains of linked tetrahedra. Some silicate minerals have single silicate strands of formula $(SiO_3)_n^{2n-}$ bonded to metal cations which balance the negative charges. Asbestos has the double-stranded structure shown in Figure S2.56.

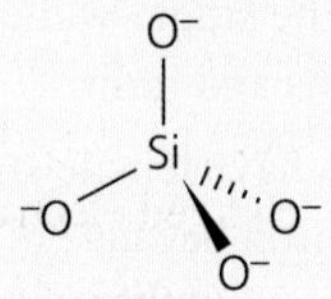

■ **Figure S2.54** The $[SiO_4]^{4-}$ ion

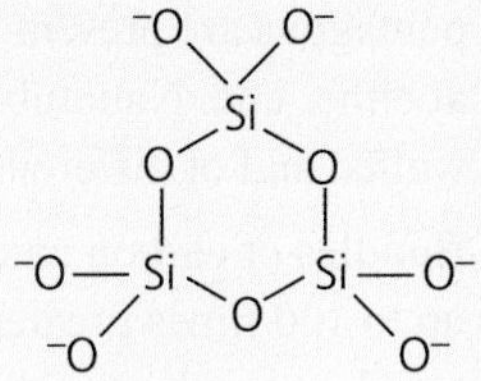

■ **Figure S2.55** The $[Si_3O_9]^{6-}$ ion

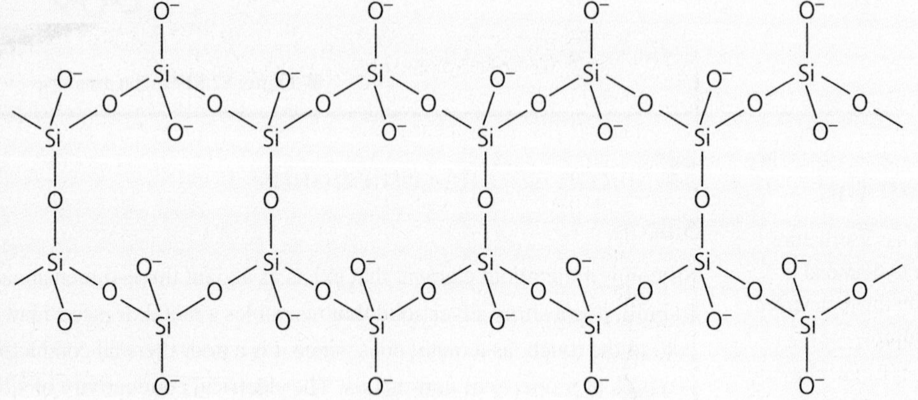

■ **Figure S2.56** Double strands of silicate tetrahedra

LINKING QUESTION

Why are silicon–silicon bonds generally weaker than carbon–carbon bonds?

◆ **Intermolecular forces:** Weak forces between molecules.

Intermolecular forces

Simple molecular compounds are formed by the covalent bonding of a relatively small number of atoms. The bonds holding the atoms together in molecules are strong covalent bonds. The molecules interact in the solid and liquid states through much weaker **intermolecular forces** (Figure S2.57). These forces, responsible for the physical properties of simple molecular compounds, include London (dispersion) forces; dipole–dipole, dipole–induced dipole and ion–dipole interactions; and hydrogen bonds.

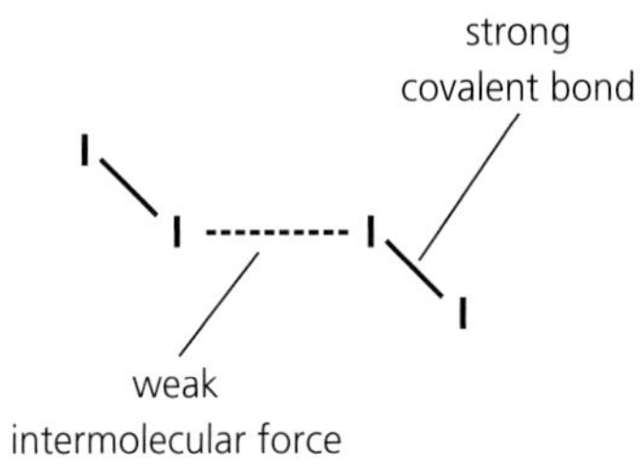

strong covalent bond

weak intermolecular force

■ **Figure S2.57** Strong covalent bonds and weak intermolecular forces

Common mistake

Misconception: *Intermolecular forces are the forces within a molecule.*

Intermolecular forces are the weak attractive forces between molecules. Those within the molecule are referred to as **intra**molecular forces.

LINKING QUESTION

How do the terms 'bonds' and 'forces' compare?

Physical properties of simple molecular compounds

- Simple molecular compounds are gases, liquids or soft solids with low melting points.
- Most simple covalent compounds whose intermolecular forces are mainly London (dispersion) forces (for example, iodine and the halogenoalkanes) are poorly soluble in water, but are soluble in less polar or non-polar solvents.
- Simple covalent compounds whose intermolecular forces are hydrogen bonds (for example, amines, carboxylic acids, amides and sugars) are often soluble in water provided they have relatively low molar mass or can form multiple hydrogen bonds.
- Generally, simple molecular compounds do not conduct electricity when molten. This is because they do not contain ions. Molecules are electrically neutral and are not attracted to charged electrodes.
- A number of soluble simple molecular compounds undergo hydrolysis – a chemical reaction with water. The molecular substance is completely or partially converted into ions. Substances that undergo hydrolysis with water include chlorine, Cl_2, ammonia, NH_3, hydrogen chloride, HCl, and many non-metal chlorides such as PCl_3, PCl_5, NCl_3 and SCl_2.

Inquiry 1: Exploring and designing

Pilot methodologies

If you have a plan for a large-scale chemistry experiment with a large range for the independent variable and many trials—such as testing the electrical conductivity of equimolar solutions of soluble ionic salts—it may be appropriate to do a pilot study first.

In a pilot study, you pre-test the methodology you plan to use in the full study, using a smaller sample and smaller range. The pilot helps you find any problems with your planned methodology (such as replacing an ammeter with a conductivity meter) so you can make changes and improvements before you carry out the full study. Your pilot can also give you an idea of what the results of your full study might be.

LINKING QUESTION

What experimental data demonstrate the physical properties of covalent compounds?

London (dispersion) forces

Noble gases can be liquefied and solidified at very low temperatures. It follows that there must be weak attractive forces between their atoms which keep the atoms together in the liquid and solid states.

◆ **London (dispersion) force:** An attractive force between atoms or molecules due to the formation of small instantaneous and induced dipoles.

London (dispersion) forces are short-range attractive forces that arise due to the formation of temporary dipoles caused by random fluctuations in the electron density in a molecule or an atom. The random and independent movement of electrons means that, at one instant, there may be more electrons on one side of the molecule or atom than the other, leading to a separation of charge.

Top tip!

London forces operate between all molecules. In non-polar molecules they are the only attractive forces acting when the substance is in the liquid or solid state.

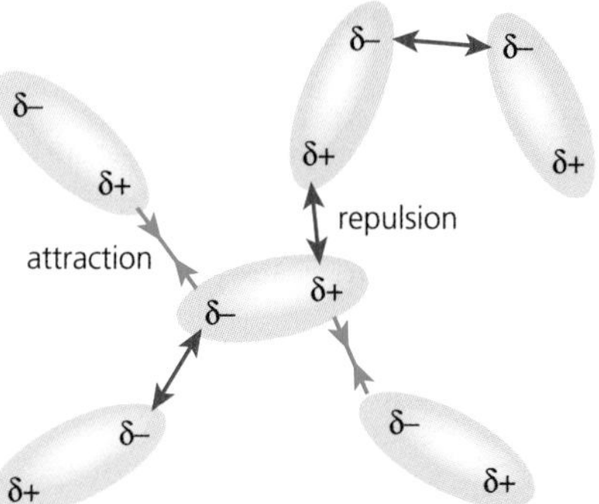

■ **Figure S2.58** Attraction and repulsion between molecules with dipoles

Top tip!

London forces account for the deviations of the noble gases and the halogens from ideal gas behaviour.

Over an averaged period of time, the electron density is spread evenly around the atom or molecule. At any instant, however, the electron density distribution may be asymmetrical, giving the atom or molecule a temporary dipole (Figure S2.59).

The formation of a temporary dipole in one atom or molecule causes electrons in the nearest atom or molecule to move by repulsion, resulting in the formation of another temporary dipole. This process is called induction (Figure S2.60) and the formation of induced dipoles is rapidly transmitted through the molecular lattice.

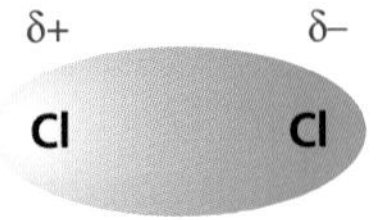

At this instant in time, more of the electron cloud is at one end of the molecule; the Cl_2 molecule has an instantaneous dipole

■ **Figure S2.59** The formation of a temporary dipole in a chlorine molecule

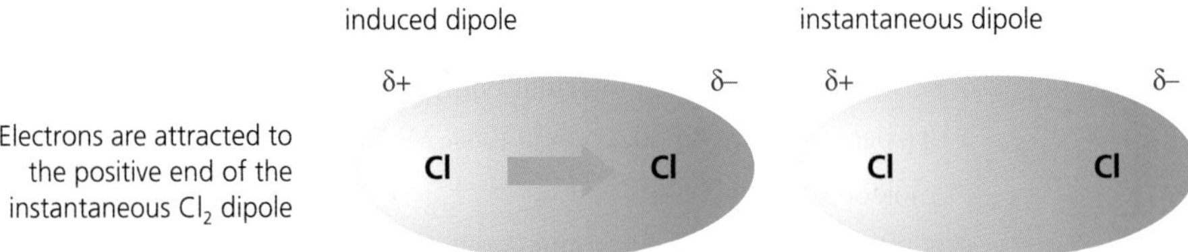

■ **Figure S2.60** An instantaneous dipole–induced dipole attraction

The strength of dispersion forces depends on the **polarizability** of the atom. The larger the molecule or atom, the greater the volume occupied by the electrons, the greater the distortion of the electron density by an electric field, and the larger the size of the temporary dipole. Hence, London forces increase with molar mass.

◆ **Polarizability**: A measure of the response of the electron cloud of an atom or molecule to an electric charge (or electric field).

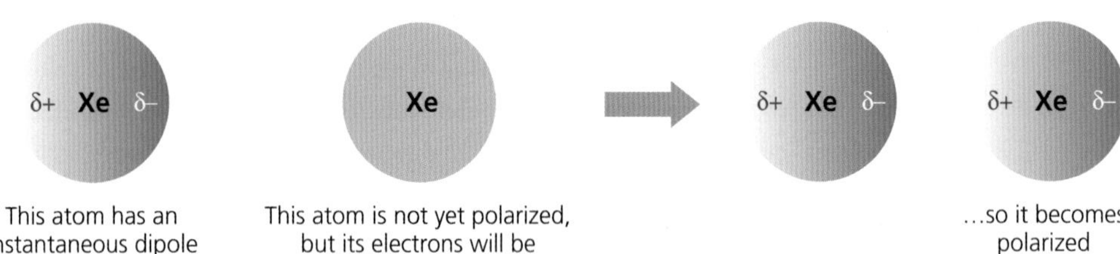

■ **Figure S2.61** The polarization of xenon atoms leads to production of induced dipole forces between atoms

Top tip!

Note that induced or temporary dipoles occur in all molecules, whether or not they have a permanent dipole and whether or not they hydrogen bond with one another.

Top tip!

The strength of the attraction between induced dipoles decreases rapidly with distance and so this is only a very short-range force. An ideal gas assumes London (dispersion) forces do not operate.

Table S2.17 shows the effects of number of electrons and molecular shape in determining the size or extent of dispersion forces.

■ **Table S2.17** Boiling points of some molecular elements and compounds where the only intermolecular forces are London (dispersion) forces

Molecule	Boiling point / °C	Comments
CH_4	−162	Both molecules are tetrahedral. The number of electrons increases from 10 in CH_4 to 18 in SiH_4.
SiH_4	−112	
F_2	−188	All these diatomic molecules are linear. The number of electrons increases down the group: 18 in F_2; 34 in Cl_2; 70 in Br_2.
Cl_2	−34	
Br_2	+58	
CH_4	−162	As the alkane molecules become longer there is a larger surface area of contact as well as more electrons (eight for each additional CH_2 group).
C_2H_6	−89	
C_3H_8	−42	
C_4H_{10}	0	
$C(CH_3)_4$ (central C bonded to four CH_3 groups)	+10	Both these molecules have the same number of electrons. The first is tetrahedral (the electron clouds around the atoms make it almost spherical) and the second is linear. The second has a greater area of contact.
$CH_3CH_2CH_2CH_2CH_2CH_3$	+36	

Dipole–dipole forces

◆ **Dipole–dipole force:** The attractive force between polar molecules due to their dipole moments (polarities).

A **dipole–dipole force** exists between polar molecules in solids and liquids because the positive end of the dipole of one molecule will electrostatically attract the negative end of the dipole of another molecule (Figure S2.62).

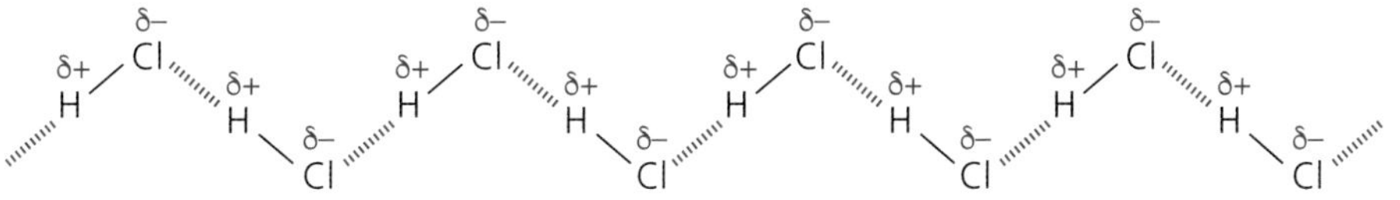

■ **Figure S2.62** Dipole–dipole forces in HCl(s)

The strength of a dipole–dipole force depends on the molecular polarity. The more polar the molecule, the greater the strength of the dipole–dipole force (as measured by the dipole moment (in debyes)). For polar substances with similar molar masses, the higher the molecular polarity, the stronger the dipole–dipole attractions and the higher the boiling point, as shown in Table S2.18.

■ **Table S2.18** Dipole moments and boiling points for molecules having similar molar masses

Name of substance	Formula	Relative molar mass (g mol^{-1})	Dipole moment / D	Boiling point / K
propane	$CH_3CH_2CH_3$	44	0.1	231
methoxymethane	CH_3OCH_3	46	1.3	249
ethanenitrile	CH_3CN	41	3.9	355

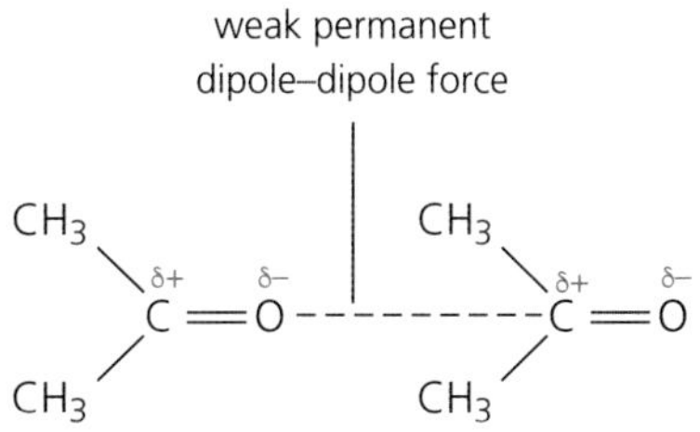

Dipole–dipole forces are often stronger than London forces. For example, consider propanone (Figure S2.63) (CH_3COCH_3, M = 58 g mol^{-1}) and butane ($CH_3CH_2CH_2CH_3$, M = 58 g mol^{-1}). More energy is needed to break the stronger intermolecular forces between polar propanone molecules than between non-polar butane molecules. The dipole–dipole forces between propanone molecules are strong enough to make this substance a liquid under standard conditions. There are only weak dispersion forces between butane molecules, so butane is a gas under standard conditions.

■ **Figure S2.63** Dipole–dipole forces between propanone molecules

In comparing the relative strengths of intermolecular forces, the following generalizations are useful:

- When molecules have very different molar masses, London forces are more significant than dipole–dipole forces. The molecule with the largest molar mass has the strongest intermolecular attractions.
- When molecules have similar molar masses, dipole–dipole forces are more significant. The most polar molecule has the strongest intermolecular attraction.

WORKED EXAMPLE S2.2F

Explain the trend in the boiling points of the hydrogen halides shown in the table.

Molecule	Molar mass (g mol^{-1})	Difference in electronegativity	Melting point / K	Boiling point / K
HCl	36.5	1.0	159	188
HBr	81.0	0.8	186	207
HI	128.0	0.5	222	238

Answer

Stronger intermolecular forces result in higher boiling points. The strength of the dipole–dipole interactions increases with the polarity of the hydrogen–halogen bond and therefore with the difference in electronegativity between the hydrogen and the halogen atoms. The strength of the dispersion forces increases with the number of electrons.

The data in the table show that electronegativity differences decrease from HCl to HI and, therefore, dipole–dipole forces decrease. This trend suggests that the boiling points should decrease from HCl to HI.

This prediction conflicts with the data so London forces need to be considered. The number of electrons in a molecule increases from HCl to HI and so the strength of the London forces increases too. Therefore, the boiling points should increase from HCl to HI, in agreement with the data. This analysis suggests that dispersion forces dominate dipole–dipole interactions for these molecules.

Ion–dipole forces and dipole–induced dipole forces

An ion–dipole attraction (Figure S2.64) is a weak attraction between an ion and a polar molecule.

For example, when an ionic substance is dissolved in water, hydrated anions and cations are formed. Ion–dipole forces then develop between these anions and cations and the partially positive or negative ends of polar water molecules.

A dipole–induced dipole attraction (Figure S2.65) is a weak attraction that results when a polar molecule induces a dipole in an atom or in a non-polar molecule, by polarizing the electrons in the non-polar species.

An example is when a hydrogen chloride molecule (polar) interacts with an iodine molecule (non-polar) in the gas phase. There is a weak interaction when they approach each other.

A non-polar molecule (for example O_2) may be polarized by the presence of an ion (for example Fe^{2+}) and becomes an induced dipole. The interactions between them are called ion–induced dipole interactions. The strength of these interactions depends upon the charge on the ion and how easily the non-polar molecule gets polarized.

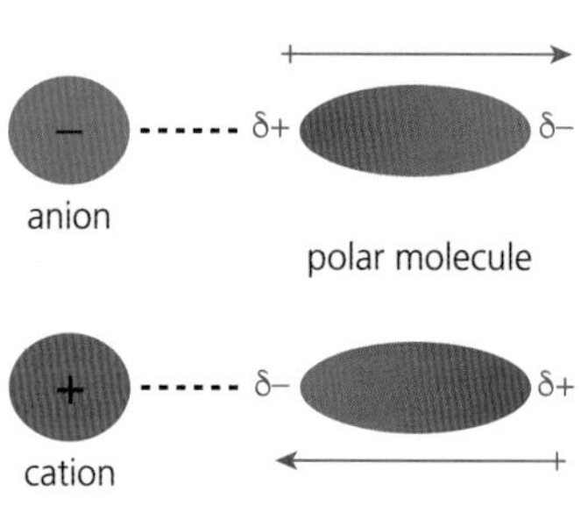

■ **Figure S2.64** Ion–dipole forces

δ+ δ−
A B
permanent dipole (polar molecule)
non-polar molecule
δ+ δ−
A B
δ+ δ−
permanent dipole (polar molecule)
induced dipole in non-polar molecule

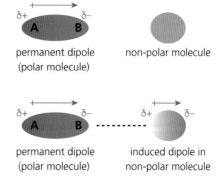

■ **Figure S2.65** Dipole–induced dipole forces

Top tip!

The term 'van der Waals forces' includes dipole–dipole, dipole–induced dipole and London (dispersion) forces.

LINKING QUESTION

To what extent does a functional group determine the nature of the intermolecular forces?

LINKING QUESTION

To what extent can intermolecular forces explain the deviation of real gases from ideal behaviour?

Hydrogen bonds

◆ **Hydrogen bond:** A type of intermolecular force involving electrostatic attraction between molecules where one molecule or both have hydrogen atoms bonded directly to nitrogen, oxygen or fluorine atoms.

A **hydrogen bond** (Figure S2.66) may be described (in a simple model) as the electrostatic attraction between a hydrogen atom bonded directly to a nitrogen, oxygen or fluorine atom and a lone pair on the nitrogen, oxygen or fluorine atom of a neighbouring molecule (in the liquid or solid state). (The lone pair on the nitrogen, oxygen or fluorine atom is also involved in the formation of the hydrogen bond by being partially shared with the hydrogen atom.)

Top tip!

Nitrogen, oxygen and fluorine are three atoms with relatively small atomic radii and relatively high values of electronegativity. The hydrogen atom is small and can closely approach the lone pair of electrons, leading to a strong attraction.

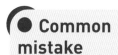

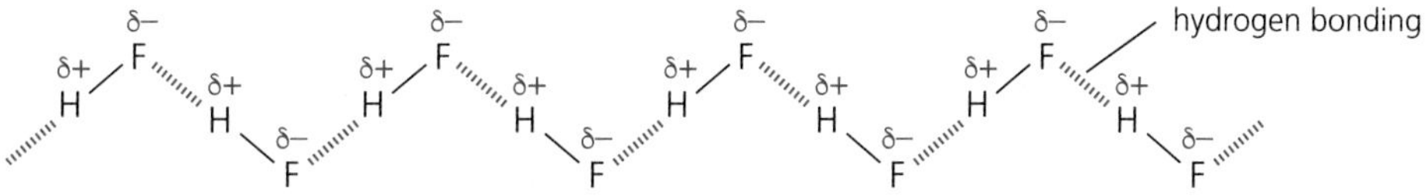

■ **Figure S2.66** Hydrogen bonding in liquid hydrogen fluoride: the dotted lines represent the intermolecular hydrogen bonds

Common mistake

Misconception: *Intermolecular forces are due to the difference in polarity of the molecules.*

Hydrogen bonds and dipole–dipole forces arise due to bond polarity (permanent dipoles).

Hydrogen bonding is (as a first approximation) a very strong permanent dipole–dipole attraction between molecules. The hydrogen bond strength varies but it is approximately ten times as strong as other dipole–dipole forces, and is only about ten times weaker than a typical covalent bond.

In the ammonia molecule, the nitrogen atom has one lone pair of electrons which means that the nitrogen in each molecule can form a single hydrogen bond (Figure S2.67).

The nitrogen atom is larger and less electronegative than the fluorine atoms so the hydrogen bonding in ammonia is weaker than that in hydrogen fluoride.

A water molecule has two lone pairs, each of which can form a hydrogen bond with another water molecule (Figure S2.68).

Since there are twice as many hydrogen bonds per molecule, the collective strength of the hydrogen bonds in water is greater than the strength of those in hydrogen fluoride.

LINKING QUESTION

How can advances in technology lead to changes in scientific definitions such as the updated IUPAC definition of the hydrogen bond?

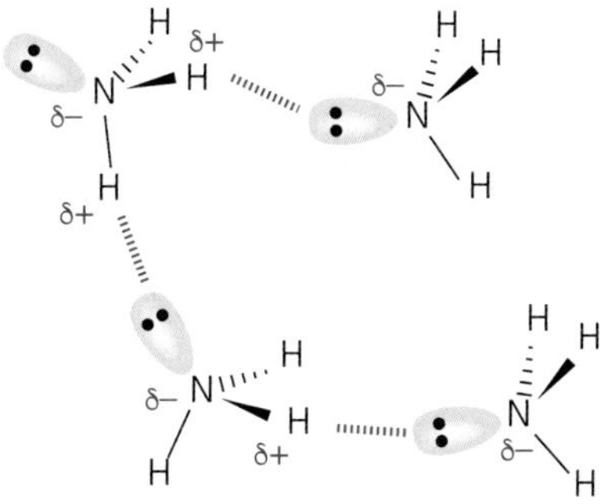

■ **Figure S2.67** Hydrogen bonding in ammonia

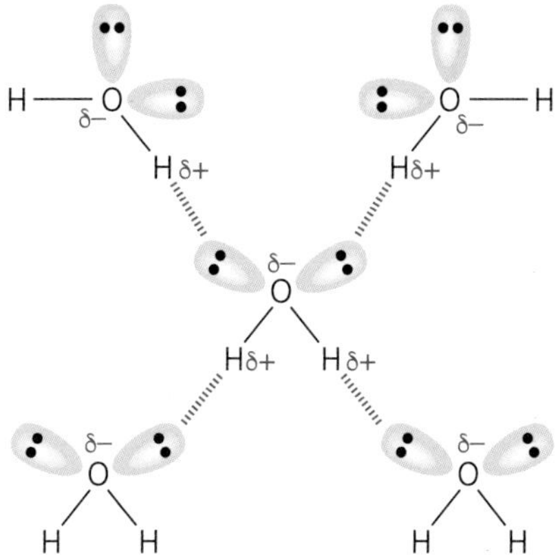

■ **Figure S2.68** Hydrogen bonding in water

What is the role of nomenclature and terminology in the chemical community?

The role of IUPAC (International Union of Pure and Applied Chemistry) is interesting to consider in relation to theory of knowledge. One of the primary roles of IUPAC is to recommend:

> *unambiguous, uniform, and consistent nomenclature and terminology for specific scientific fields, usually presented as: glossaries of terms for specific chemical disciplines; definitions of terms relating to a group of properties; nomenclature of chemical compounds and their classes; terminology, symbols, and units in a specific field; classifications and uses of terms in a specific field; and conventions and standards of practice for presenting data in a specific field.*
>
> **https://iupac.org/what-we-do/recommendations/**

We have, then, a group of individual scientists working together to establish a set of terms and definitions which all other chemists are meant to use and follow. Does this raise any questions for you about whether this might limit the individual approach of any particular chemist? Or is the uniformity necessary in order to enable an individual to take part in the community or/and for the field itself to develop over time?

Effects of hydrogen bonding

Hydrogen bonding affects the physical properties of molecular substances as follows:

- The boiling points of water, ammonia, hydrogen fluoride and other molecules are anomalously high.
- The boiling points of alcohols are considerably higher than those of alkanes and, more significantly, ketones of similar molar mass.
- The solubility of simple covalent molecules such as ammonia, methanol and ethanoic acid in water is higher than expected.
- At atmospheric pressure, liquid water can be denser than ice (greatest density at 4 °C).
- The viscosity of liquids; for example, viscosity increases from ethanol, to ethane-1-2-diol, to propane-1,2,3-triol with one, two and three –OH groups.

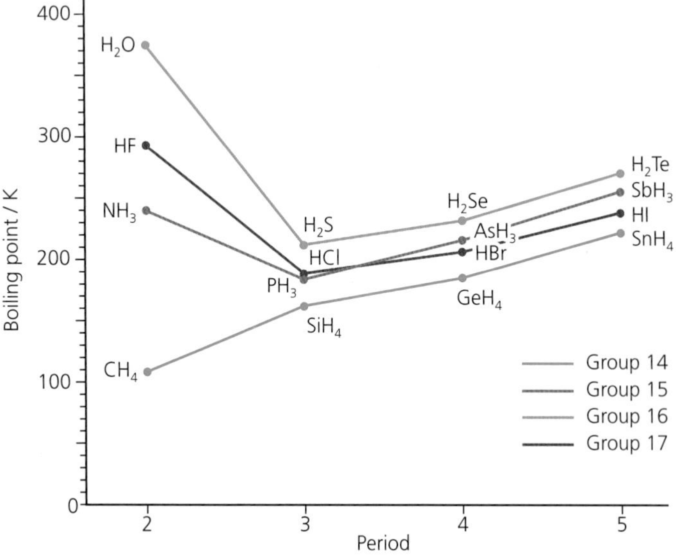

■ **Figure S2.69** The boiling points of the hydrides of elements in groups 14 to 17

Effect of hydrogen bonding on boiling point

The hydrides of group 14 elements (methane, CH_4; silane, SiH_4; germane, GeH_4; and stannane, SnH_4) have boiling points that increase regularly when the relative molecular mass increases. This is because the London forces increase as the number of electrons increases. The same pattern is not seen in groups 15, 16 and 17, even though London forces do increase with the number of electrons. The boiling points of ammonia, water and hydrogen fluoride are anomalously high due to the existence of hydrogen bonds.

Tool 3: Mathematics

Extrapolate and interpolate graphs

A line or curve of best fit is usually drawn to cover a specific range of measurements recorded in an experiment. If we want to predict other values within that range, we can usually do that with confidence by interpolation.

If we want to predict what would happen outside the range of measurements, we need to extrapolate by extending the line or curve. Lines are often extrapolated to see if they pass through the origin, or to find an intercept.

Predictions made by extrapolation should be treated with care, because it may be wrong to assume that the behaviour seen within the range of measurements also applies outside that range.

WORKED EXAMPLE S2.2G

Describe all the intermolecular forces that can occur between the molecules in each of the following liquids. Hence predict which will have the lowest boiling point and which will have the highest.

A $CH_3CH_2CH_2CH_2CH_3$

B $CH_3CH_2ClCH_2OH$

C $CH_3CH_2CHCl_2$

Answer

- Molecules of **A** will experience only London (dispersion) forces.
- Molecules of **B** will experience dipole–dipole forces (due to the –Cl and C–O polar bonds) and hydrogen bonding (due to the hydroxyl groups, –O–H), as well as London (dispersion) forces.
- Molecules of **C** will experience dipole–dipole forces (due to the dipoles of the C–Cl bonds not cancelling each other out vectorially) in addition to London (dispersion) forces.

Therefore, **A** will have the weakest intermolecular forces and hence the lowest boiling point; and **B** will have the highest boiling point.

ATL S2.2F

Hydrogen bonding results in water having some unusual physical properties including a higher than expected boiling point, high surface tension, high specific heat capacity, high latent heat of vaporization and ice being less dense than water. Outline how hydrogen bonding in ice (Figure S2.70) and liquid water accounts for these physical properties. Present your findings using software such as PowerPoint.

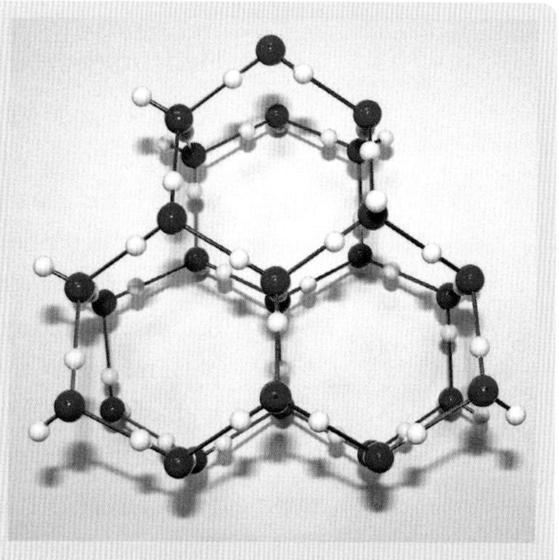

Figure S2.70 Structure of ice

Intermolecular versus intramolecular hydrogen bonding

An intermolecular hydrogen bond is formed during the interaction between two or more molecules, for example, between water and ethanol (Figure S2.71).

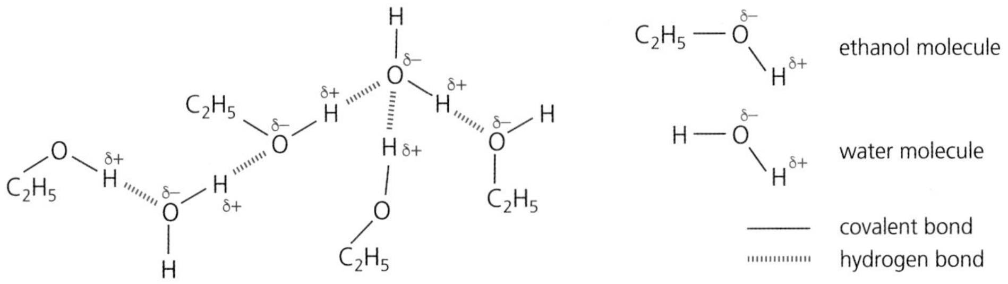

■ **Figure S2.71** Intermolecular hydrogen bonds between water and ethanol

An intramolecular hydrogen bond is formed within a single molecule. It usually occurs in organic compounds with suitable functional groups close to each other. If intramolecular hydrogen bonding is present, intermolecular hydrogen bonding will be less extensive (weaker) because the lone pairs and hydrogen atoms are involved in intramolecular hydrogen bonding.

In 2-nitrophenol, the hydrogen atom of the hydroxy group, –OH, can form an intramolecular hydrogen bond with the oxygen atom of the nitro group, $-NO_2$ (Figure S2.72).

In 4-nitrophenol, the hydrogen atom of the alcohol group cannot form an intramolecular hydrogen bond with the oxygen atom in the nitro group because they are too far apart. 4-nitrophenol forms intermolecular hydrogen bonds (Figure S2.73). The melting point of 4-nitrophenol is therefore higher than that of 2-nitrophenol, which is mainly associated via London forces.

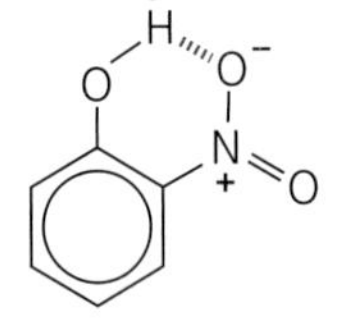

■ **Figure S2.72** Hydrogen bonding in 2-nitrophenol

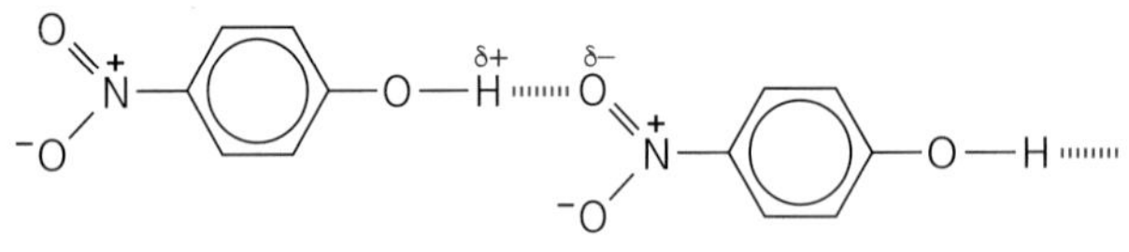

■ **Figure S2.73** Hydrogen bonding in 4-nitrophenol

ATL S2.2G

When the molar mass of organic acids, such as ethanoic acid, is measured in the gas phase or in a non-polar solvent, the molar mass is twice the theoretical value.

Use the internet to research this phenomenon and present an explanation with reference to hydrogen bonding and dimers, using diagrams to outline your arguments.

Hydrogen bonding and solubility

Water is a good solvent for liquids and gases composed of small polar molecules that can form hydrogen bonds with water molecules, such as small amines ($R–NH_2$) and small alcohols (R–OH), where R has a low molar mass. However, as the molar mass increases, the hydrocarbon portion of the molecule becomes larger and the amine ($-NH_2$) or hydroxyl functional group (–OH) is a smaller portion of the molecule. Since hydrocarbons are virtually insoluble in water, the solubility decreases as the molar mass of the amine, phenol or alcohol increases. Hydrocarbons are non-polar molecules and only weakly attracted to water molecules which form strong hydrogen bonds with each other. It is not energetically favourable to replace these strong intermolecular forces with weaker interactions.

Phenol, C_6H_5OH, has a polar hydroxyl group, –OH, which allows it to hydrogen bond to water molecules (Figure S2.74) making it slightly soluble in water. The solubility in water is limited by the bulky non-polar benzene ring. However, this part of the molecule interacts with molecules of non-polar solvents (for example, benzene), so phenol also dissolves in these. This is an example of the 'like dissolves like' principle: polar solvents dissolve polar and ionic substances; non-polar solvents dissolve non-polar substances.

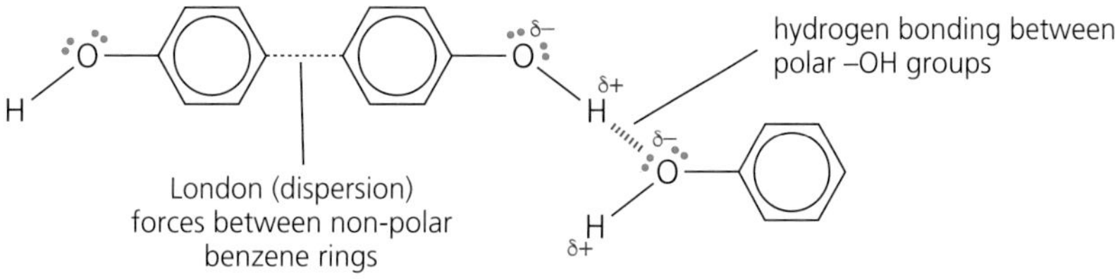

hydrogen bonding between polar –OH groups

London (dispersion) forces between non-polar benzene rings

■ **Figure S2.74** Explaining the solubility of phenol in polar and non-polar solvents

Ignoring entropy considerations, the solubility of solids in water (and other solvents) can be explained using simple energetics arguments. A solid is more likely to have a high solubility if hydration (for water) or solvation (for other solvents) is exothermic. This will be the case only if the strength of the attractions between the solvent molecules and solute ions or molecules is greater than the combined strengths of the attractions between molecules in the pure solid and between molecules in the pure solvent.

The dissolving of one liquid in another may be explained in a similar way to the dissolving of a solid in a liquid. For example, when water is added to tetrachloromethane, CCl_4, two layers separate out. The water molecules attract each other strongly, via hydrogen bonds, but have no tendency to mix with the molecules of tetrachloromethane (Figure S2.75).

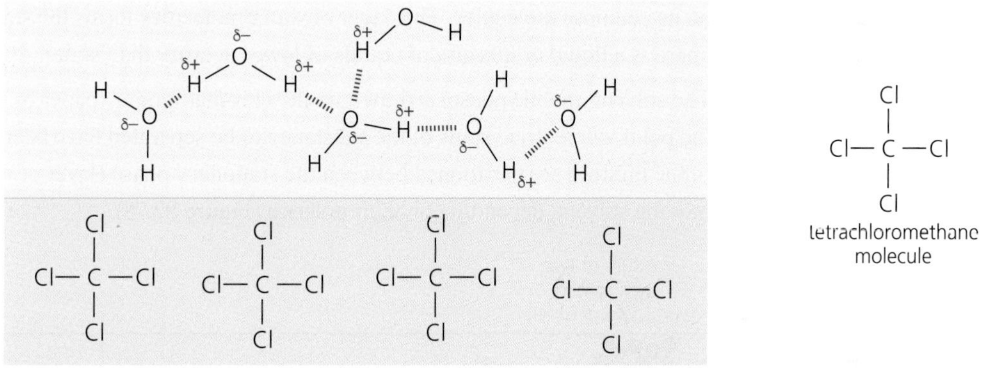

tetrachloromethane molecule

■ **Figure S2.75** The interface between water and tetrachloromethane, a pair of immiscible liquids

■ **Figure S2.76** Iodine introduced to hexane and water

It is not energetically favourable to replace the hydrogen bonds formed between water molecules with the weaker London forces formed between water and tetrachloromethane molecules. Water is a very poor solvent for substances that do not form ions or hydrogen bonds, because the attractions between its molecules are quite strong and molecules of a solvent coming between them would cause these interactions to break.

Gases—including oxygen, hydrogen, nitrogen and the noble gases—are generally only slightly soluble in water. A small number of gases (for example, SO_2, CO_2 and Cl_2) are highly soluble in water because they can form hydrogen bonds to water or because they chemically react with water to release ions.

18 Use simple energetics arguments to account for the observation that iodine (non-polar) is soluble in hexane (a non-polar solvent) but poorly soluble in water (a polar solvent).

Chromatography

Separation by **chromatography** involves placing a sample on a liquid or solid **stationary phase** and passing a liquid or gaseous **mobile phase** over it. This process is known as **elution**. The substances present in the mixture move (are eluted) at different speeds, which leads to their separation over a period of time and distance.

Chromatographic techniques can be classified by the nature of the interactions between the mobile and stationary phases. The two main forms of interaction between the **analyte** (mixture to be analysed) and the stationary and mobile phases are adsorption, used in thin-layer chromatography (TLC), and **partitioning**, used in paper chromatography (Figure S2.77).

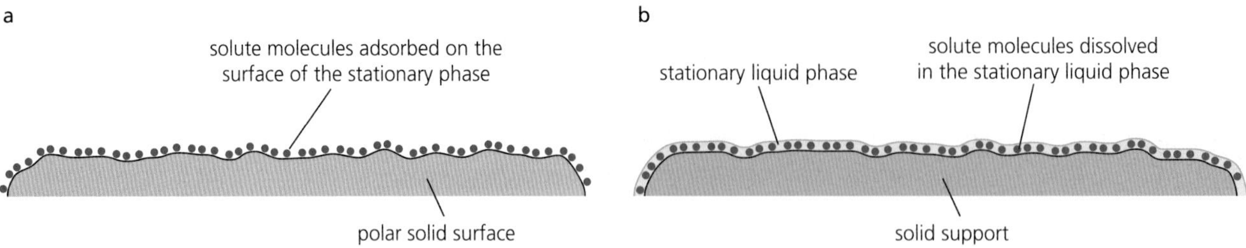

■ **Figure S2.77** Two mechanisms of separation: **a** adsorption chromatography and **b** partition chromatography

◆ **Chromatography**: The separation of a mixture by passing it through a medium in which its components move at different rates.

◆ **Stationary phase**: Solid material or liquid immobilized on an inert support, which attracts components in the mobile phase during chromatography.

◆ **Mobile phase**: Liquid (or gas) which moves through or along the stationary phase during chromatography.

◆ **Elution**: The process of washing a mixture through during chromatography.

◆ **Analyte**: The chemical species to be identified or quantified.

◆ **Partitioning**: The distribution of a solute between two immiscible solvents.

Paper chromatography

Chromatography paper contains water molecules hydrogen-bonded to the hydroxyl groups (–OH) on its cellulose molecules. This layer of water molecules forms the stationary phase and the mobile phase is a liquid or mixture of liquids of lower polarity than water.

The solvent (mobile phase) is drawn up the chromatography paper by capillary action. As it passes the point where dried spots of the substance to be separated have been applied, the substances in the mixture are partitioned between the stationary phase (layer of water molecules) and the moving solvent, depending on their polarity (Figure S2.78).

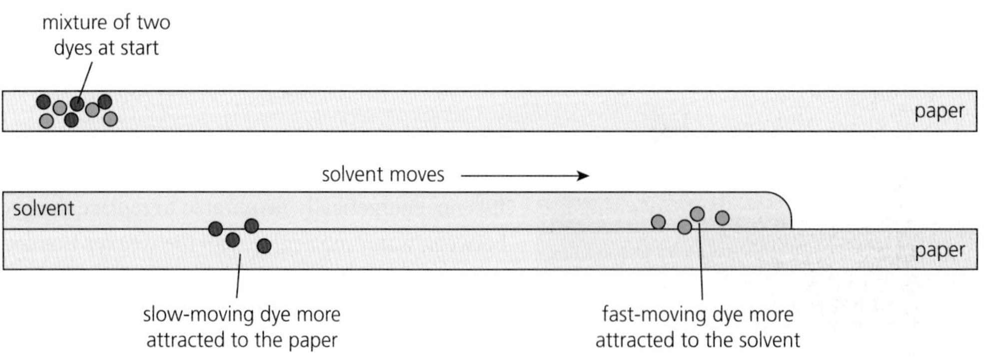

■ **Figure S2.78** The principles of paper chromatography

The process never reaches equilibrium because fresh solvent is constantly moving up the paper. The rate of elution of the spots is determined by the values of their partition coefficients (an equilibrium constant) between the solvent and water.

The greater the polarity of a substance, the smaller the value of its partition coefficient. This means it will spend a longer time dissolved in the layer of water molecules on the cellulose than in the moving solvent, and it will elute more slowly up the paper.

Thin-layer chromatography

In TLC, the stationary phase is a uniform thin layer of silica, SiO_2, or alumina, Al_2O_3, spread over the surface of a thin glass plate. The mobile phase is a solvent or solvent mixture.

The dry plate is spotted with a small sample of the mixture. The plate is then suspended in a tall vessel, such as a large beaker, with the spot just above the surface of the solvent. The beaker is often covered so that the plate is surrounded by solvent vapour. As the solvent front rises up the plate, the components separate. The plate may be removed to view the separated spots.

Top tip!

TLC has several advantages over paper chromatography: the chromatograms on the plates develop more quickly; it gives better resolution (separation) of the components of a mixture; and its results are more reproducible.

Tool 1: Experimental techniques

Paper and thin-layer chromatography

Two common methods for testing the purity of an organic product are thin-layer chromatography (TLC) and paper chromatography. TLC is similar to the familiar paper chromatography but is even more sensitive and better at separating mixtures for analysis.

A small sample of solid product is dissolved in a solvent and a few drops are placed on a glass plate covered in silica, SiO_2, or alumina, Al_2O_3, forming a spot, which is then dried. The plate is then mounted in a beaker with a small volume of solvent (or a mixture of solvents) which is able to run up the dry surface by capillary action.

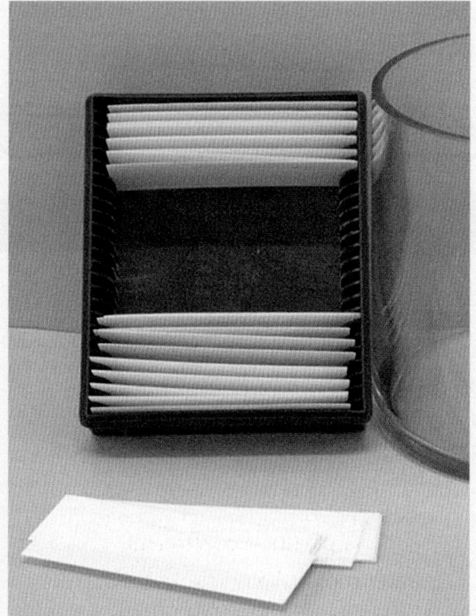

■ **Figure S2.79** TLC plates

Each compound in the sample mixture travels at a different rate due to the balance between their attraction to the stationary phase (the solid surface on the TLC plate) and their solubility in the solvent (mobile phase). The solvent will therefore carry different compounds different distances up the plate.

If the sample is pure, the spot moves up the plate at a uniform rate and will not separate.

If it is impure, more than one compound will be present and the spot will separate into two (or more) spots. Viewing the spots may require the use of UV light, which causes the spots on the silica to fluoresce or glow, and / or an indicator which reacts with the compounds to make them visible (for example, ninhydrin reacts with amino acids to give a deep blue or purple spot).

The distance the spot moves from its starting location divided by the distance moved by the solvent is known as the R_F value. The value of the retention / retardation factor (R_F) depends on the compound, solvent, temperature and nature of the TLC plate.

It is possible to use TLC (or paper chromatography) to identify a compound and this is usually done by putting a spot of pure, known compound next to the unknown one. If they move the same distance up the plate (and therefore have the same R_F), they are likely to be the same compound.

TLC can also be used to monitor the course of a reaction by taking small samples of an organic reaction mixture at regular intervals, spotting them on a TLC plate, and running alongside controls of the organic reactant and product.

Retardation factor

◆ **Retardation factor,** R_F, (in chromatography): The distance travelled by a given component divided by the distance travelled by the solvent front.

The usual way of identifying the compounds that make up the various spots on a chromatogram produced through paper chromatography or TLC is to measure their **retardation factor,** R_F, values (Figure S2.80) where:

$$R_F = \frac{\text{distance moved by component}}{\text{distance moved by solvent}}$$

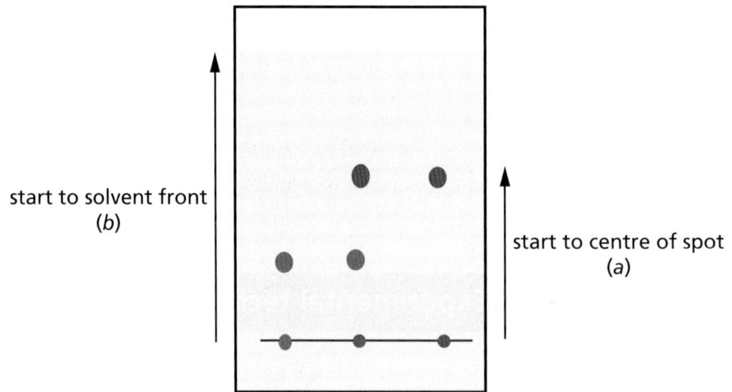

$$R_F = \frac{a}{b} = \frac{22\text{ mm}}{38\text{ mm}} = 0.58$$

■ **Figure S2.80** Calculation of the retardation factor, R_F

Note that R_F is not dependent on the distance travelled by the solvent. It does, however, depend on the nature of the mobile and stationary phases as well as the component of interest.

The R_F values of each component can be compared with those of known compounds or with those of reference compounds applied to the chromatography paper or TLC plate at the same time as the mixture.

Top tip!

Tables are available showing R_F values for a wide variety of substances obtained using particular solvents under standard conditions of temperature and solid phase. These allow the identification of unknown components.

Tool 1: Experimental techniques

Measuring length

The standard SI unit of length is the metre (m). However, a metre rule will be subdivided into centimetres (cm) and millimetres (mm). A metal or plastic rule is likely to be more accurate (better calibrated) than a wooden metre rule.

For more precise measurement of small lengths, such as the dimensions of a metal cube, a pair of vernier calipers or a micrometer screw gauge may be used.

Tool 3: Mathematics

Record uncertainties in measurements as a range (±) to an appropriate level of precision

You can never be 100% certain of any measurement you take in science. There will always be a level of uncertainty or estimation about it. We take this into account using uncertainties. It is important to remember that this method is still just an approximation, and we can never be 100% certain of the true value we are measuring.

There are two types of apparatus to be considered: analogue apparatus and digital apparatus. An analogue piece of apparatus will have a scale that you will need to read and interpret. Common examples of analogue apparatus around the laboratory will be measuring cylinders and burettes.

The accepted way of recording the precision for a reading on an analogue piece of equipment is to use half the smallest division. For example, a 25 cm^3 measuring cylinder can measure to 2 cm^3. Half the smallest division is therefore 1 cm^3. So, if you were trying to measure 22 cm^3, you could only be certain that you had between 21 cm^3 and 23 cm^3. This would be expressed as 22 cm^3 ± 1 cm^3.

For digital equipment, the uncertainty is the smallest value. If a digital balance can record to 0.001 g and you have measured 15.381 g, you can be certain that the mass is somewhere between 15.380 g and 15.382 g or 15.381 g ± 0.001 g.

Going further

Two-dimensional chromatography

To separate more complex mixtures, it is sometimes necessary to use a variation of simple chromatography. After the initial separation process, the paper is allowed to dry and then rotated 90 degrees. A different solvent is used to repeat the process after the paper has been rotated. This technique is known as two-way or two-dimensional chromatography (Figure S2.81).

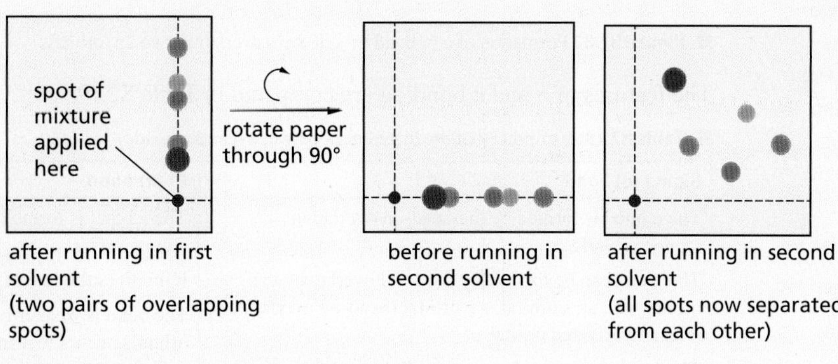

■ **Figure S2.81** Two-dimensional chromatography

Introduction to molecular orbital theory

Top tip!

MO theory explains many chemical bonding issues such as molecular shape, the existence and reactivity of double and triple bonds and why some organic molecules are coloured or magnetic.

◆ **Sigma (σ) bond**: A sigma molecular orbital is a bond formed by the head-on combination (overlap and merging) of atomic orbitals along the bond axis (an imaginary line joining the two nuclei); the electron density is concentrated between the two nuclei and this axis.

◆ **Pi (π) bond**: A pi molecular orbital is a bond formed by the lateral combination (overlap and merging) of two p orbitals; in a π bond the electron density lies to either side of a plane through the nuclei of the two atoms.

Molecular orbital (MO) theory is used to describe how atomic orbitals combine when atoms combine to produce molecules.

In a covalent bond, the atomic orbitals overlap and merge so that a new molecular orbital, containing two electrons, is formed.

Figure S2.82 shows how the s atomic orbitals of two hydrogen atoms overlap to form a covalent bond in the hydrogen molecule. This molecular orbital is known as a sigma (σ) orbital and is lower in energy than the two atomic orbitals from which it was formed. The electron density of the resultant **sigma (σ) bond** is symmetrical about a line joining the nuclei of the atoms forming the bond.

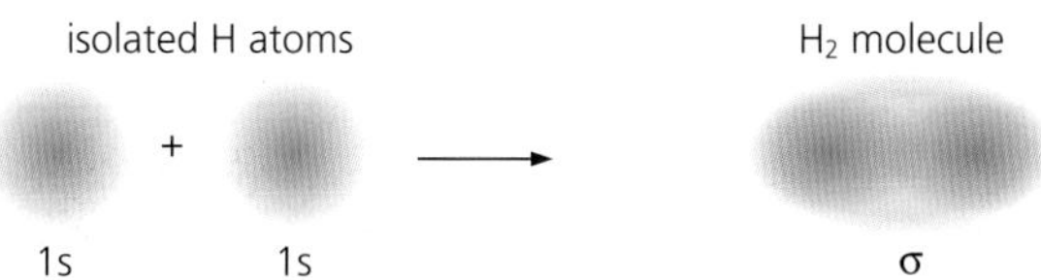

■ **Figure S2.82** Formation of a σ molecular orbital in a hydrogen molecule

Top tip!

The amount of overlap of the atomic orbitals determines the strength of the bond: the greater the overlap, the stronger the covalent bond.

The second-row atoms of the periodic table have 2s and 2p valence orbitals. The shape and orientation of p orbitals mean they can overlap (and, therefore, interact) in two ways when forming bonds: head-on to form a sigma bond and laterally to form a **pi (π) bond** (Figure S2.83). All single bonds are sigma bonds; pi bonds are only present in double and triple bonds.

Top tip!

A single π bond is often drawn as two electron clouds, one arising from each lobe of the p orbitals. You must remember, though, that the two clouds of electrons in a π bond represent one bond consisting of two electrons (a spin pair) in a single molecular orbital.

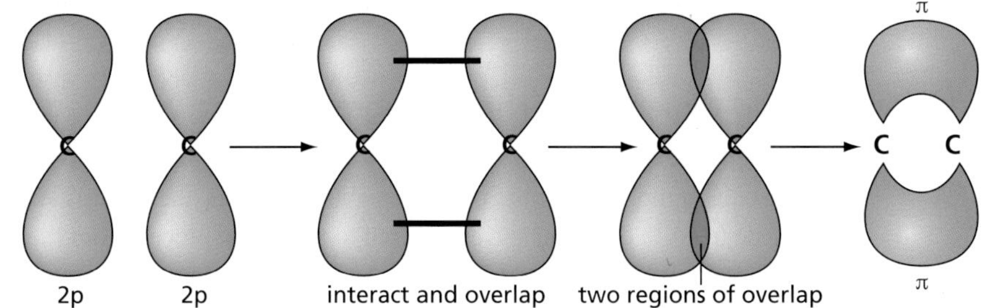

■ **Figure S2.83** Formation of a pi bond by sideways overlap of two $2p_z$ orbitals

The features of σ and π bonding are compared in Table S2.19.

■ **Table S2.19** A summary of the differences between σ and π bonds

Sigma (σ) bond	Pi (π) bond
The σ bond is formed by the head-on overlap of atomic orbitals.	The π bond is formed by the lateral overlap of p atomic orbitals.
This bond can be formed by the axial overlap of an s orbital with an s orbital, a p orbital (head-on overlap) or with a hybridized orbital.	It involves the lateral overlap of parallel p orbitals only.
The bond is stronger because overlapping of atomic orbitals can take place to a larger extent.	The bond is weaker because the overlapping of atomic orbitals occurs to a smaller extent.
The electron cloud formed by head-on overlap is symmetrical about the bond axis and consists of a single electron cloud with high electron density around the nuclei.	The electron cloud of the π bond consists of two identical electron clouds above and below the plane of the two nuclei.
There can be a free rotation of atoms around the σ bond.	Free rotation of atoms around the π bond is not possible (under standard conditions) because it involves the breaking of the π bond.
The σ bond may be present between the two atoms either alone or in combination with one or more π bonds.	The two atoms joined by a π bond will always already have a σ bond between them.
The shape of the molecule or polyatomic ion is determined by the σ framework around the central atom.	The π bonds do not contribute to the shape of the molecule.

WORKED EXAMPLE S2.2H

Describe the formation of the oxygen molecule, O_2, in terms of atomic orbitals (s and p).

Answer

The oxygen atom has the electron configuration $1s^2\ 2s^2\ 2p_x^{\ 2}\ 2p_y^{\ 1}\ 2p_z^{\ 1}$ and hence has two half-filled p orbitals in its valence shell. One of the half-filled p orbitals overlaps axially (head-on) with the half-filled p orbital of the other oxygen atom to form a σ bond. The other half-filled p orbitals of the two oxygen atoms overlap laterally (sideways) to form a π bond (p π–p π bond).

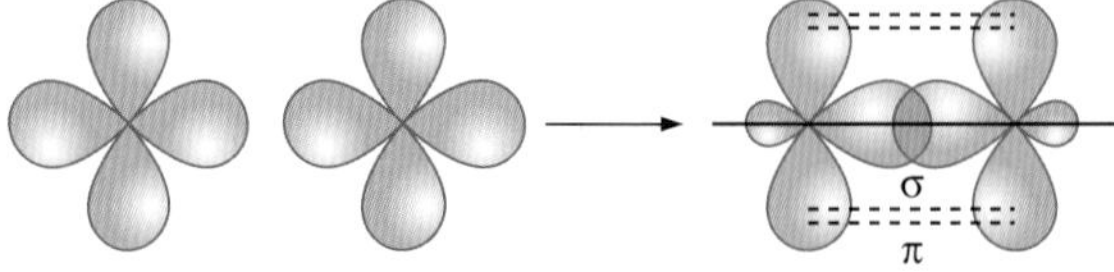

Hybridization

These two bonding types cannot completely explain the known structure and shapes of simple molecules. This deficiency led chemists to develop the theory of hybridization.

In the atoms of period 2 elements, the orbitals available for bonding are the 2s orbitals and the three 2p orbitals (along *x*-, *y*- and *z*-axes).

Figure S2.84 shows how s and p bonding atomic orbitals can overlap to form a σ bond. However, a σ molecular orbital formed from two 2p orbitals points in the same direction as the original p atomic orbitals. So molecular orbitals formed from the $2p_x$, $2p_y$ and $2p_z$ orbitals would be pointing along the *x*-, *y*- and *z*-axes; that is, at right angles to each other.

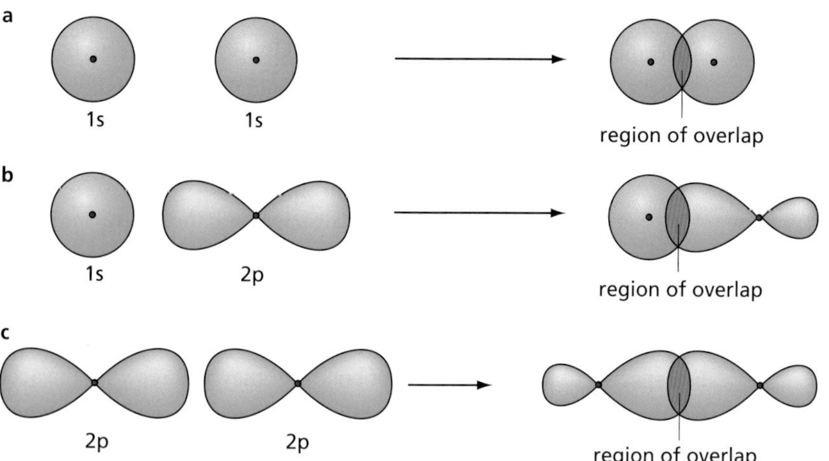

■ **Figure S2.84** The shapes of molecular orbitals formed by the overlap of 1s and 2p bonding atomic orbitals

However, VSEPR theory requires trigonal planar molecules to have three identical orbitals arranged at 120° to one another, and tetrahedral molecules to have four identical orbitals separated by 109.5°. Suitable new molecular orbitals that point in the right directions can be formed by mixing, or hybridizing, the s and p atomic orbitals (Figure S2.85).

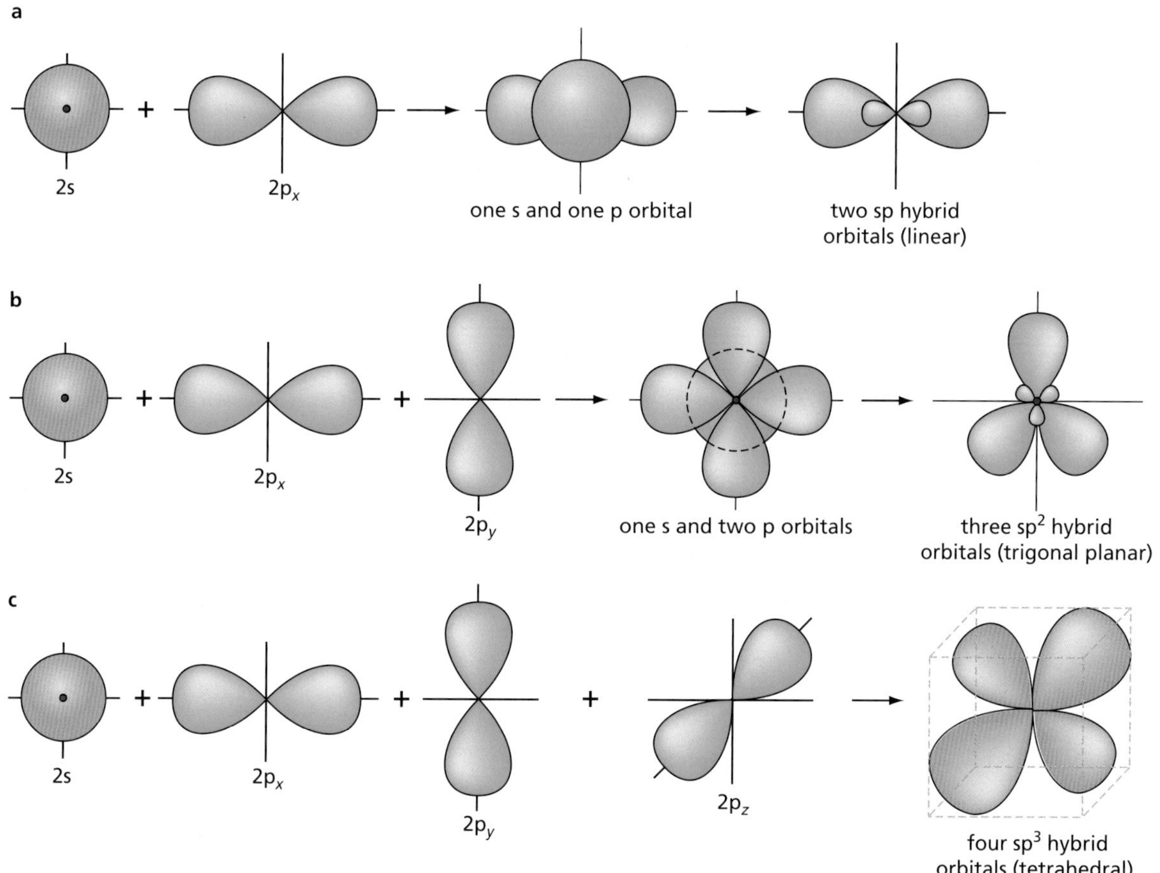

■ **Figure S2.85** sp, sp^2 and sp^3 hybrid orbitals

HL ONLY

◆ **Hybridization**: The mixing of two or more atomic orbitals to form the equivalent number of hybrid molecular orbitals of identical shape and energy which overlap and form covalent bonds.

◆ **Hybrid orbital**: An atomic orbital formed by mixing of s and p (and sometimes d or f) atomic orbitals.

- sp **hybridization** occurs when a single 2p orbital is hybridized with the 2s orbital. This produces two sp hybrid orbitals which are 180° apart, allowing two bonds to form in a linear arrangement.
- If the $2p_x$ and $2p_y$ orbitals are hybridized with the 2s orbital, the result is three identical sp^2 hybrid orbitals in a trigonal planar arrangement.
- If the 2s orbital is hybridized with all three of the 2p orbitals, four sp^3 **hybrid orbitals** are formed and these are arranged tetrahedrally. The term sp^3 hybridization indicates that the hybrid orbitals are derived from one s and three p orbitals, so that each has 25% s character and 75% p character.

Hybridization in carbon

A methane molecule has a tetrahedral shape with the central carbon atom having a valency of four and forming four single (σ) bonds with identical lengths. Consider the electronic structure of an isolated carbon atom in the ground state (Figure S2.86).

Since the 2s sublevel is full and there are two unpaired electrons in the 2p sublevel, carbon might be expected to form *two* covalent bonds (or maybe three with the acceptance of a lone pair). The p orbitals are arranged at 90° to each other (along *x*-, *y*- and *z*-axes); however, in methane, the carbon atom forms four identical bonds arranged tetrahedrally.

To explain how a carbon atom forms four bonds, a 2s electron has to be unpaired and promoted into an empty orbital of the 2p sublevel (Figure S2.87).

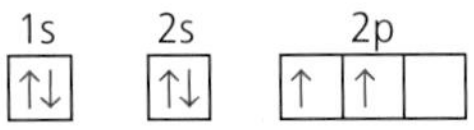

1s [↑↓] 2s [↑↓] 2p [↑][↑][ ]

■ **Figure S2.86** Carbon atom in the ground state

1s [↑↓] 2s [↑] 2p [↑][↑][↑]

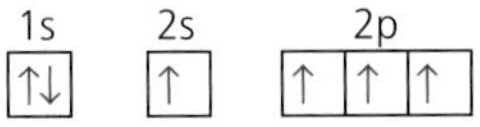

■ **Figure S2.87** Carbon atom in an excited state

This is an endothermic process, but the excess energy is regained when the two extra C–H bonds are formed (Figure S2.88). Bond formation is always exothermic and lowers the energy of the molecule, making it more stable.

The 2s orbital—which now has one unpaired electron—and the three 2p orbitals undergo hybridization to produce four identical sp^3 hybrid orbitals, each containing a single electron (Figure S2.89).

The four sp^3 hybridized orbitals of the carbon atom can then overlap with the 1s electrons of four hydrogen atoms (Figure S2.90) to form four σ bonds arranged tetrahedrally around the central carbon atom.

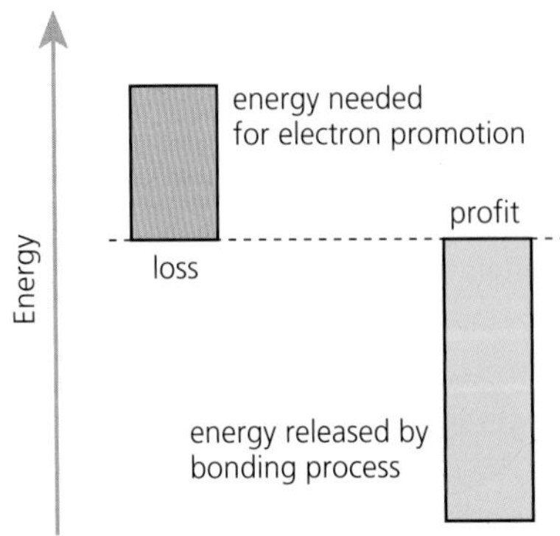

■ **Figure S2.88** The 'cost' (loss) and 'benefit' (profit)

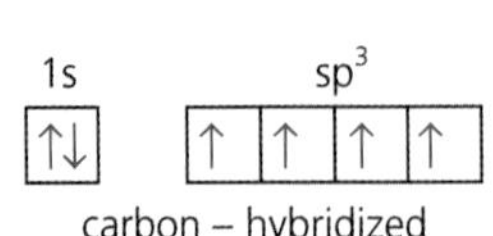

■ **Figure S2.89** Orbitals in an sp^3 hybridized carbon atom

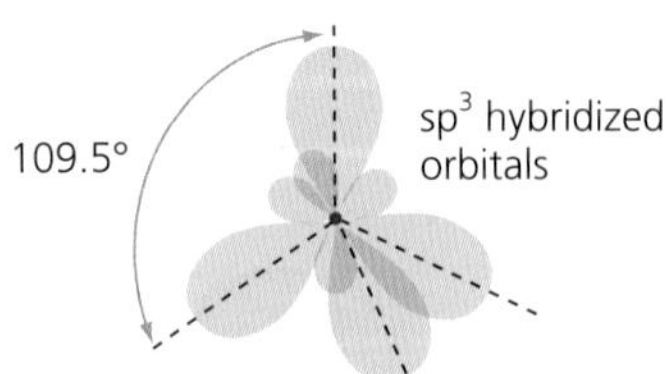

■ **Figure S2.90** Tetrahedrally arranged sp^3 hybrid molecular orbitals

HL ONLY

The process of sp^3 hybridization can also be shown as an energy diagram (Figure S2.91). The sp^3 hybrid orbitals are degenerate (have the same energy) and have identical shapes.

They overlap more effectively than p orbitals due to the presence of large lobes of electron density on one side of the hybrid orbital (Figure S2.92).

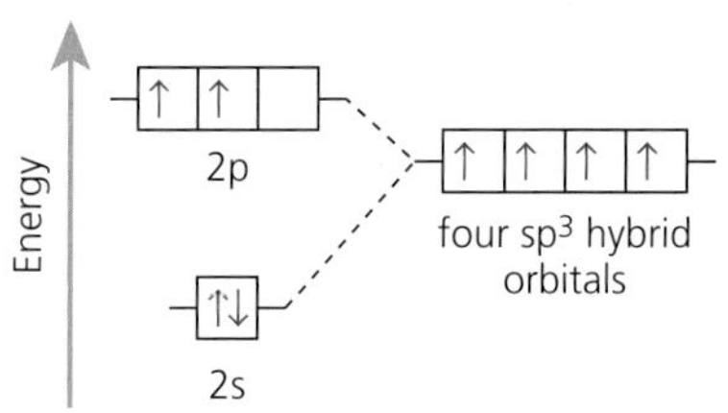

■ **Figure S2.91** Energy changes during sp^3 hybridization

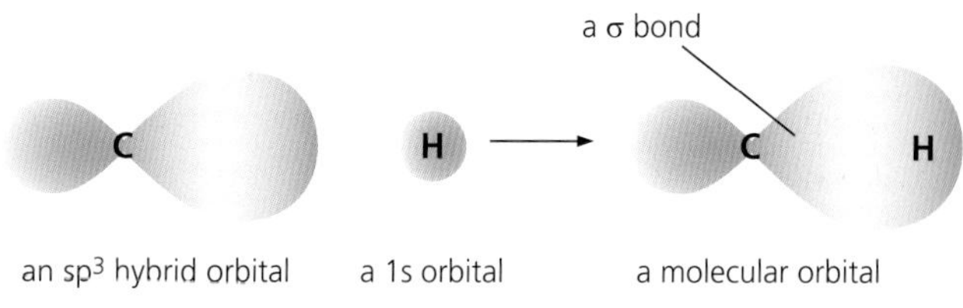

■ **Figure S2.92** Overlap between an sp^3 hybrid orbital from a carbon atom and a 1s orbital from a hydrogen atom to form a σ bond

Top tip!

Any carbon atom covalently bonded to four other atoms or groups has **sp^3** hybridized orbitals.

Hybridization also explains the shape of other alkane molecules. The ethene molecule has a trigonal planar arrangement and a carbon–carbon double bond which, unlike carbon–carbon single bonds, does not allow free rotation along the bond axis. The bond angles are 120° and all the atoms lie in the same plane.

An electron from the 2s sublevel is again promoted into the 2p sublevel. The 2s orbital hybridizes with two of the $2p^1$ hybrid orbitals to give three sp^2 hybrid orbitals (Figure S2.93). The $2p_z$ orbital does not participate in the hybridization process and is used in π bond formation.

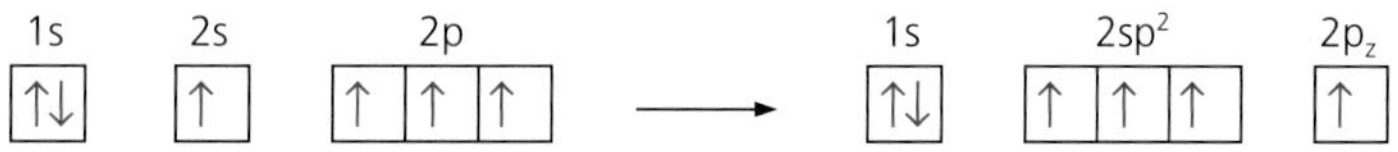

1s	2s	2p		1s	$2sp^2$	$2p_z$
↑↓	↑	↑ ↑ ↑	→	↑↓	↑ ↑ ↑	↑

■ **Figure S2.93** Excited carbon atom and an sp^2 hybridized carbon atom

The hypothetical formation of ethene from its atoms is summarized in Figure S2.94.

- The 2s orbital and the two 2p orbitals ($2p_x$ and $2p_y$) hybridize to form three equivalent sp^2 hybrid orbitals (step 1).
- One sp^2 hybrid orbital on each carbon atom overlaps and merges with that of the other carbon atom to form a carbon–carbon σ bond (step 2).
- The other two sp^2 hybrid orbitals of each carbon atom overlap with the 1s orbitals on hydrogen atoms to form carbon–hydrogen σ bonds (step 3).
- The unhybridized $2p_z$ orbital on each carbon atom overlaps sideways with that on the neighbouring carbon atom to form the carbon–carbon π bond (step 4).

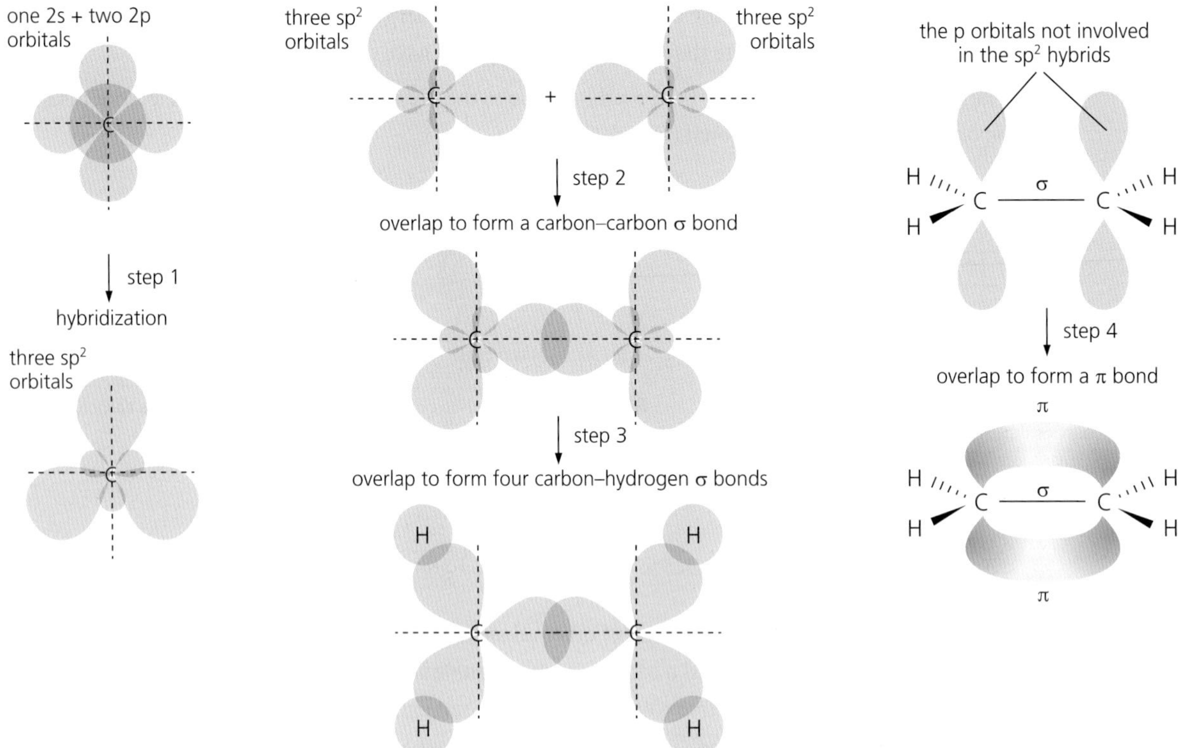

■ **Figure S2.94** The formation of the ethene molecule from sp^2 hybridized orbitals

In the ethyne molecule, C_2H_2, the two carbon atoms form sp hybrid orbitals. An electron from the 2s sublevel is again promoted into the 2p sublevel and the 2s orbital hybridizes with *one* of the 2p orbitals to give two sp hybridized orbitals. The $2p_y$ and $2p_z$ orbitals do not participate in the hybridization process (Figure S2.95).

A triple bond is formed in ethyne, C_2H_2, when two unhybridized p orbitals overlap sideways, forming two π bonds (Figure S2.96), in addition to the σ bond already present. The total electron density of the two combined p orbitals takes the shape of a cylinder that encircles the molecule.

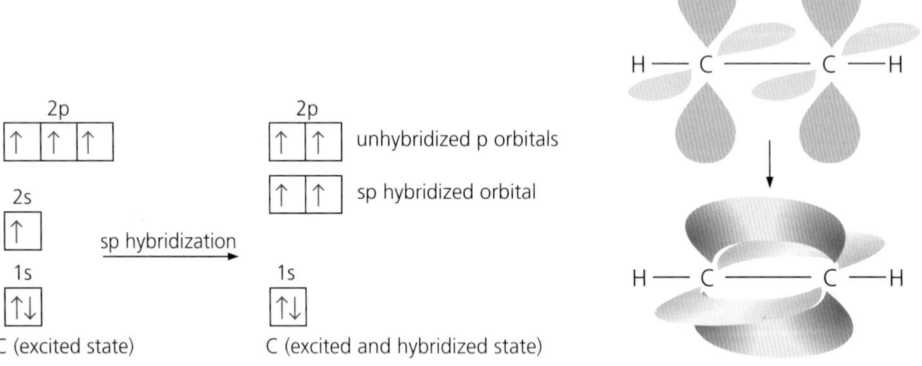

■ **Figure S2.95** Excited carbon atom and an sp hybridized carbon atom

■ **Figure S2.96** π bond formation in ethyne

Table S2.20 summarizes the three types of hybridization shown by carbon.

■ **Table S2.20** Summary of hybridization in carbon

Type of hybridized orbital	**sp^3**	**sp^2**	**sp**
Atomic orbitals used	s, p, p, p	s, p, p	s, p
Number of hybridized orbitals	4	3	2
Number of atoms bonded to the carbon atom	4	3	2
Number of σ bonds	4	3	2
Number of unhybridized p orbitals	0	1	2
Number of π bonds	0	1	2
Bonding arrangements	tetrahedral; four single bonds only	trigonal planar; two single bonds and a double bond	linear; one single bond and a triple bond or two double bonds
Example	CH_4, C_2H_6, CCl_4	$H_2C{=}CH_2$, H_2CO	$H{-}C{\equiv}C{-}H$, $H_2C{=}C{=}CH_2$

Hybridization and VSEPR theory

VSEPR theory can be used to identify the hybridization state of any central atom in a molecular structure (Table S2.21).

- sp^3 hybridization occurs where there are four electron domains around the centrally bonded atom. An electron domain is a lone pair or a covalent bond (whether single, double or triple).
- sp^2 hybridization occurs when there are three electron domains around the centrally bonded atom.
- sp hybridization occurs where there are two electron domains around the centrally bonded atom.

■ **Table S2.21** The relationship between the Lewis structure and the hybridization of the central atom

Hybridization state of central atom	**Number of electron domains**	**Number of covalent bonds**	**Number of lone pairs**	**Shape**	**Examples**
sp	2	2	0	linear	BeF_2, CO_2 (Figure S2.98) and N_2 (Figure S2.97)
sp^2	3	3	0	trigonal planar	BF_3, graphite, fullerenes, SO_3, CO_3^{2-}
sp^2	3	2	1	V-shaped or bent	SO_2, NO_2^-
sp^3	4	4	0	tetrahedral	CH_4, diamond, ClO_4^-, SO_4^{2-}
sp^3	4	3	1	trigonal pyramidal	NH_3, NF_3, PCl_3, H_3O^+
sp^3	4	2	2	V-shaped or bent	H_2O, H_2S, NH_2^-

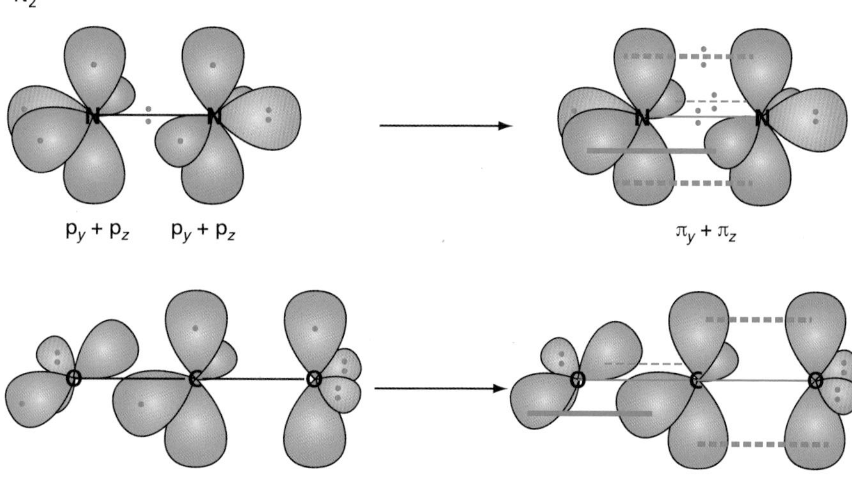

■ **Figure S2.97** Triple bond in the nitrogen molecule with the nitrogen atoms sp hybridized

■ **Figure S2.98** The two pi bonds in carbon dioxide with the carbon atoms sp hybridized

Nature of science: Theories

Explaining molecular bonds

Chemical bonding theory was introduced using Lewis formulas. The American chemist Gilbert Lewis (1875–1946) first developed these diagrams in 1902 as a teaching tool for students. These electron dot diagrams allow us to predict the number of bonds an atom forms (known as its valency) together with the formulas and structures of a large number of simple compounds which, with the introduction of VSEPR theory, leads to molecular geometries and polarities. It gives us the octet rule and a way of introducing the formation of ionic and molecular compounds.

Lewis theory is a non-mathematical and pre-quantum-mechanical model largely based on empirical patterns of valence. However, it fails as a physical theory to explain why and how bonding occurs and explain spectroscopic data.

The theory of hybridization was developed by the American chemist Linus Pauling (1901–1994) in 1931 to reconcile observed geometries of molecules with atomic orbitals of atoms in the ground state. Pauling wanted to make quantum mechanics consistent with empirically known molecular structures. He was committed to the theory of valence and bonding developed by Lewis.

It is important to remember that hybrid orbitals—and orbitals in general—are not physical, observable entities. Hybridization is not a physical process, but a mathematical procedure of combining wavefunctions (mathematical expressions involving the coordinates of an electron in space).

Resonance

◆ **Resonance structures**: The Lewis formulas that can be drawn for an ion or molecule stabilized by resonance.

◆ **Resonance**: The representation of the structure of a molecule by two or more Lewis formulas in which there is more than one possible position for pi bonds (double bonds).

◆ **Resonance hybrid**: The representation of the structure of a molecule or ion by a single structure, often using dotted lines to show partial bonds and displaying fractional charges, if appropriate.

Some molecules and ions have structures that cannot be represented by one Lewis formula only. For example, the ozone molecule can be represented by two Lewis formulas as shown in Figure S2.99.

Top tip!

The double headed arrow in Figure S2.99 represents resonance and the curly arrow shows movement of electron pairs. This is not to be confused with the double-headed equilibrium arrow which describes a physical change or chemical reaction.

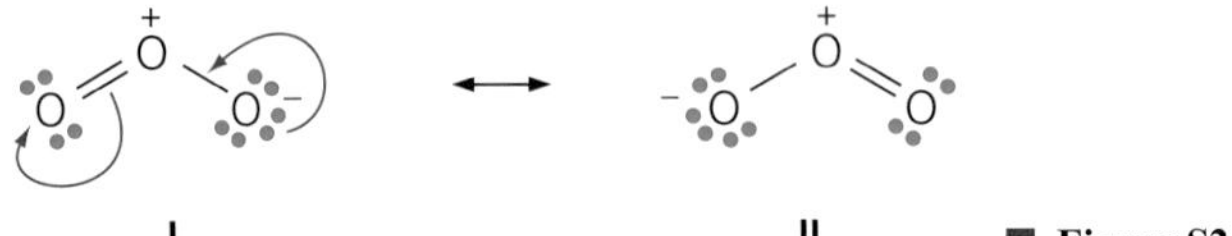

■ **Figure S2.99** Lewis formulas for ozone

The Lewis formulas I and II are known as the **resonance structures** and are equivalent except for the position of the valence electrons. **Resonance** refers to a molecule or polyatomic ion for which two or more possible Lewis formulas can be drawn.

Structures I and II in Figure S2.99 do not represent the real structure of ozone because they show that the molecule has two types of bonds: a double bond, O=O, of bond length 0.121 nm; and a single bond, O–O, with a bond length of 0.132 nm. X-ray analysis shows that both bonds have the same length of 0.128 nm and are intermediate between a single and a double bond.

The ozone molecule exists as a **resonance hybrid** (Figure S2.100) which can be regarded as an average or blend of the two structures with equal contribution from each. The dotted line indicates that the bonds have partial double bond character (bond order 1.5). This means (at a simple level) that the bond length and bond strength of oxygen–oxygen bonds is halfway between the values of single and double oxygen bonds.

An equivalent MO description of the bonding in ozone has the electrons in the double bond occupying a single delocalized π orbital combining unhybridized p orbitals on all three atoms (Figure S2.101).

Common mistake

Misconception: *Resonance structures exist and are in equilibrium.*

Resonance structures do not exist; the real molecule is described by imaginary resonance structures.

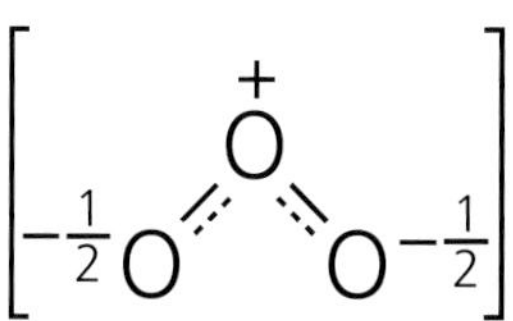

■ **Figure S2.100** Resonance hybrid structure for the ozone molecule

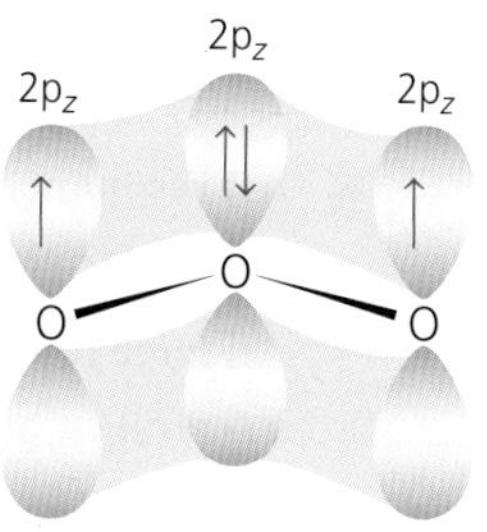

■ **Figure S2.101** Delocalized molecular orbital model of the ozone molecule

Many polyatomic ions, including the carbonate ion, can be described by the resonance and delocalization models. Figure S2.102 shows the three equivalent Lewis formulas of the carbonate ion which contribute equally to the resonance hybrid of Figure S2.103.

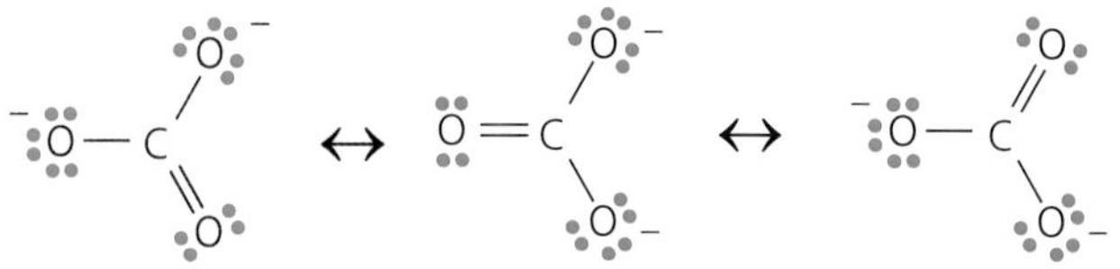

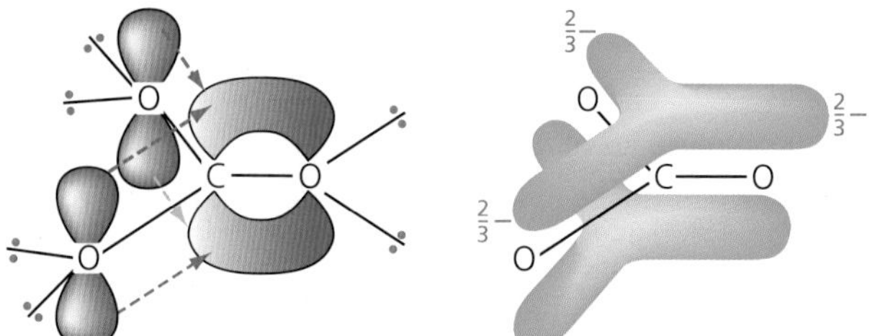

■ **Figure S2.102** Three equivalent Lewis formulas for the carbonate ion

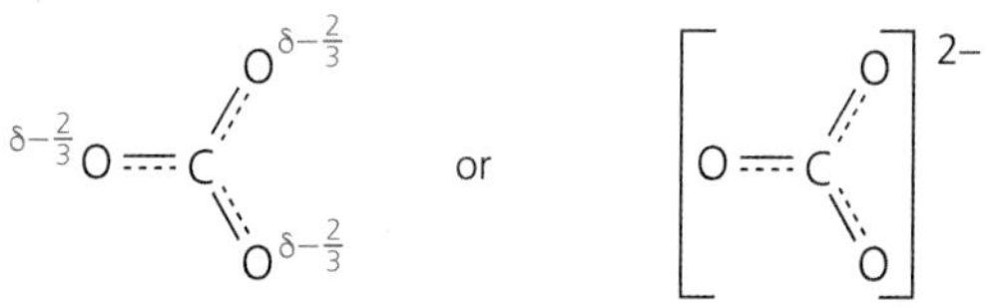

■ **Figure S2.103** Resonance hybrid for the carbonate ion

The two singly bonded oxygen atoms each have an extra electron, making them anionic (negatively charged). Each of them also contains three lone pairs, one of which will be in a 2p$_z$ orbital.

In the delocalization model, the lone pairs of the filled 2p$_z$ orbitals can overlap with the lobe of the 2p$_z$ orbital on the central carbon atom (Figure S2.104). The lone pair is thus partially donated, forming a coordinate π bond. The 2p$_z$ orbital on each oxygen atom can overlap with the 2p$_z$ orbital of the carbonyl group, >C=O, forming a four-centre delocalized π molecular orbital containing a pair of electrons.

■ **Figure S2.104** p orbital overlap leading to delocalized π bonding in the carbonate ion

HL ONLY

For significant resonance stabilization to occur within a molecule or ion, the suggested resonance structures must meet all of the following requirements, as illustrated by the carbonate ion:

- **All resonance forms must have the same distribution of atoms in space (the same molecular shape).**

  All three resonance structures of the carbonate ion have the three oxygen atoms located around a central carbon atom in a trigonal planar arrangement.

- **No resonance form may have a very high energy. No resonance form containing carbon, oxygen and nitrogen (or another atom from the second period) may have more than an octet (or, for hydrogen, a duplet).**

  The three oxygen atoms and one carbon atom of the carbonate ion each obey the octet rule.

- **All resonance forms must contain the same number of electron pairs.**

  All the resonance structures of the carbonate ion have three s pairs, one p pair and eight lone (non-bonded) pairs.

- **All resonance forms must carry the same total charge.**

  All the resonance structures of the carbonate ion have a total charge of minus two.

◆ **Resonance energy**: The difference in energy between the resonance hybrid (with delocalized electrons) and the individual resonance structures (with localized electrons).

When resonance occurs, the resonance hybrid is more stable and has a lower energy than any of the resonance structures. The difference between the energy of the most stable form and the hybrid is known as the **resonance energy**. This concept is illustrated for the carbonate ion in Figure S2.105.

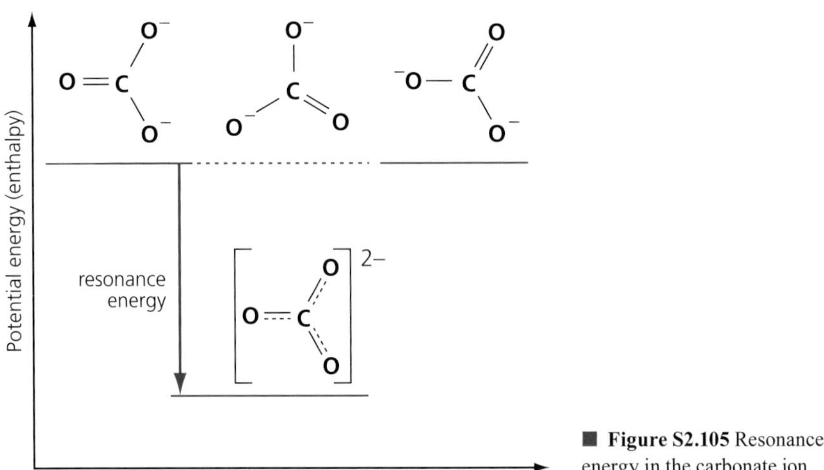

■ **Figure S2.105** Resonance energy in the carbonate ion

To understand the importance of resonance in the stabilization of a structure, and the contribution of a particular resonance form, the following points must be considered.

- In the case of ions, the most stable resonance structures have negative charges on the electronegative atoms (usually oxygen and sulfur) and positive charges on the less electronegative atoms (usually carbon and nitrogen).
- The most important resonance structures are those where as many atoms as possible obey the octet rule, have the maximum number of bonded electrons and have the fewest charges.
- Resonance effects are always stabilizing and, *generally,* the greater the number of resonance structures that can be drawn, the greater the stability of the resonance hybrid.
- Contribution is highest when the resonance structure has the lowest formal charges.

19 Draw resonance structures for the following inorganic species:
  a nitrate(V) ion, NO_3^- (three resonance structures)
  b nitrate(III) ion, NO_2^- (two resonance structures)
  c phosphate(V) ion, PO_4^{3-} (four resonance structures)
  d sulfate(IV) ion, SO_4^{2-} (six resonance structures)
  e manganate(VII) ion, MnO_4^- (four resonance structures).

A number of organic molecules, including benzene (C_6H_6), and ions exist as resonance hybrids. The ethanoate ion, CH_3COO^-, has two equivalent resonance structures making equal contributions to the resonance hybrid (Figure S2.106). The two oxygens cannot be distinguished and the two C–O bond lengths are identical.

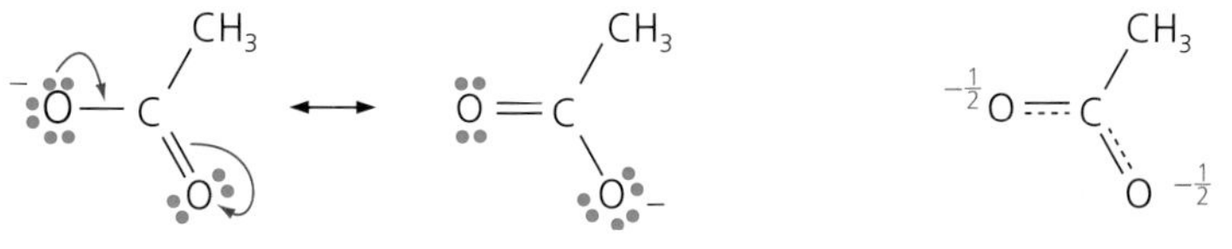

■ **Figure S2.106** Resonance structures and hybrid of the ethanoate ion

Top tip!

Resonance can be used to explain why ethanoic acid, CH_3COOH, is a stronger acid than ethanol, C_2H_5OH. Neither the ethanol molecule nor the ethoxide ion ($C_2H_5O^-$) is stabilized by resonance, hence the very low dissociation of ethanol into ions.

Propanone, the simplest ketone, can be described as a resonance hybrid of two resonance structures (Figure S2.107). The first resonance structure is known as the major resonance form and makes a larger contribution to the resonance hybrid; the second resonance structure, with the separation of charge, makes a smaller contribution to the resonance hybrid.

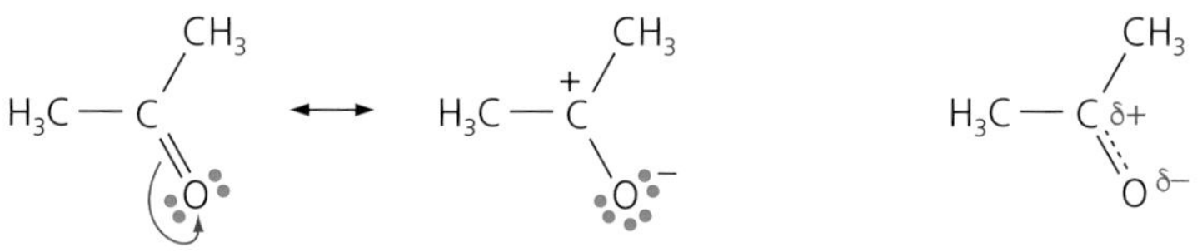

■ **Figure S2.107** Resonance structures and hybrid of the propanone molecule

Top tip!

This description for propanone explains the chemical reactions of ketones, which frequently involve attack on the electron-deficient and partially positive carbon atom of the carbonyl group by nucleophiles (electron-pair donors). The resonance model accounts for the polarity and hence dipole moment of the carbonyl carbon–oxygen bond and the observation that the carbon-oxygen double bond has a bond strength and bond length that is less than that of a double bond between carbon and oxygen.

LINKING QUESTION

Why are oxygen and ozone dissociated by different wavelengths of light?

20 Explain why the two C–O bond lengths in propanoic acid, C_2H_5COOH, are different and yet the two C–O bond lengths in the propanoate anion, $C_2H_5COO^-$, are the same length.

HL ONLY

The Montreal protocol

The detection of the depletion of the ozone layer by compounds such as chlorofluorocarbons (CFCs) led to the signing of the Montreal Protocol in 1987. This resulted in the freezing of the production of CFCs and the search for substitutes and is an example of successful international cooperation in response to an environmental problem. Predictions are that the ozone layer may be beginning to recover, though estimates now suggest that it will not be completely recovered until 2060–2070. The Montreal protocol is a positive model for the steps required in solving other, more major and significant, global environmental problems, such as global warming and loss of biodiversity.

How is scientific knowledge selected for inclusion in textbooks?

While knowing about the detection of ozone and its depletion in the stratosphere is interesting, one might ask whether this knowledge is all that important here? In a world where there is a wealth of interesting and detailed knowledge about chemistry, why should these facts find themselves in a chemistry textbook?

The knowledge that students are presented with is the result of choices. The IB has chosen what content makes it into the chemistry curriculum. Textbook writers make choices about how much detail to give, the way information is presented, and what sort of questions are included to encourage students to think about the content. Students should pay attention to these choices and recognize what role these choice-makers play in their understanding of the material.

We have a good example here with these facts about the ozone layer. We also now recognize that the planet is undergoing dangerous climate change, which makes knowledge about how carbon is stored by the environment important at political and social levels as well as the 'chemical' level. In other words, facts about the world do not exist in a vacuum: they exist in a social and political world where decisions about how human beings should live are being made every day. Knowledge about science and how the natural world works is hugely important when making these decisions. If decision-makers do not have sufficient scientific understanding of the world, then their choices might lead us further into problems rather than in the direction of solutions.

The authors and publishers of this textbook, not to mention the IB's own curriculum developers, recognize how important it is that future decision-makers have a solid understanding of scientific processes and how they act in the world. It is equally important that those who continue to study chemistry, or work in related fields, are aware of the huge impact scientific knowledge has in political and social decision-making.

Benzene

Benzene is a starting molecule in the preparation of many important chemical compounds, including medicinal drugs. The benzene molecule can be represented as a resonance hybrid of two equivalent and symmetrical resonance structures (Figure S2.108).

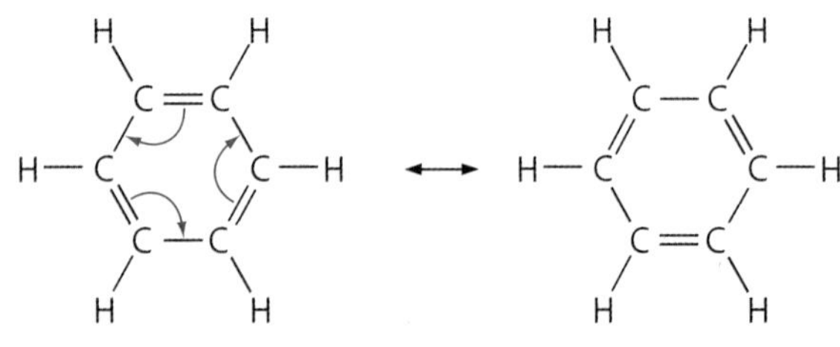

■ **Figure S2.108** Resonance structures for the benzene molecule

Each carbon–carbon bond has a bond order of 1.5 and all the equal carbon–carbon bond lengths of 140 pm lie between 134 pm (that of the carbon–carbon double bond) and 154 pm (the carbon–carbon single bond length).

Top tip!

Benzene is described as a conjugated system since it can be considered as having alternating carbon–carbon double and single bonds.

The benzene molecule can also be described by the delocalized model derived from molecular orbital theory.

Each carbon atom has $2p_z$ orbitals (with lobes above and below the ring) which overlap with those of the carbon atoms on either side to give a six-centre delocalized π molecular orbital above and below the plane of the benzene ring (Figure S2.109).

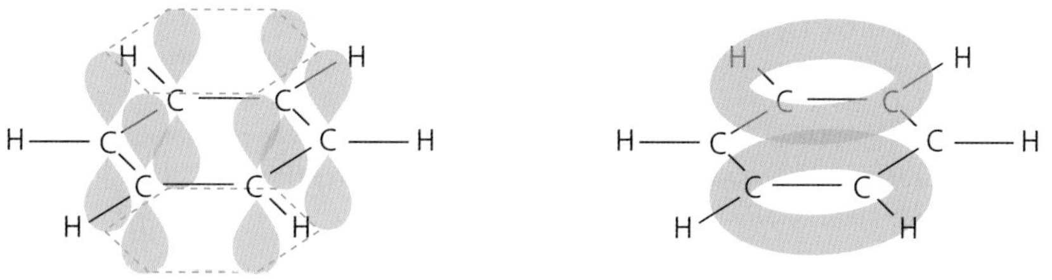

■ **Figure S2.109** Overlapping of $2p_z$ orbitals in benzene to form a delocalized cyclic π orbital

It is the resonance energy from these π electrons which is responsible for the kinetic stability of benzene, its unreactivity and its tendency to undergo substitution reactions rather than addition reactions (like alkenes).

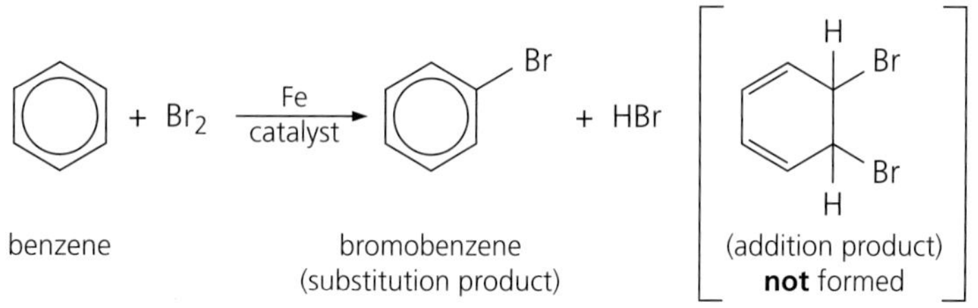

■ **Figure S2.110** Benzene reacts slowly with bromine to form bromobenzene via substitution

21 Explain why penta-1,4-diene, $H_2C{=}CH{-}CH_2{-}CH{=}CH_2$, does not contain any delocalized π electrons whereas penta-1,3-diene, $H_2C{=}CH{-}CH{=}CH{-}CH_3$, does contain delocalized π electrons.

LINKING QUESTION

How does the resonance energy in benzene explain its relative unreactivity?

LINKING QUESTION

What are the structural features of benzene that favour it undergoing electrophilic substitution reactions?

HL ONLY

What is the role of imagination and intuition in the creation of hypotheses in the natural sciences?

The German chemist August Kekulé (1829–1896) claimed that the cyclic structure (with alternating single and double bonds) he proposed for benzene in 1859 came to him in a dream, as described below in his own words:

> *I was sitting writing at my textbook, but the work did not progress; my thoughts were elsewhere. I turned my chair to the fire [after having worked on the problem for some time] and dozed. Again the atoms were gambolling before my eyes. This time the smaller groups kept modestly to the background. My mental eye, rendered more acute by repeated vision of this kind, could now distinguish larger structures, of manifold conformation; long rows, sometimes more closely fitted together; all twining and twisting in snakelike motion. But look! What was that? One of the snakes had seized hold of its own tail, and the form whirled mockingly before my eyes. As if by a flash of lightning I awoke. Let us learn to dream, gentlemen.*

However, his early training in architecture may have also played a critical role in helping him conceive his structural theories about chemical bonding. Kekulé may have therefore brought outside abilities to the field of chemistry and his architect's spatial and structural sense may have given him another approach to the problem of benzene, that his contemporaries did not possess. He was able to place a problem in his subconscious mind and turn his dreams loose on it.

When Kekulé published his suggestion that benzene had a cyclic structure with an alternating series of double and single bonds, he did not mention that he had been inspired by a dream. This was presumably to avoid ridicule from his scientific peers, since his model was not derived from experimental observations.

While imagination certainly plays a role in developing initial hypotheses or proposing interesting directions for investigation, merely imagining is not nearly enough. Scientists must back up their initial imaginings with observational evidence. That is to say, a scientist's imagination must be tested by observational and experimental evidence.

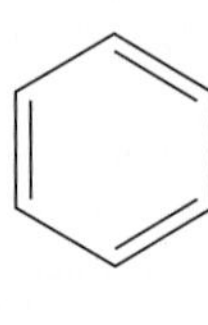

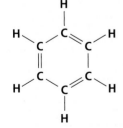

■ **Figure S2.111** Kekulé's snake and his structural and skeletal formulas for the structure of benzene

Kekulé's initial structure could not account for the existence of only one isomer of bromobenzene and the unreactivity of benzene. To overcome this disagreement between his structure and the properties of benzene, Kekulé proposed that the bonding in the benzene ring rapidly interconverted between two equivalent structures (Figure S2.112).

■ **Figure S2.112** Kekulé's oscillating model of benzene

The structure and shape of benzene was only established by the application of X-ray diffraction. The regular hexagonal shape of the benzene ring was demonstrated in 1929 by the English chemist Kathleen Lonsdale, who studied crystals of hexamethylbenzene, which crystallizes at a lower temperature than benzene.

In the 1930s Linus Pauling's resonance theory overcame many of the difficulties faced by Kekulé's equilibrium model in describing the structure and properties of benzene. Pauling used the analogy that a rhinoceros (a real animal) could be described as a blend or hybrid of a dragon and a unicorn (two imaginary animals).

It is important to distinguish between the concept of resonance and Kekulé's equilibrium model. In the resonance model, there is no interconversion between the two Kekulé structures—neither exists. To use another analogy of Pauling's, a mule (a hybrid of a donkey and a horse) is a mule every day of the week and not a horse or a donkey on alternate days.

Another early representation of the structure of benzene was proposed by the German chemist Johannes Thiele in 1899 (Figure S2.113). It can now be regarded as a resonance hybrid and clearly shows the partial double bond nature of benzene (bond order of 1.5).

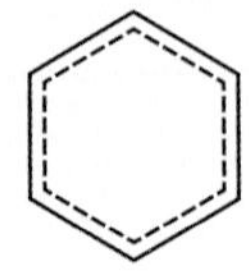

■ **Figure S2.113** The representation of the benzene molecule proposed by Johannes Thiele in 1899

The metallic model

Guiding question

- What determines the metallic nature and properties of an element?

◆ **Metals**: Chemical elements which are good conductors of heat and electricity; they form cations.

◆ **Malleable**: The ability of metals to be beaten into thin sheets without breaking.

◆ **Ductile**: The ability of metals to be stretched into wires.

◆ **Lustrous**: Having a shiny surface; a property of metals.

◆ **Metallic bonding**: At a simple level the electrostatic attraction between cations and the sea of delocalized valence electrons.

◆ **Electron sea model**: A simple model of metallic bonding with cations in fixed positions with mobile valence electrons moving in the lattice.

◆ **Main group metals**: Metals in groups 1, 2 and 13–15 in the periodic table.

SYLLABUS CONTENT

By the end of this chapter, you should understand that:

- a metallic bond is the electrostatic attraction between a lattice of cations and delocalized electrons
- the strength of the metallic bond depends on the charge of the ions and the radius of the metal ion
- transition elements have delocalized d electrons (HL only).

By the end of this chapter you should know how to:

- explain the electrical conductivity, thermal conductivity and malleability of metals
- relate characteristic properties of metals to their uses
- explain trends in melting points of s and p-block metals
- explain the high melting point and electrical conductivity of transition elements (HL only).

Metallic bonding

The properties of **metals** are very different from those of ionic and covalent compounds. Many metals are strong; many are **malleable** (can be beaten into thin sheets) and **ductile** (can be drawn into wires); and most have a high density compared to non-metals. Metals are shiny (**lustrous**) when cut or polished and are excellent electrical and thermal conductors.

Any theory of **metallic bonding** must account for all these physical properties and, ideally, any observed differences or trends in these properties.

A metal is an element whose atoms tend to lose electrons from the valence shell to form cations. The physical properties of elemental metals can be explained by the simple **electron sea model** (Figure S2.114) in which cations are in fixed positions in a relatively rigid lattice surrounded by a 'sea' of moving delocalized valence electrons.

Metallic bonding is the electrostatic attraction between the metal ions and the delocalized electrons. Metallic bonding is non-directional in the **main group metals**: all of the valence electrons are attracted to all of the metal ions but the attractions decrease with distance. The metal cations can be packed in several ways to create lattices with different unit cells, which are ignored in this model. The presence of cations and their packing arrangements are confirmed by X-ray diffraction studies.

Top tip!

Metallic bonding is present in pure metals and in most alloys that contain a mixture of metals. It is favoured by atoms of low electronegativity and low ionization energy.

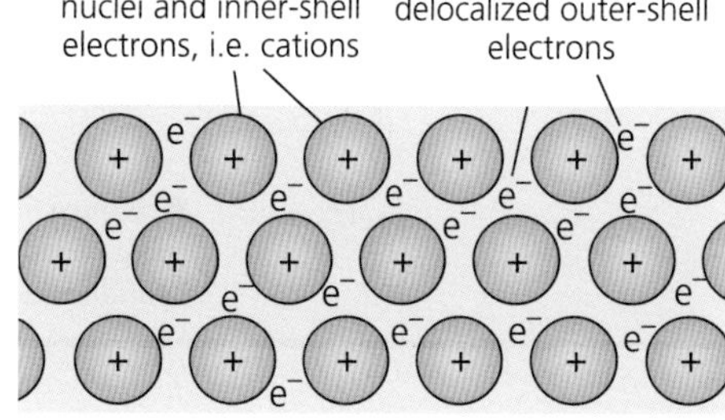

■ **Figure S2.114** Electron sea model of metallic bonding

ATL S2.3A

Work through the diagnostic questions on the presentation at **https://thescienceteacher.co.uk/wp-content/uploads/2018/10/Metallic-bonding-diagnosisv2.pptx** which have been created to see whether students have any misconceptions about metallic bonding. Answers are provided in the notes section below each slide.

Going further

Band theory

Band theory is another description of metallic bonding that leads to a sea of electrons model. This model assumes that all the valence atomic orbitals overlap with one another to form a giant molecular orbital extending over three dimensions through the entire lattice.

When the atomic orbitals of two metal atoms overlap and merge, two molecular orbitals are formed, one of lower energy and one of higher energy. When the atomic orbitals of four adjacent metal atoms overlap, four molecular orbitals are formed and so on.

The molecular orbitals of highest and lowest energy result from the overlap of orbitals from adjacent metal atoms. Intermediate energy levels arise from the overlap of the atomic orbitals of metal atoms further away in the lattice. The larger the lattice, the more intermediate energy levels are formed. Eventually these intermediate levels are so close that there is effectively a band of energies instead of discrete levels (Figure S2.115).

A magnesium atom has an electronic configuration of $1s^2\ 2s^2\ 2p^6\ 3s^2\ 3p^0$ and the 3s and 3p valence sublevels each produce energy bands. Figure S2.116 shows the energy bands in magnesium metal. The electrons in the 1s, 2s and 2p orbitals are localized on each of the five magnesium atoms. However, the delocalized molecular 3s orbitals (initially filled) and the 3p orbitals (initially empty) overlap. Electrons can easily move from the lower filled 3s valence band to the upper 3p band and move through it, forming an electric current when a voltage is applied. This upper band is therefore known as the conduction band.

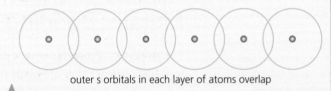

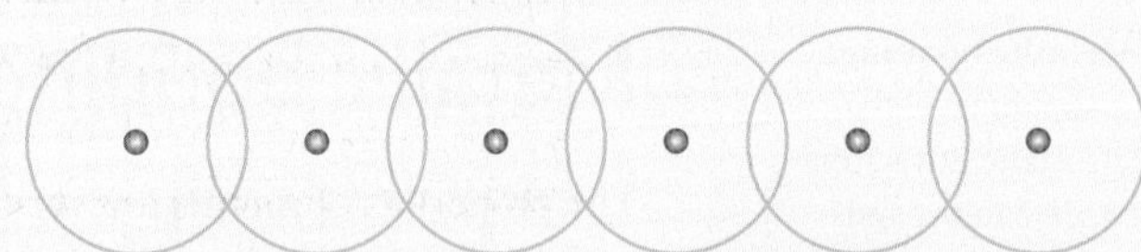

outer s orbitals in each layer of atoms overlap

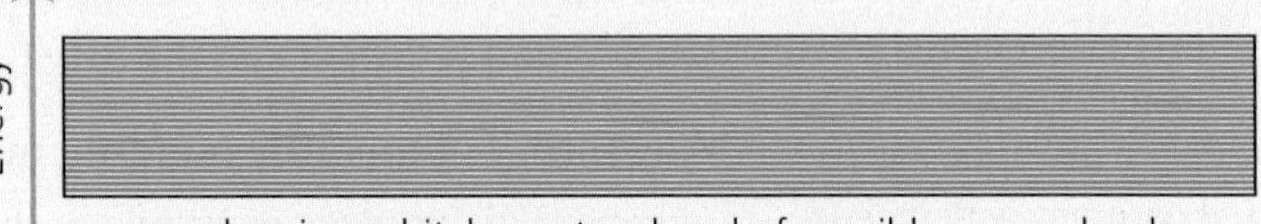

overlapping orbitals create a band of possible energy levels

■ **Figure S2.115** The origin of a band of energy levels in metals from the interaction of valence s orbitals

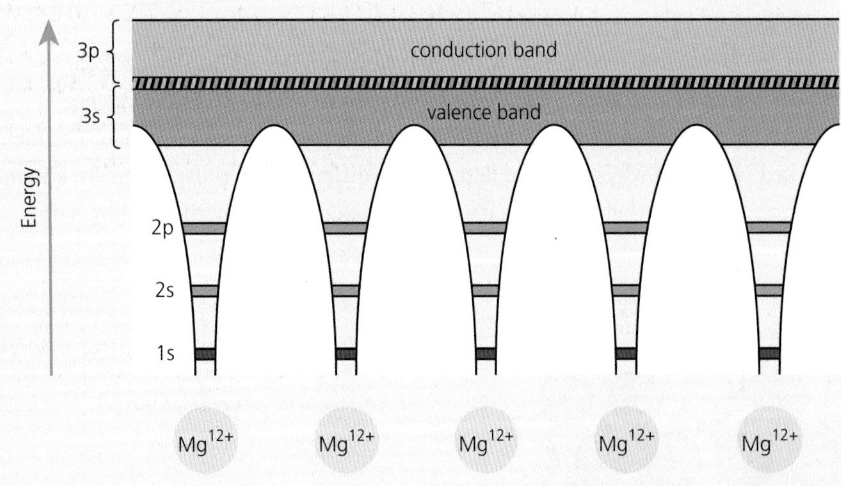

■ **Figure S2.116** Formation of energy bands in magnesium (ignoring hybridization of the atomic orbitals)

◆ **Semiconductor**: Crystalline material with a conductivity intermediate between that of a conductor and an insulator.

A number of elements, including silicon, are **semiconductors**: they are insulators at absolute zero but have a measurable conductivity when the temperature is raised. In semiconductors, the conduction and valence bands are separated by an energy gap (Figure S2.117). A small number of electrons may have enough energy to be 'promoted' to the conduction band and create a current when a voltage is applied. Heating the semiconductor leads to an increase in conductivity since more electrons have sufficient energy to enter the conduction band.

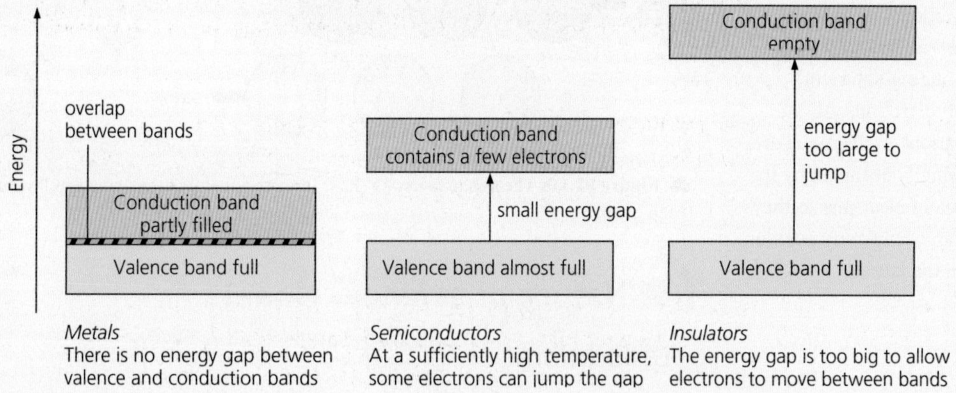

■ **Figure S2.117** The band structure of metals, semiconductors and insulators

Physical properties of metals

Table S2.22 compares the properties of solid crystalline substances with different types of bonding, including metallic.

■ **Table S2.22** Comparing and contrasting the properties of solid crystalline substances

Type of bonding	***Metallic***	***Covalent network structures***	***Simple molecular***	***Ionic***
Examples	sodium, magnesium, aluminium, iron, mercury, brass (copper and zinc)	diamond, silicon, silicon dioxide, graphene, graphite	iodine, methane, hydrogen chloride, water, benzoic acid, ethanol, ammonia, fullerenes	sodium chloride, magnesium oxide, calcium fluoride, sodium carbonate
Composition	metal atoms	non-metallic atoms	molecules	ions
Nature of bonding	cations attracted to delocalized valence electrons	atoms bonded by strong covalent bonds	covalently bonded molecules held together by weak intermolecular forces	strong electrostatic attraction between oppositely charged ions
Physical state at room temperature and pressure	solids (except mercury)	solids	gases, liquids, solids	solids
Hardness	usually hard (but group 1 metals are soft)	extremely hard	soft (if solids)	hard and brittle; undergo cleavage
Melting point	usually high (except group 1 and mercury)	very high	very low or low	high
Electrical conductivity	conductors (in solid and liquid states)	non-conductors (except graphite and graphene)	non-conductors in solid and liquid states	conductors in molten and aqueous states
Solubility	nil, but dissolve in other metals to form alloys	totally insoluble in all solvents	usually soluble in non-polar solvents; usually less soluble in polar solvents	usually soluble in polar solvents; insoluble in non-polar solvents

Common mistake

Misconception: *Because metals are malleable and ductile, metallic bonding is weak.*

Malleability is a reflection of metallic bonding acting in all directions; all the cations in the metal are equally attracted by the sea of electrons so they can occupy any position in the lattice.

Malleability and ductility

The malleability and ductility of metals can be explained by the electron sea model. If a stress is applied to the metal, planes of atoms slide over each other. This means the structure can change shape without the crystal fracturing (Figure S2.118). In contrast, stress applied to an ionic crystal will cause the crystal to shatter.

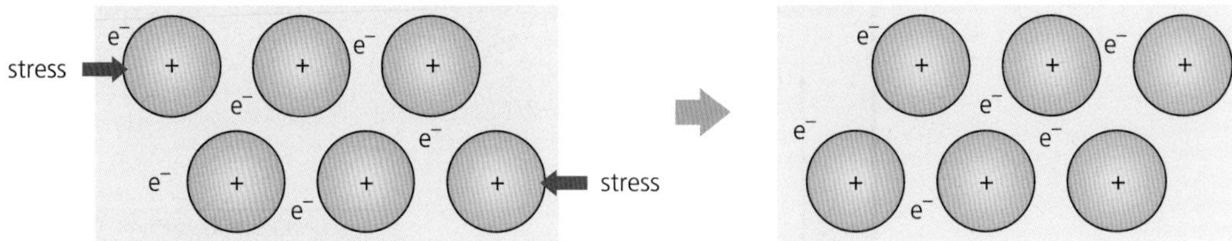

■ **Figure S2.118** The application of a shear force to a metal lattice: adjacent layers of metal atoms can slide over each other

Nature of science: Models

Dislocations

Theoretical calculations, based on the strength of metallic bonds, suggest that it should be much harder to force layers of metal atoms to slide past each other than in practice. This difference between theory and practice is explained by the presence of defects in the metal crystal known as dislocations. (Dislocations can be observed using an electron microscope.)

Figure S2.119 shows one type of dislocation: part of an extra plane of atoms inserted into the metal structure like a wedge.

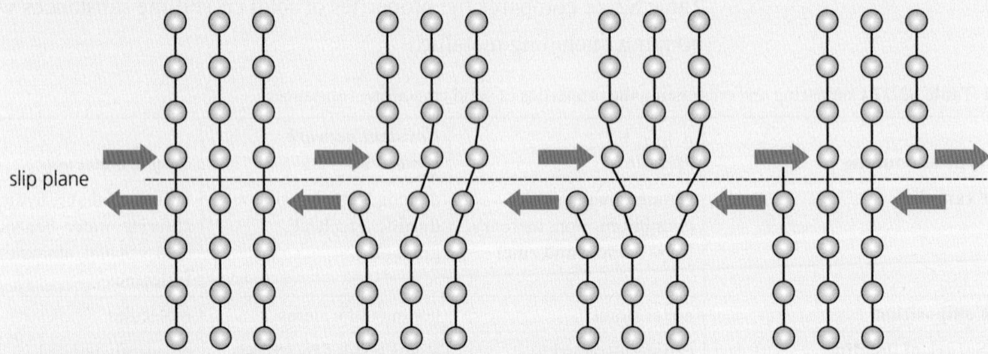

■ **Figure S2.119** The movement of a dislocation through a metal crystal

The presence of a dislocation makes it possible for the rows of metal atoms to shift along the slip plane by breaking and reforming only a small fraction of the bonds between them. As a result, slip can happen at a much lower stress than might otherwise be expected and this explains the malleability and ductility of metals.

ATL S2.3B

A bubble raft is an array of bubbles that can be made by bubbling air through soap. It models the plan of a close-packed crystal. Follow the instructions at https://thescienceteacher.co.uk/wp-content/uploads/2014/09/a-model-of-a-metallic-lattice.pdf to build a bubble-raft model of a metallic lattice.

Evaluate how well this model helps to explain the properties of metals.

Sonority

Metals are sonorous: they produce a ringing sound when struck. This makes metals useful for making bells and wires for musical instruments.

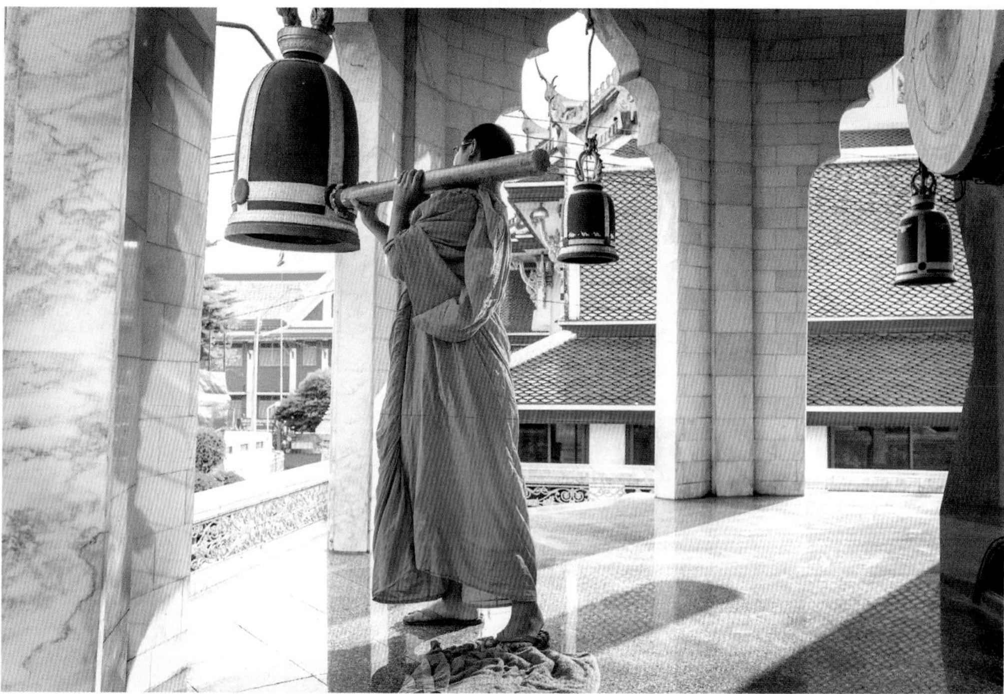

■ **Figure S2.120** A monk rings a bell in a Buddhist temple

Conductivity

The high thermal conductivity of metals is also explained by the simple metallic bonding model. When heat is supplied to a metal, the kinetic energy of the delocalized electrons is increased and these energetic electrons move through the lattice to the cold regions of the metal. The thermal conductivity is also partly due to the lattice vibrations passed on from one metal ion to the next.

The electron sea model also explains the high electrical conductivity of metals. If a potential difference (voltage) is applied across the ends of a metal sample, the delocalized electrons flow toward the positive electrode and this movement of electrons is an electric current. In practice, it is not the orderly flow shown in Figure S2.121: some electrons are scattered by the cations (hence causing resistance). However, there is a net flow of electrons in the direction shown.

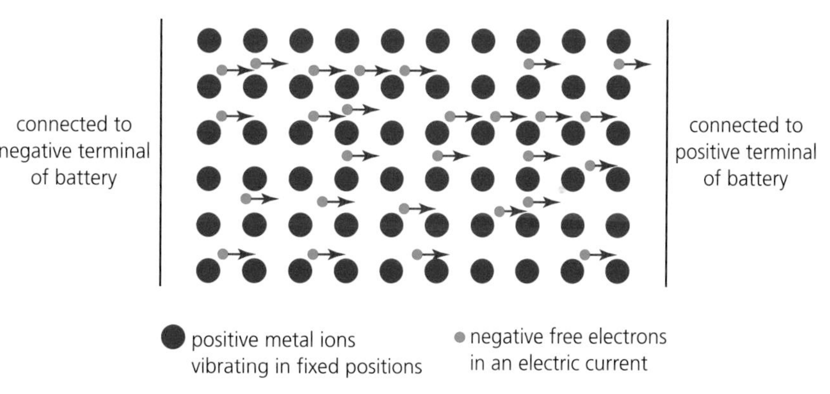

■ **Figure S2.121** The flow of current in a metal

Common mistake

Misconception: *Metals are conductors in the solid and liquid states due to the movement of cations.*

The charge carriers in both the solid and liquid states of metals are delocalized valence electrons.

TOK

Why are empirical facts important in the development of scientific models?

Earlier, we discussed how theories are a collection of ideas and beliefs that make sense of observations. This electron sea model is a good example of how we justify our acceptance of different scientific models: we think the electron sea model is useful precisely because it offers an explanation of the high electrical conductivity of metals. If it did not, we might need a better model.

Top tip!

Mercury and solid metals that have been melted are still metallically bonded, even though the ordered structure has been broken down. The bonding, between the cations and the delocalized electrons, is not fully broken or overcome until the metal boils.

ATL S2.3C

The interactive animation from the University of Colorado at **https://phet.colorado.edu/sims/cheerpj/conductivity/latest/conductivity.html?simulation=conductivity** demonstrates electrons moving through a material.

- Observe the differences in electron movement between metal and plastic when you increase the voltage of the battery.
- Explain why the metal conducted electricity while the insulator did not.
- Use the simulation to find out about the properties of photoconductors (a type of semiconductor) and research and present their uses.

Common mistake

Misconception: *Only metals and alloys can show metallic properties.*

Graphene and graphite show metallic behaviour in two dimensions due to the presence of delocalized electrons. Silicon and germanium can also show metallic properties when treated with impurities in a process known as doping.

22 The diagram shows a simple model of part of a metallic structure for sodium.

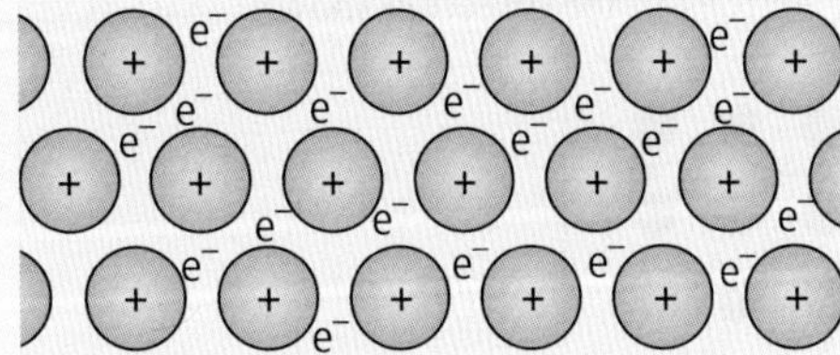

a State which atomic orbital in the sodium atom the sea of electrons was released from.

b Explain why these valence electrons are described as delocalized.

c Outline how a piece of sodium conducts electricity when a voltage is applied.

d Suggest two ways metallic bonding is similar to ionic bonding.

e State a limitation of this simple diagram as a model of sodium.

Going further

Packing of metal atoms

The strength of the metallic bonds leads to metal atoms being packed together as closely as possible.

Figure S2.122 shows **close packing** of atoms (modelled as hard spheres) in a single layer.

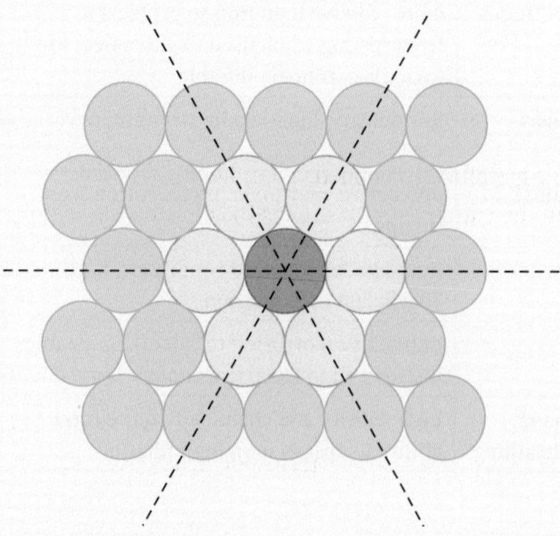

■ **Figure S2.122** Close packing of atoms

Imagine marbles that need to be packed in a box. The marbles would be placed on the bottom of the box in neat orderly rows and then a second layer started. The second layer of marbles cannot be placed directly on top of the other marbles and so the rows of marbles in this layer enter the dimples formed by a triangle of three marbles in the first layer. The first layer of marbles can be designated as A and the second layer as B, giving the two layers a designation of AB.

There are two possible ways of packing marbles in the third layer. Rows of atoms will again sit in the spaces between atoms in the second layer. If the rows of marbles in the third layer are packed so they are directly over the first layer (A) then the arrangement could be described as ABA. A packing arrangement with alternating layers—ABABAB—is known as hexagonal close packed.

If the rows of atoms in the third layer are packed so that they do not lie over atoms in either the A or B layers, then the third layer is called C. This ABCABC packing sequence is known as cubic close packing.

In both cases, the coordination number is 12, meaning each metal atom has 12 nearest neighbours: six in the same layer, three above and three below.

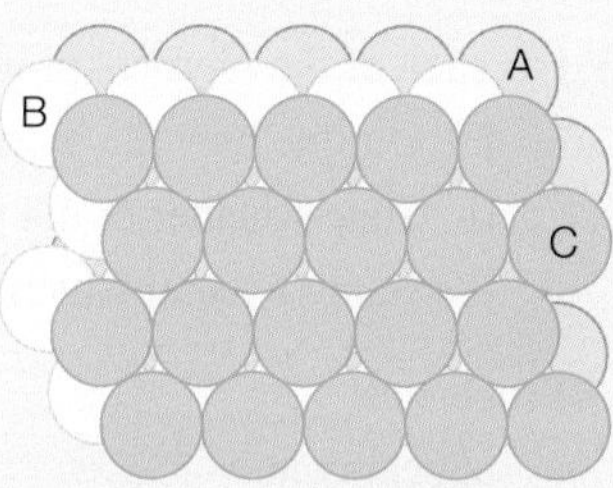

■ **Figure S2.123** Three layers in cubic close packing: A (blue), B (yellow) and C (green)

ATL S2.3D

Use the internet to find diagrams of the two structures commonly found in metals: cubic close packing and hexagonal close packing. Make models of these structures from expanded polystyrene spheres and mounting putty (such as Sticky Tack or Blu Tack). Compare the two models. Present them to your class and use them to illustrate the concepts of close packing, coordination number and unit cell.

◆ **Close packing**: The arrangement of atoms characterized by the largest number of atoms per unit volume of the crystal.

Uses of metals and alloys

Metals

Metallic properties make metals useful materials. Copper and aluminium, for example, are used for wires in circuits and electrical cables because of their high conductivity. Table S2.23 shows other common uses of a selection of metals and the chemical and physical properties that each use depends on.

■ **Table S2.23** Uses and properties of selected metals

Metal	Use as the pure metal	Properties
sodium	coolant in nuclear reactors	good conductor of heat and low melting point so that the metal flows along pipes
aluminium	airplane parts, window frames, overhead pylon cables, aluminium foil and trays	good conductor of electricity (lower cost and lower density than copper) when used as overhead cables resistant to corrosion (via spontaneous formation of a thin protective oxide layer)
zinc	electroplating to form galvanized iron, electrodes in batteries and die-casting	more reactive than iron so protects it from rusting when used to galvanize, low toxicity, extremely durable
titanium	spectacle frames, lightweight alloys and inert reaction vessels	low density, high **tensile strength**, unreactive
tin	electroplating of steel cans to form traditional 'tin' cans	unreactive, non-toxic, protects iron from rusting
mercury	liquid-in-glass thermometers, electrical switches	liquid under standard conditions, high coefficient of expansion
gold and silver	jewellery, electronics	unreactive (corrosion resistant), malleable, lustrous, high electrical conductivity, ductile
lead	car battery electrodes, weight belts for diving, radiation protection, ammunition, cable sheathing	high density and chemically quite inert, ability to absorb ionizing radiation

◆ **Tensile strength**: The maximum longitudinal force a material can withstand without breaking.

◆ **Alloy**: A mixture of a metal and one or more other elements.

23 For each of the physical properties a to d below:

i Outline how the property helps in the recognition of metals.

ii State whether or not the physical property is exclusive to metals.

iii Name a metal which does not have this typical property.

Properties:

a appearance

b melting point

c malleability and ductility

d conductivity (thermal and electrical).

LINKING QUESTION

What experimental data demonstrate the physical properties of metals, and trends in these properties in the periodic table?

ATL S2.3E

Find out which metals are used in the construction of smartphones. Choose one transition metal (d-block metal), such as nickel, and an f-block metal, such as neodymium. Outline the uses of the two metals and how their uses depend on their properties and cost. Evaluate the need to recycle the metals.

Alloys

While metals used in electrical circuits need to have high purity, pure metals are not usually used in engineering because they may easily undergo corrosion or be too soft. However, the properties of a metal may be modified by the formation of an alloy.

An **alloy** is a mixture of metals or a mixture of one or more metals and trace amounts of a non-metal (usually carbon but sometimes phosphorus, silicon or boron). Alloys usually have different physical properties from those of the component elements.

LINKING QUESTION

What are the features of metallic bonding that make it possible for metals to form alloys?

Top tip!

Atoms or ions of a different size (smaller or larger) also interrupt the passage of dislocations. Dislocations also get in each other's way, which is why a metal becomes stronger and less ductile as you deform it (think of bending a paper clip back and forth).

Going further

Alloys and the electron sea model

Alloys are often harder than the original metals because the irregularity in the structure helps to stop rows of metal atoms from slipping over each other (Figure S2.124).

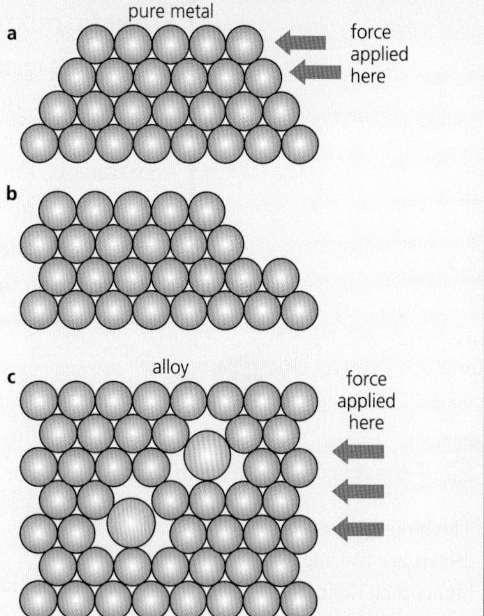

■ **Figure S2.124 a** The position of the atoms in a pure metal before a force is applied; **b** after the force is applied, slippage has taken place; **c** in an alloy, slippage is prevented because the atoms of different size cannot easily slide over each other

ATL S2.3F

Watch the Royal Society of Chemistry video at www.youtube.com/watch?v=3GOpP6L-XC0

Summarize the information it gives about the sodium–potassium alloy commonly known as NaK, which is used as a coolant in fast reactors.

Alloys are generally made by mixing the molten metal with the other elements and allowing the mixture to cool and solidify. Alloys are usually divided into **ferrous alloys**, which are mostly iron, and non-ferrous alloys that do not contain iron. **Steels** are ferrous alloys that contain carbon.

◆ **Ferrous alloy**: An alloy that is mainly iron.

◆ **Steel**: An iron-based alloy that contains a small amount of carbon.

■ **Table S2.24** Composition, uses and special properties of selected alloys

Alloy	Composition	Uses	Special properties
cupronickel	copper and nickel	'silver' coins, piping, heat exchangers	hard, silvery colour
stainless steel	iron, chromium and nickel	kitchen sinks, cutlery, surgical instruments	does not rust, remains lustrous, strong
magnalium	aluminium and magnesium	aircraft fuselages	strong
duralumin	aluminium and copper	aircraft fuselages, bicycle parts	lighter, stronger and more corrosion resistant than copper
brass	copper and zinc	musical instruments (such as trombones), door furniture, electrical connections	harder than copper, corrosion resistant
bronze	copper and tin	statues, ornaments, bells, castings, machine parts	harder than brass, corrosion resistant, sonorous
solder	tin and lead; tin, silver and copper	joining wires and pipes	low melting point
pewter	tin, lead and a small amount of antimony	plates, ornaments, mugs	resists corrosion
mild steel	iron and carbon	car bodies, large structures, machinery	easily worked, has lost most of the brittleness of hard steel
hard steel	iron and a higher percentage of carbon than mild steel	cutting tools, chisels, razor blades	tough but brittle

Going further

Eutectic mixtures

Unlike pure metals, most alloys do not have a single melting point; rather, they have a melting *range* in which the substance is a mixture of solid and liquid. However, for some alloys, such as solder (a mixture of tin and lead), there is one particular proportion of constituents, known as the eutectic mixture, at which the alloy has a unique melting point.

The reason why the eutectic composition has a unique melting point is that there is a single temperature where the three phases—one liquid and two solid—that make up the eutectic mixture are in thermodynamic equilibrium.

High-entropy alloys

High-entropy alloys (HEAs) are formed by mixing equal or relatively large proportions of (usually) five or more elements. The term 'high-entropy alloys' was coined because the entropy increase when pure metals are mixed in an alloy is significantly higher when there is a larger number of elements in the mix and their proportions are more nearly equal. HEAs have exceptional strength, ductility and toughness, especially at low temperatures. Researchers from Zhejiang University, the Georgia Institute of Technology and the University of California, Berkeley, engaged in collaborative research using atomic-resolution chemical mapping to understand the properties of HEAs.

Top tip!

The boiling points of metals are considerably higher than their melting points. This implies that most of the metallic bonding still exists in the liquid state. However, when the liquid changes into a gas (vapour), the atoms must be separated to large distances which involves breaking the metallic bonds. Hence, boiling points are a more accurate guide to the relative strength of metallic bonding.

◆ **Charge density**: The ratio of ion charge to the volume of the ion.

Melting points in s and p block metals

The melting point is an approximate measure of the strength of the metallic bonding in a metal lattice. The higher the melting point, the stronger the metallic bonding.

One factor controlling the melting and boiling point is the number of ionized valence electrons. The other factor controlling the strength of metallic bonding is the size of the metal ion. The smaller the ionic radius, the greater the **charge density** (Figure S2.125) and the stronger the metallic bonding. Thus:

$$\text{strength of metallic bond} \propto \text{charge density} \propto \frac{\text{number of valence electrons per atom}}{\text{metallic ion radius}}$$

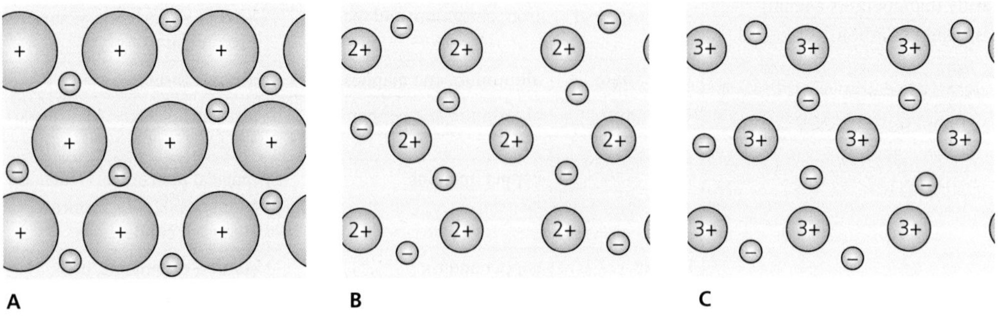

■ **Figure S2.125** Metals with small and highly charged cations form stronger metallic bonds: metallic bond strength A < B < C; charge density C > B > A

In period 3, from sodium to aluminium, the number of valence electrons per atom increases and the ionic radius decreases, as the large increase in nuclear charge outweighs the small increase in shielding. The strength of metallic bonding therefore increases from sodium (s-block), through magnesium (s-block) to aluminium (p-block). Consequently, the melting points, boiling points and electrical conductivity increase from sodium to aluminium (Table S2.25).

■ **Table S2.25** Melting points, boiling points and relative electrical conductivities of period 3 main group metals

Element	**sodium**	**magnesium**	**aluminium**
Melting point / °C	98	650	660
Boiling point / °C	883	1090	2519
Electrical conductivity (relative to aluminium)	0.55	0.61	1.00
Ionic radius / 10^{-12} m	102	72	54

The same trend is exhibited by the s-block metallic elements in period 2 (Table S2.26). The lithium ion has a charge of +1 and has a greater ionic radius than the beryllium ion. The lithium ion has a lower charge density than the beryllium ion and hence has weaker attraction to the sea of delocalized valence electrons.

■ **Table S2.26** Melting points, boiling points and relative electrical conductivities of period 2 main group metals

Element	**lithium**	**beryllium**
Melting point / °C	181	1287
Boiling point / °C	1342	2468
Electrical conductivity (relative to aluminium)	0.150	0.250
Ionic radius (10^{-12} m)	76	45

Top tip!

Boron is a metalloid and has a high melting point (2077 °C) due to the presence of a covalent network structure. Boron is a semiconductor with a low but measurable electrical conductivity at temperatures above absolute zero.

The boiling and melting points of s-block metals generally decrease as atomic number increases (Table S2.27). This is because the same number of delocalized electrons and larger cation radius lead to lower charge density which results in weaker metallic bonds.

Top tip!

Any 'breaks' in the expected trends will be due to a change in the lattice structure.

■ **Table S2.27** Melting and boiling points of s-block metals

Group 1 metal	**Boiling point / °C**	**Group 2 metal**	**Boiling point / °C**
lithium	1342	beryllium	2468
sodium	883	magnesium	1090
potassium	759	calcium	1484
rubidium	688	strontium	1377
caesium	671	barium	1845

Tin and lead are p-block metals in group 14. They have relatively low melting points which a simple theory of metallic bonding can explain as being due to the presence of large, low charge density Sn^{2+} and Pb^{2+} ions in the metallic lattice. The same theory would predict both the melting and boiling points of tin (period 5) to be higher than those of lead (period 6). This is true of the boiling point (2586 °C for tin and 1749 °C for lead) but the melting point of tin (231.9 °C) is lower than that of lead (327.5 °C) because tin and lead have different metallic structures.

Inquiry 2: Collecting and processing data

Identify and address issues that arise during data collection

You need to be able to identify where repeat measurements or experiments are necessary. You must conduct *at least one* repeat trial to ensure that the data point is reliable (that is, reproducible under the same conditions). If the second trial does not agree with the first (within experimental error), a third trial should be conducted and the outlier reading should be highlighted and excluded from the average.

If a graph is plotted from a series of values, the graph itself may give an indication of any outliers, which can be highlighted. This data should not be used when doing calculations or drawing lines or curves of best fit. If you are using a spreadsheet, you can achieve this by adding the suspect data to a separate data series in a spreadsheet.

Inquiry 2: Collecting and processing data

Identify and justify the removal or inclusion of outliers in data

You may find during an experimental investigation that one result in a set of measurements does not agree well with the other results and appears to be an outlier.

There are no hard and fast rules on how to deal with outliers; they should be treated case by case. Ideally, they should be tested statistically using a Q test (see, for example, **https://contchart.com/outliers.aspx**).

A simple checklist for experiments has two questions:

- Was the suspected outlier recorded in error?
- Was the suspected outlier recorded under different conditions to the other values?

If the answer to either of these questions is yes, then the outlier should be omitted from the data set and the mean should be calculated without this value.

HL ONLY

LINKING QUESTION

What trends in reactivity of metals can be predicted from the periodic table?

Top tip!

The units of electrical conductivity are siemens per metre, $S\,m^{-1}$. Measurements must be compared at the same temperature, typically 298 K.

Properties of transition elements

Electrical conductivity

The transition elements may have higher or lower values of electrical conductivity than p-block metals. For example, aluminium's electrical conductivity (at 25 °C) is $40.8 \times 10^6\,S\,m^{-1}$; that of silver is higher ($66.7 \times 10^6\,S\,m^{-1}$) but nickel has a lower value ($16.4 \times 10^6\,S\,m^{-1}$).

The relatively high electrical conductivity of some transition metals is due, in part, to the greater number of delocalized electrons donated by each metal atom. A simple model assumes both the 3d and 4s electrons are available since these electrons have similar energies. However, there is no systematic variation of conductivity with valence electron density across the first row of the d-block metals.

ATL S2.3G

Use the internet to find the following data about the first-row transition metals:

- valence electron configuration
- lattice structure
- number of 3d and 4s (valence) electrons
- electrical conductivity (absolute values with units or relative to a specific metal).

Present and analyse the data, then evaluate the ability of the electron sea model to account for any observed trends and correlations.

Electrical conductivity cannot be fully explained using valence charge density alone. The fact that both it and thermal conductivity depend on temperature is a clear indication that additional concepts are needed to explain any differences or trends.

Melting points

The transition metals have higher melting and boiling points than s-block metals due to the stronger metallic bonding that results from the greater number of valence electrons (3d and 4s) and smaller ionic radii. The cations have a higher charge density and hence a stronger electrostatic attraction to the delocalized sea of electrons.

LINKING QUESTION

Why is the trend in melting points of metals across a period less evident across the d-block?

■ **Table S2.28** Melting and boiling points of first-row d-block elements

Element	Sc	Ti	V	Cr	Mn	Fe	Co	Ni	Cu	Zn
Melting point / °C	1541	1670	1910	1907	1246	1538	1495	1455	1085	419.5
Boiling point / °C	2836	3287	3407	2671	2061	2861	2927	2913	2560	907

The melting points also depend to some extent on the packing arrangement of ions, which varies between metals (Figure S2.126). The lattice structure also affects conductivity.

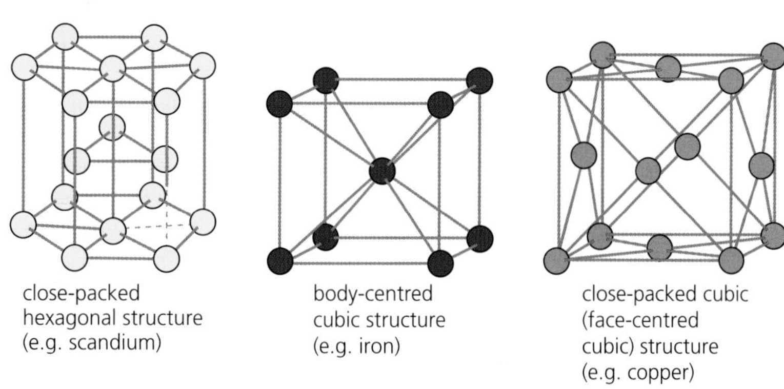

■ **Figure S2.126** Three common metal lattices: close-packed hexagonal, body-centred cubic and face-centred cubic structures (also known as cubic close-packed)

Going further

Superconductivity

Metals are good electrical conductors because of the delocalized valence electrons in the metallic lattice. At standard temperature metals have some resistance to the flow of electrons because the electrons collide with the cations in the lattice.

In 1911, Heike Kamerlingh Onnes, a Danish scientist, discovered that when mercury is cooled to about 4 K (–269 °C) it loses all resistance to electrical flow, a property known as superconductivity.

In superconductors, electrons form 'Cooper pairs' and travel through the lattice as a wave. The wavelength of this wave is such that it does not interact with the lattice and, therefore, the electrons are not scattered, leading to no resistance. This means that superconductors can carry a current indefinitely without any electrical energy transferred to heat.

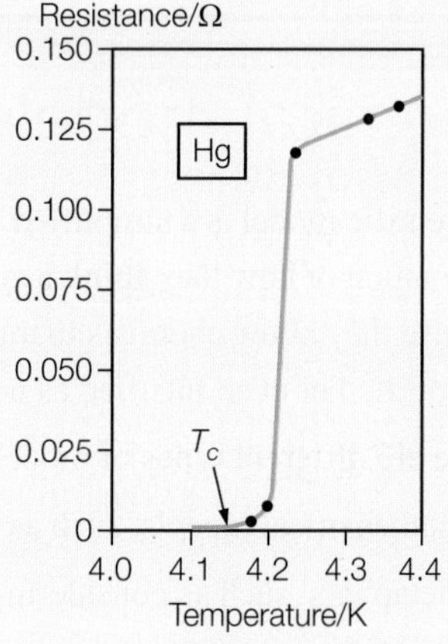

■ **Figure S2.127** Plot of resistance against temperature for mercury with T_c representing the transition temperature to superconductivity

HL ONLY

S2.4 From models to materials

Guiding question

- What role do bonding and structure have in the design of materials?

SYLLABUS CONTENT

By the end of this chapter, you should understand that:

- bonding is best described as a continuum between the ionic, covalent and metallic models, and can be represented by a bonding triangle
- the position of a compound in the bonding triangle is determined by the relative contributions of the three bonding types to the overall bond
- alloys are mixtures of a metal and other metals or non-metals; they have enhanced properties
- polymers are large molecules, or macromolecules, made from repeating subunits called monomers
- addition polymers form by the breaking of a double bond in each monomer
- condensation polymers form by the reaction between functional groups in each monomer with the release of a small molecule (HL only).

By the end of this chapter you should know how to:

- use bonding models to explain the properties of a material
- determine the position of a compound in the bonding triangle from electronegativity data
- predict the properties of a compound based on its position in the bonding triangle
- explain the properties of alloys in terms of non-directional bonding
- describe the common properties of plastics in terms of their structure
- represent the repeating unit of an addition polymer from given monomer structures
- represent the repeating units of polyamides and polyesters from given monomer structures (HL only)
- describe the hydrolysis and condensation reactions that break down and form biological molecules (HL only).

Chemical models

Link
Electronegativity is explored further in Chapter S3.1, page 234.

A scientific model is a simplified description of an idea that allows scientists to create an explanation of how they think a system or process of the physical world works. Models are useful because they allow chemists to make predictions and carry out calculations. Models may of course be modified or even falsified as new data are discovered.

There are different types of models including:

- mathematical models, such as Coulomb's law
- metaphors, such as considering electrons to be spinning particles
- teaching models, such as considering electrons to be orbiting the nucleus in circular orbits.

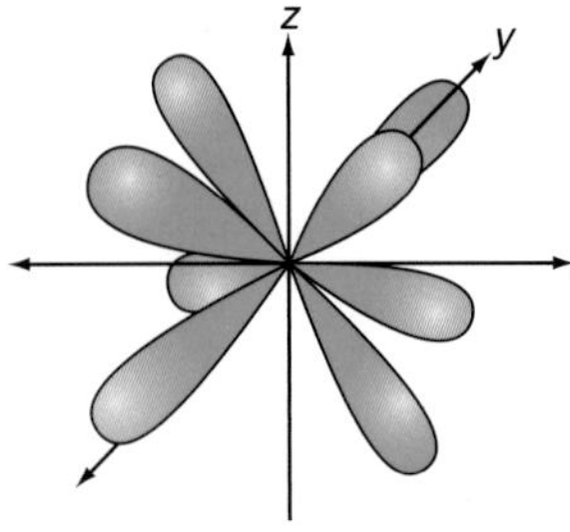

■ **Figure S2.128** Computer-generated model showing the shape of one 4f orbital. There are eight lobes pointing out

Some chemical models use precisely defined concepts; for example, the model of atomic orbitals (Figure S2.128) is grounded in quantum physics.

However, chemistry also makes use of somewhat 'fuzzy' concepts, such as aromaticity and **electronegativity**. Electronegativity is intended to be a measure of the electron-pair attracting power of an atom in a molecule. Although, in practice, this is influenced to some extent by the environment of the atom within the molecule and the oxidation state of the atom, the concept is still useful and used in the bonding triangle diagram (van Arkel diagram) used later in this chapter.

The term 'model' is potentially misleading since a scientific model is *not* the same as a scale model: atoms are not physical structures that can be seen or directly observed.

This scientific model is simply the abstract idea that subatomic particles exist and show certain behaviours. Such particles are not necessarily real in the way that macroscopic objects that we can see are real.

You should not confuse models with reality—they are meant to be useful tools. Chemical models of the atom and molecules, and hence chemical bonding models, have changed over time as new experimental or theoretical data have been obtained.

Material scientists, like chemists, are interested in the electron configurations and hence shapes and energies of orbitals in atoms because these determine the way they bond together.

The three main classes of materials can be related to the three types of strong chemical bonding.

- In metals and alloys, the bonding is metallic.
- In polymers, the strong bonding is covalent (with weak intermolecular forces such as hydrogen bonds between the polymer molecules).
- In ceramics, the bonding may be ionic, covalent or some combination of the two.

The bonding triangle (van Arkel diagram)

◆ **Bonding triangle**: A diagram (based on electronegativity difference) used to show binary compounds in varying degrees of ionic, metallic and covalent bonding.

The simple **bonding triangle** (Figure S2.129) applies to ionic or covalent binary substances (ones that contain two different elements) and to elements.

Top tip!

The bonding triangle only applies to binary compounds and alloys formed of two elements.

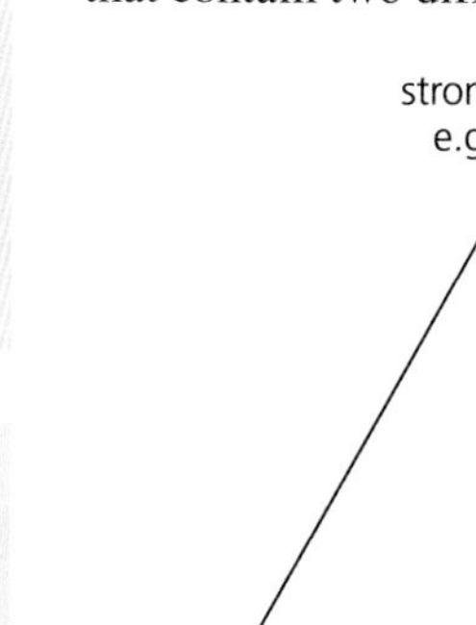

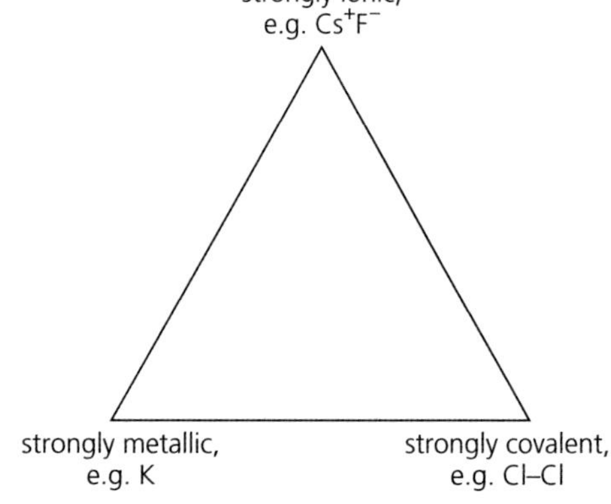

■ **Figure S2.129** Simple bond triangle diagram

TOK

Does the list of disciplines included in, or excluded from, the natural sciences change from one era to another, or from one culture or tradition to another?

In the early 20th century, most engineering schools in universities had a department of metallurgy, with a focus on steel structure and, perhaps, ceramics. The fundamental understanding of other materials was not advanced enough for them to be regarded as academic disciplines. However, after World War II, the scientific study of polymers progressed rapidly. Administrators and research scientists began to think of materials science as a new discipline in its own right, one that viewed all substances of engineering importance from a unified perspective involving both chemical and physical properties.

Metallic bonding is favoured by atoms with low values of electronegativity; covalent bonding is favoured by atoms with high values of electronegativity; and ionic bonding is favoured where one atom has a low value of electronegativity and the other a high value (Table S2.29).

- Binary ionic compounds consist of a metal and a non-metal: for example, potassium fluoride, KF.
- Binary elements include the halogens and the diatomic gaseous elements O_2, H_2 and N_2. The bonding in these is molecular covalent.
- Binary inorganic compounds which are covalent include the electron-deficient boron trichloride, BCl_3.
- Binary metallic alloys include solder (tin and lead), bronze (copper and tin) and brass (copper and zinc).

Top tip!

The bonding between atoms is determined by the electronegativities of the atoms and the differences between these electronegativities.

■ **Table S2.29** Bonding types and electronegativity

Type of bonding	Size of electronegativity values	Difference between electronegativity values
metallic	small	small
covalent	large	small
ionic	small (metal) and large (non-metal)	large

In practice, there are many substances along the three sides of the bonding triangle that have intermediate properties. For example, a substance with polar covalent bonds exhibits properties intermediate between those of substances with highly ionic and highly covalent character.

ATL S2.4A

Find out about the work of the Dutch chemists van Arkel and Ketelaar, and that of William Jensen, in developing various bond triangles. Present your research using electronic presentation software, such as PowerPoint. Outline the strengths and limitations of this approach to chemical bonding theory.

The physical properties of substances can be explained by their type of chemical bonding and structure. The triangular bonding diagram (van Arkel diagram) shown in Figure S2.130 is a useful quantitative model that helps chemists predict the structure and properties of binary compounds. This diagram is in section 17 of the IB *Chemistry data booklet*.

Top tip!

The van Arkel diagram maps bonding type to theories of chemical bonding.

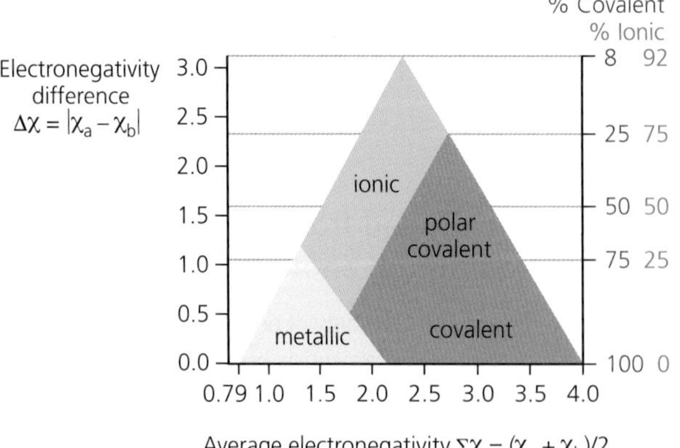

■ **Figure S2.130** The triangular bonding diagram (van Arkel diagram)

The triangular bonding diagram is a plot of the difference of the electronegativity values of the two elements in the compound (*y*-axis) against the average of the electronegativity values (*x*-axis). The electronegativity values are used without any regard for the stoichiometry (formula) of the binary compound.

The x-axis describes the degree of localization of the bonding electrons and so provides information about the degree of covalency. The most electropositive metals have delocalized valence electrons and lie at the left-hand extreme of the x-axis. At the right-hand extreme of the x-axis is the most electronegative element, fluorine (F–F).

The y-axis gives information about the degree of ionic character of the substance. Caesium fluoride, $[Cs^+ F^-]$, is found at the top of the triangle. This is the compound with the highest degree of ionic character as the two elements are the most electropositive (least electronegative) and the most electronegative.

Predictions from the bonding triangle

The triangular bonding diagram shows the three types or models of chemical bonding as part of a single scheme in which they are all related and have equal weighting. It offers a quantitative appreciation of intermediate bonding and also shows that intermediate bonding is not limited to polar covalent or ionic with covalent character. In addition, it is based on the familiar concept of electronegativity.

Plotting points on a triangular bonding diagram allows us to predict the bonding type (and hence properties).

Method one

Ignore the stoichiometry of the compound (the numbers in the formula). Calculate the average of the electronegativity values of the two elements. This is the x-coordinate. Deduce the difference between the two electronegativity values. This is the y-coordinate.

Tool 2: Technology

Generate data from models and simulations

Online tools can quickly and accurately calculate and predict the type of bonding in a binary compound or binary alloy using the values of electronegativity for the two elements.

- Visit the Metasynthesis synthlet at **www.meta-synthesis.com/webbook/38_binary/binary.php?mge_2=K&mge_1=Na** and use it to predict the bonding types of the following substances: OF_2, AlP, SnPb, CsS, NF_3, SF_6, Na_3Bi, BrCl and Mg_2Si.

WORKED EXAMPLE 2.4A

Deduce the nature of the bonding in phosphorus(V) chloride from the triangular bonding diagram. Electronegativity values of phosphorus and chlorine are 2.2 and 3.2.

Answer

$$\text{average electronegativity} = \frac{2.2 + 3.2}{2}$$

$$\text{difference in electronegativity} = 3.2 - 2.2 = 1.0$$

PCl_5 is polar covalent.

Method two

Find the electronegativities of the two elements that form the compound on the x-axis. Draw a line from the left-hand element parallel to the left-hand side of the triangle; draw a line from the right-hand element parallel to the right-hand side of the triangle. Where these lines intersect is the place to plot the compound in the triangular bonding diagram.

WORKED EXAMPLE 2.4B

Determine the position of gallium nitride, GaN, on the triangular bonding diagram. The electronegativity values for the elements are Ga = 1.8 and N = 3.0.

Answer

Using method 1

The average is 2.4 and the difference is 1.2. This places it in the polar covalent region of the triangular bonding diagram.

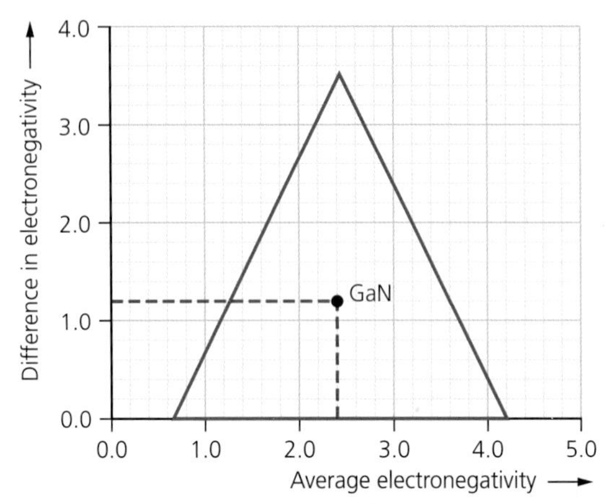

Using method 2

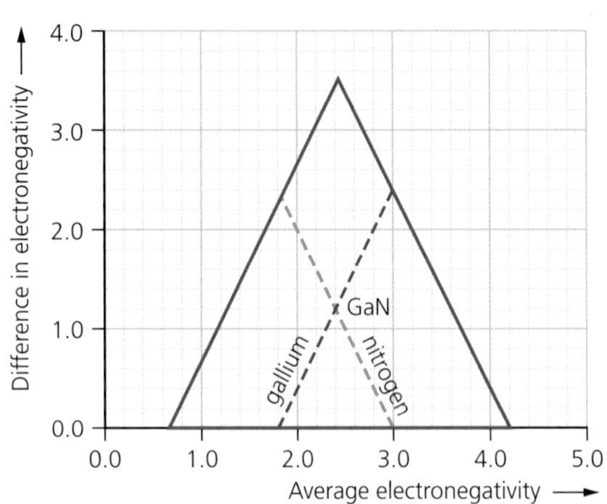

WORKED EXAMPLE 2.4C

For each of the chlorides listed below, find their coordinates on the triangular bonding diagram, predict their bonding type and, where appropriate, comment on it.

Chlorides: NaCl, $MgCl_2$, $AlCl_3$, $SiCl_4$ and SCl_2.

Answer

Formula of compound	Triangular bonding diagram coordinates	Bonding type predicted from the triangular bonding diagram	Comment
NaCl	2.05, 2.30	ionic	ionic
$MgCl_2$	2.25, 1.90	ionic	ionic
$AlCl_3$	2.40, 1.60	polar covalent	the high charge density on Al^{3+} distorts the chloride ion resulting in polar covalent bonding
$SiCl_4$	2.55, 1.30	polar covalent	simple molecular
SCl_2	2.90, 0.60	covalent	simple molecular

24 Use values of electronegativity and the triangular bonding diagram to predict the type of bonding present in:

a indium(III) chloride, $InCl_3$

b hydrogen fluoride, HF

c caesium hydride, CsH

d aluminium iodide, AlI_3

e disulfur dichloride, S_2Cl_2

f potassium chloride, KCl

g sodium–potassium alloy, NaK.

Going further

Limitations of the van Arkel diagram

All models have their limitations and tend to break down at some point. Although the van Arkel diagram provides information on bonding, it does not predict the crystal structures (lattices) of solids as these depend on stoichiometry and electron configuration.

This is illustrated by the existence of elements with allotropes, such as carbon. Graphite and diamond have very different structures and properties (consider, for example, electrical conductivity) even though both are composed of a single element with low electronegativity.

The same is also true when there are several forms of a binary compound because one of the elements can exist in several stable oxidation states. A good example of multiple oxidation states are the chlorides of lead. Lead(IV) chloride, $PbCl_4$, is essentially covalent and it hydrolyses in water, whereas lead(II) chloride, $PbCl_2$, is essentially ionic and dissolves in water.

LINKING QUESTION

What are the limitations of discrete bonding categories?

LINKING QUESTION

How do the trends in properties of period 3 oxides reflect the trend in their bonding?

LINKING QUESTION

Why do composites such as cement, which are made from both ionic and covalently bonded components, have unique properties?

Tool 1: Experimental techniques

Physical models

Making your own physical models can help you to visualize concepts more clearly, especially those displayed in three dimensions. Kinaesthetic approaches that involve making something also add variety and stimulate engagement with the concepts.

- Download and print this page from Metasynthesis: **www.meta-synthesis.com/TT.pdf** Use it to create a physical model of the truncated tetrahedron of structure, bonding and material type.

Materials science in Mesopotamia

The word 'Mesopotamia' is formed from the ancient Greek word *meso*, meaning between or in the middle of, and *potamos*, meaning river. It is in what is now southern Iraq near the Tigris and Euphrates rivers; the world's earliest civilization developed here.

The early inhabitants, known as the Sumerians, found their environment had few trees, almost no stone and no metals or their ores. Clay was the most abundant material and the Sumerians used this for pottery (for storing food), terracotta sculptures, tablets (for recording information as pictures and, later, as a written language—cuneiform) and clay cylinder seals, used to mark documents or property.

Around 3400 BC, at the city of Uruk, the Sumerians developed a type of concrete: a wet mix of gypsum (calcium sulfate dihydrate) and crushed baked bricks that dried hard. Clay is used in the production of cement, but Portland cement was first developed in 1824 and was the world's first hydraulic cement, which hardens when water is added.

Going further

Concrete

Concrete is a composite material and consists of a binder and a filler. The filler is usually gravel (small stones) and sand. The binder is cement, usually Portland cement. The three phases (sand, cement and gravel) in combination give better mechanical properties than each on its own and this is the reason why composite materials are used. The sand and gravel particles must be bound together by the cement phase so that as many as possible of the interstices (holes) are filled and the maximum strength is produced. The setting of cement upon the addition of water is a slow and complex set of physical and chemical processes.

Inquiry 1: Exploring and designing

Formulate research questions and hypotheses

A research question usually asks how an independent variable affects a dependent variable (with other variables controlled and monitored). A good chemical research question will be precisely focused on a narrow chemical topic, supported by theory with one or more models. This may allow for the generation of a hypothesis with testable predictions.

For example, 'What factors affect the setting time of concrete?' is a poor and unfocused research question. A better research question would be, 'How does the amount of water in a concrete mix of Portland cement, sand and gravel affect the tensile strength of the concrete?'. If too much water is added, the binding effect of the cement is lost; if too little water is added, the binding action will not be complete. The quality and quantity of the cement will affect the strength of the bond between the particles of aggregate and so must be controlled.

Link

The uses of alloys are discussed in Chapter S2.3, page 188. For a list of the composition, uses and special properties of selected alloys, see Table S2.24, page 189.

Top tip!

Alloys are often described as mixtures, but compounds may be formed; cementite, Fe_3C, in steel is a simple example.

Top tip!

Because the metallic radii of the d-block elements are all similar, they can often form a range of alloys with one another through an atom of one element taking the place of another.

LINKING QUESTION

Why are alloys more correctly described as mixtures rather than as compounds?

Alloys

Most pure metals are not used in engineering because of poor mechanical properties, such as low tensile strength, high malleability and softness. The properties of metals may be enhanced by the addition of other elements, usually metals but sometimes metalloids or non-metals, especially carbon. The products are known as alloys if they have the physical properties of metals.

The properties of alloys are affected by their composition, their crystalline structure and their grain structure (the size and distribution of individual crystals in the bulk material).

In homogeneous alloys, atoms of the different elements are distributed uniformly, either randomly or in regular arrays, like true compounds. Examples are some types of brass, bronze and the alloys used for making coins.

Heterogeneous alloys consist of mixtures of crystalline phases with different compositions. Examples are tin–lead solder, the mercury–silver amalgams traditionally used to fill teeth, and steel which has an iron matrix with cementite (Fe_3C) particles that make the steel strong.

Alloys are generally prepared by mixing molten liquid metals together or dissolving a solid (for example carbon) in a molten metal (for example iron), but this is not always possible. For very high melting point alloys, powder metallurgy is used: fine powder is compressed into shapes at high pressure followed by heating. Some alloys (for example, brass) can be deposited electrolytically.

Nature of science: Observations

Early metallurgy

Metals and alloys were used for many different purposes long before the development of a scientific understanding of their properties based on chemical bonding theory and experimental evidence such as X-ray crystallography.

As far as we know, the early knowledge and skill in metallurgy (study of metals and alloys) must have been reached by a process of trial and error; or, in other words, by the method of experiment and observation.

Copper is found native (pure) and was used by people over 10,000 years ago. Early metalworkers (smiths) must have observed that copper hardened when it was hammered and could be softened by heating it again (annealing).

It is not known when, where or how bronze was discovered. A curious coppersmith may have deliberately added tin to melted copper. Bronze is easier to cast than copper and much harder.

Typical properties of alloys

Increased hardness

Pure metals are soft but their alloys are hard. For example, pure gold is soft but when alloyed with copper it becomes hard. Rose gold is a gold–copper alloy widely used for specialized jewellery (Figure S2.131).

Lower melting points

Alloying can reduce the melting point of metals. For example, Wood's metal (an alloy of lead, bismuth, tin and cadmium) melts at approximately 70 °C, significantly less than its component metals. It is used as a solder (Figure S2.132).

■ **Figure S2.131** Rose-gold jewelry

■ **Figure S2.132** Wood's metal is used as a solder for connecting electrical circuits

Improved corrosion resistance

Some pure metals are sensitive to corrosion but alloying often makes them less reactive. Bronze, an alloy of copper and tin, is more resistant to corrosion than copper. Pure iron rusts easily but stainless steel (an alloy of iron, nickel and chromium) is very resistant to corrosion and is used in cutlery and medical instruments (Figure S2.133). The formation of a dense film of chromium(III) oxide prevents rusting.

Top tip!

Ferrous alloys contain iron as the major component and are known as steels.

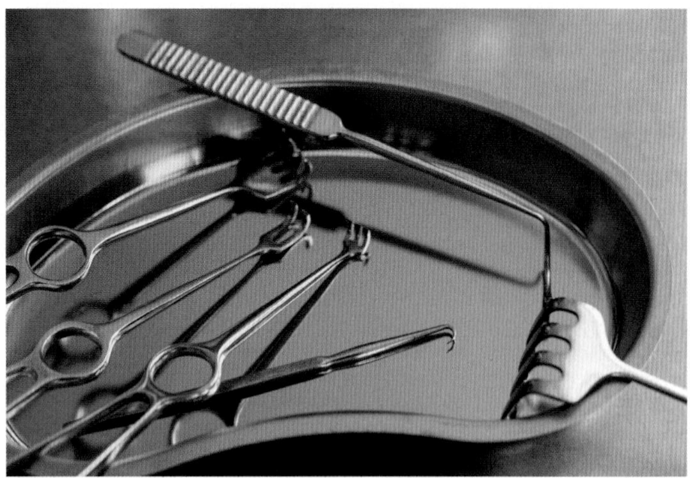

■ **Figure S2.133** Non-magnetic stainless-steel surgical instruments

Modified colour

The colour of metals can change when they are alloyed. For example, brass is yellow but its components are red (copper) and silver-white (zinc).

25 Define the term alloy and state two properties of alloys that are typically enhanced compared to the pure metals.

26 Distinguish between ferrous and non-ferrous alloys. Give one specific example of each type.

ATL S2.4B

Choose two alloys not mentioned in the text (for example: nichrome, electrum, invar, sterling silver or alnico) and develop a presentation that includes the following information: name of alloy, typical composition by mass, atomic structure, microstructure, properties that are enhanced compared to the pure metal(s) of which it is composed, and specific uses.

The properties of alloys containing metals arise from the nature of the metallic bond and the crystalline structure of metals. In a very simple model of metallic bonding, metals are giant structures in which the delocalized valence electrons are free to move.

Metals can easily change shape without breaking because the metallic bonding allows layers of cations to slip over one another when the metal is under stress (when a force is applied to it). Unlike covalent bonds, metallic bonds are not directional so, when a plane of atoms has slipped, the crystal structure is still the same as it was before. (In practice, the plastic deformation of metals involves defects known as dislocations.) In an alloy, the different sized atoms result in a disordered lattice without planes of atoms. This makes the alloy harder and less ductile.

Going further

Grains

Objects made from metals and mixtures of metals are often formed by casting: allowing molten metal or a mixture of molten metals to cool down in a mould. As the liquid cools, small nuclei of the solid appear and, as cooling continues, small crystals are formed. The crystals grow and meet to form a solid mass of small crystals, a polycrystalline solid. These crystals are called grains (Figure S2.134).

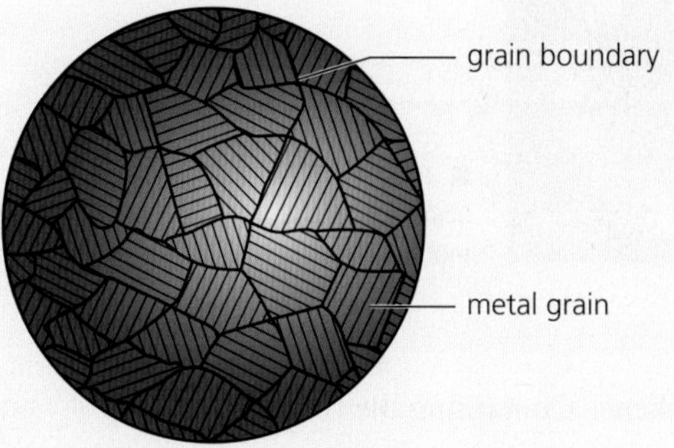

■ **Figure S2.134** Arrangement of metal grains

The size, shape and alignment of these crystal grains give rise to the microstructure of the metal or metal alloy.

The grains of a piece of polished metal can be seen easily with a microscope; however, the best place to see metallic crystal grains is on a galvanized lamp post where large grains of zinc are clearly visible (Figure S2.135).

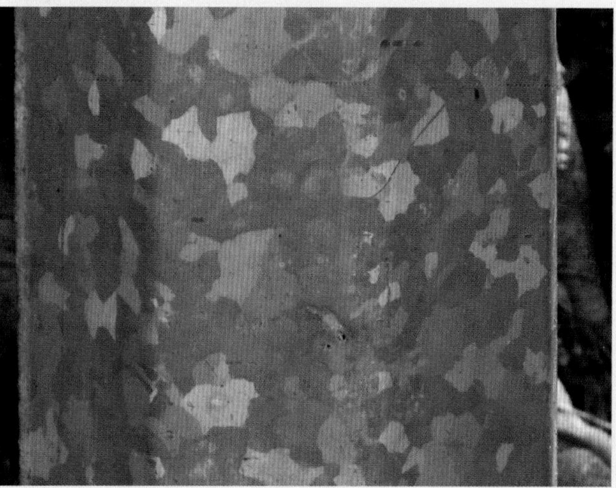

■ **Figure S2.135** The grains of zinc on a galvanized lamp post can be seen clearly

ATL S2.4C

The properties of metals and metal alloys vary with their microstructure so metallurgists have developed ways to alter the shape and size of the metal grains.

Find out about the following techniques and how grains are involved in each: hammering (forging), quenching, tempering and annealing.

Polymers

◆ **Monomers**: Small molecules that are covalently bonded to make a polymer.

◆ **Polymer**: A long-chain molecule made by covalently bonding a large number of molecules.

◆ **Polymerization**: The chemical reaction involving the formation of a polymer from its monomers.

Polymers are molecules of very high molar mass, with each molecule being formed from a large number of small molecules known as **monomers**. The process by which monomers are covalently bonded to form a polymer is known as **polymerization**. There are two major classes of polymers: addition polymers (including polyethene) and condensation polymers (including nylon and most biological polymers). Rubber, cotton, proteins, starch, cellulose and nucleic acids are natural polymers; plastics are artificial polymers.

Top tip!

Polymers are sometimes referred to as macromolecules as they have very long molecules with high values of molar mass.

Nature of science: Theories

Polymer chemistry

Up until the 1920s, there was little understanding of the molecular structure of plastics. It was generally assumed that the small molecules (monomers) from which they were made physically aggregated together into larger units. It was the German chemist Hermann Staudinger who first recognized that polymers are formed by the polymerization of monomers joining covalently to make larger molecules (Figure S2.136). He is called the 'founder of polymer chemistry' and received the 1953 Nobel Prize in Chemistry.

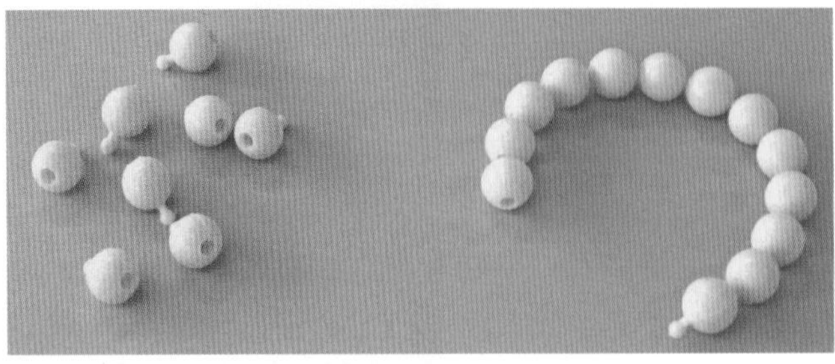

■ **Figure S2.136** The assembly of beads into a chain represents a simple model of the polymerization process

Addition polymers

◆ **Addition polymerization**: A type of polymerization that occurs with alkene-based monomers undergoing repeated addition reactions.

The simplest addition polymers are formed from alkenes (containing the functional group >C=C<) and their derivatives. These molecules can be made to join together in the presence of a suitable catalyst. High pressure is also sometimes required. The π bond breaks and the monomers join together by forming covalent bonds (new sigma bonds) (Figure S2.137). The product of this addition process is a very long hydrocarbon chain. No other product is formed and so the reaction is known as **addition polymerization**.

monomer ethene

double bonds break open (π bonds broken)

polymer poly(ethene)

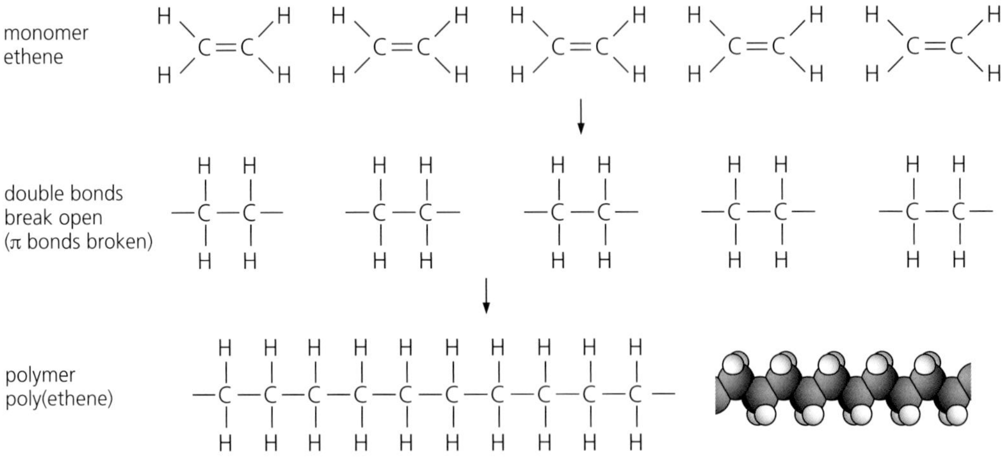

■ **Figure S2.137** A simple description of the formation of polyethene from ethene

Top tip!

Typically, there are several hundred or even thousands of monomers bound together in each polymer chain.

◆ **Repeating unit**: The fundamental recurring unit of a polymer.

The polymer is made up of a unit that repeats many times. This is known as the **repeating unit**. A single repeating unit shows the full structure of the polymer when written as shown in Figure S2.138.

$$n\ H_2C{=}CH_2 \xrightarrow{\text{polymerization}} \left(-CH_2-CH_2- \right)_n$$

■ **Figure S2.138** The formation of polyethene (where *n* represents a large number of monomer molecules) showing the repeating unit.

Nature of science: Observations

Ziegler catalysts

Some of the greatest scientific discoveries have resulted from what at first seemed to be failed experiments. The modern era of polyalkene chemistry, which has helped to make plastics an important presence in our everyday lives, has its origins in such an experiment.

Karl Ziegler (1898–1973) was a German chemist who won the Nobel Prize in Chemistry in 1963. He experimented with ethene with the aim of polymerizing it on the industrial scale in the presence of triethylaluminium, but his attempts were unsuccessful because of a competing elimination reaction. He reasoned this was due to the presence of a contaminant which he eventually found to be caused by a trace of nickel salts left after cleaning the stainless-steel reaction vessel.

Ziegler's observations conflicted with his working model based on earlier successful polymerization with triethylaluminium and he needed to develop the model he was using.

He reasoned other metal salts might delay this competing elimination reaction and promote the main polymerization reaction. Following a systematic study, he found titanium salts formed a complex in the presence of triethylaluminium that catalysed the rapid formation of stereoregular crystalline polyethene.

Polymers made from alkenes are known as polyalkenes. Polyalkenes are long-chain hydrocarbons which are saturated (only contain carbon–carbon single bonds) and non-polar.

ATL S2.4D

Find out about the two forms of polyethene: low density and high density polyethene. Describe the reaction conditions during their manufacture and explain why they have different physical properties (such as density, tensile strength, melting point and flexibility) by reference to their structures and packing arrangements. Outline their uses.

Chemists can also use substituted alkenes, such as chloroethene, as monomers since they have a reactive alkenyl functional group.

Some more examples of monomers and the addition polymers that they form are shown in Table S2.30.

■ **Table S2.30** Selected addition polymers

Monomer	Structure of polymer and IUPAC polymer name	Alternative names for polymer	Properties	Uses
ethene	poly(ethene)	polyethylene, polythene, PE	tough, durable	plastic bags, bowls, bottles, packaging
chloroethene	poly(chloroethene)	poly(vinyl chloride), PVC	strong, hard (not as flexible as polyethene)	insulation, pipes and guttering, waterproof coatings
propene	poly(propene)	polypropylene, PP	tough, durable	crates and boxes, plastic rope, carpets
tetrafluoroethene	poly(tetrafluoroethene)	polytetrafluoroethylene, Teflon, PTFE	non-stick surface, withstands high temperatures	non-stick frying pans, non-stick taps and joints, plumbing tape
phenylethene	poly(phenylethene)	polystyrene, PS	light, poor conductor of heat	insulation, packaging (foam)

WORKED EXAMPLE 2.4D

Part of the structure of an addition polymer is shown below.

a Deduce and draw the repeating unit of this polymer.

b Deduce and draw the structure of the monomer used to synthesize this polymer.

c Explain why this polymer is not a hydrocarbon.

Answer

a This polymer has four repeating units. One is drawn and placed inside square brackets and multiplied by *n*, a very large number.

b The repeating unit has its carbon–carbon single bond converted to a carbon–carbon double bond and the dangling bonds at each end are removed. The monomer is drawn showing the trigonal planar shape around the two carbon atoms.

c It is not a hydrocarbon since it contains fluorine and chlorine. A hydrocarbon is a compound that contains carbon and hydrogen atoms only.

Top tip!

The name of a polymer is the prefix *poly* followed by the name of the monomer in brackets.

ATL S2.4E

Visit **https://javalab.org/en/addition_polymerization_en/** and use the simulation to convert various monomers into their addition polymers.

27 Draw two repeating units of the polymer produced by the following alkenes:

a propene
b but-1-ene
c but-2-ene
d phenylethene
e tetrafluoroethene

28 For each of these polymers, draw the monomer from which it is made.

a $-[C(CH_3)_2-C(CH_3)_2-C(CH_3)_2-C(CH_3)_2]_n-$

b $-[CH(CH_3)-CHCl-CH(CH_3)-CHCl]_n-$

c $-[CH_2-CCl_2-CH_2-CCl_2]_n-$

Properties of addition polymers

The structure of polymers gives them a number of characteristic properties:

- Since the hydrocarbon chains are often very long, the London (dispersion) forces between the chains are often very strong and the polymers have relatively high softening and decomposition points.
- Since the chain length is variable, most polymers contain chains of a variety of different lengths. The London (dispersion) forces are therefore of variable strength and these polymers tend to melt gradually over a range of temperatures rather than at a fixed temperature.
- Since the chains are not rigidly held in place by each other, the polymers tend to be relatively soft.
- Since the chains are non-polar, polyalkenes and many other polymers are insoluble in water and are usually impermeable to water too.
- Since the intermolecular forces between the molecules are strong and the chains are often tangled, many polymers are insoluble in non-polar solvents as well.
- Since the hydrocarbon chains that form the 'backbone' of many polymers are saturated, most plastics—and polyalkenes in particular—are very unreactive and resistant to chemical attack. However, some plastics may be damaged by exposure to ultraviolet radiation (sunlight).
- Since they lack delocalized electrons, polyalkenes and many other polymers are thermal and electrical insulators.

The specific properties of a polymer are affected by the other groups that are present. For example, the presence of the polar $C^{\delta+}-Cl^{\delta-}$ bonds in poly(chloroethene) (PVC) gives it very different properties from those of poly(ethene). Every PVC molecule has a permanent dipole allowing strong dipole–dipole interactions to occur between neighbouring chains. Also, chlorine atoms are relatively large and this restricts the ability of the chains to move relative to each other.

Since the combustion of chlorinated plastics such as PVC releases hydrogen chloride and dioxins (a family of very toxic organic chlorine chemicals), these materials are not suitable for use at high temperatures.

Going further

Co-polymerization

Addition polymers can be made not only by polymerizing a single monomer, but from two or more different monomers. This is called co-polymerization and can be used to create polymers with specific properties.

Although the different monomers are likely to bond together in a relatively random order, it is usual to draw the repeat unit as containing just one of each monomer. For example, the copolymer of propene and chloroethene is drawn as:

$CH_3CH{=}CH_2 + CH_2{=}CHCl \rightarrow -[CH(CH_3)-(CH_2)_2-CHCl]-$

■ **Figure S2.139** Tapping of rubber tree

Rubber

The rubber tree (*Hevea brasiliensis*) is indigenous to South America but was brought to South East Asia in the early 18th century by European colonists. When its bark is stripped, it oozes a sticky latex, which is an emulsion of rubber in water (Figure S2.139). The purified rubber is not useful in its natural form, because it has a low melting point and low strength.

Rubber is an addition polymer (Figure S2.140) of the diene 2-methylbutadiene. This molecule, commonly known as isoprene, is a building block in a number of biomolecules, including beta-carotene, the orange pigment in carrots.

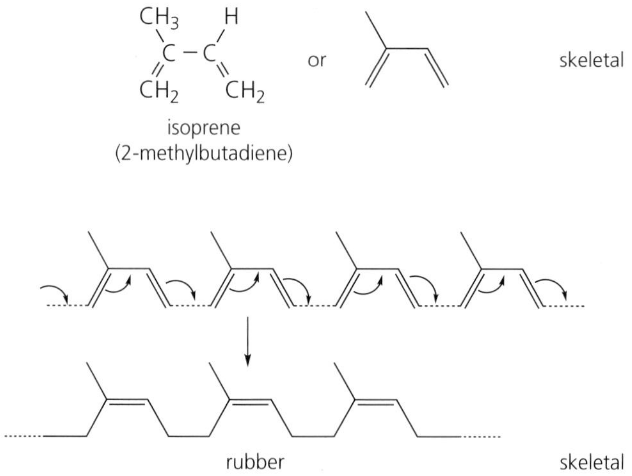

CH_3 H
C – C
CH_2 CH_2
isoprene
(2-methylbutadiene)
or
skeletal

rubber
skeletal

■ **Figure S2.140** Isoprene undergoing addition polymerization to form rubber (simplified mechanism)

Note that when a diene undergoes addition polymerization, there is still a double bond in the product. The double bond may be in the *cis* or *trans* configuration.

In natural rubber, all the double bonds are in this *cis* configuration, meaning the two $-CH_2-$ groups are on the same side of the double bond.

◆ **Thermoplastic:** A plastic that softens on heating and hardens again on cooling, allowing it to be reshaped multiple times.

The naturally occurring substance called gutta percha is an isomer of rubber in which all double bonds are *trans*. Gutta percha is harder than rubber and is **thermoplastic**. It was used in the past to make the cores of golf balls, as well as drinking vessels and other moulded objects.

LINKING QUESTION

What functional groups in molecules can enable them to act as monomers for addition reactions?

LINKING QUESTION

Why is the atom economy 100% for an addition polymerization reaction?

Biodegradable polymers

Synthetic polymers are highly stable and, although it is this property which makes them so useful, it makes disposal a problem. Waste polymers are normally buried in landfill sites but because they do not readily break down or degrade in the environment, such sites are rapidly reaching capacity. Recycling and incineration (burning) are possible solutions but another approach is to make biodegradable polymers. Many polymers will undergo pyrolysis, which is chemical decomposition occurring as a result of high temperatures. Some polymers can be made to undergo depolymerisation and reform their monomers (in a reversible reaction).

Biopolymers are natural polymers and many have similar properties to synthetic polymers. Unlike many synthetic polymers, however, they can be broken down by microorganisms in the environment; that is, they are biodegradable.

Figure S2.141 shows the structure of polyhydroxybutyrate (PHB), a biodegradable polymer manufactured by certain strains of bacteria. Although relatively expensive, it has been used as a packaging material for cosmetics and motor oils. However, its main intended applications were medical—for surgical stitches and in the controlled release of medicines into the body.

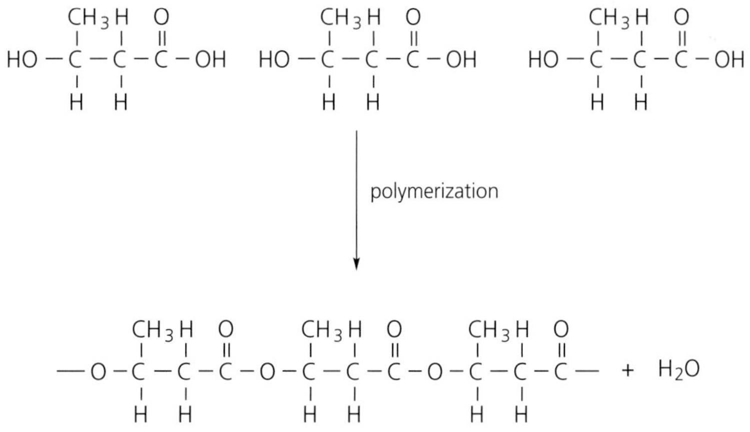

```
     CH3 H  O                CH3 H  O                CH3 H  O
     |   |  ||               |   |  ||               |   |  ||
HO - C - C - C - OH     HO - C - C - C - OH     HO - C - C - C - OH
     |   |                   |   |                   |   |
     H   H                   H   H                   H   H

                          |
                          |  polymerization
                          v

       CH3 H  O            CH3 H  O            CH3 H  O
       |   |  ||           |   |  ||           |   |  ||
 — O - C - C - C - O - C - C - C - O - C - C - C —   +  H2O
       |   |               |   |               |   |
       H   H               H   H               H   H
```

■ **Figure S2.141** Polymerization of 3-hydroxybutanoic acid to form PHB

LINKING QUESTION

What are the structural features of some plastics that make them biodegradable?

Going further

Structure of addition polymers

Many plastics, such as biopol, polyethene, polychloroethene and polyphenylethene, can be softened by heating. They will flow as viscous liquids and solidify again when cooled. Such plastics are useful because they can be remoulded. They are known as thermoplastic polymers.

Another, smaller, group of polymers can be heated and moulded only once. Such polymers are known as thermosetting polymers and include polyester resin, fibreglass, polyurethane, vulcanized rubber and Bakelite. The chains in these polymers are cross-linked to each other by permanent covalent bonds (Figure S2.142). These bonds make the structures rigid and no softening takes place on heating.

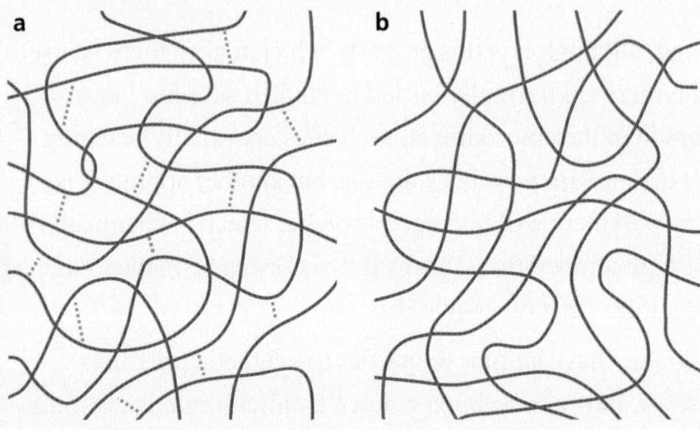

■ **Figure S2.142 a** Thermosetting and **b** thermoplastic polymers have different properties

Most polymers are amorphous—they have no regular packing of their chains. A few polymers, including polyethene and some forms of polypropene, have structures which are mixtures of crystalline regions where chains are closely packed in an ordered fashion, and amorphous regions in which the chains are further apart, randomly arranged and have more freedom to move (Figure S2.143). A single polymer chain will pass through both crystalline and amorphous regions along its length.

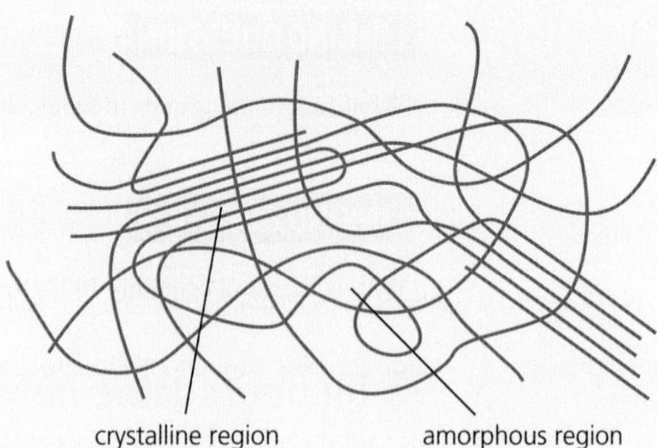

■ **Figure S2.143** Crystalline and amorphous regions of a polymer

Linear, unbranched chain structures are most likely to form crystalline regions. Poly(chloroethene) does not give rise to crystalline structures because the chlorine atoms are rather bulky and are irregularly spaced along the molecular chain. Cross-linked polymer chains cannot pack in a regular manner because of the covalent links. Although semi-crystalline polymers are relatively rare, they are among the most useful. The combination of hard, closely packed crystalline regions and relatively soft amorphous regions confers toughness, while the crystallinity reduces permeability to gases such as oxygen or CO_2 and this is important in packaging applications. Whereas amorphous polymers are typically transparent, semi-crystalline ones may be translucent because the crystals scatter light.

HL ONLY

Condensation polymers

◆ **Condensation polymerization**: A polymerization process that involves the bonding together of monomer units with elimination of small molecules, usually water or hydrogen chloride.

Condensation polymerization involves two monomers, each with a reactive functional group, bonding together to make a polymer. As the monomers join together, small molecules such as water or hydrogen chloride are released. This is why the process is known as condensation polymerization. Condensation polymerization reactions are catalysed by mineral acids and bases.

There are many examples of condensation polymers. Some, including polyesters such as terylene and polyamides such as nylon and Kevlar, are synthetic. There are also many naturally occurring condensation polymers including starch, cellulose, proteins and nucleic acids.

◆ **Polyamide:** A condensation polymer where the monomer molecules are linked by amide bonds.

Polyamides

Polyamides are formed when molecules with two carboxyl (–COOH) functional groups react with molecules with two amine ($–NH_2$) functional groups. The carboxyl groups react with the amine groups to form amide linkages (–CONH–) and release water molecules. A similar but more rapid reaction occurs between a diacylchloride and a diamine (Figure S2.144) and releases hydrogen chloride molecules.

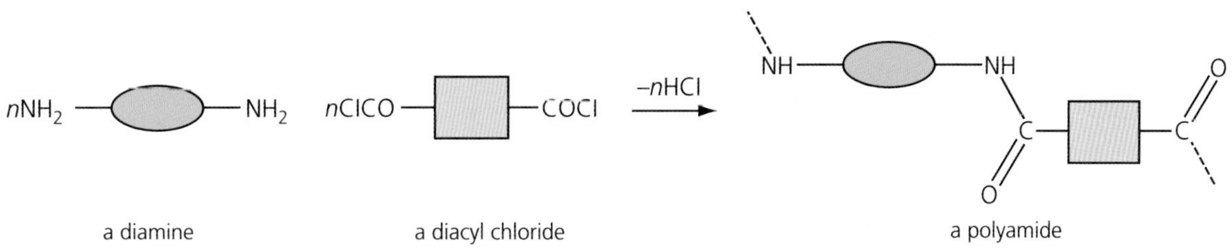

nNH$_2$ —⬭— NH$_2$ nClCO —▢— COCl $\xrightarrow{-n\text{HCl}}$ polyamide

a diamine a diacyl chloride a polyamide

■ **Figure S2.144** Generalized synthesis of a polyamide from a diamine and a diacyl chloride

Polyamides can also be formed from monomers that each contain one carboxyl group and one amine group (Figure S2.145).

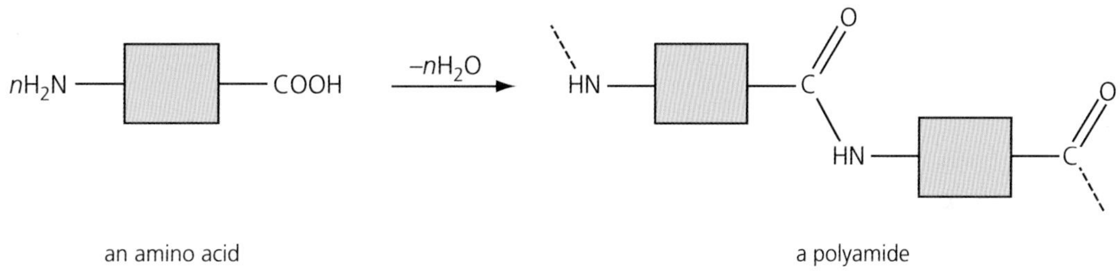

nH$_2$N —▢— COOH $\xrightarrow{-nH_2O}$ polyamide

an amino acid a polyamide

■ **Figure S2.145** Synthesis of a polyamide from one monomer with two different functional groups

29 Complete the table below contrasting addition polymerization with condensation polymerization.

Addition polymerization	**Condensation polymerization**
Generally involves one monomer	
Polymerization does not lead to elimination of smaller molecules	
Empirical formula is the same as that of monomer	
Monomers are unsaturated and usually have one reactive carbon–carbon double bond	
Generally rapid process	

WORKED EXAMPLE 2.4E

The two monomers below can react to form a polyamide.

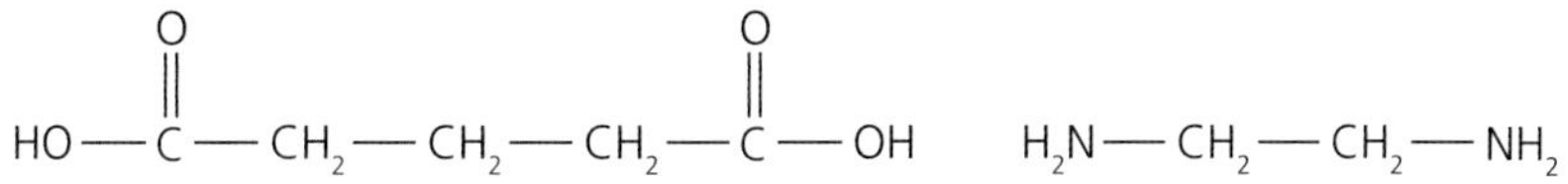

$$HO-\overset{O}{\overset{\|}{C}}-CH_2-CH_2-CH_2-\overset{O}{\overset{\|}{C}}-OH \qquad H_2N-CH_2-CH_2-NH_2$$

Deduce the repeating unit and structure of the polymer.

HL ONLY

Answer

A dimer should be formed from the reaction between the carboxyl group of the diacid and the amine group of the diamine. A water molecule is eliminated.

$$HO-\overset{O}{\overset{\|}{C}}-CH_2-CH_2-CH_2-\overset{O}{\overset{\|}{C}}-O-H \quad H-\underset{H}{\underset{|}{N}}-CH_2-CH_2-NH_2$$

$$\downarrow$$

$$HO-\overset{O}{\overset{\|}{C}}-CH_2-CH_2-CH_2-\overset{O}{\overset{\|}{C}}-\underset{H}{\underset{|}{N}}-CH_2-CH_2-NH_2 \; + \; H_2O$$

Deduce the repeating unit by removing –OH from the left end of the dimer and –H at the right end.

$$-\overset{O}{\overset{\|}{C}}-CH_2-CH_2-CH_2-\overset{O}{\overset{\|}{C}}-\underset{H}{\underset{|}{N}}-CH_2-CH_2-\underset{H}{\underset{|}{N}}-$$

Deduce the structure of the polymer by putting the repeating unit in square brackets and multiplying by *n*, a large number.

$$\left[-\overset{O}{\overset{\|}{C}}-CH_2-CH_2-CH_2-\overset{O}{\overset{\|}{C}}-\underset{H}{\underset{|}{N}}-CH_2-CH_2-\underset{H}{\underset{|}{N}}-\right]_n$$

Nylon

◆ **Nylon**: A synthetic polyamide fibre usually formed by the reaction of a diamine and a dicarboxylic acid or dicarboxylic acid chloride.

Nylon was originally made by the reaction of a diamine with a dicarboxylic acid. The two monomers used initially were 1,6-diaminohexane and hexane-1,6-dioic acid:

$$nH_2N(CH_2)_6NH_2 + nHOOC(CH_2)_4COOH \rightarrow [-OC(CH_2)_4CONH(CH_2)_6NH-]_n + nH_2O$$

The polymer chain is made up from the two monomers repeating alternately and results in the chain type: –A–B–A–B–A–B–A–

Each time a condensation reaction takes place between the two monomers, a molecule of water is lost (Figure S2.146). The new bond formed between the monomers is an amide link and nylon is a polyamide.

HL ONLY

a

$H_2N{-}(CH_2)_6{-}NH_2$ + $HOOC{-}(CH_2)_4{-}COOH$

1,6-diaminohexane   hexane-1,6-dioic acid

↓

$^{*}H_2N{-}(CH_2)_6{-}NH{-}CO{-}(CH_2)_4{-}COOH^{*}$

* further reaction can occur at both ends

b

$\left[{-}CH_2{-}CH_2{-}CH_2{-}CH_2{-}CH_2{-}CH_2{-}NH{-}CO{-}CH_2{-}CH_2{-}CH_2{-}CH_2{-}CO{-}NH{-}\right]_n$

nylon-6,6

■ **Figure S2.146 a** The polymerization of 1,6-diaminohexane and hexane-1,6-dioic acid to form nylon. **b** Because the two monomers each contain a six-carbon chain, this form of nylon is known as nylon-6,6.

Nylon can be synthesized in the laboratory by adding aqueous 1,6-diaminohexane to a solution of decanedioyl dichloride, $ClOC(CH_2)_8COCl$, in tetrachloromethane. Decanedioyl dichloride is more reactive than hexane-1,6-dioic acid and releases hydrogen chloride during the reaction.

The polymer forms at the interface of the two immiscible liquids and may be drawn out of the mixture. This process of interfacial polymerization is known as the nylon rope trick (Figure S2.147).

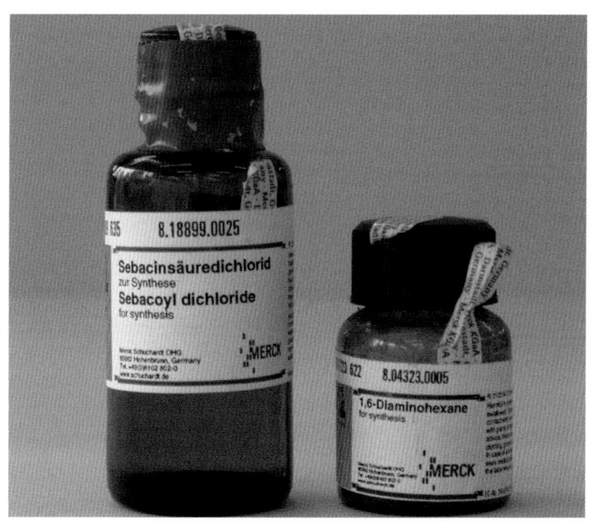

■ **Figure S2.147 a** The monomers for the 'nylon rope' experiment: decanedioyl dichloride (sebacoyl dichloride) and 1,6-diaminohexane. **b** The rope of nylon-6,10 can be continuously drawn off from the interface between the solutions of the two monomers.

Top tip!

There are several different nylons depending on the monomers used and the number of carbon atoms they contain, for example:

nylon-6,10 (two repeating units with the amine given first)

$-NH{-}(CH_2)_6{-}NH{-}CO{-}(CH_2)_8{-}CO{-}NH{-}(CH_2)_6{-}NH{-}CO{-}(CH_2)_8{-}CO-$

nylon-6,8 (two repeating units)

$-NH{-}(CH_2)_6{-}NH{-}CO{-}(CH_2)_6{-}CO{-}NH{-}(CH_2)_6{-}NH{-}CO{-}(CH_2)_6{-}CO-$

The molecular chains in nylon fibres interact through hydrogen bonding between the hydrogen atoms of the N–H groups of the amide links in one polymer chain and the C=O groups on adjacent polymer chains (Figure S2.148).

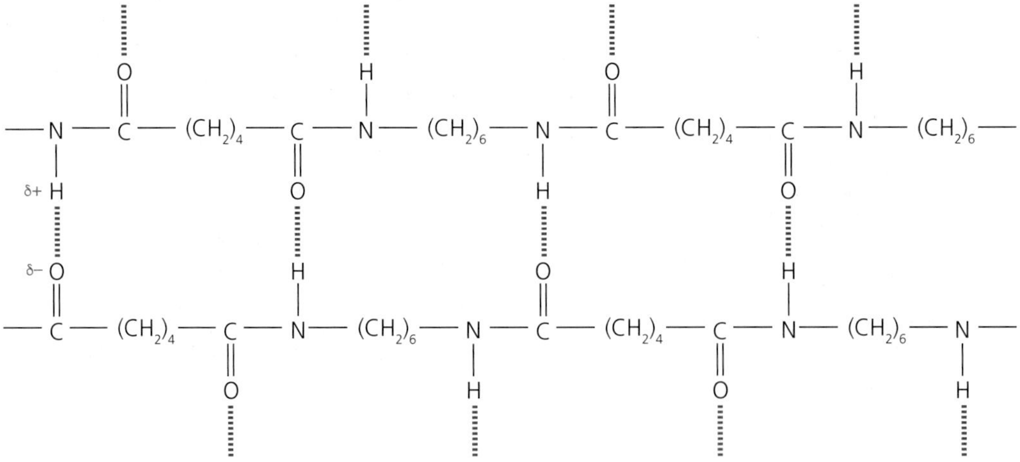

■ **Figure S2.148** Hydrogen bonding between adjacent nylon chains

Kevlar

Kevlar has a polyamide structure similar to nylon-6,6 but the amide links join benzene rings together rather than chains of carbon atoms (Figure S2.149). The two monomers are benzene-1,4-dicarbonyl chloride (an acyl chloride) and 1,4-diaminobenzene. When the amino and carboxyl groups react, they eliminate hydrogen chloride molecules and the resultant polymer chains can hydrogen bond to each other (Figure S2.150).

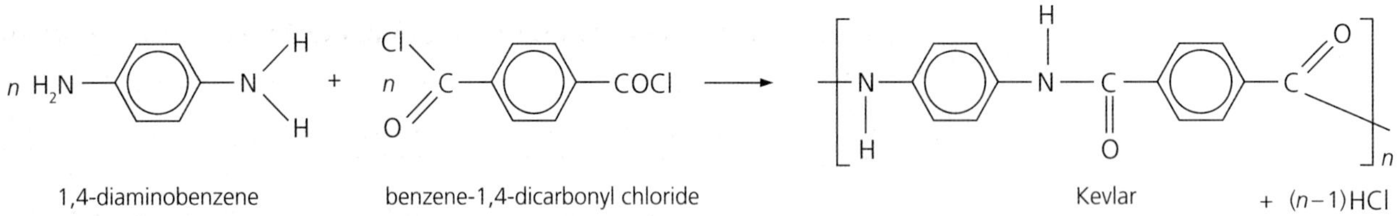

■ **Figure S2.149** The formation of Kevlar from its monomers

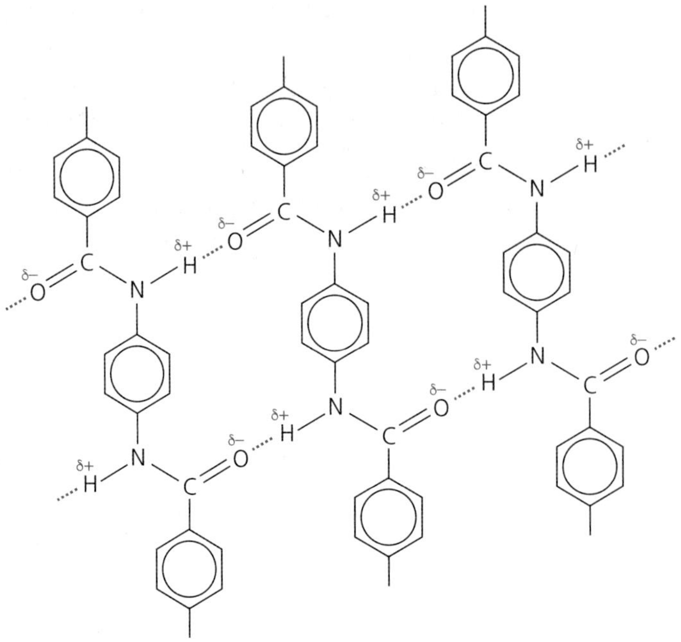

■ **Figure S2.150** The structure of Kevlar showing the hydrogen bonding between chains

HL ONLY

Kevlar is a very tough polymer which is used in protective clothing (Figure S2.151).

■ **Figure S2.151** This motorcycle helmet is reinforced with Kevlar

The Kevlar molecule is essentially flat and this is mainly due to π-electron delocalization, which involves not just the benzene rings but extends through the amide group and hence over the whole length of the molecule. Pi-electron delocalization also strengthens the covalent bonds and prevents any rotation about individual bonds within the chain. All these factors make the individual Kevlar molecules strong, stiff and rod-like.

Kevlar is processed to produce fibres in which flat sheets of molecules stack together around the fibre axis. This crystalline arrangement is shown in Figure S2.152. The very high degree of molecular alignment within the fibre is the largest contributing factor to Kevlar's very high tensile strength.

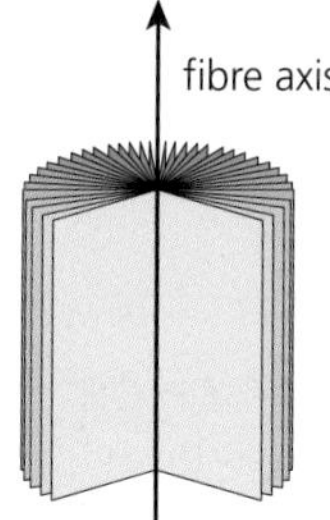

■ **Figure S2.152** The crystalline structure of Kevlar

Proteins

Amino acids are an important group of compounds that all contain the amine functional group ($-NH_2$) and the carboxyl functional group (–COOH). Naturally occurring amino acids have the amine functional group bonded to the carbon atom next to the carboxyl functional group (Figure S2.153).

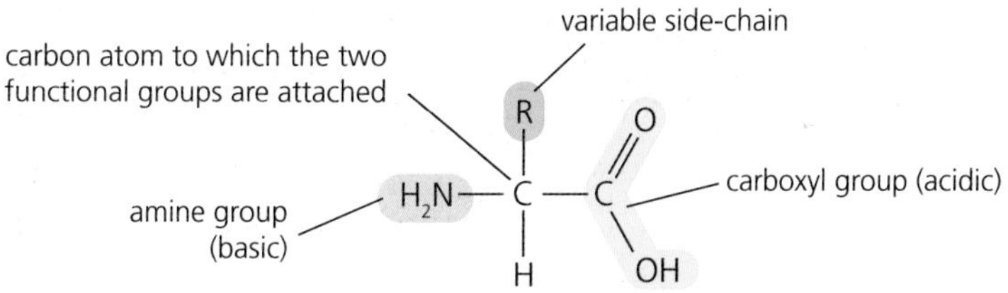

variable side-chain

carbon atom to which the two functional groups are attached

carboxyl group (acidic)

amine group (basic)

$H_2N-C(R)(H)-C(=O)-OH$

■ **Figure S2.153** Generalized structure of a naturally occurring amino acid

◆ **Amino acid**: A compound with a carboxyl (–COOH) and an amine ($-NH_2$) group bonded to the same carbon atom.

◆ **Peptide bond**: An amide bond resulting from the condensation reaction between the amine group of one amino acid and the carboxyl group of another.

The variable side-chain, R, has a different structure in each of the 20 amino acids and determines the shape and chemical and physical properties of the folded proteins in which it is found. In the simplest amino acid, glycine (Gly), R represents a hydrogen atom (–H); in alanine (Ala), R represents a methyl group, $-CH_3$.

Adjacent amino and carboxyl functional groups react to form a **peptide bond**. This reaction is a condensation process since it involves the formation of a water molecule. The reaction between two amino acids results in the formation of a dipeptide (Figure S2.154). The two amino acid residues are joined by a strong carbon–nitrogen bond in an amide functional group.

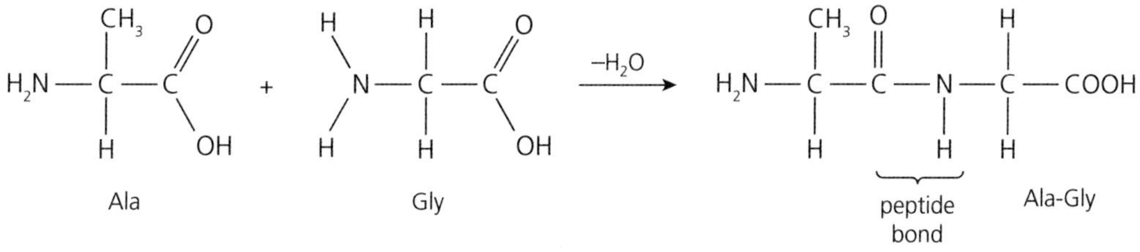

$$\underset{\text{Ala}}{H_2N-CH(CH_3)-COOH} + \underset{\text{Gly}}{H_2N-CH_2-COOH} \xrightarrow{-H_2O} \underset{\text{Ala-Gly}}{H_2N-CH(CH_3)-\underbrace{CO-NH}_{\text{peptide bond}}-CH_2-COOH}$$

■ **Figure S2.154** Peptide bond formation between alanine (Ala) and glycine (Gly)

The process of condensation polymerization is repeated until a long chain of bonded amino acids, known as a polypeptide, is formed.

Depending on which ends of the amino acids form the peptide bond, two distinct dipeptides can be formed from a given pair of amino acids. For example, the two dipeptides formed from Ala and Gly are glycyl-alanine, Gly-Ala, and alanyl-glycine, Ala-Gly (Figure S2.155).

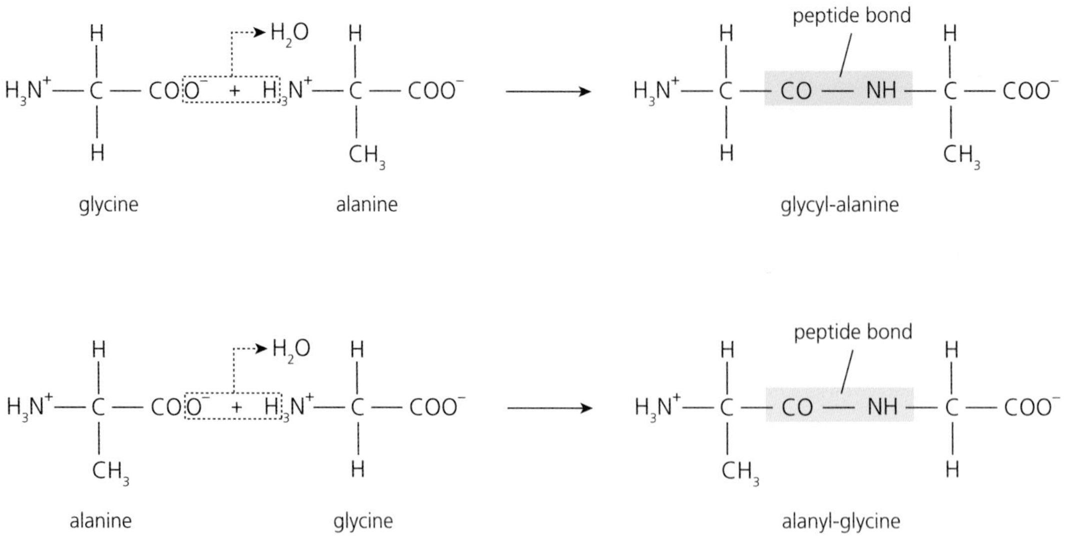

■ **Figure S2.155** The structures of the two dipeptides formed from Ala and Gly

30 Draw the structures of the two dipeptides which can form when one of the amino acids shown here reacts with the other.

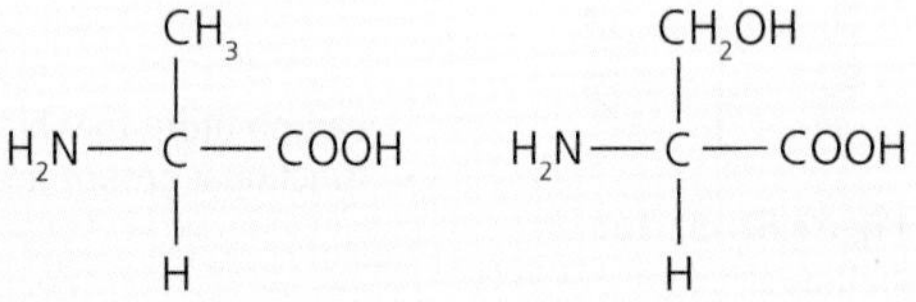

Polyesters

Polyesters are formed when molecules with two carboxyl (–COOH) functional groups react with molecules with two hydroxyl (–OH) functional groups. The carboxylic acid groups react with the alcohol groups to form ester linkages (–COO–) and release water molecules (Figure S2.156).

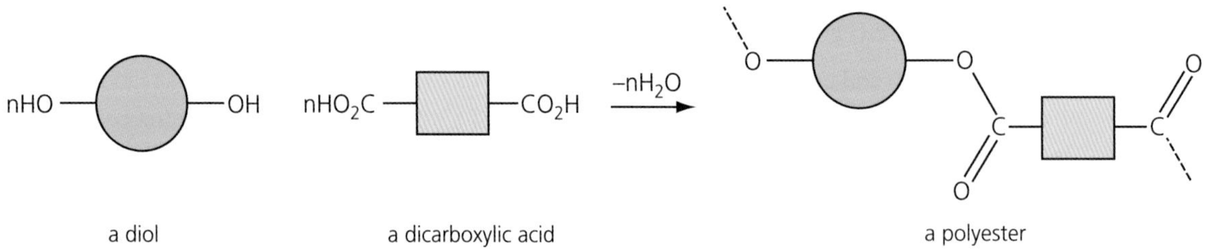

■ **Figure S2.156** Generalized synthesis of a polyester from a dicarboxylic acid and a diol

Polyesters can also be formed from monomers that each contain one carboxyl group and one hydroxyl group (Figure S2.157).

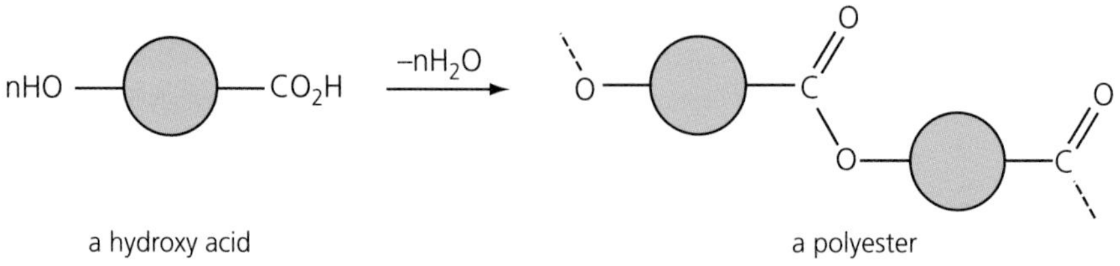

■ **Figure S2.157** Synthesis of polyester from one monomer

HL ONLY

WORKED EXAMPLE 2.4F

The two monomers below can react to form a polyester.

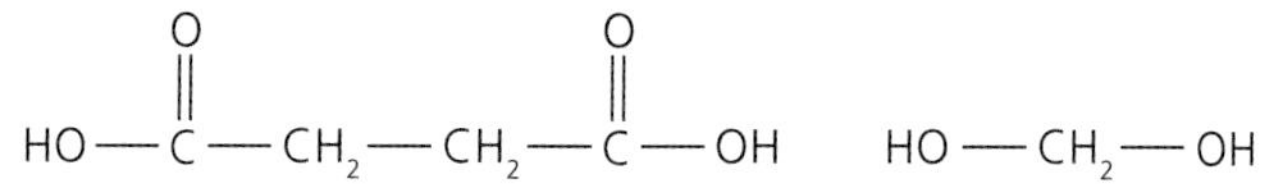

$$HO-\overset{\overset{\displaystyle O}{\|}}{C}-CH_2-CH_2-\overset{\overset{\displaystyle O}{\|}}{C}-OH \qquad HO-CH_2-OH$$

Deduce the repeating unit and structure of the polymer.

Answer

A dimer should be formed from the reaction between the hydroxy group of the diol and the carboxyl group of the diacid. A water molecule should be eliminated.

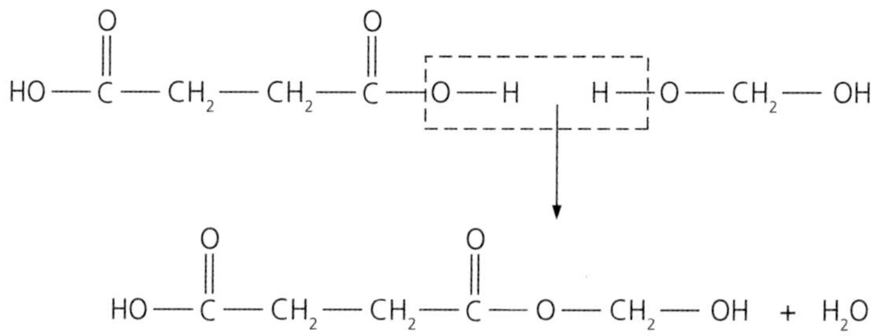

$$HO-\overset{\overset{\displaystyle O}{\|}}{C}-CH_2-CH_2-\overset{\overset{\displaystyle O}{\|}}{C}-\boxed{O-H \quad H}-O-CH_2-OH$$

$$\downarrow$$

$$HO-\overset{\overset{\displaystyle O}{\|}}{C}-CH_2-CH_2-\overset{\overset{\displaystyle O}{\|}}{C}-O-CH_2-OH \; + \; H_2O$$

Deduce the repeating unit by removing –OH from the left end of the dimer and –H from the right end.

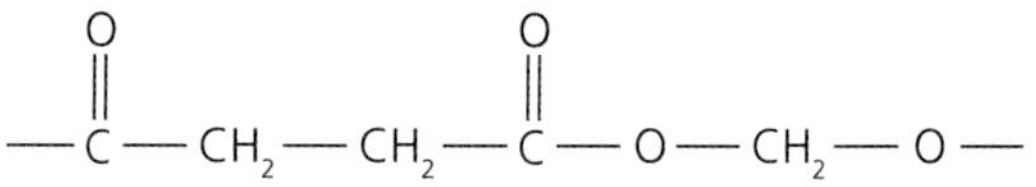

$$-\overset{\overset{\displaystyle O}{\|}}{C}-CH_2-CH_2-\overset{\overset{\displaystyle O}{\|}}{C}-O-CH_2-O-$$

Deduce the structure of the polymer by putting the repeating unit in square brackets and multiplying by *n*, a large number.

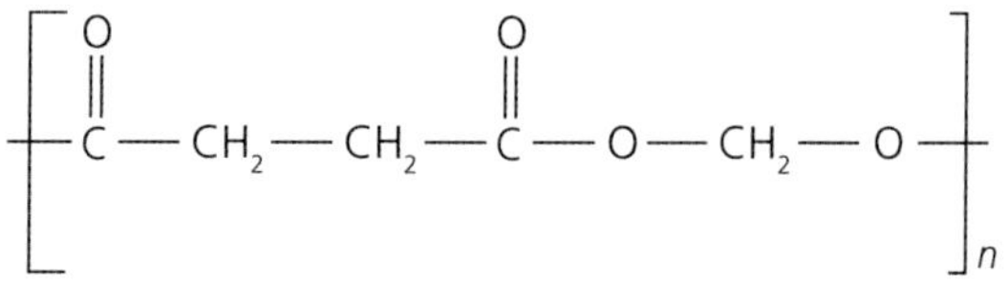

$$\left[-\overset{\overset{\displaystyle O}{\|}}{C}-CH_2-CH_2-\overset{\overset{\displaystyle O}{\|}}{C}-O-CH_2-O-\right]_n$$

Terylene

If ethane-1,2-diol is reacted with benzene-1,4-dicarboxylic acid then a polyester known as poly(ethylene terephthalate) or PET is produced (Figure S2.158).

In its fibre form, this polyester is known as Terylene in the UK and as Dacron in the USA. It can also be produced as a packaging film (Mylar, Melinex) or in a form suitable for making bottles.

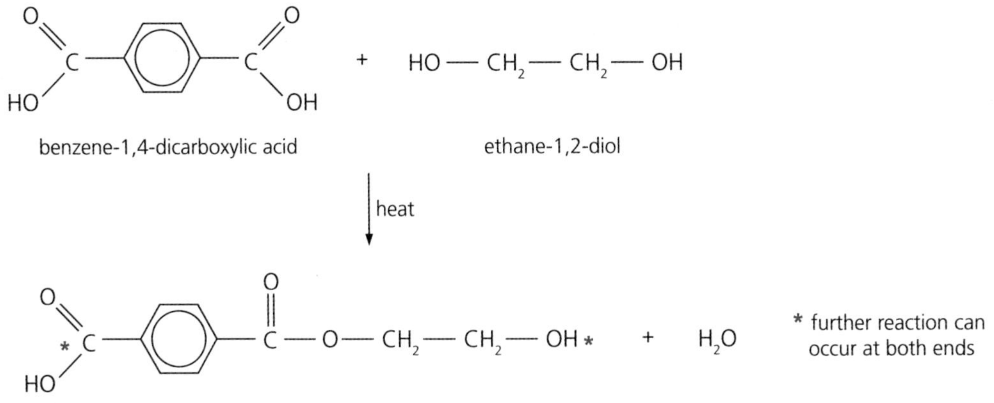

$$HOOC-C_6H_4-COOH + HO-CH_2-CH_2-OH \xrightarrow{heat} {}^{*}HOOC-C_6H_4-CO-O-CH_2-CH_2-OH^{*} + H_2O$$

■ **Figure S2.158** The formation of poly(ethylene terephthalate)

31 Draw the structures of the products of the following reactions:
  a benzene-1,4-dicarboxylic acid with 1,2-diaminoethane
  b hexane-1,6-dioyl chloride with propan-1,2-diol
  c ethanedioic acid with ethan-1,2-diol

32 State the monomers from which the following condensation polymers can be synthesized:
  a Terylene b nylon-6,6

Going further

Polymer chain growth

Polymerization occurs mainly in two ways: chain-growth polymerization and step-growth polymerization (Figure S2.159).

Chain-growth polymerization occurs in addition polymerization and is started with a small amount of a chemical known as an initiator. The monomers are covalently bonded to the chain at the active site, one monomer at a time. The growth of polymer chains occurs only at the ends. A free-radical mechanism, ionic mechanism or coordination complexes may be involved.

Step-growth polymerization typically occurs when monomer molecules have functional groups at each end, as in condensation polymerization. The reaction starts with the formation of dimers (two monomers covalently bonded together), trimers (three monomers bonded together), and tetramers (four monomers bonded together). These can react with each other to form long polymer chains: multiple units may be added to either end of the chain rather than single units to one end only as in chain-growth polymerization.

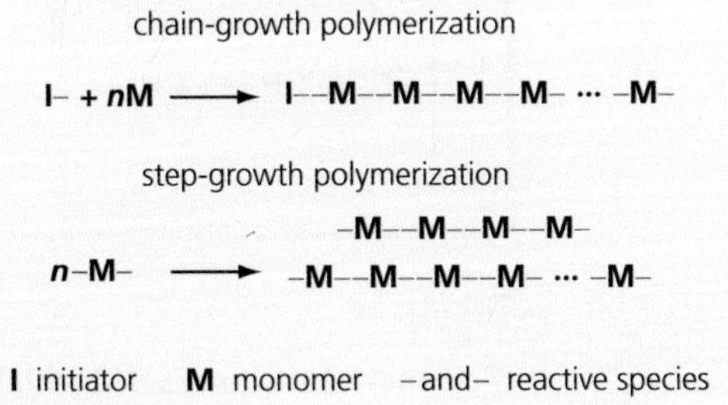

■ **Figure S2.159** Chain-growth and step-growth polymerization

◆ **Condensation reaction:** An addition reaction immediately followed by an elimination reaction in which a small molecule (such as hydrogen chloride or water) is formed each time a new bond is formed joining together two monomers.

Condensation and hydrolysis reactions

These are two extremely important reactions in the synthesis and breakdown of molecules in biological systems. Biological **condensation** is the joining together of two molecules with the formation of a covalent bond and the elimination of a water molecule.

Condensation reactions occur between amino acids to form proteins, and between carboxylic acids and alcohols to form the esters present in lipids (fats and oils). These reactions also occur between monosaccharide molecules to form disaccharides and polysaccharides (Figure S2.160).

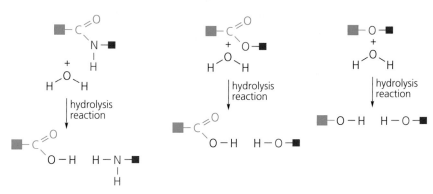

■ **Figure S2.160** The condensation reactions involved in the formation of proteins, lipids and polysaccharides

◆ **Hydrolysis:** A chemical reaction involving water as a reactant, that splits a molecule apart.

A **hydrolysis** reaction is essentially the reverse of condensation—it involves breaking a covalent bond by the addition of a water molecule. The water is split with –H and –OH attaching separately to the product molecules (Figure S2.161).

■ **Figure S2.161** Hydrolysis reactions involve the breaking of a covalent bond by the addition of the fragments of water across the bond

These reactions occur during chemical digestion. Biologically they are catalysed by enzymes, but can also be favoured by heat and acidic or alkaline conditions.

The relationship between these two key reaction types is summarized below (Figure S2.162).

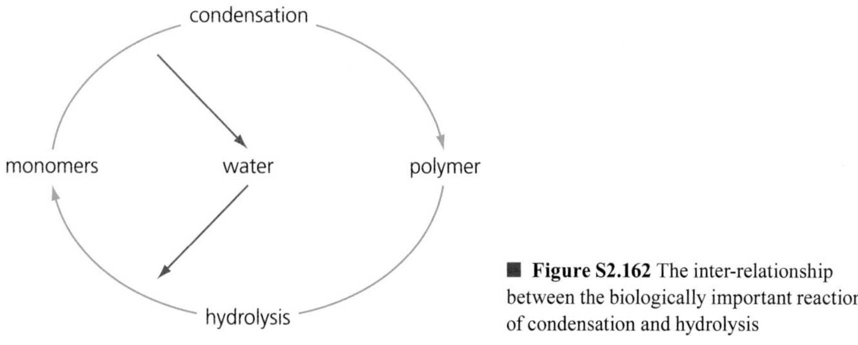

■ **Figure S2.162** The inter-relationship between the biologically important reactions of condensation and hydrolysis

33 Maltose is a disaccharide formed by a condensation reaction between two glucose molecules to form a 1-4 glycosidic bond. (Alpha and beta glucose molecules are closely related forms of glucose that differ in the arrangement of the –OH group at carbon atom 1.) Draw the resulting structure of maltose and highlight the glycosidic bond.

α-glucose

HL ONLY

Going further

Zwitterions

Amino acids have relatively high melting or decomposition points and are more soluble in water than in non-polar solvents. These observations suggest that amino acids exist as dipolar ions, known as zwitterions (Figure S2.163).

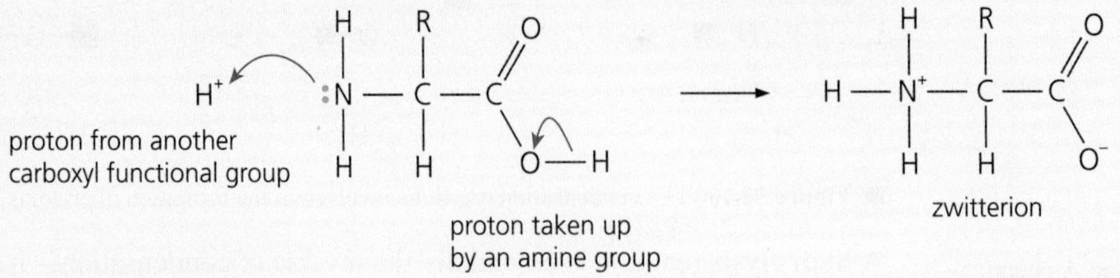

■ **Figure S2.163** Zwitterion formation by glycine

- The ionic nature of amino acids explains the fact that they are fairly soluble in water: the zwitterions of amino acids can be readily hydrated by water molecules.
- The low solubility of amino acids in non-polar solvents is because the energy needed to overcome the stronger electrostatic forces between the ions cannot be compensated for by the formation of relatively weak London (dispersion) forces between the amino acid and the solvent molecules.
- The electrostatic forces of attraction between oppositely charged functional groups account for high melting or decomposition points. Zwitterion formation can be regarded as an *internal* acid–base reaction due to the transfer of a proton (H^+) from the acid –COOH group to the basic $–NH_2$ group in the same amino acid.

LINKING QUESTION

What functional groups in molecules can enable them to act as monomers for condensation reactions?

S3.1 The periodic table: classification of elements

Guiding question

- How does the periodic table help us to predict patterns and trends in the properties of the elements?

SYLLABUS CONTENT

By the end of this chapter, you should understand that:

- the periodic table consists of periods, groups and blocks
- the period number shows the outer energy level that is occupied by electrons
- elements in a group have a common number of valence electrons
- periodicity refers to trends in properties of elements across a period and down a group
- trends in properties of elements down a group include the increasing metallic character of group 1 elements and decreasing non-metallic character of group 17 elements
- metallic and non-metallic properties show a continuum; this includes the trend from basic metal oxides through amphoteric to acidic non-metal oxides
- the oxidation state is a number assigned to an atom to show the number of electrons transferred in forming a bond; it is the charge that atom would have if the compound were composed of ions
- discontinuities occur in the trend of increasing first ionization energy across a period (HL only)
- transition elements have incomplete d-sublevels that give them characteristic properties (HL only)
- the formation of variable oxidation states in transition elements can be explained by the fact that their successive ionization energies are close in value (HL only)
- transition element complexes are coloured due to the absorption of light when an electron is promoted between the orbitals in the split d-sublevels; the colour absorbed is complementary to the colour observed (HL only).

By the end of this chapter you should know how to:

- identify the positions of metals, metalloids and non-metals in the periodic table
- deduce the electron configuration of an atom up to $Z = 36$ from the element's position in the periodic table and vice versa
- explain the periodicity of atomic radius, ionic radius, ionization energy, electron affinity and electronegativity
- describe and explain the reactions of group 1 metals with water, and of group 17 elements with halide ions
- deduce equations for the reactions with water of the oxides of group 1 and group 2 metals, carbon and sulfur
- deduce the oxidation state of an atom in an ion or compound
- explain why the oxidation state of an element is zero
- explain how discontinuities in first ionization energy across a period provide evidence for the existence of sublevels (HL only)
- recognize properties of transition elements, including: variable oxidation state, high melting points, magnetic properties, catalytic properties, formation of coloured compounds and formation of complex ions with ligands (HL only)
- deduce the electron configurations of ions of the first-row transition elements (HL only)
- apply the colour wheel to deduce the wavelengths and frequencies of light absorbed and/or observed (HL only).

Introduction to the periodic table

Nature of science: Models

Organizing elements

Chemists in the nineteenth century had a problem. Over sixty elements had been discovered and many of their compounds synthesized and studied. There was a large amount of data about elements but it was not classified and organized. The elements had to be grouped together in some way so that similarities between elements could be noted and patterns and trends could be observed. Only when chemists had managed to organize their facts could the study of chemistry advance. One crucial event was the Karlsruhe Congress of 1861, the first-ever international scientific conference. Further advances in chemistry followed the development of early forms of the periodic table such as that shown in Figure S3.1. Early versions were based on the properties of the elements; later ones were organized by atomic mass then atomic number. The modern version is explained by electron configurations.

Group	1		2		3		4		5		6		7		8
Sub-group	**A**	**B**	**A**	**B**	**A**	**B**	**A**	**B**	**A**	**B**	**A**	**B**	**A**	**B**	
1st period	H														
2nd period	Li		Be			B		C		N		O		F	
3rd period	Na		Mg			Al		Si		P		S		Cl	
4th period	K	Cu	Ca	Zn	–	–	Ti		V	As	Cr	Se	Mn	Br	Fe Co Ni
5th period	Rb	Ag	Sr	Cd	Y	In	Zr	Sn	Nb	Sb	Mo	Te	–	I	Ru Rh Pd
6th period	Cs	Au	Ba	Hg	La	Tl	–	Pb	Ta	Bi	W	–	–	–	Os Ir Pt
7th period	–		–		–		Th		–		U				

■ **Figure S3.1** A modernized version of Mendeleev's short-form periodic table

ATL S3.1A

Before Mendeleev, a number of people attempted to arrange the elements to show similarities in properties.

Work collaboratively with a partner to research and present the contributions (with text and appropriate diagrams) of the following chemists:

- Johann Döbereiner
- John Newlands
- Alexandre-Émile Béguyer de Chancourtois (the first person to list the known elements in order of increasing mass of their atoms).

Evaluate their contributions.

Suggest why Mendeleev is considered to be the 'father' of the periodic table while the contributions of these other chemists are, perhaps, overlooked.

Chemical symbols

The ‘language’ of chemistry transcends cultural, linguistic and national boundaries.

Chemical symbols were developed by the Swedish chemist Berzelius (1779–1848). They are not abbreviations, but symbols intended to be used by people of all languages and alphabets. The unique chemical symbols are based on the name of the element, but not necessarily the English name. For example, tungsten has the chemical symbol W which comes from the German *wolfram*.

■ **Figure S3.2** The Japanese kanji for sulfur translates as ‘yellow substance’

Chemical symbols are understood internationally whereas element names might need to be translated as they are sometimes language dependent. Often, the end of the name characterizes the specific language. For example, magnesium changes to *magnésium* in French and *magnesio* in Spanish. In Japanese, katakana characters are used to reproduce the sound of the English ‘magnesium’. However, an element’s name may be very different in a different language: for example, *Sauerstoff* is German for oxygen.

Arrangement of elements in the periodic table

The main purpose of the periodic table is to classify elements, which helps chemists in their study of them. The periodic table can also be used to make predictions about chemical and physical properties of the elements.

■ **Figure S3.3** Samples of period 3 elements

◆ **Periodicity**: The regular repetition of chemical and physical properties as you move across and down the periodic table.

The arrangement of elements into the periodic table leads to **periodicity**—repeating patterns of chemical and physical properties. These patterns are due to repeating changes in electron configuration. Physical properties such as melting points (Figure S3.4) and boiling points, atomic properties such as ionization energy and electronegativity, and chemical properties such as rate of reaction with water, all show periodicity.

◆ **Periodic table**: A table of elements arranged in order of increasing atomic (proton) number to show the similarities of chemical elements with related electronic configurations.

◆ **Group**: A column of the periodic table which contains elements with similar chemical properties with the same number of electrons in their outer or valence shell.

◆ **Period**: A horizontal row in the periodic table which contains elements with the same number of shells, and with an increasing number of electrons in the outer or valence shell, as the period is crossed from left to right.

The **periodic table** in the IB *Chemistry data booklet* (see also Figure S1.4 page 5), known as the medium form, has 18 **groups** (columns) and 7 **periods** (rows). The periodic table lists the elements in order of increasing atomic number (number of protons). The atomic number is above the element symbol and the relative atomic mass is below.

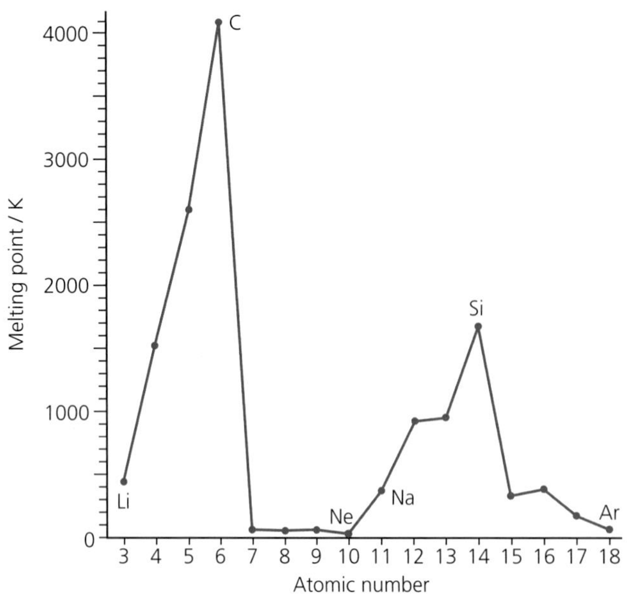

■ **Figure S3.4** Periodicity in the melting points of the first 18 elements

Going across the periodic table from left to right, the elements gradually change from being metals to non-metals (Table S3.1). The division between metals and non-metals is a 'staircase' (zig zag) that descends from the left of boron (period 2, group 13) to the right of polonium (period 6, group 16).

Top tip!

The periodic table shows relative atomic mass and not mass number. This is the weighted average mass of the isotopes of each element.

■ **Table S3.1** Properties of elements in period 3

Element	***sodium***	***magnesium***	***aluminium***	***silicon***	***phosphorus (white)***	***sulfur***	***chlorine***	***argon***
Formula	Na	Mg	Al	Si	P_4	S_8	Cl_2	Ar
Appearance	silvery	silvery	silvery	blue-grey	white solid	yellow solid	green gas	colourless gas
Bonding	metallic	metallic	metallic	covalent network structure	simple molecular	simple molecular	simple molecular	monatomic
Ion formed	Na^+	Mg^{2+}	Al^{3+}	none	P^{3-}	S^{2-}	Cl^-	none

◆ **Metallic character**: The tendency of an element to lose electrons and form cations.

◆ **Metalloids**: A group of chemical elements intermediate in properties between metals and non-metals.

Elements to the left of the staircase are metallic and electropositive (low electronegativity) and tend to form cations. Elements to the right of the staircase are non-metallic and electronegative and tend to form anions. **Metallic character** decreases across any period and increases down any group.

However, there are many elements near the staircase, such as silicon, which have metallic and non-metallic properties or intermediate properties. These elements are known as **metalloids** (Figure S3.5). Metalloids are neither strongly metallic nor strongly non-metallic in their properties.

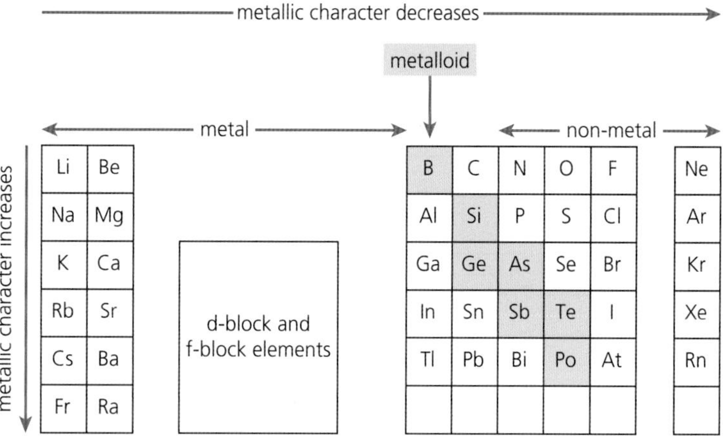

■ **Figure S3.5** The positions of metals, non-metals and metalloids in the periodic table

Table S3.2 summarizes the typical properties of the three types of elements: metals, metalloids and non-metals.

■ **Table S3.2** Typical physical properties of the three types of elements

	Metals	**Metalloids**	**Non-metals**
Appearance	lustrous (shiny)	lustrous	dull (non-lustrous)
Physical state	solids (except mercury)	solids	solids, liquids and gases
Melting and boiling points	usually high (except group 1 and mercury)	high	low (except carbon (diamond and graphite) and silicon)
Ductility and malleability	ductile (can be pulled into wires); malleable (can be beaten and shaped without breaking)	brittle (easily broken when a force is applied)	brittle if solid
Thermal conductivity	excellent	moderate	poor (except carbon)
Electrical conductivity	excellent	moderate	poor (except graphite)

Metals contain metallic bonding while metalloids are likely to have covalent network structures. Non-metals are usually simple molecular or monatomic.

1 Classify the following elements as metals, non-metals or likely metalloids based on their positions in the periodic table:
 a rubidium
 b radon
 c germanium
 d strontium
 e silicon
 f fluorine
 g copper
 h lead
 i mercury
 j lanthanum
 k boron.

◆ **Block:** A set of elements in the periodic table whose atoms have the same sublevel being filled.

Nature of science: Falsification

New elements

Science is often a long process of trial and error, in which ‘discoveries’ may, on further careful study, turn out not to be new after all. These failures can teach us as much—or more—than successes.

In the past, a number of people claimed to have found new elements. For example, in 1898, English physicist Sir William Crookes (1832–1919) announced the discovery of an element he at first called *monium*. He then renamed it *victorium*, in honour of Queen Victoria, who had recently knighted him. However, the ‘element’ was later shown to be a mixture of gadolinium and terbium.

Blocks

Based on the electron configurations of the atoms of the elements, the periodic table can be divided into four **blocks** of elements (Figure S3.6):

- s-block elements
- p-block elements
- d-block elements
- f-block elements.

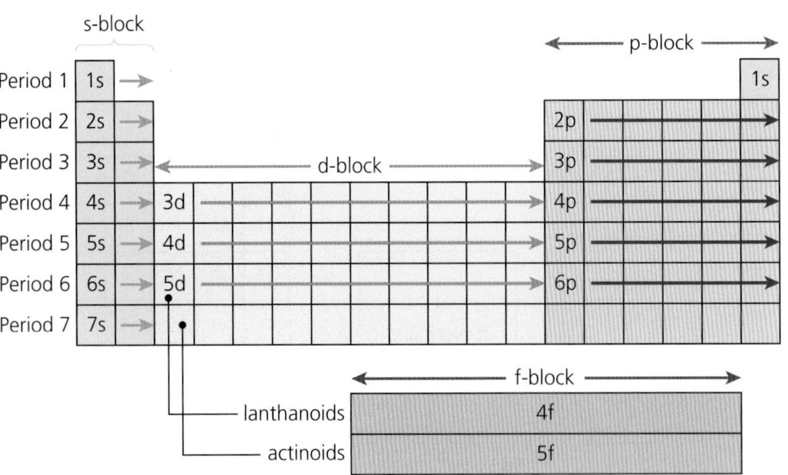

■ **Figure S3.6** Electron sublevels in periods 1 to 7

◆ **s-block**: The first two groups of the periodic table where the elements all have noble gas core configurations plus outer ns^1 (group 1) or ns^2 (group 2) electrons.

◆ **p-block**: Groups 13 to 18 of the periodic table; the outer electron configurations of all these elements have the form $ns^2\ np^x$ where $x = 1$ to 6.

◆ **d-block**: A block of elements, many of which have two s electrons as well as d electrons in their valence shell, giving an outer electron configuration of the form $(n{-}1)\ d^x\ ns^2$ where $x = 1$ to 10.

◆ **f-block**: The lanthanoid series and the actinoid series; the atoms often have two s electrons in their outer shell (n) and f electrons in their inner ($n - 2$) shell.

The **s-block** consists of hydrogen, helium and groups 1 (alkali metals) and 2 (alkaline earths). All the s-block elements have a half-filled s orbital (s^1) or a completely filled s orbital (s^2) in the outermost (valence) shell.

The **p-block** consists of groups 13 to 18. The s- and p-blocks are collectively called the main group elements. Each p-block element has an outer electron configuration which varies from $s^2\,p^1$ (group 13), $s^2\,p^2$ (group 14) through to $s^2\,p^6$ (the noble gases in group 18).

■ **Figure S3.7** British £2 coin. The gold-coloured outer ring is nickel–brass made from 76% copper, 20% zinc and 4% nickel; the inner cupronickel disc is 75% copper and 25% nickel.

The **d-block** consists of three series of metals that occupy the space between groups 2 and 13. Each series of d-block elements contains ten metals with outer electron configurations ranging from $d^1\,s^2$ to $d^{10}\,s^2$. Nearly all of the d-block elements are known as transition metals.

Unlike metals in groups 1 and 2, the transition metals have very similar chemical and physical properties. They have relatively high boiling points and melting points, form more than one stable cation, often have coloured compounds and can form alloys (Figure S3.7).

Top tip!

Chromium and copper in the first row of the d-block have $4s^1$ configurations; anomalous configurations are also found in a small number of transition elements in the second and third rows of the d-block.

The **f-block** consists of two rows of 14 metals at the bottom of the periodic table. The two rows, known as the lanthanoids and actinoids, contain elements in which f orbitals are being filled. They have some similarities to the transition metals.

Periods

The period number indicates the outer energy level (shell) that is occupied by electrons.

Top tip!

The arrangement of the periodic table reflects the filling of sublevels with electrons.

$1s^1$																	$1s^2$
$2s^1$	$2s^2$											$2p^1$	$2p^2$	$2p^3$	$2p^4$	$2p^5$	$2p^6$
$3s^1$	$3s^2$											$3p^1$	$3p^2$	$3p^3$	$3p^4$	$3p^5$	$3p^6$
$4s^1$	$4s^2$	$3d^1$	$3d^2$	$3d^3$	$3d^4$	$3d^5$	$3d^6$	$3d^7$	$3d^8$	$3d^9$	$3d^{10}$	$4p^1$	$4p^2$	$4p^3$	$4p^4$	$4p^5$	$4p^6$
$5s^1$	$5s^2$	$4d^1$	$4d^2$	$4d^3$	$4d^4$	$4d^5$	$4d^6$	$4d^7$	$4d^8$	$4d^9$	$4d^{10}$	$5p^1$	$5p^2$	$5p^3$	$5p^4$	$5p^5$	$5p^6$
$6s^1$	$6s^2$		$5d^2$	$5d^3$	$5d^4$	$5d^5$	$5d^6$	$5d^7$	$5d^8$	$5d^9$	$5d^{10}$	$6p^1$	$6p^2$	$6p^3$	$6p^4$	$6p^5$	$6p^6$
$7s^1$	$7s^2$		$6d^2$	$6d^3$	$6d^4$	$6d^5$	$6d^6$	$6d^7$	$6d^8$	$6d^9$	$6d^{10}$	$7p^1$	$7p^2$	$7p^3$	$7p^4$	$7p^5$	$7p^6$

$5d^1$	$4f^1$	$4f^2$	$4f^3$	$4f^4$	$4f^5$	$4f^6$	$4f^7$	$4f^8$	$4f^9$	$4f^{10}$	$4f^{11}$	$4f^{12}$	$4f^{13}$	$4f^{14}$
$6d^1$	$5f^1$	$5f^2$	$5f^3$	$5f^4$	$5f^5$	$5f^6$	$5f^7$	$5f^8$	$5f^9$	$5f^{10}$	$5f^{11}$	$5f^{12}$	$5f^{13}$	$5f^{14}$

■ **Figure S3.8** Periodic table showing the number of electrons in the highest level and sublevel

Top tip!

The first three periods are called short periods; the next four (which include d-block elements) are long periods.

Short periods (periods 1, 2 and 3)

In period 1, where $n = 1$, the 1s orbital is filled. Period 1 consists of hydrogen ($1s^1$) and helium ($1s^2$).

In period 2, the 2s and 2p sublevels ($n = 2$) are filled. Period 2 consists of eight elements from lithium ($1s^2\ 2s^1$) to neon ($1s^2\ 2s^2\ 2p^6$). The 2s orbital is filled first, followed by the three 2p orbitals.

In period 3, the 3s orbital ($n = 3$) is filled first, followed by the 3p orbitals. Period 3 consists of eight elements from sodium ($1s^2\ 2s^2\ 2p^6\ 3s^1$) to argon ($1s^2\ 2s^2\ 2p^6\ 3s^2\ 3p^6$).

Long periods

In period 4, the 4s, 3d and 4p orbitals are filled. Period 4 consists of 18 elements from potassium ($1s^2\ 2s^2\ 2p^6\ 3s^2\ 3p^6\ 4s^1$) to krypton ($1s^2\ 2s^2\ 2p^6\ 3s^2\ 3p^6\ 3d^{10}\ 4s^2\ 4p^6$) and includes a set of the d-block elements.

In period 5, rubidium to xenon, the 5s, 4d and 5p orbitals are filled.

In period 6, which consists of 32 elements, the 6s, 4f, 5d and 6p orbitals are filled. The lanthanoids are often removed from this period and shown at the bottom of the periodic table to allow the table to be displayed concisely.

In period 7, which also contains 32 elements, the 7s, 5f, 6d and 7p orbitals are filled. The actinoids are usually shown at the bottom of the periodic table.

■ **Table S3.3** Summary of orbitals filled across the periods

Period number	Orbitals being filled	Number of elements
1	1s	2
2	2s, 2p	8
3	3s, 3p	8
4	4s, 3d, 4p	18
5	5s, 4d, 5p	18
6	6s, 4f, 5d, 6p	32
7	7s, 5f, 6d, 7p	32

Groups

Groups in the periodic table are numbered from 1 to 18 although some groups are also commonly known by names as shown in Table S3.4.

■ **Table S3.4** Selected groups from the periodic table and their names

Group	**Name of group and type of elements**
Group 1	Alkali metals (reactive metals)
Group 2	Alkaline earth metals (reactive metals)
Group 17	Halogens (reactive non-metals)
Group 18	Noble gases (unreactive non-metals)

Atoms of elements in the same group have the same number of valence electrons. For example, all elements of group 1 (the alkali metals) and hydrogen have one valence electron in an s sublevel (Table S3.5). Hence the general valence electron configuration is ns^1, where $n = 1, 2, 3, 4$, etc.

Similarly, all the elements in group 13 have three valence electrons (Table S3.6). Hence the general electron configuration is $ns^2\ np^1$.

■ **Table S3.5** Electron configurations for the first four members of group 1

Element	**Full electron configuration**
H	$\mathbf{1s^1}$
Li	$1s^2\ \mathbf{2s^1}$
Na	$1s^2\ 2s^2\ 2p^6\ \mathbf{3s^1}$
K	$1s^2\ 2s^2\ 2p^6\ 3s^2\ 3p^6\ \mathbf{4s^1}$

■ **Table S3.6** Electron configurations for the first four members of group 13

Element	**Full electron configuration**
B	$\mathbf{1s^2\ 2p^1}$
Al	$1s^2\ 2s^2\ 2p^6\ \mathbf{3s^2\ 3p^1}$
Ga	$1s^2\ 2s^2\ 2p^6\ 3s^2\ 3p^6\ 3d^{10}\ \mathbf{4s^2\ 4p^1}$

Table S3.7 summarizes the relationship between the group numbers (for the s and p-blocks).

■ **Table S3.7** Generalized valence electron configurations for groups 1, 2 and 13 to 18

Group number	**Generalized valence electron configuration**
1	ns^1
2	ns^2
13	$ns^2\ np^1$
14	$ns^2\ np^2$
15	$ns^2\ np^3$
16	$ns^2\ np^4$
17	$ns^2\ np^5$
18	$ns^2\ np^6$

As the chemical properties of an element depend on its electron configuration, elements of the same group have similar chemical properties. A number of groups—1 and 17 especially—display trends in properties down the group. Different groups show many similar trends: for example, the ionization energy of an atom of the element decreases down a group.

Electronic configuration and location on the periodic table

The number of valence electrons of an atom can be deduced from its position in the periodic table.

- In the s-block, the number of valence electrons equals the group number. For example, sodium (Na) in group 1 has a [Ne] $3s^1$ configuration: one valence electron.
- In the p-block, the number of valence electrons is ten less than the group number. For example, selenium (Se) in group 16 has the configuration [Ar] $3d^{10}\ 4s^2\ 4p^4$: six valence electrons.
- In the d-block, the number of s and d electrons equals the group number. For example, cobalt (Co) is in group 9, with the configuration [Ar] $3d^7\ 4s^2$: nine valence electrons.

WORKED EXAMPLE S3.1A

From its position on the periodic table, predict the condensed and full electron configuration of an atom of each of the elements shown below.

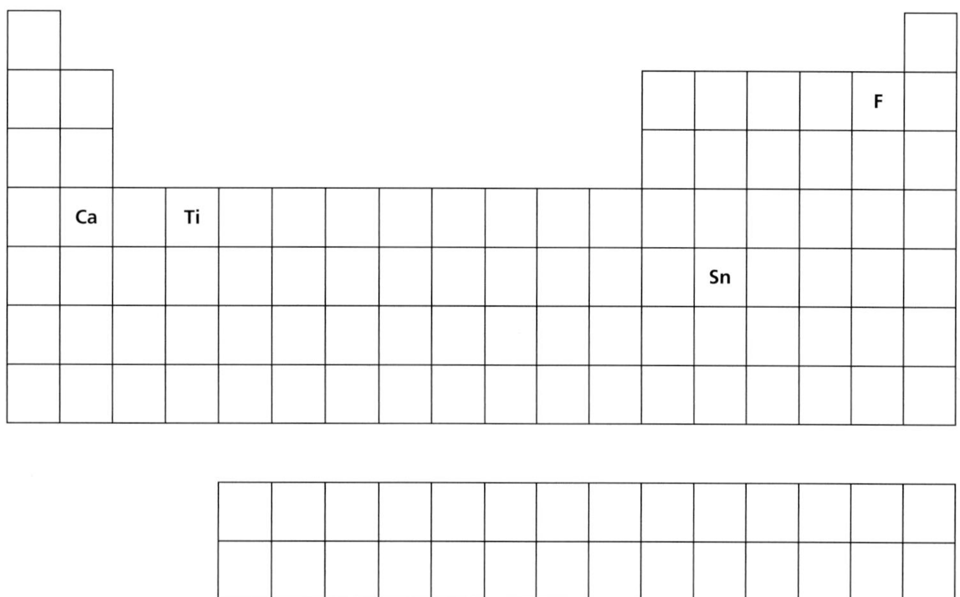

Answer

Calcium is in the fourth period and located in the second column of the s-block. Hence its electron configuration should end with $4s^2$. Argon is the nearest noble gas. The condensed electron configuration is [Ar] $4s^2$ and its full electron configuration is $1s^2\ 2s^2\ 2p^6\ 3s^2\ 3p^6\ 4s^2$.

Tin is located in the second column of the p-block and the fifth period. Hence its electron configuration should end with $5p^2$. Krypton is the noble gas at the end of the previous period. The condensed electron configuration is [Kr] $5s^2\ 4d^{10}\ 5p^2$ and the full electron configuration is $1s^2\ 2s^2\ 2p^6\ 3s^2\ 3p^6\ 4s^2\ 3d^{10}\ 4p^6\ 5s^2\ 4d^{10}\ 5p^2$ (listed here in ascending order of orbital energies).

Fluorine is located in the fifth column of the p-block and is in the second period. Hence its electron configuration should end with $2p^5$. Helium is the nearest noble gas. The condensed electron configuration is [He] $2s^2\ 2p^5$ and its full electron configuration is $1s^2\ 2s^2\ 2p^5$.

Titanium is in the fourth period and the second column of the d-block. Since the 3d sublevel is filled after the 4s sublevel, its electron configuration ends in $4s^2\ 3d^2$. The condensed electron configuration is [Ar] $4s^2\ 3d^2$ and its full electron configuration is $1s^2\ 2s^2\ 2p^6\ 3s^2\ 3p^6\ 4s^2\ 3d^2$.

ATL S3.1B

PBS Learning Media offers an interactive periodic table at **https://contrib.pbslearningmedia.org/WGBH/conv19/phy03-int-ptable/index.html**

You can click on each element to reveal its properties and the electron configuration in the form of main energy levels (shells).

Select the mystery elements activity and use the information provided to drag twelve elements into their correct positions in the periodic table.

2 The electronic configuration of the magnesium atom is $1s^2\ 2s^2\ 2p^6\ 3s^2$. Deduce its position in the periodic table.

3 Element X is in group 2 and period 3 of the periodic table. Deduce the electron configuration of this element.

4 Deduce the electron configurations of the valence shells in atoms of a gallium and b lead.

5 Deduce the electron configurations and names of the following elements:

a X in group 2 and period 3

b Y in group 15 and period 2

c Z in group 18 and period 3.

6 Identify the positions, and hence the names, of the following elements:

a X (... $3s^2\ 3p^5$)

b Y (... $4s^2\ 4p^5$)

c Z (... $3s^2\ 3p^6\ 4s^2$).

LINKING QUESTION

How has the organization of elements in the periodic table facilitated the discovery of new elements?

Tool 2: Technology

Identify and extract data from databases

A wide range of chemical databases are available on the internet. Many of them store information about the elements and provide important chemical and physical data.

WebElements (www.webelements.com) displays a wide range of properties for each element including atomic properties, physical properties, chemical reactions, crystal structure, thermochemistry, isotopes and geology.

The Orbitron (https://winter.group.shef.ac.uk/orbitron/) provides information on atomic orbitals from the first to the seventh shell.

There are many free online chemical databases that show structures. **ChemSpider** (www.chemspider.com) displays molecular structures and includes properties of compounds. It also allows you to search by structure.

Synthesis of new elements

During the Cold War, scientists in America and Russia competed to form new elements. However, element 118, oganesson, was first synthesized in 2002 at the Joint Institute for Nuclear Research (JINR) in Dubna, near Moscow by a joint team of Russian and American scientists. At the time of writing, active efforts are underway at RIKEN in Japan and JINR to synthesize element 119, known as *eka*-francium, or ununennium (Uue), the seventh alkali metal, via nuclear bombardment. This element is expected to be highly radioactive with short-lived isotopes.

Going further

The left-step periodic table

The periodic table has been extended as new elements have been synthesized and chemists are still debating the 'best' or optimum arrangement. Figure S3.9 shows an alternative known as the left-step periodic table, first proposed by Charles Janet in 1929. It has been developed from the principles of quantum mechanics.

The left-step periodic table is obtained from the long form by moving helium from the top of the periodic table to the top of group 2. The entire s-block is then moved to the right-hand side of the p-block and the f-block elements are fully incorporated into the table on the left. Physicists like this form since it displays the order of orbital filling more clearly. However, it is less useful for chemistry because it places helium into the group 2 metals.

One philosophical approach to the periodic table is that of a realist. If the approximate repetition in the properties of the elements is an objective and natural fact about the physical world, there is an optimal way to display the elements. Such an optimal periodic table would highlight fundamental empirical 'truths' about the elements and their relationships.

f-block														d-block										p-block						s-block	
																														1 H	2 He
																														3 Li	4 Be
																								5 B	6 C	7 N	8 O	9 F	10 Ne	11 Na	12 Mg
																								13 Al	14 Si	15 P	16 S	17 Cl	18 Ar	19 K	20 Ca
														21 Sc	22 Ti	23 V	24 Cr	25 Mn	26 Fe	27 Co	28 Ni	29 Cu	30 Zn	31 Ga	32 Ge	33 As	34 Se	35 Br	36 Kr	37 Rb	38 Sr
														39 Y	40 Zr	41 Nb	42 Mo	43 Tc	44 Ru	45 Rh	46 Pd	47 Ag	48 Cd	49 In	50 Sn	51 Sb	52 Te	53 I	54 Xe	55 Cs	56 Ba
57 La	58 Ce	59 Pr	60 Nd	61 Pm	62 Sm	63 Eu	64 Gd	65 Tb	66 Dy	67 Ho	68 Er	69 Tm	70 Yb	71 Lu	72 Hf	73 Ta	74 W	75 Re	76 Os	77 Ir	78 Pt	79 Au	80 Hg	81 Tl	82 Pb	83 Bi	84 Po	85 At	86 Rn	87 Fr	88 Ra
89 Ac	90 Th	91 Pa	92 U	93 Np	94 Pu	95 Am	96 Cm	97 Bk	98 Cf	99 Es	100 Fm	101 Md	102 No	103 Lr	104 Rf	105 Db	106 Sg	107 Bh	108 Hs	109 Mt	110 Ds	111 Rg	112 Cn	113 Nh	114 Fl	115 Mc	116 Lv	117 Ts	118 Og	119	120

■ **Figure S3.9** The left-step periodic table

ATL S3.1C

Black Panther is an African superhero with a body suit made from material containing the fictional metal vibranium, which has unique properties. Extraction and exploitation of this metal support a STEM economy (led by a woman of colour). If vibranium actually existed, where do you think the metal would be placed in the periodic table? Why? What would be the electron configuration? What would be the elemental symbol? Briefly explain your answers.

◆ **Atomic radius**: When the atoms are bound by a single covalent bond, or in a metallic crystal, this is half the distance between nuclei of atoms of the same element.

◆ **Nuclear charge**: The total charge of all the protons in the nucleus.

Trends in atomic properties down groups

Atomic and ionic radius

At the left of the periodic table, the **atomic radius** is defined as the radius of the atom in the metal lattice (the metallic radius). At the right of the periodic table, the atomic radius is the covalent radius: half the distance between the nuclei of two covalently bonded atoms. For the noble gases, the atomic radius is the radius of an isolated atom (the van der Waals radius) (Figure S3.10).

In general, the atomic radius of an atom is determined by the balance between two opposing factors:

1. The **nuclear charge**. The protons produce an attractive electrostatic force that pulls all the electrons closer to the nucleus. This tends to reduce the atomic radius.
2. The shielding effect. The negative charge of electrons in the inner shell(s) partially neutralizes the nuclear charge. This tends to increase the atomic radius.

When moving down a group in the periodic table, both the nuclear charge and the shielding effect increase. However, the outer electrons enter new shells and so are further away as well as being more effectively shielded. There is, therefore, an increase in the atomic radius as the nuclear charge increases (Figure S3.11).

a

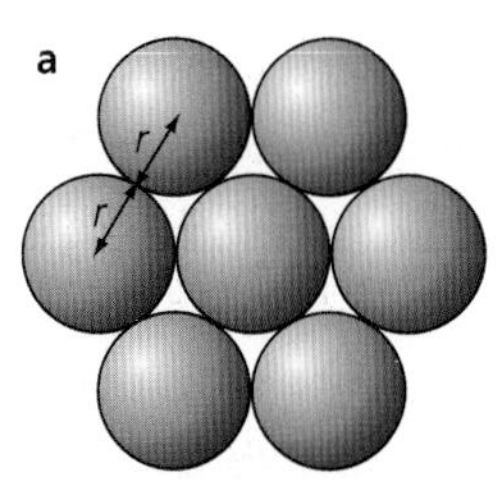

metallic radius

b

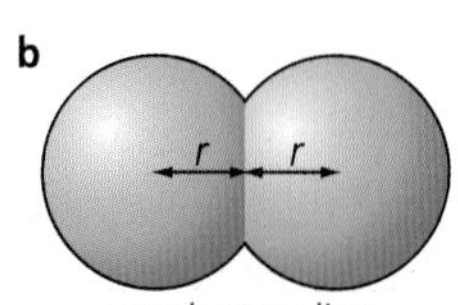

covalent radius

c

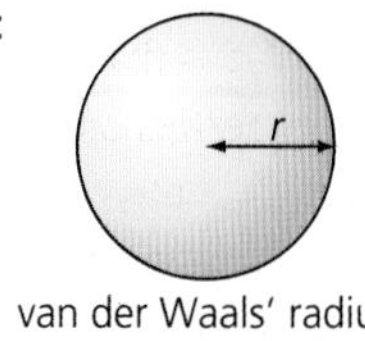

van der Waals' radius (for group 18)

■ **Figure S3.10** Metallic radius, covalent radius and van der Waals radius

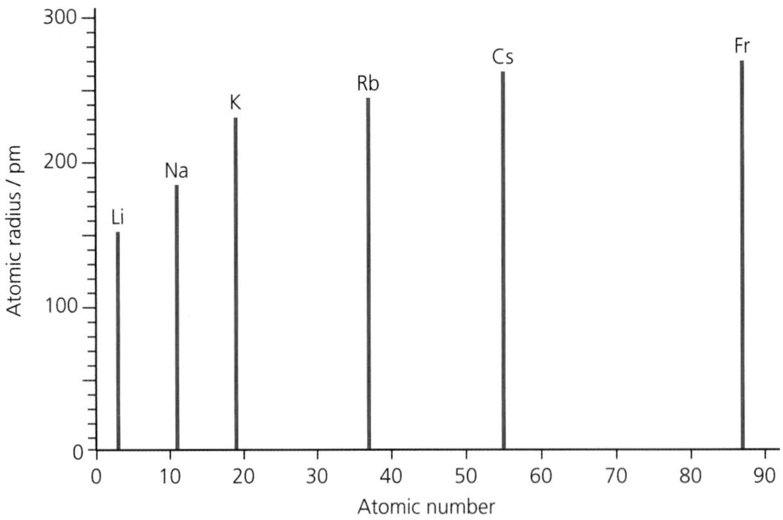

■ **Figure S3.11** Atomic radii in group 1

Inquiry 2: Collecting and processing data

Identify, describe and explain patterns, trends and relationships

Once you have collected your quantitative or qualitative data, you should look for and describe any trends, patterns or relationships you see in it.

The trend or pattern could be simply described: for example, 'As the relative atomic mass of the halogen increases, there is a decrease in the reactivity with iron wool.' However, you should also go further and explain the reasons for the trend or pattern. Using the above example, you may go on and say 'This is because as the halogen gets larger, the nucleus finds it harder to attract electrons towards it as the electrons are further from the nucleus.'

If your data is quantitative, you should plot a graph of the results as this makes it easy to spot relationships. Be prepared to describe the shape of the graph using the terms directly proportional (as one variable increases so does the other in a proportional amount) or inversely proportional (as one value increases, the other value decreases in a proportional amount). Again, you should go further than this and be prepared to give detailed scientific reasons for the relationship you have seen.

For example, a graph of temperature versus rate has been drawn for a reaction:

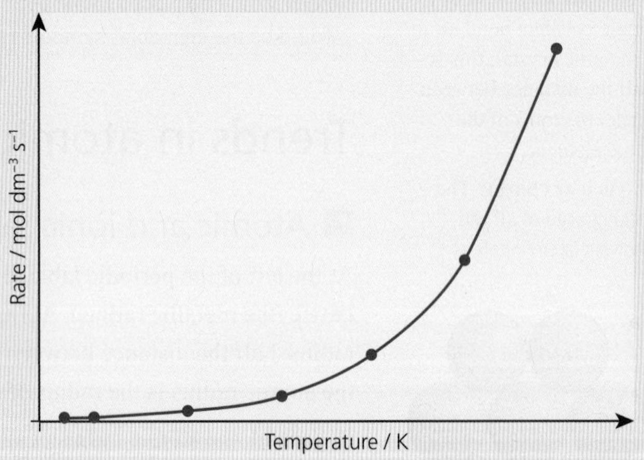

■ **Figure S3.12** A graph of temperature versus rate

You should describe this pattern – increasing the temperature increases the rate of reaction in an exponential manner (this explains the shape of the curve) – and give reasons why the curve looks like this: the particles have more kinetic energy so there are more frequent collisions, and more particles have kinetic energy that is greater than the activation energy.

Tool 2: Technology

Use spreadsheets to manipulate data

Using a spreadsheet to plot graphs allows you to update, amend or add to the data you are displaying and to process the data before creating the graph. It also allows you to easily change the type of graph to display the data in the most appropriate way.

- Use a spreadsheet, such as Excel, to plot the atomic radii of group 17 elements (halogens) as a bar chart. Explain the trend you observe.

Ionic radii are the radii of ions in a crystalline ionic compound. For anions or cations of the same charge, these radii also increase down a group and for the same reason: the increase in nuclear charge is more than outweighed by the addition of electron shells which introduce more shielding and place valence electrons further from the nucleus.

Table S3.8 Ionic radii in group 1 and group 17

Ion	Atomic number	Ionic radius / pm
Li^+	3	68
Na^+	11	98
K^+	19	133
Rb^+	37	148
Cs^+	55	167
Fr^+	87	No data

Ion	Atomic number	Ionic radius / pm
F^-	9	133
Cl^-	17	181
Br^-	35	196
I^-	53	219
At^-	85	No data

The radii of cations are smaller than those of their atoms since they have lost their outer shells and thus have one less electron shell. The nuclear charge is unchanged but acts on a smaller number of electrons which experience a greater electrostatic attraction.

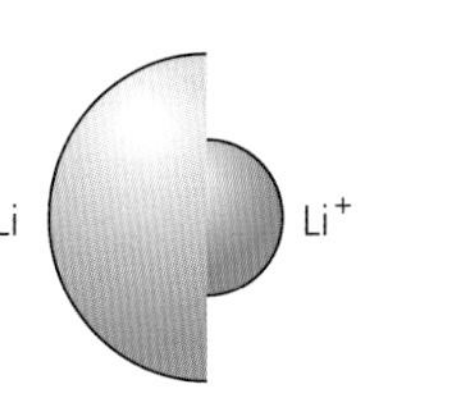

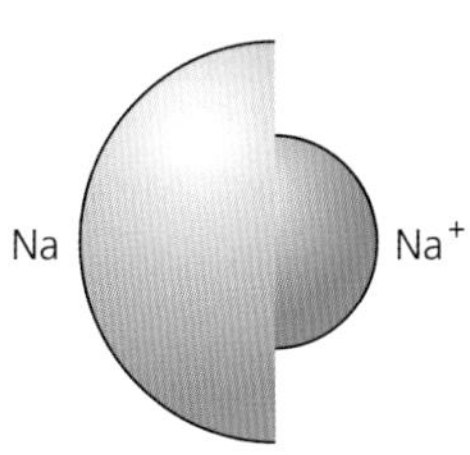

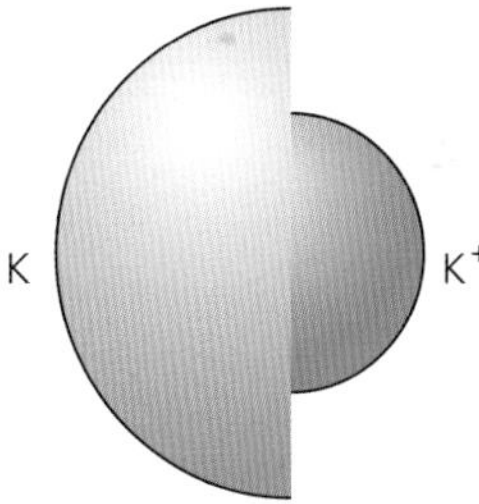

Figure S3.13 Relative sizes of atoms and ions of group 1 metals

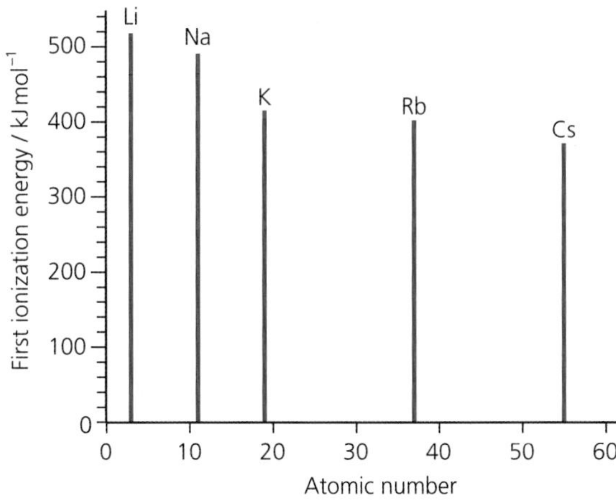

Figure S3.14 First ionization energies in group 1

First ionization energy

On moving down a group, the atomic radius increases (see above). The further the outer or valence shell is from the nucleus, the smaller the attractive force exerted by the nucleus and the more easily an outer electron can be removed. So the first ionization energies decrease down a group (Figure S3.14).

The ionization energy is a measure of how much energy is required to remove the valence electron from an atom. The first ionization energy is the enthalpy change when one mole of gaseous atoms form a mole of gaseous unipositive ions and a mole of electrons.

ATL S3.1D

Use a spreadsheet, such as Excel, to plot the first ionization energy in group 17 (halogens). Explain the trend observed.

Link

Electronegativity is discussed in relation to ionic bonding in Chapter S2.1, page 122 and in chemical models and covalent bonding in Chapter S2.4, page 194.

Since cation formation is a property of metals, the metallic character of elements can be compared in terms of first ionization energies. In general, reactive metals have low ionization energies but reactive non-metals have high ionization energies. Going down a group, the metallic character and reactivity increase as the first ionization energy decreases.

Electronegativity

Top tip!

Electronegativity values are pure numbers based on the Pauling scale with a value of 4.0 given to fluorine, the most electronegative element.

Top tip!

The trends in electronegativity can be used to explain the redox properties of groups 1 and 17. Reducing power (ability to donate electrons) increases *down* group 1; oxidizing power (ability to accept electrons) increases *up* group 17.

The electronegativity of an atom is a measure of its affinity for the bonding pair of electrons in a covalent bond.

Electronegativity values generally decrease down a group. This trend is clear in groups 1 and 17 (Table S3.9).

■ **Table S3.9** Electronegativity in groups 1 and 17

Atom	Atomic number	Electronegativity
Li	3	1.0
Na	11	0.9
K	19	0.8
Rb	37	0.8
Cs	55	0.8
Fr	87	0.7

Atom	Atomic number	Electronegativity
F	9	4.0
Cl	17	3.2
Br	35	3.0
I	53	2.7
At	85	2.2

Decreasing electronegativity indicates a decrease in non-metallic character and an increase in metallic character.

The decrease in electronegativity down the groups can also be explained by the increase in atomic radius: the increasing distance between the nucleus and shared pairs of electrons leads to the attractive force being decreased. Although the nuclear charge increases down a group, the larger electrostatic force this produces is counteracted by the increased shielding due to additional electron shells.

Electron affinity

◆ **First electron affinity**: The enthalpy change when one mole of electrons are added to one mole of gaseous atoms (under standard thermodynamic conditions) to form one mole of gaseous uninegative ions; the enthalpy change, $\Delta_{EA1}H^\ominus(X)$, in the reaction $X(g) + e^- \rightarrow X^-(g)$.

The **first electron affinity** is the enthalpy change when one mole of gaseous atoms accepts a mole of electrons to form a mole of gaseous ions of charge −1: $X(g) + e^- \rightarrow X^-(g)$. For example, $O(g) + e^- \rightarrow O^-(g)$; first electron affinity = −141 kJ mol^{-1}.

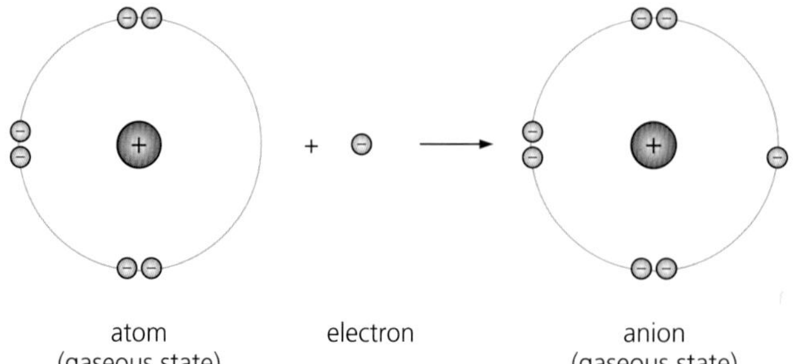

■ **Figure S3.15** The process corresponding to first electron affinity

The process of adding an electron to a gaseous atom is usually exothermic. In an exothermic process heat is released—the ion is more stable than the atom. The more negative the value, the greater the tendency for an atom of that element (in the gas phase) to accept electrons.

On moving down a group, the atomic size increases as well as nuclear charge. However, the effect of the increase in atomic size is much greater than that of the increase in nuclear charge. Hence the values of first electron affinity become less negative moving down a group.

Tool 2: Technology

Generate data from models and simulations

Many computer-generated models and simulations provide information that is not easily visualized when using physical models. They may also provide contextual information that aids understanding.

- Use **https://teachchemistry.org/classroom-resources/periodic-trends-electron-affinity-atomic-radius-ionic-radius-simulation** to examine the formation of an anion and compare the atomic radius of a neutral atom to the ionic radius of its anion for a selection of elements.

This simulation gives a clear visual comparison of the relative sizes of the various species. The electron arrangements and relevant physical data, such as atomic radius, are also automatically displayed.

7 a Write equations (with state symbols) illustrating the first ionization energy and the first electron affinity of hydrogen.
  b Explain why the first ionization energy is endothermic but the first electron affinity is exothermic.
  c State the electron configuration of H^-.
  d Explain why the hydride ion, H^-, has a greater radius than the hydrogen atom, H.

Trends in atomic properties across periods

Atomic radii

When moving from group to group across a period, left to right, the number of protons and the number of electrons increases by one. Since the electrons are added to the same shell, there is only a slight increase in the shielding effect. The effect of the increase in nuclear charge more than outweighs this small increase and, consequently, stronger electrostatic forces of attraction pull all the electrons closer to the nucleus. Hence, atomic radii decrease across a period. Figure S3.16 shows this for period 3.

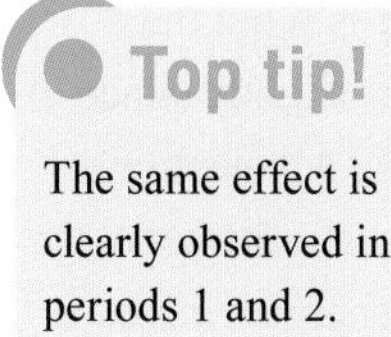

Top tip!

The same effect is clearly observed in periods 1 and 2.

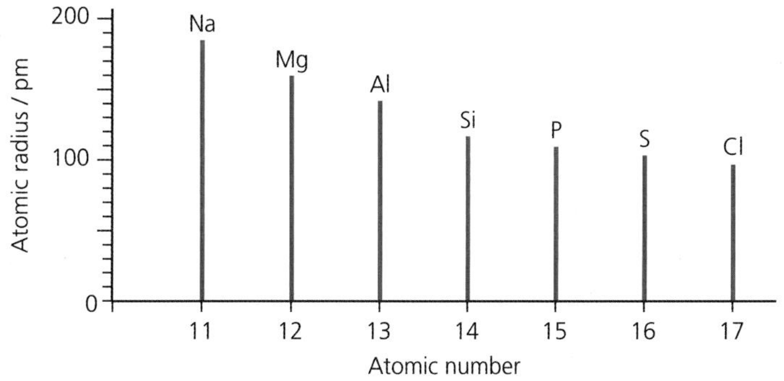

■ **Figure S3.16** Atomic radii of period 3 elements

Ionic radii

Table S3.10 shows the trend in ionic radii across period 3. Bonding in silicon is covalent: it does not form simple ions (Si^{4+} or Si^{4-}) but the radii of the two hypothetical ions can be calculated from an electrostatic model to complete the trend across the period.

- The radii of cations decrease from the sodium ion, Na^+, to the aluminium ion, Al^{3+}.
- The radii of anions decrease from the phosphide ion, P^{3-}, to the chloride ion, Cl^-.
- The ionic radii increase from the aluminium ion, Al^{3+}, to the phosphide ion, P^{3-} (if silicon is ignored).

Table S3.10 Period 3 ionic radii

Element	sodium	magnesium	aluminium	silicon	phosphorus	sulfur	chlorine
Ion	Na^+	Mg^{2+}	Al^{3+}	(Si^{4+} and Si^{4-})	P^{3-}	S^{2-}	Cl^-
Ionic radius / pm	98	65	45	(42 and 271)	212	190	181

Cations of the three metals are isoelectronic species meaning they contain the same number of electrons. The nuclear charge increases from the sodium ion to the aluminium ion so all the electrons in the aluminium ion experience a stronger electrostatic attraction and are located closer to the nucleus.

Table S3.11 Simple ions of period 3 elements

Species	Na^+	Mg^{2+}	Al^{3+}	P^{3-}	S^{2-}	Cl^-
Nuclear charge	11	12	13	15	16	17
Number of electrons	10	10	10	18	18	18
Ionic radius / pm	98	65	45	212	190	181

The nuclear charge increases from the phosphide ion to the chloride ion. The higher nuclear charge causes the electron shells in these isoelectronic species to be pulled closer to the nucleus and, hence, the anionic radii decrease.

The radii of the phosphide, sulfide and chloride anions are larger than their atoms because they have more electrons as an additional complete shell. These electrons not only repel each other, but also partially shield one another from the attractive force of the positive nucleus.

Ionization energy

The *general* trend is an increase in first (and second) ionization energy across the periodic table. Figure S3.17 shows the first and second ionization energies for the atoms of the elements in period 3.

When moving across a period from left to right, the nuclear charge increases but the shielding effect only increases slightly (since electrons enter the same shell). All electrons experience a stronger electrostatic attraction so first (and second) ionization energies increase.

The second ionization energies show a similar trend but displaced by one atomic number unit and with higher values due to the presence of an additional shell which increases the shielding of the electron being removed.

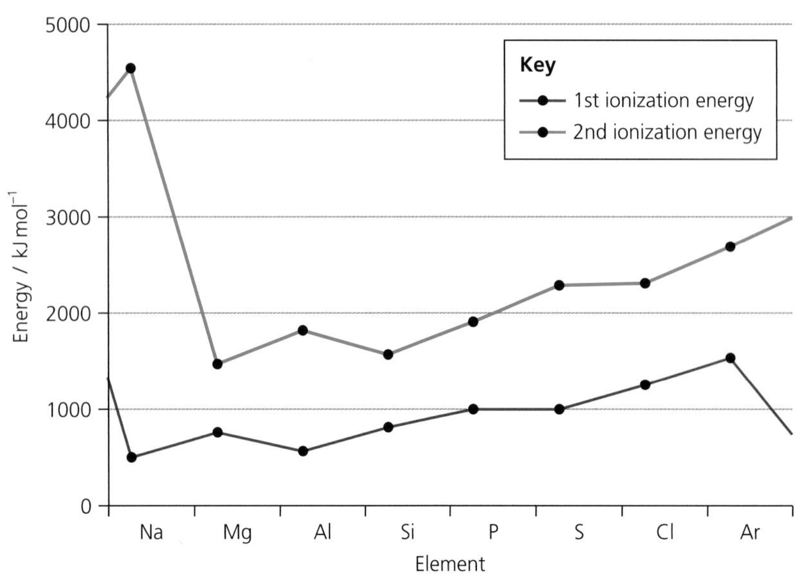

Figure S3.17 First and second ionization energies for atoms of period 3 elements

Explaining discontinuities

There are two decreases in the values of first ionization energy in period 3 (between Mg and Al and between P and S) which can only be explained by reference to sublevels and orbitals. The decrease in ionization energy is a result of the sum of two factors, the change in sublevel and the increase in nuclear charge. The change in sublevel outweighs the increase in nuclear charge.

phosphorus

sublevel			
3p	↑	↑	↑
3s	↑↓		
2p	↑↓	↑↓	↑↓
2s	↑↓		
1s	↑↓		

sulfur

sublevel			
3p	↑↓	↑	↑
3s	↑↓		
2p	↑↓	↑↓	↑↓
2s	↑↓		
1s	↑↓		

■ **Figure S3.18** Orbital diagrams for phosphorus and sulfur atoms

The decrease in first ionization energy from magnesium ($1s^2\ 2s^2\ 2p^6\ 3s^2$) to aluminium ($1s^2\ 2s^2\ 2p^6\ 3s^2\ 3p^1$) occurs because the outermost electron in aluminium enters a 3p orbital which is further away from the nucleus than the 3s electrons and so subject to additional shielding.

The first ionization energy of sulfur ($1s^2\ 2s^2\ 2p^6\ 3s^2\ 3p^4$) is less than that of phosphorus ($1s^2\ 2s^2\ 2p^6\ 3s^2\ 3p^3$) because the new electron is part of a p orbital spin pair and experiences increased inter-electron repulsion (Figure S3.18).

Common mistake

Misconception: *Half-filled (and also completely filled) sublevels of electrons have an intrinsic stability that can be used to explain the dips in first ionization energy across periods 2 and 3.*

The dips in first ionization energy across periods 2 and 3 can be explained by a change in sublevel (energy and hence distance) and additional electron–electron repulsion (Pauli exclusion principle).

Top tip!

The fact that electronegativity values of chemical elements generally increase across a period and decrease down a group can be used to assess the relative electronegativity of two elements. The further apart the two elements are in the periodic table, the larger the difference in their electronegativities.

8 a Define the term 'first ionization energy'.
  b State and explain the general trend in ionization energy across period 2 in the periodic table.
  c (HL only) Explain the reason for the small decrease in ionization energy between beryllium and boron.
  d (HL only) Explain the reason for the small decrease in ionization energy between nitrogen and oxygen.

Electronegativity

The general trend is an increase in electronegativity across a period from left to right. Between successive atoms, the nuclear charge increases but the shielding effect only increases slightly (since electrons enter the same shell). Consequently, the electron shells are pulled progressively closer to the nucleus and, as a result, electronegativity values increase.

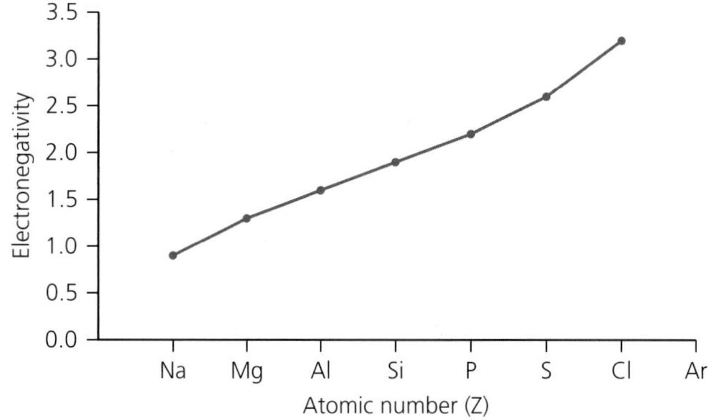

■ **Figure S3.19** Electronegativity against atomic number for period 3 elements

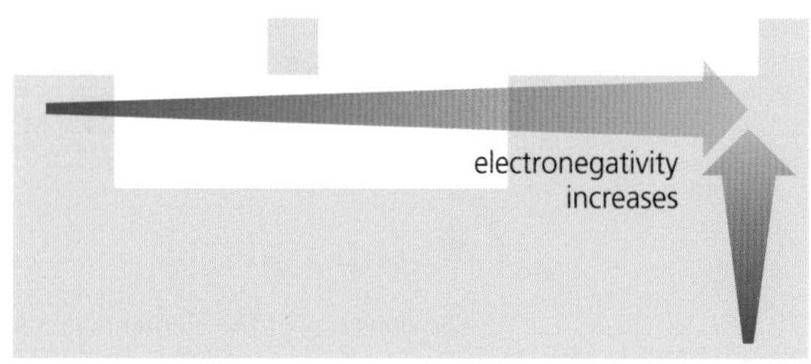

■ **Figure S3.20** Trends in electronegativity for s- and p-block elements

Electron affinity

On moving across a period, the atomic radius decreases and the nuclear charge increases. Both of these factors result in greater attraction for the incoming electron. Hence first electron affinities tend to become more negative (more exothermic) across a period from left to right.

Tool 3: Mathematics

Radar charts

A radar chart is a way of showing two or more sets of data and the correlation between them. They are often useful for making comparisons.

Each variable is given an axis that starts from the centre. All axes are arranged radially, with equal distances between them; the same scale is maintained between all axes. Grid lines that connect from axis-to-axis are used as a guide. Each variable value is plotted along its individual axis and all the variables in a single data set are connected to form a polygon.

ATL S3.1E

- Plot a radar chart showing IB *Chemistry data booklet* section 9 values of first electron affinity plotted next to electronegativity for periods 2 and 3.
- Comment on the relationship and degree of correlation between the two atomic properties.
- (HL only) Explain any unexpected drops in electron affinity by reference to electron configurations based on atomic orbitals.

Group 1 (alkali metals)

◆ **Alkali metals**: The elements of group 1; they have one outer (valence) s electron.

The main elements of group 1 are lithium, sodium, potassium, rubidium and caesium. They are known as the **alkali metals** and they are all soft silvery metals that can be cut with a knife. They are very reactive, easily forming a wide range of ionic compounds with non-metals such as the halogens and oxygen. They are stored in bottles of oil to stop them reacting with water and oxygen in the air.

The atomic and physical properties of the alkali metals are summarized in Table S3.12.

■ **Table S3.12** The atomic and physical properties of three alkali metals (electrode potential is a measure of reducing strength in aqueous solution)

Element	lithium	sodium	potassium
Electron arrangement	2,1	2,8,1	2,8,8,1
Electron configuration	$1s^2\ 2s^1$	$1s^2\ 2s^2\ 2p^6\ 3s^1$	$1s^2\ 2s^2\ 2p^6\ 3s^2\ 3p^6\ 4s^1$
Chemical symbol	Li	Na	K
First ionization energy / kJ mol^{-1}	519	494	418
Atomic radius / nm	0.152	0.186	0.231
Melting point / K	454	371	337
Boiling point / K	1600	1156	1047
Density / g cm^{-3}	0.53	0.97	0.86
Standard electrode potential, $E^\ominus$ $M^+(aq)$ | $M(s)$ / V	−3.03	−2.71	−2.92

Top tip!

The densities of the elements generally increase down groups 1 and 2 because the relative atomic masses of the elements increase more rapidly down the group than atomic volumes. Larger atoms have weaker metallic bonding and the packing arrangement also affects the density.

Top tip!

This reaction is an example of a redox reaction, in which atoms of the group 1 element act as a reducing agent and release electrons: $M(s) \rightarrow M^+(aq) + e^-$.

The alkali metals all react with water, releasing hydrogen gas and forming an aqueous solution containing a metal hydroxide. The general equation for this reaction is $2M(s) + 2H_2O(l) \rightarrow 2MOH(aq) + H_2(g)$, where M represents the group 1 metal.

Reactivity towards water increases down the group with decreasing values of first ionization energy. If there is sufficient heat, a characteristic coloured flame will be observed due to the hydrogen burning with a small amount of gaseous metal atoms (see Figure S1.47, page 42).

Going further

Reactivity with water

Although lithium reacts rather slowly with water, sodium reacts quite vigorously and the heavier alkali metals (potassium, rubidium and caesium) react so vigorously that they explode. This trend, which is not consistent with the standard reduction potentials of the atoms of the elements, is an example of the complex interplay of different forces and phenomena—in this case, kinetics and thermodynamics.

ATL S3.1F

Outline and present, using presentation software (such as PowerPoint), the life and research work of Marguerite Catherine Perey, the discoverer of francium, who named it after France.

Find out how official names of new elements are decided by IUPAC and outline one element-naming controversy.

9 Caesium is an alkali metal.

a State which group, period and block caesium is located in.

b Explain (in simple terms) why the alkali metals are the most chemically reactive metals.

c State the equation describing the reaction between caesium and cold water.

d State two observations that could be made during the reaction.

e Explain why the solution formed when caesium reacts with water has a high pH.

Caesium reacts with chlorine, oxygen and nitrogen to form ionic compounds.

f State the equations describing these redox reactions.

g Explain why caesium is more reactive than potassium.

Group 17 (halogens)

Properties of the halogens

◆ **Halogens**: The elements of group 17; they have outer (valence) electron configurations $ns^2\ np^5$.

The **halogens** are a group of very reactive non-metals that exist as diatomic molecules. Their atomic and physical properties are summarized in Table S3.13.

Table S3.13 The atomic and physical properties of the halogens

Element	**chlorine**	**bromine**	**iodine**
Chemical formula	Cl_2	Br_2	I_2
Structure	Cl–Cl	Br–Br	I–I
Electron arrangement	2,8,7	2,8,18,7	2,8,18,18,7
Outer shell configuration	$3s^2\ 3p^5$	$4s^2\ 4p^5$	$5s^2\ 5p^5$
State at room temperature and pressure	gas	liquid	solid
Colour	pale green	red–brown	black
Melting point / K	172	266	387
Boiling point / K	239	332	458 (sublimes)
Standard electrode potential, $E^\ominus$ $X_2(aq)$ | $X^-(aq)$ / V	1.36	1.09	0.54

The halogens become more reactive up the group from iodine to fluorine. This generally correlates with an increase in electronegativity from 2.7 to 4.0 and a change in first electron affinity from $-295\,kJ\,mol^{-1}$ to $-328\,kJ\,mol^{-1}$.

Halogens form uninegative anions known as halide ions (for example, fluoride, F^-) as they have seven electrons in their outer shell and gain one more to obtain eight valence electrons in reactions. Compounds such as sodium chloride that contain halogens are known as halides.

Reactions of the halogens

The halogens all react easily with metals by the transfer of electrons from the metal to the halogen. The halogens react with other non-metals by sharing pairs of electrons to form molecules.

◆ **Displacement reaction (halogens)**: a redox reaction where a halogen gains electrons and the halide ions lose electrons.

The change in reactivity of the halogens can be seen by observing **displacement reactions** between halogens and halide ions.

A more reactive non-metal can displace a less reactive non-metal from a compound. This means that a more reactive halogen can displace a less reactive halogen from a halide compound.

When chlorine water is added to an aqueous solution of potassium bromide, KBr, the solution becomes yellow-orange due to the formation of bromine:

$$Cl_2(aq) + 2Br^-(aq) \rightarrow Br_2(aq) + 2Cl^-(aq)$$

Chlorine also reacts with potassium iodide solution to form a brown solution of iodine:

$$Cl_2(aq) + 2I^-(aq) \rightarrow I_2(aq) + 2Cl^-(aq)$$

The two reactions shown above are known as displacement reactions since the more reactive halogen, chlorine, displaces or 'pushes out' a less reactive halogen , bromine or iodine, from its salt.

They are also redox reactions: there is a transfer of electrons from the iodide or bromide ions to the chlorine molecules (Figure S3.21). The halogen acts as an oxidizing agent and the halide ion acts as a reducing agent. Going down group 17, the halogens become more weakly oxidizing and the halide ions become more strongly reducing.

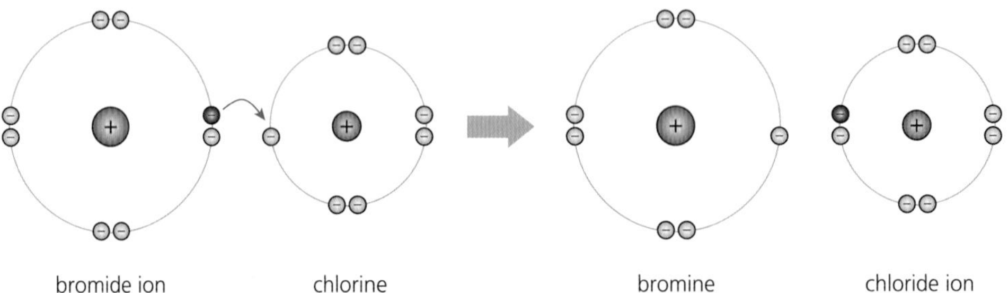

■ **Figure S3.21** The reaction between a halide ion and a halogen atom

Bromine water will give a displacement reaction with a solution of an iodide:

$$Br_2(aq) + 2I^-(aq) \rightarrow I_2(aq) + 2Br^-(aq)$$

However, as bromine is less reactive than chlorine, it is unable to displace chloride ions and no reaction occurs.

Iodine, being the most unreactive halogen, is unable to displace bromide or chloride ions and no reaction occurs.

The trends in oxidizing and reducing power for the halogens and the halide ions can be easily explained in terms of the relative sizes of the halogen atoms and halide ions.

A halide ion is oxidized by the removal of one of its outer eight electrons. In a large halide ion, the outer electrons are more easily removed as they are further from the nucleus and more effectively shielded from its attraction by the inner electrons. Small halide ions have their outer electrons located closer to the nucleus and less effective shielding occurs, hence their affinity for electrons is higher.

A similar argument explains why a small halogen atom can attract an extra electron with a greater affinity than a larger halogen atom.

LINKING QUESTION

Why are simulations and online reactions often used in exploring the trends in chemical reactivity of group 1 and group 17 elements?

10 Iodine is a halogen.
  a Distinguish between the terms 'group' and 'period'.
  b State which group, period and block iodine is located in.
  c Explain (in simple terms) why the halogens are the most chemically reactive non-metals.
  d Fluorine gas is bubbled into a solution of potassium iodide. State the colour change and state the ionic equation for the reaction (ignore any side reaction with water).
  e Explain why chlorine is more reactive than iodine.

11 a Plot a line graph of melting points versus atomic number for the alkali metals from lithium to caesium.
  b Extrapolate the curve to estimate the melting point of francium (see Top tip). Evaluate the accuracy of your extrapolation by comparing against the value for the melting point of francium from the IB *Chemistry data booklet.*
  c Explain the shape of the graph, stating any assumptions present in this simple approach.

Top tip!

Extrapolation is a graphical procedure involving extending a line or curve to estimate values that lie beyond the data points obtained. In this instance, the data is discontinuous so the graph should be a bar chart. However, a line graph is needed to enable extrapolation.

Oxides

Table S3.14 shows the formulas and properties of some of the common oxides of the period 3 elements.

■ **Table S3.14** Formulas and properties of the oxides of period 3 elements

Formula	Na_2O	MgO	Al_2O_3	SiO_2	P_4O_6 and P_4O_{10}	SO_2 and SO_3	Cl_2O and Cl_2O_7
Physical state under standard conditions	solid	solid	solid	solid	solids	gas and volatile liquid	gas and solid
Bonding	ionic	ionic	ionic (with covalent character)	giant covalent	molecular covalent	molecular covalent	molecular covalent
Acid–base nature	basic	basic	amphoteric	weakly acidic	acidic	acidic	acidic

The maximum oxidation state (see page 252) of the elements increases as the period is crossed from left to right because every period 3 element can use all the electrons in its outermost shell when bonding to oxygen (oxidation state = −2).

Top tip!

In period 3 oxides, the atoms of the period 3 element have positive oxidation states because oxygen has a higher electronegativity than any period 3 element.

Common mistake

Misconception: *The number of covalent bonds formed equals the number of electrons in the outer shell.*

The maximum number of covalent bonds often corresponds to the number of valence electrons, but other numbers of bonds may be formed: for example, sulfur can form 2, 4 and 6 bonds.

◆ **Basic oxide**: An ionic oxide, usually an oxide of a metal, that reacts with acids to form a salt and water.

As Table S3.14 shows, metals form **basic oxides**. The oxides of sodium and magnesium react with water to form hydroxides. The presence of excess aqueous hydroxide ions, $OH^-(aq)$, makes these solutions alkaline:

$$Na_2O(s) + H_2O(l) \rightarrow 2NaOH(aq)$$

$$MgO(s) + H_2O(l) \rightleftharpoons Mg(OH)_2(aq)$$

Top tip!

Metallic oxides tend to be ionic and basic. The more reactive metals form oxides that react with water to form alkaline solutions.

Magnesium oxide is not as soluble as sodium oxide because it has stronger ionic bonding and hence much lower solubility in water. As a result, reactions involving magnesium oxide are always less exothermic than those of sodium oxide, and the pH of a solution of magnesium oxide is less than that of a solution made from the same masses of sodium oxide and water.

Unlike sodium and magnesium oxides, aluminium oxide does not react with or dissolve in water, which is why a thin oxide layer can protect aluminium metal from corrosion. It does react slowly with warm dilute aqueous solutions of acids to form salts. For example:

$$Al_2O_3(s) + 6HCl(aq) \rightarrow 2AlCl_3(aq) + 3H_2O(l)$$

$$Al_2O_3(s) + 6H^+(aq) \rightarrow 2Al^{3+}(aq) + 3H_2O(l)$$

When aluminium oxide reacts with an acid, it behaves like a base or basic oxide: it forms a salt (aluminium chloride in the example with dilute hydrochloric acid above) and water.

Aluminium oxide also reacts with warm concentrated solutions of strong alkalis to form aluminates. For example:

$$Al_2O_3(s) + 2NaOH(aq) + 3H_2O(l) \rightarrow 2NaAl(OH)_4(aq)$$

$$Al_2O_3(s) + 2OH^-(aq) + 3H_2O(l) \rightarrow 2Al(OH)_4^-(aq)$$

Amphoteric oxides are likely to be formed by metals near the division between metals and non-metals.

When it reacts with an alkali, aluminium oxide behaves like an acid and forms a salt (sodium tetrahydroxoaluminate in the example with sodium hydroxide above).

This dual behaviour provides evidence that the chemical bonding in aluminium oxide is neither purely ionic nor purely covalent: it is an **amphoteric oxide**.

◆ **Amphoteric oxide**: An oxide that can react with both acids and bases.

◆ **Acidic oxide**: A covalent oxide, usually an oxide of a non-metal that reacts with bases to form a salt.

Silicon dioxide is also insoluble in water. Water cannot react with or dissolve its covalent network structure.

However, the oxide will react with and dissolve in hot, concentrated alkali:

$$SiO_2(s) + 2NaOH(aq) \rightarrow Na_2SiO_3(aq) + H_2O(l)$$

Silicon dioxide acts as an acid when it reacts with sodium hydroxide, forming a salt (sodium silicate) plus water. It reacts with strong bases but does not react with acids, so it is classified as an acidic oxide.

The pH of the solution formed by hydrolysis of a non-metallic oxide is concentration-dependent.

The remaining oxides of period 3 (from phosphorus to chlorine) have covalent or polar covalent bonding. These covalent oxides, if soluble, react and dissolve in water to form acids. They are **acidic oxides** and react with bases.

Phosphorus(V) oxide reacts vigorously with water in a hydrolysis reaction, dissolving to form an acidic solution of phosphoric(V) acid (pH $\approx$ 3):

$$P_4O_{10}(s) + 6H_2O(l) \rightarrow 4H_3PO_4(aq) \rightleftharpoons 4H^+(aq) + 4H_2PO_4^-(aq)$$

Top tip!

These are both reactions in which the oxidation state of sulfur is unchanged. In these reactions the non-metal (sulfur) is acting as a Lewis acid (see Chapter S2.2, the covalent model, page 136) and the product then undergoes dissociation in water.

Sulfur trioxide reacts with and dissolves in water, forming a very acidic solution (pH ≈ 1) of sulfuric(VI) acid:

$$SO_3(g) + H_2O(l) \rightarrow H_2SO_4(aq) \rightleftharpoons H^+(aq) + HSO_4^-(aq)$$

Sulfur dioxide reacts and dissolves in water to form a less acidic solution of sulfuric(IV) acid (sulfurous acid, a weak acid):

$$SO_2(g) + H_2O(l) \rightarrow H_2SO_3(aq) \rightleftharpoons H^+(aq) + HSO_3^-(aq)$$

Both oxides of chlorine react vigorously with water in hydrolysis reactions to form acidic solutions. For example:

$$Cl_2O(g) + H_2O(l) \rightleftharpoons 2HClO(aq)$$

12 Chlorine(VII) oxide, Cl_2O_7, is an acidic oxide and reacts violently with water to form chloric(VII) acid, $HClO_4$.

Chlorine(VII) oxide explodes to form its constituent elements.

Write balanced equations for both reactions.

Other common acidic oxides are carbon dioxide, CO_2, and nitrogen dioxide, NO_2.

Carbon monoxide, CO, is a neutral oxide: it does not react with either dilute alkali or dilute acid under standard conditions.

Going further

Oxides and electronegativity

The electronegativity difference between oxygen and hydrogen is 1.2 and so the –O–H bond has considerable ionic character, δ– O–H δ+. An element M attached to an O–H group can ionize in two different ways:

As an acid: $M\text{–}O\text{–}H \rightarrow M\text{–}O^- + H^+$

As a base: $M\text{–}O\text{–}H \rightarrow M^+ + OH^-$

Which of these takes place is determined mainly by the value of the electronegativity of M.

- If M is more electronegative than hydrogen (2.2), the M–O bond has less ionic character than the O–H bond and the substance behaves as an acid. This is generally the case when M is nitrogen, phosphorus, sulfur or chlorine in, for example, chloric(I) acid, HOCl(aq), or nitric(III) acid, HONO(aq).
- If M is less electronegative than 1.5, the M–O bond has a greater ionic character than the O–H bond and the substance behaves as a base. This is the case when M is in group 1 or group 2 (with the exception of beryllium), for example, NaOH(aq) and $Ca(OH)_2$(aq).
- If the electronegativity of M is only slightly lower than that of hydrogen (2.2), the substance is amphoteric (it can behave as an acid or a base). This happens when M is beryllium (1.6) or aluminium (1.6). Beryllium hydroxide, $Be(OH)_2$, is amphoteric, has some covalent character and is poorly soluble in water. It is also the case for tin (2.0) and lead (1.8) and for many of the elements in the d-block.

This information is summarized in Table S3.15.

Table S3.15 Acid–base character of M–O–H

Electronegativity of M	Acid–base character
< 1.5	basic
1.5–2.5	amphoteric
> 2.5	acidic

Some oxides that might be expected to be amphoteric are not (for example, iron and nickel). This may be because their oxides and hydroxides are highly insoluble—it is impossible to make them dissolve in excess alkali.

LINKING QUESTION

How do differences in bonding explain the differences in the properties of metal and non-metal oxides?

Reactions of oxides in the Earth system

Acid rain

Pure rainwater is naturally slightly acidic because carbon dioxide, a rare gas in the atmosphere (420 ppm or 0.042%), dissolves in and reacts with water to form carbonic acid which immediately dissociates into hydrogen and hydrogencarbonate ions.

$$CO_2(g) + H_2O(l) \rightleftharpoons H_2CO_3(aq) \rightleftharpoons H^+(aq) + HCO_3^-(aq)$$

◆ **Acid rain**: Rainwater with a pH less than 5.6 due to the presence of sulfuric and nitric acids.

When rain has a pH lower than that of normal rainwater (5.6), it is considered to be **acid rain**. The cause is usually an airborne pollutant, which may have a natural source.

Sulfur dioxide enters the atmosphere when a fossil fuel (especially coal) is burnt or a volcanic eruption occurs: $S(s) + O_2(g) \rightarrow SO_2(g)$. This reacts with water in moist air to form sulfurous acid and sulfuric acid as shown above.

Oxides of nitrogen (NO_x) are the second contributor to acid rain. Nitrogen monoxide is formed at high temperatures in car engines and fossil-fuelled power stations. The nitrogen monoxide is then rapidly converted to nitrogen dioxide:

$$N_2(g) + O_2(g) \rightarrow 2NO(g)$$

$$2NO(g) + O_2(g) \rightarrow 2NO_2(g)$$

Nitrogen dioxide dissolves and reacts in water to give a mixture of nitrous (nitric(III)) acid (a weak acid) and nitric(V) acid (a strong acid):

$$2NO_2(g) + H_2O(l) \rightarrow HNO_2(aq) + HNO_3(aq)$$

Animals and plants that are sensitive to changes in pH can be severely affected when acid rain flows as runoff into rivers, streams and lakes. If acidic rainwater is absorbed by soil, aluminium ions in the soil dissolve and enter streams. High concentrations of aluminium ions are toxic to fish.

ATL S3.1G

Find out about the following:

- pre-combustion and post-combustion methods of counteracting the effects of acid rain
- how acid deposition levels in your own country have changed over time, what effects they have had, and what emission controls are implemented in industry and fossil fuel power stations
- practical steps that you as an individual can take to reduce acid rain production.

Present your findings to the class, using presentation software (for example, PowerPoint). Include annotated diagrams and chemical equations.

Acid rain

The smoke and gas from coal-fired power stations is usually released into the air from very tall chimneys to reduce pollution in the surrounding area. However, the dry acidic oxides can be carried by winds and give rise to acid rain in regions far from the power station. The death of lakes and forests in Germany and Sweden was blamed on coal-fired power stations in the United Kingdom. Reducing this trans-boundary problem required international political cooperation in the 1980s to agree on research, monitoring and exchange of data.

Going further

Catalytic converters

In an ideal vehicle engine, the hydrocarbons in unleaded petrol or diesel are converted completely into carbon dioxide and water. This assumes complete combustion and no competing side reactions.

In practice, however, three other substances are present in the exhaust fumes:

- unburnt hydrocarbons
- carbon (soot)
- carbon monoxide from the incomplete combustion of the fuel

The amount of each impurity produced depends on the type, efficiency and temperature of the engine. For example, diesel engines produce higher quantities of carbon particles than petrol engines.

To reduce the amounts of these impurities, the exhaust gases are passed through a catalytic converter (Figure S3.22). This contains transition metals such as platinum and rhodium supported on an inert honeycomb support, a design that provides a large surface area on which reactions such as the following occur:

$$2CO(g) + 2NO(g) \rightarrow 2CO_2(g) + N_2(g)$$
(reduction of NO_x emissions)

$$2CO(g) + O_2(g) \rightarrow 2CO_2(g)$$

$$2C_8H_{18}(l) + 25O_2(g) \rightarrow 16CO_2(g) + 18H_2O(l)$$

Because lead poisons this catalyst by binding irreversibly to the surface, it is essential to use unleaded petrol in a car fitted with a catalytic converter.

As transport fleets worldwide begin to be replaced by electric-powered vehicles (which use a secondary power source), the generation of SO_x and NO_x in cities will be reduced. However, this may still occur at power stations.

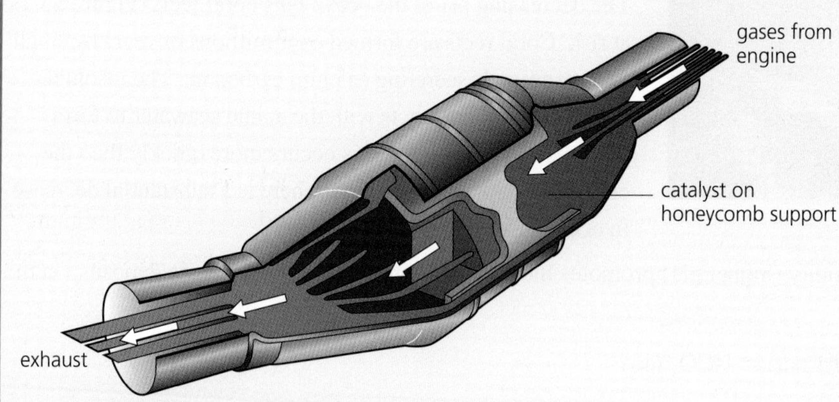

■ **Figure S3.22** A catalytic converter

Ocean acidification

Carbon dioxide dissolved in the upper layer of the oceans, and carbon dioxide in the atmosphere, exist in dynamic equilibrium:

$$CO_2(g) \rightleftharpoons CO_2(aq)$$

Top tip!

The position of this equilibrium is affected by temperature and the concentration of carbon dioxide in the air.

As the concentration of carbon dioxide in the atmosphere increases, the rate of absorption of carbon dioxide in the oceans increases until a new equilibrium is established in which the concentration of dissolved carbon dioxide is higher and the pH decreases.

The oceans act as a sink (store) of carbon dioxide: much of the additional carbon dioxide entering the atmosphere from burning fossil fuels dissolves in the oceans reversibly. This leads to the formation of carbonic acid which immediately dissociates to form hydrogencarbonate ions.

$$CO_2(aq) + H_2O(l) \rightleftharpoons H_2CO_3(aq) \qquad (1)$$

(The involvement of water molecules in equations (2) and (3) reflects the use of the Brönsted–Lowry model of acids and bases.)

$$H_2CO_3(aq) + H_2O(l) \rightleftharpoons HCO_3^-(aq) + H_3O^+(aq) \qquad (2)$$

The hydrogencarbonate ion dissociates (to a small extent) to form carbonate ions (CO_3^{2-}):

$$HCO_3^-(aq) + H_2O(l) \rightleftharpoons CO_3^{2-}(aq) + H_3O^+(aq) \qquad (3)$$

Both dissociations generate aqueous hydrogen (hydronium) ions, H_3O^+, which are responsible for making the ocean water acidic.

◆ **Acidification:** A reduction in the pH of the ocean over an extended period of time, caused by uptake of carbon dioxide.

A pair of simpler but equivalent equations can also be written, without the involvement of water molecules and hydronium (H_3O^+) ions:

$$H_2CO_3(aq) \rightleftharpoons HCO_3^-(aq) + H^+(aq)$$

$$HCO_3^-(aq) \rightleftharpoons CO_3^{2-}(aq) + H^+(aq)$$

■ **Figure S3.23** Coral reef

An increase in the concentration of dissolved carbon dioxide will result in an increased concentration of carbonic acid in equilibrium (1). Increasing the concentration of carbonic acid in equilibrium (2) leads to a shift to the right—as predicted by Le Chatelier's principle—and therefore a higher concentration of hydrogencarbonate and hydronium ions. As the reactions move to the right, the pH decreases since $pH = -\log_{10} [H_3O^+(aq)]$. This is the cause of ocean **acidification**.

The decreasing pH of the ocean puts coral reefs (Figure S3.23) at risk. Coral reefs are formed over millions of years by small marine animals secreting calcium carbonate. The insoluble calcium carbonate reacts with the acidic seawater to form aqueous calcium ions. If this occurs more quickly than the corals can replace the material, there is a substantial decrease in ocean biodiversity.

A decrease in acidity (higher pH) promotes the precipitation of calcium carbonate deposits via the reactions below:

$$CO_2(aq) + OH^-(aq) \rightleftharpoons HCO_3^-(aq)$$

$$OH^-(aq) + HCO_3^-(aq) \rightleftharpoons H_2O(l) + CO_3^{2-}(aq)$$

$$CO_3^{2-}(aq) + Ca^{2+}(aq) \rightleftharpoons CaCO_3(s)$$

TOK

To what extent do the classification systems we use in the pursuit of knowledge affect the conclusions that we reach?

Chemists classify elements based on evidence about their chemical, physical and electronic properties.

A classification system, such as classifying elements as metals, non-metals or metalloids, needs a theory (or a number of alternative theories) to explain it and justify the classification.

Chemists will attempt to test each theory or extend the application of a classification. When new observations are made, new classifications or theories may be required.

Electronegativity can be used to predict bond type (metallic, ionic or covalent) and these behaviours can be mapped to the bonding triangle. Linus Pauling (1901–1994) was the first chemist to propose a quantitative scale for the concept of electronegativity. He developed an empirical scale derived from thermochemical bond enthalpy data in the 1930s. It runs from 0.7 (francium) to 4.0 (fluorine) and in 1947, Pauling asserted that metalloids had electronegativity values close to 2. Consequently, they came to be regarded as strange in-between elements. This was reinforced as the semiconductor industry emerged in the 1950s and the development of solid-state electronics began in the early 1960s. The semiconducting properties of germanium and silicon (and boron and tellurium) supported the idea that metalloids were 'in-between' or 'half-way' elements.

■ **Figure S3.24** Semiconductor-based LEDs are used for low-energy decorative lighting

However, the non-metals at the centre of the electronegativity scale are not semiconductors but substances that lie between the metalloids and the halogens, namely: H, C, N, O, P, S and Se. It could be argued that metalloids can be classified as non-metals that are weakly metallic, so the third class of metalloids is unnecessary.

◆ **Oxidation state:** A positive, negative or zero number (usually an integer), given to indicate whether an element in a compound has been reduced or oxidized during a redox reaction.

Oxidation states

The **oxidation state** is a signed number (zero, positive or negative) which represents the number of electrons lost (or gained, if the number is negative) by an atom of an element in a compound. Oxidation states are useful in identifying and balancing redox reactions, as well as for naming inorganic compounds.

- For a cation, the oxidation state is the number of electrons which must be added to form an atom. For example, $Mg^{2+} + 2e^- \rightarrow Mg$ so the oxidation state of Mg^{2+} is +2.
- For an anion, the oxidation state is the number of electrons which must be removed to form an atom. For example, $S^{2-} \rightarrow S + 2e^-$ so the oxidation state of sulfur in the sulfide ion is −2.

There are some simple rules based on the electronegativities of elements and the bonding of a compound that we can apply to find the oxidation state of atoms of individual elements in chemical species:

1. The oxidation state of any uncombined element is zero.
   For example, in O_2(g), the oxidation state of oxygen is zero.
2. For a simple ion, the oxidation state of the ion is equal to the charge on the ion.
   For example, in Cl^- and Fe^{2+}, the oxidation states are −1 and +2, respectively.
3. The sum of the oxidation states of the elements in a compound is zero.
   For example, in NaCl [Na^+ Cl^-], the sum of the oxidation states is +1 + (−1) = 0.

Top tip!

Note the different order for the sign and number of the oxidation state compared to the charge of the ion.

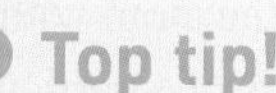

Top tip!

In manganese(II) chloride, $MnCl_2$, the oxidation state of chlorine is −1, but the sum of the two oxidation states of chlorine is −2.

By convention, the less electronegative element appears first in the formula of a binary compound, so chlorine fluoride has the formula ClF and not FCl.

4 In a polyatomic ion, the sum of the oxidation states of the elements is equal to the charge on the ion.

For example, in the sulfate ion, SO_4^{2-}, the sum of the oxidation states is +6 + (–2 × 4) = –2.

5 The oxidation state of hydrogen is +1 except where it is combined with a metal (as in, for example, sodium hydride, NaH [Na^+ H^-]) where it is –1.

6 The oxidation state of oxygen is –2 except in peroxides (where it is –1) and oxygen difluoride, OF_2 (where it is +2).

7 In a covalent bond in a molecule, the more electronegative element is given a negative oxidation state and the less electronegative element is given a positive oxidation state.

For example, the chlorine atom is less electronegative than the fluorine atom (the most electronegative element). Therefore, in chlorine fluoride, ClF, chlorine is assigned an oxidation state of +1 and fluorine that of –1.

When determining the oxidation states of elements in compounds, it helps to rewrite the formulas showing the presence of any ions.

WORKED EXAMPLE S3.1B

Determine the oxidation state of each element in the following species:

$K_2Cr_2O_7$ $Fe(OH)_3$ $Fe(OH)_2$ CO_3^{2-} CN^-

$K_3Fe(CN)_6$ C_{60} $SnCl_4$ MgH_2 $Ca(NO_3)_2$

Answer

For $K_2Cr_2O_7$, potassium is always present in compounds as the cation K^+ and hence the dichromate ion must have an overall charge of −2 to balance charges. The sum of the oxidation states in the dichromate ion $Cr_2O_7^{2-}$ must be −2 (rule 2). We know that the oxidation state of oxygen is −2 (rule 6), so all that remains is to determine the oxidation state of chromium. Labelling this as y: $2(y) + 7(-2) = -2$ so $y = +6$.

The remaining element oxidation states are listed in this table:

Formula of species (and ionic formulation where appropriate)	Oxidation states of elements in substance
$Fe(OH)_3$ [Fe^{3+} $3OH^-$]	O = −2 (rule 6) and H = +1 (rule 5) so Fe = +3 (rule 3)
$Fe(OH)_2$ [Fe^{2+} $2OH^-$]	O = −2 (rule 6) and H = +1 (rule 5) so Fe = +2 (rule 3)
CO_3^{2-}	O = −2 (rule 6) and C = +4 (rule 4)
CN^- [$^-C{\equiv}N$]	N forms three bonds and is more electronegative than C so N = −3 If oxidation state of C = x; $x + (-3) = -1$ so $x = +2$ (rule 4)
$K_3Fe(CN)_6$ [$3K^+$ Fe^{3+} $6CN^-$]	K = +1, Fe = +3, C = +2, N = −3
C_{60}	C = 0 (rule 1)
$SnCl_4$	Sn = +4, Cl = −1 (rule 3)
MgH_2 [Mg^{2+} $2H^-$]	Mg = +2, H = −1 (rule 5) (rule 3)
$Ca(NO_3)_2$ [Ca^{2+} $2NO_3^-$]	Ca = +2, N = +5, O = −2 (rule 6) (rule 3)

Top tip!

The transition metals in the d-block usually have several oxidation states.

The idea of an oxidation state is an artificial concept since it considers all compounds, even covalent ones, to be ionic. An alternative, but equivalent, model is to assume that the atom with the greater electronegativity 'owns' or 'controls' all the bonding or shared electrons of a particular covalent bond.

For example, the sulfur trioxide molecule, SO_3, is assumed to be [S^{6+} $3O^{2-}$] and the water molecule, H_2O, is assumed to be [$2H^+$ O^{2-}], as shown in Figure S3.25. The oxidation states for the central sulfur and oxygen in these two species are +6 and –2, respectively.

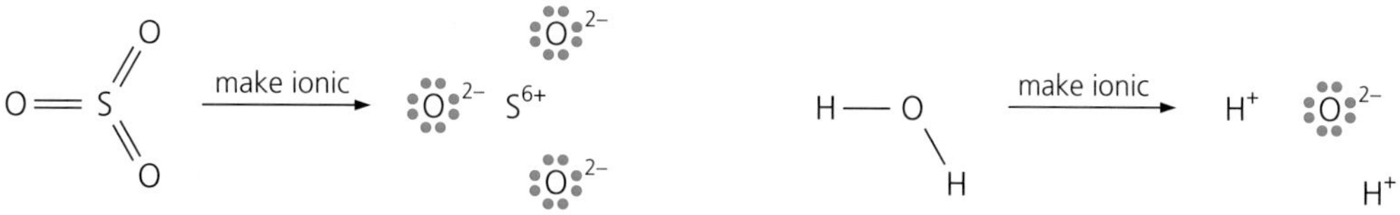

Top tip!

The highest oxidation state an element in groups 1 to 7 can have in periods 2 and 3 is its group number.

Top tip!

Metallic elements have only positive oxidation states, whereas non-metallic elements may have either positive or negative oxidation states.

$$O{=}S({=}O){=}O \xrightarrow{\text{make ionic}} O^{2-}\ S^{6+}\ O^{2-}\ O^{2-} \qquad H{-}O{-}H \xrightarrow{\text{make ionic}} H^{+}\ O^{2-}\ H^{+}$$

■ **Figure S3.25** The ionic formulations of the sulfur trioxide and water molecules

The numerical value of an oxidation state indicates the number of electrons over which control has changed in the compound or ion compared to the situation in the element. A negative sign for an oxidation state means the atom has 'gained control' of the electrons and a positive sign for an oxidation state means that the atom has 'lost control' of electrons.

So, in the examples above, SO_3 and H_2O, sulfur has lost control of six electrons present in the atom and the oxygen atom has gained control of two additional electrons.

Bonds between elements

When dealing with organic compounds and formulas with multiple atoms of the same element, it is easier to work with molecular formulas and average oxidation states.

In the hydrogen peroxide molecule (H–O–O–H), the sum of oxidation states is zero (rule 3). Hydrogen is always +1 (rule 5). The two oxygens must have equal oxidation states since they have equal electronegativity (rule 7). Therefore, the oxidation state of oxygen, x, is given by $(2 \times 1) + 2x = 0$ so $x = -1$.

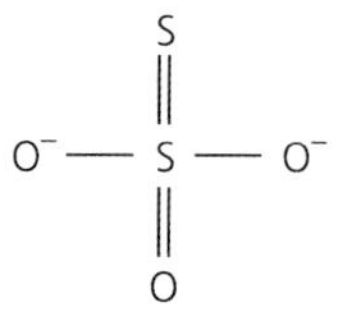

Sodium thiosulfate contains the thiosulfate ion, $S_2O_3^{2-}$. Since the oxidation state of oxygen is −2 (rule 6) and the ion has a charge of −2, the oxidation state of the two sulfurs, y, is +2 since $(-2 \times 3) + y = -2$.

$$O^{-}{-}S({=}S)({=}O){-}O^{-}$$

■ **Figure S3.26** Displayed structural formula of the thiosulfate ion

The outside sulfur atom of the thiosulfate ion (Figure S3.26) is bonded only to the other central sulfur and has an oxidation state of zero since it is in an element–element bond. The central sulfur atom forms two single bonds and one double bond to the three oxygens and so has an oxidation state of +4. The average oxidation state of sulfur in this ion is therefore +2.

Note that this leads to the possibility that the overall oxidation state of an element in an ion or molecule can be fractional. For example, in the radical superoxide anion, $\cdot O_2^-$, oxygen has an oxidation state of −½.

13 Determine the oxidation states of vanadium in the following ions and compounds:

a VO_2^+	c $[V(H_2O)_6]^{3+}$	e NH_4VO_3	g VO_3^-
b V^{2+}	d V_2O_5	f VO^{2+}	h VCl_3

ATL S3.1H

Deduce the oxidation states of atoms in the following organic molecules: ethanol, C_2H_5OH, propene, CH_3–CH=CH_2, benzene, C_6H_6, ethyne, C_2H_2 and ethanoic acid, CH_3COOH. Use the oxidation numbers calculator at www.periodni.com/oxidation_numbers_calculator.php to check your answers.

Naming inorganic compounds

The concept of oxidation state is used in a method of naming inorganic ionic substances known as Stock notation. The magnitude of the oxidation state is written in Roman numerals and inserted in brackets immediately after the name of an ion.

For example:

$FeCl_2$ [Fe^{2+} $2Cl^-$]   iron(II) chloride

$FeCl_3$ [Fe^{3+} $3Cl^-$]   iron(III) chloride

It is not usually necessary to indicate the oxidation state of metals from groups 1, 2 and aluminium in group 13 (so, for example, calcium chloride is not written as calcium(II) chloride). This notation is used for the d-block and p-block metals (with the exception of aluminium and zinc) where variable or multiple oxidation states are exhibited.

Some compounds contain two cations, for example the 'mixed' oxide of lead:

Pb_3O_4 [$2Pb^{2+}$ Pb^{4+} $4O^{2-}$] dilead(II) lead(IV) oxide

Stock names are used for non-metals in the names of the following oxyanions:

chromate(VI)	CrO_4^{2-}	dichromate(VI)	$Cr_2O_7^{2-}$
manganate(VII)	MnO_4^-	manganate(VI)	MnO_4^{2-}
chlorate(I)	ClO^-	chlorate(III)	ClO_2^-
chlorate(V)	ClO_3^-	chlorate(VII)	ClO_4^-

This is because names such as 'chlorate' and 'manganate' potentially refer to more than one species.

Oxyacid molecules contain hydrogen, oxygen and a central atom that can have various oxidation states. For example: HClO (chloric(I) acid), $HClO_2$ (chloric(III) acid), $HClO_3$ (chloric(V) acid), and $HClO_4$ (chloric(VII) acid).

Stock notation can be used for compounds between non-metals: the numbers of atoms in the molecular formula are instead shown in the name. For example, N_2O is known as dinitrogen oxide or nitrogen(I) oxide and SF_6 as sulfur hexafluoride or sulfur(VI) fluoride.

LINKING QUESTION

How can oxidation states be used to analyse redox reactions?

Transition elements

The d-block metals are a group of metals that occur in a large block between group 2 and group 13 of the periodic table. These elements have *similar* physical and chemical properties. The first row of the d-block, scandium to zinc, contains ten elements in which the 3d sublevel is being filled with electrons. This configuration is responsible for the characteristic properties of these elements.

The **transition elements** are a subset of the d-block elements that form at least one stable cation with an incomplete 3d sublevel. All members of the first row of the d-block are transition elements except zinc: the electronic configuration of the zinc ion, Zn^{2+}, is [Ar] $3d^{10}$.

Copper is a transition element since it forms the copper(II) ion, [Ar] $3d^9$, which has an incomplete d sublevel. The copper(I) ion is [Ar] $3d^{10}$ and is non-transitional.

Scandium is a transition element since it can form Sc^+, [Ar] $3d^1 4s^1$, and Sc^{2+}, [Ar] $3d^1 4s^0$, in a limited number of compounds. For example, in the compound $CsScCl_3$ [Cs^+ Sc^{2+} $3Cl^-$] scandium has oxidation state +2.

The characteristic properties of the transition elements are:

- high densities, high melting points and high boiling points
- the ability to exist in a variety of stable oxidation states in both simple ions (for example, Mn^{2+}) and **oxyanions** (for example, MnO_4^-)

◆ **Transition element:** A metal in the d-block of the periodic table which forms at least one stable cation with a partly filled d sublevel.

◆ **Oxyanion:** An anion containing oxygen.

Top tip!

Ions with a half-filled 3d sublevel ($3d^5$) or a filled 3d sublevel ($3d^{10}$) are usually relatively stable, but a number of factors (including lattice enthalpies) are involved in determining the stability of transition metal compounds in the solid state.

HL ONLY

◆ **Complex ion:** An ion with a central cation (usually transition metal) bonded by coordination bonds to one or more molecules or anions (known as ligands).

◆ **Diamagnetic:** A diamagnetic compound has all of its electron spins paired, giving a net spin of zero; diamagnetic compounds are weakly repelled by a magnetic field.

◆ **Paramagnetic:** A paramagnetic compound has one or more unpaired electrons; paramagnetic compounds are attracted by a magnetic field.

- the formation of coloured ions (both simple and complex)
- the ability to form a variety of **complex ions** where the transition metal ion forms coordination bonds to molecules or ions, known as ligands
- the ability to act as catalysts
- the ability to show magnetism in their elements and compounds.

Zinc has a relatively low melting point, boiling point and density; it has only one stable oxidation state (+2) and zinc ions are colourless. Zinc does show some catalytic properties and can form complex ions (as can lead and tin in group 14).

Table S3.16 compares the physical and chemical properties of metals in the s-, p- and d-blocks.

■ **Table S3.16** Properties of metals in the first three blocks of the periodic table

	s-block (groups 1 and 2)	**d-block (groups 3 to 12)**	**p-block (groups 13 to 18)**
Physical properties	soft, low melting points	harder, with higher melting points than p-block elements	harder, with higher melting points than s-block elements
Reaction with water	react, often vigorously (rapidly)	may not react or react only slowly with cold water, faster with steam	react only slowly with cold water, faster with steam
Type of bonding in compounds	ionic	ionic, covalent or complex ions	usually covalent or complex ions
Properties of ions	simple ions with noble gas configuration (octet)	complex ions readily formed	simple ions have a completed d sublevel; easily form complex ions
Complex ions	simple ions can be hydrated through ion–dipole forces to form colourless complex ions	formed readily; usually highly coloured	colourless complex ions are formed more readily than simple ions
Oxidation state	oxidation state = group number	oxidation states usually vary by 1 unit; +2 and +3 are common	highest oxidation state = group number − 10

14 Cobalt is a d-block element and a transition element. The most stable cation is the cobalt(III) ion.

a Draw the orbital diagram (using the arrow-in-box notation) for the Co^{3+} ion.

b State the other most common cation of cobalt and write out its full electron configuration.

c Explain why cobalt is a d-block element and a transition element.

Physical properties of transition elements

Melting and boiling points

The melting and boiling points of the first row of the d-block elements (except zinc) are much higher than those of the s-block elements. The high melting points suggest that the 4s and 3d electrons are involved in metallic bonding.

Magnetic properties

Substances can be classified according to their behaviour in a magnetic field: a **diamagnetic** substance is repelled by (moves out of) a strong magnetic field and a **paramagnetic** substance is attracted by (moves into) a strong magnetic field.

Diamagnetism indicates that all the electrons in a molecule are paired; paramagnetism indicates that a molecule has unpaired electrons.

	diamagnetic	paramagnetic
electron pairing	↑↓ ↑↓ ↑↓ no unpaired electrons	↑↓ ↑↓ ↑ at least one unpaired electron
spin alignment with magnetic field B	← B antiparallel	← B parallel
reaction to magnets	S N very weakly repelled	S N attracted
effect on magnetic field lines	field bends slightly away from the material	field bends towards the material

■ **Figure S3.27** Types of magnetism

All transition metal atoms (except zinc) have unpaired electrons and are paramagnetic (although iron, cobalt and nickel show a stronger form of magnetism, known as ferromagnetism). Paramagnetic strength increases with the number of unpaired electrons.

Variable oxidation states

When transition elements of the first row of the d-block lose electrons to form cations, the 4s electrons are lost first followed by the 3d electrons. The common and more stable oxidation states of the first-row transition elements are shown in Table S3.17.

■ **Table S3.17** Common oxidation states of the first-row transition elements

d-block metal	Sc	Ti	V	Cr	Mn	Fe	Co	Ni	Cu
Common oxidation states	+1 +2 +3	+3 +4	+2 +3 +4 +5	+2 +3 +6	+2 +4 +6 +7	+2 +3	+2 +3	+2	+1 +2
Examples of ions in these oxidation states	Sc^+ Sc^{2+} Sc^{3+}	Ti^{3+} Ti^{4+}	V^{2+} V^{3+} VO^{2+} VO_2^+	Cr^{2+} Cr^{3+} CrO_4^{2-} $Cr_2O_7^{2-}$	Mn^{2+} Mn^{4+} MnO_4^{2-} MnO_4^-	Fe^{2+} Fe^{3+}	Co^{2+} Co^{3+}	Ni^{2+}	Cu^+ Cu^{2+}

Copper and potassium both form +1 cations but have very different values of first ionization energy: potassium (419 kJ mol^{-1}) and copper (745 kJ mol^{-1}).

Copper forms Cu(I) compounds as it has a single 4s electron. The most common oxidation state for d-block metals is +2: losing two 4s electrons to form an M^{2+} ion. Chromium, iron and cobalt form M^{3+} ions as their third ionization energies are relatively low.

The maximum stable oxidation state often corresponds to the maximum number of electrons available for bonding in the 3d and 4s sublevels. For example, manganese ($3d^5\ 4s^2$) has a maximum oxidation state of +7.

Maximum oxidation state of transition metal = electrons in 4s sublevel + unpaired electrons in 3d sublevel.

To remove further electrons beyond the maximum for a transition element would involve the removal of 3p core electrons, which are much closer to the nucleus and much more tightly bound.

Transition metals in low oxidation states tend to exist as cations but are usually found as covalently bonded oxyanions in high oxidation states: for example, +6 in the chromate(VI) and dichromate(VI) ions (Figure S3.28).

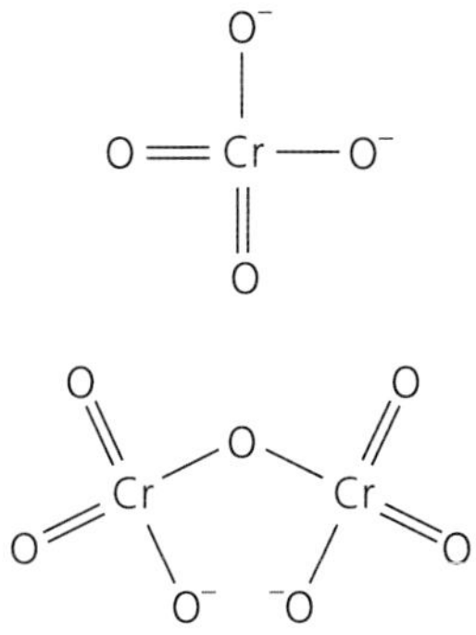

■ **Figure S3.28** Structures of chromate(VI) and dichromate(VI) ions

Top tip!

Transition elements usually show their highest oxidation states when they are bonded with fluorine or oxygen, the most electronegative elements.

Top tip!

The relative stability of the oxidation states is extremely important and is discussed in terms of standard electrode potentials which can be used to establish whether a redox reaction is feasible or spontaneous (under standard thermodynamic conditions).

The variable oxidation states are due, in part, to the relatively small energy difference between the 3d and 4s sublevels. This can be illustrated by examining the successive ionization energies in Table S3.18.

■ **Table S3.18** Successive ionization energies of selected metals

Element	First ionization energy / kJ mol^{-1}	Second ionization energy / kJ mol^{-1}	Third ionization energy / kJ mol^{-1}	Fourth ionization energy / kJ mol^{-1}
calcium	590	1150	4940	64 800
chromium	653	1590	2990	4770
manganese	716	1510	3250	5190
magnesium	736	1450	7740	10 500
iron	762	1560	2960	5400

Table S3.18 shows that to remove the third electron from a calcium atom requires about as much energy as removing the fourth electrons from chromium or manganese atoms. This is because the electron in the calcium atom is from an inner shell (main energy level) and close to the nucleus. This electron and others in the inner shells (the core electrons) never participate in ion formation.

However, assessing oxidation state stability is not as simple as comparing ionization energies. Metal ions react to form a solid compound or hydrated ions in aqueous solution. You need to consider two enthalpy changes when assessing oxidation state stability: the lattice enthalpy (for solids) and the enthalpies of hydration (for an aqueous solution).

The higher the charge on the ion, the greater the number of valence electrons that have to be removed and the greater the amount of energy that has to be supplied. Offsetting this endothermic change, however, is that the greater the charge on the ion, the greater the amount of energy released during lattice formation or as the enthalpy of hydration of the metal cation.

LINKING QUESTION

What are the arguments for and against including scandium as a transition element?

When magnesium forms an ionic compound, such as magnesium oxide, MgO [Mg^{2+} O^{2-}], the energy released during lattice formation is much greater than the energy required to form the magnesium and oxide ions (the sum of the first and second ionization energies and first and second electron affinities). However, the lattice enthalpy is *not* able to supply the energy required to remove the third electron from a magnesium atom.

TOK

How does the social context of scientific work affect the methods and findings of science?

Scientists are people with emotions who belong to a number of different cultures or communities. Therefore, when making scientific assessments, formulating a scientific hypothesis or deciding on a research direction, their culture or membership of a given community will indirectly have influence. Life outside the research laboratory may influence life inside it. Additionally, the fields of science studied, where scientists work and the extent of their research may be influenced by culture, especially political and religious cultures.

Changing oxidation states

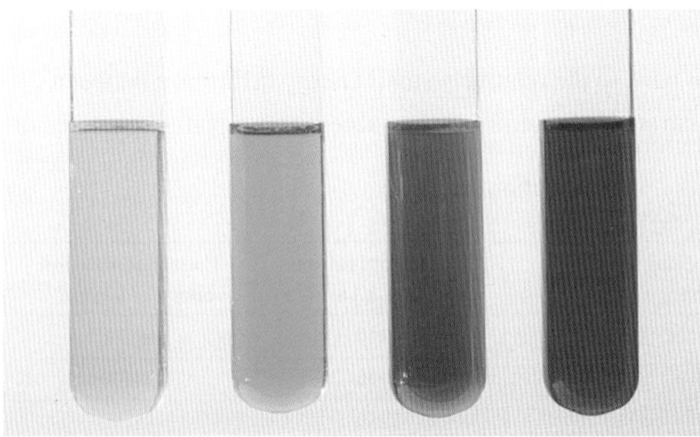

■ **Figure S3.29** Colour changes due to reduction of the acidified solution of VO_2^+ (yellow), through VO^{2+} (blue) and V^{3+} (green) to V^{2+} (lavender)

It is possible to change the oxidation state of transition elements using suitable oxidizing agents or reducing agents under appropriate conditions.

Ammonium vanadate(V), NH_4VO_3, is a common vanadium compound. In acidic conditions the trioxovanadate(V), $VO_3^-(aq)$, ions are converted to dioxovanadium(V), $VO_2^+(aq)$, ions:

$$2H^+(aq) + VO_3^-(aq) \rightleftharpoons VO_2^+(aq) + H_2O(l)$$

These dioxovanadium(V) ions react with zinc (a powerful reducing agent) and dilute sulfuric acid undergoing stepwise reduction to vanadium(II) ions, $V^{2+}(aq)$. Figure S3.29 shows the colours of the ions as the reaction proceeds.

15 Explain the following observations with reference to electronic configuration:
  a Iron has a higher melting point than magnesium.
  b Iron can form two stable cations, Fe^{2+} and Fe^{3+}, while magnesium forms only one ion, Mg^{2+}.

Formation of complex ions

Link

Ligands and complex ions are also discussed in Chapter S2.2, page 140 and Chapter R3.4, page 650.

A d-block metal complex ion consists of a d-block cation surrounded by a fixed number of ligands—molecules or negative ions that have a lone pair of electrons available. The ligands share their lone pair with empty orbitals in the central d-block metal ion and form coordination bonds. The ligands are Lewis bases (electron pair donors).

Common ligands are water molecules, H_2O; ammonia molecules, NH_3; chloride ions, Cl^-; hydroxide ions, OH^-; and cyanide ions, CN^-.

The net charge on a complex ion is the sum of the charge on the d-block metal ion and the charge on the ligands (if they are ions). The overall charge may be positive, negative or zero.

The coordination number (typically 2, 4 or 6) is the number of atoms or ligands directly bonded to the metal atom or ion. Complexes with a coordination number of two will be linear, those with four are usually tetrahedral (occasionally square planar) and those with six are octahedral (Figure S3.30). (Some octahedral complexes may have distorted shapes.)

◆ **Monodentate ligand**: A ligand in which there is only one point from which coordination can occur.

◆ **Chelating agent**: A ligand that binds to a metal ion using more than one atom to form coordination bonds and forms a ring structure.

◆ **Ligand replacement**: A reaction where one or more ligands in a complex ion are replaced, often reversibly, by another ligand.

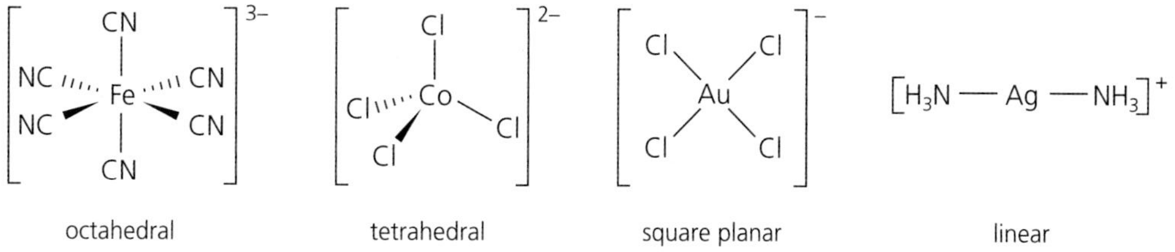

octahedral　tetrahedral　square planar　linear

■ **Figure S3.30** Common shapes of complex ions with coordination numbers of six, four (tetrahedral forms are common, square planar forms less common) and two

Polydentate ligands

Water molecules, ammonia molecules, chloride ions and cyanide ions behave as **monodentate ligands**: they form only one coordination bond with the central d-block metal ion. A number of larger ligands are polydentate: they are able to form two or more coordination bonds with the central metal cation.

Polydentate ligands are often known as **chelating agents** and can form very stable complexes. Ethanedioate ions (oxalate) and 1,2-diaminoethane molecules are bidentate ligands (Figure S3.31), while the negative ion derived from EDTA (**e**thylene**d**iamine**t**etraacetic **a**cid) is hexadentate, forming up to six coordination bonds per ion (Figures S3.31 and S3.32).

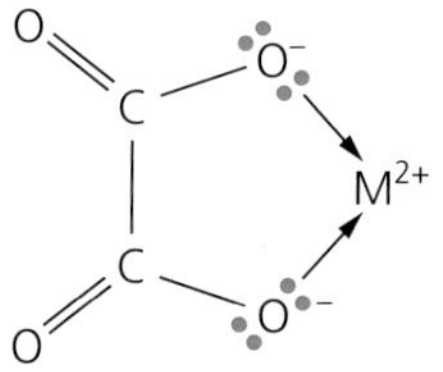

a The ethanedioate (oxalate) ion acts as a bidentate ligand via the lone pairs on its charged oxygens

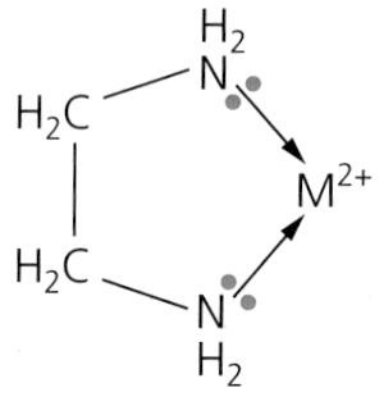

b The 1,2-diaminoethane molecule acts as a bidentate ligand via the lone pairs on the nitrogen atoms in its amine groups

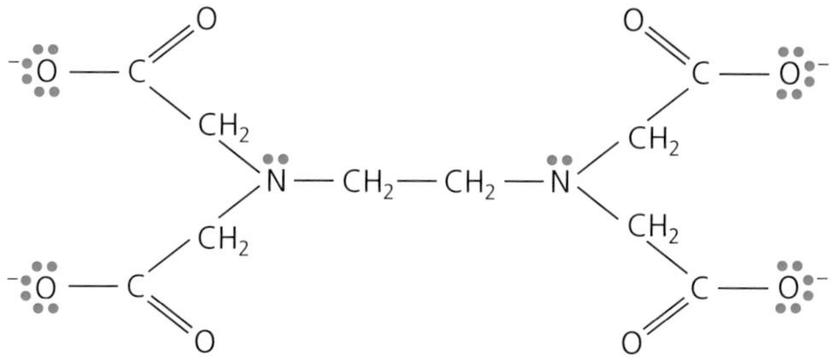

c The $EDTA^{4-}$ ion acts as a hexadentate ligand, using lone pairs on both its nitrogen atoms and its charged oxygens

■ **Figure S3.31** Formation of bidentate and polydentate complexes

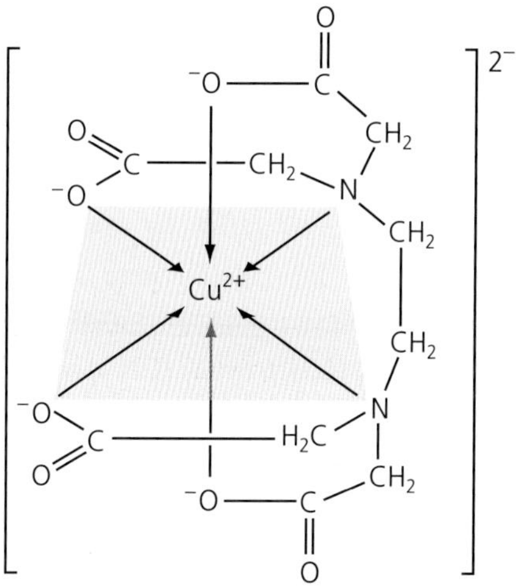

■ **Figure S3.32** Complex formed between a copper(II) ion and EDTA, a polydentate ligand

Ligand replacement reactions

If aqueous ammonia solution is added to a solution of a copper(II) salt, a process of **ligand replacement** occurs: four of the water molecules are replaced by ammonia molecules and the colour changes from light blue to dark (royal) blue.

$$[Cu(H_2O)_6]^{2+}(aq) + 4NH_3(aq) \rightarrow [Cu(H_2O)_2(NH_3)_4]^{2+}(aq) + 4H_2O(l)$$

blue　　　　　　　　　dark blue

HL ONLY

■ **Figure S3.33** Solutions containing **a** the $[Cu(H_2O)_6]^{2+}$ ion and **b** the $[CuCl_4]^{2-}$ ion

If a solution of a copper(II) salt is treated with an excess of concentrated hydrochloric acid, the water molecules are replaced in a similar ligand replacement reaction. The six water molecules are replaced by four chloride ions to form a yellow complex ion.

$$\underset{\text{blue}}{[Cu(H_2O)_6]^{2+}(aq)} + 4Cl^-(aq) \rightarrow \underset{\text{yellow}}{[CuCl_4]^{2-}(aq)} + 6H_2O(l)$$

A ligand replacement reaction also occurs when concentrated hydrochloric acid is added to a solution of cobalt(II) chloride. The colour changes from pink to blue.

$$\underset{\text{pink}}{[Co(H_2O)_6]^{2+}(aq)} + 4Cl^-(aq) \rightarrow \underset{\text{blue}}{[CoCl_4]^{2-}(aq)} + 6H_2O(l)$$

16 When excess aqueous ammonia is added to a solution of hydrated nickel ions, the blue hexaammine complex ion is formed from the hexaaqua ion.

$$\underset{\text{green}}{[Ni(H_2O)_6]^{2+}(aq)} + 6NH_3(aq) \rightarrow \underset{\text{blue}}{[Ni(NH_3)_6]^{2+}(aq)} + 6H_2O(l)$$

a State the oxidation state of nickel in both complexes.
b Identify the ligands in the reaction.
c State the type of bond formed by the ligands with the central nickel cation.
d State the coordination number of both complexes.
e State the type of reaction occurring and explain why it is not a redox reaction.

Going further

Naming complex ions and coordination complexes

When naming a complex ion, the following rules apply:

- A cation has the usual metal name; for example, copper.
- The names of anions containing oxygen are derived from the metal name and have an —*ate* suffix; for example, chromate, ferrate, manganate, vanadate.
- The oxidation state is indicated in the usual way; for example, iron(III) and copper(II).
- The ligands are given specific names: chloro (Cl^-), aqua (H_2O), hydroxo (OH^-), ammine (NH_3), cyano (CN^-), carbonyl (CO) and oxalato ($C_2O_4^{2-}$).
- The number of ligands is indicated by the prefixes di—, tetra— and hexa—.

So the complex ion $[Fe(H_2O)_6]^{3+}$ is the hexaaquairon(III) ion, and $[Cu(H_2O)_2(NH_3)_4]^{2+}$ is the tetraamminediaquacopper(II) ion.

In coordination compounds, the name of the positive counterion is given first. For example, $K_2[Co(NH_3)_2Cl_4]$ is known as potassium diamminetetrachlorocobaltate(II).

Colour in transition-element complex ions

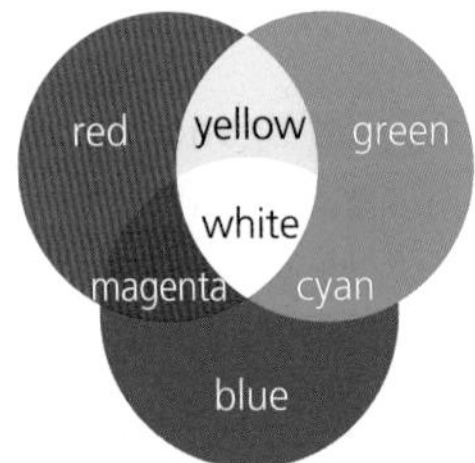

■ **Figure S3.34** Primary and secondary colours of light

When red, green and blue lights of equal intensity are combined and projected on a screen, white light is produced (Figure S3.34). Almost any colour can be made by overlapping light of three colours and adjusting the brightness (intensity) of each colour.

When white light is shone on any substance, some light is absorbed and some is reflected (from an opaque solid) or transmitted (by a liquid or solution).

- If all the light is absorbed, the substance appears black.
- If only certain wavelengths are absorbed, the compound appears coloured.
- If all the light is reflected, then the substance appears white (if solid) or colourless (if in aqueous solution).

Top tip!

Red, green and blue light are known as primary colours since all other colours can be formed from them.

Most transition-element compounds are coloured, both in solution and in the solid state. The colours of these compounds are due to the presence of incompletely filled 3d sublevels.

The following explanation is restricted to the very common octahedral complexes in which the transition element cation has an incomplete 3d sublevel containing at least one 3d electron.

In an isolated gaseous d-block metal element atom, the five 3d sublevels have different shapes but *identical* energies (they are degenerate). In a complex ion, however, the 3d sublevels are oriented differently relative to the ligands. The 3d electrons close to a ligand experience repulsion and are raised in energy. The 3d electrons further from the ligand are reduced in energy.

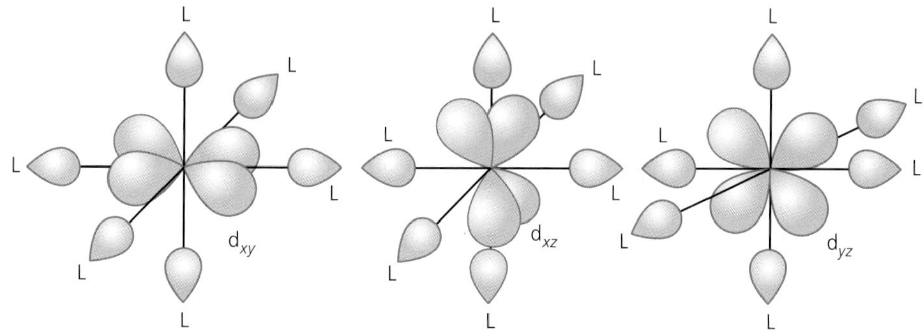

■ **Figure S3.35** Ligands at the ends of the *x*, *y* and *z* axes have weaker interactions with d_{xy}, d_{yz} and d_{xz} orbitals than with d_{z^2} and $d_{x^2-y^2}$ orbitals

The 3d sublevel has now been split into two energy levels as shown in Figure S3.36 for the hydrated titanium(III) ion: two d orbitals are raised in energy; three d orbitals are lowered in energy.

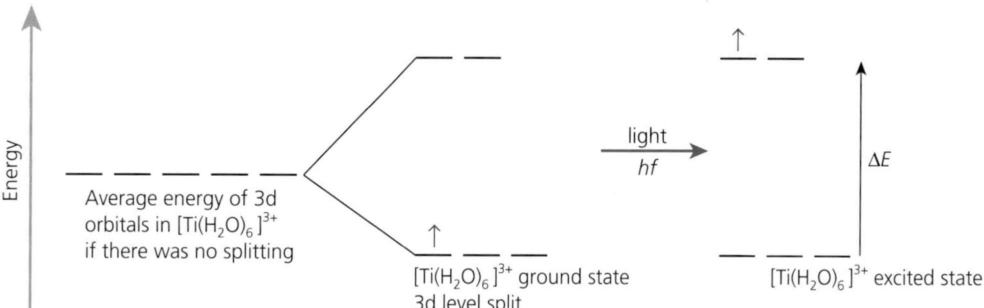

■ **Figure S3.36** Splitting of 3d sublevels in the titanium(III) ion in the octahedral $[Ti(H_2O)_6]^{3+}(aq)$ complex

Link

Planck's constant is defined in Chapter S1.3, page 40.

The energy difference, ΔE, between the two sets of 3d energy levels is directly proportional to the frequency of light that is necessary to cause an electron to be excited from the lower energy level to the higher energy level.

$\Delta E = hf$ where h is Planck's constant and f is the frequency.

◆ **d–d transition**: An electron in a d orbital on the metal is excited by a photon to another d orbital of higher energy.

When white light is absorbed by a solution of titanium(III) ions, some of the light waves (photons) have energies that correspond to the energy difference between the two groups of 3d energy levels. This light is absorbed, promoting a single 3d electron to the higher energy level. This is known as a **d–d transition**.

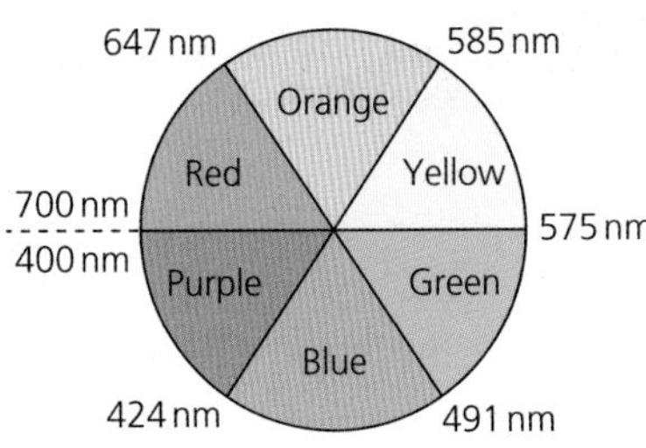

■ **Figure S3.37** Colour wheel of complementary colours

Hydrated titanium(III) ions, $[Ti(H_2O)_6]^{3+}(aq)$, absorb yellow light (of a particular frequency). Since the **complementary colour** is reflected or transmitted, substances containing these ions appear purple (Figure S3.37).

Top tip!

This explanation of colour is different from atomic emission where the light comes from electrons returning to lower energy levels after they have been excited by a flame or discharge.

◆ **Complementary colour:** One of two colours of light that, when blended together, produce white light.

The energy difference, and hence the colour of transition element complex ions, depends on three factors:

- **The ionic charge** For example, vanadium(II) ions ($3d^3$) are lavender in colour, while chromium(III) ions (also $3d^3$) are green.
- **The number of d electrons** d–d transitions occur only if there is an incomplete 3d sublevel into which a 3d electron can be promoted. Scandium(III), copper(I) and zinc(II) compounds are all colourless due to the absence of 3d electrons (Sc^{3+}) or the presence of a filled 3d sublevel (Cu^+ and Zn^{2+}).
- **The nature of the ligand** Different ligands, because of their different sizes and charges, produce different energy differences between the two groups of 3d orbitals. For example, ammonia ligands produce a larger energy gap than water molecules so light of a higher energy and frequency (shorter wavelength) is absorbed. It is this that leads to the colour change when aqueous ammonia solution is added to a solution of a copper(II) salt.

WORKED EXAMPLE S3.1C

Explain why a solution of hexaaquacobalt(III) ions, $[Co(H_2O)_6]^{3+}$, is blue but a solution of hexaamminecobalt(III) ions $[Co(NH_3)_6]^{3+}$ is yellow–orange.

Answer

The ammonia molecule, NH_3, causes greater splitting and a larger energy difference between the d orbitals; hence the wavelength of light absorbed is shorter (purple–blue). The water molecule, H_2O, causes less splitting and the complex absorbs light energy of longer wavelength (orange). The colours observed are complementary to those of the colours absorbed.

LINKING QUESTION

What is the nature of the reaction between transition element ions and ligands in forming complex ions?

Tool 1: Experimental techniques

Colorimetry

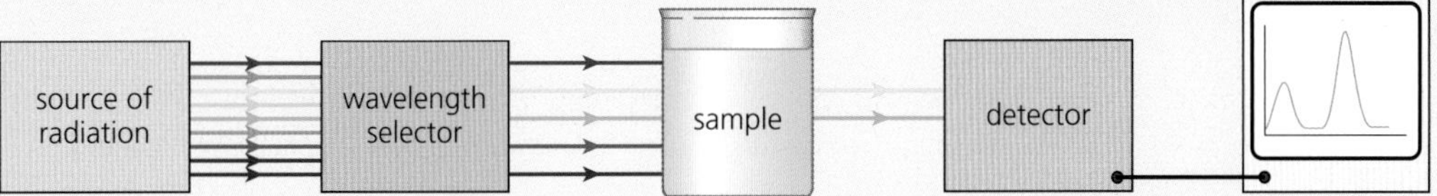

■ **Figure S3.38** Block diagram of a colorimeter

Colorimetry is the technique that helps to determine the concentration of a coloured solution.

Colorimeters (Figure S3.38) pass light of a set colour (frequency or wavelength) through a transparent sample and record the absorbance of the light. The sample is normally a solution held in a cuvette, a small plastic or quartz tube with an open top and flat sides that are transparent to visible light.

As the reaction proceeds (and the solution changes colour), the absorbance of the light passing through the solution changes and this indicates the concentration of the coloured reactant or product.

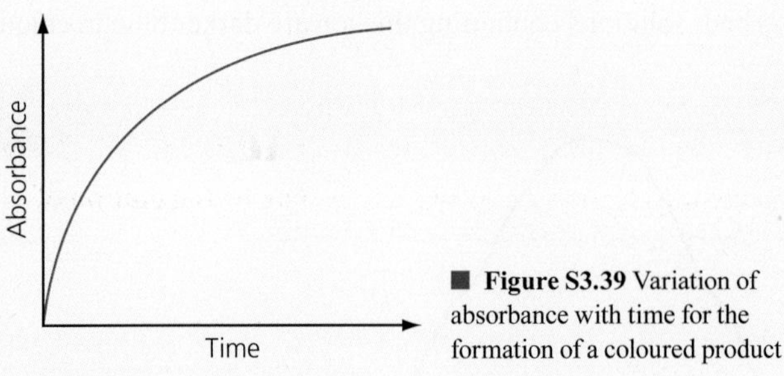

■ **Figure S3.39** Variation of absorbance with time for the formation of a coloured product

The relationship between the concentration of a coloured compound and the absorbance is assumed to be linear (Figure S3.40) and therefore the absorbance is directly proportional to the concentration (provided the concentration is low). This is known as the Beer–Lambert law.

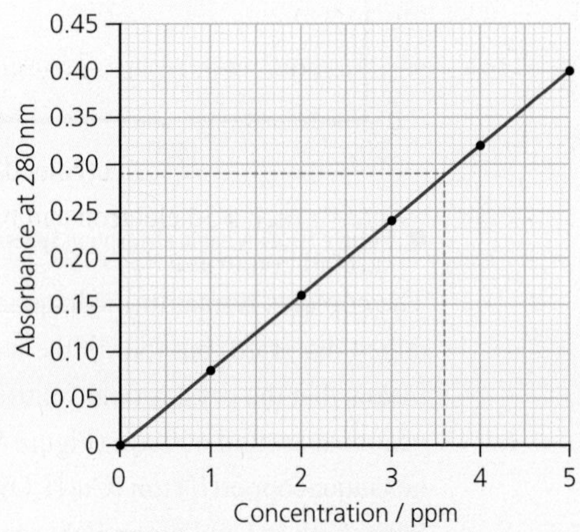

■ **Figure S3.40** Calibration line for the UV assay of proteins using absorbance at 280 nm

It is important to zero the colorimeter with a cuvette containing just water (or the solvent used to dissolve the sample) before the experiment begins, and to use a filter to select the colour of the light to match that absorbed by the compound.

For example, bromine appears orange in aqueous solution as it absorbs blue light (the complementary colour) and transmits the other wavelengths. So, to study the concentration of aqueous bromine, the colorimeter is set to measure the absorbance of blue light.

Top tip!

Ultraviolet radiation has shorter wavelengths and is of higher energy and higher frequency than visible radiation.

Ultraviolet and visible absorption spectroscopy

When an atom, molecule or ion absorbs energy, electrons are promoted from their ground state to a higher energy excited state. The energy differences in d–d transitions correspond to the ultraviolet (wavelengths 200–400 nm) and visible (wavelengths approximately 400–750 nm) regions of the electromagnetic spectrum.

A spectrometer is an instrument that measures the intensity of the radiation transmitted through a dilute solution of a sample and compares this value with the intensity of the incident radiation.

The instrument can be set to scanning mode to generate an absorption spectrum. In this mode, it will plot the proportion of light absorbed (the absorbance) at a particular frequency or wavelength against the frequency or wavelength of the radiation.

When radiation is absorbed over a certain frequency range, the spectrum has a broad absorption band.

For example, the blue hexaaquacopper(II) complex ion, $[Cu(H_2O)_6]^{2+}$(aq), has an absorption band in the red part of the visible spectrum, as shown in Figure S3.41 (λ_{max}, the wavelength corresponding to peak absorbance, is centred around 750 nm). Since red light is absorbed, blue and green light are transmitted and aqueous copper(II) salts appear blue–green (cyan) to our eyes.

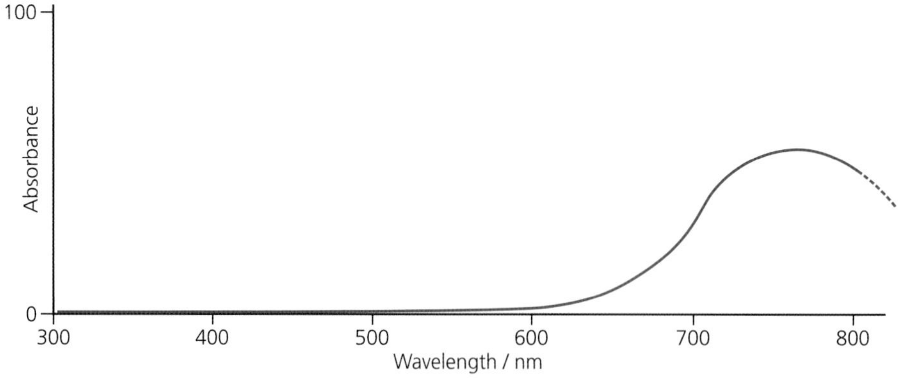

■ **Figure S3.41** Visible absorption spectrum of the hexaaquacopper(II) ion, $[Cu(H_2O)_6]^{2+}$(aq)

The complex tetraamminediaquacopper(II) ion, $[Cu(H_2O)_2(NH_3)_4]^{2+}$ is still octahedral in shape (albeit distorted) but four of the water ligands have been replaced by ammonia molecules. This means that the crystal field splitting energy has been increased and maximum absorption now occurs at around 600 nm (Figure S3.42)—that is, at a shorter wavelength than for the hexaaquacopper(II) ion $[Cu(H_2O)_6]^{2+}$(aq). The absorption band is slightly more towards green. Since both red and green light are absorbed, solutions containing this ion are darker blue in colour.

LINKING QUESTION

How can colorimetry or spectrophotometry be used to calculate the concentration of a solution of coloured ions?

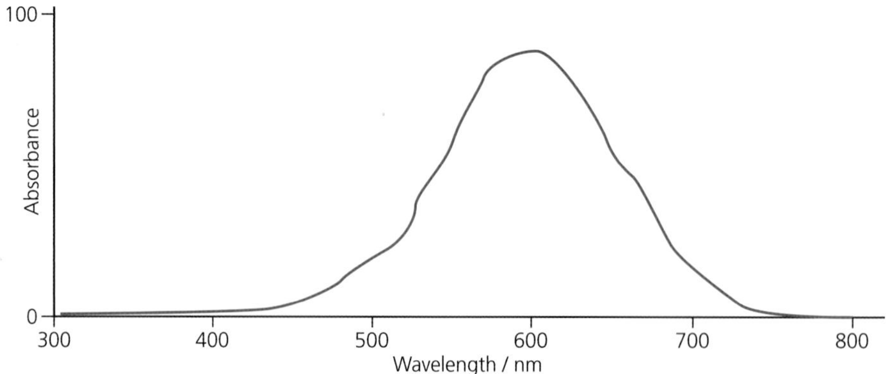

■ **Figure S3.42** Visible spectrum of the tetraamminediaquacopper(II) ion, $[Cu(H_2O)_2(NH_3)_4]^{2+}$, ($\lambda_{max}$ is centred around 600 nm)

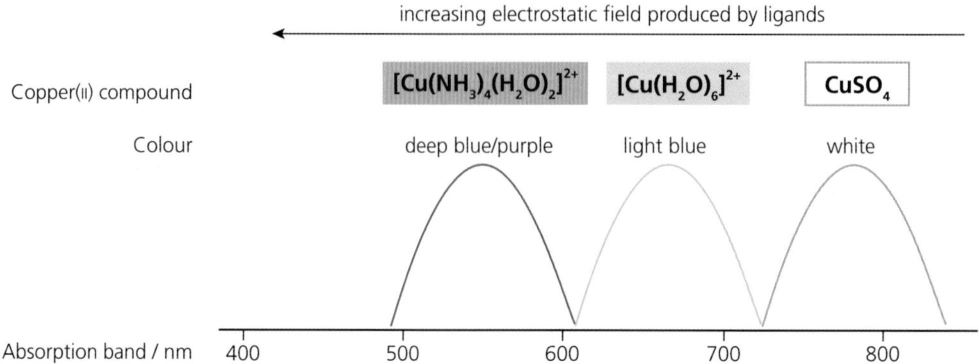

■ **Figure S3.43** Absorption spectra and colours of selected copper(II) compounds showing the effect of a stronger ligand field

Transition elements as catalysts

Catalysts can greatly increase the rate of chemical reactions by providing an alternative reaction pathway of lower activation energy. Transition elements and their compounds are often used as catalysts to increase the rates of industrial processes (see Table S3.19).

Table S3.19 Some transition metal-based industrial catalysts

Process	Reaction catalysed	Products	Catalyst
Haber	$N_2(g) + 3H_2(g) \rightleftharpoons 2NH_3(g)$	ammonia	iron, Fe
Contact	$2SO_2(g) + O_2(g) \rightleftharpoons 2SO_3(g)$	sulfuric acid	vanadium(V) oxide
Hydrogenation of unsaturated oils	$RCH{=}CHR' \rightarrow RCH_2CH_2R'$	semi-solid saturated fat	nickel, Ni
Hydrogenation	alkene to alkane	alkane	nickel, Ni
Ziegler–Natta polymerization of alkenes	$nCH_2{=}CHR \rightarrow -[CH_2{-}CHR]_n-$	stereoregular polymer	complex of $TiCl_3$ and $Al(C_2H_5)_3$

Industrial transition metal catalysts are heterogeneous: the catalyst is in a different physical state (phase) from the reactants. Typically, the catalyst is a powdered solid (often on the surface of an inert support) and the reactants are a mixture of gases.

Heterogeneous catalysis can be demonstrated in the laboratory by adding a small amount of manganese(IV) oxide, MnO_2, to a dilute solution of hydrogen peroxide. Other transition metal oxides (for example, TiO_2) and transition metals (for example, platinum) can also catalyse the decomposition of hydrogen peroxide to water and oxygen.

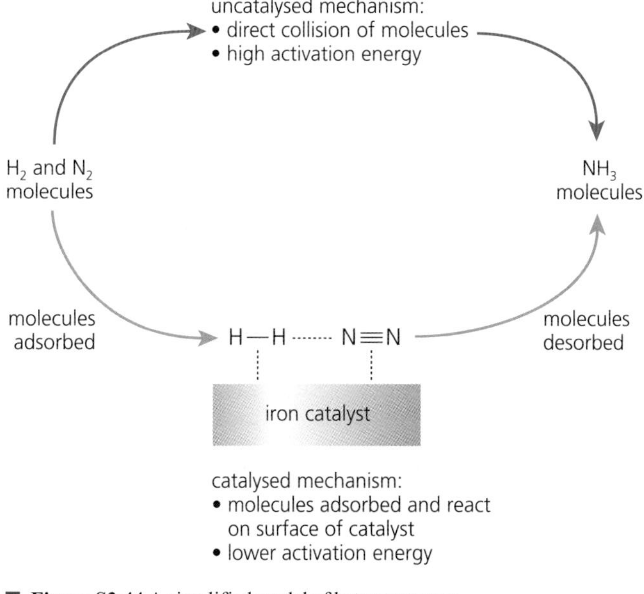

Figure S3.44 A simplified model of heterogeneous catalysis, using the Haber process as an example

During heterogeneous catalysis, gas molecules are adsorbed on the surface of the solid catalyst (Figure S3.44), where their bonds are weakened via complex formation, so the products are produced more rapidly. The adsorbed gas molecules are much closer together than in the free gas.

The strength of the adsorption helps to determine the activity of a catalyst. Some transition elements or their cations adsorb very strongly, so products are released slowly. Others adsorb weakly so the concentration of reactants on the surface is low.

The catalytic properties of heterogeneous catalysts involving transition elements or their compounds depend on the presence of empty orbitals or partially occupied d orbitals for temporary bond formation. It is believed that the presence of empty d orbitals or the presence of unpaired d electrons allows intermediate complexes to form, providing reaction pathways of lower energy than the uncatalysed reaction.

For example, passing ethene into a solution containing titanium(IV) chloride, $TiCl_4$ (molecular covalent), and triethyl aluminium, $(C_2H_5)_3Al$, catalyses its polymerization into poly(ethene). Figure S3.45, in which R represents the ethyl groups, shows how the coordination number of titanium changes as it catalyses the reaction. This is a Ziegler–Natta catalyst which allows addition polymers to be prepared at low temperatures and pressures.

$TiCl_4 + AlR_3 \rightarrow Cl_4Ti(R)(\text{vacant}) \xrightarrow{CH_2=CH_2} Cl_4Ti(R)(CH_2{=}CH_2) \rightarrow Cl_4Ti(\text{vacant})(CH_2CH_2R) \xrightarrow{nCH_2=CH_2} {+}CH_2{-}CH_2{+}_nR$

■ **Figure S3.45** Mechanism of action of Ziegler–Natta catalyst during the polymerization of ethene molecules

Transition metal ions also exhibit catalysis when they and the reactant(s) are in the same physical state (phase); usually all are in solution. For example, iron(III) ions act as a homogeneous catalyst for the reaction between peroxodisulfate and iodide ions to form sulfate ions and iodine molecules. Iron catalyses the reaction by converting between its two common oxidation states, iron(II) and iron(III). This facilitates the electron transfer processes that occur.

Another example of homogeneous catalysis happens on adding pink cobalt(II) chloride solution to a solution of hydrogen peroxide and potassium sodium tartrate at about 50 °C. Before the cobalt(II) ions are added, the reaction progresses very slowly with only a few bubbles of carbon dioxide gas being released. Almost as soon as a small mass of cobalt(II) chloride is added, there is a colour change from pink to green and rapid gas production. When the reaction is over, the pink colour returns. In terms of oxidation states, the cobalt ion has changed from +2 (pink) to +3 (green) and then back to +2 at the end.

■ **Figure S3.46** The reaction between tartrate ions and hydrogen peroxide in the presence of cobalt(II) ions: the pink solution on the left contains cobalt(II) ions, the green solution in the middle contains a temporary green intermediate containing cobalt(III) ions, and the pink solution on the right contains regenerated cobalt(II) ions

Inquiry 2: Collecting and processing data

Interpreting results: assessing validity

Validity refers to the suitability of the methodology to answer the research question. For example, an investigation to find out how the rate of a chemical reaction in solution varied with different solid catalysts would not be a valid procedure if the temperature of the reactants was not controlled and the mass and surface area of the solid catalyst were not controlled.

A valid conclusion is supported by valid experimental data, obtained from an appropriate experimental methodology and based on logical reasoning.

Functional groups: classification of compounds

Guiding question

- How does the classification of organic molecules help us to predict their properties?

SYLLABUS CONTENT

By the end of this chapter, you should understand that:

- organic compounds can be represented by different types of formulas; these include empirical, molecular, structural (full and condensed), stereochemical and skeletal
- functional groups give characteristic physical and chemical properties to a compound
- organic compounds are divided into classes according to the functional groups present in their molecules
- a homologous series is a family of compounds in which successive members differ by a common structural unit, typically CH_2; each homologous series can be described by a general formula
- successive members of a homologous series show a trend in physical properties
- 'IUPAC nomenclature' refers to a set of rules used by the International Union of Pure and Applied Chemistry to apply systematic names to organic and inorganic compounds
- structural isomers are molecules that have the same molecular formula but different connectivities
- stereoisomers have the same constitution (atom identities, connectivities and bond multiplicities) but different spatial arrangements of atoms (HL only)
- mass spectrometry (MS) of organic compounds can cause fragmentation of molecules (HL only)
- infrared (IR) spectra can be used to identify the type of bond present in a molecule (HL only)
- proton nuclear magnetic resonance spectroscopy (^{1}H NMR) gives information on the different chemical environments of hydrogen atoms in a molecule (HL only)
- individual signals can be split into clusters of peaks (HL only)
- data from different techniques are often combined in structural analysis (HL only).

By the end of this chapter you should know how to:

- identify different formulas and interconvert molecular, skeletal and structural formulas
- construct 3D models (real or virtual) of organic molecules
- identify the following functional groups by name and structure: halogeno, hydroxyl, carbonyl, carboxyl, alkoxy, amino, amido, ester, phenyl
- identify the following homologous series: alkanes, alkenes, alkynes, halogenoalkanes, alcohols, aldehydes, ketones, carboxylic acids, ethers, amines, amides and esters
- describe and explain the trend in melting and boiling points of members of a homologous series
- apply IUPAC nomenclature to saturated or mono-unsaturated compounds that have up to six carbon atoms in the parent chain and contain one type of the following functional groups: halogeno, hydroxyl, carbonyl, carboxyl
- recognize isomers, including branched, straight-chain, position and functional group isomers
- describe and explain the features that give rise to *cis–trans* isomerism; recognize it in non-cyclic alkenes and C3 and C4 cycloalkanes (HL only)
- draw stereochemical formulas showing the tetrahedral arrangement around a chiral carbon (HL only)
- describe and explain a chiral carbon atom giving rise to stereoisomers with different optical properties (HL only)
- recognize a pair of enantiomers as non-superimposable mirror images from 3D modelling (real or virtual) (HL only)
- deduce information about the structural features of a compound from specific MS fragmentation patterns (HL only)
- interpret the functional group region of an IR spectrum, using a table of characteristic frequencies (wavenumber / cm^{-1}) (HL only)

- interpret ¹H NMR spectra to deduce the structures of organic molecules from the number of signals, the chemical shifts and the relative areas under signals (integration traces)
- interpret ¹H NMR spectra from splitting patterns showing singlets, doublets, triplets and quartets to deduce greater structural detail (HL only)
- interpret a variety of data, including analytical data, to determine the structure of a molecule (HL only).

Organic molecules

◆ **Organic chemistry**: The study of carbon-containing compounds with the exception of the allotropes of the element itself, its oxides and halides, and metal carbonates.

◆ **Catenation**: The formation of chains of atoms in chemical compounds.

Organic chemistry is the study of compounds containing carbon with other elements such as hydrogen, oxygen, nitrogen and the halogens. These atoms are usually bonded to carbon by covalent bonds.

Carbon shows the property of **catenation**. In its compounds, carbon atoms are able to bond covalently together to form chains of varying lengths. These chains may be straight, branched or **cyclic**, resulting in a huge number of organic molecules (Figure S3.47).

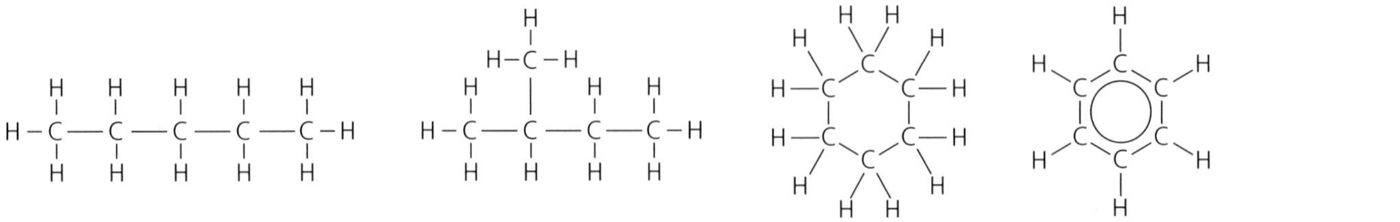

■ **Figure S3.47** Carbon compounds can form straight chains (pentane), chains with branches (2-methylbutane (methylbutane)) and rings (cyclohexane and benzene)

◆ **Cyclic**: Having atoms arranged in a ring or closed chain.

◆ **Saturated (hydrocarbons)**: Organic molecules such as alkanes that contain no carbon–carbon multiple bonds but only carbon–carbon single bonds.

◆ **Unsaturated**: Organic molecules such as alkenes that contain one or more carbon–carbon double or triple bonds.

◆ **Functional group**: The group of atoms responsible for the characteristic reactions of a compound.

◆ **Alkyl group**: A group with the general formula C_nH_{2n+1} usually represented by R.

Many organic compounds—for example, ethane (C_2H_6)—have carbon atoms covalently bonded to each other only by carbon–carbon single bonds and are known as **saturated**. Some organic compounds—for example, ethene (C_2H_4) and ethyne (C_2H_2)—have some carbon atoms joined by multiple bonds (double or triple) and are known as **unsaturated**.

Top tip!

Saturated molecules contain only carbon–carbon single bonds, but unsaturated molecules contain one or more carbon–carbon double or carbon–carbon triple bonds.

Common mistake

Misconception: *Alkenes are less reactive than alkanes due to the presence of a stronger and less reactive double bond.*

Alkenes are more reactive than alkanes due to the presence of a pi bond with a polarizable electron pair that can be attracted to electrophiles. The reactions of alkenes generally involve the breaking of this weaker pi bond.

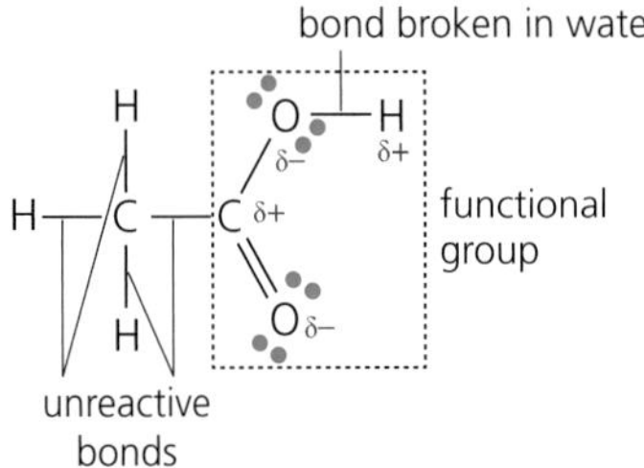

■ **Figure S3.48** An ethanoic acid molecule with a carboxyl functional group and a methyl alkyl group

Organic compounds are classified on the basis of the **functional group** present. The functional group (Figure S3.48) is the reactive part of the molecule and determines its chemical and physical properties. For example, the carboxyl group is the functional group present in all carboxylic acids. The unreactive part of the molecule is often a hydrocarbon-based **alkyl group**.

Common mistake

Misconception: *Carboxylic acids contain the C–O–O–H group.*

Carboxylic acids contain a carbon bonded to an oxygen (via a C=O double bond) and a hydroxyl group (via a C–O single bond).

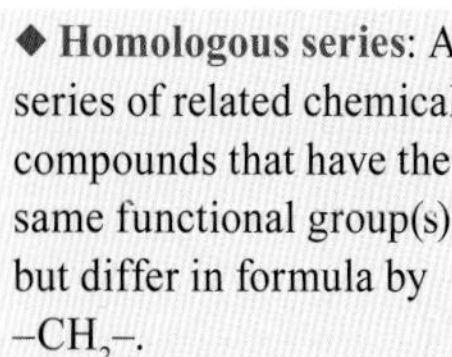

◆ **Homologous series:** A series of related chemical compounds that have the same functional group(s) but differ in formula by $-CH_2-$.

◆ **Homologues:** Successive members of a homologous series.

A group of compounds that contain the same functional group is a **homologous series** and can be represented by a general formula.

Members of a homologous series have similar chemical properties because they have the same functional group. Successive members of a homologous series are known as **homologues**.

Physical properties such as melting point, boiling point, density and solubility show a gradual change as the number of carbon atoms in the chain increases.

ATL S3.2A

Outline the vital force theory proposed by the Swedish chemist Berzelius. Present the research of Wöhler (Figure S3.49), Kolbe and Berthelot in falsifying this theory.

■ **Figure S3.49 a** Friedrich Wöhler (1800–1882) and **b** the structure of urea

LINKING QUESTION

What is unique about carbon that enables it to form more compounds than the sum of all the other elements' compounds?

Representing molecules

There are several types of formula that chemists use to show the structure of organic molecules, depending on the complexity of the chemical process or substances involved. Table S3.20 summarizes some of these.

17 Draw skeletal formulas (Table S3.20) for the following molecules:

a ethane, C_2H_6

b ethene, C_2H_4

c ethanol, CH_3-CH_2-OH

d ethanoic acid, CH_3-COOH

e propane, $CH_3-CH_2-CH_3$

f propene, C_3H_6

g propan-1-ol, $CH_3-CH_2-CH_2-OH$

h 1,2-dichloroethane, $Cl-(CH_2)_2-Cl$

i butane, $CH_3-(CH_2)_2-CH_3$

j 2-methylpropane, $HC(CH_3)_3$

k propan-2-ol, $(CH_3)_2CHOH$

l cyclohexane, C_6H_{12}

◆ **Condensed structural formula:** A formula showing groups of atoms but not the bonds between them.

◆ **Full structural formula:** A formula showing all atoms in the molecule and the bonding between them.

◆ **Skeletal formula:** A formula showing the carbon 'skeleton' of a molecule together with any functional groups.

◆ **Stereochemical formula:** A three-dimensional representation of a molecule.

■ **Table S3.20** Types of organic formula

Type of formula	Definition	Example
Molecular formula	Number and type of atoms bonded together	$C_4H_8O_2$
Empirical formula	Simplest whole number ratio of atoms of each element	C_2H_4O
Condensed structural formula	Groups of atoms, including functional groups	$CH_3CH_2CH_2COOH$ or $CH_3(CH_2)_2COOH$
Full structural formula	All atoms and bonds	H–C(H)(H)–C(H)(H)–C(H)(H)–C(=O)–O–H
Skeletal formula	Covalent bonds (shown as zig-zag lines) and symbols for atoms of elements other than carbon and hydrogen	(zig-zag)–C(=O)–O–H
Partial skeletal formula	As per skeletal formula but includes a group or groups other than the functional group(s)	H_3C–(zig-zag)–C(=O)–O–H
Stereochemical formula	The shape of the molecule in three dimensions: atoms or functional groups in the plane of the page are joined by lines, out of the page by solid wedges and into the page by tapered dashes	C_2H_5, C, H, H, CO_2H

Tool 1: Experimental techniques

Digital molecular modelling

Many topics in chemistry lend themselves to coding activities, including modelling the shapes of molecules in 3D, creating visualizations and graphs of periodic trends and modelling a mathematical relationship. Comparing such models with experimentally derived data may be an appropriate step in practical investigations.

One language that lends itself to visual modelling is VPython, which is the Python programming language plus a 3D graphics module called Visual. Online IDEs (integrated development environments) that allow you to code using Python and/or VPython without downloading any software include Cloud9 (c9.io), GlowScript (glowscript.org) and Trinket (trinket.io).

As an indication of the power of VPython, the six lines of code shown in Figure S3.51 were used to create a 3D model of a methane molecule (Figure S3.50) using the Trinket IDE. After running the code, the model can be rotated in three dimensions by right-clicking on it.

■ **Figure S3.50** Output of the code from Figure S3.51

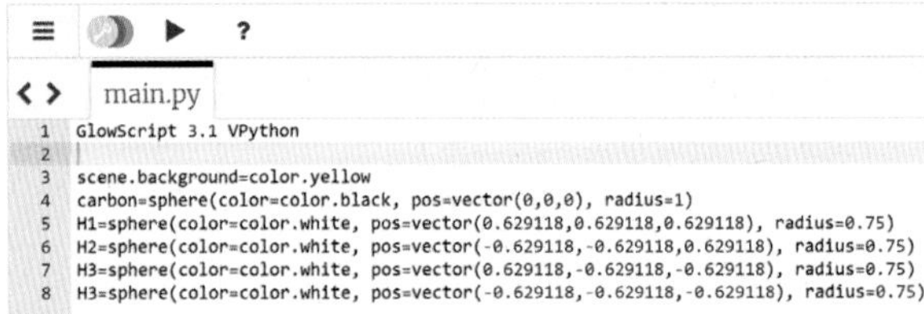

main.py

```
GlowScript 3.1 VPython

scene.background=color.yellow
carbon=sphere(color=color.black, pos=vector(0,0,0), radius=1)
H1=sphere(color=color.white, pos=vector(0.629118,0.629118,0.629118), radius=0.75)
H2=sphere(color=color.white, pos=vector(-0.629118,-0.629118,0.629118), radius=0.75)
H3=sphere(color=color.white, pos=vector(0.629118,-0.629118,-0.629118), radius=0.75)
H3=sphere(color=color.white, pos=vector(-0.629118,-0.629118,-0.629118), radius=0.75)
```

■ **Figure S3.51** Trinket code to generate a 3D model of a methane molecule

18 Expand each of the following condensed formulas into their full structural formulas.

a $CH_3CH_2COCH_2CH_3$

b $CH_3CH{=}CH(CH_2)_3CH_3$

19 For each of the following molecules, write a condensed structural formula and draw a skeletal formula.

a $HOCH_2CH_2CH_2CH(CH_3)CH(CH_3)CH_3$

b
OH
|
N≡C—CH—C≡N

20 Draw full structural formulas (showing all atoms and bonds) for the following molecules shown as skeletal formulas.

a

b
OH

21 Draw a stereochemical formula for the amino acid alanine (2-aminopropanoic acid), $H_2N\mathbf{C}H(CH_3)CO_2H$, where the carbon in bold is the central carbon atom.

TOK

How can it be that scientific knowledge changes over time?

Scientific knowledge is subject to revision and refinement as new data, or new ways to interpret existing data, are found. A paradigm is a model or theory with great explanatory power that is accepted by the majority of the chemical community.

One paradigm found in organic chemistry textbooks is that the bond length between two atoms becomes shorter if you increase the s-character of one of the two atoms. So the distance between atoms decreases $sp^3 \rightarrow sp^2 \rightarrow sp$, as shown by the C–H bond in ethane (109.5 pm), ethene (107.6 pm) and ethyne (106.4 pm).

The hybridization rule seemed plausible to organic chemists and you can often observe corresponding correlations. But whether sp hybridization is the reason for bond length trends has never been proven using quantum mechanics.

A combination of experimental data and computer modelling (based on molecular orbital theory) revealed that this paradigm was wrong and the observed effect was explained by steric repulsion.

LINKING QUESTION

What are the advantages and disadvantages of different depictions of an organic compound (structural formula, stereochemical formula, skeletal formula, 3D models, etc.)?

Tool 1: Experimental techniques

Digital molecular modelling

Using computer models to construct 3D visualizations of organic compounds can, like other interactive techniques, enhance understanding of molecular structure and the relative orientations of individual atoms or groups.

MolView (**http://molview.org/**) has two main parts: a structural formula editor and a 3D model viewer. You can draw a compound, or choose a compound from one of three databases, then click the 2D to 3D button to view a 3D model of the molecule in the viewer.

Can all knowledge be expressed in words or symbols?

The meaning of the term 'knowledge' is not universally agreed so the answer to this question may depend on how you understand it.

The Greek philosopher Plato believed there are truths to be discovered; that knowledge is possible. He defined knowledge as justified, true belief. Plato's definition is limiting, however, because it seems to include only what we might call 'propositional' knowledge, or knowledge that takes the form of propositions. Propositions are sentences which make claims about the world which can be true or false, depending on whether what they say about the world is actually true. Propositional knowledge generally takes the form of 'I know that…' as in 'I know that carbon atoms in a chain can be bonded by single, double or triple covalent carbon–carbon bonds.'

Propositional knowledge can be described as 'explicit' knowledge in that it can be represented verbally and visually. For example: the tetrahedral shape of the methane molecule can be represented by a stereochemical formula; the synthesis of an organic substance can be represented by a written methodology.

The physical chemist and philosopher Michael Polanyi suggested another form of knowledge called 'tacit' or 'implicit' knowledge. Tacit knowledge can be defined as skills, ideas and experiences that are possessed by people but are not codified and may not be easily expressed in words and symbols. With tacit knowledge, people are not often aware of the knowledge they possess, for example, the ability to study a molecular structure and devise a retrosynthesis (Figure S3.52).

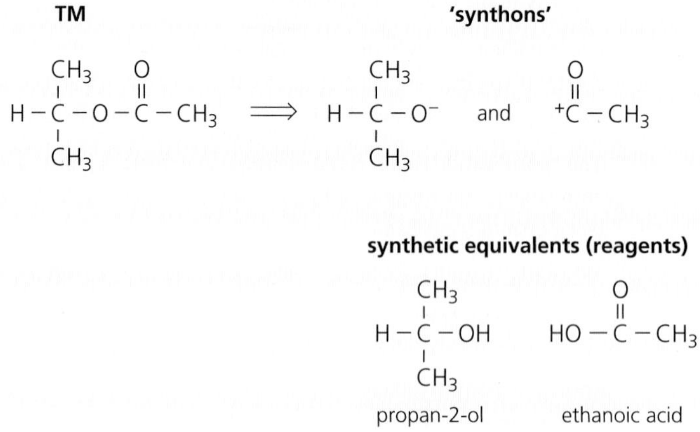

TM

$$H-C(CH_3)_2-O-\overset{O}{\overset{\|}{C}}-CH_3 \Longrightarrow$$

'synthons'

$$H-C(CH_3)_2-O^- \quad \text{and} \quad {}^{+}\overset{O}{\overset{\|}{C}}-CH_3$$

synthetic equivalents (reagents)

$$H-C(CH_3)_2-OH \qquad HO-\overset{O}{\overset{\|}{C}}-CH_3$$

propan-2-ol ethanoic acid

■ **Figure S3.52** Retrosynthesis of an ester, where TM represents the target molecule

Polanyi suggests that tacit knowledge cannot (currently) be represented visually and verbally. However, he argues for the importance of tacit or implicit knowledge in science, particularly in regard to laboratory work—for example, the skills and knowledge required to successfully synthesize and purify an organic compound in high yield. These skills might be characterized as 'ability knowledge' and take the form of 'I know how to…'

It could be argued that even implicit knowledge can be represented in some systematic way because it is represented in a human brain which at one level is a physical structure, albeit a very complex one. If we physically describe those elements of the brain which account for the skill, we might therefore be describing that implicit knowledge. Therefore, in principle—perhaps in the future—it might be possible to represent the full structure of the brain, and anything represented in it, in some form of language or symbolic system.

Another type of implicit knowledge might be knowledge that someone can only acquire by direct experience via sense perception; for example, seeing a particular shade of blue, such as copper(II) sulfate crystals, or smelling a particular odour, such as pungent ammonia gas. The mathematician and philosopher Bertrand Russell called this 'knowledge by acquaintance'. But does this type of knowledge represent a kind of knowledge that lies beyond the ability of science to describe? If so, then perhaps the scope of the natural sciences is limited; perhaps the world does contain facts and knowledge that cannot be described by science!

To explore this idea, the philosopher Frank Jackson imagined a thought experiment involving a scientist called Mary who knows everything about the neural and physical processes related to colour and colour vision, but who has been brought up in a monochrome world. That is to say she has only ever experienced the world in black and white.

Mary has a complete and detailed biological and biochemical knowledge of the structure of the human eye and knows the exact wavelengths of light that stimulate the retina when we observe a light red sky. However, would she acquire some additional knowledge on seeing a red apple for the first time? Is it possible that the knowledge she gains from this experience escapes physical explanations? If Mary did know all explicit knowledge about colour and colour vision before her experience, but learns something new when seeing the apple, then perhaps we must conclude that scientific explanations alone cannot capture and represent all there is to know.

What is the difference between the natural sciences and philosophy?

Jackson's 'Mary's Room Argument' offers an interesting comparison between the natural sciences and philosophy as areas of knowledge. Like those working in the scientific community, those in the philosophical community also use experiments. However, philosophical experiments are often hypothetical in nature: 'thought experiments'. Philosophers use them to create scenarios with slightly different variations in order to see what changes the variations create. The philosophical method is then used to interpret and explore those new outputs. In scientific experiments, data are outcomes which need interpreting and analysing. In philosophical thought experiments, the outcome is intuitions—which also need interpreting and analysing.

In the case of Mary's Room, many people's intuition is that Mary does gain new knowledge and the interpretation of this is that the scope of the natural sciences might be limited in profound ways. Jackson himself, however, continued to explore these intuitions and ultimately developed the position that while Mary did have new experiences, they did not amount to new 'explicit' knowledge of the kind that challenges the ability of the natural sciences to offer complete descriptions of the world.

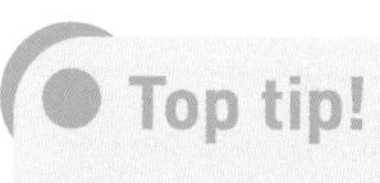

Note that a general formula assumes there is only one functional group present in the molecule.

Functional groups

The functional group responsible for the chemistry of an organic compound may consist of an atom (for example, iodo, –I) or a covalently bonded group of atoms (for example, nitro, $-NO_2$). Table S3.21 shows the functional groups for some common homologous series.

To obtain the molecular formula of a particular compound, you substitute an integer for *n* in the general formula for the relevant homologous series.

■ **Table S3.21** Common homologous series

Homologous series	Functional group and condensed structural formula	Suffix in compound name	General formula	Functional group structure
alkanes	$-CH_2-CH_2-$	—ane	C_nH_{2n+2}	*
alkenes	$-CH{=}CH-$ alkenyl	—ene	C_nH_{2n}	$>C{=}C<$
alkynes	$-C{\equiv}C-$ alkynyl	—yne	C_nH_{2n-2}	$-C{\equiv}C-$
halogenoalkanes	–X where X = F, Cl, Br, I	prefix fluoro—, chloro—, bromo—, iodo—	$C_nH_{2n+1}X$	–X where X = F, Cl, Br, I
alcohols	–OH hydroxyl	—ol	$C_nH_{2n+1}OH$ or ROH	–O–H
aldehydes	–CHO aldehyde (carbonyl)	—al	$C_nH_{2n+1}CHO$ or RCHO	$-C(=O)H$
ketones*	–CO– carbonyl	—one	$C_nH_{2n+1}COC_mH_{2m+1}$ or RCOR′	$R(R')C{=}O$
carboxylic acids	–COOH or CO_2H carboxyl	—oic acid	$C_nH_{2n+1}COOH$ or RCOOH	$-C(=O)-O-H$

*The alkane structure is the basic backbone into which the functional groups are introduced.
**R and R′ represent hydrocarbon chains (alkyl groups) attached to the group. These chains can be identical or different (as represented here).

There are several other key functional groups, one of which is fundamental to a whole separate area of organic chemistry based on aromatic hydrocarbons (arenes). Two other groups extend the range of oxygen-containing compounds, while the remaining groups are present in important nitrogen-containing compounds.

■ **Table S3.22** Other homologous series (the amine and amide shown are both primary since the nitrogen atoms are bonded to two hydrogen atoms)

Homologous series	Functional group and condensed structural formula	Suffix in compound name	General formula	Functional group structure
ethers	R–O–R′ ether	prefix —alkoxy	$C_nH_{2n+1}OC_mH_{2m+1}$	$-O-$
esters	R–COO–R′ ester	—oate	$C_nH_{2n+1}COOC_mH_{2m+1}$	$-C(=O)-O-$
amines	$R-NH_2$ amino	—amine or prefix —amino	$C_nH_{2n+1}NH_2$	$-N(H)H$
amides	$R-CONH_2$ carboxamide	—amide	$C_nH_{2n+1}CONH_2$	$-C(=O)-N(H)H$
nitriles	R–CN nitrile	—nitrile	$C_nH_{2n+1}CN$	$-C{\equiv}N$
arenes	$-C_6H_5$ phenyl—	—benzene or prefix —phenyl	C_6H_5-	$-C_6H_5$ (benzene ring)

Top tip!

A functional group behaves in nearly the same way in every molecule it is present in.

Tool 2: Technology

Use computer modelling

VSEPR, Lewis structures and MO models are all used to convert chemical theories into chemical predictions. Computer-based modelling is an ideal tool for this as software can give a much more accurate display of theoretical predictions. It can do this by creating molecular models that may display molecular orbitals, bond lengths, bond angles, electron densities, electrostatic potentials (the energy of interaction of a point positive charge with the nuclei and electrons of a molecule) and electrostatic potential maps (showing distribution of charge, Figure S3.53).

■ **Figure S3.53** The computer-generated mapping of the electrostatic potential of benzene mapped on to its pi electron density (with yellow showing high electron density and blue showing low electron density)

LINKING QUESTION

How useful are 3D models (real or virtual) to visualize the invisible?

22 Taxol is a natural product that was first isolated from the Pacific yew tree in 1967. In the late 1970s it was found to be a powerful anti-cancer agent.

Identify the functional groups present in Taxol.

Going further

Cycloalkanes

Carbon atoms can form a variety of rings and cyclic molecules such as cyclic alkanes with the general formula C_nH_{2n}. Some smaller cyclic molecules may be strained. For example, the bond angles in cyclopropane are 60° and not the usual tetrahedral bond angle of 109.5°. Cyclopropane is strained (higher in energy) making it unstable.

Figure S3.54 shows the structural and skeletal formulas of the first three cycloalkanes. The cyclopropane ring is flat but the cyclobutane and cyclopentane rings are not flat (the rings are described as puckered).

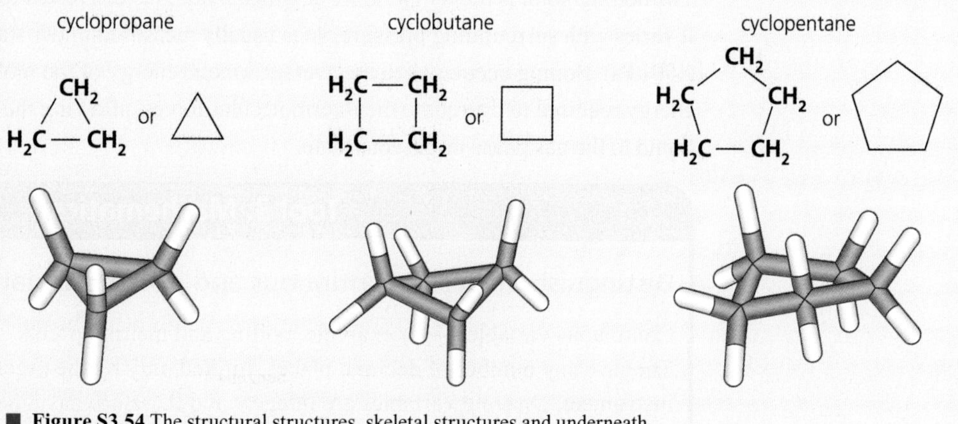

■ **Figure S3.54** The structural structures, skeletal structures and underneath, three-dimensional shapes of the first three cycloalkanes

LINKING QUESTION (HL ONLY)

What is the nature of the reaction that occurs when two amino acids form a dipeptide?

LINKING QUESTION

How can functional group reactivity be used to determine a reaction pathway between compounds; for example, converting ethene into ethanoic acid?

Going further

Natural products

A number of important medicinal drugs are either natural products or derived from compounds found in plants. An example is calanolide A, an anti-HIV drug derived from *Calophyllum lanigerum* var *austrocoriaceum* which is found only in Malaysia and the Botanic Gardens in Singapore. Calanolide A is a relatively simple molecule so it can also be synthesized from simple organic molecules.

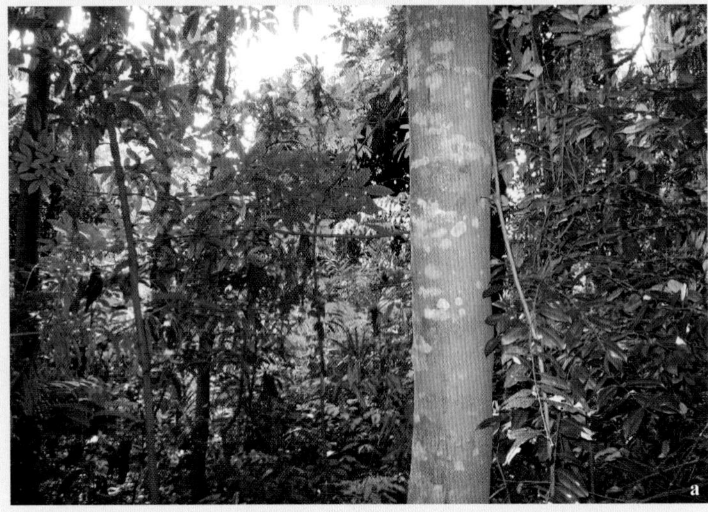

b

■ **Figure S3.55 a** Primary rainforest in the Singapore Botanic Gardens
b The structure of calanolide A, an anti-HIV drug

Trends in physical properties in a homologous series

Physical properties include melting points, boiling points or (for some organic molecules) decomposition points, viscosity of liquids and solubility in water and other solvents. These are determined largely by the intermolecular forces operating between molecules.

Boiling point

The boiling point is the temperature at which a liquid is converted to gas at constant temperature. It varies with surrounding pressure, so is usually measured under standard conditions of pressure (100 kPa). Boiling occurs when the average kinetic energy of the molecules is greater than the energy required to overcome the intermolecular forces, allowing the molecules to move from the liquid to the gas phase or gaseous state.

Tool 3: Mathematics

Distinguish between continuous and discrete variables

Continuous variables—for example, boiling and melting points—are real numbers that can have any number of decimal places, limited only by the precision of the measuring instrument. Discrete variables are integers and therefore have magnitudes limited to the whole numbers. The number of carbon atoms in an alkane is a discrete variable.

Two factors control the boiling point: the type of intermolecular forces operating in the liquid and the shape of the molecule.

Hydrogen bonding is stronger than dipole–dipole forces, which are stronger than London (dispersion) forces for molecules of similar molar mass (Figure S3.56). So if molar masses are not significantly different, hydrogen bonded liquids will have higher boiling points than liquids in which only London forces operate.

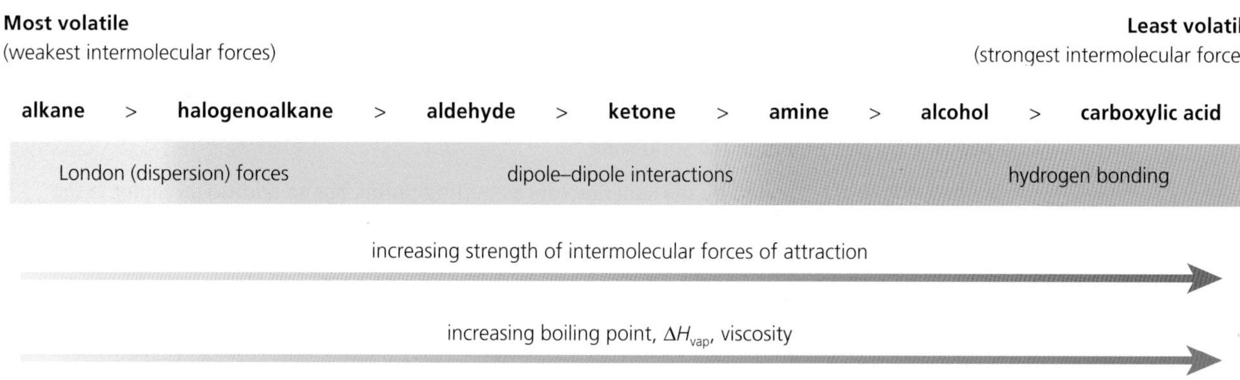

■ **Figure S3.56** The influence of the functional group on intermolecular forces and certain physical properties

This is shown with the liquids propan-1-ol, propanal and butane, all of which have approximately the same molar mass.

■ **Table S3.23** The effect of different types of intermolecular force on boiling point

Compound	M_r	Structure	Bonding	Boiling point / K
propan-1-ol	60	$CH_3CH_2CH_2$–O–H ⋯ H–O–C_3H_7 (H–C–C–C–O–H, hydrogen bond to O of second molecule)	hydrogen	371
propanal	58	H–C(H)(H)–C(H)(H)–$C^{\delta+}$(H)=$O^{\delta-}$	dipole–dipole	322
butane	58	H–C(H)(H)–C(H)(H)–C(H)(H)–C(H)(H)–H	London (dispersion) forces	135

In any series of organic compounds where the same type of intermolecular forces operate, the dispersion forces will increase with molar mass as the greater total number of electrons available for polarization results in larger temporary induced dipoles.

Butane has a higher boiling point than its structural isomer 2-methylpropane because the methyl side-chain prevents the close approach of molecules and so reduces their area of contact. Less energy is therefore needed to overcome the weaker London forces.

Common mistake

Misconception: *Long-chain alkane molecules have high boiling points due to the large number of C–H bonds.*

Long-chain alkane molecules generally have high boiling points due to the large number of electrons available for temporary dipole formation via London (dispersion) forces.

The substitution of a larger and heavier atom, such as a halogen, for a hydrogen atom in a hydrocarbon molecule will significantly increase the boiling point. It increases the molar mass and hence the number of electrons that participate in the production of dispersion forces. It also creates a relatively polar halogen–carbon bond and hence is able to form dipole–dipole forces. For example, chloromethane boils at −24 °C and propane boils at −42 °C.

Melting point

The melting point is the temperature at which a solid is converted to liquid at constant temperature. Melting points are lowered and their range widened by impurities. Melting occurs when the average kinetic energy of molecules allows them to overcome the intermolecular forces operating between them in the solid state.

Melting points are affected by the same factors as boiling points but an additional factor is the packing arrangement of molecules within the lattice. Those molecules that can pack closely and efficiently have higher melting points than molecules that pack poorly in the solid state. This is because the strength of intermolecular forces decreases rapidly with distance.

Butane and 2-methylpropane are structural isomers. Butane has the higher melting temperature (135 K, compared to 114 K for 2-methylpropane) because its straight-chain molecules can pack together more closely and hence more effectively due to their greater surface area of contact. More heat is needed to overcome the stronger London forces.

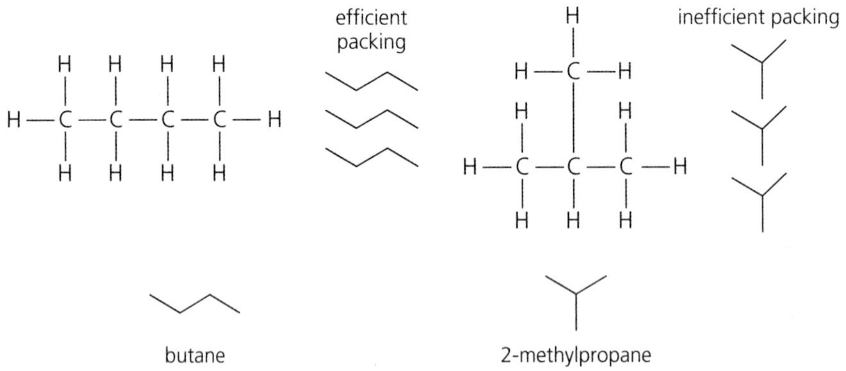

efficient packing

inefficient packing

butane

2-methylpropane

■ **Figure S3.57** The packing of butane and 2-methylpropane molecules

Tool 1: Experimental techniques

Melting point determination

A simple technique to test the purity of a compound is to accurately measure its melting point using apparatus such as that shown in Figure S3.58. Pure compounds melt at a specific temperature. Impurities normally lower the melting point by disrupting the lattice packing and cause the compound to melt over a wider temperature range. If the compound is pure, a carefully measured melting point can also be used to confirm its identity by reference to published data. However, some organic compounds decompose rather than melting.

■ **Figure S3.58** Commercial melting point apparatus

Tool 3: Mathematics

Calculate and interpret percentage change and percentage difference

A simple way to calculate the percentage change is to calculate the difference between the new quantity and the original and then express this as a percentage of the original value.

This calculation can be expressed in the formula:

$$\text{percentage change} = \frac{\text{new quantity} - \text{original quantity}}{\text{original quantity}} \times 100\%$$

Another version of this equation is used in experiments to assess accuracy.

$$\text{percentage difference} = \frac{\text{measured value} - \text{true value}}{\text{true value}} \times 100\%$$

The absolute error has the same units as the measurements. Note that the percentage difference has a sign associated with it: + if the measured value is larger than the true or literature value, and − if less than the true or literature value.

Solubility

The solubility of an organic compound in water is largely determined by two factors: the functional group and the hydrocarbon skeleton. These have opposite effects.

If the molecule has a low molar mass and the functional group is able to form hydrogen bonds (as is the case in, for example, amines, carboxylic acids and alcohols) then the compound should be soluble in water. However, a long hydrocarbon chain (alkyl group) can counteract any effect of the functional group.

In general, the solubility of organic molecules decreases with increasing chain length.

Tool 3: Mathematics

Construct and interpret tables, charts and graphs

In a data table, the independent variable is shown in the left column, the values of the dependent variable obtained in different trials are in the next column, and any derived or calculated values (often averages and, perhaps, information that has been further processed) are in columns on the far right. The table must have a title. Where appropriate, measurements must have units and a value of the random uncertainty in them.

The type of graph you plot will depend on the type of data.

Histograms are used to show the distribution of a single variable. There is no independent or dependent variable. If the single variable is continuous, there are no gaps between the bars and each bar represents a range of values. If the variable is discontinuous or categoric, there are gaps. The Maxwell–Boltzmann distribution curve is really a histogram (frequency distribution) of the kinetic energies of a very large number of particles.

Bar charts are used for plotting a numeric dependent variable against a categoric independent variable. For example, a bar chart would be used to display the specific heat capacities of different metals; the metal is the categoric independent variable.

Two numerical (continuous) variables from an experiment where the remaining variables have been controlled can be plotted on a **scatter graph**. The independent variable is always plotted on the x-axis.

If the data do not suggest a causal or simple mathematical relationship, then the points should remain unconnected. Otherwise, a line (straight or curved) of best fit may be drawn through the data points. For example, the relationship between rate of enzyme activity and temperature is shown in a curved graph.

23 The table shows the boiling and melting points and densities of alkanes.

Name	Formula	Boiling point / °C	Melting point / °C	Density / $g\,cm^{-3}$
methane	CH_4	–164	–183	0.55
ethane	C_2H_6	–89	–183	0.51
propane	C_3H_8	–42	–189	0.50
butane	C_4H_{10}	0	–138	0.58
pentane	C_5H_{12}	36	–130	0.63
hexane	C_6H_{14}	69	–95	0.66
heptane	C_7H_{16}	98	–91	0.68
octane	C_8H_{18}	126	–57	0.70
nonane	C_9H_{20}	151	–51	0.72
decane	$C_{10}H_{22}$	174	–30	0.73
dodecane	$C_{12}H_{26}$	216	–10	0.75
pentadecane	$C_{15}H_{32}$	271	10	0.77
octadecane	$C_{18}H_{38}$	317	28	0.76
eicosane	$C_{20}H_{42}$	343	37	0.79
triacontane	$C_{30}H_{62}$	450	66	0.81
tetracontane	$C_{40}H_{82}$	—	81	—
pentacontane	$C_{50}H_{102}$	—	92	—

a Plot a graph of boiling point against number of carbon atoms for the straight chain alkanes.

b Explain the shape of the graph and state what happens to the percentage difference between the boiling points.

c An unbranched alkane has a molar mass of $198\,g\,mol^{-1}$. Predict the boiling point.

d Suggest one reason why two of the alkanes do not have a boiling point.

e Explain why the alkanes are not soluble in water.

Tool 3: Mathematics

Draw and interpret uncertainty bars

The random uncertainty in the average or single measurement of a physical quantity (for example, temperature) can be shown on a line or curve graph as a bar drawn through a data point to show the possible range of values of the measurement. These bars, therefore, show the absolute uncertainty of values plotted on graphs. The length of an uncertainty bar shows the uncertainty of a data point: a short uncertainty bar shows that values are precise, indicating that the plotted average value is more likely. A long uncertainty bar indicates that the values are more spread out and less reliable.

Usually, uncertainty bars are vertical, showing the potential range of *y*-values. However, they can also be plotted for *x*-values.

LINKING QUESTION

What is the influence of the carbon chain length, branching and the nature of the functional groups on intermolecular forces?

Nomenclature of organic compounds

To clearly identify organic molecules, a systematic method of naming has been developed by IUPAC (the International Union of Pure and Applied Chemistry). In the IUPAC system of nomenclature, the names of organic molecules correlate with their structures and have three parts determined as shown in Figure S3.59.

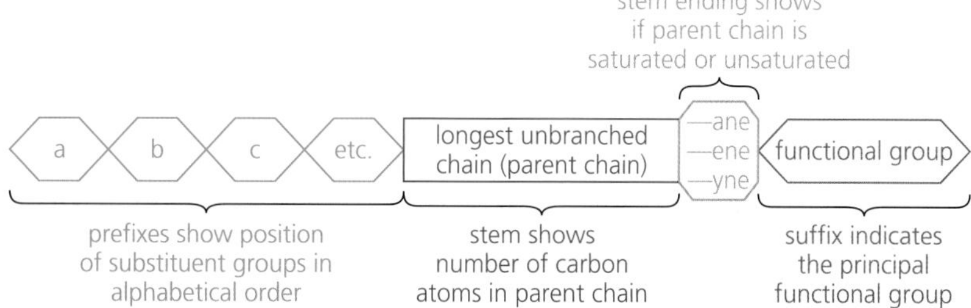

■ **Figure S3.59** IUPAC naming of organic compounds

Stem

◆ **Parent chain**: The longest continuous chain of connected carbon atoms in a hydrocarbon.

The first step is to identify the number of carbon atoms in the longest unbranched chain (**parent chain**) (Table S3.24).

■ **Table S3.24** Stem names and alkane equivalent names for parent chains with 1 to 6 carbon atoms

Number of carbon atoms	Stem name	Alkane equivalent
1	meth—	methane
2	eth—	ethane
3	prop—	propane
4	but—	butane
5	pent—	pentane
6	hex—	hexane

Top tip!

From 5 upwards, the stem is an abbreviation of the corresponding Greek word for the number (pent—, hex—, hept—, oct—, non—, dec—).

Suffix

The second step is to identify the main functional group present (Table S2.25). The suffix indicates the functional group in the molecule. An alkane is not a functional group but is included for the purposes of naming. The functional groups in bold are the functional groups that may feature in IB examination questions on the naming of organic molecules.

■ **Table S3.25** Functional groups, structures and suffixes for selected functional groups (where the numbers show the positions of functional groups (e.g., –Br and >C=C<) and alkyl side chains (e.g., –CH3), along the longest unbranched chains of the molecules)

Class	Functional group	Structure	Suffix	Example
carboxylic acids	**carboxyl**	—C(=O)—O—H	**—oic acid**	methanoic acid, HCOOH
ester	ester	—C(=O)—O—	—oate	methyl ethanoate, CH_3COOCH_3
ether	alkoxy	R–O–R′	none	methoxymethane, $CH_3–O–CH_3$
amine	amino	$–NH_2$, –NHR, $–NR_2$ —N(H)H —N(H)R —N(R)R	—amine	butan-1-amine, $CH_3(CH_2)_3NH_2$
amide	amido	—C(=O)—N(H)H	—amide	propanamide, $CH_3CH_2CONH_2$
alcohol	hydroxyl	—O—H	**—ol**	butan-2-ol, $CH_3CH(OH)CH_2CH_3$
aldehyde	carbonyl (aldehyde)	—C(=O)—H	**—al**	methanal, H_2CO
ketone	carbonyl (ketone)	>C=O R(R')C=O	**—one**	butanone, $CH_3COCH_2CH_3$
halogenoalkanes	halogeno	–X	a prefix (fluoro, etc.)	2-bromobutane, $CH_3CH_2CHBrCH_3$
alkenes	alkenyl	>C=C<	—ene	4-methyl-pent-2-ene, $(CH_3)_2CH–CH=CH–CH_3$
alkynes	alkynyl	—C≡C—	—yne	but-1-yne, H_3CCH_2CCH
alkanes	alkanes	none	—ane	heptane, $CH_3(CH_2)_5CH_3$
arenes	phenyl	(benzene ring)	benzene	benzene carboxylic acid, C_6H_5COOH (benzoic acid)

Where a number is present before the suffix in the name of an organic molecule, it indicates the position of the functional group on the longest unbranched chain (parent chain). The number is always the *lowest* number possible.

◆ **Substituent (naming):** An atom or group regarded as having replaced a hydrogen atom in a chemical derivative.

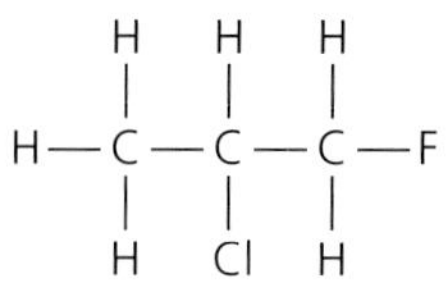

Prefix

The third step is to describe the type of substituents present on the parent chain. A **substituent** is an atom or group of atoms replacing a hydrogen atom on the parent chain. The carbon atoms in the parent chain are numbered to indicate the positions of any substituent groups in the chain.

The numbering starts from the end of the chain which gives the substituent the lowest value. For example, the molecule shown in Figure S3.60 is 2-chloro-1-fluoropropane (based on alphabetical order, c comes before f) and not 2-chloro-3-fluoropropane.

H H H
H—C—C—C—F
H Cl H

■ **Figure S3.60** Structure of 2-chloro-1-fluoropropane

The numbers and name are separated by a hyphen and the numbers are separated by a comma, for example, 2,2-dimethylbutane (Figure S3.61).

CH_3
$H_3C—C—CH_2—CH_3$
CH_3

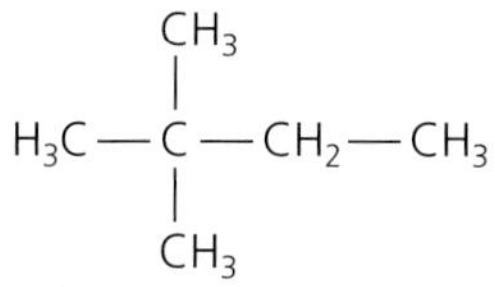

■ **Figure S3.61** Structure of 2,2-dimethylbutane

Table S3.26 shows examples of alkyl substituent groups.

■ **Table S3.26** Alkyl group names and their structural formulas for straight chain alkyl groups from one to five carbon atoms

Number of carbon atoms	Name of alkyl group	Formula
1	methyl	$-CH_3$
2	ethyl	$-CH_2-CH_3$
3	propyl	$-CH_2-CH_2-CH_3$
4	butyl	$-CH_2-CH_2-CH_2-CH_3$
5	pentyl	$-CH_2-CH_2-CH_2-CH_2-CH_3$

Table S3.27 shows examples of non-alkyl substituent groups.

■ **Table S3.27** Substituent groups and their names

Substituent name	Substituent group
fluoro	–F
chloro	–Cl
bromo	–Br
iodo	–I
nitro	$-NO_2$
cyano	–CN
hydroxyl	–O–H
oxo	>C=O
amino	$-NH_2$
amido	$-CONH_2$
alkenyl	>C=C<
alkynyl	–C≡C–
phenyl	$-C_6H_5$

If there are a number of identical substituents in the molecules, the prefixes shown in Table S3.28 are used.

■ **Table S3.28** Prefixes used to indicate number of identical substituents

Number of identical substituents	Prefix
2	di—
3	tri—
4	tetra—
5	penta—

For cyclic organic compounds, 'cyclo' is introduced into the prefix.

Going further

Order of precedence

When organic compounds contain more than one functional group, the order of precedence determines which groups are named with prefix or suffix forms. The highest precedence group takes the suffix, with all others taking the prefix form. However, double and triple bonds only take suffix form (—en and —yn) and are used with other suffixes.

For example, the compound lactic acid has the condensed formula $CH_3CH(OH)COOH$. The systematic name for lactic acid is 2-hydroxypropanoic acid; here the carboxyl group (–COOH) takes precedence in the naming, forcing the –OH group to be referred to as the hydroxyl group.

a $Cl-CH_2-CH_2-CHO$ is

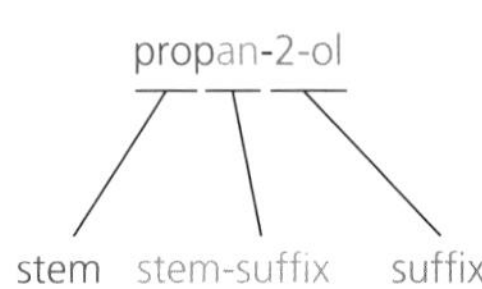

b $CH_3-CH(OH)-CH_3$ is

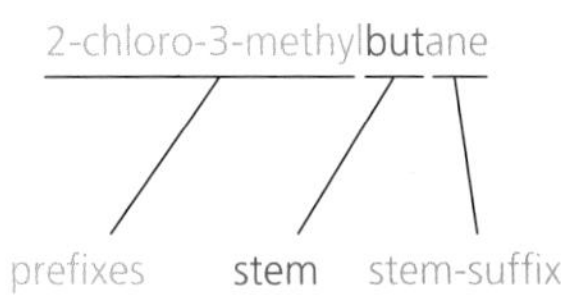

c $CH_3-CHCl-CH(CH_3)_2$ is

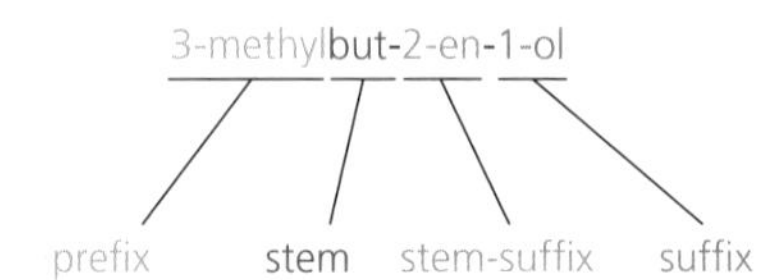

d $(CH_3)_2C=CH-CH_2OH$ is 3-methylbut-2-en-1-ol

prefix (3-methyl), stem (but), stem-suffix (-2-en), suffix (-1-ol)

■ **Figure S3.62** Examples of applying IUPAC naming rules

Nature of science: Science as a shared endeavour

Naming organic compounds

To allow international communication and understanding of chemistry, there needs to be a standardized language with which to refer to compounds. The IUPAC system of naming organic molecules names a compound unambiguously on the basis of its structure. For example, the IUPAC nomenclature for a polymer is based on the source (monomer).

This system is a relatively recent development and there is a whole set of trivial names for many compounds in the wider world of chemistry. For example: ethanamide (CH_3CONH_2) can also be referred to as acetamide; ethanenitrile as acetonitrile; phenylamine or aminobenzene ($C_6H_5-NH_2$) as aniline; and benzenecarboxylic acid (C_6H_5-COOH) as benzoic acid. Complex molecules, in particular, are often referred to by trivial names; examples include beta-carotene, chlorophyll-a and beta-D-glucose.

While the IB syllabus and examination papers use only the IUPAC names, you should be aware that other names may be found on reagent bottles in a chemistry laboratory.

WORKED EXAMPLE S3.2A

Draw the displayed structural formula for the molecule 2-bromo-1-chloro-2-methylbutane.

Answer

Step 1 Identify the parent chain (longest unbranched chain) and number it. In this example it is butane (C_4).

```
Step 1                    Step 2                          Step 3

                               CH3                       H   CH3  H   H
                          1   |2    3    4               |   |    |   |
 1   2   3   4         Cl—C—C—C—C                   Cl—C—C—C—C—H
C—C—C—C                       |                          |   |    |   |
                               Br                        H   Br   H   H
```

Step 2 Add the substituents (atoms and groups bonded to the parent chain) to the correct carbon atoms. In this case, add the bromo group to carbon 2, the chloro group to carbon 1 and the methyl group to carbon 2.

Step 3 Finally, add hydrogen atoms (keeping the numbers as low as possible) to all the remaining bonds to form C–H bonds.

ATL S3.2B

Go to www.chem.ucalgary.ca/courses/351/WebContent/orgnom/index.html. Select alkanes, review the IUPAC naming of this homologous series and then try the practice naming questions. Repeat for homologous series containing the following functional groups: halogeno, hydroxyl, carbonyl and carboxyl.

Isomerism

The existence of two or more compounds with the same molecular formula but different properties is known as isomerism. Such compounds are called **isomers**. Table S3.29 shows how different types of isomerism are related.

◆ **Isomers**: Compounds having the same molecular formula, but different structural (or spatial) arrangements of atoms.

◆ **Structural isomers**: Isomers in which the connectivity of the atoms differs.

◆ **Chain isomers**: Molecules with the same molecular formula, but with their carbon atoms joined together in different ways.

◆ **Straight-chain isomer**: A hydrocarbon with an open chain of atoms and no side-chains.

◆ **Branched-chain isomer**: A hydrocarbon which has alkyl groups bonded to its longest unbranched chain.

■ **Table S3.29** Types of isomerism

Structural isomerism: Atoms and functional groups attached in different ways.	**Chain**: Different carbon skeletons.
	Positional: Double / triple bonds and/or functional groups in different places on the carbon skeleton.
	Functional group: Different functional groups.
Stereoisomerism (HL only): Atoms and functional groups have different spatial arrangements.	***cis–trans***: Exists where there is restricted rotation around bonds.
	Optical: Exists where there is an asymmetric carbon atom.

Structural isomerism

Compounds having the same molecular formula but different structures (due to different connectivities between atoms) are classified as **structural isomers**. The connectivity of atoms in a molecule refers to the pattern in which the atoms in a molecule are bonded to one another.

Chain isomers

When two or more organic compounds have similar molecular formulas but different carbon skeletons, these are referred to as **chain isomers**. Figure S3.63 shows three chain isomers, all with the molecular formula C_5H_{12}: 2,2-dimethylpropane, 2-methylbutane and pentane. Pentane is a **straight-chain isomer** and the other two isomers are **branched-chain isomers**.

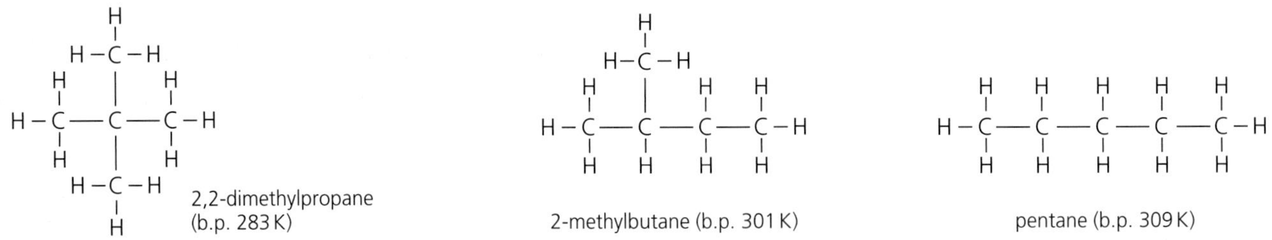

2,2-dimethylpropane (b.p. 283 K)

2-methylbutane (b.p. 301 K)

pentane (b.p. 309 K)

■ **Figure S3.63** Structural isomers of pentane

Remember that carbon chains have a 'zig zag' tetrahedral arrangement. There is completely free rotation around each carbon–carbon single (sigma) bond even though structural formulas are often drawn with 90° bond angles.

Going further

Conformations

Because sigma bonds are cylindrically symmetrical, rotation about a carbon–carbon single bond can occur without any change in orbital overlap. The different arrangements in space of atoms that result from rotation about a single bond are called conformations. A specific conformation is called a conformer.

The formulas in Figure S3.64 represent different conformers of the pentane molecule: under standard conditions, a particular molecule will rapidly change shape from one shape to another.

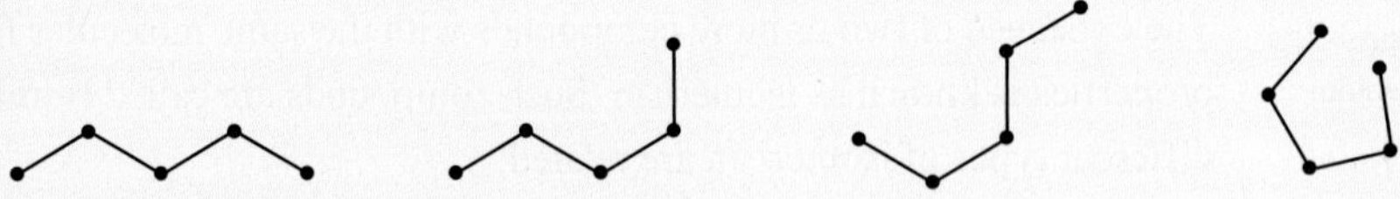

■ **Figure S3.64** Conformations of pentane

When rotation occurs about the carbon–carbon bond of ethane, two extreme conformations exist: a staggered conformer and an eclipsed conformer.

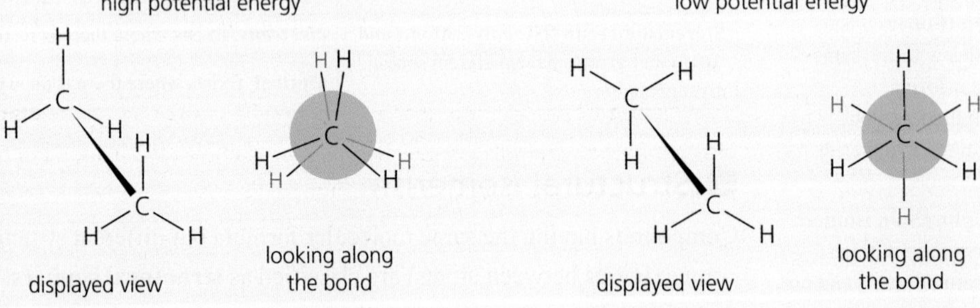

■ **Figure S3.65** Conformers of ethane

The electrons in a C–H bond repel the electrons in another C–H bond. The staggered conformation is the most stable conformation of ethane because the C–H bonds are as far away from each other as possible. The eclipsed conformation is the least stable conformation because in no other conformation are the C–H bonds as close to one another. The extra energy present in the eclipsed conformation is called torsional strain. Conformers are interconvertible and cannot be isolated at room temperature.

Positional isomers

◆ **Positional isomers:** Structural isomers that have the same carbon skeleton and the same functional groups but a different location of the functional groups on or in the carbon chain.

◆ **Functional group isomers:** Structural isomers that have different functional groups.

When two or more compounds differ in the position of substituent atoms or functional groups on the carbon skeleton, they are called **positional isomers** (Figure 3.66).

```
   H   H   H   H                            H
   |   |   |   |                            |
H—C—C—C—C—O—H                  H   H   O   H
   |   |   |   |                    |   |   |   |
   H   H   H   H                 H—C—C—C—C—H
                                    |   |   |   |
                                    H   H   H   H
     butan-1-ol                        butan-2-ol
```

■ **Figure S3.66** Positional isomers of butanol: butan-1-ol and butan-2-ol

Positional isomerism also occurs in ring systems (cyclic molecules), as shown in Figure 3.67.

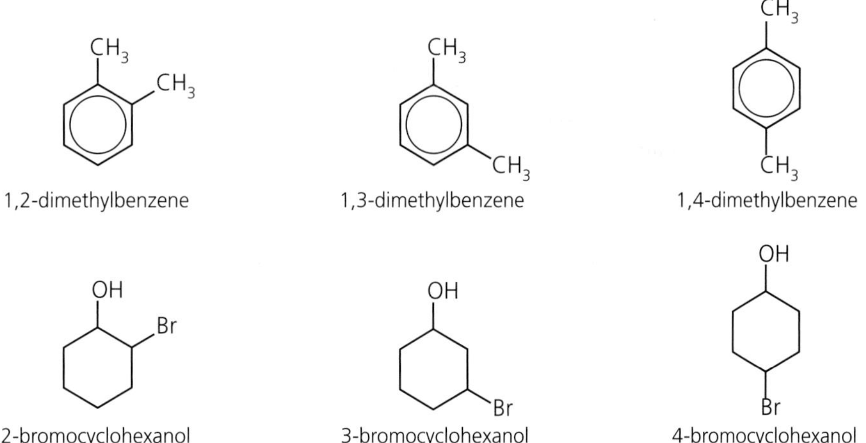

■ **Figure S3.67** Positional isomers in cyclic molecules. Note that the carbon atoms in the rings are numbered starting at one of the substituted ones, where one of the carbon atoms is bonded to carbon or bromine. In cyclohexanol, the carbon atom attached to the –OH group is carbon number 1.

Functional group isomers

Two or more compounds having the same molecular formula but different functional groups are called **functional group isomers**.

Two such functional group isomers are ethanol, C_2H_5OH, and methoxymethane, CH_3OCH_3 (Figure S3.68). Both molecules have the molecular formula C_2H_6O but one is an alcohol (with a hydroxyl functional group) and the other is an ether (with an alkoxy functional group).

Top tip!

Common pairs of functional group isomers include alkenes and cycloalkanes; ketones and aldehydes; and esters and carboxylic acids.

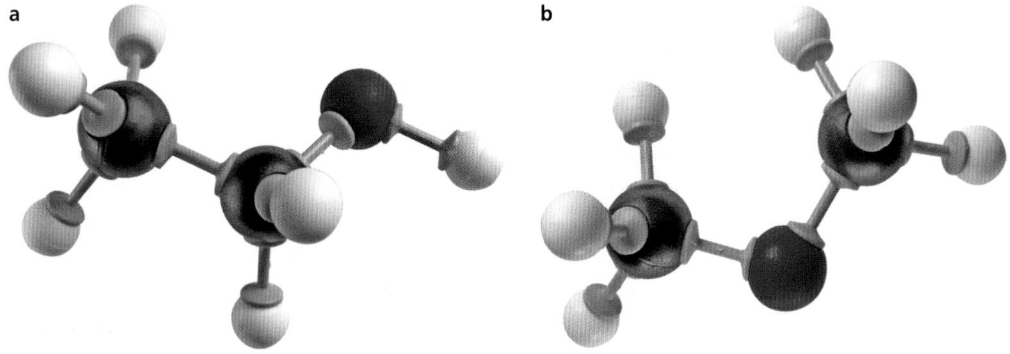

■ **Figure S3.68** Models of the functional group isomers of formula C_2H_6O: **a** ethanol (C_2H_5OH) and **b** methoxymethane $(CH_3)_2O$

Structural classification of molecules

Alcohols can be classified depending on the position of the hydroxyl group, –OH, on the chain of carbon atoms (Figure S3.69). The three classes of alcohols behave differently during oxidation.

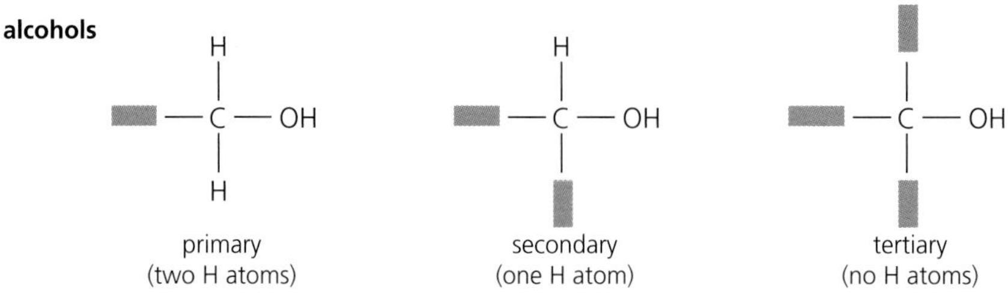

alcohols

primary (two H atoms) — secondary (one H atom) — tertiary (no H atoms)

■ **Figure S3.69** The structures of primary, secondary and tertiary alcohols. The grey blocks are alkyl or aryl groups (they may all be the same group, or different).

- In primary (1°) alcohols, the carbon bonded to the –OH group is only bonded to one alkyl group.
- In secondary (2°) alcohols, the carbon atom with the –OH group attached is bonded directly to two alkyl groups, which may be the same or different.
- In a tertiary (3°) alcohol, the carbon atom attached to the –OH group is bonded directly to three alkyl groups, which may be the same or different.

Halogenoalkanes can be classified in the same way (Figure S3.70). The three types of halogenoalkanes follow different mechanisms during substitution with nucleophiles.

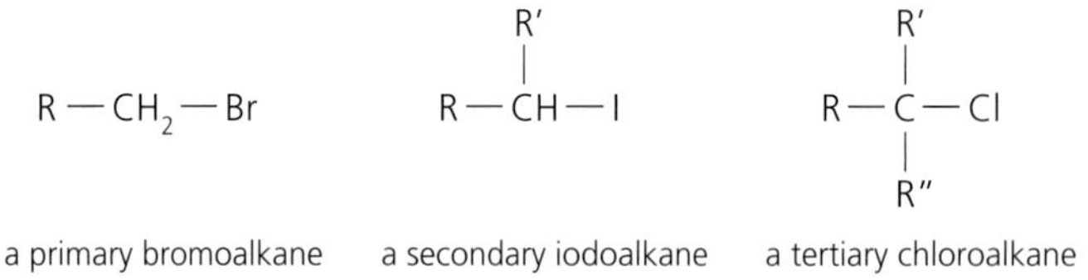

$R-CH_2-Br$ — a primary bromoalkane

$R-CH(R')-I$ — a secondary iodoalkane

$R-C(R')(R'')-Cl$ — a tertiary chloroalkane

■ **Figure S3.70** The structures of primary, secondary and tertiary halogenoalkanes. R, R′ and R″ are alkyl groups (they may all be the same group, or different)

Amines may also be classified into primary, secondary and tertiary as shown in Figure S3.71. This system is based on regarding them as being derived from ammonia.

Top tip!

Amides can also be classified as primary ($RCONH_2$), secondary (RCONHR) and tertiary ($RCONR_2$).

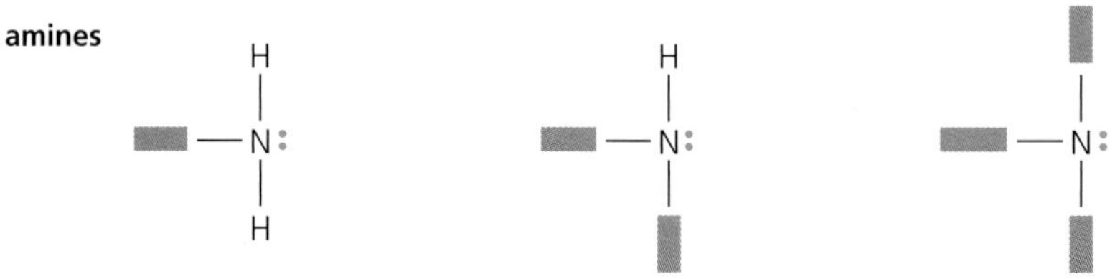

amines

■ **Figure S3.71** The structures of primary, secondary and tertiary amines (where blocks represent alkyl groups)

LINKING QUESTION (HL ONLY)

How does the fact that there are three isomers of dibromobenzene support the current model of benzene's structure?

POPs

A number of toxic chemicals—such as DDT (developed as an insecticide), dioxin (from waste incineration and metal smelting) and polychlorinated biphenyls (PCBs) (previously used as coolants in electrical equipment)—can enter water sources from industrial dumping, atmospheric emissions, agricultural use and household dumping.

Animals on farms and fish consume water and then accumulate these chemicals. Human consumption of food or water contaminated with these chemicals can lead to increased cancer risk, liver damage and central nervous system damage. Governments around the world have collaborated to ban the production of a number of these kinds of chemicals, which are known as persistent organic pollutants or POPs.

Stereoisomers

HL ONLY

◆ **Constitution**: The number and type of atoms in a molecule.

◆ **Stereoisomerism**: The form of isomerism where molecules have the same formula and functional groups but differ in the arrangement of groups in space.

◆ ***cis–trans* isomerism**: This occurs when isomers have groups in different positions with respect to a double bond or ring.

◆ ***cis* isomer**: An isomer with functional groups or atoms on the same side of the molecule or double bond as each other.

◆ ***trans* isomer**: An isomer with functional groups or atoms on opposite sides of the double bond or ring.

Compounds that have the same **constitution** and sequence of covalent bonds but differ in the relative positions of their atoms or groups in space are called stereoisomers. There are two types of **stereoisomerism**: ***cis–trans* isomerism** and optical isomerism.

cis–trans isomerism

Alkenes in which the carbons of the double bond each have two different groups attached can exist in two forms referred to as *cis–trans* isomers.

Top tip!

cis–trans isomerism occurs in alkenes because it is not possible to rotate one end of a C=C double bond with respect to the other due to the presence of a pi bond.

When two identical functional groups or atoms are on the same side of the double bond, the compound is called the ***cis* isomer**; when they are on opposite sides of the double bond, the compound is called the ***trans* isomer**.

Figure S3.72 shows the possible configurations of but-2-ene, where the two methyl groups can be on the same side of the double bond (*cis* isomer) or on opposite sides of the double bond (*trans* isomer).

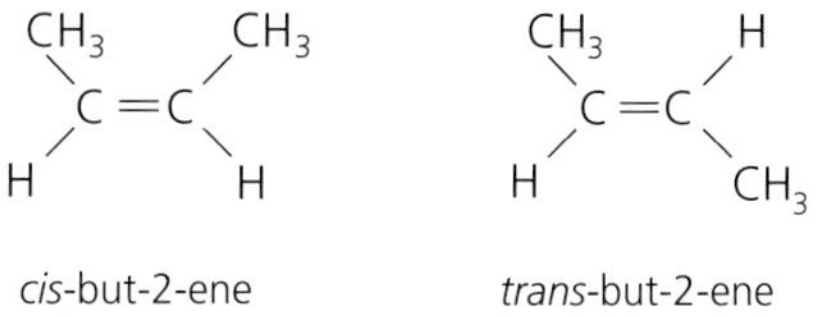

cis-but-2-ene | *trans*-but-2-ene

■ **Figure S3.72** *cis* and *trans* isomers of but-2-ene

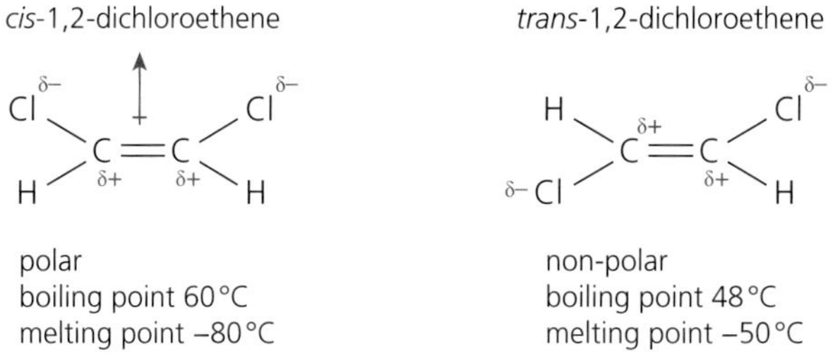

cis-1,2-dichloroethene: polar, boiling point 60 °C, melting point −80 °C

trans-1,2-dichloroethene: non-polar, boiling point 48 °C, melting point −50 °C

■ **Figure S3.73** Structures and properties of 1,2-dichloroethene

1,2-dichloroethene is another example of a molecule that exhibits *cis–trans* isomerism (Figure S3.73).

The *cis* isomer has a higher boiling point than the *trans* isomer because the bond dipoles do not cancel out and thus the molecule has a net dipole moment. The molecule is polar and has stronger dipole–dipole forces than the *trans* isomer which has only London (dispersion) forces between the molecules. More energy is needed to overcome the stronger dipole–dipole forces between molecules.

> **Common mistake**
>
> Misconception: *The boiling points of* cis–trans *isomers are the same because* cis–trans *isomers have the same chemical formulas.*
>
> The boiling points of *cis–trans* isomers are different because they have different molecular polarities and geometries (shapes).

The *trans* isomer has a higher melting point than the *cis* isomer. This is because it has a more efficient packing of molecules in the solid state, giving rise to stronger London (dispersion) forces between molecules. Hence more energy is needed to overcome the London (dispersion) forces between the molecules, resulting in a higher melting point.

Rotation about sigma bonds in small cyclic molecules is restricted. The rigid structure of the ring prevents free rotation and can give rise to isomers. For example, 1,2-dichlorocyclopropane exists as *cis* and *trans* isomers (Figure S3.74).

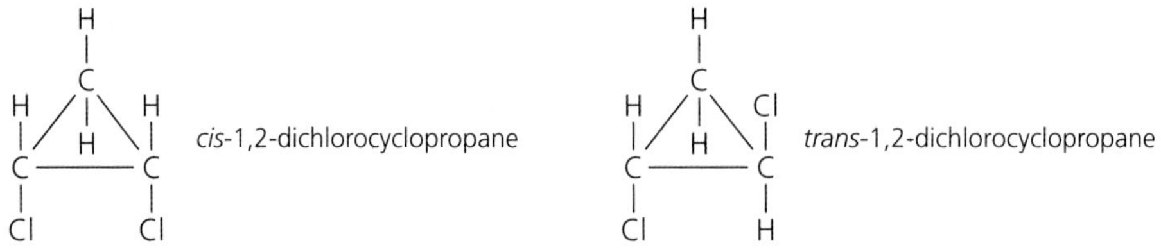

cis-1,2-dichlorocyclopropane

trans-1,2-dichlorocyclopropane

■ **Figure S3.74** Simplified representations of the *cis* and *trans* isomers of 1,2-dichlorocyclopropane

In addition to 1,1-dichlorocyclobutane, there are four isomers of dichlorocyclobutane. Firstly there are two structural isomers of this compound: 1,2-dichlorocyclobutane and 1,3-dichlorocyclobutane (Figure S3.75). Each of these structural isomers then has two *cis–trans* isomers.

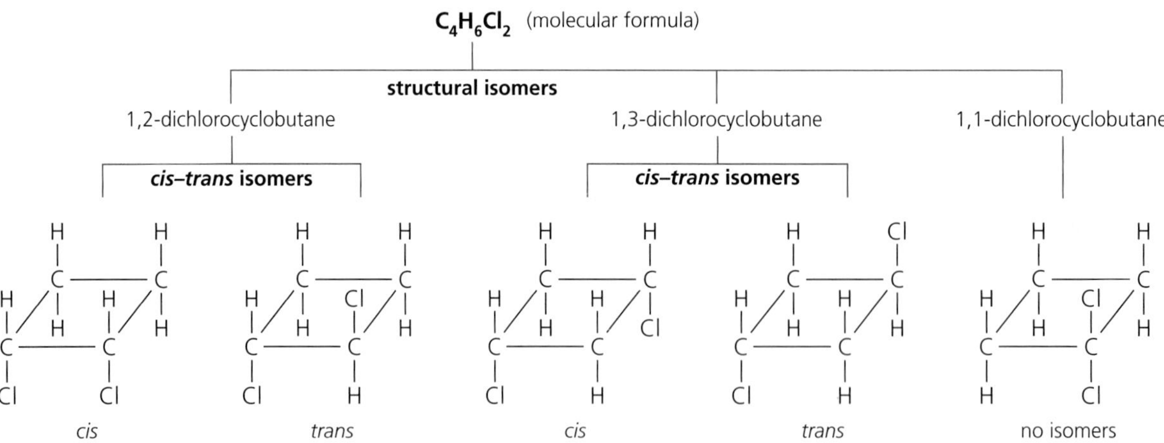

■ **Figure S3.75** The *cis–trans* isomers of dichlorocyclobutane

Going further

E–Z isomers

Geometric isomerism (*cis–trans* isomerism) can also exist when there are three or four different groups bonded to the carbon–carbon double bond. Such alkenes cannot be named using the *cis–trans* system. Instead, they require the *E–Z* system, in which the naming results from assigning priorities to the bonded groups based on the atomic number and/or molar mass of the group (Figure S3.76).

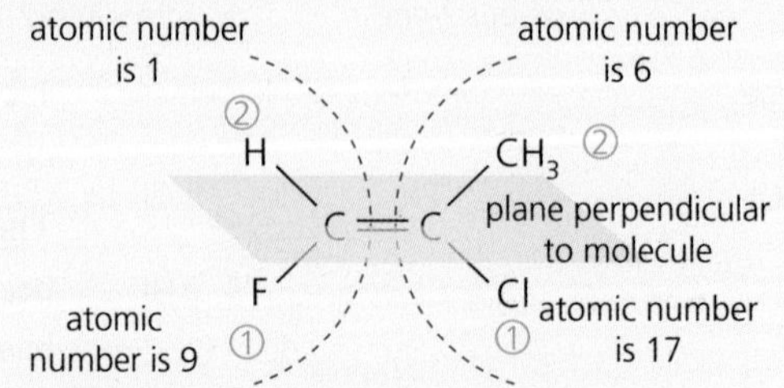

■ **Figure S3.76** The structure of (*Z*)-1-fluoro-2-chloropropene. Note that the high priority groups (labelled 1) are both on the same side of the double bond

Optical isomerism

A carbon atom which has four different atoms or functional groups bonded to it is known as a **chiral carbon**.

Stereoisomers are formed when two chiral molecules have the same molecular and structural formulas but it is not possible to superimpose the mirror image of one molecule on the other. This property is known as **optical isomerism** and the stereoisomers form a pair of **enantiomers** (Figure S3.77).

Top tip!

Chirality is a property of the entire molecule, not just the chiral carbon atom.

- ◆ **Chiral carbon**: A carbon atom bonded to four different atoms or groups of atoms.
- ◆ **Optical isomerism**: Optical isomers are molecules that have the same structural and molecular formula, but are non-superimposable mirror images of each other.
- ◆ **Enantiomers**: A pair of molecules consisting of one chiral molecule and the mirror image of this molecule.
- ◆ **Optically active**: A substance able to rotate plane-polarized light.
- ◆ **Racemic mixture**: An equimolar mixture of the pair of enantiomers of an optically active compound.
- ◆ **Polarimeter**: An instrument used to determine the angle through which the plane of polarization of plane-polarized light is rotated on passing through an optically active substance.

Top tip!

An achiral molecule such as methane (CH_4) can be superimposed on its mirror image.

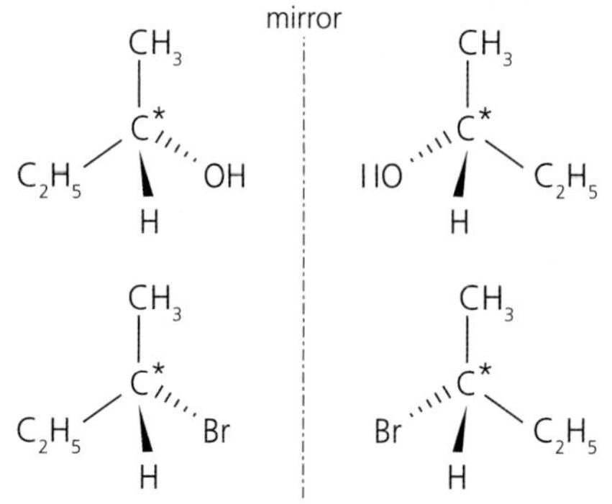

■ **Figure S3.77** Enantiomers of butan-2-ol and 2-bromobutane (* indicates a chiral carbon atom)

Enantiomers have identical physical properties (such as melting points, boiling points, density and solubility) and identical chemical properties unless the reacting reagent is chiral.

Common mistake

Misconception: *The boiling points of a pair of enantiomers are different.*

The boiling points of a pair of enantiomers are identical.

Chiral compounds that can rotate plane-polarized light are described as being **optically active**. Enantiomers are optically active. One of a pair will rotate the plane of polarized light clockwise (+) and the other anticlockwise (−) through the same angle.

An equimolar mixture of a pair of enantiomers (that is, of the same compound) will not rotate plane polarized light. The rotating powers of the molecules of the enantiomer cancel each other out. This mixture is called a **racemic mixture**.

Common mistake

Misconception: *A racemic mixture can rotate plane-polarized light.*

A racemic mixture is optically inactive. The rotation of one enantiomer is cancelled by the rotation of the other enantiomer.

A **polarimeter** (Figure S3.78) is an instrument used to distinguish optical isomers and measure optical activity. Light of a single wavelength passes through a slit to produce a thin beam of light which then passes through a polarizing filter. The polarized beam is directed through a sample of the solution being tested and then passes through another polarizing filter (analyser) which can be rotated. A photocell measures the intensity of light transmitted for each angle of the analyser.

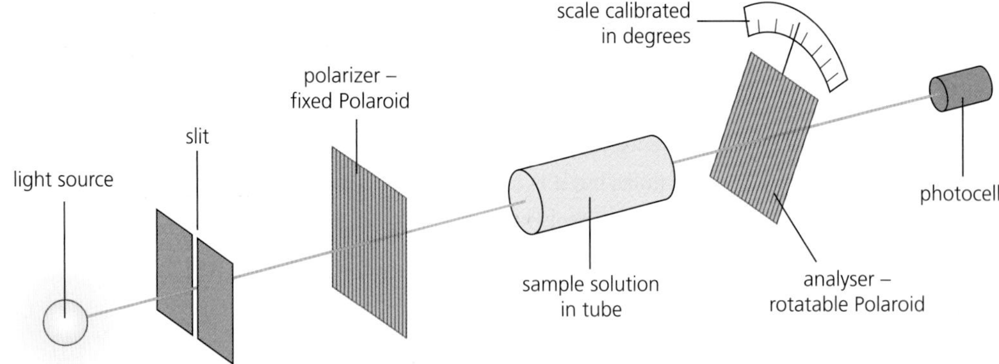

■ **Figure S3.78** A polarimeter

Nature of science: Observations

Resolving enantiomers

The French chemist Louis Pasteur (1822–1895) carefully examined crystals of salts of tartaric acid he had prepared by allowing a saturated solution to evaporate slowly. He observed two types of crystal that were mirror images of each other (Figure S3.79). Using a hand lens and tweezers, he separated the crystals into two piles.

He then dissolved each sample of crystals in water and examined the solutions using a polarimeter. He found that a solution of one sample of crystals rotated the plane of polarized light to the right but the other rotated it to the left to an equal extent. He was the first person to separate the enantiomers from a racemic mixture, a process known as resolution.

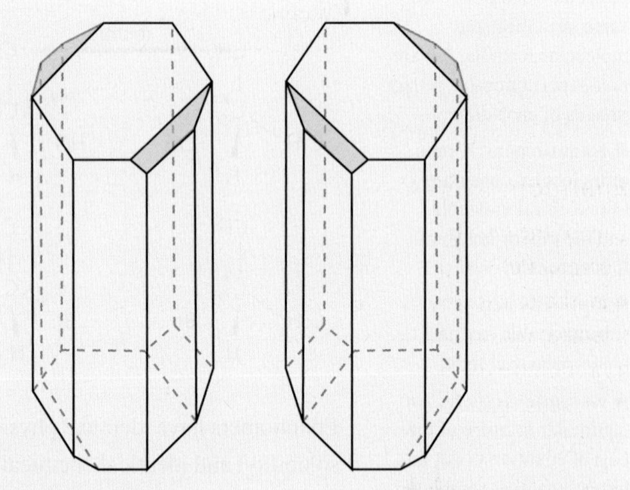

■ **Figure S3.79** Mirror image forms of sodium ammonium tartrate crystals

In the body, different enantiomers can have completely different physiological effects. For example, one of the enantiomers of the amino acid asparagine ($H_2N–CH(CH_2CONH_2)–COOH$) tastes bitter, whereas the other optical isomer tastes sweet.

A biologically active chiral molecule must fit into a chiral receptor at a target site, like a hand fits into a glove. But just as a right hand can fit only into a right-hand glove, so a particular stereoisomer can fit only into a receptor having the proper complementary shape (Figure S3.80).

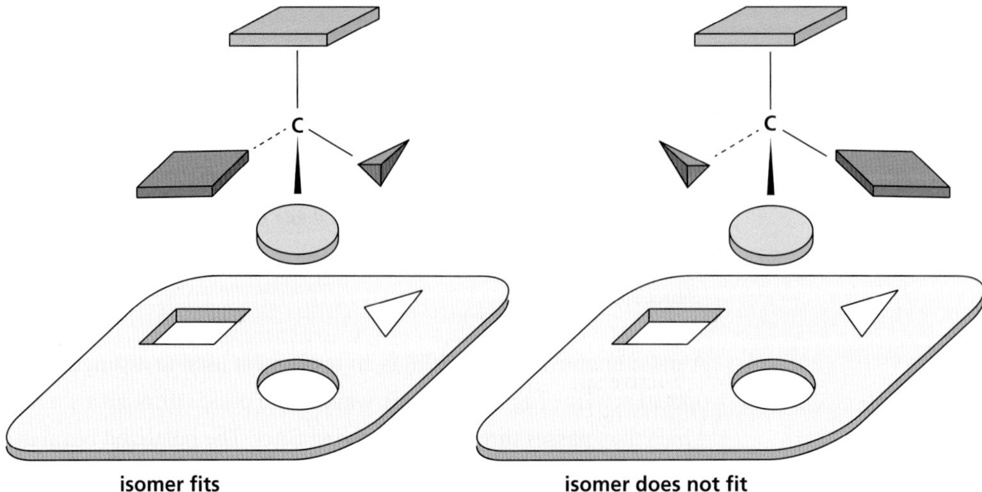

■ **Figure S3.80** Interaction of a pair of enantiomers with a chiral biological receptor

HL ONLY

Tool 2: Technology

Use computer modelling

Chemical computer simulations give you an opportunity to improve your understanding of chemical concepts by connecting the symbolic and particulate views of topics, such as molecular structures and shapes.

View the achiral acetone cyanohydrin (a bifunctional molecule with a cyano (–CN) functional group and a carbonyl (>C=O) functional group) at **www.chemtube3d.com/stturning/**. Then view the molecule at **www.chemtube3d.com/stchiralcentersrs/** which is chiral due to the presence of a single chiral carbon atom. Use this page to explore how chiral molecules are assigned an absolute configuration (labelled R or S) according to well-defined rules based on atomic number. (This will not be tested in IB examinations.)

Going further

Enantiomers in medicine

One of the enantiomers of salbutamol, used in the treatment of asthma, is significantly more effective than the other. However, it can be expensive to separate (resolve) optical isomers, so many medicines are produced as a mixture of enantiomers, only one of which is pharmacologically active.

A racemic mixture of the two enantiomers of a drug called thalidomide, developed in West Germany in the 1950s, was used to treat morning sickness (nausea) during pregnancy. One enantiomer, which was thought to be inactive, turned out to cause damage to the unborn child. Babies were born with a range of disabilities in a number of countries before it was realized the drug was responsible. Regulations were tightened significantly after the thalidomide tragedy to ensure that both enantiomeric forms of chiral medicinal drugs are tested using cultured cells before clinical trials in patients.

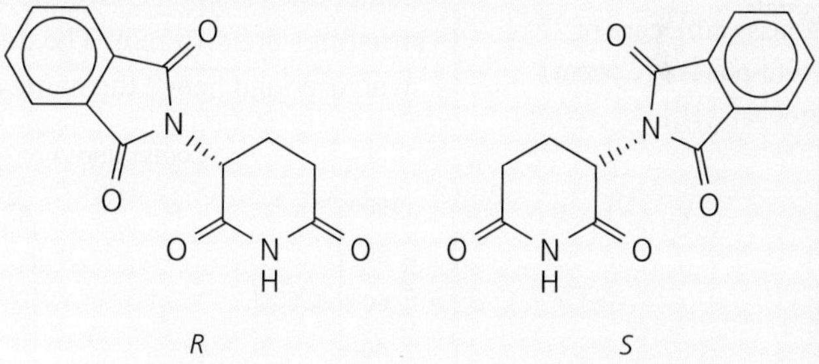

R

R-thalidomide
sedative and treatment for morning sickness

S

S-thalidomide
teratogenic agent causing birth defects

■ **Figure S3.81** The two enantiomers of thalidomide

Spectroscopy

The determination of the molecular structure of a compound, obtained either from a natural source or as the product of a chemical synthesis, is an important part of experimental organic chemistry. Without a knowledge of structure, chemists cannot explain the physical or the chemical properties of molecules. Traditionally, this was done by gradually breaking the molecule down and chemically analysing the decomposition products.

However, spectroscopic techniques, such as infrared spectroscopy (IR) and nuclear magnetic resonance spectroscopy (NMR), need only very small amounts of a compound and they require much less time than chemical methods. Chemists now determine organic structure by non-destructive spectroscopic techniques and recover the sample unchanged after they determine its spectrum.

Infrared spectroscopy (IR) shows the functional groups in an organic compound. Proton nuclear magnetic resonance spectroscopy (^{1}H NMR) provides information about the carbon–hydrogen framework of a compound.

This is another excellent example of how the development of technology helps create new opportunities for observations and experiments, which translate into new knowledge.

Link

Mass spectrometry is discussed in detail in Chapter S1.2, page 34.

Mass spectrometry

Mass spectrometry (MS) is a technique used to determine an accurate value for the molar mass of an organic compound. It can also determine structural features. High-energy electrons bombard the sample with enough energy to knock electrons out of the molecules which are ionized and broken into smaller ion fragments.

Mass spectrometry is used in the determination of the structures of organic compounds in two main ways:

- Finding the molecular formula of a compound by measuring the mass of its molecular ion to a high degree of accuracy.
- Deducing the structure of a molecule by looking at the fragments produced when a molecular ion fragments inside a mass spectrometer.

Table S3.30 (which can also be found in the IB *Chemistry data booklet*) shows common mass differences between peaks in a mass spectrum and suggests groups that correspond to the difference.

Table S3.30 Mass spectral fragments

Mass lost	Possible fragment lost
15	CH_3
17	OH
18	H_2O
28	$CH_2{=}CH_2$ or C=O
29	CH_3CH_2 or CHO
31	CH_3O
45	COOH

A mass difference of 77 corresponds to a phenyl group, $[C_6H_5]$.

WORKED EXAMPLE S3.2B

Compounds A and B have the same molecular formula, C_3H_6O, but their molecular structures are different. The diagrams show the mass spectra of A and B. Deduce the structures and hence the names of A and B.

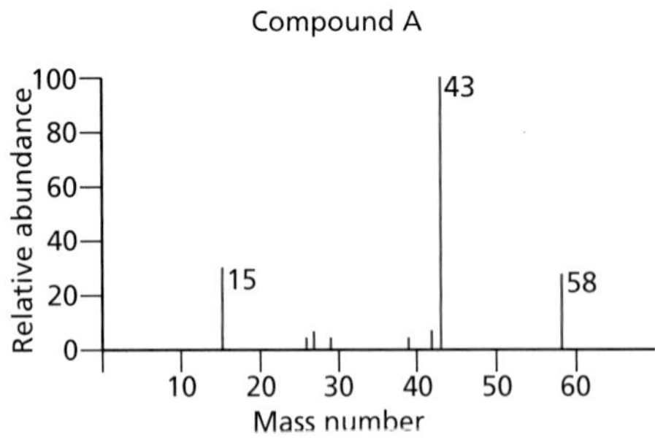

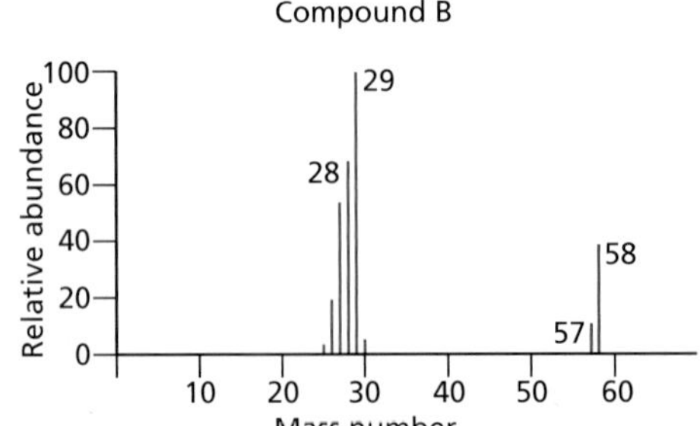

Answer

Compound A

The mass difference from 58 to 43 is 15. This indicates that the loss of a methyl group (CH_3) is likely to be responsible for this peak.

The ion of $m/z = 43$ is probably due to loss of a methyl group from the molecule, so that a fragment of formula $[C_2H_3O]^+$ would form the $m/z = 43$ ion and a likely structure for this is $[CH_3CO]^+$.

A is therefore propanone, $CH_3{-}CO{-}CH_3$, which fragments to form $[CH_3CO]^+$ and $[CH_3]^+$.

Compound B

The loss of one atomic mass unit from 58 to 57 can only be due to loss of a hydrogen atom.

The mass difference from 57 to 29 is 28, probably due to loss of a carbonyl group, >C=O.

The ion of $m/z = 28$ is the $[C{=}O]^+$ fragment; the addition of one mass unit (a hydrogen, H) to this would form a $[CHO]^+$ (aldehyde) at the $m/z = 29$ peak.

B is therefore propanal,$CH_3{-}CH_2{-}CHO$.

The peak at 43 is absent in propanal, because the molecule cannot fragment to give a CH_3CO^+ or a $C_3H_7^+$ ion.

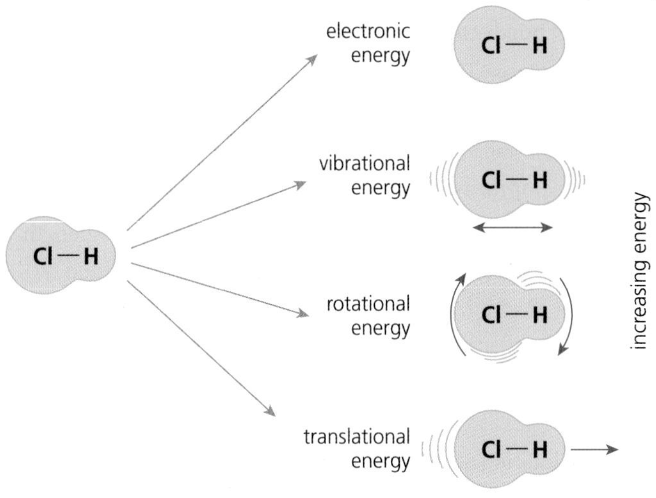

■ **Figure S3.82** The different types of energy associated with a hydrogen chloride molecule

Interaction of radiation with matter

When electromagnetic radiation interacts with molecules, the molecules absorb energy and this can cause a variety of changes, as shown for HCl in Figure S3.82. Which changes occur depends on the molecule and the energy of the electromagnetic radiation.

Molecules can absorb only certain discrete amounts of energy due to the quantization of their energy levels. There are separate energy levels, with fixed energy gaps or differences between them, for each type of change (Figure S3.83).

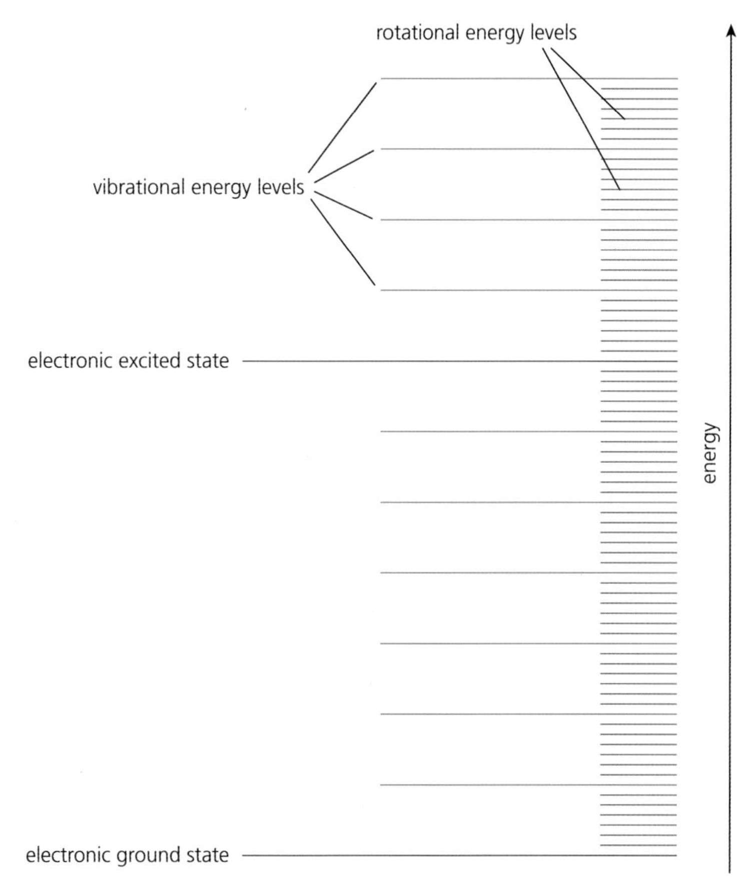

■ **Figure S3.83** The vibrational and rotational energy levels in a molecule

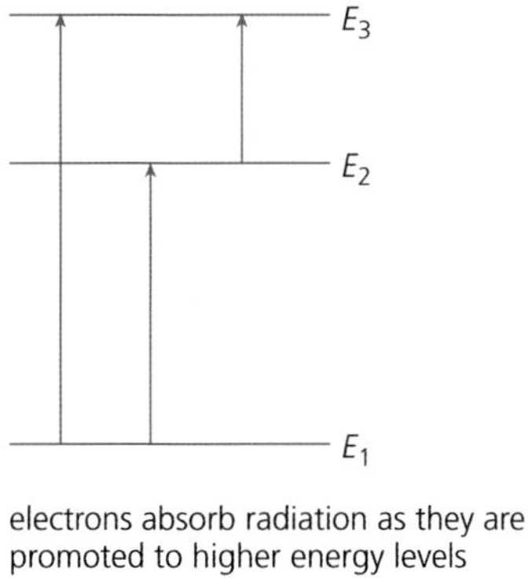

E_3

E_2

E_1

electrons absorb radiation as they are promoted to higher energy levels

■ **Figure S3.84** Electronic transitions caused by absorption of quanta of electromagnetic radiation

To change the energy of a molecule from E_1 to E_2, photons of electromagnetic radiation must provide a specific energy equal to the energy difference, $(E_2 - E_1)$. Figure S3.84 shows an increase in energy level from E_1 to E_2 resulting from absorption of a photon of energy hf.

◆ **Infrared radiation**: Electromagnetic radiation with wavelength longer than that of red light but shorter than that of radio waves.

◆ **Infrared spectroscopy**: Absorption spectroscopy carried out in the infrared region of the electromagnetic spectrum, detecting bond stretching and bending.

◆ **Infrared spectrum**: A plot of measured infrared transmittance against wavenumber.

The energy absorbed by the molecule can change its electronic, vibrational or nuclear energy. For example, **infrared radiation** causes covalent bonds to stretch and bond angles to bend. Protons in hydrogen-1 atoms absorb radio waves, which causes changes in proton spin.

In many types of spectroscopy, radiation is passed from a source through a sample that may or may not absorb certain wavelengths of the radiation. The wavelength (and thus frequency and energy) is systematically changed, and a detector measures any transmitted radiation.

The amount of light absorbed by the molecule (absorbance) is plotted as a function of wavelength. At most wavelengths the amount of radiation detected by the detector is equal to that emitted by the source—that is, the molecule does not absorb radiation. At a wavelength (frequency) that corresponds to the energy necessary for a molecular change, the radiation emitted by the source is absorbed by the molecule.

Infrared spectroscopy

Infrared radiation causes certain bonds, especially polar covalent bonds, to vibrate and undergo stretching and bending (Figure S3.85). This provides information about which functional groups are present and is the basis of **infrared spectroscopy**. It can also give information about bond strength.

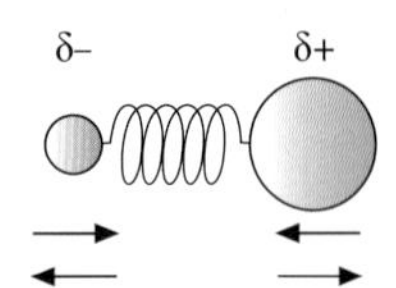

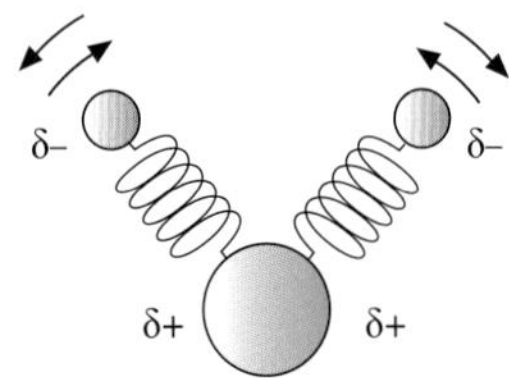

■ **Figure S3.85** Stretching and bending vibrations resulting from the absorption of infrared radiation

Most organic molecules have bonds and functional groups that absorb infrared radiation.

- The frequencies that are absorbed depend on the strength of the bonds in the molecule and the masses of the atoms at each end.
- The fraction of energy absorbed (the absorbance) depends on the change in dipole moment as the bond vibrates. Bonds to electronegative atoms such as oxygen, which are found in alcohols (–O–H) and carbonyl compounds (>C=O), show very strong absorptions.

The absorbed energies can be displayed as an **infrared spectrum**. By analysing this spectrum, we can determine details of a compound's chemical structure. In particular, the spectrum of an organic compound can allow us to identify the functional groups it contains, as each functional group has a characteristic absorption frequency or range of frequencies as shown in Table S3.31 which can also be found in the IB *Chemistry data booklet*.

■ **Table S3.31** Characteristic ranges for infrared absorption due to stretching vibrations in organic molecules

Bond	Types of molecule	Wavenumber / cm^{-1}	Intensity
C–I	iodoalkanes	490–620	strong
C–Br	bromoalkanes	500–600	strong
C–Cl	chloroalkanes	600–800	strong
C–F	fluoroalkanes	1000–1400	strong
C–O	alcohols, esters, ethers	1050–1410	strong
C=C	alkenes	1620–1680	medium-weak; multiple bands
C=O	aldehydes, ketones, carboxylic acids, esters	1700–1750	strong
C≡C	alkynes	2100–2260	variable
O–H	carboxylic acids (with hydrogen bonding)	2500–3000	strong, very broad
C–H	alkanes, alkenes, arenes	2850–3090	strong
O–H	alcohols and phenols (with hydrogen bonding)	3200–3600	strong
N–H	primary amines	3300–3500	medium; two bands

The actual frequency does vary slightly between molecules (and with solvent), so the vibration bands are in a region of the infrared spectrum, rather than at a specific frequency.

There are two main uses for infrared spectra:

- to identify the functional groups present in an organic molecule
- to identify the molecule (since every molecule has a unique infrared spectrum).

Nature of science: Models

Bonds as springs

Models are developed to explain certain phenomena that may not be observable. For example, the interpretation of IR spectra is based on the bond-vibration model.

A covalent bond behaves as if it were a vibrating spring connecting two atoms. The bond in a diatomic molecule such as hydrogen chloride can vibrate by stretching vibration, but bonds in more complex molecules may vibrate by stretching or/and bending.

ATL S3.2C

Visit **www.chemtube3d.com/category/structure-and-bonding/molecular-vibrations-ir/**

Observe the animations and draw annotated diagrams of the stretching and bending vibration modes of the following molecules: CO_2, H_2O, BF_3, C_6H_6, C_2H_4 and HCN.

Outline and present, using presentation software (such as PowerPoint), the vibrational modes of the greenhouse gases, carbon dioxide and water. State and explain clearly which vibrational modes are IR-active and IR-inactive.

◆ **Transmittance**: Ratio of the intensity of exiting light to the intensity of incident light (often expressed as a percentage).

The infrared spectrum of ethanol is shown in Figure S3.86. The horizontal scale represents the wavenumber of the infrared radiation. The vertical axis represents the **transmittance**—the percentage of infrared radiation that passes through the sample. 100% transmittance corresponds to no absorption and 0% transmittance corresponds to total absorption so high absorption shows as a dip.

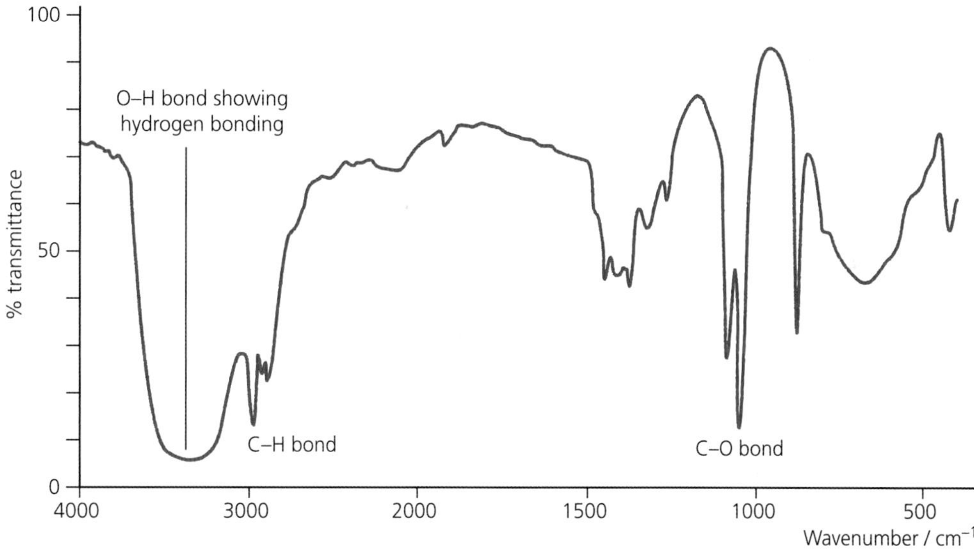

■ **Figure S3.86** The infrared spectrum of ethanol

Top tip!

The wavenumber is equal to the number of wavelengths per centimetre and the unit is referred to as the reciprocal centimetre, cm^{-1}. Larger wavenumbers correspond to higher frequencies and energies.

Top tip!

Absorption bands below 1500 cm^{-1} correspond to vibrations of the entire molecule or part of the molecule. This fingerprint region acts as a unique identifier for a molecule.

The infrared spectrum of ethanol shows the stretching vibrations of the O–H and C–O bonds. The O–H bond absorption is broad (wide) because of hydrogen bonding. The C–O bond vibration occurs at 1050 cm^{-1} and the O–H bond vibration occurs at 3350 cm^{-1}.

General approach to analysing infrared spectra:

- Examine the spectrum from left to right, starting at 4000 cm^{-1}.
- Note which are the strongest absorptions and attempt to match them from Table S3.31.
- Note the absence of peaks in important areas.
- Do not attempt to match all the peaks, especially in the fingerprint region (see Top tip).

Top tip!

The position, intensity, and shape of an absorption band are helpful in identifying functional groups.

Figures S3.87 to S3.89 show the infrared spectra of selected organic compounds. All of them have a strong absorption at around 3000 cm^{-1} due to the carbon–hydrogen bond, C–H. Alcohols have a broad absorption band between 3000 and 3500 cm^{-1} due to the presence of an oxygen–hydrogen bond, O–H. The strong absorption in propanone from approximately 1700 to 1725 cm^{-1} is due to the presence of a carbonyl functional group, >C=O.

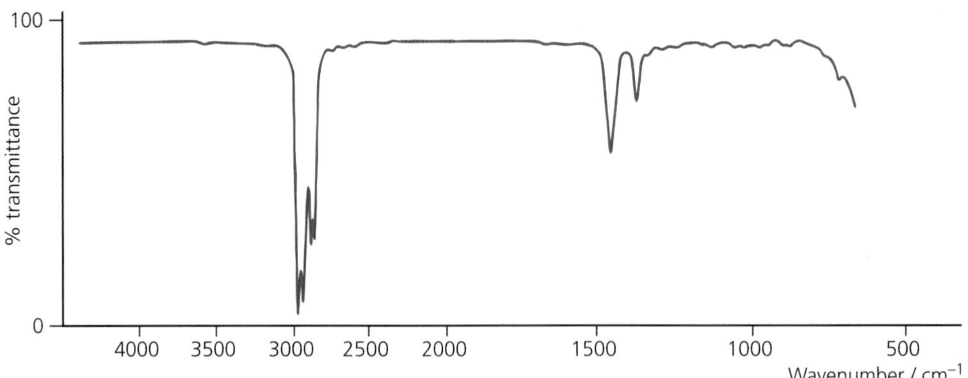

■ **Figure S3.87** IR spectrum of hexane

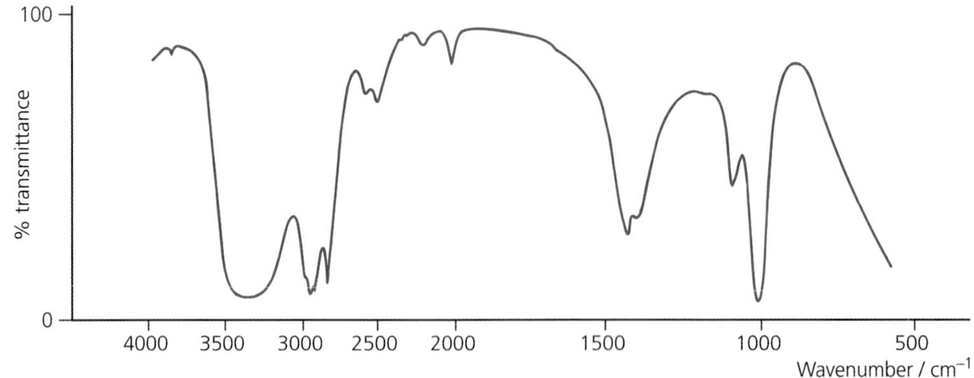

■ **Figure S3.88** IR spectrum of methanol

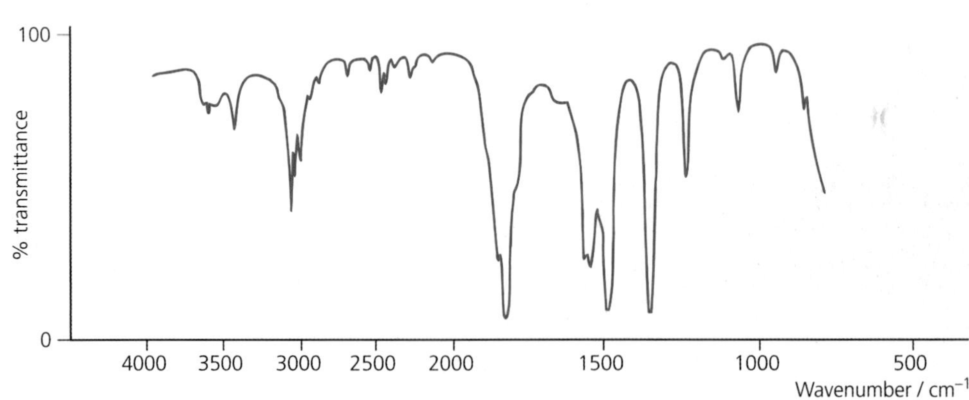

■ **Figure S3.89** IR spectrum of propanone

24 Deduce the number of structural isomers with the molecular formula $C_5H_{10}O_2$ that give infrared absorptions both at approximately 1300 cm^{-1} and at approximately 1740 cm^{-1}.

Draw skeletal formulas for the molecules, knowing there are no absorptions in the region 3000 to 3500 cm^{-1}.

Going further

Infrared spectrometers

An infrared spectrometer measures the extent to which infrared radiation is absorbed by a sample over a particular frequency range of infrared radiation. A schematic diagram of an infrared spectrometer is shown in Figure S3.90.

The source (which is essentially a glowing hot wire) produces infrared radiation over a particular frequency range. The beam of infrared radiation is then split by a beam splitter, a partially coated mirror that reflects half of the infrared radiation and passes the remaining half through a sample and a blank. The blank is identical to the sample except that the substance to be analysed is not present, only the solvent.

The detector compares the relative intensities of bands at each wavelength of the beam leaving the sample tube with those leaving the blank tube. The differences are caused by the sample absorbing infrared radiation. The results are then displayed on a screen or a chart recorder in the form of an infrared spectrum.

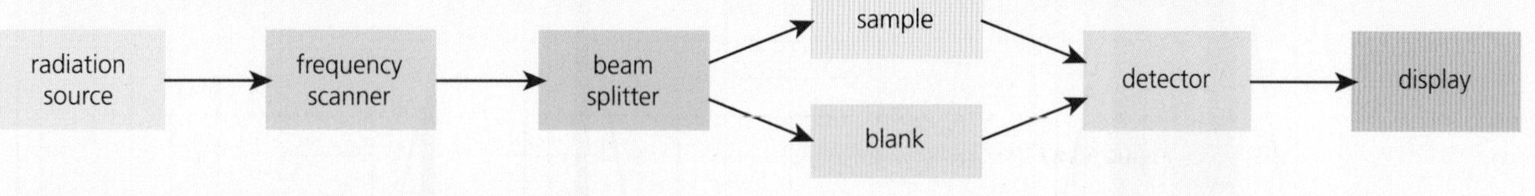

■ **Figure S3.90** Principle of a double-beam infrared spectrometer

◆ **Greenhouse effect**: A heating effect occurring in the atmosphere because of the presence of greenhouse gases that absorb infrared radiation.

◆ **Greenhouse gas**: A gas whose molecules absorb long-wavelength infrared radiation but not ultraviolet or visible light.

Nature of science: Observations

Infrared radiation

One of the first scientists to observe infrared radiation was the German astronomer William Herschel (1738–1822). In the early 19th century, he observed that when he attempted to record the temperature of each colour in visible light that had been dispersed using a prism, there was a marked increase in temperature in the area just beyond red light. However, it was only in the early 20th century that chemists started to investigate how infrared radiation interacted with matter. The first commercial infrared spectrometers were manufactured in the USA during the 1940s. Infrared spectroscopy is used in modern breath testing machines to identify individuals who drive while under the influence of alcohol.

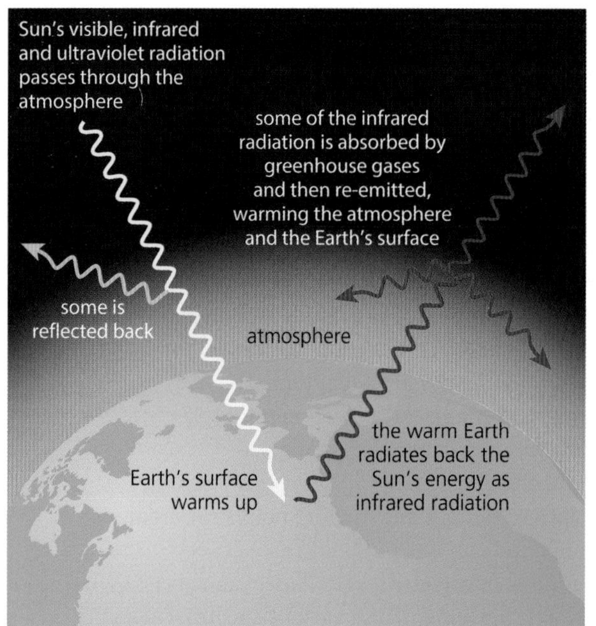

■ **Figure S3.91** The greenhouse effect

Greenhouse gases

The absorption of infrared radiation plays an important role in the atmosphere and has a direct effect on the temperature of the Earth through the **greenhouse effect** (Figure S3.91).

The Sun emits solar radiation and some of this reaches the Earth as sunlight containing electromagnetic radiation in the ultraviolet, visible and short-wavelength (near) infrared regions. Much of this radiation passes through the gases in the atmosphere and warms the surface of the Earth.

The Earth also emits electromagnetic radiation. As the Earth is cooler than the Sun, the radiation emitted by the Earth has a longer wavelength than that from the Sun and is in the long-wavelength (thermal) infrared region.

Some of the gases in the atmosphere absorb infrared radiation but do not absorb the other wavelengths of electromagnetic radiation from the Sun. These are known as **greenhouse gases** and have a significant effect on the temperature of the Earth.

LINKING QUESTION

What properties of a greenhouse gas determine its global warming potential?

Many of these gases occur naturally; they include water vapour, carbon dioxide and methane. Figure S3.92 shows the wavelength (micrometres) range of sunlight and infrared radiation from the Earth and the absorption of infrared radiation by three greenhouse gases.

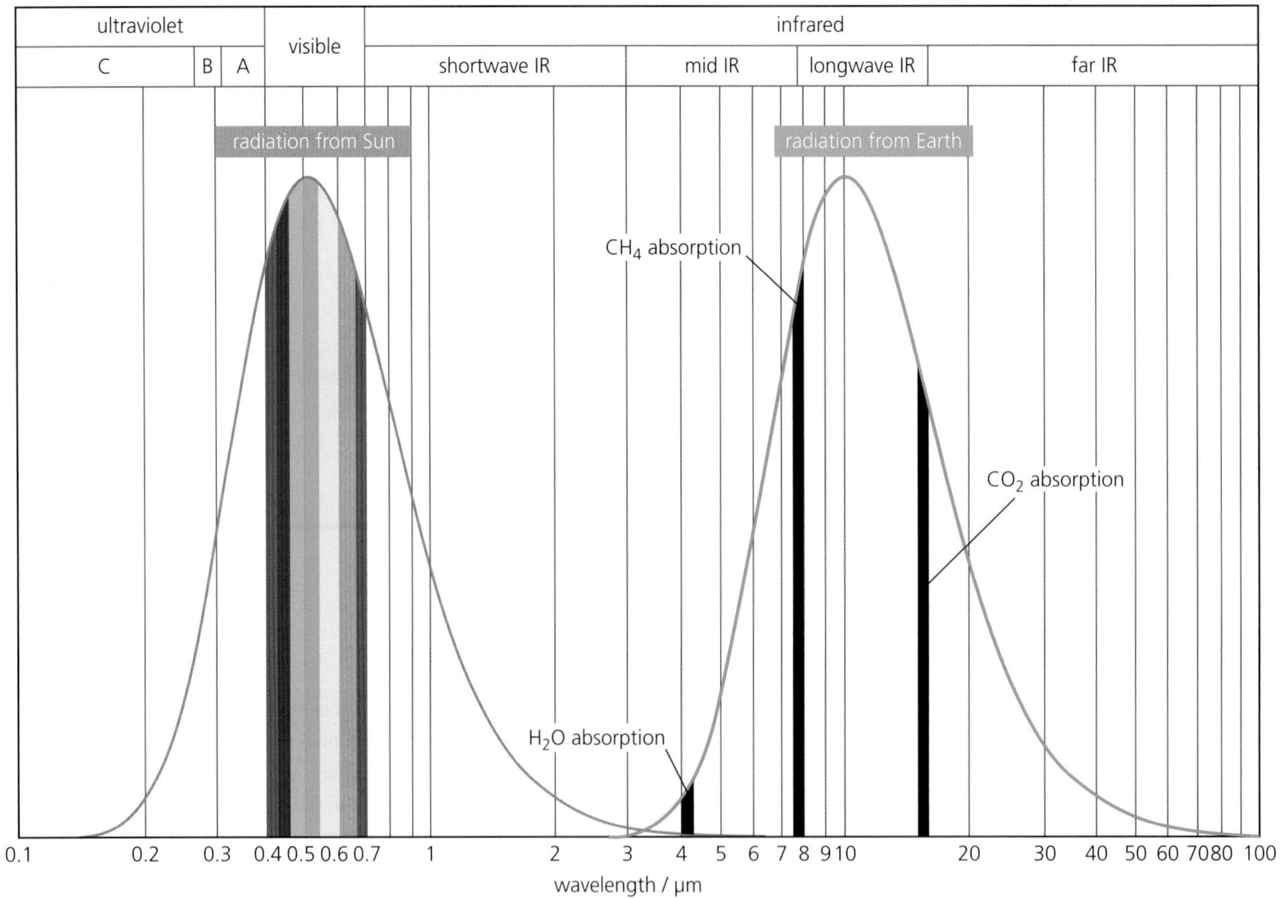

■ **Figure S3.92** The wavelengths of the sunlight from the Sun and the wavelengths of absorption for water, methane and carbon dioxide molecules

Inquiry 2: Collecting and processing data

Interpret diagrams, graphs and charts

Diagrams are often used in chemistry to show information quickly in a visual manner. The most common examples show how to set up equipment for experimental investigations, but there are many others in the IB chemistry syllabus that you may need to draw or interpret including:

- emission spectra
- orbital diagrams
- Lewis formulas
- diagrams, skeletal and structural formulas showing the structure and shapes of molecules and polyatomic ions
- the triangular bonding diagram
- the colour wheel
- reaction mechanism diagrams.

Similarly, graphs and charts are also used for more than displaying results. The most obvious example is the periodic table but others with which you should be familiar include:

- cooling curves
- graphs or charts of periodic properties (for example, ionisation energies / electronegativity) or comparing properties of related organic compounds (for example, within or between homologous series)
- energy (enthalpy) level diagrams
- Born–Haber and Hess' law energy cycles
- concentration–time plots
- Arrhenius plots
- graphs of boiling and melting points for organic compounds
- diagrams of organic reaction mechanisms (showing electron movement) and various spectra.

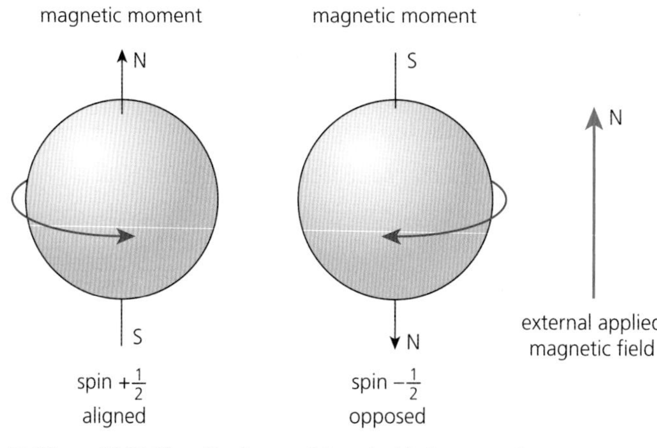

■ **Figure S3.93** The +½ spin state (aligned with the external magnetic field) is lower in energy and the −½ spin state (opposed to the external magnetic field) is higher in energy

Nuclear magnetic resonance

The subatomic particles that make up atoms may be regarded as spinning charged particles. Hydrogen or proton nuclear magnetic resonance (^{1}H NMR or just NMR) depends on the behaviour of hydrogen-1 atoms or, more specifically, their nuclei. This isotope has a single proton and an electron, with no neutrons in the nucleus.

The proton (when viewed as a particle) can spin clockwise or anticlockwise and generate a magnetic field, represented in diagrams by an arrow. When there is no applied magnetic field, the **nuclear spins** of the proton are of the same energy and a sample of hydrogen-1 atoms (at equilibrium) will have equal numbers of both spin states.

◆ **Nuclear spin (proton):** The property of the proton that generates a magnetic field.

The property of nuclear spin makes the nuclei (protons) of hydrogen-1 atoms behave like tiny bar magnets. When they are placed in a strong magnetic field they line up with the magnetic field (in an almost parallel manner) or against it (antiparallel) (Figure S3.94). Proton spins parallel to / aligned with the external magnetic field (lighter, pink spheres) have lower energy. Spins antiparallel or opposed to the magnetic field (darker, green spheres) have higher energy.

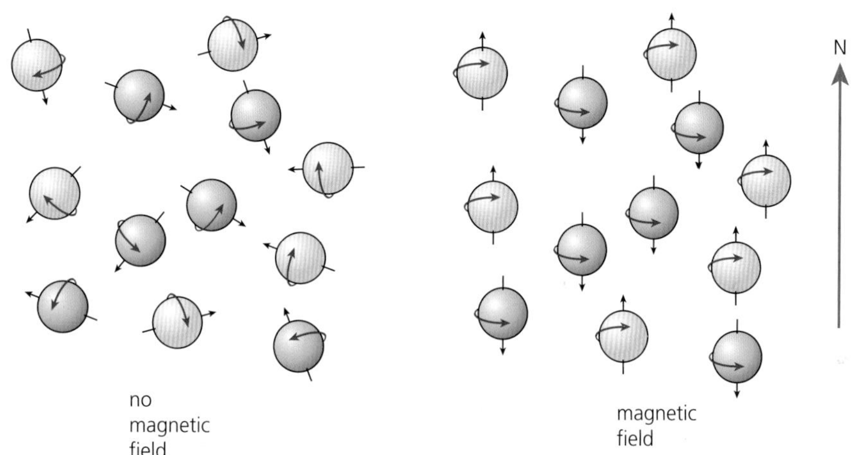

■ **Figure S3.94** The orientation of protons in the absence and presence of a strong magnetic field

The two different spin states give rise to two energy levels with a very small energy gap (Figure S3.95). The ±½ represents the spin quantum number, which distinguishes between the two spin states.

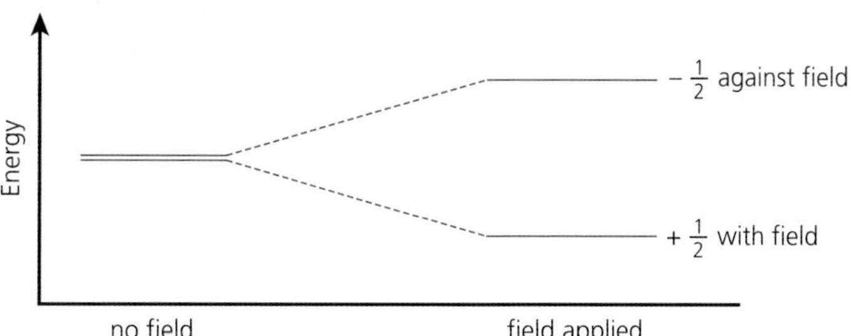

■ **Figure S3.95** The spin states of a proton in the absence and the presence of an applied magnetic field

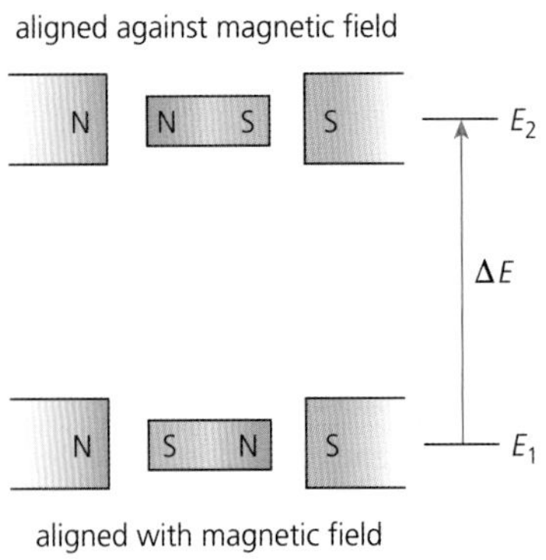

aligned against magnetic field

N N S S E_2

ΔE

N S N S E_1

aligned with magnetic field

■ **Figure S3.96** The principle of nuclear magnetic resonance

Consider the arrangement of magnets shown in Figure S3.96. The top magnet, aligned against the magnetic field, could spontaneously flip round to align itself with the field (if there are no other forces acting). However, energy must be supplied (a force applied) to flip the bottom magnet into the position of the top magnet.

Similarly, if the protons (hydrogen nuclei) in a magnetic field are irradiated with a frequency of radio waves equal to the energy gap between the two energy levels, then the hydrogen nuclei that are aligned with the magnetic field will move into the higher energy state. They are said to have 'flipped over' and undergone spin resonance. This is the basis of **nuclear magnetic resonance**.

Top tip!

^{1}H NMR is often called proton NMR as it is the hydrogen nuclei (protons) which are involved.

◆ **Nuclear magnetic resonance**: The absorption of radio waves at a precise frequency by a proton in an external magnetic field.

The basics of a traditional nuclear magnetic resonance spectrometer are shown in Figure S3.97. The sample is held in a spinning tube between the poles of a powerful electromagnet.

An NMR spectrum was traditionally acquired by varying or sweeping the magnetic field over a small range while observing the radio frequency signal from the sample. An equally effective technique is to vary the frequency of the radiation while holding the external magnetic field constant.

The sample is surrounded by radio-frequency electromagnetic radiation which is detected by a receiver. A computer or recorder linked to the receiver shows the proportion of the radiation absorbed at a given magnetic field strength and radio frequency and, hence, the strength of the magnetic field needed to make the protons in the substance undergo spin resonance (flip over).

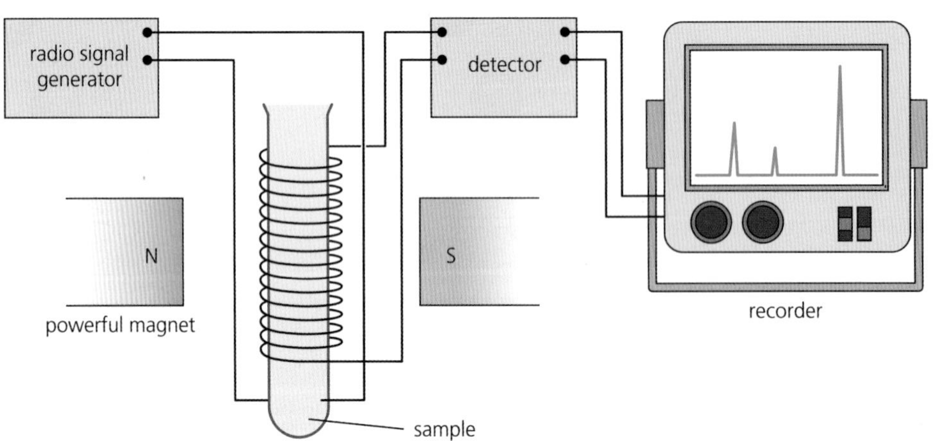

■ **Figure S3.97** The basic features of an NMR spectrometer

The greater the electron density around a hydrogen atom (proton), the stronger the magnetic field that is required for the proton in the nucleus to undergo spin resonance and 'flip' from the low energy, spin aligned state to the higher energy spin opposed state.

Nuclei with a higher electron density around them are better **shielded** from the external magnetic field.

◆ **Shielding (magnetic)**: Electrons around a nucleus create their own local magnetic fields and thereby shield the nucleus from the applied magnetic field.

◆ **Chemical environment**: The number and types of atoms to which a particular atom within a molecule is bonded.

NMR spectra

Chemical environments

This effect is illustrated in Figure S3.98 showing the NMR spectrum of ethanol. In this molecule, there are hydrogen atoms with three different levels of shielding. These different levels are dependent on the **chemical environment** of the hydrogen atom.

In benzene, all six hydrogen atoms (protons) are in the same chemical environment and therefore the NMR spectrum has only one signal (peak).

Top tip!

Butane, CH_3–CH_2–CH_2–CH_3, has hydrogen atoms in two different chemical environments, not four. This is due to symmetry: the two –CH_3 groups on the ends are in identical chemical environments, as are the two –CH_2– groups in the middle.

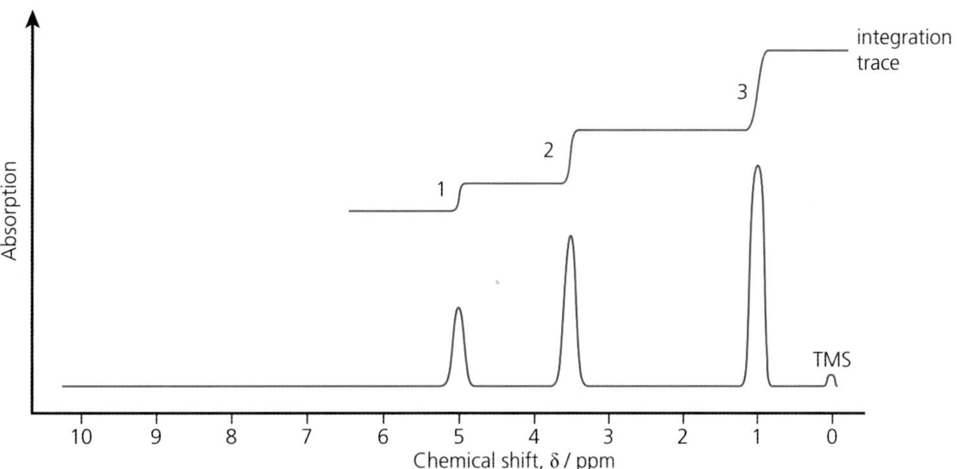

■ **Figure S3.98** The low-resolution ^{1}H NMR spectrum of anhydrous ethanol, with the integration trace. The *x*-axis shows the magnetic field increasing from left to right.

The single peak at low magnetic field represents the –O–H hydrogen which has little shielding. The peak at a slightly higher magnetic field represents the –CH_2– hydrogens, which have more shielding. The final peak represents the –CH_3 hydrogens, which have the most shielding as they are furthest away from the electronegative oxygen and so flip at highest magnetic field.

WORKED EXAMPLE S3.2C

Deduce how many chemical environments and hence how many peaks are in the NMR spectrum for the following molecules: $CH_3(CH_2)_4OH$ and $(CH_3)_3OH$. Use circles and different colours to indicate the different chemical environments.

Answer

$CH_3(CH_2)_4OH$: six chemical environments; six peaks in the NMR spectrum

The electronegative oxygen atom draws electron density towards it which deshields the adjacent –CH_2– group to a significant extent. This deshielding effect weakens along the chain from right to left, meaning the four –CH_2– groups and the terminal –CH_3 group are each in a different chemical environment.

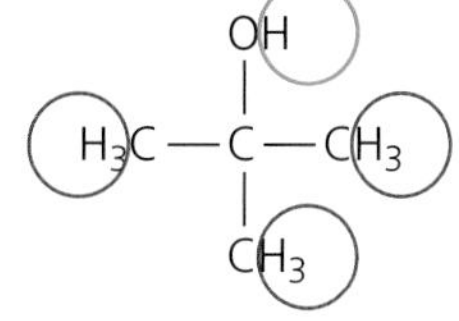

$(CH_3)_3OH$: two chemical environments; two peaks in the NMR spectrum

Due to the tetrahedral symmetry of the molecule, the three –CH_3 groups occupy identical chemical environments.

25 Draw displayed structural formulas for the molecules below.

For each molecule, determine the number of different chemical environments and hence the number of signals in the NMR spectrum.

a $CH_3CHClCH_2CH_3$
b $C(CH_3)_3CCl$
c $CHO(CH_2)_3CHO$
d $CH_3COCH_2COCH_3$
e CH_3OCH_3
f $HOCH_2COOH$

Integration traces

◆ **Integration trace:** The area under an NMR resonance peak, which is proportional to the number of hydrogen atoms which that peak represents.

◆ **Chemical shift**: The position of a signal on an NMR spectrum relative to that of the signal of tetramethylsilane (measured in parts per million).

An NMR spectrum also includes an **integration trace**. The height of each step is a measure of the area under the peak beneath; this is proportional to the number of hydrogen atoms (protons) undergoing spin resonance at this point in the NMR spectrum. In the case of ethanol (Figure S3.98), there are three steps, which are in the ratio of 1 : 2 : 3 (from left to right).

Top tip!

NMR can give structural information about the relative numbers of different types of hydrogen atoms (chemical environments) and where they are in relation to electronegative atoms.

Tool 3: Mathematics

Ratios

A ratio shows the quantitative relation between two or more amounts. For example, if the areas were 0.7 cm^2, 1.4 cm^2 and 2.1 cm^2 in a proton NMR spectrum, the ratio of the peak areas (as integers) would be 1 : 2 : 3. Ratios are similar to fractions; both can be simplified by finding common factors and then dividing by the highest common factor. Ratios are used in many calculations in chemistry, for example, in deducing empirical formulas, in calculating reacting masses and in balancing equations.

Chemical shifts

The sample undergoing NMR is referenced against a substance called tetramethylsilane (TMS). It has only one proton environment and is assigned a value of 0 ppm (parts per million). It is the peak on the far right of the NMR spectrum.

Top tip!

TMS, ($Si(CH_3)_4$), is non-toxic, unreactive and readily removed after analysis by evaporation.

The difference in magnetic field strength between that required for the protons in TMS and that of the protons in other chemical environments to undergo spin resonance is called the **chemical shift**. It is measured in ppm and given the symbol δ. Values for chemical shifts for different proton environments are given in Table S3.32 and in the IB *Chemistry data booklet*.

■ **Table S3.32** ^{1}H NMR data

Type of proton	Chemical shift / ppm
$-CH_3$	0.9–1.0
$-CH_2-R$	1.3–1.4
$-CHR_2$	1.5
$RO-C(=O)-CH_2-$	2.0–2.5
$R-C(=O)-CH_2-$	2.2–2.7
$-C_6H_4-CH_3$ (benzene ring–CH_3)	2.5–3.5
$-C\equiv C-H$	1.8–3.1
$-CH_2-Hal$	3.5–4.4
$R-O-CH_2-$	3.3–3.7
$R-C(=O)-O-CH_2-$	3.7–4.8
$R-C(=O)-O-H$	9.0–13.0
$R-O-H$	1.0–6.0
$-CH=CH_2$	4.5–6.0
$-C_6H_4-OH$ (benzene ring–OH)	4.0–12.0
$-C_6H_4-H$ (benzene ring–H)	6.9–9.0
$R-C(=O)-H$	9.4–10.0

Top tip!

A chemical shift of 1 ppm means that the radio frequency radiation required to cause spin resonance (the flip of a proton from one spin state to the other) differs in frequency from TMS by 1 part per million.

By comparing the chemical shift data for the three peaks in the NMR spectrum of ethanol with the values given in Table S3.32 it is possible to identify the structural units in the ethanol molecule (Table S3.33).

Top tip!

The higher the ppm value, the further downfield the peak, and hence the weaker the magnetic field. TMS is very upfield and hence corresponds to a strong magnetic field.

■ **Table S3.33** Low-resolution NMR chemical shift data for ethanol

Chemical shift, δ	Integration ratio	Structural feature
5.0	1	H in R–O–**H**
3.6	2	H in R–CH_2–OH
1.0	3	H in R–CH_3

26 The low-resolution ^{1}H NMR spectrum of compound A with the molecular formula $C_4H_8O_2$ is shown in the diagram. Suggest the structure of A.

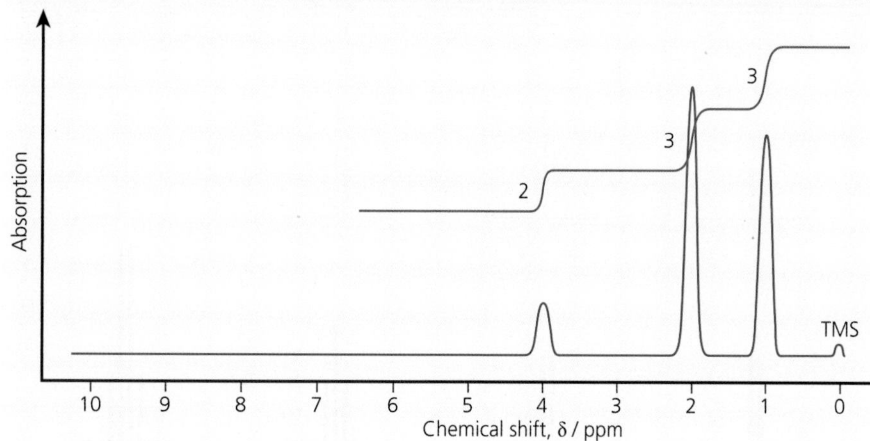

HL ONLY

High-resolution NMR

◆ **High-resolution NMR:** NMR performed in the presence of a strong and uniform magnetic field so that splitting can be observed.

◆ **Splitting:** The interaction of the magnetic fields of two or more hydrogen atoms (protons) that converts an absorption signal (peak) into a multiplet (a set of smaller signals/peaks).

◆ ***n* + 1 rule:** Absorption by *n* equivalent protons on adjacent carbon atoms produces a multiplet with *n* + 1 individual peaks.

In a **high-resolution NMR** spectrum (collected in the presence of a very strong and homogeneous magnetic field), many signals that were single in the low-resolution spectrum are split into clusters of signals.

The **splitting** occurs because the magnetic field experienced by the hydrogen atoms (protons) of one group is influenced by the spin arrangements of the hydrogen atoms (protons) in an adjacent group. (This effect involving nuclei is known as spin–spin coupling.)

The amount of splitting indicates the number of hydrogen atoms bonded to the carbon atoms immediately next to the one you are currently interested in.

The number of signals in a cluster is one more than the number of hydrogens attached to the *adjacent* carbon atom(s) (***n* + 1 rule**) (Figure S3.99) where *n* is the number of hydrogen atoms attached to the adjacent carbon atom(s):

- One adjacent hydrogen atom splits a signal (peak) into two equal signals (peaks) of the same height called a doublet.
- Two adjacent hydrogen atoms split a signal (peak) into a triplet of height ratio 1 : 2 : 1.
- Three adjacent hydrogen atoms split a signal (peak) into a quartet of height ratio 1 : 3 : 3 : 1.

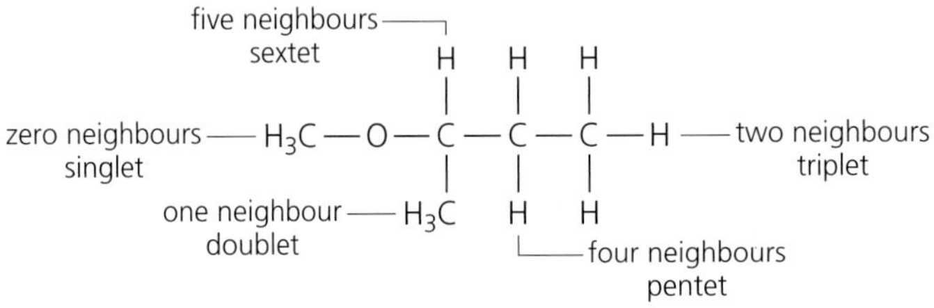

Top tip!

Peak height ratios can be obtained from the appropriate row of Pascal's triangle.

five neighbours sextet
zero neighbours singlet — H_3C—O—C—C—C—H — two neighbours triplet
one neighbour doublet — H_3C
four neighbours pentet

■ **Figure S3.99** *n* + 1 rule

Table S3.34 and Figure S3.100 summarize some commonly observed splitting patterns.

■ **Table S3.34** Peak clusters in a high-resolution ^{1}H NMR spectrum

Number of hydrogens on carbon adjacent to resonating hydrogen	Number of lines in cluster (multiplet)	Relative intensities
1	2	1:1
2	3	1:2:1
3	4	1:3:3:1
4	5	1:4:6:4:1

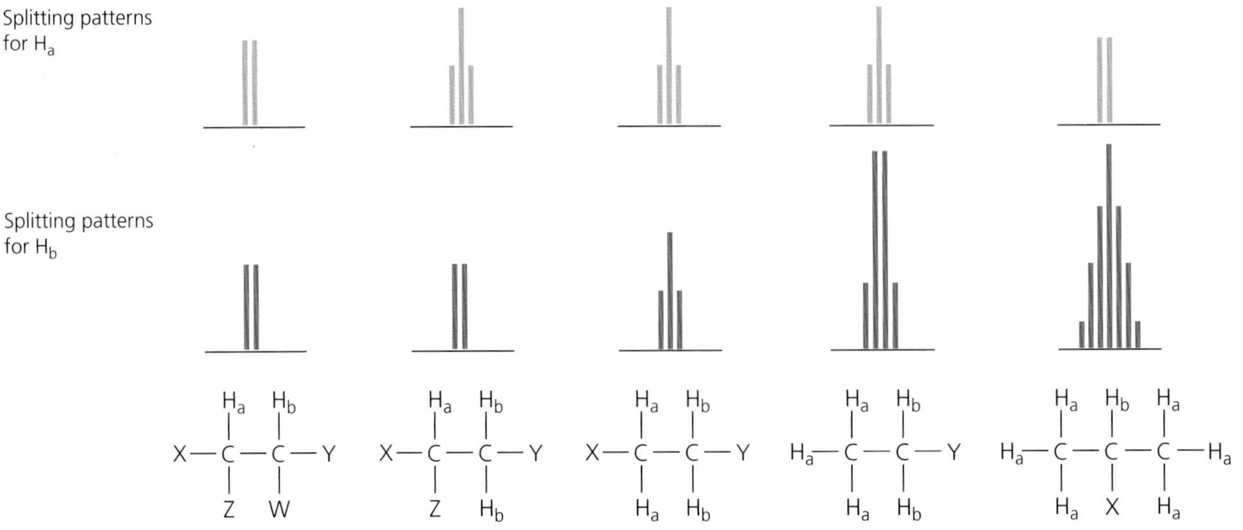

Splitting patterns for H_a

Splitting patterns for H_b

X—C(H_a)(Z)—C(H_b)(W)—Y   X—C(H_a)(Z)—C(H_b)(H_b)—Y   X—C(H_a)(H_a)—C(H_b)(H_b)—Y H_a—C(H_a)(H_a)—C(H_b)(H_b)—Y H_a—C(H_a)(H_a)—C(H_b)(X)—C(H_a)(H_a)—H_a

■ **Figure S3.100** Commonly observed splitting patterns

Figure S3.101 shows the high-resolution ^1H NMR spectrum of ethanol in the presence of a small amount of water or acid.

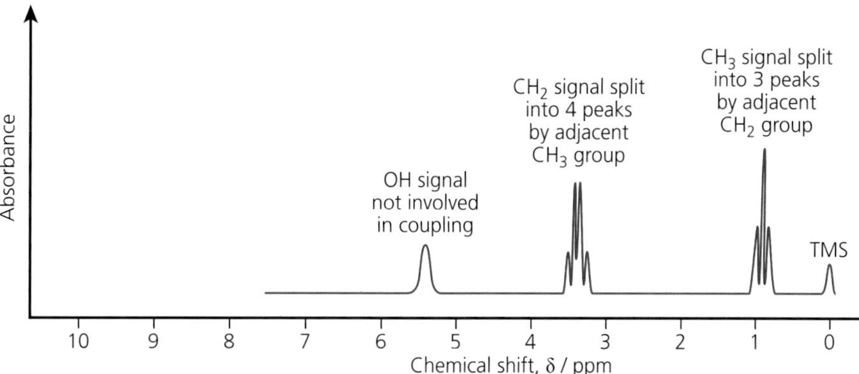

■ **Figure S3.101** High-resolution NMR spectrum of ethanol, CH_3–CH_2–OH

Summary of steps in determining structure from high resolution ^1H NMR:

- The number of signals gives the number of different chemical environments in which hydrogen atoms are present in the compound.
- The chemical shifts give possible functional groups, for example, phenyl (which has a relatively high chemical shift), and hybridization of the carbon.
- The integration of each signal gives the number of hydrogen atoms in that environment: 3 implies a $-CH_3$ group, 2 implies a $-CH_2-$ group and 1 implies CH or OH.
- The splitting of each signal gives information about what is bonded to a given carbon atom:
  - n lines corresponds to $(n - 1)$ H neighbours (when working from the NMR spectrum to structure)
  - nH neighbours corresponds to $(n + 1)$ lines (when you know the structure and you are trying to predict what the NMR spectrum should look like).

Going further

Explaining splitting

The theory which explains splitting patterns explains the intensities of the peaks in the splitting pattern. The basic principle is that the protons on adjacent carbon atoms can either add to the strength of the external magnetic field or they can oppose it, depending on their spin state (Figure S3.102).

Each adjacent proton (hydrogen atom) has its own spin state (either +½ or –½), so increasing numbers of protons mean there are more possible permutations for the distribution of those states. This in turn leads to more complex splitting patterns as the protons in different molecules experience a range of different local magnetic fields.

If a given hydrogen atom, denoted by H_a, has two sets of adjacent non-equivalent hydrogen atoms, the splitting pattern due to two different sets will operate independently as shown in Figure S3.103.

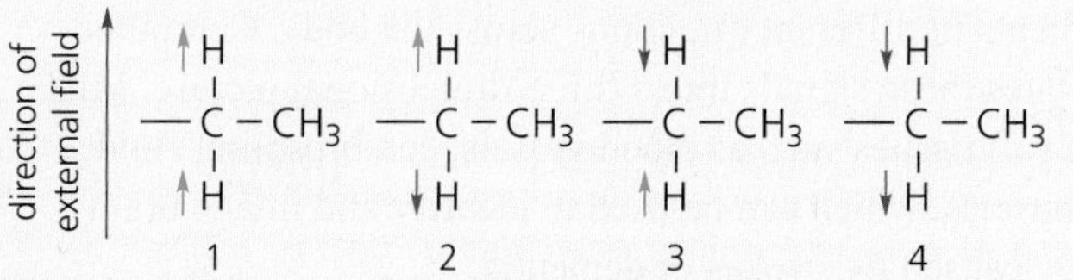

■ **Figure S3.102** The directions of the magnetic moments of the $-CH_2-$ hydrogens have an effect on the $-CH_3$ hydrogens

■ **Figure S3.103** Splitting pattern due to two different sets of adjacent hydrogen atoms

Combined techniques

Table S3.35 provides a comparison of IR spectroscopy and NMR spectroscopy showing the key features of each technique.

■ **Table S3.35** A comparison of IR spectroscopy versus NMR spectroscopy

	IR spectrosopy	**NMR spectroscopy**
Type of radiation	Infrared.	Radio waves.
Cause of absorption	Vibration of a bond with a dipole that undergoes a change in dipole moment at the same frequency as the radiation.	Transition (flip) in proton spin state.
What the key features of the spectrum represent	Bands represent the vibration of specific covalent bonds.	Peaks represent hydrogen atoms in specific chemical environments.
***x*-axis**	Shows wavenumber (cm^{-1}).	Shows chemical shift (in ppm).
Information given about an organic compound	Functional groups present.	Structure of the compound.

How do the tools that we use shape the knowledge that we produce?

Infrared spectroscopy (IR) permits the detection of certain key functional groups and can measure bond strengths. However, rather like ultraviolet spectroscopy, it has been overtaken by more precise structural analytical techniques such as mass spectrometry (MS), nuclear magnetic resonance (NMR) and X-ray crystallography.

MS gives information about molecular mass and key fragmentations give subunit information.

NMR gives structural information and information about functional groups.

^{1}H NMR is different from other spectroscopic techniques in that it gives three pieces of information. Just as with IR and UV, two measurements (the chemical shift and integration) give information about the species absorbing the electromagnetic radiation. The third measurement (multiplicity) gives information about neighbouring hydrogen atoms (protons) and this allows the connectivity of the molecule to be determined. However, NMR is less sensitive than IR which can detect molecules at lower concentrations with shorter lifetimes.

X-ray analysis allows the definitive assignment of crystalline structures in the thermodynamically most stable form. However, this is not necessarily the form in which the substance reacts in solution. X-ray crystal structure determinations are non-destructive but pure crystals are needed and some substances do not crystallize readily. Crystallization introduces new intermolecular interactions and crystal packing effects which can change the conformation of the molecule.

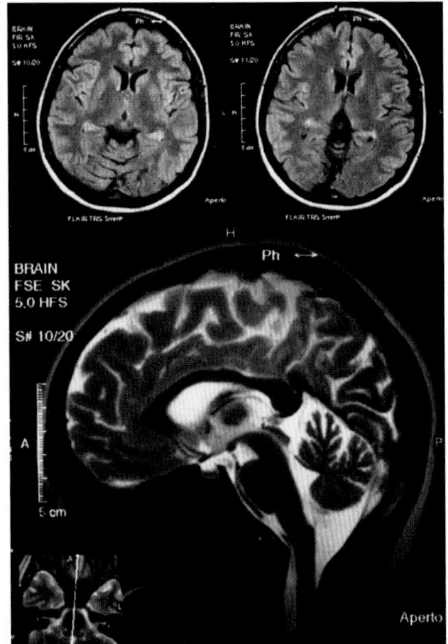

■ **Figure S3.104** MRI scans of the brain

Going further

Magnetic resonance imaging (MRI)

Magnetic resonance imaging (MRI) uses NMR for medical diagnosis. The patient is placed inside a cylinder that contains a very strong magnetic field. Radio waves then cause the protons in the hydrogen atoms in the water molecules of the body to resonate. Each type of body tissue emits a different signal, reflecting the different hydrogen density (water content) of the tissue. The three-dimensional information from MRI is derived by applying magnetic field gradients in different directions across the body. Computer software then translates these signals into a three-dimensional picture. MRI produces images of soft tissues such as blood vessels, cerebrospinal fluid, bone marrow and muscles, which can be used to identify and image brain tumours and damage caused by strokes or dementia.

NMR is a very powerful technique for determining the structure of organic compounds but is often used in conjunction with other analytical techniques to confirm the structure of an unknown organic compound.

1 The first step in determining the structure of an unknown organic compound is to establish which chemical elements are present and their percentages by mass to determine the empirical formula. This is usually done by combustion analysis.
2 The molecular (parent) ion can be identified from a mass spectrum. This can be compared with the empirical formula to establish the molecular formula.
3 The functional groups present can be established from the infrared spectrum and chemical reactions.
4 The NMR spectrum and the fragmentation pattern of the mass spectrum will establish or confirm the structural formula.

WORKED EXAMPLE S3.2D

Use the data below to deduce the formula and structure of compound X.

Elemental analysis of compound X gave the following percentage composition by mass: carbon (66.77%), hydrogen (11.11%) and oxygen (22.22%).

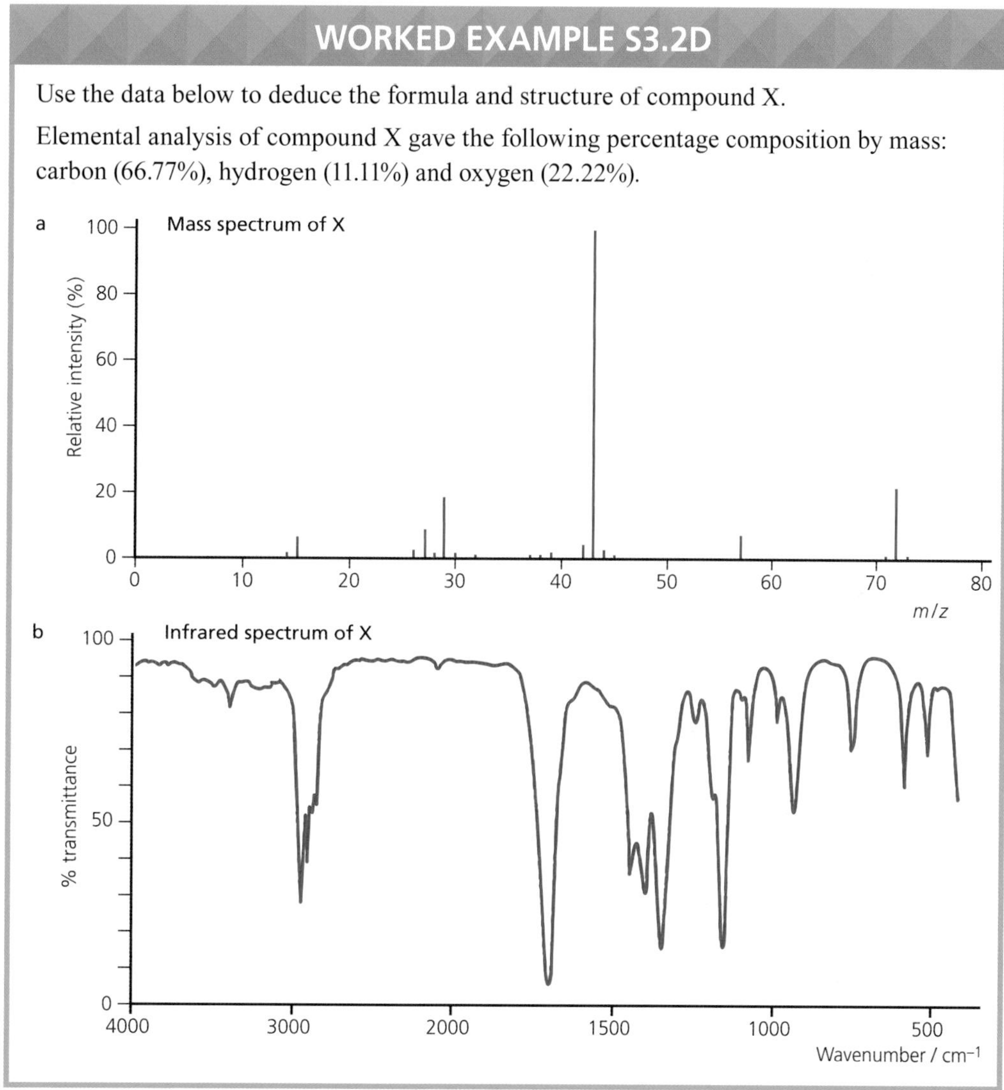

c

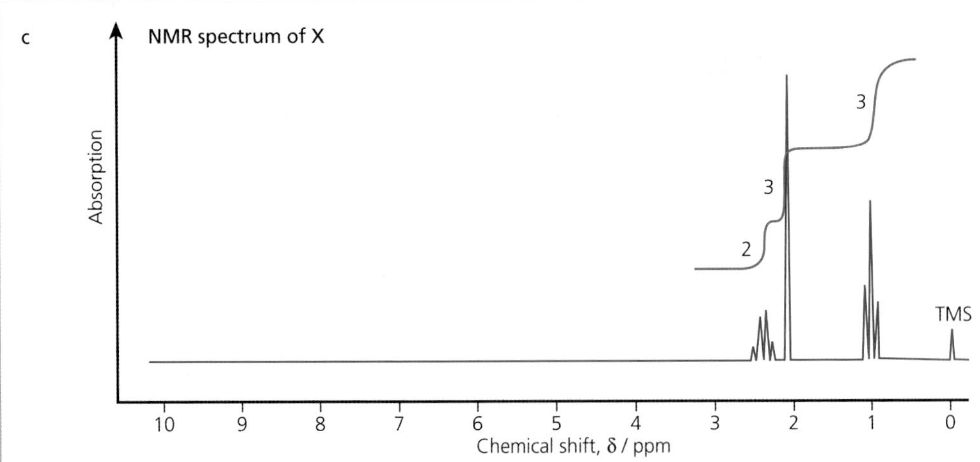

Answer

The empirical formula can be determined using the relative atomic masses.

carbon: $\frac{66.77\,g}{12.01\,g\,mol^{-1}} = 5.560\,mol$

hydrogen: $\frac{11.11\,g}{1.01\,g\,mol^{-1}} = 11.00\,mol$

oxygen: $\frac{22.22\,g}{16.00\,g\,mol^{-1}} = 1.389\,mol$

Integer ratio of amounts of atoms: 4 : 8 : 1

Hence, the empirical formula of X is C_4H_8O.

The mass spectrum shows a molecular ion at 72. This means that the molecular formula is C_4H_8O.

The infrared spectrum shows a strong absorption band centred around approximately $1720\,cm^{-1}$ and an absence of absorption bands in the range 3000 to $3500\,cm^{-1}$. This is due to the presence of a carbonyl group from an aldehyde or ketone (so only one oxygen atom is present and it cannot be an ester or carboxylic acid). The bands in the region of $2900\,cm^{-1}$ suggest the presence of several C–H bonds.

The high-resolution ^1H NMR spectrum of compound X shows three sets of signals (peaks). The triplet of peaks is from the protons of a methyl group, $-CH_3$, that has been split by the protons of an adjacent methylene group, $-CH_2-$. The single peak is due to a methyl group that is adjacent to a carbonyl group. The quartet of peaks is due to the protons of a $-CH_2-$ group split by an adjacent $-CH_3$ group.

The combined analytical information suggests that X is butanone.

```
δ = 2.1     δ = 2.5  δ = 1
   H          H     H
   |          |     |
H— C —  C  —  C  —  C —H
   |    ||    |     |
   H    O     H     H
```

Measuring enthalpy changes

Guiding question

- What can be deduced from the temperature change that accompanies chemical or physical change?

SYLLABUS CONTENT

By the end of this chapter, you should understand that:

- chemical reactions involve a transfer of energy between the system and the surroundings, while total energy is conserved
- reactions are described as endothermic or exothermic, depending on the direction of energy transfer between the system and the surroundings
- temperature changes accompany endothermic and exothermic reactions
- the relative stability of reactants and products determines whether reactions are endothermic or exothermic
- the standard enthalpy change for a chemical reaction, $\Delta H^{\ominus}$, refers to the heat transferred at constant pressure under standard conditions and states; it can be determined from the change in temperature of a pure substance.

By the end of this chapter, you should know how to:

- understand the difference between heat and temperature
- understand the temperature change (decrease or increase) that accompanies endothermic and exothermic reactions, respectively
- sketch and interpret energy profiles for endothermic and exothermic reactions
- apply the equations $Q = mc\Delta T$ and $\Delta H^{\ominus} = \frac{-Q}{n}$ in the calculation of the enthalpy change of a reaction.

There is no higher-level only material in R1.1.

◆ **Heat**: The form of energy transferred between two objects due to a temperature difference between them; the origin of heat energy is the movement of atoms, ions and molecules, that is, their kinetic energy.

◆ **Thermal equilibrium**: The condition reached when there is no overall transfer of energy, as heat, between two objects in contact with one another.

Heat and temperature

Thermodynamics is a field of study which aims to define what is happening in chemical and physical processes and, more generally, the links between energy, heat, temperature and work.

Heat

When any hot object is placed in contact with a colder object, thermal energy is transferred from the hot to the cold object. The thermal energy that is transferred from one object to another because of a temperature difference between them is called **heat**. Although heat is not a fluid, we say that heat flows from the hotter object to the cooler object until they are both at the same temperature. The two objects are then said to be in **thermal equilibrium**. The direction of spontaneous energy transfer is always from a warmer object to a cooler object (Figure R1.1).

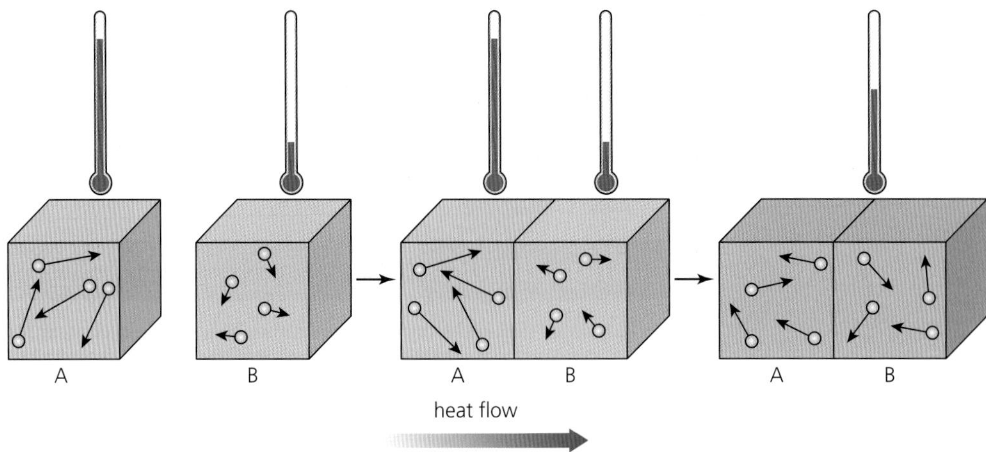

■ **Figure R1.1** Two objects reaching thermal equilibrium

Temperature

The temperature of a substance is related to the average kinetic energy of the particles that form the substance. When recording the temperature of a sample, you are indirectly measuring the average kinetic energy of its particles.

The absolute temperature scale (also known as the Kelvin or thermodynamic scale), which begins at absolute zero, was introduced in S1.1. On this scale, the average kinetic energy of the particles is directly proportional to the temperature.

LINKING QUESTION

What is the relationship between temperature and kinetic energy of particles?

Top tip!

The Fahrenheit scale of temperature (°F) is widely used in the United States. The number 32 indicates the temperature at which water freezes and the number 212 indicates the temperature at which water boils. The Fahrenheit scale is not part of the metric system. It is named after the German physicist Gabriel Fahrenheit (1686–1736).

Tool 1: Experimental techniques

Measuring temperature

Temperature is measured with a thermometer. Analogue thermometers consist of a liquid enclosed in glass, and these have a level of precision equal to half the smallest division on the scale. Digital thermometers usually have a precision of ±1 in the last digit. A temperature probe may be coupled with a data logger to record many temperature measurements.

The most appropriate analogue thermometer to use in a particular experiment depends on the range of temperatures to be measured. Thermometers that measure a smaller range will usually allow more precise readings.

If you are measuring a temperature change in an enthalpy experiment, be aware that the temperature will keep rising or falling after you have finished your experiment. This is due to the lag time in terms of the heat transferred from the surroundings to the thermometer. You should therefore always keep recording the time until the maximum or minimum temperature change is recorded.

Common mistake

Misconception: *A doubling of the temperature in degrees Celsius is a doubling of the absolute temperature.*

A doubling of the temperature in degrees Celsius is not a doubling of the absolute temperature. For example, a doubling of the temperature from 200 °C to 400 °C is a rise from (200 + 273) = 473 K to (400 + 273) = 673 K. The final temperature is only 673 K ÷ 473 K = 1.42 times the starting temperature.

Table R1.1 summarizes the differences between heat and temperature.

■ **Table R1.1** A comparison of temperature and heat

Temperature	**Heat**
Absolute temperature is directly proportional to the average kinetic energy of the particles in the measured sample.	The energy transferred from a hotter body to a colder body.
Always positive on the absolute scale; may be positive or negative on other scales.	Positive if net gain (+*Q*), negative if net loss (−*Q*).
Units include kelvin (SI base unit), Celsius, Fahrenheit.	Units include joule (SI base unit), kilojoule, calorie.
Measured directly (using a thermometer).	Measured indirectly; calculated using mass, thermal properties of material(s) and, where relevant, temperature change.

Nature of science: Theories

Caloric

At the beginning of the 18th century, a Dutch medical doctor named Boerhaave (1668–1738) described heat as a fluid that passed from one substance to another. This idea was taken up by Joseph Black (1728–1799) in 1760 when he was working in Glasgow University. Black suggested that an invisible fluid, which he called *caloric*, flowed into objects that were being heated and that the amount of caloric possessed by a substance was dependent on its mass and temperature.

An American, Benjamin Thompson (1753–1814), known as Count Rumford, suspected that there was a link between heat and work. Watching horse-powered drills being used to bore cannons, he noticed that the metal chips produced from boring the cannon became very hot. He also observed that a blunt borer did not bore the hole as quickly. He used a blunt borer to drill out cannons submerged in a barrel of water. The water boiled and, when the cannon and the chips were examined, there was no change in their total mass. The same cannon, now lighter, could be used to boil another barrel of water: it seemed to have an unlimited amount of caloric. This convinced Thompson that the caloric produced was not a substance.

The non-SI unit of energy, the calorie (cal), derives its name from this early theory on heat transfer. The energy content of food items is often given on packaging in kcal (1 cal = 4.2 joules so 1 kcal = 4200 joules).

Conservation of energy

The law of conservation of energy states that the amount of energy in the universe remains constant: energy is neither created nor destroyed but can be converted from one form to another. The amount of energy in the universe at the end of a chemical reaction is, therefore, the same as at the beginning.

In chemistry, we are interested in the transfer of energy between a chemical system and its surroundings. The system is the part of the universe that is of interest; that is, the sample of matter or reaction that is being studied. Everything outside the system is referred to as the surroundings.

Figure R1.2 shows how the chemical potential energy of zinc and hydrochloric acid is converted into heat in the surroundings (in this case, the water that is the solvent in an aqueous solution). The change in the chemical potential energy of the system equals the change in heat energy in the surroundings. This agrees with the law of conservation of energy.

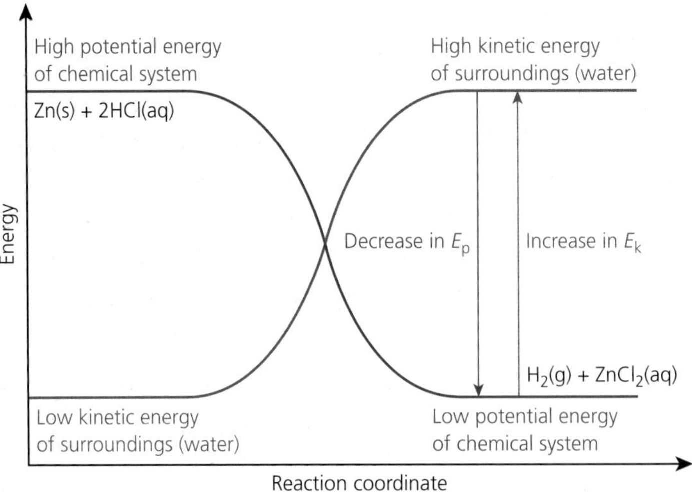

■ **Figure R1.2** The law of conservation of energy

Going further

Thermodynamic systems

Thermodynamics defines different types of system (Figure R1.3):

- An **open system** is one where energy and matter can be interchanged with the surroundings. A beaker of boiling water is an open system. When it is placed over a Bunsen burner, heat energy can enter the system from the flame, and water is slowly released into the surroundings as steam.
- A **closed system** can exchange energy, but not matter, with the surroundings. A stoppered flask is an example of a closed system. No matter can be transferred out of the flask, but heat can be conducted through the glass walls.
- An **isolated system** cannot transfer matter or energy to its surroundings. This means its internal energy (the sum of the kinetic energies and intermolecular potential energies of all the molecules inside it) remains constant. A vacuum flask containing hot coffee approximates to an isolated system. However, perfect thermal isolation is hard to achieve, and because the vacuum surrounding the hot coffee is not perfect, heat can slowly be transferred from the coffee to the surroundings. Eventually the coffee and surroundings reach thermal equilibrium (have the same temperature).

◆ **Open system:** A system which can exchange matter and energy with the surroundings.

◆ **Closed system:** A system which can exchange energy but not matter with the surroundings.

◆ **Isolated system:** A system which cannot exchange energy or matter with the surroundings.

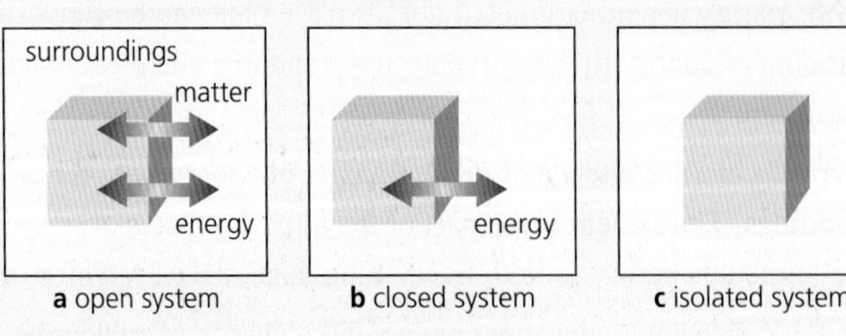

■ **Figure R1.3** Types of thermodynamic system

Nature of science: Falsification

Thermodynamics

'A theory is the more impressive the greater the simplicity of its premises, the more different the kinds of things it relates, and the more extended its area of applicability. Hence the deep impression that classical thermodynamics made upon me. It is the only physical theory of universal content which I am convinced that, within the framework of applicability of its basic concepts, will never be overthrown.'

Albert Einstein, Autobiographical notes.

The Austrian philosopher Karl Popper (1902–1994) proposed the requirement that a scientific hypothesis be falsifiable by experiment as the 'criterion of demarcation' which distinguishes scientific knowledge from other forms of knowledge. A hypothesis that is not subject to the possibility of empirical falsification does not belong in the realm of the sciences. Chemists only accept those laws and theories which have been extensively tested and, so far, have not been falsified. The laws of thermodynamics are among these.

Endothermic and exothermic reactions

◆ **Endothermic reaction:** A reaction where there is a transfer of heat energy from the surroundings to the system.

◆ **Exothermic reaction:** A reaction where there is a transfer of heat energy from the system to the surroundings.

In an **endothermic reaction**, a system gains energy because heat flows into it from the surroundings. Endothermic reactions usually result in a reduction in the temperature of the immediate surroundings.

An **exothermic reaction** releases heat energy. The released heat energy will flow into the surroundings, so exothermic reactions are usually characterized by an increase in the temperature of the immediate surroundings.

When ammonium chloride and hydrated barium hydroxide are mixed, an endothermic reaction occurs which can be approximated as:

$$Ba(OH)_2{\bullet}8H_2O(s) + 2NH_4Cl(s) \rightarrow BaCl_2{\bullet}2H_2O(s) + 2NH_3(g/aq) + 8H_2O(l)$$

The transfer of heat to the system can be demonstrated in a well-ventilated laboratory by carrying out the reaction in a flask that is wet on the outside and placed on a wooden block. As the reaction progresses, heat energy from the flask and water in contact with it (both part of the surroundings) is transferred to the system. The temperature of the surroundings drops dramatically, the water outside the flask freezes and the flask sticks to the block (Figure R1.4).

Sports cooling packs utilize an endothermic process, usually the dissolving of a solid such as ammonium nitrate or urea, to draw heat energy from an injured area of the body (Figure R1.5).

Top tip!

We can determine if a reaction is endothermic or exothermic by measuring the temperature change of the reaction system's surroundings.

■ **Figure R1.4** The endothermic reaction of hydrated barium hydroxide and ammonium chloride

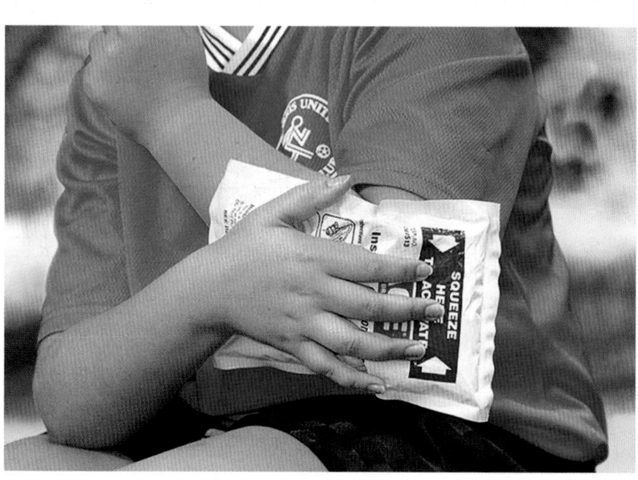

■ **Figure R1.5** A sports cooling pack being used to reduce the swelling around an injury

■ **Figure R1.6** An exothermic reaction: burning a candle

When a candle is burnt (Figure R1.6), the mixture of hydrocarbons that form the wax and oxygen from the air react to form carbon dioxide and water. During this combustion reaction, the temperature of the air around the candle increases. This demonstrates that the process transfers heat energy to the surroundings – an exothermic reaction is occurring.

LINKING QUESTION

Most combustion reactions are exothermic. How does the bonding in N_2 explain the fact that its combustion is endothermic?

LINKING QUESTION

What observations would you expect to make during an endothermic and an exothermic reaction?

Potential energy profiles

The total energy of a mixture of substances is very difficult to measure. The energy of the mixture will include the energy of its chemical bonds, the kinetic energy of vibration, rotation and translation of the molecules, and the energy associated with intermolecular forces. If the mixture undergoes a chemical reaction, the energies of the final products are similarly difficult to analyse. However, we can measure – rather easily – the energy changes that result from the reaction. These energy changes may appear as heat, work (such as expansion against atmospheric pressure) or, in special cases, light or electrical current.

◆ **Potential energy**: The energy is stored in a system. Part of this is chemical potential due to the electrostatic attractions of atomic nuclei and their electrons, the chemical bonds between atoms, and intermolecular interactions between neighbouring molecules.

◆ **Kinetic energy**: The energy due to motion, including translation (movement from place to place), vibration and rotation.

The total energy of a system (also known as its internal energy) is composed of two parts: the **kinetic energy** its particles possess, and their **potential energy**.

internal energy = kinetic energy + potential energy

At normal temperatures, the potential energy of a system is much larger than its kinetic energy.

The potential energy of a system can be compared to a bank account. When energy is transferred to the system in an endothermic reaction, the potential energy of the system increases. This is equivalent to depositing money into the account. We can also transfer energy out of the system to the surroundings, and this lowers the potential energy of the system. This is like withdrawing some money and seeing your bank balance go down.

These changes can be represented in a potential energy profile. Potential energy is shown on the *y*-axis, and the potential energies of the reactants and products are indicated by two lines at the appropriate levels (Figure R1.7).

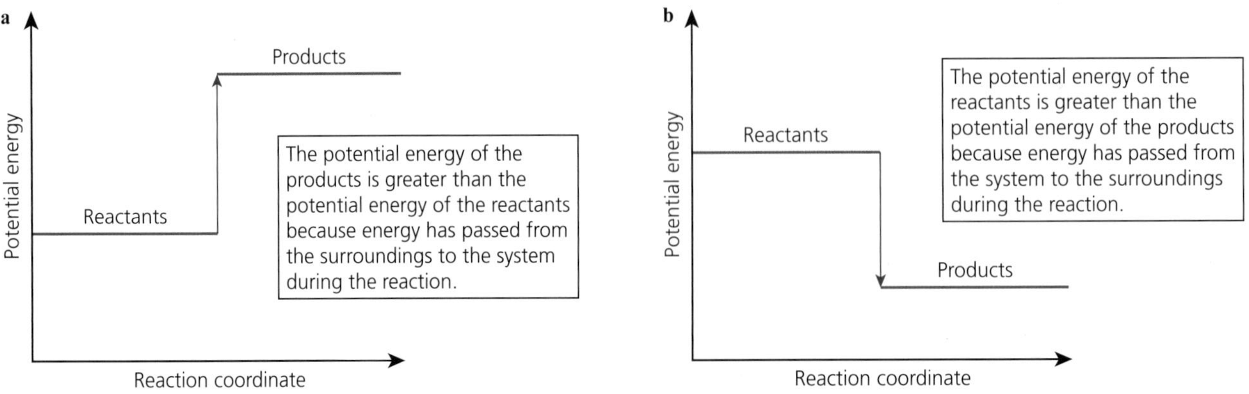

■ **Figure R1.7** The potential energy profiles of (a) endothermic and (b) exothermic reactions

In other textbooks you may see the y-axis of diagrams like the ones in Figure R1.7 given the label *enthalpy* or just *energy*. If a system expands during a reaction, it does work and the change in potential energy is not equal to the change in enthalpy. This means labelling the y-axis of a reaction profile using this term is not strictly correct.

The relative potential energy of the reactants and products determines whether a reaction is endothermic or exothermic.

Going further

Stability

The terms stability and relative stability are often used in an imprecise way. The term stability may refer to thermodynamic or kinetic stability.

If a system is said to be thermodynamically unstable, this means there are spontaneous physical or chemical processes (which happen without a continuous input of energy) that can make the system more thermodynamically stable. However, the spontaneous process will not occur at an appreciable rate if there is a large energy barrier and so the system is described as being kinetically stable.

An analogy can be drawn from a tower of wooden blocks. The tower is thermodynamically unstable because it will collapse if given enough energy to start the process, but it is kinetically stable because under usual conditions there is not enough energy to start the collapse. After the collapse of the tower, the system is in a thermodynamically more stable state. Providing it with the same stimulus that caused its collapse will not result in it reforming the tower. A number of familiar compounds (for example, benzene and ethyne) are thermodynamically unstable relative to their elements and their combustion products. These compounds do not quickly decompose or start burning when left in air at room temperature because both reactions have a large activation energy. This acts as a barrier, making the compounds kinetically stable. Thermodynamic stability is sometimes referred to as energetic stability.

1 Sketch the potential energy profile for each of the following reactions.

a The combustion of sulfur. This is an exothermic reaction.

$$S_8(s) + 8O_2(g) \rightarrow 8SO_2(g)$$

b The thermal decomposition of copper(II) carbonate. This is an endothermic reaction.

$$CuCO_3(s) \rightarrow CuO(s) + CO_2(g)$$

Does the list of disciplines included in, or excluded from, the natural sciences change from one era to another, or from one culture or tradition to another?

The Latin word *scientia* was anglicized to science and applied to any formal branch of knowledge until the beginning of the 17th century when its use gradually became restricted to the natural sciences.
At the beginning of the 19th century, the term philosophy was usually applied to all branches of knowledge including science. The rapid advances in science resulted in a separation of the two terms: natural sciences (for example, chemistry, physics and biology) and philosophy, which studies, among other topics, the nature of knowledge (epistemology – the focus of TOK), reality, and existence as well as language and ethics. Hence, the terms natural philosophy and natural science were equivalent in meaning, but the former has declined in use; today, it only survives in the job title 'chair of natural philosophy', describing a physics professor at the older universities.

Calculating heat energy transferred

The heat energy transferred into or away from a pure substance, Q, can be determined using the following equation:

$$Q = m \times c \times \Delta T$$

where:

- Q is the heat energy transferred (J)
- m is the mass of the substance undergoing the change (g)
- c is the specific heat capacity of the substance ($J\,g^{-1}\,K^{-1}$)
- ΔT is the change in temperature of the substance, $T_{final} - T_{initial}$ (K).

The specific heat capacity of a substance, c, is the energy required (in joules) to raise the temperature of 1 g of the substance by 1 K. Intuitively, it makes sense that in order to raise the temperature of 2 g of the substance by 5 K we would have to take the specific heat capacity of the substance and multiply it by 2 and 5, the mass and change in temperature respectively.

WORKED EXAMPLE R1.1A

Calculate the heat energy required to increase the temperature of 20.0 grams of nickel (specific heat capacity $0.440\,J\,g^{-1}\,K^{-1}$) from 50.0 °C to 70.0 °C.

Answer

$$Q = mc\Delta T$$

$$Q = 20.0\,g \times 0.440\,J\,g^{-1}\,K^{-1} \times (343.15 - 323.15)\,K$$

$$Q = 176\,J$$

Note that a temperature change in kelvin (20.0 K in this example) is equivalent to the change in degrees Celsius. Although in the example the change is calculated in kelvin for completeness, working out the change of temperature directly from two degrees Celsius values is the best approach.

Top tip!

Temperature change is the same in degrees Celsius and kelvin. If you are given both initial and final temperatures in degrees Celsius, you do not need to convert to kelvin.

WORKED EXAMPLE R1.1B

Determine the temperature change of $250.0\,cm^3$ of pure water when 2.00 kJ of heat energy is supplied to it. The specific heat capacity of water is $4.18\,J\,g^{-1}\,K^{-1}$.

Answer

We need the energy transferred in joules: 2.00 kJ = 2000 J

$$Q = mc\Delta T$$

$$2000\,J = 250\,g \times 4.18\,J\,g^{-1}\,K^{-1} \times \Delta T$$

$$\Delta T = 1.91\,K \text{ or } 1.91\,°C$$

Top tip!

The density of water is approximately $1\,g\,cm^{-3}$. This means that the volume of water in cm^3 is equal to the mass of the water in grams. Between about 0 °C and 30 °C, the true density of water (to 3 significant figures) is $1.00\,g\,cm^{-3}$, but the density decreases at higher temperatures, reaching $0.96\,g\,cm^{-3}$ at 100 °C.

In other chemistry texts you may encounter questions where you are given the heat capacity of a substance rather than the specific heat capacity. The heat capacity of a substance is the heat energy required to raise the whole mass of a substance by 1 K (or °C), not just 1 g.

heat capacity = specific heat capacity × mass of substance

It is usual for the heat capacity to be given the symbol *C* (uppercase) while *c* (lowercase) is used for specific heat capacity.

WORKED EXAMPLE R1.1C

Calculate the heat capacity of 80.0 g of water. The specific heat capacity of water is $4.18\,J\,g^{-1}\,K^{-1}$.

Answer

Heat capacity (C) = specific heat capacity (c) × mass (m) = $4.18\,J\,g^{-1}\,K^{-1} \times 80.0\,g = 334\,J\,K^{-1}$

Tool 3: Mathematics

Delta notation

A 'Δ' (Greek letter delta) indicates a finite change. For example, ΔT is the temperature change of a system. It is calculated by taking the final temperature (T_2) and subtracting the initial temperature (T_1).

$$\Delta T = T_2 - T_1$$

Another use of Δ is Δn. This is the total number of moles of species on the right-hand side of an equation minus the total number of moles of species on the left-hand side of the equation. For the equation:

$$Al_2O_3(s) + 6HI(g) \rightarrow 2AlI_3(s) + 3H_2O(l)$$

$$\Delta n = (2 + 3) - (1 + 6) = -2$$

2 Calculate the heat energy transferred to each of the substances in the table.

	Substance	Mass of substance / g	Specific heat capacity / $J\,g^{-1}\,K^{-1}$	Initial temperature / °C	Final temperature / °C
a	water	200	4.18	20	65
b	steel	35	0.466	20	200
c	methanol	700	2.14	20	40
d	water	500	4.18	10	37
e	gold	1500	0.129	−5	30

3 The temperature of 800 g of water is raised from 10.0 °C to 80.0 °C. How much heat energy is used to do this? The specific heat capacity of water is $4.18\,J\,g^{-1}\,K^{-1}$.

4 A 700 g sample of water loses 240 kJ of heat energy when it cools from its boiling point. What temperature has the water cooled to? The specific heat capacity of water is $4.18\,J\,g^{-1}\,K^{-1}$.

Enthalpy changes

Imagine a reaction which takes place in a closed system which is maintained at constant volume and is in good thermal contact with its surroundings. Since work = force × distance moved, and the system does not expand or contract, the system does no work and all energy absorbed or released must appear as heat. The amount of heat transferred into or out of the system, which is easily measured, thus equals the change in the internal energy of the system.

However, many chemical reactions are carried out under conditions of constant pressure which means the system is able to expand or contract into its surroundings. An example of this is the exothermic reaction of a metal carbonate with an acid in an open beaker. The reaction produces carbon dioxide gas which will expand against the surrounding atmosphere. In this scenario, where pressure acting on the system has remained constant, some of the energy change of the reaction is used to do the work needed to expand the system. The heat energy transferred from the system is less than it would be in constant volume conditions because some of the energy released is transferred as work.

◆ **Enthalpy change**: The heat transferred to or from a system under constant external pressure.

The heat absorbed or evolved in a process at constant external pressure is equivalent to the change in a property of the system known as enthalpy. Enthalpy is given the symbol H, and thus the change in enthalpy as a result of a chemical reaction is $\Delta H^{\ominus}$. Heat energy and enthalpy are both measured in the same unit, the joule (J). The difference between the internal energy and **enthalpy change** of a reaction is usually small and is only significant when evolution or absorption of a gas is involved.

Going further

Internal energy and enthalpy

The two extensive state properties, enthalpy, H, and internal energy, U, are related through the equation $H = U + PV$ where P and V refer to pressure and volume respectively.

So, at constant pressure, $\Delta H^{\ominus} = \Delta U + \Delta(PV) = \Delta U + P\Delta V$.

This suggests we need to measure the volume of gas evolved (or reacted) as well as the heat transferred (because $\Delta H^{\ominus} = -Q$) in order to determine the internal energy change associated with a reaction.

In a bomb calorimeter, the measurement is made under constant volume conditions rather than constant pressure conditions so $-Q = \Delta H^{\ominus} = \Delta U + \Delta(PV) = \Delta U + V\Delta P$.

In such constant-volume calorimetry, the heat transferred is equal to the internal energy change, ΔU, only if there is no pressure change: that is, if the amount of gas evolved in the reaction is equal to the amount of gas required for the reaction.

However, if the gases involved in the reaction can be assumed to be ideal, no further measurements are needed to determine ΔU in either case. We can use the ideal gas equation ($PV = nRT$, see S1.5) to replace the PV term, giving $H = U + nRT$ and $-Q = \Delta H^{\ominus} = \Delta U + \Delta nRT$ where Δn is the change in the amount of gaseous species during the process.

Transferring heat energy, Q, from the surroundings to the system causes the enthalpy, H, of the system to increase. In this endothermic process, the value of the enthalpy change for the system, $\Delta H^{\ominus}$, is positive.

Top tip!

Endothermic reactions have a positive $\Delta H^{\ominus}$.

Exothermic reactions have a negative $\Delta H^{\ominus}$.

Transferring heat energy, Q, from the system to the surroundings causes the enthalpy, H, of the system to decrease. In this exothermic process, the value of the enthalpy change for the system, $\Delta H^{\ominus}$, is negative.

The change in enthalpy accompanying a chemical reaction is usually quoted in kilojoules per mole, $kJ\,mol^{-1}$ and the value is stated next to the reaction equation, as shown below.

$$2H_2(g) + O_2(g) \rightarrow 2H_2O(l);\ \Delta H^{\ominus} = -572\,kJ\,mol^{-1}$$

The change in enthalpy value gives the amount of energy, in kJ, that would be absorbed or released by the system when the amounts given in the balanced equation, in moles, are reacted. From the above equation and its accompanying enthalpy change we can deduce that 572 kJ of energy will be released when 2 moles of hydrogen molecules react with 1 mole of oxygen molecules to form 2 moles of water molecules.

If the reaction is reversed, the sign for the enthalpy change is also reversed.

$$2H_2O(l) \rightarrow 2H_2(g) + O_2(g);\ \Delta H^{\ominus} = +572\,kJ\,mol^{-1}$$

When the amounts of reactants and products are halved, the enthalpy change for the reaction is also halved:

$$H_2(g) + \frac{1}{2}O_2(g) \rightarrow H_2O(l);\ \Delta H^{\ominus} = -286\,kJ\,mol^{-1}$$

Standard enthalpy changes

Enthalpy changes during a chemical reaction (or physical process) depend on the surrounding conditions, such as temperature, pressure, physical state and the amount of substance involved.

◆ **Standard enthalpy change:** The enthalpy change when the amounts of reactants in a specified stochiometric equation react together under standard conditions.

A **standard enthalpy change** is the enthalpy change when the amounts of reactants in a specified stochiometric equation react together under the following conditions:

- a pressure of 1 bar or 10^5 pascal (very close to 1 atmosphere)
- involving the most thermodynamically stable allotrope under standard conditions (for example, carbon in the form of graphite)
- using solutions (where relevant) with a concentration of $1.0\,mol\,dm^{-3}$
- at a specified temperature, typically 298.15 K.

The symbol $\ominus$ is used to indicate chemical reactions or physical processes taking place under standard conditions.

When quoting the enthalpy change of a reaction, you must always show the physical states of reactants and products in the equation. These states should be the 'standard' states of the elements or compounds involved at 298 K and 1 bar pressure, for example, $H_2O(l)$, Fe(s) and $O_2(g)$.

A change in state of a reactant or product affects the enthalpy change of the reaction. For example:

$$H_2(g) + \frac{1}{2}O_2(g) \rightarrow H_2O(l);\ \Delta H^{\ominus} = -286\,kJ\,mol^{-1}$$

$$H_2(g) + \frac{1}{2}O_2(g) \rightarrow H_2O(g);\ \Delta H^{\ominus} = -242\,kJ\,mol^{-1}$$

The difference between the two $\Delta H^{\ominus}$ values, $+44\,kJ\,mol^{-1}$ is due to the standard enthalpy change of vaporization of water.

ATL R1.1A

The stoichiometric reaction between hydrogen and oxygen described above is spectacular because it releases energy quickly and creates a shockwave. An interesting and detailed video showing this reaction can be found at www.youtube.com/watch?v=I9CI6KSV560.

Watch the video (starting at 4 minutes and ending at 14 minutes) and comment on why the reaction creates a much stronger shockwave when the reactants are pre-mixed and the mixture is ignited from the centre of the balloon.

Precise values of standard enthalpy changes for many reactions are available in reference material. Table R1.2 shows the molar enthalpies of combustion (that is, the enthalpy change when 1 mole of the substance is burnt in excess oxygen) for some elements. The molar enthalpies of combustion for a range of compounds can be found in section 14 of the IB *Chemistry data booklet.*

■ **Table R1.2** Selected standard enthalpy changes of combustion

Substance	Formula	State	$\Delta H^{\ominus}$ / kJ mol^{-1}
hydrogen	H_2	g	−286
sulfur	S	s	−297
carbon (graphite)	C	s	−394

Calculating the standard enthalpy change for a reaction

Consider the reaction between iron(III) oxide and aluminium.

$$Fe_2O_3(s) + 2Al(s) \rightarrow Al_2O_3(s) + 2Fe(l)$$

This reaction is highly exothermic and the heat given out melts the iron produced. The molten iron can be used to weld together steel railway tracks (Figure R1.8).

As stated before, the enthalpy change is that when the quantities stated in the equation, 1.0 mole of Fe_2O_3 and 2.0 moles of Al, react together. In theory, 1.0 mole of Fe_2O_3 and 2.0 moles of Al could be reacted together and the heat energy change of the surroundings could be measured; this would tell us the enthalpy change of the reaction when it occurs in the amounts specified. However, we seldom carry out reactions using such exact quantities. In a typical scenario, there will be a limiting reactant, all of which will react (be 'used up') and the other reactants will be present in excess. There is more information about limiting reactants in R2.1.

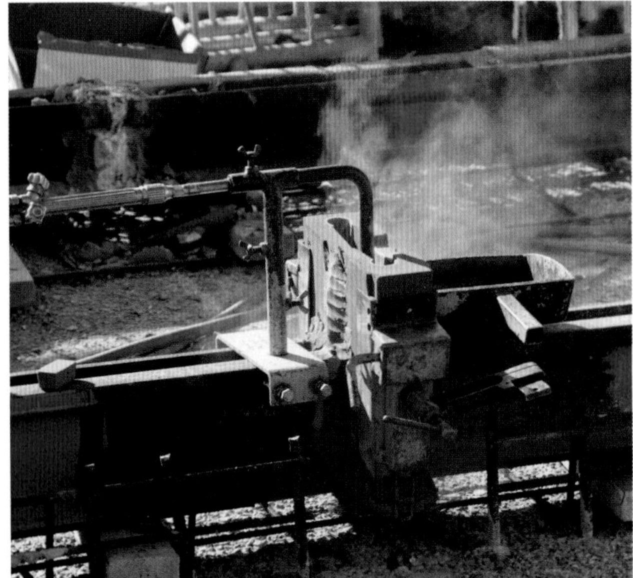

■ **Figure R1.8** The thermite reaction between aluminium and iron(III) oxide

The enthalpy change of a reaction, $\Delta H^{\ominus}$, can be calculated from the heat energy transferred to the surroundings, Q, the amount of the limiting reactant (in mol), and the stoichiometric coefficient of the reaction. To determine the enthalpy change:

$$\Delta H^{\ominus} = \frac{-Q\ (\text{kJ})}{\text{amount of limiting reactant (mol)}} \times \text{stoichiometric coefficient of limiting reactant}$$

In this case, if Al is the limiting reactant, as would be usual because Al is more expensive than Fe_2O_3, the enthalpy change of the reaction would be:

$$\Delta H^{\ominus} = \frac{-Q}{n(\text{Al})} \times 2$$

Top tip!

Q is the heat energy transferred to the surroundings of a system. The system has the opposite heat energy change and hence the change in enthalpy, $\Delta H^{\ominus}$, is proportional to $-Q$.

This value, given in $kJ\,mol^{-1}$, is the enthalpy change when 1.0 mole of Fe_2O_3 reacts with 2.0 moles of Al.

And, if we measured that the reaction transferred 2500 kJ of energy to the surroundings when 6.0 moles of aluminium were used:

$$\Delta H^{\ominus} = \frac{2500\,kJ}{6.0\,mol} \times 2 = 830\,kJ\,mol^{-1}$$

Common mistake

Misconception: *When calculating* $\Delta H^{\ominus}$, *Q is often given in the units of J because this is the unit it has when it is calculated from the equation* $Q = m \times c \times \Delta T$.

Because $\Delta H^{\ominus}$ is given in the units of $kJ\,mol^{-1}$, Q must be converted into kJ first.

WORKED EXAMPLE R1.1D

When 0.025 moles of sulfuric acid is reacted with an excess of sodium hydroxide, the heat energy change, Q, is 2.79 kJ. Calculate the enthalpy change of the reaction that produces 1 mole of water.

Answer

$$\frac{1}{2}H_2SO_4(aq) + NaOH(aq) \rightarrow \frac{1}{2}Na_2SO_4(aq) + H_2O(l)$$

$$\Delta H^{\ominus} = \frac{-2.79\,kJ}{0.025\,mol} \times \frac{1}{2} = -55.8\,kJ\,mol^{-1}$$

5 Ethanoic acid is neutralized by sodium hydroxide.

$$CH_3COOH(aq) + NaOH(aq) \rightarrow CH_3COONa(aq) + H_2O(l)$$

When 0.048 moles of ethanoic acid is mixed with 0.040 moles of sodium hydroxide, the heat energy change of the reaction, Q, is 2.09 kJ. Calculate the enthalpy change, $\Delta H^{\ominus}$, of the reaction.

6 When 0.0222 mol of decane, $C_{10}H_{22}$, is combusted in an excess of oxygen, the heat energy change of the surroundings is 145 kJ. Use this information to calculate the enthalpy change of the reaction given below:

$$2C_{10}H_{22}(l) + 31O_2(g) \rightarrow 20CO_2(g) + 22H_2O(l) \quad \Delta H^{\ominus} = ?$$

Measuring the standard enthalpy change of a chemical reaction

◆ **Calorimetry**: The act of measuring the temperature change of a body for the purpose of deriving the heat energy transfer to it.

Calorimetry experiments allow us to determine the enthalpy changes of reactions. These experiments use a calorimeter, which is a vessel used to contain and measure the quantity of heat energy produced or absorbed during a physical process or chemical change. The exact apparatus chosen for a calorimetry experiment depends on the reaction that is being investigated.

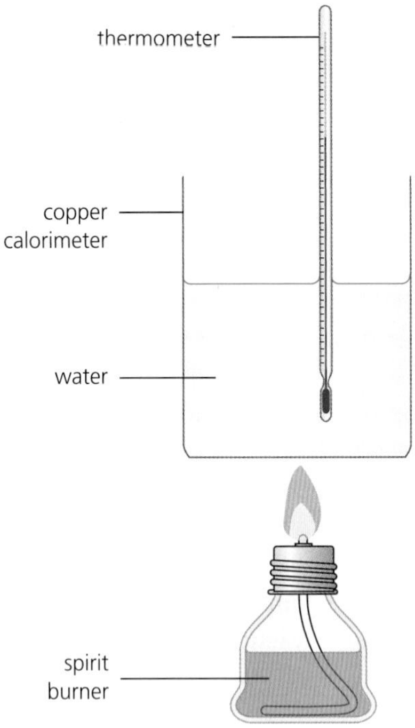

Combustion of a liquid

As an example, consider combusting 0.013 mol of ethanol in a spirit burner beneath a metal can filled with $100.0\,cm^3$ of water (Figure R1.9). The purpose of the can of water is to capture the heat energy released by the reaction so it is the calorimeter in this experiment. The equation is:

$$C_2H_5OH(l) + 3O_2(g) \rightarrow 2CO_2(g) + 3H_2O(g)$$

thermometer
copper calorimeter
water
spirit burner

■ **Figure R1.9** Simple apparatus used to measure enthalpy change of combustion of liquids

As the ethanol is burnt, the temperature of the water rises by 28 °C. The heat energy transferred to the water is:

$$Q = mc\Delta T$$

$$Q = 100\,g \times 4.18\,J\,g^{-1}\,K^{-1} \times 28\,K$$

$$= 12\,000\,J \text{ (2 s.f.)}$$

In this case, the limiting reactant is the ethanol because there is a plentiful supply of air around the spirit burner, and 0.013 mol of ethanol were used.

$$\Delta H^{\ominus} = \frac{-12\,kJ}{0.013\,mol} \times 1 = -920\,kJ\,mol^{-1} \text{ (2 s.f.)}$$

This value is less exothermic than the literature value of $-1367\,kJ\,mol^{-1}$ because several assumptions are made when calculating the heat transferred to the water, and none of these is completely fulfilled. These assumptions are as follows:

- All the heat energy released by the combustion of ethanol was transferred to the water in the copper calorimeter.
- The only reaction that took place was the one being studied – complete combustion in this case.
- The reaction and energy transfers were sufficiently rapid for the maximum temperature of the water to be reached before it began cooling to room temperature.

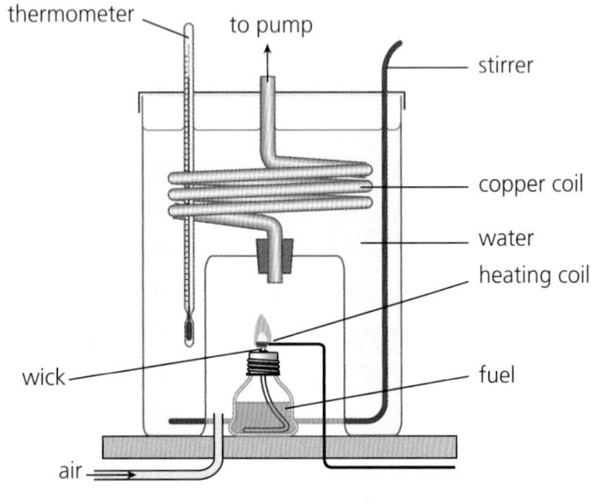

■ **Figure R1.10** A Thiemann fuel calorimeter

In reality, heat energy will have been transferred to other parts of the reaction's surroundings (the air, the copper calorimeter, the thermometer), not just the water. The warmed water will also have transferred heat energy to the air, and some of the water may have evaporated (a process which absorbs energy). Some incomplete combustion of the ethanol is likely to have occurred, too. All these things impact the accuracy of the experimentally obtained value.

More sophisticated calorimetry equipment addresses some of the deficiencies of a simple set up. A Thiemann fuel calorimeter (Figure R1.10) does this by reducing heat loss to the surroundings and ensuring complete combustion.

WORKED EXAMPLE R1.1E

A chemist carries out a calorimetry experiment to determine the enthalpy change for the combustion of propanone.

$$CH_3COCH_3(l) + 4O_2(g) \rightarrow 3CO_2(g) + 3H_2O(g)$$

The following procedure and results were recorded:

- 250.0 cm^3 of water was measured using a 250 cm^3 measuring cylinder. The water was poured into a copper calorimeter.
- The mass of the spirit burner and lid were measured using a 2 d.p. electronic balance.
- The temperature of the water was recorded with a digital thermometer.
- The copper calorimeter was clamped above the spirit burner and the wick was lit with a burning splint.
- The water was heated until it reached approximately 60 °C. The spirit burner was extinguished by replacing the lid. The maximum temperature reached by the water was recorded.
- The spirit burner was weighed again.

Mass of burner + propanone before experiment / g	142.34
Mass of burner + propanone after experiment / g	140.49
Starting temperature of water / °C	21.5
Maximum temperature of water / °C	63.7

Calculate the enthalpy change of the reaction based on these results.

Answer

$Q = 250\,g \times 4.18\,J\,g^{-1}\,K^{-1} \times (63.7 - 21.5)\,K$

$Q = 44\,099\,J$

Mass of propanone = (142.34 – 140.49) g = 1.85 g

Amount of propanone $= \dfrac{1.85\,g}{58.09\,g\,mol^{-1}} = 0.0318\,mol$

$\Delta H^{\ominus} = \dfrac{-44.1\,kJ}{0.0318\,mol} \times 1 = -1390\,kJ\,mol^{-1}$

As expected, this experimental value is less exothermic than $-1790\,kJ\,mol^{-1}$, the value given in section 14 of the IB *Chemistry data booklet.*

Tool 3: Mathematics

Calculate and interpret percentage error and percentage uncertainty

Percentage errors and percentage uncertainty are useful tools in helping to evaluate if our experimental value agrees with the accepted value.

Calculating percentage uncertainty

To calculate the percentage uncertainty, take the uncertainty value of the apparatus you are using, divide by the measured value, and multiply by 100.

For example, 20.00 cm^3 ± 0.05 cm^3 of water is measured using a glass pipette.

The percentage uncertainty is calculated:

$$\frac{0.05}{20.00} \times 100 = 0.25\% = 0.3\% \text{ (1 s.f.)}$$

Note, it is standard practice to quote percentage uncertainties to 1 s.f. The smaller the percentage error, the more precise your data will be.

Calculating percentage error

To calculate the percentage error, you need to know the literature value for a piece of data.

Percentage error is calculated from:

$$100 \times \frac{\text{experimental value} - \text{literature value}}{\text{literature value}}$$

For example, an enthalpy of solution is calculated to be $+2.65\,kJ\,mol^{-1}$. The literature value is $+3.11\,kJ\,mol^{-1}$.

The percentage error is:

$$100 \times \frac{2.65 - 3.11}{3.11} = -14.8\% \text{ (3 s.f.)}$$

The purpose of carrying out this calculation is to see how close your experimental value is to the true value (literature value) and to help you identify methodological weaknesses. The sign is also important. In this example, the sign is negative, meaning that the value is less than expected. A reason for this is that less substance was dissolved than should have been. This would result in a smaller than expected value being obtained. Your evaluation should take this into account. It would be wrong to say that too much substance had been added as this would give a result greater than expected.

Inquiry 1: Exploring and designing

Controlling variables

Scientists aim to measure and control all the variables which can affect an experiment. This means that a change in the dependent variable can be ascribed to a change in the independent variable.

Consider experiments to measure the enthalpy changes when methanol or hexanol are combusted in a spirit burner. Simple flame calorimetry experiments have many variables which can affect the calculated enthalpy change. The proportion of heat transferred to the calorimeter, and the rate and total amount of heat loss from the calorimeter, are two important variables. Ideally, these would be kept constant to allow a fair comparison of results.

The size of the flame in the two experiments is likely to differ, and this will affect the efficiency of heat transfer to the calorimeter. The two experiments will experience different rates of heat loss from the calorimeter because of differences in the air flow past the experiment; a draught shield may help reduce this.

It might appear that the enthalpy change calculated is independent of both the mass of water used and the temperature rise recorded because these values are accounted for in the calculation of heat energy change, Q. This is not truly the case. Different masses of water will lose heat to the surroundings at different rates, as will water of different temperatures, and so these do have an indirect effect on the results by affecting the efficiency of the calorimeter apparatus.

Ideally, only complete combustion would occur, but this is less likely for a less volatile and longer chain molecule of hexanol.

The variables of a reaction should be considered carefully, and it is important to plan how they can be controlled to a reasonable extent.

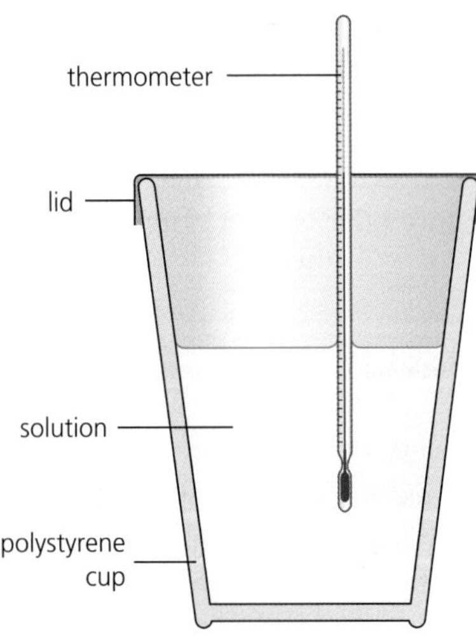

■ **Figure R1.11** A simple calorimeter for measuring the enthalpy change involving an aqueous solution

7 When 0.130 g of propyne, C_3H_4, is burnt in an excess of oxygen it raises the temperature of 400.0 g of water by 3.76 °C.

$$C_3H_4(g) + 4O_2(g) \rightarrow 3CO_2(g) + 2H_2O(l)$$

a Calculate the moles of propyne burnt.

b Calculate the heat energy change of the water heated by the reaction.

c Calculate the change in enthalpy when 1.00 mol of propyne is combusted in an excess of oxygen. You should assume that all the heat from the reaction was transferred to the water.

Calorimetry involving solutions

Many reactions have at least one aqueous reactant. In this case, the enthalpy change of the reaction can be established from the change in temperature of the solvent, which is the immediate surroundings of the reacting particles. A polystyrene coffee cup is often used as a calorimeter in these cases (Figure R1.11).

As an example, consider 50.0 cm^3 of 0.200 mol dm^{-3} copper(II) sulfate placed in an insulated container with a lid. The temperature of the solution is taken, and the lid is removed before addition of 1.20 g of zinc. The temperature of the solution rises by 9.5 °C because it absorbs the heat energy released in the displacement reaction.

$$Zn(s) + CuSO_4(aq) \rightarrow ZnSO_4(aq) + Cu(s)$$

$$Zn(s) + Cu^{2+}(aq) \rightarrow Zn^{2+}(aq) + Cu(s)$$

If we make the approximation that the density and specific heat capacity of the 0.200 mol dm^{-3} copper(II) sulfate are identical to those of water (these are valid assumptions because, although the solutes do change the density and specific heat capacity, the solution is dilute and, therefore, largely consists of water), the heat energy transferred to the surroundings by the reaction is:

$$Q = mc\Delta T$$

$$Q = 50.0\,\text{g} \times 4.18\,\text{J}\,\text{g}^{-1}\,\text{K}^{-1} \times 9.5\,\text{K}$$

$$Q = 1990\,\text{J}$$

The amount of reactants used is:

$$n(CuSO_4) = 0.200\,\text{mol}\,\text{dm}^{-3} \times 0.0500\,\text{dm}^3 = 0.0100\,\text{mol}$$

$$n(Zn) = \frac{1.20\,\text{g}}{65.38\,\text{g}\,\text{mol}^{-1}} = 0.0184\,\text{mol}$$

Taking into account the stoichiometry of the reaction, the copper(II) sulfate is the limiting reactant.

$$\Delta H^{\ominus} = \frac{-1.99\,\text{kJ}}{0.0100\,\text{mol}^{-1}} \times 1 = -199\,\text{kJ}\,\text{mol}^{-1}$$

As with the flame calorimeter experiment, the experimental result obtained using this method and apparatus will differ from the accepted literature value. The reasons for this are:

- The reaction mixture will lose some heat energy to the air and polystyrene calorimeter.
- Heat energy may be used in the evaporation of water.
- The density and specific heat capacity of the reaction mixture may be different from that of water.
- The heat capacity of the solid in the reaction mixture has not been accounted for.

WORKED EXAMPLE R1.1F

2.00 g of $NH_4NO_3(s)$ is dissolved in 30.0 g of water. The temperature of the water falls from 22.2 °C to 17.5 °C. Calculate the standard enthalpy change of this process (the enthalpy change of solution). The specific heat capacity of water is 4.18 J g^{-1} K^{-1}.

Answer

$$NH_4NO_3(s) \rightarrow NH_4^+(aq) + NO_3^-(aq)$$

$$Q = 30.0\,\text{g} \times 4.18\,\text{J}\,\text{g}^{-1}\,\text{K}^{-1} \times -4.7\,\text{K}$$

$$= -590\,\text{J (2 s.f)}$$

$$n = \frac{2.00\,\text{g}}{80.06\,\text{g}\,\text{mol}^{-1}} = 0.0250\,\text{mol}$$

$$\Delta H^{\ominus} = \frac{-(-0.59\,\text{kJ})}{0.025\,\text{mol}} \times 1 = +24\,\text{kJ}\,\text{mol}^{-1}$$

The temperature of the surroundings has decreased, so the reaction was endothermic.

LINKING QUESTION

Why do calorimetry experiments typically result in a smaller change in temperature than is expected from theoretical values?

8 50.0 cm^3 of a 0.500 mol dm^{-3} solution of iron(II) sulfate is placed in a polystyrene cup and its temperature is measured. 0.800 g of powdered magnesium is added, and the temperature is observed to rise by 20.6 °C.

a Give the chemical equation for the reaction occurring.

b Calculate the standard enthalpy change of the reaction.

c Identify three assumptions you have made in calculating the standard enthalpy change of the reaction.

Tool 1: Experimental techniques

Calorimetry

Specialist apparatus has been developed to overcome the limitations of simple calorimetry methods. A bomb calorimeter (Figure R1.12) is able to yield highly accurate values for enthalpies of combustion.

The 'bomb' is a sealed pressure vessel filled with oxygen, to ensure complete combustion. The fuel is ignited using an electrical current, and the energy change of the fuel is absorbed by a known mass of water in the calorimeter. The calorimeter is itself submerged in a bath of water which is electrically heated to match the temperature of the calorimeter, so ensuring heat is not transferred from the calorimeter to the surroundings. The bomb calorimeter is calibrated by using a reaction of known enthalpy change to determine the bomb calorimeter's heat capacity.

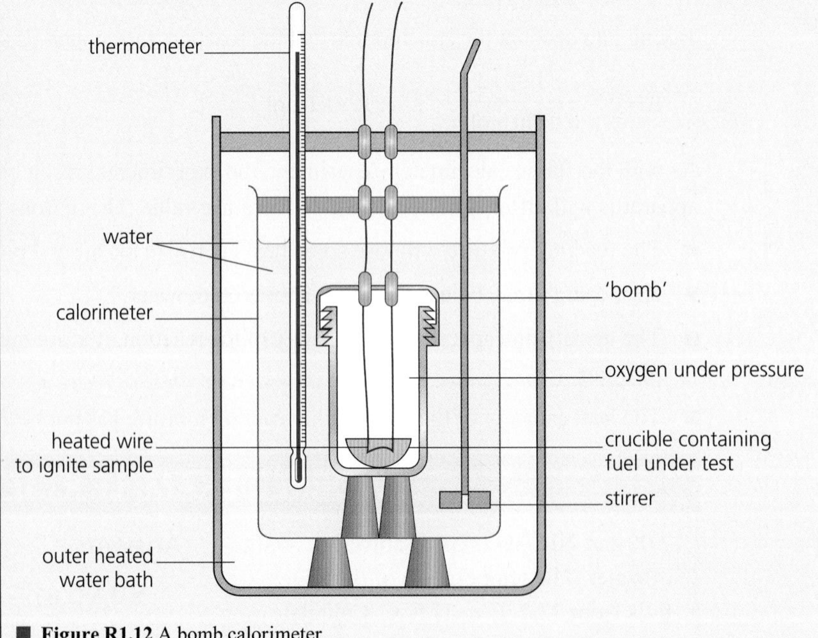

■ **Figure R1.12** A bomb calorimeter

Temperature corrections in simple calorimetry

Simple calorimeters (for example, a polystyrene cup fitted with a lid) can give accurate results for fast reactions, such as neutralizations or ionic precipitations. However, this equipment gives less accurate results for slower reactions, such as metal ion displacement. This is because the heat loss to the surroundings increases if the reaction is slow – the heat is lost over a longer period of time. Consequently, the temperature rise of the water in the calorimeter is not as large as it should be. However, an allowance can be made for this by plotting a temperature–time graph (a cooling curve).

One reagent is placed in the calorimeter and its temperature recorded at regular intervals by the investigator or, even better, by a datalogger. The second reagent is added after a period of time and the mixture is stirred continuously. The temperature is recorded until the maximum temperature (or minimum if the reaction is endothermic) is reached and for at least 5 minutes thereafter. A graph of temperature against time will have the form shown in Figure R1.13 or Figure R1.14. In these examples the temperature was recorded at 1-minute intervals, and the mixing of reactants was at 4 minutes.

The temperature given by extrapolating the final straight-line segment of the curve to the time of mixing is an estimation of the maximum (or minimum) temperature that would have been reached if the reaction had been instantaneous and there had been no heat loss to the surroundings. The approximation made is that rate of heat exchange of the reaction solution with its surroundings stays constant from the point of mixing.

Inquiry 3: Concluding and evaluating

Discuss the impact of uncertainties on the conclusions

It is impossible to obtain a result with no uncertainties – there will always be a degree of uncertainty associated with a final answer. The role of uncertainties is to determine how valid your calculated value is compared to the literature value. If the literature value falls within the range of uncertainties for your calculated value, you may conclude that most of the uncertainties in the experiment have been considered. These are the random errors. If, however, the literature value is outside the range of your uncertainties, there must be some uncertainties that you have not considered – these are systematic errors.

For example, a calculated value for the pK_a of ethanoic acid is 4.9 ± 0.2. The literature value is 4.76. This means that the calculated result is accurate as the literature value lies within the uncertainty range. We can assume that most random errors in this experiment have been measured, and that there are few systematic errors.

If, however, your value for the enthalpy of combustion of ethanol is $-810\,kJ\,mol^{-1} \pm 50\,kJ\,mol^{-1}$ but the literature value is $-1367\,kJ\,mol^{-1}$, the literature value does not lie within the range of uncertainties. In this experiment, there are systematic errors and part of the evaluation will need to address possible sources of these unaccounted errors. In this case, they are likely to be from heat loss and incomplete combustion.

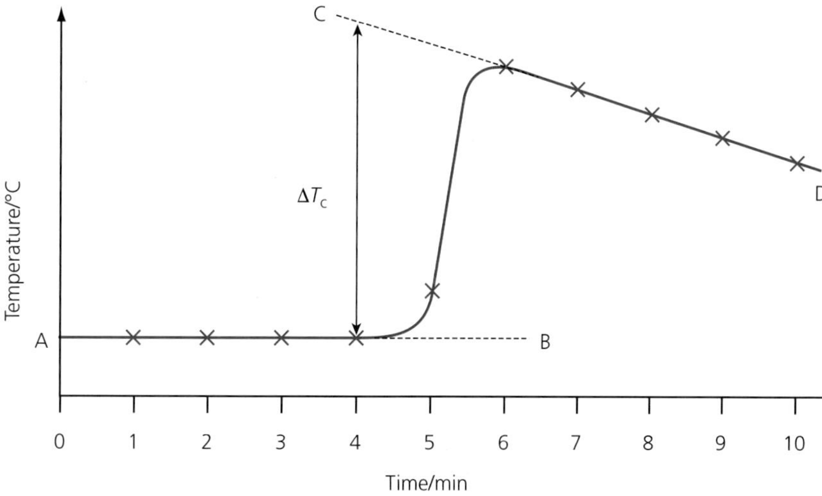

■ **Figure R1.13** A temperature correction curve for an exothermic reaction

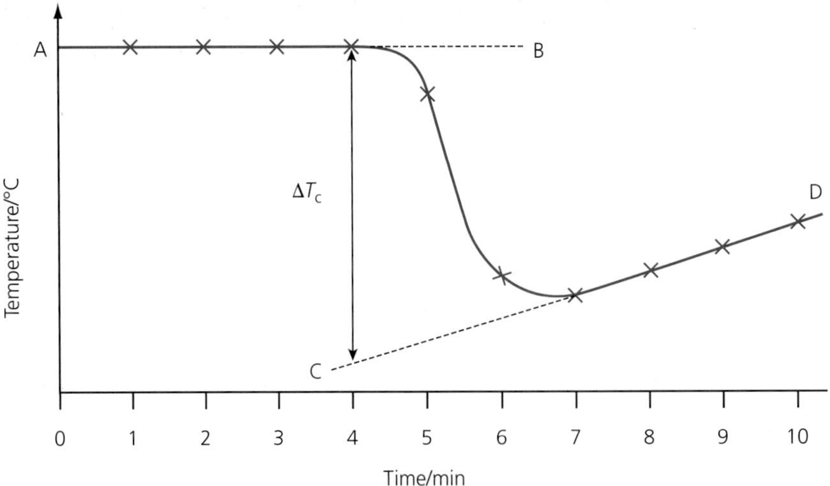

■ **Figure R1.14** A temperature correction curve for an endothermic reaction

WORKED EXAMPLE R1.1G

0.50 g of powdered magnesium was added to $50.0\,cm^3$ of $0.60\,mol\,dm^{-3}$ ethanoic acid.

$$Mg(s) + 2CH_3COOH(aq) \rightarrow Mg(CH_3COO)_2(aq) + H_2(g)$$

The temperature of the solution was recorded over a 10-minute period, and the maximum temperature rise was determined (by back extrapolation to the time of mixing) to be 30.0 °C. Calculate the standard enthalpy change of reaction. State any assumptions you have made about the solution.

Answer

$$Q = 50.0\,g \times 4.18\,J\,g^{-1}\,K^{-1} \times 30\,K$$

$$= 6300\,J \text{ (2 s.f.)}$$

$$n(Mg) = \frac{0.50\,g}{24.31\,g\,mol^{-1}} = 0.021\,mol$$

$$n(\text{ethanoic acid}) = 0.60\,mol\,dm^{-3} \times 0.05000\,dm^3 = 0.030\,mol$$

Taking into account the stoichiometry of the reaction, the ethanoic acid is the limiting reactant.

$$\Delta H^{\ominus} = \frac{-6.3\,kJ}{0.030\,mol} \times 2 = -420\,kJ$$

$$Mg(s) + 2CH_3COOH(aq) \rightarrow Mg(CH_3COO)_2(aq) + H_2(g) \qquad \Delta H^{\ominus} = -420\,kJ$$

The assumptions are that the solution has a density of $1.00\,g\,cm^{-3}$ and its specific heat capacity is $4.18\,J\,g^{-1}\,K^{-1}$.

Tool 3: Mathematics

Propagate uncertainties in processed data

Combining uncertainties given in a single unit

When adding and subtracting values, the uncertainties are also added together. For example, when calculating the volume dispensed from a burette, the initial reading, $4.55\,cm^3 \pm 0.05\,cm^3$ is subtracted from the final reading $22.80\,cm^3 \pm 0.05\,cm^3$. The volume dispensed from the burette has an uncertainty of $0.05\,cm^3 + 0.05\,cm^3 = 0.1\,cm^3$.

Combining uncertainties given in different units

In most calculations, you will be combining quantities – and hence their uncertainties – measured in different units. You cannot add these together. For example, the temperature of $50.1\,g \pm 0.1\,g$ of water (specific heat capacity $4.18\,J\,g^{-1}\,K^{-1}$) is raised by $34.6\,°C \pm 0.1\,°C$. The heat energy gained by the water is $50.1 \times 4.18\,J\,g^{-1}\,K^{-1} \times 34.6\,°C = 7250\,J$, but it would be incorrect to express the uncertainty as $0.1\,g + 0.1\,°C$ or $0.2\,g\,°C$!

We get around this problem by determining the percentage uncertainty for each value, adding them together to get an overall percentage uncertainty, and then converting this percentage uncertainty back into an absolute value.

The percentage uncertainty is calculated from:

$$\frac{\text{uncertainty}}{\text{measured value}} \times 100$$

So in this example, the percentage uncertainties are:

$$\text{mass} = \frac{0.1}{50.1} \times 100 = 0.2\%$$

$$\text{temperature} = \frac{0.1}{34.6} \times 100 = 0.3\%$$

$$\text{total \% uncertainty} = 0.2\% + 0.3\% = 0.5\%$$

The final value will therefore be expressed as $7250\,J \pm 0.5\%$

Converting the percentage uncertainty back to an absolute value, this is:

$7250\,J \pm 40\,J$

Dealing with exponents and uncertainties (HL only)

Exponents are treated by first converting the uncertainty into a percentage uncertainty and then multiplying this uncertainty by the exponent (if the exponent is 2, multiply the uncertainty by 2; if the exponent is 3, multiply the uncertainty by 3 and so on).

For example, a rate equation is rate = $k[A]^2$.

The concentration of A is given as $0.041\,mol\,dm^{-3} \pm 0.002\,mol\,dm^{-3}$.

When calculating a value for $[A]^2$ (for example, to use the value of k to determine the rate), we carry out the following operations:

$$\text{percentage uncertainty} = \frac{0.002}{0.041} \times 100 = 4.878\%$$

Exponent = 2, therefore the percentage uncertainty is $4.878\% \times 2 = 9.756\% = 10\%$

$[A]^2 = 0.001\,7\,mol^2\,dm^{-6} \pm 10\% = 0.001\,7 \pm 0.0002\,mol^2\,dm^{-6}$

WORKED EXAMPLE R1.1H

A student placed 25.0 cm^3 of 1.00 mol dm^{-3} sodium hydroxide in a glass beaker. She then added 5.0 cm^3 portions of hydrochloric acid from a burette. After each addition she measured the temperature of the mixture. Her results are plotted in the graph below. The acid and alkali had the same initial temperature.

$$NaOH(aq) + HCl(aq) \rightarrow NaCl(aq) + H_2O(l)$$

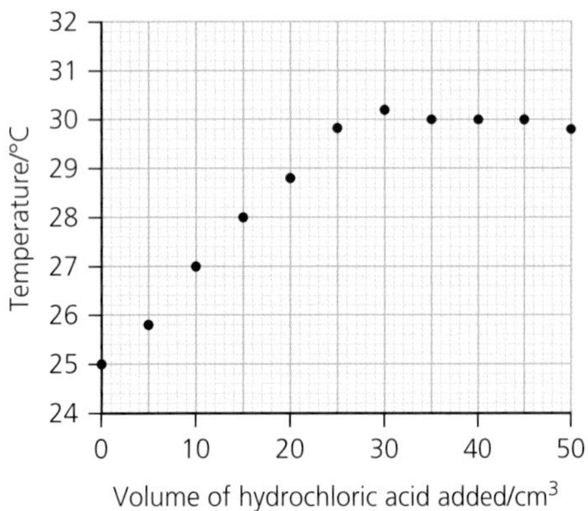

Use the results to calculate:

1 the volume of acid required to neutralize the sodium hydroxide

2 the enthalpy of neutralization for the reaction.

Answer

The graph can be split into two parts. First, heating occurs due to the exothermic nature of the reaction, and secondly the warm mixture starts to cool when the sodium hydroxide has been completely neutralized and room temperature acid is being added to the mixture. Neutralization has occurred where the two sections of the graph cross. This crossing point will also give the maximum temperature achieved. Note that this may not coincide with one of the 5.0 cm^3 portions of acid added.

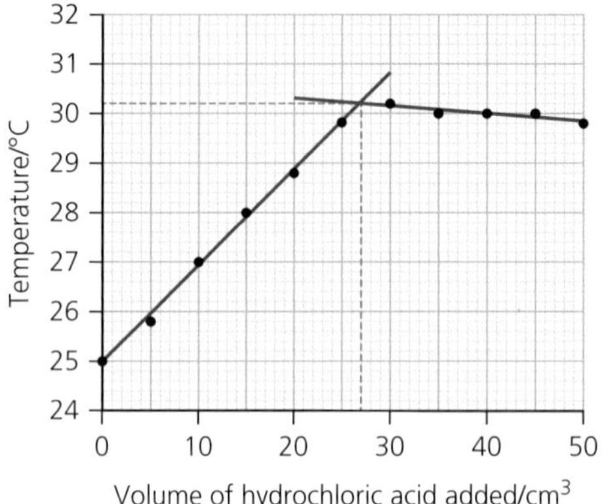

The maximum temperature achieved is 30.2 °C, and this occurs when 27.0 cm^3 of hydrochloric acid has been added.

The heat energy change is:

$$Q = (25.0 + 27.0)\,cm^3 \times 4.18\,J\,g^{-1}\,K^{-1} \times (30.2 - 25.0)\,K$$

$$Q = 1130\,J$$

$$n(NaOH) = 0.025\,dm^3 \times 1.00\,mol\,dm^{-3} = 0.025\,mol$$

$$\Delta H^{\ominus} = \frac{-1.13\,kJ}{0.025\,mol} \times 1 = -45\,kJ\,mol^{-1}$$

Inquiry 3: Concluding and evaluating

Interpret processed data to draw and justify conclusions

Interpreting processed data is a real skill and to do this you will need to draw on your best analytical skills as well as your chemistry knowledge.

Quantitative processed data will probably be in the form of a graph. To interpret this data, you need to look at the independent variable (*x*-axis) and dependent variable (*y*-axis) and try to spot trends or patterns. Once you have spotted a trend, you will then need to draw on your chemical knowledge to explain it.

Qualitative data can be a little harder to interpret as there will probably not be a graph, just observations. If this is the case, try to put the data into groups or order. Then, once you have sorted the data, use your chemistry knowledge to explain what the results are telling you.

9 Sodium hydroxide is reacted with hydrochloric acid.

$$NaOH(aq) + HCl(aq) \rightarrow NaCl(aq) + H_2O(l)$$

20.0 cm^3 of 2.00 mol dm^{-3} sodium hydroxide is measured using a pipette and placed into a polystyrene cup. The temperature of the solution is taken every minute for 3 minutes. On the fourth minute, 25.0 cm^3 of 2.00 mol dm^{-3} hydrochloric acid is added to the polystyrene cup. The temperature of the mixture is taken at 1-minute intervals until 12 minutes have passed in total. The results are recorded in the table.

Time / minutes	1	2	3	4	5	6	7	8	9	10	11	12
Temperature / °C ± 0.1 °C	23.0	23.0	23.0		34.0	34.5	33.5	33.0	32.5	32.0	32.0	31.5

a Plot a graph of temperature against time and estimate the maximum temperature change by using extrapolation to account for the cooling the mixture experiences.

b Calculate the enthalpy change for the reaction.

c The literature value for the neutralization of sodium hydroxide and hydrochloric acid is −57.9 kJ mol^{-1}. Use this accepted value to calculate the experimental error.

ATL R1.1B

Plan a calorimetry investigation into the enthalpy changes when the straight-chain alcohols containing 1–5 carbon atoms are combusted.

You should write a detailed method which could be repeated by someone who wished to replicate your results. Consider ways to mitigate the problems associated with calorimetry experiments. Include a list of the apparatus you would use.

You should include a full risk assessment of the chemicals and possible hazards. Consider any adjustments you can make to reduce risks, and to minimize environmental impacts.

Use a software package such as Microsoft *Excel* to set up a results table which is ready to process the experimental data you collect (Figure R1.15). Include formulae so you can immediately process your results. This level of planning helps experiments run smoothly and allows you to immediately spot any anomalous results.

For an extra challenge, have a graph set up so that your data is processed and plotted onto a graph as you type it in.

	A	B	C	D	E	F	G
1			Methanol			Ethanol	
2		1	2	3	1	2	3
3	Mass of water/g						
4	Mass of burner before/g						
5	Mass of burner after/g						
6	Initial temperature/K						
7	Maximum temperature/K						
8	Mass of alcohol used/g						
9	Temperature change/K						
10	Heat energy change/J						
11	Amount of alcohol used/mol						
12	Enthalpy change/kJ mol^{-1}						

■ **Figure R1.15** A spreadsheet ready to process raw data collected in a calorimetry experiment

LINKING QUESTION

How can the enthalpy change for combustion reactions, such as for alcohols or food, be investigated experimentally?

Inquiry 3: Concluding and evaluating

Explain realistic and relevant improvements to an investigation

When evaluating any investigation, you should suggest improvements, but these need to be realistic, relevant and in the correct direction.

Realistic improvements are things you could do yourself. If you were investigating gas properties over two different days and found that the air pressure was different, it would not be realistic to say that you would ensure the air pressure remained constant as this is impossible! However, if you found that the room temperature fluctuated from day to day you could control this, perhaps by using a water bath.

Improvements also need to be relevant. If you are determining the concentration of an unknown solution using a colorimeter, stating that room pressure needed to be controlled would be irrelevant as it has no effect on the experiment. However, if you stated that you used a stock solution over several different days and that you should have made a fresh solution each time (as the solution may have oxidized), this would be a relevant improvement.

The direction of the improvement can be difficult to get right. If you were using a redox titration to determine a quantity and found that your calculated value was more than expected, it would be wrong to say that this could have occurred due to the solution being less concentrated than expected due to oxidation, as this would give a result that was less than expected. You would need to look elsewhere for an error in the methodology. Perhaps you used a larger quantity of one of the reagents instead.

R1.2 Energy cycles in reactions

Guiding question

- How does application of the law of conservation of energy help us to predict energy changes during reactions?

SYLLABUS CONTENT

By the end of this chapter, you should understand that:

- bond-breaking absorbs and bond-forming releases energy
- Hess's law states that the enthalpy change for a reaction is independent of the pathway between the initial and final states
- standard enthalpy changes of combustion, $\Delta H_c^\ominus$, and formation, $\Delta H_f^\ominus$ data are used in thermodynamic calculations (HL only)
- an application of Hess's law uses enthalpy of formation data or enthalpy of combustion data to calculate the enthalpy change of a reaction (HL only)
- a Born–Haber cycle is an application of Hess's law, used to show energy changes in the formation of an ionic compound (HL only).

By the end of the chapter, you should know how to:

- calculate the enthalpy change of a reaction from given average bond enthalpy data
- apply Hess's law to calculate enthalpy changes in multistep reactions
- deduce equations and solutions to problems involving standard enthalpy changes of combustion and formation (HL only)
- calculate enthalpy changes of a reaction using $\Delta H_f^\ominus$ data or $\Delta H_c^\ominus$ data (HL only)
- interpret and determine values from a Born–Haber cycle for compounds composed of univalent and divalent ions (HL only).

Common mistake

Misconception: *Energy is released when bonds break.*

Energy is always absorbed when bonds break. Work needs to be done to overcome the attractive force in the bond or intermolecular force.

Bond-breaking absorbs and bond-forming releases energy

In a simple covalent molecule, energy is required to overcome or break the attractive forces between a shared pair or pairs of electrons and the two nuclei. If enough energy is expended, two well separated atoms with no bonding attraction between them can be formed. This concept is illustrated in Figure R1.16 using the hydrogen molecule.

Similarly, overcoming the electrostatic forces of attraction between the oppositely charged ions in an ionic compound (breaking the ionic bonding) requires an input of energy.

Energy is also required to overcome metallic bonding and vaporize a metal.

Bond breaking is always an endothermic process.

An application of Hess's law (which you will meet later in this chapter, on page 339) is that if bond-breaking is always endothermic, the reverse process, bond-making, must always be exothermic.

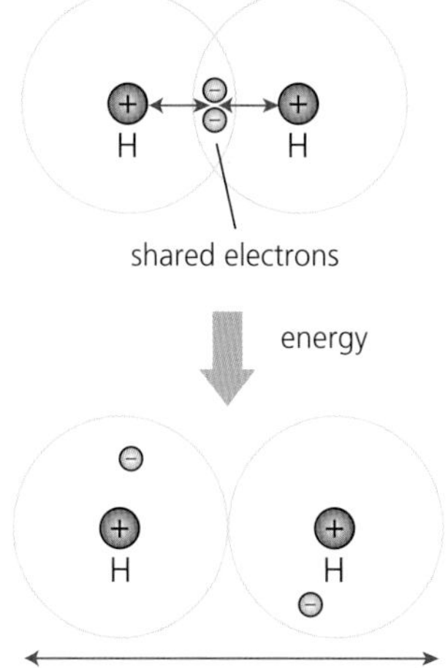

■ **Figure R1.16** To break a covalent bond between two H atoms, the attractive forces between the shared pair of electrons and the nuclei of the two atoms need to be overcome

Going further

Bond-forming and bond order

Molecular orbital (MO) theory addresses some of the deficiencies of the Lewis model in which electrons are shared between two nuclei. MO theory was developed by Friedrich Hund (1896–1997) and Robert Mulliken (1896–1986), and Mulliken received a Nobel Prize in 1966.

In MO theory, new molecular orbitals are formed from the interaction of atomic orbitals. In a hydrogen molecule, the 1s atomic orbitals from each atom interact to form a σ (bonding) and a σ* (antibonding) molecular orbital (Figure R1.17). The two electrons in the molecule fill the molecular orbitals starting with the lowest energy orbitals.

When a bond is formed between two hydrogen atoms, the electrons from the atomic orbitals enter the lower-energy σ molecular orbital. Energy is released as the bond is formed, so this is an exothermic process.

In MO theory we define a quantity known as the bond order for two interacting atoms. This is equivalent to describing a bond as single, double or triple in the Lewis model of bonding.

$$\text{Bond order} = \frac{1}{2}\left(\begin{array}{c}\text{number of electrons in bonding}\\ \text{molecular orbitals} - \text{number of electrons}\\ \text{in antibonding molecular orbitals}\end{array}\right)$$

The bond order for an H_2 molecule is 1, the equivalent of a single bond in the more familiar Lewis model.

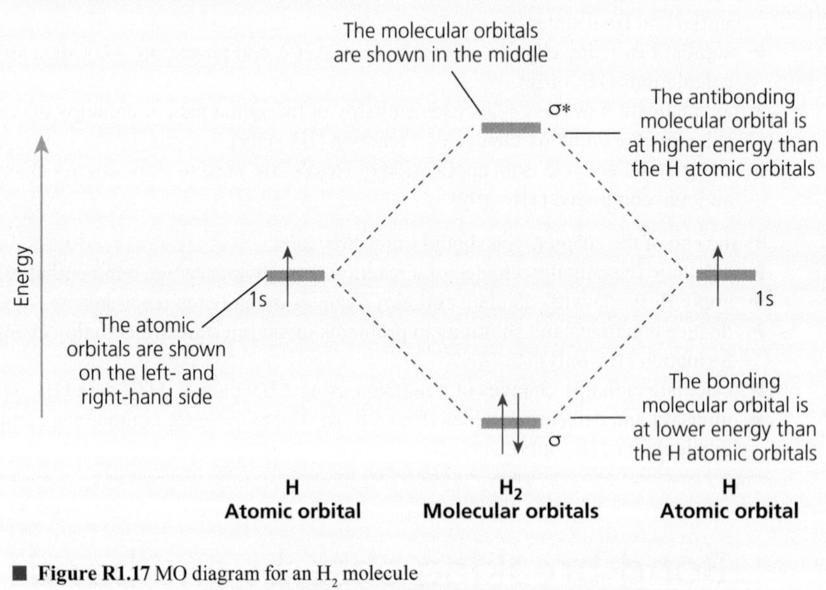
The molecular orbitals are shown in the middle
σ*
The antibonding molecular orbital is at higher energy than the H atomic orbitals
Energy
1s
1s
The atomic orbitals are shown on the left- and right-hand side
The bonding molecular orbital is at lower energy than the H atomic orbitals
σ
H
Atomic orbital
H₂
Molecular orbitals
H
Atomic orbital

■ **Figure R1.17** MO diagram for an H_2 molecule

Bond enthalpies

Bond enthalpy is the enthalpy change when one mole of a specific covalent bond is broken within gaseous molecules. For a diatomic molecule, XY, the bond enthalpy of X–Y is the enthalpy change when one mole of the molecules are separated into the gaseous constituent atoms.

$$\text{X–Y(g)} \rightarrow \text{X(g)} + \text{Y(g)}; \qquad \Delta H^{\ominus} = \text{bond enthalpy of X–Y}$$

For the hydrogen molecule, the thermochemical equation describing the bond-breaking process is:

$$H_2(g) \rightarrow 2H(g); \qquad \Delta H^{\ominus} = +436\,\text{kJ mol}^{-1}$$

The bond enthalpy for an H–H bond is $436\,\text{kJ mol}^{-1}$.

The strength of a covalent bond is indicated by the size of the bond enthalpy. The larger the bond enthalpy, the stronger the covalent bond.

LINKING QUESTION

How would you expect bond enthalpy data to relate to bond length and polarity?

Average bond enthalpies

A hydrogen-to-hydrogen bond only exists within a molecule of hydrogen, but most other covalent bonds are found in many different molecules. For example, a C–H bond is present in nearly all organic molecules, including hydrocarbons, sugars and amino acids. There is some variation in the bond enthalpies for the same type of bond in different compounds.

To break the first C–H bond in a molecule of methane requires 439 kJ mol^{-1}. The equivalent process in ethane and methanol requires 423 kJ mol^{-1} and 402 kJ mol^{-1}, respectively.

In methane, each subsequent C–H bond broken requires a different amount of energy to break.

$CH_4(g) \rightarrow CH_3(g) + H(g)$; $\Delta H^{\ominus} = +439\,kJ\,mol^{-1}$

$CH_3(g) \rightarrow CH_2(g) + H(g)$; $\Delta H^{\ominus} = +462\,kJ\,mol^{-1}$

$CH_2(g) \rightarrow CH(g) + H(g)$; $\Delta H^{\ominus} = +422\,kJ\,mol^{-1}$

$CH(g) \rightarrow C(g) + H(g)$; $\Delta H^{\ominus} = +338\,kJ\,mol^{-1}$

so $CH_4(g) \rightarrow C(g) + 4H(g)$; $\Delta H^{\ominus} = +1661\,kJ\,mol^{-1}$

Clearly, it is impractical to list the bond strength of every C–H bond in every possible compound. We can calculate an average C–H bond enthalpy in CH_4 as $\frac{+1661\,kJ\,mol^{-1}}{4} = 415\,kJ\,mol^{-1}$, but the average value for other compounds is different.

◆ **Average bond enthalpy**: The energy needed to break the specific covalent bonds in one mole of gaseous molecules, separating the molecules into gaseous atoms, averaged over similar compounds.

By looking at the bond enthalpies for a specific type of bond in a range of different molecules it is possible to determine an **average bond enthalpy**. The average bond enthalpy given in the IB *Chemistry data booklet* for a C–H bond is 414 kJ mol^{-1}. A selection of bond enthalpies and average bond enthalpies are given in Table R1.3. A more extensive table of bond enthalpies and average bond enthalpies at 298.15 K is given in section 12 of the IB *Chemistry data booklet*.

Top tip!

In the IB *Chemistry data booklet*, bond enthalpy values are given for bonds which only exist in one type of molecule, for example an F–F bond. For bonds which exist in many different molecules, an average bond enthalpy is given.

Table R1.3 A selection of average bond enthalpies at 298.15 K

Bond	H–H	C–C	C=C	C≡C	N≡N	O–O	O=O	N–H
Bond enthalpy / kJ mol^{-1}	436	346	614	839	945	144	498	391

Bond	F–F	Cl–Cl	Br–Br	I–I	C–H	O–H	C≡N	C=O
Bond enthalpy / kJ mol^{-1}	159	242	193	151	414	463	890	804

Tool 3: Mathematics

Simple mean and range

The mean is calculated by adding together all the values and dividing by the total number of values. The purpose of averaging data (provided it follows a normal distribution) is to reduce the effect of random errors.

The range is the difference between the largest and smallest data values. Since it only depends on two of the measurements, it is most useful in representing the dispersion of small data sets.

Both mean and range are measured in the same units as the data.

Calculation of the enthalpy change of a reaction from given average bond enthalpy data

Bond enthalpies can be used to estimate the enthalpy change for a particular reaction involving molecules in the gaseous state. Consider, for example, the combustion of methane:

$$CH_4(g) + 2O_2(g) \rightarrow CO_2(g) + 2H_2O(g)$$

The reaction can be regarded as occurring in two steps. In the first, all of the bonds in the reactants have to be broken to form atoms. This is an endothermic process and heat energy has to be absorbed from the surroundings. In the second step, bond formation occurs. This is an exothermic process and releases heat energy to the surroundings. The overall reaction is exothermic since the energy released during bond formation is greater than the energy absorbed during bond-breaking (Figure R1.18).

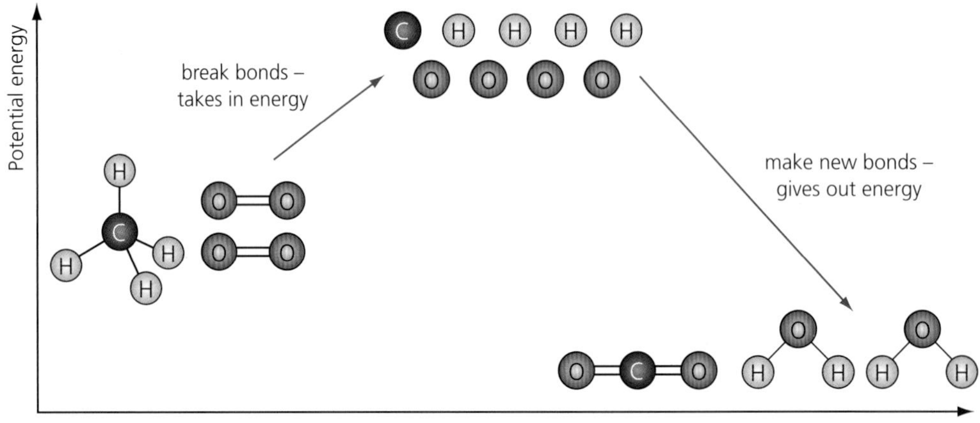

■ **Figure R1.18** Breaking and forming of bonds during the combustion of methane

In the first step, energy is required to:

- break 4 C–H bonds in a methane molecule = $4 \times 414\,kJ\,mol^{-1} = 1656\,kJ\,mol^{-1}$
- break 2 O=O bonds in two oxygen molecules = $2 \times 498\,kJ\,mol^{-1} = 996\,kJ\,mol^{-1}$.

Total amount of energy required to break all these bonds = $(1656 + 996)\,kJ\,mol^{-1} = 2652\,kJ\,mol^{-1}$.

In the second step, energy is released:

- forming 2 C=O bonds in a carbon dioxide molecule = $2 \times 804\,kJ\,mol^{-1} = 1608\,kJ\,mol^{-1}$
- forming 4 O–H bonds in two water molecules = $4 \times 463\,kJ\,mol^{-1} = 1852\,kJ\,mol^{-1}$.

Total amount of energy released to surroundings when these bonds are formed = $(1608 + 1852)\,kJ\,mol^{-1} = 3460\,kJ\,mol^{-1}$.

The enthalpy change of the reaction is calculated by considering the difference between the energy required to break bonds and the energy released when new bonds form:

$$\Delta H^{\ominus} = \sum\text{Bond enthalpy of bonds broken} - \sum\text{Bond enthalpy of bonds formed}$$

$$\Delta H^{\ominus} = (2652 - 3460)\,kJ\,mol^{-1}$$

$$\Delta H^{\ominus} = -808\,kJ\,mol^{-1}$$

Since more energy is released when the new bonds in the products are formed than is needed to break the bonds in the reactants to begin with, there is an overall release of energy in the form of heat. The reaction is exothermic. In a reaction which is endothermic, the energy absorbed by bond-breaking is greater than the energy released by bond formation.

Because bond enthalpies are often average values, enthalpy changes calculated using bond enthalpies will *not* be exactly equal to an accurate experimentally determined value. For methane, the accepted value for its complete combustion to form gaseous carbon dioxide and water is $-810\,kJ\,mol^{-1}$.

Remember, bond enthalpy values can be used for reactions which are entirely in the gaseous phase. If a component is not in the gaseous phase, the energy change associated with the required change of state should be factored into the calculation.

WORKED EXAMPLE R1.2A

The bond enthalpies for $H_2(g)$ and $HCl(g)$ are $436\,kJ\,mol^{-1}$ and $431\,kJ\,mol^{-1}$ respectively. Calculate the bond enthalpy of chlorine, given:

$$H_2(g) + Cl_2(g) \rightarrow 2HCl(g) \qquad \Delta H^{\ominus} = -184\,kJ\,mol^{-1}$$

Answer

$$\Delta H^{\ominus} = \sum \text{Bond enthalpy of bonds broken} - \sum \text{Bond enthalpy of bonds formed}$$

$$-184\,kJ\,mol^{-1} = (436\,kJ\,mol^{-1} + Cl\text{–}Cl) - (2 \times 431\,kJ\,mol^{-1})$$

$$678\,kJ\,mol^{-1} = 436\,kJ\,mol^{-1} + Cl\text{–}Cl$$

$$Cl\text{–}Cl = 242\,kJ\,mol^{-1}$$

WORKED EXAMPLE R1.2B

Use the bond enthalpy data provided to estimate the enthalpy change when cyclobutene is hydrogenated to cyclobutane. Give a reason why this estimation may differ from the actual value for the hydrogenation of cyclobutene.

Bond	H–H	C–C	C=C	C–H
Bond enthalpy / $kJ\,mol^{-1}$	436	346	614	414

Answer

$$C_4H_6(g) + H_2(g) \rightarrow C_4H_8(g)$$

```
H       H                      H       H
 \     /                        \     /
H—C — C                       H—C — C—H
  |   ||     +  H−H  ———→         |   |
H—C — C                       H—C — C—H
 /     \                        /     \
H       H                      H       H
```

In this example, it is sensible just to consider the bonds broken and formed because many within the structure of the cyclobutene remain unchanged. A C=C and an H–H bond are broken and replaced by a C–C and 2 × C–H bonds.

$$\Delta H^\ominus = \sum\text{Bond energy of bonds broken} - \sum\text{Bond energy of bonds formed}$$

$$= \{C{=}C + H{-}H\} - \{C{-}C + (2 \times C{-}H)\}$$

$$= (614\,\text{kJ mol}^{-1} + 436\,\text{kJ mol}^{-1}) - \{346\,\text{kJ mol}^{-1} + (2 \times 414\,\text{kJ mol}^{-1})\}$$

$$= (1050 - 1174)\,\text{kJ mol}^{-1}$$

$$= -124\,\text{kJ mol}^{-1}$$

The calculation gives an estimate because it uses average bond enthalpy values. The values for these bonds within cyclobutene and cyclobutane will differ from the average values given in the table.

Experimental and computational data can be used to achieve a more accurate estimation of the enthalpy change as $-32\,\text{kJ mol}^{-1}$. This highlights that in certain cases, when the actual bond enthalpies differ significantly from the average values, the use of average bond enthalpies can produce an inaccurate estimate.

Going further

Bond enthalpies in cyclic compounds

VSEPR, introduced in S2.2, is a method for predicting the shape of molecules based on the number of electron domains around an atom. Where there are three electron domains, the domains aim to achieve an angle of 120° between them; where there are four, the ideal bond angle is 109.5°. In smaller cyclic molecules, it may not be possible to achieve the angles predicted by VSEPR theory. In the case of cyclobutane (Figure R1.19), the bond angle defined by 3 carbon atoms within the ring is significantly less than 109.5°.

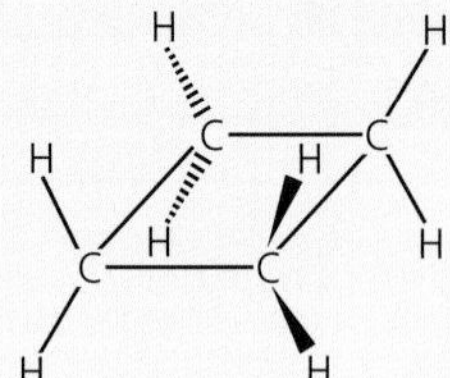

■ **Figure R1.19** The structure of cyclobutane

The atomic orbitals on each carbon in cyclobutene cannot overlap with one another effectively because of the constrained angle between the atoms. The result is a weakened bond. A C–C bond in cyclobutane has a bond enthalpy of $272\,\text{kJ mol}^{-1}$, lower than that of a typical C–C bond ($346\,\text{kJ mol}^{-1}$). Chemists describe this phenomenon as ring strain.

Ring strain is also present in other 3- and 4-membered cyclic compounds. One 4-membered ring of interest is the beta-lactam ring (cyclic amide) present in penicillin molecules (Figure R1.20). The beta-lactam ring is the key to penicillin's antibacterial properties. The ring strain means the beta-lactam ring is able to break apart relatively easily.

This occurs when penicillins meet the bacterial enzyme transpeptidase, and the construction of bacterial cell walls is disrupted as a result.

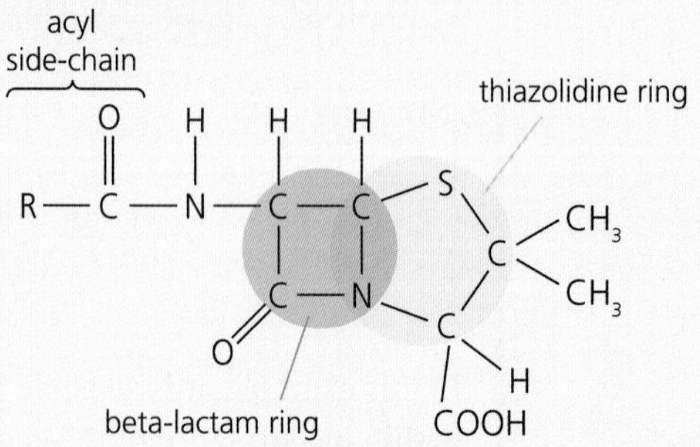

■ **Figure R1.20** The general structure of penicillins

Tool 1: Experimental techniques

Substitution reactions of halogenoalkanes

The difference in reactivity of structurally similar halogenoalkanes (R–Hal) can be attributed, in part, to the bond enthalpy of the C–Hal bond. The relative rates of hydrolysis of various halogenoalkanes (Figure R1.21) demonstrate their differing reactivities.

$$>C-Hal + H_2O \longrightarrow C-OH + HHal$$

■ **Figure R1.21** Generalized equation for the hydrolysis of a halogenoalkane to form an alcohol

To study this in practice, three test tubes (labelled R–Cl, R–Br and R–I) containing ethanol ($2\,cm^3$) to act as a solvent are placed in a water bath and heated to a temperature of 40 °C. A fourth test tube containing aqueous silver nitrate solution ($5\,cm^3$, $0.10\,mol\,dm^{-3}$) is also placed in the same water bath. The labelled test tubes are removed from the water bath and two drops of 1-chlorobutane, 1-bromobutane or 1-iodobutane are added to each tube. $1\,cm^3$ of silver nitrate solution is then added quickly to each tube and the tubes are shaken.

The tubes are observed for 10 minutes. The water acts as a nucleophile (electron pair donor) and a bonded halogen atom (–Hal) is substituted by a hydroxyl functional group (–OH). The displaced halide ion rapidly reacts with silver ions to give a silver halide precipitate:

$$Ag^+(aq) + Hal^-(aq) \rightarrow AgHal(s)$$

The mixtures slowly become cloudier as more and more precipitate forms.

The amount of silver halide that is formed in a given time is proportional to the rates of hydrolysis. The results clearly show that the rate of hydrolysis is slowest for 1-chlorobutane and fastest for 1-iodobutane. This correlates with the relative bond strength of the C–Hal bond.

LINKING QUESTION

How does the strength of a carbon–halogen bond affect the rate of a nucleophilic substitution reaction?

WORKED EXAMPLE R1.2C

Ethanol is a colourless liquid which can be combusted in a plentiful supply of oxygen.

$$C_2H_5OH(l) + 3O_2(g) \rightarrow 2CO_2(g) + 3H_2O(g)$$

Use bond enthalpy data in section 12 of the IB *Chemistry data booklet*, and the enthalpy of vaporization of ethanol given below, to estimate the enthalpy change when liquid ethanol is combusted.

$$C_2H_5OH(l) \rightarrow C_2H_5OH(g) \qquad \Delta H^\ominus = +44\,kJ\,mol^{-1}$$

Answer

Using bond enthalpies:

$$\Delta H^\ominus = \sum\text{Bond enthalpy of bonds broken} - \sum\text{Bond enthalpy of bonds formed}$$

$$\Delta H^\ominus = \{C–C + (5 \times C–H) + C–O + O–H + (3 \times O{=}O)\} - \{(4 \times C{=}O) + (6 \times O–H)\}$$

$$\Delta H^\ominus = (4731 - 5994)\,kJ\,mol^{-1}$$

$$\Delta H^\ominus = -1263\,kJ\,mol^{-1}$$

However, 44 kJ of energy is required to vaporize each mole of ethanol, so the overall enthalpy change for the reaction:

$$\Delta H^\ominus = -1263\,kJ\,mol^{-1} + 44\,kJ\,mol^{-1}$$

$$\Delta H^\ominus = -1219\,kJ\,mol^{-1}$$

10 Use the bond enthalpies in section 12 of the IB *Chemistry data booklet* to calculate the enthalpy changes associated with the following reactions:

a $CH_4(g) + Cl_2(g) \rightarrow CH_3Cl(g) + HCl(g)$

b $C_4H_{10}(g) \rightarrow C_2H_6(g) + C_2H_4(g)$

c $CH_4(g) + 2H_2O(g) \rightarrow CO_2(g) + 4H_2(g)$

11 Use the enthalpy change for the reaction of iodine with chlorine to complete the table.

$$I_2(g) + Cl_2(g) \rightarrow 2ICl \qquad \Delta H^\ominus = -70.2\,kJ\,mol^{-1}$$

Bond	I–I	Cl–Cl	I–Cl
Bond enthalpy / kJ mol^{-1}	151	242	?

Inquiry 1: Exploring and designing

State and explain predictions using scientific understanding

In many experiments you will make a hypothesis or prediction based upon existing chemical knowledge.

For example, bond enthalpies predict a linear relationship between the enthalpy of combustion of monohydric alcohols (those with one –OH group) and the number of carbon atoms. These predictions arise because each methylene group, $-CH_2-$, is responsible for a fixed increment in the enthalpy of combustion.

As the number of atoms in an alcohol molecule increases, more energy is required to break the bonds, but even larger amounts of energy are released as these atoms form carbon dioxide and water. Upon combustion, each alcohol molecule forms one more carbon dioxide molecule and one more water molecule than the previous alcohol.

■ **Table R1.4** Bonds broken and formed during the complete combustion of monohydric alcohols

	Bonds broken					Bonds formed	
Name of alcohol	**C–C**	**C–H**	**C–O**	**O–H**	**O=O**	**C=O**	**O–H**
methanol	0	3	1	1	1.5	2	4
ethanol	1	5	1	1	3	4	6
propan-1-ol	2	7	1	1	4.5	6	8
butan-1-ol	3	9	1	1	6	8	10

There is a constant difference between the alcohols as the homologous series is ascended:

1 extra C–C bond is broken = 1 × 346 = 346 kJ extra energy absorbed.

2 extra C–H bonds are broken = 2 × 414 = 828 kJ extra energy absorbed.

1.5 extra O=O bonds are broken = 1.5 × 498 = 747 kJ extra energy absorbed.

Hence, total extra energy absorbed = 346 + 828 + 747 = 1921 kJ mol^{-1}.

2 extra C=O bonds are formed = 2 × 804 = 1608 kJ extra energy released.

2 extra O–H bonds are formed = 2 × 463 = 926 kJ extra energy released.

So, total extra energy released = 1608 + 926 = 2534 kJ mol^{-1}.

The extra energy released is greater than the extra energy absorbed and thus, as the homologous series is ascended, the enthalpies of combustion become more exothermic.

Also, because the change in the structures of the alcohols is fixed, the difference in the enthalpy of combustion from one homologue to the next will also be fixed and will be: (1921 – 2534) kJ mol^{-1} = –613 kJ mol^{-1}.

Figure R1.22 shows the experimentally determined values for the combustion of alcohols in the gaseous phase. The values are derived by combining standard enthalpies of combustion and enthalpies of vaporisation. The graph shows that the trend is for the combustion reactions to become increasingly exothermic with increasing carbon chain length, as predicted by bond enthalpy data. The relationship between enthalpies of combustion and carbon chain length is close to being linear, and the line of best fit shows a jump of $-615\,kJ\,mol^{-1}$ between each member of the homologous series.

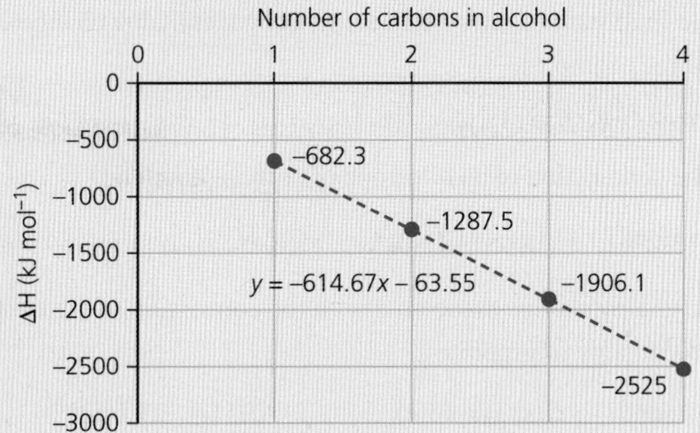

■ **Figure R1.22** Experimentally determined values for the combustion of alcohols $C_nH_{2n+1}OH(g) + O_2(g) \rightarrow nCO_2(g) + (n+1)H_2O(g)$

Hess's law

Hess's law is a statement of the first law of thermodynamics (conservation of energy) in a form useful in chemistry. It states that, if a reaction is carried out in a series of steps, the enthalpy change for the overall reaction will equal the sum of the enthalpy changes for the individual steps. The overall enthalpy change is independent of the number of steps or the path by which the reaction is carried out.

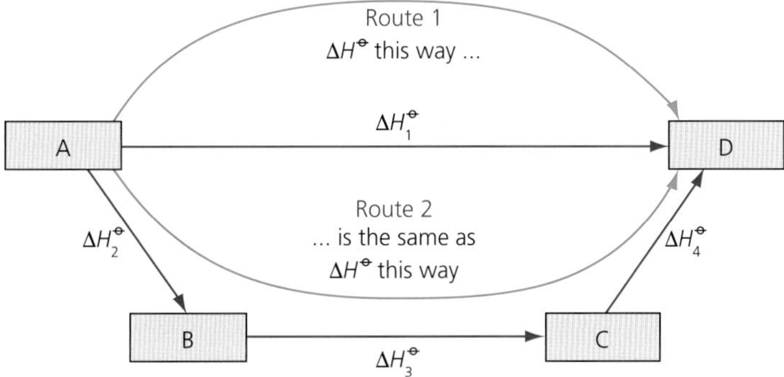

■ **Figure R1.23** Hess's law

Figure R1.23 illustrates Hess's law. The conversion of A to D is possible via two pathways: route 1 and route 2.

Hess's law tells us that the enthalpy changes of routes 1 and 2 are equal, therefore:

$$\Delta H_1^\ominus = \Delta H_2^\ominus + \Delta H_3^\ominus + \Delta H_4^\ominus$$

Hess's law provides a useful means of calculating enthalpy changes that are difficult to measure directly. Chemists can calculate $\Delta H^{\ominus}$ for any chemical reaction or physical process as long as they can find a route for which $\Delta H^{\ominus}$ is known for each step. This means that a relatively small number of experimental measurements can be used to calculate $\Delta H^{\ominus}$ for a huge number of different reactions.

Using Hess's law to calculate enthalpy changes

To calculate the enthalpy change associated with a reaction, we can construct a pathway from the initial to the final state of the system using known enthalpy changes. This is demonstrated in the following three examples.

Combustion of diamond

Consider the combustion of an allotrope of carbon, diamond.

$$C(s, diamond) + O_2(g) \rightarrow CO_2(g); \qquad \Delta H_1^{\ominus}$$

Given the enthalpy change associated with the conversion of diamond to graphite and the combustion enthalpy of graphite:

$$C(s, diamond) \rightarrow C(s, graphite); \qquad \Delta H_2^{\ominus} = -2\,kJ\,mol^{-1}$$

$$C(s, graphite) + O_2(g) \rightarrow CO_2(g); \qquad \Delta H_3^{\ominus} = -393\,kJ\,mol^{-1}$$

We can apply Hess's law and construct a Hess's law cycle (Figure R1.24).

Figure R1.24 shows that

$\Delta H_1^{\ominus} = \Delta H_2^{\ominus} + \Delta H_3^{\ominus}$

$\Delta H_1^{\ominus} = (-2 + -393)\,kJ\,mol^{-1}$

$\Delta H_1^{\ominus} = -395\,kJ\,mol^{-1}$

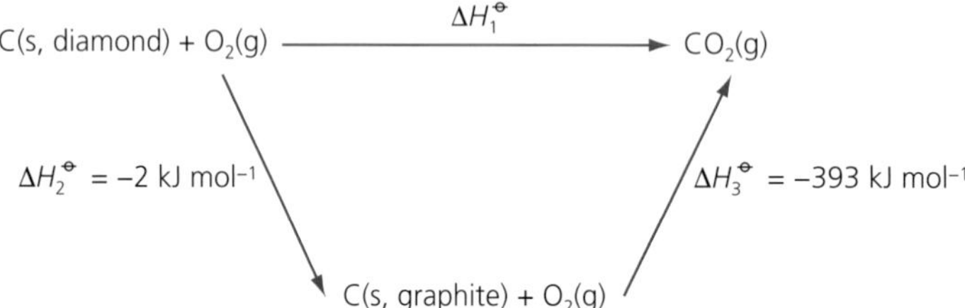

■ **Figure R1.24** Hess's law cycle for the combustion of diamond

It follows from Hess's law that chemical equations can be treated in the same way as algebraic equations: they can be added, subtracted or multiplied by a constant value. This presents the option to solve a problem algebraically, as opposed to constructing a Hess's law cycle.

Given:

$C(s, diamond) \rightarrow C(s, graphite); \quad \Delta H_2^{\ominus} = -2\,kJ\,mol^{-1}$

$C(s, graphite) + O_2(g) \rightarrow CO_2(g); \quad \Delta H_3^{\ominus} = -393\,kJ\,mol^{-1}$

Adding the two enthalpy changes together:

$[C(s, diamond) \rightarrow C(s, graphite)] + [C(s, graphite) + O_2(g) \rightarrow CO_2(g)] = \Delta H_2^{\ominus} + \Delta H_3^{\ominus}$

And grouping reactants and products together:

$C(s, diamond) + \cancel{C(s, graphite)} + O_2(g) \rightarrow \cancel{C(s, graphite)} + CO_2(g) = -2 - 393\,kJ\,mol^{-1}$

$C(s, diamond) + O_2(g) \rightarrow CO_2(g); \quad \Delta H_1^{\ominus} = -395\,kJ\,mol^{-1}$

Incomplete combustion of graphite

Consider the combustion of carbon (as graphite) to give carbon monoxide:

$$C(s) + \frac{1}{2}O_2(g) \rightarrow CO(g); \qquad \Delta H_4^{\ominus}$$

The enthalpy change of this reaction cannot be found experimentally because upon mixing carbon and oxygen, carbon monoxide is not the exclusive product, even in a limited amount of oxygen. However, by considering the known enthalpy changes for the complete combustion of carbon and carbon monoxide, a Hess's law cycle can be constructed (Figure R1.25).

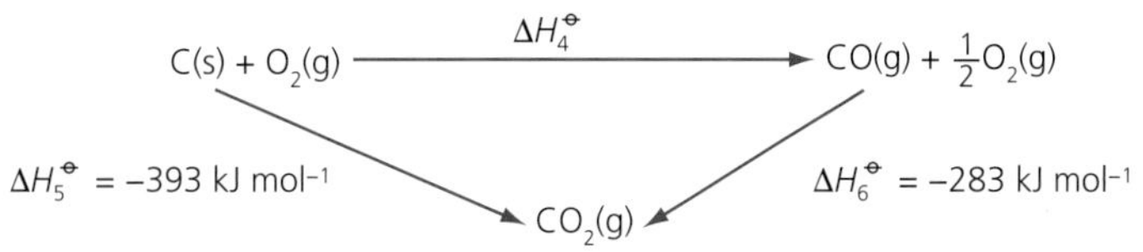

$C(s) + O_2(g)$ —— $\Delta H_4^{\ominus}$ ——→ $CO(g) + \frac{1}{2}O_2(g)$

$\Delta H_5^{\ominus} = -393\ \text{kJ mol}^{-1}$ ↘ $CO_2(g)$ ↙ $\Delta H_6^{\ominus} = -283\ \text{kJ mol}^{-1}$

■ **Figure R1.25** Hess's law cycle to determine the enthalpy change when carbon and oxygen form carbon monoxide

$$C(s) + O_2(g) \rightarrow CO_2(g); \qquad \Delta H_5^{\ominus} = -393\,\text{kJ mol}^{-1}$$

$$CO(g) + \frac{1}{2}O_2(g) \rightarrow CO_2(g); \qquad \Delta H_6^{\ominus} = -283\,\text{kJ mol}^{-1}$$

The reaction across the top of the Hess cycle in Figure R1.25, $\Delta H_4^{\ominus}$, is the conversion of C(s) and $\frac{1}{2}O_2(g)$ to form CO(g).

If a reaction is reversed, the sign for the enthalpy change is reversed. This tells us that:

$$CO_2(g) \rightarrow CO(g) + \frac{1}{2}O_2(g) \qquad -\Delta H_6^{\ominus} = +283\,\text{kJ mol}^{-1}$$

From the Hess's law cycle (Figure R1.25) we can see that $\Delta H_4^{\ominus}$ is equal to the addition of $\Delta H_5^{\ominus}$ and the negative of $\Delta H_6^{\ominus}$.

$$\Delta H_4^{\ominus} = \Delta H_5^{\ominus} - \Delta H_6^{\ominus}$$

$$\Delta H_4^{\ominus} = (-393 + 283)\ \text{kJ mol}^{-1}$$

$$\Delta H_4^{\ominus} = -110\ \text{kJ mol}^{-1}$$

Algebraically, reversing $\Delta H_6^{\ominus}$ and adding it to $\Delta H_5^{\ominus}$

$$C(s) + O_2(g) + \cancel{CO_2(g)} \rightarrow \cancel{CO_2(g)} + CO(g) + \frac{1}{2}O_2(g) \qquad = (\Delta H_5^{\ominus} - \Delta H_6^{\ominus})$$

$$C(s) + \frac{1}{2}O_2(g) \rightarrow CO(g) \qquad = -110\,\text{kJ mol}^{-1}$$

Oxidation of hydroquinone

Quinone, $C_6H_4O_2$, can be formed by the reaction of hydroquinone, $C_6H_4(OH)_2$, with hydrogen peroxide, H_2O_2. The equation for this reaction is:

$$C_6H_4(OH)_2(aq) + H_2O_2(aq) \rightarrow C_6H_4O_2(aq) + 2H_2O(l); \qquad \Delta H_7^{\ominus}$$

The following data can be used to calculate the enthalpy change of the above reaction:

$$C_6H_4\,(OH)_2(aq) \rightarrow C_6H_4O_2(aq) + H_2(g); \qquad \Delta H_8^{\ominus} = +177\,\text{kJ mol}^{-1}$$

$$H_2(g) + O_2(g) \rightarrow H_2O_2(aq); \qquad \Delta H_9^{\ominus} = -191\,\text{kJ mol}^{-1}$$

$$H_2(g) + \frac{1}{2}O_2(g) \rightarrow H_2O(l); \qquad \Delta H_{10}^{\ominus} = -286\,\text{kJ mol}^{-1}$$

The challenge is to find a pathway with the given enthalpy changes. This is presented in Figure R1.26.

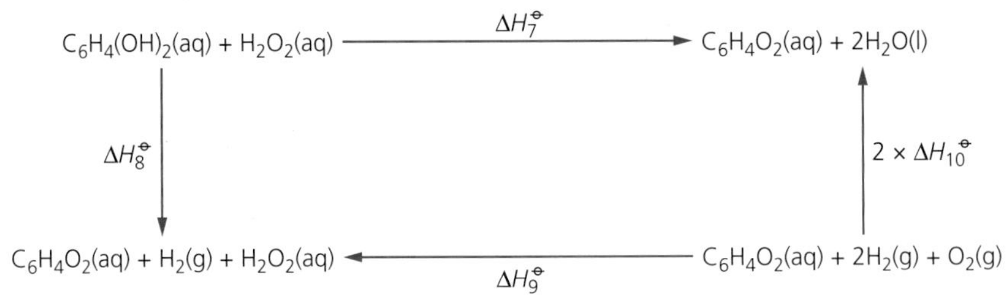

$C_6H_4(OH)_2(aq) + H_2O_2(aq)$ —$\Delta H_7^{\ominus}$→ $C_6H_4O_2(aq) + 2H_2O(l)$

$\Delta H_8^{\ominus}$ (down) $\qquad$ $2 \times \Delta H_{10}^{\ominus}$ (up)

$C_6H_4O_2(aq) + H_2(g) + H_2O_2(aq)$ ←$\Delta H_9^{\ominus}$— $C_6H_4O_2(aq) + 2H_2(g) + O_2(g)$

■ **Figure R1.26** A Hess's law cycle to produce quinone

The enthalpy change of the reaction in question, $\Delta H_7^{\ominus}$, is:

$$\Delta H_7^{\ominus} = \Delta H_8^{\ominus} - \Delta H_9^{\ominus} + (2 \times \Delta H_{10}^{\ominus})$$

$$= 177\,\text{kJ mol}^{-1} - (-191\,\text{kJ mol}^{-1}) + (2 \times -286)\,\text{kJ mol}^{-1}$$

$$= -204\,\text{kJ mol}^{-1}$$

Algebraically, the correct combination of enthalpy changes can be chosen to equal the enthalpy change of interest, $\Delta H_7^{\ominus}$.

$$\Delta H_8^{\ominus} - \Delta H_9^{\ominus} = C_6H_4(OH)_2(aq) + H_2O_2(aq) \rightarrow C_6H_4O_2(aq) + 2H_2(g) + O_2(g)$$

Then add $2 \times \Delta H_{10}^{\ominus}$

$$C_6H_4(OH)_2(aq) + H_2O_2(aq) + \cancel{2H_2(g) + O_2(g)} \rightarrow C_6H_4O_2(aq) + \cancel{2H_2(g) + O_2(g)} + 2H_2O(l)$$

$$C_6H_4(OH)_2(aq) + H_2O_2(aq) \rightarrow C_6H_4O_2(aq) + 2H_2O(l) = \Delta H_7^{\ominus}$$

So $\Delta H_7^{\ominus} = \Delta H_8^{\ominus} - \Delta H_9^{\ominus} + 2 \times \Delta H_{10}^{\ominus} = 177\,\text{kJ mol}^{-1} - 191\,\text{kJ mol}^{-1} + (2 \times 286)\,\text{kJ mol}^{-1}$

$\Delta H_7^{\ominus} = -204\,\text{kJ mol}^{-1}$

WORKED EXAMPLE R1.2D

Calcium carbonate decomposes to calcium oxide and carbon dioxide:

$$CaCO_3(s) \rightarrow CaO(s) + CO_2(g); \qquad \Delta H_1^{\ominus}$$

The reaction is slow and a high temperature is required to bring it to completion. Direct measurement of the enthalpy change of reaction is therefore not practical. Instead, two reactions that take place readily at room temperature are carried out and their enthalpy changes used to find the enthalpy of decomposition of calcium carbonate. These reactions are the reactions of calcium carbonate and calcium oxide with dilute hydrochloric acid:

$$CaCO_3(s) + 2HCl(aq) \rightarrow CaCl_2(aq) + H_2O(l) + CO_2(g) \qquad \Delta H_2^{\ominus} = -17\,\text{kJ mol}^{-1}$$

$$CaO(s) + 2HCl(aq) \rightarrow CaCl_2(aq) + H_2O(l) \qquad \Delta H_3^{\ominus} = -195\,\text{kJ mol}^{-1}$$

Use the data given to calculate the enthalpy change for the decomposition of calcium carbonate as stated in the equation.

Answer

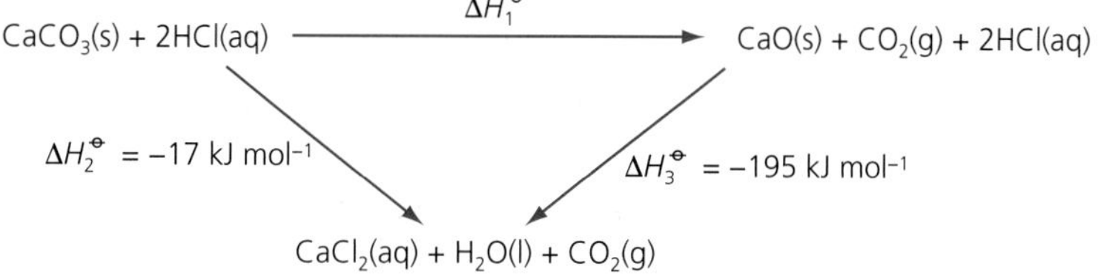

$$\Delta H_1^{\ominus} = (-17 - -195)\,kJ\,mol^{-1}$$

$$= +178\,kJ\,mol^{-1}$$

Using the algebraic approach, reversing $\Delta H_3^{\ominus}$ and adding it to $\Delta H_2^{\ominus}$:

$$CaCO_3(s) + \cancel{2HCl(aq)} + \cancel{CaCl_2(aq)} + \cancel{H_2O(l)} \rightarrow \cancel{CaCl_2(aq)} + \cancel{H_2O(l)} + CO_2(g) + CaO(s) + \cancel{2HCl(aq)}$$

$CaCO_3(s) \rightarrow CaO(s) + CO_2(g)$ $\quad \Delta H_2^{\ominus} + -\Delta H_3^{\ominus} = (-17 + 195)\,kJ\,mol^{-1}$

$CaCO_3(s) \rightarrow CaO(s) + CO_2(g)$ $\quad = +178\,kJ\,mol^{-1}$

Going further

Enthalpies of transition

It is essential to specify the physical states of the substances involved when writing thermochemical equations to represent an enthalpy change. This is because any change in physical state has its own enthalpy change (Figure R1.27).

The enthalpy of fusion of ammonia is the enthalpy change when 1 mole of solid ammonia is brought into the liquid phase.

$$NH_3(s) \rightarrow NH_3(l)$$

and the enthalpy of vaporization of ammonia is the enthalpy change when one mole of liquid ammonia is brought into the gaseous phase.

$$NH_3(l) \rightarrow NH_3(g)$$

The enthalpy of sublimation for iodine is the enthalpy change when one mole of iodine molecules are brought into the gaseous phase.

$$I_2(s) \rightarrow I_2(g)$$

Enthalpies of fusion, vaporization and sublimation represent the enthalpy changes associated with the processes commonly known as melting, boiling and subliming, respectively. These are always endothermic since intermolecular forces of attraction need to be overcome. The values of these enthalpy changes will vary with the strength of intermolecular forces.

The enthalpy changes of the reverse changes of phase have an equal magnitude but the opposite sign. Condensing, freezing and deposition are all exothermic processes. Changes of state are covered in detail in Chapter S1.1, page 18.

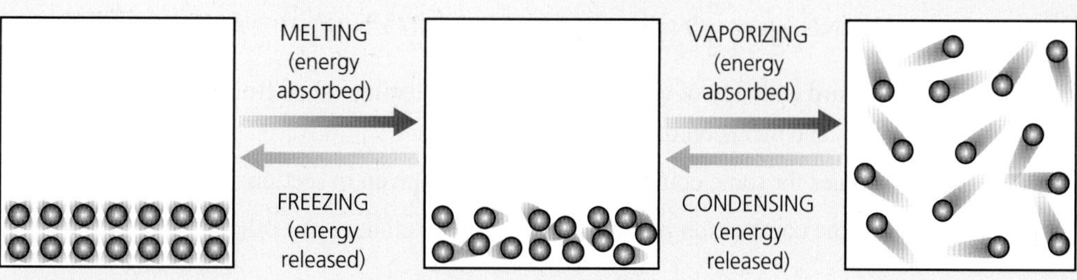

■ **Figure R1.27** Changes of state and enthalpy changes

12 Use the Hess's law energy cycle given to determine the enthalpy change when copper(II) sulfate pentahydrate is dehydrated to anhydrous copper(II) sulfate.

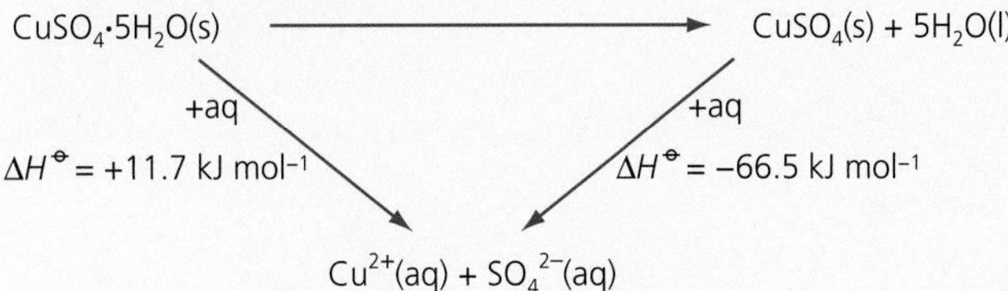

13 Given the following enthalpy changes,

$S(s) + O_2(g) \rightarrow SO_2(g) \qquad \Delta H^\ominus = -297\,kJ\,mol^{-1}$

$SO_2(g) + \frac{1}{2}O_2(g) \rightarrow SO_3(s) \qquad \Delta H^\ominus = -98\,kJ\,mol^{-1}$

what is the standard enthalpy change of the following reaction?

$S(s) + \frac{3}{2}O_2(g) \rightarrow SO_3(g)$

14 Given the following enthalpy changes,

$Fe_2O_3(s) + CO(g) \rightarrow 2FeO(s) + CO_2(g)$
$\Delta H^\ominus = -3\,kJ\,mol^{-1}$

$Fe(s) + CO_2(g) \rightarrow FeO(s) + CO(g)$
$\Delta H^\ominus = +11\,kJ\,mol^{-1}$

what is the standard enthalpy change, in $kJ\,mol^{-1}$, of the following reaction?

$Fe_2O_3(s) + 3CO(g) \rightarrow 2Fe(s) + 3CO_2(g)$

15 Given the following enthalpy changes,

$C(s) + O_2(g) \rightarrow CO_2(g) \qquad \Delta H^\ominus = -394\,kJ\,mol^{-1}$

$H_2(g) + \frac{1}{2}O_2(g) \rightarrow H_2O(l) \qquad \Delta H^\ominus = -286\,kJ\,mol^{-1}$

$3C(s) + 4H_2(g) \rightarrow C_3H_8(g) \qquad \Delta H^\ominus = -104\,kJ\,mol^{-1}$

what is the standard enthalpy change of the following reaction?

$C_3H_8(g) + 5O_2(g) \rightarrow 3CO_2(g) + 4H_2O(l)$

HL ONLY

Top tip!

Standard enthalpies, such as those of combustion and formation, are the enthalpy changes measured:

- under a gas pressure of 1 bar or 10^5 pascal (very close to 1 atmosphere)
- involving the most stable allotrope under standard conditions (for example, carbon in the form of graphite)
- using solutions (where relevant) with a concentration of $1.0\,mol\,dm^{-3}$
- at a constant specified temperature, typically 298.15 K.

Standard enthalpy of combustion, $\Delta H_c^\ominus$, and formation, $\Delta H_f^\ominus$

There are some standard reaction enthalpies which have particular significance. You need to be familiar with two of these – the standard enthalpy of combustion and the standard enthalpy of formation.

Standard enthalpy of combustion, $\Delta H_c^\ominus$

Standard enthalpy of combustion is the enthalpy change for the complete oxidation of 1 mole of a compound to carbon dioxide and liquid water. If the organic compound undergoing combustion contains nitrogen, the standard enthalpy of combustion is defined as producing N_2 in addition to the CO_2 and H_2O.

$$CH_4(g) + 2O_2(g) \rightarrow CO_2(g) + 2H_2O(l); \qquad \Delta H_c^\ominus = -891\,kJ\,mol^{-1}$$

$$CO(NH_2)_2(s) + \frac{3}{2}O_2(g) \rightarrow CO_2(g) + 2H_2O(l) + N_2(g); \qquad \Delta H_c^\ominus = -698\,kJ\,mol^{-1}$$

A standard enthalpy of combustion, $\Delta H_c^\ominus$, is distinguished from a standard enthalpy change, $\Delta H^\ominus$, by the inclusion of a subscript 'c'.

The values for some common compounds are given in section 14 of the IB *Chemistry data booklet*.

O_2 and the combustion products CO_2 and H_2O all have an enthalpy of combustion of $0\,kJ\,mol^{-1}$.

◆ **Standard enthalpy of combustion:** The enthalpy change when one mole of a substance is burnt completely in excess oxygen with all reactants and products in their standard states under standard conditions.

◆ **Standard enthalpy of formation**: The enthalpy change when one mole of a compound is formed from its constituent elements with all reactants and products in standard states under standard conditions.

Standard enthalpy of formation, $\Delta H_f^\ominus$

Standard enthalpy of formation is the enthalpy change when 1 mole of a compound is formed from its constituent elements.

A standard enthalpy of formation, $\Delta H_f^\ominus$, is distinguished from a standard enthalpy change, $\Delta H^\ominus$, by the inclusion of a subscript 'f'.

The enthalpy of formation of methane is represented by the equation:

$$C(s) + 2H_2(g) \rightarrow CH_4(g); \qquad \Delta H_f^\ominus = -74\,kJ\,mol^{-1}$$

The enthalpies of formation for potassium manganate(VII) and silver bromide are:

$$K(s) + Mn(s) + 2O_2(g) \rightarrow KMnO_4(s); \qquad \Delta H_f^\ominus = -813\,kJ\,mol^{-1}$$

$$Ag(s) + \frac{1}{2}Br_2(l) \rightarrow AgBr(s); \qquad \Delta H_f^\ominus = -99.5\,kJ\,mol^{-1}$$

LINKING QUESTION

Would you expect allotropes of an element, such as diamond and graphite, to have different $\Delta H_f^\ominus$ values?

The enthalpies of formation for elements in their standard states are zero. There is no reaction (and, therefore, no enthalpy change) in forming an element from that element. For example:

$$O_2(g) \rightarrow O_2(g) \qquad \Delta H_f^\ominus = 0\,kJ\,mol^{-1}$$

However, the enthalpies of formation for ozone ($O_3(g)$) and diamond (C(s, diamond)) are not zero since these are not the standard states of the elements oxygen and carbon.

Top tip!

One mole of substance is burnt during a standard enthalpy of combustion, and one mole of a substance is formed in a standard enthalpy of formation. This may mean that you need to balance the equation with non-integer values.

The thermochemical equation for the standard enthalpy of combustion of propan-1-ol, $\Delta H_c^\ominus$, is:

$$C_3H_7OH(l) + \frac{9}{2}O_2(g) \rightarrow 3CO_2(g) + 4H_2O(l)$$

For

$2C_3H_7OH(l) + 9O_2(g) \rightarrow 6CO_2(g) + 8H_2O(l)$ the enthalpy change is $2 \times \Delta H_c^\ominus$ for propan-1-ol.

The thermochemical equation for $\Delta H_f^\ominus$ of magnesium oxide is:

$$Mg(s) + \frac{1}{2}O_2(g) \rightarrow MgO(s)$$

For

$2Mg(s) + O_2(g) \rightarrow 2MgO(s)$; $\Delta H^\ominus$ is $2 \times \Delta H_f^\ominus$ of magnesium oxide.

IUPAC Gold book

The *Compendium of Chemical Terminology* is a book published by IUPAC that contains internationally accepted definitions for chemical terms. The book helps chemists communicate clearly with one another by setting out clear and precise definitions. Work on the first edition was initiated by Victor Gold (1922–1985), hence its common name: the Gold book.

The Gold book defines the standard enthalpy of formation $\Delta H_f^\ominus$ of a compound as the change of enthalpy during the formation of 1 mole of the substance from its constituent elements, with all substances in their standard states.

The superscript Plimsoll ($\ominus$) on this symbol indicates that the process has occurred under standard conditions. The standard pressure value $p^\ominus = 10^5$ Pa (100 kPa, which is 1 bar) is recommended by IUPAC, but prior to 1982 the value 1.00 atmosphere (101.325 kPa) was used.

HL ONLY

There is no standard temperature and so the temperature is usually specified as an additional piece of information (usually 25 °C or 298.15 K).

It is useful to note that the enthalpies of combustion of the elements hydrogen and carbon are equivalent to the enthalpies of formation of water and carbon dioxide, respectively:

$$H_2(g) + \frac{1}{2}O_2(g) \rightarrow H_2O(l); \qquad \Delta H_c^{\ominus}(H_2) = \Delta H_f^{\ominus}(H_2O)$$

$$C(s, \text{graphite}) + O_2(g) \rightarrow CO_2(g); \qquad \Delta H_c^{\ominus}(C) = \Delta H_f^{\ominus}(CO_2)$$

Most enthalpies of formation are negative, that is, the corresponding reactions are exothermic. However, some compounds have positive enthalpies of formation, for example benzene and nitrogen monoxide:

$$6C(s) + 3H_2(g) \rightarrow C_6H_6(l); \qquad \Delta H_f^{\ominus} = +49\,\text{kJ mol}^{-1}$$

$$\frac{1}{2}N_2(g) + \frac{1}{2}O_2(g) \rightarrow NO(g); \qquad \Delta H_f^{\ominus} = +90\,\text{kJ mol}^{-1}$$

A positive enthalpy of formation contributes towards making a compound energetically unstable towards decomposition into its elements. This means that decomposition of the compound to its elements can occur. Both benzene and nitrogen monoxide, whilst being energetically unstable to decomposition at 298 K, are kinetically stable – there is a large activation energy barrier to decomposition at room temperature. However, nitrogen monoxide does undergo significant decomposition in the presence of a platinum catalyst, as the catalyst lowers the activation energy barrier to the reaction.

The full criteria for assessing whether a compound is stable relative to decomposition are covered later in Chapter R1.4, page 386.

16 Give equations which represent the following enthalpy changes:

a $\Delta H_c^{\ominus}$ of $H_2(g)$
b $\Delta H_c^{\ominus}$ of $C_2H_6(g)$
c $\Delta H_c^{\ominus}$ of $C_8H_{18}(l)$
d $\Delta H_c^{\ominus}$ of $C_6H_{12}O_6(s)$
e $\Delta H_f^{\ominus}$ of $NH_3(g)$
f $\Delta H_f^{\ominus}$ of $SO_2(g)$
g $\Delta H_f^{\ominus}$ of $Na_2O(s)$
h $\Delta H_f^{\ominus}$ of $CHCl_3(l)$

Going further

Computational chemistry

In academia, computer aided modelling has largely replaced the more approximate method of estimating enthalpy changes using bond enthalpies.

Computational chemistry allows chemists to calculate the structure and properties of molecules. Techniques can be grouped into three main categories:

- ***Ab initio*** (meaning from the beginning): No, or very few, assumptions are made about bonding in a molecule. It is purely based on quantum mechanics and requires the Schrödinger equation to be solved. This method is only used for small molecules because it quickly becomes very computationally demanding as more atoms are added.
- **Semi-empirical:** This is an approximate and much faster version of *ab initio*, incorporating some experimentally derived parameters, for example bond lengths. It is useful for determining certain properties of small and medium sized molecules, such as charge distributions and molecular orbitals. It is also used as a starting point for more accurate *ab initio* calculations.
- **Empirical (molecular mechanics):** No quantum mechanics at all. Molecules are treated like animated models, with simple rules governing bond angles, lengths etc. It is mainly useful for organic and biological chemistry, and is the fastest – and often the only – way of modelling large molecules such as proteins, DNA and synthetic polymers.

ATL R1.2A

Computational calculations require a lot of processing power and specialist software, but there are websites which allow you to submit molecules for computation. ChemCompute gives free access to state of the art computational software as well as an introduction to computational chemistry.

Go to **https://chemcompute.org/** and choose the GAMESS package; here there are different experiments that you can choose to follow. Under 'General Education Chemistry' select 'Lab Experiment' under 'Computational Chemistry for Non-Majors'. This takes you to a page where you can submit molecules for computation using the GAMESS software package. There are 13 short activities to work through on the Instructions tab. These introduce how to draw molecules, and how to submit them for analysis.

Applications of Hess's law

Enthalpy changes from enthalpies of formation

The enthalpy change associated with any reaction can be determined by calculation from the enthalpies of formation of all the substances involved. A Hess cycle which expresses this is shown in Figure R1.28.

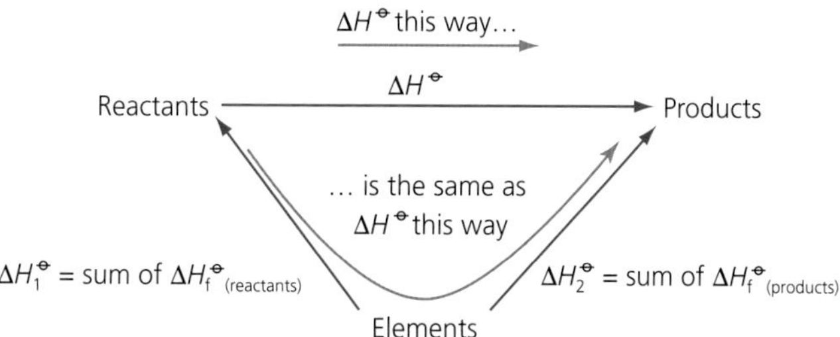

■ **Figure R1.28** Hess cycle for calculating the standard enthalpy of a reaction from standard enthalpies of formation

The elements which form the reactants in $\Delta H_1^\ominus$ are the same ones which can form the products in $\Delta H_2^\ominus$. This means that for any reaction $\Delta H^\ominus = -\Delta H_1^\ominus + \Delta H_2^\ominus$, so the enthalpy change of any reaction can be calculated using the expression:

$$\Delta H^\ominus = \sum \Delta H_{f\ (products)}^\ominus - \sum \Delta H_{f\ (reactants)}^\ominus$$

where:

- $\sum \Delta H_{f\ (products)}^\ominus$ is the sum of the enthalpies of formation for all the products multiplied by their stoichiometric coefficients in the reaction equation
- $\sum \Delta H_{f\ (reactants)}^\ominus$ is the sum of the enthalpies of formation for all the reactants multiplied by their stoichiometric coefficients in the reaction equation.

WORKED EXAMPLE R1.2E

Calculate the enthalpy change of the following reaction:

$$3CuO(s) + 2Al(s) \rightarrow 3Cu(s) + Al_2O_3(s)$$

$$\Delta H_f^{\ominus} (CuO) = -155\,kJ\,mol^{-1}$$

$$\Delta H_f^{\ominus} (Al_2O_3) = -1669\,kJ\,mol^{-1}$$

Answer

This is a redox reaction and involves a more reactive metal displacing a less reactive metal from its oxide. The enthalpies of formation of copper oxide and aluminium oxide are given, and we know that the enthalpies of formation of the elements Al and Cu = 0 kJ mol^{-1}.

Using the equation to calculate the standard enthalpy change from the given formation enthalpies:

$$\Delta H^{\ominus} = \sum \Delta H^{\ominus}_{f\ (products)} - \sum \Delta H^{\ominus}_{f\ (reactants)}$$

$$= [(3 \times 0) + (-1669)]\,kJ\,mol^{-1} - [(3 \times -155) + (2 \times 0)]\,kJ\,mol^{-1}$$

$$= -1204\,kJ\,mol^{-1}$$

The relevant Hess's law cycle for the reaction is shown below.

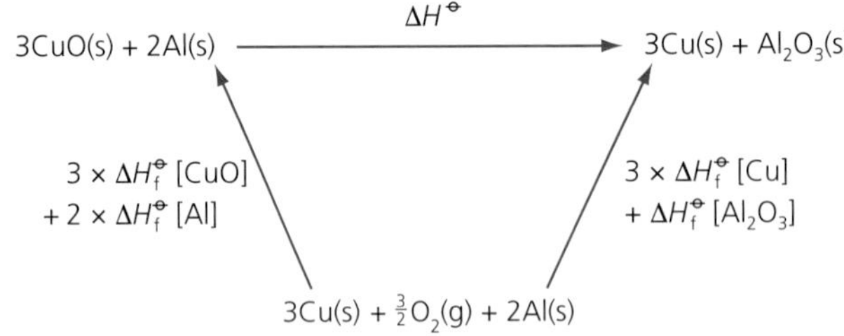

Enthalpy changes from enthalpies of combustion

The enthalpy change of any reaction can also be determined by calculation from the enthalpies of combustion of all the substances in the chemical equation. A Hess's law cycle which expresses this is shown in Figure R1.29.

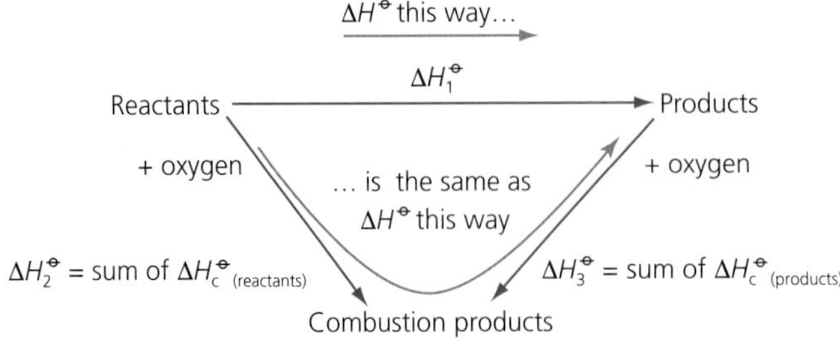

Figure R1.29 Hess's law cycle for calculating the standard enthalpy of a reaction from standard enthalpies of combustion

The pathways between reactants and products are $\Delta H_1^\ominus = \Delta H_2^\ominus - \Delta H_3^\ominus$, and more generally:

$$\Delta H^\ominus = \sum \Delta H^\ominus_{c\ (reactants)} - \sum \Delta H^\ominus_{c\ (products)}$$

where:

- $\sum \Delta H^\ominus_{c\ (reactants)}$ is the sum of the enthalpies of combustion for all the reactants multiplied by their stoichiometric ratios in the reaction equation
- $\sum \Delta H^\ominus_{c\ (products)}$ is the sum of the enthalpies of combustion for all the products multiplied by their stoichiometric ratios in the reaction equation.

Notice that the general formulae for calculating unknown enthalpy changes from enthalpy of formation or combustion values involve products minus reactants, and reactants minus products respectively. This is a consequence of the direction of the arrows from the intermediate state to the reactants and products. These arrows both point up, or both point down.

WORKED EXAMPLE R1.2F

Calculate the enthalpy change of reaction for the hydrogenation of propene to form propane given:

$$CH_3{-}CH{=}CH_2(g) + H_2(g) \rightarrow CH_3{-}CH_2{-}CH_3(g)$$

$$\Delta H_c^\ominus\ (C_3H_6) = -2058\,kJ\,mol^{-1}$$

$$\Delta H_c^\ominus\ (H_2) = -286\,kJ\,mol^{-1}$$

$$\Delta H_c^\ominus\ (C_3H_8) = -2219\,kJ\,mol^{-1}$$

Answer

Using the equation to calculate standard enthalpy change from given combustion enthalpies:

$$\Delta H^\ominus = \sum \Delta H^\ominus_{c\ (reactants)} - \sum \Delta H^\ominus_{c\ (products)}$$

$$\Delta H^\ominus = (-2058 + -286)\,kJ\,mol^{-1} - -2219\,kJ\,mol^{-1}$$

$$\Delta H^\ominus = -125\,kJ\,mol^{-1}$$

The relevant Hess's law cycle for the reaction is shown below.

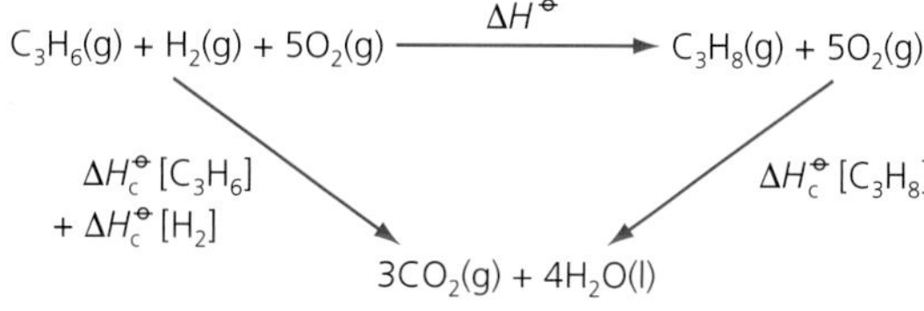

HL ONLY

Use the data from the table to answer the questions which follow.

Compound	$\Delta H_f^\ominus$ / kJ mol^{-1}
$TiCl_3(s)$	−721
$TiCl_4(l)$	−804
$PCl_3(l)$	−320
$PCl_5(s)$	−444
$POCl_3(l)$	−597
$CH_4(g)$	−74
$GeO(s)$	−212
$GeO_2(s)$	−551
$HCl(g)$	−92
$CCl_4(l)$	−128

Compound	$\Delta H_c^\ominus$ / kJ mol^{-1}
$NH_3(g)$	−46
$C_6H_6(l)$	−3268
$H_2(g)$	−286
$C_6H_{12}(l)$	−3920
$C_2H_2(g)$	−1301
$C_2H_6(g)$	−1561
$C_2H_5OH(l)$	−1367
$C_2H_4(g)$	−1411
$CH_3COOH(l)$	−874

17 Use standard enthalpies of formation to calculate the reaction enthalpies for:

a $2TiCl_3(s) + Cl_2(g) \rightarrow 2TiCl_4(l)$

b $PCl_3(l) + Cl_2(g) \rightarrow PCl_5(s)$

c $2PCl_3(l) + O_2(g) \rightarrow 2POCl_3(l)$

d $CH_4(g) + 4Cl_2(g) \rightarrow CCl_4(l) + 4HCl(g)$

e $2GeO(s) \rightarrow Ge(s) + GeO_2(s)$

18 Use standard enthalpies of combustion to calculate the reaction enthalpies for:

a $C_2H_2(g) + 2H_2(g) \rightarrow C_2H_6(g)$

b $C_6H_6(l) + 3H_2(g) \rightarrow C_6H_{12}(l)$

c $3C_2H_2(g) \rightarrow C_6H_6(l)$

d $C_2H_4(g) + H_2O(l) \rightarrow C_2H_5OH(l)$

e $C_2H_5OH(l) + O_2(g) \rightarrow CH_3COOH(l) + H_2O(l)$

19 Calculate the enthalpy change requested in each case:

a $\Delta H_f^\ominus$ of $NH_4Cl(s)$

Given $NH_3(g) + HCl(g) \rightarrow NH_4Cl(s)$; $\Delta H^\ominus = -176\,kJ\,mol^{-1}$

b $\Delta H_f^\ominus$ of $MgCl_2(s)$

Given $TiCl_4(l) + 2Mg(s) \rightarrow 2MgCl_2(s) + Ti(s)$; $\Delta H^\ominus = -478\,kJ\,mol^{-1}$

c $\Delta H_f^\ominus$ of $(CH_3CO)_2O(l)$

Given $(CH_3CO)_2O(l) + H_2O(l) \rightarrow 2CH_3COOH(l)$; $\Delta H^\ominus = -46\,kJ\,mol^{-1}$

What role do paradigm shifts play in the progression of scientific knowledge?

The phlogiston theory was a theory of combustion (burning) and rusting which hypothesized that all materials that could be burnt contained a substance known as phlogiston.

According to the theory, burning and rusting both represent the escape of phlogiston, and air is necessary for both processes because phlogiston is absorbed into it. When the air becomes saturated with phlogiston, the phlogiston has no place to go and the flame goes out or the rusting stops.

Although the theory made qualitative sense and helped explain burning and rusting, it suffered from a quantitative defect: it could not adequately account for the observed changes in mass that accompany burning and rusting. It was known as early as 1630 that, when a piece of iron rusts, the rust formed has a greater mass than the original iron.

Some chemists tried to explain this by asserting that phlogiston had negative mass. However, when a lump of charcoal (carbon) burns, again with the loss of phlogiston, its mass decreases.

The theory was later falsified (disproved) by the work of the French chemist Antoine Lavoisier (1743–1794) and the English chemist Joseph Priestley (1733–1804) who proposed a theory of burning (combustion) involving a chemical reaction with oxygen. It completely replaced the phlogiston theory in a paradigm shift. This occurs when a scientific model or way of thinking is quickly and completely replaced by a very different scientific model or way of thinking.

Thomas Kuhn (1922–1996) wrote about the development of science through paradigm shifts. More information about his theory can be downloaded from **www.hoddereducation.co.uk/media/Documents/magazine-extras/IB%20Review/IBRev%204_2/IBReview4_2_poster.pdf?ext=.pdf**

Energy changes in ionic compounds

You have already been introduced to some standard enthalpy changes in this chapter, such as enthalpies of formation and combustion, but there are many different types of enthalpy change encountered in thermodynamics; some commonly used ones are presented in Table R1.5. The IUPAC recommended symbol notation is to have the subscript before the *H*, for example $\Delta_c H^\ominus$, although it is traditional and more common to see the subscript after the *H* too, for example $\Delta H_c^\ominus$. The IB chooses to follow the second convention, and this is what has been adopted in this book.

■ **Table R1.5** Types of enthalpy change. (*There are a small number of examples that do not obey this general rule.)

Process	Transition	Symbol	Endothermic	Exothermic
combustion	compounds (s,l,g) + O_2(g) → CO_2(g), H_2O(l)	$\Delta H_c^\ominus$		✓*
formation	elements → compound	$\Delta H_f^\ominus$		✓*
transition	allotrope 1 → allotrope 2	$\Delta H_{trs}^\ominus$	✓	✓
fusion	s → l	$\Delta H_{fus}^\ominus$	✓	
vaporization	l → g	$\Delta H_{vap}^\ominus$	✓	
sublimation	s → g	$\Delta H_{sub}^\ominus$	✓	
mixing	pure → mixture	$\Delta H_{mix}^\ominus$		
solution	solute → solution	$\Delta H_{sol}^\ominus$	✓	✓
hydration	$X^{n\pm}$(g) → $X^{n\pm}$(aq)	$\Delta H_{hyd}^\ominus$		✓
atomization	species (s,l,g) → atoms (g)	$\Delta H_{at}^\ominus$	✓	
ionization	X(g) → X^+ + e^-(g)	$\Delta H_{ion}^\ominus$	✓	
electron affinity	X(g) + e^-(g) → X^-(g)	$\Delta H_{ea}^\ominus$		✓*
reaction	reactants → products	$\Delta H_r^\ominus$	✓	✓
bond-breaking	X–Y(g) → X + Y(g)	$\Delta H_{BE}^\ominus$	✓	
lattice enthalpy	MX(s) → M^+(g) + X^-(g)	$\Delta H_L^\ominus$	✓	

The standard enthalpy changes which are used to calculate the lattice enthalpies of ionic compounds are introduced or recapped below.

Lattice enthalpy

The lattice enthalpy can be represented as $\Delta H_L^\ominus$, $\Delta H_{latt}^\ominus$ or $\Delta H_{lattice}^\ominus$. It is the energy change when one mole of an ionic compound is separated into its component gaseous ions.

$$M_aX_b(s) \rightarrow aM^{b+}(g) + bX^{a-}(g); \qquad \Delta H_L^\ominus (MX)$$

For calcium fluoride:

$$CaF_2(s) \rightarrow Ca^{2+}(g) + 2F^-(g); \qquad \Delta H_L^\ominus (CaF_2) = +2620\,kJ\,mol^{-1}$$

This is an endothermic process because the electrostatic attractions between the oppositely charged ions must be overcome. The term 'separated to infinite distance' is included in the definition given in the IB *Chemistry data booklet* because this is when the attractions between the ions will have been completely overcome. The process is shown in Figure R1.30. However, note that the separated ions are not shown at an infinite distance as this would be impossible to draw!

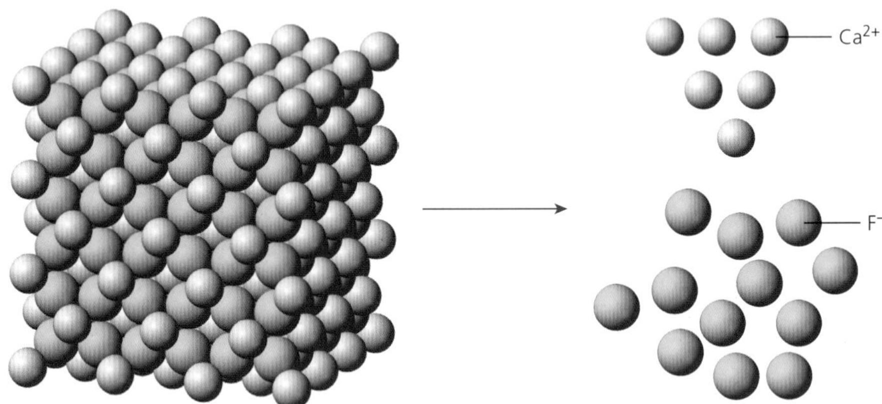

■ **Figure R1.30** The lattice enthalpy of calcium fluoride

In S2.1, the factors which affect lattice enthalpies were discussed. Large lattice enthalpies are expected when the ions are highly charged and small. Theoretical lattice enthalpies can be calculated using equations which take into account the charge of the ions, the distance between the ions and the arrangement of the ions in the lattice. If there is a difference between the experimentally determined value and the theoretically calculated value, this implies there is a degree of covalent contribution to the bonding.

The values of some lattice enthalpies are given in section 16 of the IB *Chemistry data booklet*.

Top tip!

Lattice enthalpies are always positive because energy must be absorbed to overcome the strong electrostatic forces of attraction between the oppositely charged ions that make up an ionic compound.

Top tip!

In other chemical literature you may encounter the **lattice enthalpy of formation**. This is the enthalpy change when one mole of an ionic solid is formed from its constituent gaseous ions. This enthalpy change has the same magnitude as the lattice enthalpy but the opposite sign. Ensure that you are able to give the correct definition for a lattice enthalpy and know that the value of a lattice enthalpy is positive.

Nature of science: Science as a shared endeavour

Estimating lattice enthalpies

The advancement of knowledge and understanding in a field of science is a product of collaborative work, discussion and refinement of ideas. Ideas and experimental discoveries are shared through the process of peer-reviewed publications, and this allows other scientists around the world to further contribute to an area of research. Science is a collaborative process and rarely are there paradigm shifts.

The names of scientists who contributed to the development of theoretical lattice enthalpies are attached to the equations they helped to formulate. The Born–Landé, Born–Mayer and Kapustinskii equations all predict lattice enthalpies. The basis for each of these models is the Coulomb interaction of charged particles, which arises from the work of the French physicist Charles-Augustin de Coulomb (1736–1806) published in the 18th century.

Coulomb's law states that the force, *F*, operating between two oppositely charged particles is dependent on the size of the charges and the square of the distance, *d*, that separates them:

$$F \propto \frac{\text{charge on positive ion} \times \text{charge on negative ion}}{d^2}$$

Max Born (1882–1970) and Alfred Landé (1888–1976) were German-born physicists. Joseph Edward Mayer (1904–1983) was an American scientist who, along with Born, modified and improved the Born–Landé equation. Anatoli Kapustinskii (1906–1960) was a Russian chemist who recognized the drawbacks of the previous work, and noticed he could dispense with the need for detailed prior knowledge of the structure, and so formulated the Kapustinskii equation. The Kapustinskii equation allows chemists to estimate the ionic radii of polyatomic ions from the lattice enthalpies of their compounds.

LINKING QUESTION

What are the factors that influence the strength of lattice enthalpy in an ionic compound?

Ionization energies

Ionization energies are represented by the symbols $\Delta H^{\ominus}_{IE}$, $\Delta H^{\ominus}_{i}$ or $\Delta H^{\ominus}_{ion}$ and the number 1, 2, 3 … is usually included to indicate the first, second or third ionization of the atom. If the number is omitted, the enthalpy changed quoted is assumed to be the first. The terms ionization enthalpy and ionization energy refer to the same process and they are used interchangeably.

◆ **First ionization energy**: The enthalpy change, $\Delta H^{\ominus}_{IE1}(X)$, in the reaction $X(g) \rightarrow X^{+}(g) + e^{-}$.

The **first ionization energy** is the enthalpy change when one mole of electrons is removed from one mole of gaseous atoms to form one mole of gaseous ions with a 1+ charge. Using magnesium as an example:

$$Mg(g) \rightarrow Mg^{+}(g) + e^{-}; \qquad \Delta H^{\ominus}_{IE1}(Mg) = +738\,kJ\,mol^{-1}$$

The second ionization energy is the enthalpy change when one mole of electrons is removed from one mole of 1+ gaseous ions to form one mole of gaseous ions with a 2+ charge:

$$Mg^{+}(g) \rightarrow Mg^{2+}(g) + e^{-}; \qquad \Delta H^{\ominus}_{IE2}(Mg) = +1451\,kJ\,mol^{-1}$$

As discussed in S3.1, ionizations are endothermic processes, and each successive ionization energy for atoms of the same element gets larger.

The first ionization energies of the elements are given in section 9 of the IB *Chemistry data booklet.*

Electron affinities

The electron affinity of an element is represented by $\Delta H^{\ominus}_{ea}$ or $\Delta H^{\ominus}_{EA}$. It is also represented by $\Delta H^{\ominus}_{eg}$ where 'eg' represents electron gain. As with ionization enthalpies, a number is usually included to indicate the first, second or third electron affinity of the element.

The first electron affinity is the enthalpy change when one mole of electrons is gained by one mole of gaseous atoms to form one mole of gaseous 1− ions. Using sulfur as an example:

$$S(g) + e^- \rightarrow S^-(g); \qquad \Delta H^{\ominus}_{EA1}(S) = -200\,kJ\,mol^{-1}$$

The second electron affinity is the enthalpy change when one mole of electrons is gained by one mole of singularly negative ions to form one mole of 2− ions:

$$S^-(g) + e^- \rightarrow S^{2-}(g); \qquad \Delta H^{\ominus}_{EA2}(S) = +545\,kJ\,mol^{-1}$$

If there is an unfilled orbital in an atom's outer shell, the effective nuclear charge of the atom is usually strong enough to attract an electron and energy is released in this process. Most first electron affinities are exothermic. Values for the first electron affinities of elements correlate quite well with electronegativity. A small atom with a high nuclear charge has a high electronegativity value and a relatively large exothermic electron affinity. The trends in electron affinity values were discussed in S3.1 page 234.

The second electron affinity of an element is an endothermic process because energy is required to overcome the mutual repulsion between the negatively charged ion and the electron being added.

The first (and second, when relevant) electron affinities for many elements are given in section 9 of the IB *Chemistry data booklet.*

Enthalpy of atomization

◆ **Enthalpy of atomization:** The enthalpy change when forming one mole of gaseous atoms of an element from the element in its standard state under standard conditions.

The **enthalpy of atomization** of an element is the enthalpy change for the formation of one mole of gaseous atoms. It is represented by the symbol $\Delta H^{\ominus}_{at}$ or $\Delta H^{\ominus}_{a}$.

The enthalpy of atomization for sodium is:

$$Na(s) \rightarrow Na(g); \qquad \Delta H^{\ominus}_{at} = +109\,kJ\,mol^{-1}\ (= \Delta H^{\ominus}_{sub})$$

For sodium, $\Delta H^{\ominus}_{at}$ is equal to the enthalpy of sublimation, $\Delta H^{\ominus}_{sub}$. Sublimation is a change in state (at constant temperature) from a solid to a gas. The atomization enthalpy of mercury is equal to its enthalpy change of vaporization:

$$Hg(l) \rightarrow Hg(g); \qquad \Delta H^{\ominus}_{at} = +59\,kJ\,mol^{-1}\ (= \Delta H^{\ominus}_{vap})$$

The enthalpy of atomization for chlorine is:

$$\frac{1}{2}Cl_2(g) \rightarrow Cl(g) \qquad \Delta H^{\ominus}_{at} = +121\,kJ\,mol^{-1}$$

Comparing this to the bond enthalpy of Cl–Cl:

$$Cl_2(g) \rightarrow 2Cl(g) \qquad \Delta H^{\ominus}_{BE} = +242\,kJ\,mol^{-1}$$

It can be deduced that the enthalpy of atomization for chlorine is equal to half of the Cl–Cl bond enthalpy.

$$\Delta H^{\ominus}_{at}(Cl) = \frac{1}{2}\Delta H^{\ominus}_{BE}(Cl\text{–}Cl)$$

This same principle is true for any gaseous diatomic element.

Some of the standard enthalpy changes introduced in this section, as applied to sodium and chlorine, are shown in Figure R1.31.

Top tip!

A question may not give you the enthalpy change of atomization of a diatomic element such as O_2 or Cl_2. However, you are given the bond enthalpy of O=O and Cl–Cl in Section 12 of the IB *Chemistry data booklet*, and you can divide this value by two to calculate the enthalpy change of atomization.

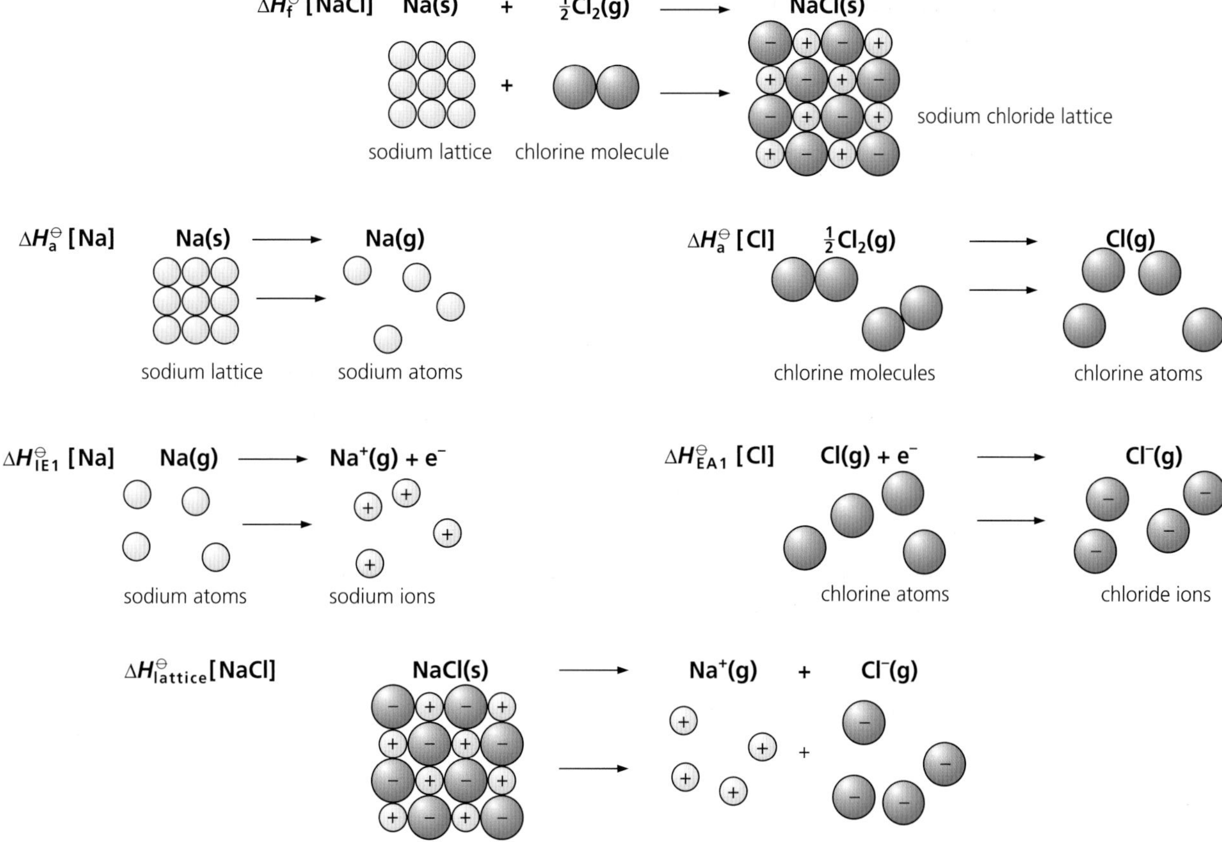

■ **Figure R1.31** Standard enthalpy changes involving sodium and chlorine

For the solid diatomic element iodine, the atomization enthalpy is a combination of its sublimation enthalpy and bond enthalpy.

$$\frac{1}{2}I_2(s) \rightarrow I(g); \qquad \Delta H^{\ominus}_{at} = \{\Delta H^{\ominus}_{sub} + \frac{1}{2}\Delta H^{\ominus}_{BE}(I\text{–}I)\} = +107\,kJ\,mol^{-1}$$

20 Give equations which represent the following enthalpy changes:

a first ionization enthalpy of iron
b enthalpy of atomization of zinc
c lattice enthalpy of magnesium oxide
d first electron affinity of chlorine
e enthalpy of atomization of bromine
f second ionization enthalpy of calcium
g second electron affinity of oxygen
h lattice enthalpy of lithium oxide.

21 Give the enthalpy changes which correspond to the following processes. Some of the processes are described by multiples of combinations of enthalpies.

a $Ca(s) + Cl_2(g) \rightarrow CaCl_2(s)$

b $CaCl_2(s) \rightarrow Ca^{2+}(g) + 2Cl^-(g)$

c $O_2(g) \rightarrow 2O(g)$

d $Zn(g) \rightarrow Zn^+(g) + e^-$

e $Zn(g) \rightarrow Zn^{2+}(g) + 2e^-$

f $Zn(s) \rightarrow Zn^{2+}(g) + 2e^-$

g $\frac{1}{2}F_2(g) + e^- \rightarrow F^-(g)$

h $O(g) + 2e^- \rightarrow O^{2-}(g)$

Born–Haber cycles

A Born–Haber cycle is formed by combining the standard enthalpy changes introduced in the previous pages. The relative heights of the lines in these cycles represent the potential energy the system has in the specified state, and a completed Born–Haber cycle can be used to calculate the lattice enthalpies, $\Delta H_L^\ominus$, of an ionic compound – this cannot be done experimentally.

The sum of the enthalpy of formation and lattice enthalpy of an ionic compound (the red enthalpy changes in Figure R1.32) are equal to the enthalpy change associated with converting the elements forming the compound into the relevant gaseous ions through atomization, ionization and electron affinity processes (the blue enthalpy changes in figure R1.32). The relative heights of each level reflect the endothermic or exothermic nature of the previous step, but there is no strict scale.

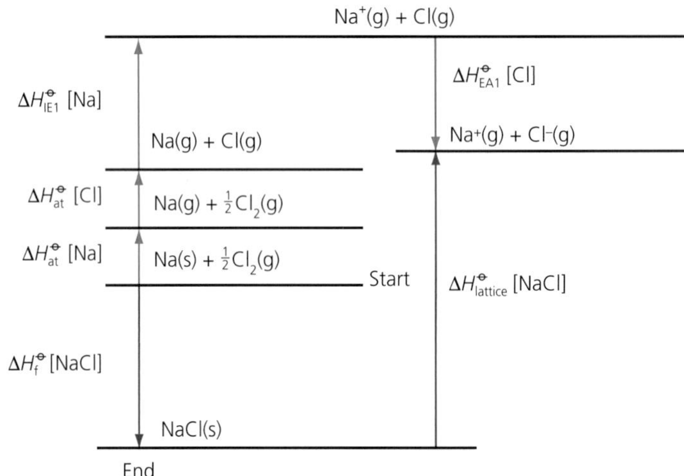

■ **Figure R1.32** A Born–Haber cycle for sodium chloride

For the Born-Haber cycle involving the formation of sodium chloride shown in Figure R1.32, the red and blue pathways start at the same point and finish at the same point. Hess's law tells us that the total enthalpy change of the two pathways must be equal, that is:

$$\Delta H_f^\ominus(\text{NaCl}) + \Delta H_L^\ominus(\text{NaCl}) = \Delta H_{at}^\ominus(\text{Na}) + \Delta H_{at}^\ominus(\text{Cl}) + \Delta H_{IE1}^\ominus(\text{Na}) + \Delta H_{EA1}^\ominus(\text{Cl})$$

The arrows going upwards represent endothermic enthalpy changes. The arrows going downwards represent exothermic enthalpy changes. It is important to include state symbols, and give each enthalpy change its own step in a Born–Haber cycle.

When you are given values for all but one of the enthalpy changes in a Born–Haber cycle, you can deduce the unknown enthalpy change using the following steps:

1 Write out the two pathways which are equivalent to one another. For sodium chloride:

$$\Delta H_f^\ominus(\text{NaCl}) + \Delta H_L^\ominus(\text{NaCl}) = \Delta H_{at}^\ominus(\text{Na}) + \Delta H_{at}^\ominus(\text{Cl}) + \Delta H_{IE1}^\ominus(\text{Na}) + \Delta H_{EA1}^\ominus(\text{Cl})$$

2 Substitute in the values given, in this case all the values for sodium chloride are known, except for the lattice enthalpy:

$\Delta H_f^\ominus(\text{NaCl}) = -411\,\text{kJ mol}^{-1}$ $\Delta H_{IE1}^\ominus(\text{Na}) = 496\,\text{kJ mol}^{-1}$

$\Delta H_{at}^\ominus(\text{Na}) = 109\,\text{kJ mol}^{-1}$ $\Delta H_{EA1}^\ominus(\text{Cl}) = -349\,\text{kJ mol}^{-1}$

$\Delta H_{at}^\ominus(\text{Cl}) = 121\,\text{kJ mol}^{-1}$ $-411\,\text{kJ mol}^{-1} + \Delta_L H^\ominus(\text{NaCl}) = (109 + 121 + 496 - 349)\,\text{kJ mol}^{-1}$

3 Rearrange to find the unknown.

$\Delta H_L^\ominus(\text{NaCl}) = +788\,\text{kJ mol}^{-1}$

Sodium chloride is composed of univalent ions. The pathway involving atomization, ionization and electron affinity involves more steps when divalent ions are present in the compound. Consider the Born–Haber cycle for magnesium chloride shown in Figure R1.33.

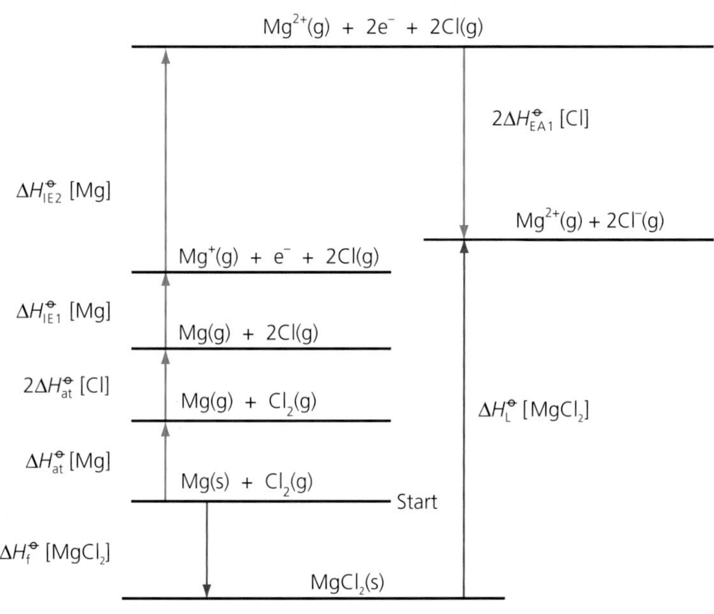

■ **Figure R1.33** A Born–Haber cycle for magnesium chloride

Top tip!

It is easy to miss that the enthalpy of atomization and electron affinity of chlorine need to be multiplied by two to account for the 2 moles of gaseous Cl^- ions needed.

Top tip!

You may be provided with the enthalpy of a Cl–Cl bond (in section 12 of the IB *Chemistry data booklet*), but not the enthalpy of atomization of Cl. Remember, $2\Delta H_{at}^\ominus(\text{Cl}) = \Delta H_{BE}^\ominus(\text{Cl–Cl})$.

The second ionization enthalpy of magnesium is needed to produce the +2 oxidation state, and because 2 moles of $Cl^-(g)$ need to be made, the atomization enthalpy and electron affinity of chlorine need to be multiplied by 2.

The two equivalent pathways for magnesium chloride are:

$$\Delta H_f^\ominus(\text{MgCl}_2) + \Delta H_L^\ominus(\text{MgCl}_2) = \Delta H_{at}^\ominus(\text{Mg}) + 2 \times \Delta H_{at}^\ominus(\text{Cl}) + \Delta H_{IE1}^\ominus(\text{Mg}) + \Delta H_{IE2}^\ominus(\text{Mg}) + 2 \times \Delta H_{EA1}^\ominus(\text{Cl})$$

WORKED EXAMPLE R1.2G

Use the values in the table to calculate the lattice enthalpy of magnesium oxide.

Enthalpy change	$\Delta H^\ominus$ / kJ mol^{-1}
enthalpy of formation of magnesium oxide	−602
enthalpy of atomization of oxygen	248
first electron affinity of oxygen	−142
second electron affinity of oxygen	753
first ionization enthalpy of magnesium	736
second ionization enthalpy of magnesium	1450
enthalpy of atomization of magnesium	150

HL ONLY

Answer

The constructed Born–Haber cycle is shown in the figure.

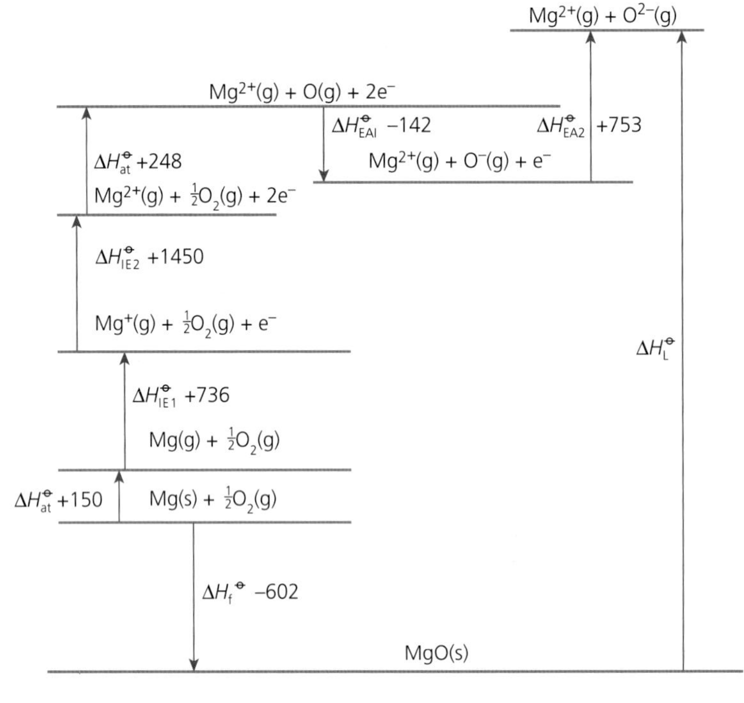

$$-602 + \Delta H^\ominus_L(\text{MgO}) = 150 + 736 + 1450 + 248 - 142 + 753$$

$$\Delta H^\ominus_L(\text{MgO}) = +3797\,\text{kJ mol}^{-1}$$

Going further

Enthalpy of solution

Many ionic compounds, such as sodium chloride, dissolve well in water to form solutions. Others, such as silver chloride, are almost insoluble. When an ionic substance dissolves, the lattice of the ionic crystal needs to be broken up (Figure R1.34). Lattice enthalpies are always endothermic, so if the compound is to dissolve, the energy needed to achieve this must be supplied from enthalpy changes that occur during dissolving.

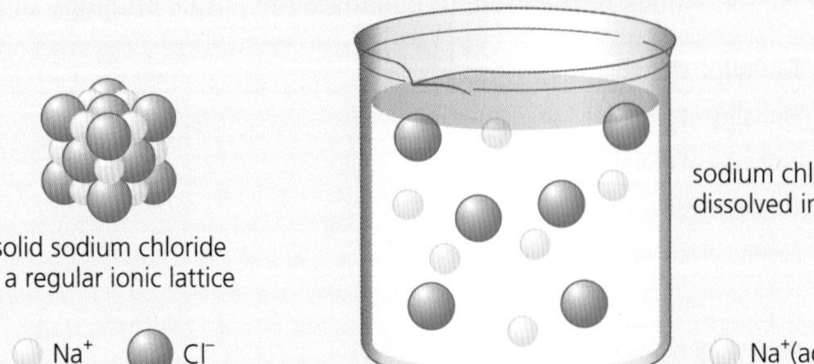

■ **Figure R1.34** The formation of a sodium chloride solution from a sodium chloride lattice

The ions in the crystal lattice become separated as they become hydrated by polar water molecules. The hydration process involves the cations and anions being surrounded by a number of water molecules (Figure R1.35). Ion–dipole bonds are formed and hence hydration is exothermic.

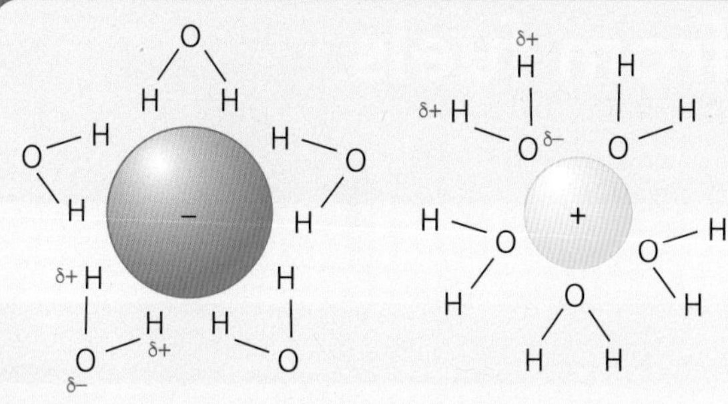

■ **Figure R1.35** A hydrated anion and cation

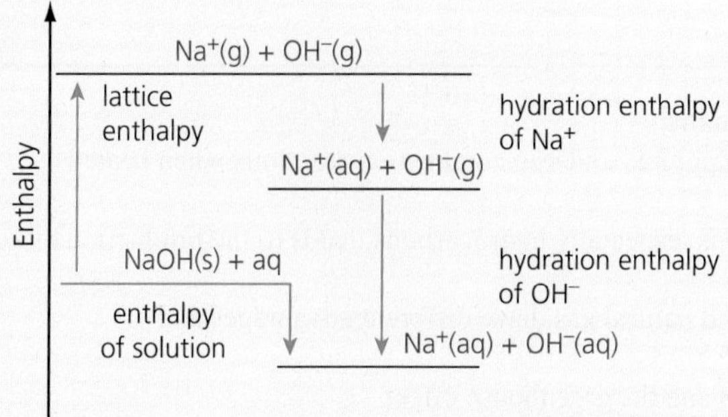

■ **Figure R1.36** An energy cycle for the dissolving of sodium hydroxide

The separation of the ions in the lattice is strongly endothermic and the hydration of the ions exothermic. The enthalpy change of solution, $\Delta H^{\ominus}_{sol}$, is the difference between these two enthalpy changes. Figure R1.36 shows an energy cycle for sodium hydroxide.

In the example of sodium hydroxide given, the enthalpy of solution is a relatively large exothermic value ($-44.5\,kJ\,mol^{-1}$). Dissolving sodium hydroxide in water makes the solution noticeably warmer as heat energy is transferred from the chemical system to the surroundings (in this case, the water acting as the solvent). Ionic substances that are soluble generally have a large negative value for the enthalpy of solution. Conversely, insoluble ionic substances generally have a large positive value for the enthalpy of solution. However, these are 'rules of thumb' and a number of exceptions occur. This is because free energy changes (not just the sign of $\Delta H^{\ominus}_{sol}$) determine solubility. Free energy changes are introduced later in Chapter R1.4.

22 Construct a Born–Haber cycle for Na_2O and use it to calculate the lattice enthalpy of Na_2O.

Enthalpy change	**kJ mol⁻¹**
$\Delta H^{\ominus}_{f}(Na_2O)$	−414
$\Delta H^{\ominus}_{at}(Na)$	109
$\Delta H^{\ominus}_{IE1}(Na)$	496
$\Delta H^{\ominus}_{at}(O)$	248
$\Delta H^{\ominus}_{EA1}(O)$	−141
$\Delta H^{\ominus}_{EA2}(O)$	+753

23 Construct a Born–Haber cycle for $MgBr_2$ and use it to calculate the bond enthalpy of a Br–Br bond.

Enthalpy change	**kJ mol⁻¹**
$\Delta H^{\ominus}_{f}(MgBr_2)$	−524
$\Delta H^{\ominus}_{at}(Mg)$	150
$\Delta H^{\ominus}_{IE1}(Mg)$	736
$\Delta H^{\ominus}_{IE2}(Mg)$	1450
$\Delta H^{\ominus}_{L}(MgBr_2)$	2420
$\Delta H^{\ominus}_{vap}(Br)$	30.9
$\Delta H^{\ominus}_{EA1}(Br)$	−342

24 Sodium chloride and magnesium chloride are two salts formed when seawater is evaporated to dryness. State which has the higher lattice enthalpy, and give a reason for your choice.

HL ONLY

R1.3 Energy from fuels

Guiding question

- What are the challenges of using chemical energy to address our energy needs?

SYLLABUS CONTENT

By the end of this chapter, you should understand that:

- reactive metals, non-metals and organic compounds undergo combustion reactions when heated with oxygen
- incomplete combustion of organic compounds, especially hydrocarbons, leads to the production of carbon monoxide and carbon
- fossil fuels, which include coal, crude oil, and natural gas, have different advantages and disadvantages
- there is a link between carbon dioxide levels and the greenhouse effect
- biofuels are produced from the biological fixation of carbon over a short period of time through photosynthesis
- renewable and non-renewable energy sources exist
- a fuel cell can be used to convert chemical energy from a fuel directly to electrical energy.

By the end of this chapter, you should know how to:

- deduce equations for reactions of combustion, including hydrocarbons and alcohols
- deduce equations for the incomplete combustion of hydrocarbons and alcohols
- evaluate the amount of carbon dioxide added to the atmosphere when different fuels burn
- consider the advantages and disadvantages of biofuels
- deduce half-equations for the electrode reactions in a fuel cell.

There is no higher-level only material in R1.3.

Combustion reactions

Combustion reactions occur when a fuel is heated in the presence of oxygen. Reactive metals, non-metals and organic compounds act as fuels that are oxidized by molecular oxygen to form combustion products.

Combustion of organic compounds

Many organic molecules readily undergo combustion when they are heated in an atmosphere containing oxygen. The type of combustion, complete or incomplete, depends on the supply of oxygen, the type of organic compound being combusted and the temperature of the combustion.

Complete combustion

◆ **Complete combustion:** The reaction of an organic fuel with a plentiful supply of oxygen to produce carbon dioxide and water.

When there is a plentiful supply of oxygen, the carbon atoms in an organic compound are oxidized to carbon dioxide and the hydrogen atoms form water. This is known as **complete combustion**.

Butane is one of the constituents of bottled gas and its complete combustion is shown in the equation:

$$2C_4H_{10}(g) + 13O_2(g) \rightarrow 8CO_2(g) + 10H_2O(g)$$

In an open system, such as when burning the gas on a stove in the atmosphere, the water produced is mainly in the gaseous form. Note that this equation does not represent the standard enthalpy of combustion of butane because all reactants and products are not in their standard states, and 2 moles of butane are being combusted in the stated equation.

The general equation for the complete combustion of a hydrocarbon has the form:

$$C_xH_y(s/l/g) + \left(x + \frac{y}{4}\right)O_2(g) \rightarrow xCO_2(g) + \frac{y}{2}H_2O(g)$$

Top tip!

You should become used to balancing combustion equations. It is helpful to balance the equation for carbon and hydrogen first, before balancing the oxygen afterwards.

If a non-integer value of oxygen is required to balance the equation, it is customary to double all coefficients to obtain the balanced equation with integer coefficients. For example, propene, C_3H_6:

$$C_3H_6(g) + \frac{9}{2}O_2(g) \rightarrow 3CO_2(g) + 3H_2O(g)$$

becomes:

$$2C_3H_6(g) + 9O_2(g) \rightarrow 6CO_2(g) + 6H_2O(g)$$

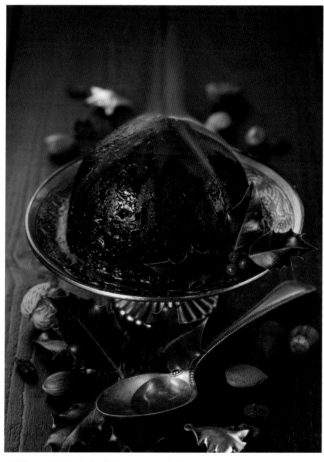

■ **Figure R1.37** In the United Kingdom, brandy, a spirit with a high ethanol content, is poured over a Christmas pudding and then lit. The combustion of the ethanol gives a light blue flame which surrounds the pudding.

If an oxygen-containing organic compound is combusted, the external supply of oxygen needed is reduced. For this reason, oxygen-containing organic compounds have a greater tendency to undergo complete combustion, but they also release a little less energy than the parent hydrocarbon because they are already partially oxidized.

The combustion of ethanol is:

$$CH_3CH_2OH(g) + 3O_2(g) \rightarrow 2CO_2(g) + 3H_2O(g)$$

Ethanol burns with a faint blue flame (Figure R1.37).

LINKING QUESTION

Which species are the oxidizing and reducing agents in a combustion reaction?

Analysis of combustion products allows the molecular formula of an organic compound containing carbon, hydrogen and oxygen to be determined (see Chapter S1.4 for more on molecular formulae).

WORKED EXAMPLE R1.3A

When 15.0 g of an organic compound is burnt in an excess of oxygen, 22.0 g of carbon dioxide and 9.0 g of water are produced as the only products of combustion. Determine the empirical formula of the organic compound.

Answer

We know that the carbon dioxide produced is as a result of the oxidation of the carbon in the organic compound:

$$n(\text{carbon dioxide}) = \frac{22.0\,\text{g}}{44.02\,\text{g mol}^{-1}} = 0.500\,\text{mol}$$

so the compound contained 0.500 mol of carbon atoms, C. This is 6.01 g of C.

The water produced was from the hydrogen in the organic compound:

$$n(\text{water}) = \frac{9.0\,\text{g}}{18.02\,\text{g mol}^{-1}} = 0.50\,\text{mol}$$

Because 2 hydrogen atoms are present in 1 molecule of water, the compound contained (2 × 0.50 mol) 1.0 mol of hydrogen atoms, H. This is 1.0 g of H.

The remaining mass of the compound must be due to oxygen because carbon dioxide and water were the only products of combustion.

Mass of oxygen atoms in compound = 15.0 g – (6.0 + 1.0) g = 8.0 g

$$\text{Amount of oxygen atoms, O, in the compound} = \frac{8.0\,\text{g}}{16.00\,\text{g mol}^{-1}} = 0.50\,\text{mol}$$

The compound contains 0.5 mol of carbon atoms, 1.0 mol of hydrogen atoms and 0.5 mol of oxygen atoms.

Hence the empirical formula is CH_2O.

This type of question regularly appears in IB Chemistry examination papers.

◆ **Incomplete combustion**: Combustion that occurs when the supply of oxygen is restricted; the products are carbon monoxide and/or carbon, and water.

Top tip!

If you are required to give an equation for the incomplete combustion of a hydrocarbon, ensure you include carbon monoxide, carbon and water as the products.

Incomplete combustion

Incomplete combustion of hydrocarbons occurs in a limited supply of oxygen and it produces carbon with a lower oxidation state than +4. This is as either carbon monoxide, CO (carbon with an oxidation state of +2), or carbon, C (carbon with an oxidation state of 0). Water is still produced from the hydrogen atoms in the hydrocarbon.

One equation for the incomplete combustion of butane is shown below:

$$C_4H_{10}(g) + 4O_2(g) \rightarrow 3CO(g) + C(s) + 5H_2O(l)$$

There are many other correct equations for incomplete combustion. The key point to notice is that carbon monoxide or carbon are included as products, and this indicates that some incomplete combustion has occurred. In practice, a mixture of carbon-containing products will form: some carbon dioxide, some carbon monoxide and some carbon. The unburnt carbon, along with solid partially oxidized products, is referred to as soot.

Going further

Charcoal

Charcoal is a porous solid produced from organic material (such as wood) in a process known as destructive distillation (pyrolysis). This process involves heating the organic material in a limited amount of oxygen to remove the water and volatile components. What is left behind is a solid, largely composed of carbon, known as charcoal. Charcoal is a useful fuel because it burns at a higher temperature than wood.

Charcoal has pores (gaps) within its structure. Heating charcoal to a high temperature with water vapour increases the number of pores, and the resulting product is activated charcoal. Activated charcoal is used for adsorbing gases and purifying and clarifying liquids. Charcoal filters are used in some gas masks.

Going further

The colour of a flame

Consider the burning of methane gas using a Bunsen burner. If the methane is premixed with air before ignition (by opening the valve at the bottom of the burner) the combustion will be complete. Each methane molecule is in close proximity to oxygen molecules from the air, so they can combust completely to form carbon dioxide and water. We observe a light blue colour (the flame) due to electronic transitions in short-lived intermediates produced in the combustion reaction of hydrocarbons.

When the valve is closed, the flame colour changes to orange. The orange colour is a result of partially oxidized solid particles, mainly carbon (soot), glowing due to their high temperature, and this is known as incandescence. The small soot particles are carried upwards by the rising hot air, where they cool quickly and stop emitting light in the visible region.

■ **Figure R1.38** Bunsen burner flames

LINKING QUESTION

What might be observed when a fuel such as methane is burnt in a limited supply of oxygen?

Larger hydrocarbons, and those with aromatic rings, are more likely to undergo incomplete combustion. This can be seen when different fractions produced by the fractional distillation of crude oil are burnt. The fractions with short-chain hydrocarbons burn with a clean flame, but the longer-chain hydrocarbon fractions have a much smokier flame – an indication that incomplete combustion is occurring.

LINKING QUESTION

Why do larger hydrocarbons have a greater tendency to undergo incomplete combustion?

◆ **Particulates**: Haze, smoke or dust. Any type of solid particle or droplet which can remain suspended in the atmosphere for extended periods of time.

For the large number of people around the world who rely on solid fuels, such as wood and charcoal, the biggest pollution concern is **particulates**. Particulates are tiny airborne particles which irritate our eyes and respiratory system. When solid fuels are burnt, the inadequate supply of oxygen means that some of the carbon in the fuel is released in tiny particles referred to as soot. The US Environmental Protection Agency states that clean air should contain fewer than

LINKING QUESTION

How does limiting the supply of oxygen in combustion affect the products and increase health risks?

Top tip!

Breathing in particulates and carbon monoxide produced by incomplete combustion poses major health risks to human beings.

15 micrograms of particles per cubic metre. An open fire can produce three hundred times as much. Among the many people worldwide who rely on such fires, respiratory illnesses are a major cause of death.

Particulate pollution from diesel engines and large ships is a worldwide problem above oceans and land. The smaller particulates and droplets (also referred to as aerosols) interfere with the chemistry of the atmosphere.

The carbon monoxide produced in incomplete combustion is a colourless and odourless gas which binds to the iron in haemoglobin more strongly than oxygen does, and therefore strongly inhibits the binding (and subsequent transportation) of oxygen. Breathing in a high dose of carbon monoxide is fatal. Many modern homes are fitted with detectors which will warn occupants of high levels of carbon monoxide in the air.

25 Give balanced equations for the complete combustion of the following organic compounds:
- a C_3H_8
- b C_3H_6
- c $C_6H_{12}O_6$
- d C_3H_6S
- e C_6H_5OH

26 a Incomplete combustion of hexane produces unburnt carbon (soot) and another carbon-containing combustion product. What is this product?
- b Explain why incomplete combustion occurs.
- c State two health hazards associated with the incomplete combustion of hexane.
- d Outline the observations you could make when observing incomplete (as opposed to complete) combustion.

■ **Figure R1.39** A freshly cut sample of lithium metal

Metals and oxygen

Lithium (Figure R1.39) has a shiny silver surface which tarnishes quickly when exposed to moist air. This high reactivity is typical of Group 1 metals, and it is why they are stored in oil.

When lithium is burnt in an atmosphere of pure oxygen, its primary reaction is to form a white solid, lithium oxide.

$$4Li(s) + O_2(g) \rightarrow 2Li_2O(s)$$

Although the analogous oxides (X_2O where X = Na, K, Rb, Cs) are a product of other Group 1 metals reacting with oxygen, these are not the primary products of combustion.

Sodium forms sodium peroxide, Na_2O_2, which contains the peroxide ion O_2^{2-} as the dominant product:

$$2Na(s) + O_2(g) \rightarrow Na_2O_2(s)$$

Moving further down the group, the formation of superoxides, compounds containing the O_2^- ion, dominates:

$$Rb(s) + O_2(g) \rightarrow RbO_2(s)$$

Top tip!

Oxides are compounds containing O^{2-} ions. Peroxides contain O_2^{2-} ions. Superoxides contain O_2^- ions.

In each reaction, the Group 1 metal has been oxidized, and the oxygen reduced. A mixture of different oxides is produced for each metal, and the relative amounts will be somewhat dependent on the conditions of the reactions.

The combustion of Group 2 metals gives oxides as the main product. Magnesium can be burnt to form magnesium oxide, and a bright white light is produced by the reaction.

$$2Mg(s) + O_2(g) \rightarrow 2MgO(s)$$

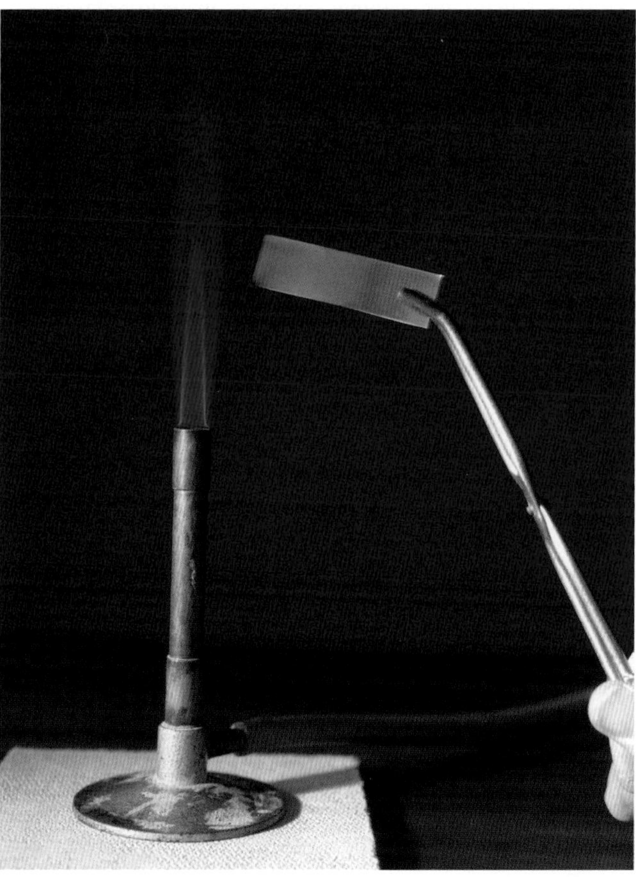

■ **Figure R1.40** Heating a copper strip in air produces copper(II) oxide.

Transition metals have variable oxidation states, and typically form many different oxides. When a copper pipe is heated in a flame it will tarnish due to the formation of copper(II) oxide (Figure R1.40).

Vanadium oxides exist for vanadium in the +2, +3, +4 and +5 oxidation states. Finely divided iron burns in oxygen to form Fe_3O_4 as the main product, although other oxides of iron can be prepared. Other metals, such as gold, resist oxidation with molecular oxygen, even when heated, because of the instability of the oxide formed.

The oxides of the first-row transition elements are summarized in Table R1.6. The most stable compounds for each metal are highlighted in blue. The main point to note is that the transition metals can have different oxidation states, and this leads to a variety of different oxides.

■ **Table R1.6** Oxides of first-row transition elements

Oxidation state	**Sc**	**Ti**	**V**	**Cr**	**Mn**	**Fe**	**Co**	**Ni**	**Cu**
>+4			V_2O_5	CrO_3	Mn_2O_7				
+4		TiO_2	VO_2	CrO_2	MnO_2				
+3	Sc_2O_3	Ti_2O_3	V_2O_3	Cr_2O_3	Mn_2O_3	Fe_2O_3			
+2		TiO*	VO*		MnO	FeO*	CoO*	NiO	CuO
+1									Cu_2O
Mixed					Mn_3O_4	Fe_3O_4	Co_3O_4		

*Forms non-stoichiometric compounds, M_xO, where x is 0.7–1.3 due to vacancies in the ionic structure.

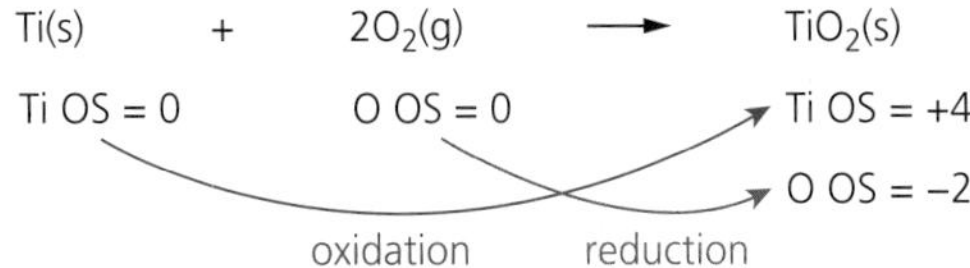

■ **Figure R1.41** The oxidation of titanium metal

■ **Figure R1.42** The combustion of phosphorus in oxygen

It is beyond the scope of this chapter to discuss the reaction of each metal with oxygen, but one aspect of their reactions is common; the elemental metal experiences an increase in its oxidation state.

Non-metals and oxygen

Most non-metals will react with oxygen in combustion reactions. Sulfur, for example, combusts in air to produce sulfur dioxide:

$$S_8(s) + 8O_2(g) \rightarrow 8SO_2(g)$$

This is often written simply as:

$$S(s) + O_2(g) \rightarrow SO_2(g)$$

However, this ignores that sulfur's most common allotropes consist of molecules with 8 atoms.

The combustion of carbon with oxygen is an incredibly important reaction to our modern world. In coal-fired power stations, carbon is reacted with oxygen in the air to form carbon dioxide:

$$C(s) + O_2(g) \rightarrow CO_2(g)$$

The heat energy released by the reaction is used to convert water into steam which drives turbines which, in turn, spin generators. The overall result is the conversion of chemical energy into heat energy into electricity. Unfortunately, the large amount of carbon dioxide, CO_2, produced in power stations is having detrimental effects on the Earth's climate.

The direct oxidation of nitrogen, N_2, to nitrogen dioxide, NO_2, is thermodynamically unfavorable under standard conditions, in part due to the great strength of the triple bond in a nitrogen molecule. Nitrogen oxides can be formed from nitrogen and oxygen at high temperatures, and their release from combustion engines into the atmosphere leads to problems such as photochemical smog and acid rain.

The reaction of phosphorus with oxygen produces phosphorus(V) oxide; this is a spectacular reaction to demonstrate because of the warm light produced (Figure R1.42).

$$P_4(s) + 5O_2(g) \rightarrow P_4O_{10}(s)$$

In the reaction, the phosphorus has an oxidation state change from 0 → +5 – it is oxidized.

In general, non-metals are oxidized in reaction with oxygen. There is one exception: molecular oxygen is oxidized, not reduced, when it reacts with fluorine as the halogen is a more powerful oxidizing agent. Fluorides of oxygen, O_nF_2 (where n = 1, 2 or 3 most commonly), in which oxygen has a positive oxidation state, have been prepared but most are unstable at normal temperatures.

Top tip!

In nearly every case, oxygen is reduced in combustion reactions and the metal or non-metal is oxidized.

Going further

Photochemical smog

Photochemical smog affects some of the world's big cities. It has an unpleasant odour and causes irritation to our respiratory systems. This is particularly serious for vulnerable groups such as babies, the elderly and people with asthma.

■ **Figure R1.43** The orange tinge of nitrogen dioxide in photochemical smog

Traditional, non-electric vehicles rely on diesel or gasoline to fuel their combustion engines. These produce carbon dioxide, particulates, nitrogen oxides (NO_x), sulfur dioxide and other volatile organic compounds (VOCs) which are mainly unburnt and partially oxidized fuel.

The VOCs and nitrogen oxides released by combustion engines react with hydroxyl radicals, •OH, naturally present in the atmosphere. This results in the production of peroxy radicals, $RCH_2O_2\bullet$, and nitrogen dioxide, NO_2:

$$RCH_3 + \bullet OH \rightarrow RCH_2\bullet + H_2O$$

$$RCH_2\bullet + O_2 \rightarrow RCH_2O_2\bullet$$

$$RCH_2O_2\bullet + NO \rightarrow RCH_2O\bullet + NO_2$$

The nitrogen dioxide gives the smog its yellow-brown colour. Once NO_2 has been formed, it can react with peroxy radicals to form relatively long-lasting peroxy nitrates, or it is photochemically decomposed back to NO and oxygen atoms by UV radiation. The oxygen atoms produced react with oxygen molecules, O_2, to form ozone, O_3:

$$RCH_2O_2\bullet + NO_2 \rightarrow RCH_2O_2NO_2 \text{ (peroxy nitrate)}$$

$$NO_2 \rightarrow NO + O$$

$$O_2 + O \rightarrow O_3$$

The secondary pollutants produced in these radical reactions (peroxy nitrates and ozone) are the main components of photochemical smog.

Because hydroxyl radicals are produced by the action of sunlight on water molecules in the atmosphere, and because the Sun is at its strongest in the summer months, photochemical smog is usually at its worst in the summer.

In the 20th century, great efforts were made to decrease localized pollution through the introduction of electronic management of vehicle engines and catalytic converters in the exhaust. This meant that motor vehicles drastically reduced the harmful emissions they released. The switch to electric vehicles has gathered pace in recent years and this will further reduce local pollution.

Although electric vehicles do not emit pollutants from their engines, it is still important to consider how the electricity required to run the vehicles is generated. Currently the generation of electricity relies heavily on fossil fuels being burnt in power stations. If carbon capture can be developed, it will be much easier to apply this to a smaller number of power stations than to individual vehicles. Also, moving to more renewable and 'green' energy sources for electricity production will help lower the overall emissions associated with driving an electric car.

Fossil fuels

◆ **Fossil fuel**: A fuel formed from the remains of living organisms over millions of years. Fossil fuels have a high carbon content and release energy when combusted.

Fossil fuels is the name given to a set of energy-rich combustible carbon compounds which have been formed from organic remains over millions of years. Coal, crude oil and natural gas are three such fossil fuels that are integral to our modern society. The energy in fossil fuels originates from solar energy which drove photosynthesis in prehistoric plants. The biological material produced by the plants and the organisms which fed on them was reduced by the action of heat, pressure and bacteria into the fossil fuels we now extract from beneath the Earth's surface.

The chemical energy fossil fuels possess can be released by combustion, and this is used to do work (for example, on a piston in a car engine), generate heat (for example, in a gas boiler) or produce steam to drive electricity generation (for example, in a coal-fired power station).

Fossil fuels are abundant but they are non-renewable.

Key fossil fuels

Coal

Coal is the most plentiful of the fossil fuels. It originated as prehistoric forests were flooded, buried and then gradually compressed by layer upon layer of soil. Most dead plant material decomposes where it falls. However, when large quantities of plant matter are buried and isolated from oxygen, it can, over millions of years, form coal.

The action of pressure and heat changed the prehistoric forests into peat, then lignite, then bituminous soft coal, and finally anthracite (hard coal). At each stage of the formation process, the carbon percentage increases and the **specific energy** of the coal also increases. Although anthracite is the most desirable form in terms of specific energy, other forms of coal are often used, mostly because they are plentiful and cheap in local situations.

◆ **Specific energy**: The energy released per unit of mass of fuel consumed; it is usually expressed in megajoules per kilogram, $MJ\,kg^{-1}$, which is equivalent to $kJ\,g^{-1}$.

◆ **Crude oil**: A viscous mixture of hydrocarbons found beneath the ground.

Crude oil

Crude oil remains one of the most important raw materials in the world today. It is a complex mixture of hydrocarbons and supplies us with fuels for a range of transport types and for electricity generation. In addition, it is an important chemical feedstock used in the production of organic polymers, pharmaceuticals, dyes and solvents.

Crude oil formed over geological periods of time (that is, millions of years) from the remains of marine animals and plants. These creatures died and accumulated as sediment at the bottom of the oceans before becoming trapped under layers of rock. Under these conditions of high temperature and high pressure, this matter decayed in the presence of bacteria and the absence of oxygen to form crude oil (petroleum) and natural gas.

Crude oil is a limited resource and eventually reserves will be so depleted that chemists will need to consider other sources of carbon, both as a fuel and as a chemical feedstock. Indeed, the balance of these two uses is an issue if we are to conserve this non-renewable resource for as long as possible.

Unrefined crude oil varies greatly in appearance depending on its composition, but usually it is black or dark brown. In underground reservoirs, it is often found in association with natural gas, which forms a gas cap over the crude oil (Figure R1.44).

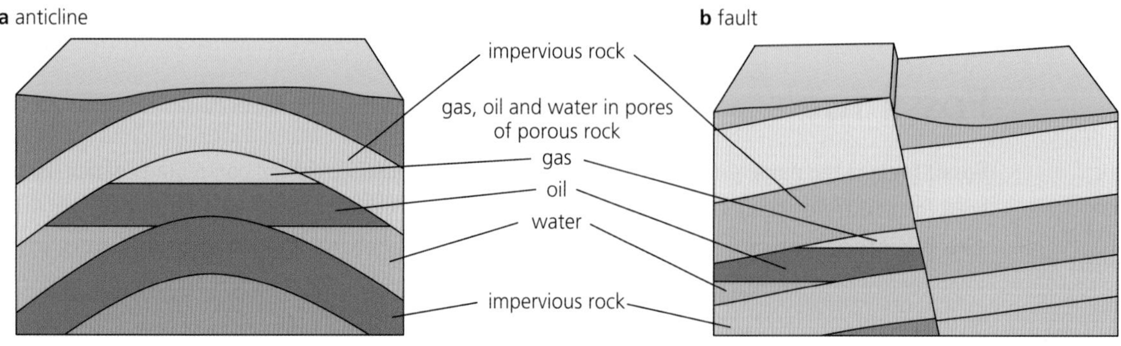

■ **Figure R1.44** Crude oil is found in underground reservoirs in certain geological situations. It permeates the rock in these reservoirs and is usually found in association with natural gas.

Crude bitumen is a semi-solid form of crude oil mixed with sand and water in sources known as oil sands or tar sands.

Crude oil needs to be refined before it is used as a fuel or chemical feedstock. This involves removing the sulfur impurities, and then separating it into groups of compounds with similar boiling points. The groups are known as fractions. The proportion of each fraction differs depending on the precise source of the oil, and fields producing oil with a high amount of short-chain hydrocarbons (with 5–10 carbon atoms, used for motor vehicle fuels) are particularly valuable.

Natural gas

◆ **Natural gas**: A naturally occurring mixture of gaseous hydrocarbons consisting primarily of methane.

Methane is the primary component of **natural gas**. Natural gas is formed in the same way as crude oil, and hence these two fossil fuels are often found together. Natural gas is stored in geological formations (porous rocks) capped by a layer of impervious rock (Figure R1.44) and can be formed from the decomposition of crude oil and coal deposits.

Natural gas is the cleanest of the fossil fuels because it can easily be treated to remove impurities, and its combustion is usually complete. This means harmful carbon monoxide and particulates are not produced.

Top tip!

The 'petroleum industry' is usually taken to include both the oil and natural gas industries. The term *petroleum* comes from the Latin *petra* meaning rock, and *oleum*, meaning oil.

Transportation of natural gas requires significant infrastructure. It may be liquefied and shipped or pumped along pipelines. Many countries rely on natural gas for their energy needs, and because there is a long lag time associated with building new supply lines, having control of the supply of this vital commodity can yield political power.

International pipelines can be used to apply political pressure to countries. In February 2022, Germany denied certification of the recently completed Nord Stream 2 gas pipeline. This stopped Russia exporting a greater volume of natural gas into northern Europe, and was a move designed to damage the Russian economy and persuade Russia not to start military action in Ukraine. Later in the year, Russia effectively shut its gas pipelines into Europe in an effort to destabilize the political leadership of the European Union because of their opposition to invasion of Ukraine.

Nature of science: Global impact of science

The importance of fossil fuels in science

As well as supplying energy, fossil fuels also provide the scientific community with many chemicals to draw upon for a wide range of other applications. Before the mid-19th century, drugs that existed were usually extracted from plants. Today's multi-billion-dollar pharmaceutical industry uses many chemicals which are ultimately derived from fossil fuels. Petroleum by-products are also used in the manufacture of dyes and plastics.

Fossil fuels have provided a 'means' to scientific progress, as they have provided raw materials for many scientific fields. In the coming years, scientists will need to use further scientific progress to solve the issues exploitation of fossil fuels has caused.

Comparing fossil fuels

One aspect to consider when comparing fossil fuels is the specific energy of the fuel.

$$\text{specific energy} = \frac{\text{energy released from the fuel (MJ)}}{\text{mass of fuel consumed (kg)}}$$

Top tip!

The specific energy is usually quoted in MJ kg^{-1}, which is equivalent to kJ g^{-1}.

The approximate specific energy of some fuels is given in Table R1.7. The values vary between different deposits, so these numbers are only a guide.

■ **Table R1.7** The characteristics of some common fuels

Fuel	Specific energy / $MJ\,kg^{-1}$	Carbon content, % by mass
natural gas (mainly methane)	55	75
gasoline (petrol)	46	82
coal	33	94
wood	16	70

The specific energy of a fuel can be calculated from its standard enthalpy of combustion.

WORKED EXAMPLE R1.3B

Use the necessary data given in sections 7 and 13 of the IB *Chemistry data booklet* to calculate the specific energy of octane, C_8H_{18} (the major component of gasoline).

Answer

Necessary data regarding octane from the IB *Chemistry data booklet*:

$\Delta H_c^{\ominus}$: $-5470\,kJ\,mol^{-1}$ M: $114.26\,g\,mol^{-1}$

The enthalpy change of combustion tells us that burning 1.00 mol of octane would release 5470 kJ of heat energy.

1.00 mol of octane has a mass of $(1.00\,mol \times 114.26\,g\,mol^{-1}) = 114.26\,g$

$$\text{specific energy} = \frac{\text{energy released from the fuel}}{\text{mass of fuel consumed}} = \frac{5470\,kJ}{114.26\,g} = 47.9\,kJ\,g^{-1} = 47.9\,MJ\,kg^{-1}$$

(Quoting the answer in the more commonly used units of $MJ\,kg^{-1}$, $47.9\,kJ\,g^{-1} = 47.9\,MJ\,kg^{-1}$)

LINKING QUESTION

Why is a high activation energy often considered to be a useful property of a fuel?

27 Biodiesel is an alternative fuel to regular diesel derived from crude oil. Given the enthalpy of combustion of a sample of biodiesel is $12.0\,MJ\,mol^{-1}$, and its molar mass is $299\,g\,mol^{-1}$, determine the specific energy, in $kJ\,g^{-1}$, of the biodiesel.

Fossil fuels such as gasoline and kerosene have a high specific energy, meaning that a comparatively small mass of fuel is needed for a given energy content. This has made these fuels very effective for transport applications because they offer very long ranges.

Battery technology has improved greatly in the past decades and electric vehicles are becoming a viable alternative to traditional petrol or diesel vehicles, especially over short distances. Many car manufacturers have shifted their focus to electric vehicles (EVs), and some countries have signalled their intentions to prohibit the sale of new diesel and petrol vehicles in the near future. Drawbacks are that large batteries which offer a comparable driving range to modern petrol vehicles are heavy and expensive to make, and these reduce the efficiency of vehicles. The majority of journeys made do not require the full range a large, heavy battery can provide, so more efficient EVs designed for many short journeys are the most common type of EVs available today. How the electricity to charge EVs is generated also needs to be considered.

Many countries are now aiming to reduce their greenhouse gas emissions in order to limit global warming, so a second aspect to consider when comparing fossil fuels is the amount of carbon dioxide added to the atmosphere when they are combusted. This is determined by the mass of fuel burnt and the carbon content of the fuel. A higher carbon content will result in more carbon dioxide being produced during combustion.

The specific energy and carbon content of some common fuels are shown in Table R1.7.

If 1.00 kg of coal is burnt, $(1.00\,kg \times 0.94) = 0.94\,kg$ of carbon is being burnt. The amount of carbon atoms in 0.94 kg is:

$$\frac{0.94\,kg \times 1000}{12.01\,g\,mol^{-1}} = 78\,\text{mol of carbon atoms}$$

Because all the carbon in the coal is oxidized to carbon dioxide (assuming complete combustion)

$$C(s) + O_2(g) \rightarrow CO_2(g)$$

it follows that 78 mol of carbon dioxide, CO_2, are formed by the complete combustion of 1.00 kg of coal.

Burning 1.00 kg of natural gas would produce less carbon dioxide (62 mol of carbon dioxide, CO_2) because the carbon content of natural gas is lower.

However, it is also important to note the specific energy values, which tell us that the same mass of each fuel does not release the same amount of heat energy. We must remember this if we want to compare the masses of carbon dioxide generated by different fossil fuels producing the same energy output.

When considering the three main fossil fuels (coal, crude oil and natural gas), natural gas produces the least amount of carbon dioxide per unit of energy released, and coal produces the most.

WORKED EXAMPLE R1.3C

Calculate the mass of carbon dioxide produced when 100 MJ of energy is released by burning wood and gasoline separately.

Specific energy of wood = $16\,MJ\,kg^{-1}$ Carbon content of wood = 70%

Specific energy of gasoline = $46\,MJ\,kg^{-1}$ Carbon content of gasoline = 82%

Answer

For wood

First calculate the mass of wood needed to produce 100 MJ of energy using:

$$\text{specific energy} = \frac{\text{energy released from the fuel}}{\text{mass of fuel consumed}}$$

$$\text{mass of wood consumed} = \frac{100\ \text{MJ}}{16\,\text{MJ}\,\text{kg}^{-1}} = 6.25\,\text{kg}$$

Then calculate the mass of carbon combusted:

$$\text{mass of carbon in the wood} = 6.25\,\text{kg} \times 0.70 = 4.38\,\text{kg}$$

Then calculate the mass of carbon dioxide produced:

$$\text{amount of carbon in the wood} = \frac{4380\,\text{g}}{12.01\,\text{g}\,\text{mol}^{-1}} = 365\,\text{mol}$$

$$\text{amount of carbon dioxide, CO}_2\text{, produced} = 365\,\text{mol}$$

$$\text{mass of carbon dioxide, CO}_2 = 365\,\text{mol} \times 44.01\,\text{g}\,\text{mol}^{-1} = 16\,064\,\text{g} = 16\,\text{kg (2 s.f.)}$$

For gasoline

$$\text{mass of gasoline} = \frac{100\,\text{MJ}}{46\,\text{MJ}\,\text{kg}^{-1}} = 2.17\,\text{kg}$$

$$\text{mass of carbon dioxide, CO}_2\text{, produced} = 2.17\,\text{kg} \times 0.82 \times \frac{44.01\,\text{g}\,\text{mol}^{-1}}{12.01\,\text{g}\,\text{mol}^{-1}} = 6.5\,\text{kg (2 s.f.)}$$

Much more carbon dioxide is produced when using wood as an energy source.

WORKED EXAMPLE 1.3D

The specific energy of a fuel can be expressed in the units $kWh\,kg^{-1}$. The specific energy of natural gas is $15.4\,kWh\,kg^{-1}$. A kilowatt-hour is a unit of energy equivalent to 3.6 MJ.

Calculate the mass of carbon dioxide produced when enough natural gas is burnt to produce 1.00 kilowatt-hour of energy. Assume that the natural gas is pure methane.

Answer

$$\text{specific energy} = \frac{\text{energy released from the fuel}}{\text{mass of fuel consumed}}$$

$$\text{mass of natural gas required} = \frac{1.00\,kWh}{15.4\,kWh\,kg^{-1}} = 0.0649\,kg = 64.9\,g$$

$$\text{methane's carbon content by mass} = \frac{12.01\,g\,mol^{-1}}{16.05\,g\,mol^{-1}} \times 100 = 74.8\%$$

$$\text{mass of carbon dioxide, } CO_2\text{, produced} = 64.9\,g \times 0.748 \times \frac{44.01\,g\,mol^{-1}}{12.01\,g\,mol^{-1}} = 178\,g\ (3\ s.f.)$$

28 The average home in the United Kingdom uses 15 000 kWh of energy in a year. Calculate the mass of carbon dioxide the production of this amount of energy creates if the following fuels are used:

a natural gas (specific energy = $15.4\,kWh\,kg^{-1}$, carbon content 75%)

b heavy fuel oil, a fraction from crude oil (specific energy = $11.6\,kWh\,kg^{-1}$, carbon content 85%).

The comparative advantages and disadvantages of each type of fossil fuel are summarized in Table R1.8.

■ **Table R1.8** The advantages and disadvantages of different fossil fuels

Fossil fuel	Advantages	Disadvantages
Coal	• Cheap to extract and plentiful throughout the world • Large infrastructure already exists for transport and burning • Coal reserves will last much longer than oil reserves – still several hundred years remaining	• Lowest specific energy of the three fossil fuels • Releases the most carbon dioxide per unit of energy produced • Sulfur impurities release sulfur dioxide when burnt, which leads to acid rain – gaseous factory emissions must be scrubbed to remove this • Coal mining is a dangerous activity and an open mine can spoil the beauty of the countryside around it
Crude oil	• Very versatile – different crude oil fractions have various physical properties making them suitable for many different fuel applications • Convenient for use in vehicles due to high volatility • Sulfur impurities can be removed easily	• Limited reserves – the crude oil we have is becoming harder to find and extract • It is not evenly distributed around the world • Its use in vehicles produces local pollution and health risks
Natural gas	• Produces fewer pollutants per unit of energy • Has the highest specific energy • Can be piped directly into homes	• Limited supply • Risk of explosions and leaks makes it impractical for some uses • Expensive to invest in gas pipeline infrastructure • Unless liquefied it occupies a much larger volume per unit of available energy than other fossil fuels (lower energy density)

Historical, geological and technological factors all play a part in determining which fossil fuel is selected for a specific use. Because different societies have different priorities, different fossil fuels may be preferred in different regions of the world.

Carbon dioxide and the greenhouse effect

Link

The greenhouse effect is also covered in Chapter S3.2, page 295 (HL only).

In S3.2 (page 295), carbon dioxide was identified as a greenhouse gas because it is able to absorb infrared radiation and cause the warming of the Earth's atmosphere. The greenhouse effect is a necessary mechanism for maintaining the Earth's temperature at a habitable level. It is estimated that, without this effect, the surface temperature would be around 30 °C lower.

Going further

Heat transfer in the atmosphere

By considering how a real-life greenhouse operates, we can see that the term 'greenhouse gas' is a misnomer. A greenhouse has warm air inside it because the Sun's visible radiation passes through its glass roof and warms the floor and air inside. Unlike the air outside the greenhouse, the air inside cannot rise to mix with cooler air above. The roof glass stops infrared radiation radiated from the floor leaving the greenhouse, but if the roof was made from NaCl (which is transparent to infrared radiation), the air inside would still warm significantly even though the infrared radiated from the floor would no longer be trapped. This shows that the primary reason a greenhouse works is the restriction of air circulation and not because it traps infrared radiation from the ground in a manner akin to a greenhouse gas. Both effects rely on trapping energy derived from the Sun and both have a warming effect, but the mechanisms by which this is achieved are different.

There are other factors that affect the transfer of energy into and around the atmosphere and from it into space. Some energy is carried by water as it vaporizes. Water vapour rises in the atmosphere by convection and then condenses in the cooler upper troposphere, producing clouds and releasing the energy at this higher altitude. The energy may then escape the Earth in the form of infrared radiation. The atmosphere is therefore heated not only by radiation, but also by convection.

LINKING QUESTION (HL ONLY)

Why is carbon dioxide described as a greenhouse gas?

In recent years the level of carbon dioxide in the atmosphere has been rising (Figure R1.45). Rising concentrations of infrared-absorbing gases decrease the amount of energy escaping from the Earth by radiation (by absorbing the energy on its journey upwards) and increase the amount of energy moving downwards towards the surface (by re-radiating this as infrared). The upwards / downwards equilibrium is therefore disturbed, so the surface temperature rises. This will continue until the upwards energy flow again equals the downwards flow.

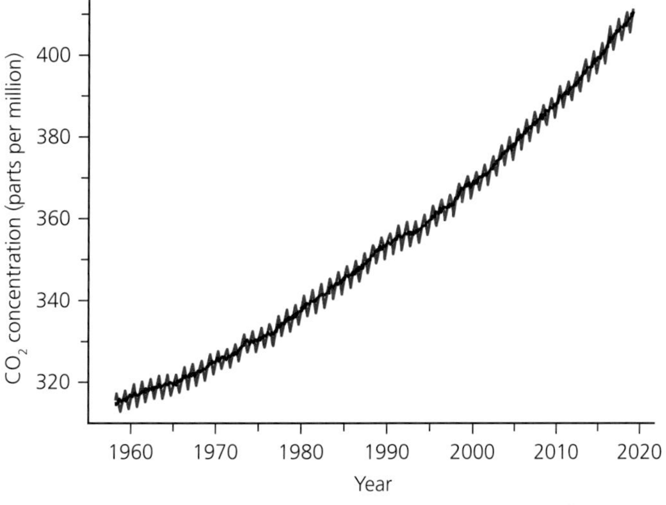

■ **Figure R1.45** The concentrations of carbon dioxide in the atmosphere (in parts per million) observed at NOAA's Mauna Loa Observatory in Hawaii have been steadily increasing over the course of 60 years

Burning fossil fuels releases the vast amounts of carbon stored in them as carbon dioxide, and scientists agree that this carbon dioxide is leading to an increase in global temperatures. The Intergovernmental Panel on Climate Change (IPCC) is the United Nations body assessing the science and impacts of climate change. The first headline statement from scientists contributing to the IPCC's sixth assessment report (2021) is that '*it is unequivocal that human influence has warmed the atmosphere, ocean and land. Widespread and rapid changes in the atmosphere, ocean, cryosphere and biosphere have occurred.*'

Politicians, the public and global organizations are becoming aware of the need to limit the warming of the Earth. The role carbon dioxide emissions have in this process was recognized in the Paris Climate Accords, and nearly all of the world's top carbon dioxide emitters signed to agree to reduce their carbon dioxide emissions. The aim of the agreement is to limit the warming of the Earth to 2 °C (ideally 1.5 °C) above pre-industrial levels.

Nature of science: Global impact of science

The impact of fossil fuels

Large-scale fossil fuel use is a relatively recent development. Underground coal mining was developed in the late 18th century, and oil drilling began in the 19th century.

The industrialization of the world was driven by fossil fuels. Coal led to steam engines, which led to steam ships and mass manufacturing in factories. Factories drove urbanization and steam ships drove international trade and migration. Coke (a high-carbon fuel made from coal) replaced charcoal in the manufacture of iron and made this process much more efficient. As oil gradually became more readily available, motor vehicles and aircraft were developed, which led to further migration, mass travel and globalization, and cities grew outwards, with suburban zones for those who could afford cars until oil overtook coal as a source of carbon dioxide.

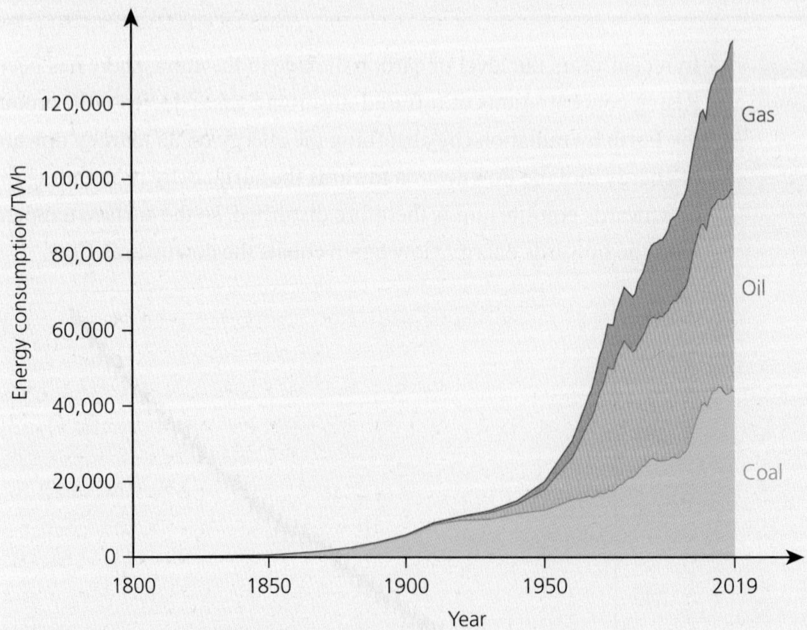

■ **Figure R1.46** Global fossil fuel consumption

This progress was driven by science and technology, as scientists around the world invented and developed more efficient ways to utilize fossil fuels for transport and power generation.

While fossil fuels have driven large societal changes, this has not been without harm to the planet and communities. One issue is climate change caused by the release of vast amounts of carbon dioxide. This is a problem affecting the whole world and is only likely to worsen in future decades. The United Nations states that climate change can lead to intense droughts, water scarcity, severe fires, rising sea levels, flooding, melting polar ice, catastrophic storms and declining biodiversity. The impacts of climate change are not felt equally around the world, and it is often communities that have made relatively little contribution to climate change who face the worst consequences. Rising sea levels and increasing temperatures are creating climate refugees, while other communities that have been able to develop their economies through exploitation of fossil fuels.

It is recognized that everyone must reduce their use of fossil fuels to limit climate change, and scientists are working to diversify our energy supplies away from fossil fuels.

LINKING QUESTION

What are some of the environmental, economic, ethical and social implications of burning fossil fuels?

Inquiry 1: Exploring and designing

Selecting sufficient and relevant sources of information

If you ask one group of people to briefly research the pros of eating chocolate and another group the cons, both groups will quickly collect plenty of information. It is not difficult to find claims that chocolate is healthy because, for example, it seems to offer an improvement in cardiovascular health. But, equally, there are numerous claims linking chocolate to poor health and cardiovascular disease. This highlights that there are a lot of contradictory claims, and we cannot assume that a statement is true just because it has been published in a book or journal, broadcast or shared on a website.

Scientists have a crucial role in assessing the validity of claims. When researching a topic, it is important to consider what you read carefully. Just some of the questions to ask are: Are the claims supported by other independent research? Are any results reliable and reproducible? What are the interests of the author? What might the vested interests be? Has the source been peer reviewed? Is the author well recognized within the field?

Primary literature is mostly found in peer-reviewed journal articles, and this adds credibility to the findings. Secondary literature compiles the findings of primary literature, and specialist books or journal-published literature reviews can be a useful starting point to gain an overview of a topic. Information found on websites should be treated with some caution because it is less likely to have been treated with the same scrutiny. Again, ask questions such as: What is the reputation of the author or website?

ATL R1.3A

Nationally determined contributions

The commitments made by signatories of the Paris Climate Accords are different. The intended nationally determined contributions (NDCs) – that is, the precise contributions each country will make – were determined by the countries themselves. These contributions are reviewed every five years. This was agreed to be the right approach because all economies are at different stages, and some are more able to move away from fossil fuels faster than others.

In many areas of the world, charities, journalists and pressure groups closely watch the latest emissions data, and they try to build public pressure in order to encourage governments and corporations to adhere to the commitments they have made.

Research the NDCs made by your government (**https://unfccc.int/NDCREG**). Search for data or articles which indicate whether your country is on track to meet the commitments made. Compile the evidence you find and write a summary on whether you believe your country will meet its NDCs. Cite the evidence you have found in your research.

TOK

What is the influence of political pressure on different areas of knowledge?

As we have discussed before, the construction of knowledge (including knowledge about the adverse effects of the consumption of fossil fuels) occurs within the context of various communities of knowers, each with their own scopes, methods and interests. These communities comprise people, who also have their own needs and interests. The knowledge around climate change is an important example of this. Chemists or physicists can focus exclusively on facts related to the impact of carbon in the atmosphere, but biologists might want to focus on the effect of this on organisms and ecosystems. Politicians might want to focus on public policy and economists might focus on the financial impact of rising energy costs or the effects of climate change on different industries.

The challenge then is to coordinate these different approaches in the hopes that the best outcome can be achieved.

Unfortunately, the public policy decisions required to meet intergovernmental agreements like the Paris Climate Accords (which recognize the scientific facts provided by chemists, physicists and biologists) are often made by politicians, who are personally affected by political pressures which have little to do with the environment. A politician might be representing a community with deep ties to fossil fuel consumption, so they see regulations as a threat to people's livelihoods. They might be financially supported by lobbyists supporting fossil fuel industries, who have no interest in seeing fossil fuel use regulated or curtailed. In some cases, these interests might result in individuals challenging well-regarded and well-supported scientific evidence. This can give the impression that there is genuine scientific debate where, in fact, there is none.

Constructing scientific facts about climate change is one thing, but weighing up scientific knowledge with the interests of political or financial communities is quite another.

29 Carbon dioxide is a greenhouse gas.

a Explain the general mechanism by which greenhouse gases lead to the warming of the Earth.

b Explain the molecular mechanism by which molecules of carbon dioxide act as a greenhouse gas.

Nature of science: Science as a shared endeavour

Climate science as a transdisciplinary endeavour

The study of global warming and its likely effects is highly transdisciplinary, encompassing aspects of biology, chemistry, physics, geology, economics, mathematics and computer science.

- Atmospheric chemists study the composition of the atmosphere. In order to do this, they use a combination of direct measurement, remote sensing, laboratory work and computer modelling.
  - Direct measurement can be carried out by using aircraft to collect samples of the atmosphere.
  - Remote sensing uses spectroscopic methods to assess the concentration of different compounds.
  - Laboratory experiments allow the kinetics of atmospheric chemical reactions to be studied.
  - Atmospheric and experimental data are used to set the parameters for a computer model which will attempt to predict future changes.
- Atmospheric physicists study how the atmosphere interacts with electromagnetic radiation, and also use statistics to develop more advanced climate models.
- Biologists explore the natural processes involved in sustaining carbon dioxide levels, and the impact of global warming on different ecosystems, including coral reefs.
- Geologists employ an understanding of the carbon cycle and volcanic emissions to determine their effects. They also collect samples to build up our understanding of historical data about the temperature and composition of the atmosphere.
- Economists provide data about how climate change, regulations or policy changes will affect industries and financial markets.
- The expertise of mathematicians and computer scientists is needed to produce complex climate models.

Experts in all these fields, from countries all over the world, contribute to our understanding of climate change.

Renewable energy sources

◆ **Non-renewable energy sources**: Finite energy resources that will no longer be available to us once they have been used up.

◆ **Renewable energy sources**: Energy sources that are naturally replaced on a timescale useful to humans.

◆ **Biofuels**: Non-petroleum based renewable fuels, such as ethanol or biodiesel, derived from biomass.

Coal, crude oil and natural gas will eventually run out; they are **non-renewable energy sources**. Reserves also tend to be concentrated in certain countries or regions, meaning that some countries have become very wealthy on the basis of an accident of geography.

Renewable energy sources are ones which cannot be used up or can be replenished at the rate they are being used. Renewable energy sources are often referred to as 'alternative' energy or 'green' energy.

Renewable sources of energy include:

- photovoltaic cells
- **biofuels**
- wind
- biomass
- ocean (tidal and wave)
- geothermal
- hydroelectric power.

Renewables are highly dispersed over the Earth, meaning they should be equally available to everyone in varying proportions. One disadvantage is that they can be seasonal and unreliable. For example, wind turbines will not generate any power if there is no wind, and solar cells will not generate power in the dark.

Generation of power in nuclear power stations relies on fission (splitting) of ^{235}U atoms. In the future, it is possible that fusion reactors may be developed that use hydrogen as a fuel. In December 2022 it was announced that, for the first time, more energy had been gained from a hydrogen fusion experiment than had been put in. Since hydrogen is much more abundant than uranium, fusion would essentially be renewable, but it will likely be many years before this technology will be released as a viable method of renewable everyday energy production.

Common mistake

Misconception: *Nuclear power is a renewable energy resource.*

In a nuclear reactor, uranium-235 atoms are split into atoms of other substances. Since sources of uranium are finite, nuclear power is usually classified as non-renewable.

Biofuels

Solar energy reaching the Earth's surface can be converted into chemical energy by plants. Plants use the solar energy to convert carbon dioxide and water into glucose and oxygen. This conversion of fully oxidized carbon in the air into carbon in organic matter is known as biological fixation. Some organic matter made through the biological fixation of carbon can be processed into biofuels. Biofuels are an alternative to fossil fuels, and they are renewable – they can be grown in a short space of time.

The first step in producing a biofuel relies on plants. Solar energy reaching the surface of the Earth is absorbed by plants and used to drive photosynthesis (Figure R1.47).

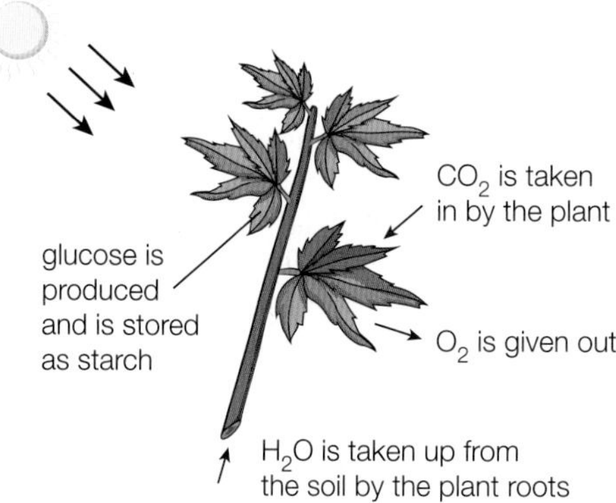

energy from sunlight is trapped by chlorophyll in green leaves
CO_2 is taken in by the plant
glucose is produced and is stored as starch
O_2 is given out
H_2O is taken up from the soil by the plant roots

■ **Figure R1.47** Photosynthesis

Photosynthesis produces glucose and oxygen from atmospheric carbon dioxide and water:

$$6CO_2 + 6H_2O \rightarrow C_6H_{12}O_6 + 6O_2$$

Top tip!

You are expected to be able to recall this equation for photosynthesis.

Photosynthesis is a complicated biological process which relies on chlorophyll molecules found within plant cells. These chlorophyll molecules (chlorophylls a and b) absorb red and blue wavelengths in the visible light regions to initiate photosynthesis. Green light is reflected.

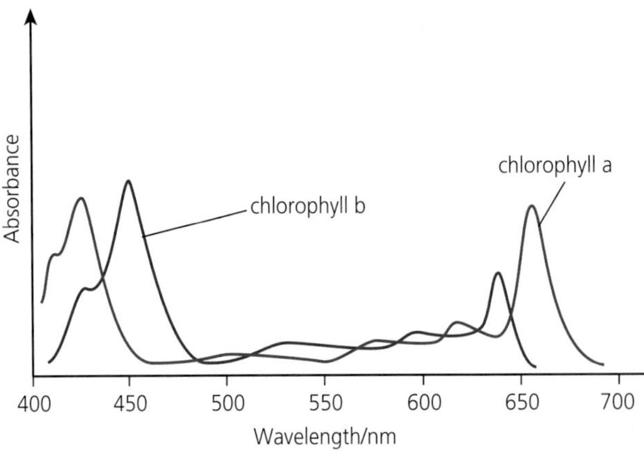

■ **Figure R1.48** Absorption spectra of chlorophyll a and b

The glucose made by plants is either stored or used in respiration. Some of the carbon in the synthesized glucose will make its way into other organic matter (such as oils) through biosynthetic pathways within plant cells. It is this organic matter which is then made into biofuels.

Ethanol

◆ **Fermentation:** A chemical change performed on carbohydrates by microorganisms in anaerobic conditions.

Bioethanol is ethanol produced by the **fermentation** of starchy or sugar-rich crops, including sugar cane, wheat, sorghum and corn. Its production is a multi-step process, involving both physical and chemical conversions. It starts with grinding of the plant material, to produce a fine powder. The powder is mixed with water to form a mash to which an enzyme (amylase) is added; this enzyme hydrolyses the liquefied starch into glucose. Yeast (a single-celled fungus) is able to ferment the glucose (via an enzyme known as zymase), forming ethanol and carbon dioxide gas. This process forms the basis of beer and wine making and can be simplified as:

$$C_6H_{12}O_6 \rightarrow 2C_2H_5OH + 2CO_2$$

Fermentation produces a solution of ethanol along with side products such as aldehydes and other alcohols. The maximum achievable ethanol concentration is limited to about 12% by mass as concentrations higher than this will poison the yeasts and stop fermentation.

The fermented mixture is distilled to separate the ethanol from the water and side products. This distillation step requires a significant input of energy. Bioethanol production is relatively inefficient since only a certain amount of the plant (about 30% in the case of sugar cane) is available for fermentation: the rest is mainly structural compounds such as cellulose and lignin which cannot be fermented.

Advantages of ethanol

- Lower greenhouse gas emissions: ethanol has been adopted in some countries as a means of lowering greenhouse gas emissions. Ethanol does produce carbon dioxide when burnt. However, this carbon was absorbed from the atmosphere as the plants used to make the ethanol grew so, theoretically, the ethanol is carbon-neutral.
- Reduced crude oil use: sugar cane is a large-scale crop in Brazil and, since 1976, all cars in the country must use gasoline blended with ethanol. Many vehicles sold in Brazil are designated 'flex-fuel', because they can run on a wide range of blends, ranging from almost pure ethanol to almost pure gasoline. In the USA, most new cars can run on gasoline containing 10% ethanol without modification or damage, which means that oil use (and so the carbon footprint) per person will decrease in areas where ethanol is widely available.

Disadvantages of ethanol

- Lower specific energy and potential damage to engines: ethanol has a lower specific energy than gasoline. Approximately 1.5 dm^3 of ethanol is required to provide the same amount of energy as 1 dm^3 of gasoline. Unlike gasoline, ethanol absorbs moisture from the atmosphere, which can lead to engine corrosion. Blended fuels, containing ethanol and gasoline, are more likely to cause engine damage than pure gasoline.
- High energy cost of distillation: although ethanol is a renewable fuel, as sugar crops can be quickly replenished, ethanol production is criticized for being energy intensive. The distillation step requires large amounts of heat energy. By some calculations, when the processing is taken into account, use of ethanol does not actually reduce greenhouse gas emissions relative to traditional gasoline.
- Food versus fuel debate: When maize (corn) is used as a source of ethanol, as in the USA, critics claim that the price of maize on world markets is forced up, making poorer countries more likely to export their staple food crops and leaving their own populations vulnerable to hunger.

ATL R1.3B

Research the arguments for and against using fertile land to grow crops for biofuels instead of crops for human consumption. Is it morally acceptable to encourage this? Does this push the price of food staples up and hurt poorer countries?

Evaluate the competing claims.

Hold a debate in your class and vote to decide the outcome.

Use of gasoline and ethanol mixtures in different countries – the E nomenclature

■ **Figure R1.49** Bioethanol

Most cars in the United States can run on 'E10' – gasoline containing 10% ethanol by volume. In corn-producing regions of the USA, higher proportions of ethanol are common: E85 is widely available.

In Brazil, some vehicles run on 100% ethanol (E100), but an ethanol content of at least 18% (E18) is mandatory for all gasoline vehicles.

In Sweden, flex-fuel vehicles are available which run on blends from E0 to E85. Such vehicles rely on electronic engine-management systems to adjust the spark and inlet valve timings, according to the fuel composition, to prevent knocking (pre-ignition) in the engine.

Biodiesel

Ethanol is not a suitable fuel for diesel-powered vehicles because it is less viscous and much more volatile than diesel, so the vehicle's fuel system would need to be extensively modified. However, plant-based fuels derived from vegetable oils (triglycerides) are much closer to diesel in their properties and are commonly called biodiesels.

Plant oils have the general structure shown in Figure R1.50.

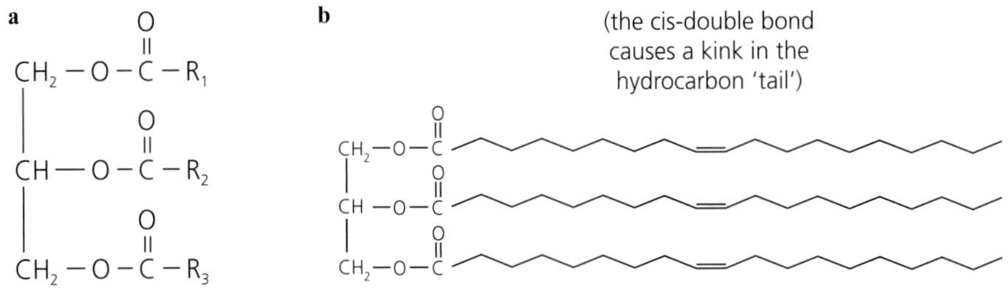

a

$CH_2-O-C(=O)-R_1$

$CH-O-C(=O)-R_2$

$CH_2-O-C(=O)-R_3$

b

(the cis-double bond causes a kink in the hydrocarbon 'tail')

$CH_2-O-C(=O)-$

$CH-O-C(=O)-$

$CH_2-O-C(=O)-$

triolein, m.p. –4 °C

■ **Figure R1.50** a) The general structure of oils (triglycerides); b) triolein, a symmetrical unsaturated oil found in olive oil

The oils are extracted by pressing the fruit or the seeds of plants, and then they are chemically modified in a process known as transesterification. In transesterification, each triglyceride molecule is broken down into three fatty acid methyl ester molecules (RCO_2CH_3) and a molecule of glycerol (propane-1,2,3-triol) through reaction with methanol and an acid or alkaline catalyst (Figure R1.51). The fatty acid esters are more volatile and less viscous than the oils they came from, making them much more suitable for use as biodiesel.

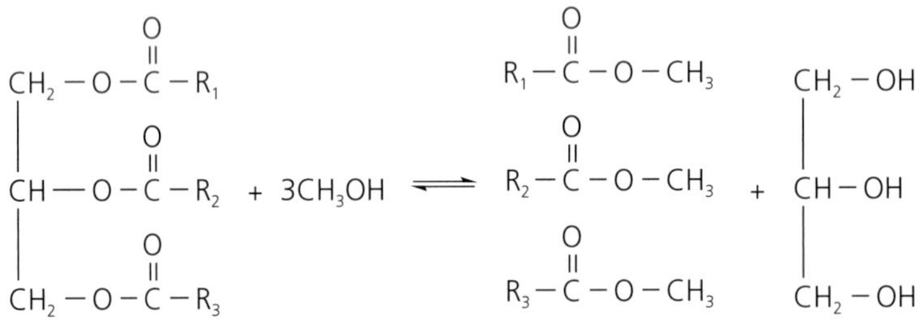

$$\begin{array}{l} CH_2-O-C(=O)-R_1 \\ CH-O-C(=O)-R_2 \\ CH_2-O-C(=O)-R_3 \end{array} + 3CH_3OH \rightleftharpoons \begin{array}{l} R_1-C(=O)-O-CH_3 \\ R_2-C(=O)-O-CH_3 \\ R_3-C(=O)-O-CH_3 \end{array} + \begin{array}{l} CH_2-OH \\ CH-OH \\ CH_2-OH \end{array}$$

■ **Figure R1.51** The transesterification of oils produces three fatty acids and a molecule of propane-1,2,3-triol. An acid or alkaline catalyst is used to speed up the rate of transesterfication, and an excess of methanol is used to push the reaction towards the products

As is the case with ethanol and gasoline, biodiesel is most often blended in various proportions with crude-oil-derived diesel.

The specific energy density (that is, the energy provided per volume of fuel) is shown in Table R1.9.

■ **Table R1.9** The energy density of fossil fuels compared with biodiesel

Fuel	**Energy density / MJ dm^{-3}**
gasoline	46
diesel	43
sunflower oil (from which biodiesel can be derived)	33
commercial biodiesel	33–35

There are advantages and disadvantages of using biodiesel; some of these are common to biofuels, and thus are shared with ethanol too:

Advantages of biodiesel

- Decreased fossil fuel use.
- Potential for lower greenhouse gas emissions because the carbon released has been recently taken from the atmosphere by plants.
- Can perform very well as fuels.
- More easily biodegradable in the event of a spill.
- Free of sulfur, so no contribution to acid rain.

> **Top tip!**
>
> Biofuels are described as carbon neutral because, it is claimed, their use leads to no overall change in the amount of atmospheric carbon dioxide. However, this is a simplistic view that fails to take account of the energy used in the transport of feedstock, production of the biofuel and its distribution.

Disadvantages of biodiesel

- Slightly lower specific energy than the fossil fuel equivalent.
- More viscous, so requires pre-warming.
- Displacement of food crops by crops grown specifically for fuel leads to higher food prices.
- Lots of energy is required to process the oils into useful biofuels.
- A lot of impure glycerol is produced as a by-product.

Biofuels and biodiversity

Another controversy surrounding biodiesel is loss of biodiversity.

In South-East Asian countries, such as Malaysia and Indonesia, large areas of primary tropical rainforest have been logged for timber and then replaced with a monoculture based on palm oil production.

Palm oil is used extensively in foods and toiletries around the world, but can also be processed for biodiesel.

Very large areas of primary tropical rainforests – home to an enormous range of trees and other plants that support a complex ecosystem of insects, birds and mammals – have been replaced with rows of palm trees in which many rainforest organisms cannot survive. As plantations grow and rainforests shrink, the numbers of organisms that can be supported will fall.

Governments are under pressure from environmental groups to protect regions of primary rainforest in order to prevent widespread extinctions. However, the palm oil industry is hugely important economically. It provides employment (not just on plantations, but in associated transport, processing and infrastructure projects) and export income for the countries concerned. The rise of biodiesel as a popular 'green' fuel in economically developed countries is, potentially, a driver for environmental harm in less developed countries.

Figure R1.52 a) Palm oil plantation in Malaysia; b) primary rainforest in Borneo, Malaysia

Nature of science: Global impact of science

Ethics of energy generation

The European Group on Ethics (EGE) in Science and New Technologies is a committee of the European Commission in Brussels which attempts to explore the ethical issues surrounding energy production. They propose that international regulatory policies should take into account three main elements: societal impact, economic impact and environmental impact. The committee tries to apply the principle of justice to ensure that all EU citizens have access to energy, that the EU maintains a secure energy supply, and that the policies implemented are sustainable. When these different considerations come into conflict, politicians try to find a balance between them. Different political parties in different countries may disagree with the balance proposed: for example, some feel that sustained economic growth should override all other concerns, while others feel that we have a moral obligation to future generations to protect our environment.

The switch from using fossil fuels is challenging because it requires new expertise and investment. At present, renewable energy sources have yet to match fossil fuels in terms of low cost and ease of production, so governments offer funds to promote investment. Governments have to justify this expenditure and the funding is subject to political pressure.

Many scientists and prominent figures have campaigned to raise the profile of issues associated with fossil fuel use and, in many countries, public opinion is moving to demand humans meet more of their energy needs through renewable resources.

Greta Thunberg is an environmental activist from Sweden, who was born in 2003. She has made a powerful international impact with her passionate calls for action against climate change. Despite her relatively young age, she has addressed the United Nations Climate Conference and memorably addressed world leaders at the 2019 UN Climate Action Summit, asking how they dared steal her childhood and dreams with empty words and no meaningful action.

■ **Figure R1.53** Brussels, Belgium, 21 February 2019. Sixteen-year-old Swedish climate activist Greta Thunberg attends the European Economic and Social Committee event

Fuel cells

When fossil fuels are burnt, the hot gases produced do work moving a piston or driving a turbine. No matter how well an engine or power station is designed, a large proportion of the energy released in the combustion reaction is not transferred into useful work. Typically, a combustion engine is 20% efficient and fossil fuel power stations are between 30% and 50% efficient, depending on the type of fuel they burn.

Fuel cells convert the chemical energy stored in a fuel directly into electrical energy. Specialized hydrogen–oxygen fuel cells have been shown to be up to 70% efficient.

The same general principles apply to all fuel cells: two reactants (typically a fuel and an oxidizing agent as for combustion reactions) are continuously fed into the cell where they separately undergo oxidation or reduction at electrodes. This results in a flow of electrons (a current) from the negative electrode (**anode**) to the positive electrode (**cathode**). If we connect an electrical component (for example, a motor) across the two electrodes, it can use the current generated to function.

◆ **Anode**: The electrode where oxidation (the loss of electrons) occurs during an electrochemical process.

◆ **Cathode**: The electrode where reduction (the gain of electrons) occurs during an electrochemical process.

Hydrogen–oxygen fuel cells

Figure R1.54 shows a fuel cell which consumes hydrogen and oxygen gas in alkaline conditions. As long as the hydrogen and oxygen continue to be supplied at a relatively high temperature and pressure, the cell will continue to generate a current.

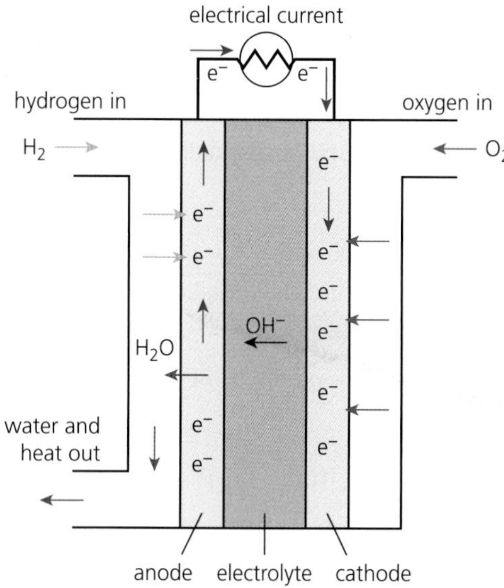

■ **Figure R1.54** An alkaline hydrogen–oxygen fuel cell

Hydrogen gas enters the fuel cell and is oxidized at the anode in the presence of hydroxide ions. Oxygen gas enters the opposite side of the fuel cell and is reduced at the cathode.

Anode: $H_2(g) + 2OH^-(aq) \rightarrow 2H_2O(l) + 2e^-$ (oxidation)

Cathode: $O_2(g) + 2H_2O(l) + 4e^- \rightarrow 4OH^-(aq)$ (reduction)

The electrons released at the anode by the oxidation reaction travel through the external circuit to the cathode where they combine with oxygen in the reduction reaction.

The overall reaction occurring in the hydrogen–oxygen fuel cell is found by combining the two half-equations:

$$2H_2(g) + O_2(g) \rightarrow 2H_2O(l)$$

The water produced leaves the fuel cell.

In this alkaline hydrogen–oxygen fuel cell, potassium hydroxide electrolyte fills the space between the electrodes and hydroxide ions move from the cathode (where they are produced) to the anode (where they are consumed). Although the electrolyte is not depleted, it usually needs to be replaced regularly as it becomes contaminated with by-products.

The electrodes are made from platinum, or from graphite with metals deposited on the surface to catalyze the reactions. In other fuel cells, the electrodes are made of conducting polymers.

Hydrogen–oxygen fuel cells can run in acidic conditions using phosphoric(V) acid as the electrolyte. With this electrolyte, the half-equations are:

Anode: $H_2(g) \rightarrow 2H^+(aq) + 2e^-$ (oxidation)

Cathode: $O_2(g) + 4H^+(aq) + 4e^- \rightarrow 2H_2O(l)$ (reduction)

The overall reaction is the same as in alkaline conditions.

Although hydrogen has a very high specific energy – almost three times that of gasoline – its energy density (measured in kJ dm^{-3}) is very low. This means large volumes are needed to ensure a continuous supply and it is usually stored under pressure in very large and heavy tanks. Hydrogen is also a highly flammable gas and so the associated explosion risks are another disadvantage.

LINKING QUESTION

What are the main differences between a fuel cell and a primary (voltaic) cell?

Research scientists are working on these problems, and a promising solution is to store the hydrogen within a stable compound which can react to release it as a gas when it is needed. Metallic hydrides such as ZrH_x release hydrogen when heated, although they are easily 'poisoned' by impurities. Another approach is to use porous materials which store hydrogen in their pores.

Methanol fuel cells

An alternative to hydrogen is methanol, which has a far higher energy density than hydrogen because it is a liquid at room temperature and pressure. It is, therefore, far more easily stored (in sealed cartridges, for example) and transported.

Methanol fuel cells rely on the catalytic oxidation of methanol to form carbon dioxide and water.

The electrode reactions are:

Anode: $CH_3OH + H_2O \rightarrow 6H^+ + 6e^- + CO_2$ (oxidation)

Cathode: $O_2 + 4H^+ + 4e^- \rightarrow 2H_2O$ (reduction)

Overall: $CH_3OH + \frac{3}{2}O_2 \rightarrow CO_2 + 2H_2O$ (redox)

The design of a methanol fuel cell (Figure R1.55) is similar to that of an acidic hydrogen–oxygen fuel cell.

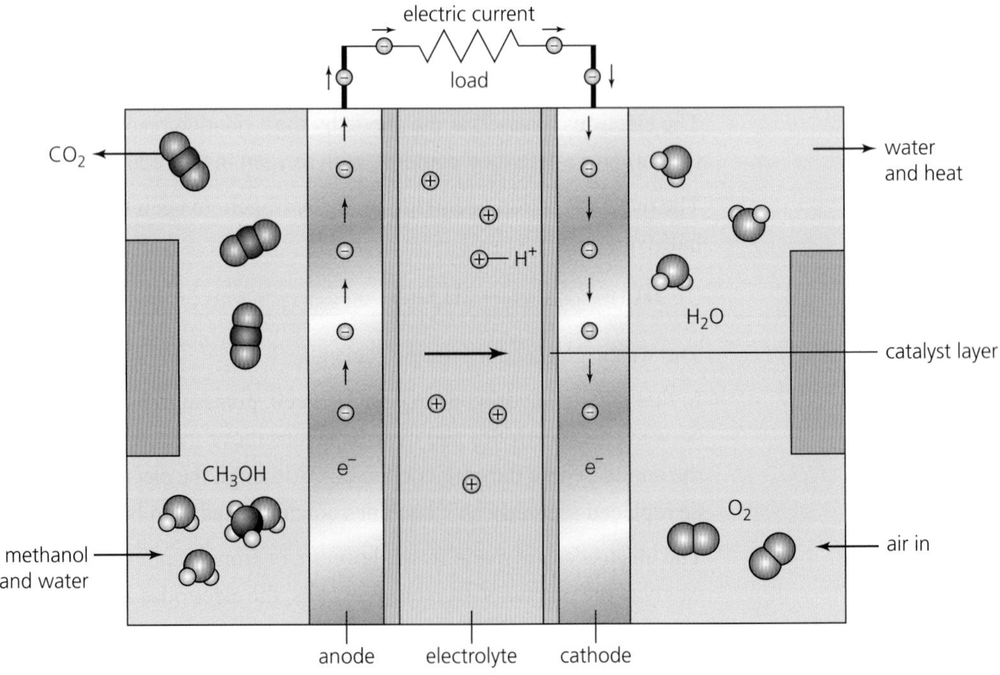

■ **Figure R1.55** A methanol fuel cell

Methanol fuel cells rely on the transition metals, platinum and ruthenium, as catalysts at both electrodes. This contributes to the high cost of these cells. There are also practical difficulties in getting methanol fuel cells to work: keeping the reactants separate from one another while still letting electrolytes move between the electrodes is challenging. Methanol fuel cells also produce the greenhouse gas carbon dioxide, albeit in small quantities.

Table R1.10 summarizes the reactions in each type of fuel cell.

■ **Table R1.10** Fuel cell reactions

Hydrogen–oxygen (alkaline electrolyte)	anode	$H_2(g) + 2OH^-(aq) \rightarrow 2H_2O(l) + 2e^-$
	cathode	$O_2(g) + 2H_2O(l) + 4e^- \rightarrow 4OH^-(aq)$
	overall	$2H_2(g) + O_2(g) \rightarrow 2H_2O(l)$
Hydrogen–oxygen (acidic electrolyte)	anode	$H_2(g) \rightarrow 2H^+(aq) + 2e^-$
	cathode	$O_2(g) + 4H^+(aq) + 4e^- \rightarrow 2H_2O(l)$
	overall	$2H_2(g) + O_2(g) \rightarrow 2H_2O(l)$
Methanol–oxygen	anode	$CH_3OH + H_2O \rightarrow 6H^+ + 6e^- + CO_2$
	cathode	$O_2 + 4H^+ + 4e^- \rightarrow 2H_2O$
	overall	$CH_3OH + \frac{3}{2}O_2 \rightarrow CO_2 + 2H_2O$

ATL R1.3C

Carbon monoxide detectors alert you to dangerous levels of carbon monoxide in the atmosphere. It is possible to buy different types of alarms, but the type which is most prevalent in the United States and Europe uses a fuel cell to act as the sensor. The fuel cell uses the carbon monoxide as a fuel to generate a current. The current generated is proportional to the concentration of carbon monoxide in the air, and if it exceeds a set value in a given time it will trigger the alarm.

Research how a carbon monoxide detector works, and what chemical reactions are occurring at the electrodes.

Prepare a short presentation to explain the chemistry behind a carbon monoxide detector.

R1.4 Entropy and spontaneity

Guiding question

- What determines the direction of chemical change?

SYLLABUS CONTENT

By the end of this chapter, you should understand that:

- entropy, S, is a measure of the dispersal or distribution of matter and/or energy in a system (HL only)
- the more ways energy can be distributed, the higher the entropy (HL only)
- under the same conditions, the entropy of a gas is greater than that of a liquid, which in turn is greater than that of a solid (HL only)
- the change in Gibbs energy, ΔG, relates the energy that can be obtained from a chemical reaction to the change in enthalpy, $\Delta H^{\ominus}$, change in entropy, ΔS, and absolute temperature, T (HL only)
- at constant pressure, a change is spontaneous if the change in Gibbs energy, ΔG, is negative (HL only)
- as a reaction approaches equilibrium, ΔG_r becomes less negative and finally reaches zero and at equilibrium $\Delta G^{\ominus} = -RT \ln K$ (HL only).

By the end of this chapter, you should know how to:

- predict whether a physical or chemical change will result in an increase or decrease in entropy of a system (HL only)
- calculate standard entropy changes, $\Delta S^{\ominus}$, from standard entropy values, $S^{\ominus}$ (HL only)
- apply the equation $\Delta G^{\ominus} = \Delta H^{\ominus} - T\Delta S^{\ominus}$ to calculate unknown values of these terms (HL only)
- interpret the sign of ΔG calculated from thermodynamic data (HL only)
- determine the temperature at which a reaction becomes spontaneous (HL only)
- perform calculations using the equations $\Delta G_r = \Delta G^{\ominus} + RT \ln Q$ and $\Delta G^{\ominus} = -RT \ln K$ (HL only).

What are spontaneous processes?

Why does mixing certain chemicals lead to a chemical reaction, yet combining others does not? Why do some reactions effectively go to completion when others reach equilibrium with both products and reactants in the system? Thermodynamics can answer these important questions, and predict the different types of behaviour we observe.

Spontaneous processes are those that, once started, will continue towards their final equilibrium state without being driven by an external influence. If work is required to make a process happen it is said to be **non-spontaneous**.

◆ **Spontaneous process:** A process which does not require work to be done to make it happen.

◆ **Non-spontaneous process:** A process which requires work to be done to make it happen.

Consider some examples:

1 The movement of heat from a hot body to a cold body through a thermally conducting barrier (Figure R1.1, page 308) is a spontaneous physical process. The reverse process, transfer of heat from the cold body to the hotter one, is non-spontaneous – no matter how long we wait, it will not happen.

Link

Heat and temperature are discussed further in Chapter R1.1, page 307.

2 Acidified water is electrolysed when a potential difference is applied (Figure R1.56).

$$2H_2O(l) \rightarrow 2H_2(g) + O_2(g)$$

However, this process will stop as soon as the power supply which is driving the process is switched off. The process is non-spontaneous. The reverse process, combination of hydrogen with oxygen, is spontaneous but, at room temperature, occurs too slowly to be observed.

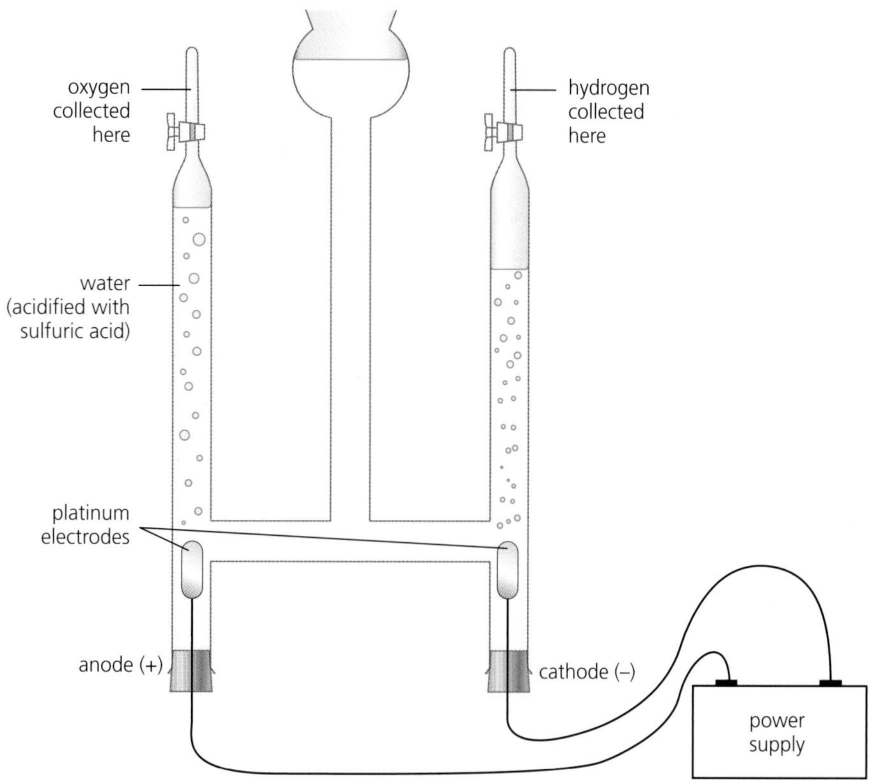

■ **Figure R1.56** The electrolysis of acidified water in a Hoffman apparatus

3 When a spirit burner is filled with methanol, and a small amount of energy is supplied by lighting the wick, a spontaneous reaction occurs between methanol and oxygen in the atmosphere. The reaction will continue until all the methanol (or oxygen in the atmosphere if the burner is in a closed system) has been used up.

Like the combination of hydrogen with oxygen, the oxidation of methanol by oxygen occurs spontaneously but is slow under standard conditions. By igniting the methanol vapour with a match, energy is supplied to the reactants, which allows them to overcome the relatively high activation energy barrier. The reaction is exothermic and, once it has started, releases enough energy to maintain the high rate of reaction, even when the match has been removed.

The last example highlights that the thermodynamic spontaneity of a reaction and its kinetic properties need to be considered separately. A reaction may be spontaneous, but this tells us nothing about the rate at which it occurs. Kinetics is the study of a reaction's transition states and intermediates and allows us to determine why a reaction is slow or fast – its kinetic properties.

- Some spontaneous processes, such as the conversion of diamond to graphite under standard conditions, are so slow that they are never realized in practice.
- Other spontaneous processes, such as the expansion of a gas into a vacuum, occur almost instantaneously.

When we consider the energy changes associated with any spontaneous process, we must consider the first law of thermodynamics.

Link

The factors that determine the rate of a chemical reaction are covered in Chapter R2.2, page 433.

The first law in its broadest sense is a statement about the universe. It tells us that there is a quantity which we call energy. If we measure its total value before and after some process – such as a chemical reaction – there is no change. In the example of a hot body transferring heat to a cooler body, the heat lost by the hot body is equal to the heat gained by the cold body. If a chemical reaction loses energy (is exothermic) and is in thermal contact with its surroundings, that same energy appears as heat in the surroundings – its form changes but the total energy does not.

We might speculate that only exothermic reactions and processes would be spontaneous, but experience quickly shows that there are many spontaneous processes which are endothermic, so we have to look beyond enthalpy changes for ways of predicting the direction of spontaneous changes. In fact, we find that a property of systems known as entropy, S, which measures the way in which energy is distributed, provides a measurable quantity which always changes in a consistent way in a spontaneous process.

◆ **Entropy**: A property of a system which reflects the number of ways the particles and energy in the system can be distributed.

Entropy

The **entropy** of a system is a measure of the dispersal or distribution of matter and energy in the system. A system that has its particles dispersed throughout the system has a higher entropy than if the particles were concentrated in one area. Likewise, a system that has its energy distributed amongst all the particles has a higher entropy than if the same amount of energy was concentrated in only a few of the particles.

Top tip!

When energy and matter are dispersed across a system, the entropy of the system increases and so does the disorder. For this reason, the disorder of a system is a useful indicator of the system's entropy.

Entropy and disorder

When a drop of ink is added to a beaker of water it will diffuse through the solution. The ink molecules spread out and become more disordered. Because the molecules are spread throughout the system after diffusion, the entropy of the system has increased. The amount of disorder in the system is an indicator of the entropy, and because of this, entropy and disorder are often associated together. More disorder is a sign of higher entropy.

Going further

Entropy and statistical thermodynamics

The concept of entropy and its relation to spontaneity come originally from a detailed analysis of the conversion of heat into work in engines. However, it is much more easily understood through statistical thermodynamics, which considers the molecular nature of a system and how energy is distributed.

Thermodynamics makes much use of the idea of the internal energy of a collection of molecules. The internal energy even of a single substance can be difficult to analyse because there are contributions from translation (moving to a different place), rotation and vibration of molecules and these are all affected by intermolecular forces. Mixtures are even more complex because there are different intermolecular forces between like and unlike molecules.

Even so, it should be easy to see that a system may have the same total energy distributed in quite different ways. For example, if we could take a 'snapshot' of a molecular gas at some instant, we would find some molecules moving rapidly (in translation, rotation and vibration) and others moving more slowly. At some later instant we would find a different distribution of kinetic energies, despite the fact that the overall internal energy was unchanged. Each of these different ways in which the energy can be distributed while keeping the overall internal energy unchanged is known as a microstate of the system.

Statistical thermodynamics tells us that the entropy of a system is determined by the number of ways the energy and matter in the system can be arranged: that is, the number of microstates it can have. The more microstates that are possible, the higher the entropy of the system. The Boltzmann formula defines the precise relationship between the two variables.

$$S = k_B \ln W$$

where:

- S = absolute entropy in $J\,K^{-1}$
- k_B = Boltzmann constant = $1.381 \times 10^{-23}\,J\,K^{-1}$
- W = number of ways of distributing the total energy over all of the molecules (the number of microstates).

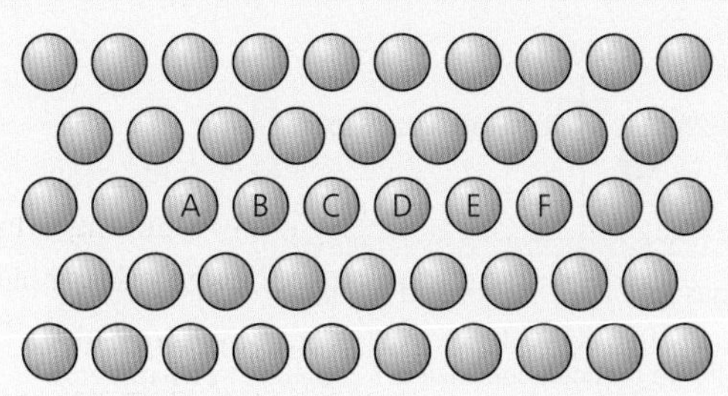

■ **Figure R1.57** A solid crystalline sample of helium, with six of the atoms labelled A to F. These atoms are identical but distinguished by their position in the lattice.

Consider a sample of frozen crystalline helium with a large number of atoms arranged in a lattice (Figure R1.57).

Six of the atoms labelled A to F form our system in this example. Each is assumed to have a set of four evenly spaced vibrational energy levels. (In reality, the vibrational energy levels converge in a similar way to electronic energy levels.) If the temperature is very low, close to absolute zero, each helium atom will have the lowest possible vibrational energy, E_0. This is shown in scenario (a) in Figure R1.58. There is only one way of distributing the energy between the atoms if they all have kinetic energy E_0.

(a)

E_3 = 3 units	
E_2 = 2 units	
E_1 = 1 unit	
E_0 = 0 unit	A B C D E F

(b)

E_3 = 3 units						
E_2 = 2 units						
E_1 = 1 unit	A	B	C	D	E	F
E_0 = 0 unit	B C D E F	A C D E F	A B D E F	A B C E F	A B C D F	A B C D E

(c)

E_3 = 3 units						
E_2 = 2 units	A	B				
E_1 = 1 unit			A B	A C	C D	B F
E_0 = 0 unit	B C D E F	A C D E F	C D E F	B D E F	A B E F	A C D E

■ **Figure R1.58** Examples of how atoms can be arranged among vibrational energy levels. Each possible arrangement is one microstate. (a) Shows the only option for a system that possesses 0 units of energy. Each microstate in (b) has 1 unit of energy, and each microstate in (c) has 2 units of energy

If the system is given one extra unit of energy (equivalent to the energy difference between the vibrational energy levels), then only one of the atoms can reach the next highest vibrational energy level, E_1. However, because there are six helium atoms, there are six ways of arranging them so that one atom has kinetic energy E_1 and five atoms have kinetic energy E_0. These are shown in scenario (b) of Figure R1.58.

If the collection of helium atoms is given another single unit of energy, the number of possible arrangements greatly increases. It is possible to have one atom with energy E_2 and the other five atoms with E_0; as before, there are six arrangements by which this could be achieved. Alternatively, two helium atoms could have kinetic energy E_1, with the other four atoms with E_0. There are 15 different ways this can be achieved, giving 21 different arrangements in total.

Some of these arrangements are shown in scenario (c) of Figure R1.58. This example demonstrates that as the total kinetic energy available to a system increases, there is a rapid increase in the number of possible arrangements and, therefore, the entropy of the system.

A second simple example to consider is that of five particles distributed between two otherwise empty gas jars. First consider the particles partitioned into one side of the system (Figure R1.59).

When the partition is removed, each particle will move in a straight line until it collides with another particle or the walls of the container. The particles move around in a random manner and, at any time in the future, each particle is equally likely to be in either of two places: the left-hand or the right-hand container.

There are five molecules, each with two possible places once the partition is removed, giving a total of $2 \times 2 \times 2 \times 2 \times 2 = 2^5 = 32$ possible arrangements (Figure R1.60). Each of these arrangements is equally likely. If a sixth molecule was added to the system, there would be $2^6 = 64$ possible arrangements, and the entropy of the system would therefore be greater.

In the case of five particles, each of the 32 possible arrangements is equally likely. However, only two of these arrangements have all five particles in one side of the container. The probability that we will find all five particles in one side of the container is $\frac{2}{32} = 0.0625$.

Having the particles split in a 3 : 2 arrangement between the two sides of the container is the most probable configuration. This accounts for 20 of the possible arrangements, and so the probability of observing this configuration is $\frac{20}{32} = 0.625$.

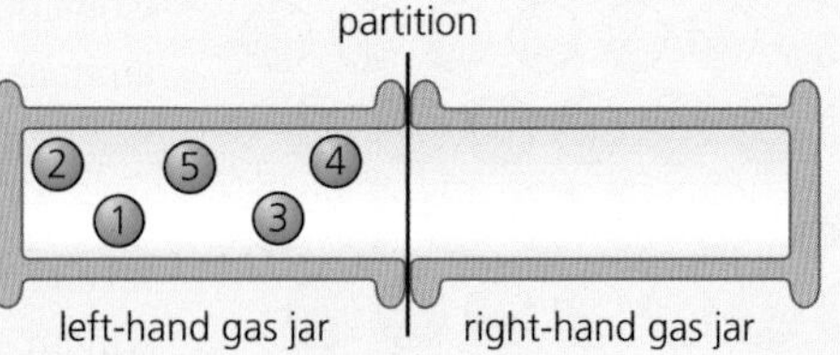

■ **Figure R1.59** Five particles confined to a container by a partition

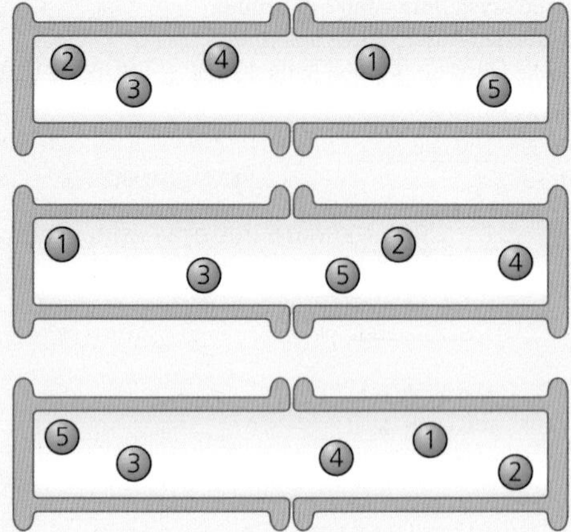

■ **Figure R1.60** Three of the possible 32 arrangements when the partition between the gas jars is removed

If we extend this analysis to one mole of a gas containing 6.02×10^{23} molecules, the probability of finding all the molecules in one container drops to less than 2×10^{-100}. There is no realistic chance of ever observing this outcome. When the numbers of molecules involved are this large, it also causes the probability of any configuration other than the most probable one to become insignificant. The molecules will adopt the configuration in which they are, within any possible limits of precision of measurement, equally distributed through the whole container. This is why we observe that a gas expands into a vacuum and, as it does, its entropy increases.

Nature of science: Observations

Defining entropy

There are many ways to define entropy. All of them are equivalent. For example:

- The thermodynamic definition, which was developed through observation of heat–work cycles and the efficiency of heat engines. The thermodynamic definition describes how to measure the entropy of an isolated system in thermodynamic equilibrium. It directly connects to physics, but makes no reference to the atomic and molecular nature of matter.
- The statistical definition was developed later by analysing the statistical behaviour of the particles in a system. Ludwig Boltzmann proposed the link between the distribution of molecules over energy levels and the entropy.

The thermodynamic definition provides the experimental definition of entropy, while the statistical definition of entropy extends the concept, providing a deeper understanding of its nature and connecting with quantum mechanics.

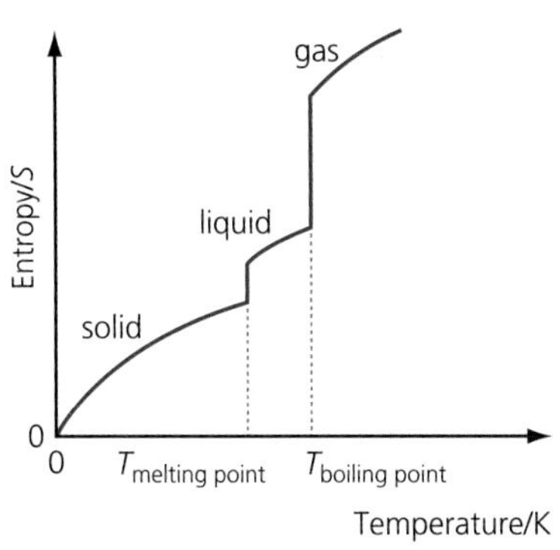

■ **Figure R1.61** Variation of entropy with temperature for a pure substance undergoing melting and boiling.

Standard entropy changes

Consider a pure, perfectly crystalline substance at absolute zero (0 K). All the molecules are identical and, at this temperature, all the atoms in each molecule are in the lowest possible energy state. The entropy of this substance is zero.

As the temperature of the crystal increases, the amount of energy the crystal has increases (recall that temperature is proportional to the kinetic energy of a substance). With increasing energy, the number of ways the energy can be distributed increases, and the disorder within the crystal does too. A schematic relationship between the entropy and temperature of a pure substance is shown in Figure R1.61.

Starting from the entropy of a perfect crystal at 0 K, the entropy of a pure substance at a higher temperature can be calculated, and this yields the **standard entropy** values, $S^{\ominus}$.

◆ **Standard entropy, $S^{\ominus}$:** The entropy of one mole of a substance at 298 K and 100 kPa.

The standard entropy values of some substances are shown in Table R1.11. The standard entropy values for more compounds are given in section 13 of the IB *Chemistry data booklet*.

■ **Table R1.11** Standard entropy values, $S^{\ominus}$, at 298 K. All values are in $J K^{-1} mol^{-1}$

Solids		**Liquids**		**Gases**	
ice ($H_2O(s)$)	45	water ($H_2O(l)$)	70	water vapour ($H_2O(g)$)	189
C(s) (diamond)	2.4	ethanol(l)	161	CO(g)	198
C(s) (graphite)	5.7	benzene(l)	173	$CO_2(g)$	214
NaCl(s)	72	propanone(l)	200	$H_2(g)$	131

Top tip!

The entropy of a system increases with increasing temperature.

Top tip!

In diamond, identical covalent bonds hold atoms in a rigid three-dimensional crystal structure. In graphite, the atoms bond together in sheets which have the freedom to slide past each other. The less constrained structure of graphite gives it a higher value of standard entropy than diamond.

Notice that the standard entropy of H_2O is highest when it is in the gaseous form and lowest as a solid. The molecules in a gas have more disorder than those in the ordered three-dimensional lattice of a solid. As a general rule, for molecules containing about the same number of atoms, gases have a higher standard entropy than liquids, which have a higher standard entropy than solids.

Common mistake

Misconception: *The standard entropy of an element is zero.*

Unlike the enthalpy of formation of an element, the entropies of elements in their standard states are not zero. Only the entropy of a perfect crystal at 0 K would be equal to zero. The entropy value of an element is positive.

The standard entropy change of a chemical reaction, $\Delta S^{\ominus}$, is the difference between the standard entropy of the products and the reactants.

$$\Delta S^{\ominus} = \Sigma S^{\ominus}_{(products)} - \Sigma S^{\ominus}_{(reactants)}$$

$\Sigma S^{\ominus}_{(products)}$ is the sum of the standard entropies of all the products.

$\Sigma S^{\ominus}_{(reactants)}$ is the sum of the standard entropies of all the reactants.

This equation is used to calculate the entropy change which accompanies a reaction or a physical process.

LINKING QUESTION

Why is the entropy of a perfect crystal at 0 K predicted to be zero?

WORKED EXAMPLE R1.4A

Calculate the entropy change that occurs during the complete combustion of one mole of ethane with all reactants and products in their standard states.

Answer

$$C_2H_6(g) + \frac{7}{2}O_2(g) \rightarrow 2CO_2(g) + 3H_2O(l)$$

$S^\ominus[C_2H_6(g)] = 230\,J\,K^{-1}\,mol^{-1}$

$S^\ominus[O_2(g)] = 205\,J\,K^{-1}\,mol^{-1}$

$S^\ominus[CO_2(g)] = 214\,J\,K^{-1}\,mol^{-1}$

$S^\ominus[H_2O(l)] = 70\,J\,K^{-1}\,mol^{-1}$

$\Delta S^\ominus = \Sigma S^\ominus_{(products)} - \Sigma S^\ominus_{(reactants)}$

$\Delta S^\ominus = [(2 \times 214) + (3 \times 70)\,J\,K^{-1}\,mol^{-1}] - [230 + (3.5 \times 205)]\,J\,K^{-1}\,mol^{-1}$

$\Delta S^\ominus = -310\,J\,K^{-1}\,mol^{-1}$

The data shown in the table are required to answer the questions that follow.

Compound	$H_2(g)$	$Cl_2(g)$	HCl(g)	Zn(s)	$ZnCl_2(s)$	$O_2(g)$
Entropy / $J\,K^{-1}\,mol^{-1}$	131	223	187	41.6	116	205

30 Calculate the standard entropy changes for the following reactions:

a $H_2(g) + Cl_2(g) \rightarrow 2HCl(g)$

b $Zn(s) + Cl_2(g) \rightarrow ZnCl_2(s)$

c $Zn(s) + 2HCl(g) \rightarrow ZnCl_2(s) + H_2(g)$

31 Calculate the entropy change when 3.00 mol of zinc chloride decomposes into its elements in their standard states.

32 The formation of zinc oxide from its elements has the standard entropy change $-100\,J\,K^{-1}\,mol^{-1}$. Calculate the standard entropy of zinc oxide.

Predicting the sign of a change in entropy

In many chemical reactions and physical processes, it is possible to predict the sign of the entropy change (of the system), ΔS, by examining the reactants and products in the balanced equation. If the products are more disordered than the reactants, then the entropy change of the system, ΔS, is positive. If the products are less disordered than the reactants, then the entropy change of the system, ΔS, is negative.

Consider the reaction of hydrogen with oxygen to produce water:

$$2H_2(g) + O_2(g) \rightarrow 2H_2O(l)$$

The reaction equation shows that 3 moles of gaseous reactants react to form 2 moles of liquid product. Because we know that gases have more disorder than liquids, it can be predicted that in this chemical reaction there will be a decrease in the entropy. The sign of ΔS will be negative.

The key factor to look for when deducing the sign of an entropy change of a reaction is a change in the number of gaseous molecules. When assessing a process such as melting or boiling, think about whether the system is becoming more or less ordered. Some common examples are summarized in Table R1.12.

■ **Table R1.12** Entropy changes of some common physical and chemical processes

Chemical reaction or physical change	Entropy change	Example
melting	increase	$H_2O(s) \rightarrow H_2O(l)$
boiling	large increase	$H_2O(l) \rightarrow H_2O(g)$
condensing	large decrease	$H_2O(g) \rightarrow H_2O(l)$
sublimation	very large increase	$I_2(s) \rightarrow I_2(g)$
vapour deposition	very large decrease	$I_2(g) \rightarrow I_2(s)$
freezing	decrease	$H_2O(l) \rightarrow H_2O(s)$
dissolving a solute to form a solution	generally an increase (except with highly charged ions)	$NaCl(s) + (aq) \rightarrow NaCl(aq)$
precipitation	large decrease	$Pb^{2+}(aq) + 2Cl^-(aq) \rightarrow PbCl_2(s)$
crystallization from a solution	decrease	$NaCl(aq) \rightarrow NaCl(s)$
chemical reaction: solid or liquid forming a gas	large increase	$CaCO_3(s) \rightarrow CaO(s) + CO_2(g)$
chemical reaction: gases forming a solid or liquid	large decrease	$2H_2S(g) + SO_2(g) \rightarrow 3S(s) + 2H_2O(l)$
increase in number of moles of gas	large increase	$2NH_3(g) \rightarrow N_2(g) + 3H_2(g)$

WORKED EXAMPLE R1.4B

Predict the sign of the entropy change for the following reactions.

a $2SO_3(g) \rightarrow 2SO_2(g) + O_2(g)$
b $H_2O(l) \rightarrow H_2O(s)$
c $Br_2(l) \rightarrow Br_2(g)$
d $H_2O_2(l) \rightarrow H_2O(l) + \frac{1}{2}O_2(g)$

Answer

a $2SO_3(g) \rightarrow 2SO_2(g) + O_2(g)$ $\Delta S > 0$ because 2 moles of gas produce 3 moles of gas
b $H_2O(l) \rightarrow H_2O(s)$ $\Delta S < 0$ because a liquid forms a solid
c $Br_2(l) \rightarrow Br_2(g)$ $\Delta S > 0$ because a liquid forms a gas
d $H_2O_2(l) \rightarrow H_2O(l) + \frac{1}{2}O_2(g)$ $\Delta S > 0$ because 1 mole of liquid produces 1 mole of liquid and $\frac{1}{2}$ mole of gas

HL ONLY

33 Three of the following reactions have a significant increase in entropy. State which three reactions show a significant increase in entropy.

A $CaCO_3(s) \rightarrow CaO(s) + CO_2(g)$

B $H_2(g) + Cl_2(g) \rightarrow 2HCl(g)$

C $C(s) + O_2(g) \rightarrow CO_2(g)$

D $2C(s) + O_2(g) \rightarrow 2CO(g)$

E $C_9H_{20}(l) + 14O_2(g) \rightarrow 9CO_2(g) + 10H_2O(g)$

34 Predict whether the following reactions and changes will result in an increase, decrease or little change in entropy. Consider the stoichiometry of the reaction and the physical state of the reactants and products.

a $H_2(g) + Br_2(g) \rightarrow 2HBr(g)$

b $H_2O_2(l) \rightarrow H_2O_2(s)$

c $Cr^{3+}(aq) + 6H_2O(l) \rightarrow [Cr(H_2O)_6]^{3+}(aq)$

d $Hg(l) \rightarrow Hg(g)$

e $AgNO_3(s) \rightarrow AgNO_3(aq)$

f $CaO(s) + CO_2(g) \rightarrow CaCO_3(s)$

g $H^+(aq) + OH^-(aq) \rightarrow H_2O(l)$

h $2HCl(g) + Br_2(l) \rightarrow 2HBr(g) + Cl_2(g)$

i $2SO_2(g) + O_2(g) \rightarrow 2SO_3(g)$

j $H_2(g) \rightarrow 2H(g)$

The second law of thermodynamics

The spontaneity of a physical or chemical change is governed by the second law of thermodynamics. One way of expressing this law is:

Spontaneous processes are those that increase the total entropy of the universe.

Or, since entropy is related to the disorder of matter or energy, spontaneous processes are those which lead to an increase in the dispersal of matter and/or energy in the universe.

The second law, as stated above, refers to the universe. To determine the spontaneity of a process, the entropy change of a system and its surroundings (the rest of the universe) must be considered together. The sum of the entropy change of a system and its surroundings is usually referred to as $\Delta S_{(total)}$, instead of ΔS of the universe.

$$\Delta S_{(total)} = \Delta S_{(system)} + \Delta S_{(surroundings)}$$

If $\Delta S_{(total)}$ is a positive value, the entropy of the universe has increased, and the process is spontaneous.

$$\Delta S_{(total)} > 0 \text{ for a spontaneous process}$$

Calculating the total entropy change in the general case is often extremely difficult. However, it can be simplified in a way which is extremely useful in chemistry if we consider only a closed system which is maintained at constant pressure and temperature. Such a system cannot change the entropy of its surroundings by an exchange of matter, only of heat or work.

A full thermodynamic analysis tells us that the entropy change of the surroundings, $\Delta S_{(surroundings)}$, of a closed system at constant pressure and temperature is given by:

$$\Delta S_{(surroundings)} = -\frac{\Delta H^{\ominus}}{T}$$

where:

- $\Delta S_{(surroundings)}$ is the change in the entropy of the surroundings in $J\,K^{-1}\,mol^{-1}$
- $\Delta H^{\ominus}$ is the enthalpy change of the chemical or physical process measured in the unusual unit $J\,mol^{-1}$
- T is the absolute temperature at which the process occurs in kelvin.

The negative sign in the formula means that, if heat flows from an exothermic system into the surroundings, the entropy of the surroundings increases and, if heat flows from the surroundings into an endothermic system, the entropy of the surroundings decreases.

Top tip!

Not altering the units of entropy and enthalpy often causes lost marks in examinations. Entropies are given in $J K^{-1} mol^{-1}$, whereas enthalpy changes are given in $kJ mol^{-1}$. When they are combined in the same equation a conversion must be made to allow the units to match.

Top tip!

According to the second law of thermodynamics, the entropy of the universe must increase. The entropy of a system may decrease during a spontaneous process, but this must be accompanied by a larger increase in entropy in the surroundings.

WORKED EXAMPLE R1.4C

Calculate the entropy change of the surroundings, and hence the total entropy change, when water condenses on a window at 25 °C.

$H_2O(g) \rightarrow H_2O(l) \qquad \Delta H^{\ominus} = -44.0\,kJ\,mol^{-1} \qquad$ and $\qquad \Delta S^{\ominus} = -118\,J\,K^{-1}\,mol^{-1}$

(where $\Delta S^{\ominus}$ represents the standard entropy change of the system)

Answer

$$\Delta S_{(surroundings)} = -\frac{\Delta H^{\ominus}}{T}$$

$$= -\frac{-44\,000\,J\,mol^{-1}}{298\,K}$$

$$= +148\,J\,K^{-1}\,mol^{-1}$$

(Note the conversion of the enthalpy change from $kJ\,mol^{-1}$ to $J\,mol^{-1}$. This is because the units of entropy and entropy changes are $J\,K^{-1}\,mol^{-1}$.)

The overall entropy change in the universe can then be calculated:

$$\Delta S_{(total)} = \Delta S_{(system)} + \Delta S_{(surroundings)}$$

$$= (-118 + 148)\,J\,K^{-1}\,mol^{-1}$$

$$= +30\,J\,K^{-1}\,mol^{-1}$$

The total entropy change is positive, so the process is spontaneous at this temperature. This corresponds to our everyday observation that water vapour will condense if it meets a surface below its boiling point.

Going further

Entropy and time

According to the second law of thermodynamics, all spontaneous processes lead to an increase in the overall entropy of the universe. Hence, the passing of time is accompanied by an increase in the entropy of the universe. You can tell that time has passed by observing a process (for example, the melting of ice cubes at room temperature) which involves an increase in entropy. Hence the expression, 'entropy is time's arrow'.

The second law predicts that the universe (if it is a closed system) will eventually reach a completely disordered state with maximum entropy. This is known as the heat death of the universe. Once this point is reached, time would have no meaning because there would be no overall entropy change to observe. The temperature would be close to absolute zero and no life would be possible.

TOK

Are there questions that science cannot explore?

For some, the impending heat death of the universe is not an attractive idea – it seems to spell the end of all of our hopes and dreams (not that any of us would be around to experience it!). However, this insight also illustrates an intriguing limit to scientific knowledge. Science explores the observable and, in this case, the observable (observing the phenomena that led to the second law) leads us to something of a dead end. For many people, this contradicts the deep human intuition that there is more to our existence than merely being witnesses of the heat death of the universe. Religious knowledge is often used as a way of making sense of other possibilities; some would suggest that beyond these laws, which seem to suggest the universe is on a one-way trip to oblivion, there are other truths about the universe which extend beyond what the scientific method might provide. The meaning or significance of human life, for instance, which is considered central to the scope of religious knowledge, would place the scientific expectations of the second law in a different context. This clearly extends beyond the ability of scientific knowledge to unpack or explore.

Going further

Entropy changes during changes of phase

Phase changes are accompanied by an enthalpy change. Using the thermodynamic definition of entropy at constant pressure, we can calculate the change in entropy of a state change:

$$\Delta S = \frac{\Delta H^{\ominus}}{T}$$

where:

- ΔS = entropy change
- $\Delta H^{\ominus}$ = enthalpy change during change of phase
- T = absolute temperature (K).

For a change of state such as vaporization:

$$X_2(l) \rightarrow X_2(g)$$

$$\Delta_{vap} S = \frac{\Delta H_{vap}}{T_b}$$

where T_b is the boiling point.

For liquid benzene, $\Delta H_{vap} = +30.8\,kJ\,mol^{-1}$ and $T_b = 353\,K$.

$$\Delta S_{vap} = \frac{30.8 \times 10^3\,J\,mol^{-1}}{353\,K} = +87.3\,J\,K^{-1}\,mol^{-1}$$

The change in entropy of the benzene is positive. This is to be expected because it is moving from a more ordered liquid phase to a less ordered gaseous phase. The total change in entropy, $\Delta S_{(total)}$, accompanying a phase change is zero. This means the entropy change of the surroundings was $-87.3\,J\,K^{-1}$ per mol of benzene vaporized.

Common mistake

Misconception: *An increase in entropy means an increase in temperature.*

There is an increase in entropy when a molecular liquid boils at constant temperature under standard conditions. This is a phase change that occurs at constant temperature.

Frederick Thomas Trouton (1863–1922) noticed that the entropy change of vaporization was similar for many liquids at their boiling points. Trouton's rule says that $\Delta S_{vap} \approx 10.5R$ (where R is the gas constant) or alternatively $\Delta S_{vap} \approx 87\,J\,K^{-1}\,mol^{-1}$ (Table R1.13).

Table R1.13 Data illustrating Trouton's rule

Substance	ΔH_{vap} / kJ mol^{-1}	T_b / K	ΔS_{vap} / J K^{-1} mol^{-1}
methane	+9.27	112	+83.0
carbon tetrachloride	+30.0	350	+85.8
cyclohexane	+30.1	354	+85.1
hydrogen sulfide	+18.8	214	+88.0
water	+40.7	373	+109

Water does not obey the rule due to the strong hydrogen bonds it forms which result in a liquid that is more structured than most others.

Gibbs energy change

It is possible to find the total entropy change associated with a reaction using only terms referring to the system. The expression is:

$$\Delta S_{(total)} = \Delta S_{(system)} + \left(-\frac{\Delta H}{T}\right)$$

This can be multiplied by $-T$ and rearranged to give:

$$-T\Delta S_{(total)} = \Delta H - T\Delta S_{(system)}$$

Top tip!

The equation, $\Delta G^{\ominus} = \Delta H^{\ominus} - T\Delta S^{\ominus}$ is a useful way of applying the second law of thermodynamics in terms of properties of the system only.

$-T\Delta S_{(total)}$ is known as the change in Gibbs energy, ΔG, and by substituting this into the previous expression we arrive at an important equation:

$$\Delta G = \Delta H - T\Delta S$$

Under standard conditions, this becomes:

$$\Delta G^{\ominus} = \Delta H^{\ominus} - T\Delta S^{\ominus}$$

where:

- $\Delta G^{\ominus}$ = standard change in Gibbs energy (kJ mol^{-1})
- $\Delta H^{\ominus}$ = standard change in enthalpy of the system (kJ mol^{-1})
- T = absolute temperature (K)
- $\Delta S^{\ominus}$ = standard change in entropy of the system (kJ K^{-1} mol^{-1}).

This equation is given in section 1 of the IB *Chemistry data booklet.*

Because the change in Gibbs energy is usually quoted in kJ mol^{-1}, the entropy change must be converted from its normal units (J K^{-1} mol^{-1}) to kJ K^{-1} mol^{-1} through multiplying by 10^{-3}.

Top tip!

The IB *Chemistry data booklet* gives both enthalpy changes and Gibbs energy changes in kJ mol^{-1}. Standard entropy values are given in J K^{-1} mol^{-1}.

HL ONLY

WORKED EXAMPLE R1.4D

Calculate the Gibbs energy change, ΔG, for the reaction of nitrogen with hydrogen to form ammonia at (a) 298 K and (b) 598 K.

$N_2(g) + 3H_2(g) \rightarrow 2NH_3(g)$ $\Delta H^{\ominus} = -92.2\,\text{kJ mol}^{-1}$ $\Delta S^{\ominus} = -199\,\text{J K}^{-1}\text{mol}^{-1}$

Answer

a $\Delta G^{\ominus} = \Delta H^{\ominus} - T\Delta S^{\ominus}$

$= -92.2\,\text{kJ mol}^{-1} - (298\,\text{K} \times -199 \times 10^{-3}\,\text{kJ K}^{-1}\text{mol}^{-1})$

$= (-92.2 + 59.3)\,\text{kJ mol}^{-1}$

$= -32.9\,\text{kJ mol}^{-1}$

b $\Delta G = \Delta H - T\Delta S$

$= -92.2\,\text{kJ mol}^{-1} - (598\,\text{K} \times -199 \times 10^{-3}\,\text{kJ K}^{-1}\text{mol}^{-1})$

$= (-92.2 + 119)\,\text{kJ mol}^{-1}$

$= +26.8\,\text{kJ mol}^{-1}$

Be aware that the enthalpy change and entropy change values for the reaction at 598 K are not identical to the given standard enthalpy and entropy change values (that is, those at 298 K). However, the standard values are an acceptable approximation to those at a different temperature if the change in temperature does not result in a change of state of a reactant or product. In this example, all species are gases at 298 K, and this would also be true at 598 K.

If more accurate values are required for the enthalpy and entropy values at temperatures other than 298 K, you need to know the standard values and the molar heat capacities of the substances involved in the reaction; this is beyond the scope of this text.

Top tip!

For calculations you will face in your IB Chemistry course, it is acceptable to assume the values for entropy and enthalpy changes are independent of temperature.

35 Calcium carbonate can be decomposed to calcium oxide and carbon dioxide.

$CaCO_3(s) \rightarrow CaO(s) + CO_2(g)$ $\Delta H^{\ominus} = +178\,\text{kJ mol}^{-1}$; $\Delta S^{\ominus} = +164\,\text{J K}^{-1}\text{mol}^{-1}$

a Calculate the total entropy change, $\Delta S_{(total)}$, when calcium carbonate is decomposed at 298 K.

b With reference to your answer in part (a), is the decomposition of calcium carbonate spontaneous under these conditions?

c Calculate the standard Gibbs energy change, $\Delta G^{\ominus}$, for the reaction at 298 K.

The significance of Gibbs energy change

A decrease in Gibbs energy corresponds to an increase in the entropy of the universe ($-T\Delta S_{(total)} = \Delta G$). Therefore:

- if $\Delta G < 0$, the reaction or process is spontaneous
- if $\Delta G > 0$, the reaction or process is non-spontaneous.

The Gibbs energy change of a reaction represents the maximum amount of (non-expansion) work that can be obtained from the reaction. For the reaction of hydrogen and oxygen to form water, $\Delta G^{\ominus} = -237\,kJ\,mol^{-1}$. This means that this reaction can do up to $237\,kJ\,mol^{-1}$ of work when it is carried out in a way which allows the energy to be used in, for example, a fuel cell. The Gibbs energy change is sometimes called the free energy because it is the energy which is free to do useful work.

Link

Fuel cells are covered in more detail in Chapter R1.3, page 382.

ATL R1.4A

It is possible to perform a simple investigation into the thermodynamics of the stretching and contracting processes of a rubber band. It is recommended that you use a wide rubber band in order to get the best observations.

Quickly stretch the band until it is near full extension and place it in contact with a sensitive part of your skin (such as your lips or forehead). Answer the following questions:

a Does the stretched rubber band feel warmer or cooler?

b Has an exothermic or endothermic process occurred?

c The stretching of the rubber band is a non-spontaneous process because work was needed to carry it out. What sign does the Gibbs energy change of a non-spontaneous process have?

d Based upon your answers to (b) and (c), what is the sign of the entropy change when a rubber band is stretched?

e Think about the organisation of the polymer strands in a contracted and stretched rubber band. Draw a diagram for each and state which one is more ordered and therefore has the lower entropy.

Tool 3: Mathematics

Use basic arithmetic and algebraic calculations to solve problems

- An equation is unaffected if the same quantity is added to or subtracted from each side.
- An equation is unaffected if each side is multiplied or divided by the same quantity.
- Any term occurring as a factor on both the top and bottom of a fraction on the same side of an equation can be cancelled.

The Gibbs equation can be rearranged to give an expression for the entropy change, ΔS:

$$\Delta G = \Delta H - T\Delta S$$

Move the terms $T\Delta S$ and ΔG to the opposite sides by subtracting ΔG on each side of the equation and adding the term $T\Delta S$ to each side of the equation:

$$\Delta G - \Delta G + T\Delta S = \Delta H - T\Delta S - \Delta G + T\Delta S$$

Cancel the terms on each side of the equation: the ΔG and $-\Delta G$ on the left-hand side and the $-T\Delta S$ and $+T\Delta S$ on the right-hand side.

$$T\Delta S = \Delta H - \Delta G$$

Divide both sides by T:

$\dfrac{T\Delta S}{T} = \dfrac{(\Delta H - \Delta G)}{T}$; cancel T on the left-hand side to get $\Delta S = \dfrac{(\Delta H - \Delta G)}{T}$

Gibbs energy change and temperature

The feasibility of a process (whether it is spontaneous or non-spontaneous) is affected by the enthalpy change of the reaction and the entropy change of the reaction:

$$\Delta G = \Delta H - T\Delta S$$

Table R1.14 shows a summary of the four possible combinations.

Table R1.14 Different combinations of enthalpy and entropy changes

Enthalpy change, ΔH	Entropy change, ΔS	Comments
+	+	endothermic reaction spontaneous at high temperatures when the magnitude of $T\Delta S > \Delta H$
+	–	endothermic reaction non-spontaneous at all temperatures
–	–	exothermic reaction spontaneous at low temperatures when the magnitude of $\Delta H > T\Delta S$
–	+	exothermic reaction spontaneous at all temperatures

If the two terms which comprise the equation for a change in Gibbs energy have the same sign, their contributions must be weighed against one another. Figure R1.62 shows the possibilities for an exothermic reaction which is accompanied by a decrease in the entropy of the system. The exothermic nature of the reaction makes a favourable contribution to spontaneity, while the decrease in entropy makes an unfavourable contribution to spontaneity (the sign of $-T\Delta S$ is positive). Under the conditions of (a), the contribution of ΔH is greater than $-T\Delta S$ and the reaction is spontaneous ($\Delta G < 0$). In (b), the temperature has been raised and the $-T\Delta S$ term now dominates, making the reaction non-spontaneous ($\Delta G > 0$).

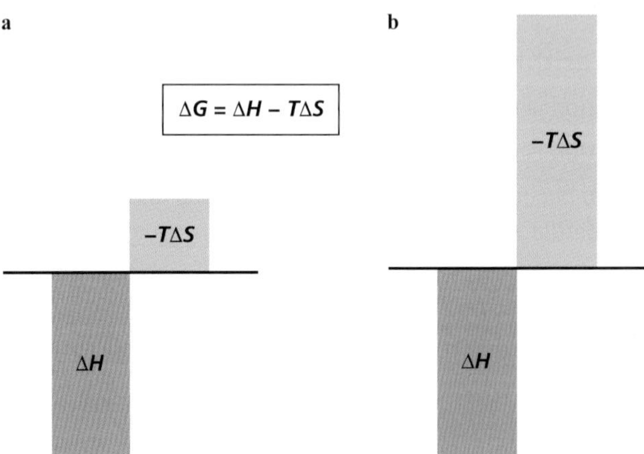

Figure R1.62 The feasibility of a reaction which has a negative enthalpy and negative entropy change. (a) At low temperatures the reaction is feasible because ΔH makes a larger negative contribution than the positive contribution of $-T\Delta S$. (b) At higher temperatures the reaction is not feasible because of the larger positive contribution of $-T\Delta S$

This information is represented graphically in Figure R1.63. At 0 K, all exothermic reactions have a negative Gibbs energy change, ΔG, and are therefore spontaneous. At 0 K, all endothermic reactions have a positive Gibbs energy change, ΔG, and are therefore non-spontaneous. As the temperature at which the reaction occurs rises, the entropy term becomes more important, and this may change the spontaneity of the reaction.

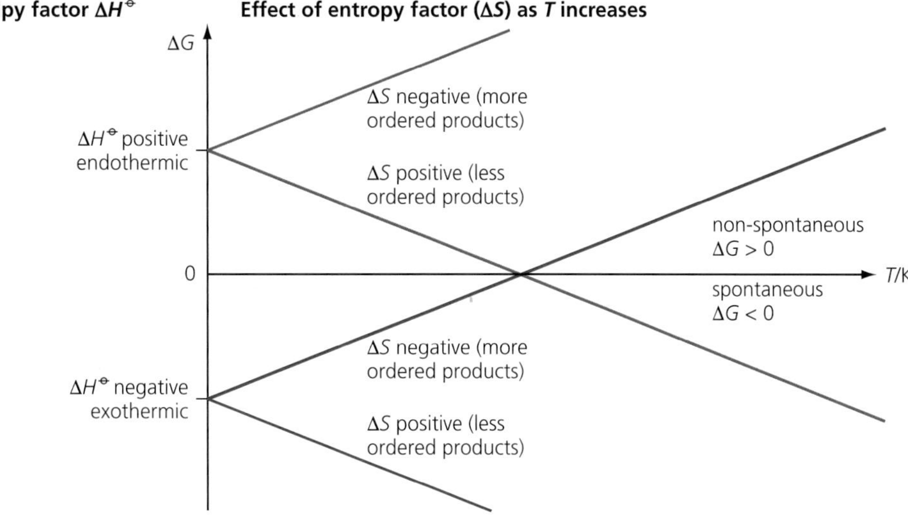

■ **Figure R1.63** A graphical summary of the conditions for spontaneity (treating the values of entropy and enthalpy changes as independent of temperature)

For the reaction of iron(III) oxide with carbon:

$$2Fe_2O_3(s) + 3C(s) \rightarrow 4Fe(s) + 3CO_2(g)$$

$\Delta H^\ominus = +468\,kJ\,mol^{-1}$; $\Delta S^\ominus = +558\,J\,K^{-1}\,mol^{-1}$

The Gibbs energy change at 298 K is:

$\Delta G^\ominus = 468\,kJ\,mol^{-1} - (298\,K \times 558 \times 10^{-3}\,kJ\,K^{-1}\,mol^{-1})$

$\Delta G^\ominus = +302\,kJ\,mol^{-1}$

The reaction is non-spontaneous at 298 K. However, if it is heated to 900 K:

$\Delta G = 468\,kJ\,mol^{-1} - (900\,K \times 558 \times 10^{-3}\,kJ\,K^{-1}\,mol^{-1})$

$\Delta G = -34.2\,kJ\,mol^{-1}$

The reaction is spontaneous under these conditions. The spontaneity of the reduction of iron(III) oxide with carbon is modelled by the blue line with a negative gradient in Figure R1.63.

For reactions where $T\Delta S$ and ΔH have the same signs, there will be a unique temperature at which the reaction changes from being spontaneous to being non-spontaneous or vice versa. This happens at the temperature which gives $\Delta G = 0$.

WORKED EXAMPLE R1.4E

Calculate the temperature at which the decomposition of calcium carbonate becomes spontaneous.

$CaCO_3(s) \rightarrow CaO(s) + CO_2(g)$ $\quad \Delta H^\ominus = +178\,kJ\,mol^{-1}$ $\quad \Delta S^\ominus = +164\,J\,K^{-1}\,mol^{-1}$

Answer

The temperature at which the reaction becomes spontaneous is given by setting $\Delta G = 0$.

$$\Delta G = \Delta H - T\Delta S$$

$$0 = 178\,kJ\,mol^{-1} - T \times 164 \times 10^{-3}\,kJ\,K^{-1}\,mol^{-1}$$

$$T = \frac{178}{0.164}\,K$$

$$= 1085\,K\ (812\,°C)$$

Top tip!

The temperature at which the spontaneity of a process changes is given by $T \approx \frac{\Delta H}{\Delta S}$.

Going further

Thermodynamics in organic chemistry

Values of $\Delta H^{\ominus}$ are relatively easy to calculate, so organic chemists frequently evaluate reactions only in terms of that quantity.

Strictly, you can only ignore the entropy term and assume $\Delta G^{\ominus} \approx \Delta H^{\ominus}$ if the reaction involves an entropy change small enough for $T\Delta S^{\ominus}$ to be much smaller than $\Delta H^{\ominus}$.

The entropy term cannot be ignored for organic reactions that occur with a significant change in entropy or at high temperatures as $T\Delta S^{\ominus}$ is then significant.

36 Carbon monoxide and hydrogen can, under the right conditions, react to form methanol.

$$CO(g) + 2H_2(g) \rightarrow CH_3OH(l) \qquad \Delta H^{\ominus} = -129\,kJ\,mol^{-1} \qquad \Delta S^{\ominus} = -218\,J\,K^{-1}\,mol^{-1}$$

a Explain why the reaction will be spontaneous at low temperatures.

b Calculate the temperature at which the reaction ceases to be spontaneous.

ATL R1.4B

An online activity which allows you to visualise how ΔG changes when the parameters of ΔH, ΔS and T are varied is available at www.geogebra.org/. Visit the website and search for 'Gibbs free energy'.

The Gibbs free energy and spontaneity application allows the Gibbs energy of a reaction to be plotted against the temperature. The sliders can be used to adjust the values of enthalpy and entropy change. The + and – keys (with the use of the shift key also) allow the sliders to be moved in smaller increments.

Use the sliders to create a line that shows the change in Gibbs energy for the hydrogenation of ethene.

$$C_2H_4(g) + H_2(g) \rightarrow C_2H_6(g) \qquad \Delta H^{\ominus} = -132\,kJ\,mol^{-1}$$
$$\Delta S^{\ominus} = -121\,J\,K^{-1}\,mol^{-1}$$

Zoom in and determine the temperature at which the reaction ceases to be spontaneous. How would you expect this temperature to change if the reaction had a more exothermic enthalpy change? Move the slider to see if you are correct.

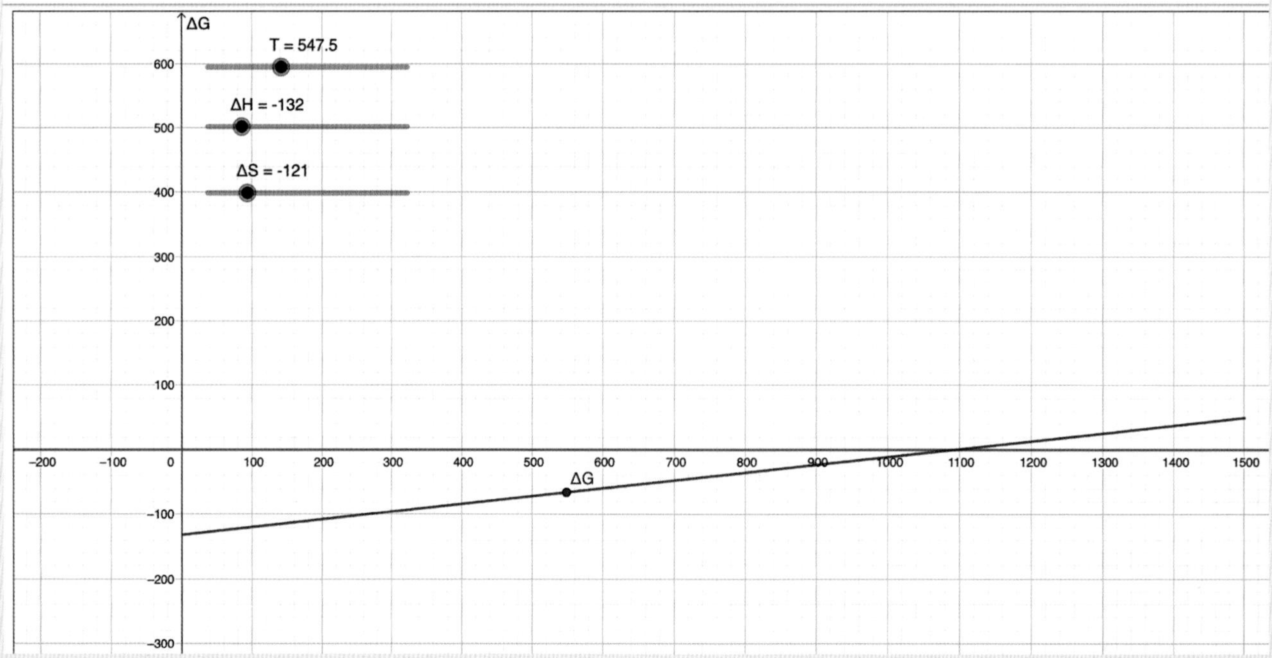

■ **Figure R1.64** The Gibbs free energy and spontaneity activity on Geogebra

HL ONLY

Gibbs energy change and equilibrium

Link

Equilibrium is covered in more detail in Chapter R2.3, page 482.

An equilibrium mixture of reactants and products is established in all reactions, but it is usual to state that a reaction has gone to completion if the equilibrium mixture is composed overwhelmingly of products ($K > 10^4$), or does not occur if it is composed overwhelmingly of reactants ($K < 10^{-4}$).

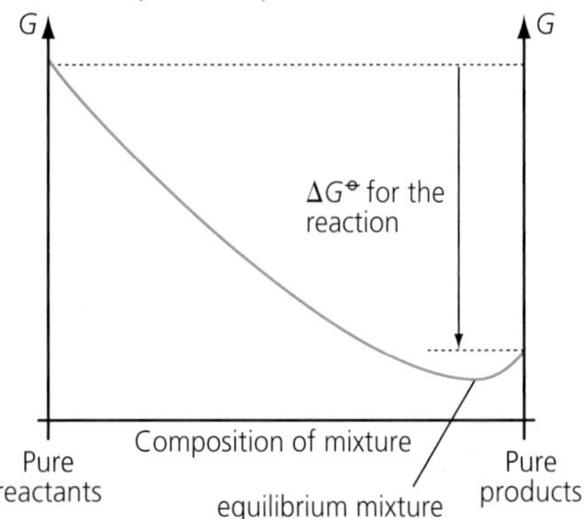

■ **Figure R1.65** The variation of Gibbs energy for a simple reaction ($A \rightleftharpoons B$) that comes to equilibrium with observable amounts of reactants and products present

Observations of the many reactions that contain appreciable amounts of reactants and products in their equilibrium mixture reveal that the equilibrium mixture will be reached regardless of whether a reaction proceeds from the starting point of entirely reactants or entirely products. This indicates that it is spontaneous for reactions to form an equilibrium mixture from either direction. The equilibrium mixture therefore represents a maximum of total entropy – the entropy of the mixture of species is greater than that of either the pure reactants or the pure products. This maximum entropy corresponds to a minimum of Gibbs energy (Figure R1.64).

Top tip!

As an equilibrium reaction proceeds from either pure reactants or pure products the Gibbs energy of the mixture decreases, reaching a minimum when equilibrium is reached.

The green line in Figure R1.65 shows the Gibbs energy for this reaction. The minimum Gibbs energy is when the reaction is at equilibrium and this is established closer to the pure products.

The standard change in Gibbs energy, $\Delta G^{\ominus}$, is the change in Gibbs energy when the reactants specified in a reaction equation are completely transformed to products under standard conditions. It is a constant (at constant temperature) and is the difference between the two dotted red lines in Figure R1.65. The standard change in Gibbs energy is different from the **reaction Gibbs energy, ΔG_r**. The reaction Gibbs energy, ΔG_r, is the difference between the Gibbs energy possessed by the products and reactants in a particular ratio of reactants to products. ΔG_r therefore varies as the ratio of products to reactants changes, but will have the value zero at equilibrium.

◆ **Reaction Gibbs energy, ΔG_r**: The Gibbs energy of the products present in a reaction mixture minus the Gibbs energy of the reactants present in a reaction mixture. Its value is 0 when the reactants and products have the same amount of Gibbs energy, and this corresponds to the equilibrium position.

The reaction Gibbs energy is represented by the gradient of the green line in Figure R1.65 and its value is given by the equation:

$$\Delta G_r = \Delta G^{\ominus} + RT \ln Q$$

where:

- ΔG_r = reaction Gibbs energy; the difference between the Gibbs energy of the products and reactants ($J\,mol^{-1}$)
- $\Delta G^{\ominus}$ = standard change in Gibbs energy ($J\,mol^{-1}$)
- R = gas constant ($J\,K^{-1}\,mol^{-1}$)
- T = absolute temperature (K)
- Q = reaction quotient.

Top tip!

Note that the standard change in Gibbs energy is in $J\,mol^{-1}$ (not the more usual $kJ\,mol^{-1}$) because the gas constant has the units $J\,K^{-1}\,mol^{-1}$.

HL ONLY

Top tip!

The subscript 'r' is sometimes omitted from ΔG_r, as is the case in the IB *Chemistry data booklet*.

ΔG_r and $\Delta G^\ominus$ are easily confused. The standard Gibbs energy, $\Delta G^\ominus$, is constant for a given reaction and is the change in Gibbs energy when the reaction, as stated in the reaction equation, goes to completion. The reaction Gibbs energy, ΔG_r, varies as equilibrium is approached as it is effectively the difference between the Gibbs energy of the products and reactants in a system at any point in time. The two are linked by the equation shown above.

As a reaction mixture gets closer to the equilibrium mixture, the difference between the Gibbs energy of the products and reactants gets smaller, and the gradient of the line showing the change in Gibbs energy with reaction composition approaches zero (Figure R1.65).

When the reaction is at equilibrium, $Q = K$, the value of ΔG_r will be zero so:

$$0 = \Delta G^\ominus + RT \ln K$$

$$\Delta G^\ominus = -RT \ln K$$

This gives the important relationship between the standard change in Gibbs energy of a reaction, $\Delta G^\ominus$, and the equilibrium constant, K. This equation is given in section 1 of the IB *Chemistry data booklet.*

- A reaction which has an equilibrium constant, $K > 1$ will have a negative standard Gibbs energy change.
- A reaction which has an equilibrium constant, $K < 1$ will have a positive standard Gibbs energy change.

LINKING QUESTION

What is the likely composition of an equilibrium mixture when $\Delta G^\ominus$ is positive?

If the standard change in Gibbs energy of a reaction is $+23\,\text{kJ}\,\text{mol}^{-1}$ (or greater), substituting the value into $\Delta G^\ominus = -RT \ln K$ shows K is very low, and the reaction will not occur to any significant extent (less than 0.01% conversion). The opposing statement can be made about reactions with $\Delta G^\ominus < -23\,\text{kJ}\,\text{mol}^{-1}$: they go effectively to completion (more than 99.99% conversion).

WORKED EXAMPLE R1.4F

$N_2O_4(g) \rightleftharpoons 2NO_2(g)$ $\qquad K = 0.15$ (at 298 K)

a Calculate the standard change in Gibbs energy, $\Delta G^\ominus$, for the decomposition of dinitrogen tetroxide into nitrogen dioxide.

b Calculate the reaction Gibbs energy, ΔG_r, when the reaction quotient, Q, has the value of 0.1.

Answer

a $\Delta G^\ominus = -RT \ln K$

$\Delta G^\ominus = -8.31\,\text{J}\,\text{K}^{-1}\,\text{mol}^{-1} \times 298\,\text{K} \times \ln 0.15$

$\Delta G^\ominus = +4700\,\text{J}\,\text{mol}^{-1}$ $(= +4.7\,\text{kJ}\,\text{mol}^{-1})$

b $\Delta G_r = \Delta G^\ominus + RT \ln Q$

$\Delta G_r = 4700\,\text{J}\,\text{mol}^{-1} + (8.31\,\text{J}\,\text{K}^{-1}\,\text{mol}^{-1} \times 298\,\text{K} \times \ln 0.1)$

$\Delta G_r = -1000\,\text{J}\,\text{mol}^{-1} = -1\,\text{kJ}\,\text{mol}^{-1}$

WORKED EXAMPLE R1.4G

Calculate the equilibrium constant for the reduction of silver ions by iron(II) ions at 298 K.

$Ag^+(aq) + Fe^{2+}(aq) \rightleftharpoons Fe^{3+}(aq) + Ag(s)$ $\Delta G^\ominus = -9.08\,kJ\,mol^{-1}$

Answer

$$\frac{\Delta G^\ominus}{-RT} = \ln K$$

$$\frac{-9080\,J\,mol^{-1}}{-8.31\,J\,K^{-1}\,mol^{-1} \times 298\,K} = \ln K$$

$$3.67 = \ln K$$

$$K = e^{3.67}$$

$$K = 39.1$$

37 Two allotropes of oxygen exist in equilibrium with one another:

$$\frac{3}{2}O_2(g) \rightleftharpoons O_3(g) \qquad \Delta G^\ominus = +163\,kJ\,mol^{-1}$$

a Calculate a value for the equilibrium constant, K, at 298 K.

b Calculate the reaction Gibbs energy, ΔG_r, when the reaction quotient, Q, has a value of 1.

c What name is given to the phenomenon achieved when the reaction Gibbs energy, ΔG_r, is zero?

Going further

Coupled reactions

It is possible to make non-spontaneous reactions occur. This is done by coupling a non-spontaneous reaction to another with a larger, negative $\Delta G^\ominus$. The non-spontaneous reaction is driven by the work of the spontaneous one, and in total, there is still a decrease in Gibbs free energy.

Figure R1.66 shows an analogy. The non-spontaneous process of lifting a weight has been achieved by coupling it with the spontaneous process of a falling heavier weight.

Many biochemical processes rely on coupled reactions. Adenosine triphosphate (ATP) in cells acts as a short-term energy store and source. The hydrolysis of ATP is coupled with non-spontaneous reactions (for example, polymerization of amino acids into proteins) and provides sufficient Gibbs energy to make them spontaneous.

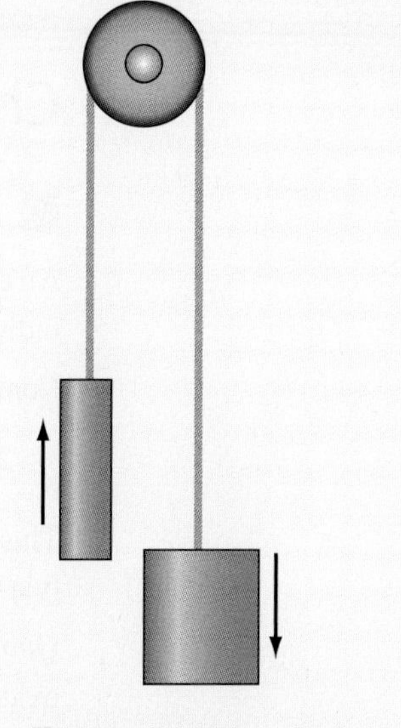

■ **Figure R1.66** A mechanical analogy to illustrate the concept of a coupled reaction

LINKING QUESTION

How can electrochemical data also be used to predict the spontaneity of a reaction?

HL ONLY

R2.1 How much? The amount of chemical change

Guiding question

- How are chemical equations used to calculate reacting ratios?

SYLLABUS CONTENT

By the end of this chapter, you should understand that:

- chemical equations show the ratio of reactants and products in a reaction
- the mole ratio of an equation can be used to determine:
  - the masses and/or volumes of reactants and products
  - the concentrations of reactants and products for reactions occurring in solution
- the limiting reactant determines the theoretical yield
- the percentage yield is calculated from the ratio of experimental yield to theoretical yield
- the atom economy is a measure of efficiency in green chemistry.

By the end of this chapter you should know how to:

- deduce chemical equations when reactants and products are specified
- calculate reacting masses and/or volumes and concentrations of reactants and products
- identify the limiting and excess reactants from given data
- distinguish between the theoretical yield and the experimental yield
- solve problems involving reacting quantities, limiting and excess reactants, theoretical, experimental and percentage yields
- calculate the atom economy from the stoichiometry of a reaction.

There is no higher-level only material in R2.1

Chemical equations

Amounts and equation coefficients

An equation is a symbolic representation of a microscopic-scale reaction (a reaction between individual colliding particles), which takes place at the macroscopic scale (in bulk).

Consider the following balanced equation:

$$2NO_2(g) \rightarrow N_2O_4(g)$$

This equation allows us to perform calculations involving moles, volumes, numbers of molecules (via the Avogadro constant) and masses (via relative molecular masses).

Qualitatively, a word equation states the names of the reactants (nitrogen dioxide or nitrogen(IV) oxide) and products (dinitrogen tetroxide) and gives their physical states (pure gases) using state symbols.

Quantitatively, it expresses the following relationships:

- The **relative number of molecules** of the reactants and products. Here, two molecules of nitrogen dioxide chemically react to form one molecule of dinitrogen tetroxide.
- The **amounts** (in moles) of the particles that form the reactants and products: 2 moles of nitrogen dioxide molecules react to form 1 mole of dinitrogen tetroxide molecules.

Link

The mole concept is defined and discussed in detail in Chapter S1.4, page 65.

Link

Avogadro's law is covered in more detail in Chapter S1.4, page 93.

- The **masses** of reactants and products. 92.02 g (2 mol) of nitrogen dioxide molecules reacts to form 92.02 g of dinitrogen tetroxide molecules (1 mol). The law of conservation of mass is obeyed.
- The **relative volumes** of gaseous reactants and products (Avogadro's law). 2 volumes of nitrogen dioxide react to form 1 volume of dinitrogen tetroxide. For example, 1.0 dm^3 of nitrogen dioxide would react to form 0.5 dm^3 of dinitrogen tetroxide, or 10 cm^3 of nitrogen dioxide would react to form 5 cm^3 of dinitrogen tetroxide.

1 Nitrogen monoxide reacts with oxygen to form nitrogen dioxide:

$$2NO(g) + O_2(g) \rightarrow 2NO_2(g)$$

Complete the table below based on the balanced equation above assuming NO and O_2 are combined in stoichiometric amounts.

NO(g)	**O_2(g)**	**NO_2(g)**
20 molecules		
2000 molecules		
1.204×10^{24} molecules		
0.2 mol		
60.02 g		

Deducing chemical equations

A chemical equation using chemical symbols can be deduced when the names of all the reactants and products are included in a word equation. First add the correct formulas for all the reactants and products and then balance the equation.

For example:

ammonia	+	nitrogen monoxide	→	nitrogen	+	water	word equation
NH_3	+	NO	→	N_2	+	H_2O	unbalanced chemical equation
$2NH_3$	+	NO	→	N_2	+	**$3H_2O$**	balancing hydrogen
$2NH_3$	+	**3NO**	→	N_2	+	$3H_2O$	balancing oxygen
$2NH_3$	+	3NO	→	$\frac{5}{2}N_2$	+	$3H_2O$	balancing nitrogen
$4NH_3$	+	**6NO**	→	**$5N_2$**	+	**$6H_2O$**	removing fractional coefficient (multiplying through by two)
$4NH_3(g)$	+	6NO(g)	→	$5N_2(g)$	+	$6H_2O(l)$	adding state symbols (for standard conditions)

2 Write balanced symbol equations for the following reactions:

a lithium hydroxide + sulfuric(VI) acid → lithium sulfate + water

b sodium + magnesium fluoride → sodium fluoride + magnesium

c copper + gold(I) nitrate → gold + copper(II) nitrate

d sucrose + oxygen → carbon dioxide + water

e magnesium carbonate + hydrobromic acid → magnesium bromide + water + carbon dioxide

f zinc + lead(II) nitrate → lead + zinc nitrate

g aluminium bromide + chlorine → aluminium chloride + bromine

h sodium phosphate(V) + calcium chloride → calcium phosphate(V) + sodium chloride

i calcium hydroxide + phosphoric(V) acid → calcium phosphate(V) + water

j manganese + sulfuric acid(VI) → manganese(II) sulfate + hydrogen

3 Pharaoh's serpent is the name given to the set of reactions that occur when mercury(II) thiocyanate, $Hg(SCN)_2$ (extremely toxic by skin absorption), undergoes thermal decomposition.

After setting fire to the mercury(II) thiocyanate a series of chemical reactions occur. Balance the following equations:

a $\_Hg(SCN)_2(s) \rightarrow \_HgS(s) + CS_2(l) + C_3N_4(s)$

b $CS_2(l) + \_O_2(g) \rightarrow CO_2(g) + \_SO_2(g)$

c $\_C_3N_4(s) \rightarrow \_(CN)_2(g) + N_2(g)$

Going further

Deducing products in an unbalanced equation

Urea, $CO(NH_2)_2$, can be used to remove nitrogen dioxide, NO_2, from the flue gases of power stations, converting it into harmless nitrogen.

$$2CO(NH_2)_2 + 3NO_2 \rightarrow 2A + bH_2O + cN_2$$

What is the identity of the product A (which does not contain hydrogen) and the values of b and c?

By comparing the number of H atoms: $2 \times 4 = b \times 2 \Rightarrow b = \mathbf{4}$

By comparing the number of N atoms:

$$(2 \times 2) + (3 \times 1) = c \times 2 \Rightarrow c = \frac{7}{2}$$

Let the formula of A be CO_a.

By comparing the number of O atoms:

$(2 \times 1) + (3 \times 2) = (2 \times a) + (b \times 1) \Rightarrow a = 2$

The identity of A is $\mathbf{CO_2}$.

Inquiry 2: Collecting and processing data

Identify and record relevant qualitative observations

It is important when recording your observations that you use the correct scientific language. Table R2.1 lists some changes you may observe during tests, recorded using the correct chemical language.

■ **Table R2.1** Recording observations during chemical reactions

Test	Example of correct recording of observation
A colour change in solution.	The orange solution turned dark green.
On mixing two solutions a solid forms.	A bright yellow precipitate rapidly formed from the mixing of two clear, colourless solutions.
A gas is given off.	There was effervescence (bubble production) and a colourless and odourless gas was released at the surface of the solution.
A solid reacts in an exothermic reaction to form a solution.	The white solid reacted to form a colourless solution; heat was released causing the flask to warm.

Going further

Types of reactions with examples

Direct combination (synthesis)

A synthesis or direct combination reaction involves a single compound that is formed from its elements (atoms or molecules).

$$Ca(s) + Br_2(l) \rightarrow CaBr_2(s)$$

Decomposition

A decomposition reaction involves a compound being converted to produce two simpler substances.

$$CaCO_3(s) \rightarrow CaO(s) + CO_2(g)$$

Redox

Redox is reduction and oxidation. The definition of redox includes all reactions where there is a transfer of electrons and a change in the oxidation state of one or more atoms. Lewis acid/base reactions are not redox reactions because there is no change in oxidation states.

$$P_4(s) + 5O_2(g) \rightarrow P_4O_{10}(s)$$

$$2PbO_2(s) \rightarrow 2PbO(s) + O_2(g)$$

Ionic precipitation

Ionic precipitation reactions occur when solutions of different salts (or a salt and a base, or a salt and an acid) react together to form an insoluble salt. The solid that is formed is known as a precipitate.

$$2AgNO_3(aq) + Na_2S(aq) \rightarrow Ag_2S(s) + 2NaNO_3(aq)$$

Acid–base reactions

Arrhenius acid–base reactions are a group of reactions that involve acids reacting with bases (often metal oxides or metal hydroxides) to form a salt and water only.

$$Fe_2O_3(s) + 6HCl(aq) \rightarrow 2FeCl_3(aq) + 3H_2O(l)$$

LINKING QUESTION

When is it useful to use half-equations?

ATL R2.1A

Can you think of any types of inorganic reaction not included here? Can you think of some examples of reactions that can have two or more different classifications?

Develop a Venn diagram to summarize types of inorganic reaction with examples.

Reacting masses and volumes

Consider the synthesis of ammonia from nitrogen and hydrogen. The balanced equation for the stoichiometric reaction is: $N_2(g) + 3H_2(g) \rightarrow 2NH_3(g)$. Figure R2.1 shows the three types of information present in the equation.

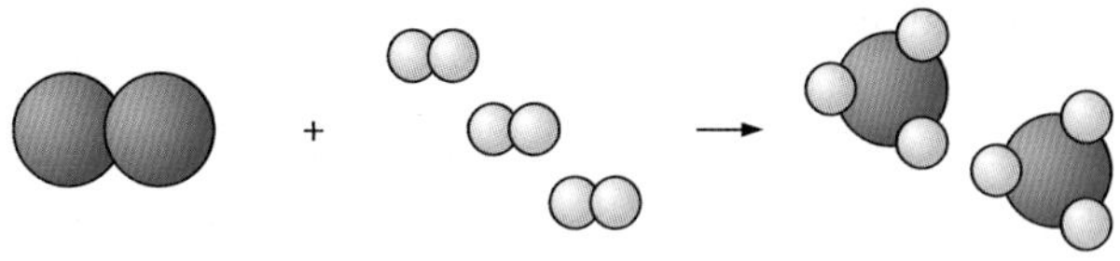

N_2	+	$3H_2$	→	$2NH_3$
1 molecule of N_2		3 molecules of H_2		2 molecules of NH_3
1 mole of N_2		3 moles of H_2		2 moles of NH_3
2 N atoms		6 H atoms		2 N atoms and 6 H atoms

■ **Figure R2.1** This representation of the production of ammonia from nitrogen and hydrogen shows several ways to interpret the quantitative information of a chemical reaction

The amounts (in mol) can be scaled up or scaled down and, hence, the reacting masses of the reactants and products are also scaled up or down.

For example, doubling the amounts leads to doubling of the masses of the reactants and products:

$$2\text{ mol of }N_2 \quad + \quad 6\text{ mol of }H_2 \quad \rightarrow \quad 4\text{ mol of }NH_3$$

$$56.04\text{ g} \quad + \quad 12.12\text{ g} \quad \rightarrow \quad 68.16\text{ g}$$

whereas halving the amounts halves the masses of the reactants and products:

$$0.5\text{ mol of }N_2 \quad + \quad 1.5\text{ mol of }H_2 \quad \rightarrow \quad 1\text{ mol of }NH_3$$

$$14.01\text{ g} \quad + \quad 3.03\text{ g} \quad \rightarrow \quad 17.04\text{ g}$$

According to Avogadro's law, there are equal numbers of molecules in equal volumes of gas at the same temperature and pressure. Therefore, gas volumes are related in the same way as amounts. Doubling the number of molecules doubles the volume of gas and halving the number of molecules halves the volume.

$N_2(g)$	+	$3H_2(g)$	→	$2NH_3(g)$
1 dm^3		3 dm^3		2 dm^3
2 dm^3		6 dm^3		4 dm^3

Tool 3: Mathematics

Understand direct and inverse proportionality, as well as positive and negative correlations between variables

The simplest relationship between two variables is direct proportionality. This means that, if one variable, say x, doubles, then the other variable, y, also doubles; if y is halved in value, then x is halved, etc. The ratio of the two variables (x/y or y/x) is constant. Proportionality is shown using the symbol $\propto$, so $y \propto x$ and x/y = constant are equivalent statements.

To check whether two variables are proportional to each other, you can either (i) calculate their ratio at different values to see whether it is constant or (ii) plot x against y to determine whether it produces a straight line through the origin.

You can do this quickly in a spreadsheet program such as *Excel* by using the 'add trend line' function. This feature allows you to view the equation of the line of best fit and also the square of the correlation coefficient (R^2), which is a measure of how well your data points fit the given straight line. The closer R^2 is to 1, the better the fit. A positive correlation means that as one variable increases, the other variable increases. A negative correlation means that as one variable increases, the other decreases.

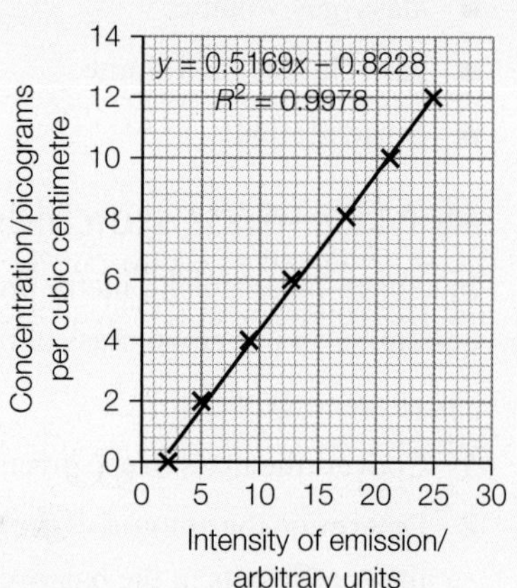

■ **Figure R2.2** Spectroscopic calibration data plotted in *Excel* showing equation of the line and coefficient of determination (square of correlation coefficient) (R^2)

If one variable increases while the other decreases, they have an inverse relationship. The simplest inverse relationship is when one variable, for example x, doubles when the other variable, y, halves and the product of the two variables xy is constant. Inverse proportionality can be expressed as $y \propto 1/x$ or xy = constant.

WORKED EXAMPLE R2.1A

Determine the amount of ammonia produced if 42.0 moles of hydrogen are reacted with an excess of nitrogen.

Answer

Note: the problem states that there is an *excess* of nitrogen so we are not concerned with any mole ratio involving nitrogen.

$$N_2(g) + 3H_2(g) \rightarrow 2NH_3(g)$$

42.0 mol

The hydrogen and ammonia are in a 3 : 2 ratio. Since $2 = 3 \times \frac{2}{3}$, the conversion factor is $\times \frac{2}{3}$.

Hence the amount of ammonia is $(42.0\,\text{mol} \times \frac{2}{3}) = 28.0\,\text{mol}$.

4 Determine the amount of aluminium oxide that can be produced from 4.00 mol of oxygen (molecules) reacting with *excess* aluminium (atoms).

Stoichiometry problems

There are four basic types of stoichiometry problem involving the mole concept:

- mass–mass
- mass–gas volume
- gas volume–gas volume
- concentration.

Mass–mass stoichiometry problems

In a mass–mass stoichiometry problem, you will use a given mass of a reactant or product to determine an unknown mass of reactant or product.

There are three steps:

1 Convert the mass of the given substance to moles using the molar mass of the given substance.

2 Determine the amount of the required substance using the amount of the given substance and the coefficients in the balanced equation.

3 Convert the amount of the required substance to a mass using the molar mass of the required substance.

WORKED EXAMPLE R2.1B

Determine the mass of mercury produced from the decomposition of 0.00125 kg of mercury(II) oxide.

$$2HgO(s) \rightarrow 2Hg(l) + O_2(g)$$

Answer

1 Convert the mass of mercury(II) oxide to the amount of mercury(II) oxide using the molar mass of mercury(II) oxide (216.59 g mol^{-1}).

$$\text{amount of mercury(II) oxide} = \frac{12.50\,\text{g}}{216.59\,\text{g mol}^{-1}} = 0.0577\,\text{mol}$$

2 Determine the amount of mercury using the amount of mercury(II) oxide and the coefficients in the balanced equation.

amount of mercury = 0.0577 mol

(because the amounts of mercury(II) oxide and mercury are in a 1 : 1 ratio).

3 Convert the amount of mercury to the mass using the molar mass.

mass of mercury = (0.0577 mol × 200.59 g mol^{-1}) = 11.6 g

5 49.00 g of potassium chlorate(V) is heated and decomposes according to the equation:

$$\mathbf{2KClO_3(s) \rightarrow 2KCl(s) + 3O_2(g)}$$

Determine the mass of oxygen formed.

Mass–gas volume stoichiometry problems

In a mass–volume stoichiometry problem, you will use a given mass of a reactant or product to determine an unknown volume of reactant or product.

There are three steps:

1. Convert the mass of the given substance to an amount using the molar mass of the given substance.
2. Determine the amount of the required substance using the amount of the given substance and the coefficients in the balanced equation.
3. Convert the amount of the required substance to cubic decimetres (dm^3) of gas using the molar gas volume at STP.

WORKED EXAMPLE R2.1C

Determine the volume of hydrogen gas produced from the reaction between 1.64 g of aluminium metal and hydrochloric acid. Assume STP.

$$2Al(s) + 6HCl(aq) \rightarrow 2AlCl_3(aq) + 3H_2(g)$$

Answer

1. Convert the mass of aluminium to the amount of aluminium using the molar mass of aluminium ($26.98\,g\,mol^{-1}$).

   $$\text{amount of aluminium (Al)} = \frac{1.64\,g}{26.98\,g\,mol^{-1}} = 0.0608\,mol$$

2. Determine the amount of hydrogen using the amount of aluminium and the coefficients in the balanced equation.

   $$\text{amount of hydrogen } (H_2) = (0.0608\,mol \times \frac{3}{2}) = 0.0912\,mol$$

   (because the amounts (mol) of aluminium and hydrogen are in a 2 : 3 ratio).

3. Convert the amount of hydrogen to cubic decimetres of hydrogen using the molar gas volume at STP.

   $$\text{volume of hydrogen gas } (H_2) = (0.0912\,mol \times 22.7\,dm^3\,mol^{-1}) = 2.07\,dm^3$$

6 Determine the volume of carbon dioxide produced from the complete combustion of 17.117 g of sucrose ($M = 342.34\,g\,mol^{-1}$). Assume STP.

$$C_{12}H_{22}O_{11}(s) + 12O_2(g) \rightarrow 12CO_2(g) + 11H_2O(l)$$

Gas volume–gas volume stoichiometry problems

In a gas volume–gas volume stoichiometry problem, you will use a given volume of a gas to determine an unknown volume of a gaseous reactant or product.

There is one step:

- Convert the given volume to the unknown volume using the mole ratio (which is also the volume ratio) from the coefficients in the balanced chemical equation.

WORKED EXAMPLE R2.1D

Determine the volume of oxygen gas that reacts with 2.75 dm³ of sulfur dioxide gas to form sulfur trioxide gas.

$$2SO_2(g) + O_2(g) \rightarrow 2SO_3(g)$$

Answer

From the balanced equation, 1 mole of oxygen molecules reacts with 2 moles of sulfur dioxide molecules to form 2 moles of sulfur trioxide molecules.

Hence, from Avogadro's law, 1 dm³ of oxygen molecules reacts with 2 dm³ of sulfur dioxide molecules to form 2 dm³ of sulfur trioxide molecules.

Hence, using ratios, the volume of oxygen gas (O_2) = $(2.75\,dm^3 \times \frac{1}{2}) = 1.38\,dm^3$

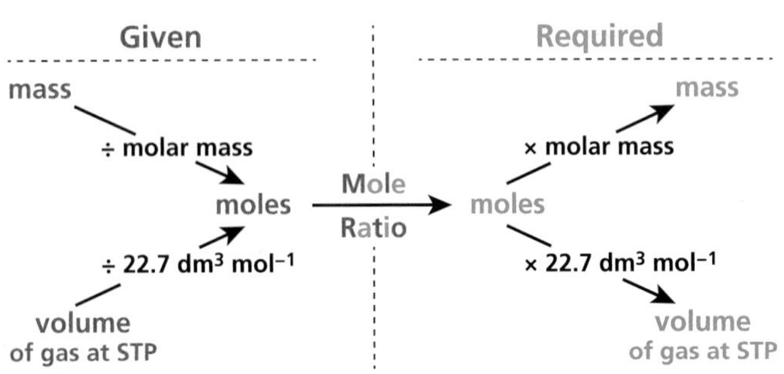

■ **Figure R2.3** Interconversion between masses and volumes of gas via the mole ratio

7 When nitrogen (N_2) and hydrogen (H_2) react they produce ammonia (NH_3). Both reactants and products are gases under standard conditions.

a State the balanced equation with state symbols.

b If there is 12.5 dm³ of nitrogen, determine how much ammonia is produced and what volume of hydrogen is used.

Figure R2.3 summarizes how to interconvert between masses and/or volumes of gas via the mole ratio given by the coefficients in a balanced equation.

WORKED EXAMPLE R2.1E

Ozone is usually made by passing oxygen gas through a tube between two highly charged electrical plates.

$$3O_2(g) \rightarrow 2O_3(g)$$

The reaction does not go to completion, so a mixture of the two gases is produced.

We can determine the concentration of ozone, O_3, in the mixture by its reaction with aqueous potassium iodide, KI.

$$O_3(aq) + 2KI(aq) + H_2O(l) \rightarrow I_2(aq) + O_2(aq) + 2KOH(aq)$$

We can determine the amount of iodine formed by its reaction with sodium thiosulfate.

$$2Na_2S_2O_3(aq) + I_2(aq) \rightarrow Na_2S_4O_6(aq) + 2NaI(aq)$$

When 500.00 cm³ of an oxygen / ozone gaseous mixture at STP was passed into an excess of aqueous KI, and the iodine titrated, 15.00 cm³ of 0.100 mol dm⁻³ $Na_2S_2O_3$ was required to discharge the iodine colour.

1 Calculate the amount in moles of iodine produced.

2 Calculate the percentage of ozone in the gaseous mixture.

Answer

1 $2S_2O_3^{2-}(aq) + I_2(aq) \rightarrow S_4O_6^{2-}(aq) + 2I^-(aq)$

$nI_2 : nS_2O_3^{2-} = 1:2$

amount of $S_2O_3^{2-} = \frac{15.00}{1000} \times 0.100 = 1.5 \times 10^{-3}$ mol

amount of $I_2 = \frac{1}{2} \times 1.5 \times 10^{-3} = 7.5 \times 10^{-4}$ mol

2 $nI_2 : nO_3 = 1:1$

amount of $O_3 = 7.5 \times 10^{-4}$ mol

volume of $O_3 = 7.5 \times 10^{-4}\,\text{mol} \times 22\,700\,\text{cm}^3\,\text{mol}^{-1} = 17.03\,\text{cm}^3$

% of $O_3 = \frac{17.03}{500} \times 100\% = 3.4\%$

Tool 1: Experimental techniques

Measuring variables: Volume

Volume of liquid or solution

The SI unit for volume is the cubic metre (m^3). Since this is a large unit, chemists usually measure volumes of liquids or solutions in cubic centimetres (cm^3). Larger volumes may be measured in cubic decimetres (dm^3). $1\,dm^3 = 1000\,cm^3$ (since $1\,dm = 10\,cm$ and $10\,cm \times 10\,cm \times 10\,cm = 1000\,cm^3$). Other common units for volume are litres (L) and millilitres (ml). There are 1000 millilitres in 1 litre. Both measurements are part of the metric system and are accepted for use in the International System of Units (SI), but are not SI units. $1\,ml = 1\,cm^3$ and $1\,L = 1\,dm^3$.

The volume of a liquid may be measured accurately using a pipette, a burette or a volumetric flask.

If high precision is not required, a measuring cylinder may be used. These are often used when the reagent is to be added in excess and so the exact volume is not required.

Gases

To measure the volume of a gas produced in a chemical reaction, we can collect it in a graduated gas syringe.

The syringe must be lightly greased to reduce friction, allowing the plunger to move when gas is produced. Gas syringes may leak where they are joined to a delivery tube (this can be minimized by using vaseline and tightly fitting rubber connectors). It is good practice to periodically twist the plunger of the syringe (without pulling or pushing) to ensure that it is moving freely.

An alternative method of collecting and measuring gas volumes is via displacement of water in an inverted burette, measuring cylinder or eudiometer (see Figure S1.113, page 108). However, this method is only suitable for gases that have a low solubility in, and do not react with water. Even then, as more gas is introduced, the pressure of the gas above the water increases, increasing its solubility and reducing its ability to displace water. This limits the accuracy of measuring the volume of gas produced.

Tool 3: Mathematics

Express measurement and processed uncertainties – absolute, fractional (relative), percentage – to an appropriate number of significant figures or level of precision

Measured uncertainties should be quoted as half the smallest division of the apparatus used to record them. Measurements made using the apparatus should therefore be given to the same precision as the uncertainty. In practice, this usually means to the same number of decimal places. For example, a burette with a scale marked every $0.02\,cm^3$ records to $0.01\,cm^3$ so $25\,cm^3$ of solution should be correctly recorded as $25.00\,cm^3 \pm 0.01\,cm^3$. $25\,cm^3$ (or $25.0\,cm^3$) would be considered incorrect as the uncertainty is given with more precision than the measured value. A volume of $25.000\,cm^3$ would also be incorrect as the recorded value is more precise than the uncertainty. This rule applies to final calculated values as well.

It is worthwhile remembering that uncertainties themselves are only approximations, so it is considered acceptable to approximate a percentage uncertainty to 1 significant figure.

Concentration stoichiometry problems

In aqueous and non-aqueous reactions, quantities of reactants and products are often specified in terms of volumes and concentrations. Once again there are three steps:

1. Convert the volume and concentration of the given substance to moles using

   amount (mol) = volume (dm^3) × molar concentration ($mol\,dm^{-3}$)
2. Determine the amount of the required substance using the amount of the given substance and the coefficients in the balanced equation.
3. Convert the amount of the required substance to a volume of gas, or mass depending on what is required using the same equation as in step 1.

We make the conversions between solution concentration, volume of gas, or mass, depending on what is required, and amounts of solute in moles using the molar concentrations of the solutions. We make the conversions between amounts in moles of A and B using the stoichiometric coefficients from the balanced chemical equation.

The neutralization reaction below shows how coefficients in a balanced equation determine the mole ratio of the reactants and products in a stoichiometric equation.

$Ba(OH)_2(aq)$	+	$2HCl(aq)$	→	$BaCl_2(aq)$	+	$2H_2O(l)$
1 mol		2 mol		1 mol		2 mol
($1\,dm^3 \times 1\,M$)		($2\,dm^3 \times 1\,M$)		($1\,dm^3 \times 1\,M$)		2 mol
($0.1\,dm^3 \times 1\,M$)		($0.2\,dm^3 \times 1\,M$)		($0.1\,dm^3 \times 1\,M$)		0.2 mol

(where M represents $mol\,dm^{-3}$)

8 Determine the maximum mass of anhydrous zinc sulfate crystals ($ZnSO_4$) that could be formed when 100.00 cm^3 of 2.00 mol dm^{-3} sulfuric(VI) acid, H_2SO_4, is reacted with an excess of metallic zinc.

WORKED EXAMPLE R2.1F

Determine the volume (dm^3) of a 0.610 M NaOH solution needed to neutralize 0.0200 dm^3 of a 0.245 M H_2SO_4 solution. (M represents mol dm^{-3}.)

$$2NaOH(aq) + H_2SO_4(aq) \rightarrow Na_2SO_4(aq) + 2H_2O(l)$$

Answer

1 Amount of sulfuric acid = $0.0200\,dm^3 \times 0.245\,mol\,dm^{-3} = 4.90 \times 10^{-3}\,mol$

2 From the stoichiometry we see that 1 mole of H_2SO_4 reacts with 2 moles of NaOH. Therefore, the amount of reacted NaOH(aq) must be $(2 \times 4.90 \times 10^{-3}\,mol) = 9.80 \times 10^{-3}\,mol$.

3 Volume of NaOH(aq) = $\dfrac{9.80 \times 10^{-3}\,mol}{0.610\,mol\,dm^{-3}} = 0.0161\,dm^3$

Nature of science: Experiments

The discovery of oxygen and a study of the combustion process

When mercury(II) oxide, a red powder once known as calx of mercury, is strongly heated it decomposes into mercury and a gas. The English chemist Joseph Priestley (1733–1804) collected this gas and found that flammable substances burnt much more strongly in it than in normal air.

In France, Antoine Lavoisier (1743–1794) carried out an experiment to find out more about Priestley's gas using the apparatus shown in Figure R2.4. He kept the mercury in the retort at a temperature just lower than its boiling point for several days.

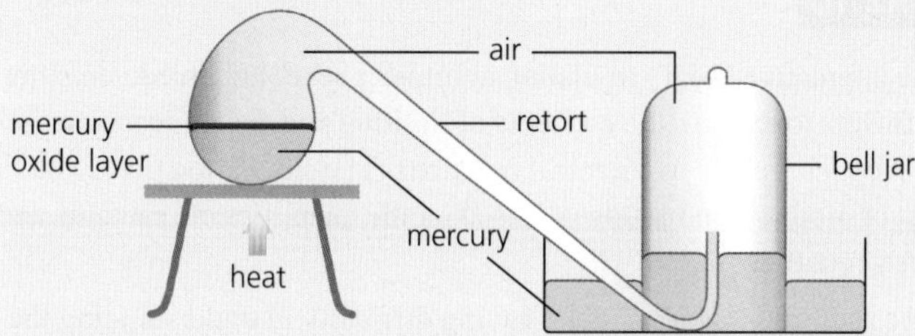

Figure R2.4 Lavoisier's preparation of oxygen

At the end of this time, he made the following observations:

- The level of the mercury in the bell jar had risen, showing the volume of air in the bell jar had been reduced by 20%.
- A layer of red powder, which he was able to show was calx of mercury, had formed on the surface of the hot mercury in the retort.
- The gas remaining in his apparatus would not support combustion.

On the basis of these observations and those of Priestley, Lavoisier proposed that the 20% of the air that supports combustion consists of a gas identical to that which Priestley produced (which we now know to be oxygen). He also proposed that, when substances burn, they chemically combine with this gas to form new substances (oxides).

LINKING QUESTION

How does the molar volume of a gas vary with changes in temperature and pressure?

LINKING QUESTION

In what ways does Avogadro's law help us to describe, but not explain, the behaviour of gases?

Should the natural sciences be regarded as a body of knowledge, a system of knowledge or a method?

It is helpful to think about the natural sciences as distinguishing between the claims they produce and the methods used to produce them. Science is more than the acquisition of an ever-increasing series of claims or statements or facts about the world (in other words, a body of knowledge composed of, for example, empirical data, models, theories and laws); it is also a methodology. It includes an understanding of the testable and provisional nature of scientific claims, and how these claims are generated, established and communicated between members of the scientific community. Scientists rely on a set of established procedures and practices associated with scientific inquiry to gather evidence and test their ideas of how the physical world works. Becoming a part of this community requires the individual to understand these methods and conventions as well as established facts; neither is more important than the other. However, there is no single method or single philosophy of science, and the real process of scientific research is often iterative (requiring many repeated observations) and complex, with researchers following many different paths. Even so, all the methods require observations and experiments and these are at the heart of what it means to be 'scientific'.

Limiting reactant

◆ **Limiting reactant**: The reactant that is used up when a reaction goes to completion.

◆ **Excess reactant**: A reactant is in excess when, after the reaction is complete, some of it remains unreacted.

In a chemical reaction, the reactants are often added in amounts which are not stoichiometric. The **limiting reactant** is the reactant that is used up first while an **excess reactant** is one that is not completely used up. The amount of product formed is determined by the limiting reactant.

ATL R2.1B

Use the interactive PhET simulation at https://phet.colorado.edu/en/simulations/reactants-products-and-leftovers to explore the concepts of limiting and excess reactants in the context of everyday experiences as well as in chemical reactions. The simulation gives you the opportunity to predict the amount of products and leftovers based on the quantities of reactants and ratios of molecules in the balanced equation.

Imagine someone in your class is having difficulty in understanding the concepts of limiting and excess reactants. Use presentation software (such as PowerPoint) to develop your own visual method for teaching this concept.

Consider the reaction between hydrogen and chlorine to form hydrogen chloride:

$$H_2(g) + Cl_2(g) \rightarrow 2HCl(g)$$

One mole of hydrogen molecules, H–H, reacts with one mole of chlorine molecules, Cl–Cl, to form two moles of hydrogen chloride molecules, H–Cl.

Table R2.2 shows the results of three experiments that involve different amounts of hydrogen and chlorine molecules as reactants. In experiment 1, the exact amounts (numbers) of the two reactants are used, resulting in a stoichiometric reaction. In experiments 2 and 3, an excess of one reactant is used, resulting in a non-stoichiometric reaction.

■ **Table R2.2** The concept of a limiting reactant applied to the synthesis of hydrogen chloride

Experiment	Number of molecules of reactants used		Number of molecules of products or unreacted reactants		
1	1	1	2	0	0
	H–H	Cl–Cl	H–Cl H–Cl	Stoichiometric reaction – no excess reactant	
2	3	1	2	2	0
	H–H H–H H–H	Cl–Cl	H–Cl H–Cl	H–H H–H Chlorine is limiting reactant; hydrogen is in excess	
3	1	2	2	0	1
	H–H	Cl–Cl Cl–Cl	H–Cl H–Cl		Cl–Cl Hydrogen is limiting reactant; chlorine is in excess

Top tip!

In order to determine the limiting reactant, divide the amount of each reactant (in moles) by its coefficient in the balanced equation. The reactant with the lowest number of moles is the limiting reactant.

WORKED EXAMPLE R2.1G

Sulfur hexafluoride is synthesized by burning sulfur in fluorine gas. The reaction is described by the equation:

$$S(s) + 3F_2(g) \rightarrow SF_6(g)$$

4 moles of sulfur (atoms) are added to 20 moles of fluorine (molecules). Deduce which will be the limiting reactant.

Answer

Amount of sulfur atoms ÷ 1 = (4 mol ÷ 1) = 4 mol; divided by 1 because the coefficient is 1.

Amount of fluorine molecules ÷ 3 = (20 mol ÷ 3) = 6.67 mol; divided by 3 because the coefficient is 3. However, they react in a 1 : 3 molar ratio.

Therefore, sulfur is the limiting reactant and fluorine is present in excess.

Common mistake

Misconception: *The limiting reagent is the reagent for which the lowest mass or lowest amount of reactant is present.*

The limiting reagent is deduced from the amounts in moles of the reagents and the stoichiometric coefficients in the balanced equation.

WORKED EXAMPLE R2.1H

0.250 mol of sulfur atoms are heated with 0.350 mol of iron atoms. They react according to the following equation:

$$Fe(s) + S(s) \rightarrow FeS(s)$$

Deduce the limiting reactant and the excess reactant.

Answer

According to the balanced equation, 1 mole of iron atoms reacts with 1 mole of sulfur atoms to produce 1 mole of iron(II) sulfide.

Hence, 0.250 mol of sulfur atoms will react with 0.250 mol of iron atoms to produce 0.250 mol of iron(II) sulfide.

Therefore, sulfur is the limiting reactant and iron is present in excess. The amount of unreacted iron atoms is (0.350 mol – 0.250 mol) = 0.100 mol.

WORKED EXAMPLE R2.1I

8.08 g of hydrogen gas reacts with 48.00 g of oxygen gas in a reaction to form water. Deduce what chemicals remain after the reaction.

Answer

The balanced equation for the reaction is:

$$2H_2(g) + O_2(g) \rightarrow 2H_2O(l)$$

$$\text{amount of hydrogen} = \frac{8.08\,\text{g}}{2.02\,\text{g mol}^{-1}} = 4.00\,\text{mol}$$

$$\text{amount of oxygen} = \frac{48.00\,\text{g}}{32.00\,\text{g mol}^{-1}} = 1.50\,\text{mol}$$

The stoichiometric molar ratio is 2 : 1 so the 1.50 mol of oxygen reacts with 3.00 mol of the hydrogen.

After the reaction there is 1 mole of unreacted hydrogen and 2 moles of water; there is no oxygen, since oxygen is the limiting reagent.

9 Calculate the mass of magnesium sulfide that can be obtained from the reaction between 4.862 g of magnesium and 3.207 g of sulfur.

$$Mg(s) + S(s) \rightarrow MgS(s)$$

Identify the limiting reactant and calculate the mass of the unreacted element present in excess.

10 Chloroethane, C_2H_5Cl, reacts with oxygen to form carbon dioxide, water and hydrogen chloride.

a Complete the following equation:

$$C_2H_5Cl + \_O_2 \rightarrow \_CO_2 + \_H_2O + HCl$$

b Deduce the limiting reagent in this reaction when 3.0 mol of chloroethane and 3.0 mol of oxygen are reacted.

c Determine the amount of carbon dioxide formed.

◆ **Theoretical yield**: The maximum amount or mass of a particular product that can be formed when the limiting reactant is completely consumed and there are no losses or side reactions.

◆ **Experimental yield**: The quantity of a product that is obtained from a chemical reaction.

◆ **Percentage yield**: The experimental yield as a percentage of the theoretical yield.

Percentage yield

The calculated mass, volume or amount of product formed when all the limiting reactant reacts is called the **theoretical yield**. The quantity of a product actually obtained in a chemical reaction is the **experimental yield**.

The **percentage yield** can be calculated from the following expression:

$$\text{percentage yield} = \frac{\text{experimental yield}}{\text{theoretical yield}} \times 100$$

Top tip!

Experimental and theoretical yields may be expressed in any appropriate terms: for example, as masses, molar amounts or, if the product is a gas, volumes. The units will cancel when percentage yield is calculated, as long as both yields are expressed using the same units.

In most reactions, not all of the reactants actually react and, when chemical reactions are carried out at industrial scales, they often form by-products as well as the desired product. Therefore, what is left at the end of a reaction is typically a mixture that includes a certain amount of the starting reactants and by-products as impurities.

Achieving high percentage yields is especially important in industries such as the pharmaceutical industry that rely on organic synthesis. Creating a drug may require ten different steps and, if each reaction has a percentage yield of 90%, the overall yield is only $(0.9 \times 0.9 \times 0.9 \times 0.9 \times 0.9 \times 0.9 \times 0.9 \times 0.9 \times 0.9 \times 0.9) = 0.35$ or 35%. In reactions such as the Haber process, where the reactants can be recycled, low yields are less of an issue.

Top tip!

Percentage yields are of particular importance in organic chemistry because many organic reactions are reversible and there are significant side reactions.

11 Sodium hydrogencarbonate, $NaHCO_3$, can be prepared from sodium sulfate by a three-step process as shown.

$$Na_2SO_4(s) + 4C(s) \rightarrow Na_2S(s) + 4CO(g)$$

$$Na_2S(s) + CaCO_3(s) \rightarrow CaS(s) + Na_2CO_3(s)$$

$$Na_2CO_3(s) + H_2O(l) + CO_2(g) \rightarrow 2NaHCO_3(s)$$

Assuming a yield of 90% in each step, determine the mass of sodium hydrogencarbonate that could be obtained from 100 kg of sodium sulfate. Give your answer to the nearest kilogram.

WORKED EXAMPLE R2.1J

In an experiment to produce a sample of hex-1-ene, 20.40 g of hexan-1-ol was heated with excess phosphoric(V) acid. The phosphoric(V) acid acted as a dehydrating agent, removing water from the alcohol to form hex-1-ene:

$$CH_3CH_2CH_2CH_2CH_2CH_2OH \rightarrow CH_3CH_2CH_2CH_2CH{=}CH_2 + H_2O$$

hexan-1-ol hex-1-ene

After purification, the hex-1-ene produced had a mass of 10.08 g. Calculate the percentage yield (to the nearest integer). State three reasons why this preparation does not produce a 100% yield.

Answer

From the equation, 1 mol of hexan-1-ol produces 1 mol of hex-1-ene.

Amount of hexan-1-ol $= \dfrac{20.40\,\text{g}}{102.20\,\text{g mol}^{-1}} = 0.1996\,\text{mol}$

Hence, the theoretical amount of hex-1-ene produced is 0.1996 mol (since there is excess phosphoric(V) acid).

Amount of hex-1-ene $= \dfrac{\text{mass (g)}}{\text{molar mass (g mol}^{-1})} = 0.1996\,\text{mol}$

Rearranging, mass of hex-1-ene $= 84.16\,\text{g mol}^{-1} \times 0.1996\,\text{mol} = 16.80\,\text{g}$.

Since only 10.08 g of hex-1-ene was produced, the percentage yield is: $\dfrac{10.08\,\text{g}}{16.80\,\text{g}} \times 100 = 60\%$

The yield is not 100% because:

- the reaction may not be complete
- side reactions may occur and other products might be formed
- some hex-1-ene may have been lost during purification.

Competing reactions

In certain circumstances, the same two chemicals can react to give different products. For example, when carbon burns in a plentiful supply of oxygen, it reacts to produce carbon dioxide, CO_2(g), according to the following chemical equation:

$$C(s) + O_2(g) \rightarrow CO_2(g)$$

However, carbon monoxide, CO(g), is also produced to a small extent even when there is excess oxygen available:

$$2C(s) + O_2(g) \rightarrow 2CO(g)$$

This is an example of a competing reaction. Since some of the carbon reacts to form carbon monoxide in the competing reaction, the experimental yield of carbon dioxide is always less than predicted.

LINKING QUESTION

What errors may cause the experimental yield to be **a** higher and **b** lower than the theoretical yield?

12 When 50.00 tonnes of iron(III) oxide were reacted with excess carbon, 30.00 tonnes of iron were produced. Determine the percentage yield. (1 tonne = 1000 kg.)

$$2Fe_2O_3(s) + 3C(s) \rightarrow 4\,Fe(l) + 3CO_2(g)$$

ATL R2.1C

A carbon footprint is defined as the mass of carbon dioxide that has the same effect as the total amount of all greenhouse gases given out over the full life cycle of a product or service.

For example, the carbon footprint of a plastic bag made from poly(ethene) could include carbon dioxide released as fuels arc burnt to provide energy to:

- drill, pump and transport the crude oil
- heat and vaporize crude oil in fractional distillation (to separate the hydrocarbons)
- heat long-chain alkanes to crack them to make ethene and shorter chain alkanes
- provide heat and pressure to polymerize ethene to make poly(ethene)
- transport the plastic bags to shops
- transport used plastic bags to waste disposal sites, where they may be incinerated or buried in landfill (burning and decomposing both release carbon dioxide).

Use an online website such as https://footprint.wwf.org.uk/#/ to calculate your personal carbon footprint and find out how you and industry can reduce carbon emissions.

Green chemistry

◆ **Green chemistry**: The design of chemical products and processes that reduce or eliminate the use and generation of hazardous substances.

Green chemistry involves the use of chemicals and chemical processes designed to reduce or eliminate impacts on the environment. This may involve using less energy, reduction of waste products, use or production of non-toxic chemicals and improved efficiency. One important principle of green chemistry is that it is better to prevent waste than to treat or clean waste after it has been produced.

In 2015, the United Nations created a universal call to action to end poverty, protect the planet, and ensure that all people enjoy peace and prosperity by 2030. This framework, comprising seventeen aspirational goals known as the Sustainable Development Goals (SDGs), has been adopted by governments, industry and many other organizations worldwide. SDG5 is concerned with good health and well-being. Medicinal chemistry and green chemistry will help achieve that goal.

ATL R2.1D

Organize and manage a discussion with a small group of your peers based on the following questions:

- What are the roles of chemists in the chemical industry?
- What are examples of products from the chemical industry?
- What do you think or visualize when you hear or see the term, 'green chemistry'?
- What is environmental science?
- What global environmental problems do you think our planet has?
- How do you think the global community will address and solve those problems?

Make a summary of the discussion using presentation software (such as PowerPoint) to present to a different group.

Nature of science: Evidence

Evidence for anthropogenic climate change

The clearest evidence for surface warming is from wide-ranging thermometric records (Figure R2.5), some of which extend back to the late 19th century. Temperatures over both the land and ocean surface are now monitored at a large number of locations. Indirect estimates of temperature change from sources such as tree rings and ice cores help to place recent temperature changes in the context of the past, in terms of the average surface temperature of Earth. These indirect estimates show that 1989 to 2019 was the warmest 30-year period in more than 800 years.

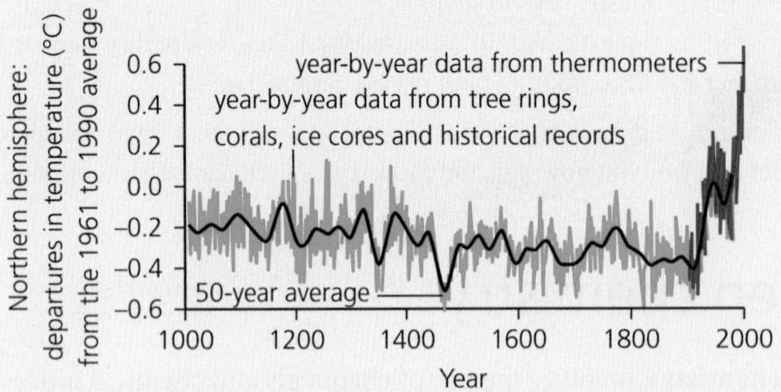

■ **Figure R2.5** Variations in the Earth's surface temperature over time

A wide range of other observations give a more comprehensive picture of warming throughout the climate system. For example, the lower atmosphere and the upper layers of the ocean have also warmed, snow and ice cover are decreasing in the Northern hemisphere, the Greenland ice sheet is shrinking and sea level is rising. These measurements are made with a variety of land-, ocean- and space-based monitoring systems, which increases confidence in the reality of global-scale warming of Earth's climate.

■ Climate change and the United Nations

The Intergovernmental Panel on Climate Change (IPCC) was established by the United Nations in 1988 to study anthropogenic (human caused) climate change, its likely impacts and possible solutions.

The IPCC is a collaborative endeavour involving a large number of scientists who assess the peer-reviewed literature on climate change and prepare reports which are then reviewed by governments. The IPCC publishes a summary report that sets out its recommendations, which is used by the UN Framework Convention on Climate Change (UNFCCC).

ATL R2.1E

If carbon dioxide is accepted to be the main cause of global warming and hence climate change, are there any viable methods to remove carbon dioxide from the air after combustion of fossil fuels? Research the various carbon capture and storage techniques, one of which is shown in Figure R2.6.

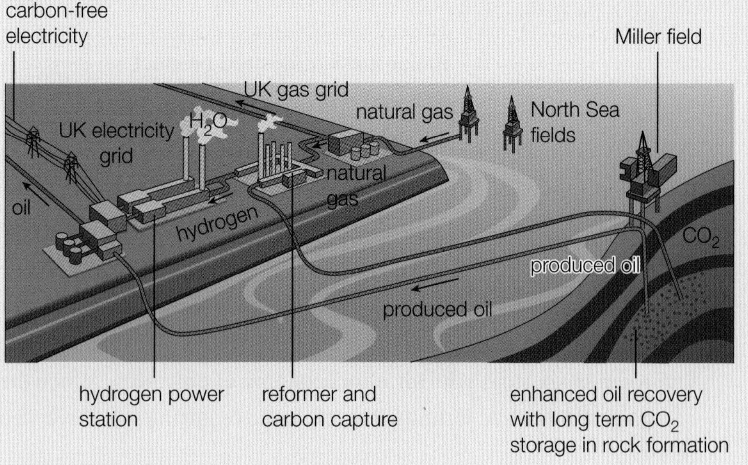

■ **Figure R2.6** Carbon removal techniques

Classify each method as biological or chemical and evaluate their cost and their potential ability to help reduce global warming and ocean acidity.

Link

Ocean acidity is covered in more detail in Chapter S3.1, page 245.

TOK

How might we, as members of the public, judge whether to accept scientific findings if we do not have detailed scientific knowledge?

When Greta Thunberg, the young Swedish climate activist, testified in the US Congress, submitting as her testimony the IPCC 1.5 °C report, she was asked by one member why we should trust the science. She simply replied, 'because it's science'.

Many members of the public lack the scientific knowledge and skills necessary to fully understand or participate in the work of the scientific community as it researches the causes and effects of climate change. They might not be aware of the nature of scientific knowledge and highlight the fact that 'even' scientists 'admit' that *all* scientific knowledge is tentative. However, despite the fact that scientific claims are, by their nature, always open to falsification, there is virtually no disagreement among experts within the scientific community about the dangers posed by climate change (although some scientists do dispute the conclusions of the majority with regard to anthropogenic climate change). The reliability of the claims made by these experts is the result of the use of various scientific methods, followed by collective peer evaluation; such methods and evaluation are the most critical aspects of generating reliable scientific knowledge. Modern society relies on trust in a whole range of 'experts' – not just professional scientists. Even if we accept that experts in any field are sometimes mistaken, they are certainly more reliable than non-experts.

Atom economy

The concept of percentage yield is useful but, from a green chemistry perspective, it is only one factor to be considered. This is because the percentage yield is calculated by considering only the desired product and one (the limiting) reactant. But many atoms in the reactants do not end up in the desired product. This can lead to waste and pollution.

For example, when calcium carbonate (limestone) is decomposed at high temperature to produce calcium oxide (quicklime), part of the calcium carbonate is, in effect, lost to the atmosphere as carbon dioxide.

◆ **Atom economy**: Ratio of the total mass of atoms in the desired product to the total mass of atoms in the reactants.

ATL R2.1F

You can explore the efficiency of different ways of synthesizing ibuprofen in more detail by working through the questions from the RSC activity at www.ch.ic.ac.uk/marshall/4l6/Ibuprofen2.pdf. (Some questions require a knowledge of organic chemistry.)

One of the key principles of green chemistry is that processes should be designed so that the maximum amount of all the raw materials ends up in the product and a minimum amount of waste is produced. The concept of **atom economy** can be used to calculate the overall efficiency of a chemical process. The atom economy of a reaction is the molar mass of the desired product expressed as a percentage of the sum of the molar masses of all the reactants as shown in the equation for the reaction.

$$\text{atom economy} = \frac{\text{molar mass of desired product}}{\text{sum of molar masses of all reactants}} \times 100$$

A higher atom economy means more of the reactant atoms are incorporated into the final useful product or products. The higher the atom economy of a reaction, the less waste is produced. Efficient processes with high atom economy are important for sustainable development as they conserve natural resources and create less waste.

The production of ibuprofen, a drug which reduces swelling and pain, is an excellent example of the application of atom economy. In the 1960s, the British pharmaceutical company Boots made ibuprofen in six steps with an atom economy of only 40%. When the patent expired, another company synthesized it through a new route requiring just three steps and with an atom economy of 99%.

TOK

In what ways have developments in science challenged long-held ethical values?

Science has played a role in changing some long-held ethical values. For example, the invention of mechanical ventilators and use of synthetic drugs, such as midazolam (which is used for sedation), changed ethical values associated with death and dying.

Ventilators and other life support systems allow doctors to keep a person 'alive' by ensuring their blood keeps receiving oxygen and continues to flow around the body, even if their brain has stopped functioning. However, recovery is not possible once the brain is no longer active, so, in many places where this technology is available, people are only considered to be dead once they are 'brain dead'. But reaching this new definition of death required the resolution of new ethical questions, and the technology has also raised new ones. Is it ethically acceptable to switch off the ventilation machine after brain death? Would this be equivalent to killing someone? Should we keep people artificially alive in order to harvest organs for the purposes of donation?

Top tip!

The atom economy can be improved by finding a use for the 'waste' products. An example of this is that carbon dioxide, a by-product of the petrochemical industry, can be used in fizzy drinks or as a substitute for hydrochloric or sulfuric acids in adjusting the pH of waste water.

WORKED EXAMPLE R2.1K

Ethanol can be produced by the following two reactions:

$$C_2H_4 + H_2O \rightarrow C_2H_5OH$$

$$C_2H_5Br + KOH \rightarrow C_2H_5OH + KBr$$

Without calculation, explain which reaction has the higher atom economy.

Answer

Hydration of ethene (by steam) has a higher atom economy (of 100%) because all of the reactants are converted into the desired product, whereas the substitution of bromoethane produces potassium bromide as a waste product, especially given that atom economy is defined in terms of mass (bromine and even potassium have high values of relative atomic mass), not number of atoms.

Top tip!

You need to write balanced equations for a reaction before you can calculate the atom economy.

WORKED EXAMPLE R2.1L

Calculate the atom economy of the reaction between copper(II) oxide and sulfuric acid to form copper(II) sulfate (desired product) and water.

Answer

$CuO(s) + H_2SO_4(aq) \rightarrow CuSO_4(aq) + H_2O(l)$

Mass of starting atoms is

$CuO = (63.55 + 16.00)\,g = 79.55\,g$

$H_2SO_4 = (2.02 + 32.07 + 64.00)\,g = 98.09\,g$

Total = 177.64 g

Mass of desired product is

$CuSO_4 = (63.55 + 32.07 + 64.00)\,g = 159.62\,g$

$$\text{atom economy} = \frac{159.62}{177.64} \times 100 = 89.9\%$$

A reaction with a low atom economy can nevertheless have a high percentage yield. Both the percentage yield and the atom economy have to be taken into account when designing a green chemical process, but there are other factors to be considered as well.

Top tip!

It is important not to confuse percentage yield and atom economy. The percentage yield is about how much product is formed compared to the maximum amount that could be formed theoretically. Atom economy measures what percentage of the mass of the reactants theoretically ends up in the desired product.

ATL R2.1G

What are the twelve principles of green chemistry? Using software such as PowerPoint, create a presentation describing these principles and outlining a relevant example for each one.

13 Calculate the atom economy of the hydrogen produced in the reaction between carbon and steam to form carbon dioxide and hydrogen.

14 Calculate the atom economy of the titanium product in the reaction between titanium(IV) chloride and magnesium to form titanium and magnesium chloride.

Going further

Catalysts and atom economy

Sometimes a catalyst is used in place of a stoichiometric reagent. A catalyst is (strictly speaking) not a reagent, so its effect on the atom economy is neutral. The alternative stoichiometric reagent is a reactant, so it will reduce the atom economy. This leads to the conclusion that a catalytic method offers a better atom economy than an alternative, non-catalytic method.

From a slightly different perspective, atom economy is one aspect of measuring the amount of waste that comes out of a reaction. A more relevant quantity is perhaps the ratio of the mass of the desired product(s) to the total mass of all of the reaction inputs (including reactants, catalysts, solvents, purification materials, etc). Since catalysts can be used in smaller amounts than stoichiometric reagents, a catalyzed reaction will involve a smaller mass of reaction inputs, which increases that ratio.

LINKING QUESTION

The atom economy and the percentage yield both give important information about the 'efficiency' of a chemical process. What other factors should be considered in this assessment?

ATL R2.1H

In 2005, the Nobel Prize in Chemistry was jointly awarded to three chemists for the discovery of a catalytic chemical process called metathesis which has broad uses in the chemical industry. Using presentation software, such as PowerPoint, outline this research and show how it is relevant to green chemistry.

Going further

Synthesis of shikimic acid

Enzyme-catalyzed biochemical reactions are highly selective and efficient, and proceed in aqueous solution under relatively mild conditions of temperature and pH. Many pharmaceutical drugs or synthetic intermediates (primary precursors) can be produced from renewable materials by genetically modified organisms. One such intermediate, shikimic acid (Figure R2.7), is a precursor to the antiviral drug for treating flu known as oseltamivir (Tamiflu).

For many years shikimic acid was extracted from the Chinese star anise plant using solvent extraction with hot water and column chromatography. Modern biosynthetic technologies allow shikimic acid to be produced on an industrial scale by genetically modified *E. coli* bacteria.

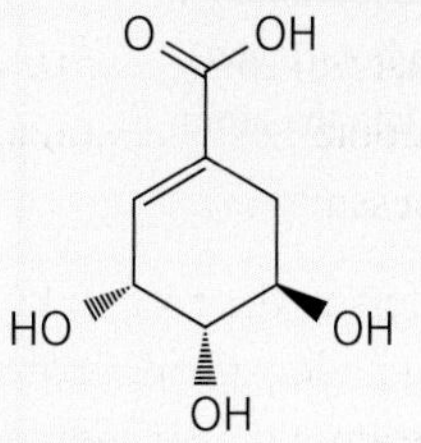

■ **Figure R2.7** Structure of shikimic acid

ATL R2.1I

Work with a small group of classmates. Divide yourselves into two teams and debate the issues outlined below referring to the principles of green chemistry. Summarize the arguments supporting each opinion.

■ **Table R2.3** Green chemistry: issues for debate

Topic	Opinion	Opinion
A large grant of money can be used either to prevent hazardous waste from entering the environment or to decrease worker exposure to hazardous chemicals.	It should be used to prevent environmental impact.	It should be used to increase occupational safety.
A company can either reduce the amount of packaging material or reduce the amount of waste generated during the product manufacturing process.	The amount of material used to produce packing should be minimized.	The waste generated during the manufacturing process should be minimized.

Going further

Solvents

Many solvents used in traditional organic syntheses are chlorinated, highly toxic and cause ozone depletion. Traditionally solvents are 'disposed' of in two ways: they are recycled through a distillation process or burnt in a waste incineration plant.

The green chemistry approach to solvents may involve the use of supercritical water or carbon dioxide, along with 'ionic liquids' (salts with a low melting point that do not evaporate). Ionic liquids are excellent solvents for many substances and they can also be recycled.

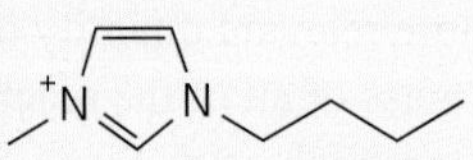

■ **Figure R2.8** Structure of ([BMIM] [PF_6]), an ionic liquid

In some reactions, a well-designed ionic solvent can lead to higher percentage yields alongside milder conditions than those used with traditional solvents. Some reactions that are traditionally carried out in organic solvents can also be carried out in the solid or gas phases.

Tool 1: Experimental techniques

Recognize and address relevant safety, ethical or environmental issues in an investigation

You must minimize the risk of harm to yourself and other people in the chemistry laboratory by performing a risk assessment.

Risk refers to the probability that a chemical might be harmful under specific conditions. Risk is often associated with the amount, concentration (if in solution) and physical state of a chemical. Risks can often be reduced by changing to a less hazardous chemical, using a more dilute solution, changing the design of the apparatus or lowering the temperature and using a safety screen or fume cupboard.

A hazard is a potential source of harm. Risk assessment aims to minimize the risk of a hazard becoming an actual source of harm. For example, exposure of the skin to $2.00\,mol\,dm^{-3}$ sulfuric acid is a hazard and blistering of the skin or damage to the eyes is a potential harm.

Your risk analysis should consider the hazards associated with the materials and chemical substances you plan to use in a practical investigation. These risks are best assessed by reference to the Material Safety Data Sheet (MSDS) appropriate to the chemical(s) in use. These sheets are generally supplied by the chemical manufacturer along with the chemical.

Safety responses to common hazards include:

- irritant – dilute acids and alkalis – wear goggles
- corrosive – stronger acids and alkalis – wear goggles
- flammable – keep away from naked flames
- toxic – wear gloves; avoid skin contact; wash hands after use
- oxidizing – keep away from flammable or easily oxidized materials.

A risk assessment should also consider what actions would need to be taken in the event of potential harm becoming actual harm. For example, acid on skin – thoroughly wash affected area with water to dilute and remove the acid; burn from flame – cool under water for 20 minutes, etc.

Record in your risk assessment procedures to be followed to deal with spillage in the laboratory and correct disposal of waste chemicals (for example, dispose to drain with water dilution; neutralize (if acidic) then dispose to drain with suitable dilution; or use a liquid waste container). Chlorinated and non-chlorinated organic solvents must be disposed of separately.

How fast? The rate of chemical change

Guiding question

- How can the rate of a reaction be controlled?

SYLLABUS CONTENT

By the end of this chapter, you should understand that:

- the rate of reaction is expressed as the change in concentration of a particular reactant / product per unit time
- species react as a result of collisions of sufficient energy and proper orientation
- factors that influence the rate of a reaction include pressure, concentration, surface area, temperature and the presence of a catalyst
- activation energy, E_a, is the minimum energy that colliding particles need for a successful collision leading to a reaction
- catalysts increase the rate of reaction by providing an alternative reaction pathway with lower E_a
- many reactions occur in a series of elementary steps and the slowest step determines the rate of the reaction (HL only)
- energy profiles can be used to show the activation energy and transition state of the rate-determining step in a multistep reaction (HL only)
- the molecularity of an elementary step is the number of reacting particles taking part in that step (HL only)
- rate equations depend on the mechanism of the reaction and can only be determined experimentally (HL only)
- the order of a reaction with respect to a reactant is the exponent to which the concentration of the reactant is raised in the rate equation (HL only)
- the order with respect to a reactant can describe the number of particles taking part in the rate determining step (HL only)
- the overall reaction order is the sum of the orders with respect to each reactant (HL only)
- the rate constant, k, is temperature dependent and its units are determined from the overall order of the reaction (HL only)
- the Arrhenius equation uses the temperature dependence of the rate constant to determine the activation energy (HL only)
- the Arrhenius factor, A, takes into account the frequency of collisions with proper orientations (HL only).

By the end of this chapter you should know how to:

- determine the rates of reactions
- explain the relationship between the kinetic energy of the particles and the temperature in kelvin, and the role of collision geometry
- predict and explain the effects of changing conditions on the rate of a reaction
- construct Maxwell–Boltzmann energy distribution curves to explain the effect of temperature on the probability of successful collisions
- sketch and explain energy profiles with and without catalysts for endothermic and exothermic reactions
- construct Maxwell–Boltzmann energy distribution curves to explain the effect of different values for E_a on the probability of successful collisions
- evaluate proposed reaction mechanisms and recognize reaction intermediates (HL only)
- distinguish between intermediates and transition states and recognize both in energy profiles (HL only)
- construct and interpret energy profiles from kinetic data (HL only)
- interpret the terms 'unimolecular', 'bimolecular' and 'termolecular' (HL only)
- deduce the rate equation for a reaction from experimental data (HL only)

- sketch, identify and analyze graphical representations of zero-, first- and second-order reactions (HL only)
- solve problems involving the rate equation, including the units of k (HL only)
- describe the qualitative relationship between temperature and the rate constant (HL only)
- analyse graphical representations of the Arrhenius equation, including its linear form (HL only)
- determine the activation energy and the Arrhenius factor from experimental data (HL only).

Common mistake

Misconception: *Exothermic reactions occur faster than endothermic reactions.*

There is a wide variety of rates (due to differences in concentration, temperature, surface area and catalysts) in both exothermic and endothermic reactions. An exothermic reaction can also speed itself up when the heat is released and hence temperature increases.

◆ **Collision theory:** A simple model which considers particles to be hard spheres that react with each other when they collide with sufficient kinetic energy.

◆ **Unsuccessful collision:** A collision between two reactant particles without the correct orientation and/or sufficient kinetic energy to overcome the activation energy.

◆ **Successful collision:** A collision between two reactant particles with the correct orientation and with sufficient kinetic energy for the reaction to occur.

◆ **Steric effect:** An effect in which the rate or path of a chemical reaction depends on the size or arrangement of groups in a molecule.

Introduction

Some chemical reactions are fast (for example, ionic precipitation and neutralization) while others are slow (for example, rusting). The rate of a reaction is a measure of how quickly concentrations of reactants or products change with time.

Reaction kinetics is the study of the rates of chemical reactions.

The study of kinetics allows chemists to:

- determine how fast a reaction will take place
- determine the conditions required for a specific reaction rate
- propose a reaction mechanism.

The reaction mechanism describes the chemical reactions behind the changes that can be observed. Reactions take place via elementary steps (bond breaking or bond forming) that happen as the reactants are converted to the products.

Collision theory

The effects of concentration for solutions and pressure for gases, temperature, surface area (for solids and immiscible liquids) and catalysts on rates of reaction are explained using **collision theory**.

Collision theory predicts that in order to react with each other, particles (atoms, ions or molecules) must collide in the correct orientation and with sufficient kinetic energy.

When reactant particles collide they may simply bounce off each other, without reacting. Such an **unsuccessful collision** will take place if the particles do not have enough kinetic energy and/or do not have the correct orientation.

If the reactant particles have the correct orientation and do have enough kinetic energy to react, they will change into product particles when they collide – a **successful collision** (Figure R2.9) results in a chemical reaction.

An example of a **steric effect** is provided by the bromination of ethene (Figure R2.10), where the bromine molecule has to approach the pi bond of the double bond 'sideways on' during electrophilic addition.

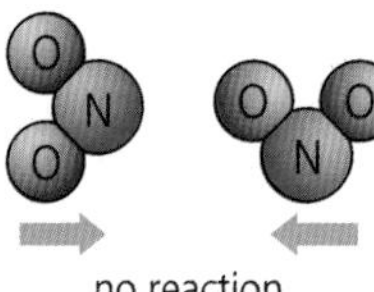

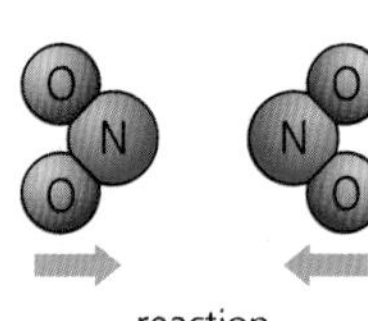

■ **Figure R2.9** Two nitrogen dioxide molecules approaching with sufficient kinetic energy to overcome the activation energy barrier must collide in the correct orientation in order to form dinitrogen tetroxide in the reaction $O_2N(g) + NO_2(g) \rightarrow N_2O_4(g)$

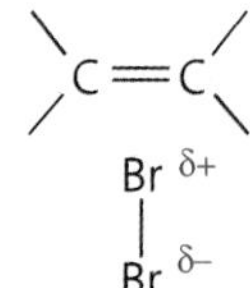

■ **Figure R2.10** A bromine molecule undergoing polarization as it approaches an ethene molecule in the reaction $C_2H_4(g) + Br_2(g) \rightarrow C_2H_4Br_2(g)$

15 The diagram shows some possible orientations for collisions between ethene and hydrogen chloride molecules in the reaction:

$$C_2H_4(g) + HCl(g) \rightarrow C_2H_5Cl(g)$$

Only one of the possibilities shown (collision 1) results in a successful or effective collision. Discuss why this is the case and the reasons for the other collisions being unsuccessful. (You should consider the orientation and polarization of the interacting molecules.)

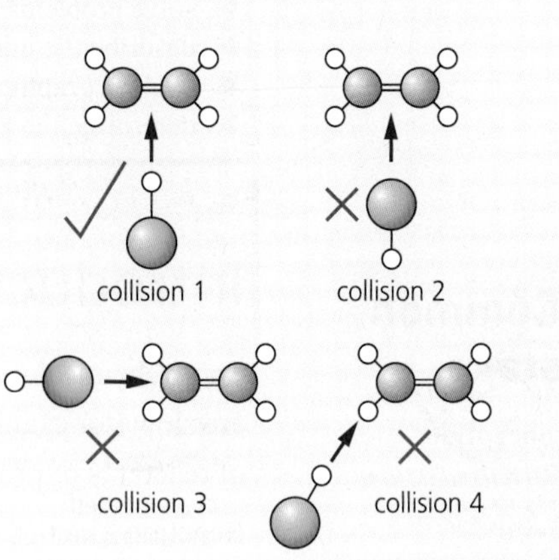

Activation energy

Not all collisions are successful, even if the particles have the correct orientation. For a reaction to occur, repulsion between electron clouds must be overcome and some bonds must be broken, which requires energy.

Each of the colliding particles must be moving fast enough so that when they collide there is enough kinetic energy to allow the reaction to occur.

This fixed amount of kinetic energy that the particles need to overcome an endothermic 'energy barrier' is known as the **activation energy** (E_a).

Fast reactions are associated with low energy barriers (and hence low activation energies) and slow reactions are associated with high energy barriers under standard conditions.

Top tip!

The value of the activation energy determines the temperature sensitivity of the reaction rate. The larger the activation energy, the greater the effect of temperature on reaction rate.

◆ **Activation energy**: The minimum amount of combined total kinetic energy required by a colliding pair of ions, atoms or molecules for a chemical reaction to occur; the energy barrier that has to be overcome to form the transition state.

◆ **Transition state**: The partially bonded, short-lived chemical species of highest potential energy located at the top of the activation energy barrier as a reaction proceeds from reactants to products.

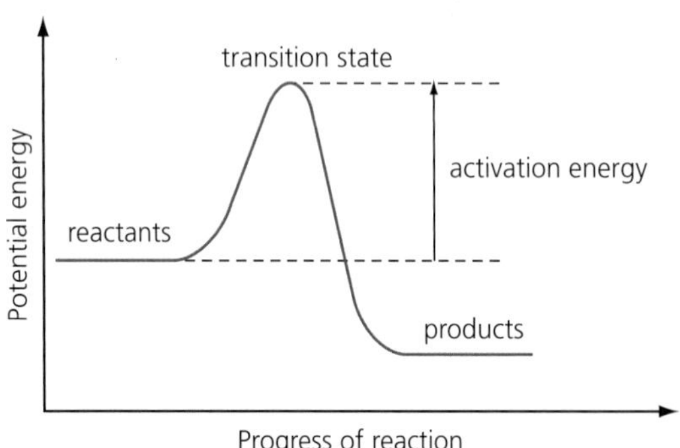

■ **Figure R2.11** Concept of barrier height and activation energy (for the forward reaction of an exothermic reaction)

The **transition state** at the top of the energy profile is a high energy species formed before the product. It cannot be easily isolated and studied because it is unstable.

Common mistake

Misconception: *Activation energy is the (total) amount of energy released in a reaction.*

Activation energy is the minimum kinetic energy needed for the reacting species to form the transition state.

Factors affecting the rate of reaction

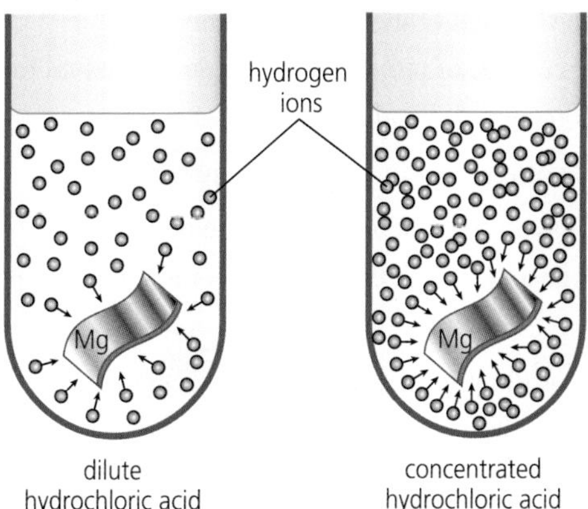

■ **Figure R2.12** The effect of concentration on the collision frequency between magnesium (magnesium atoms on surface) and hydrochloric acid (hydrogen ions)

Concentration

For reactions in solution, an increase in concentration often causes an increase in the reaction rate. In these cases, if the concentration is increased, then the frequency of collisions between reacting particles in solution also increases because the particles are closer together and there are more per unit volume (Figure R2.12).

Pressure

For reactions in which molecules collide and react in the gas phase, an increase in pressure will cause an increase in the rate of reaction. The increase in pressure forces the particles closer together, so there are more particles per unit volume (equivalent to an increase in concentration). This causes an increase in the number of collisions per unit time and hence the rate of reaction.

> **Top tip!**
>
> Since liquids and solids undergo little change in volume when the pressure is increased their reaction rates are little affected by changes in pressure.

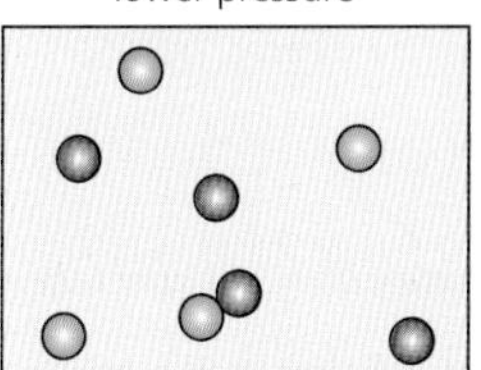

higher pressure

Both reactants are gases

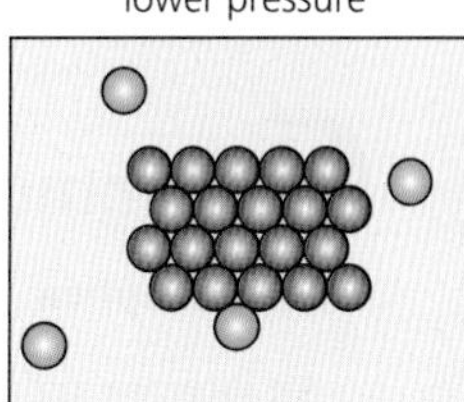

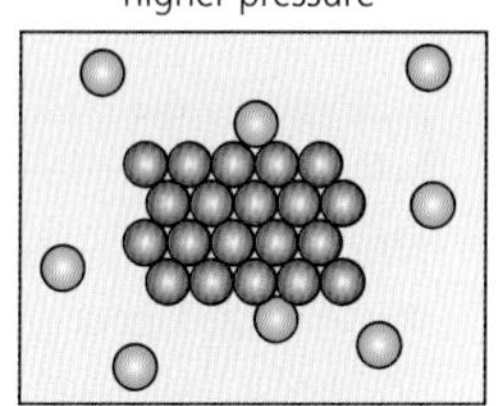

One reactant is a solid

■ **Figure R2.13** The effect of pressure on a reaction involving gas particles (green and pink spheres) and between a gas and a solid

Number of particles

E_a Kinetic energy

■ **Figure R2.14** Maxwell–Boltzmann distribution of kinetic energies in an ideal gas

Temperature

Theoretical calculations and experimental measurements show that kinetic energies of molecules in an ideal gas are distributed over a wide range known as a Maxwell–Boltzmann distribution (Figure R2.14). Similar distributions of kinetic energies are present in the particles of liquids and the particles in solutions.

The total area under the curve is directly proportional to the total number of molecules and the area under any portion of the curve is directly proportional to the number of molecules with kinetic energies in that range.

> **Top tip!**
>
> Temperature is a measure of the average kinetic energy of the particles of a substance.

When the temperature of a gas, liquid or solution is increased, a number of changes occur in the shape of the Maxwell–Boltzmann distribution (Figure R2.15).

- The peak of the curve moves to the right, so the most probable value and the average value of kinetic energy for the molecules increases.

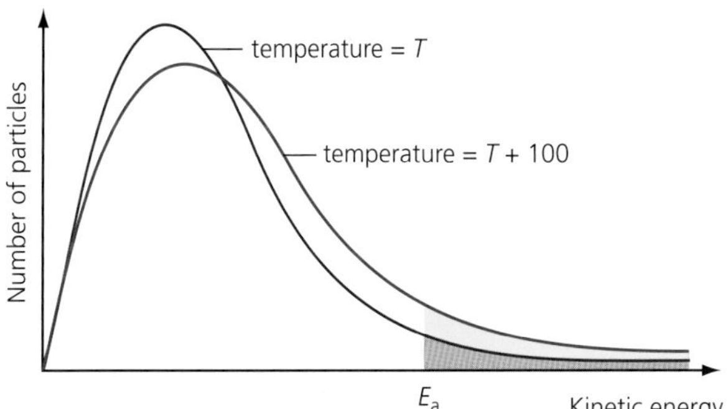

■ **Figure R2.15** Maxwell–Boltzmann distribution of kinetic energies in a solution or gas at two different temperatures

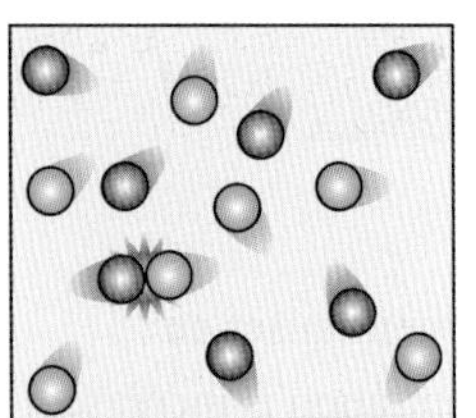

cold – slow movement, fewer collisions per unit time, little kinetic energy

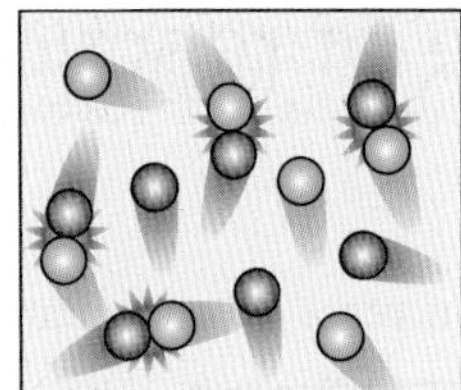

hot – fast movement, more collisions per unit time, more kinetic energy

■ **Figure R2.16** The effect of temperature on gaseous molecules

- The curve flattens, so the total area under it and, therefore, the total number of molecules is unchanged.
- The area under the curve to the right of the activation energy, E_a, increases and hence the percentage of molecules with energies equal to or greater than the activation energy, E_a, also increases.

These changes increase the rate of reaction in two ways (Figure R2.16):

- The molecules on average have a greater velocity, hence they travel a greater distance per unit time and so will be involved in more collisions per unit time.
- A larger proportion of the colliding molecules will have kinetic energies equal to or exceeding the activation energy so a larger proportion of the collisions will be successful.

Top tip!

As a rough rule of thumb, a rise of 10 °C approximately doubles the initial rate of many reactions near room temperature.

LINKING QUESTION

What is the relationship between the kinetic molecular theory and collision theory?

Common mistake

Misconception: *An increase in the initial temperature does not affect the rate of exothermic reactions.*

An increase in the initial temperature raises the rate of almost all reactions (both endothermic and exothermic).

Particle size

Some chemical reactions involve reactants (or a catalyst) that are solids. The reaction takes place on the surface of the solid and, if the solid is porous, inside the pores.

The surface area of a solid is greatly increased if it is broken up into smaller pieces or particles (Figure R2.17). The greater the surface area of a solid reactant, the greater the rate of reaction; there are more particles on the surface that can react and so there are more successful collisions per unit time.

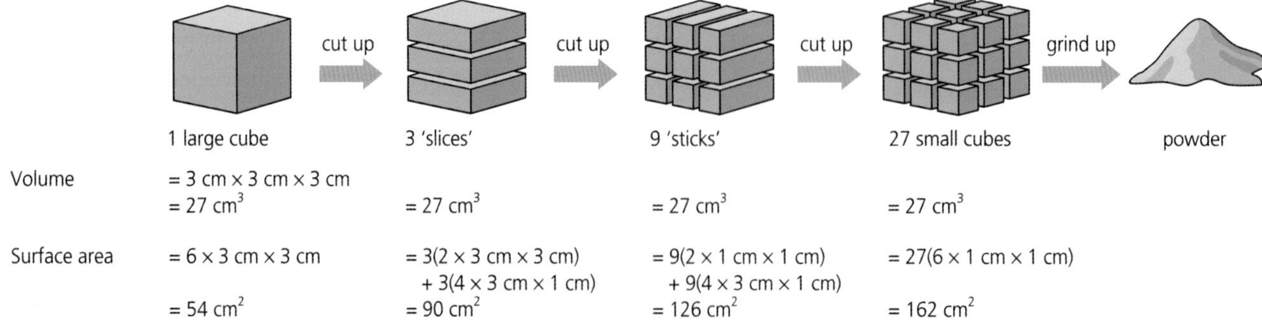

■ **Figure R2.17** The effect of particle size on the surface area of a solid reactant

Top tip!

The effectiveness of solid catalysts is increased significantly if the catalyst is a fine powder and not lumps.

Tool 1: Experimental techniques

Measuring time

Time is nearly always measured using a digital timer. Occasionally, analogue timers are used but this is very unusual. When deciding upon the uncertainty of your measurement, you should recognize that modern timers usually measure to 0.01 (one hundredth) of a second. This level of precision is faster than usual human reaction times which are between 0.15 and 0.30 s. It is therefore acceptable to estimate your own reaction time (there are plenty of experiments to do this available online) and use this as an uncertainty. If the chemical reaction is very fast, you may be able to use a digital camera to measure the speed. You can slow down the video to determine a more exact time than by watching the reaction at normal speed. If timing the formation of a precipitate, you can use a digital light gate.

Going further

Light

The rates of some reactions are greatly increased by exposure to sunlight. When some molecules absorb visible or ultraviolet light, covalent bonds in the molecules are broken and this can initiate a number of chemical reactions. The greater the intensity of light (the greater the number of photons per second) the more reactant molecules are likely to gain the energy required to initiate the process each second and so the greater the rate of reaction.

Mixtures of hydrogen and bromine, or of methane and chlorine, do not react in the dark but a very rapid reaction takes place in the presence of ultraviolet light.

Catalysts

A catalyst is a substance that increases the rate of a chemical reaction but remains chemically unchanged at the end of the reaction. The rate is often directly proportional to the concentration of the catalyst.

A catalyst works by providing a reaction pathway with a lower activation energy (Figure R2.18). This means that a greater proportion of the sample particles have sufficient kinetic energy to react when they collide (Figure R2.19).

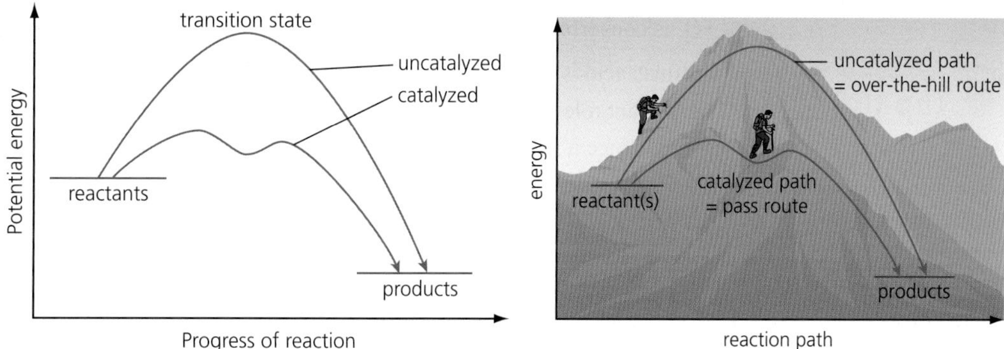

■ **Figure R2.18 a** General enthalpy level diagram for uncatalyzed and catalyzed pathways of an exothermic reaction. **b** The 'mountain pass' analogy for the mechanism of catalytic action showing the idea of the creation of an alternative reaction pathway.

Top tip!

Catalysts, which are often transition metals or their compounds, are important in many industrial processes.

◆ **Homogeneous catalyst**: A catalyst in the same phase as the reactants.

◆ **Heterogeneous catalyst**: A catalyst that is in a different phase from the reactants and provides a surface on which the reaction can occur.

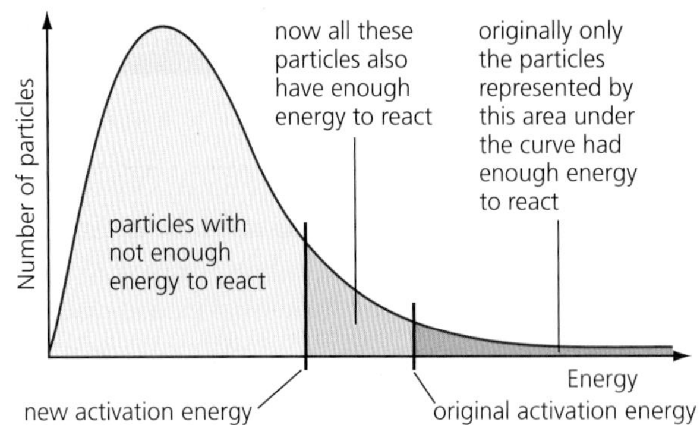

■ **Figure R2.19** Maxwell–Boltzmann distribution of kinetic energies showing the effect of the lower activation energy of the new catalyzed reaction pathway on the proportion of particles with sufficient kinetic energy to react

Homogeneous and heterogeneous catalysts

When a catalyst and the reactants in a catalyzed reaction are in the same phase, the catalyst is described as a **homogeneous catalyst**. For example, a catalyst can be described as homogeneous if it is dissolved in water and the reactants are also present as an aqueous solution.

■ **Figure R2.20** The production of an oxygen-filled foam from the manganese(IV) oxide-catalyzed decomposition of hydrogen peroxide (note that the demonstrator should be wearing a lab coat)

If the catalyst is in a different phase to the reactants, it is described as a **heterogeneous catalyst**.

Aqueous hydrogen peroxide decomposes to water and oxygen. Solid manganese(IV) oxide acts as a heterogeneous catalyst (Figure R2.20):

$$2H_2O_2(aq) \rightarrow 2H_2O(l) + O_2(g)$$

The insoluble manganese(IV) oxide can be filtered off, washed and dried before being reused as a catalyst. The decomposition of hydrogen peroxide can also be catalyzed by the enzyme catalase which can be obtained from finely chopped fresh liver or potato.

■ Sulfuric acid

The critical reaction in the production of sulfuric acid is the V_2O_5-catalyzed contact step in which SO_2 is converted to SO_3.

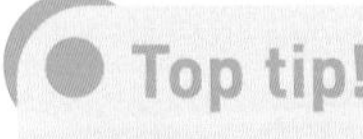

One advantage of heterogeneous catalysts is that they can be separated from the reaction products easily.

Sulfuric acid is the most-produced industrial chemical worldwide as it has many uses and plays a significant role in many chemical processes, including the production of nearly all manufactured goods. Since it is hazardous to store in large quantities, it is only produced as required; this makes it a very sensitive indicator of industrial and economic activity. Studies predict global demand for sulfuric acid will rise significantly by 2040 as a result of more intensive agriculture and a move away from fossil fuels. This trend may prevent green technology advances and threaten global food security.

16 The Haber process, used in industry to make ammonia, involves an exothermic reaction:

$$N_2(g) + 3H_2(g) \rightleftharpoons 2NH_3(g)$$

a Sketch the energy profile for the Haber process in the absence of a catalyst.

b On the same diagram, sketch the energy profile for the process in the presence of a catalyst.

c Label the activation energy on one of the energy profiles.

17 The activation energy for the uncatalyzed decomposition of ammonia to its elements is $+335\,kJ\,mol^{-1}$. The enthalpy of reaction for this decomposition is $+92\,kJ\,mol^{-1}$.

a Calculate the activation energy for the uncatalyzed formation of ammonia from nitrogen and hydrogen.

b If a catalyst, such as iron or tungsten, is introduced, the activation energy is altered. State and explain how it will change.

18 a Sketch a graph with labelled axes to show the Maxwell–Boltzmann distribution of molecular energies in a sample of ideal gas.

b State what is meant by the activation energy of a reaction.

c Shade an area on your graph to show the proportion of molecules capable of reacting.

d Mark on your graph a possible value for the activation energy for the same reaction in the presence of a catalyst.

e Shade an area on your graph showing the additional number of molecules capable of reacting because of the catalyst.

f Draw on your graph a second curve showing the distribution of kinetic energies for the same sample at a slightly higher temperature.

19 a Explain why gases react together faster at a higher pressure.

b Explain why reactants in solution react faster at higher concentration.

c Explain why finely divided solids react more quickly than lumps of the same mass of solid.

d Explain why raising the temperature increases the rate of reaction.

Going further

Active sites on heterogeneous catalysts

The Haber process is the reaction between hydrogen and nitrogen molecules to form ammonia molecules on the surface of an iron catalyst. A simplified model of the reaction (Figure R2.21) involves the adsorption of the reactant hydrogen and nitrogen molecules onto the surface of the catalyst followed by dissociation into nitrogen and hydrogen atoms.

In the presence of iron atoms (which have empty 3d orbitals), the adsorbed nitrogen and hydrogen atoms combine to form ammonia molecules and the ammonia then undergoes desorption from the surface.

However, the reaction between the molecules only takes place at 'active sites' on the surface of the iron. Due to the high pressure used in the Haber process, there are always more reactant molecules than there are active sites. If reactant molecules already occupy all of the active sites (that is, the catalyst is saturated), an increase in pressure will not increase the rate of a reaction.

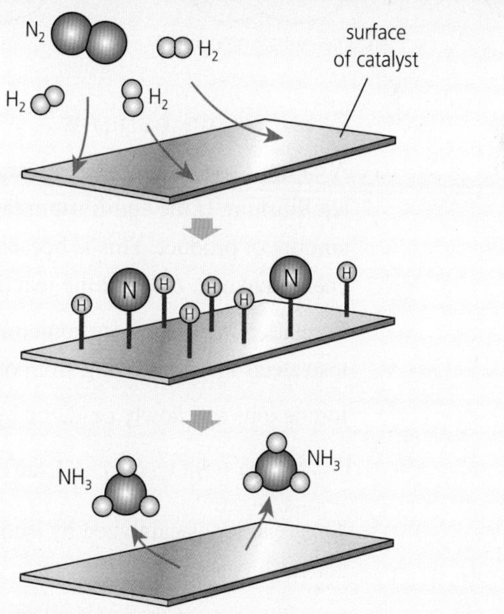

■ **Figure R2.21** Chemisorption of nitrogen and hydrogen molecules, dissociation into atoms and the subsequent reaction on the surface of the iron catalyst in the Haber process

Enzymes

◆ **Active site:** The region on the surface of an enzyme molecule that binds the substrate molecule.

◆ **Substrate:** A substance acted upon by an enzyme at its active site.

Top tip!

Remember that the active site on an enzyme molecule is actually three-dimensional and hence more complex than most diagrams showing the lock and key mechanism suggest.

Enzymes are biological catalysts that work in aqueous solution. They are protein molecules that enable biochemical reactions in living organisms to occur at a fast rate at relatively low temperatures.

Enzyme catalysis has the following specific features:

- Enzymes are more efficient than inorganic catalysts; the reaction rate is often increased by a factor of 10^6 to 10^{12}.
- Enzymes are very specific; they usually only catalyze one particular reaction.
- Enzymes generally catalyze reactions under very mild conditions (for example, 35 °C, pH 7, atmospheric pressure).
- The amount of enzyme present in a cell can be controlled.
- The activity of an enzyme can be controlled by inhibitors which slow the catalysis.

The specific substance that fits and binds into the **active site** of the enzyme molecule and is converted to product(s) is called a **substrate**. The active site is where the catalysis occurs.

In the lock and key model of enzyme activity (Figure R2.22), the products fit the active site less well, and are less strongly bound, so they are released from the enzyme, which is then free to react with another substrate molecule.

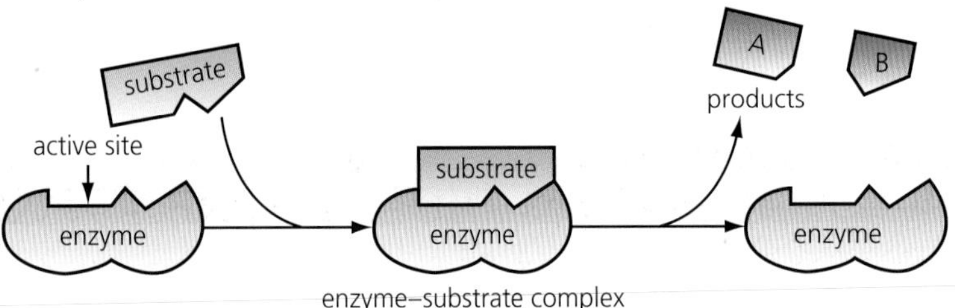

■ **Figure R2.22** The lock and key model of enzyme catalysis

At high temperatures or very low or high values of pH, the enzyme is denatured. It loses its precise three-dimensional shape, and the deformed active site is then unable to catalyze biochemical reactions.

HL ONLY

Catalytic mechanisms

A catalyst speeds up both forward and reverse reactions, but does not affect the position of the equilibrium. If the equilibrium favours the reactants, there will still not be any increase in the amount of product. This is because catalysts do not alter the enthalpy change, $\Delta H^{\ominus}$, or the Gibbs energy change, $\Delta G^{\ominus}$, of the reaction.

Transition metal ions are sometimes effective as homogeneous catalysts because they can gain and lose electrons, converting from one oxidation state to another.

Iodide ions are slowly oxidized by peroxodisulfate(VII) ions:

$$2I^-(aq) + S_2O_8^{2-}(aq) \rightarrow I_2(aq) + 2SO_4^{2-}(aq) \text{ (overall reaction)}$$

This reaction is catalyzed by iron(II) ions which provide an alternative pathway with a lower activation energy:

$$2Fe^{2+}(aq) + S_2O_8^{2-}(aq) \rightarrow 2Fe^{3+}(aq) + 2SO_4^{2-}(aq) \text{ (elementary step)}$$

$$2Fe^{3+}(aq) + 2I^- \rightarrow 2Fe^{2+}(aq) + I_2(aq) \text{ (elementary step)}$$

Note that the Fe^{2+} and Fe^{3+} ions cancel out of these two elementary steps to give the overall equation above (involving iodide and peroxodisulfate(VI) ions), indicating that the Fe^{2+} ions come through the reaction chemically unchanged, as expected for a catalyst.

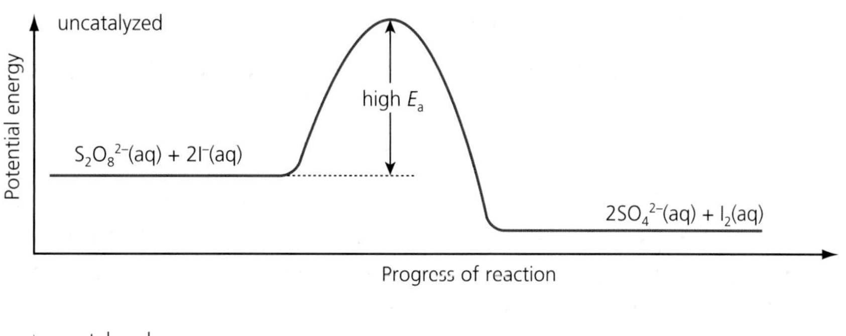

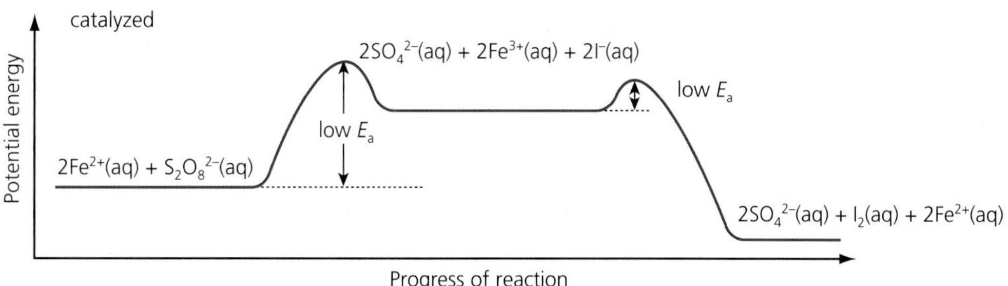

■ **Figure R2.23** Energy level diagrams for the uncatalyzed and catalyzed oxidation of iodide ions by peroxodisulfate ions

◆ **Intermediate**: A species which is neither a reactant nor a product (and so does not appear in the stoichiometric equation) but which is formed during one step of a reaction, then used up in the next step.

Homogeneous catalysis usually involves the formation of an **intermediate** species which is then depleted, forming the products and releasing the catalyst (Figure R2.24) to initiate further reactions. An intermediate is a long-lived species (usually a molecule or ion) of low chemical potential energy (a minimum on the energy profile diagram) that can be experimentally detected and studied, and sometimes even isolated. An intermediate is not the same as the transition state which is a species of high chemical potential energy (a maximum on the energy profile diagram) that is neither a reactant nor a product, but resembles the structure of both to some extent. Transition states cannot be isolated or easily studied due to their very short lifetime.

Top tip!

Transition states have partially formed bonds. Intermediates have fully formed bonds.

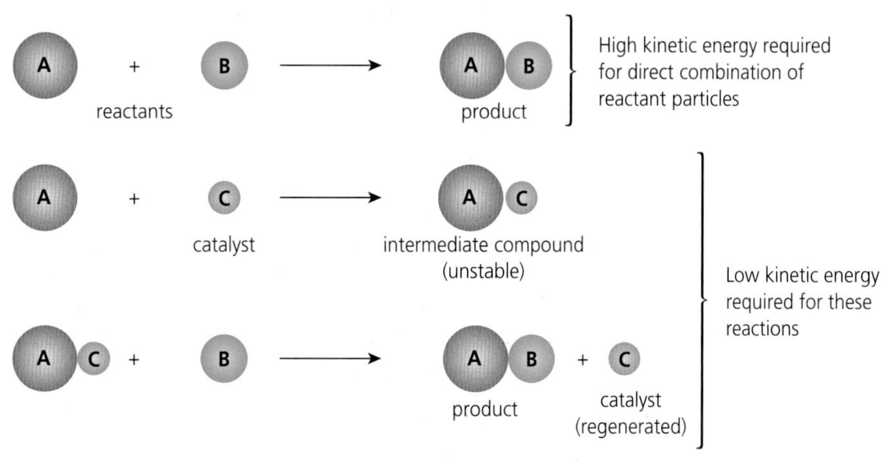

■ **Figure R2.24** The principle of homogeneous catalysis

LINKING QUESTION

What are the features of transition elements that make them useful as catalysts?

The relative effect of factors affecting rates of reaction

Most factors affecting the rate of a simple (first-order) reaction usually have a proportional effect on the rate of reaction.

For example, if the concentration of reactants in solution, the pressure of reacting gases or the surface area of a solid reactant is doubled, the reaction rate *usually* doubles. This is because if the factor is doubled, the number of reactant particles available for collisions doubles and so the frequency of successful collisions will double (Figure R2.25).

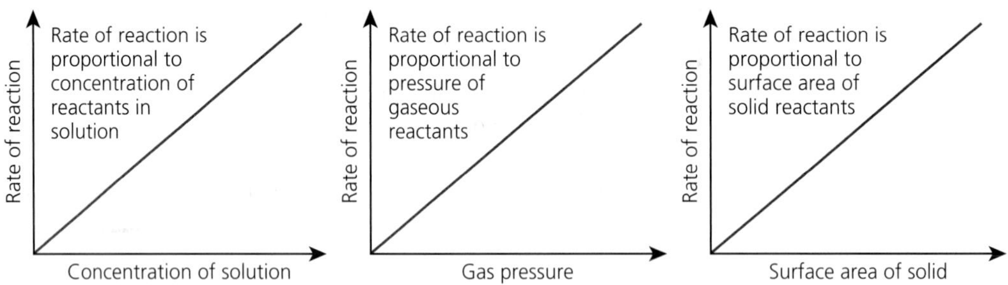

■ **Figure R2.25** Proportional relationships between concentration, gas pressure and surface area of solid and rate of a simple (first-order) reaction

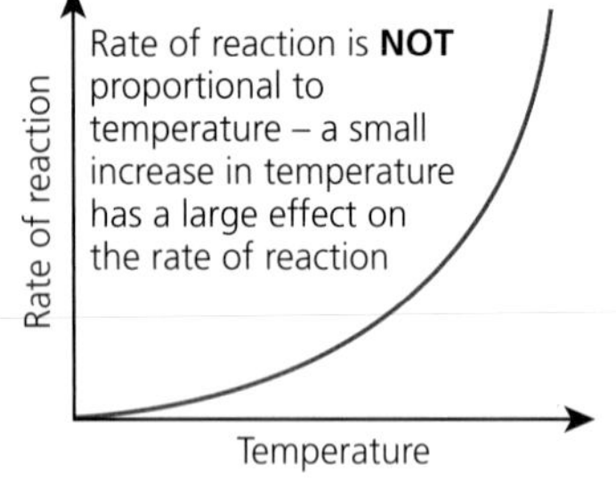

■ **Figure R2.26** Exponential relationship between temperature and rate of reaction

However, changes in temperature have a much greater effect and there is an exponential relationship between temperature and rate (Figure R2.26). This is because a small increase in temperature leads to many more particles having higher values of kinetic energy (see Figure R2.15, page 433).

Top tip!

Remember that changing pressure only affects the rate of a reaction if one or more of the reactants are gases.

Table R2.4 summarizes the effects of temperature, concentration, light and particle size on reaction rate.

■ **Table R2.4** Summary of the factors affecting rate of reaction

Factor	Reaction	Change made in conditions	Usual effect on the initial rate of reaction
temperature	all	increase	increase
		increase by 10 K (~3% increase at room temperature)	approximately doubles
concentration/gas pressure	reactions with aqueous reactants or gaseous reactants	increase	usually increases (unless zero order)
		doubling of concentration of one of the reactants	usually exactly doubles (if first order)
light	generally those involving reactions of mixtures of gases, including the halogens	reaction in sunlight or ultraviolet light	potentially very large increase
particle size	reactions involving solids and liquids, solids and gases or mixtures of solids	powdering the solid, resulting in a large increase of surface area	potentially very large increase

Tool 3: Mathematics

Determine rates of change from tabulated data

To determine whether rates of change are constant or variable, calculate the difference between consecutive data points.

■ **Table R2.5** Variation of y with x

Variable x	0.0	1.0	2.0	3.0	4.0
Variable y	0.0	3.0	5.0	8.0	10.0

For variable x the difference between consecutive data points is: +1.0, +1.0, +1.0 and +1.0.
For variable y the difference between consecutive data points is: +3.0, +2.0, +3.0 and +2.0.

To determine the rate of change, calculate the ratio of each change in y to the corresponding change in x: 3.0, 2.0, 3.0 and 2.0. In this case, the rate of change is variable.

An average rate of change can be calculated for the range of data: $\frac{10}{4} = 2.5$.

Nature of science: Experiments

Molecular beams are used to find out what happens during a molecular collision. A molecular beam consists of a stream of gas molecules moving with the same velocity (same speed and direction).

A molecular beam may be directed at a gaseous sample or into the path of a second beam consisting of molecules of a second reactant. If the molecules react when the beams collide, the experimenters can then detect the products of the collision and the direction in which the products emerge.

Researchers also use spectroscopic techniques to determine the vibrational and rotational energy of the products.

By repeating the experiment with molecules having different speeds and different states of rotational or vibrational energy, chemists can learn more about the collision itself.

LINKING QUESTION

What variables must be controlled in studying the effect of a factor on the rate of a reaction?

Calculating reaction rates

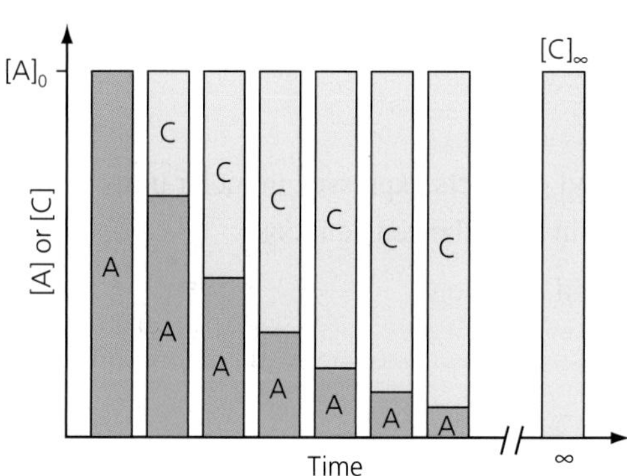

■ **Figure R2.27** The relationship between reactant and product concentrations

Reactions that are fast have a high rate of reaction and are complete after a short time. Reactions that are slow have a low rate of reaction and take a long time to reach completion. (In both cases, this assumes that the value of the equilibrium constant is high, that is, the reaction is non-reversible.)

Figure R2.27 shows a graphical method of visualizing how reactant and product concentrations change with time for the essentially irreversible reaction A → C. [A] represents reactant concentrations and [C] represents product concentrations.

There is a 1 : 1 molar ratio between the amount of A used up and the amount of C formed when the reaction has reached completion. $[C]_\infty$ represents the concentration of the product after sufficient time has passed for the reaction to have gone to completion and stopped.

The average rate of a reaction is defined as:

$$\text{rate} = \frac{\text{change in concentration of a reactant or product}}{\text{change in time}}$$

or

$$\text{rate} = \frac{\text{concentration at time } t_2 - \text{concentration at time } t_1}{\text{time } t_2 - \text{time } t_1}$$

The SI units of rate are moles per cubic decimetre per second, $\text{mol dm}^{-3}\text{s}^{-1}$.

Chemists can follow the progress of a reaction experimentally by measuring the disappearance of a reactant or the appearance of a product.

If we follow the disappearance of reactant A, then average rate $= -\frac{\Delta[A]}{t_2 - t_1}$, where [A] represents the concentration or amount of a reactant, t_2 is greater than t_1 and Δ, the Greek letter delta, indicates a change in quantity.

Reactant A is being used up so $\Delta[A]$ is negative. The negative sign in front therefore gives rise to a positive reaction rate.

WORKED EXAMPLE R2.2A

During a reaction, 0.040 mol of a substance is produced in a $2.5\,\text{dm}^3$ vessel in 20 seconds. Calculate the average rate of reaction in units of $\text{mol dm}^{-3}\text{s}^{-1}$.

Answer

Determine the concentration produced in $1.0\,\text{dm}^3$:

$$\text{concentration} = \frac{0.04\,\text{mol}}{2.5\,\text{dm}^3} = 0.016\,\text{mol dm}^{-3}$$

Determine the change in concentration per unit time (average reaction rate):

$$\text{rate} = \frac{0.016\,\text{mol}}{20\,\text{s}} = 8.0 \times 10^{-4}\,\text{mol dm}^{-3}\text{s}^{-1}$$

Common mistake

Misconception: *Reaction rate is the time for a reaction to reach completion.*

Average reaction rate is related to the reciprocal of the reaction time, $\frac{1}{\text{time}}$, since a longer reaction time implies a smaller reaction rate. Reaction rate is change in concentration of a reactant or product per unit time.

20 A reaction produces 44.02 g of carbon dioxide in 15.00 s in a vessel of volume $4.00\,\text{dm}^3$. Calculate the average rate of reaction in units of $\text{mol dm}^{-3}\text{s}^{-1}$.

21 Acidified hydrogen peroxide and aqueous potassium iodide react to form iodine. It was found that the concentration of iodine was $0.06\,\text{mol dm}^{-3}$ after allowing the reactants to react for 30 s. Calculate the average rate of formation of iodine during this time, in units of $\text{mol dm}^{-3}\text{s}^{-1}$.

The rates of change in concentrations of all reactants and products expressed in molar units are related to each other via the stoichiometric coefficients in the balanced equation.

Any reaction can be represented by the following general equation:

$$aA + bB \rightarrow cC + dD$$

reactants products

The relative rates of reaction are given by the following expression:

$$\text{rate} = -\frac{1}{a}\frac{\Delta[\text{A}]}{\Delta t} = -\frac{1}{b}\frac{\Delta[\text{B}]}{\Delta t} = +\frac{1}{c}\frac{\Delta[\text{C}]}{\Delta t} = +\frac{1}{d}\frac{\Delta[\text{D}]}{\Delta t}$$

The negative sign accounts for the decreasing concentrations of the reactants A and B, whereas the positive sign accounts for the increasing concentrations of the products C and D with time.

Consider the reaction between acidified hydrogen peroxide and aqueous potassium iodide ions to form iodine and water:

$$2H^+(aq) + H_2O_2(aq) + 2I^-(aq) \rightarrow I_2(aq) + 2H_2O(l)$$

Top tip!

The negative sign in this expression indicates a decrease in the peroxide concentration with time. The negative sign is put there to ensure that a positive rate is obtained since the change in concentration for a reactant is negative.

The rate of disappearance or consumption of hydrogen peroxide will be the same as the rate of appearance of iodine:

$$\text{rate} = \frac{-\Delta[H_2O_2(aq)]}{\Delta t} = \frac{\Delta[I_2(aq)]}{\Delta t}$$

This is because one molecule of iodine is formed for every hydrogen peroxide molecule used up.

The average rate of appearance of water will be twice the average rate of appearance of iodine because two water molecules are formed for every iodine molecule formed:

$$\text{rate} = \frac{\Delta[I_2(aq)]}{\Delta t} = \frac{1}{2}\frac{\Delta[H_2O(l)]}{\Delta t}$$

Figure R2.28 shows a typical graph of the concentration of a reactant against time.

Figure R2.29 shows a graph of the concentration of product against time for the same reaction.

The graphs show that the rate of a reaction is not constant: it usually decreases with time as the concentration of the reactants decreases and that of the products increases. The reaction rate is zero when one or more of the reactants are all used up and the reaction stops (Figure R2.30).

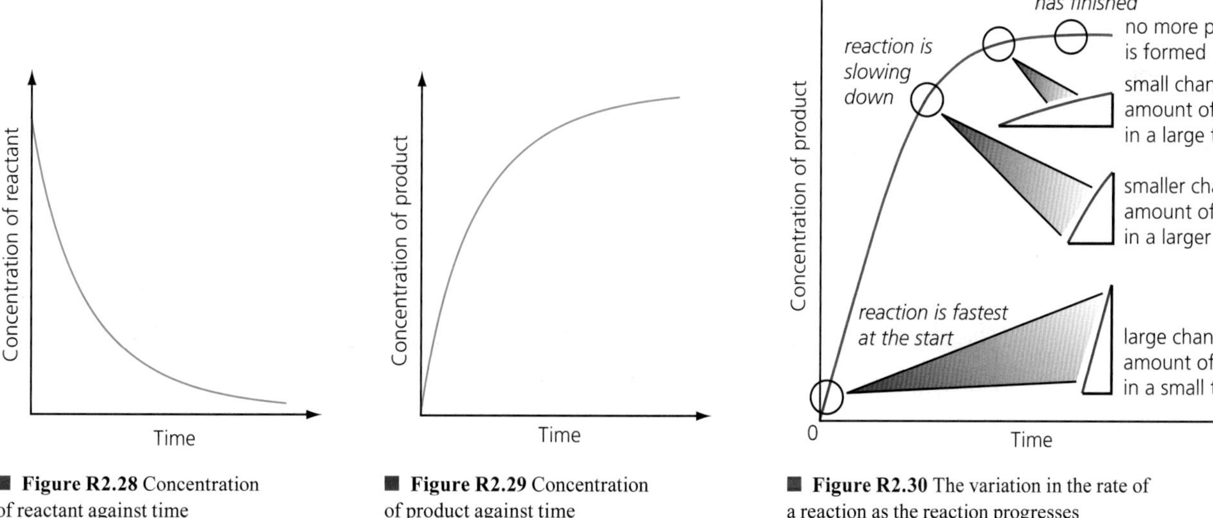

■ **Figure R2.28** Concentration of reactant against time

■ **Figure R2.29** Concentration of product against time

■ **Figure R2.30** The variation in the rate of a reaction as the reaction progresses

◆ **Instantaneous rate:** The rate of a reaction at a given point – the slope of the tangent of a graph of concentration against time at that point.

It is often useful to know the rate of a reaction at a particular time or at a particular concentration of a reactant or product. This **instantaneous rate** of reaction is equal to the gradient (slope) of a graph of product or reactant concentration at the point of interest.

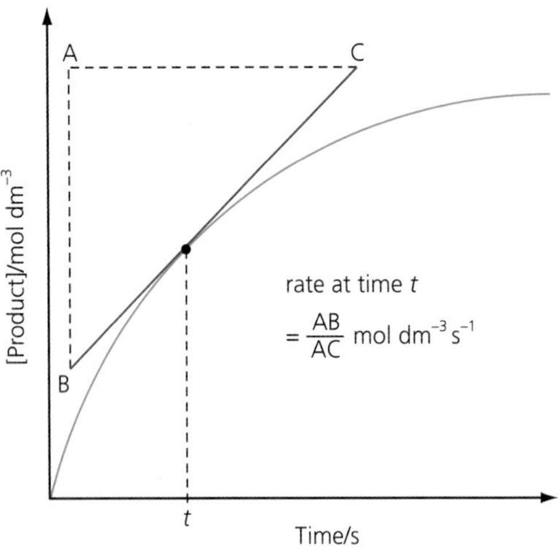

■ **Figure R2.31** The rate of formation of product at time t is the gradient of the product concentration–time curve at time t

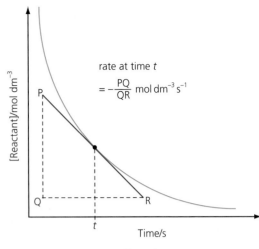

■ **Figure R2.32** The rate of loss of reactant at time t is the absolute value of the gradient of the reactant concentration–time curve at time t

The steeper the gradient of the tangent to the curve, the higher the reaction rate. When the tangent is horizontal, the gradient is zero, the rate of reaction is zero and the reaction has finished (or not begun).

A series of gradients from a concentration–time graph can be used to create a rate–concentration graph (Figure R2.33).

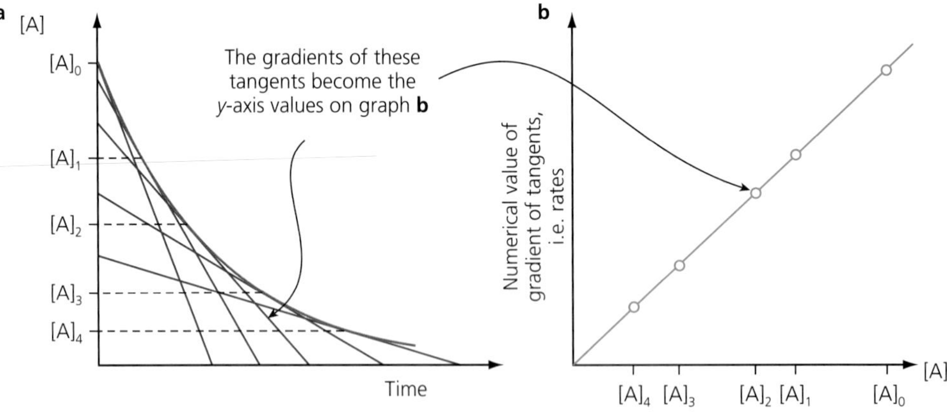

■ **Figure R2.33 a** Concentration versus time graph; **b** rate versus concentration graph

Top tip!

Many reactions show a directly proportional relationship between rate and concentration of reactant. These are known as first-order reactions.

Tool 3: Mathematics

Apply the coefficient of determination (R^2) to evaluate the fit of a trend line or curve.

R-squared (R^2) or the coefficient of determination is a statistical measure with a value between 0 and 1 that can easily be generated on a spreadsheet. The value gives you an indication of how well your line or curve fits the plotted data. The smaller the value, the worse the correlation; 0 means there is no correlation at all and 1 indicates perfect correlation – all your data points lie on the line.

22 The table shows data for the catalyzed decomposition of hydrogen peroxide solution into oxygen gas and water.

Time / min	0	0.5	1.0	1.5	2.0	2.5	3.0	3.5	4.0	4.5	5.0
Volume of oxygen gas / cm^3	0	18	35	48	58	66	72	78	81	84	87
Hydrogen peroxide concentration / $mol\,dm^{-3}$	0.159	0.129	0.101	0.080	0.063	0.050	0.040	0.030	0.025	0.020	0.018

a Plot a line graph of the concentration of hydrogen peroxide solution (y-axis) against time (x-axis).

b Use the graph to determine the instantaneous rate of the reaction at the following concentrations of hydrogen peroxide: $0.16\,mol\,dm^{-3}$, $0.12\,mol\,dm^{-3}$, $0.08\,mol\,dm^{-3}$, $0.04\,mol\,dm^{-3}$.

c Plot a line graph (with a line of best fit) of rate (y-axis) against the concentration of hydrogen peroxide solution (x-axis).

d Evaluate the accuracy of this approach to establishing the relationship between concentration and rate.

Measuring rates of reactions

In order to find out how the rate of reaction changes with time, we need to select a suitable methodology to monitor the progress of a reaction. This method will indirectly measure either the rate of disappearance of a reactant, or the rate of appearance of a product. There are two main approaches: sampling and continuous.

Sampling

This methodology involves taking small samples of the reaction mixture at various times and then chemically analysing each sample to determine the concentration of one of the reactants or products.

This approach can be used to follow the alkaline hydrolysis of a halogenoalkane:

$$R{-}X + OH^- \rightarrow ROH + X^-$$

Samples of the reaction mixture are removed at various times and 'quenched' to stop or slow down the reaction by, for example, cooling the sample in ice. The hydroxide ion concentration can be found by titration with a standard solution of a strong acid.

A sampling method can also be used to measure the rate of alkaline hydrolysis of the ester ethyl ethanoate:

$$CH_3COOC_2H_5(l) + OH^-(aq) \rightarrow CH_3COO^-(aq) + C_2H_5OH(aq)$$

At regular intervals during the reaction, a sample of the reaction mixture is taken and titrated against standardized hydrochloric acid using a suitable indicator (Figure R2.34). This allows the concentration of the sodium hydroxide remaining in the reaction mixture to be determined. The smaller the volume of hydrochloric acid solution required for neutralization, the further the reaction has progressed.

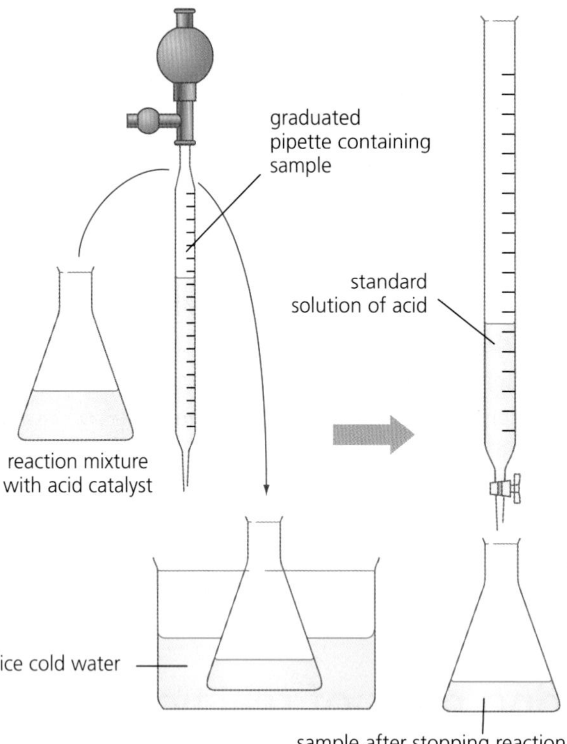

■ **Figure R2.34** Following the course of an acid-catalyzed reaction via sampling and titration with acid

LINKING QUESTION

What experiments measuring reaction rates might use time as **a** a dependent and **b** an independent variable?

Continuous

This approach involves monitoring a physical property of the reaction mixture over a period of time. The use of some of these properties is discussed below.

Inquiry 1: Exploring and designing

Justify the range and quantity of measurements

In a scientific investigation, the range describes the maximum and minimum values of the independent variable studied. It should be sufficient to reveal the relationship with the dependent variable.

Since there are some random uncertainties associated with experimental measurements, you must consider repeating some of the trials. But how much data is sufficient? The answer to this question depends on how reproducible the measurements are and on how small a change in the result of an experiment you want to detect. Obviously, the less precise the measurements (that is, the more random error present) and the smaller the change we are interested in, the more data we must collect and average to have confidence in our results. Averaging several trials under the same conditions is the best approach to ensure reliable results.

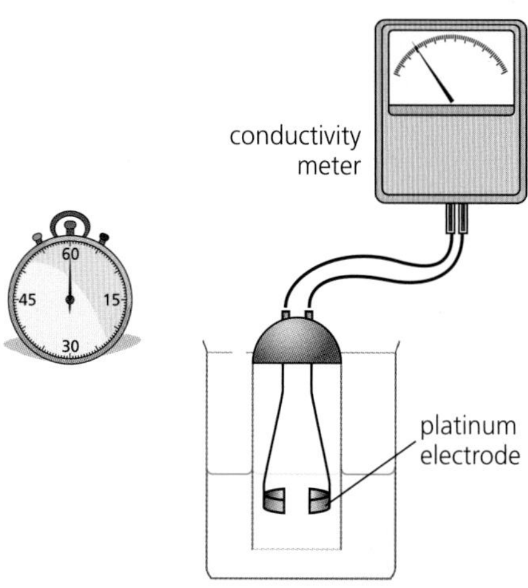

■ **Figure R2.35** A conductivity cell and meter

Conductivity

Changes in the electrical conductivity of a reaction solution can be used to follow reactions such as the hydrolysis of tertiary halogenoalkanes:

$$(CH_3)_3CBr(l) + H_2O(l) \rightarrow (CH_3)_3COH(aq) + H^+(aq) + Br^-(aq)$$

As the reaction proceeds, the electrical conductivity of the solution increases because ions are being formed.

The mobility of ions, and hence the conductivity of their solutions, varies, for example, small hydrogen ions and hydroxide ions have very high mobilities but bromide ions have a low mobility. As a result, this method can sometimes be used even if there are ions on both sides of the equation.

Inquiry 1: Exploring and designing

Calibrate measuring apparatus

Calibration is an important component of analytical chemistry and is the process of evaluating and adjusting the accuracy of equipment. It is carried out to minimize bias in the readings of an instrument but, if the calibration gradually changes (drifts), then precision is also affected.

It is important to calibrate conductivity meters as the cell constants may change with time or due to contamination. A simple way to calibrate a conductivity meter is to measure the conductivity of a standard solution and adjust the reading to match the known value.

A pH meter should be calibrated before use to ensure accurate and reliable measurements because it is possible to have small changes in the electrode output over time. Fresh buffer solutions covering the pH ranges which you will measure should be used to calibrate the meter. Generally, three buffer solutions are required: one below pH 7, one at pH 7 and one above pH 7.

The sensors used in data logging can be checked for systematic error. For example, temperature calibration may involve comparing one data logger against a reference device that is a highly precise instrument regularly checked by an accredited laboratory.

Colour changes

Spectrophotometry and colorimetry, which allow monitoring of colour changes, can be used to, for example, follow the reaction of iodine with aqueous propanone to form iodopropanone and hydroiodic acid:

$$CH_3COCH_3(aq) + I_2(aq) \rightarrow CH_3COCH_2I(aq) + HI(aq)$$

As the reaction proceeds, the colour of the iodine slowly fades.

Top tip!

The reaction between iodine and propanone can also be followed by chemical analysis. At regular time intervals, samples of the acidic reaction mixture are quenched (neutralized) with sodium hydrogencarbonate (a base) and the iodine titrated with sodium thiosulfate using starch as the indicator.

Tool 1: Experimental techniques

Spectrophotometry

Spectrophotometry is the technique of measuring how much light of a particular frequency a chemical substance in solution absorbs by measuring the reduction in the intensity of a beam as it passes through a sample of the solution.

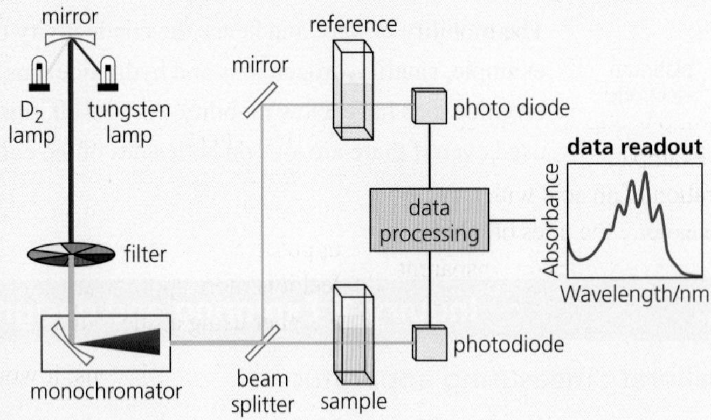

■ **Figure R2.36** A schematic diagram of a UV–Vis spectrophotometer. The combination of the filter and monochromator ensures that only light of a particular wavelength is shone through the sample. A simple colorimeter just has the filter to select the light

The key difference between colorimetry and spectrophotometry is that colorimetry uses fixed wavelengths that are only in the visible range, while spectrophotometry can use wavelengths in a wider range, including ultraviolet.

Spectrophotometry can also be used to measure the concentration of mixtures of coloured species. The spectrophotometer will detect all the species in a solution as long as they absorb light and have sufficiently different spectra: there must be regions in the measured wavelength range where the two spectra do not overlap much.

The Beer–Lambert law applies separately to each species, provided they do not react with each other and the absorbance peaks do not overlap. The Beer–Lambert law states that, for a fixed path length, the absorbance is directly proportional to concentration, provided the solution is dilute.

LINKING QUESTION

How can graphs provide evidence of systematic and random error?

HL ONLY

Optical activity

Some organic molecules rotate the plane of plane-polarized light by an amount which can be measured with a polarimeter. Changes in the concentrations of these optically active molecules cause a change in the total amount of rotation.

For example, sucrose is hydrolysed in acidic solution to form glucose and fructose:

$$C_{12}H_{22}O_{11}(aq) + H_2O(l) \rightarrow C_6H_{12}O_6(aq) + C_6H_{12}O_6(aq)$$

Although both the reactant and the two products are optically active, the sizes and directions in which they rotate plane-polarized light differ.

Link

The structures and properties of enantiomers (optical isomers) is covered in more detail in Chapter S3.2 page 287.

Volume of gas evolved

Reactions that produce gases can be investigated by collecting and measuring the total volume of gas produced in a gas syringe at different times during the reaction. The rate of increase of volume of gas (tangent to the volume–time curve) can be used as a measure of reaction rate as long as the assumption of ideal gas behaviour holds.

Tool 3: Mathematics

Construct and interpret tables, charts and graphs for raw and processed data

When you carry out an investigation you will need to record your raw data. We use a table for this purpose. Tables can hold two types of data: qualitative and quantitative data. The independent variable will be the first column of the table and the dependent variables will be put in subsequent columns.

Qualitative data contains observations, ideally in the form 'before → during → after'. For example, when methyl orange indicator is used in a titration of an acid with a base, qualitative observations would be along the lines of, 'The solution was transparent yellow, it slowly turned transparent orange before turning transparent red.' All investigations should have qualitative observations of some sort. Quantitative data will contain numbers. All quantitative data should be shown to the appropriate number of significant figures and the column headings in the table should state the precision of the instrument used. In both cases, the data table should contain headings and, if collecting quantitative data, units should be included. Tables should also have a title.

Once your raw data is collected, you may need to process it. The results of your processing should also be put in a table, using the same guidelines as above. Ensure you make it clear which table is raw data and which is processed data.

Tables should be clear and easy to follow to allow correct interpretation. Missing units or titles can hamper this process.

Once your data is processed, it will likely go into a suitable graph or chart. If the data is discrete, the appropriate form to record it would be a bar chart. A histogram is very similar to a bar chart but may refer to ranges of data as opposed to discrete pieces of data. Sometimes, you may feel it is more appropriate to plot the points and join them together using a line – this is referred to as a line graph.

If the data is continuous, it would be more appropriate to use a scatter graph with a line (or curve) of best fit as this will allow for the easiest interpretation of trends and patterns. The points are connected and presented either as a line of best fit or a curve of best fit.

In all graphs, the independent variable goes on the *x*-axis and the dependent variable on the *y*-axis. Do not forget to include a description of each variable and the units (where appropriate). Graphs should always have a title.

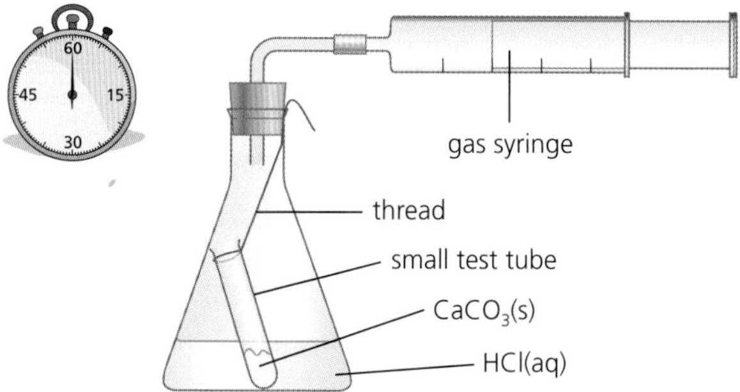

Figure R2.37 Apparatus used to study the rate of a reaction that releases a gas

Figure R2.37 shows apparatus suitable for investigating the reaction between calcium carbonate and dilute hydrochloric acid. This arrangement ensures the two reactants are kept separate while the apparatus is set up so that the start time can be accurately recorded.

When the investigator is ready to start recording measurements, they release the thread to drop the tube containing the calcium carbonate into the acid.

Line II on Figure R2.38 shows typical results. The reaction is fastest (the curve is steepest) at the beginning and the reaction finishes (the curve becomes horizontal) when one or both reactants are used up.

The other line (line I) shows data that may have been obtained from using the same amounts of calcium carbonate and hydrochloric acid at a higher temperature, *or* using acid of a higher concentration (if calcium carbonate is the limiting reactant) *or* using powdered calcium carbonate rather than lumps.

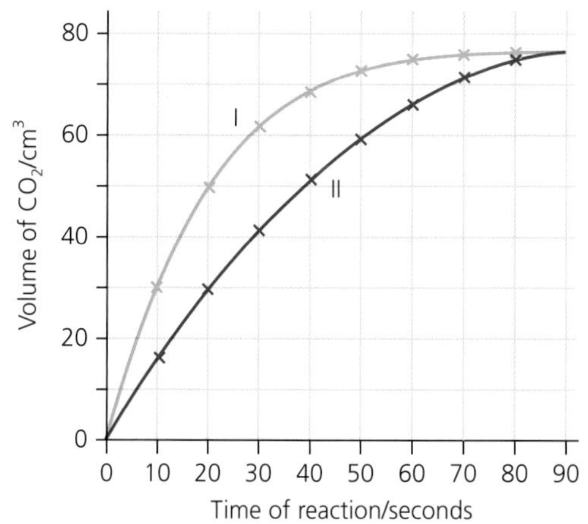

Figure R2.38 A graph of total volume of carbon dioxide collected against time

Mass of gas evolved

Alternatively, a reaction producing a gas can be performed in an open flask placed on an electronic balance. The reaction can then be monitored by recording the total loss of mass as the gas is produced (Figure R2.39).

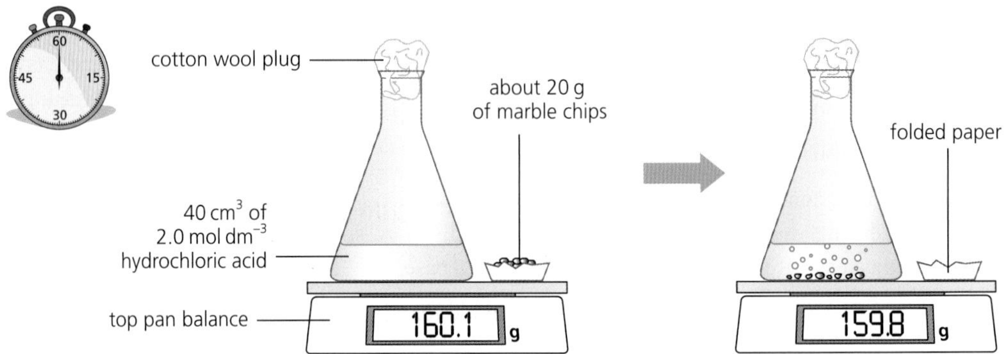

■ **Figure R2.39** Following the rate of reaction via the total loss of mass between marble chips (calcium carbonate) and dilute hydrochloric acid

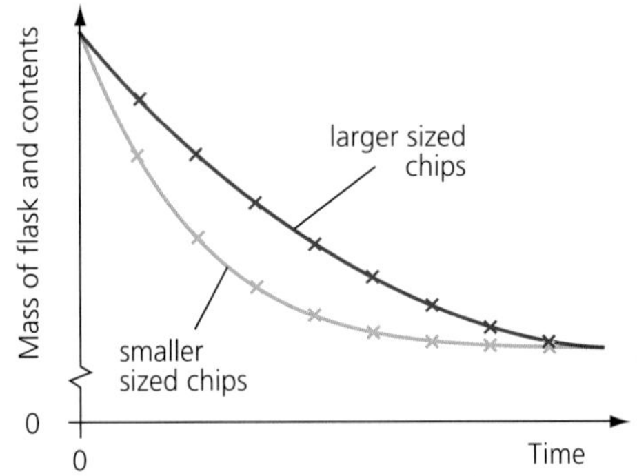

■ **Figure R2.40** The effect of chip size on the mass of flask and contents over time for the reaction between calcium carbonate and hydrochloric acid

Figure R2.40 shows the effect of chip size on a graph of mass of flask and contents versus time. The smaller sized chips have a larger surface area and hence the reaction finishes in a shorter period of time.

Top tip!

This approach is less suitable for gases with small molar masses, such as hydrogen, since the change in mass is smaller and the effect of random uncertainties is greater.

Tool 3: Mathematics

Drawing lines or curves of best fit

A line of best fit is an interpretation of the trend or pattern shown by the data collected. Any data you collect will be imperfect, due to the associated experimental uncertainties, so a line of best fit will not necessarily go through all the data points; it is a way of estimating where a point would lie if there were no uncertainties.

If possible, you should add error bars to your plotted points. The line or curve of best fit should, ideally, pass through all the error bars. Any points where this is not the case may be anomalous and should be reviewed and perhaps discarded. Error bars also allow you to draw minimum and maximum lines of best fit. These are lines with the minimum and maximum gradients possible that pass through all the error bars.

There are two ways of drawing a graph: by hand or by using software such as *Excel* or another spreadsheet application. The advantage of software is that it uses an algorithm to accurately determine the line of best fit; in contrast, if you draw by hand, the line is more of an estimation.

A line or curve of best fit also allows the data to be interpolated or extrapolated.

WORKED EXAMPLE R2.2B

Suggest methods of measuring the rates of the following reactions:

a Zinc powder reacting with excess aqueous copper(II) sulfate:

$$Zn(s) + CuSO_4(aq) \rightarrow ZnSO_4(aq) + Cu(s)$$

b Hydrogen peroxide decomposing to give water and oxygen:

$$2H_2O_2(aq) \rightarrow 2H_2O(l) + O_2(g)$$

c A tertiary halogenoalkane (2-iodo-2-methyl propane) undergoing hydrolysis to form a tertiary alcohol (2-methylpropan-2-ol) and hydroiodic acid:

$$(CH_3)_3C\text{–}I(solv) + H_2O(l) \rightarrow (CH_3)_3COH(aq) + H^+(aq) + I^-(aq)$$

d Hydrogen peroxide reacting with potassium manganate(VII):

$$2MnO_4^-(aq) + 5H_2O_2(aq) + 6H^+(aq) \rightarrow 2Mn^{2+}(aq) + 5O_2(g) + 8H_2O(l)$$

e Methyl methanoate reacting with water in a hydrolysis reaction to form methanoic acid and methanol:

$$HCOOCH_3(aq) + H_2O(l) \rightarrow HCOOH(aq) + CH_3OH(aq)$$

Answer

a Store the solution and allow the zinc to settle, then measure the change in colour of the solution using a colorimeter or spectrophotometer. If the mobilities of copper(II) and zinc ions are significantly different, it may be possible to measure a change in conductivity.

b Measure the total volume of gas released using a gas syringe or measure the decrease in mass using an electronic balance.

c Measure the electrical conductivity of the solution with a conductivity probe. Alternatively, titrate samples with a base at regular time intervals.

d Measure the volume or mass of oxygen released using a gas syringe or electronic balance; monitor absorbance of the purple colour of manganate(VII) ions using a colorimeter or spectrophotometer.

e Sample at intervals and titrate samples with standardized sodium hydroxide; monitor change in pH or conductivity since HCOOH is a weak acid and partially dissociates into ions.

Plot linear and non-linear graphs

When drawing scatter graphs, you will obtain either a straight line or a curve. A straight line may imply a directly proportional relationship – as one value increases, so does the other value, by a proportional amount. For example, as x doubles, y triples or as x doubles, y increases by 2.3 times. An inversely proportional graph will not be a straight line – it will be a curve. Inversely proportional means that as one quantity increases, the other decreases. For example, as x doubles, y is halved or as x doubles, y decreases by a factor of 3. A straight-line graph will have a fixed or unchanging gradient while a curve will have a continually changing gradient.

It should be obvious if you need to draw a line or a curve but if it is not clear you will need to collect more data points or repeat some trials.

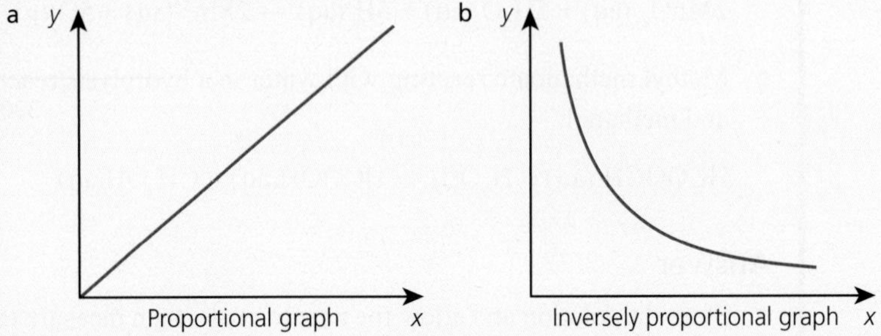

■ **Figure R2.41** Examples of **a** proportional and **b** inversely proportional graphs

During an investigation, you must make sure you collect enough data. Otherwise, it will be very difficult to identify the relationship between variables. Figure R2.42 shows a graph with too few data points.

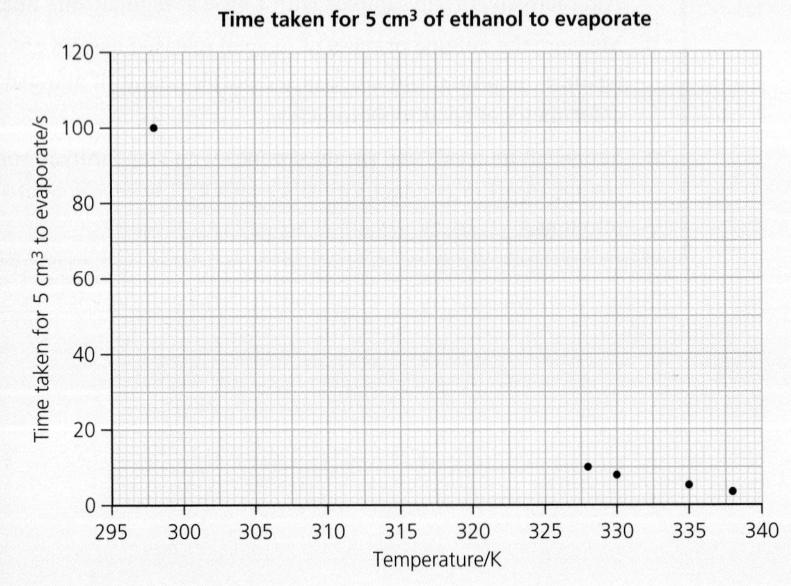

■ **Figure R2.42** A graph with insufficient data points

As shown in Figure R2.43, these points could be interpreted as a line or a curve:

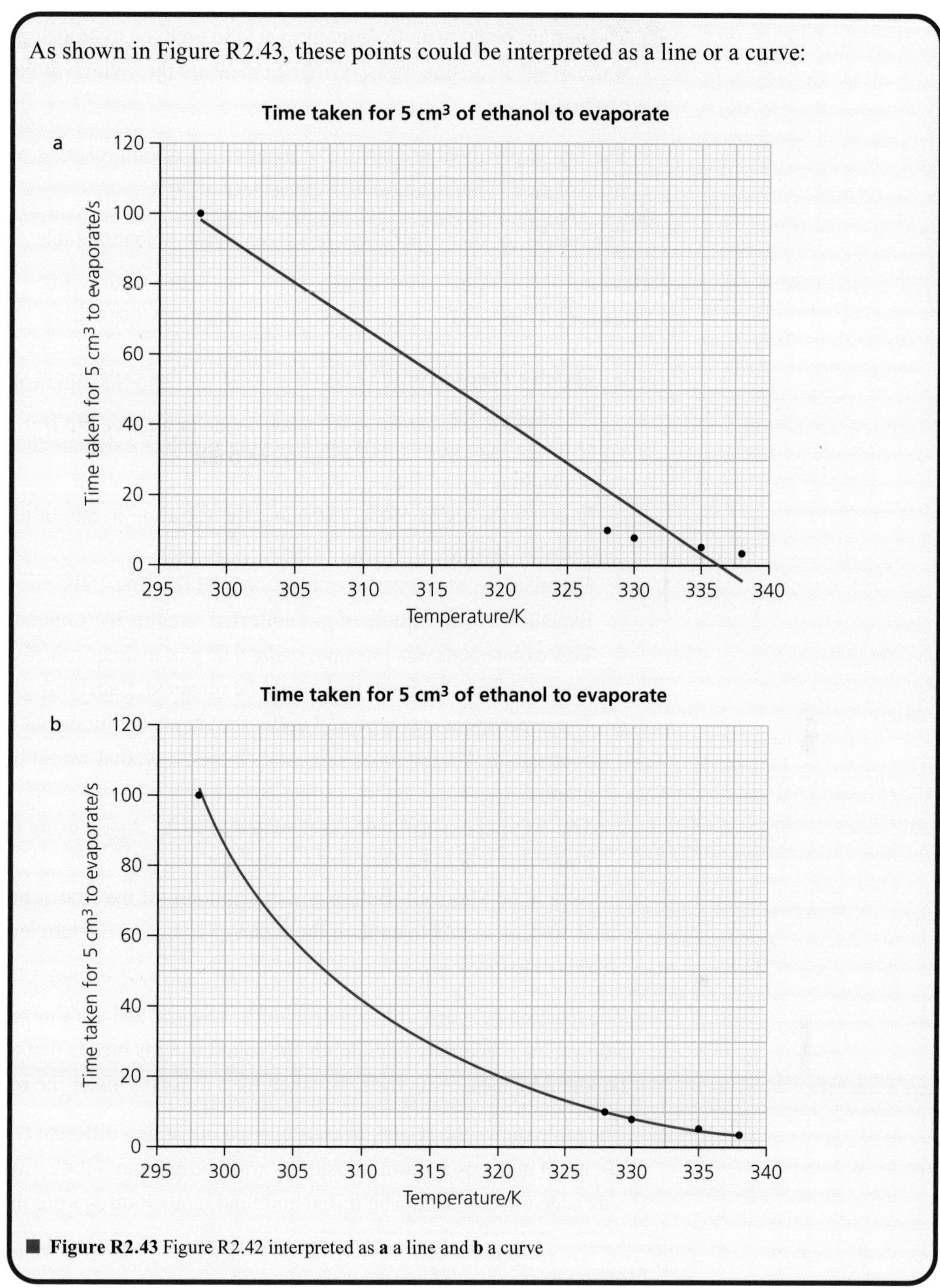

■ **Figure R2.43** Figure R2.42 interpreted as **a** a line and **b** a curve

23 Magnesium reacts with hydrochloric acid to produce hydrogen gas. A series of experiments was carried out, using a gas syringe to measure the volume of gas produced under different conditions.

The first experiment used 0.10 g of magnesium ribbon reacting with 30 cm^3 of 0.50 mol dm^{-3} HCl solution. This reaction was carried out at 20 °C.

The results for this experiment are given in the following table.

Time / s	0	15	30	45	60	75	90	105	120	135	150	165	180
Volume of gas / cm^3	0.0	18.6	32.3	44.3	54.8	62.7	68.4	72.6	74.9	75.4	75.6	75.6	75.6

a Sketch a diagram of the apparatus used for this experiment. State the balanced equation for the reaction.

b Draw a graph of this data and use your graph to calculate the initial rate (cm^3 s^{-1}) of reaction.

c Interpret the shape of the graph and the variation in rate of reaction using the collision theory.

d Calculate the average rate of reaction over the first 150 s.

e Explain why the volume of gas collected remains the same after 150 s.

f This experiment was repeated using 0.05 g of magnesium ribbon. On the same axes as before, sketch the graph that would be obtained. Label this graph B.

g The experiment was repeated under the same conditions but using 0.10 g of powdered magnesium. On the same axes, sketch the graph that would be obtained. Label this graph C.

h The original experiment was repeated at 10 °C. Again using the same axes, sketch the graph that would be obtained and label it graph D.

i Sketch the Maxwell–Boltzmann distribution for the first experiment and the experiment at 10 °C. Use this to explain the effect of changing the temperature on the rate of this reaction.

24 The reaction between dilute hydrochloric acid and sodium thiosulfate solution produces a fine yellow precipitate that clouds the solution. This means that the rate of this reaction can be found by measuring the time taken for a cross (×) under the reaction to become hidden.

The table shows the results of tests carried out at five different temperatures. In each case, 50 cm^3 of aqueous sodium thiosulfate was poured into a flask. 10 cm^3 of hydrochloric acid was added to the flask. The initial and final temperatures were measured.

Experiment	**Initial temperature / °C**	**Final temperature / °C**	**Average temperature / °C**	**Time for cross to disappear / s**
A	24	24		130
B	33	31		79
C	40	38		55
D	51	47		33
E	60	54		26

a Deduce the average temperature at which each experiment was carried out.

b Plot a graph of the time taken for the cross to disappear against the average temperature for the experiment.

c State in which experiment the rate of reaction was fastest.

d Explain why the rate was fastest in this experiment.

e Explain why the same volumes of sodium thiosulfate and hydrochloric acid solutions were used in each experiment. Explain why the conical flasks used in each test run need to be of the same dimensions.

f Deduce from the graph the time it would take for the cross to disappear if the experiment was repeated at 70 °C. Show on the graph how you found your estimate.

g Sketch on the graph the curve you would expect if all the experiments were carried out with 50 cm^3 of more concentrated sodium thiosulfate solution.

h Outline how it would be possible to carry out this experiment at a temperature of around 0 to 5 °C.

LINKING QUESTION

Concentration changes in reactions are not usually measured directly. What methods are used to provide data to determine the rate of reactions?

ATL R2.2A

There is a wide variety of software designed to simulate chemical processes or to illustrate key chemical concepts. Some of these programs are interactive, so you can change the value of variables and observe the effects on the simulated system.

Visit the online simulation of an iodine clock reaction at **http://web.mst.edu/~gbert/IClock/Clock.html** and read through the discussion.

Use the simulation to investigate the effect of changing the concentration of each solution and the temperature.

Present your findings to a peer.

(HL only) Include in your presentation a discussion of the elementary steps involved in this clock reaction.

Rate equations

HL ONLY

Chemists may investigate the effect of the concentration of the reaction participants on the rate of a reaction. They summarize the results of these investigations in the form of a **rate equation**. A rate equation shows how changes in the concentrations of reactants affect the rate of a reaction.

Take the example of a general reaction:

$$aA + bB \rightarrow \text{products}$$

where a and b are the stoichiometric coefficients in the equation.

The rate equation (sometimes referred to as the rate expression) takes this form:

$$\text{rate} = k[A]^x[B]^y$$

- [A] and [B] represent the concentrations of the reactants in moles per cubic decimetre.
- k is the **rate constant**.
- The powers x and y show the **order** of the reaction with respect to each individual reactant. The reaction above is order x with respect to A and order y with respect to B.
- The **overall order** of the reaction is $x + y$.

◆ **Rate equation**: The experimentally determined relationship between the rate of a reaction and the concentrations of the chemical species in the reaction.

◆ **Rate constant**: The constant of proportionality in a rate equation.

◆ **Order (individual)**: The power to which the concentration of a species is raised in a rate expression.

◆ **Order (overall)**: The sum of the individual orders with respect to each of the reactants in a rate expression.

Top tip!

A rate equation cannot be deduced from the balanced chemical equation: it has to be found by experiment. The values of x and y in the rate equation may or may not be the same as the values of a and b in the balanced equation for the reaction.

The rate constant, k, is only constant for a particular temperature and its units depend on the overall order of the reaction.

Table R2.6 How the dependence of the rate of a reaction on concentration of a reactant is related to order and rate equation

Order	Rate equation	Nature of dependence
zero	rate = $k[A]^0$	Rate is independent of the concentration of A; we could make any change to the concentration of A and the rate would remain the same.
first	rate = $k[A]^1$	Rate changes as the concentration of A changes; for example, if the concentration of A is doubled, the rate doubles.
second	rate = $k[A]^2$	Rate changes as the square of the change in concentration of A; if the concentration of A is tripled (threefold increase), the rate increases by nine times (ninefold, as $3^2 = 9$).

Rate equations have two main uses:

- The rate equation and the rate constant can be used to predict the rate of a reaction for a mixture of reactants of known concentrations.
- A rate equation will help to suggest a mechanism for the reaction.

Rate equations may not include all the reactants and they may include substances (such as an acid or alkali) which, although present in the reaction mixture, do not appear in the equation because they act as homogeneous catalysts.

Examples of rate equations

$2N_2O_5(g) \rightarrow 4NO_2(g) + O_2(g)$ rate = $k[N_2O_5]$

This emphasizes that the order of reaction is not obtained from the stoichiometric coefficients in the equation.

$CH_3COCH_3(aq) + I_2(aq) \rightarrow CH_2ICOCH_3(aq) + HI(aq)$ rate = $k[CH_3COCH_3][H^+]$

The reaction between propanone and iodine is catalyzed by protons (H^+). The rate equation includes propanone and the catalyst but does not include iodine, the other reactant in the equation. The concentration of iodine does not affect the rate of reaction; the reaction is zero (or zeroth) order with respect to iodine.

WORKED EXAMPLE R2.2C

The rate equation for the reaction: $2A + B + C \rightarrow 2D + E$ is rate = $k[A][B]^2$.

Deduce the overall order and the order for each individual reactant.

Answer

The overall order is $1 + 2 = 3$.

The reaction is first order with respect to A.

The reaction is second order with respect to B.

The reaction is zero (or zeroth) order with respect to C.

You may be asked to deduce the effect on the relative rate if two reactant concentrations are changed at the same time.

Common mistake

Misconception: *The concentrations of reactants in the rate equation have exponents equal to the stoichiometric coefficients of the reactants in the balanced equation.*

The individual orders of reactants must be experimentally determined. Any relation to the overall equation is coincidental (unless the reaction takes place in a single elementary step).

WORKED EXAMPLE R2.2D

The following equation represents the oxidation of bromide ions in acidic solution by bromate(V) ions:

$$BrO_3^-(aq) + 5Br^-(aq) + 6H^+(aq) \rightarrow 3Br_2(aq) + 3H_2O(l)$$

The rate equation is:

$$\text{rate} = k[BrO_3^-][Br^-][H^+]$$

Deduce the effect on the rate if the concentration of bromate(V) ions is halved, but the concentration of hydrogen ions is quadrupled (at constant temperature and bromide concentration).

Answer

The equation is first order with respect to both bromate(V) and hydrogen ions so:

- quadrupling the hydrogen ion concentration quadruples the rate
- halving the bromate(V) concentration halves the rate.

This is equivalent to the overall rate being doubled.

The rate constant

The rate of a reaction is proportional to the reactant concentrations raised to appropriate powers. Introducing the rate constant, k, allows us to write this relationship as an equation.

The value of k does not depend on the extent of the reaction or vary with the concentrations of the reactants; it is unique to a particular reaction under particular conditions such as temperature, solvent and pH (see Table R2.7).

Table R2.7 Differences between the rate and rate constant of a reaction

Rate of reaction	Rate constant, k
It is the speed at which the reactants are converted into products at a specific time during the reaction.	It is a constant of proportionality in the rate equation.
It depends upon the concentration of reactant species at a specific time.	It refers to the rate of reaction when the concentration of every reacting species is unity (one $mol\,dm^{-3}$).
It generally decreases with time.	It is constant and does not vary during the reaction.

Low values of k are associated with slow reactions, while higher values are associated with faster reactions provided the overall order and conditions are the same.

The units for the rate constant depend on the overall order of the reaction, as shown in Table R2.8.

Table R2.8 The units of the rate constant for reactions of different orders

Zeroth order	First order	Second order	Third order	nth order
rate = $k[A]^0$	rate = $k[A]$	rate = $k[A]^2$	rate = $k[A]^3$	rate = $k[A]^n$
units of rate = $mol\,dm^{-3}\,s^{-1}$	units of $\frac{\text{rate}}{\text{concentration}}$ = s^{-1}	units of $\frac{\text{rate}}{(\text{concentration})^2}$ = $mol^{-1}\,dm^3\,s^{-1}$	units of $\frac{\text{rate}}{(\text{concentration})^3}$ = $mol^{-2}\,dm^6\,s^{-1}$	units of $\frac{\text{rate}}{(\text{concentration})^n}$ = $(mol\,dm^{-3})^{1-n}\,s^{-1}$

WORKED EXAMPLE R2.2E

The rate equation for a reaction at 800 K is:

rate = $k[A]^2[B]$

The initial rate of reaction was $5.50 \times 10^{-4}\,mol\,dm^{-3}\,s^{-1}$ when the concentrations of A and B were $3.00 \times 10^{-2}\,mol\,dm^{-3}$ and $6.00 \times 10^{-2}\,mol\,dm^{-3}$, respectively. Calculate the rate constant (to 1 decimal place).

Answer

$$k = \frac{\text{rate}}{[B][A]^2}$$

$$= \frac{(5.50 \times 10^{-4}\,mol\,dm^{-3}\,s^{-1})}{(3.00 \times 10^{-2})^2\,(mol\,dm^{-3})^2 \times 6.00 \times 10^{-2}\,mol\,dm^{-3}}$$

$$= 10.2\ mol^{-2}\,dm^{6}\,s^{-1}$$

25 Deduce the overall order of the reaction and the units of the rate constant, k, for this reaction:

$3NO(g) \rightarrow N_2O(g) + NO_2(g)$; rate = $k[NO(g)]^2$

26 Propene reacts with bromine to form 1,2-dibromopropane:

propene	+ bromine	→	1,2-dibromopropane
$CH_3CH{=}CH_2(g)$	+ $Br_2(g)$	→	$CH_3CHBr{-}CH_2Br(l)$

The rate equation for this reaction is:

rate = $k[CH_3CH{=}CH_2(g)][Br_2(g)]$

and the rate constant is $30.0\,mol^{-1}\,dm^3\,s^{-1}$.

Calculate the initial rate of reaction when the concentrations of propene and bromine are both $0.040\,mol\,dm^{-3}$.

27 Hydrogen iodide decomposes according to the equation:

$2HI(g) \rightarrow H_2(g) + I_2(g)$

The rate equation for the reaction is:

rate = $k[HI(g)]^2$

At a temperature of 700 K and a concentration of $2.00\,mol\,dm^{-3}$, the rate of decomposition of hydrogen iodide is $2.50 \times 10^{-4}\,mol\,dm^3\,s^{-1}$.

Calculate the rate constant, k, at this temperature and calculate the number of hydrogen iodide molecules that decompose per second in $1.00\,dm^3$ of gaseous hydrogen iodide under these conditions. (Avogadro constant = $6.02 \times 10^{23}\,mol^{-1}$).

First-order reactions

A reaction is first order with respect to a reactant if the rate of reaction is directly proportional to the concentration of that reactant: a graph of rate against concentration gives a straight line passing through the origin. In this case, the concentration of the reactant is raised to the power of one in the rate equation:

$$\text{rate} = k[\text{A}]$$

Another approach to identifying a first-order reaction is to plot a concentration–time graph and then study the time taken for the concentration to be halved.

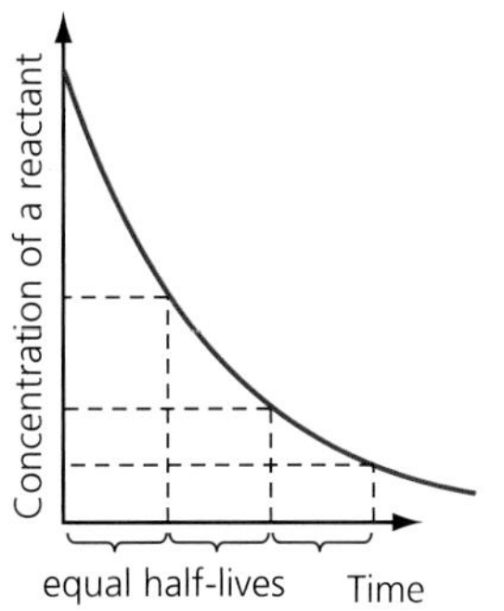

■ **Figure R2.44** Concentration–time curve for a first-order reaction, showing how successive half-lives remain constant as concentration decreases

This quantity, known as the **half-life**, $t_{\frac{1}{2}}$, is constant for a first-order reaction (at constant temperature) and is independent of the initial concentration (Figure R2.44).

The greater the value of the first-order rate constant, k, the faster the reaction, that is, the more rapid the exponential decrease in the concentration of the reactant (assuming the same initial concentration) (Figure R2.45).

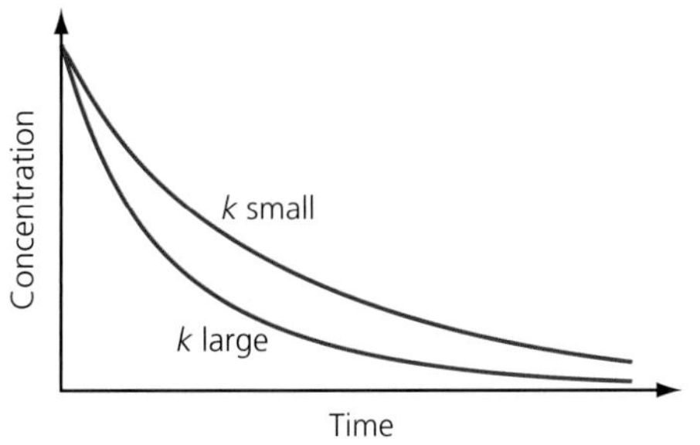

■ **Figure R2.45** First-order concentration–time graphs with different rate constant values starting from the same initial concentration

◆ **Half-life**: The time for the concentration of one of the reactants to decrease by half.

Going further

Half-life and the rate constant

It can be shown that, for a first-order equation:

$$t_{\frac{1}{2}} = \frac{\ln 2}{k} = \frac{0.693}{k}$$

where k is the rate constant.

This equation allows us to calculate the rate constant of a first-order reaction from the half-life and vice versa.

For example, if the half-life of a first-order reaction is 100 s, the rate constant is:

$$k = \frac{\ln 2}{t_{\frac{1}{2}}} = \frac{0.693}{100\,\text{s}} = 6.93 \times 10^{-3}\,\text{s}^{-1}$$

Conversely, a first-order reaction with a rate constant of $0.100\,\text{s}^{-1}$ has a half-life of:

$$t_{\frac{1}{2}} = \frac{\ln 2}{k} = \frac{0.693}{0.100\,\text{s}^{-1}} = 6.93\,\text{s}$$

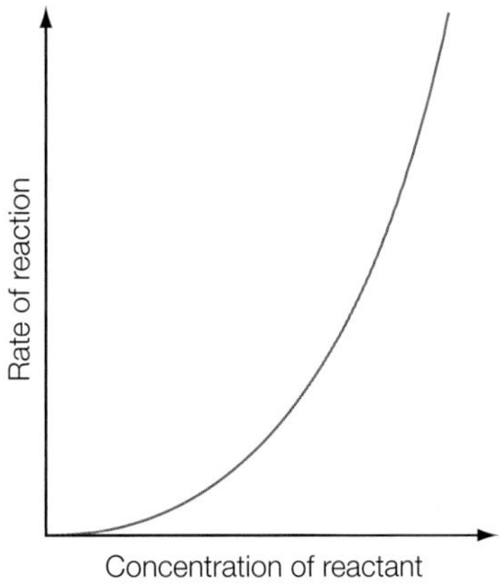

Second-order reactions

A reaction is second order with respect to a reactant if the rate of reaction is proportional to the concentration of that reactant squared. This means that doubling the concentration of A increases the rate by a factor of four and the rate–concentration graph is a parabolic (quadratic) curve rather than a straight line (Figure R2.46). At its simplest, the rate equation for a second-order reaction takes the form:

$$\text{rate} = k[\text{A}]^2.$$

Rate of reaction

Concentration of reactant

■ **Figure R2.46** The variation of reaction rate with concentration for a second-order reaction

Top tip!

A graph like the one in Figure R2.46 can be transformed into a straight line by plotting the rate of reaction against the square of the reactant concentration.

HL ONLY

A concentration–time graph of a second-order reaction has unequal half-lives (Figure R2.47). The time for the concentration to fall from its initial value c to $\frac{c}{2}$ is half the time for the concentration to decrease from $\frac{c}{2}$ to $\frac{c}{4}$.

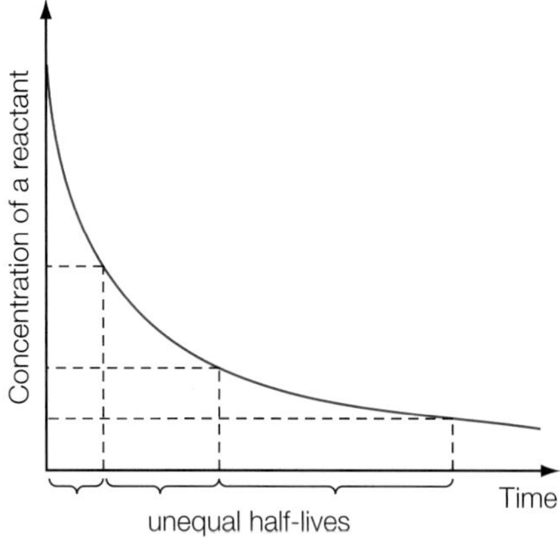

■ **Figure R2.47** The variation of concentration of a reactant plotted against time for a second-order reaction

Top tip!

The half-life of a second-order reaction is inversely proportional to the starting concentration.

Tool 3: Mathematics

Use basic arithmetic and algebraic calculations to solve problems

Apply these three principles when solving an algebraic calculation:

1. Work towards solving the variable (often given the symbol x).
2. Use the opposite mathematical operation – remove a constant or coefficient by performing the opposite operation on both sides:
   Opposite of × is ÷, opposite of + is −, opposite of x^2 is $\pm\sqrt{x}$ and opposite of $\pm\sqrt{x}$ is x^2.
3. Maintain balance: whatever is done to one side, must be done to the other side of the equation. The resulting expression is then given in its simplest form, which may involve using the power laws.
   For example, you can determine the units for a rate constant by solving the rate equation for the rate constant, k, and substituting the units for the rate and concentration(s):

$$\text{rate} = k[\text{X}]^2[\text{Y}]$$

$$\Rightarrow k = \frac{\text{rate}}{[\text{X}]^2[\text{Y}]}$$

Substituting units gives units of:

$$k = \frac{\text{mol dm}^{-3}\,\text{s}^{-1}}{(\text{mol dm}^{-3})^2(\text{mol dm}^{-3})}$$

To simplify the expression, the term 'mol dm^{-3}' can be cancelled top and bottom:

$$\text{units of } k = \frac{\text{s}^{-1}}{\text{mol}^2\,\text{dm}^{-6}} = \text{dm}^6\,\text{mol}^{-2}\,\text{s}^{-1} \text{ because } \frac{x^n}{x^m} = x^{n-m}$$

Note that the convention when writing compound units is to place the positive indices first.

Zero-order reactions

In a reaction which is zero order with respect to a reactant, the rate of reaction is unaffected by changes in the concentration of that reactant (Figure R2.48). In the rate equation, the concentration term for the reactant is raised to the power of zero, so rate = $k[\text{reactant}]^0 = k$.

In this case, the concentration–time graph for the reactant is a straight line and its gradient is constant.

Reactants which are zero order are never involved in the slowest step of a reaction mechanism, or any fast equilibrium directly before the slowest step.

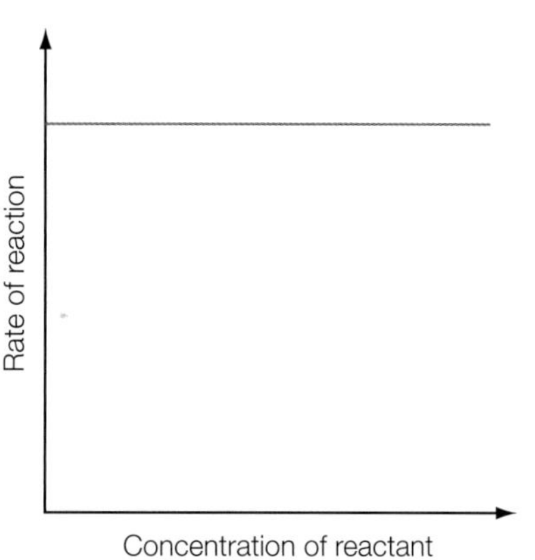

Figure R2.48 The variation of reaction rate with concentration for a zero-order reaction

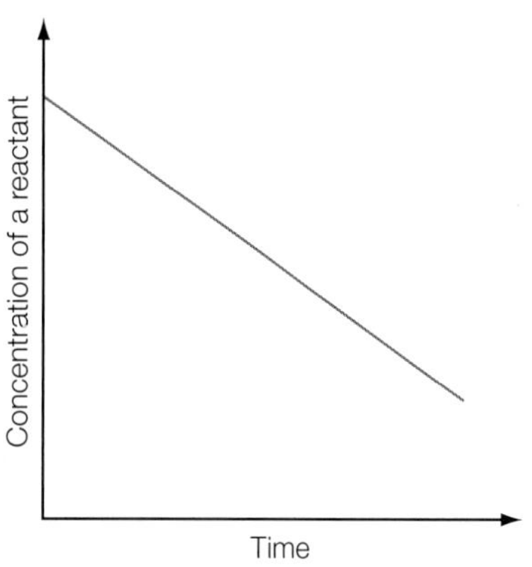

Figure R2.49 The variation of concentration of a reactant plotted against time for a zero-order reaction

Common mistake

Misconception: *An increase in the initial concentration of reactants would increase / decrease the rate of a zero-order reaction.*

If a reaction has zero-order kinetics, a change in the initial concentration of a reactant will have no effect on the rate of reaction.

Going further

Pseudo-first-order reactions

Consider the hydrolysis of the ester ethyl ethanoate to form ethanoic acid and ethanol:

$$CH_3COOC_2H_5(l) + H_2O(l) \rightarrow CH_3COOH(aq) + CH_3CH_2OH(aq)$$

From experimental studies, the rate equation is rate = $k[CH_3COOC_2H_5][H_2O]$. If water is present in excess, the amount of water used up is a small fraction of the amount of water present initially, that is, $[H_2O]$ is essentially constant:

$$\text{rate} = k[CH_3COOC_2H_5][H_2O] = k'[CH_3COOC_2H_5], \text{ where } k' = k[H_2O]$$

The reaction appears to be zero order with respect to water and first order with respect to the ester, and hence it is a pseudo-first-order reaction overall.

In general, given rate = $k[A][B]$ for a reaction which is first order with respect to each of the two reactants, the reaction will be pseudo first order if the concentration of one of the reactants is so large that it remains effectively constant during the reaction.

Experimental determination of rate equations

Initial rate method

This method is based on finding the rate at the start of a reaction when all the concentrations are known.

The investigator prepares a series of mixtures in which all the initial concentrations are the same except one. A suitable method is used to measure the change of concentration with time for each mixture and the results are used to plot concentration–time graphs. The initial rate for each mixture is then found by drawing a tangent to the curve at the start and calculating the gradient.

For a reaction: $A + B \rightarrow C + D$, where A and B represent reactants and C and D represent products, the following procedure is used to establish the rate equation:

- Carry out the reaction with known concentrations of A and B and measure the initial rate of reaction.
- Repeat the experiment using double the concentration of A, but keeping the concentration of B the same as in the first experiment. Any change in the rate of reaction must be caused by the change in the concentration of A.

If the rate is doubled in the second experiment, then the reaction is first order with respect to A; if the initial rate is increased four times, the reaction is second order with respect to A; and if there is no change in the initial rate then the reaction is zero order with respect to A.

WORKED EXAMPLE R2.2F

Iodine reacts with propanone according to the following equation:

$$I_2(aq) + CH_3COCH_3(aq) \xrightarrow{H^+} ICH_2COCH_3(aq) + HI(aq)$$

The kinetics of this reaction were investigated in four experiments carried out at constant temperature. The initial rate of reaction was measured at different concentrations of propanone, iodine and hydrogen ions as shown in the table.

Experiment number	Propanone concentration / $mol\,dm^{-3}$	Hydrogen ion concentration / $mol\,dm^{-3}$	Iodine concentration / $mol\,dm^{-3}$	Initial rate / $mol\,dm^{-3}\,s^{-1}$
1	6.0	0.4	0.04	1.8×10^{-5}
2	6.0	0.8	0.04	3.6×10^{-5}
3	8.0	0.8	0.04	4.8×10^{-5}
4	8.0	0.4	0.08	2.4×10^{-5}

Use the data to determine the individual orders and overall order of the reaction.

Answer

Comparing experiments 1 and 2, the concentration of hydrogen ions is doubled, while the other two reactant concentrations are kept constant.

The rate also doubles between experiments 1 and 2, indicating that the reaction is first order with respect to hydrogen ions.

Comparing experiments 3 and 4, there is a doubling in the concentration of iodine, while the propanone concentration is kept constant and the hydrogen ion concentration is halved. The initial rate is also halved between experiments 3 and 4, which means the reaction is zero order with respect to iodine, since the halving of the initial rate is due to the change in concentration of protons (H^+).

Comparing experiments 2 and 3, there is an increase in the concentration of propanone by a factor of $\frac{4}{3}$. There is also an increase in rate between experiments 2 and 3 by a factor of $\frac{4}{3}$. Therefore, the rate is proportional to the propanone concentration, that is, it is first order with respect to propanone.

Overall, therefore:

$$\text{rate} = k[H^+(aq)]^1[I_2(aq)]^0[(CH_3)_2CO(aq)]^1$$

or

$$\text{rate} = k[H^+(aq)]\,[(CH_3)_2CO(aq)]$$

and the reaction is second order overall.

28 Nitrogen dioxide can be produced by reacting oxygen and nitrogen monoxide:

$$O_2(g) + 2NO(g) \rightarrow 2NO_2(g)$$

A student performed a series of experiments to determine the order of reaction with respect to O_2 and NO. The results are given in the table.

Experiment	Initial $[O_2]$ / $mol\,dm^{-3}$	Initial [NO] / $mol\,dm^{-3}$	Temperature / K	Initial rate / $mol\,dm^{-3}\,s^{-1}$
1	0.02	0.013	298	0.0033
2	0.02	0.013	318	0.0376
3	0.04	0.013	298	0.0066
4	0.08	0.026	298	0.0528

Determine the order of reaction with respect to each reactant and, hence, write the rate equation.

LINKING QUESTION

What measurements are needed to deduce the order of reaction for a specific reactant?

Graphical methods

Determining order

You can also determine the order of a reaction by plotting the rate against different initial concentrations of a reactant. For reactions like the iodine clock and the reaction between thiosulfate ions and dilute acid, you can plot the relative rate $\left(\frac{1}{\text{time}}\right)$ against initial concentration. Figure R2.50 shows the distinctive shapes of the rate–concentration graphs for different orders of reaction.

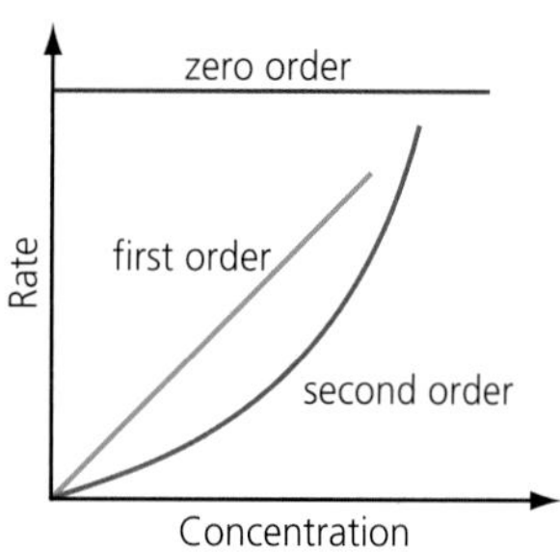

■ **Figure R2.50** Rate–concentration graphs for zero-order (zeroth order), first-order and second-order reactions

Determining *k*

Another way of calculating the rate constant for a first-order reaction is to plot a graph of the natural logarithm of the concentration against time. For a first-order reaction, the graph (Figure R2.51) is a straight line with a gradient of $-k$.

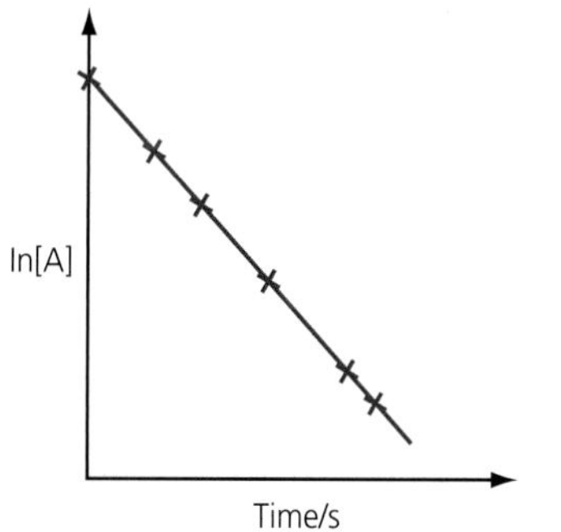

■ **Figure R2.51** Determining the rate constant for a first-order reaction by plotting the natural log of concentration against time

Graphical methods of determining *k* and the order of a reaction for zero, first and simple second-order relationships are summarized in Table R2.9.

■ **Table R2.9** Summary of graphical methods of finding orders and half-lives

Overall reaction order	zero order A → products	first order A → products	simple second order* A → products A + B → products
Rate equation	rate = k	rate = k[A]	rate = $k[A]^2$ * rate = k[A][B]
Data to plot for a straight-line graph	[A] versus t	ln [A] versus t	$\frac{1}{[A]}$ versus t
Slope or gradient equals	$-k$	$-k$	$+k$
Changes in the half-life as the reactant is consumed	$t_{\frac{1}{2}}$ becomes shorter	$t_{\frac{1}{2}}$ is constant	$t_{\frac{1}{2}}$ becomes longer
Units of k	$mol\,dm^{-3}\,s^{-1}$	s^{-1}	$dm^3\,mol^{-1}\,s^{-1}$

* A simple second-order reaction is a reaction which is second order with respect to one reactant, that is, rate = $k[A]^2$.

29 The progress of the isomerization of cyclopropane can be followed by measuring either the decrease in concentration of cyclopropane or the increase in concentration of propene.

$$\text{cyclo-}C_3H_6\,(H_2C{-}CH_2{-}CH_2\text{ ring})(g) \rightarrow CH_3{-}CH{=}CH_2(g)$$

Propene

Results for the experiment on the isomerization of cyclopropane are shown in the table.

Time / min	**[cyclopropane] / $mol\,dm^{-3}$**	**[propene] / $mol\,dm^{-3}$**
0	1.50	0.00
5	1.23	0.27
10	1.00	0.50
15	0.82	0.68
20	0.67	0.83
25	0.55	0.95
30	0.45	1.05
35	0.37	1.13
40	0.33	1.17

a Using the data in the table, calculate a value for the average rate of reaction over the first 5 minutes. (Note that you will need to convert minutes to seconds during your calculation.)

b The value calculated in part **a** is not the initial rate of reaction as – even over this relatively short time – the rate is slowing as the reactant is used up. Plot a graph of [propene] against time. Then find the initial rate of reaction by drawing the tangent to the curve at time 0 and finding its slope.

c By drawing additional tangents, calculate the rate of reaction at propene concentrations of $0.30\,mol\,dm^{-3}$, $0.60\,mol\,dm^{-3}$ and $0.90\,mol\,dm^{-3}$.

d Calculate the concentration of cyclopropane when the propene concentration is $0.00\,mol\,dm^{-3}$, $0.30\,mol\,dm^{-3}$, $0.60\,mol\,dm^{-3}$ and $0.90\,mol\,dm^{-3}$.

e Use these values to plot a graph of the rate of reaction against the concentration of cyclopropane. Then find the value of the rate constant, k, from your graph.

Going further

Saturation kinetics

A change from first-order to zero-order kinetics occurs during the decomposition of gases on the surface of a heterogeneous catalyst (for example, during the decomposition of ammonia to nitrogen and hydrogen on a tungsten catalyst).

At low pressure (that is, low concentration), gas molecules are adsorbed onto the surface and the reaction shows first-order kinetics with respect to ammonia.

However, the number of catalytic active sites on the metal surface is rate limiting; at high pressure all the active sites are occupied. The rate then becomes independent of the concentration of the reactant molecules and the reaction shows zero-order kinetics with respect to ammonia.

Enzymes show similar saturation kinetics: when the enzyme is the limiting reactant and the substrate is present in a large excess, there is a change from first to zero-order.

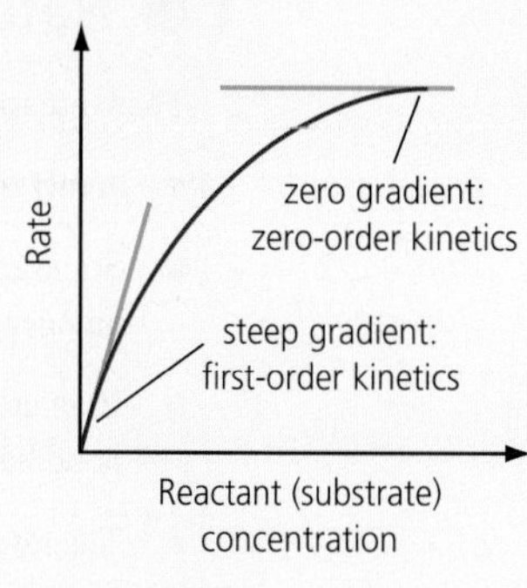

■ **Figure R2.52** Graph of the rate of an enzyme-catalyzed reaction against substrate concentration

◆ **Elementary step:** A reaction with no intermediates; a reaction that takes place in a single step with a single transition state.

◆ **Reaction mechanism:** The series of elementary steps by which an overall chemical reaction occurs.

◆ **Unimolecular:** An elementary step involving only one species (typically a molecule).

◆ **Bimolecular:** An elementary step involving two species.

◆ **Termolecular:** An elementary step involving three species.

Reaction mechanisms

Nature of science: Measurement

Rate equations of different reactions stimulated chemists to think about the mechanisms of reactions. They wanted to understand why a rate equation cannot be predicted from the balanced equation for the reaction. They could not understand why similar reactions, for example, the hydrolysis of different classes of halogenoalkanes, had different rate equations. The key to understanding the mechanism of a reaction is the idea that most reactions do not take place in one step, as suggested by the balanced equation; instead, they involve a series of chemical reactions called elementary steps.

All chemical reactions, except the very simplest, actually take place via a number of reactions called **elementary steps**. These comprise a **reaction mechanism** which is a model of what chemists believe is occurring at the molecular level.

Elementary steps are described as **unimolecular** if only one chemical species (atom, ion, radical or molecule) is involved, **bimolecular** if two chemical species are involved and **termolecular** if three species are involved.

Unimolecular steps (Figure R2.53) involve either the decomposition or dissociation of a molecule into two or more smaller molecules or ions, or the rearrangement of a molecule:

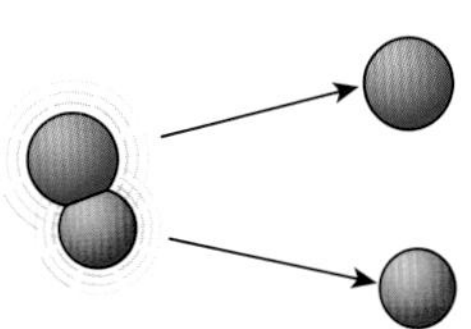

■ **Figure R2.53** Unimolecular step

$A \rightarrow B$ (rearrangement) or

$A \rightarrow B + C$ (decomposition or dissociation)

Bimolecular steps (Figure R2.54) involve two species colliding and reacting with each other:

$A + B \rightarrow$ Product or $A + A \rightarrow$ Product

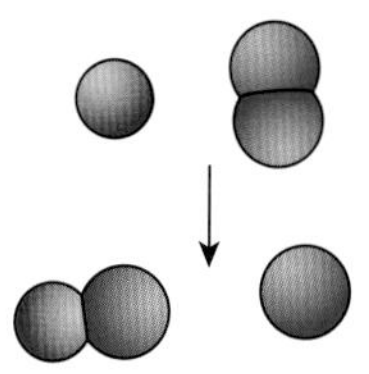

■ **Figure R2.54** Bimolecular step

The thermal decomposition of dinitrogen monoxide is described by the following equation:

$$2N_2O(g) \rightarrow 2N_2(g) + O_2(g)$$

This reaction is believed to occur via the following elementary steps:

$N_2O(g) \rightarrow N_2(g) + O(g)$ slow

$N_2O(g) + O(g) \rightarrow N_2(g) + O_2(g)$ fast

The oxygen atoms produced in the first reaction are an intermediate and they are used up in the second reaction.

Nature of science: Experiments

There are many ways of studying a reaction in order to propose a reaction mechanism. For example:

- Analytic techniques such as NMR can be used to detect the changes taking place and their sequence.
- Some intermediates are sufficiently stable for detection and isolation. For example, some halogenoalkanes form tertiary carbocations that can be isolated.
- If an atom is labelled with an isotope (not necessarily a radioactive one) the label may indicate which bond has been broken during a reaction.
- The rate of ionic reactions changes with the polarity of the solvent. For example, the rate of hydrolysis of 2-bromo-2-methylpropane is increased by the addition of sodium chloride. The sodium chloride increases the polarity of the solvent and increases the rate of ionization of the bromoalkane.

Rate equations for elementary steps

◆ **Molecularity**: The number of chemical species or particles participating in an elementary step of a mechanism.

Top tip!

Elementary steps in which three reactant particles collide, known as termolecular steps, are very rare because the probability of three particles colliding at the same time and place with sufficient kinetic energy and in the correct orientation is very small.

We describe elementary steps by their **molecularity**: the number of reactant particles involved in the elementary step.

The most common molecularities are unimolecular and bimolecular:

A → products	unimolecular
A + A → products	bimolecular
A + B → products	bimolecular

You can deduce the rate equation for an elementary step from its equation, but you cannot deduce the rate equation for a complete reaction from its overall balanced equation unless you know all the elementary steps.

Since we know that an elementary step occurs through the collision of the reactant particles, the rate is proportional to the product of the concentrations of those particles.

For example, the rate for the unimolecular step in which A reacts to form a product is proportional to the concentration of A:

A + B → products $\quad$ rate = k[A]

For example, the rate for the bimolecular elementary step in which A reacts with B is proportional to the concentration of A multiplied by the concentration of B:

A + B → products $\quad$ rate = k[A][B]

Similarly, the rate equation for the bimolecular step in which A reacts with A is proportional to the square of the concentration of A:

A + A → products $\quad$ rate = $k[A]^2$

Table R2.10 summarizes the rate equations for the common elementary steps, as well as those for the rare termolecular step (assuming they have been categorically shown not to progress through an intermediate). Notice that the molecularity of the elementary step is equal to the overall order of the step.

■ **Table R2.10** Rate equations for elementary steps

Elementary step	Molecularity	Rate equation
A → products	1	rate = k[A]
A + A → products	2	rate = $k[A]^2$
A + B → products	2	rate = k[A][B]
A + A + A→ products	3 (rare)	rate = $k[A]^3$
A + A + B → products	3 (rare)	rate = $k[A]^2[B]$
A + B + C → products	3 (rare)	rate = k[A][B][C]

◆ **Rate-determining step:** The slowest elementary step in a chemical reaction that involves a number of elementary steps.

The slowest step in a reaction mechanism is known as the **rate-determining step**. The rate of this step determines the rate of the overall reaction.

Reactants that take part in the rate-determining step will appear in the final rate equation for a reaction (after intermediates have been removed).

The rate equation of an elementary step appears in the rate equation for the overall reaction if the step:

- is the rate-determining step
- is a fast equilibrium step *directly before* the rate-determining step.

For example:

Step 1: $\mathbf{A} + \mathbf{B} + \mathbf{C} \rightleftharpoons \text{ABC}$ (fast equilibrium step)

Step 2: $\text{ABC} + \mathbf{D} \rightarrow \text{AD} + \text{BC}$ (rate-determining step)

A reactant that does not appear in one of these steps will not appear in the rate equation. This means that changing the concentration of that reactant will have no effect on the rate of the reaction. The reactant is said to have zero-order kinetics.

Top tip!

Steps with high activation energy involve:

- interaction between two molecules
- interaction between ions of the same charge
- a bond breaking homolytically to form free radicals.

Steps with low activation energy involve:

- interaction between two radicals
- interaction between ions of opposite charge
- acid–base reactions.

Top tip!

The overall reaction of a multi-step mechanism is the sum of its elementary steps.

Multi-step reactions

Multi-step mechanisms involve a sequence of elementary steps, each with its own transition state. The elementary steps may also involve the formation of intermediates.

Consider the reaction:

$$2NO_2(g) + F_2(g) \rightarrow 2NO_2F(g)$$

For this reaction to occur in a single elementary step, three molecules would need to collide simultaneously and in the correct orientation with sufficient kinetic energy. This is a highly unlikely event, particularly in the gas phase, although it is more likely to occur on the surface of a catalyst.

Experiments show that the rate equation for this reaction is:

$$\text{rate} = k[NO_2][F_2]$$

and the suggested mechanism with fluorine atoms (radicals) as an intermediate is:

Step 1 $\bullet NO_2(g) + F_2(g) \rightarrow NO_2F(g) + \bullet F(g)$ slow

Step 2 $\bullet NO_2(g) + \bullet F(g) \rightarrow NO_2F(g)$ fast

The energy profile for this reaction is shown in Figure R2.55 and it can be seen that the rate-determining step (elementary step 1) has a higher activation energy than the other step. The activation energy for the overall reaction can be regarded as being equivalent to the activation energy of the slow, rate-determining step.

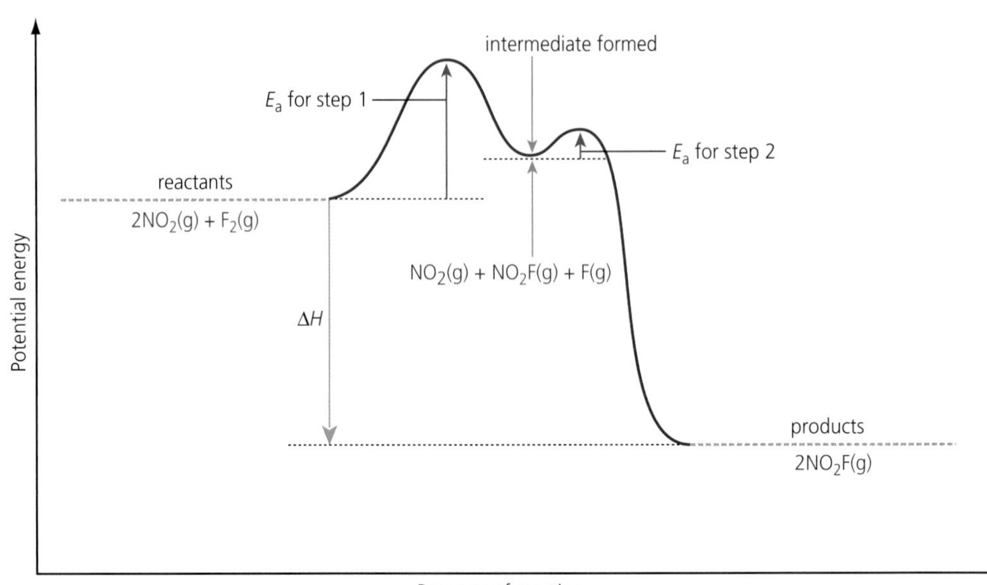

■ **Figure R2.55** Potential energy profile for the two-step reaction between NO_2 and F_2. The formation of an intermediate is represented by the potential energy well in the profile. Step 1 is the rate-determining step.

More complex situations arise where the first step of a sequence is not the rate-determining step. For example, consider the oxidation of nitrogen(II) oxide:

$$2NO(g) + O_2(g) \rightarrow 2NO_2(g)$$

for which the following reaction mechanism has been proposed:

Step 1 $NO(g) + NO(g) \rightleftharpoons N_2O_2(g)$ fast

Step 2 $N_2O_2(g) + O_2(g) \rightarrow 2NO_2(g)$ slow

Top tip!

Intermediates never appear in a rate equation – only reactants and homogeneous catalysts.

Consequently, the rate of reaction depends on step 2, for which the rate equation is:

$$\text{rate} = k[N_2O_2(g)][O_2(g)]$$

However, N_2O_2 is the intermediate product of step 1, and so the concentration of this intermediate depends on $[NO]^2$, since the first step is second order with respect to NO.

Therefore this fact must be substituted into the rate equation, which then becomes:

$$\text{rate} = k[NO(g)]^2[O_2(g)]$$

The order of this reaction is, therefore, third order overall.

This approach can be justified by the concept of equilibrium and its connection to rate constants for elementary steps:

$k_1[NO]^2 = k_{-1}[N_2O_2]$ for step 1 where k_1 and k_{-1} are the rate constants for the forward and backward reactions.

Rearranging for the concentration of N_2O_2: $[N_2O_2] = \frac{k_1}{k_{-1}}[NO]^2$. This expression is now substituted into the rate equation for the second rate-determining step: rate $= k_2 (\frac{k_1}{k_{-1}}[NO]^2) \times [O_2] = k[NO]^2 [O_2]$.

LINKING QUESTION

What are the rate equations and units of k for the reactions of primary and tertiary halogenoalkanes with aqueous alkali?

Examples of mechanisms

Kinetic studies provide much of the data required for proposing reaction mechanisms.

ATL R2.2B

Gaseous reaction mixtures can contain a complex mixture of molecular species. Nitrogen(V) oxide decomposes in the following reaction:

$$2N_2O_5(g) \rightarrow 4NO_2(g) + O_2(g)$$

This reaction is found experimentally to have the rate equation: rate $= k[N_2O_5]$.

Figure R2.56 shows a suggested mechanism for this reaction.

The rate equation suggests that a single N_2O_5 molecule is involved in the rate-determining step. This fits with the proposed mechanism, which suggests that the decomposition of N_2O_5 to form NO_2 and NO_3 is the slow rate-determining step. The steps which follow the slow step are relatively fast and so have no effect on the reaction rate.

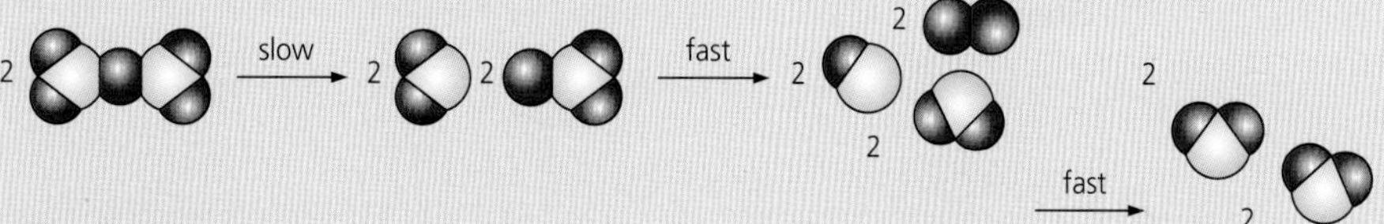

In the first step each molecule breaks down. They do not collide in pairs.

$N_2O_5 \xrightarrow{\text{slow}} NO_2 + NO_3$

$NO_2 + NO_3 \xrightarrow{\text{fast}} NO + NO_2 + O_2$

$NO + NO_3 \xrightarrow{\text{fast}} 2NO_2$

(Note that two molecules of N_2O_5 need to have reacted for subsequent steps to be completed)

■ **Figure R2.56** The proposed mechanism for the decomposition of N_2O_5 (NO_3 is a radical with an unpaired electron: NO_3•)

Make a copy of Figure R2.56 and number or label the stages to match the reaction steps with the illustrations, to create a picture of what is happening.

Note the following elementary step also fits the rate equation: $NO_3(g) + N_2O_5(g) \rightarrow O_2(g) + 3NO_2(g)$.

ATL R2.2C

The Egyptian and Arabic chemist, Ahmed Zewail (1946–2016), won the Nobel Prize in Chemistry in 1999 for his pioneering work in the US. He used an ultra-fast laser technique to study reactions such as:

$$\bullet H(g) + F_2(g) \rightarrow HF(g) + \bullet F(g)$$

Zewail closed his Nobel Prize address in Arabic with the words of the Egyptian scholar Taha Hussein (1889–1973): *Wailu li-talib al-'ilmi in radia 'an nafsihi.* 'The end will begin when seekers of knowledge become satisfied with their own achievements.'

- On a poster, outline Zewail's work in developing 'femtochemistry' and describe its importance in understanding reaction mechanisms.
- Describe Zewail's political and social work in both Egypt and the United States, giving an oral presentation to a group of your peers who are interested in global politics.

Iodine and propanone

$$CH_3COCH_3(aq) + I_2(aq) \rightarrow CH_3COCH_2I(aq) + HI(aq)$$

$$\text{rate} = k[CH_3COCH_3(aq)][H^+(aq)]$$

Link

The use of curly arrows to show electron pair movement in organic chemistry is covered in more detail in Chapter R3.3, page 627.

The accepted mechanism involves four elementary steps. These are shown below using curly arrows to show the movement of electron pairs. Curly arrows are mainly used in organic chemistry to show which bonds are broken or formed, or how charges are distributed around molecules through resonance.

Step 1

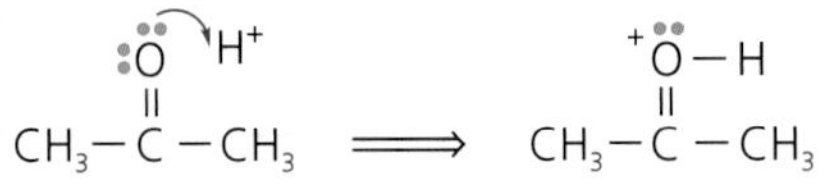

■ **Figure R2.57** Rapid protonation (gain of a proton)

Step 2

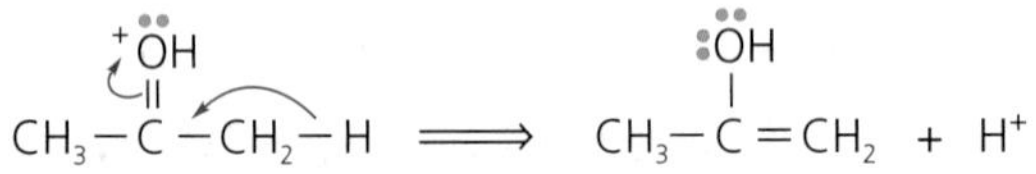

■ **Figure R2.58** Deprotonation (loss of a proton) and formation of the prop-1-en-2-ol intermediate

Step 3

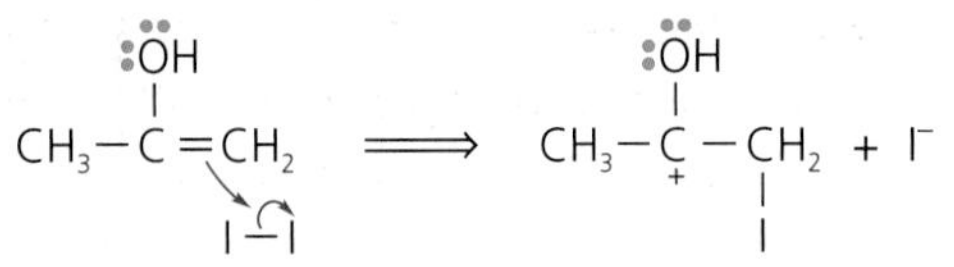

■ **Figure R2.59** Reaction of prop-1-en-2-ol with iodine to form a carbocation intermediate

Step 4

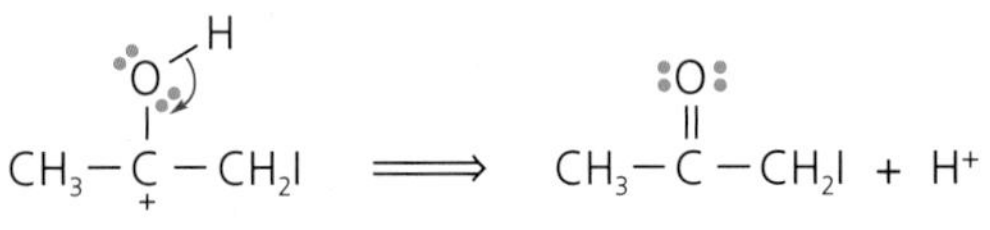

■ **Figure R2.60** Deprotonation and formation of the product, iodopropanone

Step 1 is the slowest and, hence, the rate-determining step. It has a molecularity of 2 since only propanone molecules and hydrogen ions are involved. The reaction is first order with respect to both the propanone and the acid.

The iodine is *not* involved in the rate-determining step and therefore changing its concentration has *no* effect on the overall rate.

Hydrolysis of bromoethane

Bromoethane, C_2H_5Br, undergoes rapid hydrolysis in the presence of cold dilute aqueous alkali to form ethanol and bromide ions:

$$C_2H_5Br(aq) + OH^-(aq) \rightarrow C_2H_5OH(aq) + Br^-(aq)$$

Experiments show that the reaction is first order with respect to the concentration of hydroxide ions and first order with respect to the concentration of bromoethane:

rate = $k[C_2H_5Br(aq)][OH^-(aq)]$

This means that the bromoethane molecule and hydroxide ion are both involved in the slow rate-determining step to form the high energy transition state shown below:

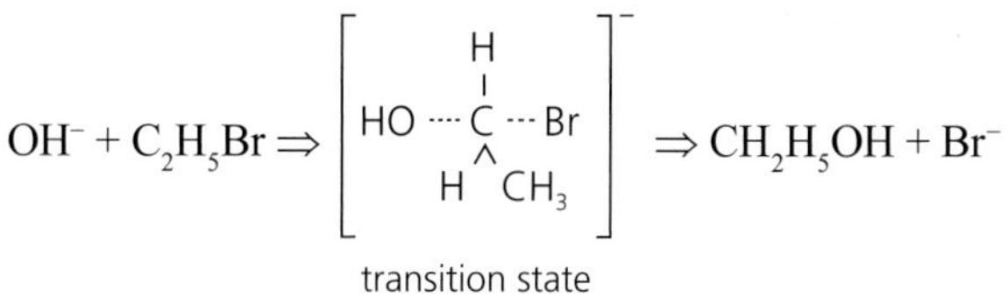

$OH^- + C_2H_5Br \Rightarrow$ [HO ···· C ··· Br (with H above C; H and CH_3 below C)]$^-$ $\Rightarrow CH_2H_5OH + Br^-$

transition state

■ **Figure R2.61** The reaction between hydroxide ions and bromoethane molecules to form ethanol molecules and bromide ions via a transition state

Link

The hydrolysis of bromoethane and nucleophilic substitution reactions of halogenoalkanes (HL only) are covered in more detail in Chapter R3.4, page 652.

This is an S_N2 reaction. The S indicates substitution – the replacement of one atom (or functional group) by another atom (or functional group). In this example, a bromine atom is replaced by the hydroxyl group. The N indicates that the organic species is attacked by a nucleophile (an electron pair donor or Lewis base – in this example the hydroxide ion). The 2 indicates a molecularity of 2.

The dotted lines in the transition state indicate partial bonds (bond order less than 1), and the negative charge is delocalized over both the partial bonds.

Two related structures, $[HO\cdots\cdots CH_2CH_3\cdots Br]^-$ and $[HO\cdots CH_2CH_3\cdots\cdots Br]^-$, appear before and after the transition state, respectively. The diagrams of these structures illustrate how the HO–C bond becomes shorter and stronger as the C–Br bond becomes longer and weaker.

Top tip!

The 2 may also indicate the order but if the reaction is carried out with a large excess of hydroxide ions, pseudo-first-order kinetics are observed.

Hydrolysis of 2-bromo-2-methylpropane

By contrast, the hydrolysis of $(CH_3)_3CBr$ shows first-order kinetics:

rate = $k[(CH_3)_3CBr]$

The initial rate is independent of the concentration of the alkali, so it shows zero-order kinetics with respect to hydroxide ions.

The kinetic data suggests a unimolecular mechanism that involves only the 2-bromo-2-methylpropane in the rate-determining step:

Step 1 $(CH_3)_3CBr \rightarrow (CH_3)_3C^+ + Br^-$ slow

Step 2 $(CH_3)_3C^+ + OH^- \rightarrow (CH_3)_3COH + H^+$ fast

This is an S_N1 reaction, where the S indicates substitution, the N indicates that the organic species is attacked by a nucleophile and the 1 indicates a molecularity of 1.

Inquiry 3: Concluding and evaluating

Evaluating hypotheses

Typically, a hypothesis is generated from some theory or model, then evaluated using data collected through some laboratory procedures. A hypothesis should either be supported by the data or falsified by the data.

Ozone reacts with nitrogen monoxide according to the following equation:

$$O_3(g) + NO(g) \rightarrow NO_2(g) + O_2(g)$$

The experimentally determined rate equation is rate = $k[O_3][NO]$. Below are several suggested mechanisms (hypotheses) consisting of elementary steps, with the rate-determining step identified and comments evaluating each mechanism.

Mechanism A

$NO(g) + O_3(g) \rightarrow NO_3(g) + O(g)$	slow
$NO_3(g) + O(g) \rightarrow NO_2(g) + O_2(g)$	fast
$O_3(g) + NO(g) \rightarrow NO_2(g) + O_2(g)$	overall

Consistent, the slow first step determines the order.

Mechanism B

$O_3(g) + NO(g) \rightarrow NO_2(g) + O_2(g)$	slow, one elementary step

Consistent, one-step reaction mechanism.

Mechanism C

$O_3(g) \rightarrow O_2(g) + O(g)$	slow
$O(g) + NO(g) \rightarrow NO_2(g)$	fast
$O_3(g) + NO(g) \rightarrow NO_2(g) + O_2(g)$	overall

Not consistent, slow first step, no dependence on [NO] is predicted by this mechanism.

Mechanism D

$NO(g) \rightarrow N(g) + O(g)$	slow
$O(g) + O_3(g) \rightarrow 2O_2(g)$	fast
$O_2(g) + N(g) \rightarrow NO_2(g)$	fast
$O_3(g) + NO(g) \rightarrow NO_2(g) + O_2(g)$	overall

Not consistent, slow first step, no dependence on $[O_3]$ is predicted by this mechanism.

LINKING QUESTION

Which mechanism in the hydrolysis of halogenoalkanes involves an intermediate?

WORKED EXAMPLE R2.2G

The mechanisms of five reactions are outlined below. State the rate equation for each reaction.

a	Overall reaction	$A + B \rightarrow D$	
	Mechanism	$A + B \rightarrow D$	
b	Overall reaction	$P + Q \rightarrow R$	
	Mechanism	$P \rightarrow S$	slow
		$S + Q \rightarrow R$	fast
c	Overall reaction	$A + B \rightarrow C + D$	
	Mechanism	$A \rightarrow M + D$	slow
		$M + B \rightarrow C$	fast
d	Overall reaction	$T + R \rightarrow P$	
	Mechanism	$T + H^+ \rightleftharpoons TH^+$	fast
		$TH^+ + R \rightarrow P + H^+$	slow
e	Overall reaction	$K + 2M \rightarrow N$	
	Mechanism	$2M \rightleftharpoons M_2$	fast
		$M_2 + K \rightarrow N$	slow

Answer

a rate = $k[A][B]$

b The rate-determining step is $P \rightarrow S$, which has a molecularity of 1, so rate = $k[P]$.

c The rate-determining step is $A \rightarrow M + D$, so rate = $k[A]$.

d The fast equilibrium step, $T + H^+ \rightleftharpoons TH^+$, occurs directly before the rate-determining step, $TH^+ + R \rightarrow P + H^+$, so rate = $k[T][H^+][R]$.

The hydrogen ions are acting as a homogeneous catalyst, so they affect the rate equation but do not appear in the overall equation.

e The fast equilibrium step, $2M \rightleftharpoons M_2$, which has a molecularity of 2, appears directly before the rate-determining step, $M_2 + K \rightarrow N$, so rate = $k[M]^2[K]$.

30 The elementary steps for a reaction mechanism are:

$XY_2 + XY_2 \rightarrow X_2Y_4$ slow

$X_2Y_4 \rightarrow X_2 + 2Y_2$ fast

a Deduce the overall equation for the reaction.
b Deduce the rate equation for the reaction.
c Deduce the units of the rate constant, k, in this rate equation.
d Sketch the energy profile for the reaction.

31 Propose a feasible mechanism for each of the following reactions on the basis of their rate equations and the given intermediate.

a $2NO_2Cl(g) \rightarrow 2NO_2(g) + Cl_2(g)$

rate = $k[NO_2Cl]$ intermediate: Cl

b $O_3(g) + 2NO_2(g) \rightarrow O_2(g) + N_2O_5(g)$

rate = $k[O_3][NO_2]$ intermediate: NO_3

c $H_2(g) + I_2(g) \rightarrow 2HI(g)$

rate = $k[I_2][H_2]$ intermediate: I

d $2NO(g) + 2H_2(g) \rightarrow N_2(g) + 2H_2O(g)$

rate = $k[NO]^2[H_2]$ intermediate: N_2O

LINKING QUESTION

Why are reaction mechanisms only considered as 'possible mechanisms?

Chain reactions

◆ **Chain reaction:** A reaction that is self-sustaining due to one elementary step producing a reactive intermediate that is used in a preceding step.

A **chain reaction** is a multi-step reaction where a reaction intermediate generated in one step reacts in such a way that this intermediate is regenerated. Chain reactions usually involve free radicals as intermediates.

The reaction between gaseous chlorine and methane in the presence of ultraviolet light (to form chloromethane and hydrogen chloride) is a chain reaction involving methyl and chlorine radicals.

$Cl\bullet + CH_4 \rightarrow HCl + CH_3\bullet$;

$CH_3\bullet + Cl_2 \rightarrow CH_3Cl + Cl\bullet$;

In the absence of an energy sink (for example, walls of the container), any two unlike radicals that collide will have enough energy to form two radicals – and more than enough energy to form chloromethane. This means they do not stick together; instead, they dissociate again.

$CH_3\bullet + Cl\bullet \rightarrow CH_3Cl \rightarrow CH_3\bullet + Cl\bullet$

Going further

Role of a third body

Only if a third body is involved in the collision can they dissipate some of their energy, which allows them to remain together and become a molecule.

$M + CH_3\bullet + Cl\bullet \rightarrow CH_3Cl + M$

The wall of the container is the most likely third body, but collisions between it and a pair of particles are rare. It is more likely that individual radicals will collide with the wall and be adsorbed, and this process tends to bring the chain reaction to an end.

Ozone depletion by chlorofluorocarbons (CFCs) (for example, dichlorodifluoromethane, CCl_2F_2) is a chain reaction that has had serious environmental consequences. In the presence of ultraviolet light, CFCs decompose to produce chlorine atoms:

$CF_2Cl_2 \rightarrow \bullet CF_2Cl + \bullet Cl$

These atoms are highly reactive radicals that catalyze the breakdown of ozone to oxygen molecules in the following elementary steps:

$\bullet Cl + O_3 \rightarrow \bullet ClO + O_2$

$\bullet ClO + O \rightarrow \bullet Cl + O_2$

The overall reaction is:

$O_3(g) + O(g) \rightarrow 2O_2(g)$

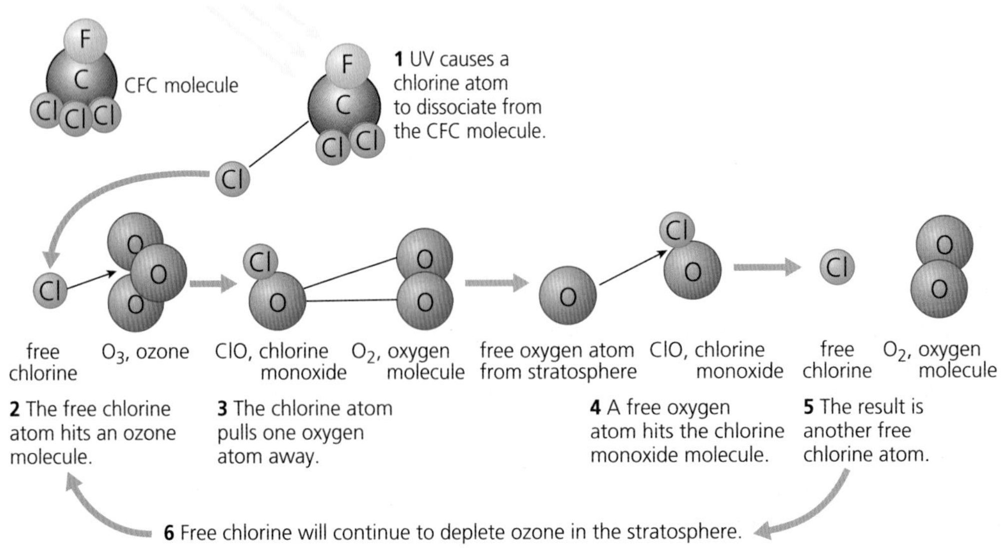
CFC molecule
1 UV causes a chlorine atom to dissociate from the CFC molecule.
free chlorine
O3, ozone
ClO, chlorine monoxide
O2, oxygen molecule
free oxygen atom from stratosphere
ClO, chlorine monoxide
free chlorine
O2, oxygen molecule
2 The free chlorine atom hits an ozone molecule.
3 The chlorine atom pulls one oxygen atom away.
4 A free oxygen atom hits the chlorine monoxide molecule.
5 The result is another free chlorine atom.
6 Free chlorine will continue to deplete ozone in the stratosphere.

■ **Figure R2.62** The catalytic cycle involved in the depletion of the ozone layer

The chlorine radical is regenerated in this catalytic cycle, which involves a chain reaction. It is estimated that a single chlorine radical can eliminate about 100 000 ozone molecules before it is physically or chemically removed from the stratosphere.

Going further

Transition state theory

According to this theory, when molecules collide, bond breaking and bond formation take place, and the interacting molecules form a high-energy and unstable species known as a transition state.

The transition state can either decompose to re-form the reactants, or it can undergo further changes to form the product molecules or an intermediate. This is illustrated by the simple bimolecular reaction between hydrogen and iodine to form hydrogen iodide.

In Figure R2.63:

- the green dotted line represents the energy of the transition state.
- **a** represents the enthalpy of the individual reactant hydrogen and iodine molecules.
- At **b** weak covalent bonds start to form between the atoms of hydrogen molecules and iodine molecules. At the same time, the hydrogen–hydrogen and iodine–iodine bonds start to lengthen and weaken.
- **c** represents the formation of the transition state.
- At **d** the two hydrogen–iodine bonds continue to shorten and strengthen, while the hydrogen–hydrogen and iodine–iodine bonds continue to lengthen and weaken; the electron density between the hydrogen and iodine atoms steadily increases.
- **e** represents the enthalpy of the two hydrogen iodide molecules.

As well as the energy considerations illustrated by the diagram, there is also a steric effect in this reaction; a successful reaction requires the hydrogen and iodine molecules to collide 'sideways'.

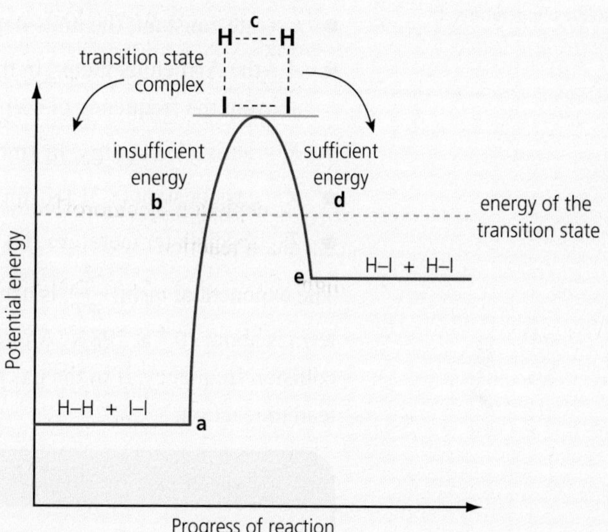

■ **Figure R2.63** Enthalpy or potential energy level diagram for the formation of hydrogen iodide

TOK

Why might some people regard science as the supreme form of all knowledge?

Some people without a deep understanding of the nature of science and the scientific method might regard science as the supreme form of knowledge. However, science is not authoritarian; instead, it tries to be objective and to identify and avoid bias.

Scientific inquiry is based on empirical evidence (data from observation and experiment) which is used to explain and predict processes in the physical world. If evidence accumulates that contradicts a belief about how the physical world operates, that belief must be re-examined – no matter how strongly it is held – and may even be relinquished. Willingness to change a belief based on evidence is one of the basic values of science.

But neither the scientific method nor the scientific community is infallible. Scientific knowledge, such as reaction mechanisms in kinetics, is always subject to change and revision in the light of new data.

For example, many pre-university text books, including this one, introduce transition state theory using the reaction between hydrogen and iodine molecules to form a four-centre transition state that then forms hydrogen iodide molecules. However, it has been demonstrated that the reaction is more likely to proceed via a mechanism involving iodine radicals which react with hydrogen molecules. One piece of evidence for an atomic mechanism is the observed increase in rate when the reaction mixture is irradiated with visible light in the region of 500 to 600 nm.

This illustrates the point that the knowledge in text books may be simplified (or even incorrect) in order to teach key concepts, rather than accurately reflecting the latest scientific research.

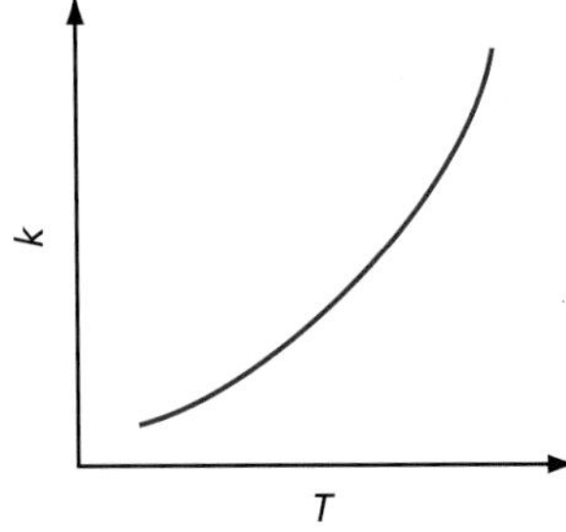

■ **Figure R2.64** Plot of rate constant, k, against absolute temperature, T

Rate constant and temperature

When the temperature increases, the rate of reaction, and hence the rate constant, increases exponentially (Figure R2.64).

This relationship between absolute temperature and the rate constant can be modelled by the Arrhenius equation:

$$k = Ae^{\frac{-E_a}{RT}}$$

where:

- k = rate constant; the units depend on the overall order of reaction
- A = the Arrhenius factor, in the same units as k (this factor allows the equation to take into account the frequency of collisions with proper orientations)
- E_a = activation energy, in J mol^{-1}
- R = gas constant ($8.31\,\text{J K}^{-1}\,\text{mol}^{-1}$)
- T = absolute temperature (K).

The exponential term, $\frac{-E_a}{RT}$, is usually very small. For example if $E_a = 50\,\text{kJ mol}^{-1}$, the exponential term is about 10^{-9} at 298 K. A rate constant is often the product of a rather large number (the collision frequency, if in the gas phase) and a very small number – the fraction of collisions which lead to reaction.

Tool 3: Mathematics

Exponential functions

Exponential functions are based on the irrational number e, which is approximately equal to 2.71828. An irrational number cannot be expressed exactly as a fraction and any decimal representation is approximate. π is another example of an irrational number.

We can use e as the base for logarithms instead of using 10, that is, if $e^b = a$ then $\log_e a = b$.

Logarithms to the base e are referred to as natural logarithms and given the symbol ln:

$$\ln x = \log_e x$$

They are very useful because they are the inverse function to raising something to the power of e:

$$e^{\ln x} = x \quad \text{for all } x > 0 \qquad \text{and} \qquad \ln(e^x) = x \text{ for all } x$$

Exponential relationships are common in the natural sciences. In IB Chemistry, they are used:

- as illustrated in this section of the book, in the Arrhenius equation, $k = Ae^{\frac{-E_a}{RT}}$ or $\ln k = \ln A - \frac{E_a}{RT}$ (used to calculate the activation energy for a reaction)
- in the equation $\Delta G = -RT\ln K$ (used to calculate the Gibbs free energy change of an equilibrium reaction (see R1.4, page 404)).

You can find powers of e and natural logarithms using your calculator, just as you work out values of 10^x and $\log_{10} x$. For example, $e^{3.72} = 41.3$, $e^{-1.36} = 0.257$ and $\ln 1570 = 7.36$.

Natural logarithms are larger numbers than the logarithms of the same number to base 10 because e < 10, so it has to be raised to a higher power to get the same number.

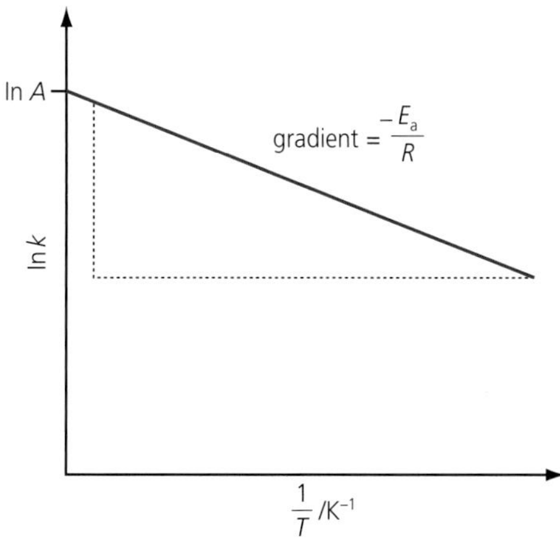

■ **Figure R2.65** An Arrhenius plot of ln k against T^{-1}

The linear form of the Arrhenius equation, $\ln k = \left(\frac{-E_a}{R}\right)\left(\frac{1}{T}\right) + \ln A$, is useful as it has the same structure as the formula for a straight line.

$$y = mx + c$$

An Arrhenius plot (Figure R2.65) shows the natural logarithm of the rate constant plotted against the reciprocal of the absolute temperature. The resulting graph is a straight line with gradient $-\frac{E_a}{R}$ and a y-axis intercept of ln$A$.

This can be used to find activation energies since $E_a = -R \times$ gradient.

Values of A from these plots are usually much less precise than E_a values because of the long extrapolation to $1/T = 0$.

Going further

Arrhenius temperature dependence

Reactions that have an activation energy of approximately 50 kJ mol^{-1} show Arrhenius temperature dependence; a rise in temperature of 10 °C will approximately double the initial rate and rate constant of the reaction over a range of temperatures (Figure R2.66).

This is sometimes stated as a general rule but, since values of activation energies vary considerably, reactions may be either much faster or much slower.

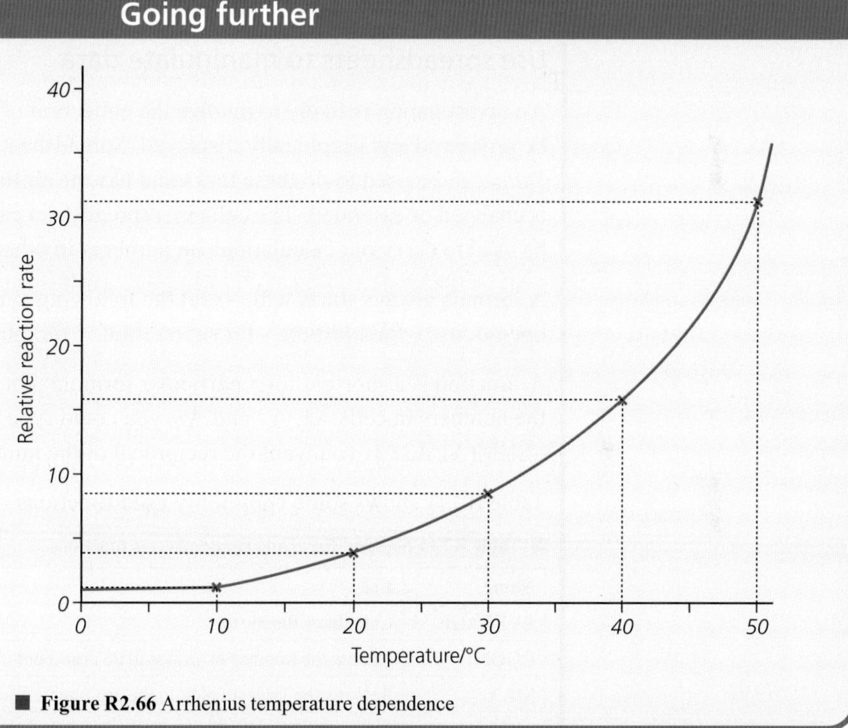

■ **Figure R2.66** Arrhenius temperature dependence

Inquiry 1: Exploring and designing

Maintain constant environmental conditions of system

Imagine you are carrying out an investigation to determine the activation energy of a reaction in aqueous solution by measuring the rate constant or relative rate at five different temperatures (as it is a logarithmic relationship, anything proportional to k can be used). The best approach is to use a thermostatically controlled water bath to bring each solution to the desired temperature. Accuracy in maintaining the temperature of the water bath is essential for obtaining precise and reliable results. You need to wait for at least 10 minutes to allow the chemicals to come to the same temperature as the water bath. You also need to measure the temperature of the reaction mixture rather than the temperature of the water bath.

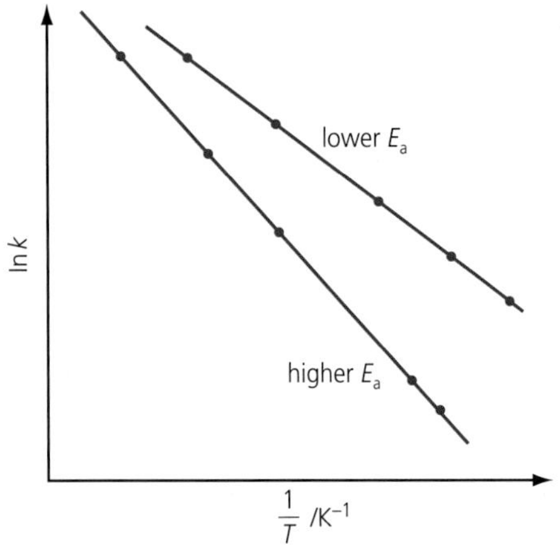

■ **Figure R2.67** Arrhenius plots for two reactions with different activation energies

If Arrhenius plots are drawn on the same axes for two reactions with different activation energies (Figure R2.67), the reaction with the higher activation energy has a steeper gradient. This means the rate constant and initial rate change with temperature much more rapidly than for the reaction with the lower activation energy.

Tool 2: Technology

Use spreadsheets to manipulate data

An investigation is likely to involve the collection of large amounts of raw data that need to be processed and graphically displayed. Spreadsheet software, such as *Excel*, *Numbers* or *Quip*, can be used to do these tasks and has the ability to update any graphs if the raw data is changed or extended. The cells in a spreadsheet can hold numbers or formulas, which can be used to carry out calculations on numbers in other cells.

A formula always starts with = and the following symbols are used for mathematical operations: + for addition, - for subtraction, * for multiplication and / for division.

A function is a shortcut for a particular formula. For example, if you wanted to add the numbers in cells A1, A2 and A3, you could type the formula: =A1+A2+A3 or =SUM(A1:A3). If you want the reciprocal of the number in cell A1, then type =1/A1.

Table R2.11 shows some commonly used functions in *Excel*.

■ **Table R2.11** Selected *Excel* and *Google Sheets* functions

Name	Use	Example
AVERAGE	Calculates the mean	=AVERAGE(A1:A3)
COUNT	Counts the number of values in an argument	=COUNT(A1:A3)
MAX	Returns the largest value in the argument	=MAX(A1:A3)
MIN	Returns the smallest value in the argument	=MIN(A1:A3)
SUM	Calculates the sum of the values in an argument	=SUM(A1:A5)
LN	Calculates the natural logarithm of a number	=LN(A1)
LOG	Calculates the logarithm to the base 10 of a number	=LOG(A1)
ROUND	Rounds the argument to the specified number of decimal places	=ROUND(A1,3)

A formula can be copied down to the cells in the same column or row by 'filling down'. In *Excel*, that means dragging the small cross in the corner of the active cell down or across to the cells where the formula needs to be copied.

Spreadsheets can also plot graphs. In *Excel*, this can be done by clicking on the Insert tab and choosing the appropriate type of graph, typically a scatter graph.

32 The rate constant, k, was determined for a reaction at various temperatures. The results are given in the table.

Temperature / °C	Second-order rate constant, k / $mol^{-1} dm^3 s^{-1}$
5	6.81×10^{-6}
15	1.40×10^{-5}
25	2.93×10^{-5}
35	6.11×10^{-5}

LINKING QUESTION

What is the relative effect of a catalyst on the rate of the forward and backward reactions?

a Plot a graph of $\ln k$ against T^{-1} where T must be expressed as an absolute temperature. (Plot the $\ln k$ axis from −12 to −8 and the T^{-1} axis from 0.0030 to 0.0038.)

b Calculate the gradient of your Arrhenius plot and use it to determine a value for the activation energy, E_a, in $kJ\,mol^{-1}$.

c Calculate an approximate value for the Arrhenius constant, A, using the Arrhenius plot.

R2.3 How far? The extent of chemical change

Guiding question

- How can the extent of a reversible reaction be influenced?

SYLLABUS CONTENT

By the end of this chapter, you should understand that:

- a state of dynamic equilibrium is reached in a closed system when the rates of forward and backward reactions are equal
- the equilibrium law describes how the equilibrium constant, K, can be determined from the stoichiometry of a reaction
- the magnitude of the equilibrium constant indicates the extent of a reaction at equilibrium and is temperature dependent
- Le Châtelier's principle enables the prediction of the qualitative effects of changes in concentration, temperature and pressure on a system at equilibrium
- the reaction quotient, Q, is calculated using the equilibrium expression with non-equilibrium concentrations of reactants and products (HL only)
- the equilibrium law is the basis for quantifying the composition of an equilibrium mixture (HL only)
- the equilibrium constant and Gibbs energy change, ΔG, can both be used to measure the position of an equilibrium reaction (HL only).

By the end of this chapter you should know how to:

- describe the characteristics of a physical and chemical system at equilibrium
- deduce the equilibrium constant expression from an equation for a homogeneous reaction
- determine the relationships between K values for reactions that are the reverse of each other at the same temperature
- apply Le Châtelier's principle to predict and explain responses to changes of systems at equilibrium
- calculate the reaction quotient, Q, from the concentrations of reactants and products at a particular time, and determine the direction in which the reaction will proceed to reach equilibrium (HL only)
- solve problems involving values of K and initial and equilibrium concentrations of the components of an equilibrium mixture (HL only).

Reversible reactions

The extent of a chemical reaction is determined by thermodynamics. The rate of a chemical reaction is determined by kinetics.

In this chapter, we focus on describing and quantifying how far a chemical reaction goes based on an experimentally measurable quantity called the equilibrium constant.

Many reactions you are familiar with are essentially **irreversible**; it is very difficult to re-form the reactants. For example, the combustion of methane is irreversible. The **forward reaction** is represented by the equation:

$$CH_4 + 2O_2 \rightarrow CO_2 + 2H_2O$$

The **backward reaction** does not occur easily. It is very difficult to form the reactants, methane and oxygen, from water and carbon dioxide.

◆ **Irreversible**: A reaction where the extent of the backward reaction is negligible.

◆ **Forward reaction**: The conversion of reactants into products in an equilibrium.

◆ **Backward reaction**: The conversion of products into reactants in an equilibrium.

◆ **Reversible:** A reaction where the products can also react to form the reactants.

◆ **Thermal dissociation:** The breakdown of a molecule or ion into smaller species due to the action of heat.

However, some chemical reactions are more readily **reversible**. One example is the dehydration (or thermal decomposition) of hydrated copper(II) sulfate, $CuSO_4.5H_2O(s)$, by strong heating; the reverse reaction is the hydration of anhydrous copper(II) sulfate, $CuSO_4(s)$.

The reaction of ammonia with hydrogen chloride (an acid–base reaction) is a reaction for which the direction of change depends on the temperature. At room temperature, the two gases react to form a white cloud of solid ammonium chloride:

$$NH_3(g) + HCl(g) \rightarrow NH_4Cl(s)$$

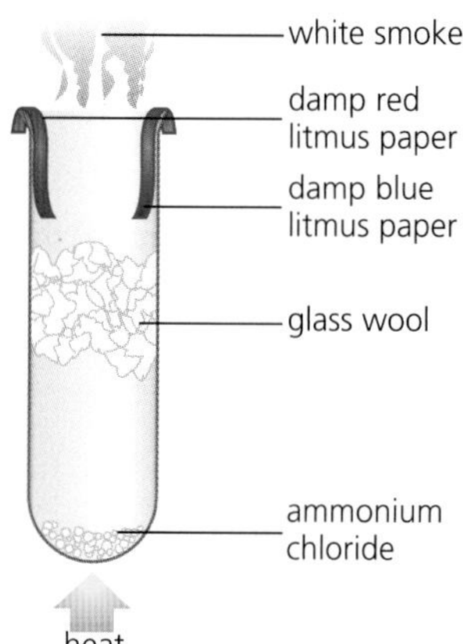

■ **Figure R2.68** Investigating the thermal decomposition of ammonium chloride

Heating makes this reversible reaction go the other way. Ammonium chloride undergoes **thermal dissociation** (decomposes) at high temperatures to form hydrogen chloride and ammonia. This can be demonstrated as shown in Figure R2.68; the red and blue litmus change colour in the presence of ammonia and hydrogen chloride, respectively, and the white smoke at the top provides evidence that the reaction reverses as the cooler gases escape the boiling tube.

$$NH_4Cl(s) \rightarrow NH_3(g) + HCl(g)$$

Changing the temperature is not the only way to alter the direction of change. Heated iron, for example, reacts with steam to form an oxide of iron in a redox reaction (Figure R2.69). Supplying excess steam and flushing away the hydrogen allows the reaction to continue until all the iron is converted to its oxide.

$$3Fe(s) + 4H_2O(g) \rightarrow Fe_3O_4(s) + 4H_2(g)$$

The forward reaction is favoured when the concentration of steam is high and the hydrogen is flushed away, keeping its concentration low.

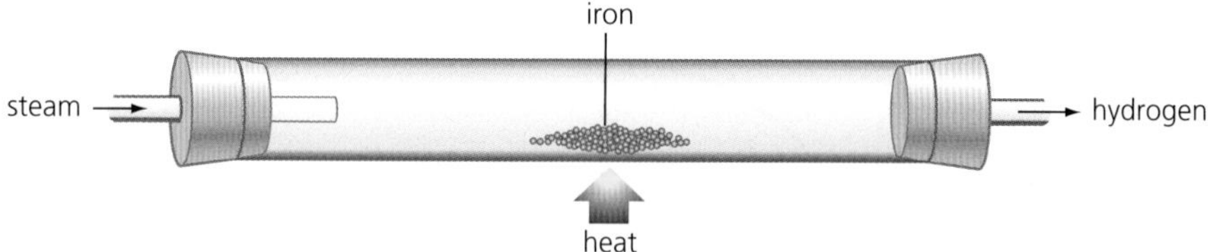

■ **Figure R2.69** The forward reaction occurs when the concentration of steam is high and the hydrogen is swept away, keeping its concentration low

Changing the conditions favours the reverse reaction. A stream of hydrogen reduces all the iron oxide to iron as long as the flow of hydrogen flushes away the steam formed (Figure R2.70).

$$Fe_3O_4(s) + 4H_2(g) \rightarrow 3Fe(s) + 4H_2O(g)$$

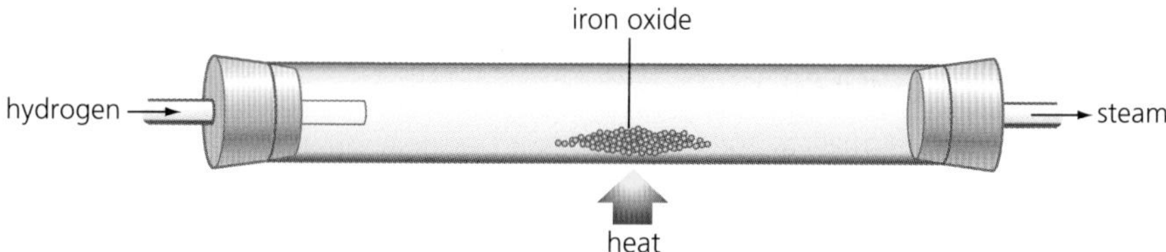

■ **Figure R2.70** The backward reaction occurs when the concentration of hydrogen is high and the steam is flushed away, keeping its concentration low

Equilibrium

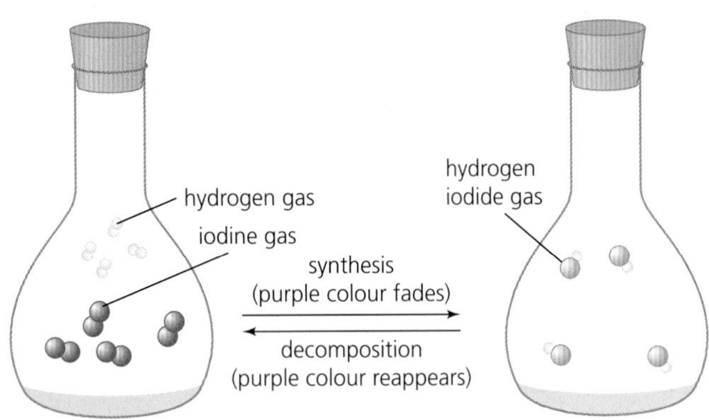

■ **Figure R2.71** Establishment of an equilibrium reaction between iodine and hydrogen to form hydrogen iodide

Consider the reversible reaction $H_2(g) + I_2(g) \rightleftharpoons 2HI(g)$ carried out in a sealed container (Figure R2.71). This is an example of a closed system – chemicals cannot leave or enter, but heat can flow in or out.

At temperatures above approximately 25 °C and below 500 °C, it is found that this reaction does not go to completion. Instead, a stoichiometric mixture of hydrogen and iodine forms a mixture containing all three components – hydrogen, iodine and hydrogen iodide – in proportions which depend on the temperature.

Hydrogen and iodine react to form hydrogen iodide and, at the same time, hydrogen iodide re-forms hydrogen and iodine.

Common mistake

Misconception: *Equilibrium reactions continue until all the reactants are consumed.*

At equilibrium reactants and products are both present.

Top tip!

Once a system is at equilibrium, it is impossible to determine whether the system was prepared by starting with reactants or products.

Figure R2.72 plots the concentrations of hydrogen and iodine (the reactants) and hydrogen iodide (the product) over time.

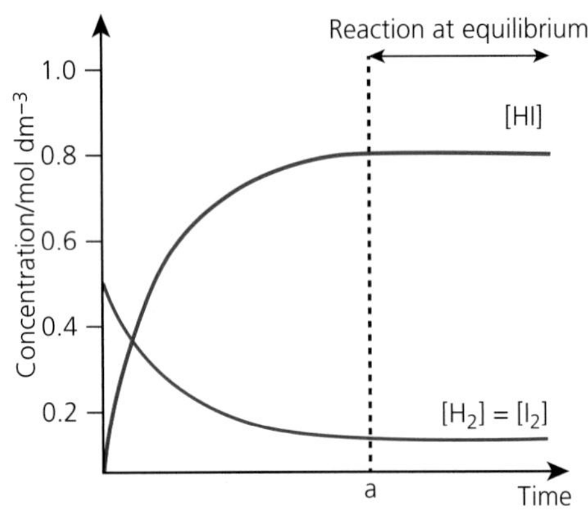

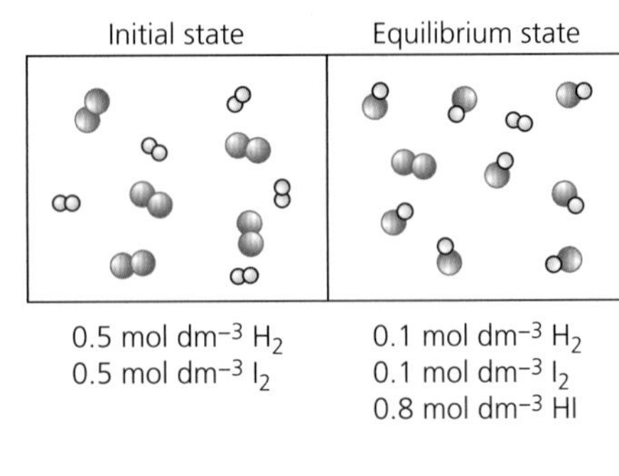

■ **Figure R2.72** A graph showing the relationship between hydrogen, iodine and hydrogen iodide concentrations as the reaction proceeds to equilibrium

- At the start of the reaction (time = 0) the concentration of hydrogen iodide [HI(g)] is zero as the reaction has not taken place yet. The concentrations of hydrogen $[H_2(g)]$ and iodine $[I_2(g)]$ are both $0.5\,mol\,dm^{-3}$.
- At times greater than zero but less than *a* (where equilibrium is established), hydrogen and iodine react to form hydrogen iodide at a fast rate (as shown by the steep gradient of the curve).
  - The concentration of hydrogen $[H_2(g)]$ and iodine $[I_2(g)]$ decreases, while the concentration of hydrogen iodide [HI(g)] increases.
  - The rate of the forward reaction decreases (as shown by the shallower gradient) as the concentrations of hydrogen $[H_2(g)]$ and iodine $[I_2(g)]$ decrease.
- At times greater than *a*, the rate of the forward reaction equals the rate of the backward reaction. The concentrations of hydrogen $[H_2(g)]$ ($0.10\,mol\,dm^{-3}$), iodine $[I_2(g)]$ ($0.10\,mol\,dm^{-3}$) and hydrogen iodide [HI(g)] ($0.80\,mol\,dm^{-3}$) remain constant.

◆ **Dynamic equilibrium**: An equilibrium is described as dynamic because, although there is no change in macroscopic properties, the forward and backward reactions are occurring at equal rates.

The chemical system is in a state of **dynamic equilibrium**. Dynamic equilibrium is established in a closed system when the rates of the forward and backward reactions are the same. At equilibrium there is no overall change in the concentration of the reactants or products (provided the temperature remains constant).

A good analogy of a system in dynamic equilibrium is a person walking up an escalator that is moving down (Figure R2.73). The person on the escalator is in a state of dynamic equilibrium if they walk up the escalator at the same rate as the escalator moves down.

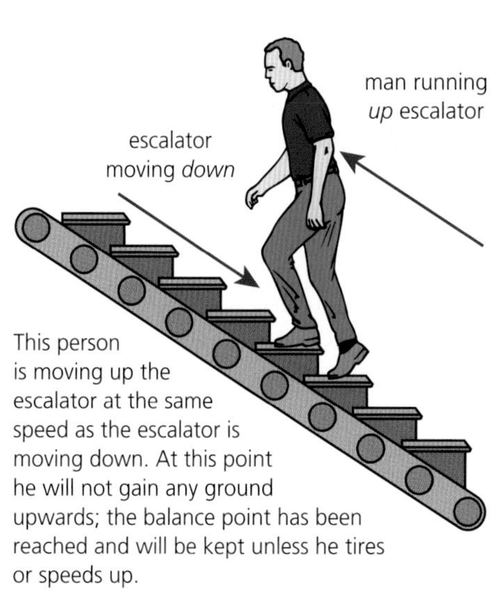

man running
up escalator
escalator
moving down
This person is moving up the escalator at the same speed as the escalator is moving down. At this point he will not gain any ground upwards; the balance point has been reached and will be kept unless he tires or speeds up.

■ **Figure R2.73** Analogy of a dynamic equilibrium

Common mistake

Misconception: *The rate of forward reaction is greater than the reverse reaction rate.*

At equilibrium the forward and backward reactions continue to occur (at the same rate) provided the system is closed.

Common mistake

Misconception: *At equilibrium, no reaction occurs.*

At equilibrium, both the forward and backward reactions are occurring at equal rates.

Top tip!

The concept of equilibrium is fundamental to the understanding of acid–base behaviour (including buffers), redox reactions (including voltaic cells), organic chemistry, metal refining and solubility (solubility product).

All reversible reactions, in a closed system, will reach dynamic equilibrium. This is represented graphically in Figure R2.74.

During the initial stage, the rate of the forward reaction is very fast and the backward reaction is taking place at a negligible rate. Gradually the forward reaction slows down and the backward reaction speeds up. Eventually, both reactions attain the same rate, and a state of dynamic equilibrium is reached.

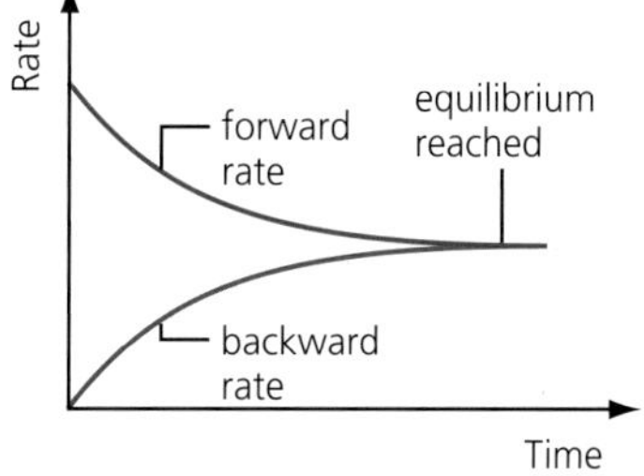

■ **Figure R2.74** The rate of the forward reaction (red): $H_2 + I_2 \rightarrow 2HI$ and backward reaction (blue): $2HI \rightarrow I_2 + H_2$ leading to an equilibrium between hydrogen, iodine and hydrogen iodide

Top tip!

Chemical equilibrium is an integrating and unifying concept because it requires connections from several areas of chemistry, such as the mole and reaction stoichiometry, gas laws, kinetics and thermodynamics.

Common mistake

Misconception: *The forward reaction goes to completion before the reverse reaction starts.*

Both reactions occur at the same time but occur at different rates until equilibrium is attained.

ATL R2.3A

One of your classmates is having problems understanding the dynamic nature of equilibrium.

- Devise a simple experiment with suitable questions using glass troughs, beakers of different sizes and coloured water to clearly illustrate this concept to your classmate.
- After your demonstration, you might want to compare your approach with the ideas shown in this website: **www.flinnsci.com/api/library/Download/adaef0ac15f64d138dd264fc8f656dac**
- Evaluate this demonstration as a 'mechanical' analogy to a chemical equilibrium.

The phosphorus cycle and eutrophication

On Earth, phosphorus is present in the form of atoms, ions and molecules such as P, PO_4^{3-} and P_4. Since matter cannot be formed or destroyed, phosphorus, like other elements, cycles through various forms as it moves through the Earth system. The phosphorus cycle includes both physical and biological environments in which compounds are interconnected through reversible chemical equilibria.

Rocks that contain phosphates are weathered and then, due to rain and erosion, phosphate ions are washed into rivers and the sea. New phosphate-containing rocks are slowly formed when organic compounds, rich in phosphorus compounds, accumulate on sea beds forming sediments that are later raised to sea level by plate movements.

Excessive levels of phosphate in waterways (from animal waste and laundry detergents) can result in uncontrolled algal growth which depletes the water of dissolved oxygen – a condition known as eutrophication, which reduces biodiversity.

Large rivers may flow through several countries, and can carry phosphate ions into coastal regions and wider marine environments including international waters. Controlling eutrophication and its consequences therefore requires international cooperation and coordination.

Conditions for dynamic equilibrium

A dynamic equilibrium (or, simply, an equilibrium) can only be achieved in a closed system at constant temperature.

Consider the thermal decomposition of calcium carbonate:

$$CaCO_3(s) \rightarrow CaO(s) + CO_2(g)$$

In an open container (Figure R2.75b), carbon dioxide will escape into the surroundings upon heating. It will not be able to react with calcium oxide to reform calcium carbonate. The system will never reach equilibrium; instead, it is essentially irreversible in the forward direction.

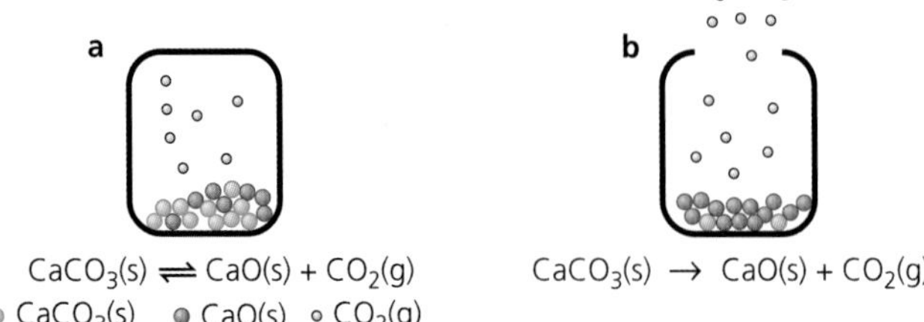

Figure R2.75 a A closed system where no carbon dioxide escapes and an equilibrium is established. **b** An open system; here the calcium carbonate is continually decomposing as the carbon dioxide is lost. The reaction goes to completion

However, if calcium carbonate is heated in a sealed container (Figure R2.75a), the carbon dioxide formed can react with calcium oxide to reform calcium carbonate. There will always be some calcium carbonate remaining, regardless of how long the solid is heated.

This system is in dynamic equilibrium and can be represented by the following equation:

$$CaCO_3(s) \rightleftharpoons CaO(s) + CO_2(g)$$

Nature of science: Models

Studying equilibria

A heavy isotope is one having an extra neutron or neutrons in the nucleus of the atom. Deuterium (^2_1H or D) is a heavier isotope of hydrogen in which the nucleus contains a neutron.

In studies of the ammonia equilibrium ($N_2(g) + 3H_2(g) \rightleftharpoons 2NH_3(g)$), some of the hydrogen is replaced by an equal amount of 'heavy hydrogen', D_2. The D_2 molecules have identical chemical behaviour to H_2 molecules (Figure R2.76).

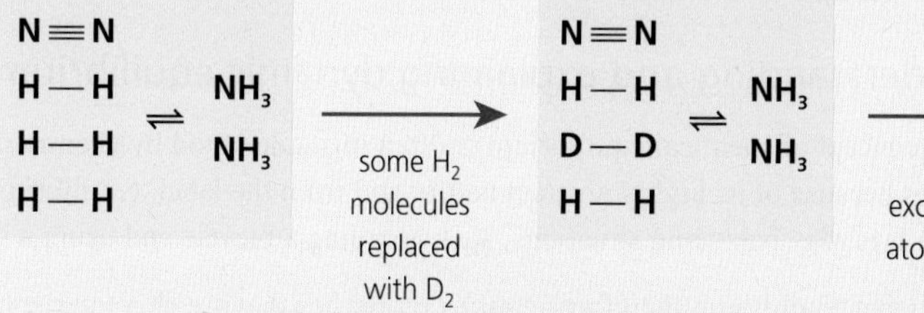

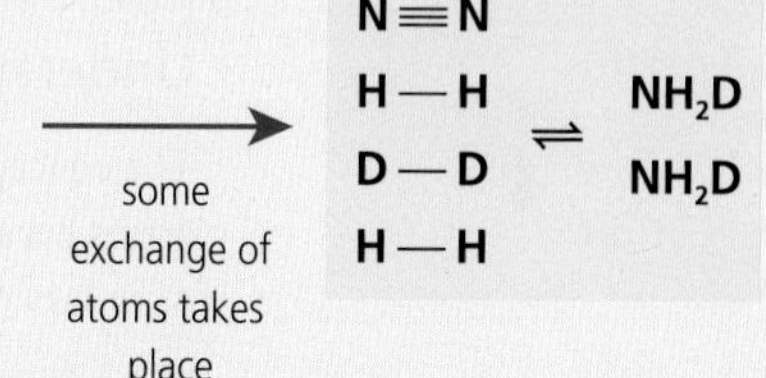

■ **Figure R2.76** Incorporation of deuterium into ammonia within an equilibrium mixture of nitrogen, hydrogen, deuterium and ammonia

When the new equilibrium mixture is analysed using a mass spectrometer, NH_2D, NHD_2, ND_3 and HD will be detected. These results are explained by an exchange of atoms between the molecules of ammonia, hydrogen and deuterium in the equilibrium mixture.

A similar experiment can be carried out with lead(II) chloride, $PbCl_2$, labelled with radioactive lead(II) ions.

Lead(II) chloride is only slightly soluble in cold water, and a saturated, filtered solution of the radioactive salt can be prepared easily. If unlabelled solid lead(II) chloride is added to the saturated solution, it precipitates to the bottom as no more solid can dissolve in the saturated solution. However, after a length of time, it is found that some of the radioactive lead is present in the solid (see Figure R2.77).

This system can be described by the following equilibrium:

$$PbCl_2(s) \rightleftharpoons Pb^{2+}(aq) + 2Cl^-(aq)$$

This shows that some of the radioactive lead(II) ions in the solution have been precipitated into the solid, and an equal number of lead(II) ions from the solid have dissolved to keep the solution saturated.

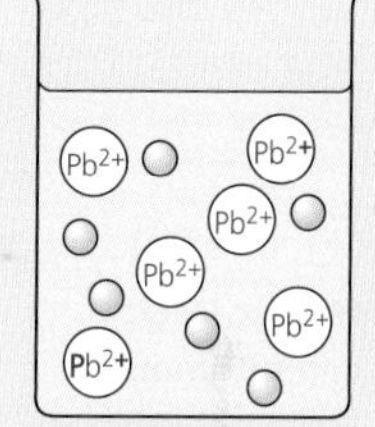

a) saturated, radioactive lead(II) chloride (filtered)

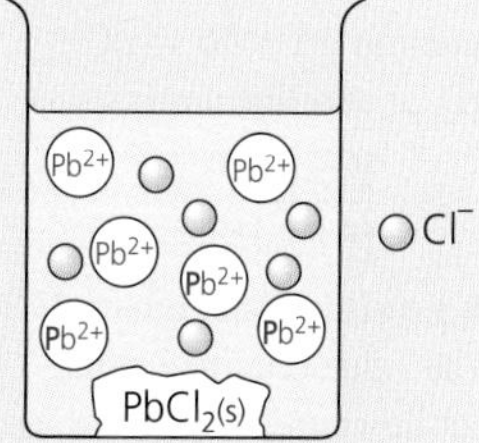

b) addition of non-radioactive $PbCl_2$ (s)

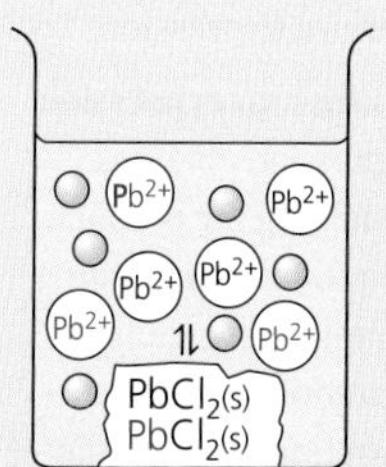

c) at equilibrium both the solution and the solid contain radioactive Pb^{2+} ions

■ **Figure R2.77** Incorporation of some of the radioactive lead(II) ions into solid lead(II) chloride

The equilibrium law

Position of equilibrium

We often use the term 'position of equilibrium' to refer to the specific state of a system – the concentration of reactants and products – at equilibrium. At equilibrium, at a particular temperature and pressure, there is a very specific relationship between the concentrations of the species present.

For the general reversible reaction:

$$A + B \rightleftharpoons C + D$$

- if the system contains a larger proportion of products (C and D) than reactants (A and B), the position of equilibrium lies to the right
- if the system contains a greater proportion of reactants (A and B) than products (C and D), the position of equilibrium lies to the left.

Nature of science: Models

Understanding and explaining dynamic equilibrium

The concept of chemical equilibrium is often misunderstood by chemistry students. This misconception arises because of its highly abstract nature and from the label 'equilibrium' being used in physics as well as in everyday balancing situations, such as riding a bicycle and using a weighing balance.

The label equilibrium therefore acquires properties that are characteristic of these situations. Hence, an equality of two sides, stability and a static and unchanging nature become associated with the concept of equilibrium.

But these attributes of equilibrium are the opposite to those of physical and chemical equilibria.

Processes that reach chemical equilibrium appear to be macroscopically stable and static systems. However, at the **microscopic level** the system is dynamic, not only because of molecular movement but also because the process of breaking and creating bonds continues for ever (provided the system is closed).

Models are important in all the natural sciences but especially in chemistry, and their development is crucial in understanding and generating chemical knowledge. In chemistry, the observed phenomenon (such as an equilibrium system) is re-conceptualized not only at the macroscopic (bulk) level but also in terms of the theoretical models of the structure of matter at the submicroscopic level.

Chemical equilibrium as a dynamic equilibrium is a relatively simple concept but it can be explained in a number of different ways: using Le Châtelier's principle (a 'rule'), by kinetics of opposing reactions and by thermodynamics. Knowledge of all three models is expected by the IB Chemistry syllabus.

◆ **Microscopic level:** Atomic and subatomic levels.

◆ **Equilibrium law**: In general, for a reversible reaction $aA + bB \rightleftharpoons cC + dD$, at equilibrium

$$K = \frac{[C]^c_{eqm}[D]^d_{eqm}}{[A]^a_{eqm}[B]^b_{eqm}}$$

◆ **Equilibrium constant:** The value obtained when equilibrium concentrations of the chemical species are substituted in the equilibrium expression.

◆ **Equilibrium expression:** The expression obtained by multiplying the product concentrations and dividing by the multiplied reactant concentrations, with each concentration raised to the power of the coefficient in the balanced equation.

Equilibrium constants

For the reversible reaction: $aA + bB \rightleftharpoons cC + dD$, at equilibrium, there is a fixed mathematical relationship between the equilibrium concentrations of reactants and products known as the **equilibrium law**. This is shown in Figure R2.78. K is the **equilibrium constant** and it is given by the **equilibrium expression** which is expressed in terms of molar concentrations.

Top tip!

Pure solids and pure liquids acting as solvents are excluded from an equilibrium expression.

> **Top tip!**
>
> An equilibrium constant is actually expressed in terms of activities (no units) (or relative concentrations) and not molar concentrations. The two values have almost the same value when the solutions are very dilute. The activity is a measure of the effective concentration since activities are always less than or equal to molar concentrations and take into account ionic interactions.

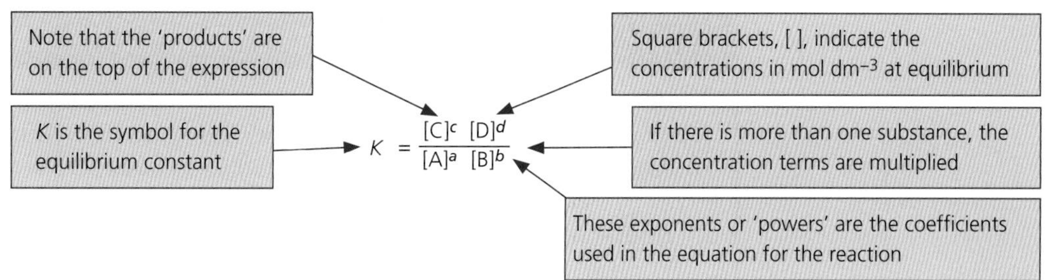

Note that the 'products' are on the top of the expression
Square brackets, [ ], indicate the concentrations in mol dm−3 at equilibrium
K is the symbol for the equilibrium constant
$K = \frac{[C]^c\ [D]^d}{[A]^a\ [B]^b}$
If there is more than one substance, the concentration terms are multiplied
These exponents or 'powers' are the coefficients used in the equation for the reaction

■ **Figure R2.78** The equilibrium law

It should be noted that the equilibrium constant, K, is constant for a given temperature. If the temperature changes, then the value of K will change.

> **LINKING QUESTION**
>
> How does the value of K for the dissociation of an acid convey information about its strength?

TOK

Why are many of the laws in the natural sciences stated using the language of mathematics?

This is a deep question, which physicists have been contemplating for many years. Many laws in the natural sciences, especially physics, are best expressed as mathematical equations. For example, the equilibrium law is a chemical law described in mathematical terms and first proposed in 1867 by the Norwegian chemist Cato Guldberg (1836–1902) and the mathematician Peter Waage (1833–1900).

One simple response to this question is that physics is so successfully described by mathematics because the physical world is completely mathematical. Another response is that the laws of nature are human inventions – nature *just is*.

The mathematical structure of a physical theory may point to further advances in that theory and even empirical predictions. For example, the theoretical physicist Paul Dirac (1902–1984) predicted the existence of the positron (a particle identical to an electron but with a positive charge) and other types of antimatter by simply analysing the mathematical structure of his equation for the electron.

Tool 3: Mathematics

Calculations involving exponents

In IB Chemistry, terms involving exponents are only multiplied or divided.

Multiplication

If the base is the same, the exponents can be added together. For example:

$$2^3 \times 2^4 = 2^{3+4} = 2^7$$

If the exponent is the same, the bases can be multiplied, and the exponent kept the same. For example:

$$2^2 \times 3^2 = (2 \times 3)^2 = 6^2$$

If both the base and the exponent are different, each value will need to be determined. For example:

$$3^2 \times 4^3 = 9 \times 64 = 576$$

Division

If the base is the same, the exponents can be subtracted. For example:

$$\frac{3^5}{3^3} = 3^{5-3} = 3^2$$

If the exponent is the same, the bases can be divided, and the exponent kept the same. For example:

$$\frac{6^4}{2^4} = \left(\frac{6}{2}\right)^4 = 3^4$$

If both base and exponent are different, as with the multiplication rule, each individual value will need to be determined. For example:

$$\frac{5^3}{3^2} = \frac{125}{9} = 13.9$$

WORKED EXAMPLE R2.3A

The table below shows the equilibrium concentrations of gases in three ammonia synthesis reactions, all carried out at 500 °C. Use the data to demonstrate that the equilibrium concentrations for a particular reaction produce a constant value of K.

$$N_2(g) + 3H_2(g) \rightleftharpoons 2NH_3(g) \qquad K = \frac{[NH_3]^2}{[N_2][H_2]^3}$$

	$[N_2]$ / mol dm^{-3}	$[H_2]$ / mol dm^{-3}	$[NH_3]$ / mol dm^{-3}
Experiment 1	0.922	0.763	0.157
Experiment 2	0.399	1.197	0.203
Experiment 3	2.59	2.77	1.82
Experiment 4	1.20	3.60	1.84
Experiment 5	0.23	0.92	0.104
Experiment 6	0.20	0.40	0.0278

Answer

Substituting values from the table:

Experiment 1 $\quad K = \frac{(0.157)^2}{0.922 \times (0.763)^3} = 0.0602$

Experiment 2 $\quad K = \frac{(0.203)^2}{0.399 \times (1.197)^3} = 0.0602$

Experiment 3 $\quad K = \frac{(1.82)^2}{2.59 \times (2.77)^3} = 0.0602$

Experiment 4 $\quad K = \frac{(1.84)^2}{1.20 \times (3.60)^3} = 0.0602$

Experiment 5 $\quad K = \frac{(0.104)^2}{0.23 \times (0.92)^3} = 0.0602$

Experiment 6 $\quad K = \frac{(0.0278)^2}{0.20 \times (0.40)^3} = 0.0602$

K is a constant value for all six experiments since all six experiments were conducted at the same temperature.

33 For each of the following reactions, write an expression for the equilibrium constant, K.

a $H_2(g) + F_2(g) \rightleftharpoons 2HF(g)$

b $O_2(g) + 2SO_2(g) \rightleftharpoons 2SO_3(g)$

c $[Ag(NH_3)_2]^+(aq) \rightleftharpoons Ag^+(aq) + 2NH_3(aq)$

d $4PF_5(g) \rightleftharpoons P_4(g) + 10F_2(g)$

e $N_2(g) + O_2(g) \rightleftharpoons 2NO(g)$

34 Write a balanced equation from the following equilibrium expression:

$$K = \frac{[O_2(g)]^3}{[O_3(g)]^2}$$

If a chemical equation is changed in some way, then the equilibrium constant for the equation changes because of the change in the chemical equation (Table R2.12).

■ **Table R2.12** The equilibrium constant for the same reaction at the same temperature can be expressed in a number of ways

Change in reaction equation	Equilibrium constant expression	Equilibrium constant
Reverse the reaction	Inverse of expression	K^{-1}
Halve the stoichiometric coefficients	Square root of the expression	$\sqrt{K}$
Double the stoichiometric coefficients	Square of the expression	K^2
Sequence of reactions	Multiply the values for the individual steps	$K = K_1 \times K_2 \times K_3 \ldots$

35 The same equilibrium system may be represented by two different equations:

$COCl_2(g) \rightleftharpoons CO(g) + Cl_2(g); K_1$

$CO(g) + Cl_2(g) \rightleftharpoons COCl_2(g); K_2$

State the equilibrium expression constants, K_1 and K_2.

Deduce the mathematical relationship between K_1 and K_2.

Tool 3: Mathematics

Carry out calculations involving reciprocals

The reciprocal of a number (also known as the multiplicative inverse) is 1 divided by the number in question; the reciprocal of $x = \frac{1}{x}$.

Any number multiplied by its reciprocal equals 1: $x \times \frac{1}{x} = 1$ (for example, $4 \times \frac{1}{4} = 1$).

Dividing by a number is the same as multiplying by the reciprocal of that number. For example, $5 \div 4$ is the same as $5 \times \frac{1}{4}$.

In scientific notation, a negative power indicates a reciprocal: $3^{-2} = \left(\frac{1}{3}\right)^2 = \frac{1}{9}$.

When written as a decimal, the reciprocal of a number should be expressed to the same number of significant figures (s.f.) as the number. For example:

$\frac{1}{9.58} = 0.104\,384 = 0.104$ (3 s.f.).

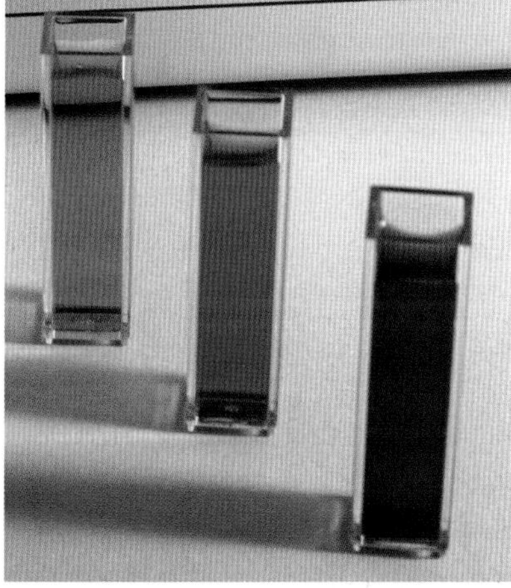

■ **Figure R2.79** The equilibrium between $Fe^{3+}(aq)$ ions and $SCN^-(aq)$ ions can be studied by colorimetry. A soluble, blood-red coloured complex is formed between the two ions

Measuring an equilibrium constant

There are several factors to consider when experimentally determining the value of an equilibrium constant.

The reactants and products must have actually reached equilibrium. This can be tested by removing small samples from the reacting mixture at different times and analysing them. If there is a colour change in the reaction, a colorimeter or spectrophotometer (Figure R2.79) can be used.

When the same result is obtained for repeated analyses, then it can be assumed that equilibrium has been reached. In a reaction involving gases, where there is a change in gas pressure, all that is required is a steady pressure reading. A steady, unchanging pressure reading indicates that equilibrium has been reached.

The temperature at which the measurement is recorded must be known, and kept constant. This is done by carrying out the reaction in a container whose temperature is kept constant using a thermostat.

Inquiry 2: Collecting and processing data

Collect and record sufficient relevant quantitative data

You should put quantitative data into a table that you have designed. The independent variable needs to go in the first column of the table and the dependent variables will be in subsequent columns. At the top of each column there should be a title for the data being collected. The title should also state the precision of the instrument used. All quantitative data should be shown to the appropriate number of significant figures.

A typical investigation will ideally involve using at least five different values of the independent variable and repeating the data collection three times for each value (sometimes referred to as the 5 × 3 approach).

Sufficient relevant quantitative data must be collected and recorded from an experimental investigation (or simulation) to give a full and detailed answer to the research question. The number and range of repeated values must be appropriate to the methodology and research question.

The most common approach to measuring concentrations when equilibrium is reached is via titration (with adequate precautions to prevent the equilibrium shifting). However, depending on the nature of the equilibrium, you can determine the concentration using a colorimeter, spectrophotometer or electrochemical cell.

Inquiry 1: Exploring and designing

Design and explain a valid methodology

An investigative methodology is valid if it is suitable to answer the research question that was formulated. Validity will be reduced if, for example, there are uncontrolled variables that may affect the value of the dependent variable.

Chemical methods, for example, titrations and gravimetric analysis, are generally accurate and precise. However, they may be time consuming and are usually limited to analysis of reasonably high concentrations of analyte.

Instrumental methods measure a signal from a sensor/detector (usually a voltage or current); this signal is related either directly or indirectly to the analyte concentration via a calibration process. Instrumental methods are generally accurate, precise and readily automated, but they require careful calibration procedures.

The reaction between ethanoic acid, ethanol, ethyl ethanoate and water is one of the few chemical equilibrium systems (other than acid–base equilibria) that is relatively easy to study in the school laboratory with simple chemical analysis:

$$CH_3COOH(l) + C_2H_5OH(l) \leftrightharpoons CH_3COOC_2H_5(l) + H_2O(l)$$

This esterification reaction is extremely slow at room temperature in the absence of a catalyst. In the presence of an acid catalyst to supply H^+ (a small amount of concentrated aqueous hydrochloric acid) the reaction mixture reaches equilibrium in about 48 hours.

If the equilibrium mixture is diluted, the rate of reaction is slow. This means you can find the equilibrium concentration of ethanoic acid by titration, without the position of equilibrium shifting significantly in the short time taken for the titration.

1 **Mix measured quantities of chemicals and allow the mixture to reach equilibrium.**

Precisely measured quantities of the chemicals are added to sample tubes. The masses of the components of the mixture can be found by weighing. The sample tubes are tightly stoppered to avoid loss by evaporation and left at constant temperature for 48 hours.

Some of the tubes at first contain just ethanol, ethanoic acid and hydrochloric acid. Others have only ethyl ethanoate, water and hydrochloric acid. The results will demonstrate that equilibrium can be reached from either side of the equation.

2 **Analyse the mixture to find the equilibrium concentration of the acid.**

Each equilibrium mixture is transferred with a pipette to a volumetric flask and diluted with water. Titration with a standard solution of sodium hydroxide determines the total amount of acid (hydrochloric and ethanoic acids) in the sample at equilibrium.

3 **Use the equation for the reaction and the information from steps 1 and 2 to determine the values for all the equilibrium concentrations.**

Some of the sodium hydroxide used in the titration reacts with the hydrochloric acid. Since the amount of hydrochloric acid does not change as the reactants reach equilibrium, it is possible to determine how much of the sodium hydroxide was used to neutralize it, given how much hydrochloric acid was added at the start.

The remainder of the alkali added during the titration reacts with ethanoic acid.

Hence the amount of ethanoic acid at equilibrium can be calculated.

The other equilibrium concentrations can be determined given the starting amounts of chemicals present and the equation for the reaction.

The flow chart in Figure R2.80 shows an approach to processing the data and calculating an equilibrium constant.

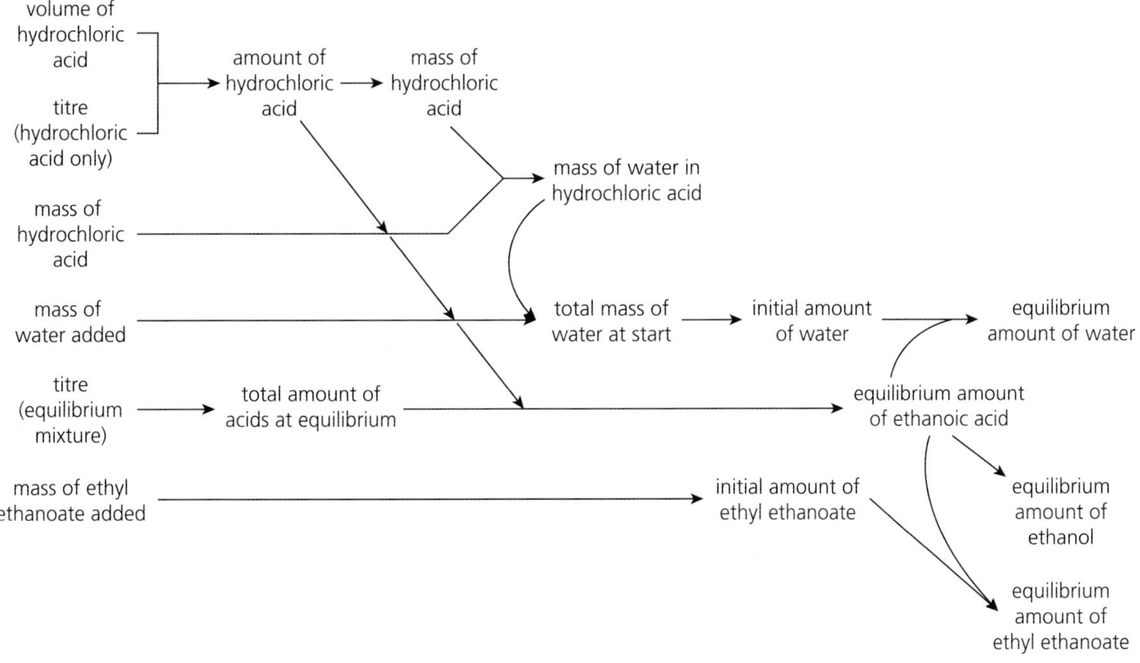

■ **Figure R2.80** Processing of raw data from ethyl ethanoate equilibrium analysis

36 The equilibrium constant for the acid hydrolysis of ethyl ethanoate can be found by experiment.

$$CH_3COOC_2H_5(l) + H_2O(l) \rightleftharpoons CH_3COOH(l) + C_2H_5OH(l)$$

A reaction mixture is set up in a sealed flask and left for 48 hours at 25 °C to reach equilibrium. Samples can then be taken from the flask and titrated with alkali to find the concentration of ethanoic acid present at equilibrium.

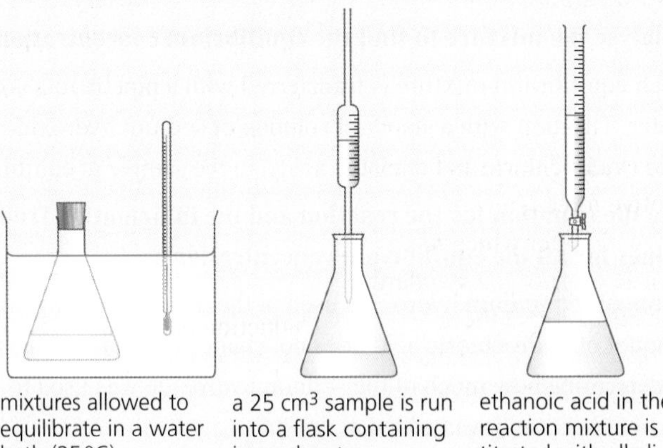

mixtures allowed to equilibrate in a water bath (25 °C) | a 25 cm³ sample is run into a flask containing ice and water | ethanoic acid in the reaction mixture is titrated with alkali

44.0 g of ethyl ethanoate was mixed with 36.0 g of water (acidified with a small amount of hydrochloric acid to act as a homogeneous catalyst) and allowed to reach equilibrium. The equilibrium mixture was then made up to 250 cm³ (3 s.f.) with distilled water.

A 25.0 cm³ sample of the diluted mixture was titrated with 1.00 mol dm^{-3} sodium hydroxide solution. After allowing for the acid catalyst present, the ethanoic acid in the equilibrium mixture required 29.5 cm³ of alkali to neutralize it.

a Determine the equilibrium constant, *K*, for the reaction by working through the following stages of calculation:

i Calculate the amounts (mol) of ethyl ethanoate and water mixed at the start of the reaction.

ii Use the titration value to calculate the number of moles of ethanoic acid present at equilibrium.

iii Complete the following table containing data on the changes taking place as the reaction reaches equilibrium.

	ethyl ethanoate	water	ethanoic acid	ethanol
At start / moles				
At equilibrium / moles				
At equilibrium* / mol dm^{-3}				

* Assume that the volume of the reaction mixture is *V* dm³

iv Write an expression for the equilibrium constant, *K*, and calculate its value using the data from the table.

b Why is it still necessary to account for the acid catalyst added at the start of the reaction when analysing the equilibrium mixture?

c Suggest a method by which you could take account of the acid catalyst added.

d Suggest a suitable indicator for the titration of the ethanoic acid produced with sodium hydroxide solution.

Tool 1: Experimental techniques

Voltaic (electrochemical) cells

Voltaic cells produce an electric potential difference (voltage) from a spontaneous redox reaction. In a voltaic cell, the two redox half-reactions occur in physically separate half-cells. The electrons flow as a current from one half-cell to the other via a wire connecting the two electrodes.

The potential difference between the two half-cells when no current is flowing is a maximum value known as the cell potential (or, in physics, electromotive force (emf) of the cell). This can be measured using a voltmeter which has a high resistance, so the current is close to zero. If a standard voltaic cell is prepared, the voltage measured at the start will be equal to the standard electrode potential. Standard conditions for electrochemical cells are 1 mol dm^{-3}, 298 K and 100 kPa.

Many voltaic cells can be set up in the laboratory using strips of cleaned metal immersed in aqueous solutions of their own ions connected by a salt bridge and a wire (and possibly other components) to form the external circuit. If the half-cell is for a transition metal that has two or more ions (for example, Fe^{2+}(aq) and Fe^{3+}(aq)) a graphite or platinum electrode is used. A platinum electrode is preferred because it is unreactive and an excellent electrical conductor.

The free-moving ions crossing the salt bridge conduct the charge and maintain electroneutrality in each half-cell. A salt bridge is usually made from a piece of filter paper (or material) soaked in a salt solution, usually potassium nitrate. It can also be a glass U-tube plugged with cotton wool containing a salt solution (usually KCL or KNO_3) in a gel. The salt should not react with the electrodes or electrode solutions. For example, potassium chloride would not be suitable for copper systems because chloride ions can form complexes with copper(II) ions.

Figure R2.81 shows a voltaic cell with copper as the anode:

$$Cu^{2+}(aq) + 2e^- \rightleftharpoons Cu(s); \qquad E^\ominus = +0.34\,V$$

$$Fe^{3+}(aq) + e^- \rightleftharpoons Fe^{2+}(aq); \qquad E^\ominus = +0.77\,V$$

Reduction occurs at the electrode with the more positive potential.

The cell potential can be varied by changing the concentration of iron(II) ions or copper(II) ions in the other half-cell. The cell potential is also affected by changes in temperature of the two solutions.

The cell potential can be used to calculate the Gibbs energy change using:

$$\Delta G^\ominus = -nFE^\ominus$$

where n represents the number of electrons transferred during the redox reaction and F represents the Faraday constant (the charge in coulombs carried by one mole of electrons), and the equilibrium constant, K, for the redox reaction under standard conditions can be calculated using:

$$\Delta G^\ominus = RT \ln K$$

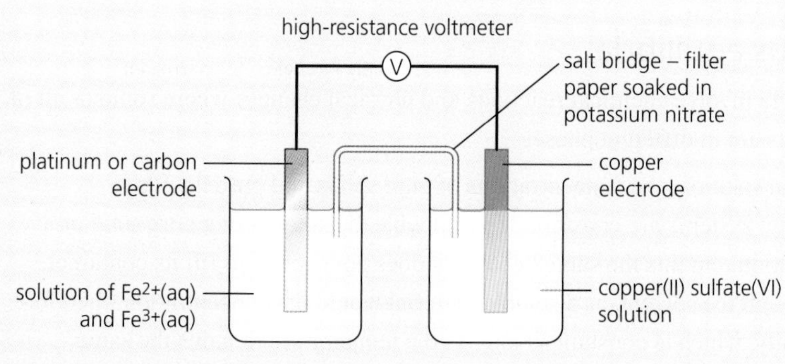

Figure R2.81 Measuring electrode potentials in a voltaic cell

Link

Voltaic cells are covered further in Chapter R3.2, from page 595.

Types of equilibrium

◆ **Phase**: A homogeneous part of a heterogeneous system that is separated from other parts by a distinguishable boundary.

◆ **Homogeneous equilibrium**: A reaction where the reactants and products are in the same phase.

◆ **Heterogeneous equilibrium**: A reaction where the reactants and products are in different phases.

In the same way as chemical equilibria involve a chemical reaction in a closed system, physical equilibria involve a physical process at equilibrium within a closed system. Some physical changes involve a change of **phase**, where a phase is a distinct part of a system which has a uniform composition.

For example, a mixture of sand and water consists of two phases: solid sand and liquid water. If glucose is shaken with water it will dissolve to form a single phase of aqueous glucose solution. Ethanol and water are another example of a single-phase system, in this case, containing two miscible liquids, ethanol and water.

Single-phase systems are also referred to as homogeneous and those consisting of two or more phases are known as heterogeneous.

Homogeneous equilibria

Homogeneous equilibria involve chemical reactions and physical changes where all the reactants and products are in the same phase.

For reactions in aqueous solution, water is excluded from the equilibrium expression as it is present in large excess and its concentration remains essentially constant. For example, in the final reaction in Table R2.13, the concentration of the water remains essentially constant at ~55.6 $mol\,dm^{-3}$.

However, when reactions take place in a non-aqueous medium and water is present in stoichiometric amounts as a reactant (not as large excess), then the water should be included in the equilibrium expression as shown in the middle line of the table.

■ **Table R2.13** Examples of homogeneous chemical equilibria

Phase	Reaction	Equilibrium constant
gaseous	$2SO_2(g) + O_2(g) \rightleftharpoons 2SO_3(g)$	$K = \frac{[SO_3]^2}{[O_2] \times [SO_2]^2}$
aqueous / liquid phase	$CH_3COOH(l) + C_2H_5OH(l) \rightleftharpoons CH_3COOC_2H_5(l) + H_2O(l)$	$K = \frac{[CH_3COOC_2H_5][H_2O]}{[CH_3COOH][C_2H_5OH]}$
	$[Cu(H_2O)_6]^{2+}(aq) \rightleftharpoons Cu^{2+}(aq) + 6H_2O(l)$	$K = \frac{[Cu^{2+}]}{[Cu(H_2O)_6]^{2+}}$

Heterogeneous equilibria

Heterogeneous equilibria involve chemical reactions and physical changes where some or all of the reactants and products are in different phases.

Unlike gases or species in solution, the concentrations of pure solids and pure liquids in heterogeneous equilibrium systems are constant. If you double the amount of a pure solid or a pure liquid, its concentration remains the same because it does not expand to fill its container, like a gas, or spread through the solvent, as a soluble material would. Its concentration, therefore, depends only on its density, which is constant as long as the temperature remains the same.

■ **Table R2.14** Examples of heterogeneous chemical equilibria

Phase	Reaction	Equilibrium constant
solid / gas	$3Fe(s) + 4H_2O(g) \rightleftharpoons Fe_3O_4(s) + 4H_2(g)$	$K = \frac{[H_2]^4}{[H_2O]^4}$
solid / solution	$AgCl(s) \rightleftharpoons Ag^+(aq) + Cl^-(aq)$	$K = [Ag^+] \times [Cl^-]$

■ **Figure R2.82** A sealed flask containing bromine demonstrates a physical equilibrium between the liquid and its vapour

◆ **Vapour**: A gas in equilibrium with the surface of its liquid.

◆ **Saturated**: A material (such as air, a solution or a heterogeneous catalyst) which contains the maximum equilibrium amount of a substance (a gas, solute or reactant, for example) under given conditions.

◆ **Saturated vapour pressure**: The pressure of the vapour over a liquid where the liquid and vapour are in equilibrium.

Physical equilibria

Solid–liquid equilibrium

When a pure solid is heated, it starts changing into liquid at the melting point. At this temperature, the solid and liquid states of the substance coexist in equilibrium at a specific pressure.

Consider ice and water in a thermos flask at 0°C. This represents a dynamic equilibrium:

ice ⇌ water; $H_2O(s) \rightleftharpoons H_2O(l)$; rate of melting = rate of freezing

Liquid–vapour equilibrium

When bromine is placed in a sealed container, the orange-brown **vapour** collects over the red-brown liquid. As the liquid evaporates, the colour of the vapour becomes more intense. Eventually, the intensity of colour of the vapour over the liquid remains constant (Figure R2.82).

A state of equilibrium has been reached between bromine liquid and bromine gas:

$$Br_2(l) \rightleftharpoons Br_2(g)$$

The rate of evaporation and the rate of condensation are the same. There is no overall change in the amounts of bromine liquid and vapour present and the air above the liquid is said to be **saturated** with the vapour.

The pressure exerted by the bromine gas is known as the vapour pressure. The vapour pressure at equilibrium depends only on the nature of the substance and the temperature; it is known as the **saturated vapour pressure**.

Figure R2.83 shows how the saturated vapour pressure of a liquid increases with temperature. It shows why two different liquids have different boiling points.

As a liquid is heated, its vapour pressure increases. When the temperature of the liquid reaches the point at which the vapour pressure equals the external pressure, vaporization can occur throughout the liquid – the liquid boils. If the container is not sealed, a liquid will boil when its vapour pressure is equal to the pressure of the atmosphere.

Gas–liquid equilibrium

When a bottle of carbonated drink is opened, it will fizz – bubbles are produced throughout the liquid, rise to the surface and break. The sealed bottle holds an equilibrium system where the rate of carbon dioxide dissolving equals the rate of degassing (Figure R2.84):

$$CO_2(aq) \rightleftharpoons CO_2(g)$$

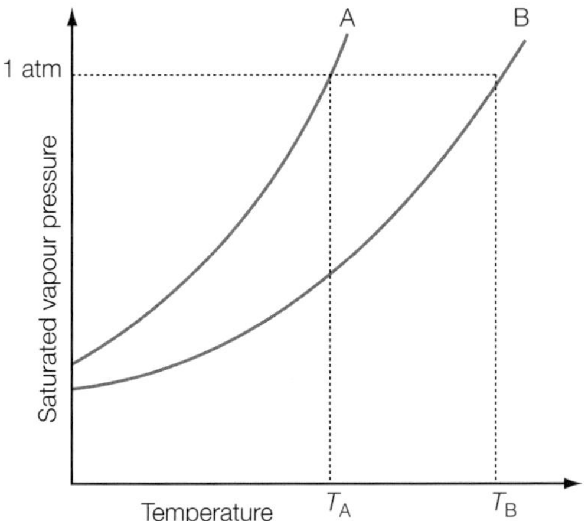

■ **Figure R2.83** Saturated vapour pressure curves for two liquids A and B, where A (e.g. ethanol, b.p. 78°C) is more volatile than B (e.g. water, b.p. 100°C)

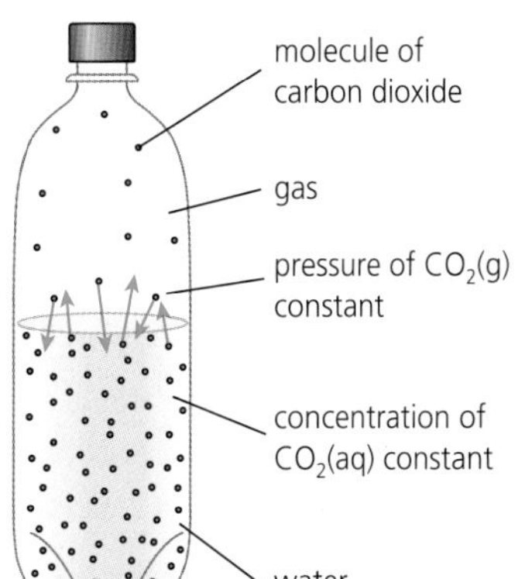

■ **Figure R2.84** Dynamic equilibrium in a sealed bottle of fizzy drink. The bottle on the left of the photograph has had its cap slightly opened, and then closed again

Once we release the pressure by slightly unscrewing the lid, the carbon dioxide starts to leave the solution. A stream of bubbles is produced and some of the gas above the liquid can leave the bottle. Retightening the cap slows down the stream of bubbles because it closes the system and allows equilibrium to be re-established.

◆ **Sparingly soluble**: An ionic substance that is partially soluble in water which results in an equilibrium between dissolved ions and undissolved ionic solid.

◆ **Solubility product**: The product of the concentrations of ions, raised to the power of their coefficients, in a saturated solution.

Going further

Equilibrium and solubility

Solubility product

Few chemicals are totally insoluble in water. Many 'insoluble' ionic substances dissolve to a small extent and are known as **sparingly soluble**.

When calcium carbonate is placed in water, a low concentration of hydrated calcium and carbonate ions is formed; these ions are in equilibrium with solid calcium carbonate:

$$CaCO_3(s) \rightleftharpoons Ca^{2+}(aq) + CO_3^{2-}(aq)$$

The equilibrium expression is therefore:

$$K = \frac{[Ca^{2+}(aq)] \times [CO_3^{2-}(aq)]}{[CaCO_3(s)]}$$

The addition of more solid calcium carbonate will not cause the position of equilibrium to shift further to the right because the solution is saturated at that particular temperature and the concentration of the solid calcium carbonate, $CaCO_3$, remains constant since the concentration is determined by the density of the solid.

Since the position of equilibrium is not changed by the addition of more solid calcium carbonate because it is a constant which can be combined with the equilibrium constant, it can be removed from the equilibrium constant expression which is now written as:

$$K_{sp} = [Ca^{2+}(aq)] \times [CO_3^{2-}(aq)]$$

and the equilibrium constant is known as the **solubility product, K_{sp}**.

At 298 K the solubility product of calcium carbonate is 5×10^{-9}. This very small value shows that there is a very low concentration of calcium and carbonate ions present in the solution. (The concentration of each ion is equal to the square root of K_{sp}, 7.07×10^{-5} mol dm^{-3}.)

A common misconception is that increasing the amount of a solid ionic substance that is at equilibrium causes more dissolved ions to be produced.

In fact, increasing the amount of a solid ionic substance at equilibrium does not cause more dissolved ions to be produced, because the solution is already saturated and contains the maximum possible amount of dissolved ions. Pure solids are not included in equilibrium expressions.

Understanding solubility products is important in many areas including the separation of metal ions by precipitation when refining an ore and the recycling of metals.

Partition

Iodine dissolves in cyclohexane, which is a non-polar hydrocarbon, to form a purple solution.

Iodine is sparingly soluble in water ($I_2(s) \rightleftharpoons I_2(aq)$) but will dissolve in a solution of potassium iodide to form a yellow, orange or brown solution (depending on the concentration). In this solution, iodine molecules react reversibly with iodide ions to form triiodide ions:

$$I_2(solv) + I^-(aq) \rightleftharpoons I_3^-(aq)$$

Figure R2.85 shows how this equilibrium can be established from both the 'reactant side' and 'product side' of the equation.

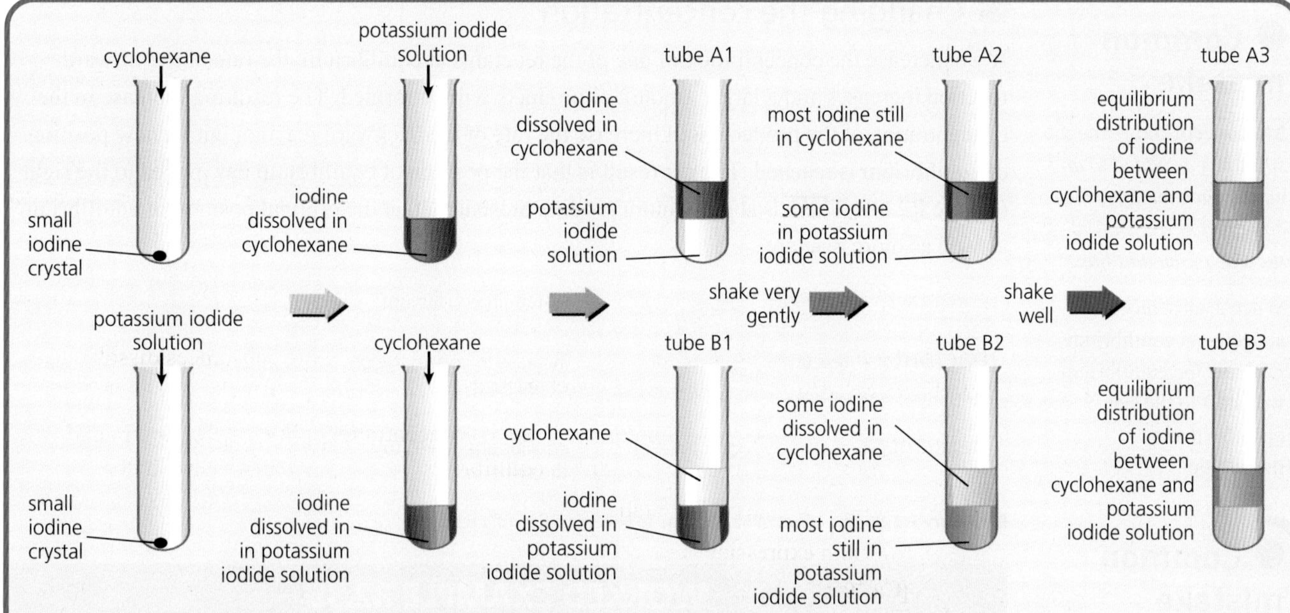

■ **Figure R2.85** Two approaches to the same equilibrium state. Note that the tubes labelled A1, A2 and A3 are the same tube at three different stages. The same is true for the tubes labelled B1, B2 and B3

Experiments show that, at a fixed temperature, the ratio of the concentrations of the iodine in the two layers is constant:

$$K = \frac{[I_2][I^-]}{[I_3^-]}$$

If iodine is dissolved in a mixture of water and tetrachloromethane, the following equilibrium is set up:

$I_2(aq) \rightleftharpoons I_2(solv)$

$$K = \frac{[I_2(solv)]}{[I_2(aq)]}$$

The equilibrium constant is known as the partition coefficient or distribution coefficient for the solute distributed between two solvents at a given temperature. The partition or distribution law states that, at a fixed temperature, a solute distributes itself between two immiscible solvents, so that the ratio of the concentrations of solute in each layer is constant.

Paper chromatography and other chromatographic techniques depend on the principle of partitioning. It is also the basis for the separation technique of solvent extraction.

Shifting the position of equilibrium

◆ **Le Châtelier's principle**: If a system is at equilibrium, any change imposed on the system tends to shift the equilibrium to counteract the effect of the applied change.

The previous chapter considered the effects of concentration, pressure of gases, temperature, particle size of solids and catalysts on the rate of reaction.

We shall now consider what effect (if any) these factors may have on the equilibrium position of a reaction. Such changes can be predicted using **Le Châtelier's principle**. This states that if the conditions of a closed system in dynamic equilibrium are changed (for example, by changing the concentration, pressure or temperature), the position of equilibrium moves so as to reduce that change.

Common mistake

Misconception: *Le Châtelier's principle can be applied in the initial state before the reaction has reached equilibrium.*

Le Châtelier's principle can only be applied to the system when it has reached equilibrium.

Common mistake

Misconception: *When a substance is added to an equilibrium mixture, the equilibrium will shift to the side of the addition.*

When a substance is added to an equilibrium mixture, the equilibrium will shift to the side opposite that of the addition.

Common mistake

Misconception: *When more products are added to an equilibrium system at constant temperature, K will increase.*

When more products are added to an equilibrium system at constant temperature, *K* will not change: the concentrations of all reactants and products will change in such a way that *K* remains constant.

Changing the concentration

If we increase the concentration of one of the reactants at equilibrium, the rate of the forward reaction increases and a larger amount of products will be formed. The resulting increase in the concentration of the products will increase the rate of the backward reaction until a new position of equilibrium is reached. The net result is that the position of equilibrium has moved to the right (Figure R2.86). There is slightly more product present than at the original position of equilibrium as well as more reactant.

$\text{reactant}_1 + \text{reactant}_2 \rightleftharpoons \text{product}$	system in equilibrium
$\text{reactant}_1 + \text{reactant}_2 \rightleftharpoons \text{product}$	increase in concentration of reactant_1
$\text{reactant}_1 + \text{reactant}_2 \rightleftharpoons \text{product}$	equilibrium moves to the right

■ **Figure R2.86** The effect of an increase in concentration of a reactant

WORKED EXAMPLE R2.3B

The ester ethyl ethanoate can be synthesized from ethanoic acid and ethanol. This is known as esterification. Predict and explain the effect of adding a small amount of concentrated sulfuric acid (a strong dehydrating agent) to the equilibrium system.

$$CH_3CO_2H(l) + C_2H_5OH(l) \rightleftharpoons CH_3CO_2C_2H_5(l) + H_2O(l)$$

ethanoic acid ethanol ethyl ethanoate water

Answer

If we remove some of the water from the equilibrium mixture then, according to Le Châtelier's principle, more carboxylic acid and alcohol will react to replace it, producing more ester.

If we add a few drops of concentrated sulfuric acid to the mixture, the sulfuric acid provides hydrogen ions, H^+. These ions act as a homogeneous catalyst for the reaction and also as a dehydrating agent, removing water from the mixture.

The addition of the acid favours the production of the ester since, by removing water, it shifts the position of equilibrium to the right.

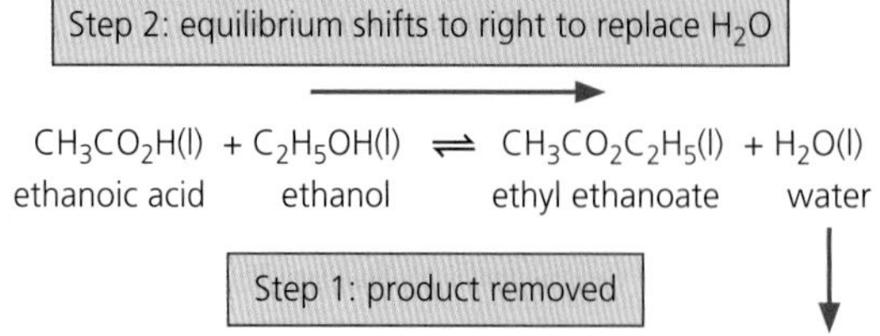

■ **Figure R2.87** The effect of removing water on the equilibrium position of an esterification reaction

Changing the pressure

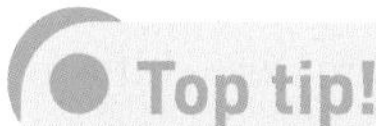

Top tip!

The more gas particles present in a fixed volume, the higher the pressure.

In an equilibrium system involving gases, a change in pressure may affect the position of equilibrium. This is similar to the way in which changing the concentration may affect the position of equilibrium in an aqueous reaction. The effect depends on whether or not there is a change in volume as the reaction proceeds.

For a mixture of gases, an increase in pressure may increase the concentration of both reactants and products to the same extent. This will mean it has no effect on the position of equilibrium.

This occurs if there are the same amounts of gas molecules on each side of the equation, so that the reaction does not change in volume as it proceeds. An example is the decomposition of hydrogen iodide into its elements:

$$2HI(g) \rightleftharpoons H_2(g) + I_2(g)$$

There are two molecules (or 2 moles of gas molecules) on each side of the equation, so changing the pressure will not change the position of equilibrium.

Now consider the synthesis of ammonia from nitrogen and hydrogen.

$$N_2(g) + 3H_2(g) \rightleftharpoons 2NH_3(g)$$

In the forward reaction, four molecules or 4 moles of gas form two molecules or 2 moles of gas, so there is a significant decrease in pressure as the reaction shifts to the right. An increase in pressure will cause the position of equilibrium to shift to the right, to reduce the overall pressure of the system. The forward reaction is favoured and so increasing the pressure results in an equilibrium mixture containing more ammonia, the product.

Dinitrogen tetroxide is a colourless gas that decomposes to brown nitrogen dioxide gas:

$$\underset{\text{colourless}}{N_2O_4(g)} \quad \rightleftharpoons \quad \underset{\text{brown}}{2NO_2(g)}$$

If an equilibrium mixture of these two gases is introduced to a gas syringe, you can change the pressure in the reaction mixture by pushing the plunger in, or by quickly pulling it out. You can then compare the colour of the gas mixture with the original colour, before the pressure change. The results are summarized in Figure R2.88.

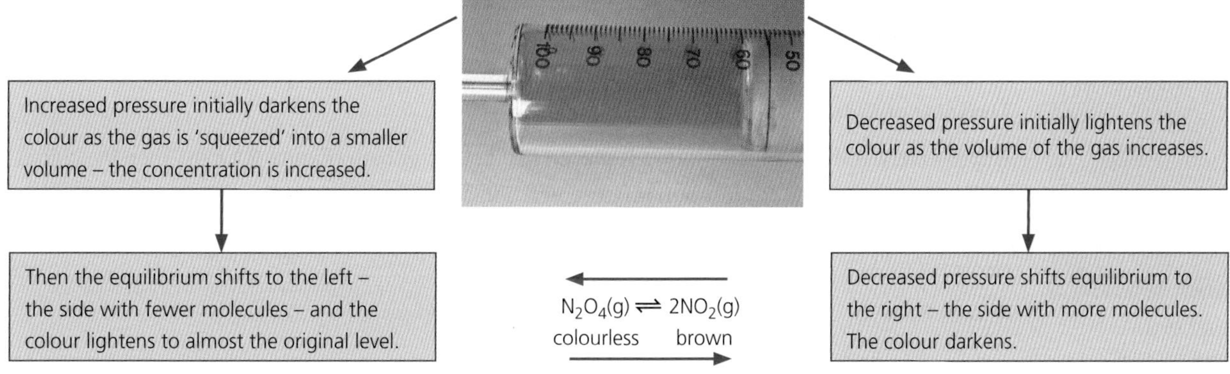

Increased pressure initially darkens the colour as the gas is 'squeezed' into a smaller volume – the concentration is increased.
Then the equilibrium shifts to the left – the side with fewer molecules – and the colour lightens to almost the original level.
Decreased pressure initially lightens the colour as the volume of the gas increases.
Decreased pressure shifts equilibrium to the right – the side with more molecules. The colour darkens.
$N_2O_4(g) \rightleftharpoons 2NO_2(g)$
colourless brown
100
90
80
70
60
50

■ **Figure R2.88** The effect of changing pressure on the gaseous equilibrium involving nitrogen dioxide

When the piston is pushed in, increasing the pressure, there is an initial darkening of the colour of the gas mixture. The darkening is due to the increase in concentration of nitrogen dioxide. However, this is quickly followed by a lightening of the colour as the equilibrium mixture shifts to its new composition in which there is a higher concentration of colourless N_2O_4. This new mixture contains fewer molecules and therefore reduces the pressure in the syringe, thus counteracting the increase in externally applied pressure.

Alternatively, if the plunger is pulled out, the pressure in the syringe decreases. After an initial lightening of the colour, the gas mixture becomes darker as the new equilibrium is established. Decreased pressure favours the side of the equation that contains more gaseous molecules, counteracting the decrease in applied pressure.

Changing the temperature

Le Châtelier's principle can also be used to predict the effect of a temperature change on the position of equilibrium. The direction of change depends on whether the forward reaction is exothermic ($\Delta H^\ominus$ negative) or endothermic ($\Delta H^\ominus$ positive) where $\Delta H^\ominus$ represents the enthalpy change. If the temperature is raised, the equilibrium will move in the direction which will tend to reduce the temperature; that is, it will favour the endothermic direction. These effects are summarized in Table R2.15.

Table R2.15 The effects of temperature changes on chemical equilibria

Nature of forward reaction (sign of $\Delta H^\ominus$)	Change in temperature	Shift in the position of equilibrium	Effect on value of K
endothermic (positive $\Delta H^\ominus$)	increase	to the right	K increases
endothermic (positive $\Delta H^\ominus$)	decrease	to the left	K decreases
exothermic (negative $\Delta H^\ominus$)	increase	to the left	K decreases
exothermic (negative $\Delta H^\ominus$)	decrease	to the right	K increases

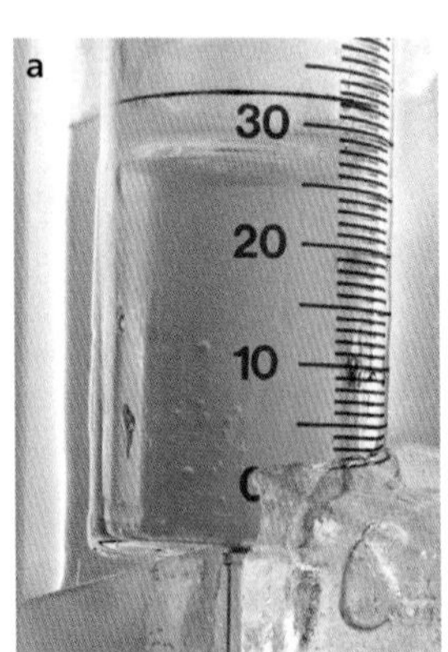

Consider the decomposition of dinitrogen tetroxide:

$$N_2O_4(g) \rightleftharpoons 2NO_2(g) \qquad \Delta H^\ominus = +57\,\text{kJ mol}^{-1}$$

colourless brown

Figure R2.89 An equilibrium mixture of N_2O_4 and NO_2 in a sealed gas syringe is placed firstly in **a** an ice bath and then in **b** a hot water bath

Any enthalpy value quoted with an equilibrium equation refers to the forward reaction so the decomposition of N_2O_4 is endothermic.

If an equilibrium mixture is set up in a sealed container at room temperature and then placed in an ice bath (Figure R2.89a), its colour will lighten as a new equilibrium mixture containing more N_2O_4 is established. A decrease in temperature causes the equilibrium position to shift to the left. The value of K decreases as a result of these changes.

Alternatively, if the original mixture is placed in a hot water bath (Figure R2.89b), then the colour will darken as the new equilibrium mixture will contain more NO_2. An increase in temperature causes the equilibrium position to shift to the right. The value of K increases as a result of these changes (Table R2.16).

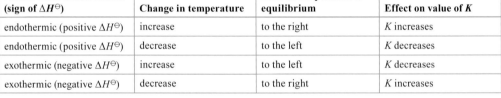

Table R2.16 The change in the equilibrium constant K for the endothermic reaction $N_2O_4(g) \rightleftharpoons 2NO_2(g)$ with temperature

Temperature / K	K
298	4.0×10^{-2}
400	1.4
500	41

increased temperature ↓   increased K ↓

Links

Solvents and saturation are covered in detail in Chapter S1.1, page 9.

The thermodynamics of solution formation (and other phase changes) is also covered in Chapter R1.4, page 393.

Le Châtelier's principle can also be used to explain the effect of temperature on the solubility of solids in liquid solvents to form solutions. If heat is absorbed **at the point where a solution becomes saturated**, then solubility increases with temperature, but if heat is released at this point, solubility decreases with an increase in temperature. If heat is absorbed during **the formation of a saturated solution**, solubility increases with temperature, but if heat is released during the formation of a saturated solution, solubility decreases with an increase in temperature.

The dissolving of gases in liquids is always exothermic and therefore the solubility of gases in liquids decreases with temperature. For gases to dissolve spontaneously ΔG (Gibbs energy change) must be <0 since ΔS (entropy change of the system) is always negative for the dissolution of a gas in a liquid. ΔH (enthalpy change) must therefore be negative (exothermic) for the dissolution to occur spontaneously, and therefore an increase in temperature leads to a decrease in solubility.

Common mistake

Misconception: *When the temperature is increased, more products form.*

When the temperature is increased, K increases to favour the forward reaction only if the forward reaction is endothermic.

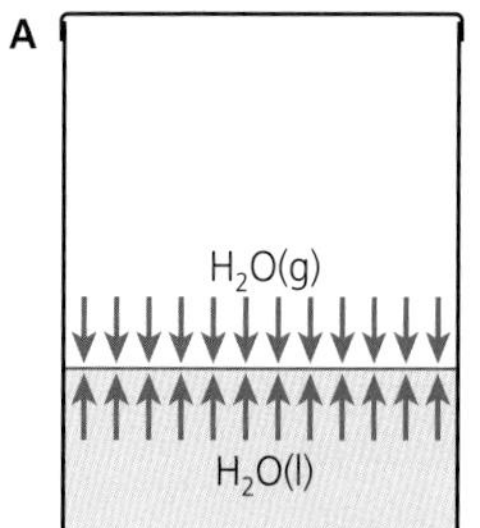

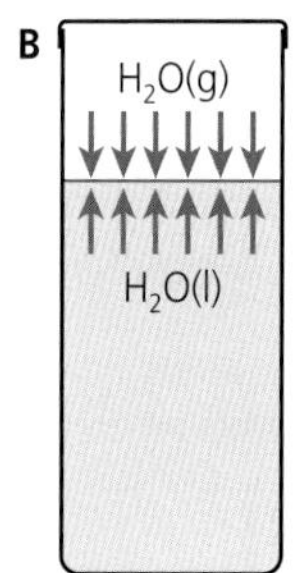

■ **Figure R2.90** The effect of changing surface area on the evaporation of water in a closed container (at a fixed temperature)

Changing the surface area of a solid or liquid

In a heterogeneous reaction, a small particle size increases the rate of reaction by increasing the surface area. If the reaction is in equilibrium, the rates of forward and backward reactions will be increased equally when the surface area is increased.

If, for example, the surface area of a liquid is doubled (at a fixed temperature), the rate at which molecules leave this surface is doubled, but so too is the rate at which they enter the surface again. The position of equilibrium is therefore unchanged (see Figure R2.90) at constant temperature.

Common mistake

Misconception: *A catalyst affects the rates of the forward and reverse reactions differently: for example, a catalyst speeds up only the forward reaction.*

A catalyst increases the rates of the forward and reverse reactions equally, since both the activation energies of the forward and backward reactions are equally lowered. A catalyst has no effect on the position of equilibrium and K remains unchanged.

Addition of a catalyst

The effect of adding a catalyst to an equilibrium mixture is similar to that of increasing the surface area of a liquid or particles of a solid. Both the forward and backward reaction rates are increased by the same amount, and the position of equilibrium is unaltered (provided temperature remains constant). However, the time to reach equilibrium is reduced. There is a greater proportional change for the forward activation energy than the reverse one.

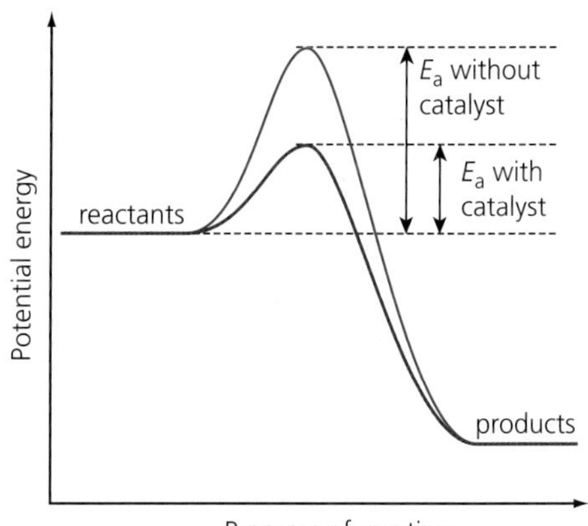

■ **Figure R2.91** Effect of a catalyst in lowering the activation energy for both forward and reverse reactions

LINKING QUESTION

Why do catalysts have no effect on the value of K or on the equilibrium composition?

Top tip!

Note that Le Châtelier's principle is a guide based on a large number of observations of reversible processes at equilibrium. The principle does not *explain* (at a deeper and fundamental level) the effect of changing conditions on systems at equilibrium; doing so requires thermodynamics and the concept of Gibbs energy.

■ **Table R2.17** A summary of the effects of changing conditions on the equilibrium constant

Change made	Effect on 'position of equilibrium'	Value of K
Concentration of one of the components of the mixture	Changes	Remains unchanged
Pressure	Changes if the reaction involves a change in the total number of gas molecules	Remains unchanged
Temperature	Changes	Changes
Use of a catalyst	No change	Remains unchanged

ATL R2.3B

Catalysts are often used in industrial processes which involve equilibrium systems. Two examples are the Haber process (Figure R2.92) for manufacturing ammonia and the Contact process for manufacturing sulfuric acid (Figure R2.93).

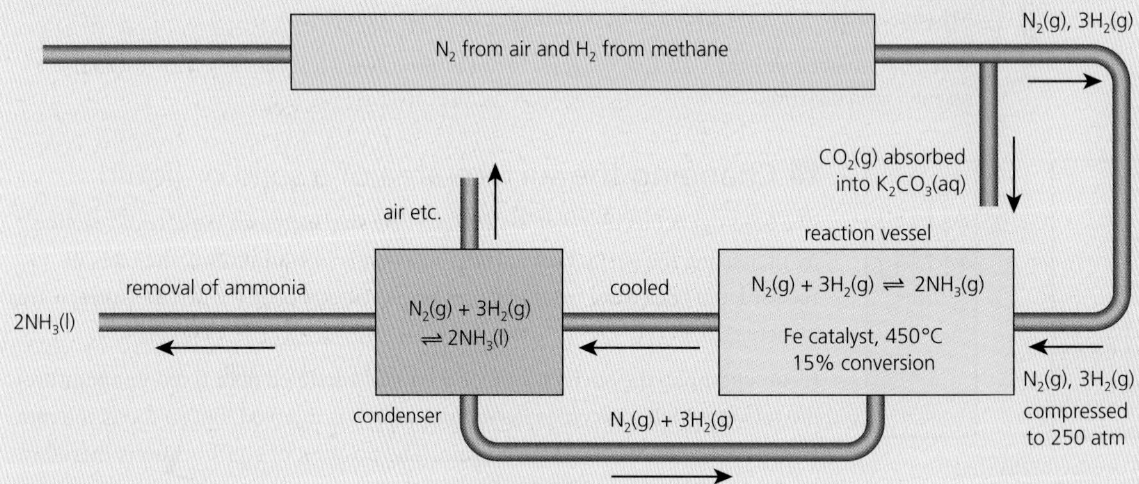

■ **Figure R2.92** Flow diagram showing the stages of the Haber process

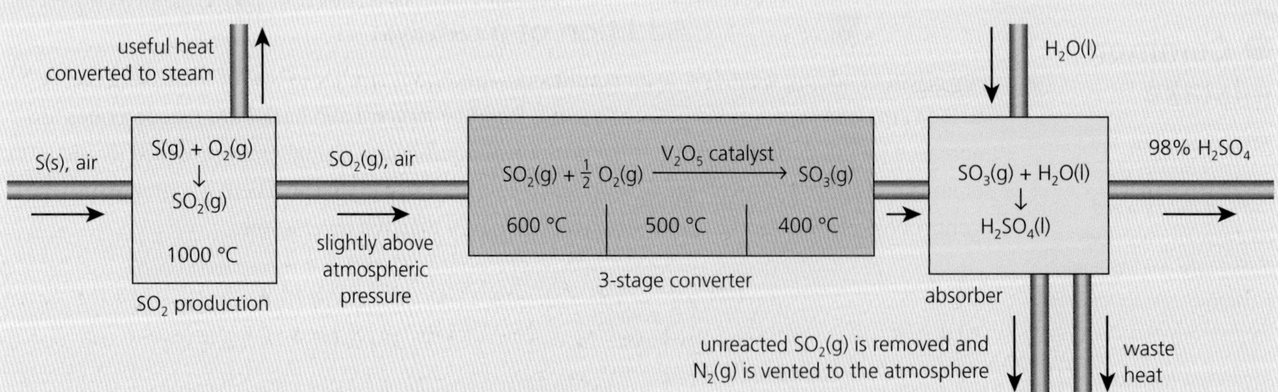

■ **Figure R2.93** Flow diagram showing the stages of the Contact process

Industrial chemists obviously want a high yield of product, and will therefore consider all the factors which shift the equilibrium position to the right. However, they must also consider the rate at which equilibrium is achieved and the impact of shifting the equilibrium position to favour the product.

- Use presentation software (such as PowerPoint) to summarize the reactions and equilibria present in the Haber process and the Contact process (Figures R2.92 and R2.93).
- In each case, state and explain:
  - the effect of a catalyst on the equilibrium system
  - the conditions of temperature and pressure (high or low) which give the highest yield of product(s) in the equilibrium mixture.

You may have realized that the conditions which give the highest yield of product may not give the highest rate of reaction.

- Explain this when relevant for one or both processes.

Graphs involving Le Châtelier's principle

Consider the synthesis of ammonia in the Haber process:

$$N_2(g) + 3H_2(g) \rightleftharpoons 2NH_3(g)$$

If more nitrogen is added at constant volume, the concentration of $N_2(g)$ increases and, using arguments based on Le Châtelier's principle, we can make the following statements about the changes:

- Hydrogen reacts with the added nitrogen to form ammonia.
- $[NH_3]$ increases significantly; $[N_2]$ suddenly goes to the new higher figure; it then drops back from there as equilibrium is reached, but even at that point it remains higher than the value before additional nitrogen was added and $[H_2]$ is decreased since K is unchanged at constant temperature.
- The position of equilibrium shifts to the right (favouring the product, NH_3).
- Both forward and backward rates will be increased.

These changes in concentration are shown in Figure R2.94a while the changes in the rates of the forward and backward reactions are shown in Figure R2.94b.

- The addition of more nitrogen increases its concentration, which causes an increase in the rate of the forward reaction.
- This uses up nitrogen and hydrogen, so the forward rate then decreases.
- At the same time, more ammonia is being made, so the backward rate increases.
- The decreasing forward rate and increasing backward rate become equal and a new equilibrium is re-established.

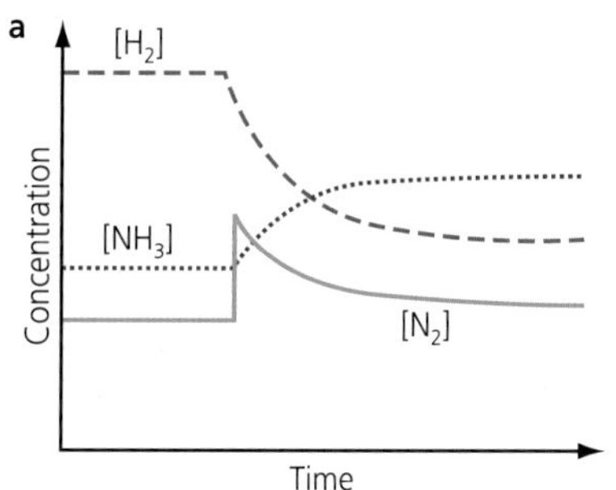

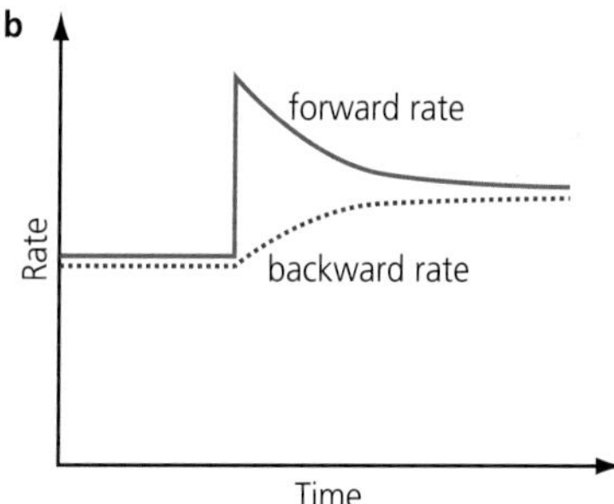

■ **Figure R2.94** The effect of the addition of nitrogen on: **a** the concentrations of the molecules present in the equilibrium mixture and **b** the rates of the forward and backward reactions

Similar arguments can be applied when considering the removal of some of the product, ammonia, from such an equilibrium mixture. Figure R2.95a shows the effect of reducing the ammonia concentration in this way, while Figure R2.95b shows the effect on the rates of the two reactions.

The consequences of removing ammonia can be summarized as a sequence of events which come from the reduction in the concentration of ammonia:

- Firstly, nitrogen and hydrogen react to make up for the lost ammonia.
- Then the concentrations of all of the components of the mixture decrease, especially hydrogen (due to the stoichiometry) since K is unchanged.
- The position of equilibrium shifts to the right (favouring the products).
- Both rates will be slower than before.

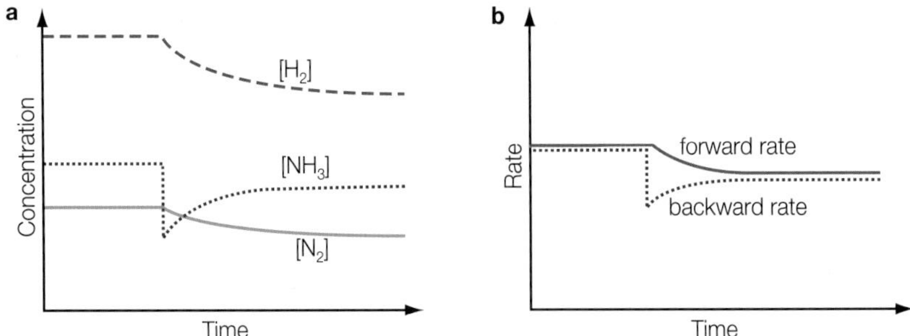

■ **Figure R2.95** The effect of the removal of ammonia on **a** the concentrations of the components of the equilibrium mixture and **b** the rates of the forward and backward reactions

The argument in terms of the rates of reactions is that the removal of some ammonia decreases the backward rate. The forward reaction continues, making more ammonia quickly. The rate of the forward reaction slows, while the backward rate increases until they are again equal.

WORKED EXAMPLE R2.3C

Consider the following reaction:

$$X(aq) + Y(g) \rightleftharpoons Z(g) \quad \Delta H^{\ominus} < 0$$

The graph shows the amounts (in moles) of gases in a reaction mixture as it changes with time.

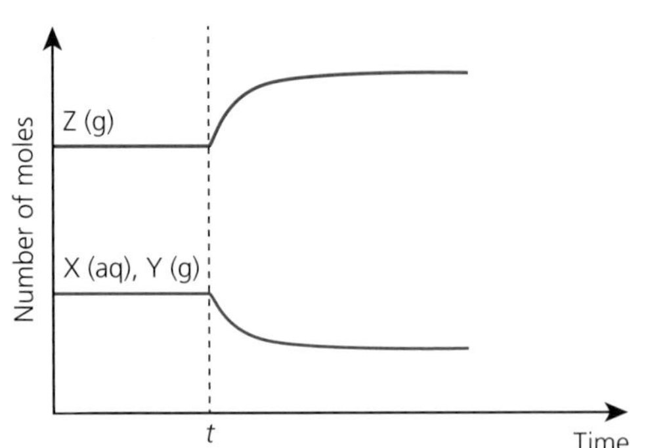

The reaction mixture is initially at equilibrium until time, t, when a disturbance is introduced to the system. Which of the disturbances at time, t, can account for the observed changes in the graph?

I an increase in pressure

II addition of water into the aqueous solution

III increase in temperature

A I only

B II only

C III only

D none of the above

Answer

Option I is incorrect. An increase in pressure will not change the position of equilibrium since the number of moles of gaseous reactants = number of moles of gaseous products.

Option II is incorrect. Addition of water results in a decrease in concentration of X. Hence, the position of equilibrium will shift to the left in order to increase the concentration of X. Hence, the concentration of Z should decrease.

Option III is incorrect. An increase in temperature will result in the position of equilibrium shifting left in order to absorb heat energy since the backward reaction is endothermic. Hence, the concentration of Z should decrease.

The answer is D.

37 Use Le Châtelier's principle to draw up a table showing how the position of equilibrium and the value of the equilibrium constant in reactions A, B and C would be affected by the following changes:

a increased temperature

b increased pressure

Reaction A: the interconversion of oxygen and ozone:

$3O_2(g) \rightleftharpoons 2O_3(g);$ $\Delta H^{\ominus} = +285\,kJ\,mol^{-1}$

Reaction B: the reaction between sulfur dioxide and oxygen in the presence of a platinum / rhodium catalyst:

$2SO_2(g) + O_2(g) \rightleftharpoons 2SO_3(g);$ $\Delta H^{\ominus} = -197\,kJ\,mol^{-1}$

Reaction C: the reaction between hydrogen and carbon dioxide:

$H_2(g) + CO_2(g) \rightleftharpoons H_2O(g) + CO(g);$ $\Delta H^{\ominus} = +41\,kJ\,mol^{-1}$

38 Using this reaction:

$CH_3COOH(l) + C_2H_5OH(l) \rightleftharpoons CH_3COOC_2H_5(l) + H_2O(l)$

explain what happens to the position of equilibrium and equilibrium constant when:

a more $CH_3COOC_2H_5(l)$ is added

b some $C_2H_5OH(l)$ is removed.

39 Using this reaction:

$Ce^{4+}(aq) + Fe^{2+}(aq) \rightleftharpoons Ce^{3+}(aq) + Fe^{3+}(aq)$

explain what happens to the position of equilibrium and equilibrium constant when:

a the concentration of $Fe^{2+}(aq)$ ions is increased

b water is added to the equilibrium mixture.

40 Predict the effect of increasing the pressure on the following gas-phase reactions:

a $N_2O_4(g) \rightleftharpoons 2NO_2(g)$

b $H_2(g) + I_2(g) \rightleftharpoons 2HI(g)$

c $CH_4(g) + H_2O(g) \rightleftharpoons CO(g) + 3H_2(g)$

41 Predict the effect of decreasing the pressure on the reaction:

$2NO_2(g) \rightleftharpoons 2NO(g) + O_2(g)$

42 Predict the effect of increasing the temperature on these reactions:

a $CO(g) + 2H_2(g) \rightleftharpoons CH_3OH(g);$ $\Delta H^{\ominus} = -90\,kJ\,mol^{-1}$

b $H_2(g) + CO_2(g) \rightleftharpoons H_2O(g) + CO(g);$ $\Delta H^{\ominus} = +41.2\,kJ\,mol^{-1}$

Going further

Addition of an inert gas

When an inert gas is added to a system in equilibrium at constant volume, the total pressure will increase. However, the concentrations of the products and reactants do not change. That is, the addition of the inert gas has no effect on the equilibrium.

Nature of science: Global impact of science

Fritz Haber

The German chemist Fritz Haber (1868–1934) is one of the most complex figures in the history of science to evaluate. His life and career were closely linked with the political upheaval in Europe that led to two world wars.

Haber devised a method for the direct synthesis of ammonia from nitrogen and hydrogen. This was crucial for the development of cheap fertilizers and revolutionized food production worldwide. He received the Nobel Prize in Chemistry in 1918 for his work on ammonia synthesis. However, this was controversial due to his involvement in the development of gas warfare during the First World War.

The Haber process also had an unintended consequence; it contributed to a global population boom with world population rising from about 1.6 billion people in 1900 to about 8 billion people today. This population growth, and the resulting growth in resource consumption and waste production, are behind the need for global sustainability.

Haber also developed a process for converting ammonia into nitric acid. Nitric acid was then used as the basis for synthesizing a variety of insecticides and producing nitrate-based high explosives. The novel production of explosives significantly helped Germany to overcome the effect of the Allied blockades of nitrates from Chile during the First World War (1914–18).

TOK

Is science, or should it be, value-free?

Many suggest that science is 'value-free'. By this, they mean that the scientific community does not try to apply the scientific method to questions of what is ethically *right* or *wrong*; it simply tries to describe what exists.

However, science cannot be absolutely objective and so cannot be value free. Scientists are part of a scientific community with shared values. The inherent values of the scientific community include the free flow of information (bearing in mind patent laws and journal pay walls), honesty, intellectual curiosity, open-mindedness and objectivity. These are 'values' in the sense that the scientific community would say that they are important to the development of knowledge and that they tend to produce better knowledge.

Scientific knowledge is the result of a social construction in which social negotiations determine what is legitimate and non-legitimate. In other words, science is conducted by human beings and no human being can construct knowledge *outside* of social conventions, context and beliefs.

Should there be any limits to our choices of which scientific areas to study? What are those limits, if any? How do we decide what they are? Should ethical values and beliefs constrain scientific research?

Is there a conflict between the inherent values of science and the scientific community, such as open-mindedness and curiosity, and the inherent values of some members in the society or culture?

Can moral disagreements be resolved with reference to empirical evidence?

The Scottish philosopher David Hume proposed a 'fact / value' or 'is / ought' distinction that many philosophers consider to be the heart of the difference between science as an area of knowledge and knowledge related to ethics and morals. The concept is best understood as a distinction between 'what *is*' (descriptions of the world of objects using empirical evidence or 'facts') and 'what *ought* to be' (moral values). The fact / value distinction is the line between what is true in an empirical sense and what is morally right or wrong. This is sometimes a source of conflict between the natural sciences and morality.

Some people argue that moral values are not 'real', because they are not objects; instead, they are purely subjective claims. In other words, moral values are a bit like taste – what is 'good' for you might not be 'good' for others – and therefore there can be no definitive moral truth.

Others, however, argue that moral truths can be derived from observations even though such moral truths do not describe *objects*. Consider differences in access to education around the world, for example. It is probably objectively true to say that a person without access to education is less able to participate in the world than those who do gain an education. It is a demonstrable, testable fact that a person who is *not allowed* to gain an education is disadvantaged, and we might say that this is *unfair* or *wrong*. Some ways of treating people do not allow them to live full, enriching lives.

Some scientific discoveries are *designed* to be used in technology that leads to human misery. This has nothing to do with one's own taste; it is just true. For example, in the 1940s Harvard professor Louis F. Fieser (1899–1977) saved countless lives after he synthesized vitamin K (helping newborn babies' blood to coagulate better). Later, however, he went to work for Dow Chemicals to develop a better Napalm (a chemical weapon used in blowtorches and incendiary bombs). At the time, he claimed that 'I have no right to judge the morality of Napalm just because I invented it' (*Time* magazine, 5 January 1968).

If you accept that scientific knowledge exists in a social context, and recognize that that social context has moral values (such as justice or fairness) at its heart, then the suggestion that science (descriptions of what *is*) and morality (claims about what *should* be) exist in different realms is far less convincing.

Reaction quotient

◆ **Reaction quotient**: A value obtained by applying the equilibrium law to any concentrations other than equilibrium concentrations.

HL ONLY

To determine the progress of a reaction relative to equilibrium, we use a quantity called the **reaction quotient** (Q). The reaction quotient is defined in the same way as the equilibrium constant, except that the reaction does not need to be at equilibrium.

For the general reversible reaction:

$$aA + bB \rightleftharpoons cC + dD$$

the reaction quotient (Q) is defined as:

$$Q = \frac{[C]^c[D]^d}{[A]^a[B]^b}$$

The equilibrium constant has only one value (at a given temperature), whereas the reaction quotient depends on the current state of the reaction and has many different values as the reaction proceeds to equilibrium. Depending on whether equilibrium is approached from the side of reactants or products, the reaction quotient may be smaller or larger than K. At equilibrium, the reaction quotient (Q) is equal to the equilibrium constant (K).

Figure R2.96 shows a plot of Q as a function of the concentrations of A and B for the simple reaction $A(g) \rightleftharpoons B(g)$, which has an equilibrium constant of $K = 1.50$.

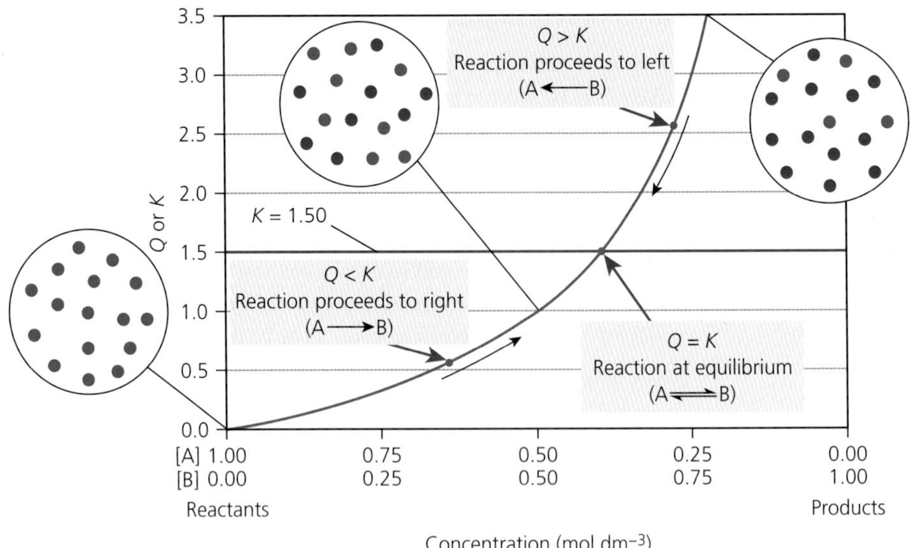

■ **Figure R2.96** The reaction quotient, Q, equilibrium constant, K, and direction in which the reaction proceeds

The far left of the graph represents pure reactant, $[A] = 1.00\,mol\,dm^{-3}$, and the far right represents pure product, $[B] = 1.00\,mol\,dm^{-3}$ (where Q becomes infinite). The midpoint of the graph represents an equal mixture of A and B, $[A] = [B] = 0.50\,mol\,dm^{-3}$.

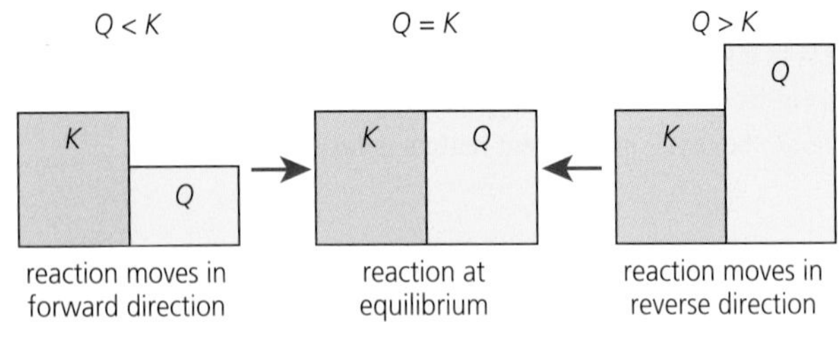

■ **Figure R2.97** The relative magnitudes of the reaction quotient and equilibrium constant indicate the direction in which a reaction mixture tends to change

By comparing the value of Q with the equilibrium constant, K, the direction of reaction can be predicted (see Figure R2.97).

- If $Q < K$, the reaction will move to the right, producing more product(s) to reach equilibrium.
- If $Q = K$, the reaction is at equilibrium.
- If $Q > K$, the reaction will move to the left, producing more reactant(s) to reach equilibrium.

Le Châtelier's principle and the reaction quotient

Use of the reaction quotient, Q, helps to explain (at a simple level) why changing the concentration of a component of an equilibrium mixture gives rise to the effect it does. Take the following reaction as an example:

$$N_2(g) + 3H_2(g) \rightleftharpoons 2NH_3(g)$$

At equilibrium:

$$Q = K = \frac{[NH_3]^2}{[N_2][H_2]^3}$$

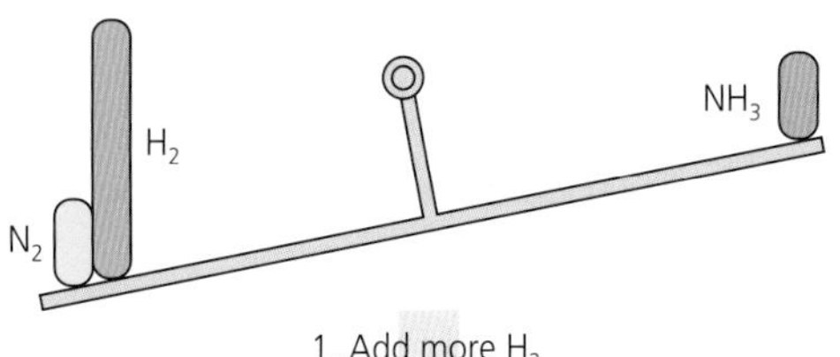
N$_2$
H$_2$
NH$_3$
1. Add more H$_2$

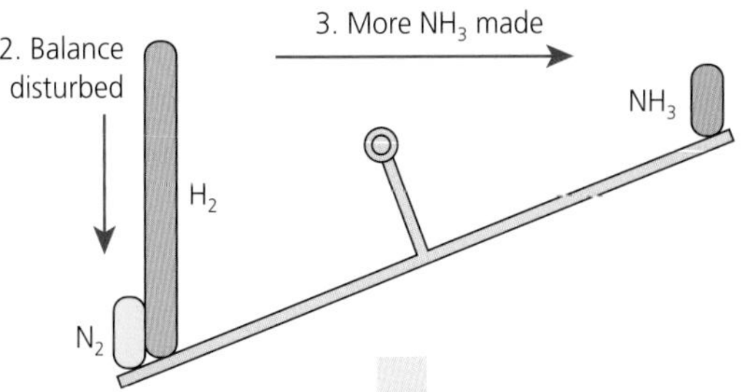
2. Balance disturbed
3. More NH$_3$ made
N$_2$
H$_2$
NH$_3$
System responds to restore position of see-saw beam – the amounts of N$_2$, H$_2$ and NH$_3$ are different, but the position is the same

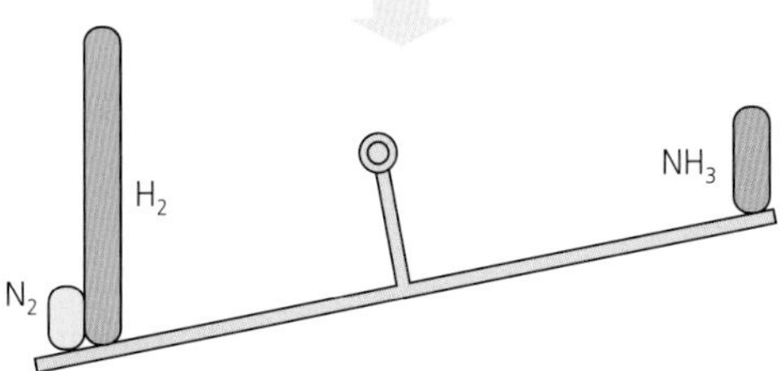
N$_2$
H$_2$
NH$_3$

■ **Figure R2.98** Diagrammatic representation of the effect of adding more hydrogen to the equilibrium mixture involved in ammonia synthesis

If more hydrogen is added to the equilibrium mixture, this will increase the value of the denominator in Q. Its value will no longer be equal to K – it will have a lower value. Therefore the reaction will adjust to increase Q by producing more ammonia.

The argument outlined here can be visualized using the analogy of a see-saw, as shown in Figure R2.98. The resting angle of the beam represents the equilibrium mixture. If the balance is disturbed, the resting angle is changed; as equilibrium is restored, so is the resting angle.

At a particular temperature, the value of K is constant. The system responds to any change in the composition of the mixture in a way that restores the angle of the beam – the value of Q – to the 'resting' angle.

Calculating equilibrium constants

Equilibrium constants are easily calculated from values of the equilibrium concentrations of all the reactants and products. The values just need to be substituted into the equilibrium expression.

WORKED EXAMPLE R2.3D

Nitrogen(II) oxide, NO, is a pollutant released into the atmosphere from car exhausts. It is also formed when nitrosyl chloride, NOCl, dissociates according to the following equation:

$$2NOCl(g) \rightleftharpoons 2NO(g) + Cl_2(g)$$

To study this reaction, different amounts of the three gases were placed in a closed container and allowed to come to equilibrium at 503 K and at 738 K.

The equilibrium concentrations of the three gases at each temperature are given in the table.

Temperature / K	Concentration / mol dm^{-3}		
	NOCl	*NO*	*Cl$_2$*
503	2.33×10^{-3}	1.46×10^{-3}	1.15×10^{-2}
738	3.68×10^{-4}	7.63×10^{-3}	2.14×10^{-4}

a Write the expression for the equilibrium constant, K, for this reaction.

b Calculate the value of K at both of the two temperatures given.

c State and explain whether the forward reaction is endothermic or exothermic.

Answer

a $K = \dfrac{[NO]^2[Cl_2]}{[NOCl]^2}$

b At 503 K:

$$K = \frac{(1.46 \times 10^{-3})^2 \times 1.15 \times 10^{-2}}{(2.33 \times 10^{-3})^2} = 4.5 \times 10^{-3}$$

At 738 K:

$$K = \frac{(7.63 \times 10^{-3})^2 \times 2.14 \times 10^{-4}}{(3.68 \times 10^{-4})^2} = 9.2 \times 10^{-2}$$

c The value of K is greater at 738 K.

K increases with temperature. This means the forward reaction is favoured, increasing the proportion of products in the equilibrium mixture, at higher temperatures. This suggests that the forward reaction is endothermic.

HL ONLY

ICE tables

In most cases, we only need to know the initial concentrations of the reactant(s) and the equilibrium concentration of any one reactant or product from experiments in order to deduce the other equilibrium concentrations from the stoichiometry of the balanced equation.

For example, consider the simple homogeneous equilibrium occurring in a fixed volume: $A(g) \rightleftharpoons 2B(g)$, where one mole of reactant A is converted to 2 moles of product B. The mole ratio is 1:2 for the forward reaction.

We have a reaction mixture in which the initial concentration of the reactant A is $1.00\,mol\,dm^{-3}$ and the initial concentration of the product B is $0.00\,mol\,dm^{-3}$. The equilibrium concentration of A is $0.75\,mol\,dm^{-3}$.

Since [A] has changed by $-0.25\,mol\,dm^{-3}$, we can deduce, based on the stoichiometry, that [B] must have changed by $2 \times (+0.25\,mol\,dm^{-3})$ or $+0.50\,mol\,dm^{-3}$.

The initial conditions, the changes and the equilibrium conditions are summarized in a table known as an ICE table:

■ **Table R2.18** ICE table for $A(g) \rightleftharpoons 2B(g)$

	[A]	**[B]**
Initial	1.00	0.00
Change	−0.25	+0.50
Equilibrium	0.75	0.50

To calculate the equilibrium constant, you use the balanced equation to write an expression for the equilibrium constant and then substitute the equilibrium concentrations from the ICE table:

$$K = \frac{[B]^2}{[A]} = \frac{(0.50)^2}{0.75} = 0.33$$

43 Ammonia is produced in the Haber process:

$$N_2(g) + 3H_2(g) \rightleftharpoons 2NH_3(g)$$

1 mole of nitrogen and 2 moles of hydrogen are placed in a reaction vessel of $2\,dm^3$. After equilibrium is reached, the amount of nitrogen remaining is 0.4 mol.

Write an expression for the equilibrium constant, K, and calculate its value.

Using measured equilibrium concentrations

You may need to calculate equilibrium concentrations of reactants or products from the equilibrium constant. These types of calculations allow you to calculate the concentration or amount of a reactant or product at equilibrium.

These types of problem can be classified into two categories:

- calculating equilibrium concentrations when given the equilibrium constant and all but one of the equilibrium concentrations of the reactants and products
- calculating equilibrium concentrations when we know the equilibrium constant and only the initial concentrations.

Calculating equilibrium concentrations from the equilibrium constant and all but one of the equilibrium concentrations

To solve this type of problem, you can substitute the concentration values into the equilibrium constant expression and rearrange.

WORKED EXAMPLE R2.3E

Consider the following reaction:

$$2COF_2(g) \rightleftharpoons CO_2(g) + CF_4(g); \quad K = 2.00 \text{ at } 1000\,°C$$

In an equilibrium mixture, the concentration of COF_2 is 0.255 mol dm^{-3} and the concentration of CF_4 is 0.120 mol dm^{-3}. Deduce the equilibrium concentration of CO_2.

Answer

$$K = \frac{[CO_2][CF_4]}{[COF_2]^2}$$

$$[CO_2] = K\frac{[COF_2]^2}{[CF_4]}$$

$$[CO_2] = 2.00\left(\frac{(0.255)^2}{0.120}\right) = 1.08\text{ mol dm}^{-3}$$

Calculating equilibrium concentrations from the equilibrium constant and initial concentrations

Top tip!

If the equilibrium constant, K, is relatively small ($< \approx 0.01$), the reaction will not proceed very far to the right and we can make the assumption that x is small relative to the initial concentration of reactant.

These kinds of problems are generally more intricate than those we just examined and require a specific procedure using an ICE table to solve them. The changes in concentration are not known and are represented by the variable x.

Suppose that, for the reaction $A(g) \rightleftharpoons 2B(g)$, we have a reaction mixture in which the initial concentration of A is 1.00 mol dm^{-3} and the initial concentration of B is 0.00 mol dm^{-3}. The equilibrium constant $K = 0.0033$, and we want to determine the equilibrium concentrations.

■ **Table R2.19** ICE table for determining equilibrium concentrations given initial concentrations and K

	[A]	[B]
Initial	1.00	0.00
Change	$-x$	$+2x$
Equilibrium	$1.00 - x$	$2x$

Note that, due to the stoichiometry of the reaction, the change in [B] must be $+2x$. As before, each equilibrium concentration is the sum of the two entries above it in the ICE table.

Since we know the value of the equilibrium constant, we can use the equilibrium constant expression to set up an equation in which x is the only variable:

$$K = \frac{[B]^2}{[A]} = \frac{(2x)^2}{1.0 - x} = 0.0033$$

$$K = \frac{4x^2}{1.0 - x} = 0.0033$$

Since x is assumed to be very small, this approximates to:

$$K \approx \frac{4x^2}{1.0} = 0.0033$$

so $x \approx 0.029$

WORKED EXAMPLE R2.3F

An organic compound X exists in equilibrium with its isomer, Y, in the liquid state at a particular temperature:

$X(l) \rightleftharpoons Y(l)$

Calculate the amount (mol) of Y that is formed at equilibrium if 1 mole of X is allowed to reach equilibrium at this temperature. K has a value of 0.020.

Answer

Let the amount of Y at equilibrium = y moles:

	X(l)	$\rightleftharpoons$	Y(l)
(I) Starting amount (moles)	1.00		0.00
(C) Change in amount (moles)	$-y$		y

From the equation, if y moles of the isomer Y are present, then y moles of X must have reacted. Therefore, $(1.00 - y)$ moles of X must remain at equilibrium. Also, if we call the volume of liquid V dm^3, then we can complete the table as follows:

	X(l)	$\rightleftharpoons$	Y(l)
(I) Starting amount (moles)	1.00		0.00
(E) Equilibrium amount (moles)	$(1.00 - y)$		y
Equilibrium concentration (mol dm^{-3})	$\frac{(1.00 - y)}{V}$		$\frac{y}{V}$

$$K = \frac{[Y]}{[X]}$$

$$0.020 = \frac{y/V}{(1.00 - y)/V}$$

The 'V' terms cancel:

$$0.020 = \frac{y}{(1.00 - y)}$$

$$0.020(1.00 - y) = y$$

$$0.020 - 0.020y = y$$

and so, $1.020y = 0.020$

therefore, $y = \frac{0.02}{1.020} = 0.0196\,\text{mol} = 0.020\,\text{mol}$ (2 s.f.)

Therefore at equilibrium the mixture contains 0.020 mol of Y.

Going further

Calculations using the quadratic formula

For the esterification reaction:

$$CH_3COOH(l) + C_2H_5OH(l) \rightleftharpoons CH_3COOC_2H_5(l) + H_2O(l)$$

calculate the amount of ethyl ethanoate that formed at equilibrium when 1.0 mole of ethanol reacted with 2.0 moles of ethanoic acid at 373 K. The value of K is 4.0 at this temperature.

Let the amount of ethyl ethanoate at equilibrium be x moles. Let the volume of the reacting mixture be $V\,dm^3$.

■ **Table R2.20** ICE table for $CH_3COOH(l) + C_2H_5OH(l) \rightleftharpoons CH_3COOC_2H_5(l) + H_2O(l)$

	$CH_3COOH(l)$	$C_2H_5OH(l)$	$CH_3COOC_2H_5(l)$	$H_2O(l)$
(*I*) Starting amount (moles)	2.00	1.00	0.00	0.00
(*C*) Change (moles)	$-x$	$-x$	x	x
(*E*) Equilibrium amount (moles)	$(2.0 - x)$	$(1.0 - x)$	x	x
Equilibrium concentration ($mol\,dm^{-3}$)	$\frac{(2.0 - x)}{V}$	$\frac{(1.0 - x)}{V}$	$\frac{x}{V}$	$\frac{x}{V}$

$$K = \frac{\left(\frac{x}{V}\right)\left(\frac{x}{V}\right)}{\left(\frac{2.0 - x}{V}\right)\left(\frac{1.0 - x}{V}\right)}$$

The V terms cancel to give:

$$K = \frac{x^2}{(2.0 - x)(1.0 - x)}$$

Therefore:

$$4.0 = \frac{x^2}{(2.0 - x)(1.0 - x)} = \frac{x^2}{x^2 - 3x + 2}$$

This rearranges to:

$$3x^2 - 12x + 8 = 0$$

The solution of this quadratic equation requires the use of the general expression:

$$x = \frac{-b \pm \sqrt{b^2 - 4ac}}{2a}$$ for the general quadratic $ax^2 + bx + c = 0$

Using this expression gives possible values for x of 0.85 moles or 3.15 moles. The second of these solutions is impossible as we only started with 1.0 mole of ethanol. Therefore the amount of ethyl ethanoate at equilibrium is 0.85 moles.

The IB syllabus states that calculations of this type (involving the solution of quadratic equations) will not be set. However, questions 44 and 45 lead to equations that include perfect squares. It is worth remembering that these can be solved by taking the square root of each side of the expression.

HL ONLY

Inquiry 1: Exploring and designing

Demonstrate independent thinking, initiative or insight

Critical thinking involves the application of logic to a problem and requires a wide range of skills. These include recording and processing data; formulating appropriate hypotheses; making conclusions; and acting on the conclusions to address the research question.

Critical thinking is especially important in chemistry, because the study of equilibria and kinetics deals with complex and dynamic systems which can be difficult to understand for a number of reasons:

- They have many different aspects, involving many interactions.
- It can be difficult to change one variable in an experiment without producing confounding variables. A confounding variable is a variable, other than the independent variable that you are interested in, that may affect the dependent variable.
- Some variables may be unmeasured or unmeasurable.
- Changes to the chemical system can lead to unexpected results.
- Conclusions drawn from the data may not be definite.

Critical thinking allows you to arrive at the most likely conclusion from the experimental results of your investigation. However, you need to reflect and consider if other conclusions might be possible.

This approach allows you to assess the possibilities and propose a hypothesis, but also to understand that your conclusions are tentative and might change when new experimental data is recorded and analysed.

44 Sulfur dioxide, SO_2, reacts with oxygen in the presence of a catalyst (vanadium(V) oxide) to form sulfur trioxide. This reaction is carried out in a sealed container of volume $3.0\,dm^3$ by mixing 2.0 mol of sulfur dioxide and 1.4 mol of oxygen and allowing equilibrium to be established. A conversion rate of 15% is achieved at 700 K. Calculate the equilibrium constant *K* at this temperature for this reaction.

45 When the reaction $SO_3(g) + NO(g) \rightleftharpoons NO_2(g) + SO_2(g)$ was carried out at a particular temperature, the equilibrium constant was found to be 6.78. The initial equimolar concentrations of SO_3 and NO were $0.030\,mol\,dm^{-3}$. Calculate the equilibrium concentration of each component in the mixture once equilibrium is established.

46 Hydrogen and carbon dioxide gases react according to the following equation: $H_2(g) + CO_2(g) \rightleftharpoons H_2O(g) + CO(g)$. The reaction was carried out in a sealed vessel of volume $10\,dm^3$. The four components of the reaction were put into the vessel in the following proportions: 2.00 mol of each of the reactants (H_2 and CO_2) and 1.00 mol of each of the products (H_2O and CO). The system was then allowed to come to equilibrium at 1200 K. The equilibrium constant, *K*, for this reaction at 1200 K is 2.10. Calculate the equilibrium concentration of each component of the reaction mixture.

LINKING QUESTION

How does the equilibrium law help us to determine the pH of a weak acid, weak base or buffer solution?

Link

The relationship between the Gibbs energy change and the position of equilibrium (as measured by the equilibrium constant) is discussed in more detail in Chapter R1.4 page 403.

Top tip!

The relationship between $\Delta G^\ominus$ and K is logarithmic – small changes in $\Delta G^\ominus$ have a large effect on K.

Top tip!

The relationship can also be written in an exponential form:

$K = e^{\frac{-\Delta G^\ominus}{RT}}$.

47 Explain how the value of the equilibrium constant, K, reflects the concentration of reactants and products at equilibrium.

Relationship between equilibrium constant, spontaneity and Gibbs energy change

The value of the equilibrium constant, K, does not give any information about the forward and backward rates of reaction. Equilibrium constants are independent of the kinetics of the reaction.

However, the chemical equilibrium constant, K, is directly related to the standard state Gibbs energy change, $\Delta G^\ominus$, by the following equation:

$$\Delta G^\ominus = -RT \ln K$$

where R represents the gas constant and T the absolute temperature in kelvin. This leads to the changes summarized in Table R2.21.

■ **Table R2.21** A summary of the relationship between $\Delta G^\ominus$ and K

$\Delta G^\ominus$	$\ln K$	K	Position of equilibrium
negative	positive	>1	to the right – products favoured
zero	zero	=1	central – neither reactants or products favoured
positive	negative	<1	to the left – reactants favoured

Figure R2.99 shows a schematic representation of the numerical relationship between the Gibbs energy change and the extent of reaction in the forward direction.

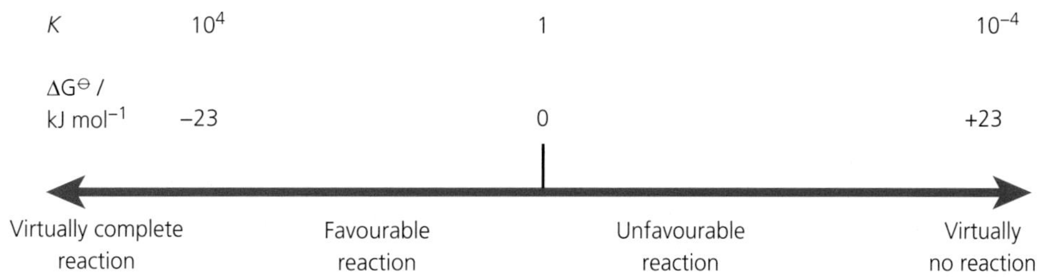

■ **Figure R2.99** The relationship between the Gibbs energy change and extent of the forward reaction

48 Given that $\Delta G^\ominus$ for a reaction is $-28.7\,\text{kJ mol}^{-1}$, calculate the equilibrium constant, K, for the reaction at 25 °C.

Relationship between thermodynamics and equilibrium

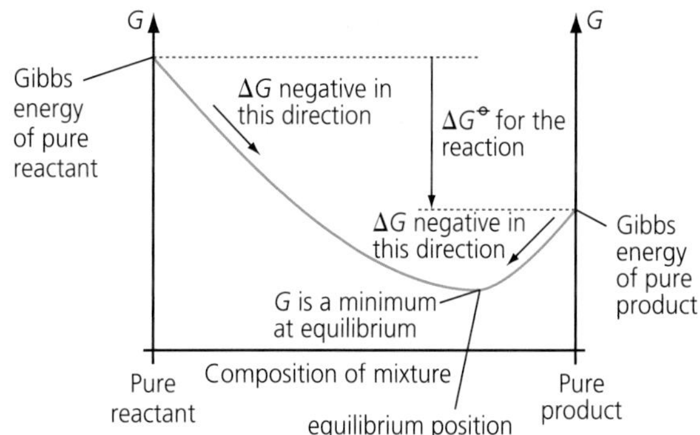

■ **Figure R2.100** The variation in the Gibbs energy for a reaction (single reactant ⇌ single product) for which the overall $\Delta G^\ominus$ is negative

The position of equilibrium, as measured by the equilibrium constant, K, corresponds to a minimum in the Gibbs energy of the system for a chemical reaction or physical change.

Figure R2.100 shows this for a simple equilibrium between a single reactant and a single product: A ⇌ B.

The conversion of either reactants or products into the equilibrium mixture is a process in which ΔG is negative, so the equilibrium mixture always has a lower Gibbs energy than the pure reactants or the pure products. G is a minimum and ΔG is zero at this point – there is no tendency to spontaneously shift in either direction away from equilibrium.

HL ONLY

Top tip!

ΔG is the 'driving force' for all chemical and physical changes. It is analogous to the difference in gravitational potential energy. If a ball is at the top of one side of a valley, it will roll down to a position where it stores less gravitational potential energy and settle in the lowest equilibrium position.

The value and sign of $\Delta G^{\ominus}$ indicate the position of equilibrium.

- If $\Delta G^{\ominus}$ is negative then the position of equilibrium will be closer to the products than the reactants (Figure R2.101a). The more negative the value of $\Delta G^{\ominus}$, the closer the position of equilibrium lies towards the products.
- If $\Delta G^{\ominus}$ is a large negative number, then the position of equilibrium lies very close to pure products – a reaction for which $\Delta G^{\ominus}$ is negative may proceed spontaneously from reactants to products and be considered to go to completion as long as its rate is high enough.
- If $\Delta G^{\ominus}$ is large and positive then the position of equilibrium lies more towards the reactants (Figure R2.101b) – the more positive the number, the closer the position of equilibrium towards pure reactants.

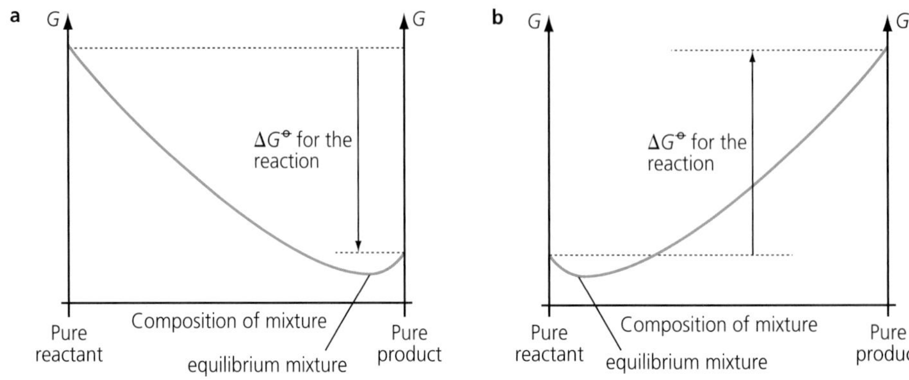

■ **Figure R2.101** The influence of the value of $\Delta G^{\ominus}$ on the position of equilibrium: **a** $\Delta G^{\ominus}$ is negative and the position of equilibrium lies closer to the products; **b** $\Delta G^{\ominus}$ is positive and the position of equilibrium lies closer to the reactants

WORKED EXAMPLE R2.3G

Determine the signs of $\Delta G^{\ominus}$ and explain the position of equilibrium for these two reactions.

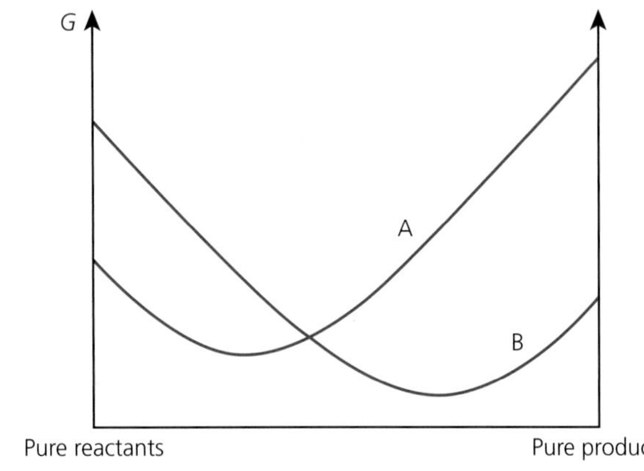

Answer

Reaction A:

$\Delta G^{\ominus} > 0$, K is expected to be <1.

The minimum value of G lies close to the reactants.

The position of equilibrium lies very much to the left and most of the reactants are left unreacted.

Reaction B:

$\Delta G^{\ominus} < 0$, K is expected to be >1.

The minimum value of G lies close to the products.

The position of equilibrium lies to the right and most of the reactants are reacted.

WORKED EXAMPLE R2.3H

The dissociation of dinitrogen tetroxide was carried out in a $1.00\,dm^3$ sealed tube at 298 K. The graph shows the change of Gibbs free energy during the reaction.

a State and explain whether the forward reaction is spontaneous.

b Determine the value of the equilibrium constant.

c Use the value of the equilibrium constant to calculate the standard Gibbs free energy change of the reaction.

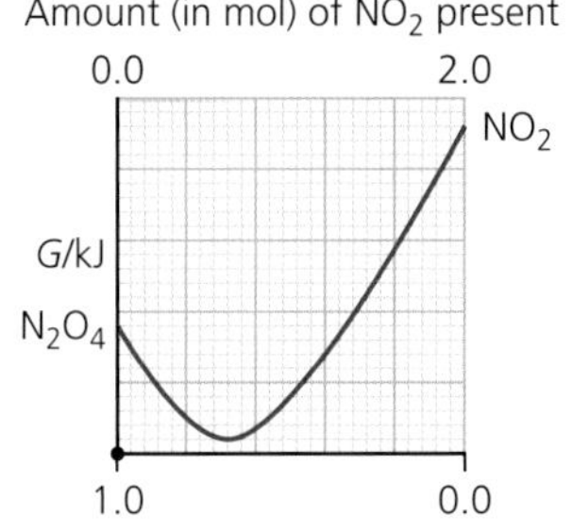

Answer

a $\Delta G^\ominus > 0$, and therefore the reaction is not spontaneous.

b $N_2O_4(g) \rightleftharpoons 2NO_2(g)$

$$K = \frac{[NO_2]^2}{[N_2O_4]}$$

$K = (0.640\,mol / 1\,dm^3)^2 \div (0.680\,mol / 1.00\,dm^3) = 0.60$

c $\Delta G^\ominus = -RT\ln K$

$= -(8.31\,J\,K^{-1}\,mol^{-1}) \times (298\,K) \times (\ln 0.60)$

$= 1260\,J\,mol^{-1} = 1.26\,kJ\,mol^{-1}$

LINKING QUESTION

How can Gibbs energy be used to explain which of the forward or backward reaction is favoured before reaching equilibrium?

Going further

The relationship between K and $\Delta H^\ominus$

To derive this relationship, you need to start from two relationships previously introduced:

$\Delta G^\ominus = \Delta H^\ominus - T\Delta S^\ominus$ and

$\Delta G^\ominus = -RT\ln K$

Combining these equations gives:

$-RT\ln K = \Delta H^\ominus - T\Delta S^\ominus$, or

$$\ln K = -\frac{\Delta H^\ominus}{RT} + \frac{\Delta S^\ominus}{R}$$

(known as the van't Hoff equation)

$\Delta H^\ominus$ and $\Delta S^\ominus$ can be treated as constants over a limited temperature range.

A plot of $\ln K$ against $1/T$ is therefore a straight line, because the equation is of the form $y = mx + c$, where the slope, $m = -\frac{\Delta H^\ominus}{R}$ and the intercept, $c = \frac{\Delta S^\ominus}{R}$. For an exothermic reaction ($\Delta H^\ominus < 0$), the magnitude of K decreases with increasing temperature, but for an endothermic reaction ($\Delta H^\ominus > 0$), the magnitude of K increases with increasing temperature, which agrees with Le Châtelier's principle.

Proton transfer reactions

Guiding question

- What happens when protons are transferred?

SYLLABUS CONTENT

By the end of this chapter, you should understand that:

- a Brønsted–Lowry acid is a proton donor and a Brønsted–Lowry base is a proton acceptor
- a pair of species differing by a single proton is called a conjugate acid–base pair
- some species can act as both Brønsted–Lowry acids and bases
- the pH scale can be used to describe the $[H^+]$ of a solution; $pH = -\log_{10}[H^+]$; $[H^+] = 10^{-pH}$
- the ion product constant of water shows an inverse relationship between $[H^+]$ and $[OH^-]$; $K_w = [H^+][OH^-]$
- strong and weak acids and bases differ in the extent of ionization
- acids react with bases in neutralization reactions
- pH curves for neutralization reactions involving strong acids and bases have characteristic shapes and features
- the pOH scale describes the $[OH^-]$ of a solution; $pOH = -\log_{10}[OH^-]$; $[OH^-] = 10^{-pOH}$ (HL only)
- the strengths of weak acids and bases are described by their K_a, K_b, pK_a or pK_b values (HL only)
- for a conjugate acid–base pair, the relationship $K_a \times K_b = K_w$ can be derived from the expressions for K_a and K_b (HL only)
- the pH of a salt solution depends on the relative strengths of the parent acid and base (HL only)
- pH curves of different combinations of strong and weak monoprotic acids and bases have characteristic shapes and features (HL only)
- acid–base indicators are weak acids, where the components of the conjugate acid–base pair have different colours; the pH of the end-point of an indicator, where it changes colour, approximately corresponds to its pK_a value (HL only)
- an appropriate indicator for a titration has an end-point range that coincides with the pH at the equivalence point (HL only)
- a buffer solution is one that resists change in pH on the addition of small amounts of acid or alkali (HL only)
- (HL only) the pH of a buffer solution depends on both:
  - the pK_a or pK_b of its acid or base
  - and the ratio of the concentration of acid or base to the concentration of the conjugate base or acid.

By the end of this chapter you should know how to:

- deduce the Brønsted–Lowry acid and base in a reaction
- deduce the formula of the conjugate acid or base of any Brønsted–Lowry base or acid
- interpret and formulate equations to show acid–base reactions of these species
- perform calculations involving the logarithmic relationship between pH and $[H^+]$
- recognize solutions as acidic, neutral and basic from the relative values of $[H^+]$ and $[OH^-]$
- recognize that acid–base equilibria lie in the direction of the weaker conjugate
- formulate equations for the reaction between acids and metal oxides, metal hydroxides, hydrogencarbonates and carbonates
- sketch and interpret the general shape of the pH curve
- interconvert $[H^+]$, $[OH^-]$, pH and pOH values (HL only)
- interpret the relative strengths of acids and bases from K_a, K_b, pK_a, pK_b data (HL only)
- solve problems involving K_a, K_b and K_w values (HL only)
- construct equations for the hydrolysis of ions in a salt, and predict the effect of each ion on the pH of the salt solution (HL only)
- interpret the general shapes of pH curves for all four combinations of strong and weak acids and bases (HL only)

- construct equilibrium expressions to show why the colour of an indicator changes with pH (HL only)
- identify an appropriate indicator for a titration from the identity of the salt and the pH range of the indicator (HL only)
- describe the composition of acidic and basic buffers and explain their actions (HL only)
- solve problems involving the composition and pH of a buffer solution, using the equilibrium constant (HL only).

Acids and bases – Arrhenius definitions

Acids

◆ **Acid (Arrhenius theory)**: A type of compound that contains hydrogen and ionizes in water to produce positive hydrogen ions (as the only cations).

◆ **Binary acid**: An acid in which the acidic hydrogen atom(s) are bonded directly to a non-metal atom other than oxygen.

◆ **Oxyacid**: An acid whose molecule contains one or more oxygen atoms.

◆ **Acid anhydride**: A molecular compound that reacts with water to form an acid.

The Arrhenius theory defines an **acid** as a substance that produces hydrogen ions, H^+, and no other cations when dissolved in water. For example:

$$HCl(aq) \rightarrow H^+(aq) + Cl^-(aq)$$

$$H_2SO_4(aq) \rightarrow 2H^+(aq) + SO_4^{2-}(aq)$$

$$CH_3COOH(aq) \rightarrow H^+(aq) + CH_3COO^-(aq)$$

In **binary acids**, the hydrogen atoms are often covalently bonded to a halogen atom, e.g. H–Cl.

Oxyacids usually consist of three elements, including hydrogen and oxygen. Common oxyacids include nitric acid, HNO_3, and sulfuric acid, H_2SO_4. These molecules also release hydrogen ions, $H^+(aq)$, when reacted with water. Organic acids such as methanoic acid, HCOOH, are also oxyacids.

Oxyacids are formed when oxides of reactive non-metals react with water in a hydrolysis reaction, for example:

$$SO_3(g) + H_2O(l) \rightarrow H_2SO_4(aq) \text{ (sulfuric(VI) acid)}$$

$$P_4O_6(s) + 6H_2O(l) \rightarrow 4H_3PO_3(aq) \text{ (phosphoric(III) acid)}$$

These oxides are known as **acid anhydrides** or acidic oxides since they are neutralized by bases, for example:

$$SO_3(g) + 2NaOH(aq) \rightarrow Na_2SO_4(aq) + H_2O(l)$$

Nature of science: Falsification

Classifying acids

The first attempt at a theoretical interpretation of acid behaviour was made by the French chemist Antoine Lavoisier (1743–1794). His theory (1776) suggested that acids contained oxygen. Many common acids (for example, H_2SO_4) are oxyacids, but this idea was falsified by the identification of acids such as hydrochloric acid, HCl, and hydrocyanic acid, HCN. In 1830, the German chemist Justus von Liebig suggested that acids are hydrogen-containing compounds in which the hydrogen may be replaced by metals.

Pure acids are often in the form of simple covalent molecules (Figure R3.1). According to the Arrhenius theory, acids only show their acidic properties in aqueous solution when the molecules undergo ionization.

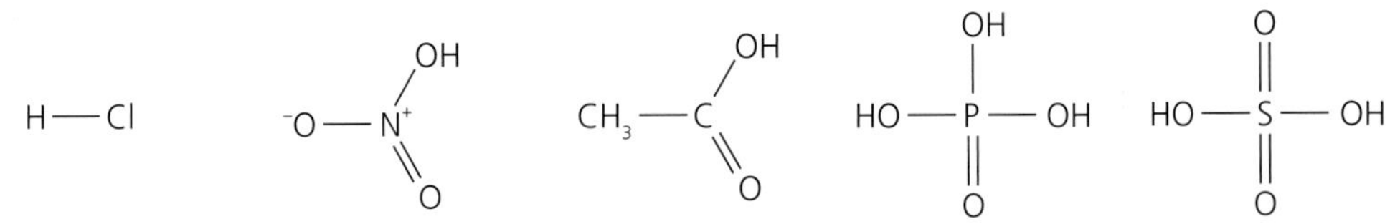

■ **Figure R3.1** Structural formulas for the molecules that form hydrochloric, nitric(V), ethanoic, phosphoric(V) and sulfuric(VI) acids in aqueous solution

When hydrogen chloride gas or solid citric acid (2-hydroxypropane-1,2,3-tricarboxylic acid) are dissolved in organic solvents, they do not exhibit acidic properties (Figure R3.2). The molecules do not ionize in organic solvents and hence it is the presence of $H^+(aq)$ ions in aqueous solutions that gives these acids their acidic properties.

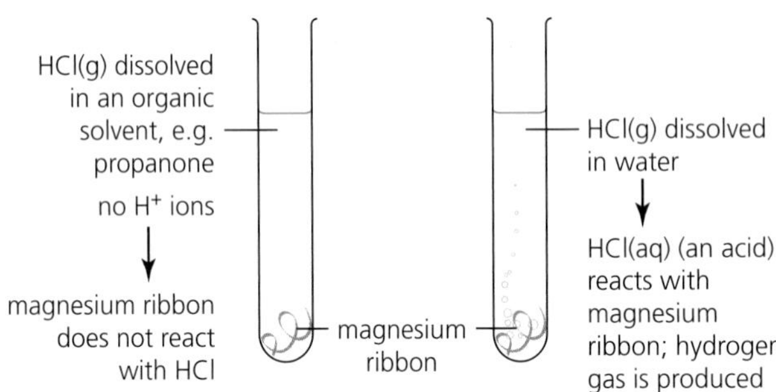

■ **Figure R3.2** The importance of water in acid formation

Acids that have a single proton to donate are said to be **monoprotic**. Common examples include hydrochloric, $HCl(aq)$, nitric, $HNO_3(aq)$, nitrous, $HNO_2(aq)$, and ethanoic, $CH_3COOH(aq)$, acids.

Acids that have two protons to donate are said to be **diprotic**. Common examples include 'carbonic acid', $H_2CO_3(aq)$, sulfuric acid, $H_2SO_4(aq)$, and sulfurous acid, $H_2SO_3(aq)$.

Top tip!

The term carbonic acid is in quotation marks because the compound H_2CO_3 does not actually exist and cannot be isolated. Attempts to isolate it result in the formation of carbon dioxide and water.

The only common inorganic triprotic acid is phosphoric(V) acid $H_3PO_4(aq)$.

The hydrogen in an acid usually has to be attached to oxygen or a halogen. This accounts for the monoproticity of ethanoic acid; only the hydrogen atom attached to the oxygen atom can be replaced by a metal ion. The other three hydrogen atoms are attached to a carbon atom and are therefore not acidic.

Bases

According to the Arrhenius theory, a **base** is any metal oxide or hydroxide that reacts with an acid to produce a salt and water only. For example:

$$Ba(OH)_2(aq) + 2HNO_3(aq) \rightarrow Ba(NO_3)_2(aq) + 2H_2O(l)$$

Common mistake

Misconception: *All compounds with hydrogen in their molecular formula are acids.*

Most compounds with hydrogen in their molecular formula are not acids. For example, the hydrogen atoms in $C_6H_{12}O_6$ are not released as ions in aqueous solution.

◆ **Monoprotic**: An acid that has only one acidic hydrogen atom in its molecules and releases one hydrogen ion (proton) per molecule in aqueous solution.

◆ **Diprotic**: An acid that has two acidic hydrogen atoms in its molecules and releases two hydrogen ions (protons) per molecule in aqueous solution.

◆ **Base (Arrhenius theory)**: A compound that reacts with an acid to give water and a salt only.

◆ **Salt**: An ionic compound formed by reaction of an acid with a base, in which the hydrogen of the acid has been replaced by metal or other cations.

◆ **Neutralization**: A chemical reaction between an acid and a base to produce a salt and water only.

A **salt** is the substance formed when the hydrogen in an acid is partially or wholly replaced by a metal cation or ammonium (NH_4^+) ion.

The reaction between acids and bases to form a salt and water only is known as **neutralization**.

Common mistake

Misconception: *Mixing an acid with a base (without regard to quantities) neutralizes the base resulting in a neutral solution.*

There is a need for equimolar quantities (stoichiometric amounts). For example, mixing equal volumes of barium hydroxide and hydrochloric acid (of the same concentration) will result in an alkaline solution since the base (hydroxide ions) is present in excess.

Ammonia, NH_3, and amines (for example, propylamine, $C_3H_7NH_2$) are also bases.

Many bases, such as copper(II) hydroxide, are insoluble in water.

The oxides and hydroxides of barium, calcium and Group 1 metals produce hydroxide ions (OH^-) in aqueous solution; for example:

$$Ba(OH)_2(aq) \rightarrow Ba^{2+}(aq) + 2OH^-(aq)$$

◆ **Alkali**: A base that dissolves in water to give hydroxide ions as the only anions.

These soluble bases are **alkalis**.

Ammonium hydroxide is very unstable and exists in equilibrium with ammonia and water (equilibrium lies to the right):

$$NH_4OH(aq) \rightleftharpoons NH_3(aq) + H_2O(l)$$

Top tip!

The hydrogen ion is responsible for the properties of acids; the hydroxide ion is responsible for the properties of alkalis.

■ **Table R3.1** Common alkalis

Name	Formula	Description	Ions produced in aqueous solutions	
sodium hydroxide	NaOH	strong alkali	Na^+	OH^-
potassium hydroxide	KOH	strong alkali	K^+	OH^-
calcium hydroxide	$Ca(OH)_2$	strong alkali (but only slightly soluble in water)	Ca^{2+}	OH^-

Common mistake

Misconception: *All compounds with a hydroxyl group in their molecular formula are bases.*

Many compounds with a hydroxyl group in their molecular formula are neutral (for example, C_2H_5OH) or acidic (for example, sulfuric acid, $S(=O)_2(OH)_2$).

According to the Arrhenius theory, acid–base reactions can be divided into three main types:

- between an acid and an alkali:

$$2HCl(aq) + Ba(OH)_2(aq) \rightarrow BaCl_2(aq) + 2H_2O(l)$$

$$H^+(aq) + OH^-(aq) \rightarrow H_2O(l)$$

- between an acid and an insoluble metal oxide:

$$2HCl(aq) + CuO(s) \rightarrow CuCl_2(aq) + H_2O(l)$$

$$2H^+(aq) + CuO(s) \rightarrow H_2O(l) + Cu^{2+}(aq)$$

- between an acid and an insoluble metal hydroxide:

$$3HCl(aq) + Fe(OH)_3(s) \rightarrow 3H_2O(l) + FeCl_3(aq)$$

$$3H^+(aq) + Fe(OH)_3(s) \rightarrow 3H_2O(l) + Fe^{3+}(aq)$$

Common mistake

Misconception: *Water is neither an acid nor a base.*

Water molecules can act as either an acid or a base: they can both accept and lose protons.

Water

Water is a molecular covalent compound but it is slightly ionized and in equilibrium with low concentrations of its ions. Hence water can act as both an acid and a base:

$$H_2O(l) \rightleftharpoons H^+(aq) + OH^-(aq)$$

However, water is neutral at all temperatures since the concentrations of the two ions are always equal (this is the basis for water being neutral). All aqueous solutions have a low concentration of hydrogen and hydroxide ions from the dissociation of water molecules.

Going further

Uses of acids and bases

A number of acids and bases are used in the home as cleaning agents (Figure R3.3). Rust removers often contain phosphoric acid, which forms a protective layer of iron(III) phosphate to help prevent further rusting. Hard water contains a high concentration of calcium ions and forms deposits of calcium carbonate in kettles and hot water pipes. The carbonate deposits can be removed with acid, for example, vinegar (ethanoic acid).

Oven cleaners (Figure R3.4) usually contain sodium hydroxide, which converts oils and fats into water-soluble products by producing propane-1,2,3-triol and carboxylate ions. Ammonia and sodium carbonate are also present in many liquid cleaners. They are both weaker bases and hence less corrosive to the skin and eyes. Sodium carbonate is present in dishwasher crystals (this is a base because carbonate ions undergo hydrolysis with water molecules to release excess hydroxide ions).

■ **Figure R3.3** Denture cleaning tablets. Active ingredients include the salts sodium hydrogencarbonate, sodium perborate and citric acid

■ **Figure R3.4** Oven cleaner pads

ATL R3.1A

Find out how ant bites, nettle stings, bee stings and wasp stings can be treated with acids and bases found in the home or pharmacy. Which bite or string would you treat with an acid and which with a base and why?

Create an infographic to present to your findings to your peers. Include appropriate chemical names, equations, acid–base concepts and terminology.

Ionic equations

◆ **Net ionic equation**: The equation for a reaction involving ionic substances, including only those ions that actually participate in the reaction.

◆ **Spectator ions**: Ions present in solution that do not participate directly in a reaction.

When a soluble ionic substance is dissolved in water, the ions separate and behave independently. When ionic solutions react, the reaction usually only involves some of the ions and not others. A **net ionic equation** shows only the ions involved in the reaction. The other ions are **spectator ions**.

For example, consider the following reaction:

$$H_2SO_4(aq) + 2NaOH(aq) \rightarrow Na_2SO_4(aq) + 2H_2O(l)$$

This can be written:

$$2H^+(aq) + SO_4^{2-}(aq) + 2Na^+(aq) + 2OH^- \rightarrow 2Na^+(aq) + SO_4^{2-}(aq) + 2H_2O(l)$$

Removing the ions that do not take part in the reaction gives:

$$2H^+(aq) + 2OH^-(aq) \rightarrow 2H_2O(l)$$

So the net ionic equation is:

$$H^+(aq) + OH^- \rightarrow H_2O(l)$$

The only particles that actually react are the OH^- ions from the sodium hydroxide and the H^+ ions from the sulfuric acid. Net ionic equations must always have the same net charge on each side of the equation; in this case, the net charge on each side of the equation is zero. Net ionic equations may be written for reactions in aqueous solution in which some of the ions originally present are removed from solution, or ions not originally present are formed. Ions are removed from solution by the following processes:

- formation of an insoluble precipitate
- formation of molecules containing only covalent bonds
- formation of a new ionic chemical species
- formation of a gas.

The ions in a solution undergo their characteristic reactions regardless of which other ions are present. For example, barium ions in solution react with sulfate ions in solution to form a white precipitate of barium sulfate. If a solution of barium chloride, $BaCl_2$, and a solution of sodium sulfate, Na_2SO_4, are mixed, a white precipitate of barium sulfate, $BaSO_4$, is rapidly produced. The following equations describe the precipitate formation:

$$BaCl_2(aq) + Na_2SO_4(aq) \rightarrow BaSO_4(s) + 2NaCl(aq)$$

$$Ba^{2+}(aq) + 2Cl^-(aq) + 2Na^+(aq) + SO_4^{2-}(aq) \rightarrow BaSO_4(s) + 2Na^+(aq) + 2Cl^-(aq)$$

The sodium and chloride spectator ions can be removed from the equation to generate a net ionic equation:

$$Ba^{2+}(aq) + SO_4^{2-}(aq) \rightarrow BaSO_4(s)$$

This shows that any soluble barium salt will react with any soluble sulfate to produce barium sulfate.

■ **Table R3.2** Solubility rules

Soluble	Insoluble
all nitrates	
most chlorides / iodides / bromides	lead(II) and silver chlorides / iodides / bromides
sodium, potassium and ammonium carbonates	all other carbonates
sodium, potassium and ammonium hydroxides, calcium and barium hydroxides (sparingly soluble)	all other hydroxides
most sulfates	calcium, lead(II) and barium sulfates

Top tip!

All sodium, potassium and ammonium salts are soluble.

ATL R3.1B

Create a rhyme or write lyrics to a song to help you remember the solubility rules.

Record a video of people you know performing your rhyme or song.

1 a Use the solubility rules to complete the following solubility grid. Write a ✓ in the boxes where a precipitate will form and a cross (✗) if no precipitate forms when the two aqueous solutions are mixed.

	$NH_4OH(aq)$ or $NH_3(aq)$	$AgNO_3(aq)$	$FeCl_3(aq)$	$Pb(NO_3)_2(aq)$	$CuSO_4(aq)$
$Na_2CO_3(aq)$					
$Fe(NO_3)_2(aq)$					
$BaCl_2(aq)$					
$K_2SO_4(aq)$					
$NaOH(aq)$					

b Write a net ionic equation for each precipitate that forms (note that silver(I) hydroxide spontaneously dehydrates to the oxide).

Going further

Salt solubility

Strong ionic bonding in a crystal lattice is favoured by the presence of small and highly charged ions. The strength of ionic bonding is measured by the lattice enthalpy which has to be overcome for the salt to dissolve in water and form an aqueous solution. However, small ions and high charges are the same factors which favour a large hydration enthalpy and hence the hydration process. The ions interact with water molecules and are stabilized. It is therefore difficult to deduce from simple principles which salts are soluble and which are insoluble.

Reactions of acids and bases

Neutralization of acids

Acids react with bases to form a salt and water only, in a process known as neutralization. For example:

nitric acid + copper(II) oxide → copper(II) nitrate + water

$$2HNO_3(aq) + CuO(s) \rightarrow Cu(NO_3)_2(aq) + H_2O(l)$$

Copper(II) oxide is a black solid that reacts with nitric acid to form copper(II) nitrate, a salt that dissolves in water to form a blue solution. The reaction is exothermic; heat is released into the solution and its temperature rises.

Another example:

nitric acid + sodium hydroxide → sodium nitrate + water

$$HNO_3(aq) + NaOH(aq) \rightarrow NaNO_3(aq) + H_2O(l)$$

$$H^+(aq) + OH^-(aq) \rightarrow H_2O(l)$$

LINKING QUESTION

Neutralization reactions are exothermic. How can this be explained in terms of bond enthalpies?

All solutions involved in this reaction are clear and colourless. This reaction is also exothermic; heat is released into the solution when the hydrogen ions form covalent bonds with hydroxide ions.

Acids and metals

Acids react with reactive metals to form a salt and hydrogen gas. This is a redox reaction.

For example:

sulfuric acid + magnesium → magnesium sulfate + hydrogen

$$H_2SO_4(aq) + Mg(s) \rightarrow MgSO_4(aq) + H_2(g)$$

or (net ionic equation):

$$2H^+(aq) + Mg(s) \rightarrow Mg^{2+}(aq) + H_2(g)$$

The metal is used up and converted to soluble metal ions – a colourless solution is formed. A colourless and odourless gas is evolved, resulting in effervescence. The gas extinguishes a lighted splint with a 'pop' sound, confirming the presence of hydrogen.

LINKING QUESTION

How could we classify the reaction that occurs when hydrogen gas is released from the reaction between an acid and a metal?

Common mistake

Misconception: *More hydrogen gas is released from a solution of a strong acid since it has more hydrogen ions than a solution of a weak acid.*

If a metal reacts with equal volumes of weak and strong acids of the same concentration, the same volume of hydrogen gas is formed. The reaction is slower with the weaker acid.

Acids and carbonates and hydrogencarbonates

Acids react with metal carbonates and metal hydrogencarbonates to form a salt, water and carbon dioxide gas.

For example:

hydrochloric acid + calcium carbonate → calcium chloride + water + carbon dioxide

$$2HCl(aq) + CaCO_3(s) \rightarrow CaCl_2(aq) + H_2O(l) + CO_2(g), \text{ or}$$

$$2H^+(aq) + CaCO_3(s) \rightarrow Ca^{2+}(aq) + H_2O(l) + CO_2(g)$$

The white solid is used up and converted to soluble metal ions – to form a colourless solution. The colourless and odourless gas released forms a white precipitate of calcium carbonate in limewater (aqueous calcium hydroxide), confirming the formation of carbon dioxide.

Bases with ammonium and alkylammonium (amine) salts

When warmed, alkalis react with ammonium salts to produce a salt, water and ammonia gas. For example:

sodium hydroxide + ammonium sulfate → sodium sulfate + water + ammonia

$$2NaOH(aq) + (NH_4)_2SO_4(aq) \rightarrow Na_2SO_4(aq) + 2H_2O(l) + 2NH_3(g)$$

$$OH^-(aq) + NH_4^+(aq) \rightarrow H_2O(l) + NH_3(g)$$

When the mixture is warmed, a pungent gas is produced that turns moist red litmus blue. The ammonia gas dissolves in the water dampening the litmus paper to form aqueous ammonia, which is a weak alkali due to its reaction with water molecules to form excess hydroxide ions:

$$NH_3(g) + H_2O(l) \rightleftharpoons NH_4^+(aq) + OH^-(aq)$$

When salts of amines are heated with an alkali, they produce water and an amine. For example:

$$NaOH(aq) + C_2H_5NH_3Cl(aq) \rightarrow C_2H_5NH_2(g) + NaCl(aq) + H_2O(l)$$

$$OH^-(aq) + C_2H_5NH_3^+(aq) \rightarrow H_2O(l) + C_2H_5NH_2(g)$$

Amines often have a 'fishy' smell. Like ammonia, they will turn moist red litmus blue because they also react with water molecules to form excess hydroxide ions:

$$C_2H_5NH_2(g) + H_2O(l) \rightleftharpoons C_2H_5NH_3^+(aq) + OH^-(aq)$$

2 Write equations, including state symbols, for the following reactions:

a sulfuric acid and copper(II) carbonate
b hydrobromic acid and calcium hydrogencarbonate
c phosphoric(V) acid and sodium carbonate
d ethanoic acid and calcium
e ammonium phosphate solution and aqueous barium hydroxide
f aminoethane (ethylamine) and hydrochloric acid

Naming salts

Salts are a class of chemical substance produced when acids react with bases, metals, metal oxides, carbonates or hydrogencarbonates. Different acids form different salts (see Table R3.3) and each acid can have its hydrogen replaced by different metal ions.

Ammonium salts can be formed either from unstable ammonium hydroxide (NH_4OH) (also known as ammonia solution, $NH_3(aq)$) or from ammonia gas ($NH_3(g)$):

$$NH_4OH(aq) + HNO_3(aq) \rightarrow NH_4NO_3(aq) + H_2O(l)$$

$$NH_3(g) + HCl(g) \rightarrow NH_4Cl(s)$$

■ **Table R3.3** Selected acids and their salts

Basicity	Acid	Salt	
1	nitric acid HNO_3	zinc nitrate potassium nitrate copper(II) nitrate	$Zn(NO_3)_2$ KNO_3 $Cu(NO_3)_2$
	hydrochloric acid HCl	aluminium chloride ammonium chloride	$AlCl_3$ NH_4Cl
	CH_3COOH ethanoic acid	potassium ethanoate magnesium ethanoate	CH_3COOK $(CH_3COO)_2Mg$
2	sulfuric acid H_2SO_4	lithium sulfate lead(II) sulfate ammonium sulfate	Li_2SO_4 $PbSO_4$ $(NH_4)_2SO_4$
3	phosphoric acid H_3PO_4	calcium phosphate iron(III) phosphate ammonium phosphate	$Ca_3(PO_4)_2$ $FePO_4$ $(NH_4)_3PO_4$

◆ **Normal salt:** A salt formed when all of the replaceable hydrogens of an acid have been replaced by metal (or ammonium) ions.

◆ **Acid salt:** Salt of a polyprotic acid (an acid having two or more acidic hydrogens) in which not all the hydrogen ions have been replaced by cations.

◆ **Parent acid:** The acid with which a particular salt was formed.

◆ **Parent base:** The base with which a particular salt was formed.

Acids containing more than one replaceable hydrogen atom can form salts in which all or some of the hydrogen is replaced. Salts formed by replacing all of the hydrogens are termed **normal salts**; those formed by replacing only some of the hydrogen are termed **acid salts**.

Acid salts are products of incomplete neutralization of acids. One such salt, potassium hydrogencarbonate ($KHCO_3$), is formed when one hydrogen ion in carbonic acid is replaced with a potassium ion:

$$H_2CO_3(aq) + KOH(aq) \rightarrow KHCO_3(aq) + H_2O(l)$$

Acids with three hydrogen atoms, such as phosphoric(V) acid, can form two types of acid salt, with one or two hydrogen atoms in the acid anion:

$$H_3PO_4(aq) + KOH(aq) \rightarrow KH_2PO_4(aq) + H_2O(l)$$

$$H_3PO_4(aq) + 2KOH(aq) \rightarrow K_2HPO_4(aq) + 2H_2O(l)$$

Table R3.4 gives examples of the sodium salts formed by common acids, including some acid salts.

■ **Table R3.4** Examples of sodium salts formed by common acids

Acid	Salt	Example
hydrochloric acid, HCl	chlorides	sodium chloride, NaCl
nitric acid, HNO_3	nitrates	sodium nitrate, $NaNO_3$
ethanoic acid, CH_3COOH	ethanoates	sodium ethanoate, CH_3COONa
sulfuric acid, H_2SO_4	sulfates (normal salts) and hydrogensulfates (acid salts)	sodium sulfate, Na_2SO_4, sodium hydrogensulfate, $NaHSO_4$
carbonic acid, H_2CO_3	carbonates (normal salts) and hydrogencarbonates (acid salts)	sodium carbonate, Na_2CO_3, sodium hydrogencarbonate, $NaHCO_3$
phosphoric acid, H_3PO_4	phosphates (normal salts), hydrogenphosphates (acid salts) and dihydrogenphosphates (acid salts)	sodium phosphate, Na_3PO_4, disodium hydrogenphosphate, Na_2HPO_4, sodium dihydrogenphosphate, NaH_2PO_4.

Parent acids and bases

To identify the **parent acid** and **parent base** in a salt, you apply the generalized equation:

acid + base → salt + water

If we write the salt as M_mX_n, where M^{n+} is the metal cation and X^{m-} is the anion:

- To find the parent base, add an appropriate number of hydroxide ions, OH^-, to M^{n+} to form a neutral compound, $M(OH)_n$.
- To find the parent acid, add an appropriate number of H^+ ions to form a neutral compound, typically the hydride of a non-metal anion H_mX.

WORKED EXAMPLE R3.1A

Identify the parent components of the following salts:

a ammonium phosphate, $(NH_4)_3PO_4$

b lithium sulfate, Li_2SO_4

Answer

a Separate the formula unit into the cation and anion:

NH_4^+ and PO_4^{3-}

Add OH^- to NH_4^+: $NH_4^+ + OH^- \rightarrow NH_4OH(aq)$, ammonium hydroxide, or aqueous ammonia ($NH_3(aq)$)

Add H^+ to PO_4^{3-}: $3H^+ + PO_4^{3-} \rightarrow H_3PO_4$, phosphoric(V) acid

b Separate the formula unit into the cation and anion:

Li^+ and SO_4^{2-}

Add OH^- to Li^+: $Li^+ + OH^- \rightarrow LiOH(aq)$, lithium hydroxide

Add H^+ to SO_4^{2-}: $2H^+ + SO_4^{2-} \rightarrow H_2SO_4$, sulfuric acid

3 Identify the parent acid and base of the following salts:

a $Ca(CH_3COO)_2$

b KNO_3

c Na_2SO_3

d $FeSO_4$

e K_2CO_3

f $BaCl_2$

Inquiry 2: Collecting and processing data

Interpret qualitative and quantitative data

Qualitative data is non-numerical and can be used to derive descriptive conclusions; the colour and physical properties of a salt can be used as a preliminary step in identification. For example, a green salt may be iron(II) sulfate or copper(II) carbonate.

The solubility of a soluble or sparingly soluble salt is quantitative data; the solubility of iron(II) sulfate is $25.6\,g / 100\,cm^3$.

Heating these two salts provides further qualitative data (or quantitative data if you measure the percentage loss when heated to constant mass). When strongly heated, iron(II) sulfate decomposes to form a reddish-brown solid and a pungent gas that moist blue litmus paper shows is acidic. This data is consistent with the following equation:

$$2FeSO_4(s) \rightarrow Fe_2O_3(s) + SO_2(g) + SO_3(g)$$

Copper(II) carbonate will also undergo thermal decomposition to form a black solid and a colourless gas that turns limewater (calcium hydroxide solution) cloudy. This data is consistent with the following equation:

$$CuCO_3(s) \rightarrow CuO(s) + CO_2(g)$$

Preparation of salts

The preparation of a salt depends on whether it is soluble or insoluble (Table R3.2, page 524).

- If the salt is insoluble, ionic precipitation has to be used.
- If the salt is soluble and it can be prepared from a solution of an alkali and an acid, then a titration may be used.
- If the base is insoluble then excess of the solid reagent is reacted with dilute acid (Figure R3.5).

LINKING QUESTION

How can the salts formed in neutralization reactions be separated?

◆ **Hydronium ion**: The ion, H_3O^+, formed when a water molecule forms a coordination bond with a proton.

◆ **Acid (Brønsted–Lowry)**: A proton donor.

◆ **Base (Brønsted–Lowry)**: A proton acceptor.

◆ **Dissociation (of acids)**: The release of hydrogen ions from molecules by bond breaking in aqueous solution.

◆ **Ionization**: A chemical reaction involving the formation of one or more ions.

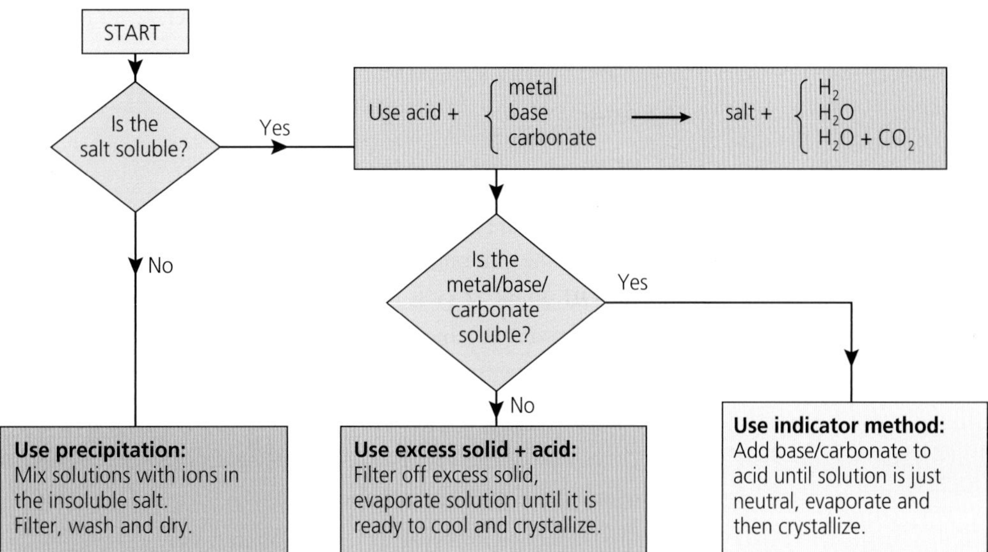

■ **Figure R3.5** Summary of the preparation of salts

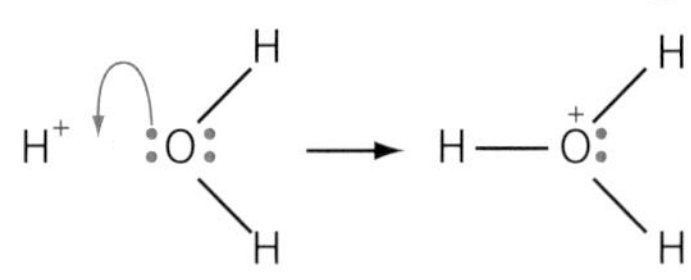

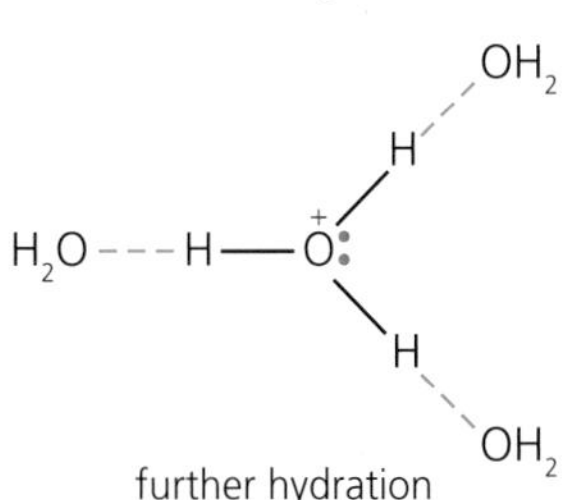

■ **Figure R3.6** Formation of the hydronium ion from the reaction between a proton and a water molecule (and its further hydration by water molecules (note: knowledge not required by the IB Guide))

Brønsted–Lowry theory

The Arrhenius model of acids assumes that aqueous solutions of acids contain hydrogen ions (protons), H^+. However, protons have a high charge density and will immediately react with water molecules to form **hydronium ions**, H_3O^+ (Figure R3.6). Brønsted–Lowry theory is also not restricted to aqueous solutions but includes reactions between gases and non-aqueous solutions.

The Brønsted–Lowry model takes this into consideration and proposes that a more accurate way of describing the reaction of hydrogen chloride with water to form hydrochloric acid is:

$$H–Cl(g) + H_2O(l) \rightarrow H_3O^+(aq) + Cl^-(aq)$$

The water molecule acts as a **base** (proton acceptor) and the hydrogen chloride molecule acts as an **acid** (proton donor). The formation of ions is known as **dissociation** or **ionization**.

The formation of aqueous ammonia solution can also be described by the Brønsted–Lowry model:

$$NH_3(g) + H_2O(l) \rightarrow NH_4^+(aq) + OH^-(aq)$$

Here, the water molecule acts as an acid (proton donor) and the ammonia molecule acts as a base (proton acceptor).

The common feature of both reactions is proton (H^+) transfer from the acid to the base (Figure R3.7).

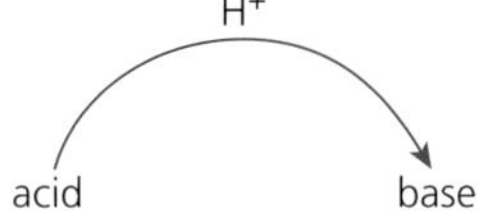

■ **Figure R3.7** Brønsted–Lowry model

The Brønsted–Lowry model of acid–base interactions includes reactions that do not require water. For example, ammonia gas and hydrogen chloride gas react rapidly to form solid ammonium chloride: $NH_3(g) + HCl(g) \rightarrow [NH_4^+(s)\ Cl^-(s)]$ with proton transfer from the hydrogen chloride molecule to the ammonia molecule.

WORKED EXAMPLE R3.1B

Chloric(VII) acid, $HClO_4$, acts as an acid in water.

Write an equation showing its ionization in water.

Identify the Brønsted–Lowry acid and base. Explain your answer.

Answer

$$HClO_4(aq) + H_2O(l) \rightleftharpoons H_3O^+(aq) + ClO_4^-(aq)$$

$HClO_4$ is the acid (proton donor) because in the forward reaction it has lost its proton (H^+) and formed the chlorate(VII) ion, ClO_4^-.

A water molecule, H_2O, is the base (proton acceptor) because in the forward reaction it has gained a proton (H^+) and formed the hydronium cation, H_3O^+.

WORKED EXAMPLE R3.1C

Phenylamine, $C_6H_5NH_2$, is monoprotic.

Write an equation showing its ionization in water.

Identify the Brønsted–Lowry acid and base. Explain your answer.

Answer

$$C_6H_5NH_2(aq) + H_2O(l) \rightleftharpoons OH^-(aq) + C_6H_5NH_3^+(aq)$$

The phenylamine molecule is the base (proton acceptor) because in the forward reaction it has accepted a proton (H^+) and formed $C_6H_5NH_3^+$.

H_2O is the acid (proton donor) because in the forward reaction it has lost a proton (H^+) and formed the hydroxide ion, OH^-.

4 Identify the Brønsted–Lowry acids and bases in the following reactions:

a $H_2O + [Al(H_2O)_6]^{3+} \rightleftharpoons [Al(H_2O)_5OH]^{2+} + H_3O^+$

b $H_2O + HCO_3^- \rightleftharpoons CO_3^{2-} + H_3O^+$

c $H_2NCONH_2 + H_2O \rightleftharpoons H_3N^+CONH_2 + OH^-$

d $HSO_4^- + H_3O^+ \rightleftharpoons H_2SO_4 + H_2O$

e $NH_4^+ + OH^- \rightleftharpoons NH_3 + H_2O$

Common mistake

Misconception: *A base is an OH^- donor.*

A Brønsted–Lowry base is a proton acceptor.

LINKING QUESTION

Why has the definition of acid evolved over time?

TOK

Maps as a metaphor for scientific knowledge

It is possible to think about theories of acids and bases in the light of the TOK map metaphor: the various ways we characterize the world are like different maps of the same landscape. In this case, the theories about bases and acids are like three different maps – each theory has its own benefits and drawbacks, but they all describe the same world.

The theories all make certain assumptions about what the map *needs* to show (levels of generality, specificity of claims around ions) and they all guide our reasoning. For example, the Arrhenius theory might lead us to think that reactions in water are the *only* way to describe the behaviour of acids and bases in solution.

There might not be a 'best' map of a given area, because the definition of best depends on what you *need* from the map. In the same way, there might be various, fully appropriate ways of characterising acids and bases but no one theory that is objectively *best*.

Amphiprotic species

◆ **Amphiprotic:** A chemical species capable of accepting and donating protons and able to behave as both a Brønsted–Lowry acid and a Brønsted–Lowry base.

Water is **amphiprotic**: it can act as both an acid and a base. In the reaction with hydrogen chloride, it acts as a base (proton acceptor), but with ammonia it acts as an acid (proton donor).

The conjugate base of a polyprotic acid (an acid having two or more acidic hydrogens) is amphiprotic: it can act either as an acid or a base because it can either donate its remaining acidic hydrogen atom as a proton or accept a proton and revert to the original acid.

For example, $HC_2O_4^-$ is amphoteric: it can gain a proton: $HC_2O_4^- + H^+ \rightarrow H_2C_2O_4$, or it can lose a proton: $HC_2O_4^- \rightarrow H^+ + C_2O_4^{2-}$.

Amino acids are also amphiprotic species. For example, the simplest amino acid, glycine, $H_2N–CH_2–COOH$, has two functional groups. The amino group, $–NH_2$, is basic due to the presence of a non-bonded pair of electrons on the nitrogen atom; the carboxyl functional group is acidic due to the presence of an acidic or ionizable hydrogen atom. In solution and in the solid state, there is an internal acid–base transfer of a proton from the carboxylic acid group to the amino group. A dipolar ion or zwitterion, $H_3N^+–CH_2–COO^-$, is formed. In aqueous solution at pH values greater than 6.0, the average charge on glycine molecules becomes negative (the anion or basic form predominates); at pH values less than 6.0, the molecules become positively charged (the cation or acidic form predominates) (Figure R3.8).

LINKING QUESTION

What is the periodic trend in the acid–base properties of metal and non-metal oxides?

$$\underset{\text{at pH} > 6.0}{NH_2CH_2COO^-} \xleftarrow{-H^+} \underset{\text{at pH} = 6.0}{{}^+NH_3CH_2COO^-} \xrightarrow{+H^+} \underset{\text{at pH} < 6.0}{{}^+NH_3CH_2COOH}$$

■ **Figure R3.8** Behaviour of glycine molecules at different pH values

Going further

Factors required for Brønsted–Lowry acidity

In order for a compound to be considered a Brønsted–Lowry acid, the ion formed during ionization must be energetically stable. For this to happen, it should be possible for the charge on the anion to be 'spread out' and its potential energy decreased by delocalization. This explains why organic compounds such as ethanoic acid are acidic in water. The negative charge on the ethanoate ion, CH_3COO^-, is spread over two oyxgen atoms, as shown in Figure R3.9.

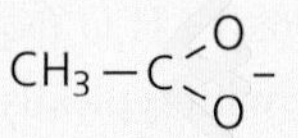

■ **Figure R3.9** The delocalization of charge in the ethanoate ion, CH_3COO^-

Ions in aqueous solution are surrounded by a layer of water molecules that prevent the ions from interacting (Figure R3.10).

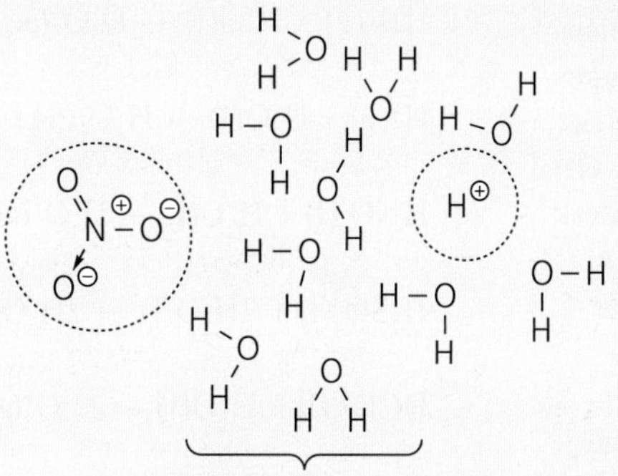

The water molecules between nitrate and hydrogen ions act as a layer of electrical insulation. The layer keeps the ions apart.

■ **Figure R3.10** Water hydrating the nitrate and hydrogen ions of nitric acid

LINKING QUESTION

Why does the release of oxides of nitrogen and sulfur into the atmosphere cause acid rain?

◆ **Strong acid**: An acid that is completely ionized when dissolved in water.

◆ **Weak acid**: An acid that is only partially ionized when dissolved in water.

Strong and weak acids and bases

Strong and weak acids

Acids are often classified into **strong** and **weak acids**.

When a strong acid dissolves, all the acid molecules react with the water to produce hydrogen ions or form hydronium ions.

In general, for a strong monoprotic acid, HA:

$$\underset{0\%}{HA(aq)} \rightarrow \underset{100\%}{H^+(aq) + A^-(aq)} \quad \text{or}$$

$$\underset{0\%}{HA(aq) + H_2O(l)} \rightarrow \underset{100\%}{H_3O^+(aq) + A^-(aq)}$$

One mole of a strong monoprotic acid forms one mole of protons (H^+) or one mole of hydronium ions, H_3O^+.

This is illustrated graphically in Figure R3.11.

■ **Figure R3.11** Graphical representation of the behaviour of a strong acid in aqueous solution

The common strong acids are hydrochloric, hydrobromic, hydroiodic, nitric, sulfuric and chloric(VII) acids:

$$HCl(g) + H_2O(l) \rightarrow H_3O^+(aq) + Cl^-(aq)$$

$$HBr(g) + H_2O(l) \rightarrow H_3O^+(aq) + Br^-(aq)$$

$$HI(g) + H_2O(l) \rightarrow H_3O^+(aq) + I^-(aq)$$

$$HNO_3(l) + H_2O(l) \rightarrow H_3O^+(aq) + NO_3^-(aq)$$

$$H_2SO_4(l) + 2H_2O(l) \rightarrow 2H_3O^+(aq) + SO_4^{2-}(aq) \text{ (simplified)}$$

$$HClO_4(l) + H_2O(l) \rightarrow H_3O^+(aq) + ClO_4^-(aq)$$

Sulfuric acid is actually only strong in its first ionization ($H_2SO_4(l) + H_2O(l) \rightarrow HSO_4^-(aq) + H_3O^+(aq)$) and is weak in its second ionization ($HSO_4^-(aq) + H_2O(l) \rightleftharpoons SO_4^{2-}(aq) + H_3O^+(aq)$), with typically only about 10% converted to sulfate ions.

Common mistake

Misconception: *A concentrated acid / base solution is strong and a dilute acid / base solution is weak.*

Concentration refers to the amount of dissolved acid or base per unit volume, commonly expressed as a molarity ($mol\,dm^{-3}$). A concentrated solution has a high concentration and a dilute solution has a low concentration. The acid or base can be weak or strong. In everyday language 'strong' can incorrectly refer to a solution with a high concentration of solute and 'weak' can refer to a solution with a low concentration of solute.

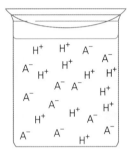

Monoprotic organic acids are usually weak. When a weak acid dissolves in water, only a small percentage of its molecules (typically 1%) react with water molecules to release hydrogen or hydronium cations.

An equilibrium is established with most of the acid molecules not undergoing ionization or dissociation. In other words, the equilibrium lies on the left-hand side of the equation.

In general, for a weak acid, HA:

$$HA(aq) \rightleftharpoons H^+(aq) + A^-(aq)$$ or

99% 1%

$$HA(aq) + H_2O(l) \rightleftharpoons H_3O^+(aq) + A^-(aq)$$

99% 1%

This is illustrated graphically in Figure R3.12.

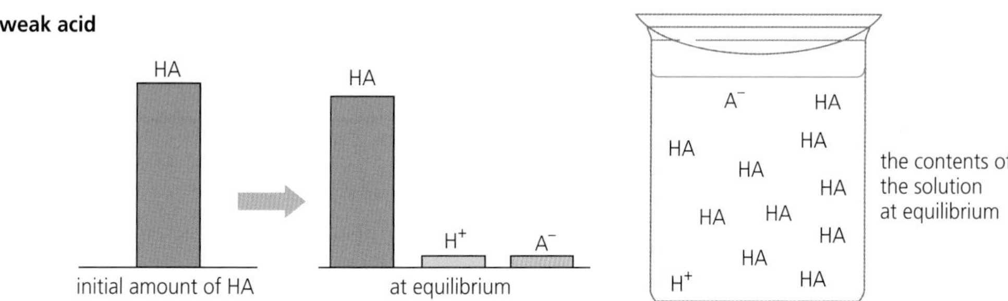

■ **Figure R3.12** Graphical representation of the behaviour of a weak acid in aqueous solution

Examples of common weak acids are ethanoic acid:

$$CH_3COOH(l) + H_2O(l) \rightleftharpoons CH_3COO^-(aq) + H_3O^+(aq)$$

and sulfurous acid:

$$SO_2(g) + H_2O(l) \rightleftharpoons H_2SO_3(aq)$$

$$H_2SO_3(aq) + H_2O(l) \rightleftharpoons HSO_3^-(aq) + H_3O^+(aq)$$

Acid drainage

LINKING QUESTION

Why does the acid strength of the hydrogen halides increase down Group 17?

Acid drainage occurs when the iron sulfides brought to the surface by mining activity react with water and air and oxidize. The process creates sulfuric acid, which breaks down surrounding rocks. This can cause toxic metals (for example, lead and iron) and metalloids (for example, arsenic) to enter water in nearby watercourses or aquifers. The United Nations (UN) has labelled it the second biggest environmental problem facing the world after global warming, and is promoting international cooperation on reducing and treating acid drainage.

WORKED EXAMPLE R3.1D

Compare the following two solutions: 2.00 mol dm^{-3} solution of methanoic acid, HCOOH(aq), and 0.50 mol dm^{-3} solution of hydrochloric acid, HCl(aq).

a State and explain which one is more concentrated.

b State and explain which one is a strong acid.

c State and explain which one has the higher concentration of hydrogen ions.

Answer

a Methanoic acid is more concentrated as it has a concentration of 2.00 mol dm^{-3}, whereas the hydrochloric acid is only 0.5 mol dm^{-3}.

b Hydrochloric acid is a strong acid because all the molecules dissociate into ions when added to water. Methanoic acid is a weak acid as only a small percentage of the molecules ionize when added to water.

c Hydrochloric acid will contain a higher concentration of hydrogen ions because all the molecules ionize. Even though the methanoic acid is four times as concentrated, only a small percentage of the molecules ionize. Therefore, as hydrochloric acid contains a higher concentration of hydrogen ions, it will have a lower pH.

Strong and weak bases

A strong base undergoes almost 100% ionization when in dilute aqueous solution. Strong bases have high pH values and high conductivities (Table R3.5). The higher the value of pH, the higher the concentration of hydroxide ions.

Table R3.5 Comparison of a weak and strong base of the same concentration

	0.1 mol dm^{-3} NaOH(aq)	**0.1 mol dm^{-3} NH_3(aq)**
[OH^-(aq)]	0.1 mol dm^{-3}	≈ 0.0013 mol dm^{-3}
pH	13	≈ 11
electrical conductivity	high	low

In general for a strong ionic base, BOH:

$$BOH(aq) \rightarrow OH^-(aq) + B^+(aq)$$

0% 100%

The three common strong bases are sodium hydroxide, potassium hydroxide and barium hydroxide:

$$Na^+(s)\ OH^-(s) + (aq) \rightarrow Na^+(aq) + OH^-(aq)$$

$$[K^+(s)\ OH^-(s)] + (aq) \rightarrow K^+(aq) + OH^-(aq)$$

$$[Ba^{2+}(s)\ 2OH^-(s)] + (aq) \rightarrow Ba^{2+}(aq) + 2OH^-(aq)$$

Weak bases are composed of molecules that react with water molecules to release hydroxide ions. In general, for a weak molecular base, B:

$$B(aq) + H_2O(l) \rightleftharpoons OH^-(aq) + BH^+(aq)$$

An equilibrium is established, with the majority of the base molecules not undergoing ionization. In other words, the equilibrium lies on the left-hand side of the equation. Weak bases have low pH values (above 7 and below 12) and low conductivities.

Examples of weak bases are aqueous ammonia and amines, such as aminoethane (ethylamine):

$$H_2O(l) + NH_3(aq) \rightleftharpoons NH_4^+(aq) + OH^-(aq)$$

$$H_2O(l) + C_2H_5NH_2(aq) \rightleftharpoons C_2H_5NH_3^+(aq) + OH^-(aq)$$

Calcium hydroxide is weak because it is only slightly soluble in water:

$$[Ca^{2+}(s)\ 2OH^-(s)] + (aq) \rightleftharpoons Ca^{2+}(aq) + 2OH^-(aq)$$

Tool 1: Experimental techniques

Dilutions

Dilutions can be a quick way to make a less concentrated solution. For example, you have a standard solution that is 0.800 mol dm^{-3} but it is too concentrated for your reaction. Instead of making a new standard solution, you can dilute your original solution to 0.200 mol dm^{-3}. In this example you are diluting the original solution to 25% of its original concentration.

This is called the dilution factor and is calculated from:

$$\frac{\text{concentration of desired solution}}{\text{concentration of original solution}} \times 100$$

or

$$\frac{0.200}{0.800} \times 100 = 25.0\%$$

Serial dilutions are used when the concentration needs to be significantly reduced; 'significantly' is often taken as meaning to less than 10% of the original concentration. Successive amounts of the same volume are diluted several times. This can be thought of as a logarithmic dilution as three dilutions can dilute the solution by 10^3 (or 1000 times). A serial dilution is far more accurate than trying to carry out the dilution in one step.

LINKING QUESTION

How would you expect the equilibrium constants of strong and weak acids to compare?

LINKING QUESTION

What physical and chemical properties can be observed to distinguish between weak and strong acids or bases of the same concentration?

◆ **Conjugate acid**: The chemical species formed when a proton or hydrogen ion is accepted by a base.

◆ **Conjugate base**: The chemical species formed when an acid loses a proton or hydrogen ion.

◆ **Conjugate acid–base pair**: Two chemical species related to each other by the loss or gain of a single proton or hydrogen ion.

Conjugate acid–base pairs

All acid–base reactions are reversible to some degree and so there is competition for protons, $H^+(aq)$. We can therefore think of the equilibrium between an acid and a base in terms of pairs of related, or conjugated, species. On the left-hand side of the equation there is an acid and base; on the right-hand side of the equation, there is a conjugate acid (of the base) and a conjugate base (of the acid).

For example, in the dissociation of nitric acid, the nitrate ion, NO_3^-, is the **conjugate base** of nitric acid, while the hydronium cation, H_3O^+, is the **conjugate acid** of water which acts as the base. Nitric acid and nitrate ions are a **conjugate acid–base pair**, while water and hydronium cations form another conjugate acid–base pair (Figure R3.13).

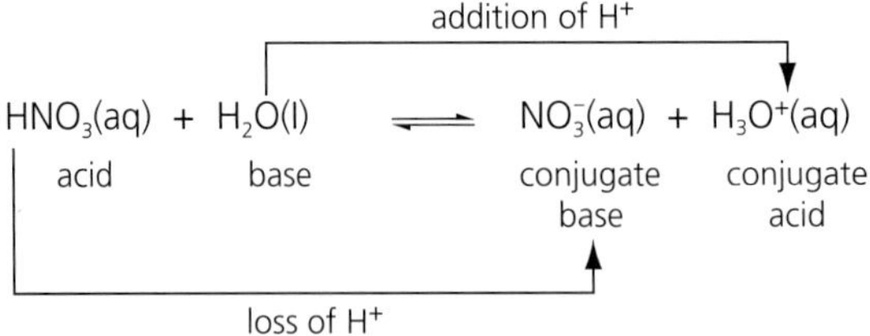

■ **Figure R3.13** Equilibrium reaction between nitric acid and water

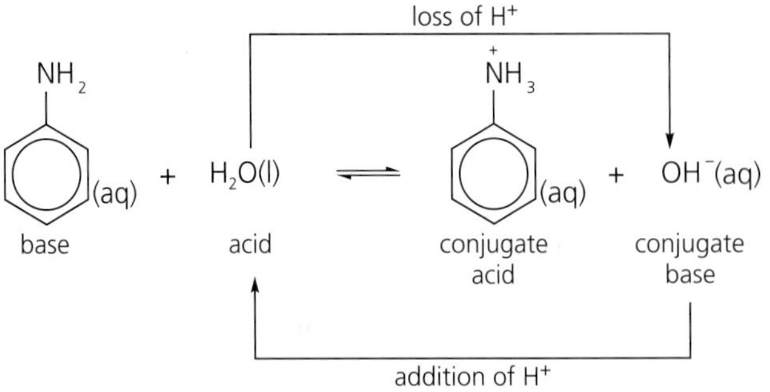

■ **Figure R3.14** Equilibrium reaction between phenylamine and water

In the ionization of phenylamine (Figure R3.14), $C_6H_5NH_2$, the phenylammonium ion, $C_6H_5NH_3^+$, is the conjugate acid of phenylamine, while the hydroxide ion, OH^-, is the conjugate base of water which acts as the acid. Phenylamine and phenylammonium ions are a conjugate acid–base pair and water and hydroxide ions form another conjugate acid–base pair.

WORKED EXAMPLE R3.1E

In pure ethanol, C_2H_5OH, the following equilibrium can exist with ammonium ions:

$$NH_4^+(solv) + C_2H_5OH(l) \rightleftharpoons NH_3(solv) + C_2H_5OH_2^+(solv)$$

where (solv) indicates the ions are surrounded by ethanol molecules. Identify the conjugate acid–base pairs.

Answer

A conjugate acid is located on the right-hand side of the equilibrium and acts as an acid (proton donor). $C_2H_5OH_2^+(solv)$ is a conjugate acid because in the backward reaction it loses a proton (H^+) and forms $C_2H_5OH(l)$. $NH_3(solv)$ is a conjugate base because in the backward reaction it gains a proton to form $NH_4^+(solv)$.

5 Identify the acid, base, conjugate acid and conjugate base in the following reactions:

a $CO_3^{2-}(aq) + H_2O(l) \rightleftharpoons HCO_3^-(aq) + OH^-(aq)$

b $2H_2SO_4(aq) + 2H_2O(l) \rightleftharpoons 2H_3O^+(aq) + 2HSO_4^-(aq)$

Acids and their conjugates

It is important to note that the competition for a proton, H^+, is between the base and its conjugate.

In the case of sulfuric acid, the water molecule is a much stronger base than the hydrogensulfate ion; in other words, the water molecule has a much greater tendency to accept a proton, $H^+(aq)$, than does the hydrogensulfate ion. The position of the equilibrium will lie on the right and nearly all of the sulfuric acid molecules will be ionized:

$$\underset{\text{acid}}{H_2SO_4(l)} + \underset{\text{base}}{H_2O(l)} \rightleftharpoons \underset{\text{conjugate acid}}{H_3O^+(aq)} + \underset{\text{conjugate base}}{HSO_4^-(aq)}$$

In the case of ethanoic acid, the ethanoate ion is a much stronger base than the water molecule; in other words, the ethanoate ion has a much greater tendency to accept a proton, $H^+(aq)$, than does the water molecule:

$$\underset{\text{acid}}{CH_3COOH(l)} + \underset{\text{base}}{H_2O(l)} \rightleftharpoons \underset{\text{conjugate acid}}{H_3O^+(aq)} + \underset{\text{conjugate base}}{CH_3COO^-(aq)}$$

Common mistake

Misconception: *The anion in solutions of strong acids is a strong base.*

The anion (conjugate base) present in solutions of strong acids is a very weak base, since the equilibrium strongly favours the products. For example, Cl^- (from HCl) has no effect on the pH of water.

In general, weak acids in aqueous solutions produce relatively strong conjugate bases (Figure R3.15).

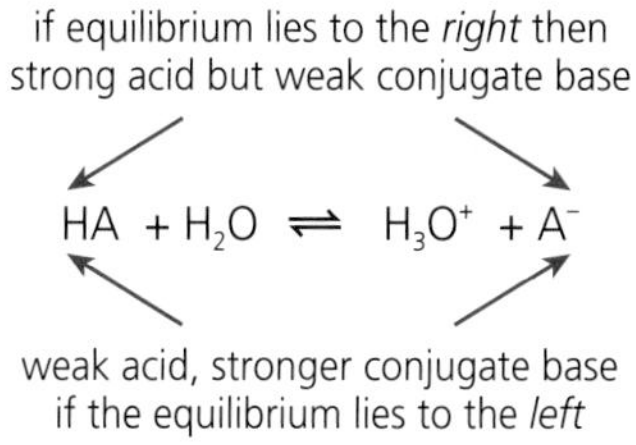

if equilibrium lies to the *right* then
strong acid but weak conjugate base

$$HA + H_2O \rightleftharpoons H_3O^+ + A^-$$

weak acid, stronger conjugate base
if the equilibrium lies to the *left*

■ **Figure R3.15** The relationship between acid and conjugate base

■ **Table R3.6** Some common acids and conjugate bases in order of their strengths

Conjugate acid	Strength	Conjugate base	Strength
H_2SO_4	very strong	HSO_4^-	very weak
HCl		Cl^-	
HNO_3		NO_3^-	
H_3O^+	fairly strong	H_2O	weak
HSO_4^-		SO_4^{2-}	
CH_3COOH		CH_3COO^-	
H_2CO_3	weak	HCO_3^-	less weak
NH_4^+		NH_3	
HCO_3^-		CO_3^{2-}	
H_2O	very weak	OH^-	fairly strong

Top tip!

In general, strong acids in aqueous solutions produce very weak conjugate bases. Weak acids produce relatively strong conjugate bases.

WORKED EXAMPLE R3.1F

Deduce the formula of the conjugate base of the hydrogen selenide molecule, H_2Se.

Answer

Hydrogen selenide behaves as an acid as shown in this equation:

$$H\text{–}Se\text{–}H \rightarrow H\text{–}Se^- + H^+$$

so HSe^- is the conjugate base.

WORKED EXAMPLE R3.1G

Deduce the formula of the conjugate acid of the butanoate ion, $C_3H_7–COO^-$.

Answer

The butanoate ion behaves as a base as shown in this equation:

$$C_3H_7–COO^- + H^+ \rightarrow C_3H_7–COOH$$

so butanoic acid, $C_3H_7–COOH$, is the conjugate acid.

6 State the formulas of the conjugate bases of the following acids:

a HI
b HNO_2
c H_2SO_4
d HSO_4^-
e HS^-

7 State the formulas of the conjugate acids of the following bases:

a Br^-
b HS^-
c CO_3^{2-}
d HSO_4^-
e H_2O
f N_2H_4

LINKING QUESTION

What are the conjugate acids of the polyatomic anions listed in Structure 2.1?

Care must be taken with the term 'conjugate' when referring to diprotic or triprotic acids. For example, consider the ionization or dissociation of the weak acid, sulfurous acid, $H_2SO_3(aq)$:

$H_2SO_3(aq)$	+	$H_2O(l)$	$\rightleftharpoons$	$H_3O^+(aq)$	+	$HSO_3^-(aq)$
acid		base		conjugate acid		conjugate base

$HSO_3^-(aq)$	+	$H_2O(l)$	$\rightleftharpoons$	$H_3O^+(aq)$	+	$SO_3^{2-}(aq)$
acid		base		conjugate acid		conjugate base

In the first equation, the hydrogensulfite ion, $HSO_3^-(aq)$, is the conjugate base of sulfurous acid, but in the second equation it is the conjugate acid of the sulfite ion, $SO_3^{2-}(aq)$.

The two equations show that conjugate is a relative term and only links a specific pair of acids and bases. The hydrogensulfite ion, like the water molecule, is an amphiprotic species.

The terms 'acid' and 'base' are also relative, for example, if two concentrated acids are reacted together, then the weaker acid of the two will be 'forced' to act as a base. For example, when concentrated nitric and sulfuric acids are reacted together in a 1 : 2 molar ratio, a nitrating mixture is formed which contains a cation known as the nitronium ion, NO_2^+. This ion is involved in the nitration of benzene.

The first equilibrium to be established in the nitrating mixture is shown below:

$HNO_3(aq)$	+	$H_2SO_4(aq)$	$\rightleftharpoons$	$H_2NO_3^+(aq)$	+	$HSO_4^-(aq)$
base		acid		conjugate acid		conjugate base

Bases and their conjugates

It is important to understand that, in the equilibria for the weak bases ammonia and carbonate ions, the competition for a proton is between the base and the conjugate base of the water molecule (the hydroxide ion):

$NH_3(g)$	+	$H_2O(l)$	$\rightleftharpoons$	$NH_4^+(aq)$	+	$OH^-(aq)$
base		acid		conjugate acid		conjugate base
$CO_3^{2-}(aq)$	+	$H_2O(l)$	$\rightleftharpoons$	$HCO_3^-(aq)$	+	$OH^-(aq)$
base		acid		conjugate acid		conjugate base

The hydroxide ion is a much stronger base than the ammonia molecule and has a much greater tendency to accept a proton, $H^+(aq)$, than the ammonia molecule. Hence the position of the equilibrium will lie on the left and few of the ammonia molecules will be ionized.

The carbonate ion is a stronger base than the ammonia molecule (although both are classified as weak). In other words, the carbonate ion has a much greater tendency to accept a proton, $H^+(aq)$, than does the ammonia molecule.

Top tip!

In general, weak bases in aqueous solution produce relatively strong conjugate acids, although note that the conjugate acid of a weak base is itself a weak acid.

Top tip!

In general, strong bases in aqueous solution produce very weak conjugate acids.

◆ **pH**: The negative logarithm (to the base 10) of the hydrogen ion concentration in an aqueous solution (in $mol\,dm^{-3}$).

8 Complete the following equations, identifying the conjugate acid–base pairs:

a $H_2O + H_2O \rightleftharpoons$

b $NH_3 + H_3O^+ \rightleftharpoons$

c $HCO_3^- + OH^- \rightleftharpoons$

d $NH_3 + HCl \rightleftharpoons$

9 Complete the following equations, assuming that the first reactant given is acting as a Brønsted–Lowry acid:

a $NH_4^+ + OH^- \rightleftharpoons \quad +$

b $H_3O^+ + \quad \rightleftharpoons \quad + HSO_4^-$

c $\quad + H^- \rightleftharpoons OH^- +$

d $HCO_3^- + \quad \rightleftharpoons \quad + H_2O$

The pH and pOH scales

pH is a measure of how acidic or alkaline a solution is in water. The symbol ‘p’ means ‘$-\log_{10}$’. The pH scale (Figure R3.16) typically ranges from 0 to 14 and its values are related to the concentration of hydrogen ions (and hydroxide ions).

$$pH = -\log_{10}[H^+(aq)]$$

This rearranges to: $[H^+(aq)] = 10^{-pH}$

This is an inverse relationship: as pH decreases, $H^+(aq)$ increases.

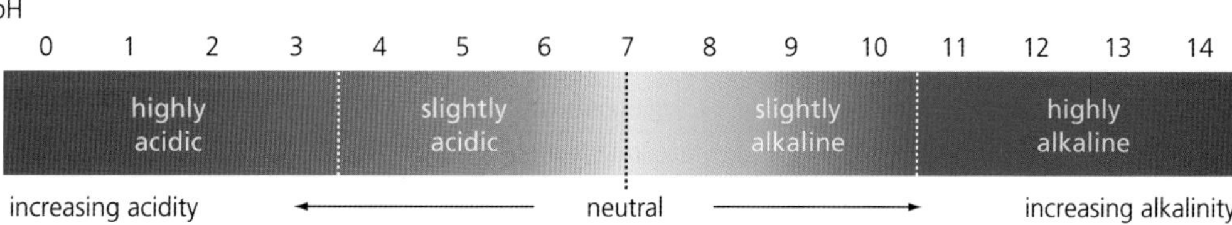

■ **Figure R3.16** The pH scale

Common mistake

Misconception: *The pH scale runs from 0 to 14 and values below 0 or greater than 14 are not possible.*

The pH scale typically runs from 0 to 14. Lower (that is, negative) and higher pH values are possible but, in these cases, ideal behaviour for ions is no longer observed and the relationship between pH and hydrogen ion concentration no longer holds.

Tool 3: Mathematics

Carry out calculations involving logarithmic functions

Logarithms to the base 10 are widely used in chemistry and form the basis for the pH scale. The base 10 logarithm ($\log_{10}$) of a number is the power to which ten must be raised to give the number. A logarithm is the inverse of an exponent.

Number	Exponent expression	Logarithm (to base 10)
100	10^2	2
10	10^1	1
1	10^0	0
1/10 = 0.1	10^{-1}	–1
1/100 = 0.01	10^{-2}	–2

Note that the logarithm of a number less than one is negative and the logarithm of a number greater than one is positive. The logarithm of a negative number is not defined.

You can use a calculator to find the logarithms of numbers that are not multiples of 10. For example, $\log_{10}$ of the numbers 62, 0.872 and 1.0×10^{-7} are 1.79, −0.0595 and –7.0, respectively.

Sometimes, as in the case of pH calculations, it is necessary to obtain the number whose logarithm is known. This procedure is known as taking the antilogarithm. For example, the antilogarithm of 1.46 is 28.8.

The pH scale is logarithmic to the base 10. This means that every change of one unit on the pH scale equates to a change in the hydrogen ion concentration by an order of magnitude (factor of 10) (Table R3.7). For example, an aqueous solution with a pH of 3 is 10 times more acidic than an aqueous solution with a pH of 4 and 100 times (10 × 10) more acidic than an aqueous solution with a pH of 5.

LINKING QUESTION

What is the shape of a sketch graph of pH against $[H^+]$?

■ **Table R3.7** The relationship between pH and $[H^+](aq)$

pH	Concentration of hydrogen ions, $H^+(aq)$ / mol dm^{-3}
0	$1 \times 10^0 = 1.0$
1	$1 \times 10^{-1} = 0.1$
2	$1 \times 10^{-2} = 0.01$
3	$1 \times 10^{-3} = 0.001$
4	$1 \times 10^{-4} = 0.0001$
5	$1 \times 10^{-5} = 0.00001$
6	$1 \times 10^{-6} = 0.000001$
7	$1 \times 10^{-7} = 0.0000001$
8	$1 \times 10^{-8} = 0.00000001$
9	$1 \times 10^{-9} = 0.000000001$
10	$1 \times 10^{-10} = 0.0000000001$
11	$1 \times 10^{-11} = 0.00000000001$
12	$1 \times 10^{-12} = 0.000000000001$
13	$1 \times 10^{-13} = 0.0000000000001$
14	$1 \times 10^{-14} = 0.00000000000001$

Common mistake

Misconception: *pH = 0 means that no acid or base is present.*

A solution with pH 0 would be very acidic and the solution would have a hydrogen ion concentration of 1 mol dm^{-3} (pH = $-\log_{10}(1) = 0$).

The smaller the pH, the more acidic the solution and the greater the concentration of H^+ ions. There is an inverse relationship between hydrogen ion concentration and pH.

A stronger alkali will have a higher pH than a weaker alkali with the same concentration at the same temperature.

Another consequence of the scale being logarithmic is that, although a pH of 5.5 appears to be 'halfway' between pH 5 and pH 6, it is not. For pH 5.5, $[H^+]$ is $10^{-5.5}$ ($= 3.16 \times 10^{-6}$), which is not halfway between 10^{-5} and 10^{-6}.

Table R3.8 indicates that freshly distilled water (pH 7) and alkalis such as 1 mol dm^{-3} aqueous sodium hydroxide (pH 14) – despite being neutral and alkaline – contain hydrogen ions at very low concentrations.

■ **Table R3.8** Hydrogen and hydroxide concentrations for aqueous solutions (at 25 °C)

Solution	**$[H^+(aq)]$ / mol dm^{-3}**	**$[OH^-(aq)]$ / mol dm^{-3}**
acidic	$>10^{-7}$	$<10^{-7}$
neutral	10^{-7}	10^{-7}
basic	$<10^{-7}$	$>10^{-7}$

Hydrogen ions are present in neutral and alkaline aqueous solutions because water itself is very slightly dissociated into hydrogen ions and hydroxide ions:

$$H_2O(l) \rightleftharpoons H^+(aq) + OH^-(aq)$$

$$2H_2O(l) \rightleftharpoons H_3O^+(aq) + OH^-(aq)$$

In distilled water (pH 7) the concentrations of hydroxide and hydrogen ions are equal, that is, $[H^+(aq)] = [OH^-(aq)]$.

Acidic solutions

An acidic solution has a pH < 7. In an acidic solution, the concentration of hydrogen ions, $H^+(aq)$, is greater than the concentration of hydroxide ions, $OH^-(aq)$. A pH < 7 does not mean that $OH^-(aq)$ ions are absent; some will be present from the ionization of water molecules.

Alkaline solutions

An alkaline solution has a pH > 7. In an alkaline solution, the concentration of $H^+(aq)$ is less than the concentration of $OH^-(aq)$. pH > 7 does not mean that $H^+(aq)$ ions are absent in the alkaline solution: some will be present from the dissociation of water. The greater the value of pH, the more alkaline the solution and the greater the concentration of $OH^-(aq)$ ions.

10 Classify each of the following aqueous solutions as acidic, neutral or alkaline:

a a solution with pH 9.5
b a solution with pH 3.0
c a solution with pH 0.0
d a solution with pH 7

11 Three solutions had pH values of 9, 12 and 14. State which one was the most alkaline.

12 Three solutions had pH values of –1, 3 and 6. State which one was the most acidic.

ATL R3.1C

Use the introduction at https://phet.colorado.edu/sims/html/acid-base-solutions/latest/acid-base-solutions_en.html to see how the simulation can show particles, graphs and tools such as a pH meter, pH paper and a conductivity tester.

Use the 'My solution' option to explore the properties of acids and bases in solution, comparing the effects of the strength and concentration.

One of your classmates claims, 'Solutions of strong acids always have a lower pH than weak acids.' Using evidence from the simulation to support your reasoning, demonstrate to the student that their claim is incorrect.

Top tip!

When calculating pH from $[H^+(aq)]$, the number of decimal places in pH should be equal to the number of significant figures in the number $[H^+(aq)]$ of which you are taking the logarithm.

Top tip!

Calculated values of pH (from $pH = -\log_{10}[H^+]$) for very dilute solutions are not accurate since they fail to include the contribution of hydrogen ions from the ionization of water molecules.

WORKED EXAMPLE R3.1H

$10.00\,cm^3$ of an aqueous solution of a monoprotic strong acid (for example, HCl(aq)) is added to $990.00\,cm^3$ of water. Calculate the change in pH.

Answer

Dilution factor is $\frac{1000}{10} = \times 100$; this will be an increase in pH of 2, since a change of one pH unit corresponds to a change in H^+ concentration of 10.

ATL R3.1D

Use the expression: $pH = -\log_{10}[H^+(aq)]$ to calculate the theoretical pH of an aqueous solution of hydrochloric acid of concentration $1.0 \times 10^{-8}\,mol\,dm^{-3}$ at 25 °C.

The experimental value is 6.98. Explain the differences between the two values.

13 Calculate the pH of $0.01\,mol\,dm^{-3}$ hydrochloric acid, HCl(aq).

14 Calculate the pH of a $0.100\,mol\,dm^{-3}$ HCl(aq) solution.

15 Calculate the pH of a $3.00 \times 10^{-7}\,mol\,dm^{-3}$ $H^+(aq)$ solution.

16 Calculate the pH of a $1.00\,mol\,dm^{-3}$ $HNO_3(aq)$ solution.

17 Calculate the pH of $0.01\,mol\,dm^{-3}$ sulfuric acid, $H_2SO_4(aq)$. It is a diprotic acid and you can assume it is fully ionized in water.

18 An aqueous solution of hydrochloric acid contains 3.646 g of hydrogen chloride (HCl) in every $250\,cm^3$ of solution. Calculate the concentration of hydrogen ions, $H^+(aq)$, and the pH of the solution.

19 Calculate the pH of a mixture of $25.00\,cm^3$ of $0.1000\,mol\,dm^{-3}$ HCl and $15.00\,cm^3$ of $0.1000\,mol\,dm^{-3}$ NaOH.

Tool 3: Mathematics

Calculations involving logarithmic functions

It is worth remembering that the purpose of a logarithmic function is to display a wide range of data in a meaningful way. The logarithmic scale works around the number system we are familiar with: that of base 10.

If we think of the number 1000 as being 10^3, the $\log_{10}$ of 1000 = 3. The general equation can be given as:

$$\log_{10}x = y \text{ or } 10^y = x$$

A doubling of the logarithmic integer will lead to a 10-fold increase in linear value. For example, on the logarithmic scale, if the value increases from 2 to 3 to 4, we have a real increase from 100 to 1000 to 10 000. This means we can represent large changes in linear numbers in a way that is easier to interpret.

In chemistry, logarithmic functions are most commonly used in determining pH, as the change in concentration of H^+ can vary dramatically, for example, from $1\,mol\,dm^{-3}$ to $1 \times 10^{-14}\,mol\,dm^{-3}$. The numbers 0–14 on the pH scale make it much easier to visualize and interpret these concentrations of H^+.

◆ **pOH**: The negative logarithm (to the base 10) of the hydroxide ion concentration in an aqueous solution (in $mol\,dm^{-3}$).

HL ONLY

pOH

The symbol 'p' can also be used with the hydroxide ion concentration to define **pOH**. The equation for pOH is given here and in section 1 of the IB *Chemistry data booklet.*

$$pOH = -\log_{10}([OH^{-}(aq)]\,mol\,dm^{-3})$$

The logarithmic pH and pOH scales both reduce a wide range (typically 1 to 10^{-14}) to a narrower range (typically 0 to 14).

20 Calculate the pOH of a solution with a hydroxide ion concentration of $0.100\,mol\,dm^{-3}$.

21 Calculate the hydroxide ion concentration of a solution with a pOH of 2.

Measuring pH

Indicators

An indicator is a chemical that has one colour when it is in an acidic solution and another colour when it is in an alkaline solution. Some common indicators are shown in Table R3.9.

■ **Table R3.9** Common indicators and their properties

	Colour of indicator in		
Indicator	***Highly acidic solution***	***pH at which indicator changes colour***	***Highly alkaline solution***
litmus	red	4.5–8.3	blue
methyl orange	red	3.1–4.4	yellow
bromothymol blue	yellow	6.0–7.6	blue
phenolphthalein	colourless	8.3–10.0	pink

Universal indicator contains a mixture of dyes and gives different colours in solutions of different pH values. It is available in paper or solution form. The pH values and corresponding colours of universal indicator with the pH values of some common substances are shown in Figure R3.17.

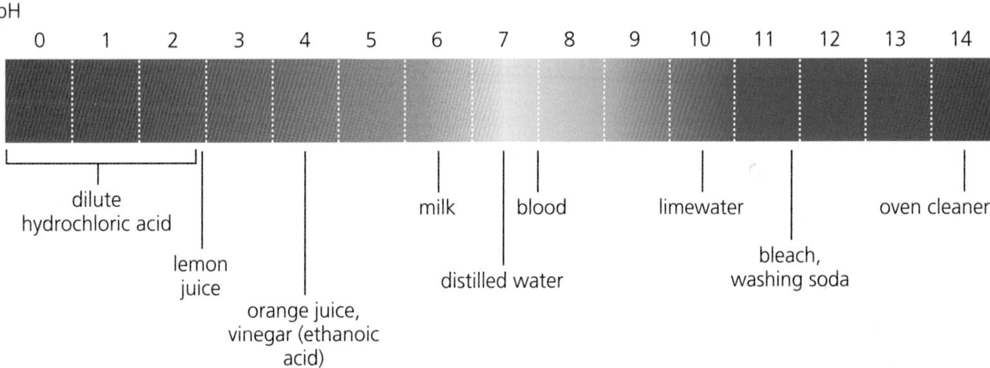

■ **Figure R3.17** The pH values of common substances measured using universal indicator

Litmus changes colour over a wide range of pH values. For this reason, it is not used in titrations, but only for detecting acidic and alkaline solutions.

pH meters

A pH meter is an electronic instrument which is more accurate than indicators. It requires calibration and is dipped into the solution directly to measure pH.

The change in pH during a neutralization reaction of an acid with an alkali can be studied using a data logger and a computer.

Tool 1: Experimental techniques

Measuring variables: pH

The pH of a solution can be measured electrochemically with a device called a pH meter (Figure R3.18). The pH probe contains a very thin glass bulb filled with acid of known concentration.

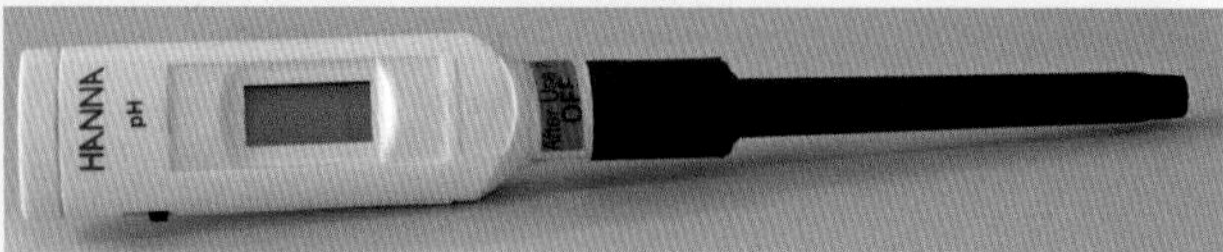

■ **Figure R3.18** pH probe

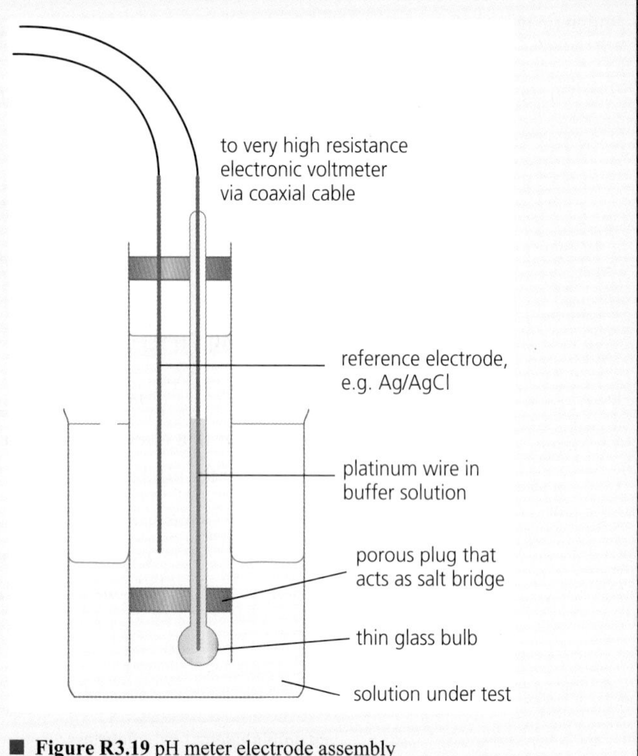

■ **Figure R3.19** pH meter electrode assembly

The technique makes use of a cell in which one electrode is sensitive to the hydrogen ion concentration, $[H^+(aq)]$ (or $[H_3O^+(aq)]$), and another electrode serves as a reference (Figure R3.19). An electrode sensitive to the concentration of a particular ion is called an ion-selective electrode.

It is often necessary to control the pH of a solution, especially if hydrogen ions, $H^+(aq)$, are being formed or used up. It may also be necessary to control pH if the nature of the species changes with pH, as happens with weak organic acids and when enzymes are being studied. pH is often controlled by performing the reaction in a buffer solution.

22 a State two ways in which the pH of a solution can be measured.

b State which method will give the more accurate value.

LINKING QUESTION

When are digital sensors (for example, pH probes) more suitable than analogue methods (for example, pH paper or solution)?

◆ **Ion product constant of water**: The product of the concentrations of hydrogen and hydroxide ions in water under standard conditions.

Top tip!

This is a key equation in acid–base equilibria.

Ion product constant of water

When water is purified by repeated distillation, its electrical conductivity falls to a very low, constant value. This is evidence that pure water ionizes to a very small extent:

$$H_2O(l) \rightleftharpoons H^+(aq) + OH^-(aq) \quad \text{or} \quad 2H_2O(l) \rightleftharpoons H_3O^+(aq) + OH^-(aq)$$

If the equilibrium law is applied to the first equation: $K_w = [H^+(aq)] \times [OH^-(aq)]$

where K_w represents a constant known as the **ion product constant of water**. At 298 K (25 °C), the measured concentrations of $H^+(aq)$ and $OH^-(aq)$ in pure water are $1.00 \times 10^{-7}\,mol\,dm^{-3}$; therefore:

$$K_w = [H^+(aq)] \times [OH^-(aq)] = (1.00 \times 10^{-7}) \times (1.00 \times 10^{-7}) = 1.00 \times 10^{-14}$$

Note that the product of $[H^+(aq)]$ and $[OH^-(aq)]$ is constant at a given temperature. Hence as the hydrogen ion concentration of a solution increases, the hydroxide ion concentration decreases (and vice versa).

WORKED EXAMPLE R3.1I

Calculate the pH of a 0.500 mol dm^{-3} $OH^-(aq)$ solution. Assume $K_w = 1.00 \times 10^{-14}$.

Answer

$K_w = [H^+(aq)] \times [OH^-(aq)]$

$1.00 \times 10^{-14} = [H^+(aq)] \times 0.500$

$$[H^+(aq)] = \frac{1.00 \times 10^{-14}}{0.500} = 2.00 \times 10^{-14}\,\text{mol dm}^{-3}$$

$pH = -\log_{10}[H^+(aq)] = -\log_{10} 2.00 \times 10^{-14} = 13.7$

23 Calculate the H^+ concentration and pH of a 0.01 mol dm^{-3} solution of KOH(aq).

24 Calculate the pH of a 2.50×10^{-3} mol dm^{-3} $OH^-(aq)$ solution.

25 Calculate the pH of 0.0200 mol dm^{-3} aqueous sodium hydroxide, NaOH(aq).

26 Calculate the pH of 0.010 mol dm^{-3} aqueous barium hydroxide, $Ba(OH)_2(aq)$.

WORKED EXAMPLE R3.1J

At 60 °C, the ion product constant of water is 9.55×10^{-14}. Calculate the pH of a neutral solution at this temperature.

Answer

$K_w = [H^+(aq)] \times [OH^-(aq)]$

$9.55 \times 10^{-14} = [H^+(aq)] \times [OH^-(aq)]$

Since it is a neutral solution:

$[H^+(aq)] = [OH^-(aq)]$

$[H^+(aq)] = \sqrt{9.55 \times 10^{-14}} = 3.09 \times 10^{-7}\,\text{mol dm}^{-3}$

$pH = -\log_{10} 3.09 \times 10^{-7} = 6.5$

The solution is described as neutral when the concentration of hydrogen ions equals the concentration of hydroxide ions, so that at 298 K (25 °C) the value of K_w is 1.00×10^{-14}.

Experimental measurements demonstrate that water becomes increasingly dissociated and hence more acidic as the temperature rises. The pH decreases with an increase in temperature, but the solution is still described as chemically neutral since the concentrations of hydroxide and hydrogen ions remain equal.

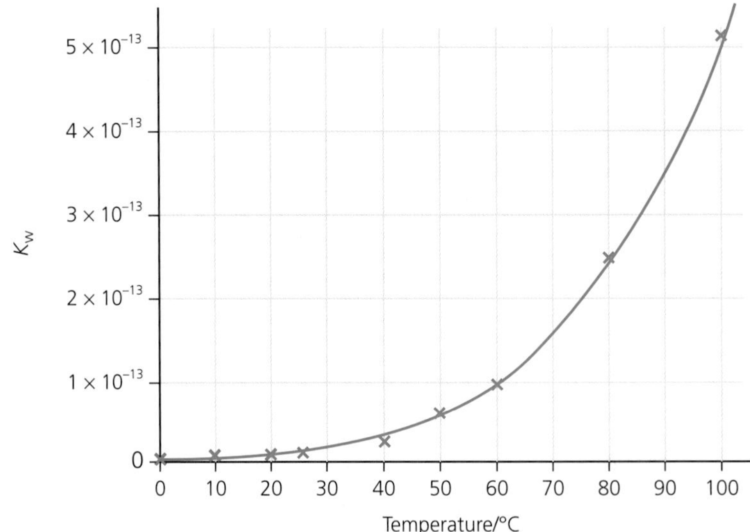

■ **Figure R3.20** The relationship between temperature and the ion product constant of water

The dissociation of water is expected to be an overall endothermic process because energy is needed to break the –OH bond and separate oppositely charged protons and hydroxide ions, which are then hydrated. It can also be deduced that the dissociation of water is an endothermic process since applying Le Châtelier's principle predicts that increasing the temperature will favour the reaction that absorbs heat – that is, the endothermic reaction. The increasing values of K_w also show that increasing the temperature favours the dissociation of water molecules, thereby producing more hydrogen ions and hydroxide ions.

Common mistake

Misconception: *Water always has a pH of 7 and is always neutral.*

Pure water is always neutral because hydrogen and hydroxide ion concentrations are equal: $H_2O(l) \rightleftharpoons H^+(aq) + OH^-(aq)$. Temperature shifts the position of equilibrium. The process is endothermic (because it involves bond breaking), so the forward reaction absorbs heat. Increasing the temperature favours the forward reaction so more hydrogen ions and hydroxide ions are formed. This raises the value of K_w and lowers the pH.

LINKING QUESTION

Why does the extent of ionization of water increase as temperature increases?

WORKED EXAMPLE R3.1K

The ion product constant of water, K_w, for the ionization of water

$$H_2O(l) \rightleftharpoons H^+(aq) + OH^-(aq)$$

at different temperatures is given in Figure R3.20.

What can be deduced from this information?

A Only at 25 °C are $[H^+]$ and $[OH^-]$ equal.

B The equilibrium lies furthest to the right at 0 °C.

C The forward reaction is exothermic.

D The pH of water decreases as temperature increases.

Answer

Option A is wrong as $[H^+]$ and $[OH^-]$ are equal at all temperatures for pure water.

Option B is wrong as the equilibrium lies furthest to the right at 50 °C (as shown by the K_w value).

Option C is wrong as the forward reaction is endothermic. When temperature increases, K_w increases. This shows that the equilibrium lies to the right.

Option D is correct. When temperature increases, $[H^+]$ increases and this leads to a decrease in pH.

pOH and K_w

Start from:

$$K_w = [H^+(aq)] \times [OH^-(aq)] = 1.00 \times 10^{-14}$$

Taking negative logarithms to the base 10 of both sides gives:

$$-\log_{10} K_w = -\log_{10}([H^+(aq)] \times [OH^-(aq)]) = -\log_{10}(1.00 \times 10^{-14})$$

$$= -\log_{10}[H^+(aq)] - \log_{10}[OH^-(aq)] = -\log_{10} 10^{-14}$$

$$pK_w = pH + pOH = 14$$

A wide variety of acid–base equilibrium problems can be solved using the relationships between pH, hydrogen ion concentration and hydroxide ion concentration in conjunction with the ion product constant of water.

27 Calculate the pH of a solution with a pOH of 1.

28 Calculate the pOH of a solution with a pH of 2.

29 Calculate the pH of a 0.100 mol dm^{-3} $Ba(OH)_2$(aq) solution.

HL ONLY

Titration curves

We can find quantitative information about acid–base reactions by titration in the presence of a suitable indicator. Typically, dilute acid is added to the burette and run into alkali in the presence of an indicator (Figure R3.21).

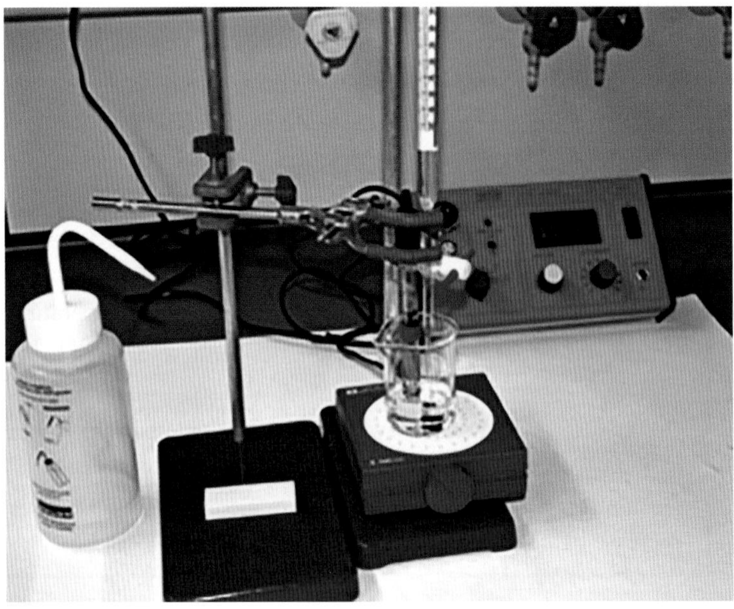

■ **Figure R3.21** Apparatus to record titration curves during neutralization; the reaction mixture is stirred magnetically and the change in the pH is recorded on the meter

◆ **Equivalence point**: The point in a titration at which a stoichiometric ratio of reagents has reacted and the reaction is complete.

The end-point, when the indicator changes colour, should correspond with the **equivalence point** at which the acid has neutralized the alkali. The indicator used should change colour over a small range of pH values that includes the equivalence point of the reaction: in this case the end-point and equivalence points are effectively the same, so the volume at which neutralization occurs can be determined accurately.

Common mistake

Misconception: *Indicators 'help' with neutralization.*

Indicators are not needed for neutralization; the molecules simply change their structure, and hence their colour, during the reactions.

◆ **Titration curve**: A plot showing the pH of a solution being analysed as a function of the amount of acid or base added.

If the pH during a titration is plotted, the resulting **titration curve** is centred around the equivalence point where there is a salt and water only.

The pH of the neutralized solution depends on the relative strength of the acid and base, as shown in Figure R3.22. In these examples, the acid is ten times as concentrated as the alkali to ensure there is very little pH change due to dilution, but it will also affect the shape of the buffer zone region with a weak acid.

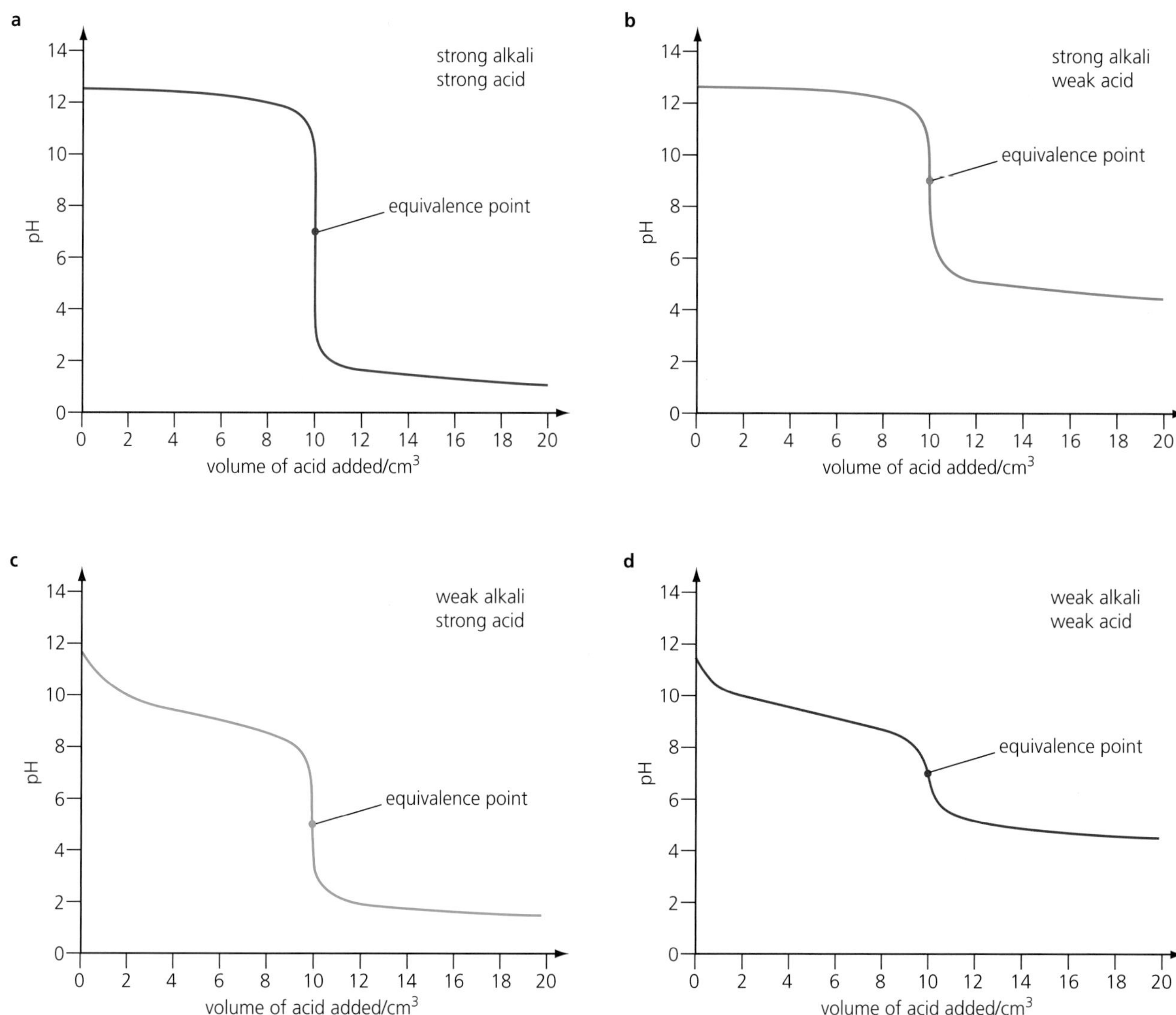

■ **Figure R3.22** pH changes as $1.0\,mol\,dm^{-3}$ acid is run into $100\,cm^3$ of $0.10\,mol\,dm^{-3}$ alkali; the equivalence point occurs after $10\,cm^3$ of acid has been added

Tool 3: Mathematics

Draw and interpret uncertainty bars

When a range of experimental values is plotted, each point can have error bars drawn on it. The size of the bar is calculated from the uncertainty due to random errors. Any line or curve of best fit that is drawn should pass through the error bars of every point. If it is not possible to draw a line of best fit within the error bars, then the systematic errors are greater than the random errors.

Strong acid–strong base

The net ionic equation for a strong acid–strong base titration is:

$$H^+(aq) + OH^-(aq) \rightarrow H_2O(l)$$

A strong base is completely ionized in aqueous solutions. Therefore, the major species present before addition of HCl will be OH^-, the cation of the base, and water. The pH is determined by the $[OH^-]$ ion from the base.

LINKING QUESTION

Why is the equivalence point sometimes referred to as the stoichiometric point?

As acid is added, the additional hydrogen ions are quickly removed by neutralization to form water so there is only a gradual change in pH.

When enough hydrogen ions have been added to react exactly with the hydroxide ions from the base, the stoichiometric or equivalence point of the titration has been reached. This occurs at pH 7: the solution is neutral.

HL ONLY

Strong acid–weak base

Compared with the curve for a strong acid and strong base, the initial pH on this curve is lower and the pH shift is more pronounced at first.

In the first part, where the curve is relatively flat, the solution contains a mixture of weak base and its salt (the conjugate acid). The pH change is gradual because the solution is a buffer mixture. This means its pH changes very little when acid is added to it.

After the equivalence point, the shape of the curve is identical to that of a strong acid–strong base reaction, as it is now the concentration of hydrogen ions, $[H^+(aq)]$, which controls the pH in both titrations.

The pH of the equivalence point of a titration of a strong acid with a weak base is always less than 7 (acidic).

Weak acid–strong base

The equivalence point occurs at pH > 7 (basic) since the anion of the salt is the relatively strong conjugate base of the weak acid.

After the equivalence point, where there is an excess of weak acid, the solution contains a weak acid and its salt (conjugate base). Again, this creates a buffer mixture so pH change is resisted after the equivalence point.

Weak acid–weak base

The equivalence point is close to 7. The anion of the salt is the relatively strong conjugate base of the weak acid and hydrolyses water by accepting protons from water molecules. The cation of the salt is the relatively strong conjugate acid of the weak base, which hydrolyses water by donating protons to water molecules. The salt solution is slightly basic, neutral or slightly acidic depending on the relative strengths of the conjugates.

Inquiry 1: Exploring and designing

Develop investigations that involve hands-on laboratory experiments, databases, simulations, modelling

Consider the research question: How does changing the strength of the weak acid reacting with the strong base affect the shape of the titration curve?

During the neutralization process, the pH, electrical conductivity and temperature of the system change. These changes can be monitored and used to detect the equivalence point of the reaction during hands-on laboratory experiments. The equivalence point of the reaction can also be determined by the use of indicators, which change colour as the pH changes.

A titration curve can be modelled mathematically using *Excel* or investigated using online simulations.

Values of K_a and pK_a for weak acids can be found in a variety of online databases. You can use these values to deduce the answer to the question above, using methods outlined elsewhere in this chapter.

30 In an experiment, a student slowly added solution Y in $0.50\,cm^3$ portions from a burette to solution X in a conical flask and swirled the solution. The pH after each addition was measured and recorded and a graph of pH against volume of Y added was drawn.

a Explain why the flask was swirled after each addition.

b Using the graph, state and explain whether solution X is acidic, alkaline or neutral.

c Using the graph, state and explain whether solution Y is acidic, alkaline or neutral.

d Describe what happens to the pH of the mixture in the conical flask as solution Y is slowly added.

e Describe what would happen to the shape of the graph if solution X was in the burette and solution Y was in the conical flask.

f Deduce the volume of Y needed to react with all of solution X to reach the equivalence point, where only salt and water are present.

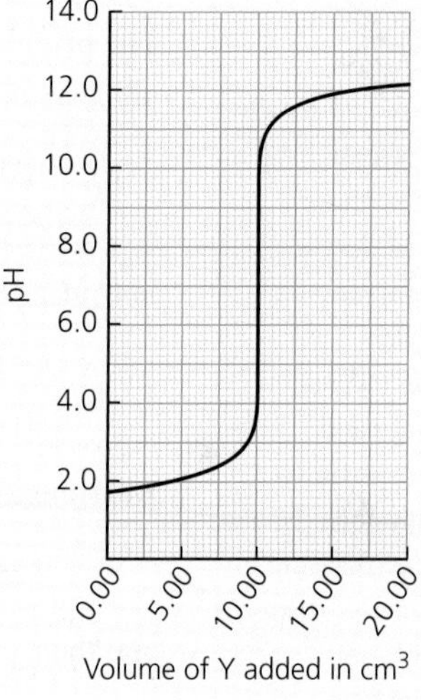

LINKING QUESTION

How can titration be used to calculate the concentration of an acid or base in solution?

Nature of science: Measurements

Measurement of pH

In 1889, the German chemist Walther Hermann Nernst (1864–1941) gave the theoretical foundation for the use of electrode potential to measure the concentration of an ion in solution. This led the Danish chemist SPL Sørensen (1868–1939) to develop the pH scale and an early pH meter, in 1900. At that time, pH was measured using known concentrations of coloured indicator dyes which changed colour by drop-wise addition of solutions. This allowed standardized measurements to be made. Litmus paper and other indicator strips were also used, with comparisons against colour charts to determine pH.

Dissociation constants

Acid dissociation constants

A weak monoprotic acid, HA, reacts reversibly with water according to the equation:

$$H_2O(l) + HA(aq) \rightleftharpoons H_3O^+(aq) + A^-(aq)$$

◆ **Acid dissociation constant:** The equilibrium constant for the reaction in which an acid ionizes and loses a proton.

The equilibrium constant for this reaction is known as the **acid dissociation constant, K_a**.

$$K_a = \frac{[H^+(aq)] \times [A^-(aq)]}{[HA(aq)]}$$

The acid dissociation constant is a measure of the strength of a weak acid. The larger the value of K_a, the greater the extent of ionization or dissociation and the stronger the acid.

Since acid dissociation constants, K_a, tend to be small and vary considerably, they are often expressed as pK_a values where:

$$pK_a = -\log_{10} K_a$$

Values of pK_a are also a measure of acid strength: the smaller the value of pK_a, the stronger the acid. A change of 1 in the value of pK_a means a change in acid strength by a factor of 10 (compare this with pH and $[H^+(aq)]$).

The pH of a solution of a weak acid can only be calculated if the acid dissociation constant, K_a (or pK_a), is known.

$$K_a = \frac{[H^+(aq)] \times [A^-(aq)]}{[HA(aq)]}$$

but since $[H^+(aq)] = [A^-(aq)]$ in a solution where only the acid is present:

$$K_a = \frac{[H^+(aq)]^2}{[HA(aq)]}$$

Rearranging:

$$[H^+(aq)] = \sqrt{[HA(aq)] \times K_a}$$

and then:

$$pH = -\log_{10}[H^+(aq)]$$

You can also use this approach to calculate the K_a (and hence pK_a) of a weak acid if you know the pH of the solution and its concentration.

Top tip!

Acid dissociation constants are not usually quoted for strong acids because these effectively undergo complete ionization in water. Their dissociation constants are very large and tend towards infinity in dilute solutions. It is difficult to measure them accurately because the concentration of unionized acid molecules is so low.

Top tip!

Values of K_a and pK_a are equilibrium constants. Like other equilibrium constants, they are not affected by changes in concentration, only by changes in temperature. This means that acid strengths vary with temperature and that the order of acid strengths can vary with temperature.

31 The pH of 0.01 mol dm^{-3} benzenecarboxylic acid solution, $C_6H_5COOH(aq)$, is 3.10. Calculate the acid dissociation constant, K_a, at this temperature.

32 Calculate the pH value of a 0.1 mol dm^{-3} solution of ethanoic acid, $CH_3COOH(aq)$, given that its K_a value is 1.8×10^{-5}.

Common mistake

Misconception: *The strength of an acid is determined by pH.*

The strength of an acid is determined by its value of K_a/pK_a, which varies with temperature. pH is related to hydrogen ion concentration; it is not a direct measure of acid strength. However, measuring the pH values of equimolar solutions of different acids will give relative acid strengths.

Nature of science: Theories

Acidity of oxyacids

The value of K_a can be related to the molecular structure of the acid. For example, chloric(I) acid, HOCl, is a stronger acid than iodic(I) acid, HOI (Figure R3.23). The acidity of oxyacids containing a halogen increases with the increasing electronegativity of the halogen atom. As the electronegativity of the atom bonded to an O–H group increases, the ease with which the hydrogen is ionized or dissociated increases. The movement of electron density towards the electronegative atom further polarizes the O–H bond, which favours ionization. In addition, the electronegative atom stabilizes the conjugate base, which also leads to a stronger acid.

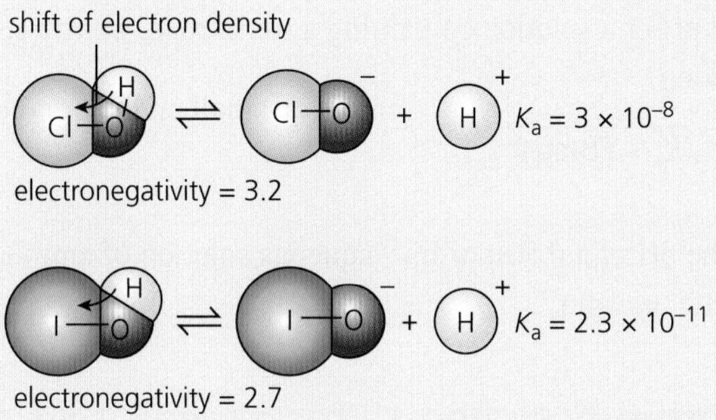

shift of electron density

$$Cl{-}O{-}H \rightleftharpoons Cl{-}O^- + H^+ \quad K_a = 3 \times 10^{-8}$$

electronegativity = 3.2

$$I{-}O{-}H \rightleftharpoons I{-}O^- + H^+ \quad K_a = 2.3 \times 10^{-11}$$

electronegativity = 2.7

■ **Figure R3.23** The acidity of oxyacids increases with increasing electronegativity of the halogen atom

Tool 3: Mathematics

Appreciate when some effects can be ignored and why this is useful

There are many examples in chemistry when effects are so small they can be ignored, either to simplify the development of a mathematical model or to enable a calculation to be carried out. For example:

- Kinetic molecular theory assumes that, in a gas, the volume of molecules and the strength of their interactions are negligible. These simplifications allowed the development of the ideal gas equation and the gas laws.
- When gases are not involved in a reaction at equilibrium (for example, in a reaction in aqueous solution), pressure changes have a negligible effect, since solids and liquids are essentially incompressible.
- The mass of the electron is negligible compared with that of an ion, so the ion mass is taken as being the same as the relative atomic mass.
- In acid–base theory, we make two assumptions to allow acid–base problems to be solved with relatively simple calculations:
  - The dissociation of a weak acid or weak base is small and therefore the concentration of acid or base is assumed to be equal to the concentration of the solution.
  - The amount of hydrogen ions contributed by the autoionization of water is even smaller and therefore negligible.

Base dissociation constants

A weak base may be either a molecule, such as ammonia, or an anion, such as ethanoate. A weak base forms a slightly basic solution:

$$B(aq) + H_2O(l) \rightleftharpoons BH^+(aq) + OH^-(aq)$$

$$B^-(aq) + H_2O(l) \rightleftharpoons BH(aq) + OH^-(aq)$$

◆ **Base dissociation constant**: The equilibrium constant for the reaction in which a base gains a proton from a water molecule in an aqueous solution.

The equilibrium constant for this reaction is known as the **base dissociation constant,** K_b:

$$K_b = \frac{[BH^+(aq)] \times [OH^-(aq)]}{[B(aq)]} \qquad K_b = \frac{[BH(aq)] \times [OH^-(aq)]}{[B^-(aq)]}$$

To calculate the pH of a solution containing a weak base, we use a similar approximation to that of a weak acid, giving:

$$[OH^-(aq)] = \sqrt{K_b \times [B(aq)]}$$

33 Calculate the pH of a 0.50 mol dm^{-3} aqueous solution of ammonia for which $K_b = 1.8 \times 10^{-5}$ mol dm^{-3}.

Relationship between K_a for a weak acid and K_b for its conjugate base

The relationship between K_a for a weak acid, HA, and K_b for its conjugate base, A^- is:

$$K_a \times K_b = K_w = 1.0 \times 10^{-14}$$

This relationship is derived as follows. Let BH^+ be the conjugate acid of a base. The expression for the *acid dissociation* constant K_a for the conjugate acid can be written as:

$$K_a = \frac{[B][H^+]}{[BH^+]} = \frac{[B][H^+][OH^-]}{[BH^+][OH^-]} = \frac{[B]}{[BH^+][OH^-]}[H^+]\,[OH^-]$$

$$K_a = \frac{1}{K_b}K_w;\ K_a \times K_b = K_w$$

Since $pK_a = -\log_{10} K_a$ and $pK_b = -\log_{10} K_b$, the logarithmic form of the equation above is:

$$pK_a + pK_b = pK_w = 14.00$$

The stronger the acid, the larger the value of K_a and the smaller the value of pK_a. Likewise the stronger the base, the larger the value of K_b and the smaller the value of pK_b.

These equations show that, as the value of K_a increases (and the value of pK_a decreases), the value of K_b decreases (and the value of pK_b increases).

LINKING QUESTION

How can we simplify calculations when equilibrium constants K_a and K_b are very small?

Analogous equations can be written to describe the relationship between K_b for a weak base B and K_a for its conjugate acid BH^+:

$$K_b \times K_a = K_w = 1.00 \times 10^{-14}$$

$$pK_b(B(aq)) + pK_a(BH^+(aq)) = pK_w = 14.00$$

HL ONLY

Salt solutions

Normal soluble salts (as opposed to acid salts) often dissolve to form neutral solutions; for example, sodium chloride ionizes to release sodium and chloride ions, neither of which react with water:

$$NaCl(s) + (aq) \rightarrow Na^+(aq) + Cl^-(aq)$$

This occurs when the salt has been formed by the neutralization of a strong acid by a strong base.

◆ **Salt hydrolysis**: When one or both ions derived from a soluble salt undergo a chemical reaction with water molecules, leading to formation of an acidic or an alkaline solution.

However, some normal salts dissolve in water to form acidic or alkaline solutions. This is because one of the ions reacts with the water to release hydrogen ions or hydroxide ions. This phenomenon is called **salt hydrolysis** and occurs when the salts are formed from weak acids or weak bases.

An example of a salt of a weak acid and a strong base is sodium carbonate, Na_2CO_3, which undergoes hydrolysis as follows:

$$CO_3^{2-}(aq) + H_2O(l) \rightleftharpoons HCO_3^-(aq) + OH^-(aq)$$

The sodium ions are spectator ions. The resulting solution contains an excess of hydroxide ions and hence has a pH > 7. (It is slightly alkaline.)

Similar reactions occur with a salt formed from a weak acid and a strong base. Consider sodium hydrogencarbonate:

$$HCO_3^-(aq) + H_2O(l) \rightleftharpoons H_2CO_3(aq) + OH^-(aq)$$

Ammonium chloride, NH_4Cl, is the salt of a strong acid and a weak base. The hydrolysis reaction is:

$$NH_4^+(aq) + H_2O(l) \rightleftharpoons NH_3(aq) + H_3O^+(aq)$$

The chloride ions are spectator ions. The resulting solution contains an excess of hydronium ions and is acidic (pH < 7).

Ammonium ethanoate, CH_3COONH_4, is the salt of a weak acid and a weak base. Both ions are hydrolysed:

$$NH_4^+(aq) + H_2O(l) \rightleftharpoons NH_3(aq) + H_3O^+(aq)$$

$$CH_3COO^-(aq) + H_2O(l) \rightleftharpoons CH_3COOH(aq) + OH^-(aq)$$

$$NH_4^+(aq) + CH_3COO^-(aq) + 2H_2O(l) \rightleftharpoons NH_3(aq) + H_3O^+(aq) + CH_3COOH(aq) + OH^-(aq)$$

Common mistake

Misconception: *Neutralization always results in a neutral solution.*

Salts formed by reacting a strong acid and a strong base are neutral. However, salt hydrolysis results in other salt solutions being acidic (if formed from a strong acid and a weak base) or basic (if formed from a weak acid and a strong base).

The final pH of the solution depends on the equilibrium constants for the two reactions. In this example, the two values are approximately the same so the two processes cancel each other out and the solution is close to neutral.

■ **Table R3.10** Summary of salt hydrolysis

	Strong acid	**Weak acid**
Strong base	both conjugate acid and conjugate base are weak and cannot hydrolyse water → neutral salt	conjugate base is stronger than conjugate acid $pK_b << pK_a$ → basic salt
Weak base	conjugate acid is stronger than conjugate base $pK_a << pK_b$ → acidic salt	depends on the relative strength of the conjugate acid and conjugate base $pK_a < pK_b$ → weakly acidic salt $pK_b < pK_a$ → weakly basic salt $pK_a \approx pK_b$ → neutral salt

WORKED EXAMPLE R3.1L

Sodium fluoride, NaF, is a salt formed when a strong base, NaOH, is reacted with hydrofluoric acid, HF. Predict and explain the pH of the NaF solution.

Answer

Sodium fluoride releases ions in aqueous solution:

$$NaF(aq) \rightarrow Na^+(aq) + F^-(aq)$$

The sodium ion, $Na^+(aq)$, is a spectator ion and does not hydrolyse.

The fluoride ion, F^-, hydrolyses because it is the conjugate base of the weak acid, HF, and is basic enough to remove a proton from a water molecule:

$$F^-(aq) + H_2O(l) \rightleftharpoons HF(aq) + OH^-(aq)$$

Excess hydroxide ions are produced, so the solution is basic and the pH is greater than 7.

34 Determine if the following salts are acidic or alkaline:

a $CH_3COONa(aq)$
b $NH_4Br(aq)$
c $KCN(aq)$
d $CH_3CH_2CH_2CH_2COOK(aq)$

35 Deduce if the solutions below have pH > 7, pH < 7 or pH ≈ 7:

a $Na_2SO_4(aq)$
b $NH_4CH_3COO(aq)$
c $Na_2S(aq)$
d $KNO_2(aq)$

◆ **Acid–base indicator:** A substance that changes colour reversibly, depending on whether the solution is acidic or basic; they are usually weak acids in which the unionized acidic form has a different colour from the anion base form.

Indicators

Nature of indicators

An **acid–base indicator** is a soluble dye or a mixture of soluble dyes that changes colour over a specific pH range.

The colour changes for two common acid–base indicators, methyl orange and phenolphthalein, are shown in Figure R3.24.

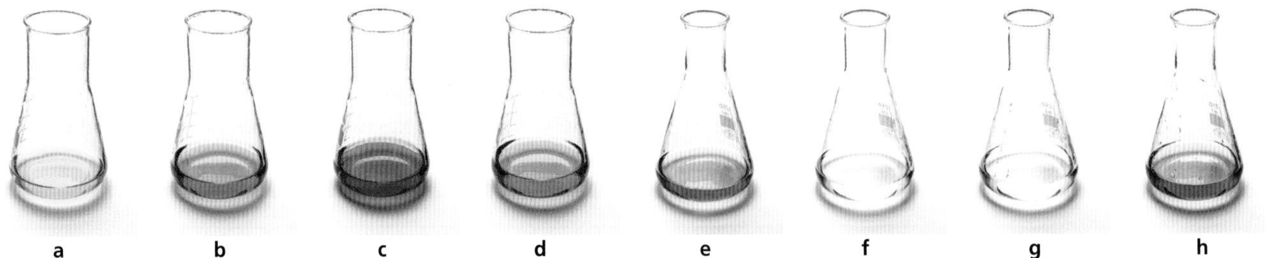

■ **Figure R3.24** The colour changes of methyl orange (a → b and c → d) and phenolphthalein (e → f and g→ h) at the end-point

Phenolphthalein is slightly unusual for an indicator as one of its two forms is colourless. Figure R3.25 shows the structural formulas of the dissociated (ionized) and unionized forms of phenolphthalein. The wavelength of the light absorbed by the weakly acidic indicator changes significantly when a proton is lost to form the conjugate base.

Top tip!

Indicators are brightly coloured, so only small quantities are needed. This means their addition to a solution will not significantly alter the concentration of hydrogen ions already present.

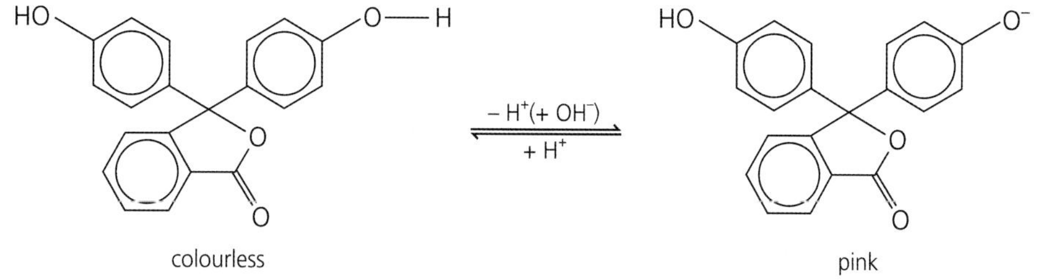

■ **Figure R3.25** Phenolphthalein is colourless in acid, but pink in alkaline conditions

Many other indicators are also weak acids in which the acid (HIn) and its conjugate base (In^-) have different colours.

Using HIn(aq) for the acidic form of an indicator and In^-(aq) for its conjugate base, the equilibrium for the ionization of the indicator can be generalized to:

$$HIn(aq) \rightleftharpoons In^-(aq) + H^+(aq)$$

or

$$HIn(aq) + H_2O(l) \rightleftharpoons In^-(aq) + H_3O^+(aq)$$

Consider a solution of the indicator bromothynol blue, which is yellow when unionized, HIn(aq), and blue when ionized, In^-(aq).

- In a neutral solution, very few of the acid molecules ionize. This means nearly all the indicator molecules will exist as the unionized yellow form, HIn(aq). The solution therefore appears yellow.
- If excess acid is added, the increase in hydrogen ion concentration will shift the equilibrium above to the left (according to Le Châtelier's principle). This means that the concentration of HIn(aq) is very high. This means almost all the indicator molecules will exist as the unionized yellow form and the solution turns yellow.
- In the presence of excess alkali, the hydroxide ions will combine with the hydrogen ions to form water, so they are removed from the equilibrium. The hydrogen ions removed will be partly replaced by the dissociation of HIn(aq). The concentration of In^-(aq) will hence be relatively high and the solution will be blue.

Since indicators are generally weak acids, an equilibrium expression for the acid dissociation constant, K_a, can be written for them. In general:

$$K_a = \frac{[H^+(aq)] \times [In^-(aq)]}{HIn(aq)}$$

or

$$K_a = \frac{[H_3O^+(aq)] \times [In^-(aq)]}{HIn(aq)}$$

The acid dissociation constant, K_a, is sometimes known as the dissociation constant for the indicator and given the symbol K_{In}.

The equation can be rearranged to make the ratio of the concentrations of the two coloured forms the subject:

$$\frac{[HIn(aq)]}{[In^-(aq)]} = \frac{[H^+(aq)]}{K_a}$$

This equation shows that the colour of an indicator depends not only on the hydrogen ion concentration (that is, the pH), but also on the value of the acid dissociation constant, K_a (or K_{In}). This means that different indicators change colour over different pH ranges.

$$K_a = \frac{[H^+(aq)] \times [In^-(aq)]}{HIn(aq)}$$

The equilibrium expression above can also be rearranged to give the following equation:

$$\frac{1}{[H^+(aq)]} = \frac{1}{K_a} \times \frac{[In^-(aq)]}{[HIn(aq)]}$$

Taking logarithms to the base 10 of both sides:

$$pH = pK_a + \log_{10} \frac{[In^-(aq)]}{[HIn(aq)]}$$

This equation allows us to calculate any of the four variables, given the other three.

The colour change of an indicator usually takes place over a range of about 2 pH units: from pH = $pK_a - 1$ to pH = $pK_a + 1$ (Table R3.11). This generally corresponds to the change described above – that is, going from 10% of one form of the indicator to 10% of the other form (Figure R3.26).

■ **Table R3.11** The pH ranges of some common acid–base indicators

Indicator	'Acid colour'	'Alkaline colour'	pH range and pK_a
methyl orange	red	yellow	3.1–4.4 and 3.7
bromothymol blue	yellow	blue	6.0–7.6 and 7.0
bromophenol blue	yellow	blue	3.0–4.6 and 4.2
phenolphthalein	colourless	pink	8.3–10.0 and 9.6
thymol blue	red	yellow	1.2–2.8 and 1.6
	yellow	blue	8.0–9.6 and 8.9
methyl red	red	yellow	4.4–6.2 and 5.1
litmus	red	blue	5.0–8.0 and 6.5

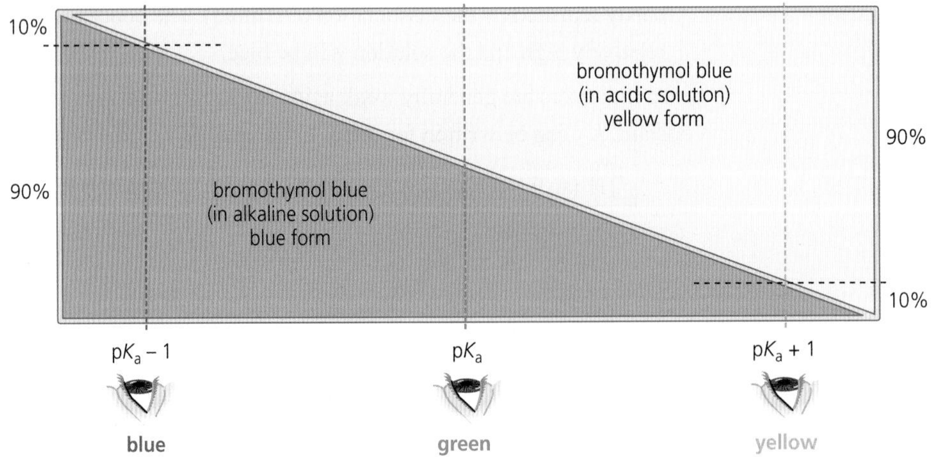

■ **Figure R3.26** A diagram illustrating the behaviour of a typical acid–base indicator: bromothymol blue is placed into a transparent plastic box diagonally divided into halves

LINKING QUESTION

What are some of the similarities and differences between indicators used in acid–base titrations and in redox titrations?

Selecting an indicator for an acid–base titration

The indicator used for any acid–base titration should ideally change colour at the pH corresponding to the mid-point of the almost vertically straight portion of the titration curve.

For a strong acid–strong base titration, any of the indicators could be used: all of them change colour within the almost vertically straight portion of the titration curve, between about pH 4 and pH 11. In other words, the pK_a values of suitable indicators must lie between 4 and 11, and preferably be centred around 7.

Methyl orange changes colour over the pH range 3.2 to 4.4 and phenolphthalein changes over the pH range 8.2 to 10.0. Both indicators are commonly used for titrations involving a strong acid and a strong base (Figure R3.27).

> **Top tip!**
>
> There is little loss in accuracy if the indicator changes colour anywhere within the range of the almost vertically straight portion of the curve, since the pH change is relatively large for the addition of a relatively small amount of acid or base.

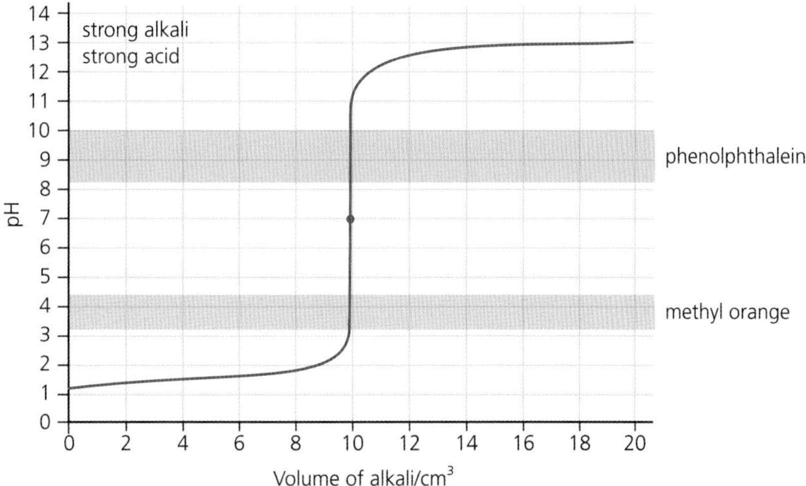

Figure R3.27 Titration curve starting with 100 cm^3 of 0.100 mol dm^{-3} strong acid and adding 1.0 mol dm^{-3} strong alkali

However, the choice of indicator is more limited if a weak acid or a weak base is used in the titration, since the pH range of the almost straight portion is much smaller and fewer indicators change colour completely over this range.

For a strong acid–weak base titration, such as that between 0.1 mol dm^{-3} aqueous ammonia and 0.1 mol dm^{-3} hydrochloric acid, the indicator needs to change between pH values 4 and 7. Methyl orange is a suitable indicator, but phenolphthalein is not.

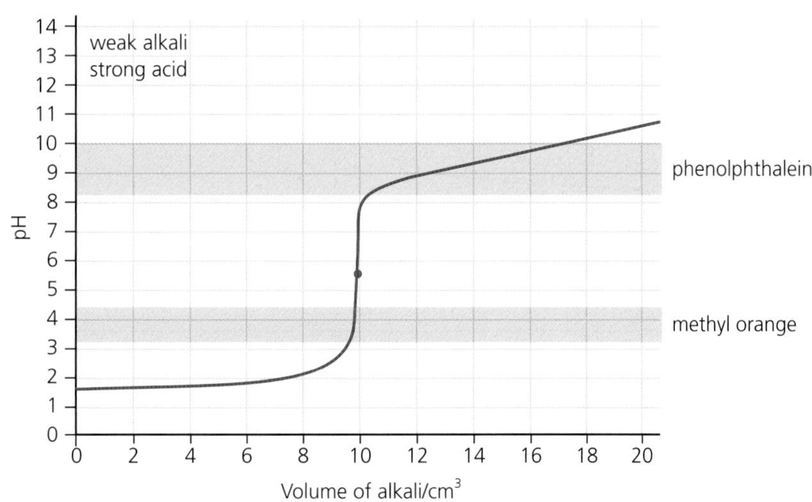

Figure R3.28 Titration curve starting with 100 cm^3 of 0.100 mol dm^{-3} strong acid and adding 1.0 mol dm^{-3} weak alkali

As Figure R3.28 shows, phenolphthalein would not be a suitable indicator because it changes colour at the wrong volume (not at the equivalence point) and over a large volume change of aqueous ammonia solution. It would therefore be impossible to find the equivalence point accurately using phenolphthalein as the indicator.

For a weak acid–strong base titration, such as that between 0.1 mol dm^{-3} ethanoic acid and 0.1 mol dm^{-3} sodium hydroxide, the indicator needs to change between pH values 6 and 10. Phenolphthalein is a suitable indicator, but methyl orange is not. Figure R3.29 shows that methyl orange will change colour very slowly over a relatively large volume of sodium hydroxide so that it would be very difficult to locate the end-point (and hence the equivalence point) accurately. In addition, the colour change would occur at the wrong volume.

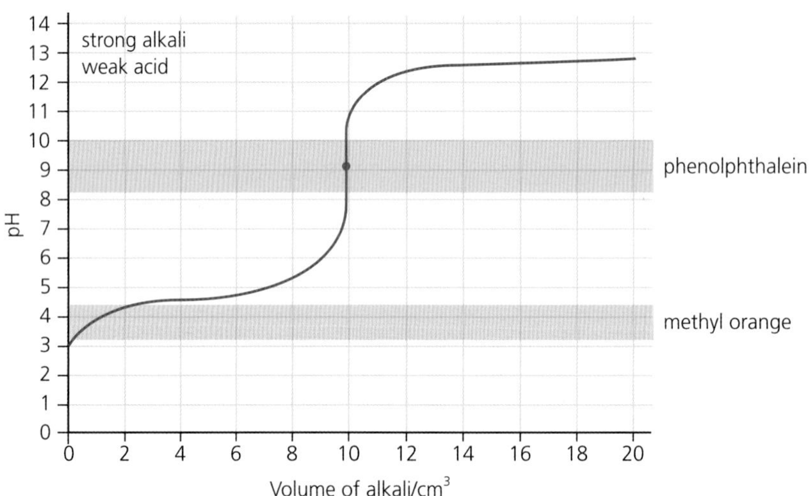

■ **Figure R3.29** Titration curve starting with 100 cm^3 of 0.100 mol dm^{-3} weak acid and adding 1.0 mol dm^{-3} strong alkali

No indicator is suitable for the titration of a weak acid with a weak base, such as that between 0.1 mol dm^{-3} ethanoic acid and 0.1 mol dm^{-3} aqueous ammonia, since there is no almost vertically straight portion present in the titration curve (Figure R3.30). In other words, the pH changes gradually throughout the titration.

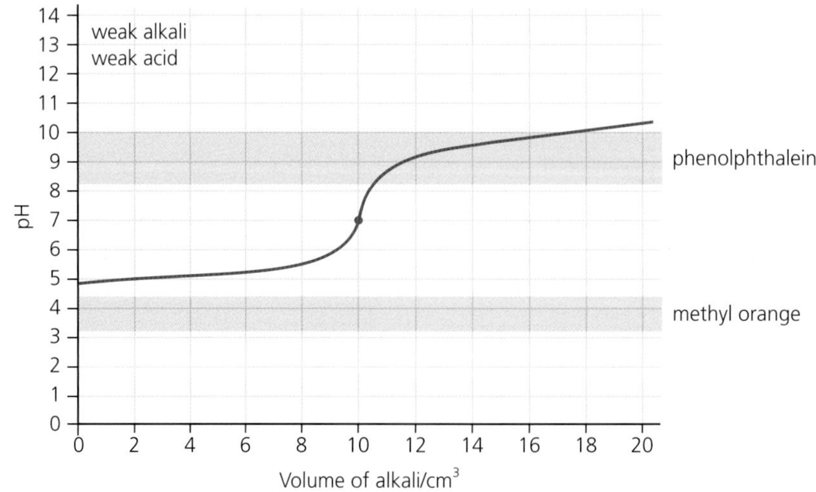

■ **Figure R3.30** Titration curve starting with 100 cm^3 of 0.100 mol dm^{-3} weak acid and adding 1.0 mol dm^{-3} weak alkali

Even if bromothymol blue, whose pK_a is approximately 7, is used as an indicator, it will change colour over a relatively large volume of ammonia. Hence it is not possible to determine the end-point (and hence the equivalence point) accurately.

Table R3.12 summarizes the use of phenolphthalein and methyl orange as acid–base indicators. The principles described here can be used to select other suitable indicators for acid–base titrations.

■ **Table R3.12** The suitability of methyl orange and phenolphthalein as indicators

Alkali	Acid	Indicator
strong	strong	methyl orange or phenolphthalein
strong	weak	phenolphthalein
weak	strong	methyl orange
weak	weak	none

Nature of science: Models

A mathematical model of neutralization

A pH curve is described by a series of equations based on acid–base equilibria. For example, consider the titration of a strong acid (0.100 mol dm^{-3} HCl, in the flask) against a strong base (0.100 mol dm^{-3} NaOH, in the burette).

The pH of the solution in the flask at the beginning of the titration will be $-\log_{10} 0.100$, that is, 1.

Let us now calculate the pH after 22.5 cm^3 of 0.100 mol dm^{-3} aqueous sodium hydroxide has been added to 25.0 cm^3 of 0.100 mol dm^{-3} aqueous hydrochloric acid.

The addition of the alkali has removed $\frac{22.5}{25.0}$ (or, if we divide the numerator and denominator by 2.5, $\frac{9}{10}$ or 90%) of the hydrogen ions. In other words, $\frac{1}{10}$ (10%) of the original number of moles of hydrogen ions is left. This ratio can be termed the reaction factor.

An approximate pH for the resulting solution is:

$$[H^+(aq)] = 0.100 \times \frac{1}{10} = 0.0100\ \text{mol dm}^{-3}$$

$$pH = -\log_{10} 0.0100 = 2.00$$

When 90% of the hydrogen ions have been neutralized, the pH has only changed by one unit: from 1 to 2.

However, this simple calculation ignores the dilution effect: the addition of aqueous sodium hydroxide not only adds hydroxide ions but also changes the volume, and hence the concentration, of the resulting solution.

We can modify the previous equation to include the dilution factor and slightly improve the accuracy of the calculation:

'new' $H^+(aq)$ concentration = 'old' $H^+(aq)$ concentration × reaction factor × dilution factor

$$[H^+(aq)] = 0.100 \times \frac{1}{10} \times \frac{25.0}{(25.0 + 22.5)} = 5.26 \times 10^{-3}\ \text{mol dm}^{-3}$$

$$pH = 2.28$$

Repeating the above calculation, without the dilution factor, for a higher volume of NaOH(aq):

- 24.75 cm^3 of 0.100 mol dm^{-3} sodium hydroxide added to 25.0 cm^3 of 0.100 mol dm^{-3} hydrochloric acid gives a pH of 3.00:

$$\text{reaction factor} = \frac{(25.0 - 24.75)}{25.0} = 0.0100$$

$$[H^+(aq)] = 0.100 \times 0.0100 = 0.001\,00\ \text{mol dm}^{-3}$$

$$pH = -\log_{10} 0.001\,00 = 3.00$$

- 24.975 cm^3 of 0.100 mol dm^{-3} sodium hydroxide added to 25.0 cm^3 of 0.100 mol dm^{-3} hydrochloric acid gives a pH of 4.00:

$$\text{reaction factor} = \frac{(25.0 - 24.975)}{25.0} = 0.001\,00$$

$$[H^+(aq)] = 0.100 \times 0.001\,00 = 0.000\,100\ \text{mol dm}^{-3}$$

$$pH = -\log_{10} 0.000\,100 = 4.00$$

- 24.9975 cm^3 of 0.100 mol dm^{-3} sodium hydroxide added to 25.0 cm^3 of 0.100 mol dm^{-3} hydrochloric acid gives a pH of 5.00:

$$\text{reaction factor} = \frac{(25.0 - 24.9975)}{25.0} = 0.000\,100$$

$$[H^+(aq)] = 0.100 \times 0.000\,100 = 0.000\,0100\ \text{mol dm}^{-3}$$

$$pH = -\log_{10} 0.000\,0100 = 5.00$$

- 24.999 75 cm^3 of 0.100 mol dm^{-3} sodium hydroxide added to 25.0 cm^3 of 0.100 mol dm^{-3} hydrochloric acid gives a pH of 6:

$$\text{reaction factor} = \frac{(25.0 - 24.999\,75)}{25.0} = 0.000\,0100$$

$$[H^+(aq)] = 0.100 \times 0.000\,0100 = 0.000\,001\,00\ \text{mol dm}^{-3}$$

$$pH = -\log_{10} 0.000\,001\,00 = 6.00$$

These simplified calculations show that the pH rises very rapidly near the equivalence point.

LINKING QUESTION

When collecting data to generate a pH curve, when should smaller volumes of titrant be added between measurements?

ATL R3.1E

Visit http://faculty.concordia.ca/bird/javascript/titration/titration-js.html and run the simulation for several combinations of strong and weak acid and base (all at 1.00 mol dm^{-3}). Use an appropriate indicator and try placing the acid in the burette and then the flask.

Repeat with lower and higher concentrations for the base and acid.

Interpret each pH curve in terms of intercept with the pH axis, equivalence point, buffer region and points where pH = pK_a or pOH = pK_b.

Going further

Conductometric titrations

The end-point of an acid–base titration can be found by monitoring the conductivity of the solution as the alkali is progressively neutralized by the addition of acid (Figure R3.31).

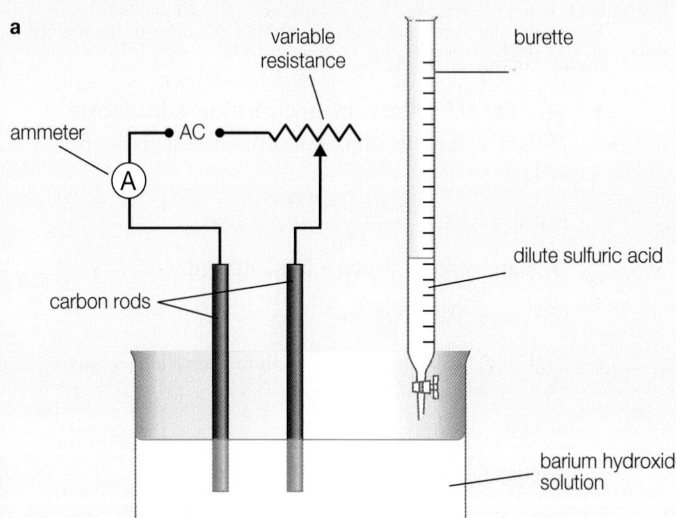

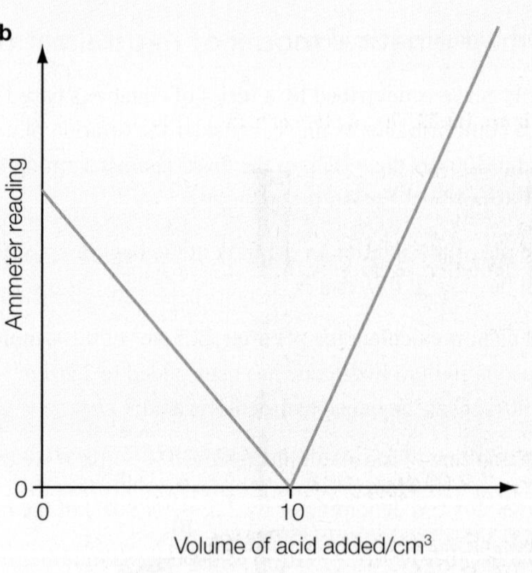

■ **Figure R3.31** Neutralization of barium hydroxide with dilute sulfuric acid: **a** apparatus to monitor the neutralization by measuring conductivity changes and **b** a graph of sample results

For example, consider the titration of barium hydroxide and dilute sulfuric acid.

$Ba(OH)_2(aq) + H_2SO_4(aq) \rightarrow BaSO_4(s) + 2H_2O(l)$

or ionically:

$OH^-(aq) + H^+(aq) \rightarrow H_2O(l)$

At the equivalence point, the electrical conductivity is close to zero (Figure R3.31b) because the ions of barium hydroxide and sulfuric acid are replaced by insoluble barium sulfate and water molecules.

Buffer solutions

◆ **Buffer solution:** A solution that resists change in pH when small amounts of a strong acid or alkali are added (provided its buffer capacity is not exceeded) or when the solution is diluted.

A **buffer solution** is an aqueous solution whose pH (and hence hydrogen ion concentration) remains unchanged by dilution with water or when relatively small amounts of acid or base are added to it. Buffers *resist* changes in pH.

ATL R3.1F

Find out about the uses of buffers in blood, electrophoresis and in products such as shampoos; research the names and actions of specific buffers. Present what you discover to your peers in the form of a poster.

There are two types of buffer solution:

- An **acidic buffer** consists of an aqueous solution of a weak acid and one of its soluble salts (conjugate base), for example, ethanoic acid and sodium ethanoate.
- A **basic buffer** consists of an aqueous solution of a weak base and one of its soluble salts (conjugate acid), for example, ammonia and ammonium chloride.

To resist changes in pH, a buffer must be able to neutralise any extra acid or alkali added to the system.

■ **Figure R3.32** The addition of hydrogen ions to an acidic buffer

■ **Figure R3.33** The addition of hydroxide ions to an acidic buffer

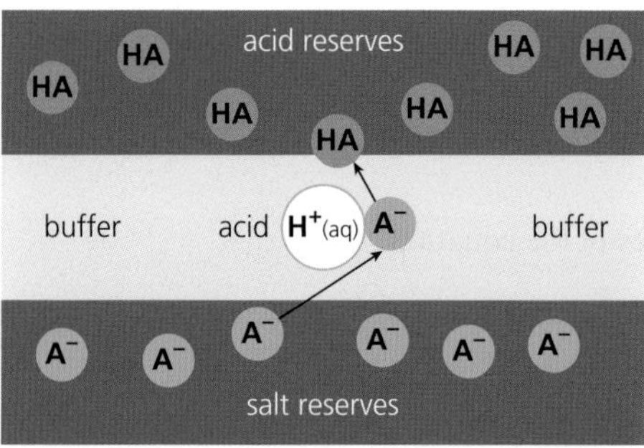

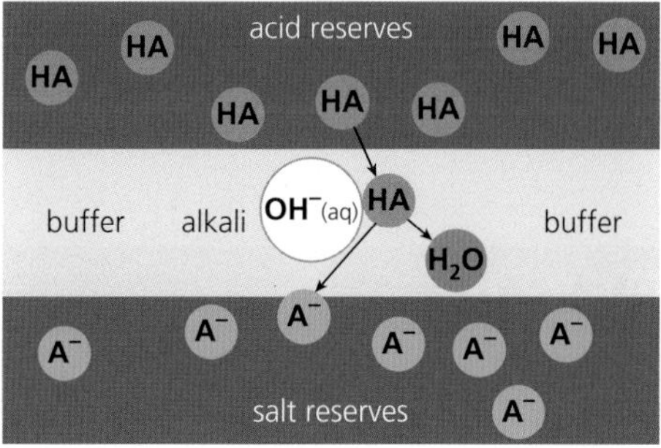

■ **Figure R3.34** The composition of an acidic buffer (HA and A^-) and its reactions upon the addition of acid or alkali

Actions of buffer solutions

If extra hydrogen ions (from an acid) are added to an acidic buffer system, the equilibrium system responds according to Le Châtelier's principle (Figure R3.32). The position of equilibrium shifts and the large reserves of A^- ions (the conjugate base) from the salt allow the H^+ ions to be removed.

If an alkali is added to the buffer, the additional hydroxide ions react with H^+ ions from the weak acid to form water molecules. The equilibrium of the weak acid shifts in the opposite direction (Figure R3.33).

Calculating pH and buffer composition

Acidic buffers

Consider the equilibrium for a weak monoprotic acid, HA, in the presence of a salt, MA:

$$HA(aq) \rightleftharpoons H^+(aq) + A^-(aq)$$

$$K_a = \frac{[H^+(aq)] \times [A^-(aq)]}{[HA(aq)]}$$

Rearranging:

$$[H^+(aq)] = \frac{K_a \times [HA(aq)]}{[A^-(aq)]}$$

Taking negative logarithms to the base 10 of both sides:

$$pH = pK_a - \log_{10}\frac{[HA(aq)]}{[A^-(aq)]}$$

or

$$pH = pK_a + \log_{10}\frac{[A^-(aq)]}{[HA(aq)]}$$

This equation is often called the Henderson–Hasselbalch equation. It indicates that the pH of a buffer solution depends on the K_a of the weak acid and the *ratio* of the concentrations of the acid and its conjugate base – not on their actual concentrations.

The following assumptions are made in deriving the Henderson–Hasselbalch equation:

- In a weak acid like HA, which is only very slightly dissociated, the concentration of HA at equilibrium is approximately the same as the molar concentration of the solution.
- The salt MA completely dissociates. Therefore $[A^-]$ is effectively the concentration supplied by the salt. The cation, M^+, is a spectator ion.
- The ionization of water molecules is ignored in this model of buffer behaviour.

Two important points can be seen from the Henderson–Hasselbalch equation:

- Since $\frac{[\text{acid}]}{[\text{salt}]}$ is a ratio, adding water to a buffer will not affect the ratio, because it will dilute both equally. Therefore, it will not affect $[H^+(aq)]$, which determines the pH.
- If [acid] = [salt] when the buffer is made up, the pH is the same as the pK_a (or $[H^+] = K_a$).

The Henderson–Hasselbalch equation allows calculation of pH of an acid buffer from its composition and acid dissociation constant. It also allows calculation of composition from the other two values.

LINKING QUESTION

Why must buffer solutions be composed of weak acid or base conjugate systems, not of strong acids or bases?

There are two methods of forming a buffer – mixing a weak acid/base with its salt and partially neutralizing a weak acid/base. For example:

- Mixing equal volumes of 1.00 mol dm^{-3} ethanoic acid with 1.00 mol dm^{-3} sodium ethanoate would form an acidic buffer and mixing equal volumes of 1.00 mol dm^{-3} aqueous ammonia solution with 1.00 mol dm^{-3} ammonium chloride would form a basic buffer.
- 100.00 cm^3 of 1.00 mol dm^{-3} phosphoric acid can be partially neutralised by 33.33 cm^3 of 1.00 mol dm^{-3} sodium hydroxide to form a buffer solution containing H_3PO_4 and $H_2PO_4^-$.

Basic buffers

The Henderson–Hasselbalch equation can be readily applied to basic buffers since $pK_w = pK_a + pK_b$.

$$B(aq) + H_2O(l) \rightleftharpoons BH^+(aq) + OH^-(aq)$$

$$K_b = \frac{[BH^+(aq)] \times [OH^-(aq)]}{[B(aq)]}$$

Taking negative logarithms (to the base 10) of both sides of the equation:

$$-\log_{10} K_b = -\log_{10}[OH^-(aq)] - \log_{10}\frac{[BH^+(aq)]}{[B(aq)]}$$

$$pK_b = pOH - \log_{10}\frac{[BH^+(aq)]}{[B(aq)]}$$

$$pOH = pK_b + \log_{10}\frac{[BH^+(aq)]}{[B(aq)]}$$

Calculating the pH of a buffer system

WORKED EXAMPLE R3.1M

Calculate the pH of a buffer containing 0.20 moles of sodium ethanoate in 500 cm^3 of 0.10 mol dm^{-3} ethanoic acid. K_a for ethanoic acid is 1.8×10^{-5}.

(Assume complete dissociation of sodium ethanoate and that the dissociation of ethanoic acid is insignificant, so that the equilibrium concentration of ethanoic acid is the same as the initial concentration.)

Answer

$$[CH_3COO^-(aq)] = 0.20 \times \frac{1000}{500} = 0.40\,\text{mol dm}^{-3}$$

$$\frac{[H^+(aq)] \times [CH_3COO^-(aq)]}{[CH_3COOH(aq)]} = 1.8 \times 10^{-5}$$

$$1.8 \times 10^{-5} = \frac{x \times 0.40}{0.10}$$

$$x = 4.5 \times 10^{-6} = [H^+(aq)]$$

$$pH = -\log_{10}[H^+(aq)] = -\log_{10}(4.5 \times 10^{-6}) = 5.3$$

Calculating the mass of a salt required to give an acidic buffer solution with a specific pH

WORKED EXAMPLE R3.1N

Calculate the mass of sodium propanoate ($M = 96.07\,\text{g mol}^{-1}$) that must be dissolved in $1.00\,\text{dm}^3$ of $1.00\,\text{mol dm}^{-3}$ propanoic acid ($\text{p}K_a = 4.87$) to give a buffer solution with a pH of 4.5.

(Let x represent the concentration of propanoate ions and y represent the amount of sodium propanoate.)

Answer

$[H^+(aq)] = 10^{-\text{pH}} = 1 \times 10^{-4.5} = 3.16 \times 10^{-5}\,\text{mol dm}^{-3}$

$K_a = 1 \times 10^{-4.87} = 1.35 \times 10^{-5}$

$$K_a = \frac{[H^+(aq)][CH_3CH_2COO^-(aq)]}{[CH_3CH_2COOH(aq)]} = 1.35 \times 10^{-5}$$

$$1.35 \times 10^{-5} = \frac{(3.16 \times 10^{-5})x}{1.00}$$

$x = 0.427\,\text{mol dm}^{-3}$

$$0.427\,\text{mol dm}^{-3} = \frac{y}{1.00\,\text{dm}^{-3}}$$

$y = 0.427\,\text{mol}$

$96.07\,\text{g mol}^{-1} \times 0.427\,\text{mol} = 41.0\,\text{g}$ of sodium propanoate is required.

36 Hydrofluoric acid is a weak acid ($K_a = 7.20 \times 10^{-4}$). Calculate the mass of solid sodium fluoride required to be added into a $250\,\text{cm}^3$ solution of $0.00420\,\text{mol dm}^{-3}$ HF to form a buffer solution of pH 5.6.

Buffer regions and titration curves

The relatively flat portions of titration curves where the pH changes most slowly on addition of acid or alkali are the buffer regions.

A titration curve for a weak acid, such as ethanoic acid, allows the $\text{p}K_a$ (4.76) and hence K_a to be calculated graphically (Figure R3.35) since the pH of the half-neutralized acid in the middle of the buffer region corresponds to the $\text{p}K_a$ of the acid.

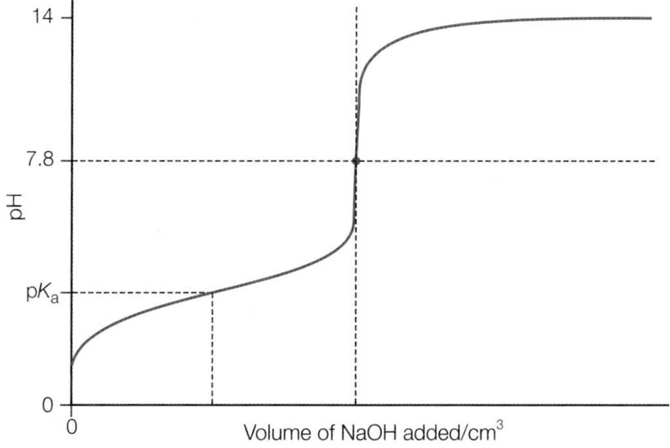

■ **Figure R3.35** The determination of the value of the $\text{p}K_a$ of ethanoic acid (a weak acid), from its titration curve with sodium hydroxide (a strong base)

During the titration of ethanoic acid with sodium hydroxide, the hydroxide ions gradually convert ethanoic acid molecules into ethanoate ions, so halfway to the equivalence point half of the ethanoic acid molecules will have been converted to ethanoate ions.

So specifically for the half-neutralized solution:

$$[CH_3COOH(aq)] = [CH_3COO^-(aq)]$$

However, in general:

$$K_a = [H^+(aq)] \times \frac{[CH_3COO^-(aq)]}{[CH_3COOH(aq)]}$$

Hence, $K_a = [H^+(aq)] \times \frac{1}{1}$ since the two concentrations are equal.

So, $K_a = [H^+(aq)]$.

Taking negative logarithms to the base 10 of both sides:

$$pK_a = pH; \ pK_a = 4.76$$

A similar approach can be used for the titration of a weak base (B) with a strong acid. The pH at the half-neutralized point (before the equivalence point) is equal to the pK_a of the weak acid, BH^+.

To determine the pK_b of the weak base (B), subtract the pK_a from 14, because pOH equals the pK_b:

$$pK_b = 14 - pK_a$$

Buffering capacity

◆ **Buffering capacity**: The amount (mol) of acid or base that must be added to one cubic decimetre of the buffer solution to decrease or increase the pH by one unit.

Buffer solutions have a limited capacity to resist pH changes. If too much strong acid or strong base is added, no more buffering action is possible. The **buffering capacity** is the ability of a buffer to resist changes in pH.

The buffering capacity increases as the concentration of the buffer salt / acid solution increases. The closer the buffered pH is to the pK_a, the greater the buffering capacity.

For example, consider the addition of a strong acid, such as HCl, to a buffer composed of ethanoic acid and sodium ethanoate.

- Initially, the HCl donates its proton to the conjugate base, CH_3COO^-, through the reaction:

  $$CH_3COO^- + HCl \rightarrow CH_3COOH + Cl^-$$

  This changes the pH by lowering the ratio $[CH_3COO^-]/[CH_3COOH]$. As long as there is still a lot of CH_3COO^- present, the change in pH will be relatively small.

- If we keep adding HCl, the conjugate base CH_3COO^- will eventually run out. Once the CH_3COO^- is gone, any additional HCl will donate its proton to water:

  $$HCl + H_2O \rightarrow H_3O^+ + Cl^-$$

  $[H_3O^+]$ dramatically increases and so the pH decreases significantly.

This is termed 'breaking the buffer solution', and we call the amount of acid a buffer can absorb before it breaks the 'buffering capacity for addition of strong acid'. A solution of this buffer with more conjugate base (higher $[CH_3COO^-]$) has a higher buffering capacity for addition of strong acid.

Similarly, a buffer will 'break' when the amount of strong base added is so large it consumes all the weak acid through the reaction:

$$CH_3COOH + OH^- \rightarrow CH_3COO^- + H_2O$$

Common mistake

Misconception: *Buffer solutions maintain pH.*

Buffer solutions have a finite capacity to resist pH changes. In an acidic buffer solution, it is the amount (in moles) of weak acid and the amount (in moles) of its conjugate base that determine the extent to which a buffer can neutralize added acid or base.

A solution with more weak acid (higher $[CH_3COOH]$) has a higher buffer capacity for addition of a strong base.

Although the pH of this buffer is determined only by the ratio $[CH_3COO^-]/[CH_3COOH]$, the ability of the buffer to absorb strong acid or base is determined by the individual concentrations of $[CH_3COO^-]$ and $[CH_3COOH]$.

WORKED EXAMPLE R3.10

Propanedioic acid, $HOOCCH_2COOH$, is a weak diprotic acid which undergoes step-wise ionization in water:

$$HOOCCH_2COOH(aq) \rightleftharpoons HOOCCH_2COO^-(aq) + H^+(aq)$$

$$HOOCCH_2COO^-(aq) \rightleftharpoons {}^-OOCCH_2COO^-(aq) + H^+(aq)$$

The following pH–volume curve is obtained when propanedioic acid is added to NaOH(aq).

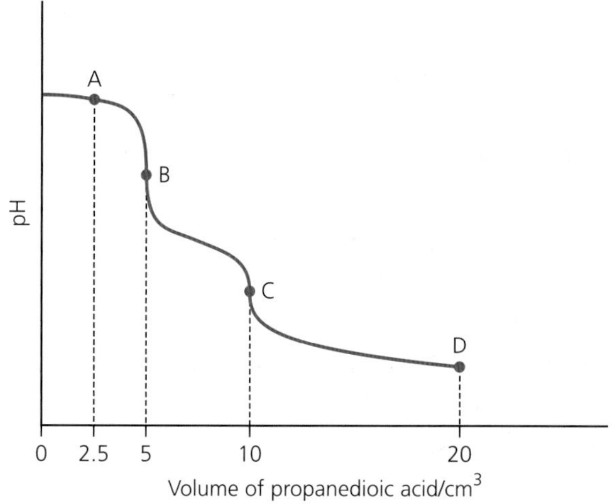

At which point on the titration curve is the mixture most able to resist pH change upon addition of a small amount of aqueous acid or base?

Answer

A NaOH will be present in excess. No buffer present.

B NaOH still present in excess.

C First equivalence point. $HOOCHCHCOO^-$ and water present.

D $HOOCHCHCOO^-$ (weak acid) and ${}^-OOCHCHCOO^-$ (salt) are present, so maximum buffering capacity.

HL ONLY

Tool 2: Technology

Represent data in a graphical form

Excel is spreadsheet software that allows the user to perform mathematical and statistical analyses on numerical data. In addition, *Excel* can also plot the data (Figure R3.36) in a variety of formats once it has been entered into the cells of the spreadsheet.

Go to the menu bar, choose 'Insert' and select 'Chart'. The Chart Wizard box appears on the screen, giving you a choice of graphs you can plot. In many cases, the most appropriate option is 'XY (Scatter)'. Click on 'Next'.

Excel also allows you to add a line or curve of best fit (a trend line) to a scatter graph and display its equation and the value of R^2.

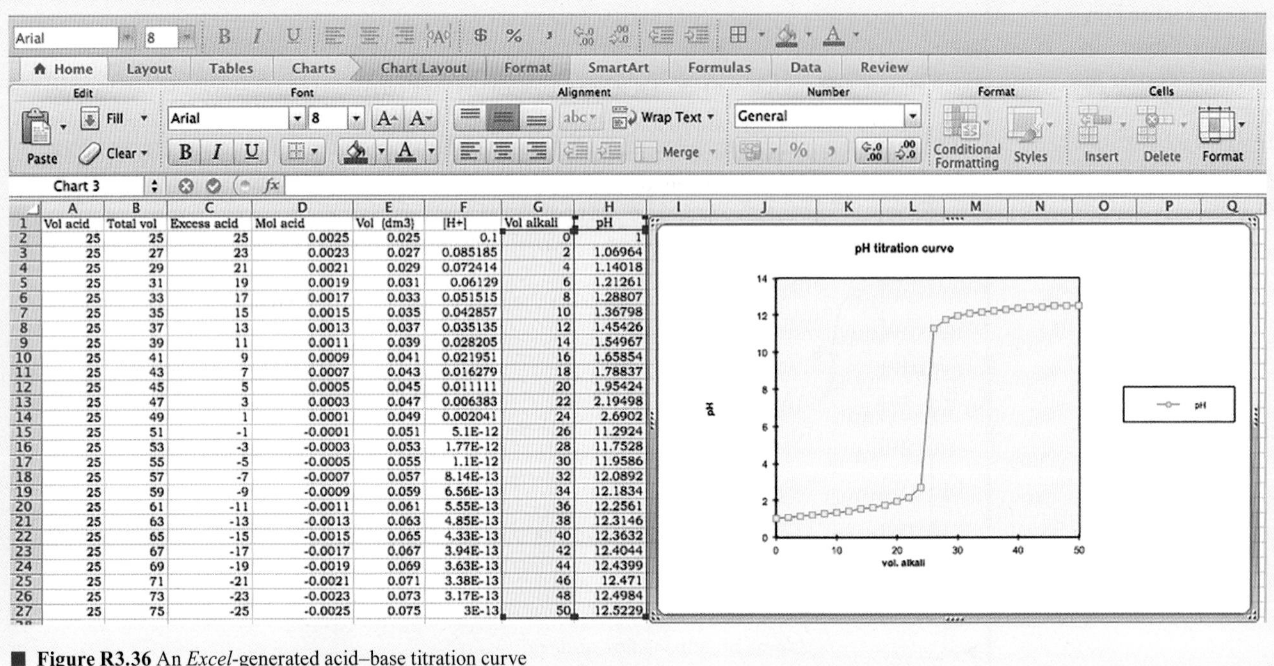

	A	B	C	D	E	F	G	H
1	Vol acid	Total vol	Excess acid	Mol acid	Vol (dm3)	[H+]	Vol alkali	pH
2	25	25	25	0.0025	0.025	0.1	0	1
3	25	27	23	0.0023	0.027	0.085185	2	1.06964
4	25	29	21	0.0021	0.029	0.072414	4	1.14018
5	25	31	19	0.0019	0.031	0.06129	6	1.21261
6	25	33	17	0.0017	0.033	0.051515	8	1.28807
7	25	35	15	0.0015	0.035	0.042857	10	1.36798
8	25	37	13	0.0013	0.037	0.035135	12	1.45426
9	25	39	11	0.0011	0.039	0.028205	14	1.54967
10	25	41	9	0.0009	0.041	0.021951	16	1.65854
11	25	43	7	0.0007	0.043	0.016279	18	1.78837
12	25	45	5	0.0005	0.045	0.011111	20	1.95424
13	25	47	3	0.0003	0.047	0.006383	22	2.19498
14	25	49	1	0.0001	0.049	0.002041	24	2.6902
15	25	51	-1	-0.0001	0.051	5.1E-12	26	11.2924
16	25	53	-3	-0.0003	0.053	1.77E-12	28	11.7528
17	25	55	-5	-0.0005	0.055	1.1E-12	30	11.9586
18	25	57	-7	-0.0007	0.057	8.14E-13	32	12.0892
19	25	59	-9	-0.0009	0.059	6.56E-13	34	12.1834
20	25	61	-11	-0.0011	0.061	5.55E-13	36	12.2561
21	25	63	-13	-0.0013	0.063	4.85E-13	38	12.3146
22	25	65	-15	-0.0015	0.065	4.33E-13	40	12.3632
23	25	67	-17	-0.0017	0.067	3.94E-13	42	12.4044
24	25	69	-19	-0.0019	0.069	3.63E-13	44	12.4399
25	25	71	-21	-0.0021	0.071	3.38E-13	46	12.471
26	25	73	-23	-0.0023	0.073	3.17E-13	48	12.4984
27	25	75	-25	-0.0025	0.075	3E-13	50	12.5229

■ **Figure R3.36** An *Excel*-generated acid–base titration curve

LINKING QUESTION

How does Le Châtelier's principle enable us to interpret the behaviour of indicators and buffer solutions?

Electron transfer reactions

Guiding question

- What happens when electrons are transferred?

SYLLABUS CONTENT

By the end of this chapter, you should understand that:

- ▶ oxidation and reduction can be described in terms of electron transfer, change in oxidation state, oxygen gain/loss or hydrogen loss/gain
- ▶ half-equations separate the processes of oxidation and reduction, showing the loss or gain of electrons
- ▶ the relative ease of oxidation and reduction of an element in a group can be predicted from its position in the periodic table
- ▶ the reactions between metals and aqueous metal ions demonstrate the relative ease of oxidation of different metals
- ▶ acids react with reactive metals to release hydrogen
- ▶ oxidation occurs at the anode and reduction occurs at the cathode in electrochemical cells
- ▶ a primary (voltaic cell) is an electrochemical cell that converts energy from spontaneous redox reactions to electrical energy
- ▶ secondary (rechargeable) cells involve redox reactions that can be reversed using electrical energy
- ▶ an electrolytic cell is an electrochemical cell that converts electrical energy to chemical energy by bringing about non-spontaneous reactions
- ▶ functional groups in organic compounds may undergo oxidation
- ▶ functional groups in organic compounds may undergo reduction
- ▶ reduction of unsaturated compounds by the addition of hydrogen lowers the degree of unsaturation
- ▶ the hydrogen half-cell $H_2(g) \rightleftharpoons 2H^+(aq) + 2e^-$ is assigned a standard electrode potential of zero by convention. It is used in the measurement of standard electrode potential, $E^\ominus$ (HL only)
- ▶ standard cell potential, $E^\ominus_{cell}$, can be calculated from standard electrode potentials. $E^\ominus_{cell}$ has a positive value for a spontaneous reaction (HL only)
- ▶ the equation $\Delta G^\ominus = -nFE^\ominus_{cell}$ shows the relationship between standard change in Gibbs energy and standard electrode potential for a reaction (HL only)
- ▶ during electrolysis of aqueous solutions, competing reactions can occur at the anode and cathode, including the oxidation and reduction of water (HL only)
- ▶ electroplating involves the electrolytic coating of an object with a thin metallic layer (HL only).

By the end of this chapter, you should know how to:

- ▶ deduce oxidation states of an atom in a compound or ion
- ▶ identify the oxidized and reduced species and the oxidizing and reducing agents in a chemical reaction
- ▶ deduce redox half-equations and equations in acidic or neutral solutions
- ▶ predict the relative ease of oxidation of metals
- ▶ predict the relative ease of reduction of halogens
- ▶ interpret data regarding metal and metal ion reactions
- ▶ deduce equations for reactions of reactive metals with dilute HCl and H_2SO_4
- ▶ identify electrodes as anode and cathode, and identify their signs / polarities in voltaic cells and electrolytic cells, based on the type of reaction occurring at the electrode
- ▶ explain the direction of electron flow from anode to cathode in the external circuit, and ion movement across the salt bridge
- ▶ deduce the reactions of the charging process from given electrode reactions for discharge, and vice versa
- ▶ discuss the advantages and disadvantages of fuel cells, primary cells and secondary cells
- ▶ explain how current is conducted in an electrolytic cell
- ▶ deduce the products of the electrolysis of a molten salt

- deduce equations to show changes in the functional groups during oxidation of primary and secondary alcohols, including the two-step reaction in the oxidation of primary alcohols
- deduce equations to show reduction of carboxylic acids to primary alcohols via the aldehyde, and reduction of ketones to secondary alcohols
- deduce the products of the reactions of hydrogen with alkenes and alkynes
- interpret standard electrode potential data in terms of ease of oxidation / reduction (HL only)
- predict whether a reaction is spontaneous in the forward or reverse direction from $E^\ominus$ data (HL only)
- determine the value for $\Delta G^\ominus$ from $E^\ominus$ data (HL only)
- deduce from standard electrode potentials the products of the electrolysis of aqueous solutions (HL only)
- deduce equations for electrode reactions during electroplating. (HL only)

Oxidation and reduction

◆ **Oxidation:** (simple definition) The gain of oxygen or loss of hydrogen.

◆ **Reduction:** (simple definition) The loss of oxygen or gain of hydrogen.

These reactions are described as redox reactions since they involve both oxidation and reduction.

Oxidation may involve gain of oxygen by an element. For example, when magnesium reacts with oxygen to form magnesium oxide, the magnesium gains oxygen and is oxidized:

$$2Mg(s) + O_2(g) \rightarrow 2MgO(s)$$

Oxidation may involve the loss of hydrogen. For example, when manganese(IV) oxide reacts with hydrochloric acid, the hydrogen chloride loses hydrogen and is oxidized:

$$MnO_2(s) + 4HCl(aq) \rightarrow MnCl_2(aq) + 2H_2O(l) + Cl_2(g)$$

Reduction may involve loss of oxygen. Copper(II) oxide reacts with hydrogen to form copper and water:

$$CuO(s) + H_2(g) \rightarrow Cu(s) + H_2O(l)$$

Copper(II) oxide has lost oxygen and is reduced. The hydrogen gains oxygen to form water and is oxidized. Reduction and oxidation have occurred together in a redox reaction.

Reduction may involve the addition of hydrogen to a compound. For example, ethyne reacts with hydrogen to form ethene. The ethyne gains hydrogen and is reduced:

$$C_2H_2(g) + H_2(g) \rightarrow C_2H_4(g)$$

Nature of science: Observations

Redox processes

Oxidation and reduction processes have been in use for about 7000 years in the extraction of metals from their ores (usually metal oxides or metal sulfides). One of the first 'textbooks' about the extraction of metals, *De re metallica*, was written (in Latin) by the German scientist Georgius Agricola (1494–1555) and published in 1556, after his death. As long ago as 1741, the word '*reduce*' was used to describe the process of extracting metals from their ores. '*Reduction*' referred to the observation that the mass decreased when the ore was converted into its metal by heating with coke (impure carbon).

In the eighteenth century, the French chemist Antoine Lavoisier (1743–1794) recognized that these ores were metal oxides. By the beginning of the nineteenth century, the idea that reactive metals were easily oxidized had become established. These observations formed the basis of the original historical definitions of oxidation as the addition of oxygen and reduction as the removal of oxygen.

Oxidation
I s
L oss

Reduction
I s
Gain

■ **Figure R3.37** OILRIG mnemonic for redox reactions and electron transfer

◆ **Half-equation:** An equation that shows one half of a redox reaction. Half-equations show either oxidation or reduction in terms of electron transfer.

When electrolysis was discovered and modern theories of atomic structure and chemical bonding were developed, the terms oxidation and reduction were redefined in terms of electrons.

Oxidation was defined as the loss of electrons from a substance; reduction was defined as the gain of electrons (Figure R3.37).

Note that these definitions include all of the oxidation and reduction reactions previously defined in terms of loss and gain of oxygen and hydrogen. For example:

$$2Mg(s) + O_2(g) \rightarrow 2MgO(s)$$

can be rewritten to show the loss and gain of electrons that occur during this reaction. The magnesium atom loses two electrons to form a magnesium ion, Mg^{2+}; each oxygen atom gains two electrons to become an oxide ion, O^{2-}.

These two processes can be described by the following **half-equations**:

Oxidation: $Mg \rightarrow Mg^{2+} + 2e^-$

Reduction: $O_2 + 4e^- \rightarrow 2O^{2-}$

Whenever there is a reduction in which one substance gains electrons, there must be an oxidation where another substance loses electrons. Such processes are called redox (reduction–oxidation) reactions. The two equations are known as half-equations since they only describe one of the two reactions that must occur together.

The net ionic equation to describe the redox reaction is obtained by adding the two half-equations together and cancelling the electrons that appear on both sides of the equation:

Half-equations: $Mg \rightarrow Mg^{2+} + 2e^-$

$O_2 + 4e^- \rightarrow 2O^{2-}$

Top tip!

These definitions of oxidation and reduction, as electron loss and electron gain, respectively, also include many examples of redox reactions which do not involve oxygen or hydrogen.

The first half-equation has to be multiplied through by 2 so that the number of electrons is the same as in the second half-equation:

$2Mg \rightarrow 2Mg^{2+} + 4e^-$

Sum of the two half-equations: $2Mg + O_2 + 4e^- \rightarrow 2Mg^{2+} + 4e^- + 2O^{2-}$

Cancelling of electrons: $2MgO + O_2 \rightarrow 2MgO[Mg^{2+}O^{2-}]$

Going further

Redox reactions in biology

Oxidation occurs when hydrogen is removed from a molecule:

$$AH_2 + B \rightarrow A + BH_2$$

Substance A has been oxidized by transferring hydrogen to a second substance, B, which acts as a hydrogen carrier. Oxidation steps like this are important to respiration because they allow hydrogen atoms to be removed from the glucose molecule, at the same time releasing energy in a useful form.

The most important hydrogen carrier in cells is nicotinamide adenine dinucleotide (NAD). NAD is usually present in solution in the form of NAD^+, and the reduced form NADH is produced according to the following equation:

$$NAD^+ + H^+ + 2e^- \rightarrow NADH$$

NAD passes on the hydrogen atoms to a system of carrier molecules located in the mitochondria. At the end of the electron transport chain, hydrogen ions and electrons react with oxygen molecules to form water:

$$4H^+ + 4e^- + O_2 \rightarrow 2H_2O$$

The electron transfer approach to oxidation (electron loss) and reduction (electron gain) is useful, but it does have some limitations. For example, consider the combustion of sulfur in air or oxygen to form sulfur dioxide:

$$S(s) + O_2(g) \rightarrow SO_2(g)$$

According to the historical definition of oxidation, the sulfur has been oxidized. However, none of the species involved is ionic, so there is no obvious transfer of electrons.

Link

Oxidation states are covered in more detail in Chapter S3.1, page 247.

◆ **Oxidation:** (complete definition) An increase in oxidation state or the loss of electrons.

◆ **Reduction:** (complete definition) A decrease in oxidation state or the gain of electrons.

The concept of an oxidation state allows chemists to avoid the problems associated with using two separate and sometimes conflicting definitions for oxidation and reduction.

Chemists use oxidation states to keep track of the electrons transferred or shared during chemical changes. With the help of oxidation states it becomes much easier to recognize redox reactions. Oxidation is now defined as an increase in oxidation state and reduction is defined as a decrease in oxidation state.

In Figure R3.38, movement up the diagram involves **oxidation**: the loss of electrons and/or a change to a more positive oxidation state. Movement down the diagram involves **reduction**: the gain of electrons and a shift to a less positive, or more negative, oxidation state. Usually, the higher the oxidation state of the element, the stronger its oxidizing power (as measured by its standard electrode potential).

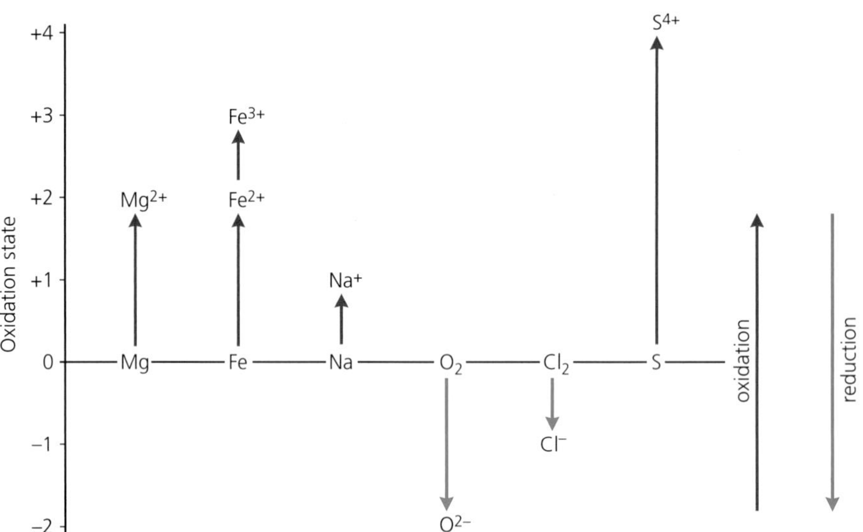

■ **Figure R3.38** Oxidation states of atoms and ions

LINKING QUESTION

What are the advantages and limitations of using oxidation states to track redox changes?

Chemists base the names of some inorganic compounds on oxidation states; for example, sulfur dioxide can be named sulfur(IV) oxide $[S^{4+}2O^{2-}]$ to show that sulfur has an oxidation state of +4.

37 Reduction is now defined in terms of change in oxidation state. Discuss how an earlier definition of oxidation and reduction may have led to conflicting answers for the conversion of aluminium hydroxide, $Al(OH)_3$, to aluminium, Al.

Common mistake

Misconception: *An element in a redox reaction cannot be both oxidized and reduced.*

In disproportionation reactions, a single element undergoes both oxidation and reduction. These reactions involve elements that can have more than one oxidation state.

◆ **Comproportionation**: A reaction in which an element in a higher oxidation state reacts with the same element in a lower oxidation state to give the element in an intermediate oxidation state.

◆ **Disproportionation**: A type of chemical reaction in which the same compound is simultaneously reduced and oxidized.

◆ **Reducing agent**: A substance that brings about reduction in other substances; it achieves this by being oxidized itself.

Going further

Comproportionation and disproportionation

If an element has three or more oxidation states, then it *may* be able to act as its own oxidizing agent and reducing agent. If a compound that contains the element in a high oxidation state is reacted with a compound that contains the element in a low oxidation state, then the intermediate oxidation state is formed. For example, if vanadium(V) is mixed with vanadium(III), the result is a compound containing vanadium(IV). The vanadium ions (III and V) have undergone simultaneous oxidation and reduction in a type of redox reaction known as **comproportionation** (the reverse of disproportionation).

$$\mathbf{VO_2^+}(aq) + \mathbf{V^{3+}}(aq) \rightarrow \mathbf{2VO^{2+}}(aq)$$

$+5 \quad\quad +3 \quad\quad +4$

Occasionally, the intermediate oxidation state is a more powerful oxidizing agent than the higher oxidation state. This is because the intermediate state has a structure that makes it unstable relative to the higher and lower oxidation states. The intermediate state then undergoes **disproportionation** and forms substances in the higher and lower oxidation states.

An example is the decomposition of hydrogen peroxide, H_2O_2, which contains oxygen in an oxidation state of −1 due to the presence of a weak –O–O– bond. This spontaneously breaks down to molecular oxygen (oxidation state zero) and water, which contains oxygen in an oxidation state of −2.

$$2H_2O_2(l) \rightarrow O_2(g) + 2H_2O(l)$$

Reducing and oxidizing agents

A **reducing agent** is a substance that brings about the reduction of another substance by undergoing oxidation (Figure R3.39).

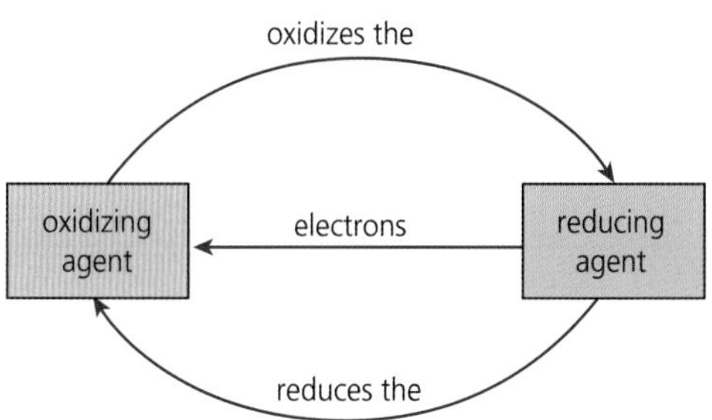

■ **Figure R3.39** The relationship between oxidizing and reducing agents

Some common reducing agents are:

Iodide ions: for example, with concentrated sulfuric acid:

$$H_2SO_4(aq) + 8H^+(aq) + 8I^-(aq) \rightarrow 4I_2(g) + H_2S(g) + 4H_2O(l)$$

Hydrogen sulfide: for example with oxygen:

$$2H_2S(g) + 3O_2(g) \rightarrow 2SO_2(g) + 2H_2O(l)$$

Hydrogen: for example, with lead(II) oxide:

$$PbO(s) + H_2(g) \rightarrow H_2O(l) + Pb(s)$$

Carbon: for example, with lead(II) oxide:

$$PbO(s) + C(s) \rightarrow Pb(s) + CO(g)$$

Top tip!

The strengths of reducing and oxidizing agents in aqueous solution are described by their standard electrode potentials (expressed as a voltage). Strong reducing agents have negative values and strong oxidizing agents have positive values.

■ **Figure R3.40** Zinc displacing copper from copper(II) sulfate solution. The copper appears as a reddish solid. Colourless zinc ions replace the copper(II) ions in solution

Top tip!

Oxygen is a very weak oxidizing agent for species in aqueous solution.

◆ **Displacement reaction:** A redox reaction in which a more reactive element displaces a less reactive element from a solution of its ions or salt, often in aqueous solution.

◆ **Oxidizing agent:** A substance that brings about oxidation in other substances; it achieves this by being reduced itself.

Carbon monoxide: for example, with iron(III) oxide:

$$Fe_2O_3(s) + 3CO(g) \rightarrow 2Fe(l) + 3CO_2(g)$$

Metals: a more reactive metal can reduce the ions of a less reactive metal. For example, copper atoms act as a reducing agent with aqueous silver nitrate solution:

$$2AgNO_3(aq) + Cu(s) \rightarrow Cu(NO_3)_2(aq) + 2Ag(s)$$

In terms of (non-spectator) ions, this can be written as:

$$2Ag^+(aq) + Cu(s) \rightarrow Cu^{2+}(aq) + 2Ag(s)$$

after removing the 'spectator' nitrate ions.

This and similar reactions (see Figure R3.40) involving metals and metal ions are known as **displacement reactions**.

An **oxidizing agent** is a substance that brings about the oxidation of other substances by undergoing reduction.

Some common oxidizing agents are:

Oxygen: oxygen gas will oxidize many elements when they are heated in a stream of air or oxygen. For example, with barium and sulfur:

$$2Ba(s) + O_2(g) \rightarrow 2BaO(s)$$

$$S(s) + O_2(g) \rightarrow SO_2(g)$$

Halogens: the halogens are strong oxidizing agents both in aqueous solution and in elemental form. For example:

$$2FeCl_2(s) + Cl_2(g) \rightarrow 2FeCl_3(s)$$

$$2Fe^{2+}(aq) + Cl_2(aq) \rightarrow 2Fe^{3+}(aq) + 2Cl^-(aq)$$

The oxidizing strength increases up the group from iodine to fluorine.

Iron(III) ions: the iron(III) ion is a mild oxidizing agent and is reduced to the iron(II) ion.

Potassium dichromate(VI): dichromate(VI) ions are a powerful oxidizing agent, usually used in acidic solution. The orange dichromate(VI) ion is reduced to the blue-green chromium(III) ion. For example:

$$K_2Cr_2O_7(aq) + 7H_2SO_4(aq) + 3Zn(s) \rightarrow Cr_2(SO_4)_3(aq) + 3ZnSO_4(aq) + K_2SO_4(aq) + 7H_2O(l)$$

$$Cr_2O_7^{2-}(aq) + 14H^+(aq) + 3Zn(s) \rightarrow 2Cr^{3+}(aq) + 3Zn^{2+}(aq) + 7H_2O(l)$$

Potassium manganate(VII): manganate(VII) ions are an even more powerful oxidizing agent, usually used in acidic solution. The purple manganate(VII) ion is reduced to the very pale pink manganese(II) ion. For example:

$$5NaNO_2(aq) + 2KMnO_4(aq) + 3H_2SO_4(aq) \rightarrow 5NaNO_3(aq) + 2MnSO_4(aq) + K_2SO_4(aq) + 3H_2O(l)$$

$$5NO_2^-(aq) + 2MnO_4^-(aq) + 6H^+(aq) \rightarrow 5NO_3^-(aq) + 2Mn^{2+}(aq) + 3H_2O(l)$$

Hydrogen peroxide: hydrogen peroxide is a moderately strong oxidizing agent that is reduced to water. For example:

$$H_2O_2(aq) + 2H^+(aq) + 2e^- \rightarrow 2H_2O(l)$$

Identifying oxidizing and reducing agents

The equations below describe some typical redox reactions of the elements in Period 3.

Reducing agents are found on the left-hand side of the periodic table and oxidizing agents are found on the right-hand side. Oxidizing strength generally increases across a period left to right and up a group (e.g., halogens) and reducing properties generally decrease across a period and down a group (e.g., alkali metals).

Oxidation is a reaction that results in the loss of electrons. Ease of oxidation follows the same trends as ionization energy. That is because the smaller the ionization energy, the less energy is required to remove an electron, thus ease of oxidation increases along the period.

Reduction is a reaction that results in the gaining of an electron. Ease of reduction follows the same trend as the electron affinity. That is because the larger the negative electron affinity, the more energetically favourable it is to gain an electron and thus it decreases along the period.

Sodium

$$2\underline{Na}(s) + 2\underline{H_2}O(l) \rightarrow 2\underline{Na}OH(aq) + \underline{H_2}(g)$$

0 +1 +1 0

Sodium is oxidized; hydrogen in water is reduced. Sodium is the reducing agent; water is the oxidizing agent.

Magnesium

$$2\underline{Mg}(s) + \underline{C}O_2(g) \rightarrow 2\underline{Mg}O(s) + \underline{C}(s)$$

0 +4 +2 0

Magnesium is oxidized; carbon in carbon dioxide is reduced. Magnesium is the reducing agent; carbon dioxide is the oxidizing agent.

Phosphorus

$$4\underline{P}(s) + 8H_2\underline{S}O_4(aq) \rightarrow 4H_3\underline{P}O_4(aq) + \underline{S}(s) + 7\underline{S}O_2(g) + 2H_2O(l)$$

0 +6 +5 0 +4

Phosphorus is oxidized; sulfur in the sulfuric acid is reduced to sulfur and sulfur dioxide. Phosphorus is the reducing agent; sulfuric acid is the oxidizing agent.

$$\underline{P}(s) + 3\underline{Na}(s) \rightarrow \underline{Na_3P}(s)$$

0 0 +1 −3

Phosphorus is reduced; sodium is oxidized. Phosphorus is the oxidizing agent and sodium is the reducing agent.

Chlorine

$$8\underline{N}H_3(aq) + 3\underline{Cl_2}(g) \rightarrow \underline{N_2}(g) + 6NH_4\underline{Cl}(aq)$$

−3 0 0 −1

Chlorine is reduced and some of the nitrogen in ammonia is oxidized. Chlorine is the oxidizing agent and ammonia is the reducing agent.

Chlorine (continued)

$$\underline{Cl}_2(g) + \underline{F}_2(g) \rightarrow 2\underline{ClF}(g)$$

$$0 \quad\quad 0 \quad\quad\quad +1\ -1$$

Chlorine is oxidized; fluorine is reduced. Chlorine is the reducing agent and fluorine is the oxidizing agent.

Link

Trends in atomic properties across periods are also covered in Chapter S3.1, page 235.

Trends in the redox properties of the elements

The redox properties of the elements in Period 3 are summarized in Table R3.13. There is a trend from strong reducing agents on the left to strong oxidizing agents on the right.

Table R3.13 Summary of redox properties of the elements in Period 3

Property / Element	Sodium, Na	Magnesium, Mg	Aluminium, Al	Silicon, Si	Phosphorus, P	Sulfur, S	Chlorine, Cl
Oxidation states (with the exception of zero)	+1 only	+2 only	+3 only	+4 (–4 rarely)	+5, +3, –3	+6, +4, +2, –2 [and more]	+7, +5, +3, +1, –1
Examples of compounds in these oxidation states	NaBr	$MgSO_4$	Al_2O_3	SiO_2, SiH_4	PCl_5 PCl_3 PH_3	SO_3 SO_2 SCl_2 H_2S	$HClO_4$ $NaClO_3$ $NaClO_2$ NaClO HCl
Redox properties (all reactions of elements are redox)	Strong reducing agent; chemistry is summarized by $Na \rightarrow Na^+ + e^-$	Strong reducing agent; chemistry is summarized by $Mg \rightarrow Mg^{2+} + 2e^-$	Strong reducing agent; chemistry is summarized by $Al \rightarrow Al^{3+} + 3e^-$	Usually a reducing agent except SiO_2	Usually a reducing agent except PCl_5	A reducing agent, but can be an oxidizing agent with hydrogen and reactive metals	An oxidizing agent, especially in solution. Can be a reducing agent with fluorine and water
Standard electrode potential, $E^\ominus$/ V	–2.71	–2.37	–1.66				+1.36
Electronegativity of element	0.9	1.2	1.5	1.8	2.1	2.5	3.0
Type of element	metal	metal	metal	metalloid	non-metal	non-metal	non-metal

Top tip!

Whether a compound is an oxidizing or reducing agent depends on the starting oxidation state and what the other reagent is.

There are also clear trends in Groups 1 and 2, with reducing strength increasing down the groups, and Group 17, with oxidizing strength of the halogens increasing up the group.

These trends usually correlate with a change in electronegativity or ionization energy.

The standard electrode potential is a measurement of the reducing or oxidizing strength of an element in aqueous solution under standard conditions. The more negative the value, the greater the reducing power of the element and the less energy is needed for it to donate electrons. The more positive the value, the greater the oxidizing power of the element and the less energy is needed for it to accept electrons.

LINKING QUESTION

Why does metal reactivity increase, and non-metal reactivity decrease, down the main groups of the periodic table?

Going further

Substances that can behave as both oxidizing and reducing agents

The terms oxidizing agent and reducing agent, like the terms acid and base, are relative terms. A weak reducing agent may be 'forced' to act as an oxidizing agent in the presence of a more powerful reducing agent. Conversely, a weak oxidizing agent may be forced to act as a reducing agent in the presence of a more powerful oxidizing agent.

For example, with acidified aqueous potassium iodide, hydrogen peroxide acts as an oxidizing agent and converts iodide ions to iodine:

$$H_2O_2(aq) + 2H^+(aq) + 2I^-(aq) \rightarrow 2H_2O(l) + I_2(aq)$$

The hydrogen peroxide is reduced to water during the reaction:

$$H_2O_2(aq) + 2H^+(aq) + 2e^- \rightarrow 2H_2O(l)$$

However, in the presence of acidified potassium manganate(VII), a stronger oxidizing agent than hydrogen peroxide, hydrogen peroxide instead acts as a reducing agent:

$$5H_2O_2(aq) + 2MnO_4^-(aq) + 6H^+(aq) \rightarrow 5O_2(g) + 2Mn^{2+}(aq) + 8H_2O(l)$$

The hydrogen peroxide is oxidized to oxygen:

$$H_2O_2(aq) \rightarrow O_2(g) + 2H^+(aq) + 2e^-$$

Substances such as hydrogen peroxide, that are able to act as both oxidizing and reducing agents, can be converted to stable compounds that have higher and lower oxidation states, that is, they can undergo by acting simultaneously as both an oxidizing agent and reducing agent.

Clean drinking water

In 2010, 122 countries formally acknowledged the human right to a clean water supply as part of an adequate standard of living in a United Nations General Assembly resolution. Although recognized in international law, the right to water is not enforceable until it is incorporated into national legislation. Today, it is estimated that one billion people lack clean water.

Disinfection of water supplies commonly uses oxidizing agents such as chlorine or ozone to kill bacteria and viruses. In water, chlorine forms chlorate(I) ions, ClO^-; these ions have germicidal properties. However, there are concerns about the use of chlorine as a disinfectant, because it can also oxidize other species. This can lead to the formation of harmful by-products such as trichloromethane, $CHCl_3$.

Redox titrations

◆ **Redox titration:** A titration used to determine the concentration of a solution of an oxidizing agent or of a reducing agent or to deduce the ionic equation of a redox reaction.

A **redox titration** is a type of titration based on a redox reaction, which involves the transfer of one or more electrons from a reducing agent to an oxidizing agent. They often involve reacting an analyte with a measured volume of titrant solution of known concentration and finding the equivalence point when the titrant and analyte have reacted stoichiometrically (a reaction where all the reactants are consumed and none remain after the chemical reaction has finished), as shown by a colour change. This may involve a redox indicator.

The overall stoichiometric equation for a redox reaction can be obtained by combining the two half-equations, so that the number of electrons lost by the reducing agent equals the number of electrons gained by the oxidizing agent.

Redox titrations are used for two main reasons:

- to find the concentration of a solution
- to determine the stoichiometry of a redox reaction and hence to suggest a likely equation for the reaction.

An example is provided by the reaction of an oxidizing agent with excess potassium iodide solution to form iodine. The iodine is then titrated with sodium thiosulfate solution, using starch as an indicator (Figure R3.41). This type of redox titration is known as iodometry.

The overall equation for the reaction is:

$$2S_2O_3^{2-}(aq) + I_2(aq) \rightarrow S_4O_6^{2-}(aq) + 2I^-(aq)$$

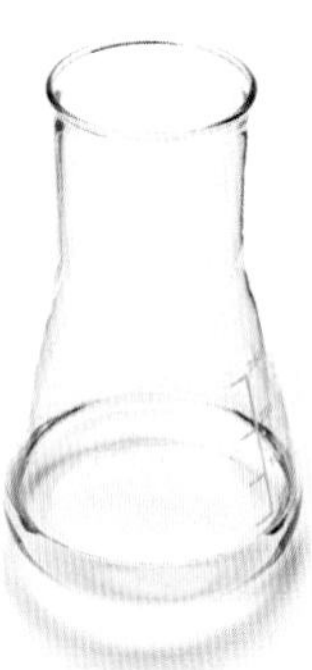

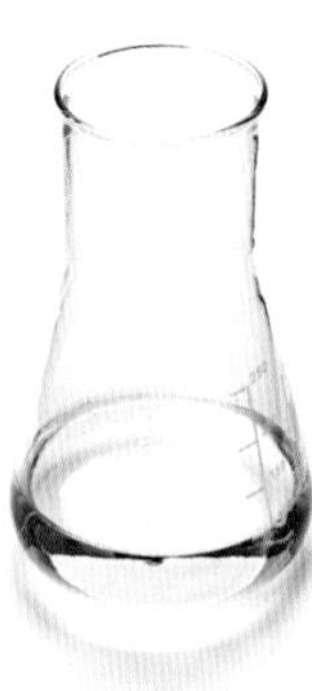

■ **Figure R3.41** The end-point of an iodine titration is made much sharper by adding starch. The flask on the left contains a high concentration of brown-coloured iodine. A few drops of starch solution are added when the solution is pale yellow (and the iodine concentration is low); the end-point occurs when the intense blue colour of the starch–iodine complex just disappears to form a colourless solution (flask on the right)

Common reagents for redox titrations are described in Table R3.14.

■ **Table R3.14** Important half-equations for redox reactions

Reduction reaction (for oxidizing agent in bold)	Comment	Oxidation reaction (for reducing agent in bold)	Comment
$\mathbf{MnO_4^-} + 8H^+ + 5e^- \rightarrow Mn^{2+} + 4H_2O$	purple to colourless in acidic conditions (not a primary standard); can only be acidified with dilute H_2SO_4 or H_3PO_4	$\mathbf{C_2O_4^{2-}} \rightarrow 2CO_2 + 2e^-$	This reaction is carried out at 80 °C since the reaction is relatively slow at room temperature. Ethanedioic acid and its salts (ethanedioates) are primary standards.
$\mathbf{Cr_2O_7^{2-}} + 14H^+ + 6e^- \rightarrow 2Cr^{3+} + 7H_2O$	orange to blue-green (primary standard)	$\mathbf{2S_2O_3^{2-}} \rightarrow S_4O_6^{2-} + 2e^-$	
$\mathbf{I_2} + 2e^- \rightarrow 2I^-$	yellow-brown to colourless	$\mathbf{2I^-} \rightarrow I_2 + 2e^-$	colourless to yellow-brown
$\mathbf{Fe^{3+}} + e^- \rightarrow Fe^{2+}$	brown to pale green	$\mathbf{Fe^{2+}} \rightarrow Fe^{3+} + e^-$	pale green to brown (iron(II) salts are not primary standards)
$\mathbf{H_2O_2} + 2H^+ + 2e^- \rightarrow 2H_2O$	can oxidize SO_3^{2-} to SO_4^{2-} and NO_2^- to NO_3^- and Fe^{2+} to Fe^{3+}	$\mathbf{H_2O_2} \rightarrow O_2 + 2H^+ + 2e^-$	

The oxidizing strengths of manganate(VII) ions, dichromate(VI) ions and hydrogen peroxide vary with pH.

Tool 1: Experimental techniques

Redox titrations

Redox titrations are carried out using potassium manganate(VII), $KMnO_4$, or potassium dichromate(VI), $K_2Cr_2O_7$. Both these reagents are strong oxidizing agents and show distinct colour changes. The manganate(VII) ion is a very deep purple colour. It is reduced to the manganate(II) ion, which is very pale pink in colour (when dilute, it looks colourless). The dichromate(VI) ion is orange in colour. When reduced, it forms the chromium(III) ion which is blue-green (this is used to oxidize alcohols, see page 602).

Both oxidizing agents need to be acidified. An advantage of potassium dichromate(VI) is that it has a slightly lower $E^\ominus_{cell}$ value (+1.36 V) than potassium manganate(VII) ($E^\ominus_{cell}$ = +1.51 V). This means that it can be acidified using hydrochloric acid, whereas potassium manganate(VII) would reduce the chloride ions present to chlorine gas ($E^\ominus_{cell}$ = +1.36 V). This is both dangerous (chlorine is toxic) and inaccurate (manganate(VII) ions will be used up in this reaction as well as in the reaction you are carrying out).

A disadvantage to using potassium dichromate(VI) is that it is a known carcinogen, which needs to be handled with extreme care and disposed of safely. It can also be inaccurate as the end-point is difficult to determine due to colour mixing.

LINKING QUESTION

Why are some redox titrations described as 'self-indicating'?

WORKED EXAMPLE R3.2A

Sodium nitrate(III), $NaNO_2$, is used as a preservative in processed meat. In an acidic solution, nitrate(III) ions, NO_2^-, are converted to nitric(III) acid, HNO_2, which reacts with the manganate(VII) ion.

A 1.00 g sample of a water-soluble solid containing $NaNO_2$ was dissolved in dilute H_2SO_4 and titrated with 0.0100 mol dm^{-3} aqueous $KMnO_4$ solution. In the reaction, NO_2^- is oxidized to NO_3^-. The titration required 12.15 cm^3 of the $KMnO_4$ solution.

Calculate the percentage by mass (to 2 s.f.) of $NaNO_2$ in the 1.00 g sample.

Answer

$NO_2^-(aq) \rightarrow NO_3^-(aq)$

$NO_2^-(aq) + H_2O(l) \rightarrow NO_3^-(aq) + 2H^+(aq) + 2e^-$ (×5)

$MnO_4^-(aq) \rightarrow Mn^{2+}(aq)$

$MnO_4^-(aq) + 8H^+(aq) + 5e^- \rightarrow Mn^{2+}(aq) + 4H_2O(l)$ (×2)

$\mathbf{5NO_2^-(aq) + 2MnO_4^-(aq) + 6H^+(aq) \rightarrow 5NO_3^-(aq) + 2Mn^{2+}(aq) + 3H_2O(l)}$

Amount of $MnO_4^- = \frac{12.15}{1000}$ dm^3 × 0.0100 mol dm^{-3} = 0.000 1215 mol

Amount of NO_2^- = (0.000 121 5 mol × $\frac{5}{2}$) = 3.0375 × 10^{-4} mol

Mass of $NaNO_2$ = 3.0375 × 10^{-4} mol × 69.00 g mol^{-1} = 0.02096 g

Percentage by mass of $NaNO_2 = \frac{0.02096\text{ g}}{1.00\text{ g}} \times 100 = 2.1\%$

Tool 3: Mathematics

Express quantities and uncertainties to an appropriate number of significant figures or decimal places

Addition / subtraction and decimal places

The rule when adding or subtracting numbers is to record the final value to the smallest number of decimal places in the quantities used. For example, a temperature of 25.84 °C was recorded. This dropped to 10.3 °C. The temperature change is 25.84 – 10.3 = 15.54 °C which, to 1 decimal place (the smallest number of decimal places in the values used) is correctly written as 15.5 °C. (Note: It would be unusual to use thermometers with different precisions in an experiment.)

Multiplication / division and significant figures

In chemistry, multiplication and division are much more common operations than addition and subtraction. The final answer should be recorded to the smallest number of significant figures used in the data from which it is calculated. For example, a heat energy change of 50.02 g of water and a temperature change of 7.8 °C would be calculated (using $4.18\,J\,g^{-1}\,K^{-1}$) as being 50.02 × 7.8 × 4.18 = 1630.852… J. This is correctly recorded as 1600 J to 2 s.f. as the smallest number of significant figures comes from the temperature change (recorded to only 2 s.f.).

Uncertainties and significant figures

Uncertainties from equipment or experiments should be quoted to 1 significant figure. For example, a measuring cylinder records a value of $9.85\,cm^3 \pm 0.05\,cm^3$.

Rounding of values

Only round your final answer at the end of all the calculation steps. Never do this from step to step as repeated rounding of numbers may produce an error in your final answer.

The final, calculated, answer should be expressed with a level of precision that matches that of the total uncertainty, expressed to one significant figure. For example, $0.0656\,mol\,dm^{-3} \pm 0.008\,mol\,dm^{-3}$ should be written as $0.066\,mol\,dm^{-3} \pm 0.008\,mol\,dm^{-3}$. Both the calculated value and the uncertainty are to 3 decimal places.

Going further

Primary standards and redox

A primary standard is a very pure stable reagent that is easily weighed. Primary standards are typically used to produce standard solutions for titrations and other analysis techniques. Ideally a primary standard satisfies the following requirements:

- easy to obtain, purify, dry and preserve in pure form
- inert to the air, that is, neither hygroscopic (absorbing moisture from the air) nor susceptible to aerial oxidation (being oxidized by molecular oxygen or water) or reaction with carbon dioxide
- capable of being tested for impurities by tests of known sensitivity
- high molar mass to reduce the effect of weighing errors
- readily soluble under the conditions in which it is used
- reactions are stoichiometric and practically instantaneous.

Primary redox standards include potassium dichromate(VI), $K_2Cr_2O_7$, iodine, I_2, and potassium iodate(V), KIO_3. High oxidation state oxyanions require acid to function properly, to facilitate the formation of water from the complexed oxide when it is released from the high oxidation state element in its reduction.

Solutions of potassium manganate(VII) are *not* primary standards because pure potassium manganate(VII) is difficult to prepare and it reacts slowly with water to form manganese(IV) oxide, especially in the presence of light.

38 A solution of potassium manganate(VII) can be standardized by titration under suitable conditions with a solution of arsenic(III) oxide, As_2O_3.

5 moles of arsenic(III) oxide are oxidized by 4 moles of manganate(VII) ions.

Calculate the oxidation state to which the manganate(VII) is reduced.

39 Hydrated iron(II) sulfate has the formula $FeSO_4.xH_2O$.

An experiment was performed to determine x, the amount of water of crystallization in hydrated iron(II) sulfate.

101.2 g of hydrated iron(II) sulfate ($FeSO_4.xH_2O$) was dissolved in 500 cm^3 of water. 20.00 cm^3 of this solution reacts completely with 24.00 cm^3 of 0.100 mol dm^{-3} potassium dichromate(VI) solution.

Calculate the value of x. The equation for the reaction is shown below.

$$Cr_2O_7^{2-}(aq) + 6Fe^{2+}(aq) + 14H^+(aq) \rightarrow 2Cr^{3+}(aq) + 6Fe^{3+}(aq) + 7H_2O(l)$$

40 Potassium manganate(VII), $KMnO_4$, oxidizes potassium iodide, KI, to liberate iodine, I_2. The iodine liberated can be titrated with aqueous sodium thiosulfate, $Na_2S_2O_3$. From the results obtained, the concentration of potassium manganate(VII) solution can be calculated. The equations for the reactions are shown below.

$$16H^+(aq) + 2MnO_4^-(aq) + 10I^-(aq) \rightarrow 2Mn^{2+}(aq) + 8H_2O\ (aq) + 5I_2(l)$$

$$2S_2O_3^{2-}(aq) + I_2(aq) \rightarrow S_4O_6^{2-}(aq) + 2I^-(aq)$$

In the experiment, the iodine liberated from 25.00 cm^3 of potassium manganate(VII) requires 26.20 cm^3 of 0.500 mol dm^{-3} sodium thiosulfate for complete reaction.

Using the equations and the titration results, calculate the concentration of the potassium manganate(VII) solution.

41 Wines often contain a small amount of sulfur dioxide that is added as a preservative. The sulfur dioxide content of a wine is found by the following method:

1 A 50.00 cm^3 sample of white wine is reacted with 40.00 cm^3 of 0.01 mol dm^{-3} aqueous iodine (an excess).

2 The sulfur dioxide in the wine is oxidized to sulfate ions in the process.

$$SO_2(aq) + I_2(aq) + 2H_2O(l) \rightarrow SO_4^{2-}(aq) + 2I^-(aq) + 4H^+(aq)$$

3 The unreacted iodine requires exactly 23.60 cm^3 of 0.02 mol dm^{-3} sodium thiosulfate for complete reaction:

$$I_2(aq) + 2S_2O_3^{2-} \rightarrow 2I^-(aq) + S_4O_6^{2-}(aq)$$

Determine the concentration of sulfur dioxide, in mol dm^{-3}, in the wine.

Inquiry 2: Collecting and processing data

Assess accuracy, precision, reliability and validity

Do not get accuracy and precision confused. Precision is a measure of how close results are to each other. The smaller the resolution of the apparatus, the more precise the results will be. For example, there are two sets of results from a titration: one set of results is 23.45 cm^3, 23.40 cm^3, 23.50 cm^3, 23.55 cm^3 and the second set is 23.90 cm^3, 25.00 cm^3, 22.10 cm^3, 22.45 cm^3. The range of the first set of results is much smaller than the second, so the values are more precise.

Accuracy describes how close your value is to the true value. An accurate thermometer will record the temperature of pure boiling water as 100 °C but an inaccurate thermometer may record it as being 96 °C. It is possible to improve the accuracy of your data by using more precise apparatus or apparatus with a more detailed scale (assuming it is correctly calibrated). For example, if measuring 10 cm^3 of water, a 100 cm^3 measuring cylinder will be less accurate than a 10 cm^3 measuring cylinder. The more accurate a measurement is, the smaller the systematic error.

Precise results are not necessarily accurate. For example, using the titration results above, if the burette has been consistently read incorrectly (for example, above eye level) the results may be close together and precise but not accurate at all. This would be an example of a systematic error.

Your data should also be reliable. This means that somebody else will be able to repeat your method and produce the same results.

To understand the term 'validity', you must first understand the term 'accurate': for a method to be valid, it must be accurate and able to measure whatever it was intended to measure with little variation. Non-valid results could be due, for example, to a piece of equipment failing or not working correctly. For example, if you pipetted 25 cm^3 of solution into a beaker and poured this into a burette, you would expect the burette to show a consistent volume each time. If the volume was not consistent, this could be due to a leaky pipette or burette. This would mean your results would not be valid as different volumes would have been used.

Reactions of metals with metal ions in solution

◆ **Reactivity series:** A series of elements (usually metals) ranked by their degree of reactivity, made for comparison of reactions of elements with other substances, for example, cations, acids and oxygen.

A group of metals can be readily sorted into a **reactivity series** showing their order of reactivity (or reducing power), via simple experiments involving the metals and aqueous solutions of their ions.

Establishing a reactivity series

1 Test tubes are filled with a small volume of the following aqueous solutions (each with the same concentration): copper(II) nitrate, lead(II) nitrate, iron(II) sulfate, magnesium nitrate, zinc nitrate and tin(II) chloride.

2 A small piece of freshly cleaned magnesium ribbon is placed in each solution. The surface of the magnesium is observed for several minutes and any colour changes indicative of a chemical reaction are recorded.

3 The process is then repeated, in turn, with fresh solutions and pieces of the other metals in the place of magnesium. The results are tabulated as shown in Table R3.15, where a tick indicates a reaction has occurred and a cross indicates no observable reaction has taken place.

Top tip!

The nitrate, sulfate and chloride ions are spectator ions and do not participate in any reactions that occur. These solutions are aqueous solutions of the metal ions.

■ **Table R3.15** Summary of results for a series of reactions between selected metals and their ions

Ion in solution / Metal	$Cu^{2+}(aq)$	$Pb^{2+}(aq)$	$Fe^{2+}(aq)$	$Mg^{2+}(aq)$	$Zn^{2+}(aq)$	$Sn^{2+}(aq)$
copper	✗	✗	✗	✗	✗	✗
lead	✓	✗	✗	✗	✗	✗
iron	✓	✓	✗	✗	✗	✓
magnesium	✓	✓	✓	✗	✓	✓
zinc	✓	✓	✓	✗	✗	✓
tin	✓	✓	✗	✗	✗	✗

4 By summing the number of reactions that each metal has produced, as shown in Table R3.16, a reactivity series can be constructed.

■ **Table R3.16** A reactivity series for selected metals based on displacement reactions

Metal	Number of displacement reactions
Mg	5
Zn	4
Fe	3
Sn	2
Pb	1
Cu	0

As we move up the reactivity series, metals become increasingly chemically reactive and their reducing power, or ability to donate electrons, increases.

Equations can be written for all the displacement reactions, for example:

Formula equation: $Mg(s) + CuSO_4(aq) \rightarrow MgSO_4(aq) + Cu(s)$

Rewriting in terms of ions: $Mg(s) + Cu^{2+}(aq) + SO_4^{2-}(aq) \rightarrow Mg^{2+}(aq) + SO_4^{2-}(aq) + Cu(s)$

Cancelling spectator ions: $Mg(s) + Cu^{2+}(aq) \rightarrow Mg^{2+}(aq) + Cu(s)$

Rewriting in terms of half-equations: $Mg(s) \rightarrow Mg^{2+}(aq) + 2e^-$
$Cu^{2+}(aq) + 2e^- \rightarrow Cu(s)$

WORKED EXAMPLE R3.2B

In order to determine the position of three metals in a reactivity series, the metals were placed in different aqueous solutions of metal ions. The table summarizes whether or not a reaction occurred.

	$Ag^+(aq)$	$Pb^{2+}(aq)$	$Ni^{2+}(aq)$
Ag(s)	—	no reaction	no reaction
Pb(s)	reaction	—	no reaction
Ni(s)	reaction	reaction	—

Deduce the ionic equations for the three reactions that take place. Use this information to place the metals silver, lead and nickel in a reactivity series, with the strongest reducing agent first. Explain your reasoning.

Answer

$Ni(s) + Pb^{2+}(aq) \rightarrow Ni^{2+}(aq) + Pb(s)$

$Ni(s) + 2Ag^+(aq) \rightarrow Ni^{2+}(aq) + 2Ag(s)$

$Pb(s) + 2Ag^+(aq) \rightarrow Pb^{2+}(aq) + 2Ag(s)$

Nickel is a stronger reducing agent than lead and silver.

Nickel is most reactive as it can reduce/displace both lead(II) ions, Pb^{2+}, and silver(I) ions, Ag^+.

Lead is a stronger reducing agent than silver but not nickel.

Lead is in the middle (of the three) as it can reduce/displace silver(I) ions, Ag^+, but not nickel ions, Ni^{2+}.

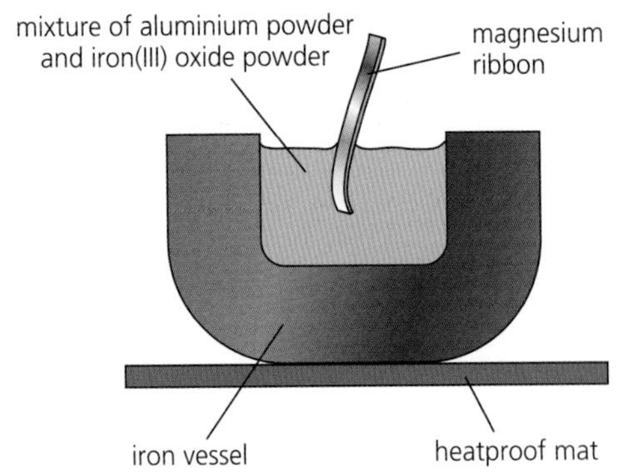

■ **Figure R3.42** Thermite reaction mixture apparatus; the magnesium ribbon acts as a fuse (which is set on fire to initiate the reaction and overcome the activation energy barrier)

Displacement reactions can also be carried out in the solid state using powdered samples of metals and metal compounds. For example, if iron(III) oxide and aluminium are heated together, a very exothermic reaction known as the thermite reaction occurs. This results in the formation of aluminium oxide and molten iron (Figure R3.42):

$$Fe_2O_3(s) + 2Al(s) \rightarrow 2Fe(l) + Al_2O_3(s)$$

The thermite reaction occurs because aluminium is a more powerful reducing agent than iron and has a stronger tendency to lose its electrons.

Using the reactivity series

Figure R3.43 shows a full reactivity series for metals. Including carbon and hydrogen extends the usefulness of the reactivity series.

Lithium
Caesium
Rubidium
Potassium
Barium
Strontium
Calcium
Sodium
Magnesium
Beryllium
Aluminium
(Carbon)
Zinc
Chromium
Iron
Cadmium
Cobalt
Nickel
Tin
Lead
(Hydrogen)
Antimony
Arsenic
Bismuth
Copper
Silver
Palladium
Mercury
Platinum
Gold

■ **Figure R3.43** A reactivity series

Reactions with water

Some metals react with water to form hydrogen.

For example, the reaction between sodium and water is:

$$2Na(s) + 2H_2O(l) \rightarrow 2NaOH(aq) + H_2(g)$$

The relevant half-equations are:

$$2Na(s) \rightarrow 2Na^+(aq) + 2e^-$$

and

$$2H_2O(l) + 2e^- \rightarrow 2OH^-(aq) + H_2(g)$$

However, some metals that should react with water (such as aluminium and lead) do not because of the presence of a thin but unreactive surface layer of metal oxide.

Reactions with acids

Metals above hydrogen, for example zinc, will replace hydrogen from dilute acids. Metals below hydrogen, for example copper, will not displace hydrogen from dilute acids. Hydrogen is behaving like a metal, since aqueous solutions of acids contain positively charged hydrogen ions, H^+.

Equation: $Zn(s) + 2HCl(aq) \rightarrow ZnCl_2(aq) + H_2(g)$

Ionic equation: $Zn(s) + 2H^+(aq) \rightarrow Zn^{2+}(aq) + H_2(g)$

Half-equations: $Zn(s) \rightarrow Zn^{2+}(aq) + 2e^-$

$2H^+(aq) + 2e^- \rightarrow H_2(g)$

Reactions with carbon

Metals above carbon in the reactivity series, such as sodium and aluminium, cannot be produced by reduction of metal oxides with carbon; instead electrolysis has to be used. Metals below carbon, such as iron and zinc, can be produced by reduction of metal oxides with carbon.

$$ZnO(s) + C(s) \rightarrow Zn(s) + CO(g)$$

ATL R3.2A

Run tests of various metals in aqueous solutions using the simulation at https://teachchemistry.org/classroom-resources/metals-in-aqueous-solutions-simulation

Interpret the data you collect to construct a reactivity series of the elements used in the simulation.

Using presentation software (such as PowerPoint), outline and present your findings. You should include appropriate equations and reference to redox concepts.

The reactions of selected metals with dilute acid, oxygen and water are summarized in Table R3.17.

■ **Table R3.17** Reactivity series of selected metals

Reactivity series	Reaction with dilute acid	Reaction with air / oxygen	Reaction with water	Ease of extraction
potassium (K)	Produce H_2 with decreasing vigour	Burn very brightly and vigorously	Produce H_2 with decreasing vigour with cold water	Difficult to extract
sodium (Na)				
calcium (Ca)		Burn to form an oxide with decreasing vigour		Easier to extract
magnesium (Mg)				
aluminium (Al)				
iron (Fe)				
lead (Pb)		React slowly to form the oxide		
copper (Cu)	Do not react with dilute acids			
silver (Ag)			Do not react with cold water or steam	Found as the element (native)
gold (Au)				
platinum (Pt)		Do not react		

WORKED EXAMPLE R3.2C

Deduce equations for the reactions of barium with oxygen, water, chlorine, dilute hydrochloric acid and sulfuric acid.

Answer

$$2Ba(s) + O_2(g) \rightarrow 2BaO(s)$$

$$Ba(s) + 2H_2O(l) \rightarrow Ba(OH)_2(aq) + H_2(g)$$

$$Ba(s) + Cl_2(g) \rightarrow BaCl_2(s)$$

$Ba(s) + 2HCl(aq) \rightarrow BaCl_2(aq) + H_2(g)$ or $Ba(s) + 2H^+(aq) \rightarrow Ba^{2+}(aq) + H_2(g)$

$Ba(s) + H_2SO_4(aq) \rightarrow BaSO_4(s) + H_2(g)$; limited reaction due to insolubility of barium sulfate

Link

The halogens (Group 17) are covered in more detail in Chapter S3.1, page 239.

LINKING QUESTION

What observations can be made when metals are mixed with aqueous metal ions, and solutions of halogens are mixed with aqueous halide ions?

Non-metal displacement reactions

Displacement reactions also occur with non-metals, in particular the halogens. A reactivity series can be written for the halogens, which corresponds to the positions of the elements of the periodic table.

As you move up the reactivity series from iodine to fluorine, the halogens become increasingly chemically reactive and their oxidizing power, or ability to receive electrons and form halide ions, increases.

◆ **Corrosion**: Chemical or electrochemical attack on the surface of a metal.

◆ **Rusting**: Corrosion of iron (or steel) to form hydrated iron(III) oxide $Fe_2O_3.xH_2O$.

Corrosion

All reactive metals undergo **corrosion**; the atoms on the surface react with oxygen and water in the air. Some, for example aluminium and chromium, form a thin layer of oxide that protects them from further attack. Iron readily corrodes to form rust, which is non-adhesive, porous and flakes off the iron, unlike the aluminum and chromium oxide layers, exposing the iron to more corrosion by oxygen and water.

Rusting is a complex electrochemical process which requires the presence of liquid water and oxygen.

An approximate overall equation describing rusting is:

$$4Fe(s) + 3O_2(g) + xH_2O(l) \rightarrow 2Fe_2O_3.xH_2O(s)$$

where the number of water molecules present in the rust is variable. Rusting can be described by the following half-equations:

$$Fe(s) \rightarrow Fe^{2+}(aq) + 2e^-$$

$$O_2(g) + 2H_2O(l) + 4e^- \rightarrow 4OH^-(aq)$$

Combining these two half-equations and multiplying the first by two gives the following ionic equation:

$$2Fe(s) + O_2(g) + 2H_2O(l) \rightarrow 2Fe^{2+}(aq) + 4OH^-(aq)$$

The iron(II) hydroxide formed by precipitation is then rapidly oxidized under basic conditions to form iron(III) hydroxide then red-brown hydrated iron(III) oxide.

$$4Fe(OH)_2(s) + O_2(g) + 2H_2O(l) \rightarrow 4Fe(OH)_3(s)$$

This is rust and can be represented by the formula $Fe_2O_3.xH_2O$. The rust flakes off the surface of iron, partly because it has a lower density than iron and partly because it only weakly adheres to the underlying iron. This exposes more iron surface and the rusting continues.

A waterproof layer protects iron from rusting to some extent. However, the protection provided by paints and other widely used materials is of limited duration, because of factors such as physical damage, chemical deterioration and weathering.

Alternative methods of rust protection use electrochemical principles. Iron pipes, ships and tanks in the ground are often protected using a method called sacrificial protection. A block of magnesium (or zinc) is electrically attached to the iron object (Figure R3.44). Magnesium (or zinc) is higher up the reactivity series than iron, so it oxidizes in preference to the iron.

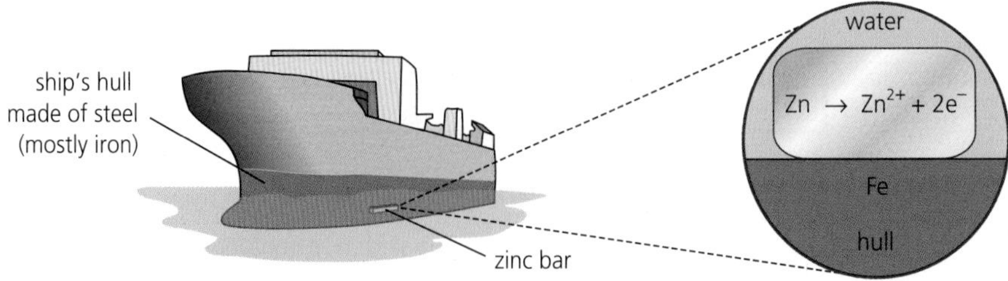

■ **Figure R3.44** Sacrificial protection of an iron ship

For example, if zinc is used, it is oxidized to form zinc ions:

$$Zn(s) \rightarrow Zn^{2+}(aq) + 2e^-$$

LINKING QUESTION

The surface oxidation of metals is often known as corrosion. What are some of the consequences of this process?

The electrons released reduce dissolved oxygen molecules to form hydroxide ions, $OH^-(aq)$:

$$O_2(g) + 2H_2O(l) + 4e^- \rightarrow 4OH^-(aq)$$

ATL R3.2B

Tin-plating and galvanizing are examples of electrochemical protection. Research the reactions used and use presentation software (such as PowerPoint) to present and explain these forms of electrochemical protection to a student in your class. Include ionic equations and diagrams and refer to the relative reactivities (reducing strength) of iron, tin and zinc in your presentation.

More half-equations

Half-equations can be used to balance equations for redox reactions because the number of electrons lost when an atom, molecule or ion is oxidized in one half-equation has to equal the number of electrons gained by the reduction of another species in the second half-equation.

When chemicals react in acidic solutions, the half-equations are balanced by including hydrogen ions and water molecules.

Constructing half-equations

The steps below are illustrated by considering the oxidizing agent $Cr_2O_7^{2-}$:

1 Write down the formulas of the reactant and products:

$$Cr_2O_7^{2-} \rightarrow Cr^{3+}$$

2 Balance with respect to the non-hydrogen or oxygen atoms (in this case, chromium):

$$Cr_2O_7^{2-} \rightarrow 2Cr^{3+}$$

3 Balance the oxygen atoms with water molecules:

$$Cr_2O_7^{2-} \rightarrow 2Cr^{3+} + 7H_2O$$

4 Balance the hydrogen atoms of the water with hydrogen ions:

$$14H^+ + Cr_2O_7^{2-} \rightarrow 2Cr^{3+} + 7H_2O$$

5 Determine the total charges on both sides of the almost completed half-equation:
LHS: $(+14 + -2) = +12$; RHS: $(2 \times +3) = +6$

6 Balance the two charges by adding electrons to the side of the equation with the more positive value:

$$6e^- + 14H^+(aq) + Cr_2O_7^{2-}(aq) \rightarrow 2Cr^{3+}(aq) + 7H_2O(l)$$

LHS: $(+14 + -2 + -6) = +6$; RHS: $(2 \times +3) = +6$

An identical process is used to construct half-equations for reducing agents that operate in an aqueous acidic solution. The one difference is that the electrons will appear on the right-hand side of the half-equation.

1 Write down the formulas of the reactant and products, for example:

$$HNO_2 \rightarrow NO_3^-$$

2 The equation is already balanced with respect to the nitrogen.

3 Balance the oxygen of the nitric(III) (nitrous) acid with a water molecule:

$H_2O + HNO_2 \rightarrow NO_3^-$

4 Balance the hydrogen present in the water and nitrous acid with hydrogen ions:

$H_2O + HNO_2 \rightarrow NO_3^- + 3H^+$

5 Determine the total charges on both sides of the almost completed half-equation:

LHS: = 0;   RHS: $-1 + (3 \times +1) = +2$

6 Balance the two charges by adding electrons to the side of the equation with the more positive value:

$H_2O(l) + HNO_2(aq) \rightarrow NO_3^-(aq) + 3H^+(aq) + 2e^-$

LHS: 0   RHS: $-1 + (3 \times +1) + -2 = 0$

Balancing using the ion-electron method

Redox equations are written by combining two half-equations: one describing the reduction of an oxidizing agent and the other describing the oxidation of a reducing agent. Often one or both of the two half-equations must be multiplied by a suitable coefficient so that the number of electrons gained by the oxidizing agent equals the number of electrons lost by the reducing agent. The electrons can then be cancelled from both sides of the equations and, if necessary, the number of water molecules and hydrogen ions (if present) simplified.

WORKED EXAMPLE R3.2D

Write a redox equation for the reduction of acidified manganate(VII) ions and the oxidation of methanol using the balanced half-equations below:

$H_2O(l) + CH_3OH(l) \rightarrow CO_2(g) + 6H^+(aq) + 6e^-$

$MnO_4^-(aq) + 8H^+(aq) + 5e^- \rightarrow Mn^{2+}(aq) + 4H_2O(l)$

Answer

Multiplying through the top half-equation by 5 and the bottom half-equation by 6:

$5H_2O(l) + 5CH_3OH(l) \rightarrow 5CO_2(g) + 30H^+(aq) + 30e^-$

$6MnO_4^-(aq) + 48H^+(aq) + 30e^- \rightarrow 6Mn^{2+}(aq) + 24H_2O(l)$

Adding the two half-equations together:

$5H_2O(l) + 5CH_3OH(l) + 6MnO_4^-(aq) + 48H^+(aq) + 30e^- \rightarrow 5CO_2(g) + 30H^+(aq) + 30e^- + 6Mn^{2+}(aq) + 24H_2O(l)$

Cancelling electrons:

$5H_2O(l) + 5CH_3OH(l) + 6MnO_4^-(aq) + 48H^+(aq) \rightarrow 5CO_2(g) + 30H^+(aq) + 6Mn^{2+}(aq) + 24H_2O(l)$

Simplifying the number of water molecules on the right-hand side of the equation:

$5CH_3OH(l) + 6MnO_4^-(aq) + 48H^+(aq) \rightarrow 5CO_2(g) + 30H^+(aq) + 6Mn^{2+}(aq) + 19H_2O(l)$

Simplifying the number of protons (H^+) on the left-hand side of the equation:

$5CH_3OH(l) + 6MnO_4^-(aq) + 18H^+(aq) \rightarrow 5CO_2(g) + 6Mn^{2+}(aq) + 19H_2O(l)$

Balancing using the oxidation state method

This method allows us to determine the reacting ratio of the species more quickly than using the ion–electron half-equation method.

1 Identify the elements that have undergone a change in oxidation state.

2 Balance the atoms which have undergone a change in oxidation state. These are the atoms that have undergone oxidation or reduction. For example, if dichromate(VI) ions are used as an oxidizing agent and reduced to chromium(III) ions, then the chromium must be balanced first.

3 Find the change in oxidation state and the number of electrons transferred for each redox species.

4 Balance the charge by adding the correct number of protons, $H^+(aq)$, for acidic aqueous solution.

5 Balance the oxygen by adding water molecules, H_2O, and the rest of the atoms not involved in redox.

In a redox reaction, the sum of the *increases* in the oxidation state of oxidized species equals the sum of the *decreases* in the oxidation state of the reduced species.

WORKED EXAMPLE R3.2E

Balance the redox reactions between acidified potassium manganate(VII) and iron(II) sulfate and ammonia to form nitrogen monoxide or nitrogen(II) oxide (NO) using the oxidation state method.

Answer

Oxidation: Fe^{2+} (+2) → $-e^-$ → Fe^{3+} (+3); 5 × (−e)

Reduction: MnO_4^- (+7) → +5e– → Mn2+ (+2); 1 × (+5e)

5 electrons transferred

Balanced equation:

$5Fe^{2+} + 8H^+ + MnO_4^- \rightarrow 5Fe^{3+} + 4H_2O + Mn^{2+}$

Oxidation: NH_3 (−3) → Increase in oxidation state, −5e → NO (+2); 4 × (−5e)

Reduction: O_2 (0) → Decrease in oxidation state, +2e → H_2O (−2); 10 × (+2e)

10 electrons transferred

Balanced equation:

$4NH_3 + 5O_2 \rightarrow 4NO + 6H_2O$

ATL R3.2C

Photochromic lenses used in sunglasses use chemical changes to compensate for changes in brightness.

Find out how the glass of photochromic glasses is made and how redox reactions account for the changes in darkness of the glasses.

Create an explanation using the following concepts: redox, oxidation, reduction, oxidation state, oxidizing agent and reducing agent.

Present your explanation to your peers using presentation software such as PowerPoint.

42 Write half-equations for each of the following conversions.

a $I^- \rightarrow I_2$
b $Fe^{2+} \rightarrow Fe^{3+}$
c $Fe \rightarrow Fe^{2+}$
d $Cl^- \rightarrow Cl_2$
e $SO_4^{2-} \rightarrow SO_2$
f $MnO_4^- \rightarrow Mn^{2+}$
g $SO_4^{2-} \rightarrow H_2S$
h $H_2O_2 \rightarrow O_2$
i $Cr^{3+} \rightarrow Cr_2O_7^{2-}$
j $C_2O_4^{2-} \rightarrow CO_2$
k $Hg_2^{2+} \rightarrow Hg$
l $IO_3^- \rightarrow I_2$

43 Write redox equations for each of the following reactions.

a $H^+/MnO_4^- + Cl^-$
b $H^+/MnO_4^- + C_2O_4^{2-}$
c $H^+/SO_4^{2-} + I^-$
d $H^+/MnO_4^- + H_2O_2$
e $I^- + H^+/IO_3^-$
f $Fe^{2+} + H^+/Cr_2O_7^{2-}$

44 Identify the oxidizing and reducing agents in the following redox reaction. Split the reaction into two half-reactions and show the changes in oxidation state.

$$3N_2H_4 + 2BrO_3^- \rightarrow 3N_2 + 2Br^- + 6H_2O$$

45 Balance the following ionic equations.

a $H_2O_2 + I_2 \rightarrow I^- + O_2$
b $S_2O_3^{2-} + Cl_2 \rightarrow SO_4^{2-} + Cl^-$

Redox equilibria

HL ONLY

◆ **Redox equilibrium:** An equilibrium involving a redox (electron transfer) reaction.

A **redox equilibrium** exists between two chemically related species that are in different oxidation states, such as metal atoms and their hydrated cations.

For example, when a copper electrode is placed in contact with an aqueous solution of copper(II) ions, the following equilibrium is established:

$$Cu(s) \rightleftharpoons Cu^{2+}(aq) + 2e^-$$

There are two opposing reactions in this equilibrium.

Copper atoms from the rod enter the solution as copper(II) ions. This leaves electrons behind on the surface of the rod, which thus ends up with a negative charge (and potential) relative to the solution. This charge cannot be measured directly – any instrument for measuring voltage measures the potential *difference* between two electrodes.

$$Cu(s) \rightarrow Cu^{2+}(aq) + 2e^-$$

Copper(II) ions in solution accept electrons from the metal rod and are deposited as copper atoms on the surface of the rod:

$$Cu^{2+}(aq) + 2e^- \rightarrow Cu(s)$$

The redox equilibrium is established when the rate of electron gain equals the rate of electron loss.

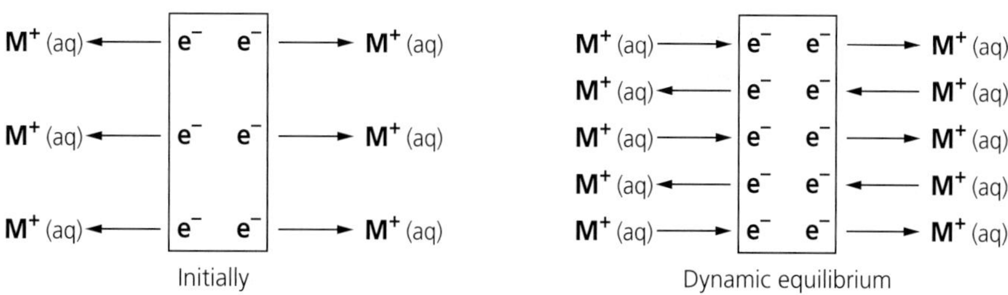

■ **Figure R3.45** The establishment of a redox equilibrium

The position of equilibrium differs for different combinations of metals placed in solutions of their ions (Figure R3.46).

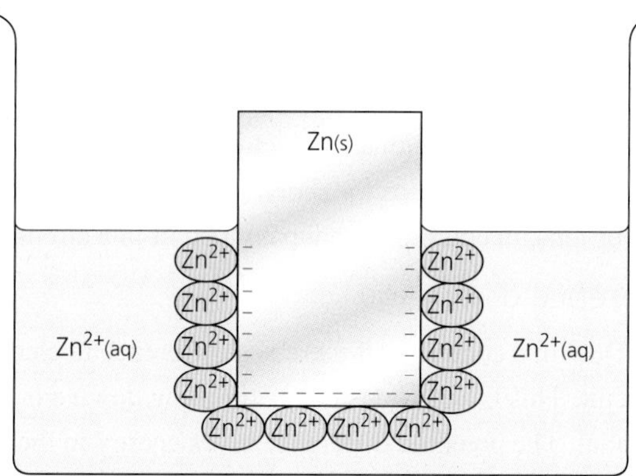

■ **Figure R3.46** The equilibrium established by zinc metal in contact with its ions. A double layer of charges is formed at the surface of the metal

For unreactive metals such as copper, the equilibrium that is established lies further to the right and favours the reduced metal over aqueous cations. This means that copper atoms are favoured over copper(II) ions.

Copper(II) ions, Cu^{2+}(aq), are therefore relatively easy to reduce. They gain electrons readily to form atoms of copper metal:

$$Cu(s) \rightleftharpoons Cu^{2+}(aq) + 2e^-$$

For reactive metals such as zinc, the equilibrium lies further to the right:

$$Zn(s) \rightleftharpoons Zn^{2+}(aq) + 2e^-$$

Zinc ions, Zn^{2+}(aq), are therefore relatively difficult to reducc. They gain electrons much less readily by comparison than copper(II) ions.

Electrochemical cells

Electricity basics

Conductors and insulators

For a substance to conduct electricity, it must contain electrically charged particles (charge carriers) that are free to move when the substance has a potential difference (voltage) applied across it. In metals, in both the solid and liquid states, the charged particles are the valence electrons whose flow through the metal forms an electric current (Figure R3.47). In electrolytes, cations and anions act as charge carriers and form an electric current when a potential difference is applied.

HL ONLY

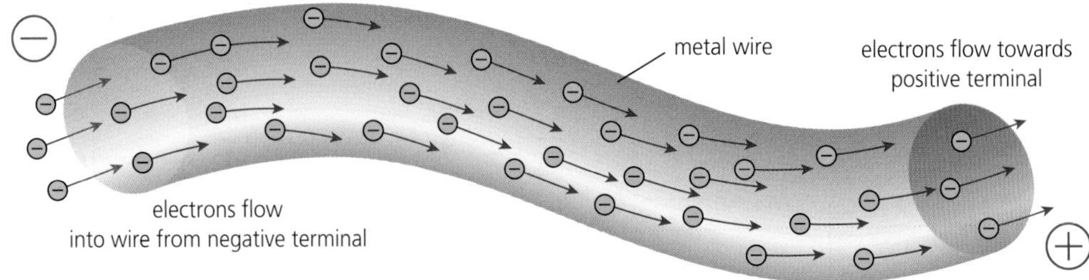

■ **Figure R3.47** Electrons flowing along a metal wire form an electric current

Ionic solids do not conduct electricity, because the ions are firmly held in the lattice by powerful electrostatic forces and cannot move. Only when the ionic substance is molten or dissolved in water are the ions released from the lattice and free to move.

Electric current

The flow of charged particles through a conductor is an electric current. The size of the electric current is measured in amperes (amps for short) (A) and is a measure of the number of single charges (electrons or ions; in coulombs) that pass a point in a circuit in one second.

Potential difference or voltage

Charged particles flow in a conductor if there is a difference in electric potential between two points in a circuit. This is analogous to a ball rolling down a hill due to a difference in gravitational potential. The potential difference gives energy to the electric charge (electrons). It is measured in volts and also termed voltage.

Electrolytic cells

◆ **Electrolyte**: Liquid or aqueous solution that conducts electricity as a result of the presence of cations and anions.

◆ **Electrolysis**: A chemical reaction caused by passing an electric current through an electrolyte.

A substance that conducts electricity and is decomposed by the passage of an electric current is known as an **electrolyte**. The process of decomposing an electrolyte with an electric current is called **electrolysis**. This is a non-spontaneous process and needs a constant input of electrical energy.

When electricity is passed through an electrolyte, the electricity enters and leaves via electrodes (usually made of graphite or an inert metal).

The electrode connected to the positive terminal of the battery or direct current (d.c.) power supply is known as the anode. The negative electrode is known as the cathode.

Negative ions, or anions, are attracted towards the anode; positive ions, or cations, are attracted towards the cathode. When the ions reach the surface of the electrodes they undergo redox reactions. Cations will gain electrons (reduction) and anions will lose electrons (oxidation) on the surfaces of the electrodes.

LINKING QUESTION

Under what conditions can ionic compounds act as electrolytes?

The simplest form of electrolysis is the electrolysis of a molten binary salt, such as lead(II) bromide, $PbBr_2$, [Pb^{2+} $2Br^-$].

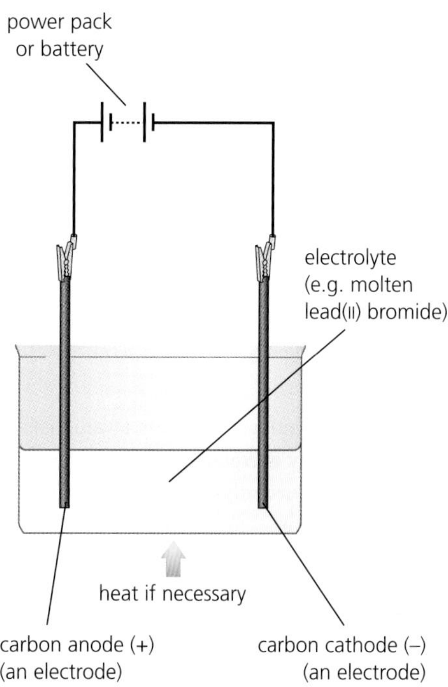

■ **Figure R3.48** An electrolytic cell used for electrolysis

Inert (chemically unreactive) graphite or metal electrodes are used (Figure R3.48). The decomposition products are molten lead (at the cathode) and bromine vapour (at the anode). The overall reaction is:

$$PbBr_2(l) \rightarrow Pb(l) + Br_2(g).$$

At the anode: Negatively charged bromide ions are electrostatically attracted towards the positively charged anode. At the anode they lose electrons and form bromine molecules:

$$2Br^-(l) \rightarrow Br_2(g) + 2e^-$$

or

$$2Br^-(l) \rightarrow 2Br(g) + 2e^-; 2Br(g) \rightarrow Br_2(g)$$

At the cathode: Positively charged lead(II) ions are attracted towards the negatively charged cathode. At the cathode they gain electrons and form lead atoms:

$$Pb^{2+}(l) + 2e^- \rightarrow Pb(l)$$

During electrolysis, each lead(II) ion accepts two electrons from the cathode. At the same time, two bromide ions each release an electron to the anode. The overall effect of these two processes is equivalent to two electrons flowing through the liquid lead(II) bromide from the cathode to the anode, however the particles flowing through the electrolyte are lead(II) and bromide ions, not electrons.

Common mistake

Misconception: *Electrons can flow through aqueous solutions without ions.*

The flow of electric current in a cell involves the flow of anions and cations and a redox reaction (reduction at the cathode and oxidation at the anode).

All ionic compounds undergo electrolysis in the molten state and obey two simple rules:

- Metals always form cations, which migrate to the cathode and are discharged as atoms.
- Non-metals always form anions, which migrate to the anode and are discharged as molecules.

Examples of the products of the electrolysis of molten (fused) electrolytes are shown in Table R3.18.

■ **Table R3.18** Examples of electrolysis of molten electrolytes

Electrolyte	Overall decomposition	Cathode half-equation	Anode half-equation
sodium chloride, $NaCl$	$2NaCl(l) \rightarrow 2Na(l) + Cl_2(g)$	$Na^+ + e^- \rightarrow Na$	$2Cl^- \rightarrow Cl_2 + 2e^-$
potassium iodide, KI	$2KI(l) \rightarrow 2K(l) + I_2(g)$	$K^+ + e^- \rightarrow K$	$2I^- \rightarrow I_2 + 2e^-$
copper(II) chloride, $CuCl_2$	$CuCl_2(l) \rightarrow Cu(l) + Cl_2(g)$	$Cu^{2+} + 2e^- \rightarrow Cu$	$2Cl^- \rightarrow Cl_2 + 2e^-$
aluminium oxide, Al_2O_3	$2Al_2O_3(l) \rightarrow 4Al(l) + 3O_2(g)$	$Al^{3+} + 3e^- \rightarrow Al$	$2O^{2-} \rightarrow O_2 + 4e^-$

46 Deduce the products of electrolysing the following molten salts:
  a lithium bromide, $LiBr$
  b caesium nitride, Cs_3N
  c iron(II) bromide, $FeBr_2$

TOK

What is the role of inductive and deductive reasoning in scientific inquiry, prediction and explanation?

The most common forms of logic used in science are inductive reasoning and deduction.

Inductive reasoning allows a chemist to move from specific instances to draw a general conclusion.

For example, consider the results of electrolysing a variety of molten binary salts, shown in Table R3.19.

■ **Table R3.19** Electrolysis of selected molten binary compounds

Binary salt	Product at cathode	Product at anode
aluminium oxide	aluminium	oxygen
sodium chloride	sodium	chlorine
lead(II) bromide	lead	bromine
magnesium nitride	magnesium	nitrogen
lithium hydride	lithium	hydrogen

The application of inductive logic in this situation leads to the general conclusion that metals are discharged at the cathode and non-metals are discharged as gases at the anode.

However, inductive reasoning can never give *certainty*. We also cannot be sure that the generalizations made in the past will continue to hold into the future. The impossibility of reaching certainty through induction is known as the 'problem of induction'. Inductive generalizations, including chemical theories or chemical laws, may be overtaken by new data. A single counter-example falsifies an inductive conclusion.

Deduction is the opposite process and allows chemists to move from general conclusions to specific predictions. The conclusions must be true if the general statements upon which they are based are true.

For example, chemists in the nineteenth century were aware that aluminium oxide in bauxite was very energetically stable. They realized that new discoveries about electrolysis might lead to a viable method for refining aluminium.

This is deductive reasoning: an electric current can be used to make a difficult (non-spontaneous) redox reaction involving an ionic compound occur; extracting aluminium from its oxide is a difficult reaction, so an electric current may be able to decompose aluminium oxide. Electrolysis would not be appropriate to reactions that are not redox and do not involve substances with ions.

Experiments demonstrated this reasoning was correct although further work was necessary to make the process energetically and economically efficient.

HL ONLY

Electroplating

◆ **Electroplating**: Method of plating one metal with another by electrodeposition. The article to be plated is made the cathode of an electrolytic cell and a rod or bar of the plating metal is made the anode.

Metals are electroplated to improve their appearance or to prevent corrosion. The most commonly used metals for **electroplating** are gold, copper, chromium, silver and tin. Familiar examples of electroplated objects include chromium-plated car bumpers and kettles, jewellery (for example, gold-plated bracelets), silver-plated cutlery and tin-plated cans.

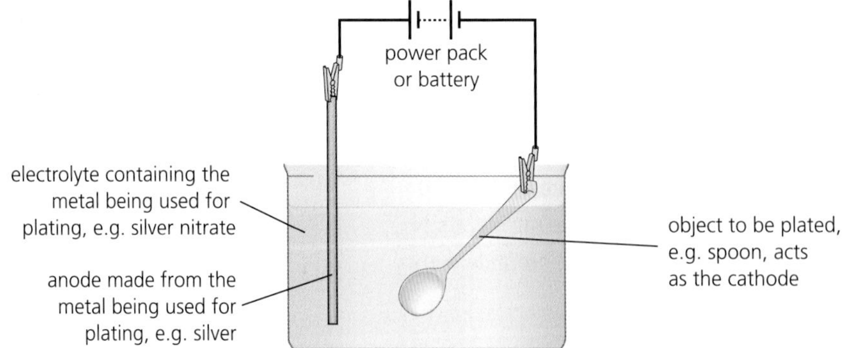

■ **Figure R3.49** An electrolytic cell used to perform silver plating

The object to be electroplated must be made the cathode (Figure R3.49). The anode must be the metal used for the plating process (unless of course you continually replenish the electrolyte). The electrolyte solution must contain cations of the metal for plating. As the current flows through the circuit, the anode slowly disappears and replaces the metal ions in the electrolyte.

HL ONLY

In chrome plating the object to be plated forms the cathode and it is placed in a solution containing a chromium species in the +6 (e.g. CrO_3) or +3 (e.g. $Cr_2(SO_4)_3$) oxidation state. The plating of chromium onto the surface of the cathode is a complex multi-step process proceeding through Cr(VI) → Cr(III) → Cr(II) → Cr(0) oxidation states. The anode is made from lead (mixed with antimony or tin) and oxidation of water to oxygen gas occurs at the anode.

Cathode: $Cr^{3+}(aq) + 3e^- \rightarrow Cr(s)$

Anode: $2H_2O(l) \rightarrow O_2(g) + 4H^+(aq) + 4e^-$

Overall: $4Cr^{3+}(aq) + 6H_2O(l) \rightarrow 3O_2(g) + 4Cr(s) + 12H^+(aq)$

In silver plating, a solution containing the complex ion $[Ag(CN)_2]^-(aq)$ is often used. This is a stable complex and produces a very low concentration of silver(I) ions, $Ag^+(aq)$.

$[Ag(CN)_2]^-(aq) \rightleftharpoons Ag^+(aq) + 2CN^-(aq)$

At the cathode the silver(I) cations undergo reduction to form silver atoms:

$Ag^+(aq) + e^- \rightarrow Ag(s)$

At the anode the silver atoms undergo oxidation to form silver(I) cations:

$Ag(s) \rightarrow Ag^+(aq) + e^-$

LINKING QUESTION (HL ONLY)

How is an electrolytic cell used for electroplating?

In order to obtain a good coating of metal during electroplating:

- the object to be plated must be clean and free of grease
- the object should be rotated to give an even coating
- the current must not be too large or the 'coating' will form too rapidly and flake off.

WORKED EXAMPLE R3.2F

A major use of tin is to make 'tin plate' – thin sheets of mild steel electroplated with tin – for use in the manufacture of food and drink cans. A tin coating of 1.0×10^{-5} m thickness is often used.

Deduce the ionic equation describing the reaction at the cathode.

Calculate the volume of tin needed to coat one side of a sheet of steel 1.0 m × 1.0 m to this thickness.

Calculate the amount (mol) of tin that this volume represents. (The density of tin is $7.3\,g\,cm^{-3}$.)

Answer

$Sn^{2+}(aq) + 2e^- \rightarrow Sn(s)$

Volume $= 1 \times 1 \times 1.0 \times 10^{-5} = 1.0 \times 10^{-5}\,m^3$ or $10\,cm^3$

Mass = volume × density $= 10\,cm^3 \times 7.3\,g\,cm^{-3} = 73\,g$

Amount of tin $= \dfrac{73\,g}{118.71\,g\,mol^{-1}} = 0.61\,mol$

Primary voltaic cells

A voltaic cell is an electrochemical cell that converts energy from spontaneous redox reactions to electrical energy.

A simple voltaic cell, known as the Daniell cell (Figure R3.50), can be made by placing a zinc electrode in a solution of zinc sulfate and a copper electrode in a solution of copper(II) sulfate. The two electrodes are connected using wires to create an external circuit which allows electrons to flow.

The Daniell cell is a primary voltaic cell, meaning it is used once and discarded. It is not recharged and reused like a secondary cell.

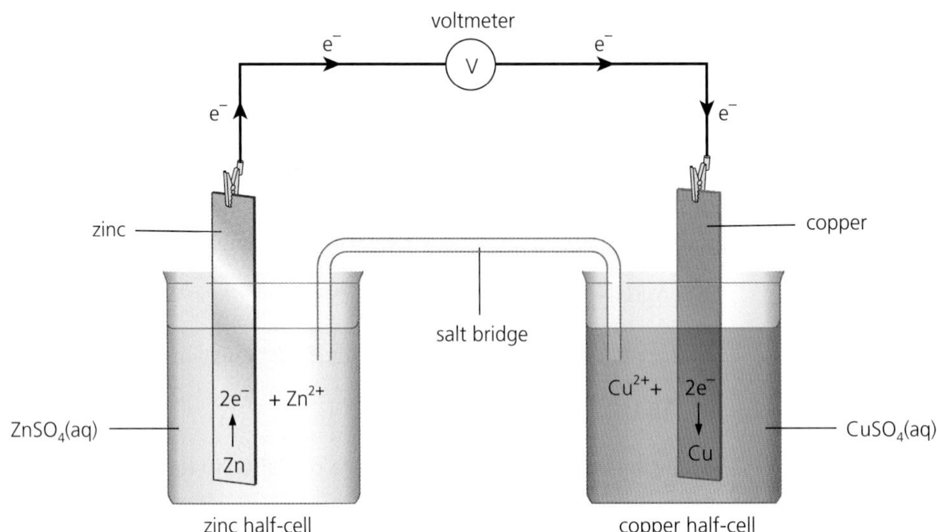

■ **Figure R3.50** A Daniell cell (a primary voltaic cell)

Including a voltmeter in the external circuit allows the potential difference across the cell to be measured.

◆ **Salt bridge**: An electrical connection made between two half-cells that allows cations and anions to flow to complete the circuit and maintain electroneutrality.

The circuit is completed by a **salt bridge** which allows ions to flow between the two half-cells. A simple salt bridge consists of a filter paper soaked in saturated potassium nitrate solution, KNO_3(aq).

Since zinc is higher than copper in the reactivity series it will undergo oxidation and release electrons onto the surface of the zinc electrode (making it negative). The zinc ions produced dissolve into the water (Figure R3.51).

The electrons flow from the surface of the zinc electrode through the external circuit to the surface of the copper electrode. Copper(II) ions on the surface of the copper electrode accept the electrons and undergo reduction (Figure R3.52).

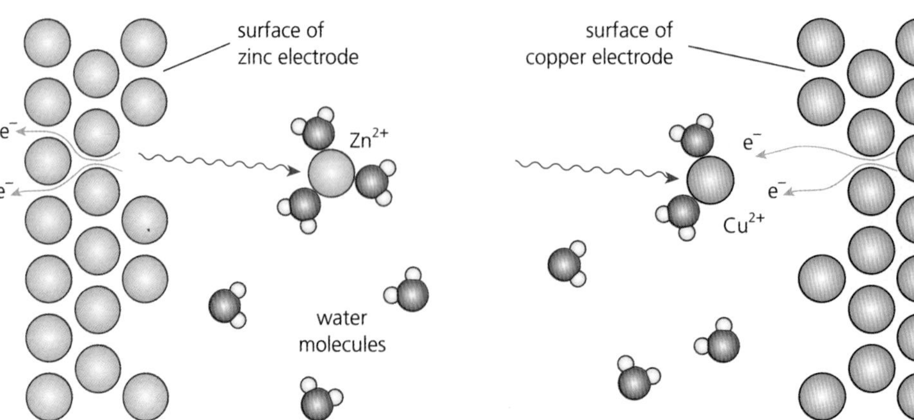

■ **Figure R3.51** Zinc atoms forming hydrated zinc ions on the surface of the zinc electrode of a Daniell cell

■ **Figure R3.52** Hydrated copper(II) ions forming copper atoms on the surface of the copper electrode of a Daniell cell

The process continues until a cell is depleted (if non-stoichiometric) or until the voltage is zero (at which point equilibrium has been attained). The zinc atoms act as a reducing agent and copper(II) ions as an oxidizing agent.

The relevant half-equations are:

Anode (oxidation): $Zn(s) \rightarrow Zn^{2+}(aq) + 2e^-$

Cathode (reduction): $Cu^{2+}(aq) + 2e^- \rightarrow Cu(s)$

Overall ionic equation: $Zn(s) + Cu^{2+}(aq) \rightarrow Cu(s) + Zn^{2+}(aq)$

The overall chemical change is the same when zinc is placed in copper(II) sulfate solution. Heat energy is released, but the arrangement in the voltaic cell, where the two reactions are physically separated, allows the production of electrical energy.

◆ **Standard cell potential:** The potential difference, measured in V, generated when two half-cells are connected under standard conditions (1 mol dm^{-3} and 298 K).

There is a flow of electrons from the anode to the cathode because there is an electric potential difference between the electrodes. This is measured in volts and known as the **standard cell potential**. A standard cell potential is measured under standard conditions of 298 K and electrolyte concentrations of 1 mol dm^{-3}. Under such conditions, the cell potential of the Daniell cell is +1.10 volts.

Similar voltaic cells can be made from different pairs of metals in contact with an aqueous solution of their ions connected by a salt bridge (which allows the flow of ions into the half-cells to balance the charges of the ions) and an external circuit. In each case, the more reactive metal forms the anode (negative electrode and where oxidation occurs) and the less reactive metal forms the cathode (positive electrode and where reduction occurs).

Inquiry 1: Exploring and designing

Identify and justify the choice of dependent, independent and control variables

What is an independent variable?

The independent variable is the variable that you choose to change. For example, if you are investigating the evaporation of liquids, you may choose to investigate how the temperature of a liquid affects its evaporation, or how the type of liquid affects its evaporation.

If you are confused, think: 'The **I**ndependent variable is the one that '**I**' choose to change.'

What is the dependent variable?

The dependent variable is the variable you measure. For example, if your independent variable is the current in an electroplating experiment, the dependent variable would be the mass of metal deposited.

What are control variables?

Control variables are essentially all the other variables in an experiment: those that are not the independent or dependent variables. However, you only need to control variables that are relevant to your experiment. Do not list or make efforts to control variables that are irrelevant.

Using the above electroplating example, the control variables would be things such as the temperature of the solution, the metal ion being deposited, the concentration of the solution and so on. These are all relevant to the experiment. The exact model of power pack or battery you use to supply your current is irrelevant.

Justification

Do not just list the independent, dependent and control variables. You must also explain your reasons for choosing them. You should state the individual values you will use for the independent variable and state how you will keep control variables constant.

Of the metal pairs shown in Table R3.20, copper and magnesium are furthest apart in the reactivity series. This combination of electrodes gives the highest cell potential. Lead and iron are the closest in the reactivity series and give the lowest voltage. Hence, the further apart the two metals are in the reactivity series, the higher the cell potential.

■ **Table R3.20** Selected voltaic cells and standard cell potentials

Metal electrodes	Standard potential / V
copper and magnesium	+2.70
copper and iron	+0.78
lead and zinc	+0.64
lead and iron	+0.32

Common mistake

Misconception: *The function of the salt bridge is to supply electrons to complete the circuit in a voltaic cell.*

The salt bridge completes the circuit so that charge (in the form of ions) can flow from one half-cell to the other and maintain electroneutrality.

The voltage of a cell depends not only on the nature of the electrodes and ions, but also on the concentrations and the temperatures of the electrolytes.

The removal of zinc from the zinc electrode in the Daniell cell will result in an increase in the concentration of zinc ions in the zinc sulfate solution. The deposition of copper(II) ions from the copper(II) sulfate solution as copper atoms will cause a decrease in the concentration of copper(II) ions in the copper(II) sulfate solution. These processes will lead to a *surplus* of positive ions in the zinc sulfate and a deficiency of positive ions in the copper(II) sulfate. Unless the concentrations of these positive ions are kept constant, then the two redox reactions will gradually slow and stop, and the current will drop to zero.

The imbalances in the concentrations of the two positive ions are restored by flows of ions from the salt bridge. For each zinc ion that enters the oxidation half-cell, two nitrate ions enter the half-cell from the salt bridge. For every two nitrate ions that enter the zinc sulfate, one zinc ion enters the salt bridge. In the copper(II) sulfate solution, positive potassium ions leave the salt bridge and are replaced by sulfate ions. For every two potassium ions that enter the copper(II) sulfate solution, one sulfate ion enters the salt bridge (Figure R3.53) to replace the two nitrate ions lost at the other end.

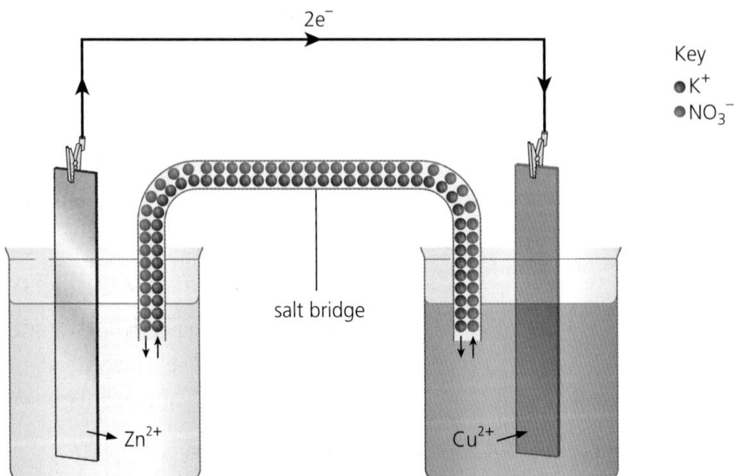

■ **Figure R3.53** Flow of ions in a salt bridge containing potassium and nitrate ions

Top tip!

The flow of ions across the salt bridge maintains an overall neutral charge in each half cell. Potassium nitrate is chosen for the salt bridge because neither potassium ions nor nitrate ions react chemically with the other ions present in the two solutions. Potassium and nitrate ions move at similar speeds to each other, meaning they enter their respective half-cells at a similar rate, maintaining charge neutrality in each half-cell.

Table R3.21 compares a voltaic cell with an electrolytic cell. They are both electrochemical cells involving redox reactions but have a number of important differences and similarities.

Table R3.21 Comparing a voltaic cell with an electrolytic cell

A voltaic cell	An electrolytic cell
Oxidation occurs at the anode (negative electrode).	Oxidation occurs at the anode (positive electrode).
Reduction occurs at the cathode (positive electrode).	Reduction occurs at the cathode (negative electrode).
It uses a redox reaction to produce a voltage.	It uses electricity to carry out a redox reaction.
A voltaic cell involves a spontaneous redox reaction $\Delta G < 0$.	An electrolytic cell involves a non-spontaneous redox reaction $\Delta G > 0$.
The voltaic cell converts chemical energy to electrical energy.	The electrolytic cell converts electrical energy to chemical energy.
There are two separate aqueous solutions connected by a salt bridge and an external circuit.	There is one electrolyte (solution or molten salt).

◆ **Cell diagram:** A diagram showing the electrodes and electrolytes in a voltaic cell.

Going further

Cell diagrams

We can represent a Daniell cell consisting of zinc and copper half-cells using the following cell diagram:

$Zn(s) \mid Zn^{2+}(aq) \parallel Cu^{2+}(aq) \mid Cu(s)$

The first half-cell (on the left) shows zinc atoms losing electrons to give zinc ions and the second half-cell (on the right) shows the reduction half-reaction. Notice that:

- the anode is written on the left and the cathode on the right
- a single vertical line represents a phase boundary between the solid electrode and the aqueous solution of ions
- the salt bridge is shown by two vertical bars.

◆ **Fuel cell:** An electrochemical cell in which the chemical energy of a fuel is converted directly into electrical energy.

Link

Fuel cells are covered in further detail in Chapter R1.3, page 382.

Fuel cells

Fuel cells and primary voltaic cells have similar basic operating principles, but there are some differences. A hydrogen fuel cell is an open system, the anode and cathode are in contact with gases and the reactants are externally supplied. A primary voltaic cell is a closed system where the anode and cathode are metals and the reactants are internally consumed.

A hydrogen fuel cell (Figure R3.54) uses the reaction between hydrogen and oxygen to produce water. The energy is released as electrical energy and not as heat. As the hydrogen and oxygen reactants are converted to products, they have to be replenished so a fuel cell can give a continuous supply of electricity.

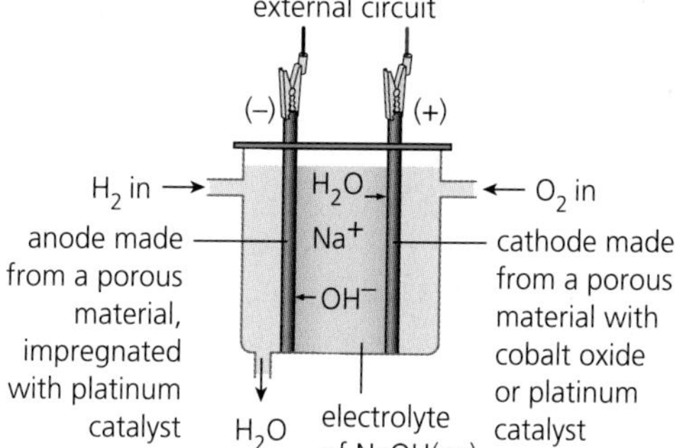

Figure R3.54 Hydrogen–oxygen fuel cell (in cross-section)

The electrolyte is a strong base, usually aqueous sodium hydroxide. It is contained within the fuel cell using porous electrodes, which allow the movement of gases and water molecules.

Oxidation: loss of electrons (negative electrode):

$$H_2(g) + 2OH^-(aq) \rightarrow 2H_2O(l) + 2e^-$$

Reduction: gain of electrons (positive electrode):

$$O_2(g) + 2H_2O(l) + 4e^- \rightarrow 4OH^-(aq)$$

ATL R3.2D

Find out about the use of fuel cells in vehicles and the current barriers to the development and widespread use of fuel cells in cars and buses. Present your findings to your peers using presentation software such as PowerPoint.

LINKING QUESTION

Electrical energy can be derived from the combustion of fossil fuels or from electrochemical reactions. What are the similarities and differences between these reactions?

◆ **Secondary cell**: A type of voltaic cell or battery that can be recharged by passing a current through it from an external d.c. supply. The charging current, which is passed in the opposite direction to that in which the cell supplies current, reverses the chemical reactions in the cell.

Fuel cells can also run on hydrocarbons and alcohols, which are more easily stored and transported than hydrogen. However, this type of cell does release carbon dioxide.

The following reactions occur in an ethanol fuel cell:

Anode: $C_2H_5OH(l) + 3H_2O(l) \rightarrow 12H^+(aq) + 12e^- + 2CO_2(g)$

Cathode: $3O_2(g) + 12H^+(aq) + 12e^- \rightarrow 6H_2O(l)$

Overall equation: $C_2H_5OH(l) + 3O_2(g) \rightarrow 3H_2O(l) + 2CO_2(g)$

Secondary voltaic cells

A **secondary cell** can be used many times and store electricity, rather than simply generating it. They differ from primary cells because a current can be passed through them to reverse the discharge chemical reactions. Secondary cells are more commonly known as rechargeable batteries.

Table R3.22 compares the advantages and disadvantages of primary cells, fuel cells and rechargeable batteries.

■ **Table R3.22** Advantages and disadvantages of primary cells, fuel cells and rechargeable batteries

	Advantages	**Disadvantages**
Primary cells	Cheap and widely available Small and highly portable Low toxicity compared to most secondary cells Long shelf life	Wasteful – they are used once and thrown away
Fuel cells	Do not require recharging – they can operate continuously as long as new fuel is supplied Fuel cells are well suited to remote locations	Expensive catalysts required (e.g. platinum in methanol fuel cells) Some fuel cells produce carbon dioxide Cannot be used in enclosed spaces due to fire risk
Rechargeable batteries (secondary cells)	Highly convenient for use in portable devices Potentially long lifetime with many charge cycles	Often contain toxic substances that can enter the environment if thrown away

TOK

How do values affect production of scientific knowledge?

Remember that nothing a chemist does, experiments on or uses exists in a vacuum. The wider social, political and environmental contexts of the processes a chemist uses to construct knowledge are always present and might constrain the actual *doing* of chemistry. Are the constraints imposed on the production or construction of chemical knowledge warranted? Do you think a chemist *should* be worried about his or her work in this wider context?

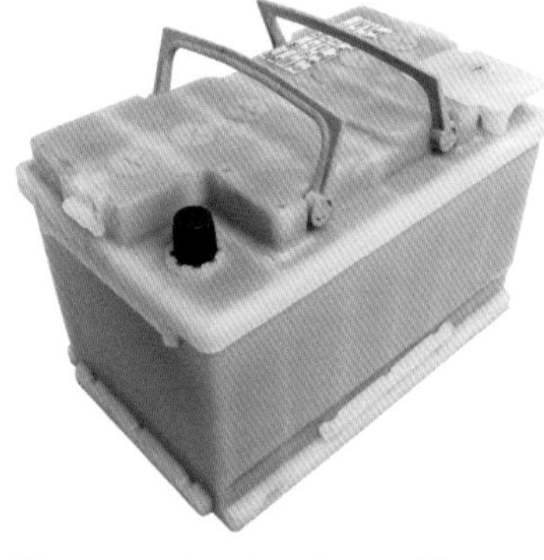

■ **Figure R3.55** A car battery (six lead–acid cells connected in series)

Lead–acid batteries

Lead–acid batteries are very widely used as car batteries and consist of six cells connected in series (Figure R3.55). Each cell generates a potential difference of approximately 2 V, leading to an overall voltage of 12 V for the battery.

The car battery must be able to generate a very high current for a short period of time, since it drives the starter motor which spins the engine to enable the internal combustion reaction to begin. It also generates the 'spark' for ignition in a petrol engine car. Once the car is running, the battery is recharged by a generator (an alternator) attached to the engine.

Both electrodes are made of lead but the cathode is also coated with lead(IV) oxide. The plates are separated by the 6 mol dm^{-3} sulfuric acid electrolyte. The lead anode is oxidized to lead(II) sulfate, and the lead(IV) oxide cathode is reduced to lead(II) sulfate.

During discharging:

Anode: $Pb(s) + SO_4^{2-}(aq) \rightarrow PbSO_4(s) + 2e^-$

Cathode: $PbO_2(s) + 4H^+(aq) + SO_4^{2-}(aq) + 2e^- \rightarrow PbSO_4(s) + 2H_2O(l)$

During recharging:

Cathode: $PbSO_4(s) + 2e^- \rightarrow Pb(s) + SO_4^{2-}(aq)$

Anode: $PbSO_4(s) + 2H_2O(l) \rightarrow PbO_2(s) + 4H^+(aq) + SO_4^{2-}(aq) + 2e^-$

Lead–acid cells have a low power-to-mass ratio due to the high density of lead. Lead(II) sulfate, which is the product of the discharge reaction at both electrodes, is very insoluble and a poor conductor. It adheres to the surfaces of the anode and cathode plates in the form of very small crystals; these crystals easily reform lead, lead(IV) oxide and sulfuric acid when the battery recharges.

There is an electrolytic side reaction that limits the life of a traditional lead–acid battery to around 1000 recharging cycles. Water in the electrolyte solution is electrolysed to hydrogen and oxygen gas. These gases diffuse out of the battery so the fluid level in the battery gradually drops. The battery can be topped up with distilled water but, as the sulfuric acid decreases, so does the voltage.

Lithium-ion batteries

In a lithium-ion battery (Figure R3.56) lithium ions (not electrons) move from the positive to the negative electrode during discharge via the polymer electrolyte (comparable to a salt bridge), while electrons flow in the external circuit.

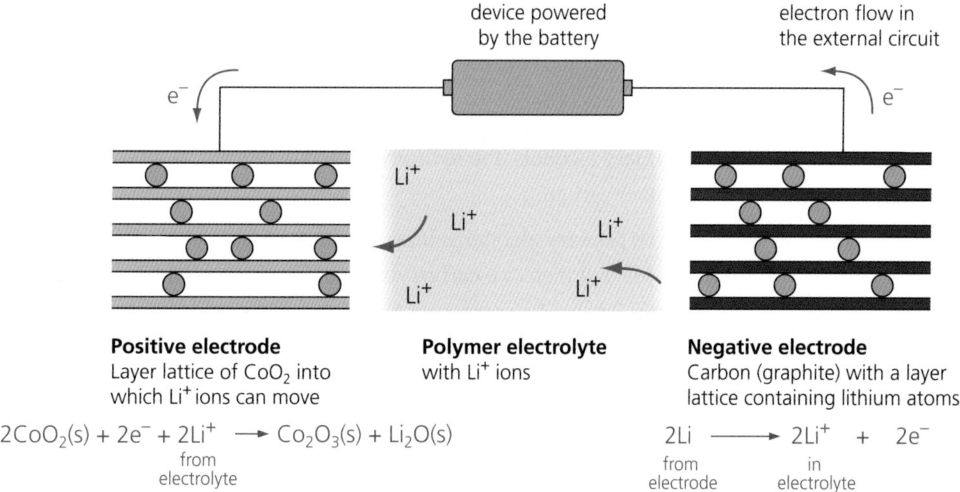

■ **Figure R3.56** Schematic diagram of a lithium-ion polymer battery showing the battery discharging. The electrode processes are reversible so the battery can be recharged

The anode is made of graphite and lithium atoms are inserted (intercalated) into the spaces between the graphite layers. During the cell reaction, lithium ions migrate out of the graphite electrode via the electrolyte (an organic polymer) while their electrons enter the external circuit.

$2Li$ (graphite) $\rightarrow 2Li^+$ (organic polymer electrolyte) $+ 2e^-$

The cathode is often cobalt(IV) oxide, CoO_2, which has a layer structure into which the lithium ions can move from the electrolyte. Electrons from the external circuit enter this lattice and reduce cobalt(IV) to cobalt(III).

$$2CoO_2(s) + 2e^- + 2Li^+ \text{ (organic polymer electrolyte)} \rightarrow Co_2O_3(s) + Li_2O(s)$$

During recharging, the lithium ions are driven back through the polymer electrolyte to the graphite anode.

Tesla motors use the lithium iron phosphate battery ($LiFePO_4$ battery) or LFP battery (lithium ferrophosphate) in some of their vehicles. These are a type of lithium-ion battery using lithium iron as the cathode, and graphitic carbon with a metallic backing as the anode.

The reaction can go both forwards and backwards (a feature of all secondary cells) and the products are stable:

Anode: $LiC_6 \rightleftharpoons Li^+ + C_6 + e^-$

lithiated graphite   graphite   (Li^+ dissolves in electrolyte)

Cathode: $FePO_4 + Li^+ + e^- \rightleftharpoons LiFePO_4$ (LFP)

LINKING QUESTION

Secondary cells rely on electrode reactions that are reversible. What are the common features of these reactions?

Oxidation of functional groups in organic compounds

LINKING QUESTION

How does the nature of the functional group in a molecule affect its physical properties, such as boiling point?

Most organic molecules contain functional groups which are the sites at which the molecule will react. The chemical characteristics of a molecule are determined by the functional groups it possesses.

Alcohols (excluding tertiary alcohols) and aldehydes, which contain hydroxyl and carbonyl functional groups respectively, are particularly susceptible to oxidation by oxidizing reagents. The transition metal ions in sodium dichromate(VI), $Na_2Cr_2O_7$, and potassium manganate(VII), $KMnO_4$, are often used in acidic conditions as oxidizing agents.

Oxidation of alcohols

Primary alcohols

In primary alcohols, the hydroxyl functional group is attached to a carbon which is bonded to only one other carbon.

Link

The structural classification of alcohols is also covered in Chapter S3.2, page 284.

When a primary alcohol is treated with a suitable oxidizing agent, it is oxidized first to an aldehyde and then to a carboxylic acid. These two oxidation steps are shown in Figure R3.57 where [O] is used to represent an appropriate oxidizing agent.

LINKING QUESTION

What is the difference between combustion and oxidation of an alcohol?

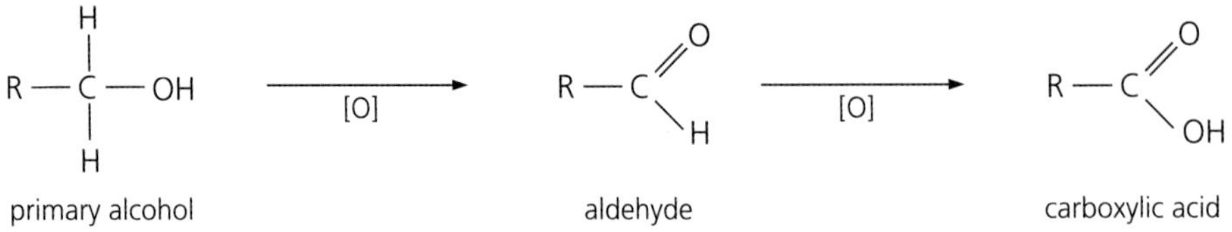

■ **Figure R3.57** The two stages in the oxidation of a primary alcohol

The oxidation of ethanol into ethanoic acid occurs when a bottle of wine is exposed to oxygen in the air. This causes the wine to smell, then to become more acidic and eventually unpleasant to drink. The first stage produces ethanal.

$$CH_3CH_2OH \rightarrow CH_3CHO + 2H^+ + 2e^-$$

The ethanal produced is then oxidized to ethanoic acid.

$$CH_3CHO + H_2O \rightarrow CH_3COOH + 2H^+ + 2e^-$$

In the laboratory, the apparatus used to oxidize primary alcohols depends on the product desired.

Producing aldehydes

If the aldehyde is required, it can be removed from the reaction mixture as it forms, by simple distillation. This is possible because aldehydes boil at a lower temperature than their related alcohol and carboxylic acid. Once distilled into a receiving flask, the aldehyde is not being exposed to the oxidizing agent in the reaction mixture, so it will not undergo the second oxidation step into the carboxylic acid. Using an excess of the alcohol and a limiting amount of oxidizing agent gives the best yield of an aldehyde.

Tool 1: Experimental techniques

Distillation

Distillation is a technique used to separate two liquids which have different boiling points. Simple distillation is used when the liquids have significantly different boiling points, such as ethanol and ethanal, but fractional distillation (Figure R3.58) is needed when liquids have much closer boiling points, for example, dodecanal and dodecanol.

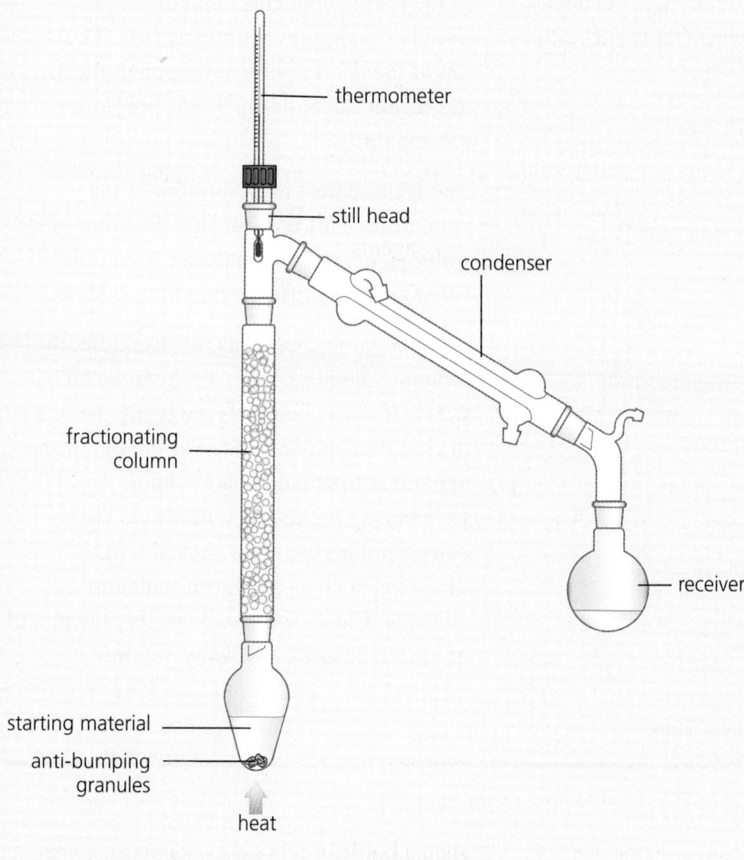

■ **Figure R3.58** Distillation using a fractionating column

Anti-bumping granules are small beads of aluminium oxide or silica glass which do not react but provide a rough surface with a large surface area on which small bubbles can form (nucleate). They are added to the reaction mixture to prevent the liquid boiling violently ('bumping') as it is heated.

When heated, the liquid in the flask produces a vapour which is rich in the most volatile liquid of the mixture. This vapour rises upwards through the fractionating column and condenses on the glass beads which fill the fractionating column. The cycle of evaporation and condensation occurs many times as the vapour rises through the column; each time, the vapour becomes richer in the most volatile component of the mixture. At the top of the fractionating column, the vapour meets the cool condenser and drips down into the collection vessel (receiver).

The condensation point of the vapour is the same value as the boiling point of the liquid. A thermometer is set in the still head, so the bulb of the thermometer is level with the exit to the condenser. This means the temperature of the vapour leaving, which condenses to form the organic product, can be monitored and recorded.

For mixtures which have a very large difference in boiling points, distillation can be done without including the fractionating column. This is known as simple distillation.

Producing carboxylic acids

If the carboxylic acid is the desired product, the aldehyde must be left in contact with the oxidizing reagent in the reaction flask by heating under reflux. An excess of the oxidizing agent is used to ensure all the alcohol and aldehyde are used up.

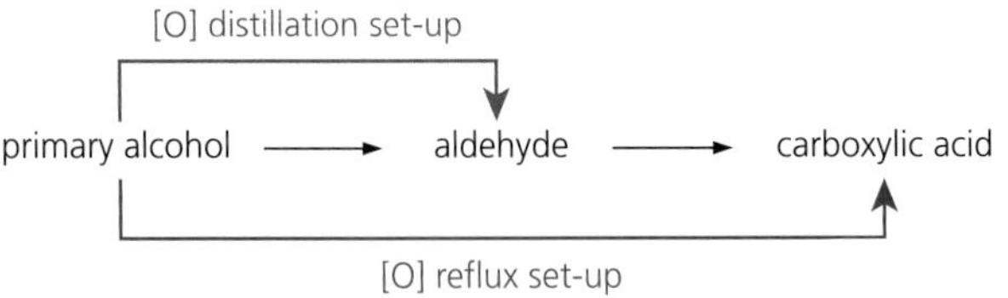

■ **Figure R3.59** Effect of set-up on product of oxidation of a primary alcohol

Tool 1: Experimental techniques

Refluxing

When an organic reaction mixture needs to be heated for a long period of time (several minutes to hours), it is heated under reflux. This allows the reaction to be heated at the boiling point of the chosen solvent without loss of the solvent. The apparatus consists of a flask connected directly to a vertically mounted Liebig condenser (Figure R3.60).

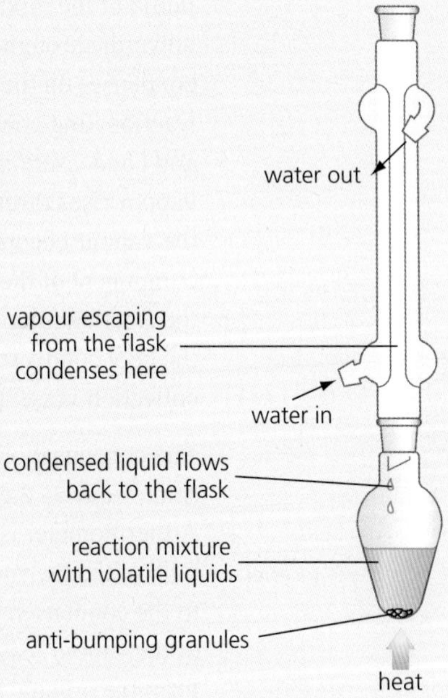

■ **Figure R3.60** Heating under reflux

Since many organic liquids are flammable, an electric heating mantle is used rather than a Bunsen burner. As with distillation, inert anti-bumping granules are added to the reaction mixture so the liquid boils smoothly. The vapour produced condenses in the Liebig condenser and drips back into the flask, preventing any loss. The top of the condenser is left open to the air to prevent the build-up of gas pressure in the apparatus.

The water enters at the bottom of the condenser. This ensures that the outer glass water jacket of the condenser is completely filled and is an effective cooling system.

Some organic chemicals (for example, the reducing agent $LiAlH_4$) are hydrolysed by water. If such chemicals are being used, a drying tube may be added to the top of the apparatus to prevent water vapour from the air entering the reaction mixture. This consists of a glass tube packed with a desiccant such as anhydrous calcium chloride, $CaCl_2$, which allows the movement of air but absorbs any water vapour.

Top tip!

Using distillation equipment for the oxidation of a primary alcohol produces an aldehyde. Using reflux equipment for the oxidation of a primary alcohol produces a carboxylic acid.

Secondary alcohols

Secondary alcohols are oxidized to ketones by a suitable oxidizing agent. The ketone will not be oxidized further, and this reaction usually works very well using acidified potassium dichromate. Propan-2-ol, the simplest secondary alcohol, is oxidized to propanone (Figure R3.61).

Tertiary alcohols cannot easily be oxidized.

> **LINKING QUESTION (HL ONLY)**
>
> Why is there a colour change when an alcohol is oxidized by a transition element compound?

$$CH_3CH(OH)CH_3 \quad + \quad [O] \longrightarrow CH_3COCH_3 \quad + \quad H_2O$$

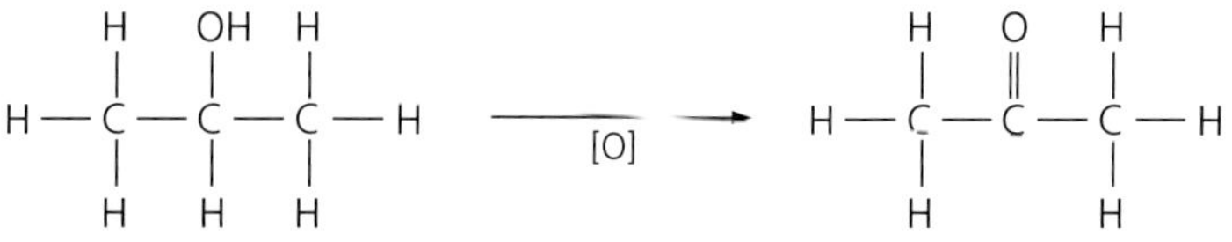

propan-2-ol (secondary alcohol) $\xrightarrow{[O]}$ propanone (ketone)

■ **Figure R3.61** The oxidation of the secondary alcohol, propan-2-ol

Going further

Breathalysers

As blood passes through the arteries in the lungs, an equilibrium is established between any ethanol in the blood plasma and the ethanol in the breath. So, if the concentration of one is known, the concentration of the other can be estimated.

The first breathalyser used to estimate a person's blood alcohol level was based on the oxidation of breath ethanol by acidified sodium dichromate(VI). The test used a sealed glass tube that contained an inert material impregnated with the oxidizing agent.

The ends of the tube were broken off; one end of the tube was attached to a mouthpiece and the other to a balloon-type bag. The person thought to be drunk blew into the mouthpiece until the bag was filled with air.

Any ethanol in the breath was oxidized as it passed through the column. When ethanol is oxidized, the orange oxidizing agent ($Cr_2O_7^{2-}$) is reduced to blue-green chromium(III) ions (Cr^{3+}). The greater the concentration of alcohol in the breath, the further the blue-green colour spreads through the tube.

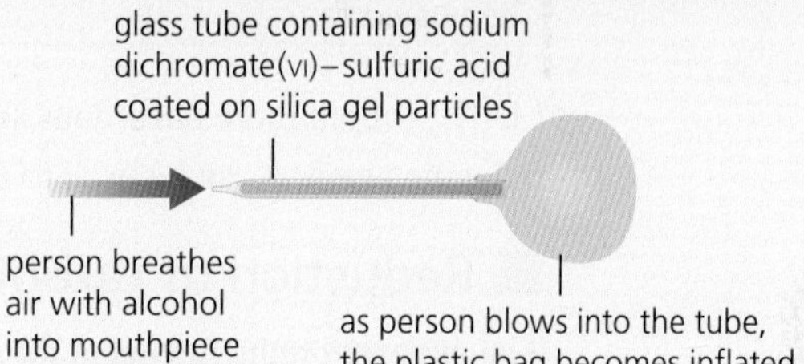
glass tube containing sodium dichromate(VI)–sulfuric acid coated on silica gel particles
person breathes air with alcohol into mouthpiece
as person blows into the tube, the plastic bag becomes inflated

■ **Figure R3.62** A simple breathalyser

Chemical breathalysers are no longer used. Instead, the amount of ethanol in the breath is now accurately measured by a fuel cell in the breathalyser. The ethanol blown over the fuel cell is oxidized at the anode of the fuel cell, and the current produced is proportional to the concentration of ethanol present.

Other organic molecules such as propanone may act as a fuel in the fuel cell and their presence can cause inaccurate results. The most sophisticated breathalysers use infrared spectroscopy to determine the concentration of ethanol based upon the characteristic absorptions across a range of wavelengths. This easily allows the elimination of any contribution due to propanone.

Reduction of functional groups

It should not be a surprise to learn that the sequence of oxidation from a primary alcohol through an aldehyde to a carboxylic acid can be reversed by using suitable reducing agents.

In these reactions, [H] is often used to represent a suitable reducing agent. No single reducing agent works brilliantly for all the steps, so the reagent chosen depends on the reaction desired. For example, borane bonded to a solvent such as ethoxyethane, $BH_3O(CH_2CH_3)_2$, works particularly well for carboxylic acids.

The most common and easy to use reducing agents are the complex hydrides lithium aluminium hydride, $LiAlH_4$, and sodium borohydride, $NaBH_4$. They are ionic compounds which contain the tetrahedral anion, $[MH_4^-]$. Table R3.23 lists some of their properties and how they are used.

■ **Table R3.23** Some properties of complex inorganic metal hydrides used as reducing agents in organic chemistry

Common name	sodium borohydride	lithium aluminium hydride
Systematic name	sodium tetrahydridoborate(III)	lithium tetrahydridoaluminate(III)
Formula	$NaBH_4$	$LiAlH_4$
Solvent used and reaction conditions	methanol–water mixture, room temperature or warming	totally dry ether, room temperature or at reflux (35 °C)
How the product is separated	adding acid and extracting the product from the aqueous mixture with an organic solvent such as ether	adding water (with care), then acid, and separating the ether solution of the product from the aqueous layer
Hazards during use	relatively safe in alkaline solution, but evolves hydrogen gas with acids	can catch fire with water or when wet, and ether solvents form explosive mixtures with air
Functional groups that can be reduced, and what they are reduced to	carbonyl compounds (to alcohols)	**carbonyl compounds (to alcohols)** esters (to alcohols) **carboxylic acids (to alcohols)** amides (to amines) nitriles (to amines)

$LiAlH_4$ is both more hazardous and more expensive than $NaBH_4$ but, as the more powerful reducing agent, it can be used to reduce carboxylic acids and their derivatives, unlike $NaBH_4$.

Reduction of carbonyl compounds and carboxylic acids

The preferred reducing agent for aldehydes and ketones is $NaBH_4$ but it is not a powerful enough reducing agent to reduce carboxylic acids.

Ketones are reduced to secondary alcohols. Carboxylic acids are reduced to primary alcohols via an aldehyde. It is not possible to stop the reduction of a carboxylic acid at the halfway stage (the aldehyde) because the powerful reducing reagent used will very quickly reduce the aldehyde.

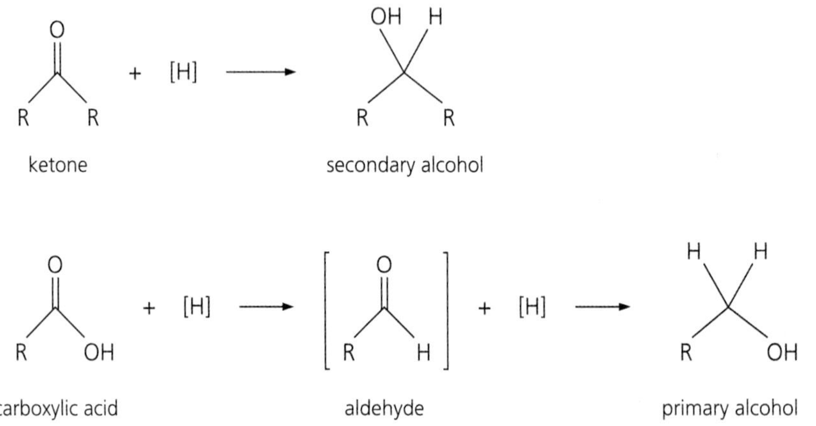

O, R, R + [H] → OH, H, R, R

ketone — secondary alcohol

O, R, OH + [H] → [O, R, H] + [H] → H, H, R, OH

carboxylic acid — aldehyde — primary alcohol

■ **Figure R3.63** Reduction of ketones and carboxylic acids to secondary and primary alcohols

Each of the reducing agents used to reduce carbonyl-containing molecules can be equated to a source of hydride, H^-, ions. The H^- ions form a new bond to the carbon of a carbonyl group, breaking the pi bond. The hydrogen ions, H^+, needed to complete the reaction are provided in a second 'work-up' step of the reaction (Figure R3.64).

■ **Figure R3.64** Reduction of propanone to propanol-2-ol via a charged intermediate and work-up with dilute acid

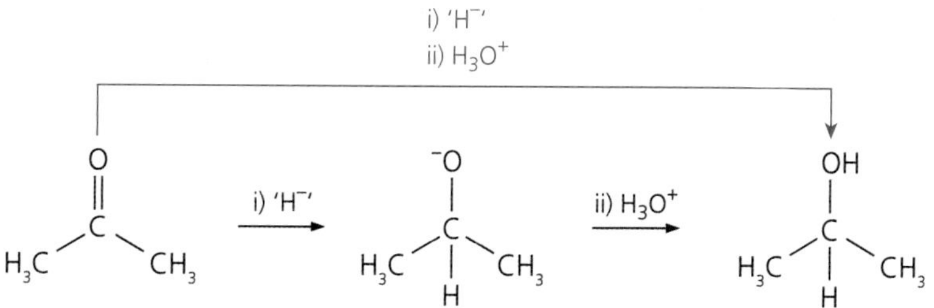

Link

Nucleophiles (electron pair donors) and electrophiles (electron pair acceptors) are covered in Chapter R3.4, page 638.

It is easy to tell whether an organic compound has been oxidized or reduced simply by looking at the change in the structure of the compound. If the reaction increases the number of C–H bonds or decreases the number of C–O, C–N, or C–X bonds (where X indicates a halogen), the compound has been reduced.

a
δ+ δ–
C═O
carbonyl group

b
H:⁻
nucleophile
δ+ δ–
C═O
electron-deficient carbon

■ **Figure R3.65 a** A carbonyl group contains an electron-deficient carbon atom. **b** Reduction begins with the nucleophilic attack of a hydride ion, H^-

In the reduction reaction, the reducing agent is providing a source of H^- ions which act as nucleophiles on the electron-deficient carbonyl carbon (Figure R3.65). Because these reagents need a partially positive carbon atom to react with, they do not reduce carbon–carbon double bonds.

47 Identify the alcohols X, Y and Z.

Alcohol	Molecular formula	Oxidized by acidified sodium dichromate(VI)?	Additional notes
X	C_3H_8O	yes	The product of oxidation evolves carbon dioxide when Na_2CO_3 is added.
Y	$C_4H_{10}O$	no	
Z	$C_5H_{12}O$	yes	The alcohol has 3 signals in its low-resolution ^1H NMR spectrum.

48 Give the structural formulas of the compounds produced when the following are reacted with an excess of lithium aluminium hydride:

a pentan-2-one

b 2-methylpropanal

c ethanoic acid

d $(C_6H_5)CH_2COCOOH$ (a keto acid)

LINKING QUESTION

How can oxidation states be used to show that the following molecules are given in increasing order of oxidation: CH_4, CH_3OH, HCHO, HCOOH, CO_2?

Reduction of unsaturated compounds

◆ **Degree of unsaturation:** The total number of pi bonds and rings possessed by a molecule.

Acyclic (non-cyclic) alkanes cannot react with hydrogen; they are said to be already saturated with hydrogen. Other hydrocarbons which contain pi bonds and/or rings can react with hydrogen and are unsaturated. These ideas lead to the concept of **degree of unsaturation** of a molecule; this can be thought of as the number of molecules of hydrogen it could potentially react with to produce a saturated molecule (Figure R3.66).

Formula	$CH_3(CH_2)_3CH_3$	$CH_3C(CH_3)CH_2$			
Name	pentane	2-methylpropene	cyclopentene	pent-1-yne	propanone
Degree of unsaturation	0	1	2	2	1

■ **Figure R3.66** Selected organic molecules with their formulas and degree of unsaturation

The degree of saturation of a hydrocarbon is easily calculated from the molecular formula. A saturated hydrocarbon will have the general formula C_nH_{2n+2}. Hexene, C_6H_{12}, would need two more hydrogen atoms (one hydrogen molecule) to have the formula of a saturated hydrocarbon, so its degree of unsaturation is 1.

WORKED EXAMPLE R3.2G

Deduce the degree of unsaturation in a molecule of the common analgesic (pain-killer) ibuprofen.

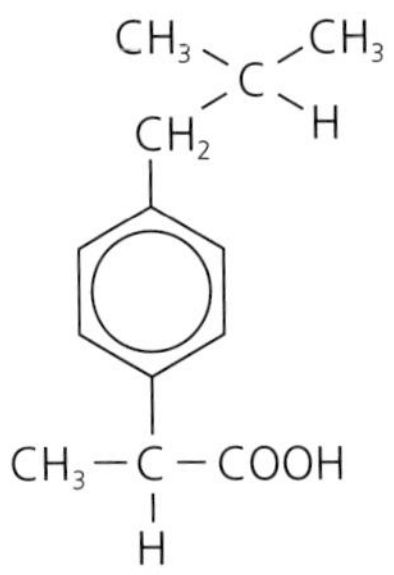

Answer

Each pi bond and ring contributes 1 to the degree of unsaturation. The arene ring in the molecule has three pi bonds and one ring, and there is one more pi bond between carbon and oxygen in the carbonyl functional group. The total degree of unsaturation is 5.

49 Deduce the degree of unsaturation of the following molecules:

a $CH_3CH{=}CH(CH_2)_3CH{=}CH_2$

b $CH_3(CH_2)_5C{\equiv}CH$

c 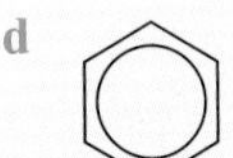

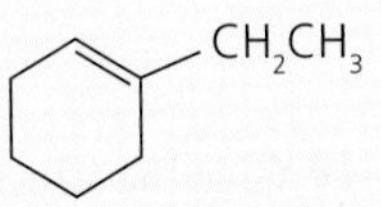

d

e

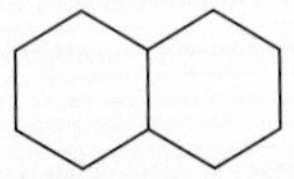

◆ **Hydrogenation**: The addition of hydrogen across a multiple bond, often performed in the presence of a heated transition metal catalyst.

Alkenes and alkynes have degrees of unsaturation of 1 and 2, respectively. When these are reacted with hydrogen in the presence of a suitable transition metal catalyst, such as palladium (Pd) or nickel (Ni), the corresponding alkane is produced (Figure R3.67). In these **hydrogenation** reactions, the degree of unsaturation of the reactant is reduced.

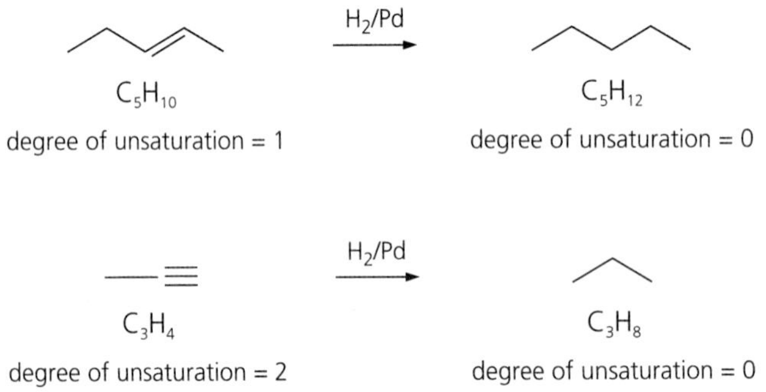

C_5H_{10} degree of unsaturation = 1 —H_2/Pd→ C_5H_{12} degree of unsaturation = 0

C_3H_4 degree of unsaturation = 2 —H_2/Pd→ C_3H_8 degree of unsaturation = 0

■ **Figure R3.67** The hydrogenation reaction of alkenes and alkynes

LINKING QUESTION

Why are some reactions of alkenes classified as reduction reactions while others are classified as electrophilic addition reactions?

The hydrogenation of alkenes is used to transform plant oils into margarine. Most plant oils have many *cis* double bonds which disrupt the packing of the alkyl chains. Hydrogenating the double bonds transforms the oils into a more suitable butter substitute with a higher melting point, and they become more stable to oxidation.

Types of half-cell

Any two different half-cells can be combined to form a voltaic cell that allows electrons to flow from the reducing agent to the oxidizing agent. The current in the external circuit allows useful work to be done (for example, lighting a bulb or driving an electric motor). It also allows chemists to measure the tendency for a redox reaction to occur.

There are three types of commonly encountered half-cell:

1 metal immersed in an aqueous solution of its own cations
2 inert electrode (for example, graphite or platinum) immersed in an aqueous solution containing two ions of the same element in *different* oxidation states (for example, $Fe^{3+}(aq)/Fe^{2+}(aq)$).
3 gas bubbling over an inert electrode immersed in an aqueous solution containing ions formed by oxidation/reduction of the gas, for example, the standard hydrogen electrode discussed below.

◆ **Hydrogen half-cell:** The standard hydrogen electrode is a type of half-cell used in measuring standard electrode potentials, which uses a platinum foil with a $1.0\,mol\,dm^{-3}$ solution of hydrogen ions, hydrogen gas at 100 kPa pressure, and a temperature of 298 K.

Hydrogen half-cell

The voltage of a single metal electrode in the half-cell of a voltaic cell cannot be measured in isolation. The solution to this problem is to choose a standard reference electrode. The internationally agreed reference electrode is the **hydrogen half-cell** (Figure R3.68) in which hydrogen gas is in equilibrium with aqueous hydrogen ions.

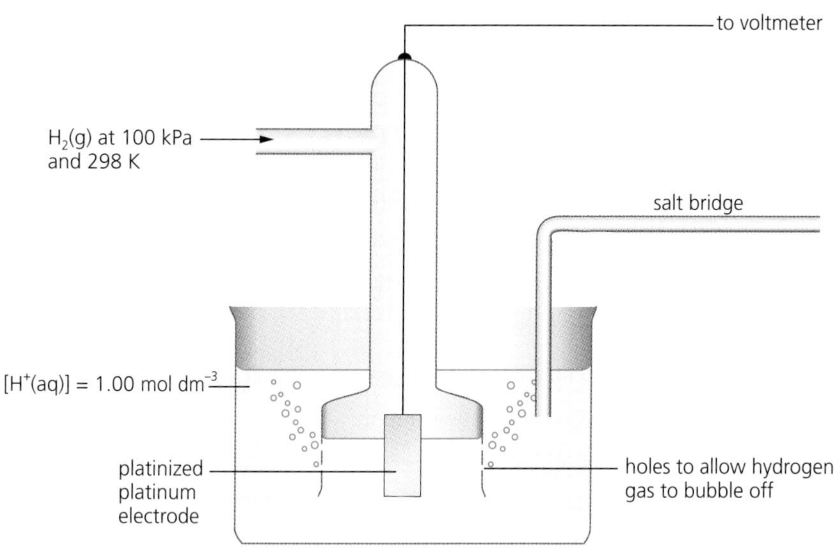

■ **Figure R3.68** A hydrogen half-cell

The hydrogen half-cell is maintained by a stream of pure hydrogen gas bubbling over a platinum electrode coated with platinum black (finely divided platinum) immersed in a solution of hydrochloric acid. The platinum electrode acts as an inert electrode; there is almost no tendency for the very unreactive platinum atoms to ionize. The platinized surface adsorbs hydrogen gas and acts as a heterogeneous catalyst, providing a surface or base for oxidation of H_2 and reduction reactions of H^+. This allows standard electrode potentials to be measured quickly.

An equilibrium is set up between the gas adsorbed on the electrode and the hydrogen ions in the acidic solution. The redox half-equation for the hydrogen half-cell is:

$$2H^+(aq) + 2e^- \rightleftharpoons H_2(g);\ E^\ominus = 0.00\,V$$

The electrode potential of this electrode is defined as 0.00 V under the following standard conditions:

- temperature at 298 K (25 °C)
- pressure of hydrogen gas at standard atmospheric pressure (100 kPa)
- hydrogen ion concentration at one mole per cubic decimetre ($1\,mol\,dm^{-3}$).

HL ONLY

Nature of science: Science as a shared endeavour

Hydrogen half-cell reference electrode

The hydrogen half-cell is the internationally agreed IUPAC reference for reporting standard electrode potentials. It was widely used in early studies as a reference electrode, and as an indicator electrode for the determination of pH values.

It is, nonetheless, an arbitrary reference: in principle any metal / metal ion system could be used as a reference electrode. To use an analogy, heights are measured using sea level as an arbitrary zero. Any level above or below sea level could be used as the reference point. The voltage taken to be zero is arbitrary; it is the difference in potential which has practical consequences.

Tabulating all electrode potentials with respect to the same hydrogen half-cell provides a practical working framework for a wide range of calculations and predictions.

All measurements are arbitrary – they are ratios of a measurement to a defined standard. The basic physical or chemical laws do not change but the values of the constants do. The important issue is not what reference we choose but that this reference is precisely defined and accepted by scientists; this emphasizes the nature of 'science' as a community of people collaborating, as well as a body of knowledge produced by this community. The knowledge is a direct consequence of people making decisions and deciding to agree to those decisions.

As a result, it is important to ask: *who* is in the scientific community? Does that community represent an adequate variety of voices? What institutional and interpersonal power structures are involved in maintaining that community and how do they manage the diversity of voices? What is the *effect* of those power or political structures on the knowledge that is produced?

Standard electrode potentials

◆ **Standard electrode potential**: The potential difference between a hydrogen half-cell and an electrode which is immersed in a solution in a half-cell containing metal ions at 1 mol dm^{-3} concentration at 298 K (25 °C) and, if a gas is present, at a pressure of 100 kPa.

Top tip!

Note that the amount of metal present does *not* influence the electrode potential of a metal.

If a standard half-cell is connected to a hydrogen half-cell electrode to form a voltaic cell, the measured voltage is the **standard electrode potential** of that half-cell.

The electrode potential of a metal depends on three factors:

- the reactivity of the metal
- the concentration of its cations in solution
- the temperature of the solution.

The concentrations and temperature of the electrolytes therefore have to be stated when comparing the electrode potentials of different elements.

Figure R3.69 shows the arrangement used to measure the standard electrode potential (−0.76 V) of a zinc half-cell.

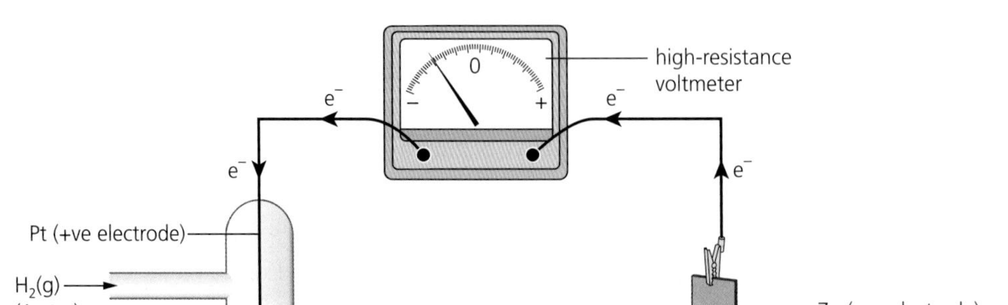

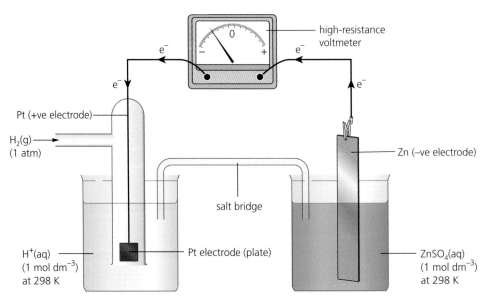

■ **Figure R3.69** Measuring the standard electrode potential of a zinc half-cell

Top tip!

The negative sign of the electrode potential is used to show that the zinc electrode is the negative terminal if it is connected to a hydrogen half-cell.

The voltmeter shows that the electrons flow from the zinc electrode to the hydrogen half-cell in the external circuit. This means that at the zinc electrode, the following reaction occurs:

$$Zn(s) \rightarrow Zn^{2+}(aq) + 2e^-$$

Oxidation occurs and the zinc electrode acts as the negative terminal, the anode.

The hydrogen half-cell electrode acts as the positive terminal, the cathode, and the overall cell reaction is:

$$Zn(s) + 2H^+(aq) \rightarrow Zn^{2+}(aq) + H_2(g)$$

By convention, the oxidized species is written first when a half-equation and its standard electrode potential are referred to. The half-equation is written as a *reduction* process:

oxidized species + ne^- ⇌ reduced species

Thus, $Ag^+(aq)/Ag(s)$, $E^\ominus = +0.80\,V$ means that the silver half-cell reaction has a standard electrode potential of +0.80 V:

$$Ag^+(aq) + e^- \rightarrow Ag(s) \qquad E^\ominus = +0.80\,V$$

oxidized species reduced species

The IB *Chemistry data booklet* contains standard reduction potentials of half-cells recorded in this format in section 18.

Table R3.24 Standard reduction potentials for some common metals (all data taken from the IB *Chemistry data booklet*)

Oxidized species	⇌	**Reduced species**	$E^\ominus$ / V
$Li^+(aq) + e^-$	⇌	$Li(s)$	−3.04
$K^+(aq) + e^-$	⇌	$K(s)$	−2.93
$Ca^{2+}(aq) + 2e^-$	⇌	$Ca(s)$	−2.87
$Na^+(aq) + e^-$	⇌	$Na(s)$	−2.71
$Mg^{2+}(aq) + 2e^-$	⇌	$Mg(s)$	−2.37
$Al^{3+}(aq) + 3e^-$	⇌	$Al(s)$	−1.66
$Mn^{2+}(aq) + 2e^-$	⇌	$Mn(s)$	−1.18
$Zn^{2+}(aq) + 2e^-$	⇌	$Zn(s)$	−0.76
$Fe^{2+}(aq) + 2e^-$	⇌	$Fe(s)$	−0.45
$Ni^{2+}(aq) + 2e^-$	⇌	$Ni(s)$	−0.26
$Sn^{2+}(aq) + 2e^-$	⇌	$Sn(s)$	−0.14
$Pb^{2+}(aq) + 2e^-$	⇌	$Pb(s)$	−0.13
$H^+(aq) + e^-$	⇌	$H_2(g)$	0.00
$Cu^{2+}(aq) + 2e^-$	⇌	$Cu(s)$	+0.34
$Cu^+(aq) + e^-$	⇌	$Cu(s)$	+0.52
$Ag^+(aq) + e^-$	⇌	$Ag(s)$	+0.80

In summary:

If $E^{\ominus} > 0$, then the electrode is the cathode: *reduction* takes place at this electrode (electrons are used up) when it is connected to the hydrogen half-cell (Figure R3.70).

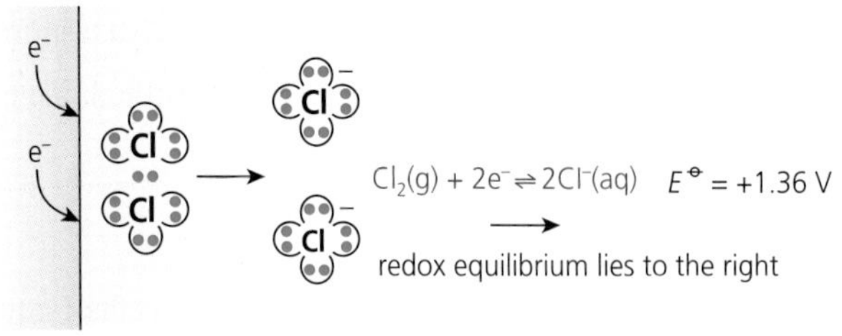

$Cl_2(g) + 2e^- \rightleftharpoons 2Cl^-(aq) \quad E^{\ominus} = +1.36\ V$

redox equilibrium lies to the right

■ **Figure R3.70** Reduction occurs at the cathode

If $E^{\ominus} < 0$, then the electrode is the anode: *oxidation* takes place at this electrode (electrons are produced) when it is connected to the hydrogen half-cell (Figure R3.71).

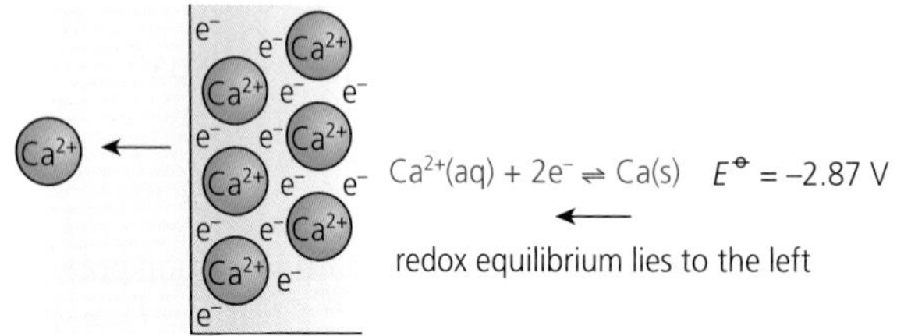

■ **Figure R3.71** Oxidation occurs at the anode

The electrochemical series

◆ **Electrochemical series:** An arrangement of elements and ions (which can undergo redox reactions) arranged in order of their standard reduction potentials, with the most negative (that is, most reducing) at the top of the series.

The arrangement of metals and hydrogen in order of increasing standard electrode potential is known as the **electrochemical series**.

Metals towards the top of the electrochemical series, with large negative electrode potentials, are very reactive and readily give up electrons in solution. In other words, they are powerful reducing agents.

Top tip!

Note that the half-cell reactions are written as reduction processes; the metal ions are gaining electrons. The standard electrode potentials are therefore sometimes known as standard reduction potentials. If the equations are reversed, we get standard oxidation potentials – numerically the same but with the opposite sign.

Towards the bottom of the electrochemical series, the metals become progressively weaker reducing agents but their oxidizing power of their cations increases. The cations of unreactive metals at the very bottom of the electrochemical series behave as weak oxidizing agents (Figure R3.72).

■ **Figure R3.72** Summary of the trends in the electrochemical series

WORKED EXAMPLE R3.2H

Use the standard electrode potentials below to arrange the following oxidizing agents in increasing order of oxidizing strength (under standard conditions):

- potassium manganate(VII) (in acidic solution)
- iodine
- iron(III) ions
- oxygen (in acidic solution).

The standard electrode potentials are:

$MnO_4^-(aq) + 8H^+(aq) + 5e^- \rightarrow Mn^{2+}(aq) + 4H_2O(l)$ $\quad E^\ominus = +1.51\text{ V}$

$\frac{1}{2}I_2(s) + e^- \rightarrow I^-(aq)$ $\quad E^\ominus = +0.54\text{ V}$

$Fe^{3+}(aq) + e^- \rightarrow Fe^{2+}(aq)$ $\quad E^\ominus = +0.77\text{ V}$

$\frac{1}{2}O_2(g) + 2H^+(aq) + 2e^- \rightarrow H_2O(l)$ $\quad E^\ominus = +1.23\text{ V}$

Answer

Oxidizing agents undergo reduction (gain of electrons). The more positive the value of the standard electrode potential, the greater the oxidizing power of the chemical species on the left-hand side of the reduction equation.

Hence, the order of increasing oxidizing power (under standard conditions) is:

- iodine
- iron(III) ions
- oxygen (in acidic solution)
- potassium manganate(VII) (in acidic solution).

The redox series

The electrochemical series has been extended to include the standard electrode potentials of redox systems in which transition metals are present in different oxidation states (Table R3.25).

Table R3.25 Part of the electrochemical series

Oxidized species	$\rightleftharpoons$	**Reduced species**	$E^\ominus$ **/ V**
$Pb^{2+}(aq) + 2e^-$	$\rightleftharpoons$	$Pb(s)$	−0.13
$H^+(aq) + e^-$	$\rightleftharpoons$	$H_2(g)$	0.00
$Cu^{2+}(aq) + e^-$	$\rightleftharpoons$	$Cu^+(aq)$	+0.15
$SO_4^{2-}(aq) + 4H^+(aq) + 2e^-$	$\rightleftharpoons$	$H_2SO_3(aq) + H_2O(l)$	+0.17
$Cu^{2+}(aq) + 2e^-$	$\rightleftharpoons$	$Cu(s)$	+0.34
$\frac{1}{2}O_2(g) + H_2O(l) + 2e^-$	$\rightleftharpoons$	$2OH^-(aq)$	+0.40
$Cu^+(aq) + e^-$	$\rightleftharpoons$	$Cu(s)$	+0.52
$\frac{1}{2}I_2(s) + e^-$	$\rightleftharpoons$	$I^-(aq)$	+0.54
$Fe^{3+}(aq) + e^-$	$\rightleftharpoons$	$Fe^{2+}(aq)$	+0.77

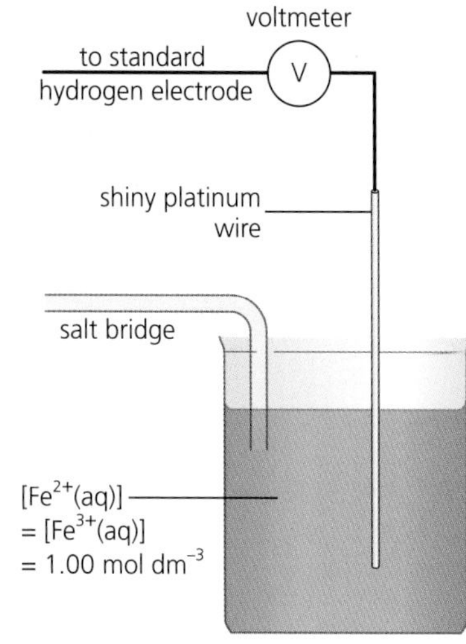

■ **Figure R3.73** The half-cell system used to measure the standard electrode potential for the $Fe^{3+}(aq)/Fe^{2+}(aq)$ system (redox couple)

$Fe^{3+}(aq) + e^- \rightarrow Fe^{2+}(aq)$ is a half-cell formed by dipping a platinum wire into an aqueous solution containing a mixture of 1 mol dm^{-3} iron(II) ions and 1 mol dm^{-3} iron(III) ions (Figure R3.73).

Electrical contact is made with the mixture of two ions by means of the platinum wire, which acts as an inert conductor. A redox equilibrium is established:

$Fe^{2+}(aq) \rightarrow Fe^{3+}(aq) + e^-$ and $Fe^{3+}(aq) + e^- \rightarrow Fe^{2+}(aq)$

$Fe^{3+}(aq) + e^- \rightleftharpoons Fe^{2+}(aq)$

Electrons produced by the backward reaction are transferred to the surface of the platinum, making it negatively charged, while the forward reaction removes electrons from the surface of the platinum wire. The resultant charge therefore depends on the relative balance between these two opposing processes.

WORKED EXAMPLE R3.21

Construct a cell diagram in which the following ionic reaction takes place:

$2Fe^{3+}(aq) + 2I^-(aq) \rightarrow 2Fe^{2+}(aq) + I_2(s)$

Answer

From the ionic reaction, it is obvious that the iodide ions are oxidized to iodine molecules and iron(III) ions are reduced to iron(II) ions. This shows that the electrode forming the oxidation terminal (the anode) will be:

$I^-(aq) \mid I_2(s)$

The electrode constituting the reduction terminal (the cathode) will be:

$Fe^{3+}(aq), Fe^{2+}(aq) \mid Pt$

The voltaic cell, representing the redox reaction will be:

$I^-(aq) \mid I_2(s) \parallel Fe^{3+}(aq), Fe^{2+}(aq) \mid Pt$

Ox-electrode Red-electrode

(anode) (cathode)

The corresponding half-equations are:

oxidation half-reaction (at anode): $2I^-(aq) \rightarrow I_2(s) + 2e^-$

reduction half-equation (at cathode): $2Fe^{3+}(aq) + 2e^- \rightarrow 2Fe^{2+}(aq)$

overall ionic equation: $2Fe^{3+}(aq) + 2I^-(aq) \rightarrow 2Fe^{2+}(aq) + I_2(s)$

(Note: there is a practical difficulty of using iodine as an electrode since it will slowly dissolve in aqueous solutions. Iodine-based batteries use an organic solvent and often incorporate the iodine into graphite.)

Standard cell potentials

Predictions of reactions at the electrodes and standard cell potentials for spontaneous redox reactions can be easily deduced and calculated.

The electrode with the more positive (or less negative) $E^\ominus$ value is more oxidizing and so will undergo reduction. This is the half-cell is on the right (the cathode, Figure R3.74).

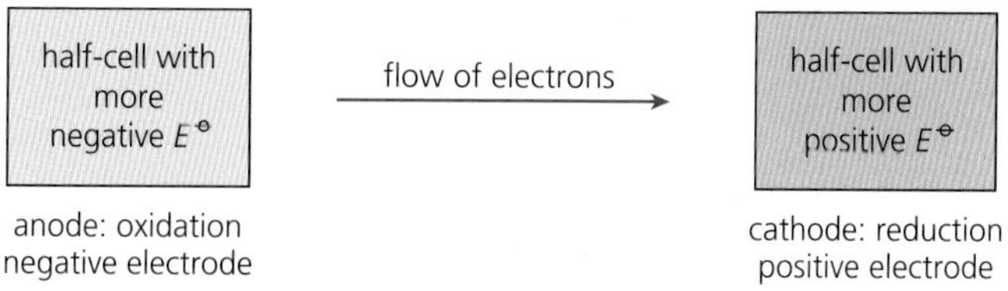

Top tip!

Reduction always takes place at the cathode and oxidation always takes place at the anode in any electrochemical cell (both voltaic and electrolytic cells).

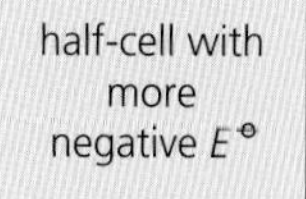

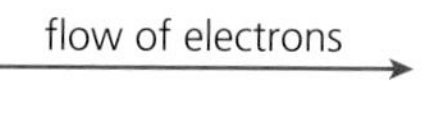

half-cell with more positive $E^\ominus$

anode: oxidation
negative electrode

cathode: reduction
positive electrode

■ **Figure R3.74** Flow of electrons in a voltaic cell

The cell potential can be determined from the cell diagram of the voltaic cell:

$$E^\ominus_{cell} = E^\ominus_C - E^\ominus_A$$

where $E^\ominus_C = E^\ominus$ of the cathode (reduction)

$E^\ominus_A = E^\ominus$ of the anode (oxidation).

Top tip!

Do not change the sign: the negative sign in the formula for $E^\ominus_{cell}$ changes the sign of the LHE for you.

If a conventional cell diagram has been written for the cell, this equates to $E^\ominus_{cell} = E^\ominus_{RHE} - E^\ominus_{LHE}$, where RHE/LHE stand for right-hand electrode/left-hand electrode.

Top tip!

The standard electrode potential of a half-cell is not changed if the stoichiometry is changed.

WORKED EXAMPLE R3.2J

A voltaic cell is constructed using magnesium and copper electrodes. Deduce the ionic equation and standard cell potential for the spontaneous reaction (under standard conditions).

The electrode potentials are:

$Mg^{2+}(aq) + 2e^- \rightarrow Mg(s)$ $\quad E^\ominus = -2.37\,V$

$Cu^{2+}(aq) + 2e^- \rightarrow Cu(s)$ $\quad E^\ominus = +0.34\,V$

Answer

Ionic equation: $Mg(s) + Cu^{2+}(aq) \rightarrow Mg^{2+}(aq) + Cu(s)$

$$E^\ominus_{cell} = E^\ominus_C - E^\ominus_A$$

$$E^\ominus_{cell} = (0.34\,V) - (-2.37\,V) = +2.71\,V$$

The magnesium electrode acts as the anode (negatively charged) and the copper electrode acts as the cathode (positively charged).

HL ONLY

HL ONLY

WORKED EXAMPLE R3.2K

A voltaic cell is made from a Co^{2+}(aq)/Co(s) half-cell and a Cu^{2+}(aq)/Cu(s) half-cell. The cell potential is +0.62 V with the copper half-cell positive. Calculate the standard electrode potential of the cobalt half-cell. The Cu half-cell standard electrode potential can be found in the IB *Chemistry data booklet*.

Answer

$$E^{\ominus}_{cell} = E^{\ominus}_{C} - E^{\ominus}_{A}$$

$$+0.62\,V = 0.34\,V - x$$

$$x = -0.28\,V$$

WORKED EXAMPLE R3.2L

Represent the cell obtained by combining copper and silver half-cells: Cu | Cu^{2+}(aq), Ag | Ag^{+}(aq). (The standard reduction potentials of these electrodes are: $E^{\ominus}_{Cu^{2+}|Cu} = +0.34\,V$, $E^{\ominus}_{Ag^{+}|Ag} = +0.80\,V$.)

Answer

For the pairs of electrodes (i) Cu(s) | Cu^{2+}(aq), (ii) Ag(s) | Ag^{+}(aq)

Given the standard reduction potentials of:

$$E^{\ominus}_{Ag^{+}|Ag}\ (= +0.80\,V) > E^{\ominus}_{Cu^{2+}|Cu}\ (= +0.34\,V)$$

the Ag | Ag^{+} electrode will be the cathode (right-hand electrode) and Cu(s) | Cu^{2+}(aq) will act as the anode (left-hand electrode). Thus, the voltaic cell obtained with these two electrodes can be represented as:

Cu(s) | Cu^{2+}(aq) || Ag^{+}(aq) | Ag(s)

anode cathode

50 Calculate cell potentials and write ionic equations for the spontaneous reactions from voltaic cells made from the following half-cells:

a Fe^{2+}(aq)/Fe(s) and Ni^{2+}(aq)/Ni(s)

b I_2(s)/I^-(aq) and MnO_4^-(aq)/Mn^{2+}(aq)

c F_2(aq)/F^-(aq) and $Cr_2O_7^{2-}$(aq)/Cr^{3+}(aq)

d Cu^{2+}(aq)/Cu(s) and Ag^{+}(aq)/Ag(s)

51 Represent the voltaic cell obtained by coupling a lead half-cell, Pb(s) | Pb^{2+}(aq), and a silver half-cell, Ag(s) | Ag^{+}(aq). The standard reduction potentials of these electrodes are: $E^{\ominus}_{Pb^{2+}|Pb} = -0.13\,V$, $E^{\ominus}_{Ag^{+}|Ag} = +0.80\,V$.

Going further

Rusting

Rusting involves the formation of hydrated iron(III) oxide from iron, water and oxygen. The first step of the reaction involves the formation of iron(II) hydroxide and can be derived from the following half-equations:

$$Fe^{2+}(aq) + 2e^- \rightarrow Fe(s);\ E^\ominus = -0.45\,V$$

$$\frac{1}{2}O_2(g) + H_2O(l) + 2e^- \rightarrow 2OH^-(aq);\ E^\ominus = +0.40\,V$$

$$\frac{1}{2}O_2(g) + H_2O(l) + Fe(s) \rightarrow Fe^{2+}(aq) + 2OH^-(aq)$$

$$E^\ominus_{cell} = E^\ominus_C - E^\ominus_A$$

$$E^\ominus_{cell} = (0.40\,V) - (-0.45\,V) = +0.85\,V$$

The positive value of the cell potential indicates that rusting is spontaneous under standard conditions. The relatively large value of the cell potential implies a large value of the equilibrium constant; this shows that the forward reaction – the formation of iron(II) and hydroxide ions – is favoured.

In practice, the iron metal is oxidized to iron(II) ions at the centre of a water drop, where the oxygen concentration is low (due to slow diffusion). The electrons released reduce the oxygen molecules at the surface of the water, where the oxygen concentration is high (Figure R3.75). The air is effectively acting as the second electrode of a cell. The iron(II) and hydroxide ions formed diffuse away from the surface of the iron object. Further oxidation by dissolved oxygen in the air results in the formation of rust: hydrated iron(III) oxide.

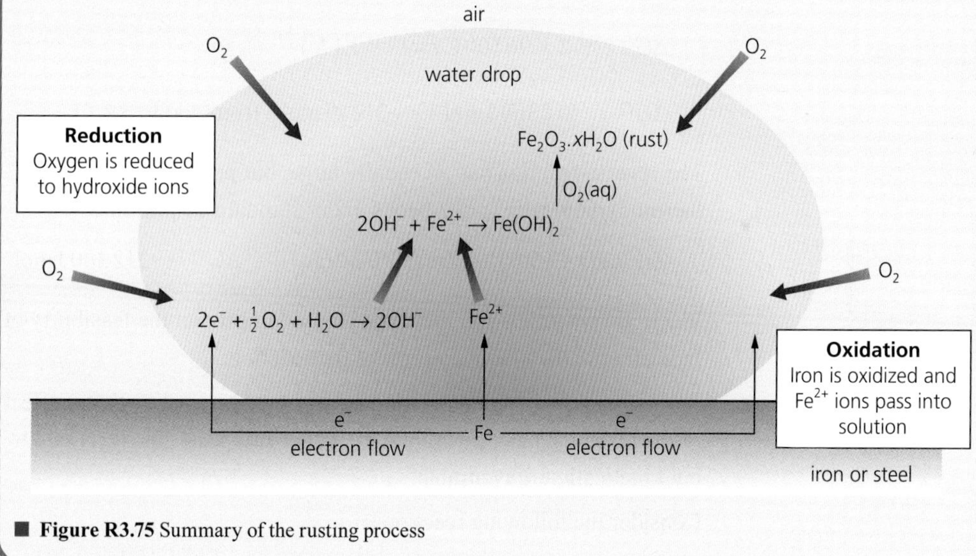

■ **Figure R3.75** Summary of the rusting process

Cell spontaneity

As long as the overall redox reaction is not at equilibrium, the oxidation reaction in a voltaic cell produces and 'pushes' electrons into the external circuit, and the reduction reaction at the cathode 'pulls' them out. The cell does work (which can be used to, for example, light a bulb or drive an electric motor) since it produces a force that moves electrons around the external circuit (produces a current). The amount of work which can be done by an electrochemical cell depends on the cell potential; the greater the cell potential, the greater the amount of work the cell can do.

HL ONLY

A cell in which the overall reaction is at equilibrium can do no work, its cell potential is zero and there is no current in the external circuit.

Link

Gibbs energy in relation to enthalpy is covered in detail in Chapter R1.4, from page 397.

The maximum amount of electrical work that can be done by an electrochemical cell is equal to the Gibbs energy change $\Delta G^\ominus$ (provided the temperature and pressure remain constant). The equation below describes the relationship between the Gibbs energy change and cell potential:

$$\Delta G^\ominus = -nFE^\ominus_{cell}$$

where:

- n = amount of electrons (in moles) transferred between the electrodes as shown in the balanced cell reaction
- $F = 96\,500\,C\,mol^{-1}$ (this is the Faraday constant, the amount of electrical charge carried by one mole of electrons)
- $E^\ominus_{cell}$ = standard cell potential.

For example, in the Daniell cell:

$$Zn(s) \rightarrow Zn^{2+}(aq) + 2e^-$$

$$Cu^{2+}(aq) + 2e^- \rightarrow Cu(s)$$

$$Zn(s) + Cu^{2+}(aq) \rightarrow Zn^{2+}(aq) + Cu(s)$$

$n = 2$ because two moles of electrons are transferred from the zinc atoms to the copper(II) ions in the above equation:

$$\Delta G^\ominus = -nFE^\ominus$$

$$\Delta G^\ominus = (-2 \times 96\,500\,C\,mol^{-1} \times 1.1\,V)$$

$$\Delta G^\ominus = -212\,300\,J\,mol^{-1} = -212\,kJ\,mol^{-1} \text{ (rounded to 3 s.f.)}$$

The reverse reaction has an equally large, but positive, value for $\Delta G^\ominus$ and is not thermodynamically spontaneous under standard conditions.

$$Zn^{2+}(aq) + Cu(s) \rightarrow Zn(s) + Cu^{2+}(aq) \qquad \Delta G^\ominus = +212\,300\,J\,mol^{-1} = +212\,kJ\,mol^{-1} \text{ (3 s.f.)}$$

Electrode potentials can therefore be used to predict the feasibility of a reaction but give no indication of the kinetics or rate of the reaction.

A reaction with a high activation energy may be so slow that it effectively does not occur. Such a reaction, which has a positive $E^\ominus_{cell}$ yet occurs very slowly, is said to be energetically favourable but kinetically unfavourable.

Consider the following reaction:

$$H_2(g) + Cu^{2+}(aq) \rightarrow Cu(s) + 2H^+(aq) \qquad E^\ominus_{cell} = +0.34\,V$$

The positive value suggests that hydrogen gas should displace copper from copper(II) ions in aqueous solution under standard conditions. In practice, the rate of reaction is so slow that the reaction is kinetically non-feasible. This is because a relatively large amount of energy is needed to break the strong hydrogen–hydrogen covalent bond before the reaction can start.

Table R3.26 summarizes the relationship between $\Delta G^\ominus$ and $E^\ominus$.

■ **Table R3.26** Summary of the relationship between $\Delta G^\ominus$ and $E^\ominus$

$\Delta G^\ominus$	$E^\ominus$	**Position of equilibrium**
negative	positive	Forward reaction is spontaneous, formation of products favoured, $K > 1$
positive	negative	Forward reaction is non-spontaneous, formation of reactants favoured, $K < 1$
zero	zero	Reactants and products favoured equally, $K = 1$

LINKING QUESTION

How can thermodynamic data also be used to predict the spontaneity of a reaction?

52 a Use the standard electrode potential data in the IB *Chemistry data booklet* to calculate the $E^\ominus_{cell}$ and $\Delta G^\ominus$ values for the following reaction:

$$4Ag(s) + O_2(g) + 4H^+(aq) \rightarrow 4Ag^+(aq) + 2H_2O(l)$$

b State whether the reaction is spontaneous under standard conditions.

c Calculate and comment on the values of $E^\ominus_{cell}$ and $\Delta G^\ominus$ for the related reaction:

$$2Ag(s) + \frac{1}{2}O_2(g) + 2H^+(aq) \rightarrow 2Ag^+(aq) + H_2O(l)$$

Non-standard conditions

Electrode potential values can only be used to predict the feasibility of a redox reaction under standard conditions.

Electrode potentials for oxidizing agents in acidic conditions refer to 1.0 mol dm^{-3} concentrations of hydrogen ions, $H^+(aq)$ (pH = 0). Increasing the $H^+(aq)$ concentration increases the oxidizing strength of the oxidizing agent and hence increases the electrode potential of the half-cell.

Consider the standard laboratory preparation of chlorine using the following reaction:

$$MnO_2(s) + 4H^+(aq) + 2Cl^-(aq) \rightarrow Mn^{2+}(aq) + 2H_2O(l) + Cl_2(g)$$

$$E^\ominus_{cell} = (1.23 - 1.36)\text{V} = -0.13\text{ V}$$

Since the cell potential, $E^\ominus_{cell}$, is negative, the reaction is not spontaneous under standard conditions. However, when concentrated hydrochloric acid is heated with manganese(IV) oxide, the cell potential becomes positive, the reaction can occur and chloride ions are oxidized.

An additional factor is the shifting of the equilibrium to the right by the loss of chlorine gas. In general, for a redox equilibrium where an oxidized species (of higher oxidation state), Ox is reduced to a reduced species, Red (of lower oxidation state):

$$\text{Ox} + ne^- \rightleftharpoons \text{Red}$$

Increasing the concentration of the oxidizing agent, [Ox], or decreasing the concentration of the reduced species, [Red], will shift the position of the equilibrium to the right. This will make the cell potential more positive. Conversely, the cell potential will become more negative if the concentration of the oxidized species, [Ox], is decreased, or the concentration of the reduced species, [Red], is increased. These shifts can be predicted by applying Le Châtelier's principle.

If the concentration of zinc ions in the zinc half-cell of a Daniell cell is decreased, then the equilibrium:

$$Zn^{2+}(aq) + 2e^- \rightleftharpoons Zn(s);\ E^\ominus = -0.76\text{ V}$$

is shifted to the left and the negative charge on the electrode is increased. This can also be predicted from Le Châtelier's principle: the removal of zinc ions will cause some of the zinc atoms to ionize, increasing the voltage of the Daniell cell to a value above 1.10 V.

If the concentration of zinc ions in the half-cell of the Daniell cell is increased, then the equilibrium is shifted to the right and the negative charge on the electrode is decreased. The addition of zinc ions will cause some of the zinc ions to gain electrons, decreasing the voltage to a value below 1.10 V.

Going further

The Nernst equation

The Nernst equation allows chemists to calculate the cell potentials of non-standard half-cells where the concentrations of ions are not 1 mol dm^{-3} or the temperature is not 298 K. This equation describes the relationship between cell potential and concentration at constant temperature, and the relationship between cell potential and temperature at constant concentration.

$$E_{cell} = E^{\ominus}_{cell} + \frac{2.3RT}{nF} \times \log_{10} \frac{\text{oxidized form}}{\text{reduced form}}$$

is a generalized form of the Nernst equation that can be used to calculate the cell potentials of voltaic cells under non-standard conditions. It provides the basis for a large range of electrochemical measurements of ion concentrations, including pH measurements.

- R = the gas constant, 8.31 J mol^{-1} K^{-1}
- F = the Faraday constant, 96 500 C mol^{-1}
- T = absolute temperature (in kelvin)
- n = number of electrons transferred.

The Nernst equation can be used to show that a half-cell potential will change by 59 mV per 10-fold change in the concentration of a substance involved in a one-electron oxidation or reduction; for two-electron processes, the variation will be 29.5 mV per 10-fold change in the concentration.

Electrolysis of aqueous solutions

Products of electrolysis of solutions

Electrolysis in an aqueous solution is complicated by the possibility of the electrolysis of water itself. Water molecules may be either oxidized or reduced according to the following half-equations:

Oxidation (anode): $H_2O(l) \rightarrow \frac{1}{2}O_2(g) + 2H^+(aq) + 2e^-$

Reduction (cathode): $H_2O(l) + e^- \rightarrow \frac{1}{2}H_2(g) + OH^-(aq)$

The water molecules are in contact with the electrodes and compete with the ions of the electrolyte to accept or release electrons at the cathode and anode, respectively.

Consider the electrolysis of aqueous sodium chloride using inert (graphite or platinum) electrodes. If the solution is concentrated, chlorine is produced at the anode and hydrogen is produced at the cathode:

Anode	Cathode
$2Cl^-(aq) \rightarrow Cl_2(aq) + 2e^-$	$2H_2O(l) + 2e^- \rightarrow H_2(g) + 2OH^-(aq)$

Common mistake

Misconception: *Water does not react during the electrolysis of an aqueous solution.*

Water molecules may gain or lose electrons at electrodes during electrolysis of aqueous solutions.

If the solution is dilute, hydrogen is produced at the cathode and oxygen is the main product at the anode:

Anode	Cathode
$2H_2O(l) \rightarrow O_2(g) + 4H^+(aq) + 4e^-$	$2H_2O(l) + 2e^- \rightarrow H_2(g) + 2OH^-(aq)$

In neither case are sodium ions discharged as sodium atoms. (However, sodium will be discharged if a flowing mercury cathode is used with concentrated sodium chloride.)

■ **Table R3.27** Examples of electrolysis of solutions containing salts or acids (examples in bold are specified in the syllabus)

Electrolyte	Electrodes	Cathode half-equation	Anode half-equation
potassium bromide, KBr(aq)	graphite / platinum	$2H_2O(l) + 2e^- \rightarrow H_2(g) + 2OH^-(aq)$	$2Br^- \rightarrow Br_2 + 2e^-$
magnesium sulfate, $MgSO_4$(aq)	graphite / platinum	$2H_2O(l) + 2e^- \rightarrow H_2(g) + 2OH^-(aq)$	$2H_2O(l) \rightarrow O_2(g) + 4H^+(aq) + 4e^-$
concentrated hydrochloric acid, HCl(aq)	graphite / platinum	$2H^+(aq) + 2e^- \rightarrow H_2(g)$	$2Cl^- \rightarrow Cl_2 + 2e^-$
dilute sulfuric acid, H_2SO_4(aq)	graphite / platinum	$2H_2O(l) + 2e^- \rightarrow H_2(g) + 2OH^-(aq)$	$2H_2O(l) \rightarrow O_2(g) + 4H^+(aq) + 4e^-$
dilute sodium hydroxide, NaOH(aq)	graphite / platinum	$2H_2O(l) + 2e^- \rightarrow H_2(g) + 2OH^-(aq)$	$4OH^-(aq) \rightarrow O_2(g) + 4H^+(aq) + 4e^-$
copper(II) sulfate, $CuSO_4$(aq)	**graphite / platinum**	$\mathbf{Cu^{2+} + 2e^- \rightarrow Cu}$	$2H_2O(l) \rightarrow O_2(g) + 4H^+(aq) + 4e^-$
copper(II) sulfate, $CuSO_4$(aq)	**copper**	$\mathbf{Cu^{2+} + 2e^- \rightarrow Cu}$	$\mathbf{Cu \rightarrow Cu^{2+} + 2e^-}$
copper(II) chloride, $CuCl_2$(aq)	graphite / platinum	$Cu^{2+} + 2e^- \rightarrow Cu$	$2Cl^- \rightarrow Cl_2 + 2e^-$
potassium iodide, KI(aq)	graphite / platinum	$2H_2O(l) + 2e^- \rightarrow H_2(g) + 2OH^-(aq)$	$2I^- \rightarrow I_2 + 2e^-$
concentrated sodium chloride, NaCl(aq)	**graphite / platinum**	$2H_2O(l) + 2e^- \rightarrow H_2(g) + 2OH^-(aq)$	$\mathbf{2Cl^- \rightarrow Cl_2 + 2e^-}$
dilute sodium chloride, NaCl(aq)	**graphite / platinum**	$2H_2O(l) + 2e^- \rightarrow H_2(g) + 2OH^-(aq)$	$2H_2O(l) \rightarrow O_2(g) + 4H^+(aq) + 4e^-$

These, and other results summarized in Table R3.27, suggest the following 'rules' regarding electrolysis of aqueous solutions:

- Metals, if produced, are discharged at the cathode.
- Hydrogen is produced at the cathode only. (For very acidic solutions the formation of hydrogen can also be described by the following equation: $2H^+(aq) + 2e^- \rightarrow H_2(g)$.)
- Non-metals, apart from hydrogen, are produced at the anode.
- Reactive metals, that is, those above hydrogen in the reactivity series, are not discharged (unless special cathodes are used or unless the molten (rather than aqueous) salt is electrolysed).
- The products can depend upon the concentration of the electrolyte in the solution and the nature of the electrode. If chloride ions are present at high concentrations, they will be oxidized (and discharged) more readily than water molecules. If no chloride ions are present or they are at very low concentrations, water molecules are oxidized (and discharged) more readily than chloride ions.

The electrolysis of water

When very dilute sulfuric acid is electrolysed, one volume of oxygen gas is collected over the anode, and two volumes of hydrogen gas are collected over the cathode (Figure R3.76).

At the anode, water molecules are oxidized (and discharged) in preference to sulfate ions. They give up electrons and form hydrogen ions and oxygen molecules. At the cathode, water molecules are reduced (and discharged) by accepting electrons to form hydrogen molecules and hydroxide ions.

Anode (oxidation): $2H_2O(l) \rightarrow O_2(g) + 4H^+(aq) + 4e^-$

Cathode (reduction): $4H_2O(l) + 4e^- \rightarrow 2H_2(g) + 4OH^-(aq)$ (or $4H^+(aq) + 4e^- \rightarrow 2H_2(g)$)

Overall reaction: $6H_2O(l) \rightarrow 2H_2(g) + O_2(g) + 4(H^+ + OH^-)$

or $2H_2O(l) \rightarrow 2H_2(g) + O_2(g)$

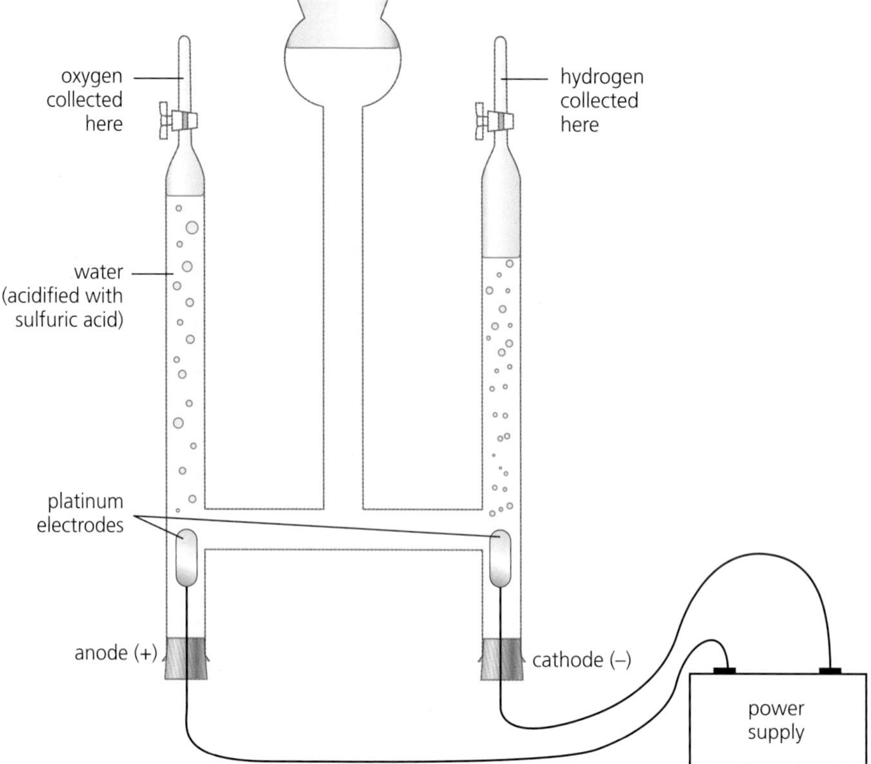

■ **Figure R3.76** Apparatus for the decomposition of acidified water (Hofmann voltameter)

Top tip!

Although the quantity of sulfuric acid is unchanged, its concentration increases as the water is consumed.

The last equation shows that the ratio of amounts or volumes of oxygen molecules to hydrogen molecules is 1 : 2 so, in effect, water is being electrolysed to its elements.

The products of electrolysis depend on standard electrode potentials of the different oxidizing and reducing species present in the electrolyte in the electrolytic cell. Out of the possible reduction reactions taking place, the reduction reaction which has highest (most positive) value of standard electrode potential takes place at the cathode (negative electrode). Similarly, out of the possible oxidation reactions, the oxidation reaction which has the lowest (most negative) value of standard electrode potential takes place at the anode (positive electrode).

Going further

Reversing the cell reaction

If the electrodes of a voltaic cell are connected, electrons flow from the negative electrode to the positive terminal. If the two electrodes are instead connected to a power supply and the external potential difference, V, is steadily increased:

- electrons will flow as before if the external voltage $V < E^{\ominus}_{cell}$
- when $V = E^{\ominus}_{cell}$, no current flows
- once $V > E^{\ominus}_{cell}$, current flows in the opposite direction and electrolysis takes place (Figure R3.77).

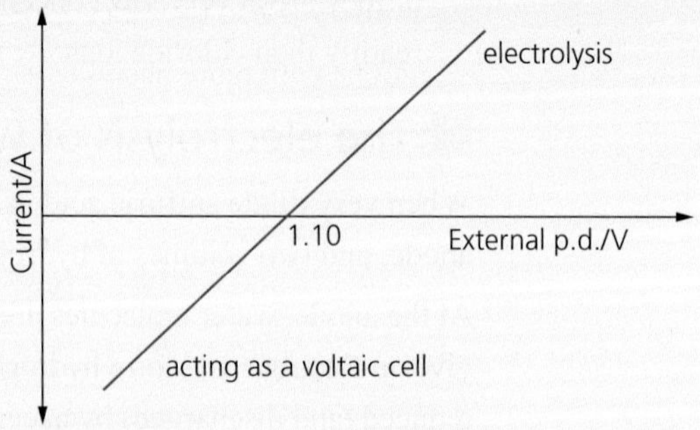

electrolysis
Current/A
1.10
External p.d./V
acting as a voltaic cell

■ **Figure R3.77** Voltage–current graph for a Daniell cell (copper / zinc voltaic cell)

For example, when an external voltage greater than 1.10 V is applied to a Daniell cell, the reaction runs in reverse:

$$Zn^{2+}(aq) + Cu(s) \rightarrow Zn(s) + Cu^{2+}(aq)$$

Standard electrode potentials and electrolysis

A more rigorous approach to predicting and explaining electrolysis products uses standard electrode potentials.

During electrolysis, cations are discharged at the cathode:

$$M^{n+}(aq) + ne^- \rightarrow M(s)$$

If hydrogen ions are discharged, then hydrogen gas is produced:

$$2H^+(aq) + 2e^- \rightarrow H_2(g)$$

Since discharge at the cathode involves reduction, ions that accept electrons readily – strong oxidizing agents with more positive standard electrode potentials – will be reduced and discharged first.

For example, it is easier to discharge copper(II) ions than zinc ions or water molecules at the cathode:

$$Zn^{2+}(aq) + 2e^- \rightarrow Zn(s);\ E^\ominus = -0.76\ V$$

$$Cu^{2+}(aq) + 2e^- \rightarrow Cu(s);\ E^\ominus = +0.34\ V$$

$$H_2O(l) + e^- \rightarrow \frac{1}{2}H_2(g) + OH^-(aq);\ E^\ominus = -0.83\ V$$

Anions are discharged at the anode during electrolysis:

$$2X^{n-}(aq) \rightarrow X_2(g) + 2ne^-$$

Since this is an oxidation reaction, ions that lose electrons most readily will be oxidized first. An anion with a more negative standard electrode potential will be discharged instead of one with a less negative standard electrode potential.

For example, it is easier to discharge bromide ions than chloride ions or water molecules:

$$Br_2(aq) + 2e^- \rightarrow 2Br^-(aq);\ E^\ominus = +1.09\ V$$

$$Cl_2(aq) + 2e^- \rightarrow 2Cl^-(aq);\ E^\ominus = +1.36\ V$$

$$H_2O(l) \rightarrow \frac{1}{2}O_2(g) + 2H^+(aq) + 2e^-;\ E^\ominus = -1.23\ V$$

In summary, when inert electrodes are used during electrolysis:

- cations with more positive $E^\ominus_{reduction}$ values will be discharged first at the cathode
- anions with more negative $E^\ominus_{reduction}$ values will be discharged first at the anode.

Nature of science: Experiments

Faraday's electrolysis experiments

When a solution of aqueous copper(II) sulfate is electrolysed using copper electrodes (Figure R3.78), the copper anode slowly disappears and the copper cathode slowly gains a deposit of copper.

Cathode: $Cu^{2+} + 2e^- \rightarrow Cu$

Anode: $Cu \rightarrow Cu^{2+} + 2e^-$

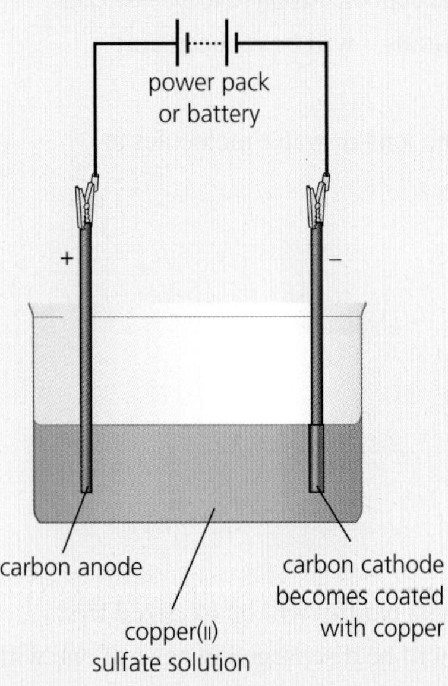

Figure R3.78 Apparatus for the electrolysis of copper(II) sulfate solution

Michael Faraday (1791–1867) performed careful measurements of changes in mass of cathodes in experiments such as these. In 1833, he summarized the results in his two laws of electrolysis.

- Faraday's first law states that the mass of an element produced during electrolysis is directly proportional to the quantity of electricity (charge) passed during the electrolysis.
- Faraday's second law states that the masses of different elements produced by the same quantity of electricity form simple whole number ratios when divided by their relative atomic masses.

Here is some experimental data that supports Faraday's second law. During an electrolysis experiment, 2.16 g of silver are deposited. When the experiment is repeated using the same current flowing for the same time but a different cell, 0.6355 g of copper are deposited.

The quantity of electricity that flowed through each cell is the same because the total charge, Q (in coulombs, C), carried by an electric current is given by $Q = I \times t$ where I = current (in amps, A) and t = time (in seconds, s).

	Silver	Copper
Amount (mol)	$\frac{2.16}{107.87} = 0.02$	$\frac{0.6355}{63.55} = 0.01$
Divide through by smallest	$\frac{0.02}{0.01} = 2$	$\frac{0.01}{0.01} = 1$

The results can be accounted for in terms of the relevant half-equations and molar quantities of ions, atoms and electrons.

$Ag^+(aq)$	+	e^-	→	$Ag(s)$
0.02 mol		0.02 mol		0.02 mol
$Cu^{2+}(aq)$	+	$2e^-$	→	$Cu(s)$
0.01 mol		0.02 mol		0.01 mol

The amount of copper formed is half the amount of silver formed because each mole of copper(II) ions needs two moles of electrons for discharge, whereas each mole of silver ions needs only one mole of electrons for discharge.

Therefore, a modern statement of Faraday's law is that the number of moles of electrons required to discharge one mole of an ion at an electrode equals the charge on the ion.

WORKED EXAMPLE R3.2M

Aqueous solutions of gold(I) nitrate, $AuNO_3$, copper(II) sulfate, $CuSO_4$, and chromium(III) nitrate, $Cr(NO_3)_3$, are electrolysed using the same quantity of electricity. How do the amounts of metal atoms (in mol) compare?

$$Au^+(aq) + e^- \rightarrow Au(s)$$

$$Cu^{2+}(aq) + 2e^- \rightarrow Cu(s)$$

$$Cr^{3+}(aq) + 3e^- \rightarrow Cr(s)$$

Answer

Au > Cu > Cr

1 mol, $\frac{1}{2}$ mol and $\frac{1}{3}$ mol of metal atoms, respectively.

Going further

Anodizing

In anodizing, a metal is coated with a layer of oxide by making the metal the anode in an electrochemical cell. When sulfuric acid is electrolysed using an aluminium anode, water molecules are reduced. Hydrogen gas is formed at the carbon cathode.

Cathode: $2H^+(aq) + 2e^- \rightarrow H_2(g)$

Anode: $2H_2O(l) \rightarrow O_2(g) + 4H^+(aq) + 4e^-$

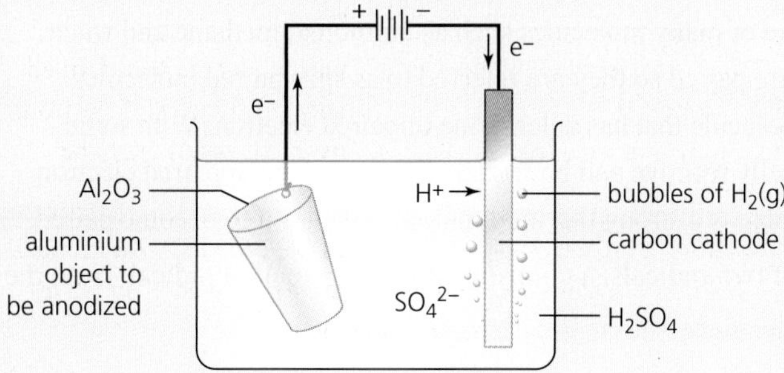

■ **Figure R3.79** Anodizing

If the anode itself is made of reactive metal, such as aluminium, then a relatively thick film of metal oxide can grow on its surface until the insulating oxide layer stops conduction.

- The thickness of the coating can be controlled by varying the current flow during electrolysis.
- The porosity of the aluminium oxide coating can also be controlled by varying the electrolysis conditions.
- If an organic dye is added to the electrolyte, dye molecules can permeate the spongy oxide layer as it forms, becoming trapped as the oxide layer hardens.

The overall effect of anodizing is to produce an attractive coloured protective coating on the surface of the anode. Aluminium bicycle parts come in a spectrum of colours as a result of anodizing.

R3.3 Electron sharing reactions

Guiding question

- What happens when a species has an unpaired electron?

SYLLABUS CONTENT

By the end of this chapter, you should understand that:

- a radical is a chemical entity that has an unpaired electron
- most radicals are highly reactive
- radicals are produced by homolytic fission
- the homolytic fission of halogens occurs in the presence of ultraviolet (UV) light or heat
- radicals take part in substitution reactions with alkanes, producing a mixture of products.

By the end of this chapter, you should know how to:

- identify and represent radicals, e.g. $\bullet CH_3$ and $\bullet Cl$
- explain, including with equations, the homolytic fission of halogens, known as the initiation step in a chain reaction
- explain, using equations, the propagation and termination steps in the reactions between alkanes and halogens.

There is no higher-level only material in R3.3.

Introduction to radicals

◆ **Radical:** A reactive species with one or more unpaired electrons.

You will be familiar with the structure of many molecules such as ammonia, methane and water. In these molecules, all the electrons are paired so they are referred to as spin-paired molecules. In contrast, a **radical** is an atom or molecule that has at least one unpaired electron. With some exceptions, these radicals are chemically reactive and have short lifetimes. The unpaired electron gives rise to this reactivity because there is a strong thermodynamic drive for it to become paired.

Figure R3.80 shows the structures of two radicals: a chlorine atom and a methyl radical. Note the unpaired electron in each.

Top tip!

The single unpaired electron in a radical is indicated in the formula by a dot. For example, •Cl represents a chlorine atom which is a radical.

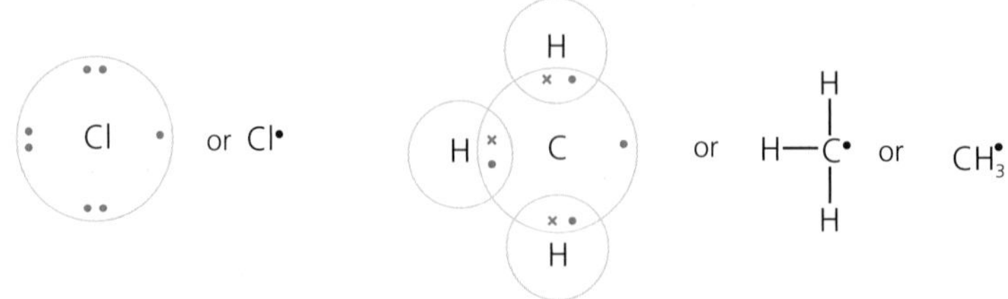

■ **Figure R3.80** The Lewis formulas of a chlorine atom and a methyl radical

53 A chlorine dioxide molecule, ClO_2, has an unpaired electron on the chlorine atom. Draw the Lewis formula of ClO_2.

54 Draw the Lewis formulas of an iodine atom (a radical) and an ethyl radical.

Common mistake

Misconception: *Radicals are always neutral atoms or molecules.*

Radicals can be neutral or charged species.

LINKING QUESTION

How is it possible for a radical to be an atom, a molecule, a cation or an anion. Consider examples of each type.

Reactions involving radicals are common in the gas phase. Radicals are often referred to as free radicals because they exist independently, free of any stabilization from other species. In solutions, ionic species are more common because they can be effectively stabilized by interactions with the solvent, particularly when the solvent is polar.

Radicals can be formed from spin-paired molecules by the unpairing of a pair of electrons in a process called homolytic fission. Heat and light are two common ways to achieve this. To understand homolytic fission, it is useful to first understand the conventions used to show the movement of electrons in organic chemistry: curly arrows.

Nature of science: Theories

The discovery of radicals

Moses Gomberg (1866–1947) was a Russian-born American chemist who, while trying to synthesize hexaphenylethane, prepared triphenylmethyl, a reactive and unstable substance. His breakthrough was recognizing that he had made a carbon radical (Figure R3.81).

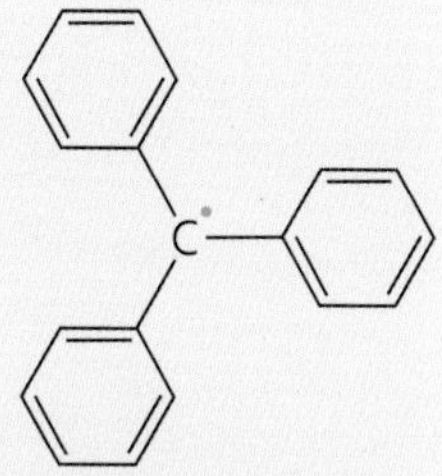

The prevailing paradigm at the time was that carbon was always tetravalent and would only be found with four bonds. Gomberg published his research in 1900, but the existence of triphenylmethyl and other organic radicals remained a contentious hypothesis for nearly a decade. Later research revealed that radicals are intermediates in a number of chemical and biochemical reactions.

This underscores the role of continued research in the sciences, even when the prevailing ideas suggest a question has been solved.

The American Chemical Society (ACS) presented a plaque marking the event to the University of Michigan on 25 June 2000, during a celebration of the 100th anniversary of Gomberg's seminal discoveries.

■ **Figure R3.81** Structure of the triphenylmethyl radical with the dot indicating the unpaired electron on the central carbon atom

Top tip!

A full-headed curly arrow shows the movement of a pair of electrons. A half-headed curly arrow (also referred to as a single-barbed arrow or fish hook) shows the movement of a single electron.

Curly arrow notation

Organic chemists use curly arrows to show the movement of electrons in reactions. The tail of the curly arrow shows where the electrons start from, and the head of the arrow points to where the electrons finish.

Heterolytic fission and homolytic fission

When a covalent bond is broken, the products that are formed depend on how the electrons in the bond are split between the atoms involved.

The carbon to bromine bond in 2-bromo-2-methylpropane can break when the substance is dissolved in an appropriate solvent. The pair of electrons in the covalent bond between the two atoms moves to become localized on the bromine atom. This is shown by the full-headed red curly arrow in Figure R3.82. The result is two ionic species: a **carbocation**, $(CH_3)_3C^+$, and a bromide ion, Br^-. This is **heterolytic fission** (heterolysis). A bond is broken, and both the electrons go to one of the atoms.

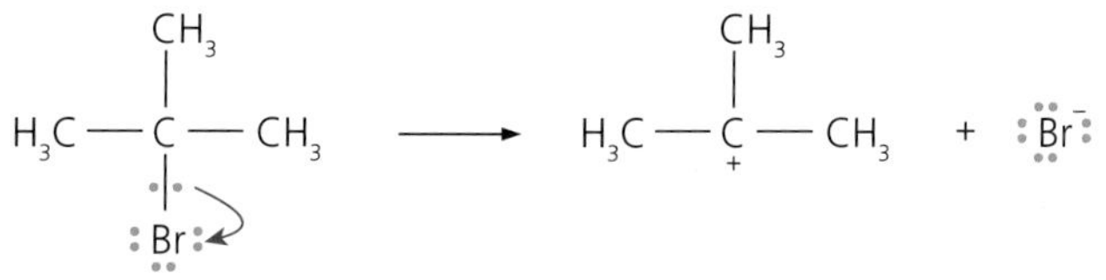

◆ **Carbocation**: An electron-deficient positively charged ion which has the charge located on a carbon atom.

◆ **Heterolytic fission**: When a bond is broken and the two electrons in the bond move to only one atom. Two ions are produced.

■ **Figure R3.82** Heterolytic fission of a carbon–bromine bond in 2-bromo-2-methylpropane

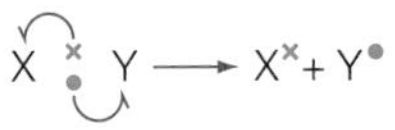

$X \overset{\times}{\bullet} Y \longrightarrow X^{\times} + Y^{\bullet}$

■ **Figure R3.83** Homolytic fission of a covalent bond

◆ **Homolytic fission:** When a bond is broken and the two electrons move to different atoms.

When a bond is broken and the pair of bonding electrons is split equally between the atoms involved in the bond, **homolytic fission** (homolysis) has occurred. This is shown by the single-headed red curly arrows in Figure R3.83. For a generic molecule, X–Y, two radical species X• and Y• are produced.

Forming radicals

Two ways of forming radicals, which are relevant to the *IB Chemistry course*, are presented below.

Radicals from homolytic fission

As you have already seen, radicals can be formed by the homolytic fission of a covalent bond. Heat, and sometimes UV light, is able to cause homolytic fission of weak sigma bonds. Most sigma bonds will undergo some homolysis when heated above 200 °C. Light of sufficient energy is able to cause homolytic fission in molecules which have a mechanism for absorbing the energy in a way that ends up being concentrated in the vibration of the bond, which leads to bond breakage.

The halogens are examples of molecules which can undergo homolytic fission when exposed to ultraviolet (UV) light. The homolytic fission of a chlorine molecule is shown in Figure R3.84. One of the original bonding electrons ends up on one chlorine radical and the second bonding electron ends up on the other chlorine radical.

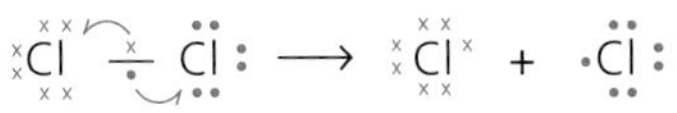

$Cl\!:\!Cl \longrightarrow Cl^{\times} + {}^{\bullet}Cl$

■ **Figure R3.84** The homolytic fission of a chlorine molecule under UV light

Some molecules very readily break down by homolytic fission, often because they contain a particularly weak sigma bond. The molecules are used as radical initiators in chemical reactions. The radicals which they form go on to make other radicals in chain reactions.

Dibenzoyl peroxide and **azo**iso**b**utyro**n**itrile (AIBN) are two radical initiators (Figure R3.85). Dibenzoyl peroxide, like all peroxides, has a weak oxygen–oxygen single bond. Unlike many organic peroxides, it is stable enough to be handled safely at normal temperatures. However, it will decompose at convenient rates when heated to temperatures in the range 50–90 °C, initially producing two benzoyloxy radicals. AIBN is similarly stable at room temperature and decomposes to give a nitrogen molecule and two cyanoisopropyl radicals. Both are widely used as initiators of polymerization reactions.

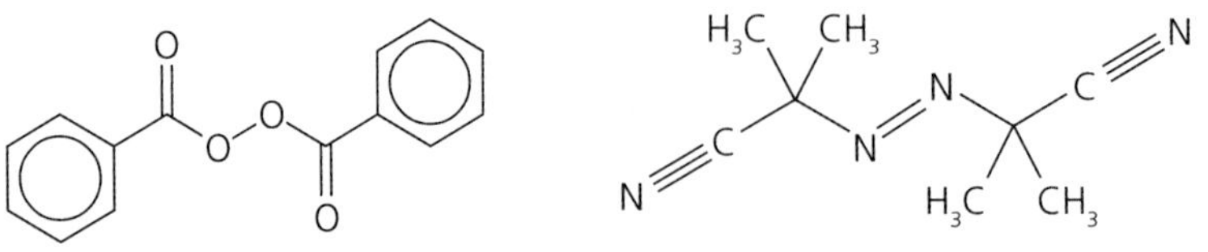

■ **Figure R3.85** The structures of the radical initiators dibenzoyl peroxide and AIBN

Radicals from radical abstraction reactions

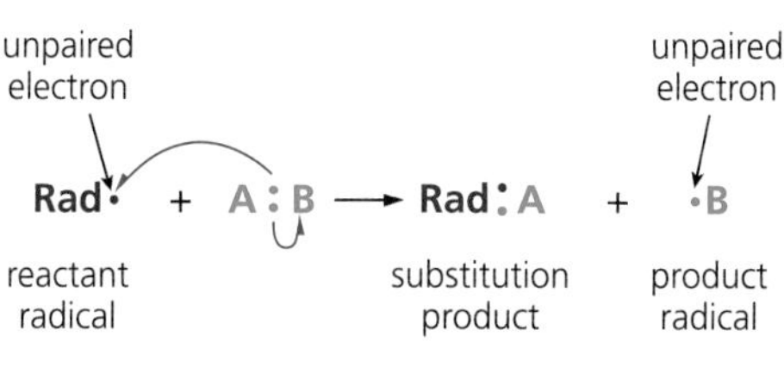

■ **Figure R3.86** A radical abstraction reaction forms a new radical

Many radicals are formed by the reaction of an existing radical with a spin-paired molecule. A radical is reactive because it contains an atom with an odd number of electrons (usually seven) in its valence shell rather than an octet. It may be able to achieve an octet by abstracting an atom or ion from a spin-paired molecule (Figure R3.86). The atom abstracted takes with it a single electron and leaves a new product radical. **Radical abstraction** reactions are also called radical transfer reactions because they transfer the radical reactivity from one species to another.

◆ **Radical abstraction**: A reaction in which a radical removes an atom or group of atoms from a spin-paired molecule. The result is a new radical on the previously spin-paired molecule.

The thermodynamic drive for radical abstraction reactions to occur is a result of the relative bond strengths of the bonds broken and formed. If, in each radical abstraction reaction, a weak bond is broken at the expense of a stronger one, the reaction is favourable.

The abstraction of hydrogen is a common feature in radical reactions. An O-based radical, RO•, will abstract hydrogen from hydrogen bromide, HBr, to form a bromine radical, Br• (Figure R3.87).

R—O• H—Br ⟶ ROH + Br•

■ **Figure R3.87** The radical abstraction of hydrogen by a peroxide radical

A radical abstraction reaction occurs in the propagation steps of radical substitution reactions.

Chemists have developed other ways of forming radicals. One commercially important way is through the addition of an existing radical to an alkene. This is the basis of roughly half of all polymer production. Another way is via single-electron oxidation or single-electron reduction of suitable precursors using an energetically excited photocatalyst. This second method is an expanding and important area in organic chemistry research.

Radical abstraction steps drive radical reactions by forming stronger bonds from weaker bonds.

55 Draw curly arrows to show the movement of electrons when a bromine radical, Br•, abstracts a hydrogen atom from a methane molecule to form a methyl radical, $•CH_3$, and HBr.

Going further

Dental composites

Dentists use dental composites to fill cavities in their patients' teeth. A composite is a material made of two or more substances with different characteristics, where the combination has improved properties compared to its component parts.

Composite resins used in dentistry have several components, each with a specific goal. Important parts of the composite are inorganic fillers, resins and photoinitiators. The inorganic fillers, typically silica based, are responsible for mechanical properties – especially the strength and abrasion resistance of the material. The resin is made from monomers which polymerize through radical intermediates to harden the composite when it has been placed in the cavity. Photoinitiators start the radical polymerization reaction when exposed to light. Other components of the composite help to bind the inorganic fillers to the resin or are present to improve the aesthetics of the composite.

Once the composite is placed into the tooth cavity, the dentist will use a small device to expose the composite to blue light. The light causes the photoinitiators to form radical species. The most commonly used photoinitiator is camphorquinone mixed with a tertiary amine. This mixture, when exposed to light, forms radicals (Figure R3.88). The camphorquinone has an unpaired electron on each of the carbons which bear an oxygen, and the tertiary amine forms a radical when the camphorquinone abstracts a hydrogen from one of its alkyl groups.

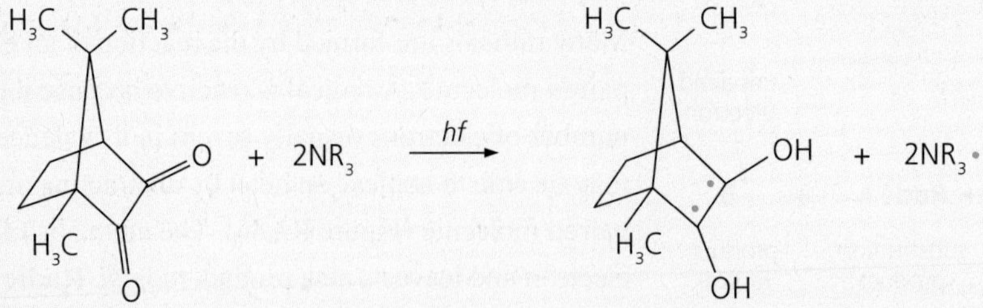

■ **Figure R3.88** Formation of radicals from the photoinitiator mixture in a dental composite

The radicals then undergo a radical addition reaction with an alkene on the end of '*bis*-GMA' monomer (Figure R3.89). This starts the polymerization process which hardens the composite and fixes it into the tooth. The mechanism for radical polymerization of alkenes is explained in the next Going further box.

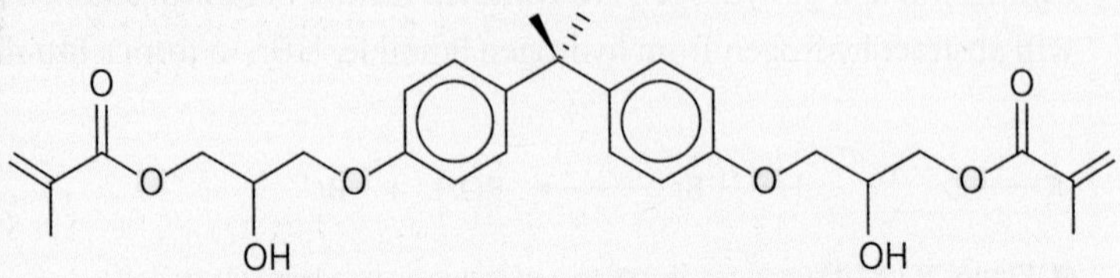

■ **Figure R3.89** The most commonly used monomer in dental resins, bisphenol A-glycidyl methacrylate (bis-GMA)

Is the depiction of the 'scientific method' traditionally found in many school science textbooks an accurate model of scientific activity?

The scientific method described in science textbooks often follows this linear sequence:

- Observe and describe some phenomenon.
- Formulate a hypothesis to explain the phenomenon, often via a relationship between an independent variable and a dependent variable.
- Use the hypothesis to make predictions.
- Test those predictions by controlled experiments to establish if the hypothesis is supported by the data.
- If it is not supported, then reject or modify the hypothesis.

This representation of the process of science emphasizes the importance of testing a hypothesis with empirical evidence (data). However, this version of the scientific method is simplified and rigid and it fails to accurately describe how real scientific research is carried out. It more accurately describes how science is summarized after the process – in textbooks and journal articles – than how the scientific method actually works.

Scientists may approach their search for new scientific knowledge differently, but they will always seek evidence that can be recorded and processed in different ways. The scientific method combines rational thought (logic), creativity and imagination to predict and explain phenomena, and the research findings of scientists are always open to scrutiny, criticism and debate.

Going further

Addition polymers

Many addition polymers form through radical chain reactions. An initiator, such as dibenzoyl peroxide shown in Figure R3.90, is mixed with the alkene monomer and homolytic fission produces two radicals.

A radical, simplified as X•, then performs a radical addition reaction with an alkene to give a new radical. The new radical can repeat the addition process with a further monomer. This is shown in Figure R3.91 with the monomer ethenylbenzene, $CH_2CH(C_6H_5)$, which is also known as styrene.

By this method, a polymer chain consisting of many thousands of monomers can be built up. Eventually the chain will stop growing when the radical abstracts a hydrogen from a nearby chain.

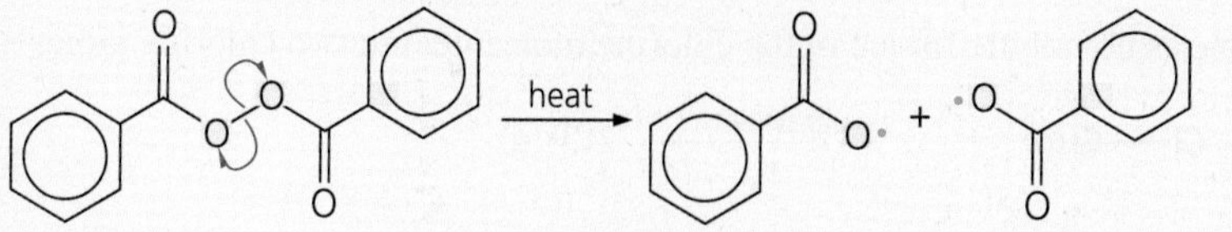

■ **Figure R3.90** The homolytic fission of dibenzoyl peroxide. Ph represents a phenyl ring, C_6H_5

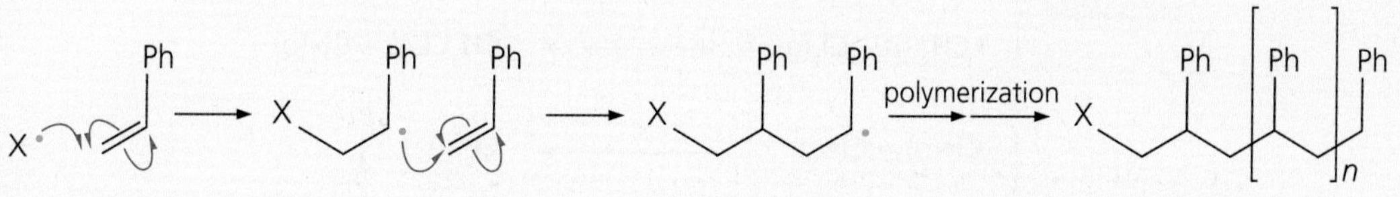

■ **Figure R3.91** Repeated radical additions form a polymer from alkene monomers. Ph represents a phenyl ring, $–C_6H_5$

Link

Addition polymers are also discussed in Chapter S2.4, page 204.

Halogenation of alkanes in radical substitution reactions

Alkanes are relatively stable molecules that do not easily undergo many chemical reactions (oxidation, reduction, reaction with electrophiles and nucleophiles) that typically occur with molecules bearing other functional groups. Their stability is due to their strong and non polar sigma bonds, and their lack of pi bonds.

Alkanes have only strong sigma bonds due to the relatively small size of carbon and hydrogen atoms which leads to effective orbital overlap. Because the carbon and hydrogen atoms of an alkane have approximately the same electronegativity (H = 2.2 and C = 2.6), the electrons in the C–H and C–C bonds are shared almost equally by the bonding atoms.

Consequently, none of the atoms in an alkane has any significant charge and their bonds have low polarity. This means that neither nucleophiles nor electrophiles are strongly attracted to them.

Alkanes are relatively unreactive compounds, but they will react with radicals, due to the chemical reactivity of the radical and low activation energy for many reactions.

LINKING QUESTION

Why are alkanes described as kinetically stable but thermodynamically unstable?

Chlorination of methane

◆ **Photochemical reaction:** A chemical reaction caused by light or ultraviolet radiation.

Alkanes react with chlorine or bromine to form halogenoalkanes in a gas-phase reaction. These halogenation reactions take place only at high temperatures or in the presence of sunlight (which contains ultraviolet radiation). When initiated by sunlight, these halogenation reactions are known as **photochemical reactions.**

Consider the reaction between methane and chlorine with a stoichiometry of 1 : 1.

In the presence of ultraviolet light, a mixture of methane and chlorine reacts to form chloromethane and hydrogen chloride (Figure R3.92). A substitution reaction has occurred. One hydrogen atom in the methane molecule has been substituted for a chlorine atom in the molecule of chlorine.

Top tip!

Radical substitution occurs in three stages: initiation, propagation and termination.

◆ **Radical substitution reaction**: A type of substitution reaction involving radical species as intermediates. One or more of the atoms or groups present in the reactant is substituted for different atoms or groups.

◆ **Initiation stage**: The stage of a radical substitution reaction that produces two radicals by homolytic fission.

◆ **Propagation stage**: The stage of a radical chain reaction that maintains the number of radicals in the reaction and produces the products of the reaction.

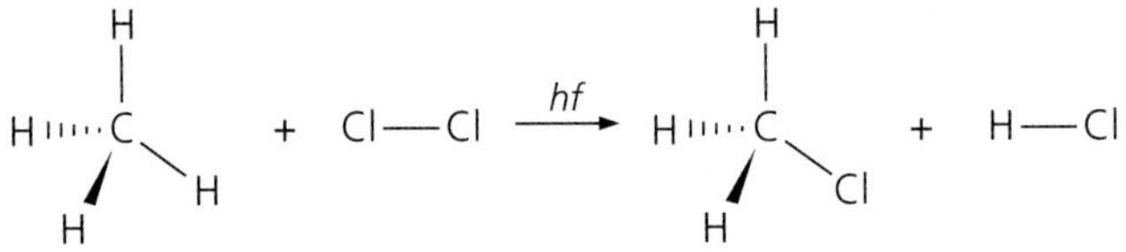

$$CH_4 + Cl{-}Cl \xrightarrow{hf} CH_3Cl + H{-}Cl$$

■ **Figure R3.92** The radical substitution reaction between methane and chlorine

This reaction is known as a **radical substitution reaction** and it happens in three stages, all of which involve radical intermediates.

Initiation stage

Figure R3.93 summarizes the steps in the radical substitution reaction between methane and chlorine in the presence of ultraviolet radiation. The first stage is **initiation** – the formation of radicals. The UV light causes homolytic fission of the chlorine molecule into two chlorine radicals.

initiation: $Cl{-}Cl(g) \xrightarrow{\text{ultraviolet}} 2Cl\bullet(g)$

propagation:

$Cl\bullet(g) + CH_4(g) \rightarrow \bullet CH_3(g) + HCl(g)$

$\bullet CH_3(g) + Cl_2(g) \rightarrow CH_3Cl(g) + Cl\bullet(g)$

possible termination steps:

$Cl\bullet(g) + Cl\bullet(g) \rightarrow Cl_2(g)$

$Cl\bullet(g) + \bullet CH_3(g) \rightarrow CH_3Cl(g)$

$\bullet CH_3(g) + \bullet CH_3(g) \rightarrow C_2H_6(g)$

overall reaction: $Cl_2(g) + CH_4(g) \xrightarrow{\text{ultraviolet}} CH_3Cl(g) + HCl(g)$

■ **Figure R3.93** Summary of the steps in the monosubstitution reaction between methane and chlorine in the gas phase

Propagation stage

The next two steps make up the **propagation** stage of the reaction. A chlorine radical abstracts hydrogen from methane and forms a methyl radical. This methyl radical then abstracts a chlorine atom from a chorine molecule to give chloromethane. The combinations of these two steps give the overall equation for the reaction. A chlorine radical is used in the first propagation step but reforms in the second. This allows the propagation steps to repeat many thousands of times. Because the product of one step, a chlorine radical, goes on to cause another methane and chlorine molecule to react together, the reaction is referred to as a chain reaction. The introduction of one initiating radical (a halogen atom in this case) can lead to the formation of very many product molecules in a chain reaction.

ATL R3.3A

Animations can be useful in helping to visualize the bond breaking and bond formation in a reaction. Animations of the propagation steps in the reaction between methane and chlorine can be found at www.chemtube3d.com/a-level-radical-reactions-propagation-steps/

Although you would not be expected to recall this, notice how in the animation, the molecular geometry of methane changes as it undergoes a radical abstraction reaction with a chlorine radical. The geometry moves from tetrahedral (for methane) to trigonal planar (for the methyl radical). This indicates that a trigonal planar geometry, in which the carbon has sp^2 hybridization, is the most stable option for the methyl radical. The unpaired electron is held in a p-orbital. Computer modelling shows that this arrangement is slightly lower in energy than the methyl radical holding a tetrahedral geometry, although the energy barrier to inversion is very low.

Make your own animation of the initiation step involved in the radical substitution of chlorine and methane. Use a free gif maker which can be found online (such as ezgif.com) to sequence still images which show UV radiation striking a molecule of chlorine and the homolytic fission which results.

Termination stage

◆ **Termination stage:** The stage of a radical chain reaction when two radicals react to form a spin-paired molecule.

Usually, the concentration of radicals in the reaction remains quite low, so the probability that two will react with one another is low. The radicals are much more likely to participate in the abstraction reactions which characterize the propagation part of the reaction, because they are surrounded by spin-paired molecules. However, the radicals will occasionally come into contact with each other and this results in **termination** steps. A termination step is the opposite of an initiation step. Two radicals react to form a spin-paired molecule. The possible termination steps depend on the radicals present in a reaction. Those for the reaction of chlorine and methane are shown in Figure R3.93.

Free radical substitution of halogens is not suitable for synthesizing specific halogenoalkanes because a mixture of substitution products is always formed. In the reaction between methane and chlorine, the products can include traces of dichloromethane, trichloromethane and tetrachloromethane, in addition to chloromethane. These other products result from propagation steps in which a chlorine radical abstracts a hydrogen from a previously produced halogenoalkane molecule (rather than methane).

For example:

$$CH_3Cl + Cl\bullet \rightarrow \bullet CH_2Cl + HCl$$

followed by:

$$\bullet CH_2Cl + Cl_2 \rightarrow CH_2Cl_2 + Cl\bullet$$

The more chlorine gas in the reaction mixture to start with, the greater the proportions of CH_2Cl_2, $CHCl_3$ and CCl_4 formed as products. Tetrachloromethane, CCl_4, can be favoured by using a large excess of chlorine.

Top tip!

When possible, convention dictates the • which signifies a radical is placed on the atom which has the unpaired electron. For the carbon-centred methyl radical, it is preferred to write $\bullet CH_3$, not $CH_3\bullet$.

Bromination of methane

Free radical substitution of methane with bromine proceeds in an analogous way to that with chlorine described above. The substitution with iodine is not energetically favourable ($\Delta G^\ominus > 0$) under standard conditions. The reaction with chlorine is faster than that with bromine because chlorine is a more electronegative element resulting in stronger bonds formed with hydrogen.

Bond enthalpy of H–Cl = 431 kJ mol^{-1}

Bond enthalpy of H–Br = 366 kJ mol^{-1}

This results in a lower activation energy in the propagation step in which chlorine radicals abstract a hydrogen.

LINKING QUESTION

What is the reverse process of homolytic fission?

LINKING QUESTION

Why do chlorofluorocarbons (CFCs) in the atmosphere break down to release chlorine radicals but typically not fluorine radicals?

LINKING QUESTION

Chlorine radicals released from CFCs are able to break down ozone, O_3, but not oxygen, O_2, in the stratosphere. What does this suggest about the relative strengths of bonds in the two allotropes?

Radical substitution of other alkanes

Radical substitution reactions between halogens and larger alkanes proceed through the same initiation, propagation and termination steps to produce a mixture of halogenoalkane products.

In the radical substitution of ethane with bromine to produce bromoethane:

$$Br_2 + C_2H_6 \rightarrow CH_3CH_2Br + HBr$$

56 Give equations that show the three stages of the radical substitution reaction between cyclohexane, C_6H_{12}, and chlorine, Cl_2.

the propagation steps are:

$$Br\bullet + C_2H_6 \rightarrow \bullet CH_2CH_3 + HBr$$

$$\bullet CH_2CH_3 + Br_2 \rightarrow CH_3CH_2Br + Br\bullet$$

The reaction mixture will also contain traces of further substituted bromoalkanes.

In most alkanes, the hydrogens are not all equivalent and substitution reactions result in halogenoalkanes which are isomers.

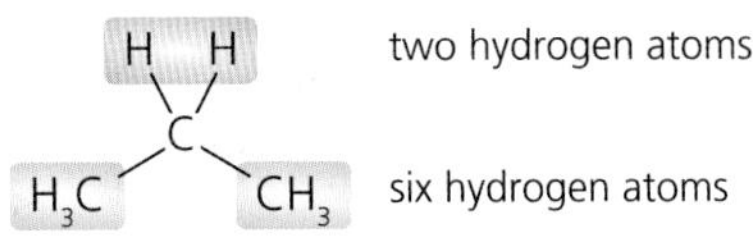

■ **Figure R3.94** Hydrogen atoms in propane

Consider the monosubstitution reaction of propane with chlorine in the gas phase in the presence of ultraviolet radiation. Two structural isomers, 1-chloropropane and 2-chloropropane, can be produced. This is because the chlorine radical can abstract one of the six primary (end-carbon) hydrogen atoms or one of the two secondary (middle-carbon) hydrogen atoms (Figure R3.94).

Because the ratio of primary hydrogen atoms to secondary hydrogen atoms is 3 : 1, the mathematical probability is that the ratio of 1-chloropropane to 2-chloropropane in the product should also be 3 : 1. However, this assumes that the abstraction of a primary hydrogen atom and of a secondary hydrogen atom are equally likely.

In practice, the ratio of 1-chloropropane to 2-chloropropane is nearly 1 : 1. This suggests that the secondary hydrogen atoms are three times more likely to be abstracted.

Top tip!

The radical substitution of alkanes with non-equivalent hydrogens results in a mixture of isomers, as well as products with further substitution.

This observation can be explained by the inductive stabilizing effect of the two alkyl groups (Figure R3.95). The secondary prop-2-yl radical is more thermodynamically stable than the primary prop-1-yl radical, and is therefore more likely to form.

$CH_3 \rightarrow \dot{C}H \leftarrow CH_3$

prop-2-yl radical: two alkyl groups give a double inductive effect

$CH_3—CH_2 \rightarrow \dot{C}H_2$

prop-1-yl radical: one alkyl group

■ **Figure R3.95** Inductive effects in propyl radicals

In general, the stability of radicals increases from primary through secondary to tertiary (Figure R3.96). This will have some degree of influence on the mixture of products formed in radical substitution reactions.

most stable: $R_3C\bullet$ (tertiary radical) > $R_2HC\bullet$ (secondary radical) > $RH_2C\bullet$ (primary radical) > $H_3C\bullet$ (methyl radical) :least stable

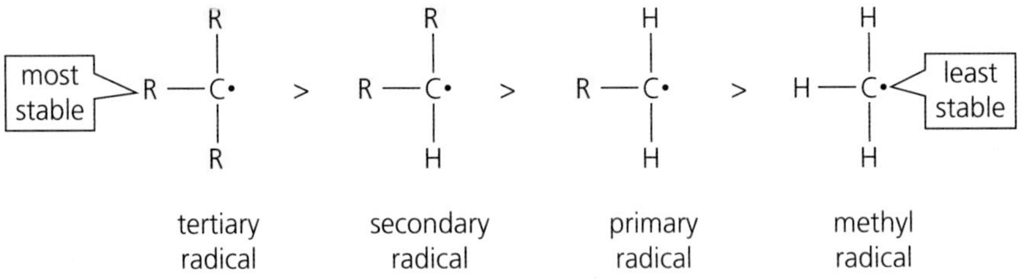

■ **Figure R3.96** Relative stability of alkyl radicals

Inquiry 3: Concluding and evaluating

Relate the outcomes of an investigation to the stated research question or hypothesis

The data and conclusion from an investigation should be related to the research question and, if appropriate, the hypothesis. Ideally, the research question should be answered, and the hypothesis supported or falsified, by the investigation.

The Kolbe reaction involves electrolysing solutions of carboxylate salts and obtaining alkanes via a free radical process. If sodium ethanoate is electrolysed, ethane is obtained via the production of methyl radicals at the anode:

$$2CH_3COO^- \rightarrow 2H_3C\bullet + 2CO_2 + 2e^-$$

$$2H_3C\bullet \rightarrow C_2H_6$$

Given this information, we can hypothesize that, if we use only this single salt, this method will only produce hydrocarbons with an even number of carbon atoms because the radicals produced couple together in the second stage of the reaction. We might propose the research question: Does the Kolbe electrolysis reaction on single salts only produce alkanes with an even number of carbons?

If an investigation into the Kolbe reaction with longer chain carboxylate salts is undertaken, the results disprove the stated hypothesis. With longer chain alkanes it is found that alkanes containing an odd number of carbons can be formed. This is because it is possible for two radicals to perform a termination step via disproportionation, producing an alkane and an alkene.

$$2C_3H_7\bullet \rightarrow C_3H_8 + C_3H_6$$

Radical reactions with other organic molecules

The reaction of alkenes with halogen radicals usually leads to a mixture of products because a radical-mediated addition reaction competes with the radical substitution process. In the case of benzene derivatives, such as methylbenzene, the radical substitution with chlorine or bromine in the presence of ultraviolet radiation proceeds well because the competing addition reaction will not occur on the more stable benzene ring (Figure R3.97). As with alkanes, more than one chlorine atom can be substituted. However, this too occurs only in the side chain and not in the benzene ring.

CH_3 + Cl_2 UV light / or boiling CH_2Cl + HCl

■ **Figure R3.97** The radical substitution reaction of methyl benzene

WORKED EXAMPLE R3.3A

3,3-dimethylpentane, C_7H_{16}, reacts with bromine to form monobromo compounds with the molecular formula $C_7H_{15}Br$.

3,3-dimethylpentane

How many possible structural isomers, each with the molecular formula $C_7H_{15}Br$, could be produced by 3,3-dimethylpentane and bromine gas in the presence of UV light?

Answer

There are three different sets of chemically inequivalent hydrogen atoms that can each be substituted to give three different monobromo products. Equivalent hydrogens are shown in the same colour. The inequivalent hydrogens are shown in different colours.

Substitution is more likely for the four red hydrogens because these are attached to a secondary carbon.

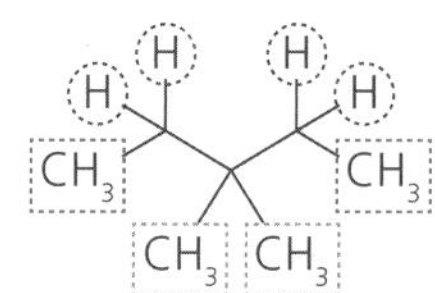

57 How many carbon-containing (organic) products could potentially be formed when the following hydrocarbons are mixed with chlorine and irradiated with ultraviolet light? (Exclude any products arising from termination reactions.)

a methane, CH_4

b methylbenzene, $C_6H_5CH_3$

c ethane, C_2H_6

Going further

Butylated hydroxyarenes and vitamin E-radical scavengers

Butylated hydroxyarenes, with the general structure shown in Figure R3.98, are added to many fatty foods to preserve them. They act by stopping the propagation steps in radical chain reactions because the radical they form is stable to further reactions. The bulky four-carbon groups bonded to the aromatic ring hinder the approach of potential reactants, and the radical is stabilized due to delocalization of the unpaired electron into the aromatic ring.

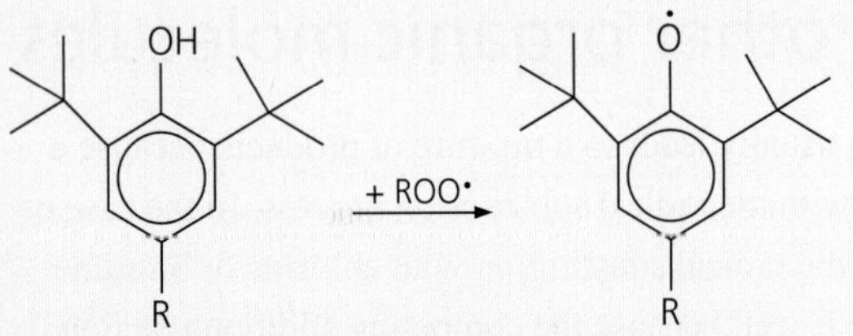

OH

+ ROO•

R

R

■ **Figure R3.98** A butylated hydroxyarene reacts with a radical to produce a stable hindered phenoxy radical

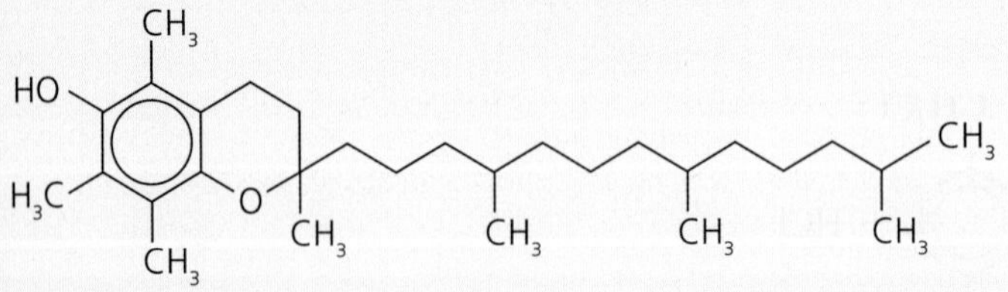

CH_3 HO H_3C O CH_3 CH_3 CH_3 CH_3 CH_3 CH_3

■ **Figure R3.99** The structure of vitamin E

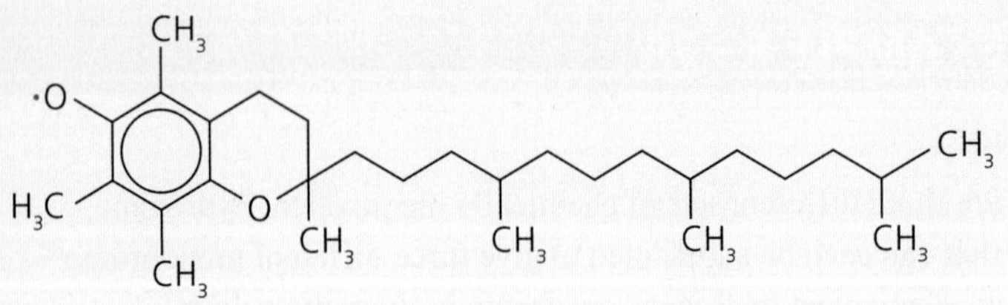

CH_3 •O H_3C O CH_3 CH_3 CH_3 CH_3 CH_3 CH_3

■ **Figure R3.100** The structure of the relatively stable radical formed by vitamin E

These compounds are also used in cosmetics, pharmaceuticals, jet fuels, rubber, petroleum products, and embalming fluid.

In our body, vitamin E performs a scavenger role similar to that of the butylated hydroxyarenes added to foods. Many molecules in our skin cells, including DNA, are susceptible to homolysis into radicals when exposed to intense light. The body uses vitamin E (Figure R3.99), to protect against the reactive radicals which form in skin cells. Vitamin E is relatively insoluble in water and lodges itself in the cell membranes where it is most effective. In the presence of radicals, vitamin E loses a hydrogen atom by abstraction from its single –OH group to form a phenoxy radical (Figure R3.100). This radical on vitamin E is relatively stable and does no further damage.

R3.4 Electron pair sharing reactions

Guiding question

- What happens when reactants share their electron pairs with others?

SYLLABUS CONTENT

By the end of the chapter, you should understand that:

- ▶ a nucleophile is a reactant that forms a bond to its reaction partner (the electrophile) by donating both bonding electrons
- ▶ in a nucleophilic substitution reaction, a nucleophile donates an electron pair to form a new bond, as another bond breaks producing a leaving group
- ▶ heterolytic fission is the breakage of a covalent bond when both bonding electrons remain with one of the two fragments formed
- ▶ an electrophile is a reactant that forms a bond to its reaction partner (the nucleophile) by accepting both bonding electrons from that reaction partner
- ▶ alkenes are susceptible to electrophilic attack because of the high electron density of the carbon-carbon double bond. These reactions lead to electrophilic addition
- ▶ a Lewis acid is an electron-pair acceptor and a Lewis base is an electron-pair donor (HL only)
- ▶ when a Lewis base reacts with a Lewis acid, a coordination bond is formed; nucleophiles are Lewis bases and electrophiles are Lewis acids (HL only)
- ▶ coordination bonds are formed when ligands donate an electron pair to transition element cations, forming complex ions (HL only)
- ▶ nucleophilic substitution reactions include the reactions between halogenoalkanes and nucleophiles (HL only)
- ▶ the rate of substitution reactions is influenced by the identity of the leaving group (HL only)
- ▶ alkenes readily undergo electrophilic addition reactions (HL only)
- ▶ the relative stability of carbocations in the addition reactions between hydrogen halides and unsymmetrical alkenes can be used to explain the reaction mechanism (HL only)
- ▶ electrophilic substitution reactions include the reactions of benzene with electrophiles (HL only).

By the end of the chapter, you should know how to:

- ▶ recognize nucleophiles and electrophiles in chemical reactions
- ▶ deduce equations for nucleophilic substitution reactions
- ▶ describe and explain the movement of electron pairs in nucleophilic substitution reactions
- ▶ explain, with equations, the formation of ions by heterolytic fission
- ▶ deduce equations for the reactions of alkenes with water, halogens, and hydrogen halides
- ▶ apply Lewis acid–base theory to inorganic and organic chemistry to identify the role of the reacting species (HL only)
- ▶ draw and interpret Lewis formulas of reactants and products to show coordination bond formation in Lewis acid–base reactions (HL only)
- ▶ deduce the charge on a complex ion, given the formula of the ion and ligands present (HL only)
- ▶ describe and explain the mechanisms of the reactions of primary and tertiary halogenoalkanes with nucleophiles (HL only)
- ▶ predict and explain the relative rates of the substitution reactions for different halogenoalkanes (HL only)
- ▶ describe and explain the mechanisms of the reactions between symmetrical alkenes and halogens, water and hydrogen halides (HL only)
- ▶ predict and explain the major product of a reaction between an unsymmetrical alkene and a hydrogen halide or water (HL only)
- ▶ describe and explain the mechanism of the reaction between benzene and a charged electrophile, E^+ (HL only).

Nucleophiles and electrophiles

When molecules react, electrons move from one species to another. At a simple level, the chemical interaction between organic molecules can be characterized as the interaction between an electron-rich species and an electron-deficient species, with the electrons moving from the former to the latter. When a pair of electrons in an orbital of an electron-rich species moves into an unfilled orbital of an electron-deficient species, a new bond is formed.

◆ **Nucleophile**: A reactant that forms a bond to its reaction partner (the electrophile) by donating a pair of electrons.

◆ **Electrophile**: A reactant that forms a bond to its reaction partner (the nucleophile) by accepting a pair of electrons from that reaction partner.

The term given to an electron-rich species which donates a pair of electrons to its reaction partner in an organic reaction is a **nucleophile**. The electron-deficient species which accepts the pair of electrons is the **electrophile**. These are very important terms in organic chemistry and it is helpful to recognize which species is the nucleophile and which is the electrophile in all reactions.

Figure R3.101 shows how we use curly arrows to communicate the movement of electrons from a nucleophile to an electrophile.

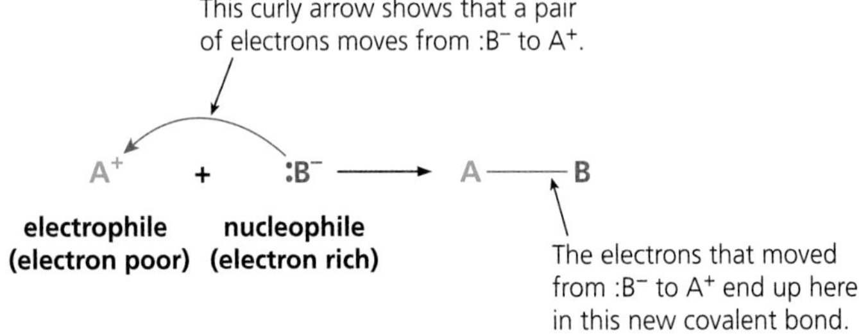

■ **Figure R3.101** The reaction between an electrophile and a nucleophile

Link

Curly arrow notation is also covered in Chapter R3.3, page 627.

Top tip!

A full-headed curly arrow shows the movement of a pair of electrons.

Of course, molecules must come into contact with one another before they can pass electrons between them and react. The combined kinetic energy of the two molecules must be sufficient to overcome the energy barrier to reaction (the activation energy) and the orientation of the collision must allow the relevant orbitals to interact. The electron-rich nature of the nucleophile and the electron-deficient nature of the electrophile may result in electrostatic attractions which initially help to draw the two reactants together. However, this is not a necessary condition and chemical reactions between two non-polar molecules are possible.

Nucleophiles

Nucleophiles can be either neutral or negatively charged species with a pair of electrons in a high-energy orbital. There is a vast number of organic molecules, many of which can act as nucleophiles with the right reaction partner present. You should be familiar with the common nucleophiles shown in Table R3.29.

■ **Table R3.29** Selected nucleophiles

Name	Molecular formula	Structural formula
water molecule	$:OH_2$	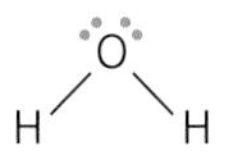
ammonia molecule	$:NH_3$	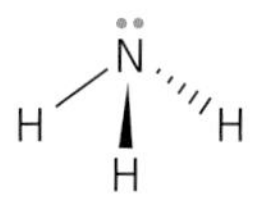
hydroxide ion	$HO:^-$	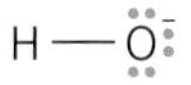
cyanide ion	$^-:CN$	
chloride ion	$:Cl^-$	
amines	$R_3N:$	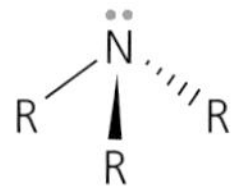
alcohols	ROH	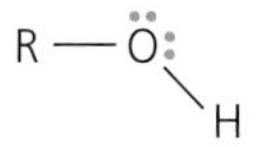
alkoxide ion	$RO:^-$	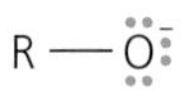
alkene	$R_2C{=}CR_2$	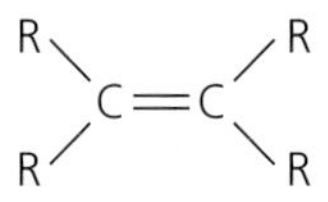

There are three classes of nucleophile: lone-pair (non-bonding pair) nucleophiles, pi-bond nucleophiles and sigma-bond nucleophiles.

Lone-pair nucleophiles

Lone-pair nucleophiles are the type you will meet most often in the *IB* course. They contain atoms with one or more non-bonding pairs available for bond formation. The non-bonding pair, which is often located on an electronegative atom, is used to make a new bond to an electrophilic atom. Alcohols (ROH), alkoxides (RO^-), amines (R_3N), water (H_2O), hydroxide (HO^-) and halides (X^-) are all examples of lone-pair nucleophiles. When these nucleophiles are shown reacting with an electrophile, convention dictates that the presence of the non-bonding pair of electrons is explicitly represented. This can be seen in Figure R3.102 and is why hydroxide ions are shown as $HO:^-$ (not HO^-).

■ **Figure R3.102** A hydroxide ion acting as a nucleophile. E^+ represents an electrophile (electron-pair acceptor)

Pi-bond nucleophiles

Pi-bond nucleophiles use the pair of electrons in a pi bond. The pi bonds of simple alkenes and arenes are weakly nucleophilic. An alkenyl functional group acting as a nucleophile is shown in Figure R3.103.

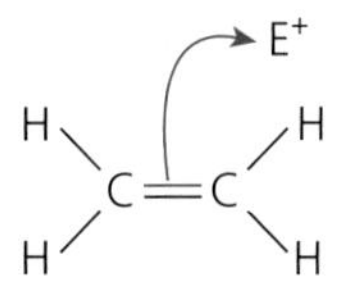

■ **Figure R3.103** The pi bond in an alkenyl functional group acting as a nucleophile. E^+ represents the electrophile

Sigma-bond nucleophiles

Sigma-bond nucleophiles are not as common as the previous two types of nucleophiles. The electrons come from a sigma bond, and this existing bond is broken as the new bond to the electrophilic centre is made. An example of a sigma-bond nucleophile is sodium borohydride, $NaBH_4$ (Figure R3.104). The borohydride ion acts as a nucleophile and is used to reduce organic compounds.

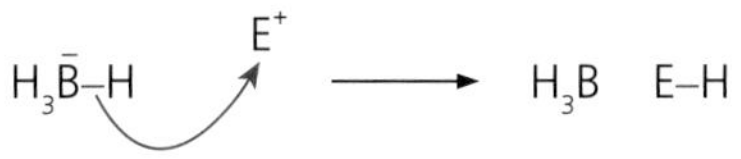

$$H_3\bar{B}{-}H \quad E^+ \longrightarrow H_3B \quad E{-}H$$

■ **Figure R3.104** A borohydride ion acting as a nucleophile. E^+ represents an electrophile

Going further

Ambident nucleophiles

The cyanide ion (Figure R3.105) has non-bonded pairs located on both the nitrogen and carbon atoms. Nucleophiles which can potentially attack from two sites are called ambident nucleophiles. These nucleophiles can attack from either of the sites depending on the reaction conditions and the reagent used.

$:\bar{C}\equiv N:$

■ **Figure R3.105** Lewis formula for the cyanide ion

For example, when halogenoalkanes are treated with an alcoholic solution of potassium cyanide, KCN (ionic), a nitrile (RCN) is the major product because attack takes place through the negatively charged carbon. However, when silver cyanide, AgCN (covalent), is used as a reagent, an alkylisocyanide (RNC) is the major product due to attack from the non-bonding pair on the nitrogen atom.

Enolate ions are another ambident nucleophile. The resonance structures of an enolate anion show the two sites it can react from: a negative charge on the carbon or the oxygen atom (Figure R3.106). Like all molecules with resonance structures, the true structure has characteristics of both resonance structures, although the second structure with a carbon-carbon double bond is closer to the structure we observe because the negative charge is more stable when carried on the electronegative oxygen atom.

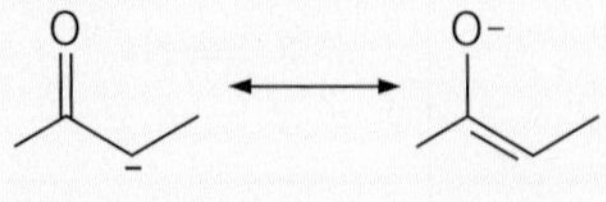

■ **Figure R3.106** The resonance structures of an enolate anion

Electrophiles

Electrophiles can be either neutral or positively charged. Some examples of electrophiles are given in Table R3.30.

■ **Table R3.30** Selected electrophiles

Name	Molecular formula
halogenoalkanes	R_3CX
halogens	X_2
carbonyl-containing compounds	$R_2C{=}O$
hydronium ion	H_3O^+
nitronium ion	NO_2^+
boron trifluoride	BF_3
aluminium chloride	$AlCl_3$
carbocations	R_3C^+

Any molecule accepting a pair of electrons to form a new bond is acting as an electrophile. All electrophiles have a relatively low-energy unfilled orbital in which they accept a pair of electrons from the nucleophile.

Figure R3.107 shows boron trifluoride, BF_3, acting as an electrophile in the presence of the nucleophile ammonia, NH_3. Boron trifluoride is electron deficient because it only has six valence electrons around the boron atom. The presence of an empty p-orbital on the boron atom means it willingly accepts a pair of electrons.

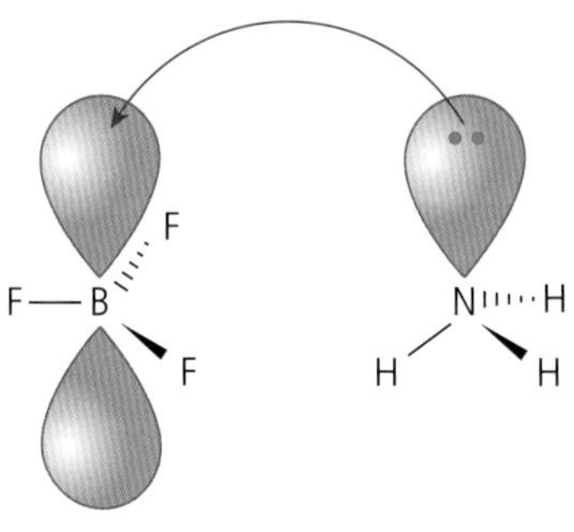

■ **Figure R3.107** The orbitals involved in the interaction of boron trifluoride, BF_3, and ammonia, NH_3

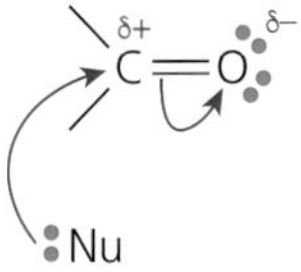

■ **Figure R3.108** A nucleophile forms a new bond with the electrophilic carbon on a carbonyl group

Many molecules act as electrophiles at sites which have a positive or partial positive (δ+) charge. Understanding which parts of a molecule are partially positive can help us to identify where a molecule may accept the pair of electrons from a nucleophile. A carbonyl functional group is polarized with the electrons lying closer to the more electronegative oxygen atom; this leaves the carbon with a partial positive charge. Carbonyl-containing molecules will act as electrophiles and accept electrons from a nucleophile at the partially positive carbon atom (Figure R3.108). This causes the pi bond between carbon and oxygen to break; the pair of electrons in this bond moves to the oxygen atom at the same time that the new bond to the nucleophile is formed.

The observation that non-polar symmetrical molecules, such as bromine, can act as electrophiles shows that a positive charge is not an essential feature of an electrophile. In fact, the only thing an electrophile must have is a low-energy unfilled orbital.

WORKED EXAMPLE R3.4A

Identify the nucleophile and electrophile in the reaction shown.

$^+CH_2CH_3$

benzene

Answer

Benzene is providing a pair of electrons which are moving towards the ethyl carbocation. Benzene is acting as the nucleophile and the ethyl carbocation is acting as the electrophile.

58 Identify the nucleophile and electrophile for the reactions shown.

a

b

c

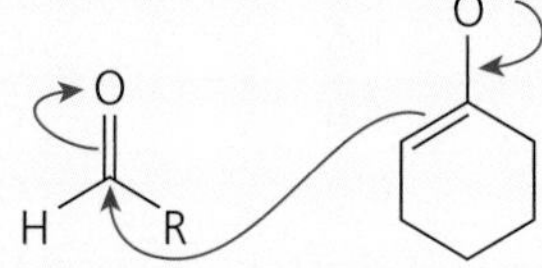

59 Identify the sites where this molecule may act as a nucleophile.

60 Identify the sites where this molecule may act as an electrophile.

Nucleophilic substitution reactions

◆ **Leaving group**: The group leaving a molecule in a substitution or elimination reaction.

A nucleophilic substitution reaction is a type of reaction in which a nucleophile replaces a **leaving group** present on an electrophile.

Consider the reaction between a hydroxide ion, OH^- (the nucleophile), and iodomethane, CH_3I (the electrophile). The hydroxide ion nucleophile replaces the iodide ion in the electrophile.

$$OH^- + CH_3I \rightarrow CH_3OH + I^-$$

The mechanism of this nucleophilic substitution reaction – that is, how the electrons move to bring about the chemical transformation – is outlined in Figure R3.109.

The charged hydroxide nucleophile donates a pair of electrons from the oxygen atom to form a new bond to the carbon of the iodomethane electrophile. The iodine atom leaves, taking with it the electrons bonding it to carbon. An iodide ion is the leaving group in this reaction.

■ **Figure R3.109** Mechanism of the nucleophilic substitution reaction between a hydroxide ion and iodomethane

All nucleophilic substitution reactions involve the substitution of a nucleophile for a leaving group, but the mechanism depends on the reactants. More details are given later in the chapter.

More generally, a nucleophilic substitution reaction can be summarized as shown in Figure R3.110. The leaving group will form an anion, and the charge of the product depends on whether the nucleophile used has a negative or neutral charge.

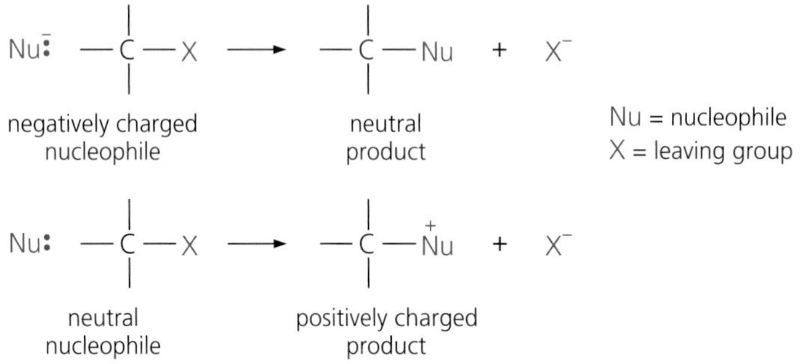

■ **Figure R3.110** Nucleophilic substitution reactions

Common leaving groups are stable anions such as the halide ions Cl^-, Br^- and I^-. Thus, halogenoalkanes often undergo nucleophilic substitution.

Table R3.31 illustrates how some common inorganic and organic nucleophiles react with a halogenoalkane such as bromoethane, C_2H_5Br.

■ **Table R3.31** Common nucleophiles that react with halogenoalkanes

Nucleophile		
Name	**Formula**	**Products when reacted with bromoethane**
water	$H_2\ddot{O}$	$CH_3CH_2OH + HBr$
hydroxide ion	$H\ddot{O}^-$	$CH_3CH_2OH + Br^-$
ammonia	$\ddot{N}H_3$	$CH_3CH_2NH_2 + HBr$
cyanide ion	$^-\ddot{C}\equiv N$	$CH_3CH_2CN + Br^-$
methoxide ion	$CH_3\text{–}\ddot{O}^-$	$CH_3CH_2\text{–O–}CH_3 + Br^-$
methylamine	$\ddot{N}H_2\text{–}CH_3$	$CH_3CH_2\text{–NH–}CH_3 + HBr$

A hydride ion, H^-, is very unstable and is therefore a poor leaving group; it will never leave in a nucleophilic substitution.

The important points to note about nucleophilic substitution reactions are that:

- a nucleophile donates a pair of electrons to form a new bond
- a bond in the electrophile breaks to produce a leaving group
- the nucleophile is substituted in place of the leaving group.

WORKED EXAMPLE R3.4B

Given the structures of sodium thiophenolate, C_6H_5SNa, and chloromethyl benzene, $C_6H_5CH_2Cl$, complete the equation for the nucleophilic substitution reaction between the two.

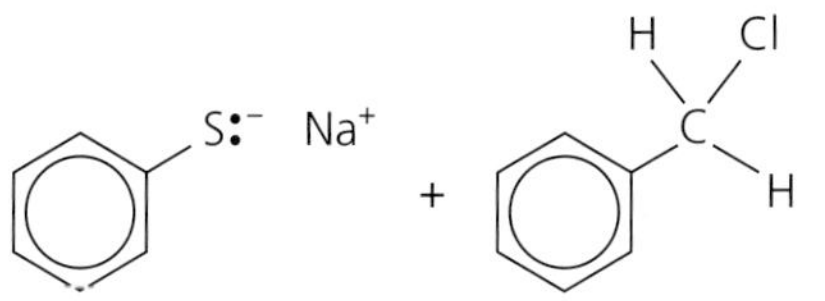

Answer

The best approach is to identify the nucleophile, the electrophile and the leaving group. Sodium thiophenolate is an ionic compound and so the anion, the thiophenolate ion, must be the nucleophile.

Chlorobenzene is the electrophile, and the polar C–Cl bond makes it electrophilic at the carbon bearing the chlorine atom.

A chloride ion will be the leaving group.

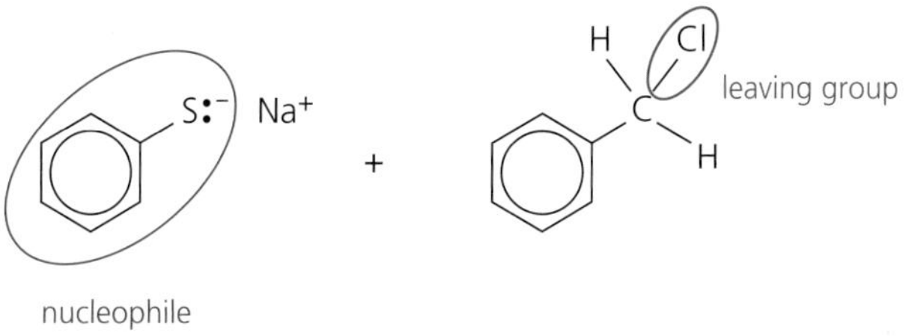

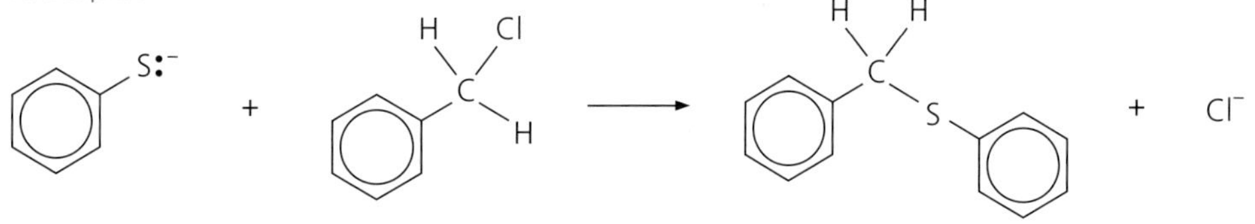

61 Indicate whether the following reactions are radical substitution reactions or nucleophilic substitution reactions.

a $C_4H_{10} + 2Br_2 \rightarrow C_4H_8Br_2 + 2HBr$

b $NaOH + C_4H_9Cl \rightarrow C_4H_9OH + NaCl$

c $NH_3 + C_2H_5I \rightarrow C_2H_5NH_3^+ + I^-$

d $NaI + C_2H_5Cl \rightarrow C_2H_5I + NaCl$

e $Cl_2 + CH_4 \rightarrow CH_3Cl + HCl$

Heterolytic fission

Link

Heterolytic fission is defined and discussed in Chapter R3.3, page 627.

Heterolytic fission involves the splitting of a covalent bond giving an unequal share of bonding electrons to each fragment formed. In Figure R3.111, the chlorine atom in chloromethane takes both the bonding electrons in the C–Cl bond, leaving the carbon atom with none.

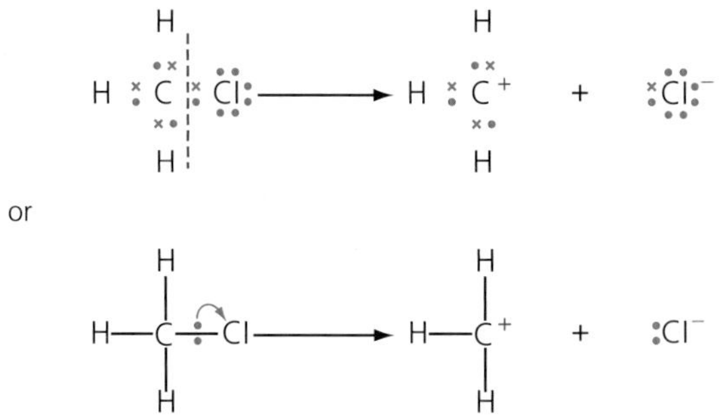

or

■ **Figure R3.111** Heterolytic fission of chloromethane to form a methyl carbocation and a chloride ion

The full-headed curly arrow in Figure R3.111 describes the movement of the bonded pair of electrons to the chlorine atom. When heterolytic fission of a bond occurs, the electrons will move to the atom with the higher electronegativity value. In this example, chlorine, being more electronegative than carbon, receives the bonding electrons.

This movement of the electron pair leaves the carbon atom in the methyl carbocation electron deficient. The carbon has only six electrons in its outer shell, and this combined with its positive charge makes it a strong electrophile (electron-pair acceptor).

TOK

Does chemistry have its own language?

In many parts of this textbook, in addition to learning facts about chemistry, you are also expected to learn the *language* of chemistry in the guise of chemical notation and various conventions used by the community of chemists.

Having the ability to understand and draw skeletal formulas is a fundamental skill if you wish to learn about organic chemistry. Additionally, curly arrows are often considered to be the language of organic chemistry because they succinctly describe the movement of electrons and help us to communicate much more effectively than words alone.

To what extent is it possible to suggest that experts in the community of chemists speak a different language? Are some meanings, denotations or concepts only available to communities who know how to 'speak' the language?

Electrophilic addition reactions with alkenes

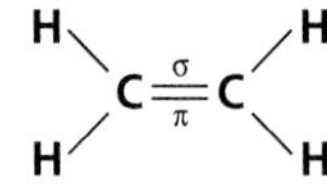

■ **Figure R3.112** Bonds in the ethene molecule

Alkenes are described as unsaturated since they have carbon atoms joined by a double bond. The double bond consists of two types of bond: a sigma bond and a pi bond.

The sigma bond, formed by two electrons, is relatively strong. The pi bond, also formed by two electrons, is relatively weak. There is a high electron density around a carbon-carbon double bond, and this allows it to act as a nucleophile when an appropriate electrophile is present.

Alkenes react by using the pair of electrons in the pi bond to form a sigma bond to an electrophile. The original sigma bond between the two carbon atoms is left intact. Normally, this results in only one molecule of product being produced from two reacting molecules during an addition reaction.

Many electrophiles can be added across the double bond of an alkene in place of the pi bond. Important products, formed by the addition of halogens, hydrogen halides and water, are described below.

Electrophilic addition of halogens

◆ **Electrophilic addition reaction:** A reaction in which an electrophile attacks a pi bond and two new sigma bonds are formed.

Alkanes react with the halogens in **electrophilic addition reactions** to form dihalogenoalkanes. This is an addition reaction because the two reactants are adding together to make a single product. It is electrophilic because the reagent used, in this case a halogen, is initially accepting a pair of electrons from the pi bond of the alkene.

For example, liquid bromine can rapidly react when excess ethene gas is bubbled into it to form colourless 1-2-dibromoethane (Figure R3.113).

ethene $H_2C{=}CH_2 + Br_2 \longrightarrow H{-}CHBr{-}CHBr{-}H$ 1,2-dibromoethane

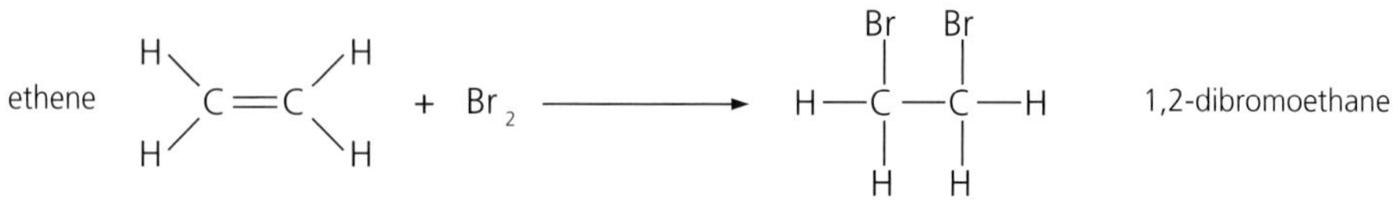

Figure R3.113 Reaction between ethene and bromine to form 1-2-dibromoethane

This reaction proceeds well with other alkenes too, and two bromine atoms become bonded to the carbons which had the pi bond between them in the starting alkene.

An aqueous solution of bromine, $Br_2(aq)$, is called bromine water and provides a simple chemical test for an alkene. The dark orange colour of the bromine water fades when an alkene is added (Figure R3.114) because the molecular bromine dissolved in the solution undergoes an electrophilic addition reaction with the alkene.

Figure R3.114 Alkenes decolourize bromine water

LINKING QUESTION

Why is bromine water decolourized by alkenes but not by alkanes?

Electrophilic addition of hydrogen halides

The hydrogen halides HCl, HBr and HI, all react with alkenes to form halogenoalkanes. The gaseous hydrogen halide may be bubbled directly through the liquid alkene, or glacial ethanoic acid may be used as a solvent. In these reactions, a hydrogen atom and halogen atom bond to the two carbons which had the pi bond in the starting alkene (Figure R3.115). The reaction mechanism is electrophilic addition, and the products are halogenoalkanes.

The reactivity of the hydrogen halides is in the order HI > HBr > HCl because of the decreasing strength of the hydrogen halide bond going down Group 17. The hydrogen halide bond is broken during the reaction, so hydrogen iodide, HI – which has the weakest and longest H–X bond – reacts the fastest.

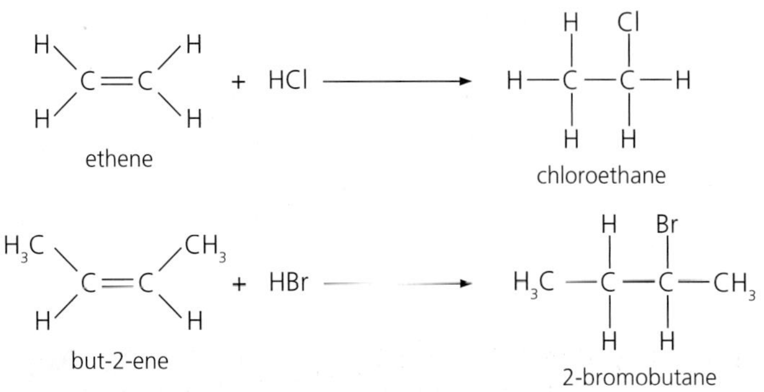

■ **Figure R3.115** The reaction between hydrogen chloride and ethene to form chloroethane, and between but-2-ene and hydrogen bromide to form 2-bromobutane

Electrophilic addition of water

Water alone is not a strong enough electrophile to react directly with alkenes. However, the reaction will proceed in the presence of a strong acid, usually concentrated sulfuric acid. The strong acid is needed to form an appreciable concentration of H_3O^+ which acts as the electrophile. The product of the acid-catalyzed electrophilic addition reaction between an alkene and water (also known as a hydration reaction) is an alcohol. Ethene will react with steam at a high temperature and pressure in the presence of an acid catalyst to form ethanol (Figure R3.116).

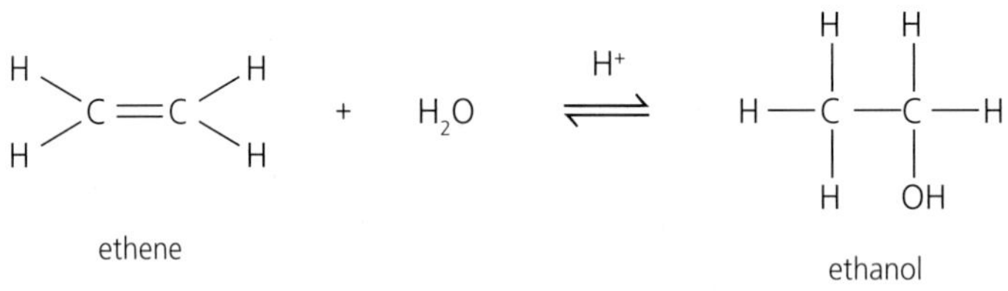

■ **Figure R3.116** The acid-catalyzed reaction between ethene and water

Figure R3.117 summarizes the products formed by the electrophilic addition reactions of ethene.

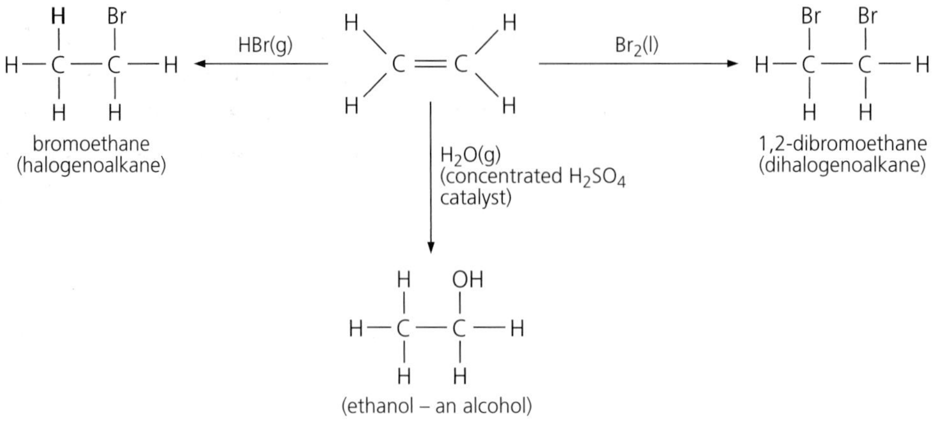

■ **Figure R3.117** A summary of the electrophilic addition reactions of ethene

WORKED EXAMPLE R3.4C

Suggest the product formed when hydrogen bromide gas is bubbled through cyclohexene, C_6H_{10}, and state the name of this type of reaction.

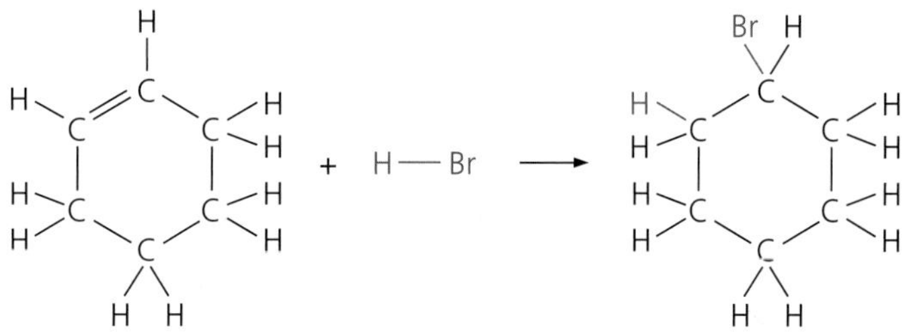

Answer

The carbon–carbon pi bond will react with the hydrogen bromide electrophile to produce bromocylohexane. The type of reaction is electrophilic addition.

Tool 1: Experimental techniques

Recrystallization

Solid organic products, for example, aspirin, are purified by recrystallization. This is done by identifying a solvent or solvent mixture (from data tables or experiments) in which the pure product is insoluble at low temperatures but soluble at high temperatures. Aspirin can be recrystallized using a 1 : 3 volume ratio of ethanol and water.

The minimum amount of hot solvent is added to dissolve the impure (crude) product. Any insoluble impurities remain solid, so hot filtration (where the filter funnel and paper are pre-heated) allows the solid impurities to be separated.

The mixture is then allowed to cool slowly and large crystals of the pure product form. As the crystals grow, they exclude molecules of solvent and any soluble impurities. The pure solid is then separated by vacuum or suction filtration (Figure R3.118), washed with ice-cold solvent, and allowed to dry.

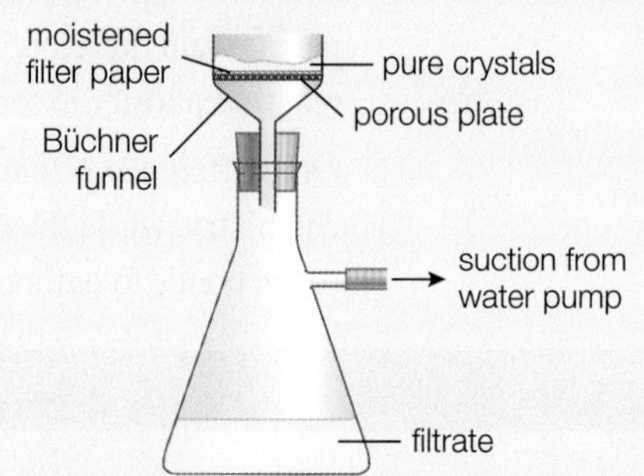

■ **Figure R3.118** Filtration under reduced pressure

LINKING QUESTION

Why are alkenes sometimes known as 'starting molecules' in industry?

Link

The Brønsted–Lowry theory is also covered in Chapter R3.1, page 529.

◆ **Lewis acid**: An electron-pair acceptor.

◆ **Lewis base**: An electron-pair donor.

HL ONLY

Lewis acids and Lewis bases

The Brønsted–Lowry theory of acids and bases states that acids are proton donors and bases are proton acceptors. A more general definition of acids and bases, related to electron pair acceptors and donors, was proposed by Gilbert Lewis (1875–1946).

Consider the reaction of a free hydrogen ion with a water molecule, which proceeds as shown in Figure R3.119.

$$H_2\ddot{O}: \quad H^+ \rightarrow H_3O^+$$

■ **Figure R3.119** The formation of a hydronium ion from a hydrogen ion (proton) and water.

Using the Brønsted–Lowry definitions, water is a base in this reaction, because it accepts a proton, and the hydrogen ion is the acid (proton donor). In the Lewis model, the hydrogen ion, H^+, is accepting a pair of electrons and hence it is a **Lewis acid**. The water molecule is donating the pair of electrons and hence it is a **Lewis base**.

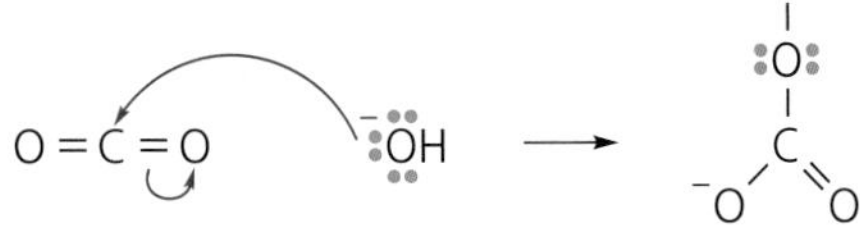

■ **Figure R3.120** The first stage of the reaction between a hydroxide ion and carbon dioxide

In the reaction of carbon dioxide, CO_2, with a hydroxide ion, ^{-}OH, to form a hydrogencarbonate ion, HCO_3^-, the carbon dioxide molecule accepts a pair of electrons (Figure R3.120). Carbon dioxide is the Lewis acid and the hydroxide ion is the Lewis base. There is no proton transfer, so it is not possible to identify the acid and base using only the Brønsted–Lowry theory.

LINKING QUESTION

What is the relationship between Brønsted–Lowry acids and bases and Lewis acids and bases?

ATL R3.4A

Acid–base theories developed as scientists tried to identify acidic and basic properties in all the substances we observe. Traditionally, acids were substances in aqueous solutions which had a sour taste and turned litmus red. However, other ideas – proposed by Arrhenius, Brønsted, Lowry, Lewis, Lux, Flood and Usanovich – have become accepted.

Research the different acid–base theories and summarize them in a table.

There are close links between the definitions of Lewis acids and bases and those of nucleophiles and electrophiles. Nucleophiles donate a pair of electrons to form a new bond to their reaction partner and therefore behave as Lewis bases. Electrophiles accept a pair of electrons from their reaction partner to form a new bond and therefore behave as Lewis acids.

One difference is that the terms acid and base are used broadly in both inorganic and organic chemistry, while the terms nucleophile and electrophile are used primarily in organic chemistry when bonds to carbon are involved.

Going further

Strengths of Lewis acids and Lewis bases

The strength of the interaction between a Lewis acid and a Lewis base is controlled by electronic and steric factors.

The electronic factor is demonstrated by the effect of electron-donating and electron-withdrawing groups. When electron-donating groups are attached to an atom, they can increase the Lewis basicity of that atom, while electron-withdrawing groups can increase the Lewis acidity.

Trimethylamine, $N(CH_3)_3$, is a Lewis base. It can be predicted to be a stronger base than NH_3 because of the electron-donating methyl groups. The pK_a values of the protonated bases, 9.8 for $N(CH_3)_3H^+$ and 9.2 for NH_4^+, show that trimethylamine (with a higher pK_a value) has a greater preference to exist in its protonated form; it is, therefore, a stronger base than ammonia. This is attributed to the increased electron density on the nitrogen atom.

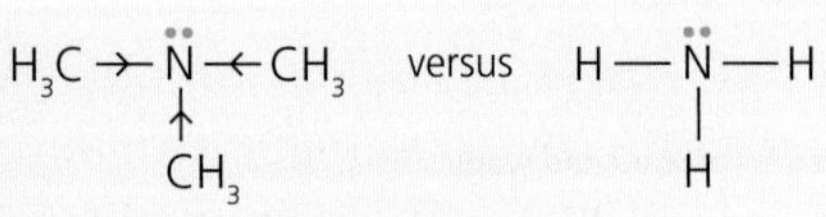

■ **Figure R3.121** Structures of the Lewis bases trimethylamine and ammonia

Similarly, one would expect BF_3 to be a stronger Lewis acid than $B(CH_3)_3$ due to the electron-withdrawing effects of the fluorine atoms which make the boron more willing to accept a pair of electrons from a Lewis base (Figure R3.122).

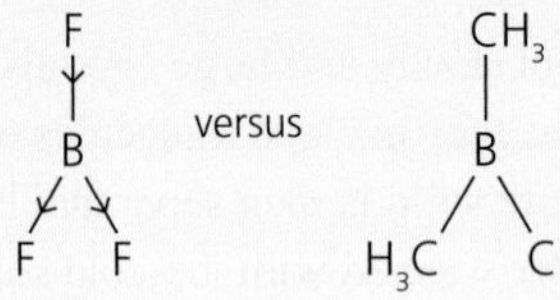

■ **Figure R3.122** Structures of the Lewis acids boron trifluoride and trimethylborane

Steric factors also play a role in determining Lewis acidity and Lewis basicity.

A trivalent nitrogen compound, NR_3, has a trigonal pyramidal shape and the nitrogen has sp^3 hybridization. Large, bulky groups on the nitrogen can act to hinder the approach of Lewis acids and stop the formation of adducts. The bulky groups make it harder for the nitrogen to act as a Lewis base, and hence make the compound less basic.

A trivalent boron compound is trigonal planar with an sp^2-hybridized boron atom. When a Lewis acid–base adduct is formed (for example, $H_3N \rightarrow BF_3$) the boron atom is rehybridized to sp^3 and the three groups on boron are forced closer together in a tetrahedral arrangement. Large, bulky groups on boron will not favour being forced closer together and this will reduce the Lewis acidity of the trivalent boron compound.

The products of Lewis acid–base reactions

When a Lewis acid reacts with a Lewis base, a coordination bond forms between the two reactants (Figure R3.123).

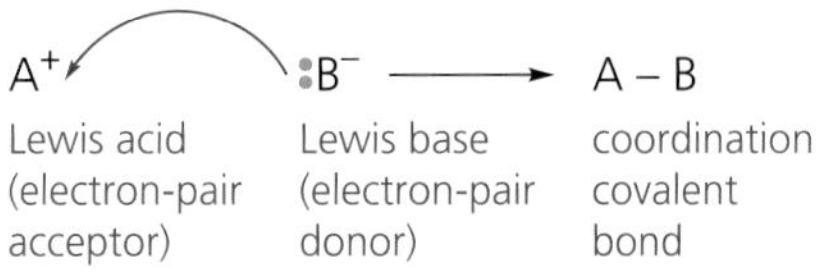

A^{+} :B^{-} → A – B
Lewis acid (electron-pair acceptor) Lewis base (electron-pair donor) coordination covalent bond

Figure R3.123 The generic reaction of a Lewis acid and Lewis base to form a coordination bond

Link

Coordination bonds are introduced on page 135.

Coordination bonds (also called dative covalent bonds) are covalent bonds in which the electrons are derived from only one of the bonded atoms. They are indistinguishable from any other covalent bond because the electrons have no memory of where they came from.

LINKING QUESTION

Do coordination bonds have any different properties from other covalent bonds?

WORKED EXAMPLE R3.4D

Fluoride ions, F^{-}, react with silicon tetrafluoride molecules, SiF_4, to form a single product, SiF_6^{2-}.

$SiF_4 + 2F^{-} \rightarrow SiF_6^{2-}$

a Identify the Lewis acid and Lewis base.

b Draw curly arrows to show the movement of the electrons in this reaction.

Answer

a Lewis acid: SiF_4. Lewis base: F^{-}. The silicon tetrafluoride accepts a pair of electrons from each fluoride ion.

b

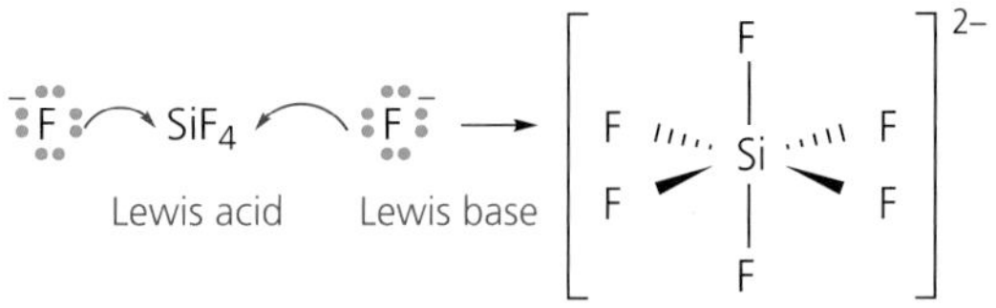

:F:⁻ SiF_4 :F:⁻ → $[SiF_6]^{2-}$
Lewis acid Lewis base

62 Draw curly arrows to show the movement of electrons in the reaction shown below. Is propanone the Lewis acid or Lewis base?

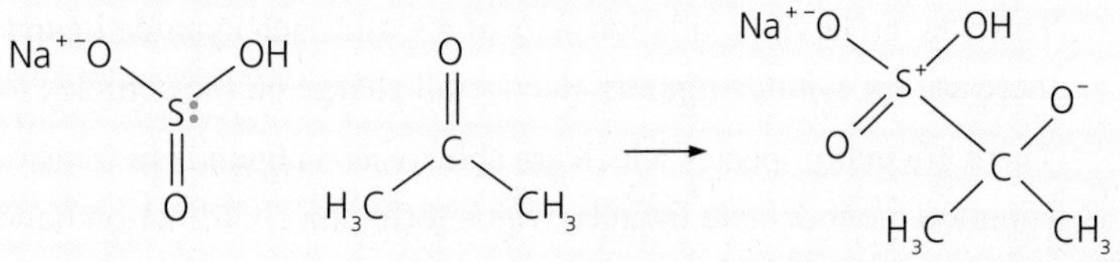

Na^{+} $^{-}$O–S(=O)–OH + H_3C–C(=O)–CH_3 → Na^{+} $^{-}$O–S^{+}(=O)(OH)–C(O^{-})(H_3C)(CH_3)

If the product of a Lewis acid–base reaction contains all the atoms from both the acid and the base (as is the case in Figure R3.123), it is often referred to as an adduct. The product does not always contain all the atoms from the acid and base because the Lewis acid can often fragment when it accepts a pair of electrons.

The nucleophilic substitution reaction between hydroxide ions and iodomethane (see Figure R3.109) is an example of a Lewis acid–base reaction in which a coordination bond is formed but the product does not contain all the atoms from the reactants. Instead, the Lewis acid has fragmented producing a leaving group, I^{-}. It is still valid to classify these reactions as involving Lewis acids and Lewis bases, or indeed, nucleophiles and electrophiles.

WORKED EXAMPLE R3.4E

Draw the Lewis formulas of the reactants and products in the reaction of an ammonia molecule, NH_3, with a hydronium ion, H_3O^+. Add curly arrows to show the movement of pairs of electrons.

Answer

63 Identify the Lewis acid and Lewis base in the forward reactions of each of the following:

a $NH_4^+ + OH^- \rightleftharpoons NH_3 + H_2O$

b $H_2O + HCO_3^- \rightleftharpoons CO_3^{2-} + H_3O^+$

c $H_2NCONH_2 + H_2O \rightleftharpoons H_3N^+CONH_2 + OH^-$

d $HSO_4^- + H_3O^+ \rightleftharpoons H_2SO_4 + H_2O$

e $H^- + H_2O \rightarrow OH^- + H_2$

64 Draw Lewis structures for the reactants and products in parts a–e of question 63. Draw curly arrows to show the movement of a pair of electrons between the Lewis acid and Lewis base.

Link

Complex ions are also discussed in Chapter S2.2, page 140 and Chapter S3.1, page 254.

Complex ions

Complex ions are formed when Lewis bases, known as ligands in this context, form coordination bonds with transition metal cations. Transition metal cations have unfilled orbitals which can accept electrons from the neutral or negatively charged ligands. (Some metals in the p-block, such as Pb, Al and Sn, can also form complex ions, but this discussion will be limited to those of transition metals.)

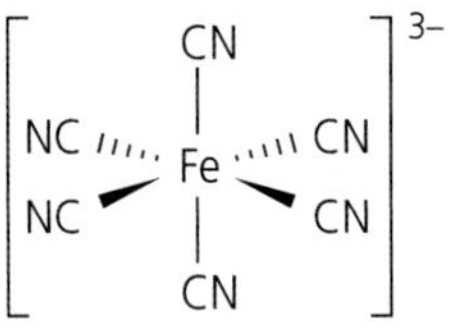

■ **Figure R3.124** The complex ion formed between iron(III) and six cyanide ligands has an octahedral shape

The complex ion formed between iron(III) and cyanide ligands, CN^-, is shown in Figure R3.124. The iron ion has an oxidation state of +3, and each cyanide ligand has a charge of −1. Because there are six cyanide ligands, the overall charge on the complex ion is −3.

There are many species which are able to act as ligands in transition metal complexes. Some common monodentate ligands, which form one coordination bond to the central metal ion, are shown in Table R3.32.

When a complex ion forms, the metal cation acts as a Lewis acid and the ligands act as Lewis bases. Each bond between a ligand and the cation is a coordination bond.

■ **Table R3.32** Common monodentate ligands

Ligand formula	Unbonded ligand name
I^-	iodide
Br^-	bromide
OH^-	hydroxide
Cl^-	chloride
F^-	fluoride

Ligand formula	Unbonded ligand name
H_2O	water
SCN^-	thiocyanate
NH_3	ammonia
CN^-	cyanide
CO	carbon monoxide

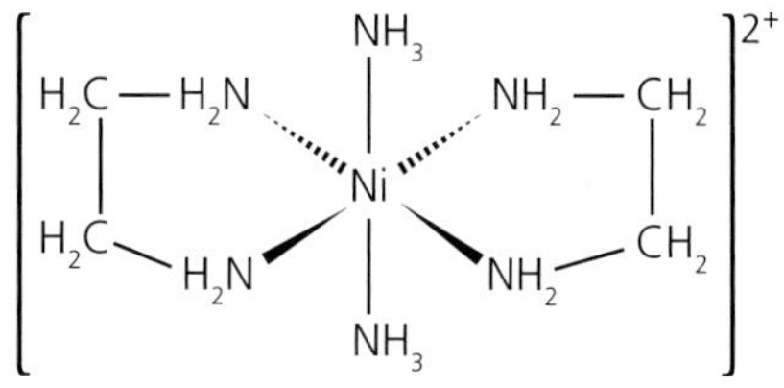

■ **Figure R3.125** 1,2-diaminoethane ligands in a complex ion with nickel and ammonia

◆ **Coordination number**: In a complex ion it is the number of donor atoms that form coordination bonds to a central metal atom or ion.

HL ONLY

Multidentate ligands can form more than one coordination bond to the metal ion through multiple binding sites. For example, 1,2-diaminoethane molecules, $H_2NCH_2CH_2NH_2$, bind to a central metal ion through two different sites as shown in Figure R3.125.

The total number of binding sites occupied by the ligands which surround a metal ion is known as the **coordination number**. The coordination number depends on the size of the ligands and how many can fit around the metal ion, as well as the identity of the metal ion.

In equations, complex ions are shown in square brackets with the charge outside, just as in the displayed formula.

- $[Fe(CN)_6]^{3-}$ shows an iron ion surrounded by six cyanide ligands. Because the overall charge on the complex is −3, and each cyanide ligand has a −1 charge, we can deduce that the iron must have a +3 oxidation state.
- In $[Cu(H_2O)_6]^{2+}$, a copper(II) ion is surrounded by six water ligands. As water ligands are neutral, the oxidation state of the copper ion must be +2 to achieve the +2 total charge on the complex ion.

When complex ions are part of an ionic compound, they must – like any other ions – exist with counter ions to make them electrically neutral. An ionic compound containing a complex ion is known as a coordination compound.

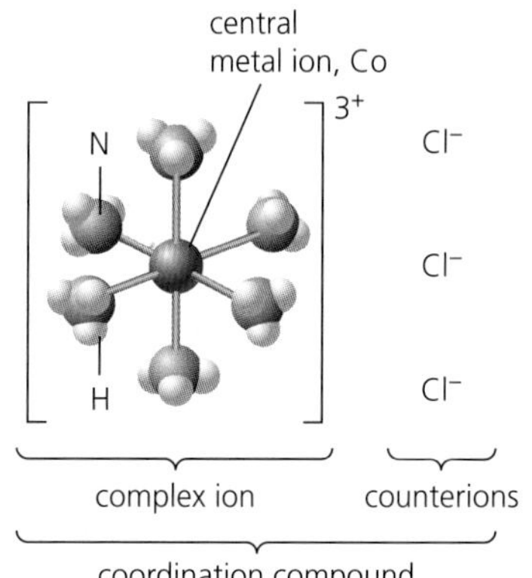

■ **Figure R3.126** Structure of hexaamminecobalt(III) chloride

Figure R3.126 shows the coordination compound hexaamminecobalt(III) chloride. Six ammonia molecules form coordination bonds with a central cobalt(III) ion. There is ionic bonding between the positively charged complex ion and three chloride ions. The formula of this compound is written as $[Co(NH_3)_6]Cl_3$. The chloride ions are outside the square bracket to show they are not coordinated to the transition metal cation.

If we know the formula of a coordination compound, we can calculate the charge of complex ions and the oxidation state of the transition metal at the centre of them.

WORKED EXAMPLE R3.4F

Calculate the oxidation state of iron in the coordination compound $K_3[Fe(CN)_4(OH)_2]$.

Answer

Potassium forms a 1+ ion, so the complex ion has a charge of 3−, $[Fe(CN)_4(OH)_2]^{3-}$.

Both CN^- and OH^- ligands have a charge of −1 and there are six of these in total; therefore the oxidation state of iron is +3.

65 Calculate the oxidation state of the metal in each of the following:

a $[Zn(OH)_4]^{2-}$
b $[Zn(H_2O)_4]^{2+}$
c $[Fe(CN)_6]^{4-}$
d $[Co(NH_3)_4Cl_2]Cl$
e $[Cu(NH_3)_4(H_2O)_2]SO_4$
f $[Fe(H_2O)_6]Cl_3$
g $[CuCl_4]K_2$
h $[CrCl_2(NH_3)_4]_2SO_4$

Nucleophilic substitution reactions of halogenoalkanes

Earlier in the chapter, halogenoalkanes were identified as electrophiles which can undergo nucleophilic substitution. The halogen atom is a suitable leaving group (because it forms a stable anion), and the presence of a halogen makes a halogenoalkane more electrophilic than an alkane.

When a cold aqueous solution of dilute sodium hydroxide is added to 1-iodopropane, a slow nucleophilic substitution reaction takes place. The iodine atom in the 1-iodopropane is replaced by a hydroxyl group (–OH) and an alcohol is formed:

$$CH_3CH_2CH_2I(l) + NaOH(aq) \rightarrow CH_3CH_2CH_2OH(aq) + NaI(aq)$$

This reaction can also be described by an ionic equation which omits the spectator Na^+ ion:

$$CH_3CH_2CH_2I(l) + OH^-(aq) \rightarrow CH_3CH_2CH_2OH(aq) + I^-(aq)$$

The aqueous hydroxide ion behaves as a nucleophile (Lewis base), because it is donating a lone pair of electrons to the carbon atom bonded to the iodine in the 1-iodopropane.

Mechanisms of nucleophilic substitution in halogenoalkanes

In all halogenoalkane nucleophilic substitution reactions, a new bond is formed between the nucleophile and electrophile, and the carbon–halogen bond is broken. Because it is not possible for carbon to form more than four complete bonds, there are only two ways in which the nucleophilic substitution reaction may proceed:

1 The carbon–halogen bond is broken first, leaving a trivalent carbon, and the nucleophile then comes in to form a nucleophile–carbon bond.
2 The carbon–halogen bond breaks at the same time as the new nucleophile to carbon bond is made.

The first route is known as the S_N1 mechanism and the second as the S_N2 mechanism. Both options are possible for any halogenoalkane, but one pathway is usually more energetically favourable and hence happens at a much faster rate than the other. The structure of the halogenoalkane and the reaction conditions are the main factors which dictate which pathway is favoured.

S_N1 mechanism

In an S_N1 reaction, the carbon–halogen bond breaks via heterolytic fission to give a carbocation and a halide ion. The carbocation produced is a very good electrophile because it has an empty orbital, and it quickly reacts with any nucleophile in close proximity. The S_N1 mechanism for the reaction of 2-bromo-2-methylpropane with hydroxide ions is shown in Figure R3.127.

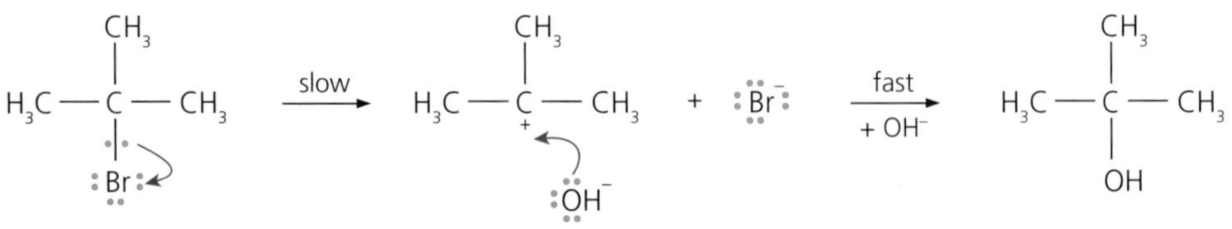

■ **Figure R3.127** The S_N1 mechanism for the reaction of 2-bromo-2-methylpropane with hydroxide ions

The first step, in which the carbocation is formed, is relatively slow and is the rate-determining step:

$$(CH_3)_3C\text{–}Br \rightarrow (CH_3)_3C^+ + Br^-$$

In the second step, the carbocation acts as a Lewis acid and forms a bond with a nearby hydroxide ion – a Lewis base:

$$(CH_3)_3C^+ + OH^- \rightarrow (CH_3)_3COH$$

The molecularity of the first elementary step is unimolecular, and because this is the slow step of the reaction, the rate equation for this reaction is:

$$\text{Rate} = k[(CH_3)_3CBr]$$

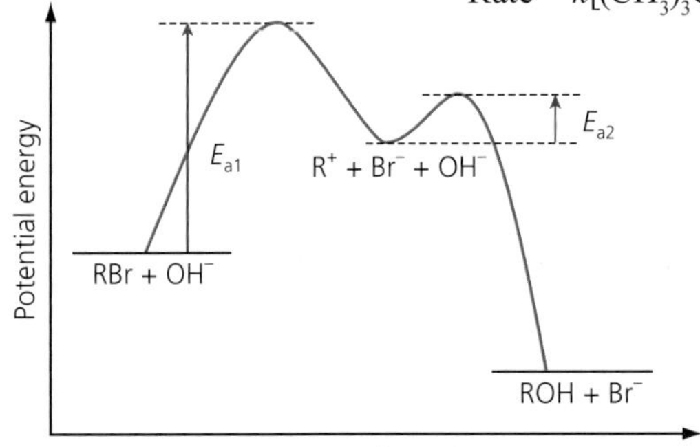

■ **Figure R3.128** The energy profile for an S_N1 mechanism

Note that the concentration of hydroxide ions has no effect on the rate of the reaction. Using hydroxide, a strong nucleophile, is not necessary in S_N1 reactions because water performs the same role!

The name of this mechanism of substitution, S_N1, is derived from S = substitution; $_N$ = nucleophilic; 1 = unimolecular rate determining step.

An energy profile for an S_N1 reaction is shown in Figure R3.128.

In the energy profile the two separate elementary steps can be seen, each with their own activation energies (E_{a1} and E_{a2}). The activation energy of the first step is larger, which makes this step the slower elementary step.

S_N2 mechanism

The S_N2 mechanism is a one-step mechanism which involves the simultaneous (concerted) attack of the nucleophile on the carbon atom and loss of the halide ion (Figure R3.129).

Nu:⁻ C—X (δ+, δ−) ⟶ Nu—C + X:⁻

■ **Figure R3.129** An S_N2 mechanism generalized for an anionic nucleophile

There is only one elementary step in the reaction and it is bimolecular. The rate equation is given by the reactants in the bimolecular elementary step:

$$\text{Rate} = k[\text{Nu:}][\text{CX}]$$

The concentrations of both reactants affect the rate of the reaction. The name of this mechanism of substitution, S_N2, is derived from: S = substitution; $_N$ = nucleophilic; 2 = bimolecular rate determining step.

An energy profile for the reaction of a hydroxide ion with a halogenoalkane in an S_N2 reaction is shown in Figure R3.130.

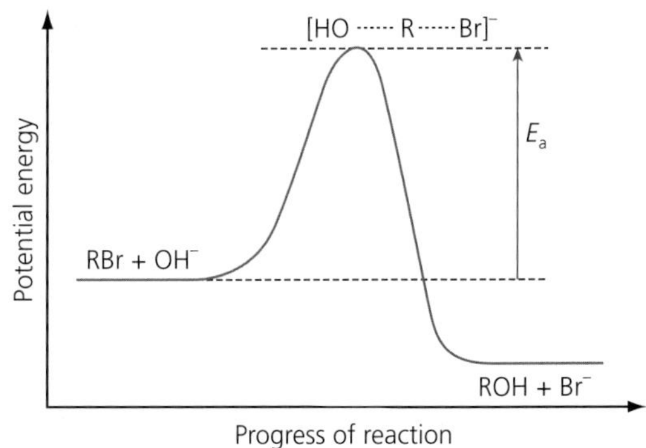

■ **Figure R3.130** The energy profile for an S_N2 mechanism

Common mistake

Misconception: *The S_N1 mechanism has one step; the S_N2 mechanism has two steps.*

The S_N1 mechanism has two steps to it; the 1 refers to the unimolecular nature of the slowest elementary step (the rate determining step). The S_N2 mechanism has just one step and the 2 refers to the bimolecular nature of this elementary step.

HL ONLY

Although there is only one elementary step in an S_N2 reaction, we can imagine what the reaction components will look like as they pass through a maximum of potential energy on the way to becoming the products. A transition state like that shown in square brackets in Figure R3.131 is usually included when drawing the mechanism of an S_N2 reaction. (In many cases, the structure of the transition states has been confirmed by computer modelling studies.)

◆ **Stereospecific:** A property of a reaction which reflects the observation that a single isomer of reactant will go on to form a single isomer of product; it is the result of the mechanism of the reaction.

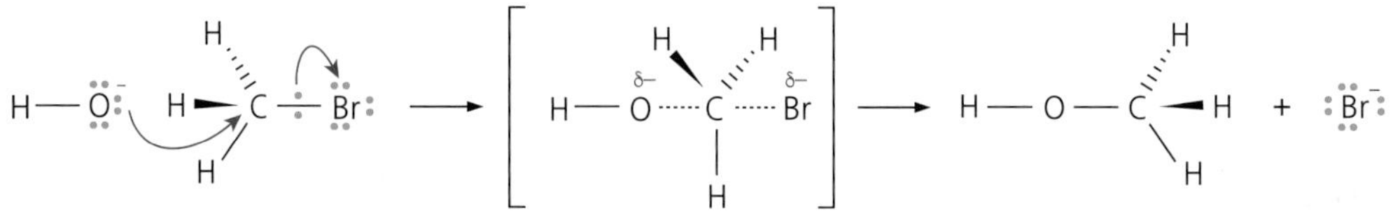

■ **Figure R3.131** The S_N2 hydrolysis of bromoethane

The dashed lines in the transition state indicate partial bonds. The oxygen-carbon bond is partially formed, and the carbon-bromine bond is partially broken. The nucleophile attacks the electron-deficient carbon atom on the opposite side from the halogen atom. This 'back-side attack' is required to correctly align the molecular orbitals involved in the reaction. The reaction is only possible in this orientation. As a result of the restricted approach of the nucleophile, S_N2 reactions are **stereospecific**. This is revisited in greater detail on page 657.

LINKING QUESTION

What differences would be expected between the energy profiles for S_N1 and S_N2 reactions?

LINKING QUESTION

What are the rate equations for S_N1 and S_N2 reactions?

Unlike an intermediate produced as the result of an elementary step within a reaction, the structure shown in a transition state cannot be isolated. A transition state represents a maximum in energy for an elementary step. An intermediate is a local minimum in energy on the pathway between the initial reactants and products. Figure R3.128 shows the energy profile of an S_N1 reaction and the intermediate formed is indicated as a local minimum. Figure R3.130 shows the energy profile of an S_N2 reaction and the transition state between the reactant and product is indicated at the maximum of energy.

S_N1 or S_N2?

The structure of the halogenoalkane is the greatest factor in determining whether nucleophilic substitution of halogenoalkanes proceeds via an S_N1 or S_N2 mechanism.

If the halogenoalkane can form a relatively stable carbocation, the S_N1 pathway will operate at a faster rate than the S_N2. If the halogenoalkane reactant is unable to form a stable carbocation, and the electrophilic carbon of the halogenoalkane is relatively unhindered by bulky substituents, the S_N2 pathway will operate at a faster rate than the S_N1 pathway.

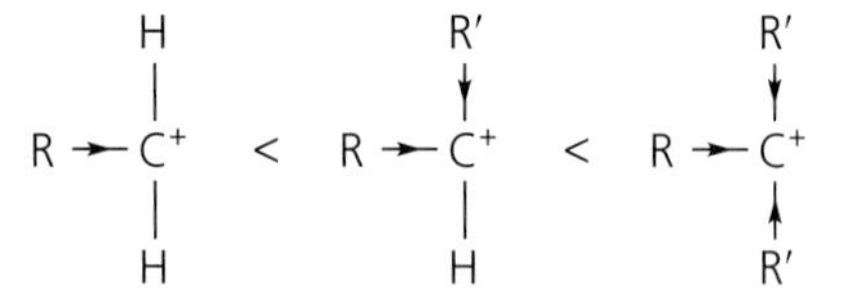

primary carbocation least stable < secondary carbocation < tertiary carbocation most stable

■ **Figure R3.132** Stability of different carbocations

Looking specifically at primary, secondary and tertiary halogenoalkanes, the stability of the carbocations formed by these is primary < secondary < tertiary (Figure R3.132).

The alkyl groups are electron-donating (as indicated by the arrows on the R–C bonds in Figure R3.132), so each one has a stabilizing effect on the carbocation.

The nucleophilic substitution pathways followed by different halogenoalkanes, as a result of the stability of their carbocation intermediates, are summarized below:

- Methyl and primary halogenoalkanes undergo nucleophilic substitution predominantly by the S_N2 pathway. They are unable to form stable carbocation intermediates, and nucleophiles are relatively free to approach their electrophilic carbon atoms.

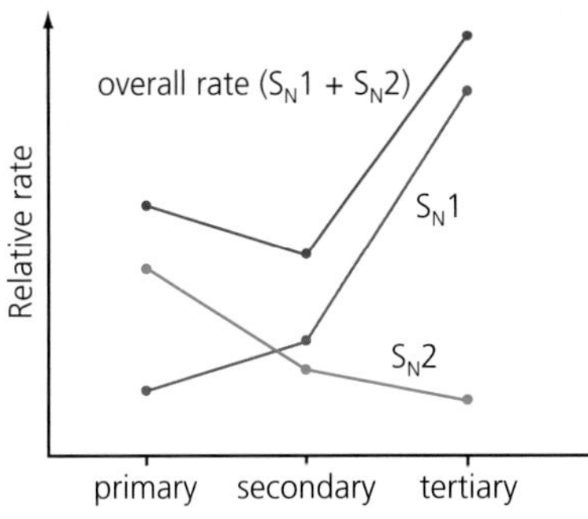

■ **Figure R3.133** Illustrative relative rates of S_N1 and S_N2 hydrolysis reactions for primary, secondary and tertiary halogenoalkanes

- Secondary halogenoalkanes are not very good at reacting by either pathway, and evidence shows that both the S_N1 and S_N2 pathways operate at a similar rate.
- Tertiary halogenoalkanes react predominantly via the S_N1 pathway owing to stabilization of the carbocation intermediate. The back-side approach of nucleophiles needed in the S_N2 pathway is usually hindered by the same bulky, stabilizing alkyl groups, making this pathway even less likely.

The relative rates are summarized in Figure R3.133.

HL ONLY

Nature of science: Hypotheses

How does your garden grow?

In his book *Elegant Solutions*, Philip Ball writes about ten beautiful experiments in chemistry. One of the experiments that Ball covers is by 17th-century Flemish chemist and physician Jan Baptist van Helmont. Van Helmont believed that all vegetables grew from water only, and therefore they, and everything else higher up the food chain, were ultimately derived entirely from water. He designed an experiment to test this hypothesis. Van Helmont took 200 pounds of earth, which he had dried in a furnace, and moistened it with rainwater. He placed the earth in a pot and planted into it a willow sapling weighing five pounds. He watered the sapling for five years, but carefully excluded all other sources of matter including dust.

At the end of the time, van Helmont weighed the tree and the dried soil again. He found that the tree had gained 160 pounds (not including the weight of the fallen leaves over the previous four autumns), and the 200 pounds of soil was only short of 2 ounces. He concluded that the mass gain must have been derived solely from the water given to the tree. It seemed illogical to suggest anything else.

We now know, of course, that the mass gain comes from water combining with carbon dioxide extracted from the air. The carbon dioxide is absorbed through leaves and turned into carbon-containing products such as cellulose through photosynthesis. To suggest that this great gain of mass came from the air would have seemed absurd in the era of van Helmont. The experiment is simple and persuasive in design and conclusion and it acts as a warning: if an important part of our understanding is missing, we can be convinced of a fallacy, just as van Helmont believed all matter was ultimately derived from water.

Inquiry 3: Concluding and evaluating

Evaluating the implications of methodological weaknesses, limitations and assumptions on conclusions

A conclusion is a general statement summarizing what your results (data) show in relation to the research question. An evaluation does more than list real or potential issues with the experiment – you need to assess the impact of each factor you identify on the investigation.

It may be useful to regard weaknesses as referring to the design of the investigation and limitations as referring to problems with the data.

Common methodological weaknesses that might undermine the conclusion are uncontrolled variables (was an effort made to control the temperature if a reaction is highly exothermic?) and measurements not being recorded with sufficient precision due to the selection of unsuitable apparatus.

Limitations could be an independent variable with a narrow range, or the accuracy of the instruments available.

It is also important to identify and discuss the various theories and models relevant to your chemical reaction or system. You should outline the assumptions you have made based on them and state why predictions derived from them have deviated from accurate experimental results.

LINKING QUESTION

How useful are mechanistic models such as S_N1 and S_N2?

Going further

Stability of carbocations

Alkyl groups are electron-donating when they are attached to a carbocation. The remarkable stabilizing effect they have on carbocations is best explained by a phenomenon known as hyperconjugation. Hyperconjugation is the mixing of filled σ-orbitals on the alkyl substituents with the empty p-orbital of the carbocation. The electrons in the sigma bonds of an alkyl group (shown on the left side of Figure R3.134) can be weakly donated into the empty p-orbital of the cation – this is not possible with a hydrogen substituent in place of the alkyl group.

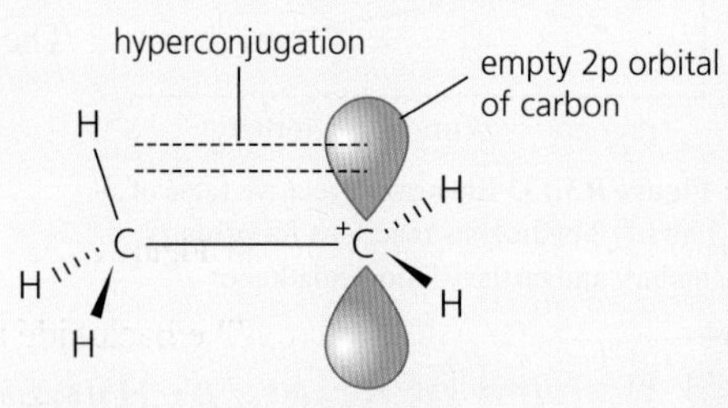

■ **Figure R3.134** Hyperconjugation of a carbocation by a methyl substituent

Hyperconjugation leads to the stabilizing of the carbocation. In extreme cases, the filled sigma bond, along with its attached hydrogen atom, moves completely over to an adjacent carbocation and a new carbocation is created one carbon away from the original one. This is known as a 1,2-hydride shift.

Many of the most stable carbocations are not tertiary carbocations stabilized solely by surrounding alkyl groups; instead, they are carbocations where there is resonance with other functional groups present in the molecule. Although it is a primary halogenoalkane, 3-bromopropene (Figure R3.135) can form a relatively stable carbocation in this way.

A phenyl group can have the same effect. The resonance structures show that the positive charge is not localized on one carbon, but is shared out across the molecule (Figure R3.136).

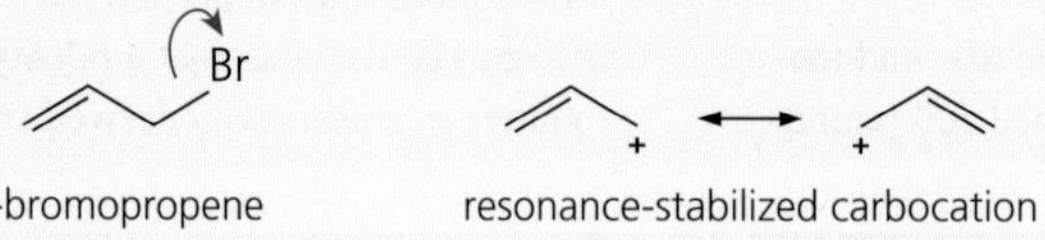

3-bromopropene resonance-stabilized carbocation

■ **Figure R3.135** The resonance-stabilized carbocation formed by 3-bromopropene

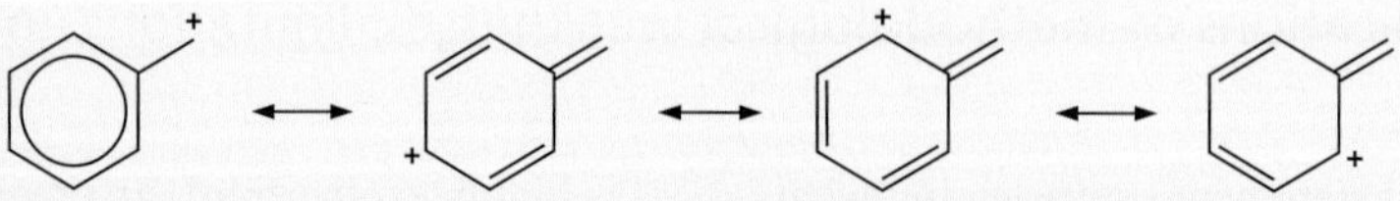

■ **Figure R3.136** The resonance structures of a benzyl carbocation

66 State which mechanism you would expect to be dominant in the nucleophilic substitution of the following halogenoalkanes.

a $CH_3CH_2CH_2CH_2I$

b $CH_3CH_2C(CH_3)_2Cl$

c $C_6H_5CH_2CH_2Br$

67 a State the type of reaction that will occur between bromomethane molecules, CH_3Br, and ethoxide ions, $CH_3CH_2O^-$.

b Draw the mechanism for the reaction.

Stereochemistry of nucleophilic substitution

Consider the reaction between bromoethane and cold dilute aqueous sodium hydroxide solution. The reactant is a primary halogenoalkane and thus the substitution proceeds almost entirely by the S_N2 mechanism. The mechanism and transition state are shown in Figure R3.137.

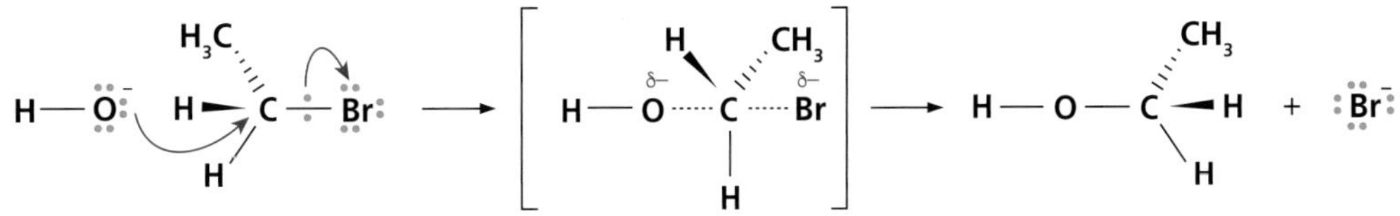

$$HO^- + CH_3CH_2Br \rightarrow [HO^{\delta-}\cdots C \cdots Br^{\delta-}] \rightarrow HO-CH_2CH_3 + Br^-$$

■ **Figure R3.137** An S_N2 reaction between bromoethane and hydroxide ions

The back-side attack of the nucleophile, which is required for the correct orbital interaction in S_N2 reactions, causes the inversion of the arrangement of the other groups or atoms around the carbon atom. The pattern is like that made when a strong wind blows an umbrella inside-out (Figure R3.138).

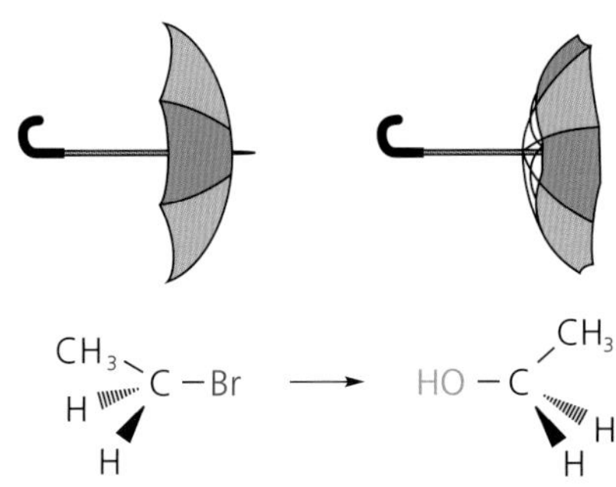

■ **Figure R3.138** The inversion of the structure of groups around the carbon atom during an S_N2 reaction is like an umbrella being blown inside-out in the wind

If an S_N2 reaction occurs at the chiral carbon of a single optical isomer (enantiomer), a single optical isomer of the product will be formed. The reaction is said to be stereospecific because the stereochemistry of the product relies on the stereochemistry of the reactant.

S_N1 reactions are not stereospecific. When a single optical isomer of a compound where the chiral carbon bears the halogen undergoes substitution, an optically inactive (racemic) mixture of the two possible products forms. This is because the intermediate carbocation formed is planar – it has the same structure regardless of the optical isomer from which it was formed. The nucleophile is equally likely to bond to the carbocation from either face of the carbocation and so the two possible optical isomers are produced in equal amounts (Figure R3.139).

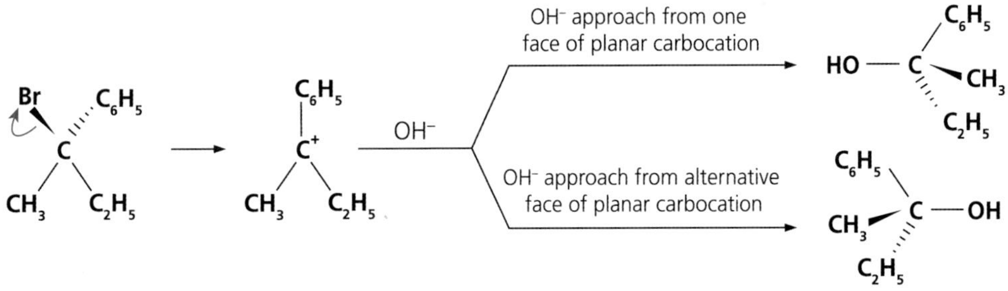

OH⁻ approach from one face of planar carbocation

OH⁻ approach from alternative face of planar carbocation

$$C(Br)(C_6H_5)(CH_3)(C_2H_5) \rightarrow C^+(C_6H_5)(CH_3)(C_2H_5) \xrightarrow{OH^-} HO-C(C_6H_5)(CH_3)(C_2H_5) \text{ and } (C_6H_5)(CH_3)(C_2H_5)C-OH$$

■ **Figure R3.139** The racemization of an optically active halogenoalkane during nucleophilic substitution (S_N1)

Top tip!

Understanding and visualizing the stereochemistry of reactants and products, and how the mechanism of substitution plays a role, is a challenge. Making models or viewing animations can be a useful way to aid your understanding.

Top tip!

The S_N2 mechanism is stereospecific. If a single optical isomer of the reactant is used, a single optical isomer of the product is produced. The S_N1 mechanism results in the loss of any stereochemical information at that centre. A racemic mixture of optical isomers is produced.

Nature of science: Experiments

Nucleophilic substitution

In 1937, Edward Hughes (1906–1963) and Christopher Ingold (1893–1970) at University College London proposed a concerted mechanism for an S_N2 reaction involving a single step with no intermediates formed. The nucleophile attacks the carbon bearing the leaving group and displaces the leaving group.

A reaction mechanism is a model that is in agreement with the experimental evidence that has been accumulated during a study of the reaction. Hughes and Ingold based their mechanism for an S_N2 reaction on the following experimental data.

- The rate of the reaction depends on the concentration of the halogenoalkane and on the concentration of the nucleophile (hydroxide ions). This means that both reactants are involved in the rate-determining step (slowest step in the mechanism).
- The reaction of a halogenoalkane in which the halogen is bonded to an asymmetric carbon leads to the formation of only one enantiomer; the configuration of the asymmetric carbon is inverted relative to its configuration in the reacting halogenoalkane.

Link

Bond enthalpies are introduced on page 332, and more information about the experimental procedure used when hydrolyzing halogenoalkanes is on page 337.

Common mistake

Misconception: *Chloroalkanes undergo nucleophilic substitution faster than bromoalkanes because the carbon bearing the more electronegative chlorine is a better electrophile.*

Although it is true that a carbon bearing a more electronegative chlorine atom will have a greater partial positive charge than a carbon bearing a bromine atom, experimental data shows that the bromoalkene undergoes substitution at the faster rate. From this we can conclude that the effect of carbon-halogen bond strength seems to predominate over the relative charge the carbon has.

Rate of substitution reactions

The nucleophile

The identity of the nucleophile has no effect on the rate of reaction if an S_N1 pathway dominates. S_N2 reactions proceed at a quicker rate with small, charged nucleophiles (such as hydroxide and alkoxide ions) which can easily access the electrophilic carbon.

The halogenoalkane

Structural features of the halogenoalkane (for example, whether it is primary, secondary or tertiary) influence the activation energy barrier, and therefore the rate of reaction, of the competing S_N1 and S_N2 pathways. The quickest S_N1 reactions occur when stable carbocations can be formed, and these reactions are orders of magnitude quicker than S_N2 reactions. If the structure of a halogenoalkane is kept constant and the halogen leaving group is altered, this has a large effect on the rate of substitution reactions.

The rate of substitution increases from fluoroalkanes through to iodoalkanes, correlating well with the strength of the bond to the leaving group (Figure R3.140).

	most reactive, best leaving group			least reactive, worst leaving group
	C–I >	C–Br >	C–Cl >	C–F
bond enthalpy/kJ mol^{-1}	228	285	346	492

■ **Figure R3.140** The bond enthalpy of the carbon–halogen bond determines how easily the halogen atom leaves the halogenoalkane molecule

The substitution reaction involves the heterolytic breaking of the carbon–halogen bond and this will proceed at a faster rate if the bond is weaker. The C–I bond is the weakest and longest and so the least amount of energy is needed to break it during the substitution reaction. The C–F bond is the strongest and shortest and so the greatest amount of energy is needed to break it during the substitution reaction.

Top tip!

The rate of substitution reactions is influenced by the identity of the leaving group. The weaker the bond to the leaving group, the easier it is to break and the faster the substitution will occur.

LINKING QUESTION

Why is iodide a better leaving group than the chloride ion?

Formation of halide ions can be used to qualitatively assess the relative rates of nucleophilic substitution reactions involving halogenoalkanes. When aqueous silver nitrate is added to a halogenoalkane dissolved in ethanol, water acts as the nucleophile and a silver halide precipitate forms as the halide ion is liberated from the halogenoalkane.

$$R\text{–}X(aq) + H_2O(l) \rightarrow R\text{–}OH(aq) + H^+(aq) + X^-(aq)$$

$$X^-(aq) + Ag^+(aq) \rightarrow AgX(s)$$

The silver halides formed are white (silver chloride), cream (silver bromide) and pale yellow (silver iodide). The longer it takes for the test tube to become cloudy due to insoluble halide formation, the slower the reaction.

If the experiment is done without silver nitrate, a quantitative assessment of the rate can be made by measuring conductivity. The increasing concentration of halide ions as the reaction progresses leads to an increase in conductivity.

Electrophilic addition of alkenes

Addition of a halogen

The electrophilic addition of a halogen to alkenes is a two-step process.

In the first step the high electron density on the alkene induces a dipole in a nearby halogen molecule. The nucleophilic alkene donates a pair of electrons from its pi bond to the halogen which is acting as an electrophile. Concurrently, the halogen–halogen sigma bond is broken by heterolytic fission. A carbocation intermediate and a halide ion are formed. This first step is the slow rate-determining step in the electrophilic addition reaction of alkenes. Figure R3.141 shows the first step of the reaction between ethene and bromine.

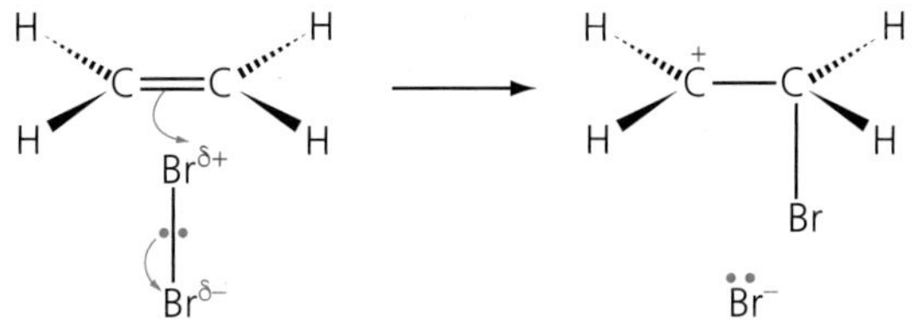

Figure R3.141 The formation of a carbocation intermediate in the reaction of ethene with bromine

In the second step (Figure R3.142), the positively charged carbocation intermediate (a powerful electrophile) reacts rapidly with the halide ion. This forms the di-halogenated product.

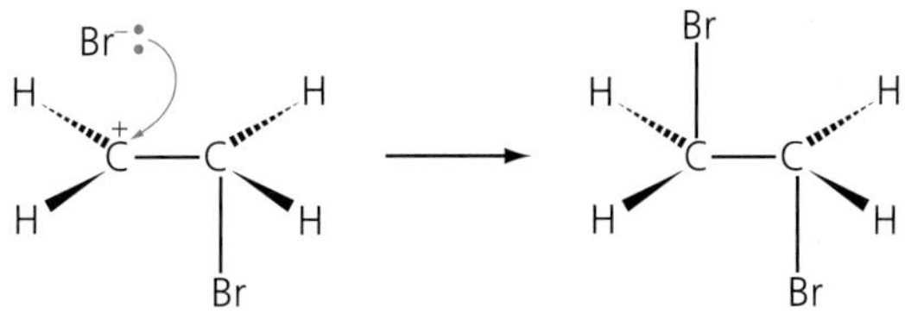

Figure R3.142 The reaction of the carbocation intermediate with a bromide ion

If the di-halogenated molecule is the desired product, care must be taken to eliminate any other nucleophiles which could react with the carbocation intermediate. For example, if bromine water is used, water molecules can react with the carbocation to give the by-product 2-bromoethanol (step 3 in Figure R3.143). If 1,2-dibromoethane is desired, bromine dissolved in a non-polar solvent would be a more suitable choice because the non-polar solvent will not react with the carbocation intermediate.

HL ONLY

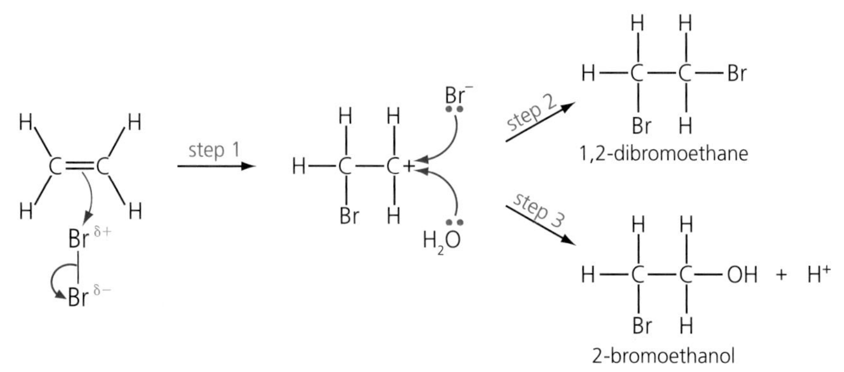

■ **Figure R3.143** The reaction of ethene and bromine water

Going further

Electrophilic addition with bromine

There is evidence to suggest the bromination of alkenes does not proceed through a carbocation intermediate. In the proposed alternative mechanism, bromine forms a bond to both carbon atoms which are part of the alkene. The intermediate formed has a positive charge on the bromine atom and is known as a bromonium ion (Figure R3.144). This ion then reacts with the nucleophile to yield the product.

The evidence for this mechanism comes from the stereospecific nature of some bromination reactions, a result that is not possible if a carbocation intermediate is present in the mechanism. In addition, some exceptionally stable bromonium ions have been characterized by X-ray crystallography and nuclear magnetic resonance (NMR) spectroscopy.

The mechanism involving a bromonium ion can be found at **www.chemtube3d.com/electrophilic-addition-to-alkenes-ethylene-and-bromine/** where the steps are animated to help visualize the process.

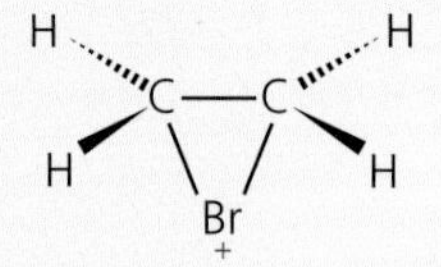

■ **Figure R3.144** A cyclic bromonium ion intermediate in a bromination reaction of an alkene

Addition of a hydrogen halide

The mechanism for the electrophilic addition of a hydrogen halide is similar to that for the addition of a halogen but, in this case, the electrophile has a permanent dipole. The electron deficient hydrogen atom of the polar hydrogen halide acts as the electrophile. Figure R3.145 shows the electrophilic addition reaction between ethene and hydrogen bromide.

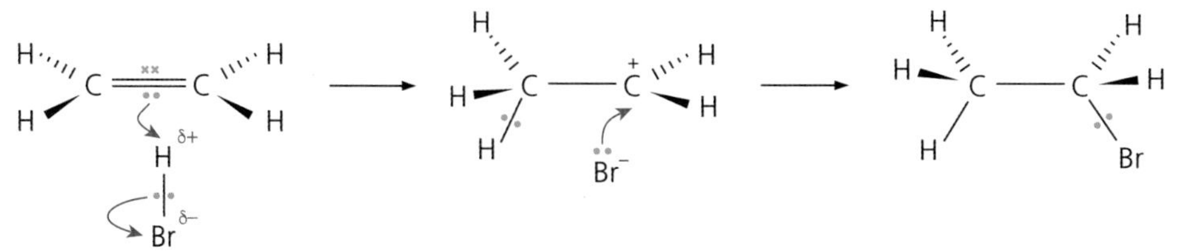

■ **Figure R3.145** The electrophilic addition of hydrogen bromide to ethene

An H–Br molecule attacks the pi bond of ethene in the initial step. The mechanism is completed by a rapid reaction between the bromide ion and the carbocation intermediate. The product of this reaction is bromoethane, CH_3CH_2Br.

Nature of science: Experiments

Supplementary nucleophiles in the electrophilic addition reactions of alkenes

Further evidence supporting the two-step mechanism of electrophilic addition is the incorporation of 'foreign' competing anions when bromination is carried out in an aqueous solution containing a mixture of salts.

For example, if ethene is reacted with bromine water that also contains dissolved sodium chloride and sodium nitrate, a mixture of products is formed (Figure R3.146).

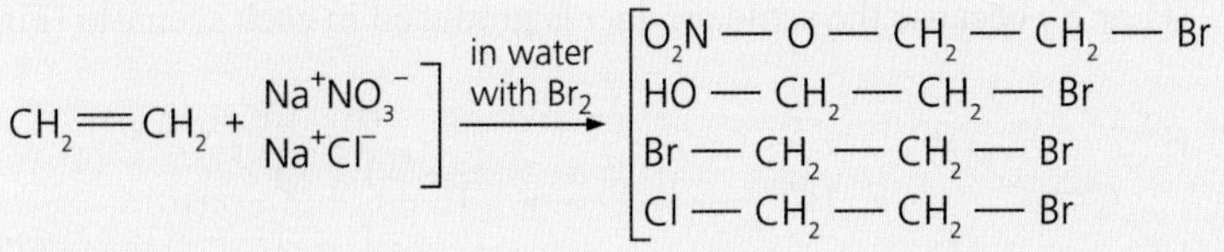

$$CH_2{=}CH_2 + \left[\begin{array}{l} Na^+NO_3^- \\ Na^+Cl^- \end{array}\right] \xrightarrow[\text{with } Br_2]{\text{in water}} \left[\begin{array}{l} O_2N-O-CH_2-CH_2-Br \\ HO-CH_2-CH_2-Br \\ Br-CH_2-CH_2-Br \\ Cl-CH_2-CH_2-Br \end{array}\right.$$

■ **Figure R3.146** The products formed when ethene reacts with bromine water containing salts

The ratio $(ClCH_2CH_2Br):(O_2NOCH_2CH_2Br)$ in the mixture is found to correspond to the $[Cl^-]:[NO_3^-]$ ratio in the original solution. This indicates that the reactive carbocation reacts with the first anion it successfully collides with.

It should also be noted that 1,2-dichloroethane is not detected in the products, indicating that the initial attacking species is not the chloride ion (Cl^-). This, and the fact that all the products contain bromine, is consistent with bromine being involved in the first step of the mechanism and the anions present reacting with the carbocation produced.

Addition of water

In certain structures, it is possible to hydrate an alkenyl functional group and so transform an alkene into an alcohol. As in the reaction with a halogen or hydrogen halide, the mechanism is electrophilic addition (Figure R3.147).

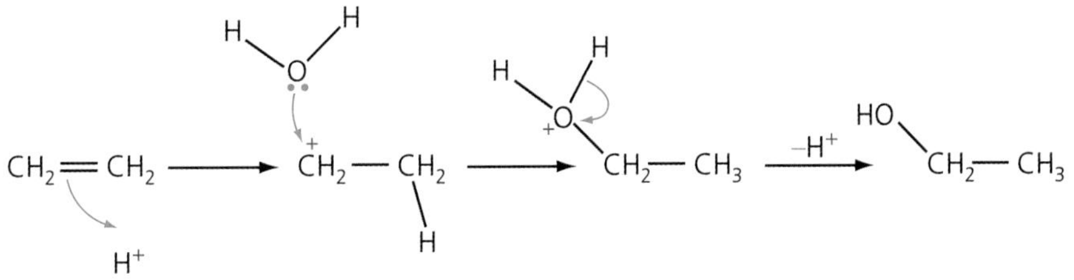

$$CH_2{=}CH_2 \xrightarrow{H^+} \overset{+}{C}H_2-CH_2-H \xrightarrow{H_2O} H_2\overset{+}{O}-CH_2-CH_3 \xrightarrow{-H^+} HO-CH_2-CH_3$$

■ **Figure R3.147** Electrophilic addition: hydration of ethene

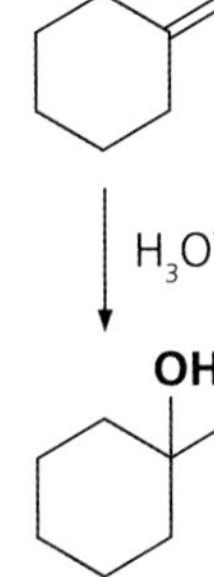

■ **Figure R3.148** The hydration of methylenecyclohexane

Aqueous acid (for example, sulfuric(VI) acid or phosphoric(V) acid) provides the protons (or more accurately, hydronium ions, H_3O^+), which act as the electrophile in the first step. Water then reacts with the carbocation intermediate and, in a further step, the $-OH_2^+$ group loses a hydrogen ion to give a neutral product. The acid is catalytic because H^+ is regenerated in the reaction.

Under standard conditions, the reaction only works well for a few alkenes which form stable tertiary carbocation intermediates that can easily be trapped by water. One example is hydration of the alkene methylenecyclohexane (Figure R3.148).

To make ethanol by this method, ethene and steam need to be subjected to non-standard conditions: they are passed over a concentrated phosphoric(V) acid catalyst absorbed onto porous pumice (to create a large surface area) at a pressure of 70 atm and a temperature of 300 °C.

Because the steps of a hydration reaction are reversible, an alcohol subjected to the same conditions will often dehydrate into the corresponding alkene.

Addition of hydrogen halides to unsymmetrical alkenes

The electrophilic addition reaction between a hydrogen halide and an alkene results in a hydrogen atom and a halogen atom forming new bonds to the carbons in the alkenyl functional group. With symmetrical alkenes, such as ethene, it does not matter which carbon received which atom (H or X) because the same product is produced in each scenario (Figure 3.149).

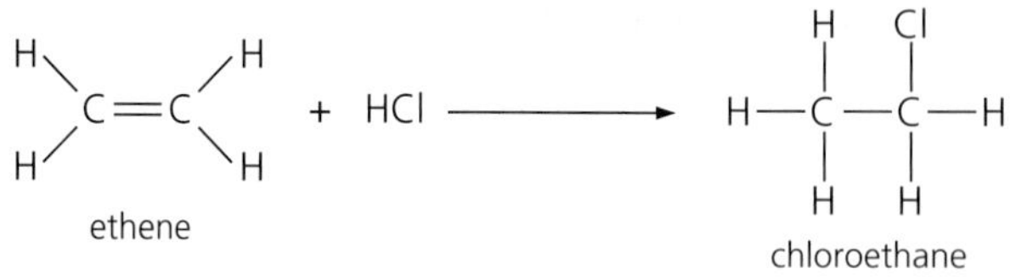

$$H_2C{=}CH_2 + HCl \longrightarrow CH_2Cl{-}CH_3$$

ethene → chloroethane

■ **Figure R3.149** The electrophilic addition reaction of ethene and hydrogen chloride

However, most alkenes are not symmetrical, so different halogenoalkane isomers are produced depending on which carbon the hydrogen and halogen atom bond with.

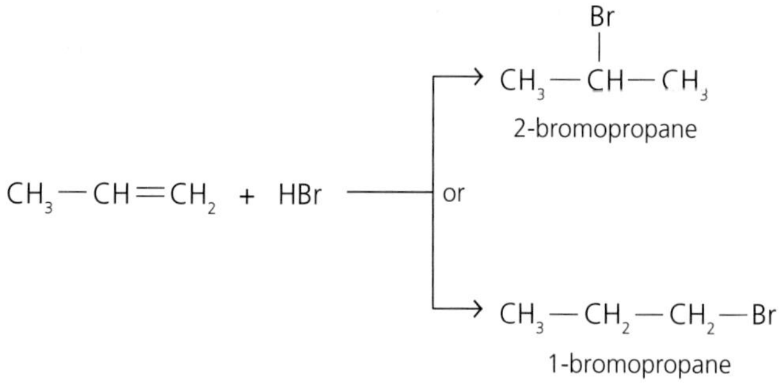

$$CH_3{-}CH{=}CH_2 + HBr \longrightarrow CH_3{-}CHBr{-}CH_3 \text{ (2-bromopropane) or } CH_3{-}CH_2{-}CH_2{-}Br \text{ (1-bromopropane)}$$

■ **Figure R3.150** The reaction between hydrogen bromide and propene

$$CH_3{-}CH{=}CH_2 + H^{\delta+}{-}Br^{\delta-} \longrightarrow CH_3{-}\overset{+}{C}H{-}CH_2{-}H + Br^- \text{ (secondary carbocation)}$$

$$CH_3{-}CH{=}CH_2 + H^{\delta+}{-}Br^{\delta-} \not\longrightarrow CH_3{-}CH(H){-}\overset{+}{C}H_2 + Br^- \text{ (primary carbocation)}$$

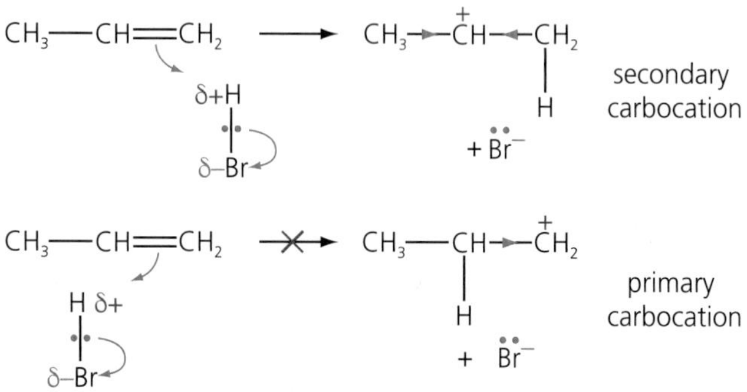

■ **Figure R3.151** Forming the two possible carbocation intermediates in the reaction between propene and hydrogen bromide

Consider the reaction between hydrogen bromide and propene. Both 2-brompropane and 1-bromopropane can be formed (Figure R3.150).

When the reaction is carried out, it is found that 2-bromopropane is the major product; almost no 1-bromopropane is formed (it is a minor product). To understand this result, consider the reaction mechanism. In the initial step of the mechanism, the pair of electrons in the pi bond moves to the hydrogen atom of the electrophilic hydrogen bromide. The carbocation formed is either a secondary carbocation, with the hydrogen atom attached to the carbon on the end of the chain, or a primary carbocation, with the hydrogen attached to the carbon in the middle of the chain (Figure R3.151).

Because a secondary carbocation is more stable than a primary carbocation, the activation energy barrier to the formation of the secondary carbocation is lower than that to the formation of the primary carbocation. The result is that more of the molecules react through the pathway involving the more stable secondary carbocation.

The bromide group in the halogenoalkane product will end up where the carbocation charge lies so 2-bromopropane, the product which arises from the secondary carbocation intermediate, is found to be the major product.

Top tip!

When adding a hydrogen halide to an unsymmetrical alkene, determine which carbon of the alkenyl functional group would form the most stable carbocation. This is where the halogen will end up.

WORKED EXAMPLE R3.4G

Predict the major and minor product in the electrophilic addition reaction between 2-methylpropene and hydrogen bromide.

$$(CH_3)_2C{=}CH_2 + HBr \longrightarrow$$

Answer

2-methylpropene can form a carbocation intermediate on the '2' carbon (the middle carbon) or the '1' carbon (the carbon on the end of the propene molecule). The carbocation formed on the 2 carbon is a stable tertiary carbocation. The carbocation formed on the 1 carbon is an unstable primary carbocation.

$(H_3C)_2C^+{-}CH_3$ versus $H_3C{-}CH(CH_3){-}\overset{+}{C}H_2$

tertiary carbocation | primary carbocation

↓ | ↓

$H_3C{-}CBr(CH_3){-}CH_2{-}H$ | $H_3C{-}CH(CH_3){-}CHBr{-}H$

2-bromo-2-methylpropane | 1-bromo-2-methylpropane

2-bromo-2-methylpropane will be the major product, 1-bromo-2-methylpropane will be the minor product.

The Russian chemist Vladimir Markovnikov (1837–1904) proposed a rule to predict which isomer is the major product. Markovnikov's rule states that, when hydrogen halides add to unsymmetrical alkenes, the hydrogen atom adds to the carbon atom that has the greatest number of hydrogen atoms bonded to it. The (hydrogen) rich get richer – is a useful way to remember Markovnikov's rule.

Although the rule is useful, Markovnikov did not understand why his rule allowed him to correctly predict the major product. The addition of hydrogen halides to unsymmetrical alkenes is best summed up as follows: the major product is the one arising from the pathway with the most stable carbocation intermediate.

Going further

Major and minor products in hydrobromination reactions

Two different halogenoalkanes are formed when but-1-ene is reacted with hydrogen bromide (Figure R3.152).

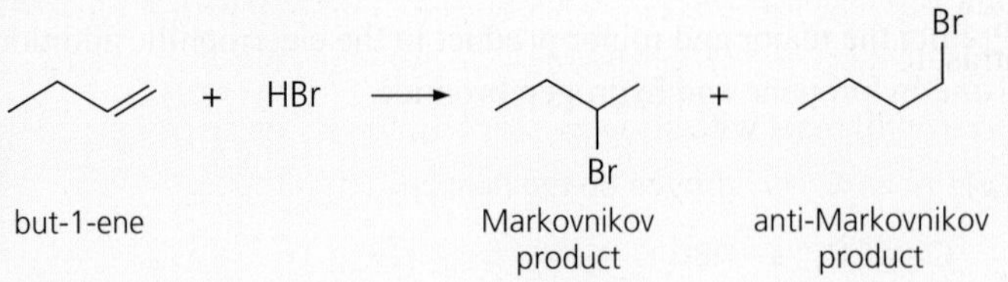

■ **Figure R3.152** The addition products of but-1-ene and hydrogen bromide

As you have seen, 2-bromobutane (sometimes known as the Markovnikov product) is the major product when the reaction proceeds via an electrophilic addition mechanism.

Curiously, the major product switches to 1-bromobutane (sometimes known as the anti-Markovnikov product) if the radical initiator benzoyl peroxide is included in the reaction mixture. The inclusion of the radical initiator changes the reaction mechanism to a chain reaction which progresses through radical intermediates.

Upon heating, the benzoyl peroxide molecules undergo homolytic fission to produce radicals which abstract the hydrogen from hydrogen bromide molecules. The resulting bromine radicals, Br•, perform an addition reaction with but-1-ene (Figure R3.153).

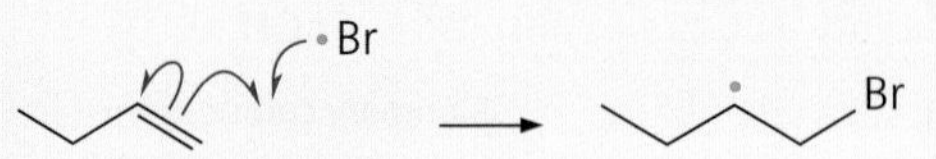

■ **Figure R3.153** Radical addition of bromine radicals to but-1-ene

The bromine radicals add selectively to the least sterically hindered end of the alkene. Since radicals are stabilized in a similar way to carbocations, this results in a secondary radical which is more stable than a corresponding primary radical.

The next step is for the carbon-centred radical to abstract hydrogen from a hydrogen bromide to produce 1-bromobutane and a bromine radical which will go on and propagate the reaction.

LINKING QUESTION

What is the difference between the bond breaking that forms a radical and the bond breaking that occurs in nucleophilic substitution?

Electrophilic substitution reactions of benzene

Benzene, C_6H_6, is a relatively unreactive molecule in which six carbon and hydrogen atoms all sit in the same plane, and each bond between the carbon atoms is of an equal strength and length. The non-hybridized p-orbitals on each of the six carbons overlap in a sideways manner to form molecular pi-orbitals, and the six electrons occupy these orbitals which spread across the structure (Figure R3.154).

HL ONLY

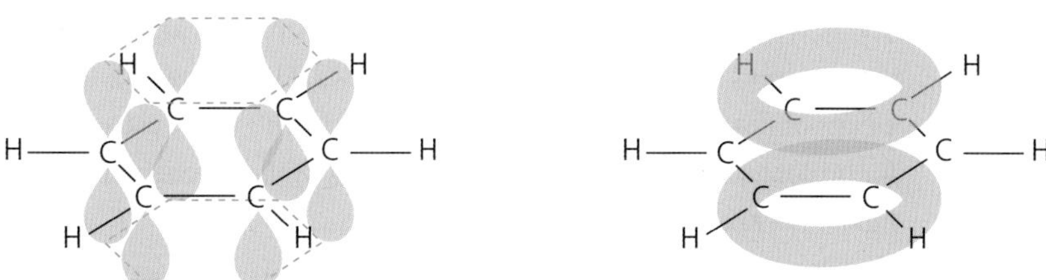

■ **Figure R3.154** The planar structure of benzene with delocalized electrons shared around its cyclic structure

LINKING QUESTION

What are the features of benzene, C_6H_6, that make it not prone to undergo addition reactions, despite being highly unsaturated?

Benzene has an area of high electron density above and below the plane of its atoms, but it is not a strong nucleophile because it is reluctant to give away the six delocalized pi electrons and lose the stability that arises from extensive delocalization. Evidence for this is that benzene will not, in contrast to alkenes, react with weak electrophiles such as aqueous bromine molecules.

Benzene will react with stronger electrophiles in a substitution reaction in which an electrophilic reagent replaces a hydrogen on the benzene ring (Figure R3.155).

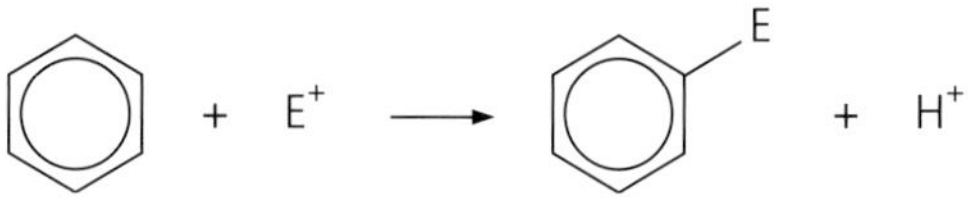

■ **Figure R3.155** The general form of an electrophilic substitution reaction involving benzene

Electrophilic substitution reactions always take place by the same two-step mechanism. In the first step, benzene reacts with the electrophile, E^+, forming a carbocation intermediate. The carbocation intermediate is stabilized by resonance of the electrons in the structure and this delocalizes the positive charge across the structure (Figure R3.156).

Step 1

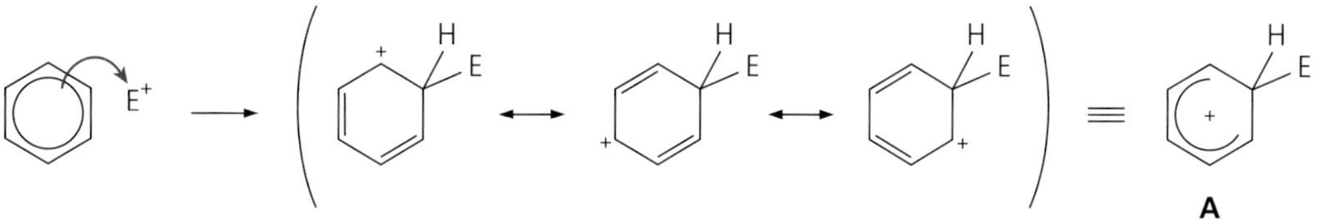

Step 2

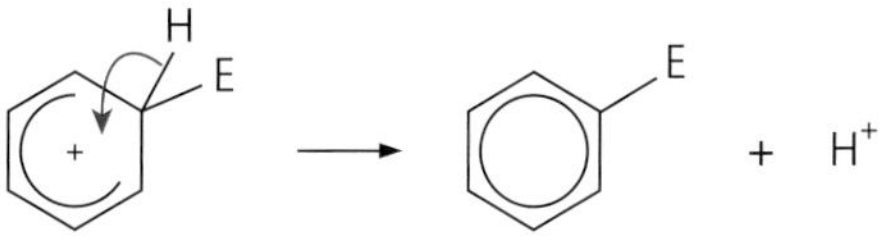

■ **Figure R3.156** The generalized mechanism of electrophilic substitution involving benzene

There are many different ways to represent the carbocation intermediate formed by this initial addition step. One way, which indicates that the positive charge is shared around the ring and that there are some remaining delocalized electrons in the structure, is labelled as A in Figure R3.156. This is the preferred way to show this structure.

In the second step of the reaction, H^+ is lost and the electrons in the C–H bond on the carbon now bearing the electrophile move into the ring to re-establish the benzene ring and a hydrogen ion is lost.

The first step is the relatively slow rate-determining step because benzene, a stable molecule, is being converted into a much less stable intermediate; this step has a high activation energy barrier via a high energy transition state. The second step is fast because it has a low activation energy barrier and it goes on to restore the delocalized pi electrons.

While weak electrophiles will react with electron rich arenes (for example, phenol), only strong electrophiles such as those listed in Table R3.33 will react in this way with benzene. These electrophiles are often formed in the reaction (*in situ*) because they are not stable enough to exist for prolonged periods of time. The products of these reactions are used in a large number of industries and are commercially important.

■ **Table R3.33** Electrophiles used in electrophilic substitution reaction with benzene

Electrophile	Reactants used to form electrophile	Product of reaction with benzene
NO_2^+	HNO_3, H_2SO_4	C_6H_5–NO_2 nitrobenzene
Cl^+	Cl_2, $AlCl_3$	C_6H_5–Cl chlorobenzene
SO_3H^+	conc. H_2SO_4	C_6H_5–SO_2OH benzenesulfonic acid
R^+	RX, $AlCl_3$	C_6H_5–R
RCO^+	RCOCl, $AlCl_3$	C_6H_5–C(=O)–R

Reactions which substitute an electron-donating substituent onto the ring have to be performed with care using a limiting amount of electrophile because the product made may be more reactive than the starting benzene. This leads to multiple sites of electrophilic substitution.

LINKING QUESTION

Nitration of benzene uses a mixture of concentrated nitric and sulfuric acids to generate a strong electrophile, NO_2^+. How can the acid / base behaviour of HNO_3 in this mixture be described?

Acknowledgements

The Publishers would like to thank the following for permission to reproduce copyright material:

p.373 Figure R1.45, Dr. Pieter Tans, NOAA/GML (gml.noaa.gov/ccgg/trends/) and Dr. Ralph Keeling, Scripps Institution of Oceanography (scrippsco2.ucsd.edu/); **p.374** Figure R1.46, Hannah Ritchie, Max Roser and Pablo Rosado (2022) – 'Energy'. Published online at OurWorldInData.org. Retrieved from: 'https://ourworldindata.org/energy' [Online Resource].

Photo credits

All photos by kind permission of Cesar Reyes except:

p.2 © Andrew Lambert Photography/Science Photo Library; **p.10** © GIPHOTOSTOCK/SCIENCE PHOTO LIBRARY; **p.36** © Dr Jon Hare; **p.69** Photograph Is Reproduced With Permission Of The BIPM, Which Retains Full Internationally Protected Copyright (Photograph Courtesy Of The BIPM); **p.98** © Andrew Lambert Photography/Science Photo Library; **p.185** © Chuchawan/Shutterstock.com; **p.201** *l* © 3djewelry/stock.adobe.com, *r* © Dmitrii/stock.adobe.com; **p.202** © Olha Kho/Shutterstock.com; **p.203** *r* © David Talbot; **p.204** © David Talbot; **p.208** © THETAE/Shutterstock.com; **p.213** *r* © Charles D. Winters/Science Photo Library; **p.215** © David Talbot; **p.223** © Andrew Lambert Photography/Science Photo Library; **p.246** © MI -_-. KI/Shutterstock.com; **p.247** © Hubb67/stock.adobe.com; **p.304** © Mikhail Basov - Fotolia; **p.311** *l* © SCIENCE PHOTO LIBRARY, *r* © Phil Degginger/Alamy Stock Photo; **p.312** © miroslavmisiura/iStock/Thinkstock; **p.318** © Lowefoto/Alamy Stock Photo; **p.361** © Sanulchik/stock.adobe.com; **p.363** © SCIENCE PHOTO LIBRARY; **p.364** © sciencephotos/Alamy Stock Photo; **p.365** © ANDREW LAMBERT PHOTOGRAPHY/SCIENCE PHOTO LIBRARY; **p.366** © Wellington College; **p.367** © Simon Fraser/Science Photo Library; **p.381** *l* © Aisyaqilumar – Fotolia, *r* © Fotos 593 - Fotolia; **p.382** © Alexandros Michailidis/Shutterstock.com; **p.436** © Dr Colin Baker; **p.522** *l* © Helen Sessions/Alamy Stock Photo, *r* © Martin Lee/Alamy Stock Photo; **p.600** © George Doyle/Stockbyte/Getty Images

Index

C

D

E

F

J

K

L

M

N

Q

R

S

T

U